Electrochemical
ENERGY
Advanced Materials and Technologies

ELECTROCHEMICAL ENERGY STORAGE AND CONVERSION

Series Editor: Jiujun Zhang
National Research Council Institute for Fuel Cell Innovation
Vancouver, British Columbia, Canada

Published Titles

Electrochemical Supercapacitors for Energy Storage and Delivery: Fundamentals and Applications
Aiping Yu, Victor Chabot, and Jiujun Zhang

Proton Exchange Membrane Fuel Cells
Zhigang Qi

Graphene: Energy Storage and Conversion Applications
Zhaoping Liu and Xufeng Zhou

Electrochemical Polymer Electrolyte Membranes
Yan-Jie Wang, David P. Wilkinson, and Jiujun Zhang

Lithium-Ion Batteries: Fundamentals and Applications
Yuping Wu

Lead-Acid Battery Technologies: Fundamentals, Materials, and Applications
Joey Jung, Lei Zhang, and Jiujun Zhang

Solar Energy Conversion and Storage: Photochemical Modes
Suresh C. Ameta and Rakshit Ameta

Electrochemical Energy: Advanced Materials and Technologies
Pei Kang Shen, Chao-Yang Wang, San Ping Jiang, Xueliang Sun, and Jiujun Zhang

Forthcoming Titles

Electrochemical Reduction of Carbon Dioxide: Fundamentals and Technologies
Jinli Qiao, Yuyu Liu,and Jiujun Zhang

Metal-Air and Metal-Sulfur Batteries: Fundamentals and Applications
Vladimir Neburchilov and Jiujun Zhang

Electrochemical ENERGY
Advanced Materials and Technologies

Edited by
Pei Kang Shen
Chao-Yang Wang
San Ping Jiang
Xueliang Sun
Jiujun Zhang

CRC Press is an imprint of the
Taylor & Francis Group, an **informa** business

CRC Press
Taylor & Francis Group
6000 Broken Sound Parkway NW, Suite 300
Boca Raton, FL 33487-2742

First issued in paperback 2017

ISBN-13: 978-1-4822-2727-7 (hbk)
ISBN-13: 978-1-138-74892-7 (pbk)

Visit the Taylor & Francis Web site at
http://www.taylorandfrancis.com

and the CRC Press Web site at
http://www.crcpress.com

Contents

SECTION I Overview of Electrochemical Energy Storage and Conversion

SECTION II Advanced Materials and Technologies for Lithium-Ion Rechargeable Batteries

SECTION III Advanced Materials and Technologies for Metal–Air Rechargeable Batteries

SECTION IV Advanced Materials and Technologies for Lead–Acid Rechargeable Batteries

SECTION V Advanced Materials and Technologies for Fuel Cells

SECTION VI Advanced Materials and Technologies for Supercapacitors

SECTION VII Advanced Materials and Technologies for Liquid–Redox Rechargeable Batteries

SECTION VIII Advanced Materials and Technologies for Water Electrolysis Producing Hydrogen

Series Preface

The goal of the *Electrochemical Energy Storage and Conversion* book series is to provide comprehensive coverage of the field, with titles focusing on fundamentals, technologies, applications, and the latest developments, including secondary (or rechargeable) batteries, fuel cells, supercapacitors, CO_2 electroreduction to produce low-carbon fuels, electrolysis for hydrogen generation/storage, and photoelectrochemistry for water splitting to produce hydrogen, among others. Each book in this series is self-contained, written by scientists and engineers with strong academic and industrial expertise who are at the top of their fields and on the cutting-edge of technology. With a broad view of various electrochemical energy conversion and storage devices, this unique book series provides essential reads for university students, scientists, and engineers and allows them to easily locate the latest information on electrochemical technology, fundamentals, and applications.

Jiujun Zhang
National Research Council of Canada
Richmond, British Colombia

Series Preface

Preface

In today's world, clean energy technologies, including storage and conversion, play the most important role in the sustainable development of human society, and are becoming the most critical elements in overcoming fossil fuel reliance and global pollution. Among possible clean energy solutions, electrochemical devices are considered the most feasible, sustainable, and resilient. However, these technologies, and the materials used to develop them, still face significant challenges in terms of cost and performance. In this regard, a comprehensive book covering important subjects in the technology, materials, and applications of electrochemical energy storage and conversion should be highly useful.

This book provides a comprehensive description of electrochemical energy conversion and storage technologies such as batteries, fuel cells, supercapacitors, and hydrogen generation, as well as their associated materials, with emphasis on existing and emerging solutions. A variety of topics such as electrochemical processes, materials, components, assembly and manufacturing, degradation mechanisms, as well as challenges and perspectives are discussed.

The authors of this book are those experts who participated in the International Conference on Electrochemical Materials and Technologies for Clean Sustainable Energy (ICES-2013) held during July 5–9, 2013, in Guangzhou, China. The conference was co-organized by Sun Yat-sen University (China), Pennsylvania State University (United States), and Yancheng Institute of Technology (China). This conference was co-chaired by Dr. Pei Kang Shen (professor at Sun Yat-sen University, China), Dr. Jiujun Zhang (principal research officer at the National Research Council of Canada), Dr. Andy X Sun (professor at University of Western Ontario, Canada), Dr. Chao-Yang Wang (professor at Pennsylvania State University, USA), Dr. San Ping Jiang (professor at Curtin University, Australia), and Dr. Baolin Wang (professor at Yancheng Institute of Technology, China). The objectives of the ICES-2013 conference were to update the latest development and progress on advanced materials and technologies for clean sustainable energy, to accelerate technology commercialization, and to enhance research collaboration among members of the electrochemical energy R&D community. More than 300 attendees from across the globe participated in ICES-2013, and gave presentations in six major themes: (i) fuel cells and hydrogen energy, (ii) lithium batteries and advanced secondary batteries, (iii) green energy for a clean environment, (iv) photo-electrocatalysis, (v) supercapacitors, and (vi) electrochemical clean energy applications and markets. Furthermore, high-quality papers presented in the conference were selected and published in a special section of the *International Journal of Hydrogen Energy* after a rigorous and regular peer-reviewed process (Volume 39, 2014).

This book contains eight sections with 25 chapters covering all aspects of synthesis, characterization, and performance evaluation of the advanced materials for electrochemical energy. Section I is an overview of electrochemical energy storage and conversion and associated materials and technologies. Section II focuses on advanced materials and technologies for Li-ion rechargeable batteries. Section III describes metal–air rechargeable batteries including working principles and materials. Section IV discusses materials and recent advances for lead–acid rechargeable batteries. Section V focuses on the working principle and material requirements for fuel cells. Section VI describes the materials and technologies for supercapacitors. Section VII describes liquid–redox rechargeable batteries and their associated capabilities and materials. Finally, Section VIII focuses on water electrolysis including catalysts and recent advances in hydrogen generation.

We would like to thank the authors/coauthors who have contributed high-quality chapters to this book. Without their tremendous efforts, it would not be possible to produce such a book. We would also like to thank CRC acquisition editor, Allison Shatkin, for her professional assistance and strong support during this project. It is our intention that this book will be a valuable reference for students, materials engineers, and researchers interested in advanced energy technologies in general.

Finally, the aforementioned conference (ICES-2013) marked the 60th birthday of this book's lead editor, Professor Pei Kang Shen. Professor Shen has worked on the cutting edge of electrochemical energy for more than 30 years and has made significant contributions to the field. It is our strong desire that this book be dedicated to his great achievements in the development of advanced materials and technologies for electrochemical energy conversion and storage, and we wish him continued success in the future.

Pei Kang Shen
Guangzhou, China

Chao-Yang Wang
University Park, Pennsylvania

San Ping Jiang
Perth, Western Australia, Australia

Xueliang Sun
London, Ontario, Canada

Jiujun Zhang
Richmond, British Columbia, Canada

Preface

Editors

Pei Kang Shen earned his BSc in electrochemistry at Xiamen University in 1982, and he continued to carry out his research and teaching at the same university for seven years before he became a visiting researcher in the United Kingdom. He earned his PhD in chemistry at Essex University in 1992. From then on, he has worked at Essex University, Hong Kong University, the City University of Hong Kong, and the South China University of Technology. Since 2001, he has been the director of and professor at the Advanced Energy Materials Research Laboratory at the Sun Yat-sen University (Guangzhou, P. R. China). He is the author of more than 300 publications in peer-reviewed journals or specialized books, with a total number of 6000 citations, 30 patents, and more than 100 meeting presentations. He has served as chairman for several international conferences. His research interests include fuel cells and batteries, electrochemistry of nanomaterials and of nanocomposite functional materials, and electrochemical engineering.

Chao-Yang Wang is the William E. Diefenderfer Chair in mechanical engineering and distinguished professor of mechanical engineering, chemical engineering, and materials science and engineering at Pennsylvania State University. He has been the founding director of the Penn State Electrochemical Engine Center since 1997. Dr. Wang holds more than 50 patents (United States, China, European Union, and Japan) and has published two books, *Modeling and Diagnostics of Polymer Electrolyte Fuel Cells* by Springer and *Battery Systems Engineering* by Wiley. He has more than 12,000 Science Citation Index citations and an *h*-index of 65 (Web of Science), and he is one of the 187 highly cited researchers in engineering named by Thomas Reuter in 2014. Dr. Wang's research interests cover the transport, materials, manufacturing, and modeling aspects of batteries and fuel cells.

San Ping Jiang earned his BEng in ceramic materials from South China University of Technology in 1982 and a PhD in electrochemistry from The City University, London, in 1988. He is a professor in the Department of Chemical Engineering and deputy director of the Fuels and Energy Technology Institute, Curtin University, Australia, and adjunct professor of the University of Sunshine Coast University, Australia. Before joining Curtin University in 2010, Dr. Jiang worked at Essex University in the United Kingdom, CSIRO Materials Science and Manufacturing Division, Ceramic Fuel Cells Ltd. in Australia, and Nanyang Technological University in Singapore. Dr. Jiang has authored and coauthored 10 book chapters, three books on fuel cells, and published ~270 journal papers, which have accrued ~10,000 citations and an *h*-index of 53. His research interests encompass solid oxide fuel cells, proton-exchange membrane and direct alcohol fuel cells, water electrolysis, solid-state ionics, electrocatalysis, and nanostructured functional materials.

Xueliang Sun is a professor and Canada Research Chair (Tier I) for the development of nanomaterials for clean energy at the University of Western Ontario, Canada. Dr. Sun earned his PhD degree in materials chemistry at the University of Manchester, United Kingdom, in 1999. Then, he worked as a postdoctoral fellow in the University of British Columbia, Canada, during 1999–2001. He was a research associate at the National Institut de la Recherche Scientifique, Quebec, Canada, during 2001–2004. Dr. Sun's research is focused on advanced nanostructured materials for energy conversion and storage, including fuel cells and Li batteries. Dr. Sun is an author and coauthor of more than 200 refereed journal articles (e.g., *Nature Communications, Advanced Materials, Journal of the American Chemical Society, Angewandte Chemie, Advanced Functional Materials*, and *Energy & Environmental Science*) and book chapters with a total of 6000 citations. He holds 10 US patents. Dr. Sun has served as the associate editor for *Frontier of Energy Storage* from 2013 to the present. His current research interests are associated with synthesis of nanomaterials for electrochemical energy storage and conversion.

Jiujun Zhang is a principal research officer at the National Research Council of Canada (NRC). He is also a fellow of the International Society of Electrochemistry (ISE). Dr. Zhang's technical expertise areas include electrochemistry, photoelectrochemistry, spectroelectrochemistry, electrocatalysis, fuel cells (PEMFC, SOFC, and DMFC), batteries, and supercapacitors. He earned his BS and MSc degrees in electrochemistry from Peking University in 1982 and 1985, respectively, and his PhD degree in electrochemistry from Wuhan University in 1988. Starting in 1990, he carried out three terms of postdoctoral research at the California Institute of Technology, York University, and the University of British Columbia. Dr. Zhang holds more than

10 adjunct professorships, including one at the University of Waterloo, one at the University of British Columbia, and one at Peking University. To date, he has approximately 400 publications with more than 16,000 citations, including 230 refereed journal papers with an *h*-index of 60, 13 edited/coauthored books, 36 book chapters, 110 conference oral and keynote/invited presentations, as well as 10 US/EU/WO/JP/CA patents, and has produced in excess of 90 industrial technical reports. Dr. Zhang serves as the editor/editorial board member for several international journals as well as the editor for a book series (Electrochemical Energy Storage and Conversion, CRC Press). He is an active member of the Electrochemical Society (ECS), the International Society of Electrochemistry (ISE), and the American Chemical Society (ACS), as well as the Canadian Institute of Chemistry (CIC).

Contributors

David Aili
Department of Energy Conversion and Storage
Section for Proton Conductors
Technical University of Denmark
Lyngby, Denmark

Jon L. Anderson
C&D Technologies Inc.
Blue Bell, Pennsylvania

Suddhasatwa Basu
Department of Chemical Engineering
Indian Institute of Technology Delhi
New Delhi, India

Mei Cai
General Motors Research and Development Center
Warren, Michigan

Zhongwei Chen
Department of Chemical Engineering
University of Waterloo
Waterloo, Ontario, Canada

Fang Dai
General Motors Research and Development Center
Warren, Michigan

Yueping Fang
Department of Chemistry
Kansas State University
Manhattan, Kansas

Jing Fu
Department of Chemical Engineering
University of Waterloo
Waterloo, Ontario, Canada

Kuan-Zong Fung
Department of Materials Science and Engineering
National Cheng Kung University
Tainan, Taiwan

Jun Gou
Energy Management
E/E Systems Engineering
Chrysler Group LLC
Auburn Hills, Michigan

Shaojie Han
Ningbo Institute of Industrial Technology
Chinese Academy of Sciences
Ningbo, Zhejiang, China

Jens Oluf Jensen
Department of Energy Conversion and Storage
Section for Proton Conductors
Technical University of Denmark
Lyngby, Denmark

Anson Lee
Energy Management
E/E Systems Engineering
Chrysler Group LLC
Auburn Hills, Michigan

Dong Un Lee
Department of Chemical Engineering
University of Waterloo
Waterloo, Ontario, Canada

Yongjun Leng
Electrochemical Engine Center
Department of Mechanical and Nuclear Engineering
Pennsylvania State University
University Park, Pennsylvania

Bing Li
Clean Energy Automotive Engineering Center
Tongji University
Shanghai, China

Jun Li
Department of Chemistry
Kansas State University
Manhattan, Kansas

Qingfeng Li
Department of Energy Conversion and Storage
Section for Proton Conductors
Technical University of Denmark
Lyngby, Denmark

Qizheng Li
Department of Mechanical and Materials Engineering
University of Western Ontario
London, Ontario, Canada

Yongliang Li
College of Chemistry and Chemical Engineering
Shenzhen University
Shenzhen, Guangdong, China

Huagen Liang
HySA Systems Competence Centre
SAIAMC Innovation Centre
Faculty of Natural Science
University of the Western Cape
Bellville, South Africa

Zijing Lin
Hefei National Laboratory for Physical Sciences at Microscales
Department of Physics
University of Science and Technology of China
Hefei, China

Hao Liu
Centre for Clean Energy Technology
University of Technology
Sydney, New South Wales, Australia

Hao Liu
Chengdu Development Center of Science and Technology of CAEP
Chengdu, Sichuan, China

Jianhong Liu
College of Chemistry and Chemical Engineering
Shenzhen University
Shenzhen, Guangdong, China

Li-Min Liu
Beijing Computational Science Research Center
Beijing, China

Zhaoping Liu
Ningbo Institute of Industrial Technology
Chinese Academy of Sciences
Ningbo, Zhejiang, China

Zhe Lü
Department of Physics
Harbin Institute of Technology
Harbin, China

Yuezhong Meng
The Key Laboratory of Low-Carbon Chemistry and Energy Conservation of Guangdong Province
State Key Laboratory of Optoelectronic Materials and Technologies
Sun Yat-sen University
Guangzhou, China

Gaind P. Pandey
Department of Chemistry
Kansas State University
Manhattan, Kansas

Sivakumar Pasupathi
HySA Systems Competence Centre
SAIAMC Innovation Centre
Faculty of Natural Science
University of the Western Cape
Bellville, South Africa

Bruno G. Pollet
HySA Systems Competence Centre
SAIAMC Innovation Centre
Faculty of Natural Science
University of the Western Cape
Bellville, South Africa

Laifen Qin
Ningbo Institute of Industrial Technology
Chinese Academy of Sciences
Ningbo, Zhejiang, China

Jingmei Shen
Optimal CAE Inc.
Plymouth, Michigan

Junli Shi
Ningbo Institute of Industrial Technology
Chinese Academy of Sciences
Ningbo, Zhejiang, China

Surya Singh
Department of Chemical Engineering
Indian Institute of Technology Delhi
New Delhi, India

Huaneng Su
HySA Systems Competence Centre
SAIAMC Innovation Centre
Faculty of Natural Science
University of the Western Cape
Bellville, South Africa

Bing Sun
Centre for Clean Energy Technology
University of Technology
Sydney, New South Wales, Australia

Xueliang Sun
Department of Mechanical and Materials Engineering
University of Western Ontario
London, Ontario, Canada

Zetian Tao
Key Laboratory for Advanced Technology in Environmental Protection of Jiangsu Province
Yancheng Institute of Technology
Yancheng, Jiangsu, China

Anil Verma
Department of Chemical Engineering
Indian Institute of Technology Delhi
New Delhi, India

Chao-Yang Wang
Electrochemical Engine Center
Department of Mechanical and Nuclear Engineering
Pennsylvania State University
University Park, Pennsylvania

Guoping Wang
Department of Chemistry and Chemical Engineering
University of South China
Hengyang, China

Guoxiu Wang
Centre for Clean Energy Technology
University of Technology
Sydney, New South Wales, Australia

Hongqing Wang
Department of Engineering
University of South China
Hengyang, China

Shuanjin Wang
The Key Laboratory of Low-Carbon Chemistry and Energy Conservation of Guangdong Province
State Key Laboratory of Optoelectronic Materials and Technologies
Sun Yat-sen University
Guangzhou, China

Wei Wang
Ningbo Institute of Industrial Technology
Chinese Academy of Sciences
Ningbo, Zhejiang, China

Yufei Wang
The Key Laboratory of Low-Carbon Chemistry and Energy Conservation of Guangdong Province
State Key Laboratory of Optoelectronic Materials and Technologies
Sun Yat-sen University
Guangzhou, China

Zhihong Wang
Department of Physics
Harbin Institute of Technology
Harbin, China

Bo Wei
Department of Physics
Harbin Institute of Technology
Harbin, China

Zhen Wei
Ningbo Institute of Industrial Technology
Chinese Academy of Sciences
Ningbo, Zhejiang, China

David P. Wilkinson
Department of Chemical and Biological Engineering
University of British Columbia
Vancouver, British Columbia, Canada

Gang Wu
Department of Chemical and Biological Engineering
University at Buffalo
The State University of New York
Buffalo, New York

Jinfeng Wu
UniEnergy Technologies LLC
Mukilteo, Washington

Lan Xia
Ningbo Institute of Industrial Technology
Chinese Academy of Sciences
Ningbo, Zhejiang, China

Yonggao Xia
Ningbo Institute of Industrial Technology
Chinese Academy of Sciences
Ningbo, Zhejiang, China

Min Xiao
The Key Laboratory of Low-Carbon Chemistry and Energy Conservation of Guangdong Province
State Key Laboratory of Optoelectronic Materials and Technologies
Sun Yat-sen University
Guangzhou, China

Qiangfeng Xiao
General Motors Research and Development Center
Warren, Michigan

Hossein Yadegari
Department of Mechanical and Materials Engineering
University of Western Ontario
London, Ontario, Canada

Li Yang
General Motors Research and Development Center
Warren, Michigan

Jianlu Zhang
Palcan Energy Corporation
Vancouver, British Columbia, Canada

Jiujun Zhang
Energy, Mining and Environment Portfolio
National Research Council of Canada
Vancouver, British Columbia, Canada

Lei Zhang
Energy, Mining and Environment Portfolio
National Research Council of Canada
Vancouver, British Columbia, Canada

Benhe Zhong
College of Chemical Engineering
Sichuan University
Chengdu, China

Hong Zhu
Department of Organic Chemistry
School of Science
Beijing University of Chemical Technology
Beijing, China

Section I

Overview of Electrochemical Energy Storage and Conversion

1 Introduction to Electrochemical Energy Storage and Conversion

Jingmei Shen, Fang Dai, Mei Cai, and Hong Zhu

CONTENTS

1.1 ELECTROCHEMISTRY: FUNDAMENTAL AND TECHNICAL OVERVIEW

Electrochemistry deals with chemical reactions taking place at the interface of an electrode and the electrolyte, and studies conversions between the electrical energy and chemical changes [1–4]. Electrons involved in these chemical reactions move between the interface and the electrodes. The basic element of each battery is the electrochemical cell, which consists of a positive electrode, a negative electrode, and an electrolyte. The electrode reactions occur at both electrodes, releasing or taking electrons. The electrode half-reactions can be written as follows:

$$\text{Negative electrode:} \quad S_{red} \rightarrow S_{ox} + ne^- \tag{1.1}$$

$$\text{Positive electrode:} \quad S_{ox} + ne^- \rightarrow S_{red} \tag{1.2}$$

where S_{red} and S_{ox} indicate the reduced and oxidized states of the chemical reactants and n is the number of electrons involved in the process. The electrons involved in the reaction are collected as a current flow.

Many devices based on electrochemistry have been built, including batteries, fuel cells, electrolyzers, and others. Figure 1.1 illustrates the schematics of electrochemical storage systems. Electrolyzers convert electrical energy into chemical energy. The energy is stored in chemical compounds. Fuel cells perform an opposite process, converting chemical energy into electrical energy. In secondary electrochemical batteries, this energy conversion process is reversible. The charging process converts electrical energy into chemical energy, which is stored in the storage component. During discharging, the chemical energy converts back into electrical energy. The converters determine the charging and discharging power, and the storage component determines the energy capacity of the systems. Generally, the active materials for converters and storages cannot be separated in secondary electrochemical batteries with internal storage. Therefore, power and capacity cannot be designed independent of each other.

Electrochemical storage systems with separate converters and storage are also attractive because practical applications sometimes desire high energy capacity rather than high power, or the other way around. The electrolyzer/hydrogen storage/fuel cell system is suitable for that purpose. Herein, we introduce several current electrochemical conversion and storage systems, including primary and secondary batteries, supercapacitors, water electrolyzers, and fuel cells.

1.2 PRIMARY BATTERIES

1.2.1 Introduction

It has been more than 100 years since the first primary battery was built [1,5]. Among all existing designs, the zinc–carbon battery was the only one in wide use up to 1940. Significant advances were made during World War II and the postwar period, not only with the zinc–carbon system but also with

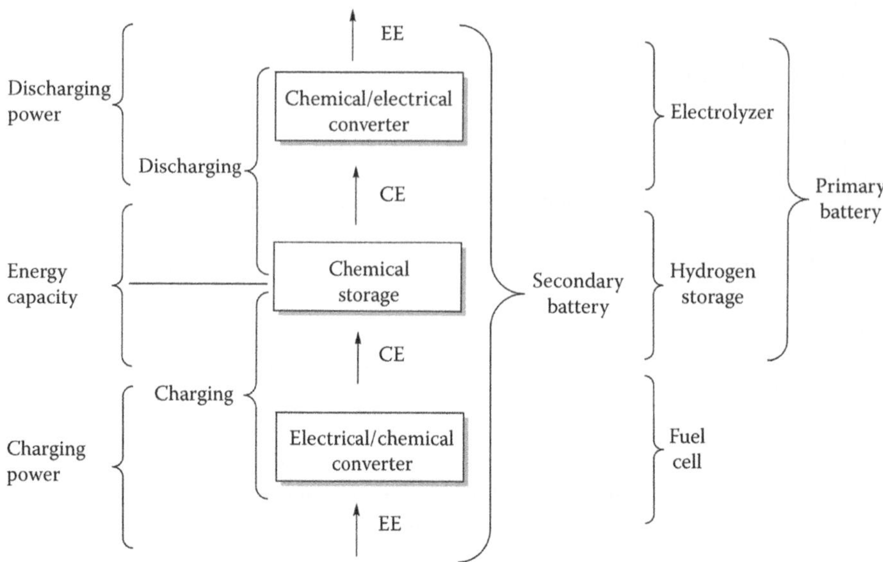

FIGURE 1.1 Schematics of electrochemical storage and conversion systems.

novel and superior types of batteries. The capacity of the batteries was much improved from <50 to >500 Wh/kg with recent advanced batteries such as lithium and zinc–air systems. Novel batteries also show excellent shelf life as high as 10 years, with a capability of storage at temperatures as high as 70°C. The operation temperature limit has been extended from 0°C to −40°C.

There are a variety of primary battery system designs using different anode–cathode combinations [5,6]. However, only a relatively few have successfully been utilized in practical applications. For example, zinc has been recognized as the most popular anode material for primary batteries because of its good electrochemical properties, compatibility with aqueous electrolytes, low cost, and availability. In contrast, although aluminum is attractive because of its high electrochemical potential and electrochemical equivalence and availability, it has not been developed successfully into a practical active primary battery system due to the passivation issue and limited electrochemical performance. Magnesium has also been used successfully in an active primary battery, particularly for military applications, because of its high energy density and good shelf life. However, commercial interest has been limited and the use in US military has been terminated. In recent years, there is an increasing focus on lithium because of its highest gravimetric energy density. The primary battery systems utilizing lithium as anode material can offer the opportunity for higher energy density and power density in the performance characteristics of primary systems.

Unlike anode materials, metal oxides such as manganese oxide, silver oxide, etc., are the dominant materials for cathode use for many types of primary batteries. Only a few other cathode materials, such as oxygen, which is used for specific battery configurations (metal–air batteries), were utilized in the primary batteries. The electrolytes for the primary batteries can be aqueous or nonaqueous systems according to the electrochemistry of the battery. Detailed demonstrations of the battery configurations, electrochemistry, and materials by different battery types are given in the following subsections.

1.2.2 Zinc–Carbon and Alkaline–Manganese Dioxide Batteries

Zinc–carbon (Zn–C or Zn–MnO_2) batteries have been well developed for more than a hundred years. The other name of "dry battery" was given to zinc–manganese dioxide (Zn/MnO_2) when it was first introduced in the middle of the 19th century by the fact that its electrolyte is immobilized in an inert support. Two types of configuration, namely Leclanché and zinc chloride systems, have become the dominant designs of the modern Zn–C batteries. The Leclanché type uses an aqueous electrolyte containing NH_4Cl (26%) and $ZnCl_2$ (8.8%), while the zinc chloride type contains only $ZnCl_2$ (15–40%). The electrodes are basically the same in both systems. One disadvantage of the Leclanché cell is the formation of sparingly soluble Zn salts, which greatly limits ion diffusion due to their accumulation near the electrode. With the $ZnCl_2$ solution, this phenomenon is much reduced to allow faster diffusion and enhanced rates of discharge. However, the better performance of the $ZnCl_2$ cell, especially at high currents and moderately low temperature (down to −10°C), is counterbalanced by a higher cost. In terms of performance and cost, this cell lies between the common Leclanché and the alkaline cell. Another advantage of the $ZnCl_2$ cell is given by its lower self-discharge rate.

In principle, the electrochemistry of the Zn–C cell is rather simple with Zn oxidation to Zn^{2+} and Mn^{4+} reduction to Mn^{3+} (MnOOH or Mn_2O_3; see Equations 1.3 through 1.8). In practice, the reactions are rather complicated and depend on several factors, such as electrolyte composition, temperature, rate, and depth of discharge.

Simplified overall reaction:

$$Zn + 2MnO_2 \rightarrow ZnO + Mn_2O_3 \tag{1.3}$$

Leclanché cell reaction with ammonium chloride electrolyte:

$$Zn + 2MnO_2 + 2NH_4Cl \rightarrow 2MnOOH + Zn(NH_3)_2\,Cl_2 \tag{1.4}$$

$$Zn + 2MnO_2 + NH_4Cl + H_2O \rightarrow 2MnOOH + Zn(OH)\,Cl + NH_3 \tag{1.5}$$

$$Zn + 6MnOOH \rightarrow 2Mn_3O_4 + ZnO + 3H_2O \tag{1.6}$$

Zinc chloride cell reaction with zinc chloride electrolyte:

$$Zn + 2MnO_2 + 2H_2O + ZnCl_2 \rightarrow 2MnOOH + 2Zn(OH)Cl \tag{1.7}$$

$$Zn + 6MnOOH + Zn(OH)Cl \rightarrow 2Mn_3O_4 + ZnCl_2 \cdot ZnO \cdot 4H_2O \tag{1.8}$$

The anode is Zn, while the cathode is a mix of electrochemically active MnO_2 and carbon.

Modern cells mostly use either chemical MnO_2 or electrolytic MnO_2, whose percentage of active material is 90–95% (the remainder is mostly H_2O plus several impurities). The highly porous acetylene black is utilized as conductive material. Zinc metal is the anode material. Owing to the acidic character of both the NH_4Cl and the $ZnCl_2$ solution, corrosion is a major issue for the Zn anode. To reduce corrosion, the Zn anode was alloyed with Cd (0.3%) and Pb (0.6%) to improve its metallurgical properties in the past. However, those toxic contents are abandoned in modern cells. Similarly, the use of Hg as the main corrosion inhibitor has been eliminated. Other materials now considered as inhibitors include Ga, Sn, Bi, glycols, or silicates.

The theoretical specific capacity of the Zn–C battery is 224 Ah/kg, based on Zn and MnO_2 and the simplified cell reaction. On a more practical basis, the electrolyte, carbon black, and water are ingredients that cannot be omitted from the system. Therefore, a specific capacity of the system is about 96 Ah/kg after content adjustment. This is the highest specific capacity a general-purpose cell can have. As a matter of fact, this number can only be approached by some of the larger Leclanché cells under certain discharge conditions. The actual specific capacity of a practical cell ranges from 75 to 35 Ah/kg, depending on different working loads considering all the cell components and the efficiency of discharge.

The alkaline battery, otherwise known as the alkaline–manganese dioxide ($Zn/KOH/MnO_2$), was introduced in the early 1960s. On the other hand, the alkaline cell, named for its use of a basic or alkaline electrolyte, was commercially produced in 1959. It became widely recognized as being superior to the Zn–C-type primary battery since the 1980s.

The overall reaction of the alkaline cell during a continuous discharge is a one-electron process, which is shown below:

$$2MnO_2 + Zn + 2H_2O \rightarrow 2MnOOH + Zn(OH)_2 \quad (1.9)$$

Since water is a reactant in reaction, the amount of water in a cell is rather important, especially in high-rate discharge applications. However, at the low- or intermittent-drain rates, the overall reaction of the cell is a 1.33-electron process, which is shown below:

$$3MnO_2 + 2Zn \rightarrow Mn_3O_4 + 2ZnO \quad (1.10)$$

Therefore, the total water management becomes an important variable in order to provide stable performance over a wide range of discharge conditions. Some battery manufacturers have included additives to the cell, such as TiO_2 and $BaSO_4$, in order to better manage this important characteristic.

The anode of an alkaline battery is made of high-purity Zn powder. The higher surface area of the Zn powder can afford higher discharge rates, while the electrolyte is more uniformly distributed. In addition, the combination of a porous anode and a conductive electrolyte reduces the extent of accumulation of reaction products near the electrode, resulting in a higher rate capability. As the Zn–C system, alloying elements such as Ir, Bi, Pd, and Al are needed as gassing suppressors. Zn powder is obviously rather reactive and can decompose H_2O with the production of hydrogen, which can cause MnO_2 self-discharge and generate an overpressure. As mentioned above, reducing the impurity level in the Zn powder greatly limits gassing. Otherwise, additives for the anode are necessary, such as ZnO (or other oxides) or organic inhibitors (polyethylene oxide compounds). Besides, a gelling agent is necessary to suspend the zinc and to maintain the conductivity and improve electrode performance. Starch, cellulose derivatives, polyacrylates, and ethylene maleic anhydride copolymers are often used as the gel agent. The anode also contains the electrolyte that is an aqueous KOH solution (35–52%).

The cathode is based on electrolytic MnO_2, which can grant high power and long shelf life. Graphite is used as the conductive carbon material, and some acetylene black is also utilized to enhance the contact surface area. The separator, which has to be chemically stable in the concentrated alkaline solution, is normally a nonwoven fabric, such as cellulose, vinyl polymer, polyolefin, or a combination thereof.

The alkaline batteries can provide a wide coverage of capacity by different constructions. For example, button cells have capacities of 25–60 mAh and the prismatic batteries can provide a capacity of 0.16–44 Ah.

1.2.3 Zinc–Air Battery

Zinc–air battery utilizes O_2 from the air as the active cathode material. O_2 diffuses through the cathode. Once the O_2 reaches the cathode interface with the alkaline electrolyte, it is catalytically reduced. The capacity of the zinc–air cell is about twice that of the $Zn–MnO_2$ or $Zn–Ag_2O$ cell with the same battery volume, since the capacity only depends on the anode. The energy of the zinc–air cell is very high both on a gravimetric and a volumetric basis.

The zinc–air battery consists of three main electrochemical components: zinc powder anode, aqueous alkaline electrolyte, and a catalytic cathode structure that converts oxygen to OH^- ions when an external circuit is established. A typical button-type Zn–air battery is shown in Figure 1.2. Unlike the anode part, the cathode region is rather complex: the holes allowing air access, the air diffuser layer distributing O_2 uniformly over the cathode, and the hydrophobic Teflon layer that is O_2 permeable but limits water vapor access. The air cathode is formed by a metallic mesh supporting the catalyst layer (carbon blended with Mn oxides and Teflon powder).

The anode reaction is shown below:

$$Zn + 2OH^- \rightarrow Zn(OH)_2 + 2e^- \quad (1.11)$$

$$Zn(OH)_2 \rightarrow ZnO + 2H_2O \quad (1.12)$$

The cathode reaction, occurring first at the catalytic site and then in the electrolyte where the ionic conduction takes place, is shown below:

$$O_2 + 2H_2O + 4e^- \rightarrow 4OH^- \quad (1.13)$$

An overall reaction may be simply written as

$$Zn + 1/2O_2 \rightarrow ZnO \quad (1.14)$$

The maximum current the cell can support depends on the O_2 availability. The cell can provide a high discharge rate

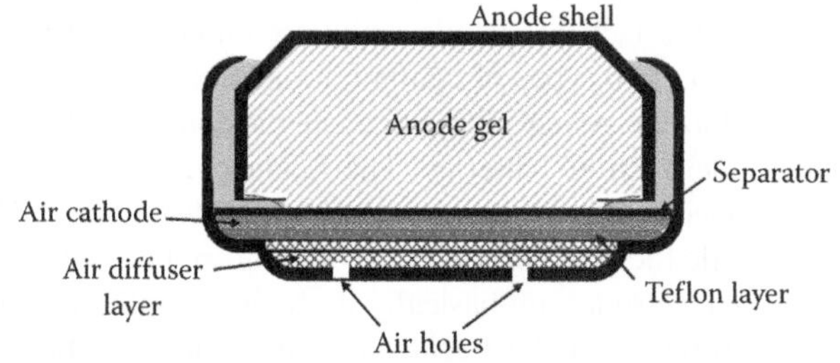

FIGURE 1.2 Cross-section scheme of zinc–air button cell.

with excess O_2 if air access was not regulated. However, strict regulation is necessary to limit the inlet of H_2O vapor and other gases that would degrade the cell. In addition, the holes and diffusion membrane are necessary, since the electrolyte concentration will change with the environmental conditions.

The button cell construction of zinc–air battery has capacity in the range of 40–600 mAh, which is delivered at low rates (0.4–2 mA). At these rates, the cell shows flat discharge curves with excellent capacity retention even at 0°C. The typical operating life is 1–3 months since the cells are in contact with the atmosphere, which favors direct Zn oxidation, carbonation of the electrolyte, and gas transfer. Therefore, the zinc–air batteries are better used in continuous-drain applications.

1.2.4 Silver Oxide Battery

Silver oxide primary batteries (Zn–Ag_2O) were first introduced in the early 1960s as power sources for electronic watches, with currents ranging from a few microamperes (LCD displays) to hundreds of microamperes (LED displays). These cells are also used in pocket calculators, hearing aids, cameras, glucometers, etc.

The Zn–Ag_2O system offers several advantages such as high capacity, a steady discharge voltage, and good storage capacity retention. The theoretical capacity of monovalent silver oxide is 231 Ah/kg. The Zn–Ag_2O battery shows a flat, constant discharge voltage profile between 1.5 and 1.6 V at both high and low discharge rates. The battery has long storage life: it can maintain >95% of its initial capacity after 1 year of room-temperature storage. It also has good low-temperature performance with delivery of about 70% of its nominal capacity at 0°C and 35% at −20°C.

The overall reaction of the Zn–Ag_2O battery is shown below:

$$Zn + Ag_2O \rightarrow ZnO + 2Ag \quad (1.15)$$

As other zinc batteries, the anode of zinc–silver oxide battery is zinc powder. The cathode is monovalent silver oxide, Ag_2O, and the electrolyte is a KOH or NaOH aqueous solution (20–45%). The Zn powder has to be highly pure, as already pointed out for alkaline cells. In commercial cells, the Zn powder is amalgamated with a low percentage of Hg to keep corrosion under control. The maximum capacity of the cell is 165 mAh. Gelling agents, such as polyacrylic acid, are added to the anode to facilitate electrolyte accessibility.

Ag_2O is now preferred as a cathode in commercial cells. Unlike AgO, used until the early 1990s, it has a stable potential and does not need to be stabilized by heavy metals, as its reactivity with alkalis is low. Since Ag_2O is a poor semiconductor, some graphite (<5%) is added as the conductive enhancer. Furthermore, the reduction of Ag_2O produces Ag, which helps decrease the cathode resistance.

The alkaline-based electrolyte contains some zincate to control gassing. KOH is preferred over NaOH in button cells because of its higher conductivity. Also, the cells with KOH electrolyte show a low working temperature as low as −28°C.

1.2.5 Magnesium/Manganese Dioxide Battery

Magnesium (Mg) is considered an attractive candidate as anode material in primary batteries. It has a high standard potential. Also, the low atomic weight and multivalence change result in a high electrochemical equivalence on both a gravimetric and a volumetric basis. In addition, it is abundant and relatively inexpensive.

The Mg primary battery uses a magnesium alloy for the anode and manganese dioxide (MnO_2) as the active cathode material. The aqueous electrolyte consists of magnesium perchlorate ($Mg(ClO_4)_2$), with barium and lithium chromate ($BaCrO_4$, Li_2CrO_4) as corrosion inhibitors, as well as magnesium hydroxide ($Mg(OH)_2$) as a buffering agent to improve storability (pH ~8.5). The amount of water is critical, as water participates in the anode reaction and is consumed during the discharge. The overall reaction is shown below:

$$Mg + 2H_2O \rightarrow Zn(OH)_2 + H_2 \quad (1.16)$$

The Mg primary battery shows superior low-temperature performance to that of the Zn–C battery, operating at temperatures of −20°C and below. However, the low-temperature performance is affected by the heat generated during discharge and is dependent on the discharge rate, battery size, battery configuration, and other such factors.

1.2.6 Lithium Primary Batteries

The development of high-energy-density battery systems was started in the 1960s and concentrated on nonaqueous primary batteries using lithium as the anode. The lithium (Li) batteries were first used in the early 1970s in selected military applications. However, the application of Li primary batteries was limited due to cell designs, formulations, and safety considerations. Li primary batteries have been designed in a variety of sizes and configurations, using a number of different chemistries. Although the aqueous primary batteries are still dominant in the market owing to their lower cost, the Li primary batteries are the best choice in some applications requiring long operation/storage times, extreme temperatures, or high power. Li primary batteries use organic solvents for the electrolyte, such as acetonitrile (CH_3CN), propylene carbonate (PC), and dimethoxyethane (DME), and inorganic solvents such as thionyl chloride ($SOCl_2$). A compatible solute is added to provide the necessary electrolyte conductivity (solid-state and molten-salt electrolytes are also used in some other Li primary cells). Different materials were utilized as the active cathode material, such as sulfur dioxide (SO_2), thionyl chloride ($SOCl_2$), manganese dioxide (MnO_2), iron disulfide (FeS_2), and carbon monofluoride (CF). Therefore, the Li primary batteries can be classified into several categories, based on the type of electrolyte (or solvent) and cathode material used.

1.2.6.1 Lithium–Sulfur Dioxide Batteries

Lithium–sulfur dioxide (Li–SO_2) cells were the first commercialized type of lithium cells. The cathode is Teflon-bonded acetylene black supported on Al screen, which provides high values of conductivity, surface area, and porosity. A cylindrical construction of Li–SO_2 battery is shown in Figure 1.3. The overall reaction of Li–SO_2 batteries is shown in below:

$$2Li + 2SO_2 \rightarrow Li_2S_2O_4 \quad (1.17)$$

The final product, lithium dithionite ($Li_2S_2O_4$), precipitates into the pores of the carbon cathode during the discharge process. The clogging issue is highly reduced owing to the highly porous structure of the cathode part. The stability of this type of cell (and of other cells with a soluble cathode) is related to the formation of a solid electrolyte interface (SEI) layer on the Li surface as soon as Li is exposed to SO_2. The growth rate of the SEI layer increases during storage of partially discharged cells.

A special construction, namely hermetic seal, is used to prevent SO_2 loss. The cell is pressurized (2 atm) to keep the electrolyte in the liquid state. A safety vent is incorporated in the cell to cope with pressure values exceeding certain limits (e.g., 24 atm).

The electrolyte contains 70% SO_2 and has a high conductivity even at −50°C ($2.2 \times 10^{-2}\ \Omega^{-1}\ cm^{-1}$). This feature allows the use of Li–SO_2 batteries in applications prohibited to other chemistries. The working voltage for Li–SO_2 batteries is 2.7–2.9 V. It also shows a relatively high practical energy density of ~260 Wh/kg. With the cylindrical construction, the Li–SO_2 can provide a high capacity of 34 Ah.

Early-design cells had an Li–SO_2 ratio as high as 1.5:1. However, it was ascertained that this ratio greatly impaired the cell safety. However, the reaction of Li with CH_3CN occurs in deeply discharged cells, when the SO_2 concentration is below 5% and the passivating film is removed. Therefore, cells with an Li–SO_2 ratio close to 1 are now preferred in order to keep Li passivated.

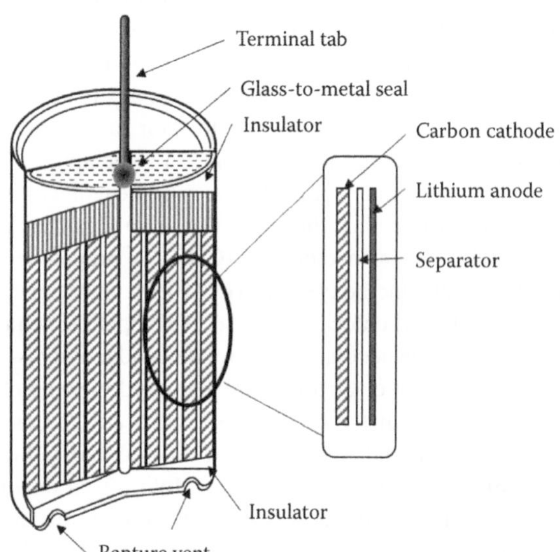

FIGURE 1.3 Lithium–sulfur dioxide battery.

1.2.6.2 Lithium–Thionyl Chloride Batteries

Lithium–thionyl chloride (Li–$SOCl_2$) batteries have a very high energy density of 450–600 Wh/kg and a long service life of 15–20 years. Moreover, these cells can be stored for long (with capacity losses of 1–2% per year at room temperature) and can be operated in an exceptionally wide temperature range (−80°C [with a special electrolyte] to 150°C). Low-rate Li–$SOCl_2$ cells are built in the bobbin-type cylindrical configuration, while moderately high-power cells are built in the spirally wound configuration. The overall cell reaction is shown below:

$$4Li + 2SOCl_2 \rightarrow 4LiCl + S + SO_2 \quad (1.18)$$

Similar as the Li–SO_2 batteries, porous carbon is the cathode support, and $SOCl_2$ is both the active cathode material and the solvent for the electrolyte salt (usually $LiAlCl_4$). LiCl precipitates on the carbon surface and stops cell operation when pore clogging occurs in cathode-limited cells. SO_2 is soluble in the electrolyte, while S is soluble up to 1 mol/L and can precipitate on the cathode toward the end of discharge. LiCl is also the main component of the passivating film formed on the Li anode.

The cell capacity could be improved by adding excess $AlCl_3$ to the electrolyte. In this case, soluble $LiAlCl_4$ is formed instead of LiCl, which helps eliminate the clogging issue. However, the use of $AlCl_3$ is limited only in high-rate cells, as $AlCl_3$ dissolves the LiCl passivating film on Li, thus allowing corrosion.

1.2.6.3 Lithium–Manganese Dioxide Batteries

Lithium–manganese dioxide (Li–MnO_2) batteries were commercialized since 1975. This widely used type of Li primary batteries shows some advanced features such as high working voltage, relatively high energy density, wide operating temperature range, long shelf life, safety, and low cost. The construction of Li–MnO_2 batteries cover coin, cylindrical, and prismatic forms (the last one including thin cells), thus allowing their use in a variety of applications.

Unlike previous examples, the Li–MnO_2 batteries are solid-cathode primary batteries. The heat-treated MnO_2 with a residual H_2O content of 1% is used as the cathode. Heat treating of electrochemical-grade MnO_2 at 350–400°C is fundamental, as the nontreated form has a poor performance. A commonly used electrolyte is $LiClO_4$–PC–DME.

The overall reaction is

$$x\text{Li} + \text{Mn}^{IV}\text{O}_2 \rightarrow \text{Li}_x\text{Mn}^{III}\text{O}_2 \quad (1.19)$$

During discharge, Li ions are gradually inserted into the channels of the cathode structure, thus giving rise to a so-called solid solution. The formula Li_xMnO_2, $0 < x < 1$, best explains the gradual accommodation of Li ions into the host structure.

1.2.6.4 Lithium–Carbon Monofluoride Batteries

Another interesting example of a solid-cathode primary battery is the so-called lithium–carbon monofluoride (Li–CF_x) battery, in which polycarbon fluoride $(CF_x)_n$ is used as a cathode. This compound is synthesized by direct fluorination of carbon with fluorine gas. The x value of the formula CF_x falls in the range of 0.9–1.2. However, only the materials with $x \geq 1$ can be used for the Li–CF_x battery. In the following, this cathode material will simply be written as CF_x (where $x = 1$ normally).

The overall reaction is

$$x\text{Li} + \text{CF}_x \rightarrow x\text{LiF} + x\text{C} \quad (1.20)$$

The formation of carbon enhances the electronic conductivity of the cathode. The typical electrolytes used for this system are $LiAsF_6$ in butyrolactone or $LiBF_4$ in PC–DME.

Apart from the energy density of 200–600 Wh/kg, the Li–CF_x system has a number of advanced features such as flat operating voltage profile (2.8 V), high capacity, and wide temperature range (−40°C to 85°C, and up to 125°C for some special designs). In addition, the Li–CF_x batteries also show the lowest self-discharge rate among all Li primary batteries. At ambient temperature, the annual capacity loss of the Li–CF_x cell is <0.5%, while it is <4% at 70°C. However, this system is not suitable for high power and is better used at low rates due to its electrochemistry limitation.

1.2.6.5 Other Lithium Primary Batteries

There are some other types of lithium primary batteries using different cathode materials, such vanadium pentoxides (V_2O_5), silver vanadium oxide ($Ag_2V_4O_{11}$), copper oxide (CuO), and iron disulfide [1,5,6].

The lithium–vanadium pentoxide battery has a high volumetric energy density (about 660 Wh/L), but with a two-step discharge profile. The lithium–silver vanadium oxide battery is capable of relatively high-rate discharge. Thus, it can be used in medical applications, such as defibrillators, that have pulse load requirements. The other solid–cathode lithium batteries operate in the range of 1.5 V and are developed to replace conventional 1.5 V button or cylindrical cells. The lithium–copper oxide cell, which was first introduced in the 1990s, shows capability at high working temperature (up to 150°C). However, it shows a significant voltage drop between open circuit voltage and working voltage. Now it has been replaced by lithium–iron sulfide batteries. The lithium–iron disulfide cell has similar advantages as other lithium primary batteries over the conventional cells, plus the advantage of high-rate performance.

1.3 SECONDARY BATTERIES

1.3.1 Introduction

It has been more than 150 years since the initial introduction of the secondary batteries. The lead–acid battery was developed in 1859 by Plante [5,7–9]. Later, the nickel–iron alkaline battery was introduced by Edison in 1908 as a power source for the early electric automobile [10]. However, it gradually lost its market share because of the high cost, maintenance requirements, and lower specific energy. The pocket-plate nickel–cadmium battery has been manufactured since 1909 primarily for heavy-duty industrial applications. The application of the nickel–cadmium batteries later extended to engine-starting and communications applications during the 1950s due to increased power capability and energy density with the sintered-plate design. The following development of the sealed nickel–cadmium battery led to its widespread use in portable and other applications. Recently, the appearance of lithium-ion batteries has changed the portable rechargeable market, in which nickel–metal hydride batteries were previously dominant, owing to their higher specific energy and energy density [1,6,10,11].

During the past few decades, significant improvements have been made with the other secondary battery systems, as well as some new types such as silver–zinc, nickel–zinc, nickel–hydrogen, and lithium-ion batteries, and the thermal system. The gravimetric and volumetric energy densities of portable rechargeable nickel–cadmium batteries are 35 Wh/kg and 100 Wh/L, respectively. The nickel–metal hydride batteries show improved gravimetric and volumetric energy densities of 100 Wh/kg and up to 430 Wh/L, respectively, by the utilization of new hydrogen-storage alloys. In contrast, lithium-ion batteries now provide a specific energy of 200 Wh/kg and an energy density of 570 Wh/L in the small cylindrical construction for consumer electronic applications. Advanced lithium-ion batteries currently provide 240 Wh/kg and 640 Wh/L.

1.3.2 Lead–Acid Batteries

As previously introduced, the lead–acid batteries have been developed for more than 150 years. Comparing with other secondary batteries, lead–acid batteries show many advantages, such as low price, good reliability, high voltage of cell (2 V), high electrochemical effectivity, and long cycle life. Thanks to these characteristics, they are now the most widely used rechargeable electrochemical energy source and represent about 60% of installed power from all types of secondary batteries. However, lead–acid batteries show relatively lower specific energy in the range 30–50 Wh/kg due to the weight of lead. Construction of lead–acid battery depends on usage. The main parts of lead–acid battery are electrodes, separators, electrolyte, vessel with lid, ventilation, and some other elements.

The electrode of the lead–acid batteries consists of grid and active mass. The positive electrode grid must be corrosion proof, since corrosion converts lead alloy to lead oxides with lower mechanical strength and electric conductivity. Grids are made from lead alloys (pure lead would be too soft), which are mostly Pb–Ca or Pb–Sb alloys with a mixture of additives such as Sn, Cd, and Se to improve corrosion resistance and achieve higher mechanical strength.

The active material is made from lead oxide, PbO, pasted onto a grid. It is then electrochemically converted into reddish-brown lead dioxide, PbO_2, on the positive electrode and on gray spongy lead, Pb, on the negative electrode. Two electrodes are electrically separated by separators. The materials used for separators can be wood veneers, cellulose paper stiffened with a phenol–formaldehyde resin binder, and those made from synthetic materials, e.g., rubber, polyvinyl chloride, polyethylene, and glass microfiber. The electrolyte is an aqueous solution of H_2SO_4. The electrolyte is mostly liquid phase covering the battery plates. Sometimes it is transformed to the gel form, or completely absorbed in separators. Some major types of battery construction are prismatic construction with grid or tubular plates, cylindrical construction (spiral wound or disc plates), or bipolar construction.

The overall reaction of lead–acid battery is

$$Pb + PbO_2 + 2H_2SO_4 \rightarrow 2PbSO_4 + 2H_2O \quad (1.21)$$

In its original design, the lead–acid battery works with its plates immersed in a liquid electrolyte. The hydrogen and the oxygen produced during overcharge are released into the atmosphere. The lost gases reflect the loss of water from the electrolyte. Thus, external water must be filled in during maintenance operation. Problems with water replenishing were overcome by invention of valve-regulated lead–acid batteries, which are designed to operate with the help of an internal oxygen cycle. Oxygen is liberated during the later stages of charging (and overcharging) on the positive electrode. The as-formed oxygen then travels through a gas space in a separator to the negative electrode where it is reduced to water. As a result, no external water is needed and part of the electrical energy delivered to the cell is consumed by the internal oxygen recombination cycle, and it is converted into heat.

1.3.3 Nickel–Cadmium Battery

Nickel–cadmium (Ni–Cd) batteries belong to nickel rechargeable batteries, which is a category of five rechargeable batteries, namely Ni–Cd, Ni–MH, Ni–H_2, Ni–Zn, and Ni–Fe batteries, having an Ni-based positive electrode and an alkaline solution in common [10]. Ni–Cd batteries have two major configurations: vented and sealed. In all types, Cd is used as the negative electrode and β-NiOOH as the positive electrode material. The electrolyte is aqueous KOH solution containing some LiOH (8–20 g/L), which improves cycle life and high-temperature performance.

The overall reaction for vented and sealed types of Ni–Cd batteries is

$$Cd + 2NiOOH + 2H_2O \rightarrow 2Ni(OH)_2 + Cd(OH)_2 \quad (1.22)$$

The reversible reduction of NiOOH to $Ni(OH)_2$ involves the intercalation of H^+ into the layered hydroperoxide structure of the electrode, thus giving rise to a solid solution.

The sealed Ni–Cd utilized a construction similar to that of sealed lead–acid batteries, which allows recombination of oxygen generated during overcharge. In this cell construction, O_2 is formed and diffuses through the separator to the negative electrode where it is reduced to OH^-. The $Cd(OH)_2$ formed is then converted into metallic Cd during cell charge.

The construction of the positive electrode is of utmost importance in determining its performance. Three types that differ for the substrate are utilized, which can be based on sintered Ni grid, Ni foam, or Ni-plated fibers. The substrates are then impregnated with the active material in its reduced state.

The negative electrode can be sintered or nonsintered type. For the sintered type, the micropores of the substrate are filled with $Cd(OH)_2$. The nonsintered type is prepared by coating an Ni-plated steel grid with a CdO-based paste. For both types, metallic Cd will be formed during the initial charge process.

The advantages of Ni–Cd batteries are high-rate capability and good shelf life in a wide temperature range (−40°C to 70°C, depending on the specific construction). Some batteries have been stored for up to 10 years and could still deliver, after recharge, almost 100% of their original capacity. On the other hand, these batteries have a high self-discharge (up to 20% loss per month) and rapidly lose their capacity even at ambient temperature.

1.3.4 Nickel–Metal Hydride Battery

Nickel–metal hydride (Ni–MH) battery was first studied in the 1960s as a derivative and replacement of both the Ni–Cd and Ni–H_2 batteries. The Ni–MH cell shows the environmental advantages associated with higher energy, lower pressure, and lower cost compared with NiCd. Since it was commercially introduced in 1989, nickel–metal hydride batteries have become a dominant commercial rechargeable battery system in high-volume production for multiple consumer applications. In recent years, the application of Ni–MH battery also expanded from portable use, such as portable personal computers and other electronic devices, to hybrid vehicle propulsion applications [11].

In the Ni–MH battery, the negative electrode is a hydrogen-storing alloy, while the positive electrode is β-NiOOH. The electrolyte is the same as that used in Ni–Cd cells. The overall process consists in the reversible transfer of a proton from an electrode to the other, as indicated by the overall reaction shown below:

$$MH + NiOOH \rightarrow Ni(OH)_2 + M \quad (1.23)$$

The proton released from the storage alloy during discharge is transferred to NiOOH to form $Ni(OH)_2$. This process is reversed upon charge. Therefore, homogeneous solid-state processes occur on both electrodes. H_2O is consumed on discharge and reformed on charge at variance with the overall reaction of the Ni–Cd system. Unlike Ni–Cd cells, there is no electrode reaction involving H_2O in the Ni–MH system. Therefore, the concentration and conductivity of the electrolyte remains the same during charge and discharge processes. During overcharge, O_2 is evolved at the positive and diffuses to the negative to form H_2O by reaction with MH. During overdischarge, H_2 is evolved at the positive and again gives rise to H_2O at the negative. Therefore, both H_2 and O_2

recombine to form H_2O, thus assuring true sealed operation of the Ni–MH battery. The capacity of the cell is limited by the positive electrode, with a negative to positive ratio of 1.5–2 to 1.

The negative electrode materials are either AB_5- or AB_2-type alloys. An example of the AB_5 type is $LaNi_5$, whereas ZrV_2 exemplifies the latter. The advantages of both types of alloys cover wide operating temperature range, high capacity, long cycle life, and high hydrogen diffusion rate. In the $LaNi_5$ alloy, using a naturally occurring mixture of rare-earth elements, La, Nd, Pr, and Ce, instead of pure La, enhances the resistance to alkali and reduces costs. Also, Ni could be partially substituted by Co, Mn, and Al. Co substitution prevents alloy pulverization. Mn substitution increases the capacity, and Al substitution improves the resistance to oxidation during the manufacture. AB_5-type alloys show capacities of about 300 mAh/g.

AB_2-type alloys (A: Zr, Ti; B: V, Ni plus minor amounts of Cr, Co, Mn, Fe) have higher specific capacities than the AB_5-type ones. However, AB_5 alloys show better performance in wider working temperature range with demanding discharge rates. AB_5 alloys are also less prone to self-discharge and, as they are cheaper and easier to use, are still preferred.

The positive electrode of Ni–MH battery is basically the same as that of Ni–Cd batteries. High-density, spherical $Ni(OH)_2$ is used (cells are assembled in the discharged state), which also contains Co and Zn. Both Co and Zn help limit the formation of γ-NiOOH on overcharge, which will result in the memory effect and cause morphological changes in the electrode.

The Ni–MH battery shows little capacity lost at elevated current density, although at the expense of the mean voltage. However, the cycling life of the cell will be much reduced with repeating discharges at high currents. The suitable discharge rate range of 0.2–0.5 C delivers the best overall electrochemical performance. At the 0.2 C rate, the Ni–MH battery delivers 40% more capacity than Ni–Cd, and a correspondingly higher energy density. In optimized cells, the energy density can approach 100 Wh/kg and 300 Wh/L.

In comparison with Ni–Cd cells, however, the rate capability of Ni–MH cell is generally poorer. The self-discharge is faster especially at high temperatures, and tolerance to overcharge is lesser. In addition, permanent damages to seals and separator will occur with long-term storage at high temperatures. Some tests have shown that continuous exposure to 45°C reduces the cycle life by 60%.

1.3.5 Lithium-Ion Battery

Development of high-energy-density lithium-ion battery started in the 1970s [12,13]. The lithium-ion (Li-ion) battery contains no metallic lithium and is therefore much safer on recharge than the earlier primary lithium–metal design of cell.

The lithium ions travel between anode and cathode during charge and discharge processes. The cathode material is typically a layered structure such as lithium cobalt oxide ($LiCoO_2$), or a tunneled structure such as lithium manganese oxide ($LiMnO_2$) [14]. The first commercialized Li-ion battery used $LiCoO_2$ as the cathode material by Sony. Besides, a number of other metal oxide–based materials are utilized as the cathode material, such as $LiMn_2O_4$, $LiFePO_4$, and $LiNi_{1-x-y}Mn_xCo_yO_2$. The corresponding cathode reaction is shown below with a general form:

$$LiMO_2 \rightarrow Li_xMO_2 + xLi^+ + xe^- \tag{1.24}$$

When the first commercial Li-ion battery was introduced, a coke-based material was used for the negative electrode. In the mid-1990s, graphite-based anode materials were utilized, since the exfoliation issue was solved by replacing PC solvent with ethylene carbonate. The graphitic carbon materials offer higher capacity (350 Ah/kg) over original coke materials (180 Ah/kg). In addition, a much lower surface area and higher density provide even higher energy density and better packing efficiency. When paired with a metal oxide as the positive electrode, the graphitic anode materials provide a relatively high voltage (from 4 V in the fully charged state to 3 V in discharged state).

The electrolyte is composed of an organic-based solvent (such as PC, ethylene carbonate, or diethyl ether) and dissolved salt (such as $LiPF_6$, $LiBF_4$, or $LiClO_4$). The positive and negative active mass is coated to both sides of thin metal foils (Al on the positive and Cu on the negative). The microporous polymer sheet between the positive and negative electrode works as the separator.

The most important advantages of lithium-ion cell are high energy density from 150 to 200 Wh/kg (from 250 to 530 Wh/L), high voltage (3.6 V), good charge–discharge characteristics, and relatively good long cycling life with acceptable low self-discharge (<10% per month). Comparing with Ni-based secondary batteries, Li-ion battery shows no memory effect, as well as better high-rate charging performance.

However, for Li-ion battery, the major disadvantage is the high price. There also must be a controlled charging process, especially close the top of charge voltage 4.2 V. Overcharging or heating above 100°C causes the decomposition of the positive electrode and the electrolyte with liberation of gas.

1.3.6 Lithium–Sulfur and Lithium–Air Batteries

The rechargeable lithium–sulfur (Li–S) has different chemistry mechanism from Li-ion battery [15,16]. The S cathode is reduced upon discharge to form a series of soluble polysulfide species (Li_2S_x $4 \le x \le 8$) that combine with Li to produce solid Li_2S_2 and Li_2S at the end of the discharge, as illustrated in Figure 1.4. On charging, Li_2S_2 and Li_2S are converted back to S via similar soluble polysulfide intermediates presented in the discharge process and lithium plates to the nominal anode, making the cell reversible. This contrasts with conventional Li-ion cells, where the lithium ions are intercalated in the anode and cathode, and consequently the Li–S system, which allows for a much higher lithium storage density.

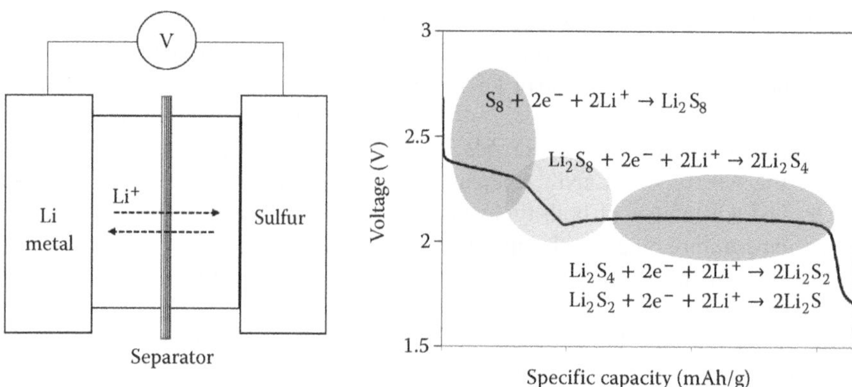

FIGURE 1.4 Scheme of Li–S battery and corresponding electrochemical reactions.

The Li–S battery, when based on the overall reaction $S_8 + 16\,Li = 8Li_2S$, shows an average voltage of 2.15 V with a theoretical specific capacity of 1675 mAh/g. This leads to a theoretical energy density of 2600 Wh/kg (2800 Wh/L), which is five times higher than that of the conventional Li-ion battery based on intercalation mechanism. Another advantage of sulfur is its abundance in nature as well as its low cost as a side product of petroleum and mineral refining. Furthermore, the unique features of Li–S chemistry greatly reduce the formation of lithium dendrites, which enhances safety particularly for high-capacity multicell battery packs.

Sulfur-based electrochemical cells had been reported in 1962; however, many severe issues, such as the electronically insulating nature of sulfur and the solubility of intermediately formed polysulfides in common liquid organic electrolytes, are still not solved satisfactorily until now. In addition, the as-formed polysulfides in the electrolyte migrate to the lithium metal anode and are electrochemically reduced, which is well known as a "shuttle reaction." This shuttle effect results in low coulombic efficiency and rapid capacity fade in Li–S batteries [16].

Recently, interest in Li–S-based rechargeable batteries has been raised due to the emerging demands of the high-energy storage systems. The recent development of new nanostructured material architectures also provides opportunities to overcome issues related to the bulk material's conductivity. Moreover, the development of novel electrolytes, binder materials, and even cell design have led to significant advances in the field of Li–S batteries within the last few years [17–26]. Therefore, the Li–S batteries will remain attractive over the longer term because of their inherently high energy content, high power capability, and potential for low cost, although they are still in the development stage.

Lithium–air systems, which are actually lithium–oxygen systems, are the most promising Li batteries in terms of energy density, since the cathode active mass (which is usually the limiting factor in batteries) is not included in the battery system [16]. While the Li-ion batteries provide an energy density of 150–200 Wh/kg (of the 900 Wh/kg theoretically possible value) at the cell level, Li–air batteries have the potential to achieve 3620 Wh/kg (when discharged to Li_2O_2 at 3.1 V) or 5200 Wh/kg (when discharged to Li_2O at 3.1 V), which is over an order of magnitude higher than that of Li-ion batteries. If the oxygen supplied is not included in the calculation, Li–air cells offer an energy density of ~11,000 Wh/kg. This is basically identical to the value for gasoline (octane) at ~13,000 Wh/kg if the external oxygen supply is also neglected. Therefore, unlike other battery technologies, Li–air is competitive with liquid fuels.

Aqueous and nonaqueous Li–air batteries were first described by Visco and Abraham [15], respectively. Although both cells involve O_2 reduction, the differences between aqueous and nonaqueous Li–air are significant enough to merit description in separate sections. A schematic illustration of both aqueous and nonaqueous Li–air batteries is shown in Figure 1.5.

For nonaqueous Li–air battery, the solid Li_2O_2 accumulates in the pores of the porous substrate cathode on discharge. Porous carbon materials are the predominant cathode materials owing to their excellent conductivity and low cost. The pore size is critical since small pores may easily become blocked by Li_2O_2 before becoming completely filled, thus limiting the discharge capacity, while large pores may compromise rate and rechargeability because of their low surface area. Therefore, it is anticipated that pore sizes in the region of 10–200 nm are likely to be optimal. Good wetting of the pore surfaces by the electrolyte is important, as is a high pore volume.

There is an unusually large separation between charge and discharge voltages, even at low charging/discharging rates. Unfortunately, this voltage separation leads to low energy storage efficiency, although the columbic efficiency is close to 100%. Thus, various catalysts have been utilized in order to reduce the voltage separation as well as sustain cycling. On the other hand, if a catalyst is used, controlling its size, morphology, and distribution within the pores are key challenges for the porous substrate materials.

The rate capability of the cell on discharge depends on the O_2 solubility and its diffusion in the electrolyte to the electrode surface. It has also been shown that as a film of insulating Li_2O_2 grows on the electrode surface, it will slow the rate and soon passivate the electrode. In addition, the electrolyte

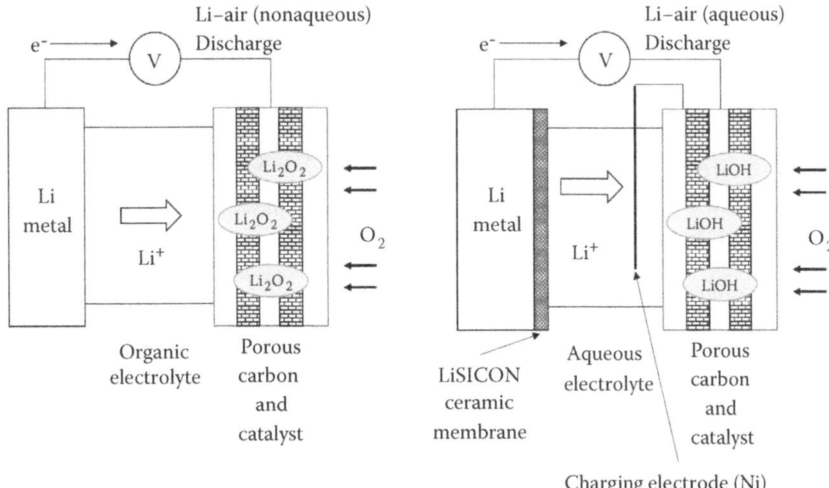

FIGURE 1.5 Scheme of nonaqueous and aqueous Li–air batteries.

must be inert to strong nucleophiles such as O^{2-}. Thus, identifying electrolyte materials that satisfy all the requirements of solubility, wettability, and stability is a formidable challenge.

If the cell is to operate in ambient air, then ingress of CO_2 and H_2O must be avoided by protecting the cathode with a membrane that blocks these species, allowing only O_2 to enter the cell.

For aqueous Li–air battery, the system can only function with a solid electrolyte layer, which should be an ionic conductor, covering the lithium anode, since lithium reacts violently with water. As a matter of fact, this innovation by Visco made the Li water cell a reality. A thin film of a nonaqueous liquid or gel electrolyte is placed between the lithium metal and the ceramic ionic conductor in an effort to address this problem with providing an acceptable cycling stability. On the other hand, the solid electrolyte/aqueous electrolyte interface also suffers from corrosion, leading to increasing cell impedance. Although an attraction of the aqueous system is that, by definition, H_2O does not have to be excluded from the cathode compartment of the cell, it is necessary to exclude CO_2 to avoid formation of Li_2CO_3 instead of LiOH. Therefore, an additional membrane protecting the cathode by preventing CO_2 ingress is also required. Therefore, aqueous Li–air battery faces many challenges similar to the nonaqueous Li–air battery for practical implementation.

1.3.7 Other Secondary Batteries

Besides the secondary batteries introduced above, there are few other specialized aqueous secondary batteries, such as nickel–hydrogen battery, silver–zinc battery, and nickel–zinc battery [1,5,6,10,27].

1.3.7.1 Nickel–Hydrogen Battery

The nickel–hydrogen (Ni–H_2) cell utilizes H_2 as a negative electrode, while the positive electrode is NiOOH that is the same as other Ni-based rechargeable batteries. This system has specifically been developed for aerospace applications and has been in continuous development since the early 1970s.

During normal operation, the overall reaction is

$$1/2H_2 + NiOOH \rightarrow Ni(OH)_2 \quad (1.25)$$

One of the major advantages of this system is tolerance to overcharge and reversal. On overcharge (and in the final charge stages), O_2 is generated at the Ni electrode and recombines at the H_2 electrode to form H_2O. During reversal, H_2 is released at the positive electrode (in a positive-limited cell) and consumed at the negative electrode at the same rate. Thus, there is no pressure buildup or change in electrolyte concentration.

The H_2 electrode consists of a thin film of Pt black catalyst supported on an Ni foil substrate backed by a gas diffusion membrane. The Ni electrode, which is similar to those used in Ni–Cd and Ni–MH batteries, consists of a porous sintered Ni powder substrate, supported by an Ni screen, electrochemically impregnated with $Ni(OH)_2$. The separator is a thin and porous ZrO_2 ceramic cloth supporting a concentrated KOH solution.

1.3.7.2 Silver–Zinc Battery

The zinc–silver oxide (Zn–AgO) battery has one of the highest energy of aqueous cells. The theoretical energy density is 300 Wh/kg (1400 Wh/L), and practical values are in the range 40–130 Wh/kg (110–320 Wh/L) depending on the specific cell design. The Zn–AgO battery shows a very low internal impedance. Its high energy density also makes them very useful for aerospace and even military purposes, although it has poor cycle life. This battery can be discharged at high currents such as 6 C rate even at low temperatures. However, it suffers a self-discharge of 20% per month at 25°C.

The overall cell reaction is shown below:

$$AgO + Zn + H_2O \rightarrow Zn(OH)_2 + Ag \quad (1.26)$$

The positive electrode is prepared by placing the silver positive active materials, which are sintered at temperatures between 400°C and 700°C, on silver or silver-plated copper grids or perforated sheets.

The Zn negative electrode is composed of a mixture of Zn, ZnO, and binders. Additives are also involved in order to minimize dendritic growth, increase the hydrogen overvoltage of the Zn electrode (reduce gassing during charge), and thus reduce corrosion.

The separator has multifunctions in the Zn–AgO battery. For example, it should prevent silver migration to the negative electrode and to control zincate migration to preserve the integrity of the Zn electrode. Typical separators used in Zn–AgO battery are of cellophane (regenerated cellulose) and synthetic fiber mats of nylon, polypropylene, and nonwoven rayon fiber mats. Synthetic fiber mats are placed next to the positive electrode to protect the cellophane from oxidation. In most cell designs, the zinc electrodes are enclosed in the separator envelope.

1.3.7.3 Nickel–Zinc Battery

The nickel–zinc (Ni–Zn) battery has NiOOH as the positive and Zn as the negative electrode. The overall reaction is shown below:

$$2NiOOH + 2H_2O + Zn \rightarrow Zn(OH)_2 + 2Ni(OH)_2 \quad (1.27)$$

The Ni–Zn battery has higher energy density than that of the Ni–Cd system, since Zn has a higher specific energy than the Cd electrode. Moreover, it also has other advantages such relatively low cost, good rate capability, and good cycle life. However, the Zn anode has few issues such as solubility in the concentrated KOH solution and dendrite growth on charge and migration, thus limiting the applications of this system. To reduce the Zn and ZnO solubility, less concentrated KOH solution is utilized. In addition, electrode additives are also helpful. $Ca(OH)_2$ is especially useful when coupled to a ZnO electrode (cells are normally assembled in the discharged state). In addition, the use of a microporous polypropylene separator, which is chemically stable in concentrated alkaline solution, proves an excellent barrier to zinc migration toward the Ni electrode.

1.3.8 Stationary Systems—Flow Batteries

The typical flow batteries store and release electrical energy based on reversible electrochemical reactions in two liquid electrolytes [1,5,6,28]. The cell has two flow loops physically separated by an ion or proton exchange membrane (PEM). Electrolytes flow through separate loops and undergo chemical reaction inside the cell with ion or proton exchange through the membrane. The electron exchange occurs through the external electric circuit. A schematic illustration of the redox flow battery system is shown in Figure 1.6. The flow battery has some advantages compared with conventional secondary batteries. For example, the capacity of the system can be greatly increased by increasing the amount of solution in electrolyte tanks. In addition, the battery can be fully discharged with little loss of electrolyte during cycling. Furthermore, the flow battery has a low self-discharge, as the electrolytes are stored separately. However, the disadvantage of the flow battery is its low energy density due to the liquid-phase active materials and the cell construction (i.e., the pumps and electrolyte reservoir), which increase the effective mass and volume of the battery.

1.3.8.1 Vanadium Redox Battery

One example of the flow battery is vanadium redox battery, in which electrolytes in two loops are separated by a PEM. The electrolyte is prepared by dissolving vanadium pentoxide (V_2O_5) in sulfuric acid (H_2SO_4). The electrolyte in the positive electrolyte loop contains $(VO_2)^+$ and $(VO)_2^+$, while the electrolyte in the negative electrolyte loop contains V^{3+} and V^{2+}. The redox reactions proceed on the carbon electrodes.

In the vanadium redox cell, the following half-cell reactions are involved during discharge:

$$\text{Negative electrode reaction:} \quad V^{2+} \rightarrow V^{3+} + e^- \quad (1.28)$$

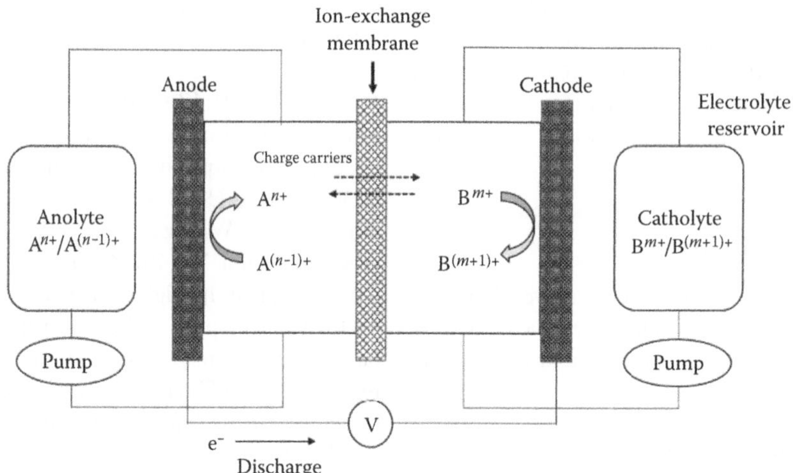

FIGURE 1.6 Scheme of a typical redox flow system.

Positive electrode reaction: $VO_2^+ + 2H^+ + e^- \rightarrow VO^{2+} + H_2O$ (1.29)

An open circuit voltage of 1.4 V is observed at 50% charge, while a fully charged cell shows >1.6 V open circuit voltage. When the cell is fully discharged, it shows an open circuit voltage of 1.0 V.

The vanadium redox flow battery can provide an extremely large capacity, which allows its application on current regulation of highly unstable power sources such as wind or solar power. Moreover, the extremely rapid response time makes it suitable for uninterruptible power supply–type applications, where they can be used to replace lead–acid batteries. On the other hand, the vanadium redox flow battery has few disadvantages, such as a low energy density of about 25 Wh/kg of electrolyte, low charge efficiency (necessity using of pumps), and a high price.

1.3.8.2 Zinc–Bromine Battery

The zinc–bromine (Zn–Br) battery is an example of the so-called semiflow battery or hybrid flow battery. The difference between this system and other typical flow batteries is that the zinc reaction does not only involve dissolved species in the aqueous phase.

The Zn–Br cell is made from the bipolar electrodes. The bipolar electrode is from a lightweight, carbon–plastic composite material. A microporous plastic separator between electrodes allows the ions to pass through it. Cells are series connected, and the battery has a positive and a negative electrode loop. The electrolyte in each storage tank is circulated through the appropriate loop. At the positive electrode, bromide ions are transformed to bromine and back to bromide ions. It is important to note that the bromide ions can combine with bromine molecules to generate the tribromide ion. A scheme illustration of a Zn–Br battery is shown in Figure 1.7. The overall cell reaction is shown below:

$$Zn + Br_2 \rightarrow ZnBr_2 \quad (1.30)$$

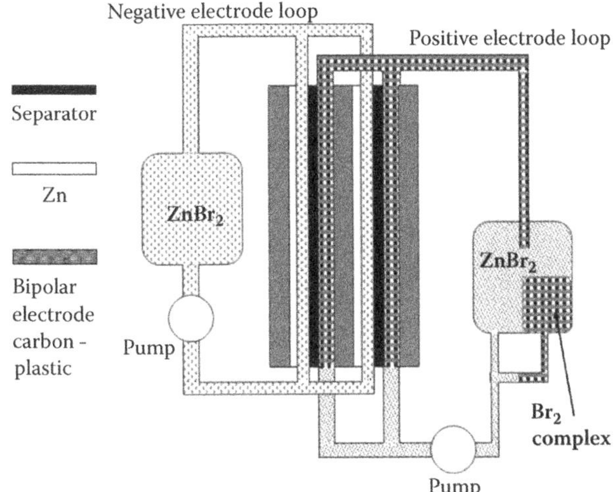

FIGURE 1.7 Scheme of zinc–bromine battery.

During the discharging process, $ZnBr_2$ is formed and becomes dissolved in the electrolyte solution. During the charging process, Br_2 is liberated on the positive electrode while Zn is deposited on the negative electrode. Br_2 is then cooperated with an organic agent, namely *N*-methyl-*N*-ethylmorpholinium bromide, to form a dense, oily liquid polybromide complex. The polybromide droplets are separated from the aqueous electrolyte on the bottom of the tank in the positive electrode loop. During discharge, bromine in the positive electrode loop is again returned to the cell electrolyte in the form of a dispersion of the polybromide oil. Moreover, diffusion of liquid-phase Br_2 toward the Zn deposit is impeded by the microporous separator.

1.3.9 Stationary Systems—Thermal Batteries

1.3.9.1 Sodium–Sulfide Batteries

Sodium (Na) has been considered as a candidate anode material owing to its high negative potential and high specific capacity. These properties allow making a battery with a high specific energy (100–200 Wh/kg). Sodium salts are cheap, nontoxic, and abundant in nature.

Sodium–sulfur (Na–S) battery uses sodium as a negative electrode, sulfur as a positive electrode, and β''-Al_2O_3 as an Na^+ conducting solid electrolyte. The operating temperature of this system is between 300°C and 350°C. In this temperature range, both Na and S are liquid, while the solid electrolyte has a high Na^+ conductivity, thus ensuring good kinetics. The energy density of the Na–S system is very high and by far exceeds that of aqueous systems (Pb–acid and Ni–Cd) used for stationary energy storage.

During discharge, Na^+ migrates from Na to S and forms polysulfides. The first stage of discharge forms Na_2S_5 at 2.07 V and the second stage forms Na_2S_3 at 1.78 V. At the C/3 rate, the average voltage is 1.9 V. The cell discharge reactions are shown below:

$$5S + 2Na \rightarrow Na_2S_5 \quad (1.31)$$

$$3Na_2S_5 + 4Na \rightarrow 5Na_2S_3 \quad (1.32)$$

During charge, the reactions reverse from Na_2S_3 to elemental S and Na forms. In the final stages, there is a marked resistance increase due to the insulating character of sulfur. Therefore, the charge process has to be stopped before complete Na plating. Thus, the subsequent discharges provide 85–90% of the theoretical capacity. The Na–S system has an excellent cycling stability (up to 5000–6000 cycles). This is mainly because the typical aging mechanism based on morphological changes of the electrodes does not operate here, owing to the liquid state of both reactants and products. An essential prerequisite for the Na electrode is high purity since contaminants tend to concentrate at the interface with the electrolyte, thus reducing the electrode-active area or even causing its failure.

The S electrode is prepared by impregnating elemental sulfur into a layer of carbon or graphite. The carbon framework maintains a good electronic conductivity, since sulfur is an insulator for both electrons and ions. Fortunately, Na polysulfides are good ionic conductors.

The β''-Al_2O_3 electrolyte is electron insulator, while it is impermeable to molten Na and S. The idealized composition of β''-Al_2O_3 is $Na_2O_{5.33}Al_2O_3$. Pure β''-Al_2O_3 phases are difficult to prepare, and it has to be stabilized with Mg or Li ions that substitute for Al ions. The ionic conductivity of this electrolyte is ~0.5 Ω^{-1} cm^{-1} at 350°C for the polycrystalline form. However, β''-Al_2O_3 is rather sensitive to moisture, which leads to deterioration of its mechanical properties. Therefore, some β''-Al_2O_3 (idealized formula, $Na_2O_{11}Al_2O_3$) is utilized as additive in the mixture, in spite of its lower conductivity, as it is less hygroscopic. A practical electrolyte shows a conductivity of 0.2 Ω^{-1} cm^{-1}.

Production of Na–S batteries for stationary applications is particularly active in Japan. Very large energy storage systems, which can offer a capacity of as high as 57 MWh, can be built with Na–S modules, providing a cycle life as long as 15 years.

1.4 SUPERCAPACITORS

1.4.1 Introduction

In recent years, electrochemical supercapacitors (ES), or ultracapacitors, have attracted significant attention, mainly owing to their high power density and long cycle life [1,29]. Moreover, the supercapacitors show a unique bridging function for the power/energy gap between traditional dielectric capacitors (which have high power output) and batteries/fuel cells (which have high energy storage). The earliest supercapacitor patent was filed in 1957. However, it was not until the 1990s that supercapacitor technology began to draw some attention, in the field of hybrid electric vehicles. The function of a supercapacitor is to boost the battery or fuel cell in a hybrid electric vehicle to provide the necessary power for acceleration. Further developments have led to the recognition that ES can play an important role in complementing batteries or fuel cells in their energy storage functions by providing backup power supplies to protect against power disruptions. As a result, the US Department of Energy has designated ES to be as important as batteries for future energy storage systems.

1.4.2 Energy Storage in Supercapacitors

Supercapacitors store energy by charge separation. The simplest capacitors store the energy in a thin layer of dielectric material that is supported by a metal current collector. The energy stored in a capacitor can be calculated by 1/2 CV^2, where C is its capacitance (farad) and V is the voltage between the terminal plates. The maximum voltage of the capacitor is dependent on the dielectric material, as well as the electrolyte. The total charge Q (coulomb) stored in the capacitor is calculated by CV. The capacitance of the dielectric capacitor depends on the dielectric constant (K) and the thickness (th) of the dielectric material and its geometric area (A)

$$C = K\frac{A}{\text{th}} \tag{1.33}$$

The construction of a supercapacitor device is much like a battery, which has two electrodes immersed in an electrolyte with a separator in between. The electrodes are fabricated from a high-surface-area microporous material. The surface area of the electrode materials used in an electrochemical capacitors is much greater than that used in battery electrodes, being 500–3000 m^2/g. Charge is stored in the micropores and given by the same expressions as cited previously for the simple dielectric capacitor. However, calculation of the capacitance of the electrochemical capacitor is much more difficult as it depends on complex phenomena occurring in the micropores of the electrodes.

The ES has two basic types, namely electrostatic or electrical double-layer supercapacitors (EDLS) and faradaic supercapacitors (FS).

1.4.2.1 Electrostatic Double-Layer Supercapacitors

The energy storage of EDLS is accomplished by an electrode-potential-dependent accumulation of electrostatic charge at the interface. The surface electrode charge generation is due to the surface dissociation as well as ion adsorption from both the electrolyte and crystal lattice defects. The electrical double-layer capacitance comes from the accumulation of charges at the interface between the electrode material and the electrolyte, while the counterbalancing charges are built up on the electrolyte side in order to meet electroneutrality. During the charging process, cations move toward the negative electrode while anions move toward the positive electrode in the electrolyte, while the electrons travel from the negative electrode to the positive electrode through the external circuit. The reverse process happens during the discharging process. In the EDLC type of supercapacitor, no charge transfers across the electrode/electrolyte interface, and no net ion exchanges occur between the electrode and the electrolyte.

1.4.2.2 Faradaic Supercapacitors

Faradaic supercapacitors (FS) or pseudocapacitors are different from electrostatic or EDLS. Fast and reversible redox reactions occur on the active materials on the electrode when a potential is applied to the device. A passage of charge across the double layer, which is similar to the charging and discharging processes that occur in batteries, is also involved, resulting in faradaic current passing through the supercapacitor. There are three types of faradaic processes: reversible adsorption, redox reactions, and reversible electrochemical doping–dedoping process.

1.4.3 Chemistry and Materials for Supercapacitors

The electrode materials of supercapacitors can be categorized into three types: carbon materials with high specific surface area, conducting polymers, and metal oxides [29–32].

Carbon materials show many advantages such as low cost, easy processing, nontoxicity, higher specific surface area, good electronic conductivity, high chemical stability, and wide operating temperature range. Therefore, carbon materials are considered promising electrode materials for industrial production. Most carbon-based electrochemical capacitors are of the EDLS type, since carbon materials normally store charges in an electrochemical double layer formed at the interface between the electrode and the electrolyte, rather than storing them in the bulk of the capacitive material. The rectangular shape of the cyclic voltammetry curves and triangular symmetrical charge–discharge voltage profile suggest their excellent capacitive properties for supercapacitor use. The important factors influencing their electrochemical performance are specific surface area, pore-size distribution, pore shape and structure, electrical conductivity, and surface functionality. Among these, specific surface area and pore-size distribution are the two most important factors affecting the performance of carbon materials. The measured specific capacitances of carbon materials in real supercapacitors are with values in the range of 75–175 F/g for aqueous electrolytes and 40–100 F/g for organic electrolytes [31,33–35].

Conducting polymers (CPs) have many unique advantages such as high conductivity in a doped state, high voltage window, high storage capacity/porosity/reversibility, and adjustable redox activity through chemical modification. Unlike carbon materials, conducting polymers are faradaic materials. When oxidation takes place, ions are transferred to the polymer backbone, and when reduction occurs, the ions are released from this backbone into the electrolyte. The redox reactions take place not only on the surface but also throughout the entire structure without any structure and phase changing, thus allowing the redox process to be fast and highly reversible. CPs can also be positively or negatively charged to balance the injected charge, thus inducing electronic conductivity by generating delocalized "n" electrons on the polymer chains. An early description of such a process is so-called doping. The positively charged polymers, introduced by oxidation on the repeating units of polymer chains, are defined as "p-doped," while negatively charged polymers are termed as "n-doped." The potentials of these doping processes are determined by the electronic state of p electrons.

Generally, metal oxides can provide higher energy density for supercapacitors than conventional carbon materials and better electrochemical stability than polymer materials. The metal oxides can both store charges through the double layer like the electrostatic carbon materials, and exhibit electrochemical faradaic reactions on electrode materials within appropriate potential windows. Some common metal oxides utilized in the supercapacitors are RuO_2, IrO_2, MnO_2, NiO, Co_2O_3, SnO_2, V_2O_5, and MoO. Besides these pure phases, mixed-oxide composites involving the oxides of Sn, Mn, Ni, V, Ti, Mo, W, and Ca are also introduced into studies in order to replace the pure RuO_2 phase to reduce the cost [30,36].

1.5 WATER ELECTROLYSIS

1.5.1 Introduction

Coal, considered as one of the earliest fuel sources, has been used for hundreds or even thousands of years. Before the liquid fuel–powered engines of today's cars, coal-fueled steam engines in trains were dominantly in use. In the mid-1800s, it was found that kerosene can be distilled from petroleum and used as a cheap alternative to burning whale oil in oil lamps. More products were invented and developed since then. The development of combustion engines for automobiles was also on a fast track. Today, more than two-thirds of US petroleum consumption is used for transportation. Natural gas is another important type of fossil fuel. Nowadays, natural gas is one of major home energy sources and used in power plants and also in industries. The investigation and development of natural gas–powered vehicles are also under way. Up to around 2020, fossil fuel will continue to be the predominant energy source. However, other major nonfossil sources, including nuclear fission, biomass, and hydropower, will probably see their shares in overall supply decline. This transition of the world's energy sector could well accelerate during the period 2020–2050 [37].

Hydrogen, a clean energy source with the highest specific energy density, could be the best alternative to fossil fuels, and has in recent years become more popular [38–40]. Currently, the hydrogen productions are mainly the reformation of natural gas, gasification of coal and petroleum coke, and gasification and reformation of heavy oil. However, the possibility of combining renewable energy to produce hydrogen, such as wind or sun, is attractive. The use of hydrogen production by renewable energy and fuel cells together could be one solution to the sustainable energy supply and to achieve the transition from the current "hydrocarbon economy" to a "hydrogen economy" [41–43]. Water electrolysis could be one simple process using renewable electricity to produce hydrogen. The application of water electrolysis to produce hydrogen and oxygen are still limited to the small scale or in unique situations currently, even though, this technology, considered as one clean method to produce extremely pure hydrogen, has been known for about 200 years. Only 4% of the world hydrogen productions are from water electrolysis.

The idea of the combination of water electrolysis technology using sustainable energy and fuel cells to approach "hydrogen economy" is illustrated in Figure 1.8 [40]. The electricity produced by renewable energy is mainly for end use, while the excess electricity can be used in water electrolysis to produce hydrogen and oxygen. Part of the hydrogen is transported to serve for industry, traffic, and household. Moreover, the other part is used to power up fuel cells to generate electricity. The remaining problems to achieve the attractive idea include the storage of hydrogen, the fuel cells' technical issues, and low gas evolution rate and the high energy consumption of water electrolysis. It is necessary to improve water electrolysis technology in order to satisfy the requirement of sustainable hydrogen production [37].

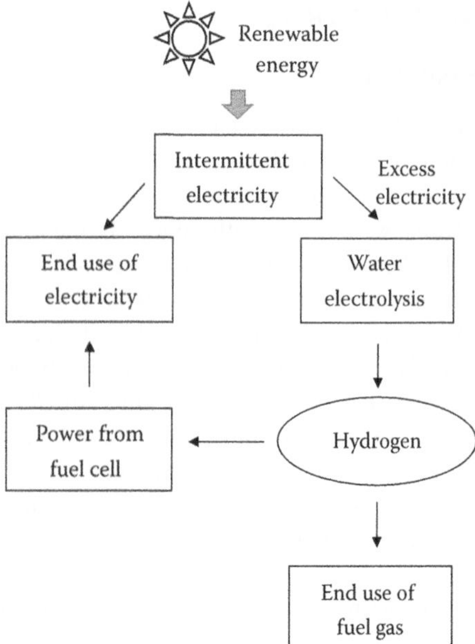

FIGURE 1.8 Schematic illustration of a conceptual distributed energy system with water electrolysis playing an important role in hydrogen production as a fuel gas and energy storage mechanism.

1.5.2 Fundamentals of Water Electrolysis

In 1789, Jan Rudolph Deiman and Adriaan Paets Van Troostwijk first produced electricity from water using an electrostatic machine. The electricity was discharged on gold electrodes in a Leyden jar with water. In 1800, the voltaic pile was designed by Alessandro. William Nicholson and Anthony Carlisle later used it and discovered the water electrolysis. In 1869, Zénobe Gramme invented the Gramme machine and made water electrolysis a simple and cheap process for the production of hydrogen. Dmitry Lachinov developed a method of industrial synthesis of hydrogen and oxygen through electrolysis in 1888. By 1902, there have been more than 400 industrial water electrolysis units in operation. The first large water electrolysis plant with a capacity of 10,000 Nm^3 H_2/h went into operation in 1939. Zdansky and Lonza developed the first pressurized industrial electrolyzer in 1948, and in 1966, General Electric built the first solid polymer electrolyte system. In 1972, the first solid oxide water electrolysis system was started. In 1978, the first advanced alkaline system was developed. Later, a considerable amount of attention was given on developments in the field of high-temperature solid oxide technology and the optimization of alkaline water electrolyzers [42–47].

Figure 1.9 illustrates a schematic of an electrochemical cell for electrolysis. The core of an electrolysis unit consists of two electrodes (anode and cathode), an external power supply, and an electrolyte. At a certain voltage, which is called critical voltage, between both electrodes, electrons flow from the negative terminal of the power supply to the cathode where the protons are reduced by the electrons to generate hydrogen gas. In the meantime, hydroxide ions transfer through the electrolyte solution to anode in order to keep the electrical charge in

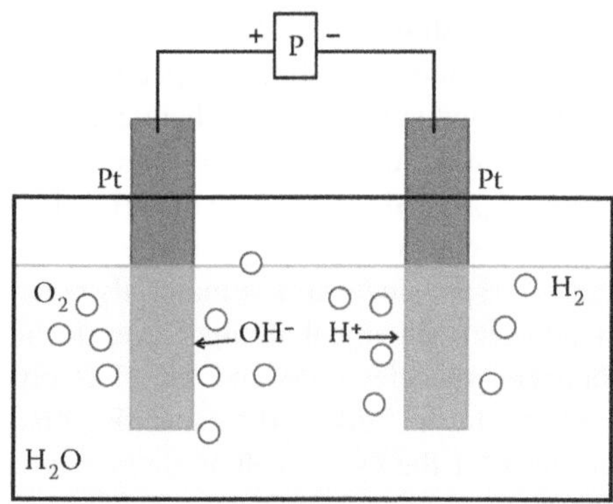

FIGURE 1.9 Schematic illustration of the structure of an electrochemical cell.

balance. At the anode, the hydroxide ions are oxidized to form oxygen gas and give away electrons, which then flow back to the positive terminal of the power source. The amount of gases produced per unit time is directly related to the current that passes through the electrochemical cell [3,4].

The half-reactions occurring on the cathode and anode, respectively, are represented here as

$$\text{Cathode:}\quad 2H^+ + 2e^- \rightarrow H_2 \tag{1.34}$$

$$\text{Anode:}\quad 2OH^+ \rightarrow \frac{1}{2}O_2 + H_2O + 2e^- \tag{1.35}$$

$$\text{Overall:}\quad H_2O \rightarrow H_2 + \frac{1}{2}O_2 \tag{1.36}$$

To split water to form hydrogen and oxygen molecules, a minimum necessary cell voltage for the start-up of electrolysis, $E°$, has to be achieved. The $E°$, also called "electromotive force" under standard conditions (pressure, temperature), is defined by the following equation:

$$E^0 = -\frac{\Delta G}{nF} \tag{1.37}$$

where ΔG is the change in the Gibbs free energy under standard conditions and n is the number of electrons transferred. F is the Faraday constant.

When there is no cell current between the two different electrode reactions and the system is stable and reversible, the equilibrium cell voltage is the equilibrium potential difference between the respective anode and cathode, as described by the following equation:

$$E^0 = E^0_{\text{anode}} - E^0_{\text{cathode}} \tag{1.38}$$

In the overall water electrolysis cell reaction, $E°$ (25°C) is 1.23 V and the Gibbs free energy change of the reaction is +237.2 kJ/mol for an open cell system, which is the minimum amount of electrical energy required to produce hydrogen.

The corresponding cell voltage is known as reversible potential. This thermodynamically unfavorable water split process at room temperature requires an overpotential, η, above the equilibrium cell voltage to get the reaction started. Besides, a number of barriers have to be overcome as shown in Equation 1.40, including external electrical circuit resistances (R_1 and R_1'), resistances from electrolyte and membrane, resistances from the partial coverage of the electrode by the oxygen bubble and hydrogen bubble, and activation energies of the electrochemical reactions occurring on the surfaces of the electrodes. Therefore, the cell potential E_{cell} can be calculated using Equations 1.39 through 1.41. In industry water electrolysis, the cell potential is always 1.8–2.0 V at the current density of 1000–300 A/m².

$$E_{cell} = E_{anode} - E_{cathode} + \Sigma\eta + iR_{cell} \quad (1.39)$$

$$R_{cell} = R_1 + R_{anode} + R_{bubble,\,O_2} + R_{electrolyte} + R_{membrane} + R_{bubble,\,H_2} + R_{cathode} + R_1' \quad (1.40)$$

$$\Sigma\eta = |\eta_{anode}(j)| + |\eta_{cathode}(i)| \quad (1.41)$$

where j is the current density at which the system operates, and Ση is the sum of the overpotentials (activation overpotential at the two electrodes, and the concentration overpotential due to the electrolyte concentration difference and the mass transport of the gaseous products away from the anode and cathode surfaces). If the system has a mild condition under which the mass transport and concentration differences can be neglected, the Ση can be calculated using Equation 1.41.

In general, the voltage efficiency of an electrochemical cell can be calculated using Equation 1.42.

$$\%\ \text{Voltage efficiency} = \frac{(E_{anode} - E_{cathode})100}{E_{cell}} \quad (1.42)$$

The Faradic efficiency and the thermal efficiency are the two other efficiency calculation methods based on the energy changes of the water electrolysis reaction, as shown in Equations 1.43 and 1.44. The Gibbs free energy change as the theoretical energy requirement and enthalpy change of water decomposition reaction are used in the calculations, respectively. The sum of the Gibbs free energy change and the energy losses are considered as the energy input in both equations.

$$\eta_{Faradic} = \frac{\Delta G}{\Delta G + \text{losses}} = \frac{E_{\Delta G}}{E_{cell}} \quad (1.43)$$

$$\eta_{Thermal} = \frac{\Delta H}{\Delta G + \text{losses}} = \frac{E_{\Delta H}}{E_{cell}} \quad (1.44)$$

To simplify the calculation, we can use the cell potential and total cell voltage in the following equations:

$$\eta_{Faradic}(25\ ^\circ\text{C}) = \frac{1.23(V)}{E_{cell}} \quad (1.45)$$

$$\eta_{Thermal}(25^\circ\text{C}) = \frac{1.48(V)}{E_{cell}} \quad (1.46)$$

In reality, if the potential drop caused by electrical resistance is 0.25 and 0.6 V for the cathode and anode overpotentials at 25°C, respectively, the faradaic efficiency is (1.23 × 100%)/(1.23 + 0.25 + 0.6) = 59%, and the thermal efficiency is (1.48 × 100%)/(1.23 + 0.25 + 0.6) = 71%.

The third way to evaluate the efficiency of a water electrolysis system is to calculate the percentage of the output of hydrogen in the total electrical energy used in the cell, as shown in Equations 1.47 and 1.48. Equation 1.47 represents the hydrogen production rate per unit electrical energy input, where the U is the cell voltage, i is the current, and t is time. $r_{H_2\ production\ rate}$ is the hydrogen production rate at unit volume electrolysis cell. On the other hand, Equation 1.48 based on the heating value of hydrogen is to calculate the energy efficiency, where 283.8 kJ is the high heating value of 1 mol hydrogen; t is the time needed for 1 mol hydrogen produced; and U and I are the cell voltage and current, respectively.

$$\eta_{H_2 production\ rate} = \frac{r_{H_2 production\ rate}}{E} = \frac{V(\text{m}^3\text{m}^{-3}\text{h}^{-1})}{Uit(\text{kJ})} \quad (1.47)$$

$$\eta_{E_2 yield} = \frac{E_{Useable}}{E} = \frac{283.8(\text{kJ})}{Uit} \quad (1.48)$$

As discussed above, many system parameters such as electrode materials, electrolyte, and system operation conditions can affect the performance and efficiency of an electrochemical cell [41,42,44,45,47]. Two broad ways to improve the energy efficiency of a cell are summarized. One is to reduce the energy thermodynamically for water split, such as by increasing the operation temperature or pressure. The other is to reduce the energy losses by reducing the resistances from each dominant component of the system.

1.5.3 Alkaline Water Electrolyzers

The resistance of the electrolyte is one important factor when considering the overpotential and the energy efficiency of a water electrolysis cell. Pure water is a bad current conductor with 0.055 μS/cm; therefore, addition of an electrolyte is necessary. Strong acids, strong base, or some solid polymer can all be used. Although water electrolysis was first discovered with strong acid as the electrolyte, in industrial plants the alkaline medium is much easier to handle, and cheaper construction materials can be used compared with acidic electrolysis technology [39,46].

Figure 1.10 illustrates the principle of alkaline water electrolysis, which is similar to that of the water-splitting

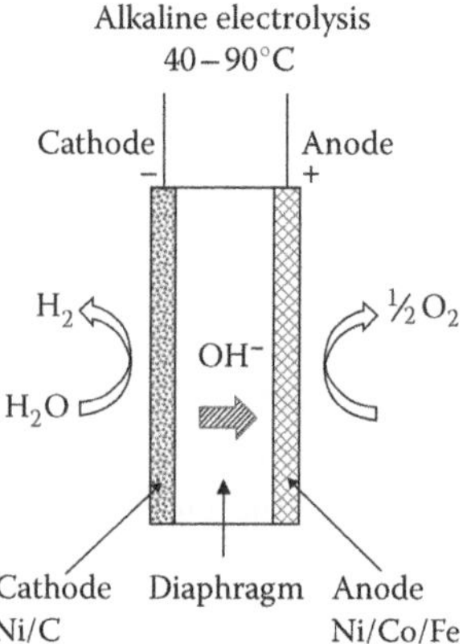

FIGURE 1.10 Schematic of the operating principle of an alkaline water electrolysis cell.

process mentioned above [48]. There are two alkaline electrolysis cell configurations: monopolar and bipolar, as shown in Figure 1.11. In the monopolar configuration, a number of individual cells in parallel are connected to one direct current power supply. The total voltage applied to the whole electrolysis system is the same as that applied to the individual pairs of the electrodes in the cell. In the bipolar configuration (Figure 1.11b), a number of individual cells are connected in series with one another via the electrolyte solution as the conducting media. The total cell voltage is the sum of the individual unit cell voltages. It is simple to fabricate and maintain using the monopolar arrangement; however, this cell configuration has large ohmic losses due to the high electrical currents at low voltages. The advantage of using the bipolar configuration is having less ohmic losses on the electrical circuit connecters; however, the design and manufacturing have to be greater to prevent electrolyte and gas leakage between cells, compared with the monopolar configuration [40].

There have been newly developed cells and plants allowing operation temperatures up to 150°C and variable pressures ranging from 5 to 30 bar, in order to reduce the required energy for water to split thermodynamically. The new cells can use current densities up to 10 kA/m^2, which is more than three times higher than the usual values of conventional low-pressure alkaline electrolysis plants. The cell efficiencies can be improved to more than 80% under those conditions. The major concern of the pressurized electrolyzers is the safety issue due to the relatively high operation pressure.

More investigations have been focusing on the development of materials for membranes and electrodes [36,45,49,50]. In the early stage of water electrolyzers, asbestos was used as the membrane to separate the O_2 and H_2; however, it is not resistant to the strong alkaline environment at high temperatures. Polymers such as perfluorosulfonic acid, arylene ether, and polytetrafluoroethylene have become common membrane materials since the 1970s. The considerations for choosing the electrode materials include the corrosion resistance, conductivity, catalytic effect, and price. Stainless steel is a cheap electrode and has low overpotential; however, it could not resist strong alkaline solutions. Lead and noble metals as electrode materials are either costly or easy to be corroded in the alkaline environment. Nickel currently is a popular electrode material owing to its better corrosion resistance to the alkaline environment compared with other transition metals. Nickel also exhibits a reasonably high hydrogen generation activity. Many researchers have been working on the physical properties and the effects of the nickel-based alloys as electrode materials [51,52].

Since the first large water electrolysis plant rated at 100 MW was developed in Canada in the 1920s, electrolyzer manufacturers worldwide have built many energy systems for different purposes. Well-known water electrolyzer developers include De Nora (Italy), Norsk Hydro (Norway), Electrolyser Corp. Ltd. (Canada), Teledyne Energy systems (USA), General Electric (USA), Stuart Cell (Canada), Hamilton Sundstrand (USA), Proton Energy Systems (USA), Shinko Pantec (Japan), and Wellman-CJB (UK). Generally, alkaline electrolysis has the voltage efficiency of 62–82%. Low partial load range, limited current density, and low operating pressure are still common issues associated with alkaline electrolyzers currently. First, the diaphragm does not completely separate the product gases and reduces the efficiency of the electrolyzer. The second drawback is the high ohmic losses across the liquid electrolyte and diaphragm. The maximum achievable current

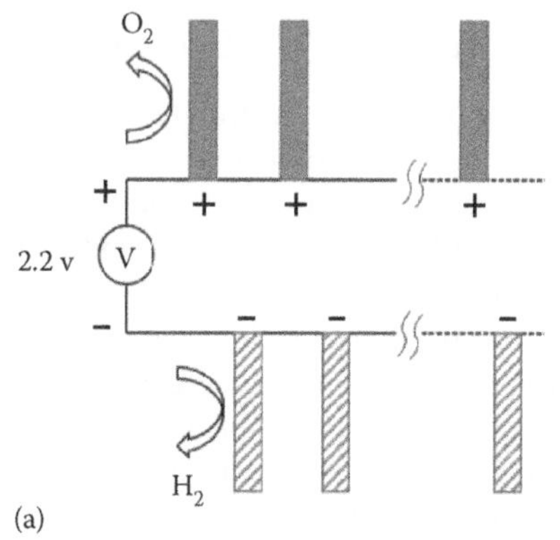

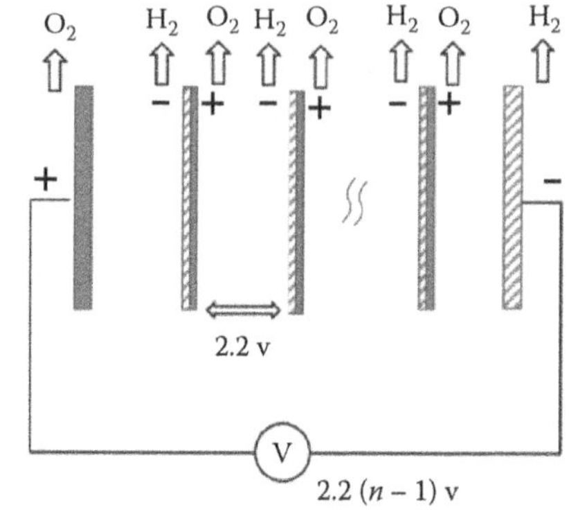

FIGURE 1.11 Schematics of the cell configuration of (a) monopolar and (b) bipolar electrolyzers.

density is usually low. Last, the operation at high pressure is difficult using the liquid electrolyte [41,44,45,53].

1.5.4 PEM Electrolyzers

PEM water electrolysis is based on the use of a polymeric PEM as the solid electrolyte [39,46,54,55]. The PEM was first proposed by General Electric for fuel cell and, later, electrolyzer applications in the 1960s to overcome the issues of alkaline electrolyzers. Grubb [54] idealized the concept using a solid sulfonated polystyrene membrane as an electrolyte. The initial performances yielded 1.88 V at 1.0 A/cm^2, which was rather efficient, compared with the alkaline water electrolysis technology at that time. The typical structure of a PEM water electrolyzer is shown in Figure 1.12. The central component is the membrane electrode assembly with a polymeric PEM. On each side of the membrane, a layer of catalyst and electrode material is coated. Ultrapure water is fed to the anode of the electrolysis cell, at which oxygen gas is produced. The membrane conducts hydrated protons from the anode to the cathode side, where hydrogen gas is generated. Appropriate swelling procedures have led to low ohmic resistances enabling the high current density of the cells. The polymer electrolyte technology has the following advantages over the alkaline water electrolysis: (i) it is a simple and compact design with greater safety and reliability, and could work at a high pressure; (ii) the low gas crossover rate of solid electrolyte membrane allows for the electrolyzer to work under a wide range of power input; and (iii) the possibility of operating cells at high current densities (>2 A/cm^2) with typical thickness of a few millimeters is afforded, therefore reducing the operational costs. Currently, PEM electrolyzers have efficiencies of up to ~85%.

There are several problems associated with PEM electrolysis. First, the costs of PEM and electrocatalysts are high. The most common membrane in PEM electrolysis is a thin (~100 mm) perfluorosulfonate (PFS) polymer membrane [52,54,55]. The standard commercial PFS membrane material is Nafion TM 117 manufactured by DuPont, with good chemical and thermal stability, mechanical strength, and high proton conductivity. Both the costs for production and for disposal of this PFS are high. There have been many studies about alternative hydrocarbon membrane materials to cut down the cost. However, these alternative membranes have either low current densities or low durability compared with Nafion TM 117 membranes. Many researchers also have been working on the development of composite or reinforced membranes that would allow a harsher operation condition for the electrolyzer.

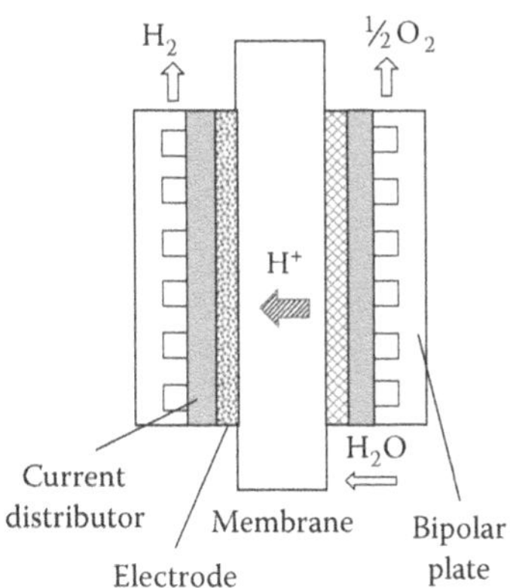

FIGURE 1.12 Structure of a PEM electrolyzer.

The high costs of the electrocatalysts used in PEM electrolyzers have been noticed since the beginning of this research area. To date, the catalysts for hydrogen evolution reaction are usually supported or unsupported platinum, while on the oxygen side, iridium, ruthenium, their oxides, or mixtures are used as the catalysts. In fact, RuO_2 was discovered to be the most active material for oxygen evolution reaction but was not very stable under the operation conditions. Iridium oxide is a material compromising the activity and stability of water electrolyzers. The catalyst loadings for electrolyzers also decreased significantly, compared with that in the early stages. The catalyst loadings for the hydrogen evolution reaction have been reduced to a range of 0.5–1 mg/cm^2, and the catalysts loadings for the oxygen evolution reaction are about 1–2 mg/cm^2.

The second drawback of the PEM electrolyzer is still related to higher operational pressures. At high pressure, the gas cross-permeation phenomenon becomes more significant. Thicker membranes have to be used in order to keep the concentrations (H_2 in O_2) of the internal gas recombiners under the safety threshold (4 vol.% in O_2).

1.5.5 Solid Oxide Electrolyzers

Both alkaline and PEM electrolysis operate at relatively low temperatures with voltage efficiency up to 85% [39]. Steam electrolysis is a technology that reaches higher total energy efficiency at higher operation temperatures [53,56–58]. Operating at high temperatures (800–1000°C) thermodynamically reduces the energy required, including heat and electricity, for water splitting to generate hydrogen and oxygen. The high temperature adds heat as a substantial part of the energy into the process and accelerates the reaction kinetics. The first solid oxide electrolyzers (SOECs) were reported in the 1980s by Dönitz and Erdle [53], using a supported tubular electrolyzer. The result was part of the Hotelly project at Dornier System GmbH. Later in this program, there have been results about single cells operating with current densities of 0.3 A/cm^2 achieving 100% of Faraday efficiency at a voltage of 1.07 V. Researchers have shown great interest in this field since then, although this technology is still under development. The Relhy project is another attractive project supported by the European Commission, including seven European partners. They focus on the development of materials for SOECs, manufacturing process, and the integration in an efficient and durable SOEC. Another advantage of the SOECs is their huge potential in the electrolysis of CO_2 to CO, or co-electrolysis of H_2O/CO_2 to H_2/CO (syngas). The drawbacks of the SOECs are mainly related to the durability of the ceramic materials at high temperatures.

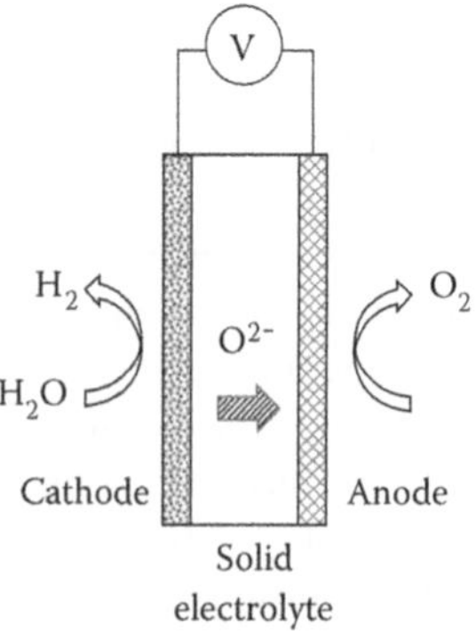

FIGURE 1.13 Scheme of a solid oxide electrolyzer cell.

Figure 1.13 illustrates the basic structure of an SOEC. The central component employs oxygen ion in conducting ceramics as the electrolyte. The oxygen ion instead of proton is the charge carrier in this system. Porous hydrogen and oxygen electrodes are on each side of the electrolyte layer. A steam heated to 800–1000°C flows on the cathode side and is split to hydrogen gas and O^{2-} ions. The oxygen ions are transported through the ceramic electrolyte to the anode, where they are oxidized and form oxygen gas. The two most common configurations for SOEC are tubular and planar cells. In the early stage of development of SOECs, tubular cells have been studied extensively. Planar cells have higher energy density and lower cost, compared with the tubular cells, and have become more popular in recent years [56,58].

SOEC uses similar materials as solid oxide fuel cell (SOFC) does in general [59,60]. ZrO_2 doped with 8 mol% of Y_2O_3 (YSZ) is the most common electrolyte material as a dense ionic conductor in use. This electrolyte has high ionic conductivity, thermal stability, and chemical stability at the operation temperatures (800–1000°C). Other materials include scandia-stabilized zirconia (ScSZ), ceria-based electrolytes (fluorite structure), or the lanthanum gallate (LSGM, perovskite structure) materials. Zirconium oxide–based electrolytes are generally used for SOECs operating at high temperatures. Ceria-based electrolytes and lanthanum gallate are suitable for intermediate-temperature operation. Besides oxygen ion conducting electrolytes, proton conducting electrolyte materials are also under investigation, since pure hydrogen can be generated with this configuration. The first reported result of steam electrolysis using proton conductors was over 30 years ago [53]. They studied the performance of the $SrCe_{1-x}MxO_{3-\delta}$ system ($x = 0.05$–0.10, M = Yb, Mg, Sc, Y, In, Zn, Nd, Sm, Dy) using Pt electrodes. Other alternate proton conducting materials include yttrium-doped barium cerate (BCY), yttrium-doped barium zirconate (BZY), and the cerate–zirconate $BaCe_{0.91-x}ZrxY_{0.1}O_{3-\delta}$ (BZCY).

The materials for the cathode have to be thermally and chemically stable, with porous structures to ensure the flow of the reactants and products gases. They also need to have good electron and oxygen ion conductivities, as well as good catalytic activities to split water. The most commonly used material is a porous cermet composed of YSZ and metallic nickel. Other materials have been considered, including samaria-doped ceria (SDC) with nickel dispersed nanoparticles, Ni–GDC ($Ce_{0.8}Gd_{0.2}O_{1.9}$), titanate/ceria composites, or lanthanum strontium chromium manganite (LSCM). For the anode, the most common material used to date is the lanthanum strontium manganite (LSM)/YSZ composite. Alternate materials include $La_{0.8}Sr_{0.2}FeO_3$ (LSF) and $La_{0.8}Sr_{0.2}CoO_3$ (LSCo), lanthanum strontium cobalt ferrite (LSCF), $Sm_{0.5}Sr_{0.5}CoO_{3-\delta}$ (SSA), $Ba_{0.5}Sr_{0.5}Co_{0.8}Fe_{0.2}O_{3-\delta}$ (BSCF), and lanthanum strontium copper ferrite (LSCuF).

To date, SOEC electrolysis is still at the laboratory stage. The following issues need to be overcome: (i) the durability of the ceramic materials at high temperature is not good enough for long-term operation; (ii) the costs of the materials and bulk system manufacture are not clear yet; and (iii) the safety issues associated with hydrogen generation and storage are still considerable. The possibility of operating cells at high current densities (>2 A/cm^2) with typical thickness of a few millimeters is afforded; therefore, the operational costs can be reduced. However, SOEC water electrolysis has huge potential for the future large-scale production of hydrogen. More investigations and development of the structure and electrochemistry of the materials are essential to the possibility of commercializing SOEC devices.

1.5.6 Photoelectrochemical and Photocatalytic H_2 Generation

For water electrolysis, the minimum required theoretical potential is 1.23 eV in dark conditions. This amount of energy equals the energy of a photon with a wavelength around 1010 nm [61]. Visible light is energetically sufficient to split water. The first results about the photodecomposition of water using semiconducting photoelectrolysis cells were reported by of Honda and Fujishima. Since then, many studies in this field have been carried out [62–67]. Figure 1.14 illustrates the structure of a photoelectrolysis cell consisting of a TiO_2 electrode and Pt electrode. Photoirradiation of the TiO_2 electrode with a small current led to the generation of H_2 and O_2 gases at

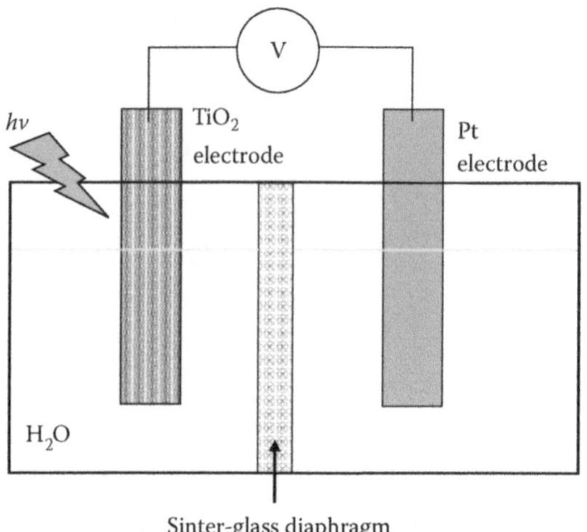

FIGURE 1.14 Electrochemical cell in which the TiO_2 electrode is connected with a Pt electrode.

the surface of the Pt electrode and TiO_2 electrode, respectively. Bard's concept has been applied to the design of photocatalytic systems since 1979. Semiconductor particles/powders was then investigated as photocatalysts. The Pt-loaded TiO_2 particle system (Pt/TiO_2), as a "short-circuited" photoelectrochemical cell, is a typical well-known example [63,68].

The first complete water split to produce H_2 and O_2 was achieved in acidic water solution using Pt/TiO_2. In 1980, [66,67] reported complete gas-phase water split at room temperature using Pt/TiO_2 coated with NaOH. Figure 1.15 represents the chemical process photocatalyzed by a semiconductor particle system, Pt/TiO_2 in this case. TiO_2 is an n-type semiconductor with a band gap of 3.0 eV. The electrons inside of the semiconductor are excited to jump from the valence band to the conduction band by the irradiation of light with enough energy, and leave holes in the valence band. Water can be electrolyzed into H_2 and O_2 with such free electron–hole pairs. Pt loaded on the surface of TiO_2 can speed up the transportation of free electron from the conduction band to the surface and accelerate the generation of H_2 gas. It is important to design photocatalysts that promote the formation of photoexcited electrons and holes as well as the transfer of these charge carriers to the adsorbed reactants. Thus far, most photocatalytic systems work under the irradiation of UV light. Some systems could split water as well under visible light. Besides TiO_2-based systems, various kinds of solid photocatalysts have been reported to split pure water into H_2 and O_2. $BaTiO_4$ loaded with RuO_2 ($RuO_2/BaTiO_4$) has been reported as an efficient photocatalyst for pure water splitting to produce H_2 and O_2 stoichiometrically under UV irradiation. $SrTiO_3$-based systems have also been extensively studied.

Materials with a layered structure have been found to be good candidates for water photoelectrolysis. For example, $K_4Nb_6O_{17}$ has been reported to have a characteristic activity for the photocatalytic splitting of water into H_2 and O_2, with a perovskite structure.

The photocatalytic splitting of water under visible light irradiation is a greatly desired reaction system for the production of H_2. Many researchers have developed the chemical doping of TiO_2 with transition metal ions or oxides as the visible light-responsive metal oxide photocatalysts. Although TiO_2 chemically doped with metal ions could induce visible light response, these catalysts showed limitations in sufficient reactivity for practical applications. The problems are related to the recombination between the photoformed electrons and holes. Usually, the addition of an oxidant or a reductant as the sacrificing reagent is necessary. Other alternative materials for water splitting under visible light include dyes (Ru complexes) and quantum dots (CdS, CdSe, InP, CdTe, Bi_2S_3)–based materials. Recently, many studies have focused on the anion-doped TiO_2, such as N-, S-, or C-doped materials [66,69–71].

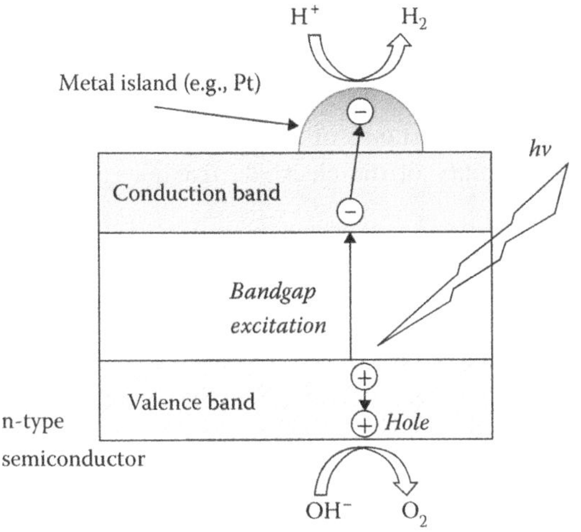

FIGURE 1.15 Illustration of semiconductor particulate systems for heterogeneous photocatalysis.

1.6 FUEL CELLS

1.6.1 Introduction

A fuel cell is an electrochemical device designed to convert the energy of a chemical reaction directly to electrical energy. The differences between fuel cells and other electrochemical power sources, the primary and secondary batteries, include the following: (i) fuel cells use a supply of gaseous or liquid reactants for the reactions rather than metals and metal oxides; (ii) it is rather easy to have a continuous supply of the reactants provided, and continuous elimination of the reaction products to keep a long operation time [72–77].

Fuel cells are named because the possible reactants or fuels could be natural gas, petroleum products, hydrogen, or water gas (syngas) produced by treating coal with steam. The biggest advantage of fuel cells is that the pollution-free electrical energy comes directly from the devices. There are no combustion reactions in fuel cells to produce environmentally unfriendly products, such as CO_2, SO_2, oxides of nitrogen, or particulate matters. In addition, fuel cells also feature good energy efficiency, light weight, silent operation, and vibration-free operation.

In 1839, Sir William Grove performed the first fuel cell experiment [78,79]. As shown in Figure 1.16, a current was

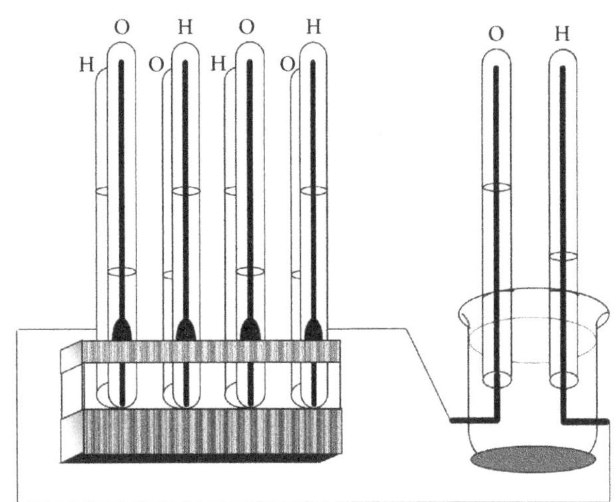

FIGURE 1.16 Scheme of Grove's gas battery (1839).

passed through the connecting wire between the anode and cathode to electrolyze water to first generate H_2 and O_2. Then, the power source was removed. He found that the devices reversely produced electricity on the electrodes. The device was named gas voltaic battery [72]. After more than 100 years, Francis Thomas Bacon further developed fuel cells in 1959. The fuel cells were used by the National Aeronautics and Space Administration as auxiliary power sources for space vehicles, as they are three times as effective as any other method of providing energy on board. In the middle of the 20th century, the use of numerous portable and other small devices for civil and military purposes required an autonomous power supply over extended periods of use. Today, numerous fuel cell–based power plants on a scale of both tens of megawatts and tens or hundreds of kilowatts have been built. Many small fuel cell units are also in use, with output between a few milliwatts and a few watts.

Many types of fuel cells have been investigated. There are mainly five types based on the characteristics of the electrolytes and operation temperature: alkaline fuel cells (low temperature), PEM fuel cells (PEMFCs, low temperature), direct methanol fuel cells (DMFCs, low temperature), phosphoric acid fuel cells (PAFCs, intermediate temperature), molten carbonate fuel cells (MCFCs, high temperature), and solid oxide fuel cells (SOFCs, high temperature) [80].

1.6.2 Fundamentals of Fuel Cells

Figure 1.17 illustrates the structure of a fuel cell in a simplest case. Two metallic (e.g., platinum) electrodes are dipped into an electrolyte solution. The anode produces electrons by "burning" a fuel, which flows on the anode. The cathode absorbs electrons in reducing an oxidizing agent. An electronically insulating porous separator is placed between the two electrodes, to prevent the mix of the reactants on each electrode and exclude accidental contact between anode and cathode. Theoretically, if a continuous supply of the reactants and continuous elimination of the reaction products can be provided, the fuel cell continues to work indefinitely. In general, the working voltage of an individual fuel cell is less than 1 V. To have a higher voltage, such as 6, 12, or 24 V, to meet the needs in real applications, stacks, which are numbers of connected individual cells, are generally used [76].

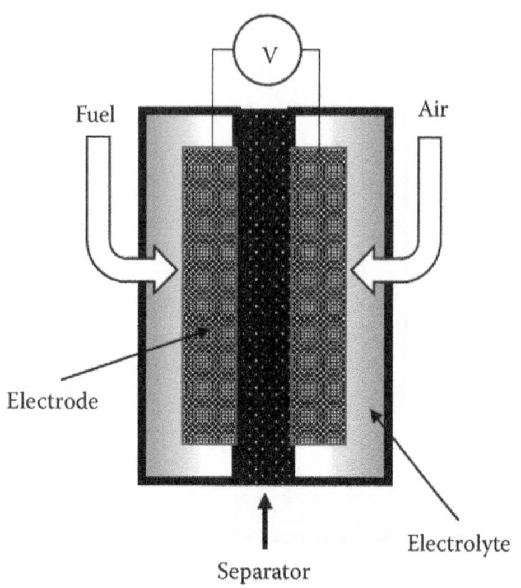

FIGURE 1.17 Schematic structure of an individual fuel cell.

When using hydrogen as the fuel and oxygen as the oxidizing reagent, the half-reactions occurring on the cathode and anode, respectively, are represented here as

$$\text{Anode:}\quad 2H_2 \rightarrow 4H^+ + 4e^- \tag{1.49}$$

$$\text{Cathode:}\quad O_2 + 4H^+ + 4e^- \rightarrow H_2O \tag{1.50}$$

The protons formed on the anode are transferred through the electrolyte toward the cathode, where oxygen are charged and react with protons to generate water. In the electrolyte, an (positive) electrical current flows from the anode to the cathode; in the external circuit, it flows in the opposite direction, from the cathode to the anode.

$$\text{Overall:}\quad 2H_2 + O_2 \rightarrow 2H_2O \tag{1.51}$$

The potentials of electrodes can be reversible (equilibrium) or irreversible (nonequilibrium). An electrode's equilibrium potential (denoted $E°$) reflects the thermodynamic properties of the electrode reaction occurring at it. On the standard hydrogen electrode (SHE) scale, the equilibrium potential of the anode, $E_{h.e.}$, is close to zero. The oxygen electrode, cathode, usually has irreversible potential. The thermodynamic potential, $E°_{o.e.}$, of the cathode is 1.229 V (relative to the SHE). In the experiment, the potential actually established at it, $E_{o.e.}$, is 0.8–1.0 V. Since the electrode potential of the cathode is more positive than that of the anode, the potential difference existing between them, being the voltage U of the fuel cell, can be calculated using the following equation:

$$U = E_{o.e.} - E_{h.e.} \tag{1.52}$$

The value of the voltage of an idle cell is called the open-circuit voltage (OCV), U_o, of this cell. The working voltage of an operating fuel cell, U_i, is lower than OCV, because the internal ohmic resistance of the cell, electrode polarization, and irreversibility of the electrode reactions shall be taken into account.

As mentioned above, there are many types of fuel cells with different fuels and electrode reactions. To have a comparison, several parameters of fuel cells need to be considered. OCV U_o and discharge or operating voltage U_i as observed of each fuel cell under given conditions (at a given discharge current) are usually used for comparisons of different cells. Second, discharge current and discharge power are used in the evaluation, calculated using Equations 1.53 through 1.55.

$$I = \frac{U_i}{R_{ext}} \tag{1.53}$$

$$I = \frac{U_0}{R_{app} + R_{ext}} \quad (1.54)$$

$$P = \frac{U_i^2 R_{app}}{(R_{app} + R_{ext})^2} \quad (1.55)$$

where R_{app} is the apparent internal resistance and R_{ext} is the resistance of the external load.

Finally, the operating efficiencies of a fuel cell, which is its efficiency in transforming a fuel's chemical energy to electrical energy, are very important parameters. Commonly defined operating efficiencies include theoretical (thermodynamic) efficiency, voltage efficiency, columbic efficiency, design efficiency (related to the fuel cell systems design), and overall efficiency [74,81,82].

1.6.3 Alkaline Fuel Cells

The alkaline fuel cell (AFC) was the first fuel cell technology used in practical service to generate electricity from hydrogen gas [75,76,83]. In the late 1950s, Allis Chalmers AFCs were equipped on a farm tractor, as the basis for vehicular applications of fuel cells. The famous Austin A40 operated by Karl Kordesch was developed in the early 1970s. Generally, AFCs can provide a power of 10–100 kW at the operation temperature of 80°C, with a cell efficiency of 60–70%. They use an aqueous solution of potassium hydroxide as the electrolyte, with typical concentrations of about 30%. The overall chemical reactions are as follows:

$$\text{Anode reaction:} \quad 2H_2 + 4OH^- \rightarrow 4H^+ + 4e^- \quad (1.56)$$

$$\text{Cathode reaction:} \quad O_2 + 2H_2O + 4e^- \rightarrow 4OH^- \quad (1.57)$$

$$\text{Overall cell reaction:} \quad O_2 + 2H_2 \rightarrow 2H_2O \quad (1.58)$$

The cells generate heat and electricity. To remove water and heat, the electrolytes are recirculated as the coolant liquid, while water is removed by evaporation.

A schematic of an AFC is shown in Figure 1.18. The electrodes consist of a double-layer structure: an active electrocatalyst layer and a hydrophobic layer that prevents the electrolyte from leaking into the reactant gas flow channels and ensures the diffusion of the gases to the reaction site. The faster kinetics of the oxygen reduction reaction in an AFC allows the use of non-noble metal electrocatalysts, such as nickel [84].

A major issue of AFCs is the requirement for low carbon dioxide concentrations in the feed oxidant stream. Carbonates form and precipitate with CO_2 present.

$$CO_2 + 2OH^- \rightarrow CO_3^{2-} + H_2 \quad (1.59)$$

The carbonates can block the electrolyte pathways and electrode pores, and therefore reduce the lifetime of the cells. The current practical solution to the problem is the use of soda lime for removing CO_2 from the air stream. In addition, liquid

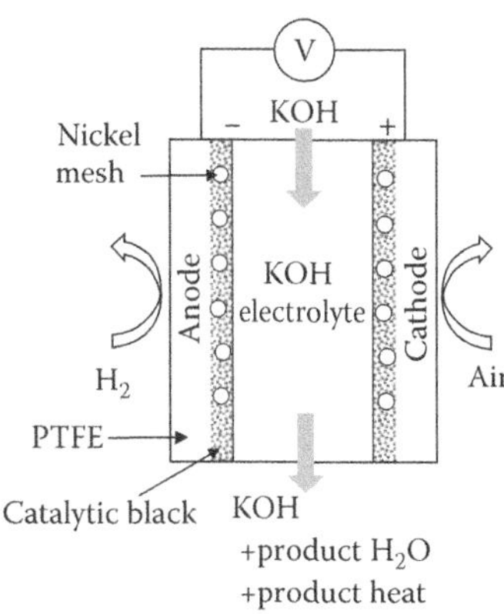

FIGURE 1.18 Schematic structure of an alkaline fuel cell composition.

electrolyte management systems used in AFCs require ongoing maintenance. Thus far, AFCs have been able to meet the 5000-h lifetime required for the design of a passenger vehicle.

1.6.4 PEM Fuel Cells

The PEM fuel cell has been considered as a preferred technology for low-temperature, moderate-power applications [82]. As shown in Figure 1.19, in a single cell, the middle component is the polymeric PEM serving as a solid electrolyte. A gas diffusion layer and electrode are on each side. The electrochemical reactions occur at the electrodes of PEMFCs [85]. Hydrogen is oxidized to protons on the anode surface, and the protons transfer through the electrolyte membrane toward the cathode, where oxygen is reduced, in the presence of protons, to water. The electrode reactions can be written as

$$\text{Anode:} \quad 2H_2 \rightarrow 4H^+ + 4e^- \quad (1.60)$$

$$\text{Cathode:} \quad O_2 + 4H^+ + 4e^- \rightarrow H_2O \quad (1.61)$$

$$\text{Overall:} \quad 2H_2 + O_2 \rightarrow 2H_2O \quad (1.62)$$

The electromotive force is 1.23 V at 25°C.

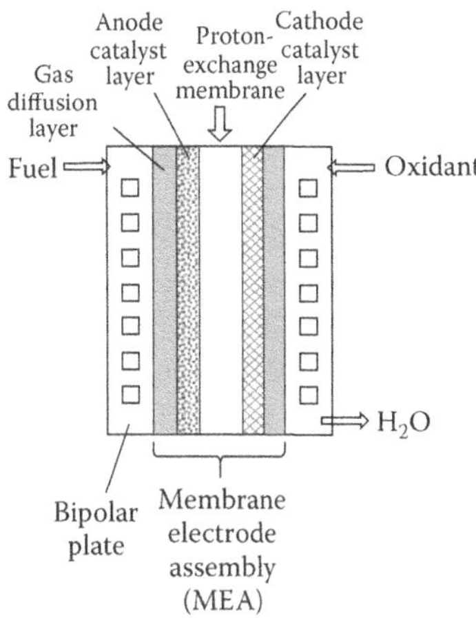

FIGURE 1.19 Basic components of a proton exchange membrane fuel cell.

In the early 1960s, the first PEMFC battery with a power of 1 kW was built by General Electric for the Gemini spacecraft. This early version of PEMFC consists of a sulfated polystyrene ion-exchange membrane as the electrolyte and about 4 mg/cm^2 of platinum catalyst in the electrode layer. Its current density was below 100 mA/cm^2, with a voltage of about 0.6 V for an individual cell. The specific power density of the cells is about 60 mW/cm^2. The major problem of this fuel cell is its high cost, which limited its application fields. In the period from the 1960s to the year 2000, dramatic improvements in PEMFC properties were achieved: (i) the specific power density of PEMFCs is 600–800 mW/cm^2; (ii) <0.4 mg/cm^2 of platinum catalysts are used; and (iii) the lifetime has been improved from 2000 h to several tens of thousands of hours. The development of Nafion proton-exchange membranes, the improved efficiency of platinum catalysts in the electrodes, and the design of the membrane-electrode assemblies are the three main factors that contributed to the fast progress [77,82,86,87].

Between 1960 and 1980, the US chemical company Du Pont de Nemours began to produce a perfluorinated sulfonic acid polymer with a continuous hydrophobic skeleton of $-(CF_2)_n-$ groups, to which a certain number of hydrophilic segments containing sulfonic acid groups $-SO_3H$ are attached. The $-(CF_2)_n-$ skeleton has a very high chemical stability. When the membrane is wetted appropriately, the sulfonic acid groups dissociate and provide a relatively high protonic membrane conductivity of about 0.1 S/cm at 80°C, three to four times higher than that of the previously used ion-exchange membranes. The current issue of this type of membrane is its high cost. A Nafion membrane costs about $700 per square meter, or $120 per kilowatt of designated power (at an average power density of 600 mW/cm^2).

Platinum, a noble metal, is the basic catalyst for electrochemical reactions at hydrogen and oxygen electrodes. Many studies have been focusing on the methods to improve the utilization efficiency and reduce the metal loading. One way is to deposit platinum onto highly dispersed carbon carriers. In various reports, the amount of platinum in the platinized carbon black can be between 10% and 40% by mass. Addition of 30–40% of proton conducting ionomer into the catalysts layer deceases the loading of platinum without any departure from the high electric parameters. The ionomer leads to a considerable increase in the contact area between catalyst and electrolyte.

For further improvement in the catalyst–membrane contact, the so-called membrane electrode assembly (MEA) is widely utilized. The following two methods have been applied to manufacture the MEA. One is applying catalysts to the carbon sheet that works as a diffusion layer. The PEM is sandwiched between two catalyst-loaded sheets, and then all the components are hot-pressed together, resulting in a catalyst-coated diffusion layer. The other technology is to paste all components of the catalytic layer as a thin layer directly on the membrane, which results in a catalyst-coated membrane. With all these developments, it was possible to reduce the loading of platinum metals to 0.4 mg/cm^2 without sacrificing performance in the mid-1990s, and by 2001, to 0.05 mg/cm^2 for the hydrogen electrode and to 0.1 mg/cm^2 for the oxygen electrode.

Light electric vehicles and portable electronic devices are the two potential applications using fuel cells as the power sources [88]. To date, PEMFCs are able to produce 100–500 kW power at an operating temperature of 50–120°C (Nafion) or 120–220°C (polybenzimidazole), with a cell efficiency of 40–70%. The cost is at $30–35 per W. However, there are still several problems that needed to be solved before the wide use of PEMFCs, including the relatively short lifetime of the membranes and catalysts, the high costs of production, and the instability of the catalysts with CO impurities in the hydrogen.

1.6.5 Direct Methanol Fuel Cells

Methanol has larger energy density compared with hydrogen. It is a promising type of fuel for fuel cells, with specific energy content of about 6 kWh/kg. The DMFCs are developed using methanol as the fuel, and expected to be one of the candidates for portable devices and even automobiles [81,89,90]. The overall cell reactions and the corresponding thermodynamic values of equilibrium electrode potentials $E°$ and EMF $\varepsilon°$ of the DMFCs are as follows:

$$\text{Anode:}\quad CH_3OH + H_2O \rightarrow CO_2 + 6H^+ + 6e^- \quad E° = 0.02V \tag{1.63}$$

$$\text{Cathode:}\quad \frac{3}{2}O_2 + 6H^+ + 6e^- \rightarrow 3H_2O \quad E° = 1.23\text{ V} \tag{1.64}$$

$$\text{Overall: } CH_3OH + \frac{3}{2}O_2 \rightarrow CO_2 + 2H_2O \qquad \varepsilon^0 = 1.23V - 0.02V = 1.21 \tag{1.65}$$

The thermodynamic parameters of the overall reaction are as follows:

Reaction enthalpy (or heat Q_{react}):

$$\Delta H = -726 \text{ kJ/mol} = 1.25 \text{ eV} \tag{1.66}$$

Gibbs reaction energy (or maximum work W_e):

$$\Delta G = -702 \text{ kJ/mol} = 1.21 \text{ eV} \tag{1.67}$$

The first studies of anodic methanol oxidation at platinum electrodes started in the 1960s. Both alkaline and acidic aqueous solutions as the possible electrolyte were used in the early test models of methanol–oxygen or methanol–air fuel cells. In the early 1980s, the first models of methanol fuel cells employed sulfuric acid aqueous solutions as the electrolyte and platinum catalysts (up to 10 mg/cm^2). The well-known pioneer works were done by Shell in England and Hitachi in Japan. The specific power attained was considered small, nearly 20 mW/cm^2.

In the mid-1990s, the development of PEMFCs speeded up the studies of DMFCs as well. The Nafion-type membranes, used in PEMFCs, were also transferred into the models of DMFCs. One problem of adopting PEM in DMFCs is the gradual penetration of methanol to the oxygen electrode by diffusion through the membrane. That causes the waste of methanol and the decrease of the oxygen electrode potential. Pt–Ru catalysts at the anode and a catalyst of pure platinum at the cathode are commonly in use. Ru adsorbs OH species on the surface, and the adsorbed OH oxidizes CO species as shown below:

$$\mathrm{Pt-Ru + H_2O \rightarrow RuOH + H^+ + e^-} \quad (1.68)$$

$$\mathrm{Pt-RuOH + Pt-CO \rightarrow Pt-Ru + Pt + CO_2 + H^+ + e^-} \quad (1.69)$$

Thus far, DMFCs are able to generate a power of 100 mW to 1 kW and 20–30% efficiency at low operating temperature (50–120°C) [76,80,91]. They have not been in commercial production or in wide practical use. Besides the methanol penetration problem, there are still several improvements related to the operation of DMFCs that needed to be done: (i) the lifetime of the DMFCs needs improvement, which is related to the lifetime of the catalysts and the dissolution of Ru ions; (ii) better operating efficiency, which associates with the methanol crossover problem, is desired; and (iii) the costs of the DMFCs need to be brought down since the current membrane and catalysts are expensive.

1.6.6 Phosphoric Acid Fuel Cells

The investigations of phosphoric acid fuel cells started from the idea of utilizing methane or other hydrocarbons as fuels in aqueous solutions for fuel cells. In the early 1960s, fuel cells with concentrated solutions of phosphoric acid with an operating temperature of 150°C were introduced. The results showed that the platinum catalysts are not active enough for a practical hydrocarbon fuel cell, but led to the invention of hydrogen–oxygen phosphoric acid fuel cells.

Phosphoric acid (H_3PO_4) in aqueous solutions dissociates into ions according to

$$\mathrm{H_3PO_4 \rightarrow H^+ + H_2PO_4} \quad (1.70)$$

In concentrated phosphoric acid solutions, the hydrogen ions exist as ions solvated by phosphoric acid molecules, $H_+ \cdot nH_3PO_4$. Phosphoric acids also dimerize to $H_4P_2O_7$ in aqueous solutions at a concentration of about 85 wt.% and high temperatures as the equation shown below:

$$\mathrm{2H_3PO_4 \rightarrow H_4P_2O_7 + H_2O} \quad (1.71)$$

This dimerization process helps maintain the concentration of $H_3P_2O_7^-$, which does not adsorb much on the platinum catalyst surface as $H_2PO_4^-$ does, and reduces the polarization of the electrodes. In addition, the water vapor pressure decreases with increasing acid concentration. The concentrated phosphoric acid solution can be, therefore, loaded into a porous solid matrix and dried. The elimination of water as a reaction product from the fuel cell's cathode space is also simplified, since water elimination can self-regulate.

In a PAFC, the common catalyst system is also Pt–Ru catalysts for the anode and Pt for the cathode. Similar to the developments in PEMFCs, the catalysts are now dispersed on a carbon support to reduce the loading from 4 to 0.4 and then to 0.25 mg/cm^2. The concentrated phosphoric acid solution as electrolyte is absorbed into the pores of a porous matrix with fine pores and a total thickness of about 50 μm. Kynar poly(vinylidene fluoride) was a matrix material in early studies, but it is not chemically stable enough at high temperatures. Recently, more researches are focusing on the silicon carbide–based composite materials for this purpose.

Nowadays, PAFCs are using 95–100% phosphoric acid solution as electrolyte with a working temperature of 150–220°C. They are able to produce <100 MW power and cost $4–4.5 per watt. Because of those features mentioned above, the first models of large-scale power plants producing a power of hundreds of kilowatts and more were built from PAFCs. Some international companies, such as United Technologies Corporation, the Japanese company Toshiba, Fuel Cells International, Japan's Fuji Electric Corporation, and Mitsubishi Electric Corporation, have made a large effort toward mass production of these fuel cells and large-scale PAFC-based power plants. These power plants are designated for the combined on-site heat and power supply of individual residential and municipal structures such as hospitals. The electrical efficiency was 35–40%. Including the thermal energy produced, the total energy conversion efficiency was as high as 85%. The major issues regarding the power plants using PAFCs are still the high costs and their reliability and lifetime [72,73,77].

1.6.7 Molten Carbonate Fuel Cells

Molten carbonate fuel cells (MCFCs) are a type of high-temperature fuel cells that have a working temperature of more than 600°C. Compared with the low-temperature fuel cells, the potential advantages of these high-temperature fuel cells include possible higher efficiency, faster electrode reactions and less electrode polarization, non-requirement for platinum catalysts, and potential of using other fuels (CO, hydrocarbons). The main disadvantage is the decrease of thermodynamic efficiency of the reaction $\eta_{thermod} = -\Delta G/-\Delta H$ with increasing temperature. The Gibbs free energy $-\Delta G$ of hydrogen oxidation by oxygen decreases with increasing temperature. It decreases to 1.06 eV at 600°C and to 0.85 eV at 1000°C from 1.23 eV at 25°C. While the reaction enthalpy,

$-\Delta H$ = 1.48 eV at 25°C, does not change significantly at higher temperatures [72,92].

At temperatures of 25°C, 600°C, and 1000°C, the thermodynamic efficiency is 0.83, 0.72, and 0.57, respectively. In addition, choosing materials with good chemical and mechanical stabilities at such high operating temperatures is challenging.

Figure 1.20 illustrates the basic structure of an MCFC, similar to that of a PEMFC. The electrolyte is generally a mixed melts with 62–70 mol% Li_2CO_3 and 30–38 mol% K_2CO_3. Sometimes, Na_2CO_3 and other salts are added to these melts. These melts are immobilized in the pores of a ceramic matrix (MgO or $LiAlO_2$). Porous metallic gas-diffusion electrodes are in use. Nickel alloy with 2–10% chromium as anode and lithiated nickel oxide as cathode are currently popular electrode systems for MCFCs.

With its fine pores filled with the carbonate melts, the matrix prevents the mix of gases for each electrode.

In a hydrogen–oxygen MCFC, the electrode reactions are as follows:

$$\text{Anode: } H_2 + CO_3^{2-} \rightarrow H_2O + CO_2 + 2e^- \quad E° = 0 \text{ V} \tag{1.72}$$

$$\text{Cathode: } \frac{1}{2}O_2 + CO_2 + 2e^- \rightarrow CO_3^{2-} \quad E° = 1.06 \text{ V} \tag{1.73}$$

$$\text{Overall: } 2H_2 + O_2 \rightarrow 2H_2O \quad E° = 1.06\text{V} \tag{1.74}$$

Depending on the partial pressures of the reactant gases, the open-circuit voltage of hydrogen–oxygen MCFCs has values of 1.00–1.06 V. The interesting part is that the cathodic reactions consume not only oxygen but also carbon dioxide. Another attractive feature of MCFCs is the possibility of using carbon monoxide as a reactant fuel. Water gas, a mixture of CO and H_2, is obtained readily by the steam gasification of coal:

$$C + H_2O \rightarrow CO + H_2 \tag{1.75}$$

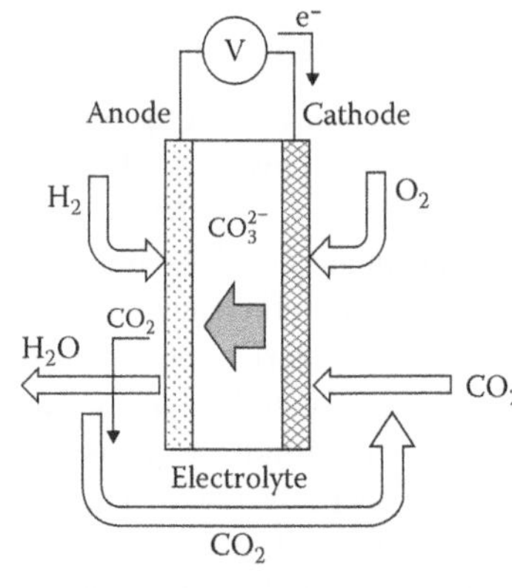

FIGURE 1.20 Schematic of reactant flow in a molten carbonate fuel cell.

The possibility of a direct electrochemical oxidation of carbon monoxide exists as well [93]. Following the equation below, it is possible that MCFCs do not need an external converter to provide hydrogen because "internal reforming" is occurring at the operating temperature.

$$CO + H_2O \rightarrow CO_2 + H_2 \tag{1.76}$$

Pure CO can be oxidized electrochemically as well according to the reaction, although the reaction is very slow.

$$CO + CO_3^{2-} \rightarrow 2CO_2 + 2e^- \tag{1.77}$$

There have been two types of internal-reforming fuel cells (IRFCs)—direct internal reforming fuel cells and indirect internal reforming fuel cells—developed, in order to utilize carbon monoxide more efficiently.

The investigations of MCFCs started from 1910s, and the concept of "solid electrolyte" (a mixture of monazite sand and other components, including alkali metal carbonates) was proposed by O. Davtyan in the 1930s. At present, MCFCs are able to provide 100 MW of power with 50–60% efficiency at 600–650°C. Industrial organizations started to show interests in MCFCs in about the mid-1960s.

Starting in 2000, more than 40 power plants in sizes from 250 kW to 1 MW have been built in several countries [93,94].

The major drawback of MCFCs thus far is the insufficiently long period of trouble-free operation. At present, it is possible to have an individual unit working for several hundreds and thousands of hours. It is still far behind the minimum requirement of 40,000 h (4.5–5 years) for a power plant. The causes of the short lifetime of MCFCs include the dissolution of nickel oxide from the oxygen electrode and the breakdown of the materials of components because of the corrosive nature of the electrolyte at high temperatures.

1.6.8 Solid Oxide Fuel Cells

Solid oxide fuel cells (SOFCs) convert chemical energy to electricity directly at the operating temperature of 600–1000°C. The efficiencies of such systems are estimated to be highest among the fuel cell systems thus far (>50%). Conventional SOFCs exist in several design variants, including basic tubular and planar cells. Monolithic cells appeared around 1990 [72,95].

Figure 1.21 is a schematic diagram for a typical solid oxide fuel cell. The important components are the anode, cathode, and electrolyte. These three components of the fuel cell—anode, cathode, and electrolyte—form an MEA. The central part is a unipolar O^{2-} ion conductor as the solid electrolyte. The most commonly known electrolyte material for this kind of fuel cells is yttria-stabilized zirconia (YSZ), that is, zirconium dioxide doped with the oxide of trivalent yttrium: ZrO_2 + 10% Y_2O_3 or $(ZrO_2)_{0.92}(Y_2O_3)_{0.08}$. The conductivity of YSZ-type electrolytes increases to about 0.15 S/cm only at temperatures above 900°C. On the two sides of electrolyte layers are electrode layers. The anode consists of

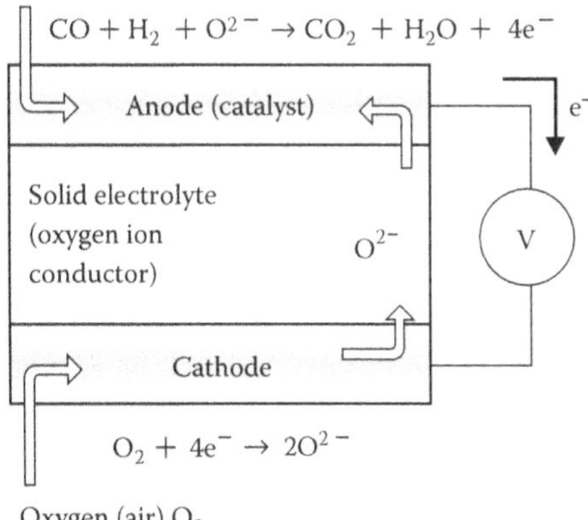

FIGURE 1.21 Schematic diagram showing the operating principles of a solid oxide fuel cell running on natural gas.

a cermet (ceramic–metal composite) of nickel and the YSZ electrolyte. Nickel particles are distributed uniformly in the solid electrolyte as the anode catalyst. The YSZ material in the anode improves the contact between the nickel catalyst and the fuel cell's electrolyte layer. The cathode layer consists of manganites or cobaltites of lanthanum doped with divalent metal ions (e.g., $La_{1-x}Sr_xMnO_3$ [LSM] or $La_{1-x}Sr_xCoO_3$ [LSC], where $0.15 < x < 0.25$). These cathode materials have both O^{2-} conductivity and some electronic conductivity [80].

When using hydrogen and carbon monoxide as the reactant fuels in SOFCs, the electrode reactions at 900°C are listed below:

When hydrogen is the fuel,

$$\text{Anode:} \quad H_2 + O^{2-} \rightarrow H_2O + 2e^- \quad E^\circ = 0 \text{ V} \tag{1.78}$$

$$\text{Cathode:} \quad \frac{1}{2}O_2 + 2e^- \rightarrow O^{2-} \quad E^\circ = 0.89 \text{ V} \tag{1.79}$$

$$\text{Overall:} \quad H_2 + \frac{1}{2}O_2 \rightarrow H_2O \quad \varepsilon^\circ = 0.89 \text{ V}, \quad \Delta H^\circ = -248.8 \text{ kJ/mol} \tag{1.80}$$

When carbon monoxide is the fuel,

$$\text{Anode:} \quad CO + O^{2-} \rightarrow CO_2 + 2e^- \quad E^\circ = 0.2 \text{ V} \tag{1.81}$$

$$\text{Cathode:} \quad \frac{1}{2}O_2 + 2e^- \rightarrow O^{2-} \quad E^\circ = 0.89 \text{ V} \tag{1.82}$$

$$\text{Overall:} \quad CO + \frac{1}{2}O_2 \rightarrow CO_2 \quad \varepsilon^\circ = 0.87 \text{ V}, \quad \Delta H^\circ = -283.0 \text{ kJ/mol} \tag{1.83}$$

The OCV U_0 in hydrogen–oxygen SOFCs at a temperature of 900°C is about 0.9 V.

At present, SOFC technologies for high-temperature operation (~1000°C) are relatively mature. Large-scale systems of ten 100 kW to MW classes can be achieved. The most critical drawback is the high cost of SOFCs. In addition, numerous problems associated with the development, manufacture, and practical applications of SOFCs are caused by the high working temperature. Many researchers have been working on decreasing the operating temperature of SOFCs to about 600°C to 700°C. Such fuel cells are called interim-temperature SOFCs (IT-SOFCs). Low-temperature SOFCs (LT-SOFCs) with operating temperatures below 600°C are also under development. The two main concerns about lowering the operating temperature are the low conductivity of current electrolyte and slow rates of the electrochemical reactions occurring at the electrodes. There have been many studies of new materials as the electrolyte to meet the requirements at lower operating temperatures [96]. Doped ceria (CeO_2) and doped lanthanum gallate ($LaGaO_3$) with higher conductivity at the temperature range of 400°C to 800°C have been particularly attractive. Both cathode and anode materials have similar conductivity issues as electrolyte does. The conductivity of LSM, a cathode material, decreases notably with decreasing temperature. Besides the development of new materials, many researchers also focus on reducing the thickness of electrolyte layers [74,96].

REFERENCES

1. Krivik, P. and P. Baca. *Electrochemical Energy Storage, Energy Storage—Technologies and Applications*, edited by A.F. Zobaa, Croatia: InTech, Rijeka, 2013.
2. Bockris, J.O'M., B.E. Conway, E. Yeager, and R.E. White. *Comprehensive Treatise of Electrochemistry: Electrochemical Processing*, Vol. 2, 2. Boston: Springer US, 1981.
3. Wendt, H. and G. Kreysa. *Electrochemical Engineering: Science and Technology in Chemical and Other Industries*. New York: Springer, 1999.
4. Zoski, C.G. *Handbook of Electrochemistry*. Amsterdam: Elsevier, 2007.
5. Reddy, T. *Linden's Handbook of Batteries*, 4th Edition. New York: McGraw-Hill Education, 2010.
6. Pistoia, G. "Chapter 2—Battery Categories and Types." In *Battery Operated Devices and Systems*, edited by G. Pistoia, 17–73. Amsterdam: Elsevier, 2009.
7. Bode, H., R.J. Brodd, and K.V. Kordesch. *Lead–Acid Batteries*. New York: Wiley, 1977.
8. Bullock, K.R. "Lead/Acid Batteries." *Journal of Power Sources* 51, no. 1–2 (1994): 1–17.
9. Huggins, R.A. (ed.) "Lead–Acid Batteries." In *Energy Storage*, 237–50. New York: Springer, 2010.
10. Shukla, A.K., S. Venugopalan, and B. Hariprakash. "Nickel-Based Rechargeable Batteries." *Journal of Power Sources* 100, no. 1–2 (2001): 125–48.
11. Zimmerman, A.H. *Nickel–Hydrogen Batteries: Principles and Practice*. El Segundo, CA: Aerospace Press, 2009.
12. Goodenough, J.B. "Evolution of Strategies for Modern Rechargeable Batteries." *Accounts of Chemical Research* 46, no. 5 (2012): 1053–61.
13. Goodenough, J.B. and Y. Kim. "Challenges for Rechargeable Li Batteries." *Chemistry of Materials* 22, no. 3 (2009): 587–603.
14. Whittingham, M.S. "Lithium Batteries and Cathode Materials." *Chemical Reviews* 104, no. 10 (2004): 4271–302.

15. Amine, K., R. Kanno, and Y. Tzeng. "Rechargeable Lithium Batteries and Beyond: Progress, Challenges, and Future Directions." *MRS Bulletin* 39, no. 5 (2014): 395–401.
16. Bruce, P.G., L.J. Hardwick, and K.M. Abraham. "Lithium–Air and Lithium–Sulfur Batteries." *MRS Bulletin* 36, no. 7 (2011): 506–12.
17. Barchasz, C., F. Molton, C. Duboc, J.C. Lepretre, S. Patoux, and F. Alloin. "Lithium/Sulfur Cell Discharge Mechanism: An Original Approach for Intermediate Species Identification." *Analytical Chemistry* 84, no. 9 (2012): 3973–80.
18. Fu, Y., Y.-S. Su, and A. Manthiram. "Highly Reversible Lithium/Dissolved Polysulfide Batteries with Carbon Nanotube Electrodes." *Angewandte Chemie International Edition* 52, no. 27 (2013): 6930–35.
19. Demir-Cakan, R., M. Morcrette, Gangulibabu, A. Gueguen, R. Dedryvere, and J.-M. Tarascon. "Li–S Batteries: Simple Approaches for Superior Performance." *Energy & Environmental Science* 6 (2013): 176–82.
20. Zhang, S.S., D. Foster, and J. Read. "A High Energy Density Lithium/Sulfur–Oxygen Hybrid Battery." *Journal of Power Sources* 195, no. 11 (2010): 3684–88.
21. Zhang, S.S. and J.A. Read. "A New Direction for the Performance Improvement of Rechargeable Lithium/Sulfur Batteries." *Journal of Power Sources* 200 (2012): 77–82.
22. Aurbach, D., E. Pollak, R. Elazari, G. Salitra, C. Scordilis Kelley, and J. Affinito. "On the Surface Chemical Aspects of Very High Energy Density, Rechargeable Li–Sulfur Batteries." *Journal of the Electrochemical Society* 156, no. 8 (2009): A694–702.
23. Zhang, S.S. "Effect of Discharge Cutoff Voltage on Reversibility of Lithium/Sulfur Batteries with Lino3-Contained Electrolyte." *Journal of the Electrochemical Society* 159, no. 7 (2012): A920–23.
24. Su, Y.-S. and A. Manthiram. "Lithium–Sulphur Batteries with a Microporous Carbon Paper as a Bifunctional Interlayer." *Nature Communications* 3 (2012): 1166.
25. Chen, S., F. Dai, M.L. Gordin, and D. Wang. "Exceptional Electrochemical Performance of Rechargeable Li–S Batteries with a Polysulfide-Containing Electrolyte." *RSC Advances* 3, no. 11 (2013): 3540–43.
26. Seh, Z.W., W. Li, J.J. Cha, G. Zheng, Y. Yang, M.T. McDowell, P.-C. Hsu, and Y. Cui. "Sulphur–TiO_2 Yolk-Shell Nano-architecture with Internal Void Space for Long-Cycle Lithium–Sulphur Batteries." *Nature Communications* 4 (2013): 1331.
27. McBreen, J. "Nickel/Zinc Batteries." *Journal of Power Sources* 51, no. 1–2 (1994): 37–44.
28. Weber, A.Z., M.M. Mench, J.P. Meyers, P.N. Ross, J.T. Gostick, and Q. Liu. "Redox Flow Batteries: A Review." *Journal of Applied Electrochemistry* 41, no. 10 (2011): 1137–64.
29. Yu, A. *Electrochemical Supercapacitors for Energy Storage and Delivery: Fundamentals and Applications*. Green Chemistry and Chemical Engineering. Boca Raton, FL: Taylor & Francis, 2013.
30. Wang, G., L. Zhang, and J. Zhang. "A Review of Electrode Materials for Electrochemical Supercapacitors." *Chemical Society Reviews* 41, no. 2 (2012): 797–828.
31. Zhang, L.L. and X.S. Zhao. "Carbon-Based Materials as Supercapacitor Electrodes." *Chemical Society Reviews* 38, no. 9 (2009): 2520–31.
32. Wang, Y. and Y. Xia. "Recent Progress in Supercapacitors: From Materials Design to System Construction." *Advanced Materials* 25, no. 37 (2013): 5336–42.
33. Jha, N., P. Ramesh, E. Bekyarova, M.E. Itkis, and R.C. Haddon. "High Energy Density Supercapacitor Based on a Hybrid Carbon Nanotube-Reduced Graphite Oxide Architecture." *Advanced Energy Materials* 2, no. 4 (2012): 438–44.
34. Tang, Z., C.-H. Tang, and H. Gong. "A High Energy Density Asymmetric Supercapacitor from Nano-Architectured Ni(Oh)2/Carbon Nanotube Electrodes." *Advanced Functional Materials* 22, no. 6 (2012): 1272–78.
35. Ghaffari, M., Y. Zhou, H. Xu, M. Lin, T.Y. Kim, R.S. Ruoff, and Q.M. Zhang. "High-Volumetric Performance Aligned Nano-Porous Microwave Exfoliated Graphite Oxide-Based Electrochemical Capacitors." *Advanced Materials* 25, no. 35 (2013): 4879–85.
36. Arico, A.S., P. Bruce, B. Scrosati, J.-M. Tarascon, and W. van Schalkwijk. "Nanostructured Materials for Advanced Energy Conversion and Storage Devices." *Nature Materials* 4, no. 5 (2005): 366–77.
37. Turner, J.A. "A Realizable Renewable Energy Future." *Science* 285, no. 5428 (1999): 687–89.
38. Amirabad, M.M., A.M. Amirabad, J. Khodagholizadeh, and A.A. Naeimi. "Producing Hydrogen through Electrolysis." *Applied Mechanics and Materials* 110–6 (2011): 2296.
39. Carmo, M., D.L. Fritz, J. Mergel, and D. Stolten. "A Comprehensive Review on PEM Water Electrolysis." *International Journal of Hydrogen Energy* 38, no. 12 (2013): 4901–34.
40. Zeng, K. and D. Zhang. "Recent Progress in Alkaline Water Electrolysis for Hydrogen Production and Applications." *Progress in Energy and Combustion Science* 36, no. 3 (2010): 307–26.
41. Mazloomi, S.K. and N. Sulaiman. "Influencing Factors of Water Electrolysis Electrical Efficiency." *Renewable and Sustainable Energy Reviews* 16, no. 6 (2012): 4257–63.
42. Wang, M., Z. Wang, X. Gong, and Z. Guo. "The Intensification Technologies to Water Electrolysis for Hydrogen Production—A Review." *Renewable and Sustainable Energy Reviews* 29 (2014): 573–88.
43. Bockris, J.O'M., F. Gutmann, and H. Bloom. *Electrochemistry, the Past Thirty Years and the Next Thirty Years: A Volume in Honor of J. O'M. Bockris*. New York: Plenum Press, 1977.
44. Ursua, A., L.M. Gandia, and P. Sanchis. "Hydrogen Production from Water Electrolysis: Current Status and Future Trends." *Proceedings of the IEEE* 100, no. 2 (2012): 410–26.
45. Bowen, C.T., H.J. Davis, B.F. Henshaw, R. Lachance, R.L. LeRoy, and R. Renaud. "Developments in Advanced Alkaline Water Electrolysis." *International Journal of Hydrogen Energy* 9, no. 1–2 (1984): 59–66.
46. Kreuter, W. and H. Hofmann. "Electrolysis: The Important Energy Transformer in a World of Sustainable Energy." *International Journal of Hydrogen Energy* 23, no. 8 (1998): 661–6.
47. de Levie, R. "The Electrolysis of Water." *Journal of Electroanalytical Chemistry* 476, no. 1 (1999): 92–3.
48. Bard, A.J. and L.R. Faulkner. *Electrochemical Methods: Fundamentals and Applications*, Vol. 2. New York: Wiley, 1980.
49. Steele, B.C.H. "Materials for Electrochemical Energy Conversion and Storage Systems." *Ceramics International* 19, no. 4 (1993): 269–77.
50. Rosa, V.M., M.B.F. Santos, and E.P. da Silva. "New Materials for Water Electrolysis Diaphragms." *International Journal of Hydrogen Energy* 20, no. 9 (1995): 697–700.
51. Suffredini, H.B., J.L. Cerne, F.C. Crnkovic, S.A.S. Machado, and L.A. Avaca. "Recent Developments in Electrode Materials for Water Electrolysis." *International Journal of Hydrogen Energy* 25, no. 5 (2000): 415–23.

52. Agbossou, K., R. Chahine, J. Hamelin, F. Laurencelle, A. Anouar, J.M. St-Arnaud, and T.K. Bose. "Renewable Energy Systems Based on Hydrogen for Remote Applications." *Journal of Power Sources* 96, no. 1 (2001): 168–72.
53. Wendt, H. *Electrochemical Hydrogen Technologies: Electrochemical Production and Combustion of Hydrogen.* New York: Elsevier, 1990.
54. Goñi-Urtiaga, A., D. Presvytes, and K. Scott. "Solid Acids as Electrolyte Materials for Proton Exchange Membrane (PEM) Electrolysis: Review." *International Journal of Hydrogen Energy* 37, no. 4 (2012): 3358–72.
55. Grigoriev, S.A., V.I. Porembsky, and V.N. Fateev. "Pure Hydrogen Production by PEM Electrolysis for Hydrogen Energy." *International Journal of Hydrogen Energy* 31, no. 2 (2006): 171–5.
56. Laguna-Bercero, M.A. "Recent Advances in High Temperature Electrolysis Using Solid Oxide Fuel Cells: A Review." *Journal of Power Sources* 203 (2012): 4–16.
57. Doenitz, W., R. Schmidberger, E. Steinheil, and R. Streicher. "Hydrogen Production by High Temperature Electrolysis of Water Vapour." *International Journal of Hydrogen Energy* 5, no. 1 (1980): 55–63.
58. Isenberg, A.O. "Energy Conversion via Solid Oxide Electrolyte Electrochemical Cells at High Temperatures." *Solid State Ionics* 3–4 (1981): 431–7.
59. Antonucci, V., A. Di Blasi, V. Baglio, R. Ornelas, F. Matteucci, J. Ledesma-Garcia, L.G. Arriaga, and A.S. Aricò. "High Temperature Operation of a Composite Membrane-Based Solid Polymer Electrolyte Water Electrolyser." *Electrochimica Acta* 53, no. 24 (2008): 7350–56.
60. Hauch, A., S.D. Ebbesen, S.H. Jensen, and M. Mogensen. "Highly Efficient High Temperature Electrolysis." *Journal of Materials Chemistry* 18, no. 20 (2008): 2331–40.
61. Bard, A.J. "Photoelectrochemistry." *Science* 207, no. 4427 (1980): 139–44.
62. Fischer, M. "Review of Hydrogen Production with Photovoltaic Electrolysis Systems." *International Journal of Hydrogen Energy* 11, no. 8 (1986): 495–501.
63. Matsuoka, M., M. Kitano, M. Takeuchi, K. Tsujimaru, M. Anpo, and J.M. Thomas. "Photocatalysis for New Energy Production: Recent Advances in Photocatalytic Water Splitting Reactions for Hydrogen Production." *Catalysis Today* 122, no. 1–2 (2007): 51–61.
64. Bard, A.J. and M.A. Fox. "Artificial Photosynthesis: Solar Splitting of Water to Hydrogen and Oxygen." *Accounts of Chemical Research* 28, no. 3 (1995): 141–5.
65. Hoffmann, M.R., S.T. Martin, W. Choi, and D.W. Bahnemann. "Environmental Applications of Semiconductor Photocatalysis." *Chemical Reviews* 95, no. 1 (1995): 69–96.
66. Kaneko, M. and I. Okura. *Photocatalysis: Science and Technology.* Tokyo: Kodansha, 2002.
67. Lewerenz, H.-J., N. Dietz, R. Collazo, A.J. Bard, T.S. Zhao, L. Peter, K. Domen, H. Frei, and F. Schuth. *Photoelectrochemical Water Splitting: Materials, Processes and Architectures RSC Energy and Environment Series 9.* Great Britain: Royal Society of Chemistry, 2013.
68. Fujishima, A. and K. Honda. "Electrochemical Photolysis of Water at a Semiconductor Electrode." *Nature* 238, no. 5358 (1972): 37–8.
69. Kraeutler, B. and A.J. Bard. "Heterogeneous Photocatalytic Preparation of Supported Catalysts. Photodeposition of Platinum on Titanium Dioxide Powder and Other Substrates." *Journal of the American Chemical Society* 100, no. 13 (1978): 4317–18.
70. Domen, K., A. Kudo, A. Shinozaki, A. Tanaka, K.-i. Maruya, and T. Onishi. "Photodecomposition of Water and Hydrogen Evolution from Aqueous Methanol Solution over Novel Niobate Photocatalysts." *Journal of the Chemical Society, Chemical Communications* no. 4 (1986): 356–7.
71. Katakis, D.F., C. Mitsopoulou, J. Konstantatos, E. Vrachnou, and P. Falaras. "Photocatalytic Splitting of Water." *Journal of Photochemistry and Photobiology A: Chemistry* 68, no. 3 (1992): 375–88.
72. Bagotsky, V.S. and Corporation Ebooks. *Fuel Cells: Problems and Solutions.* Hoboken, NJ: Wiley, 2012.
73. Hoogers, G. *Fuel Cell Technology Handbook.* Boca Raton, FL: CRC Press, 2002.
74. Revankar, S.T. and P. Majumdar. *Fuel Cells: Principles, Design, and Analysis.* Boca Raton, FL: CRC Press, 2014.
75. Srinivasan, S. *Fuel Cells: From Fundamentals to Applications.* New York: Springer, 2006.
76. Bagotsky, V.S. "The Long History of Fuel Cells." In *Fuel Cells*, 27–42. John Wiley & Sons Inc., Hoboken, NJ, 2008.
77. Perry, M.L. and T.F. Fuller. "A Historical Perspective of Fuel Cell Technology in the 20th Century." *Journal of the Electrochemical Society* 149, no. 7 (2002): S59–67.
78. Grove, W.R. "XXIV. On Voltaic Series and the Combination of Gases by Platinum." *The London and Edinburgh Philosophical Magazine and Journal of Science* 14, no. 86 (1839): 127–30.
79. Grove, W.R. "On a Gaseous Voltaic Battery." *Philosophical Magazine* 92, no. 31 (2012): 3753–56.
80. Sorrell, C.C., S. Sugihara, and J. Nowotny. *Materials for Energy Conversion Devices.* Boca Raton, FL, CRC Press, 2005.
81. Glazebrook, R.W. "Efficiencies of Heat Engines and Fuel Cells: The Methanol Fuel Cell as a Competitor to Otto and Diesel Engines." *Journal of Power Sources* 7, no. 3 (1982): 215–56.
82. Prater, K.B. "Polymer Electrolyte Fuel Cells: A Review of Recent Developments." *Journal of Power Sources* 51, no. 1–2 (1994): 129–44.
83. Gottesfeld, S. "Fuel Cell Techno-Personal Milestones 1984–2006." *Journal of Power Sources* 171, no. 1 (2007): 37–45.
84. McLean, G.F., T. Niet, S. Prince-Richard, and N. Djilali. "An Assessment of Alkaline Fuel Cell Technology." *International Journal of Hydrogen Energy* 27, no. 5 (2002): 507–26.
85. Cheng, X., Z. Shi, N. Glass, L. Zhang, J. Zhang, D. Song, Z.-S. Liu, H. Wang, and J. Shen. "A Review of PEM Hydrogen Fuel Cell Contamination: Impacts, Mechanisms, and Mitigation." *Journal of Power Sources* 165, no. 2 (2007): 739–56.
86. Smitha, B., S. Sridhar, and A.A. Khan. "Solid Polymer Electrolyte Membranes for Fuel Cell Applications—A Review." *Journal of Membrane Science* 259, no. 1 (2005): 10–26.
87. Li, X. and I. Sabir. "Review of Bipolar Plates in PEM Fuel Cells: Flow-Field Designs." *International Journal of Hydrogen Energy* 30, no. 4 (2005): 359–71.
88. Kordesch, K.V. "Power Sources for Electric Vehicles." In *Modern Aspects of Electrochemistry*, edited by J. O'M. Bockris and B. E. Conway, 339–443. Springer, New York, 1975.
89. Schultz, T., S. Zhou, and K. Sundmacher. "Current Status of and Recent Developments in the Direct Methanol Fuel Cell." *Chemical Engineering & Technology* 24, no. 12 (2001): 1223–33.
90. Hampson, N.A., M.J. Willars, and B.D. McNicol. "The Methanol–Air Fuel Cell: A Selective Review of Methanol Oxidation Mechanisms at Platinum Electrodes in Acid Electrolytes." *Journal of Power Sources* 4, no. 3 (1979): 191–201.
91. Aricò, A.S., S. Srinivasan, and V. Antonucci. "DMFCS: From Fundamental Aspects to Technology Development." *Fuel Cells* 1, no. 2 (2001): 133–61.

92. Bischoff, M. "Molten Carbonate Fuel Cells: A High Temperature Fuel Cell on the Edge to Commercialization." *Journal of Power Sources* 160, no. 2 (2006): 842–5.
93. Watanabe, T., Y. Izaki, Y. Mugikura, H. Morita, M. Yoshikawa, M. Kawase, F. Yoshiba, and K. Asano. "Applicability of Molten Carbonate Fuel Cells to Various Fuels." *Journal of Power Sources* 160, no. 2 (2006): 868–71.
94. Bischoff, M. and G. Huppmann. "Operating Experience with a 250 kWel Molten Carbonate Fuel Cell (MCFC) Power Plant." *Journal of Power Sources* 105, no. 2 (2002): 216–21.
95. Ormerod, R.M. "Solid Oxide Fuel Cells." *Chemical Society Reviews* 32, no. 1 (2003): 17–28.
96. Steele, B.C.H. and A. Heinzel. "Materials for Fuel-Cell Technologies." *Nature* 414, no. 6861 (2001): 345–52.

2 Advances in Electrochemical Energy Materials and Technologies

Hao Liu, Bing Sun, and Guoxiu Wang

CONTENTS

2.1 INTRODUCTION

Greenhouse gas emissions from consumption of fossil fuels by traditional vehicles are major causes of global warming and worldwide climate change. Rechargeable batteries are widely considered as the promising power source for the next generation of electric vehicles in order to relieve our reliance on fossil fuels. The lithium-ion battery is well recognized as the best choice among all different electrochemical power sources, such as fuel cells, solar cells, lead–acid, nickel–cadmium, and nickel metal hydride batteries. The research and development on the lithium-ion batteries has progressed rapidly since it was first commercialized in the 1990s. Rechargeable lithium-ion batteries have revolutionized portable electronic devices and have become the dominant power source for mobile phones, laptop computers, and digital cameras because of their high energy density [1,2]. However, the charge–discharge process in lithium-ion batteries at high current rates can cause a high level of polarization for bulk materials and degrade the electrochemical properties of the batteries. The development of electric vehicles or hybrid electric vehicles demands high-power batteries that can operate under high current conditions. In the following sections, we will briefly introduce advances in materials and technologies for lithium-ion batteries and lithium oxygen batteries.

2.2 ADVANCES IN MATERIALS AND TECHNOLOGIES FOR LITHIUM-ION BATTERIES

The fundamental mechanism of the lithium-ion battery relies on lithium-ion insert/extract between the electrodes during the charge–discharge processes. A schematic of the fundamental mechanism of the lithium-ion battery is shown in Figure 2.1, demonstrating that the lithium-ion battery working principle is related to the reversible extraction and insertion of lithium ions between the two electrodes. Simultaneously, to retain charge neutrality in the electrodes, removal and addition of electrons occur in the outside circuit. The lithium cobalt oxide ($LiCoO_2$) and graphite are usually used as cathode and anode materials, respectively. Although the battery has a high energy density, it has a low power rate (slow charge–discharge). The intrinsic diffusion of lithium ions in the solid state (~10^{-8} cm^2/s) inevitably limits the rate of intercalation/de-intercalation. It is crucial to develop new materials or

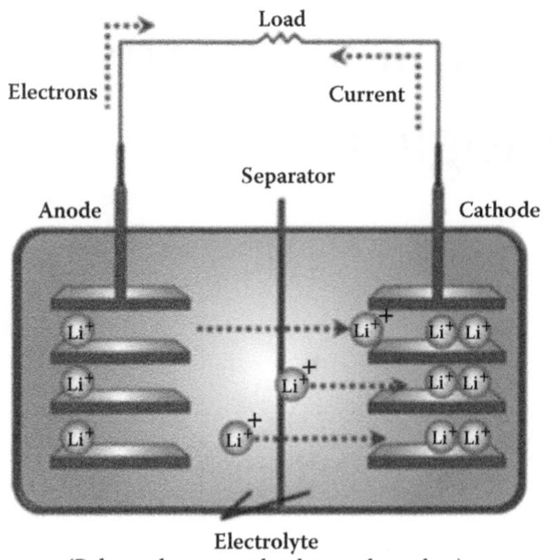

FIGURE 2.1 Schematic representation of a lithium–ion battery. Negative electrode, graphite; positive electrode, $LiCoO_2$.

technologies to improve the electrochemical performance of lithium-ion batteries at high-current charge–discharge.

2.2.1 Advances in Electrode Materials

In the 1990s, the Sony company combined an $LiCoO_2$ cathode with a carbon anode to make the first successful lithium-ion battery [3,4], which currently dominates the lithium battery market. However, because of the drawbacks of current commercial electrode materials (e.g., low capacity, high cost, and toxicity), researchers are seeking substitutes to replace them. Figure 2.2 shows voltage versus capacity for positive- and negative-electrode materials presently used or under serious considerations for the next generation of rechargeable lithium-ion batteries [1].

2.2.1.1 Cathode Materials

In lithium-ion batteries, the cathode is the positive electrode that is associated with reductive chemical reactions and gains electrons from the external circuit. The key requirements for a material to be successfully used as a cathode in a rechargeable lithium battery is that the material reacts with lithium with a high free energy of reaction (e.g., high capacity, high voltage, and high energy storage).

For commercial Li-ion batteries, because the carbon as negative electrode is lithium free, the positive electrode must act as a source of lithium ions. This requires the use of air-stable Li-based intercalation compounds to facilitate the cell assembly. On the basis of these considerations, a series of materials, including $LiCoO_2$, $LiNiO_2$, $LiMn_2O_4$, and $LiFePO_4$, have been investigated as cathode materials for lithium-ion batteries. Conductive oxide aerogels such as V_2O_5 and MnO_2 also have the potential to boost the field of energy storage, using metallic Li as the negative electrode.

The current commercial cathode material dominating the small lithium-ion cells is layered $LiCoO_2$. Although the theoretical capacity of $LiCoO_2$ is as high as 274 mAh/g, the material can only achieve about 140 mAh/g in practical applications. Many attempts have been carried out to develop new cathode materials with high capacities, good cyclability, and excellent high-rate performance. In the following sections, a brief review on advanced cathode materials such as $LiFePO_4$ and V_2O_5 will be presented.

2.2.1.1.1 $LiFePO_4$ Cathode Materials

$LiFePO_4$ is the most attractive cathode material for large-scale lithium-ion batteries, such as for electric vehicle and hybrid

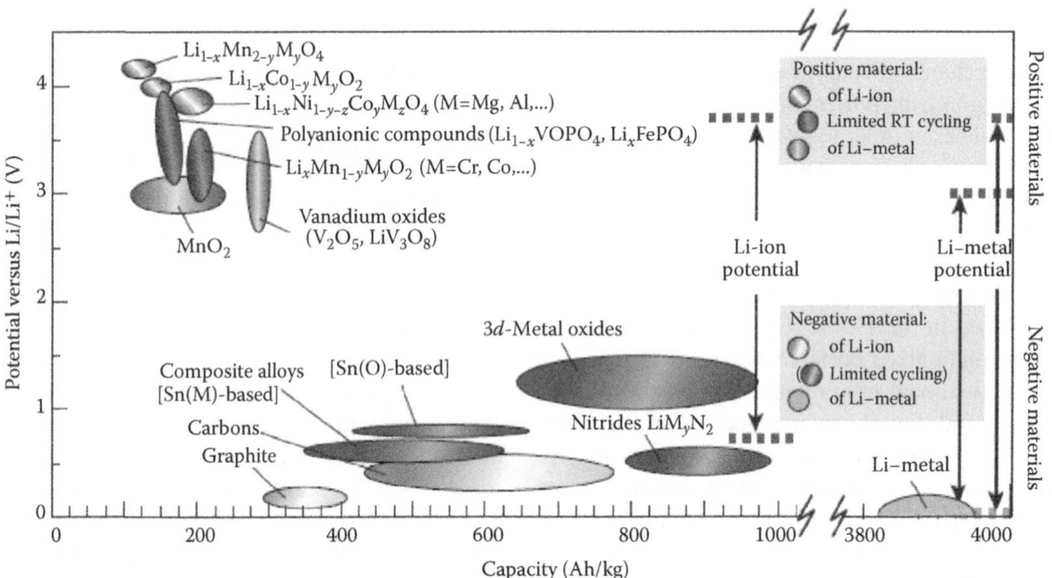

FIGURE 2.2 Voltage versus capacity for electrode materials presently used or under serious consideration for the next generation of rechargeable Li-based cells. (Reproduced by permission from Macmillan Publishers Ltd. *Nature*, Tarascon, J.M. and Armand M., Issues and challenges facing rechargeable lithium batteries, 414, 359–367, Copyright 2001.)

electric vehicle applications, owing to its low cost, abundant raw materials, and environmental friendliness. Since the first report in 1997 by Goodenough et al. [5,6], $LiFePO_4$ has been widely investigated as a promising cathode material to replace $LiCoO_2$ in lithium-ion batteries [7–12]. $LiFePO_4$ cathode material demonstrates reversible lithium intercalation, with a theoretical capacity of 170 mAh/g. The discharge potential is about 3.4 V versus lithium, and there is no obvious capacity fading even after several hundred cycles. $LiFePO_4$, with the crystal parameters of a = 10.333 Å, b = 6.011 Å, and c = 4.696 Å, belongs to the orthorhombic olivine structure with the space group P_{nma}. In the lattice of $LiFePO_4$, the polyhedrals essentially block diffusion of the lithium ions, as the diffusion is fast only along the one-dimensional (1D) tunnel, which has been investigated from first-principles calculations [13].

Because $LiFePO_4$ material has a very low electronic conductivity at room temperature, it can only achieve its theoretical capacity at a very low current density or at elevated temperatures, due to the low lithium diffusion at the interface. The loss in capacity is a diffusion-limited phenomenon associated with the two-phase character of the intercalation process [5]. There has been a surge of interests in this $LiFePO_4$ material and in understanding how to enhance the electronic conductivity. Thus far, most reports have focused on carbon coating to enhance the electronic conductivity because carbon is the most abundant and efficient conductive agent. It is reported that by increasing the carbon (10, 15, and 20 wt.%) mixed with $LiFePO_4$, an increase in the overall electrochemical performance was observed [14]. Impedance spectroscopy has been used to investigate the ohmic and kinetic contributions to the overvoltage of the cell. It was found that increasing carbon content could reduce the cell impedance because of the reduction of the charge-transfer resistance. A further decrease of the charge-transfer resistance in high-carbon-content cathodes (20 wt.% carbon) was also obtained. The carbon coating also significantly increases the tap density [15]. Wang et al. showed that an $LiFePO_4$/C nanocomposite significantly improves the electrochemical performance of $LiFePO_4$ material [16]. An ordered mesoporous $LiFePO_4$/C nanocomposite has been developed, in which $LiFePO_4$ nanoparticles are embedded in mesoporous carbon networks. This mesoporous nanoarchitecture ensures not only intimate contact between the organic electrolyte and $LiFePO_4$ nanoparticles but also high electronic conductivity for both facile mass transfer and charge transfer. Figure 2.3a shows the cyclability of mesoporous $LiFePO_4$/C cathode at a low current rate of 0.1 C. The initial discharge capacity is 166 mAh/g, which is very close to the theoretical capacity of $LiFePO_4$ cathode (170 mAh/g). After 100 cycles, the cell still delivers a capacity of 162 mAh/g. The charge and discharge capacity curves are overlapped due to the ultrahigh coulombic efficiency with an average efficiency of almost 100% (as shown in the inset of Figure 2.3a). Figure 2.3b presents the charge–discharge profiles of mesoporous $LiFePO_4$/C cathodes at different current rates in the first cycle. Although the specific capacity gradually decreases with increasing current rate, a high initial capacity of 156 mAh/g capacity was still achieved at the 1 C rate. At the high current rate of 10 C, a capacity of 118 mAh/g was obtained, demonstrating that the as-prepared mesoporous $LiFePO_4$/C nanocomposite can endure high rate charge and discharge. These results have illustrated that the electrochemical properties of $LiFePO_4$ electrode, for both long-term and high-rate performance, were improved after carbon coating.

Besides carbon species, other materials have also been used for $LiFePO_4$ coatings. Conductive metal particles (copper and silver) were dispersed on the surface of $LiFePO_4$ material to enhance the electronic conductivity [17,18]. Conductive polymers such as polypyrrole, polyethylene glycol, polyacene, poly(3,4-ethylenedioxythiophene), and polyaniline were coated on the $LiFePO_4$ surface, to achieve excellent capacity retention and rate capability [19–24]. Moreover, metal oxides have been used as coating agents to enhance the electrochemical

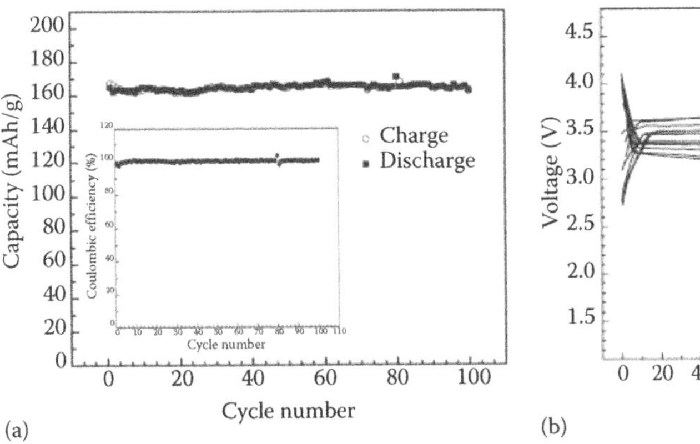

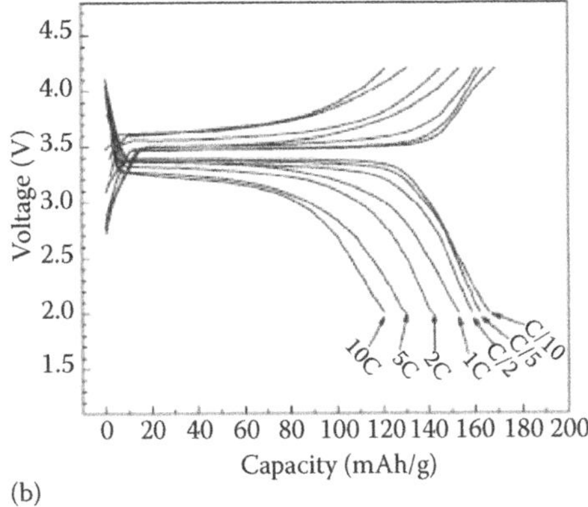

FIGURE 2.3 (a) Cycling performance of mesoporous $LiFePO_4$/C nanocomposite electrode at 0.1 C. The inset shows the coulombic efficiency. (b) First discharge and charge profiles of $LiFePO_4$/C nanocomposite electrodes at different current rates. (Based on Wang, G.X., Liu, H., Liu, J., Qiao, S.Z., Lu, G.Q.M., Munroe, P., and Ahn, H: Mesoporous $LiFePO_4$/C nanocomposite cathode materials for high power lithium ion batteries with superior performance. *Advanced Materials*. 2010. 22. 4944–4948. Copyright Wiley-VCH Verlag Gmbh & Co. KGaA. Reproduced with permission.)

properties. The coating layer can protect $LiFePO_4$ from volume change during the charge–discharge process, and hence improve the electrochemical performance of coated $LiFePO_4$ [25–28].

A report in 2002 showed excellent electrochemical behavior when the $LiFePO_4$ was "doped" with small amounts of supervalent elements such as Nb, Zr, Ti, Al, W, and Mg, due to the increase of electronic conductivity of $LiFePO_4$ material [29]. The conductivity of doped $LiFePO_4$ could be increased by 8 orders of magnitude. Liu et al. doped $LiFePO_4$ with Zn atoms [30]. During de-intercalation and intercalation of lithium ions, the doped zinc atoms in the lattice protect the $LiFePO_4$ crystal from shrinking. This kind of "pillar" effect provides larger space for the movement of lithium ions. The diffusion coefficient of lithium ions depends on the radius of the heteroatoms. As the radius of the heteroatom becomes larger, the diffusion coefficient increases [31].

The nanostructure, including morphology and crystalline orientation, has significant effect on the electrochemical performance of $LiFePO_4$ cathode. Kang and Ceder reported that $LiFePO_4$ nanoparticles can achieve high lithium bulk mobility, by creating a fast ion-conducting surface phase through controlled off-stoichiometry [32]. At a charge–discharge rate of 2 C, the nano-$LiFePO_4$ material still discharges to its theoretical capacity of 166 mAh/g. Even at a highest testing rate of 50 C, corresponding to a time of 72 s to fully discharge the capacity, the material achieved about 80% of its theoretical capacity. Mesoporous and hollow $LiFePO_4$ can be obtained by using a hard-template method [33]. Electrochemical cycling of nanowire and hollow $LiFePO_4$ cathodes demonstrates excellent rate capability even at high rate of 10 C, where they show more than 89% capacity retention of the initial capacity. Owing to the higher Brunauer–Emmett–Teller surface area, the rate capability of the hollow cathode is 6% higher than that of the nanowire cathode at 15 C. These results clearly demonstrate that mesoporous and hollow $LiFePO_4$ have higher power density and better high-rate performance than those of previously reported solid counterparts. Wang et al. reported a large exposed face of $LiFePO_4$ nanoplates synthesized by solvothermal reaction in glycol-based medium [34]. The electrochemical results illustrate that $LiFePO_4$ nanoplates with the crystal orientation along the ac facet present superior electrochemical performances, such as high 148 mAh/g at a 10 C rate. Meanwhile, the $LiFePO_4$ nanoplates with crystal orientation along the bc facet present 28 mAh/g at 10 C rate. The results demonstrated that the crystal orientation greatly affects the electrochemical properties of $LiFePO_4$ nanoplates. The high-percentage exposure of (010) planes improves the rate performance.

2.2.1.1.2 *V_2O_5*

Layered vanadium pentoxide (V_2O_5) is a typical intercalation compound. Considering that vanadium oxides possess versatile redox-dependent properties and high redox potential, it can be applied as cathode material for lithium-ion batteries, using metallic Li or Li-contained compound as the negative electrode. Electrochemical lithium intercalation occurs together with compensating electrons, resulting in the formation of vanadium bronzes.

There have been extensive interests in the preparation of vanadium oxides by the sol–gel method [35–37]. The dried gel material reacts with lithium in a single continuous discharge curve around 3.1 V and a total lithium uptake of 4.1 Li by 2.8 V, thus giving a higher capacity than crystalline V_2O_5. The xerogel vanadium pentoxides contain sheets composed of two vanadium oxide layers. The vanadyl bonds lead to a distorted octahedral coordination around the lithium crystalline V_2O_5. The double layers of vanadium oxides found in the xerogel have been described in a number of other vanadium oxides. However, their rate capabilities are still limited. Recently, a large variety of 1D nanostructured V_2O_5, including nanowires, nanobelts, nanorolls, ordered arrays of nanorods, nanotubes, and nanocables, have been developed and investigated as electrode materials for lithium-ion batteries [38–45]. All these nanostructured electrode materials exhibit excellent storage capacity and rate performance, owing to the large specific surface area and the short diffusion pathway. Lutta et al. reported a new method to prepare V_2O_5 nanofibers with dimensions less than 140 nm by coating vanadium oxide on polylactide fibers. Electrochemical measurement reveals that the V_2O_5 nanofibers are electrochemically active and undergo reversible reactions with lithium ions [46]. Cao et al. developed a mediated polyol process to synthesize V_2O_5 with highly ordered superstructures, in which hollow microspheres consist of nanorods in shells [47]. The physical state of the nanorods on the surface and the formed microspheres are tunable by changing the concentration of reactants. The prepared V_2O_5 exhibits desirable electrochemical properties such as high capacity and remarkable reversibility, as cathode material for lithium-ion battery.

A strategy to enhance the specific capacity and rate capability of layered V_2O_5 is to increase interlayer distance or decrease the interaction forces between the layers. Therefore, the intercalation dynamics of lithium ions can be boosted [48–50].

2.2.1.2 Anode Materials

The anode is the negative electrode of a cell and associated with oxidative chemical reactions that release electrons into the external circuit. The ideal anode material always has a low operating voltage and a high specific capacity. Lithium has received much attention as a promising anode material for a long time, owing to its unique properties: (i) it is the most electronegative metal (–3.04 V versus the standard hydrogen electrode), and (ii) it is the lightest metal (0.534 g/cm^3). The first aspect leads to high cell voltage, and the second aspect generates high specific capacity (3.86 Ah/g). The failure of lithium metal as an anode is attributed to the formation of lithium dendrite during the cycling. The lithium dendrite may penetrate the separator and cause safety issues such as a short circuit or even explosion. As a result, "host–guest" chemistry was considered, with such electrodes also known as "intercalation"-type electrodes.

Currently, graphite is used as the commercial negative electrode of lithium-ion batteries. The graphite anode materials have many advantages. These include the following: (i) they exhibit both higher specific charges and more negative redox potentials than most metal oxides, chalcogenides, and polymers (~0.5 V versus Li/Li^+) and (ii) its dimensional stability results in better cycling performance than that of Li alloys. The insertion of lithium into graphite is called intercalation [51]. During the electrochemical reduction (charge) process of the carbon host, lithium ions from the electrolyte penetrate into the carbon and form a lithium/carbon intercalation compound, LiC_6. Thus, graphite delivers a theoretical capacity of 372 mAh/g. Many materials with higher specific capacities have been developed as anode materials for lithium-ion batteries during the past decades. Herein, some advanced materials such as tin oxides and silicon are summarized as anode materials for lithium-ion batteries.

2.2.1.2.1 Tin Oxides

Tin-based oxides (SnO and SnO_2) have been extensively examined since the report in 1997 by Idota and coworkers [52]. However, the mechanism for these metal oxides used in lithium-ion batteries is different from that of the transition metal oxides. The tin oxides host lithium ions according to a mechanism combined with an oxide decomposition and an alloy formation. From the reactions, the Sn–O bonding is irreversibly dissociated by the introduction of lithium ions and electrons at the first lithiation step, forming Li_2O and metallic Sn. In the second step, the Sn is responsible for the reversible capacity of the Sn-based oxide anode through the alloying/de-alloying reaction with lithium up to the theoretical limit of $Li_{4.4}Sn$. The detailed mechanism of Li_2O formation in Sn oxides determined by electrochemical measurements and Auger electron spectroscopy suggested that Li_2O was formed at 0 V [53]. The Sn-based oxide anode showed improved capacity retention compared with metallic Sn, which was mainly ascribed to the resulting composite structure of electrochemically active Sn and an inactive lithium phase, such as Li_2O. Li_2O acts as a matrix where the reduced Sn phase is finely dispersed and thereby prevents the aggregation of Sn. However, upon charge–discharge cycles, Sn phases surrounded by the Li_2O matrix arising from delithiation of the Li–Sn alloy phase showed a tendency to aggregate with each other into large particles [54]. These have been directly observed by high-resolution transmission electron microscopy [55,56]. The aggregation of Sn particles after prolonged cycling caused a large volume change and destruction of the Li_2O matrix phase, which eventually degraded the capacity retention of the Sn oxide anode.

To improve the capacity retention of tin oxide anodes, it is crucial to suppress the aggregation of Sn atoms into larger particles during cycling. The control of the cutoff voltage for delithiation is highly effective in suppressing the coarsening of Sn particles, which mainly occurred in the delithiation process from an Li–Sn intermetallic phase to Sn. It should be noted that the physical modification of Sn-based oxide anode materials can provide free volume within the materials as a buffer against volume changes during cycling. Lou et al. reported a nanoarchitecture of coaxial SnO_2@carbon hollow nanospheres for lithium-ion batteries [57]. The novel architecture was prepared by a two-step coating of SnO_2 and carbon layers on a spherical silica template. The coaxial SnO_2@carbon hollow spheres exhibit a good cycle life for at least several hundred cycles and excellent rate capability. The discharge capacity can retain a capacity of 450 mAh/g after 100 cycles at 0.8 C. A stable capacity of 520 mAh/g was achieved at the current rate of 0.32 C. Afterward, the rate is increased stepwise up to 12 C and returned to 0.32 C. A stable high capacity of 500 mAh/g was maintained when the current rate is reversed back to 0.32 C after more than 200 cycles. Liu et al. prepared an ordered mesoporous SnO_2/C composite consisting of SnO_2 crystalline core and amorphous carbon shell, by a facile vacuum-assisted impregnation method. The SnO_2/C material achieved an excellent electrochemical performance as an anode material for lithium-ion batteries [58]. In Figure 2.4, the discharge capacity of mesoporous SnO_2/C

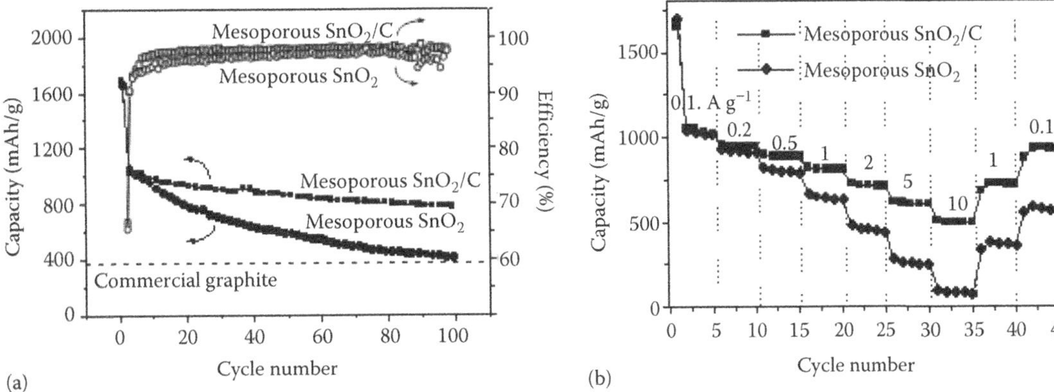

FIGURE 2.4 (a) Lithium storage performance of mesoporous SnO_2 and SnO_2/C at the galvanostatic current of 0.1 A/g. (b) Multirate tests of mesoporous SnO_2 and SnO_2/C, in order of 0.1, 0.2, 0.5, 1, 2, 5, 10 A/g, and then return to low currents of 1 and 0.1 A/g, with five cycles for each step. (Liu, H., Chen, S., Wang, G.X., and Qiao, S.Z: Ordered mesoporous core/shell SnO_2/C nanocomposite as high-capacity anode material for lithium-ion batteries. *Chemistry—A European Journal*. 2013. 19. 16897–16901. Copyright Wiley-VCH Verlag Gmbh & Co. KGaA. Reproduced with permission.)

at 0.1 A/g after 100 cycles is 780 mAh/g, which is more than double the theoretical capacity of commercial graphite material. The capacity at discharge current of 10 A/g is 510 mAh/g, illustrating that this mesoporous SnO_2/C could be potentially used for high-power batteries. The improved electrochemical performance of this core–shell mesoporous SnO_2/C could be attributed to its high specific surface area, large pore volume, and narrow pore-size distribution, which are beneficial to lithium-ion intercalation and electrolyte diffusion. Furthermore, the SnO_2/C electrode exhibits a lower resistance and higher charge-transfer dynamics than that of uncoated mesoporous SnO_2, owing to the improvement of carbon coating.

2.2.1.2.2 Silicon-Based Materials

Silicon (Si) has been considered as one of the most attractive anode materials for lithium-ion batteries because it has the highest gravimetric capacity among anodes. Si can alloy up to 4.4 lithium atoms per Si unit, corresponding to a theoretical capacity of more than 4200 mAh/g ($4.4Li + Si \leftrightarrow Li_{4.4}Si$), which is more than 10 times the capacity of commercial graphite anode material (372 mAh/g). Moreover, silicon is also abundant, cheap, and environmentally benign. Silicon swells by more than 300% volume expansion when 4.4 units of lithium ions are intercalated to form the alloy. Therefore, it presents dramatic capacity fading due to pulverization of the active material caused by the large volume expansion/shrinkage during the discharge–charge process [59].

To overcome the issue associated with volume change, many approaches have been applied to develop novel Si electrodes with high electrochemical performance. By designing the morphology of Si in the nanoscale, the pulverization issue can be suppressed due to the unique size effect of nanomaterials. Ma et al. developed nest-like Si nanospheres, which were prepared by a solvothermal method [60]. They found that the size and morphology of the products are tunable by the experimental conditions. The galvanostatic charge–discharge cycling of nest-like Si nanospheres was conducted in the potential region of 1.6–0.02 V at different current densities. The electrodes with the nest-like Si nanospheres could exhibit a large specific capacity of 3628, 3291, and 3052 mAh/g at the current density of 400 (0.1 C), 800 (0.2 C), and 2000 mA/g (0.5 C), respectively, revealing the excellent high-rate capability of this material. Chan et al. grew one-dimensional silicon nanowires by the chemical vapor deposition (CVD) method [61]. The silicon nanowires can accommodate large strain without pulverization and provide short lithium-ion diffusion pathways. The Si nanowire electrode material can maintain nearly 75% of its theoretical capacity, with almost no fading during cycling.

Another established strategy to depress the volume change is surface coating. With carbon coating, the volume change of Si particles can be restrained by the carbon layer. Additionally, the carbon layer can increase the conductivity of Si materials and therefore enhance the electrochemical performance at high rates. A novel design of carbon–silicon core–shell nanowires for high-power and long-life lithium battery electrodes has been developed. Amorphous silicon was coated onto carbon nanofibers to form a core–shell structure. The resultant core–shell nanowires showed good performance as anode materials in lithium-ion batteries. Since carbon has a much lower capacity compared with silicon, the carbon core experiences less structural stress or damage during lithium cycling and can act as a mechanical support and an efficient electron conducting media. These nanowires have a high charge-storage capacity close to 2000 mAh/g and a good cycling life. They also have a high coulombic efficiency of 90% for the first cycle and 98–99.6% in the following cycles [62]. Silicon crystalline–amorphous core–shell nanowires were grown directly on stainless-steel current collectors by a simple one-step synthesis. Amorphous Si shells can be selected to be electrochemically active instead of crystalline Si cores owing to the difference in their lithiation potentials. Therefore, crystalline Si cores function as a stable mechanical support and an efficient electrically conducting pathway, while amorphous shells store lithium ions. These core–shell nanowires achieved high charge-storage capacity (close to 1000 mAh/g, three times that of carbon) with a 90% capacity retention over 100 cycles. They also showed excellent electrochemical performance at high-rate charging–discharging (6.8 A/g, close to 20 times that of carbon at the 1 C rate) [63].

The electrochemical performance of silicon electrode is also affected by the binder used for the fabrication. Kovalenko et al. [64] reported that mixing Si nanopowder with alginate, a natural polysaccharide extracted from brown algae, yields a stable anode possessing eight times reversible capacity compared with that of the graphitic anodes. By using alginate as a binder, the initial efficiency and cycle stability were improved at a current density of 4200 mA/g. The reversible specific capacity of an alginate-based Si anode is about 2000 mAh/g, which is more than five times higher than the theoretical capacity of graphite and higher than the electrodes using poly(vinylidene fluoride) and carboxymethylcellulose as binders. The contribution of Si nanopowder alone could be calculated as ~4000 mAh/g, which is consistent with observations on other nano-Si materials but is noticeably higher than previously observed for micron-size Si. The volumetric anode capacity was determined to be about 1520 mAh/cm^3 at a current density of 140 mA/g, which is 2.5 times higher than that of graphite anodes (~620 mAh/cm^3).

2.2.2 Alternative Technologies for Lithium-Ion Batteries

The preparation technology is critical for the commercialization of electrode materials in practical applications. Currently, the solid-state method is widely applied because it is a facile route for large-scale production. However, the solid-state method consumes huge energy with low efficiency. The particle size of the product is large and not homogeneous. Therefore, researchers are seeking for other methods to prepare scalable nanomaterials with high efficiency. Thus far, many promising technologies have been developed to synthesize large-quantity electrode materials.

2.2.2.1 Spray Pyrolysis Method

Spray pyrolysis is a widely applied technology in chemical, pharmaceutical, and food industries. It enables simple, continuous, and scalable production of fine particles. Typically, the solvent serves as the fuel; thus, cost and solubility issues lead to use of solvents to dissolve the precursors. The mixture aerosol undergoes rapid combustion within milliseconds, oxidizing all the components at high temperatures. Spray pyrolysis provides a robust, versatile route to produce single-phase or composite powders in the nanometer size range with varying morphology from relatively low-cost inorganic/organic precursors.

Zhou et al. developed a facile method for the low-cost and scalable synthesis of α-Fe_2O_3 multishelled hollow spheres (MSHSs) by spray pyrolysis [65]. The synthesis involves simple spray-drying the two cheap precursors, iron nitrate and sucrose, followed by annealing in air. The successful preparation of MSHSs is affected by many factors, such as the amount of sucrose and ramp rate in calcination. More important, the α-Fe_2O_3 MSHSs exhibit high specific capacity, excellent rate capability, and good cycling stability. Chou et al. reported that large-scale hollow-structured α-Fe_2O_3/carbon (HIOC) nanocomposite with a high surface area of around 260 m^2/g was synthesized by a one-step, *in situ*, and industrially oriented spray pyrolysis method using iron lactate and sucrose solution as the precursors [66]. Electrochemical measurements showed that the carbon played an important role in affecting both the cycle life and the rate capability of the electrode. The rate capability becomes better when the carbon content in the composite increases. The HIOC composite shows the best electrochemical performance in terms of high capacity (1210 mAh/g at 0.1 C), enhanced rate capability, and excellent cycle stability with 720 mAh/g at 2 C over 220 cycles. The high performance can be ascribed to the novel high-surface-area carbon composite structure that can facilitate the contact between active materials and the electrolyte, enhance lithium and electron transport, and buffer the volume change. In addition, the HIOC can also be used for other potential applications, such as gas sensors, catalysts, and biomedical applications, because it can be easily dispersed in water and has a high surface area. The method presented here could also be adopted to synthesize other high-surface-area metal oxide. Ng et al. prepared spheroidal carbon-coated Si nanocomposite, by a spray-pyrolysis method in air [67]. This Si/C composite is a promising candidate for use as an anode material in lithium-ion batteries, as it has excellent retention of specific capacity, high coulombic efficiency, and low cost. The specific capacity of Si in the composite electrode was calculated to be 3257 mAh/g, which amounts to an impressive 77% of the theoretical value.

2.2.2.2 Electrospinning

Electrospinning is an advanced technology that uses an electrical charge to draw scalable fibrous structured materials from a solution. This preparation process does not require high temperatures to produce a solid product from a solution, and is particularly suitable for the production of fibers using large and complex molecules. Typically, liquid becomes charged when a sufficiently high voltage is applied to a liquid droplet. Then, the electrostatic repulsion counteracts the surface tension and stretches the droplet. At a critical point, a stream of liquid erupts from the surface. If the liquid has sufficiently high molecular cohesion, the breakup of stream does not occur and a charged liquid jet is formed. The jet is then elongated by a whipping process and then deposited on the grounded collector. The elongation and the thinning of the fiber lead to the formation of uniform fibers with a diameter in the nanometer scale.

Zhu et al. reported single-crystalline carbon-coated $LiFePO_4$ nanowires synthesized by using the electrospinning method [68]. These nanowires not only differ from reported analogues in crystallinity, but they are also substantially thinner. The nanowires grow along the c-direction and feature a uniform and continuous carbon coating of about 5 nm thickness. In this way, an efficient conductive network is formed, with very short diffusion lengths along the b-axis, leading to very good rate performance and cycling capability. The unique structure of carbon-coated $LiFePO_4$ nanowires results excellent discharge capacities and rate performance; for example, 169 (0.1 C), 162 (0.5 C), 150 (1 C), 114 (5 C), and 93 (10 C) mAh/g at room temperature can be achieved. Furthermore, it shows excellent cycling stability at elevated temperatures. It can achieve almost the theoretical capacity at 1 C, and 98% capacity retention was obtained after 100 cycles at an elevated temperature of 60°C. Yu et al. prepared an Sn/C composite structure, with Sn@carbon encapsulated in bamboo-like hollow carbon nanofibers, from pyrolysis of TBT/PAN mixture via a coaxial electrospinning method [69]. This composite displays a high reversible capacity of 737 mAh/g after 200 cycles at 0.5 C, demonstrating a potential to be applied as an anode material for lithium-ion batteries. It also exhibits a reversible discharge capacity of 480 mAh/g at a high rate of 5 C. The particular Sn@carbon encapsulated in hollow carbon nanofibers structure has a high content of Sn (close to 70 wt.%) and provides appropriate void volume to cushion the volume change and to prevent pulverization of the Sn nanoparticles. These above-mentioned results demonstrate that the electrospinning method is effective in preparing advanced materials for battery applications.

2.2.2.3 Large-Scale Preparation of Graphene

Graphene has been attracting enormous attention due to its unique properties. Graphene is a one-atom-thick planar sheet of sp^2-bonded carbon atoms that are densely packed in a two-dimensional honeycomb crystal lattice. The carbon–carbon bond length in graphene is about 0.142 nm. Graphene is the basic structural element of some carbon allotopes, including graphite, carbon nanotubes (CNTs), and fullerenes. It can also be considered as an infinitely large aromatic molecule.

Recently, graphene and graphene-based materials have been considered as promising alternative electrode materials in energy-related devices, because these materials have superior electrical conductivities, high surface area, chemical tolerance, high transparency, and a broad electrochemical window. The advantages of graphene-based electrodes have been demonstrated for applications in energy-related electrochemical devices, such as lithium-ion batteries, supercapacitors, and solar cells. Currently, graphene-based materials, including graphene nanosheets (GNSs) [70–73], graphene paper [74], and graphene hybrid nanostructures [75–79], have attracted extensive attention as high-capacity anode materials in lithium-ion batteries, owing to their excellent lithium storage properties. The electrochemical mechanism of the enhancement has been studied by several research groups. Paek et al. demonstrated enhanced cycling performance and lithium storage capacity of SnO_2/graphene nanoporous electrodes with three-dimensionally delaminated flexible structure [76]. In their report, SnO_2/graphene nanoporous electrode materials with delaminated structure were prepared by the reassembly of GNSs in the presence of SnO_2 nanoparticles. It was demonstrated that the obtained SnO_2/GNS electrode exhibited an enhanced reversible capacity as well as superior cycling performance in comparison with that of the bare SnO_2 electrode. Wang et al. investigated the improvement by first-principles calculations [80]. When the graphene layers are completely separated by nanoparticles such as Sn, each graphene layer can host lithium ions on both sides of its surface. The calculation results indicate that lithium can be stably stored on both sides. Therefore, the theoretical capacity of single-layer graphene should be 744 mAh/g, with the atomic model of lithiated graphene of LiC_3. This specific capacity is two times that of graphite.

Large-scale preparation methods for the synthesis of graphene material are requested for practical applications. Cheng's group developed a template-directed CVD technique for the fabrication of macroscopic 3D graphene structure using nickel foams as a template [81]. The graphene sheets are seamlessly interconnected into a 3D flexible network. The high quality of the graphene sheets and the 3D connection lead to an outstanding electrical conductivity. The unique network structure, high specific surface area, and outstanding electrical and mechanical properties of graphene sheets could enable many applications, including high-performance electrically conductive polymer composites, elastic and flexible conductors, electrode materials for lithium-ion batteries and supercapacitors, and catalyst and biomedical supports. Moreover, this template-directed CVD technique is versatile and scalable, and can be a general strategy for fabricating scalable graphene structures. The repeatable growth of graphene with large single-crystal grains on Pt substrate and its nondestructive transfer may enable various applications [82].

Chemical processes cannot prepare defect-free graphene. Paton et al. reported that high-shear mixing of graphite in suitable stabilizing liquids results in large-scale exfoliation of defect-free GNSs [83]. This was achieved using high-shear mixing. This technology is mature, scalable, and widely accessible. It is shown that exfoliation occurs whenever the local shear rate exceeds a critical value. The shear rates can be achieved in a range of mixers, including simple kitchen blenders. Furthermore, high-shear mixing can also be applied to exfoliate other 2D layered materials, including the development of novel MoS_2 and BN.

In summary, the advance of lithium-ion batteries requires large quantities of high-quality electrode materials. There are many challenges, including the synthesis, characterization, electrochemical performance and safety of lithium-ion battery systems.

2.3 ADVANCED NANOCATALYSTS FOR RECHARGEABLE LITHIUM–OXYGEN BATTERIES

2.3.1 Background

Reducing greenhouse gas emissions to decelerate the global warming and worldwide climate change is one of the major challenges for our modern society. The use of renewable energy to reduce fossil fuel consumption becomes increasingly important nowadays than at any time in the past. The petroleum consumed by automobiles and light trucks represents 34% of the world's total primary energy sources [84]. Batteries are widely considered as the best power sources to electrify transport and to be used in stationary electricity storage. Since the first commercialization in the 1990s, rechargeable lithium-ion batteries have dominated the power source market for portable electronic devices, owing to their high energy density, which is two to three times higher than conventional rechargeable batteries, such as lead–acid, Ni–Cd, and Ni–MH batteries. However, even fully developed, the specific energy (energy per unit mass) and energy density (energy per unit volume) for state-of-the-art lithium-ion batteries are still much too low to meet today's requirements as the power source for electric vehicles. New chemistry, materials, and technologies are required to explore new energy storage systems beyond the horizon of current lithium-ion batteries [85,86].

To replace the gasoline in the current transport system, we need to find an alternative power source system that has a gravimetric energy density as high as gasoline does. The energy density of gasoline is 13,000 Wh/kg. Note that the usable energy density of gasoline for automotive applications is approximately 1700 Wh/kg, assuming an average tank-to-wheel efficiency of 12.6%. There is no expectation that current rechargeable battery systems such as the lithium-ion battery will meet this target. Even with the recent intensive development of high-capacity electrode materials, the specific energy of lithium-ion batteries could be eventually pushed up to 400 Wh/kg. However, even with such high capacity, the driving range is only about 140 miles. The US Advanced Battery Consortium set the goals for electric vehicles at a calendar life of 15 years and operating temperature from −30°C to +52°C with a driving range of 300 miles per single charge, which is far beyond the electrochemical performance of today's lithium-ion batteries. Since the first report by Abraham et al., lithium–oxygen ($Li–O_2$) batteries have been considered as the most promising candidate being developed as an energy source for electric vehicles, rivaling gasoline in terms of usable energy density [87,88]. On the basis of the oxidation of 1 kg of lithium metal, the theoretical energy density of a lithium oxygen battery is calculated to be 11,680 Wh/kg, which is not much lower than that of gasoline (Figure 2.5). An energy density of 1700 Wh/kg only accounts for 14.5% of the theoretical energy of a fully charged $Li–O_2$ battery, so it is not inconceivable that such a high energy density could be achievable at the

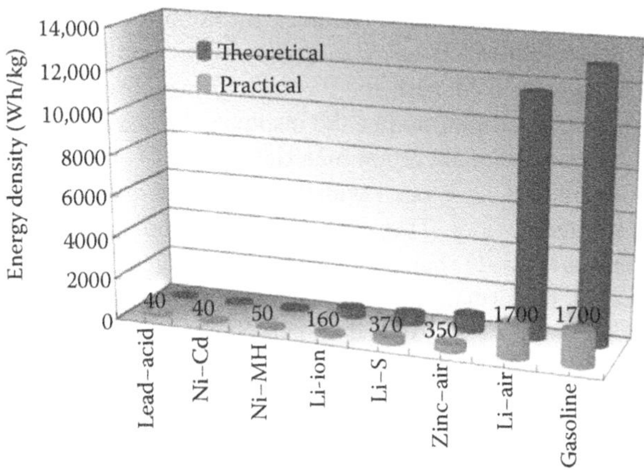

FIGURE 2.5 Practical and theoretical gravimetric energy densities (Wh/kg) for various types of rechargeable batteries compared with gasoline. (Reproduced with permission from Girishkumar, G., McCloskey, B., Luntz, A.C., Swanson, S., and Wilcke, W. Lithium–air battery: Promise and challenges. *The Journal of Physical Chemistry Letters* 1, 2193–2203. Copyright 2010 American Chemical Society.)

cell level, given the intensive research effort and long-term development.

2.3.2 Li–O$_2$ Battery Configuration and Electrochemical Reaction

The basic configuration of an Li–O$_2$ battery consists of a porous air electrode as the cathode, a lithium metal foil as the anode, and a lithium-ion conducting solvent or membrane as the electrolyte. In an aprotic type Li–O$_2$ battery, during the discharge process, the oxygen is drawn from the outside atmosphere and reduced by lithium ions from the electrolyte to form discharge products (Li_2O_2), which is denoted as oxygen reduction reaction. During the charge process, the discharge products electrochemically decompose to lithium ions and oxygen, which is denoted as an oxygen evolution reaction. Porous conductive substrates must be added to the air electrode as the reservoir for the insoluble discharge products. The amount of insoluble discharge products stored in the air electrode will ultimately determine the overall energy density. The catalysts in the porous oxygen electrode are essential to promote the oxygen reduction reactions and oxygen evolution reactions during the cell discharge and charge. However, the development of lithium oxygen batteries is still constrained by several serious challenges. The major problems of aprotic Li–O$_2$ batteries are listed below:

- Development of porous electrode with optimized pore-size distribution to accommodate discharge products, facilitate the electrolyte impregnation and oxygen diffusion
- Development of new catalysts that can promote the oxygen reduction reactions and oxygen evolution reactions in aprotic electrolyte systems
- Understanding the catalytic mechanism toward the chemical reactions in lithium oxygen batteries
- Development of new electrolytes with high stability, good lithium-ion conductivity, low volatility, and high oxygen solubility
- Development of new lithium metal anode architectures to prevent lithium dendrite formation in order to improve the cycling efficiency and safety

2.3.3 Electrochemical Reaction of Aprotic Li–O$_2$ Batteries

The electrochemical reaction in rechargeable Li–O$_2$ batteries with an aprotic electrolyte was first proposed by Bruce's group [89]. They used an *in situ* mass spectrometry measurement to demonstrate two essential prerequisites for the successful operation of a rechargeable Li–O$_2$ battery. During the discharge process, the discharge product Li_2O_2 should be formed first on discharging in the porous oxygen-breathing electrode with or without a catalyst. During the charging process, the Li_2O_2 must be decomposed to Li^+ and O_2, with or without a catalyst. Later, Bruce's group used an *in situ* spectroscopic method to study the oxygen reduction and oxygen evolution reactions in a nonaqueous solvent (acetonitrile solution), in the presence and absence of Li^+ ions [90]. Direct evidence has been provided that LiO_2 is an intermediate on oxygen reduction, and then disproportionates to Li_2O_2 during the discharge process. During the charging process, *in situ* spectroscopic studies reveal that Li_2O_2 decomposes to release oxygen and does not pass through LiO_2 as an intermediate. The electrochemical reactions are shown below:

Oxygen reduction reaction (ORR):

$$O_2 + e^- \rightarrow O_2^- \tag{2.1}$$

$$O_2^- + Li^+ \rightarrow LiO_2 \tag{2.2}$$

$$2LiO_2 \rightarrow Li_2O_2 + O_2 \tag{2.3}$$

Oxygen evolution reaction (OER):

$$Li_2O_2 \rightarrow O_2 + 2Li^+ + 2e^- \tag{2.4}$$

However, the real reactions in practical Li–O$_2$ batteries differ from the above-described ones. In early research, most Li–O$_2$ batteries use organic carbonate-based electrolytes, e.g., $LiPF_6$ in propylene carbonate. However, such electrolytes are not stable in Li–O$_2$ batteries and will decompose during the discharge process. The real redox electrochemical reactions in Li–O$_2$ batteries have attracted enormous worldwide attention [91–93]. Bruce's group used Fourier transform infrared spectroscopy and nuclear magnetic resonance (NMR) to analyze the discharge products in the air electrode after discharge and could not detect any of the desired Li_2O_2 in the discharge products. The actual discharge products observed were $C_3H_6(OCO_2Li)_2$, Li_2CO_3, HCO_2Li, CH_3CO_2Li, CO_2, and

H_2O at the cathode, due to electrolyte decomposition. The charging process of Li–O_2 batteries involves the oxidation of $C_3H_6(OCO_2Li)_2$, Li_2CO_3, HCO_2Li, and CH_3CO_2Li accompanied by CO_2 and H_2O evolution [94]. The different pathways for discharge and charge are consistent with the widely observed large charge–discharge voltage gap in Li–O_2 batteries. The battery cycling involves repeated decomposition of the electrolyte in the discharge process and the oxidation of the decomposition products in the charge process. The capacity fading of Li–O_2 batteries is associated with the starvation of the electrolyte and the accumulation of the discharge products in the cathode, such as Li_2CO_3, CH_3CO_2Li, and HCO_2Li.

Later, the electrochemical reactions in ether-based electrolytes in Li–O_2 batteries were also investigated by Bruce's group [95]. They combined electrochemical measurements with powder x-ray diffraction (XRD), Fourier transform infrared (FTIR), and NMR spectroscopy, and demonstrated that ether-based electrolytes are more stable than carbonate electrolytes and Li_2O_2 was detected in the discharge products. Recently, Li–O_2 batteries with ether-based electrolytes were demonstrated to be capable of operating for 100 cycles with capacity and rate values as high as 5000 mAh/g_{carbon} and 3 A/g_{carbon}, respectively [96]. The lithium diffusion coefficient calculated from cyclic voltammetry measurement is comparable to the values commonly reported for most cathodes currently used in lithium–ion batteries. The authors suggested that the unique behavior originates from the stable electrolyte and oxygen electrode design, which could be attributed to the configuration of Li–O_2 cells with a stable electrolyte and an advanced oxygen electrode structure.

2.3.4 Catalysts for Lithium–Air Batteries

One of the major challenges for the aprotic Li–O_2 battery is to develop a high-performance oxygen electrode. The oxygen electrode consists of a porous conductive substrate that can accommodate the insoluble discharge products and facilitate the oxygen diffusion and electrolyte impregnation. It should also have good catalytic activity to catalyze the oxygen reduction reaction and oxygen evolution reaction to minimize the discharge and charge overpotential. Therefore, we will focus on the development of nanostructure catalysts in aprotic Li–O_2 battery systems.

2.3.4.1 Nanostructured Carbon Catalysts

Carbon materials have significant advantages as electrode materials for Li–O_2 batteries, such as light weight, low cost, and abundant resources. Various carbon materials have been systematically studied as the cathode materials for Li–O_2 batteries. The electrochemical performance of commercial carbon powders as cathode materials for lithium air batteries was first investigated by Xiao et al. [97]. The results indicated that Ketjen black (KB) carbon with the highest mesopore volume among all of carbon materials demonstrated the highest specific capacity in dry air atmosphere. The uniformity of the pore sizes plays an important role in determining the electrochemical performances of lithium–air batteries. Furthermore, the KB-based electrode significantly expanded after being soaked with the electrolyte, which can provide extra space to accommodate the insoluble discharge products.

One-dimensional carbon materials such as CNTs and carbon nanofibers have been widely used in energy conversion and storage devices. They have also been studied as cathode materials in Li–O_2 batteries by several groups. The electrochemical performance of CNTs and nitrogen-doped CNTs (N-CNTs) were investigated as cathode materials for Li–O_2 battery by Li et al. The N-CNTs were synthesized by a floating catalyst CVD method [98]. The electrochemical performance of Li–O_2 batteries with N-CNTs catalysts were studied and compared with pristine CNTs in alkyl carbonate electrolyte. Li–O_2 batteries with N-CNTs delivered a discharge capacity of 866 mAh/g, which is much higher than that of pristine CNTs. The improved electrochemical performance indicated that atomic doping is a promising strategy to improve the catalytic activity of carbon materials toward oxygen reduction reactions.

All-carbon-fiber electrodes were also used for Li–O_2 batteries [99]. The porous air electrode was synthesized by growing hollow carbon fibers on a ceramic porous substrate. Li–O_2 batteries with this novel cathode exhibited high gravimetric energies (up to 2500 $Wh/kg_{discharged}$) that are about four times greater than those of lithium-ion batteries with $LiCoO_2$ as the cathode material (~600 $Wh/kg_{electrode}$). The improved high gravimetric energy is due to the unique structure of the all-carbon-fiber electrodes, which can increase the utilization of the available void volume for discharge products.

Two-dimensional GNSs have also been studied as cathode materials for Li–O_2 batteries. Wang et al. prepared GNSs from a modified Hummers' method and investigated them as cathode catalysts for Li–O_2 batteries with alkyl carbonate electrolytes [100]. The Li–O_2 battery with GNS catalysts exhibited higher specific capacity, lower charge–discharge overpotential, and better cycle stability than those of the Vulcan carbon electrode (Figure 2.6). The increased catalytic activity of GNSs may arise from the presence of many carbon vacancies and defects on the surface of GNSs.

The advantage of graphene for Li–O_2 batteries could be further strengthened by the assembly of 2D GNSs to a 3D porous architecture. Li–O_2 batteries with a novel air electrode consisting of a hierarchical arrangement of functionalized graphene sheets can deliver a high discharge capacity of 15,000 mAh/g (Figure 2.7) [101]. The excellent performance should be attributed to the unique bimodal porous structure of the electrode, which consists of multisize porous channels facilitating rapid O_2 diffusion. Meanwhile, the highly connected nanoscale pores provide a high density of reactive sites for ORRs.

Porous graphene nanoarchitectures with different pore sizes were also prepared through hard template methods and applied in Li–O_2 batteries (Figure 2.8) [102]. Li–O_2 batteries with porous graphene materials exhibited significantly higher discharge capacities than that of nonporous graphene in dimethyl sulfoxide (DMSO)-based electrolytes. Furthermore,

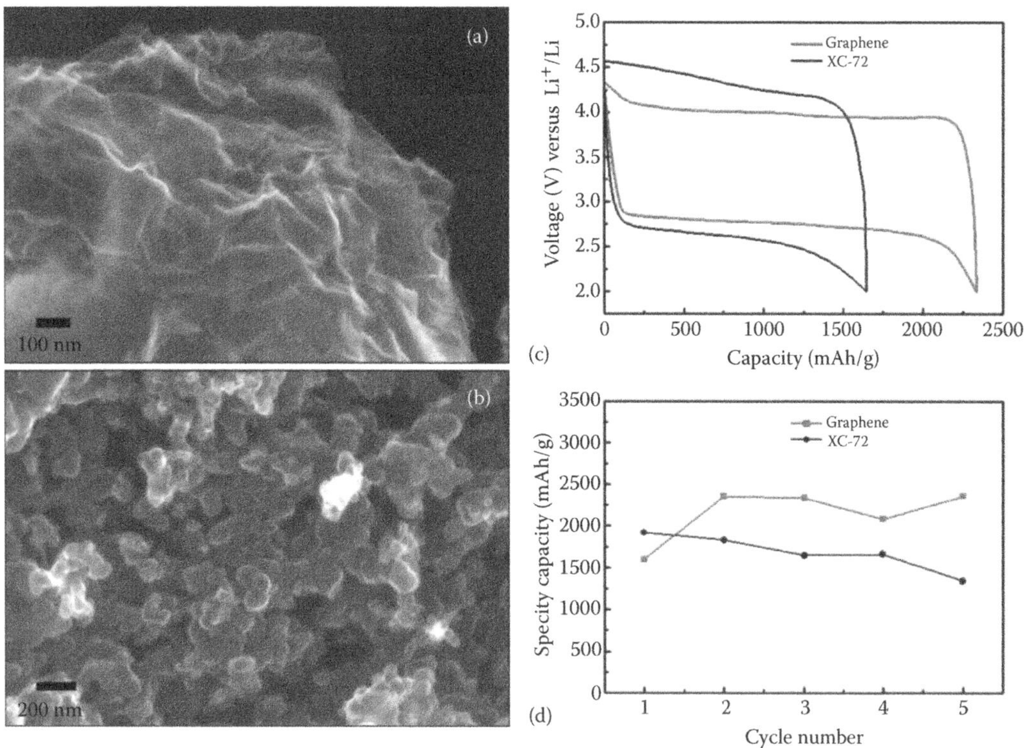

FIGURE 2.6 SEM images of (a) graphene nanosheets and (b) Vulcan XC-72 carbon. (c) Charge–discharge voltage profiles (third cycle) and (d) cycling performance of graphene nanosheets and Vulcan XC-72 carbon. (Reproduced from *Carbon*, 50, Sun, B., Wang, B., Su, D.W., Xiao, L.D., Ahn, H., and Wang, G.X., Graphene nanosheets as cathode catalysts for lithium–air batteries with an enhanced electrochemical performance, 727–733, Copyright 2012, with permission from Elsevier.)

porous graphene with a pore diameter of around 250 nm showed the highest discharge capacity among the porous graphene with the small pores (about 60 nm) and large pores (about 400 nm). It has been proved that the porous structure is an important factor influencing the overall electrochemical performance of Li–O_2 batteries. The novel porous graphene architecture inspires the development of high-performance Li–O_2 batteries.

Carbon catalysts with 3D meso- or macroporous architectures were intensively investigated by many groups. Mesocellular carbon foam with narrow pore-size distributions, centered at 4.3 and 30.4 nm, was prepared through nanocasting

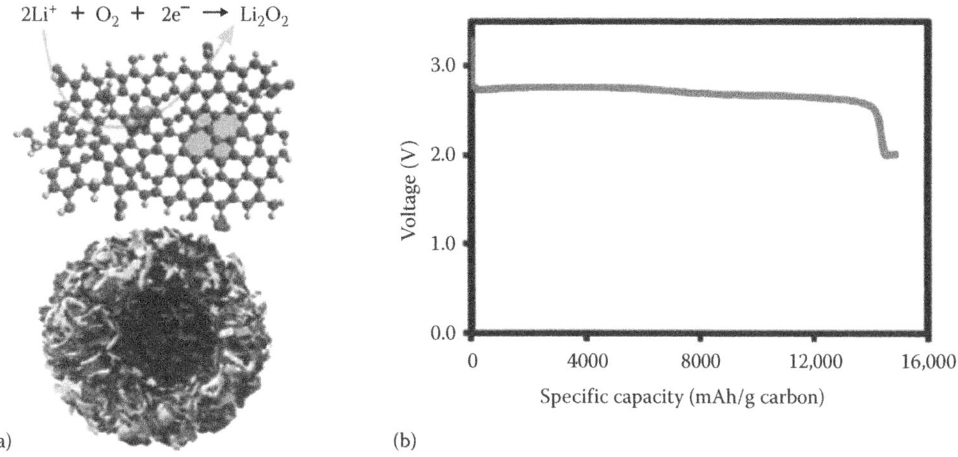

FIGURE 2.7 (a) Schematic structure of functionalized graphene sheets (upper image) with an ideal bimodal porous structure (lower image), which is highly desirable for lithium–air battery operation. (b) Discharge curve of a lithium–air battery using functionalized graphene sheets as the air electrode. (Reproduced with permission from Xiao, J., Mei, D.H., Li, X.L., Xu, W., Wang, D.Y., Graff, G.L., Bennett, W.D., Nie, Z.M., Saraf, L.V., Aksay, I.A., Liu, J., and Zhang, J.G. Hierarchically porous graphene as a lithium–air battery electrode. *Nano Letters* 11, 5071–5078. Copyright 2011 American Chemical Society.)

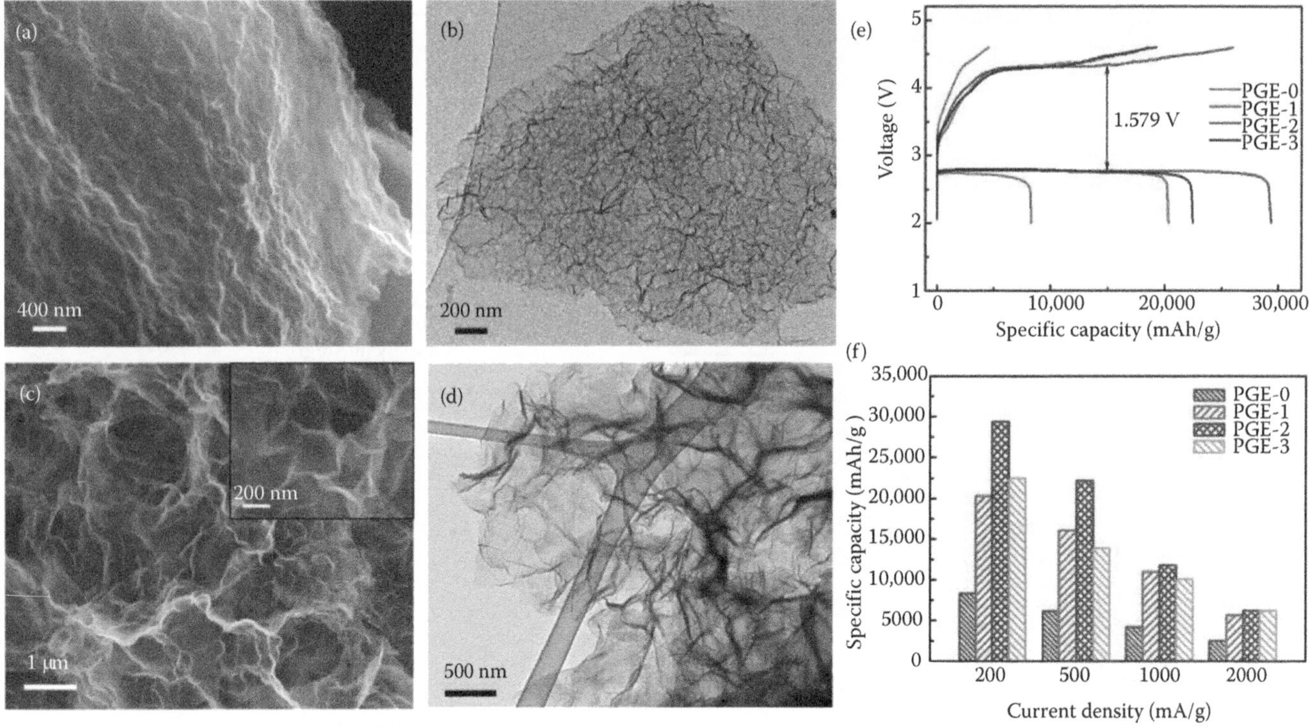

FIGURE 2.8 (a) SEM image and (b) TEM image of porous graphene (pore size about 60 nm). (c) SEM image and (d) TEM image of porous graphene (pore size about 250 nm). (e) Charge–discharge profiles of Li–O_2 batteries with different pore-size distribution graphene catalysts at the current density of 200 mA/g. (f) The discharge-specific capacities of Li–O_2 batteries with different pore-size distribution graphene catalysts at different current densities. (Reproduced with permission from Sun, B., Huang, X., Chen, S., Munroe, P., and Wang, G. Porous graphene nanoarchitectures: An efficient catalyst for low charge-overpotential, long life, and high capacity lithium–oxygen batteries. *Nano Letters* 14, 3145–3152. Copyright 2014 American Chemical Society.)

technology using mesocellular foam silica hard templates [103]. Li–O_2 batteries with mesocellular carbon foam as cathode catalysts showed an increased discharge capacity, about 40% higher than that of commercial carbon black. Three-dimensional ordered mesoporous/macroporous carbon sphere arrays (MMCSAs) were synthesized and used as the catalyst in Li–O_2 batteries by Guo et al. [104]. The ordered mesoporous channels and hierarchical mesoporous/macroporous structure of the MMCSAs facilitated the electrolyte immersion and O_2 diffusion and provided an effective space for discharge products accommodation, which effectively enhanced the electrochemical performance of the Li–O_2 batteries [104].

2.3.4.2 Nanostructured Metal Oxide Catalysts

Transition metal oxides have been widely employed as catalysts in Li–O_2 batteries. Debart et al. first explored the use of several transition metal oxides as cathode catalysts for Li–O_2 batteries with aprotic electrolytes [105]. These catalysts have already been widely investigated as electrocatalysts for O_2 electrochemistry in aqueous media, including Fe_2O_3, Fe_3O_4, NiO, CuO, and $CoFe_2O_4$. The preliminary study demonstrated that Fe_2O_3 and NiO do not perform well as catalysts for Li–O_2 batteries with aprotic electrolytes. However, Fe_3O_4, CuO, and $CoFe_2O_4$ delivered much improved capacity retentions. Co_3O_4 achieved the best performance, exhibiting reduced charge potential, the highest discharge capacity, and the best cycling performance. Later, Debart et al. examined different forms of manganese oxides for their use as the catalysts in lithium–oxygen batteries [106]. These included electrolytical manganese dioxide; commercial Mn_3O_4; bulk Mn_2O_3; bulk α-, β-, λ-, and γ-MnO_2; α-MnO_2 nanowires; and β-MnO_2 nanowires. The air cathode with α-MnO_2 as the catalyst showed the highest capacity (about 3000 mAh/g_{carbon}) and the best capacity retention compared with air cathodes using other MnO_2 polymorphs. The authors concluded that the enhanced performance of the α-MnO_2 nanowires was due to the crystal structure and the nanowire morphology with high surface area. They also discovered that the capacity retention could be improved by avoiding a high-level depth of discharge.

Ternary spinel nickel cobalt oxides with different nanostructures, such as nanorods, nanosheets, and nanoplates, have also been studied as cathode catalysts in Li–O_2 batteries [107–109]. $NiCo_2O_4$ nanorods were first reported by Sun et al. and showed an improved catalytic activity compared with pristine carbon black (Figure 2.9) [107]. Later, the same group reported the synthesis of hierarchical macroporous/mesoporous $NiCo_2O_4$ nanosheets through a hydrothermal method, followed by low-temperature calcination (Figure 2.10) [108]. When used as the cathode catalyst in Li–O_2 batteries, the as-prepared $NiCo_2O_4$ nanosheets exhibited better

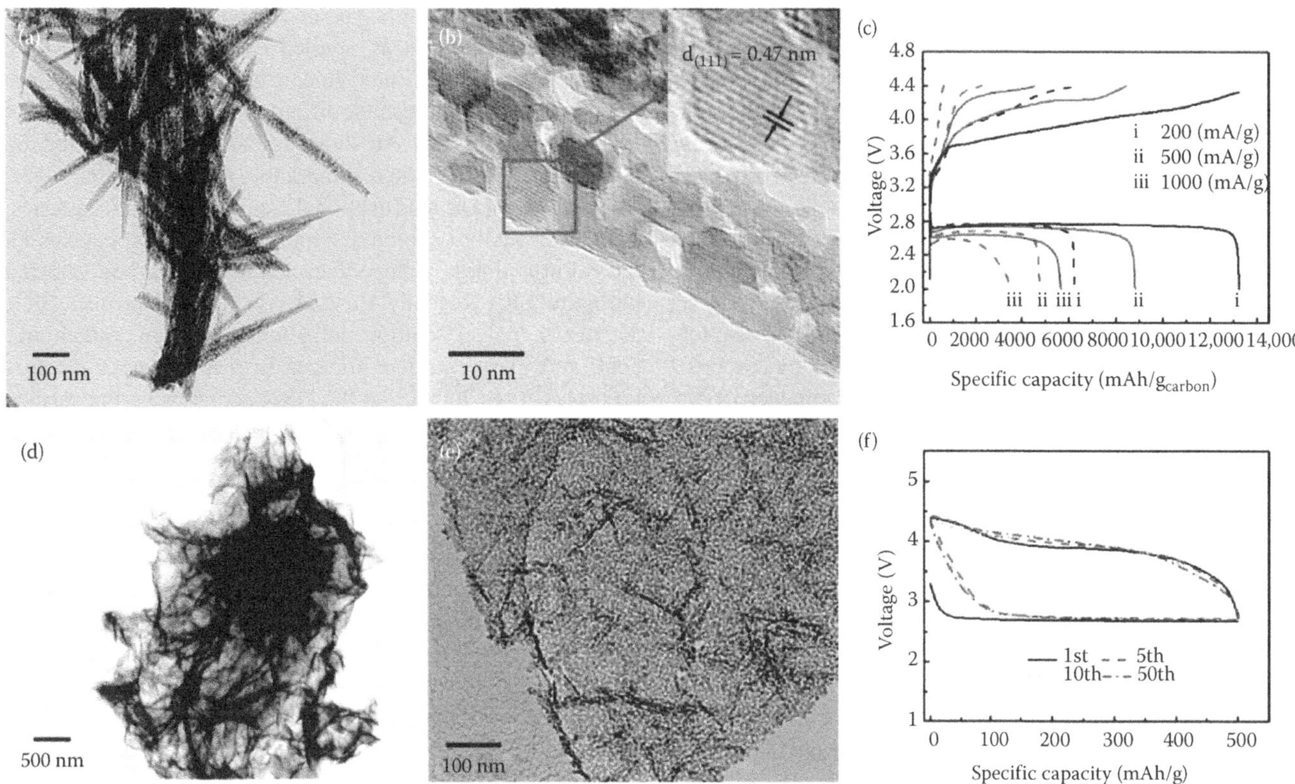

FIGURE 2.9 (a, b) TEM images of $NiCo_2O_4$ nanorods. (c) Charge–discharge voltage curves of $NiCo_2O_4$ nanorod electrodes (solid line) and carbon black electrodes (dash line) at different current densities. (Reproduced from *Electrochemistry Communications*, 13, Li, Y.L., Wang, J.J., Li, X.F., Liu, J., Geng, D.S., Yang, J.L., Li, R.Y., and Sun, X.L., Nitrogen-doped carbon nanotubes as cathode for lithium-air batteries, 668–672, Copyright 2011, with permission from Elsevier.) (d, e) TEM images of $NiCo_2O_4$ nanosheets at different magnifications. (f) Charge–discharge curves of the Li–O_2 battery with $NiCo_2O_4$ nanosheets catalysts at 200 mA/g by curtailing the capacity to 500 mAh/g in the voltage range of 2.0–4.6 V. (Reproduced from Sun, B., Huang, X., Chen, S., Zhao, Y., Zhang, J., Munroe, P., and Wang, G. Hierarchical macroporous/mesoporous $NiCo_2O_4$ nanosheets as cathode catalysts for rechargeable Li–O_2 batteries. *Journal of Materials Chemistry A*, 2, 12053–12059, 2014. Reproduced by permission of The Royal Society of Chemistry.)

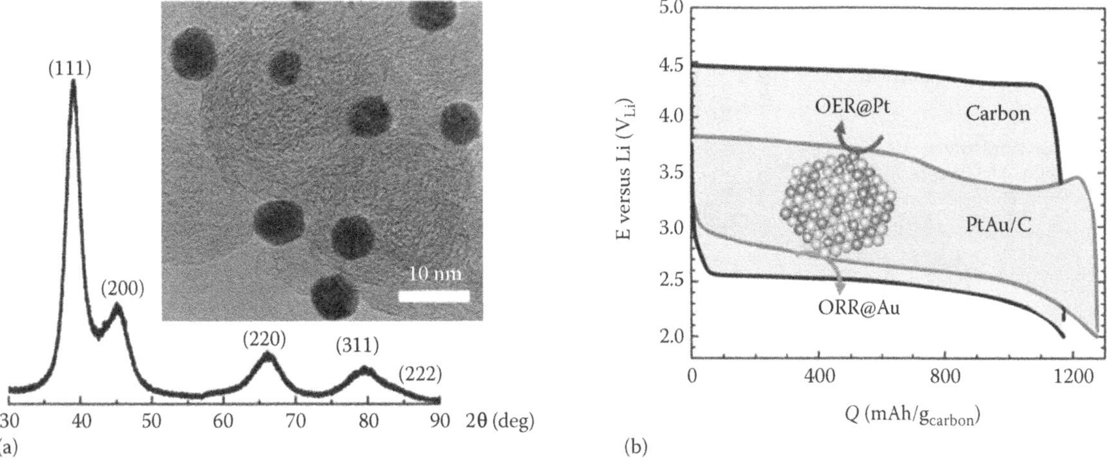

FIGURE 2.10 (a) Representative TEM image (top right) and XRD data of PtAu/C. (b) Li–O_2 cell discharge–charge profiles of carbon (black, 85 mA/g_{carbon}) and PtAu/C (gray, 100 mA/g_{carbon}) in the third cycle at 0.04 mA/cm^2. (Reproduced with permission from Lu, Y.C., Xu, Z.C., Gasteiger, H.A., Chen, S., Hamad-Schifferli, K., and Shao-Horn, Y. Platinum-gold nanoparticles: A highly active bifunctional electrocatalyst for rechargeable lithium-air batteries. *Journal of the American Chemical Society*, 132, 12170–12171. Copyright 2010 American Chemical Society.)

catalytic activity than pure carbon black toward ORRs and OERs. The Li–O_2 batteries with the $NiCo_2O_4$ catalyst showed a high reversible capacity of about 11,860 mAh/g, a low charge potential of ~4 V, and a good cycling performance up to 50 cycles (by curtailing the capacity to 500 mAh/g). The mesopores in $NiCo_2O_4$ nanosheets can facilitate the oxygen diffusion and electrolyte impregnation throughout the oxygen electrode. The macropores formed between the interconnected $NiCo_2O_4$ nanosheets can provide a large effective space to accommodate the insoluble discharge products, which can not only increase the specific capacity but also tolerate the volume change during the charge–discharge process. Furthermore, it is reported that the electronic conductivity of $NiCo_2O_4$ is higher than that of Co_3O_4 and NiO.

Owing to the insulating properties of most transition metal oxides, metal oxides/carbon nanocomposites were synthesized to increase the conductivity of the cathode catalysts. MnO_2 nanorods grown *in situ* on GNSs showed an improved catalytic activity compared with pristine graphene, MnO_2, and mechanically mixed graphene/MnO_2 [110]. Cobalt oxide nanoparticles supported on reduced graphene oxide (Co_3O_4/RGO) were mixed with KB with a weight ratio of 3:7 [111]. Li–O_2 batteries with this catalyst showed a charge plateau between 3.5 and 3.75 V, about 400 mV lower than that of KB at a current density of 140 mA/g_{carbon}. $MnCo_2O_4$–graphene hybrid materials were synthesized by direct nucleation and growth of $MnCo_2O_4$ nanoparticles on reduced graphene oxide. Li–O_2 batteries with the as-prepared hybrid materials showed an excellent catalytic activity, which led to lower overpotentials and longer cycle life than other catalysts such as platinum [112].

2.3.4.3 Noble Metal Catalysts

The electrocatalytic activities of noble metal and alloy catalysts have been widely studied in aqueous electrolyte systems. Recently, they were employed as cathode catalysts in aprotic Li–O_2 batteries. Lu et al. studied Pt, Au, and Pt–Au alloy nanoparticles loaded onto Vulcan carbon (XC-72) as electrocatalysts for rechargeable lithium–oxygen batteries [113–116]. They found that Au showed very high activity for the ORR, and Pt/C exhibited remarkable activity for the OER. Pt–Au/C can act as a bifunctional catalyst for Li–O_2 batteries. This novel catalyst exhibited a higher discharge voltage and a much lower charge voltage than that of pure carbon (showed in Figure 2.10). Recently, Lu et al. reported the intrinsic ORR activity of polycrystalline palladium, platinum, ruthenium, gold, and glassy carbon surfaces in a nonaqueous electrolyte [117]. They found that the ORR activities on the surface of these polycrystallines correlate to the oxygen adsorption energy, and the order of ORR activity is Pd > Pt > Ru ≈ Au > GC on bulk surfaces.

Pd nanocrystal combined with various carbon substrates were also applied as cathode catalysts in Li–O_2 batteries. A dramatic reduction of charge overpotential of about 0.2 V has been achieved by using a novel cathode architecture [118]. The cathode utilizes atomic layer deposition of palladium nanoparticles on a carbon surface with an alumina coating for passivation of carbon defect sites (Figure 2.11). The cathode promotes the growth of a nanocrystalline form of lithium peroxide with electronic transport properties that are needed to lower the charge potential. The combination of palladium nanoparticles attached to the carbon cathode surface, a nanocrystalline form of lithium peroxide with grain boundaries, and the alumina coating prevent electrolyte decomposition on carbon enable a low-charge-overpotential and high-energy-efficiency Li–O_2 battery. Later, Xu et al. designed and synthesized a freestanding honeycomb-like palladium-modified hollow spherical carbon deposited onto carbon paper, as a cathode for Li–O_2 batteries (Figure 2.12) [119]. The Li–O_2 batteries with the as-prepared cathode can deliver a high

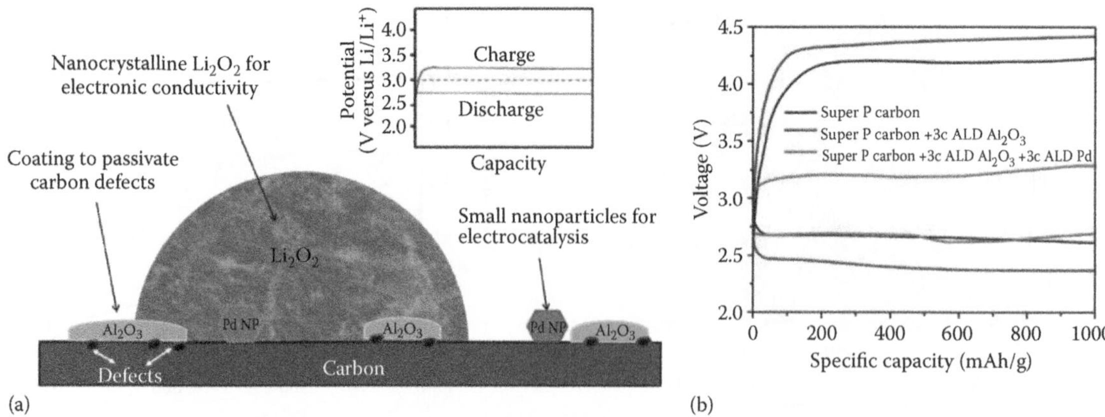

FIGURE 2.11 (a) Schematic of the nanostructured cathode architecture. The inset shows a hypothetical charge–discharge voltage profile versus capacity. The figure shows the Al_2O_3 coating, the palladium nanoparticles, and the nanocrystalline lithium peroxide, all of which contribute to lowering the overpotential. (b) Voltage profile during discharge–charge of cells (to 1000 mAh/g) based on super P carbon; super P carbon coated with Al_2O_3; and Al_2O_3-coated super P carbon further coated with Pd nanoparticles. (Reproduced by permission from Macmillan Publishers Ltd. *Nature Communications*, Lu, J., Lei, Y., Lau, K.C., Luo, X., Du, P., Wen, J., Assary, R.S., Das, U., Miller, D.J., Elam, J.W., Albishri, H.M., Abd El-Hady, D., Sun, Y.-K., Curtiss, L.A., and Amine, K. A nanostructured cathode architecture for low charge overpotential in lithium–oxygen batteries, 4, 2383, Copyright 2013.)

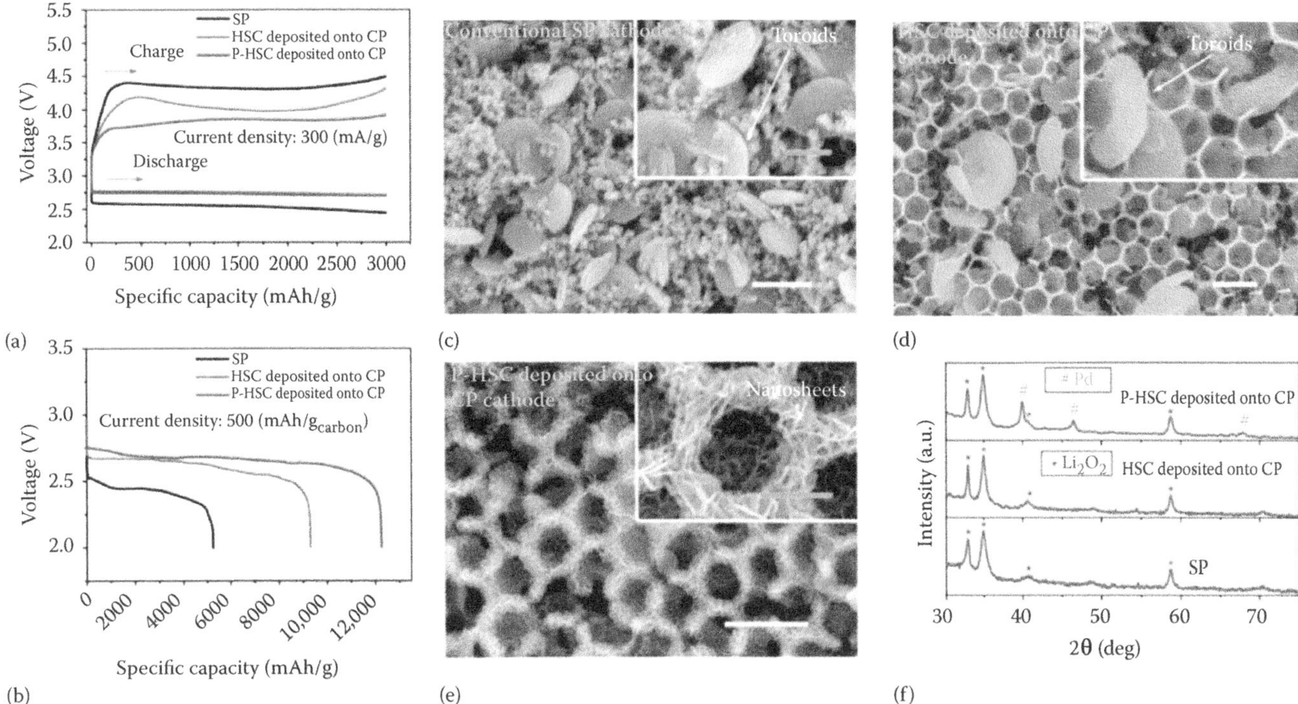

FIGURE 2.12 (a) First cycle charge–discharge profiles of Li–O_2 cells at a current density of 300 mA/g and a specific capacity limit of 3000 mAh/g. (b) The rate capability of the Li–O_2 cells with the three kinds of cathodes at the current density of 500 mA/g. FESEM images of the discharged (c) Super-P carbon black cathode, (d) HSC deposited onto CP cathode and (e) P-HSC deposited onto CP cathode with a capacity limitation of 3000 mAh/g. (f) Corresponding powder XRD patterns of the three kinds of discharged cathodes. The XRD results indicate the dominant discharge products are all Li_2O_2 for the three different electrodes. White scale bars, 1 mm. Green scale bars, 400 nm in (c–e). (Reproduced by permission from Macmillan Publishers Ltd. *Nature Communications*, Xu, J.-J., Wang, Z.-L., Xu, D., Zhang, L.-L., and Zhang, X.-B. Tailoring deposition and morphology of discharge products towards high-rate and long-life lithium–oxygen batteries, 4, 2438, Copyright, 2013.)

discharge capacity at a high current density (5900 mAh low charge overpotential/g at a current density of 1.5 A/g) and long-term cycle performance (100 cycles at a current density of 300 mA/g and a specific capacity limit of 1000 mAh/g). The improved performances were explained by the tailored deposition and morphology of the discharge products as well as the alleviated electrolyte decomposition compared with the conventional carbon cathodes.

Recently, Ru and RuO_2 have been reported as active catalysts toward OERs. Li–O_2 cells with Ru-based catalysts showed a high reversible capacity, low charge overpotential, and good cycling stability. Ru and RuO_2 were first investigated as cathode catalysts for Li–O_2 batteries with ether-based electrolytes by Jung et al. [120]. Sun et al. also reported Ru nanocrystal functionalized carbon black (Ru–CB), which showed superior electrochemical performance in DMSO-based electrolytes (Figure 2.13) [121]. Li–O_2 batteries with the as prepared Ru–CB catalysts exhibited a high reversible capacity of about 9800 mAh/g, a low charge–discharge overpotential (about 0.37 V), and an outstanding cycle performance up to 150 cycles (with a curtaining capacity of 1000 mAh/g). Later, the same group loaded Ru nanocrystals onto porous graphene and achieved further improved electrochemical performance in Li–O_2 battery applications [102]. The Ru nanocrystal-decorated porous graphene (Ru@PGE) exhibited an excellent catalytic activity as cathode in Li–O_2 batteries with a high reversible capacity of 17,700 mAh/g, a low charge–discharge overpotential (about 0.355 V), and a long cycle life up to 200 cycles (under the curtaining capacity of 1000 mAh/g).

The above reports demonstrated that ruthenium nanocrystals can significantly reduce the charge potential compared with carbon black catalysts. This indicates that ruthenium-based nanomaterials can be effective cathode catalysts for high-performance Li–O_2 batteries. A comparison of charge–discharge curves of Ru@PGE catalyst in the first cycle with different capacity limitations is shown in Figure 2.14. It is observed that the charge voltage curves of Ru@PGE electrodes showed three stages. The charge voltage differences may be owing to the different Li_2O_2 decomposition mechanisms. When the capacity is limited to 500 or 1000 mAh/g, the initial charge stage showed a sloping profile at low voltage (<3.6 V), corresponding to the Li_2O_2 decomposition through the delithiation pathway. The delithiation pathway is associated with the delithiation of Li_2O_2 to form LiO_2-like species in the first step via a solid-solution route ($Li_2O_2 \rightarrow Li^+ + e^- + LiO_2$). In the second step, the metastable LiO_2 disproportionates to evolve O_2, yielding an overall $2e^-/O_2$ oxygen evolution process ($LiO_2 + LiO_2 \rightarrow Li_2O_2 + O_2$). The second charge

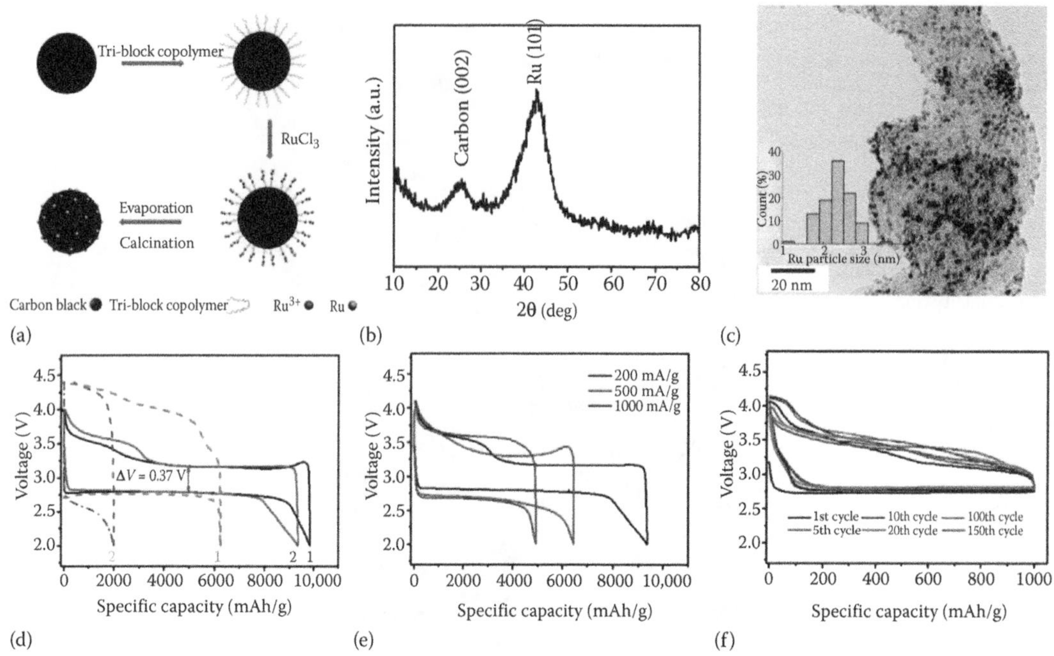

FIGURE 2.13 (a) Schematic illustration of the synthesis process of Ru–CB. (b) XRD pattern of Ru–CB. (c) TEM image and crystal size distribution of Ru nanocrystals. (d) Charge–discharge curves of the first two cycles for Ru–CB (solid line) and CB electrode (dash line) at 200 mA/g. (e) The second cycle charge–discharge voltage curves of Ru–CB at different current densities. (f) Cycling performance of the Li–O_2 cells with Ru–CB catalyst at 200 mA/g with curtailing the capacity to 1000 mAh/g. (Reproduced with permission from Macmillan Publishers Ltd. *Scientific Reports*, Sun, B., Munroe, P., and Wang, G.X. Ruthenium nanocrystals as cathode catalysts for lithium-oxygen batteries with a superior performance, 3, 2247, Copyright 2013.)

stage showed a voltage plateau at about 3.6 V, corresponding to the direct electrochemical decomposition of Li_2O_2 to form Li^+ and O_2 via a two-phase transition ($Li_2O_2 \rightarrow 2Li^+ + 2e^- + O_2$). The third charge stage is the voltage sloping higher than 3.6 V, corresponding to the decomposition of by-products and electrolyte. When increasing the charge–discharge capacity limitation, it was observed that the initial low voltage slope extended to a plateau at about 3.1 V. This indicates that the ratio of Li_2O_2 decomposed through the delithiation pathway increased.

2.3.4.4 Nanostructured Noncarbon Catalysts for Li–O_2 Batteries

Recent reports suggested that carbon can be decomposed at certain potentials in the charging process, promoting electrolyte decomposition in both charge and discharge processes. The use of alternative noncarbon materials with good electrical conductivity, high porosity, and both ORR and OER activities are expected to promote the development of Li–O_2 batteries. The first attempt was made by Peng et al. by using a nanoporous gold (NPG) electrode with a DMSO-based electrolyte [122]. Li–O_2 battery with NPG electrode achieved 95% capacity retention from the 1st to the 100th cycle at a current density of 500 mA/g under voltage limitation. Through the characterization by FTIR spectroscopy, NMR, and surface-enhanced Raman spectroscopy, the group also showed that such electrodes were particularly effective in promoting the decomposition of Li_2O_2, with all the Li_2O_2 being decomposed below 4.0 V and about 50% decomposed below 3.3 V, at a relative-rate current density. Later, the same group incorporated a redox-mediating molecule (tetrathiafulvalene, TTF) in the electrolyte of the rechargeable nonaqueous Li–O_2 battery and demonstrated that it is possible to recharge such a battery at high current densities (1 mA/cm^2) that are impossible without the redox mediator in the same cell [123]. The TTF molecule is oxidized to TTFt at the positive electrode and then, in turn, oxidizes the insulating solid Li_2O_2 in the process where TTFt is reduced back to TTF. Li–O_2 cells that contain TTF were cycled 100 times with complete reversibility of Li_2O_2 formation/decomposition on each cycle. TiC nanoparticles have also been applied as an alternative cathode for Li–O_2 batteries by the same group [124]. The good stability of the TiC cathode was explained by the formation of a TiO_2 thin layer on the surface of TiC.

The conductive indium tin oxide (ITO) electrode embedded with Ru nanoparticles was first introduced as a carbon-free cathode in Li–O_2 batteries by Li's group [125]. Li–O_2 batteries with ITO electrodes showed good cycling stability. However, owing to the high atomic weight of ITO, the specific capacity is low. Later, the same group used Sb-doped tin oxide (STO, ca. 6 nm in size) supported Ru nanoparticles (Ru/STO) as cathode materials for Li–O_2 batteries [126]. A large specific capacity of 750 mAh/g and low discharging and charging overpotentials have been obtained.

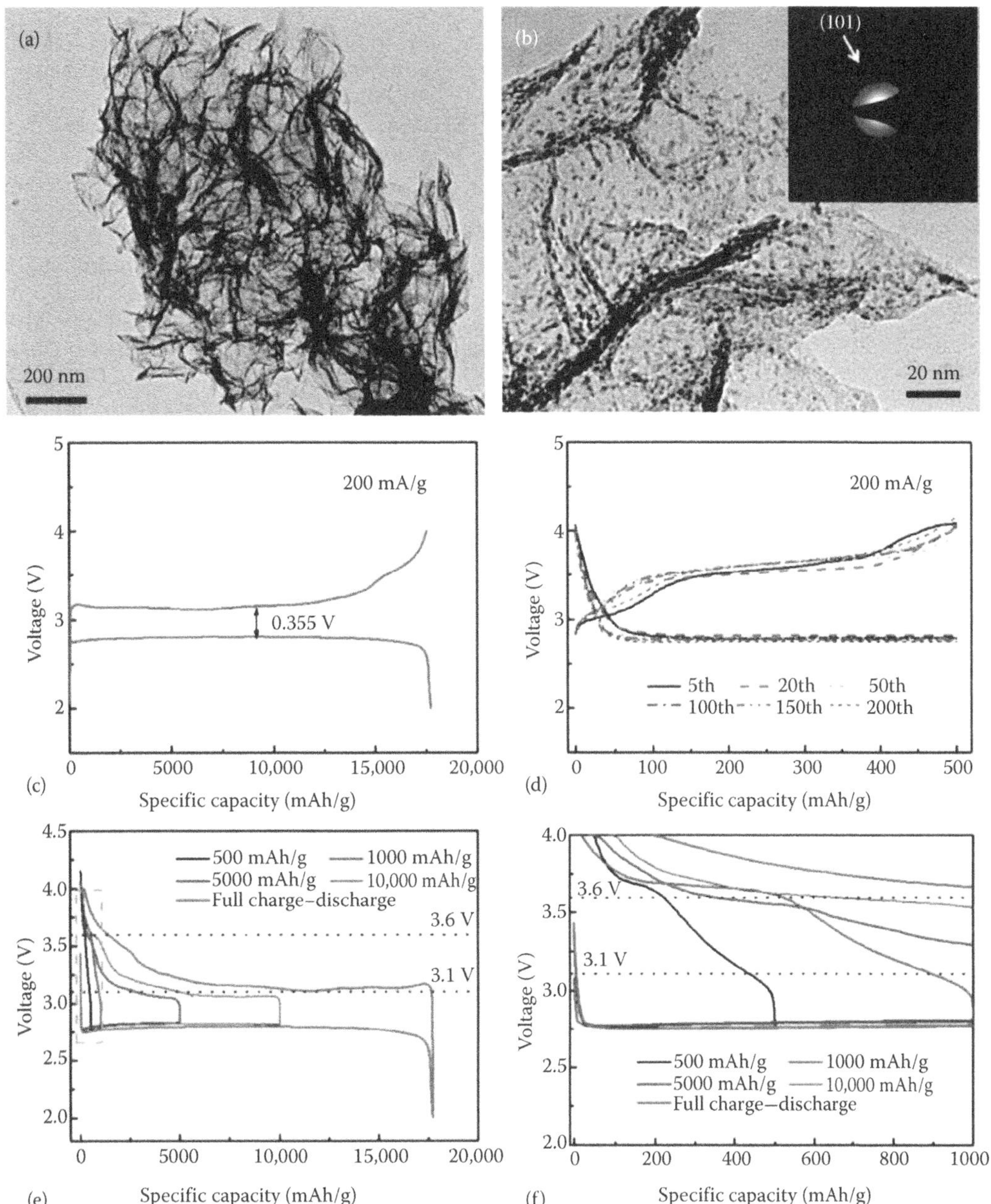

FIGURE 2.14 (a, b) TEM images of Ru@PGE and the SAED pattern (inset). (c) First cycle charge–discharge profiles of Li–O_2 batteries with Ru@PGE catalysts at 200 mA/g in the voltage range of 2.0–4.4 V. (d) Charge–discharge profiles at different cycles of Li–O_2 batteries with Ru@PGE catalysts at 200 mA/g by curtailing the capacity to 500 mAh/g in the voltage range of 2.0–4.4 V. (e) Comparison of charge–discharge curves of electrodes with Ru@PGE catalysts in the first cycle with different capacity limitations at the current density of 200 mA/g. (f) Enlarged figures of the selected area in (e). (Reproduced with permission from Sun, B., Huang, X., Chen, S., Munroe, P., and Wang, G. Porous graphene nanoarchitectures: An efficient catalyst for low charge-overpotential, long life, and high capacity lithium–oxygen batteries, *Nano Letters*, 14, 3145–3152. Copyright 2014 American Chemical Society.)

2.4 CONCLUSION

Aprotic Li–O_2 batteries have received great attention as potential power sources for future electric vehicle applications. Although this new battery system could achieve much higher energy density than any other rechargeable battery system, it is still in the early developmental stage. Many challenges need to be overcome before its consideration in practical applications. The design of novel nanostructured cathode materials (catalysts) with improved and optimized porosity to facilitate oxygen diffusion and Li ion transport toward the active surface of the electrode is critical for the overall electrochemical reactions and the resulting high capacity of the battery. More intensive research on catalyst materials is required to develop this advanced electrochemical power source.

REFERENCES

1. Tarascon, J.M., and Armand, M. Issues and challenges facing rechargeable lithium batteries. *Nature* 2001; 414:359–367.
2. Bruce, P.G., Scrosati, B., and Tarascon, J.M. Nanomaterials for rechargeable lithium batteries. *Angewandte Chemie—International Edition* 2008; 47:2930–2946.
3. Mizushima, K., Jones, P.C., Wiseman, P.J., and Goodenough, J.B. $LiCoO_2$—A new cathode material for batteries of high-energy density. *Materials Research Bulletin* 1980; 15:783–789.
4. Nagaura, T., and Tozawa, K. Lithium ion rechargeable battery. *Progress in Batteries and Solar Cells* 1990; 9:209–217.
5. Padhi, A.K., Nanjundaswamy, K.S., and Goodenough, J.B. Phospho-olivines as positive-electrode materials for rechargeable lithium batteries. *Journal of the Electrochemical Society* 1997; 144:1188–1194.
6. Padhi, A.K., Nanjundaswamy, K.S., Masquelier, C., Okada, S., and Goodenough, J.B. Effect of structure on the Fe^{3+}/Fe^{2+} redox couple in iron phosphates. *Journal of the Electrochemical Society* 1997; 144:1609–1613.
7. Yamada, A., Chung, S.C., and Hinokuma, K. Optimized $LiFePO_4$ for lithium battery cathodes. *Journal of the Electrochemical Society* 2001; 148:A224–A229.
8. Yang, S.F., Song, Y.N., Zavalij, P.Y., and Whittingham, M.S. Reactivity, stability and electrochemical behavior of lithium iron phosphates. *Electrochemistry Communications* 2002; 4:239–244.
9. Huang, H., Yin, S.C., and Nazar, L.F. Approaching theoretical capacity of $LiFePO_4$ at room temperature at high rates. *Electrochemical and Solid State Letters* 2001; 4:A170–A172.
10. Prosini, P.P., Zane, D., and Pasquali, M. Improved electrochemical performance of a $LiFePO_4$-based composite cathode. *Electrochimica Acta* 2001; 46:3517–3523.
11. Herle, P.S., Ellis, B., Coombs, N., and Nazar, L.F. Nano-network electronic conduction in iron and nickel olivine phosphates. *Nature Materials* 2004; 3:147–152.
12. Liu, H., Fu, L.J., Zhang, H.P., Gao, J., Li, C., Wu, Y.P., and Wu, H.Q. Effects of carbon coatings on nanocomposite electrodes for lithium-ion batteries. *Electrochemical and Solid State Letters* 2006; 9:A529–A533.
13. Ouyang, C.Y., Shi, S.Q., Wang, Z.X., Huang, X.J., and Chen, L.Q. First-principles study of Li ion diffusion in $LiFePO_4$. *Physical Review B* 2004; 69. No. 104303.
14. Doeff, M.M., Hu, Y.Q., McLarnon, F., and Kostecki, R. Effect of surface carbon structure on the electrochemical performance of $LiFePO_4$. *Electrochemical and Solid State Letters* 2003; 6:A207–A209.
15. Chen, Z.H., and Dahn, J.R. Reducing carbon in $LiFePO_4$/C composite electrodes to maximize specific energy, volumetric energy, and tap density. *Journal of the Electrochemical Society* 2002; 149:A1184–A1189.
16. Wang, G.X., Liu, H., Liu, J., Qiao, S.Z., Lu, G.Q.M., Munroe, P., and Ahn, H. Mesoporous $LiFePO_4$/C nanocomposite cathode materials for high power lithium ion batteries with superior performance. *Advanced Materials* 2010; 22:4944–4948.
17. Croce, F., D'Epifanio, A., Hassoun, J., Deptula, A., Olczac, T., and Scrosati, B. A novel concept for the synthesis of an improved $LiFePO_4$ lithium battery cathode. *Electrochemical and Solid State Letters* 2002; 5:A47–A50.
18. Park, K.S., Son, J.T., Chung, H.T., Kim, S.J., Lee, C.H., Kang, K.T., and Kim, H.G. Surface modification by silver coating for improving electrochemical properties of $LiFePO_4$. *Solid State Communications* 2004; 129:311–314.
19. Wang, G.X., Yang, L., Chen, Y., Wang, J.Z., Bewlay, S., and Liu, H.K. An investigation of polypyrrole–$LiFePO_4$ composite cathode materials for lithium-ion batteries. *Electrochimica Acta* 2005; 50:4649–4654.
20. Wang, J.Z., Chou, S.L., Chen, J., Chew, S.Y., Wang, G.X., Konstantinov, K., Wu, J., Dou, S.X., and Liu, H.K. Paper-like free-standing polypyrrole and polypyrrole–$LiFePO_4$ composite films for flexible and bendable rechargeable battery. *Electrochemistry Communications* 2008; 10:1781–1784.
21. Fedorkova, A., Orinakova, R., Orinak, A., Talian, I., Heile, A., Wiemhofer, H.D., Kaniansky, D., and Arlinghaus, H.F. PPy doped PEG conducting polymer films synthesized on $LiFePO_4$ particles. *Journal of Power Sources* 2010; 195: 3907–3912.
22. Xie, H.M., Wang, R.S., Ying, J.R., Zhang, L.Y., Jalbout, A.F., Yu, H.Y., Yang, G.L., Pan, X.M., and Su, Z.M. Optimized $LiFePO_4$–polyacene cathode material for lithium-ion batteries. *Advanced Materials* 2006; 18:2609–2613.
23. Murugan, A.V., Muraliganth, T., and Manthiram, A. Rapid microwave–solvothermal synthesis of phospho-olivine nanorods and their coating with a mixed conducting polymer for lithium ion batteries. *Electrochemistry Communications* 2008; 10:903–906.
24. Huang, Y.H., and Goodenough, J.B. High-rate $LiFePO_4$ lithium rechargeable battery promoted by electrochemically active polymers. *Chemistry of Materials* 2008; 20: 7237–7241.
25. Liu, H., Wang, G.X., Wexler, D., Wang, J.Z., and Liu, H.K. Electrochemical performance of $LiFePO_4$ cathode material coated with ZrO_2 nanolayer. *Electrochemistry Communications* 2008; 10:165–169.
26. Chang, H.H., Chang, C.C., Su, C.Y., Wu, H.C., Yang, M.H., and Wu, N.L. Effects of TiO_2 coating on high-temperature cycle performance of $LiFePO_4$-based lithium-ion batteries. *Journal of Power Sources* 2008; 185:466–472.
27. Cui, Y., Zhao, X.L., and Guo, R.S. High rate electrochemical performances of nanosized ZnO and carbon co-coated $LiFePO_4$ cathode. *Materials Research Bulletin* 2010; 45: 844–849.
28. Cui, Y., Zhao, X.L., and Guo, R.S. Enhanced electrochemical properties of $LiFePO_4$ cathode material by CuO and carbon co-coating. *Journal of Alloys and Compounds* 2010; 490:236–240.
29. Chung, S.Y., Bloking, J.T., and Chiang, Y.M. Electronically conductive phospho-olivines as lithium storage electrodes. *Nature Materials* 2002; 1:123–128.
30. Liu, H., Cao, Q., Fu, L.J., Li, C., Wu, Y.P., and Wu, H.Q. Doping effects of zinc on $LiFePO_4$ cathode material for lithium ion batteries. *Electrochemistry Communications* 2006; 8:1553–1557.
31. Liu, H., Li, C., Cao, Q., Wu, Y.P., and Holze, R. Effects of heteroatoms on doped $LiFePO_4$/C composites. *Journal of Solid State Electrochemistry* 2008; 12:1017–1020.
32. Kang, B., and Ceder, G. Battery materials for ultrafast charging and discharging. *Nature* 2009; 458:190–193.
33. Lim, S.Y., Yoon, C.S., and Cho, J.P. Synthesis of nanowire and hollow $LiFePO_4$ cathodes for high-performance lithium batteries. *Chemistry of Materials* 2008; 20:4560–4564.
34. Wang, L., He, X.M., Sun, W.T., Wang, J.L., Li, Y.D., and Fan, S.S. Crystal orientation tuning of $LiFePO_4$ nanoplates for high rate lithium battery cathode materials. *Nano Letters* 2012; 12:5632–5636.

35. Livage, J. Interface properties of vanadium pentoxide gels. *Materials Research Bulletin* 1991; 26:1173–1180.
36. Livage, J. Vanadium pentoxide gels. *Chemistry of Materials* 1991; 3:578–593.
37. Livage, J., Baffier, N., Pereiraramos, J.P., and Davidson, P. Vanadium Pentoxide Gels from Liquid Crystals to Lithium Batteries. *Solid State Ionics IV*, Vol. 369. *Materials Research Society Symposium Proceedings* (eds. G.A. Nazri, J.M. Tarascon, and M. Schreiber) 179–190 (1995).
38. Kim, G.T., Muster, J., Krstic, V., Park, J.G., Park, Y.W., Roth, S., and Burghard, M. Field-effect transistor made of individual V_2O_5 nanofibers. *Applied Physics Letters* 2000; 76: 1875–1877.
39. Chang, Y.J., Kang, B.H., Kim, G.T., Park, S.J., and Ha, J.S. Percolation network of growing V_2O_5 nanowires. *Applied Physics Letters* 2004; 84:5392–5394.
40. Spahr, M.E., Stoschitzki-Bitterli, P., Nesper, R., Haas, O., and Novak, P. Vanadium oxide nanotubes—A new nanostructured redox-active material for the electrochemical insertion of lithium. *Journal of the Electrochemical Society* 1999; 146:2780–2783.
41. Takahashi, K., Limmer, S.J., Wang, Y., and Cao, G.Z. Synthesis and electrochemical properties of single-crystal V_2O_5 nanorod arrays by template-based electrodeposition. *Journal of Physical Chemistry B* 2004; 108:9795–9800.
42. Takahashi, K., Limmer, S.J., Wang, Y., and Cao, G.Z. Growth and electrochemical properties of single-crystalline V_2O_5 nanorod arrays. *Japanese Journal of Applied Physics Part 1—Regular Papers Brief Communications and Review Papers* 2005; 44:662–668.
43. Takahashi, K., Wang, Y., and Cao, G.Z. Growth and electrochromic properties of single-crystal V_2O_5 nanorod arrays. *Applied Physics Letters* 2005; 86:053102.
44. Wang, Y., Takahashi, K., Shang, H.M., and Cao, G.Z. Synthesis and electrochemical properties of vanadium pentoxide nanotube arrays. *Journal of Physical Chemistry B* 2005; 109:3085–3088.
45. Takahashi, K., Wang, Y., and Cao, G.Z. Ni–V_2O_5 center dot nH_2O core–shell nanocable arrays for enhanced electrochemical intercalation. *Journal of Physical Chemistry B* 2005; 109:48–51.
46. Lutta, S.T., Dong, H., Zavalij, P.Y., and Whittingham, M.S. Synthesis of vanadium oxide nanofibers and tubes using polylactide fibers as template. *Materials Research Bulletin* 2005; 40:383–393.
47. Cao, A.M., Hu, J.S., Liang, H.P., and Wan, L.J. Self-assembled vanadium pentoxide (V_2O_5) hollow microspheres from nanorods and their application in lithium-ion batteries. *Angewandte Chemie—International Edition* 2005; 44:4391–4395.
48. Wang, Y., Takahashi, K., Lee, K., and Cao, G.Z. Nanostructured vanadium oxide electrodes for enhanced lithium-ion intercalation. *Advanced Functional Materials* 2006; 16:1133–1144.
49. Wang, J.X., Curtis, C.J., Schulz, D.L., and Zhang, J.G. Influences of treatment temperature and water content on capacity and rechargeability of V_2O_5 xerogel films. *Journal of the Electrochemical Society* 2004; 151:A1–A7.
50. West, K., Zachauchristiansen, B., Jacobsen, T., and Skaarup, S. Thin-film vanadium-oxide electrodes for lithium batteries. *Journal of Power Sources* 1993; 43:127–134.
51. Winter, M., Besenhard, J.O., Spahr, M.E., and Novak, P. Insertion electrode materials for rechargeable lithium batteries. *Advanced Materials* 1998; 10:725–763.
52. Idota, Y., Kubota, T., Matsufuji, A., Maekawa, Y., and Miyasaka, T. Tin-based amorphous oxide: A high-capacity lithium-ion-storage material. *Science* 1997; 276:1395–1397.
53. Kim, Y.J., Lee, H., and Sohn, H.J. Lithia formation mechanism in tin oxide anodes for lithium-ion rechargeable batteries. *Electrochemistry Communications* 2009; 11:2125–2128.
54. Courtney, I.A., McKinnon, W.R., and Dahn, J.R. On the aggregation of tin in SnO composite glasses caused by the reversible reaction with lithium. *Journal of the Electrochemical Society* 1999; 146:59–68.
55. Brousse, T., Retoux, R., Herterich, U., and Schleich, D.M. Thin-film crystalline SnO_2–lithium electrodes. *Journal of the Electrochemical Society* 1998; 145:1–4.
56. Kim, J.H., Jeong, G.J., Kim, Y.W., Sohn, H.J., Park, C.W., and Lee, C.K. Tin-based oxides as anode materials for lithium secondary batteries. *Journal of the Electrochemical Society* 2003; 150:A1544–A1547.
57. Lou, X.W., Li, C.M., and Archer, L.A. Designed synthesis of coaxial SnO_2@carbon hollow nanospheres for highly reversible lithium storage. *Advanced Materials* 2009; 21:2536–2539.
58. Liu, H., Chen, S., Wang, G.X., and Qiao, S.Z. Ordered mesoporous core/shell SnO_2/C nanocomposite as high-capacity anode material for lithium-ion batteries. *Chemistry—A European Journal* 2013; 19:16897–16901.
59. Cho, J. Porous Si anode materials for lithium rechargeable batteries. *Journal of Materials Chemistry* 2010; 20:4009–4014.
60. Ma, H., Cheng, F.Y., Chen, J., Zhao, J.Z., Li, C.S., Tao, Z.L., and Liang, J. Nest-like silicon nanospheres for high-capacity lithium storage. *Advanced Materials* 2007; 19:4067.
61. Chan, C.K., Peng, H.L., Liu, G., McIlwrath, K., Zhang, X.F., Huggins, R.A., and Cui, Y. High-performance lithium battery anodes using silicon nanowires. *Nature Nanotechnology* 2008; 3:31–35.
62. Cui, L.F., Yang, Y., Hsu, C.M., and Cui, Y. Carbon–silicon core–shell nanowires as high capacity electrode for lithium ion batteries. *Nano Letters* 2009; 9:3370–3374.
63. Cui, L.F., Ruffo, R., Chan, C.K., Peng, H.L., and Cui, Y. Crystalline–amorphous core–shell silicon nanowires for high capacity and high current battery electrodes. *Nano Letters* 2009; 9:491–495.
64. Kovalenko, I., Zdyrko, B., Magasinski, A., Hertzberg, B., Milicev, Z., Burtovyy, R., Luzinov, I., and Yushin, G. A Major Constituent of Brown Algae for Use in High-Capacity Li-Ion Batteries. *Science* 2010; 334:75–79.
65. Zhou, L., Xu, H.Y., Zhang, H.W., Yang, J., Hartono, S.B., Qian, K., Zou, J., and Yu, C.Z. Cheap and scalable synthesis of α-Fe_2O_3 multi-shelled hollow spheres as high-performance anode materials for lithium ion batteries. *Chemical Communications* 2013; 49:8695–8697.
66. Chou, S.L., Wang, J.Z., Wexler, D., Konstantinov, K., Zhong, C., Liu, H.K., and Dou, S.X. High-surface-area α-Fe_2O_3/carbon nanocomposite: One-step synthesis and its highly reversible and enhanced high-rate lithium storage properties. *Journal of Materials Chemistry* 2010; 20:2092–2098.
67. Ng, S.H., Wang, J.Z., Wexler, D., Konstantinov, K., Guo, Z.P., and Liu, H.K. Highly reversible lithium storage in spheroidal carbon-coated silicon nanocomposites as anodes for lithium-ion batteries. *Angewandte Chemie—International Edition* 2006; 45:6896–6899.
68. Zhu, C.B., Yu, Y., Gu, L., Weichert, K., and Maier, J. Electrospinning of highly electroactive carbon-coated single-crystalline $LiFePO_4$ nanowires. *Angewandte Chemie—International Edition* 2011; 50:6278–6282.

69. Yu, Y., Gu, L., Wang, C.L., Dhanabalan, A., van Aken, P.A., and Maier, J. Encapsulation of Sn@carbon nanoparticles in bamboo-like hollow carbon nanofibers as an anode material in lithium-based batteries. *Angewandte Chemie—International Edition* 2009; 48:6485–6489.
70. Yoo, E., Kim, J., Hosono, E., Zhou, H., Kudo, T., and Honma, I. Large reversible Li storage of graphene nanosheet families for use in rechargeable lithium ion batteries. *Nano Letters* 2008; 8:2277–2282.
71. Pan, D.Y., Wang, S., Zhao, B., Wu, M.H., Zhang, H.J., Wang, Y., and Jiao, Z. Li storage properties of disordered graphene nanosheets. *Chemistry of Materials* 2009; 21:3136–3142.
72. Wang, G.X., Shen, X.P., Yao, J., and Park, J. Graphene nanosheets for enhanced lithium storage in lithium ion batteries. *Carbon* 2009; 47:2049–2053.
73. Guo, P., Song, H.H., and Chen, X.H. Electrochemical performance of graphene nanosheets as anode material for lithium-ion batteries. *Electrochemistry Communications* 2009; 11:1320–1324.
74. Wang, C.Y., Li, D., Too, C.O., and Wallace, G.G. Electrochemical properties of graphene paper electrodes used in lithium batteries. *Chemistry of Materials* 2009; 21:2604–2606.
75. Wang, D.H., Choi, D.W., Li, J., Yang, Z.G., Nie, Z.M., Kou, R., Hu, D.H., Wang, C.M., Saraf, L.V., Zhang, J.G., Aksay, I.A., and Liu, J. Self-assembled TiO_2–graphene hybrid nanostructures for enhanced Li-ion insertion. *ACS Nano* 2009; 3:907–914.
76. Paek, S.M., Yoo, E., and Honma, I. Enhanced cyclic performance and lithium storage capacity of SnO_2/graphene nanoporous electrodes with three-dimensionally delaminated flexible structure. *Nano Letters* 2009; 9:72–75.
77. Yao, J., Shen, X.P., Wang, B., Liu, H.K., and Wang, G.X. *In situ* chemical synthesis of SnO_2–graphene nanocomposite as anode materials for lithium-ion batteries. *Electrochemistry Communications* 2009; 11:1849–1852.
78. Ji, F., Li, Y.L., Feng, J.M., Su, D., Wen, Y.Y., Feng, Y., and Hou, F. Electrochemical performance of graphene nanosheets and ceramic composites as anodes for lithium batteries. *Journal of Materials Chemistry* 2009; 19:9063–9067.
79. Yang, S.B., Feng, X.L., Wang, L., Tang, K., Maier, J., and Mullen, K. Graphene-based nanosheets with a sandwich structure. *Angewandte Chemie—International Edition* 2010; 49:4795–4799.
80. Wang, G.X., Wang, B., Wang, X.L., Park, J., Dou, S.X., Ahn, H., and Kim, K. Sn/graphene nanocomposite with 3D architecture for enhanced reversible lithium storage in lithium ion batteries. *Journal of Materials Chemistry* 2009; 19:8378–8384.
81. Chen, Z.P., Ren, W.C., Gao, L.B., Liu, B.L., Pei, S.F., and Cheng, H.M. Three-dimensional flexible and conductive interconnected graphene networks grown by chemical vapour deposition. *Nature Materials* 2011; 10:424–428.
82. Gao, L.B., Ren, W.C., Xu, H.L., Jin, L., Wang, Z.X., Ma, T., Ma, L.P., Zhang, Z.Y., Fu, Q., Peng, L.M., Bao, X.H., and Cheng, H.M. Repeated growth and bubbling transfer of graphene with millimetre-size single-crystal grains using platinum. *Nature Communications* 2012; 3. No. 669.
83. Paton, K.R., Varrla, E., Backes, C., Smith, R.J., Khan, U., O'Neill, A., Boland, C., Lotya, M., Istrate, O.M., King, P., Higgins, T., Barwich, S., May, P., Puczkarski, P., Ahmed, I., Moebius, M., Pettersson, H., Long, E., Coelho, J., O'Brien, S.E., McGuire, E.K., Sanchez, B.M., Duesberg, G.S., McEvoy, N., Pennycook, T.J., Downing, C., Crossley, A., Nicolosi, V., and Coleman, J.N. Scalable production of large quantities of defect-free few-layer graphene by shear exfoliation in liquids. *Nature Materials* 2014; 13:624–630.
84. Lu, J., Li, L., Park, J.-B., Sun, Y.-K., Wu, F., and Amine, K. Aprotic and aqueous Li–O_2 batteries. *Chemical Reviews* 2014; 114:5611–5640.
85. Armand, M., and Tarascon, J.M. Building better batteries. *Nature* 2008; 451:652–657.
86. Bruce, P.G., Freunberger, S.A., Hardwick, L.J., and Tarascon, J.M. Li–O_2 and Li–S batteries with high energy storage. *Nature Materials* 2012; 11:19–29.
87. Abraham, K.M., and Jiang, Z. A polymer electrolyte-based rechargeable lithium/oxygen battery. *Journal of the Electrochemical Society* 1996; 143:1–5.
88. Girishkumar, G., McCloskey, B., Luntz, A.C., Swanson, S., and Wilcke, W. Lithium–air battery: Promise and challenges. *The Journal of Physical Chemistry Letters* 2010; 1:2193–2203.
89. Ogasawara, T., Debart, A., Holzapfel, M., Novak, P., and Bruce, P.G. Rechargeable Li_2O_2 electrode for lithium batteries. *Journal of the American Chemical Society* 2006; 128:1390–1393.
90. Peng, Z.Q., Freunberger, S.A., Hardwick, L.J., Chen, Y.H., Giordani, V., Barde, F., Novak, P., Graham, D., Tarascon, J.M., and Bruce, P.G. Oxygen reactions in a non-aqueous Li^+ electrolyte. *Angewandte Chemie—International Edition* 2011; 50:6351–6355.
91. Xu, W., Xu, K., Viswanathan, V.V., Towne, S.A., Hardy, J.S., Xiao, J., Hu, D.H., Wang, D.Y., and Zhang, J.G. Reaction mechanisms for the limited reversibility of Li–O_2 chemistry in organic carbonate electrolytes. *Journal of Power Sources* 2011; 196:9631–9639.
92. Christensen, J., Albertus, P., Sanchez-Carrera, R.S., Lohmann, T., Kozinsky, B., Liedtke, R., Ahmed, J., and Kojic, A. A critical review of Li/air batteries. *Journal of the Electrochemical Society* 2012; 159:R1–R30.
93. Xiao, J., Hu, J.Z., Wang, D.Y., Hu, D.H., Xu, W., Graff, G.L., Nie, Z.M., Liu, J., and Zhang, J.G. Investigation of the rechargeability of Li–O_2 batteries in non-aqueous electrolyte. *Journal of Power Sources* 2011; 196:5674–5678.
94. Freunberger, S.A., Chen, Y.H., Peng, Z.Q., Griffin, J.M., Hardwick, L.J., Barde, F., Novak, P., and Bruce, P.G. Reactions in the rechargeable lithium–O_2 battery with alkyl carbonate electrolytes. *Journal of the American Chemical Society* 2011; 133:8040–8047.
95. Freunberger, S.A., Chen, Y.H., Drewett, N.E., Hardwick, L.J., Barde, F., and Bruce, P.G. The lithium–oxygen battery with ether-based electrolytes. *Angewandte Chemie—International Edition* 2011; 50:8609–8613.
96. Jung, H.G., Hassoun, J., Park, J.B., Sun, Y.K., and Scrosati, B. An improved high-performance lithium–air battery. *Nature Chemistry* 2012; 4:579–585.
97. Xiao, J., Wang, D.H., Xu, W., Wang, D.Y., Williford, R.E., Liu, J., and Zhang, J.G. Optimization of air electrode for Li/air batteries. *Journal of the Electrochemical Society* 2010; 157:A487–A492.
98. Li, Y.L., Wang, J.J., Li, X.F., Liu, J., Geng, D.S., Yang, J.L., Li, R.Y., and Sun, X.L. Nitrogen-doped carbon nanotubes as cathode for lithium–air batteries. *Electrochemistry Communications* 2011; 13:668–672.
99. Mitchell, R.R., Gallant, B.M., Thompson, C.V., and Shao-Horn, Y. All-carbon-nanofiber electrodes for high-energy rechargeable Li–O_2 batteries. *Energy and Environmental Science* 2011; 4:2952–2958.

100. Sun, B., Wang, B., Su, D.W., Xiao, L.D., Ahn, H., and Wang, G.X. Graphene nanosheets as cathode catalysts for lithium–air batteries with an enhanced electrochemical performance. *Carbon* 2012; 50:727–733.
101. Xiao, J., Mei, D.H., Li, X.L., Xu, W., Wang, D.Y., Graff, G.L., Bennett, W.D., Nie, Z.M., Saraf, L.V., Aksay, I.A., Liu, J., and Zhang, J.G. Hierarchically porous graphene as a lithium–air battery electrode. *Nano Letters* 2011; 11:5071–5078.
102. Sun, B., Huang, X., Chen, S., Munroe, P., and Wang, G. Porous graphene nanoarchitectures: An efficient catalyst for low charge-overpotential, long life, and high capacity lithium–oxygen batteries. *Nano Letters* 2014; 14:3145–3152.
103. Yang, X.H., He, P., and Xia, Y.Y. Preparation of mesocellular carbon foam and its application for lithium/oxygen battery. *Electrochemistry Communications* 2009; 11:1127–1130.
104. Guo, Z., Zhou, D., Dong, X., Qiu, Z., Wang, Y., and Xia, Y. Ordered hierarchical mesoporous/macroporous carbon: A high-performance catalyst for rechargeable Li–O_2 batteries. *Advanced Materials* 2013; 25:5668–5672.
105. Debart, A., Bao, J., Armstrong, G., and Bruce, P.G. An O_2 cathode for rechargeable lithium batteries: The effect of a catalyst. *Journal of Power Sources* 2007; 174:1177–1182.
106. Debart, A., Paterson, A.J., Bao, J., and Bruce, P.G. α-MnO_2 nanowires: A catalyst for the O_2 electrode in rechargeable lithium batteries. *Angewandte Chemie—International Edition* 2008; 47:4521–4524.
107. Sun, B., Zhang, J.Q., Munroe, P., Ahn, H.J., and Wang, G. Hierarchical $NiCo_2O_4$ nanorods as an efficient cathode catalyst for rechargeable non-aqueous Li–O_2 batteries. *Electrochemistry Communications* 2013; 31:88–91.
108. Sun, B., Huang, X., Chen, S., Zhao, Y., Zhang, J., Munroe, P., and Wang, G. Hierarchical macroporous/mesoporous $NiCo_2O_4$ nanosheets as cathode catalysts for rechargeable Li–O_2 batteries. *Journal of Materials Chemistry A* 2014; 2:12053–12059.
109. Zhang, L., Zhang, S., Zhang, K., Xu, G., He, X., Dong, S., Liu, Z., Huang, C., Gu, L., and Cui, G. Mesoporous $NiCo_2O_4$ nanoflakes as electrocatalysts for rechargeable Li–O_2 batteries. *Chemical Communications* 2013; 49:3540–3542.
110. Cao, Y., Wei, Z.K., He, J., Zang, J., Zhang, Q., Zheng, M.S., and Dong, Q.F. α-MnO_2 nanorods grown *in situ* on graphene as catalysts for Li–O_2 batteries with excellent electrochemical performance. *Energy and Environmental Science* 2012; 5:9765–9768.
111. Black, R., Lee, J.-H., Adams, B., Mims, C.A., and Nazar, L.F. The role of catalysts and peroxide oxidation in lithium–oxygen batteries. *Angewandte Chemie—International Edition* 2013; 52:392–396.
112. Wang, H.L., Yang, Y., Liang, Y.Y., Zheng, G.Y., Li, Y.G., Cui, Y., and Dai, H.J. Rechargeable Li–O_2 batteries with a covalently coupled $MnCo_2O_4$–graphene hybrid as an oxygen cathode catalyst. *Energy and Environmental Science* 2012; 5:7931–7935.
113. Lu, Y.C., Gasteiger, H.A., Parent, M.C., Chiloyan, V., and Shao-Horn, Y. The influence of catalysts on discharge and charge voltages of rechargeable Li–oxygen batteries. *Electrochemical and Solid-State Letters* 2010; 13:A69–A72.
114. Lu, Y.C., Gasteiger, H.A., and Shao-Horn, Y. Method development to evaluate the oxygen reduction activity of high-surface-area catalysts for Li–air batteries. *Electrochemical and Solid-State Letters* 2011; 14:A70–A74.
115. Lu, Y.C., Xu, Z.C., Gasteiger, H.A., Chen, S., Hamad-Schifferli, K., and Shao-Horn, Y. Platinum–gold nanoparticles: A highly active bifunctional electrocatalyst for rechargeable lithium–air batteries. *Journal of the American Chemical Society* 2010; 132:12170–12171.
116. Lu, Y.C., Gasteiger, H.A., Crumlin, E., McGuire, R., and Shao-Horn, Y. Electrocatalytic activity studies of select metal surfaces and implications in Li–air batteries. *Journal of the Electrochemical Society* 2010; 157:A1016–A1025.
117. Lu, Y.C., Gasteiger, H.A., and Shao-Horn, Y. Catalytic activity trends of oxygen reduction reaction for nonaqueous Li–air batteries. *Journal of the American Chemical Society* 2011; 133:19048–19051.
118. Lu, J., Lei, Y., Lau, K.C., Luo, X., Du, P., Wen, J., Assary, R.S., Das, U., Miller, D.J., Elam, J.W., Albishri, H.M., Abd El-Hady, D., Sun, Y.-K., Curtiss, L.A., and Amine, K. A nanostructured cathode architecture for low charge overpotential in lithium–oxygen batteries. *Nature Communications* 2013; 4:2383.
119. Xu, J.-J., Wang, Z.-L., Xu, D., Zhang, L.-L., and Zhang, X.-B. Tailoring deposition and morphology of discharge products towards high-rate and long-life lithium–oxygen batteries. *Nature Communications* 2013; 4:2438.
120. Jung, H.G., Jeong, Y.S., Park, J.B., Sun, Y.K., Scrosati, B., and Lee, Y.J. Ruthenium-based electrocatalysts supported on reduced graphene oxide for lithium–air batteries. *ACS Nano* 2013; 7:3532–3539.
121. Sun, B., Munroe, P., and Wang, G.X. Ruthenium nanocrystals as cathode catalysts for lithium–oxygen batteries with a superior performance. *Scientific Reports* 2013; 3:2247.
122. Peng, Z.Q., Freunberger, S.A., Chen, Y.H., and Bruce, P.G. A reversible and higher-rate Li–O_2 battery. *Science* 2012; 337: 563–566.
123. Chen, Y., Freunberger, S.A., Peng, Z., Fontaine, O., and Bruce, P.G. Charging a Li–O_2 battery using a redox mediator. *Nature Chemistry* 2013; 5:489–494.
124. Thotiyl, M.M.O., Freunberger, S.A., Peng, Z., Chen, Y., Liu, Z., and Bruce, P.G. A stable cathode for the aprotic Li–O_2 battery. *Nature Materials* 2013; 12:1049–1055.
125. Li, F., Tang, D.-M., Chen, Y., Golberg, D., Kitaura, H., Zhang, T., Yamada, A., and Zhou, H. Ru/ITO: A carbon-free cathode for nonaqueous Li–O_2 battery. *Nano Letters* 2013; 13:4702–4707.
126. Li, F., Tang, D.-M., Jian, Z., Liu, D., Golberg, D., Yamada, A., and Zhou, H. Li-O_2 battery based on highly efficient Sb-doped tin oxide supported Ru nanoparticles. *Advanced Materials* 2014; 26:4659–4664.

3 Future Catalyst Approaches for Electrochemical Energy Storage and Conversion

Gang Wu and David P. Wilkinson

CONTENTS

3.1 INTRODUCTION

Widespread deployment of renewable energy has been the indispensable strategy for addressing the issues resulting from the constant use of fossil fuel, such as global climate change, energy security, and sustainability. Among the available renewable energy sources, solar and wind are probably the most abundant and readily accessible, which have been considered as essential components of the future global energy portfolio [1]. However, the nature of solar and wind energy is random and intermittent. Efficient and economic energy storage and conversion technologies are required to harvest and utilize inexpensive renewable energy [2]. Among others, electrochemical energy technologies such as fuel cells and metal–air batteries can provide a temporary medium to store and release electricity when and where it is needed. Importantly, this electrochemical processes could be reversed via the oxidation or reduction of active species so as to convert chemical energy into electrical energy. Polymer electrolyte fuel cells (PEFCs) represent one of the most promising energy conversion technologies for a wide variety of applications (e.g., transportation, portable, and stationary applications), including several advantages over gasoline combustion, such as better overall fuel efficiency and reduction in CO_2 and other emissions. Meanwhile, metal–air batteries can provide significantly enhanced energy densities over traditional lithium-ion batteries. Unlike the traditional intercalation electrodes used in Li-ion batteries, the porous oxygen cathode in the metal–air cell is capable of taking reactant O_2 from the atmosphere, instead of storing bulky reactants in the electrode. As a result, the battery has significantly improved specific energy density. For example, the theoretical energy density of Li–O_2 batteries reach 5200 Wh kg^{-1}, the highest value today among studied electrochemical energy devices [3]. Thus, it has high promise to meet and exceed the battery targets set for automotive applications (1700 Wh kg^{-1}, derived from the practical energy density of gasoline) [4,5].

As shown in Figure 3.1, both fuel cells and metal–air batteries have similar oxygen cathodes and share the same operating principles for the oxygen reduction reaction (ORR), and in the case of reversible fuel cells and rechargeable metal–air batteries, require the oxygen evolution reaction (OER) to be facilitated as well [6]. The ORR and OER in acidic and alkaline aqueous electrolytes are described by the following electrochemical reactions and standard reaction potentials ($E°$):

$$O_2 + 4H^+ + 4e^- \underset{\text{OER}}{\overset{\text{ORR}}{\rightleftharpoons}} 2H_2O \quad E° = 1.229\text{ V} \tag{3.1}$$

$$O_2 + 2H_2O + 4e^- \underset{\text{OER}}{\overset{\text{ORR}}{\rightleftharpoons}} 4OH^- \quad E° = 0.401\text{ V} \tag{3.2}$$

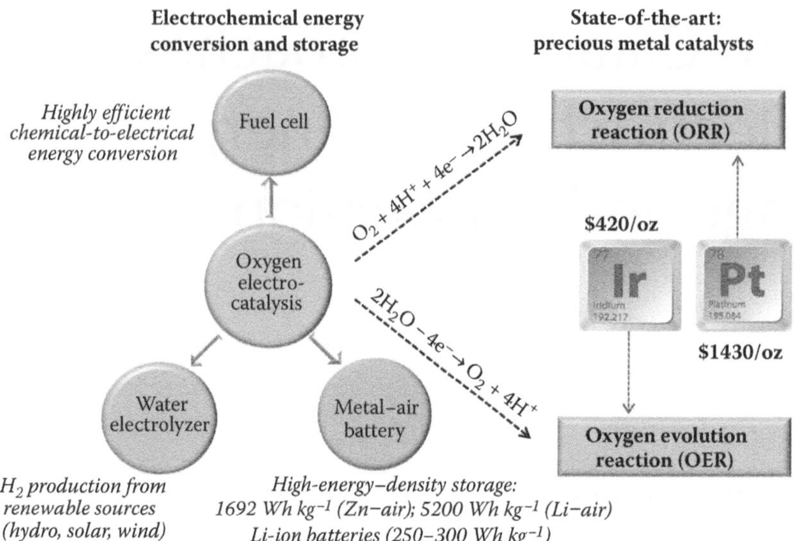

FIGURE 3.1 Electrochemical energy storage and conversion technologies and their associated oxygen reactions requiring precious metal catalysts.

The use of nonaqueous media in an Li–air battery has proven to be more feasible than an aqueous electrolyte to alleviate the parasitic corrosion on the Li metal anode [5,7]. The standard potential of the Li^+–O_2 couple in a nonaqueous solution is around 2.96 V versus Li/Li^+ according to the reaction [8]

$$2Li^+ + 2e^- + O_2 \underset{OER}{\overset{ORR}{\rightleftharpoons}} Li_2O_2 \text{(ORR discharge; OER charge)}$$

$$E° = 2.96 \text{ V} \quad (3.3)$$

It was found that the ORR overpotential in both aqueous and nonaqueous electrolytes depends first and foremost on the intrinsic activity of catalysts [9,10]. H_2O or H_2O_2 are produced during the ORR in a traditional aqueous electrolyte; however, insoluble lithium–oxygen species (Li_2O or Li_2O_2) are formed in nonaqueous media and passivate the catalysts, eventually leading to a termination of the discharge process. The catalysts can be exposed again during the charge process (OER, when an external potential is applied $E > E_0$), accompanied by the decomposition of these solid-state lithium–oxygen compounds. Thus, an efficient OER catalyst to catalyze a fast decomposition is needed to recover active sites for the ORR. The ORR and OER mechanisms in nonaqueous Li^+ electrolytes are different to traditional aqueous electrolytes, wherein various solid reactants/products and reaction intermediates are often present.

Owing to the high overpotentials of ORR and OER and inherently slow reaction kinetics, the performance of these electrochemical energy technologies is greatly limited by ORR and OER activities. Expensive and scarce precious metals such as Pt- and Ir-based catalysts represent the state of the art for the ORR and OER, respectively, in terms of their highest activity and durability. Thus, the development of low-cost, highly efficient, robust, and environmentally friendly electrochemical energy technologies is greatly dependent on new concepts and new materials in the development of nonprecious metal catalysts. In this chapter, we provide a better understanding of the current situation with electrochemical energy technologies, along with the future approach to addressing the current problems and challenges and helping move toward large-scale commercialization [11].

3.2 ADVANCED NONPRECIOUS METAL CATALYST TECHNOLOGIES

3.2.1 Challenges and Progress

Unfortunately, the high cost and limited supply of these precious metals has become a grand challenge for the widespread commercial success and implementation of clean energy technology [12,13]. The development of nonprecious metal catalysts for these oxygen-based reactions has become a hot topic in the field of electrochemical energy storage and conversion [14–16]. During several decades of searching nonprecious metal catalysts (NPMCs) for oxygen reduction, significant progress has been made in the synthesis, performance improvement, and understanding of the mechanism [13,17]. As shown in Figure 3.2, among the many studied catalyst formulations, when compared with traditional macrocyclic compounds containing metal–nitrogen coordination structures, much improved ORR catalytic activity and stability had been achieved after heat treating these macrocycle compounds at high temperatures (>700°C). This suggests that a new type of active site is generated during the pyrolysis step [18]. In an effort to further explore such heat-treated nonprecious catalysts, no longer the expensive macrocycles but much cheaper N-containing chemicals, such as ammonium and amine as well as inorganic transition metal salts like nitrate, sulfate, and chloride, have been used to develop nonprecious metal

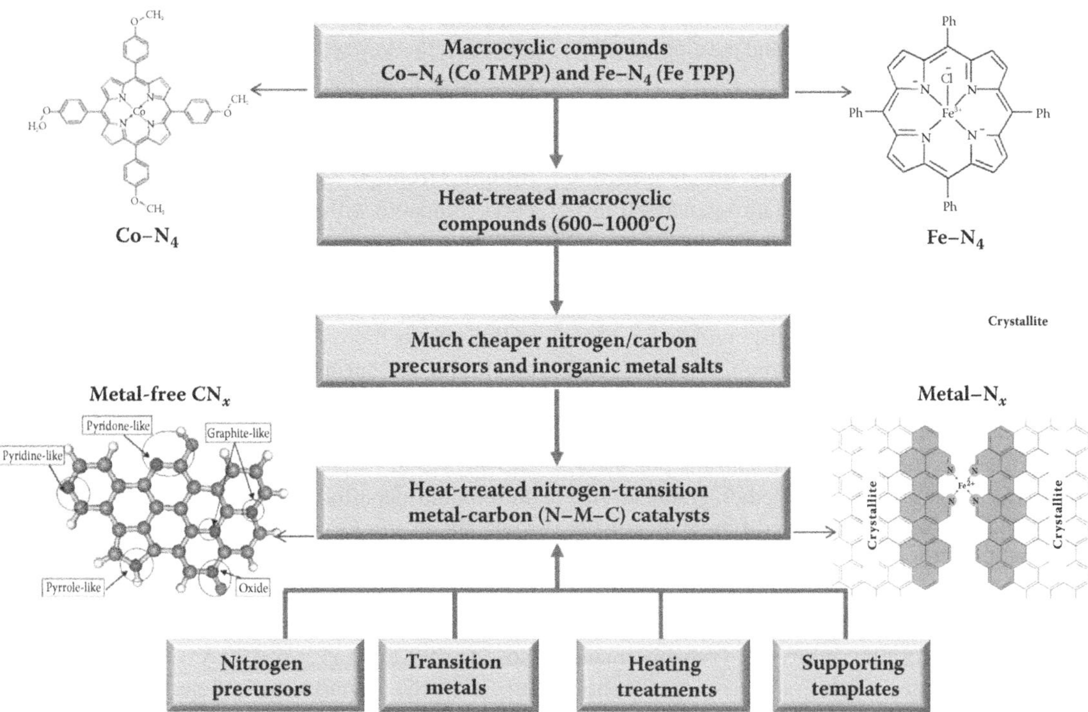

FIGURE 3.2 Development of nonprecious metal catalysts for oxygen reduction reaction.

catalysts for oxygen reduction [15,17–25]. Thus far, the nature of the nonprecious active sites in such heat-treated catalysts is still under a debate that concentrates on the role of transition metal during the catalyst synthesis. Some people believed that metals just catalyze nitrogen doping in graphitized carbon structures, not a part of the active site [26]. However, there is increasing evidence showing that metals, especially Fe, will directly participate in active sites by coordinating with nitrogen or carbon [18,27].

Despite the controversy, the catalytic activity was found to be strongly dependent on the synthetic chemistry of catalysts, including the structure of the nitrogen precursors, transition metals, heating temperatures, and supporting materials. Especially, recent breakthroughs in the synthesis of high-performance metal–nitrogen–carbon (M–N–C; M: Fe and/or Co) catalysts by simultaneously heat-treating precursors of nitrogen, carbon, and transition metals at 800–1000°C make replacement of Pt in ORR electrocatalysts with earth-abundant elements a realistic possibility. A designed approach to making active and durable materials, focusing on the catalyst nanostructure, was recently developed [13]. The approach consists of nitrogen doping, *in situ* carbon graphitization, and the use of graphitic structures as supporting templates. Various forms of nitrogen, particularly pyridinic and graphitic, can act as n-type carbon dopants in M–N–C catalysts, assisting in the formation of disordered carbon nanostructures and creating electron deficiency on carbon. While nitrogen doping at the graphene edge appears to be a key to the ORR activity, the importance of graphitic nitrogen cannot be ruled out as well.

The CN_x structures are likely a crucial part of the ORR active site(s). Noteworthy, the ORR activity is not necessarily governed by the amount of nitrogen, but by how the nitrogen is incorporated into the nanostructures. More important, highly graphitic carbon nanostructures formed during the heat treatment of nitrogen/carbon precursor in the presence of a transition metal represent an important component of M–N–C catalysts. Those structures likely act as a host for the ORR active sites. Their composition and morphology greatly depend on the synthetic chemistry, including the nature of all the precursors, type of support, and heat-treatment temperature.

3.2.2 Approaches to Advancing Nonprecious Metal Catalysts

3.2.2.1 Controlled Synthesis of Nitrogen-Doped Carbon Composite Catalysts

Future effort in the synthesis of M–N–C catalysts is likely to focus primarily on the precise control of interactions between nitrogen/carbon and metal precursors during the heat treatment in order to produce catalysts with optimum chemical composition and morphology, as well as to maximize the population of ORR active sites. Recently, the physical and chemical properties of carbon materials, such as crystal structure, morphology, and defect chemistry, have been found to have a significant impact on their catalytic properties. Thus, from both a fundamental and applied research perspective, it is of importance to study the controlled synthesis of nanocarbon catalysts with improved activity and durability, and to understand their synthesis–structure–property correlations. To realize the controlled synthesis, we developed several approaches to finely tune carbon nanostructures, morphologies, and heteroatom doping in terms of achieving maximum catalyst performance. These include using novel template materials, exploring new reaction precursors (e.g., transition metals and nitrogen–carbon precursors), and developing innovative heating and processing strategies.

The *in situ* formed graphitized carbon nanostructures in catalysts likely can link to the oxygen reduction activity and may be critical to active sites. Importantly, different carbon nanostructures are derived from different nitrogen and transition metal precursors. For example, as shown in Figure 3.3 [17], when heat-treating polyaniline (PANI) without adding any metal, no special carbon nanostructure was observed, indicating that transitional metal is indispensable for catalyzing the graphitization of nitrogen–carbon precursors and for forming highly graphitized nanocarbon. When ethylene diamine (EDA) and Co were used, onion-like carbons were formed. When another type of nitrogen carbon precursor, dicyanamide, and Fe were heat treated together, graphene-like tubes appeared [28]. We also studied other different precursors including polypyrrole (PPy) and melamine, and the formation of graphene sheet–like morphology is only observed with the PANI-derived catalysts thus far. Thus, the aromatic structure in PANI may be a key to forming graphene, probably owing to their structural similarities. Such graphene-rich morphology in the catalysts may benefit the ORR electrocatalysis by improving electron conductivity and corrosion resistance [29].

Recently, high-surface-area PANI–iron–carbon (PANI–Fe–C) catalysts with a Brunauer–Emmett–Teller (BET) surface area of 1100 $m^2\ g^{-1}$ were prepared at Los Alamos National Laboratory (LANL) (Figure 3.4) [30,31]. Compared with conventional low-surface-area catalysts (200–300 $m^2\ g^{-1}$) [19], using high iron content (30 wt.% Fe) during the catalyst synthesis by employing a three-step heating synthesis protocol are the key to generating high surface areas. The bulk Fe content in the final catalyst was determined by inductively coupled plasma and thermogravimetric analysis for these three catalysts. Unexpectedly, the Fe content in the catalyst resulting from 30 wt.% Fe was the lowest, around 2 wt.% in bulk, compared with the other two (3 and 10 wt.%) Fe catalysts [30]. In good agreement with the BET surface area, the much more porous structure was observed in the 30 wt.% Fe catalyst. This is probably because the *in situ* formed iron sulfide in this case is more efficient to leach out without enough carbon protection during the acid leaching step. As evidenced by X-ray diffraction patterns, unlike in 3 and 10 wt.% Fe catalyst, nearly no FeS was left in the 30 wt.% Fe-based catalyst after chemical acid leaching. The sulfur comes from the oxidant ammonium persulfate for polymerizing aniline in synthesis. Thus, the *in situ* formed iron sulfide during catalyst synthesis can act as an effective pore-forming agent to improve catalyst morphology. In a standard fuel cell single cell test, the high-surface-area PANI–Fe–C catalyst obtained a current density close to 190 mA cm^{-2} at 0.80 V (*i*R-free). This significantly surpasses the Department of Energy September 2014 ORR Catalyst Development Milestone that specifies a current density of 150 mA cm^{-2} at 0.8 V (*i*R-corrected) for nonprecious metal catalysts. Thus, on the basis of our current understanding of the existing catalysts, we believe that highly efficient nonprecious catalysts for PEFCs can be further improved by increasing the density of active sites with high volumetric activity, by properly selecting the precursors and optimizing solution reaction and thermal conditions.

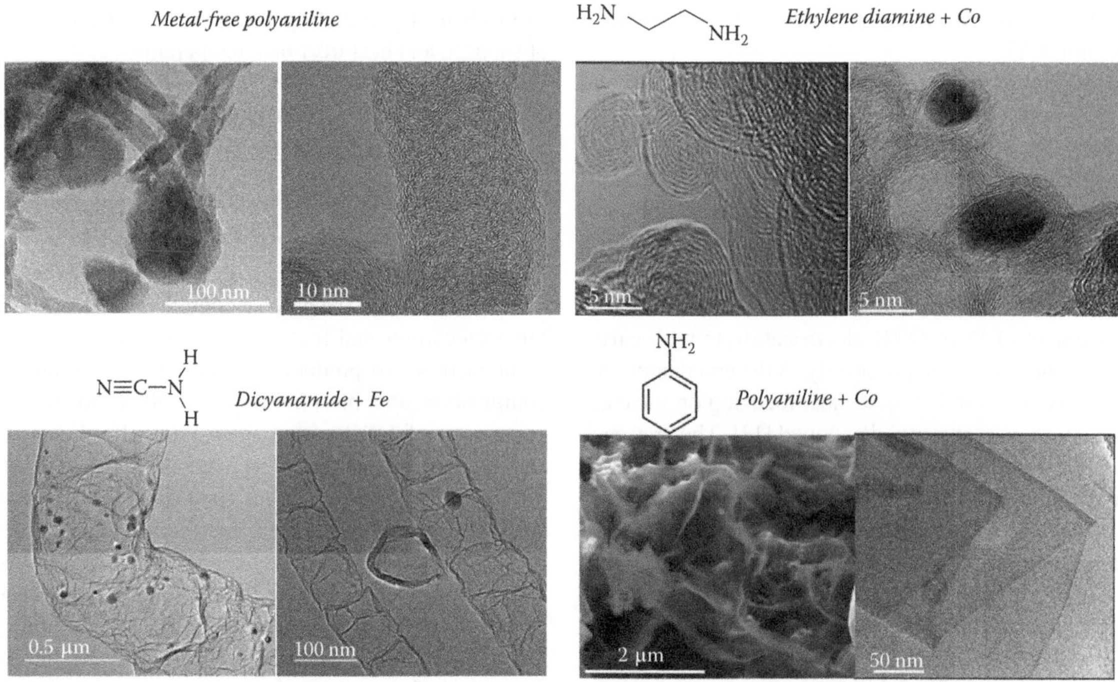

FIGURE 3.3 Diversified carbon nanostructures observed in heat-treated M–N–C catalysts. (Li, Q., R. Cao, J. Cho, and G. Wu: Nanocarbon electrocatalysts for oxygen reduction in alkaline media for advanced energy conversion and storage. *Advanced Energy Materials.* 2014. 4. 1301415. Copyright Wiley-VCH Verlag GmbH & Co. KGaA. Reprinted with permission.)

"Standard" PANI–Fe–C, 10 wt.% Fe
Two-step heat treatment

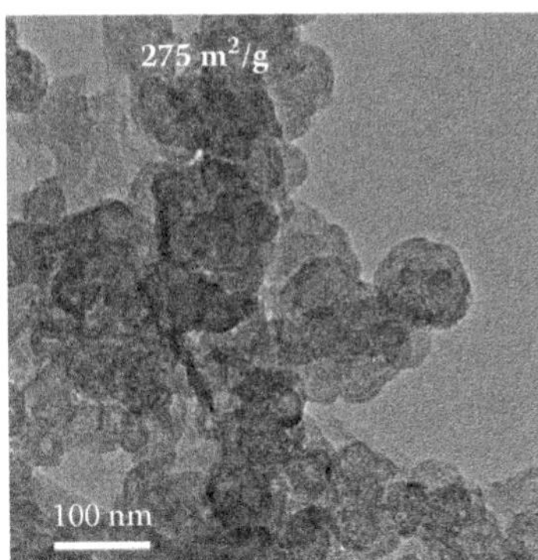

PANI–Fe–C, 30 wt.% Fe
Two-step heat treatment

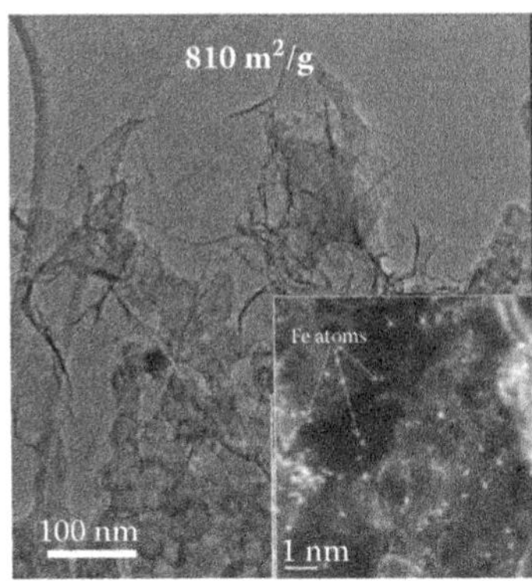

PANI–Fe–C, 30 wt.% Fe
Three-step heat treatment

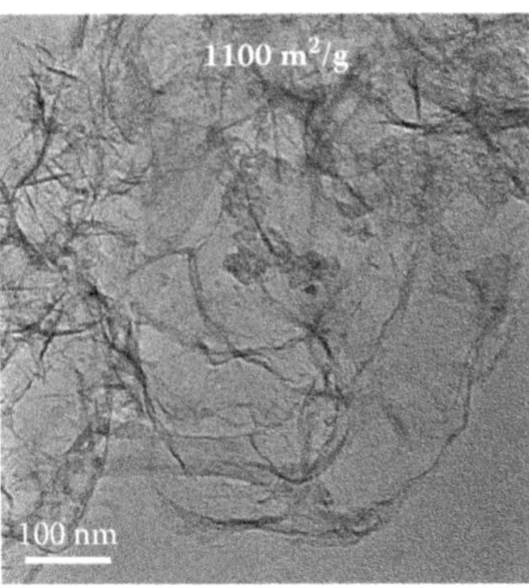

FIGURE 3.4 High-surface area Fe–N–C catalysts with much enhanced fuel cell performance. (Reprinted with permission from Li, Q., G. Wu, D.A. Cullen, K.L. More, N.H. Mack, H.T. Chung, and P. Zelenay. Phosphate-tolerant oxygen reduction catalysts. *ACS Catalysis* 4, 3193–3200. Copyright 2014 American Chemical Society.)

3.2.2.2 Graphene Catalysts with Transition Metal Incorporation

Another effective way to design advanced nonprecious metal catalysts is to explore new support materials that are able to control the hydrophobicity, morphologies, and porosity of the catalysts. Among various nanocarbon supports, a particular interesting field is to explore the possibility of incorporating graphene into ORR NPMCs. Graphene, a single atomic layer of graphite with sp^2-bonded carbon atoms arranged in a honeycomb structure, has attracted great interest since its discovery by Novoselov et al. in 2004 [32]. The promising application of graphene in NPMCs is primarily due to its unique physical and chemical properties, such as high surface area (theoretical value, ~2600 $m^2\ g^{-1}$), high electrochemical stability, excellent electrical conductivity, unique graphitic basal plane structure, and the ease of functionalization [33]. Importantly, chemical doping with heteroatoms (e.g., N, B, P, or S) into the graphene planes can tune the electronic and structural properties, provide active sites for oxygen electrocatalysis, and enhance the interactions between the carbon structures and oxygen molecules.

Currently, there are two major methods to prepare graphene. One is from reduction of graphene oxide (GO) that is usually made through oxidation of graphite powder under harsh chemical conditions. The most effective method currently used is Hummers method involving different strong oxidants (such as $KMnO_4$, sulfuric acid, and potassium persulfate). As a result, different oxygen-containing groups, especially hydroxyl, epoxy, and carboxyl, are formed on both sides and at the edges. Importantly, GO flakes can be exfoliated into single-layer and multilayer sheets when dissolved in water owing to the strong coulombic repulsion between hydrolyzed sheets. Another way to prepare graphene sheets is to use chemical vapor deposition (CVD) on Cu substrates [34]. However, the CVD method to prepare continuous graphene sheets is relatively expensive (~$100 per cm^2) and time consuming. Moreover, it is also very difficult to scale up for mass production in energy applications.

During the catalyst development through the high-temperature approach, graphene-like morphologies were found to dominate in high-performance ORR NPMCs, which is derived from carbon/nitrogen precursors via a graphitization process [17]. These graphene-like nanostructures likely link to the oxygen reduction activity and may be critical to active sites. In our recent efforts, cost-effective nitrogen-doped graphene composites, including graphene sheets or graphene tubes, were prepared from properly selected nitrogen-containing carbon precursors by using transition metals Fe or Co as catalysts and multiwalled carbon nanotubes (MWCNTs) or metal–organic frames (MOFs) as templates (Figure 3.5) [28,35]. The prepared graphene nanocomposite M–N–C catalysts (M: Co or Fe) exhibited a universally high activity for oxygen reduction in various media, including acid, alkaline, and nonaqueous electrolytes. Importantly, unlike the traditional synthetic approaches for graphene materials, the *in situ* formation of graphene nanocomposite directly through a graphitization process may open a new route for graphene catalyst preparation with controlled morphology and nitrogen functionality.

Compared with other carbon nanostructures, the formation of graphene-like morphology in catalysts is more closely associated with the enhancement of catalyst performance (Figure 3.6) [36,37]. For example, the graphene-rich TiO_2-supported catalyst has higher activity than the graphene-free Ketjen black–supported catalysts. Meanwhile, the graphene-containing MWCNT-supported catalysts exhibited much enhanced performance durability in acid media, compared with other catalysts not containing graphene. Thus, the existence of graphene-like morphologies seems to have a promotional role in improving NMPC performance in PEFCs. Furthermore, a Co–N–C catalyst derived from PANI for oxygen reduction was dominant by nitrogen-doped graphene sheets (Figure 3.7) [38]. The Co-based catalyst shows improved ORR activity and stability in 0.1 M NaOH electrolyte relative to a Pt/C reference catalyst and an Fe catalyst derived from the same nitrogen precursor [38].

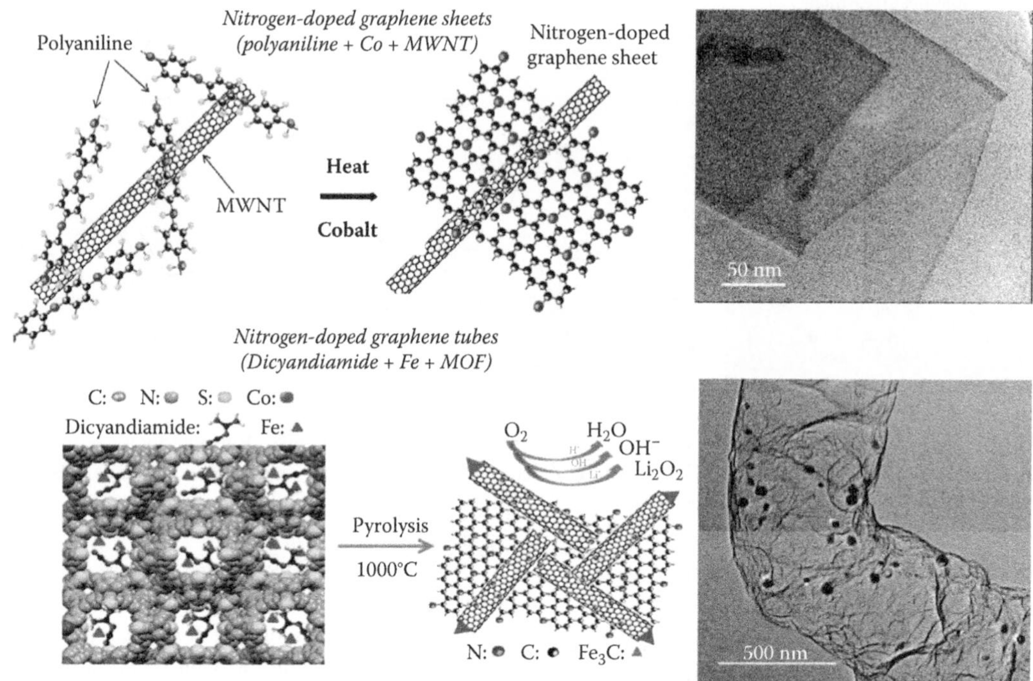

FIGURE 3.5 New approach to prepare graphene composites through graphitization processes of nitrogen–carbon precursors in the presence of transition metals. Top: graphene sheets. (Reprinted with permission from Ref. Wu, G., N.H. Mack, W. Gao, S.G. Ma, R.Q. Zhong, J.T. Han, J.K. Baldwin, and P. Zelenay. Nitrogen doped graphene-rich catalysts derived from heteroatom polymers for oxygen reduction in nonaqueous lithium–O_2 battery cathodes. *ACS Nano* 6, 9764–9776. Copyright 2014 American Chemical Society.) Bottom: graphene tubes. (Li, Q., P. Xu, W. Gao, S.G. Ma, G.Q. Zhang, R.G. Cao, J. Cho, H.L. Wang, and G. Wu: Graphene/graphene tube nanocomposites templated from cage-containing metal–organic frameworks for oxygen reduction in Li–O_2 batteries. *Advanced Materials*. 2014. 26. 1378–1386. Copyright Wiley-VCH Verlag GmbH & Co. KGaA. Reprinted with permission.)

It is worth noting that no such graphene-sheet structures were found in similar catalysts synthesized using different nitrogen–carbon precursors (e.g., EDA and cyanamide). The unique graphene morphology observed in the PANI-derived catalysts implies a significant effect of the nitrogen precursor on the structure of catalysts during the heat treatment of carbon–nitrogen precursors. Such graphene-rich morphology in the catalysts may benefit the ORR electrocatalysis by assuring high electron conductivity and providing corrosion resistance [39].

3.2.2.3 Bifunctional ORR/OER Nonprecious Metal Catalysts

During nanocarbon-based catalyst development, special emphasis has been placed on bifunctional catalysts, consisting of different active species to catalyze different electrochemical reactions [40–43]. These effects are especially important for reversible fuel cells and electrochemically rechargeable metal–air battery technologies relying on both ORR and OER processes occurring during regular operation. Usually, as the optimal active-site structures for the ORR and the OER are not similar, bifunctional catalysts consist of two or more constituents. As shown in Figure 3.8 [39,43], recently, two type of configurations were proposed to design the robust bifunctional catalysts. In particular, several metal oxides exhibit high OER activity and excellent electrochemical stability, and can be deposited onto ORR-active nitrogen-doped nanocarbon supports such as graphene or nanotubes. In our recent effort, Mn_3O_4 nanoparticles have been successfully deposited onto graphene, showing promising charge and discharge performances in Li–air batteries [44]. Further investigations on different metal oxides and nanocarbon supports, along with interface engineering, will yield novel nanocomposite catalysts with significantly improved activity and durability. On the other hand, core–shell structures can also provide optimal properties that aid in the design of robust bifunctional catalysts. ORR-active graphene-rich shells are porous and highly graphitic, so they will provide efficient electron transfer pathways and mass transport channel to the OER-active cores. Also, these metal oxide cores can serve as substrates for the synthesis of N-doped graphene, as demonstrated in our previous work using TiO_2 [36]. Following these concepts, the possible synergistic effects occurring between the ORR and OER components may play a promotional role in maximizing activity and durability.

3.2.3 Elucidating the Active-Site Structures

Despite the significant progress in synthesis and performance improvement of NMPCs, the exact structure of active sites in such N-C/M-N-C catalysts remains unknown, limiting the ability to follow a rational design paradigm for catalyst

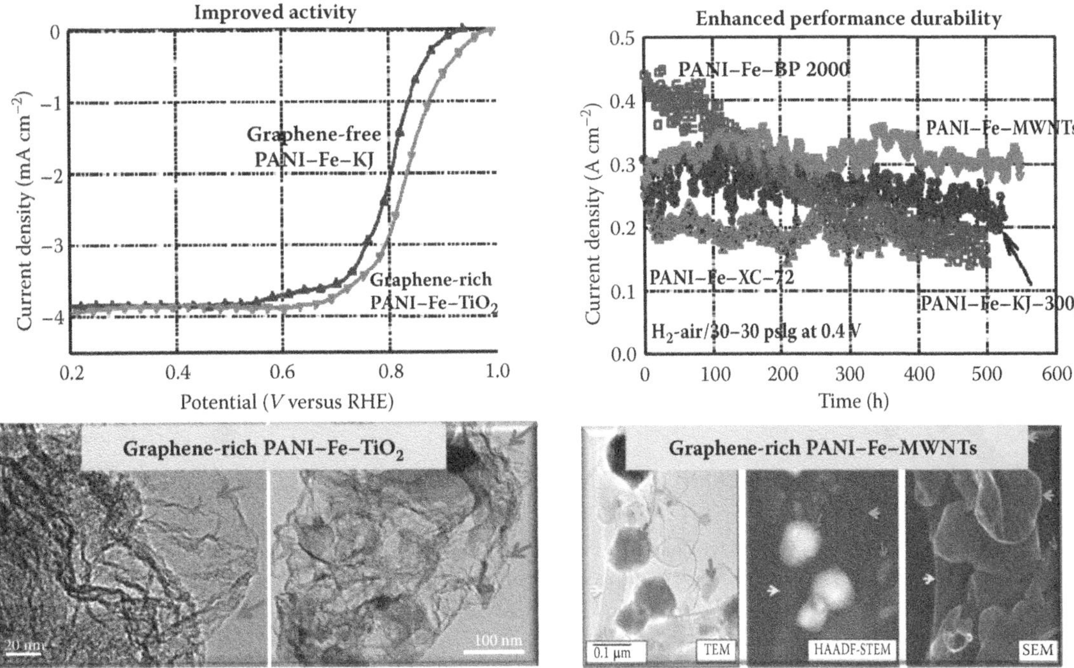

FIGURE 3.6 Promotional role of graphene-like morphology in catalyst, evidenced by enhanced activity (left) and durability (right). (From Wu, G., M.A. Nelson, N.H. Mack, S.G. Ma, P. Sekhar, F.H. Garzon, and P. Zelenay. Titanium dioxide-supported non-precious metal oxygen reduction electrocatalyst. *Chemical Communications* 46, 7489–7491, 2010. Reproduced by permission of The Royal Society of Chemistry; and Wu, G., K.L. More, P. Xu, H.L. Wang, M. Ferrandon, A.J. Kropf, D.J. Myers, S.G. Ma, C.M. Johnston, and P. Zelenay: A carbon-nanotube-supported graphene-rich non-precious metal oxygen reduction catalyst with enhanced performance durability. *Chemical Communications*, 49, 3291–3293, 2013. Reproduced by permission of The Royal Society of Chemistry.)

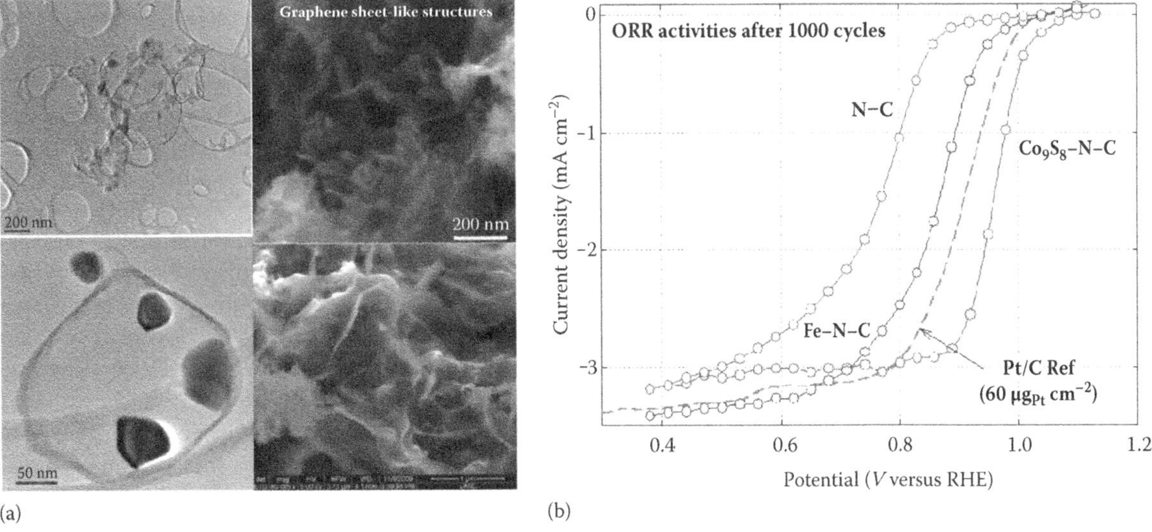

FIGURE 3.7 (a) Graphene-like morphologies observed in Co–N–C catalysts evidenced by transmission electron (TEM) and scanning electron microscopy (SEM) images. (b) RDE testing of the ORR activity in 0.1 M NaOH solution for various catalysts after 1000 potential cycles (−0.6 to 0.2 V versus 3.0 M Ag/AgCl reference; scan rate 50 mV s^{-1}). (Reproduced with permission from Wu, G., H.T. Chung, M. Nelson, K. Artyushkova, K.L. More, C.M. Johnston, and P. Zelenay. Graphene-enriched Co_9S_8–N–C non-precious metal catalyst for oxygen reduction in alkaline media. *ECS Transactions*, 41, 1709–1717. Copyright 2011, The Electrochemical Society.)

improvement. Most notably, it is unknown whether the transition metal participates in the active site or simply catalyzes its formation. In either case, the nitrogen species embedded within the carbon structures are likely critical to the active-site performance, and the optimal nitrogen–carbon structures (e.g., in-plane, edge, subsurface) are yet to be conclusively identified. Further advances in electrocatalysis require a fundamental understanding of the active-site structure, a major challenge given the M–N–C's highly heterogeneous nature. Detailed knowledge of the atomistic configuration of the

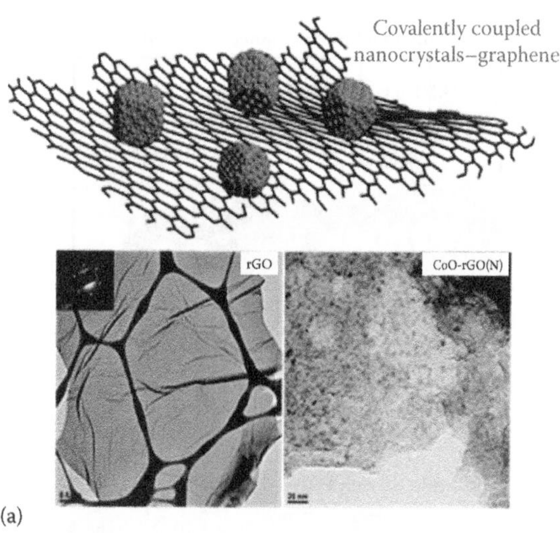

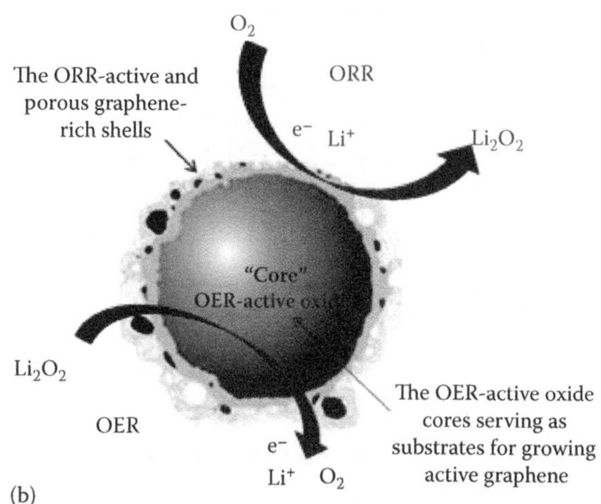

FIGURE 3.8 (a) Proposed scheme and TEM images of nanocrystal–graphene composite materials. (Reproduced from Chen, Z., A.P. Yu, R. Ahmed, H.J. Wang, H. Li, and Z.W. Chen. Manganese dioxide nanotube and nitrogen-doped carbon nanotube based composite bifunctional catalyst for rechargeable zinc–air battery. *Electrochimica Acta* 69, 295–300, 2012. by permission of the Royal Society of Chemistry; Reprinted with permission from He, Q.G., Q. Li, S. Khene, X.M. Ren, F.E. Lopez-Suarez, D. Lozano-Castello, A. Bueno-Lopez, and G. Wu. High-loading cobalt oxide coupled with nitrogen-doped graphene for oxygen reduction in anion-exchange-membrane alkaline fuel cells. *The Journal of Physical Chemistry C* 117, 8697–8707, 2013. Copyright 2014 American Chemical Society.) (b) Scheme of ORR/OER bifunctional catalyst in a "core–shell" configuration.

active sites and their associated reaction pathways for ORR and OER is essential to the rational design of highly active, selective, and stable NPMCs, which is still an elusive target even after several decades of effort. Further complicating the analysis is a lack of experimental probes for identifying active sites on the catalyst surface, defined as the topmost atomic layer. The development of such probes represents an important challenge for the proposed research. The exact nature of the active site notwithstanding, physicochemical characterization data offer strong evidence that the doped nitrogen functionality, metal bonding environment, and graphitized carbon nanostructures (e.g., graphene sheets, carbon nanofibers, and nanoshells) in the catalysts strongly depend on the nitrogen, carbon, and transition metal precursors. *In situ* generated carbon nanostructures serve as a matrix for nitrogen and metal moieties.

First-principles modeling has been applied aggressively to assist in the elucidation of active-site structure and property relationships, and the identification of thermokinetic and electronic-structure-based descriptors for catalyst activity and stability. This is particularly true for the case of Pt-based catalyst systems, where it was shown that ORR catalysts follow a Sabatier relation, in which catalytic activity is linked strongly to the properties of the d-band electronic structure [45]. It is likely that the active sites of NPMCs are covalently embedded within the planar structure of the graphitic matrix. Nitrogen has been shown to coordinate the nonprecious metal atoms directly; however, the nature of the active M_xN_y complexes is actively debated in the literature. In fact, the stability of any particular M_xN_y complex will be dependent on the M- and N-chemical potentials realized during synthesis conditions, a fact that has not been addressed yet. It has been suggested that the active site is composed of multiple metal atoms in close association, and that this facilitates the multielectron reduction steps in the ORR. Recently, at LANL, scientists explored the optimal configuration for FeN_x moieties under operating conditions. As shown in Figure 3.9 [46], using density functional theory (DFT), a range of potential NPMC ORR active-site/graphene nanoribbon edge defect geometries are, for the first time, compared in a self-consistent fashion. On the basis of the formation energies of these Fe_xN_y sites, it is found that they are thermodynamically driven to cluster. Furthermore, depending on synthesis conditions, either the 4N or 3N structures are the most stable Fe-containing defects. It is postulated that both these structures coexist in the catalysts. The 2N interedge structures are so much higher in energy than the other defects studied that their existence is postulated to be either thermodynamically prohibited or metastable due to the kinetics of their decomposition. Clustered FeN_3 structures (Fe_2N_5) seem to excel at cleaving the O_2 bond with zero barrier and thus are likely to follow a dissociative pathway for ORR. This pathway is expected to be more selective, avoiding formation of H_2O_2 at the cost of potentially overbinding ORR intermediates. *Ab initio* molecular dynamics indicates that this spontaneous reaction is likely to be unaffected by solvation. The solvent does not appear to affect the stability of these edge defects. Future work will consider the full ORR pathway.

Overall, novel design and synthesis of nonprecious metal catalysts are still required to further enhance catalytic activity and stability. For example, development and optimization of a multi-nitrogen precursor is likely to further enhance such heat-treated ORR catalysts with high activity and four-electron selectivity. Synthesis of NMPCs supported on highly graphitic carbon(s) may be a way of enhancing active-site density and improving performance durability. Preparation of

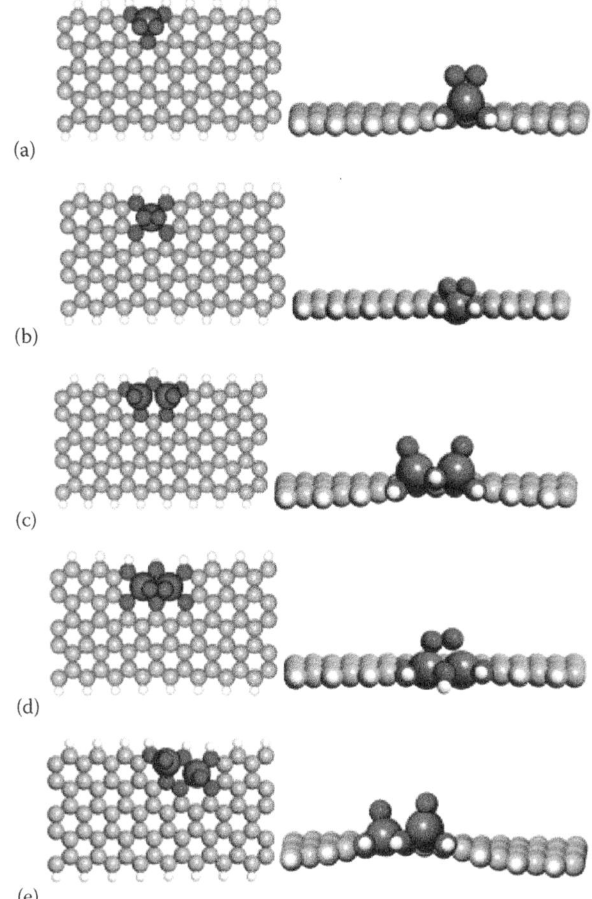

FIGURE 3.9 Optimized geometries for (a) FeN_3 structure with adsorbed O_2; (b) FeN_4 structure with adsorbed O_2; (c) bimetallic FeN_3 cluster structure with adsorbed O_2; (d) bimetallic FeN_4 cluster structure with adsorbed O_2; (e) bimetallic FeN_3/FeN_4 cluster structure with adsorbed O_2. Each structure is shown from above and from the side. (Reprinted with permission from Holby, E.F., G. Wu, P. Zelenay, and C.D. Taylor. Structure of Fe–N_x–C defects in oxygen reduction reaction catalysts from first-principles modeling. *The Journal of Physical Chemistry C*, 118, 14388–14393. Copyright 2014, American Chemical Society.)

nonprecious metal cathode with optimized distribution of ionomer (such as Nafion®) is a key to increasing the utilization of catalyst with minimum water-flooding influence. Meanwhile, the development of durability and stress-test cycling protocols specific to NMPC catalysts (including a realistic potential/voltage window under specific environmental conditions of humidity, reagent stoichiometry, etc.) is still demanded.

3.3 HYBRID NPMC–Pt CATALYSTS

Despite the significant progress in NPMCs, Pt remains the most effective catalyst to facilitate the ORR in both acidic and alkaline solutions for these electrochemical energy technologies [47]. To significantly increase Pt utilization in the catalysts, Pt nanoparticles are supported on high-surface-area carbon blacks such as Vulcan VX-72 (Pt/C); however, a fast and significant loss of electrochemically active surface area, and thus performance degradation, are often observed with traditional Pt/C catalysts over long periods of operation. As shown in Figure 3.10 [48], a variety of studies have indicated that corrosion of Pt/C catalysts occur via different mechanisms. At first, dissolved platinum is redeposited on larger platinum particles, significant particle growth can occur, and the corresponding degradation mechanism is called Ostwald ripening. However, also in the first possible case of agglomeration and coalescence due to migration, carbon corrosion may be involved and lead to a weakening of the interactions between Pt particles and supports. Alternatively, a preferential local corrosion of the support in the surrounding of the Pt particles may facilitate particle movement. A weakening of the interaction between particles and supports due to carbon corrosion is also believed to be the cause for the observed detachment of whole Pt particles from the support. In this context, the ability of Pt to catalyze the oxidation of carbon was suggested to play a decisive role. Finally, severe carbon corrosion can lead to a loss of the structural integrity of the catalyst layer, which reduces the porosity and thus can result in mass transport limitations for the reactants [49,50].

To overcome activity and stability limitations, advanced carbon materials with ordered graphitic structures, including CNTs [51–54], carbon nanofibers [55], onion-like carbon [36,56], and graphene [57,58], have been extensively studied as unique supports for Pt-based electrocatalysts owing to their unique surface structures, high electric conductivity, corrosion resistance, and large surface areas. Noteworthy, there has been increasing evidence showing that the electrochemical and physical properties of carbon materials are extremely sensitive to heteroatom (N, B, S, and P) doping into the carbon structure [56,59]. This doping thereby serves as an effective way to maintain the robust framework and intrinsic structural properties of the nanostructured carbon [59], thereby leading to enhanced electronic properties and catalytic performance. Recently, nitrogen-doped CNTs (N-CNTs) prepared via the floating catalyst CVD method were prepared as a support for Pt nanoparticles. The resulting Pt/N-CNT catalysts exhibited much higher stability than Pt/CNTs during accelerated durability tests [52]. In our recent efforts in developing carbon-based ORR transition M–N–C catalysts (M: Fe or Co) [15,17,23,25,28,30,39,60–62], *in situ* formation of various carbon nanostructures with nitrogen doping can be realized by catalyzing the decomposition of the nitrogen/carbon precursor at high temperatures (800–1000°C) [13], which was found to result in catalysts with superior activity for the ORR is comparison with other types of NPMCs. Importantly, the formation of different nanostructures (e.g., CNT, onion-like carbon, or graphene) during the catalyst synthesis is controllable by optimizing the transition metals, nitrogen/carbon precursors, and templates [13]. Apart from the obvious advantage of high electronic conductivity and enhanced corrosion resistance, the highly graphitized carbon nanostructures present in the Fe–N–C materials may serve as a matrix for supporting Pt nanoparticles.

Herein, as shown in Figure 3.11 [63], we recently developed a method to prepare highly active and stable ORR

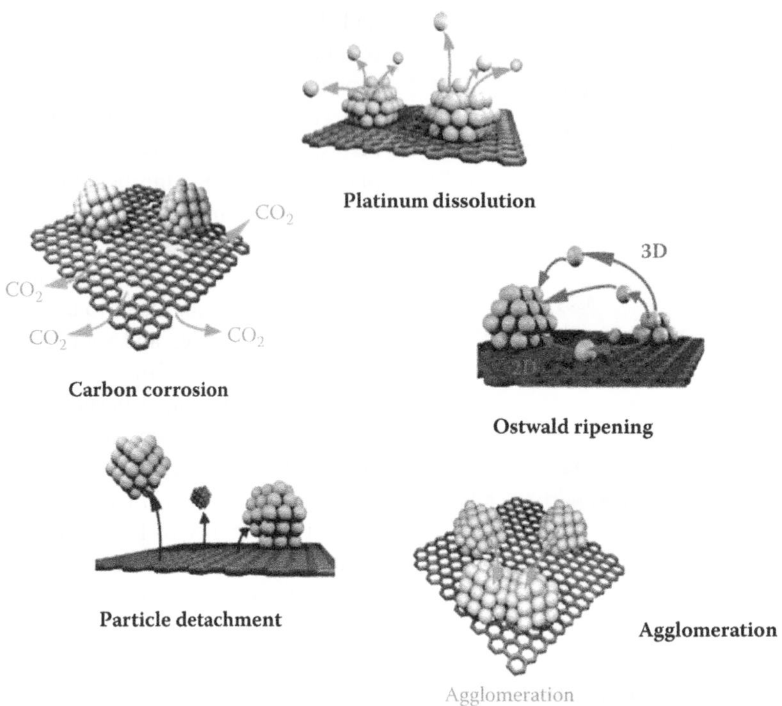

FIGURE 3.10 Simplified representation of suggested degradation mechanisms for platinum particles on a carbon support in fuel cells. (Reprinted from Meier, J.C., C. Galeano, I. Katsounaros, J. Witte, H.J. Bongard, A.A. Topalov, C. Baldizzone, S. Mezzavilla, F. Schüth, and K.J.J. Mayrhofer. Design criteria for stable Pt/C fuel cell catalysts. *Beilstein Journal of Nanotechnology* 5, 44–67, 2014, under the terms of the Creative Commons Attribution License, which permits unrestricted use, distribution, and reproduction in any medium, provided the original work is properly cited.)

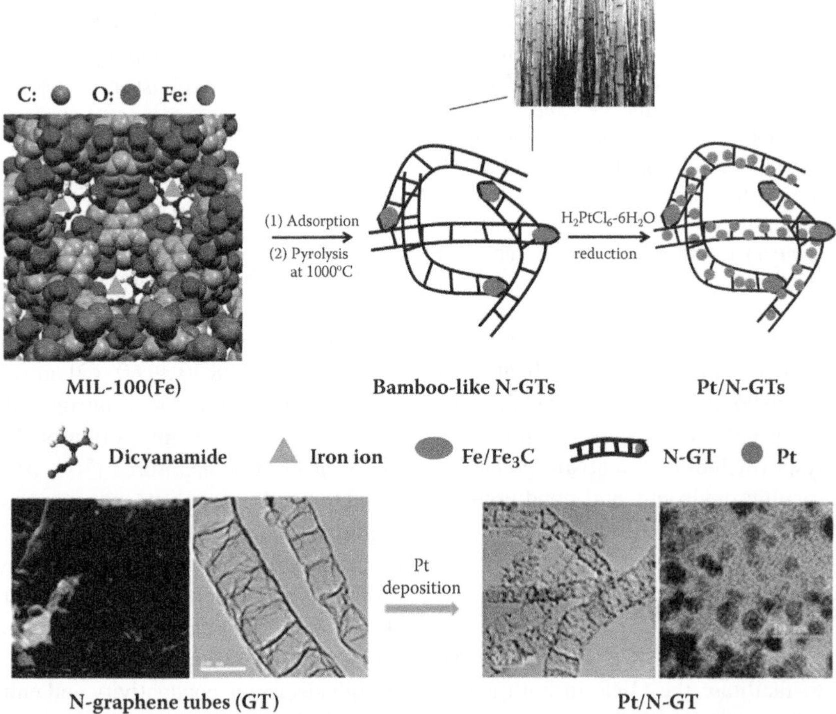

FIGURE 3.11 Synthetic scheme and microscopic images for hybrid Pt catalysts supported on nitrogen-doped graphene tubes. (Li, Q., H. Pan, D. Higgins, R. Cao, G. Zhang, H. Lv, K. Wu, J. Cho, and G. Wu: Metal–organic framework-derived bamboo-like nitrogen-doped graphene tubes as an active matrix for hybrid oxygen-reduction electrocatalysts. *Small*. 2014. doi: 10.1002/smll.201402069. Copyright Wiley-VCH Verlag GmbH & Co. KGaA. Reprinted with permission.)

catalysts by innovatively coupling Pt nanoparticles and highly active Fe–N–C materials in a unique hybrid configuration. In doing so, the ORR active carbon support materials with nitrogen doping not only offers remarkable support effects by geometrically and electronically modifying the deposited Pt particles, but also provides a large amount of complementary nonprecious ORR active sites, thereby significantly improving the ORR activity and durability of the Pt catalysts. Importantly, large-diameter nitrogen-doped graphene tubes (N-GTs) derived from dicyandiamide, iron acetate, and MOFs, MIL-100(Fe), via a high temperature method, were prepared as novel supports for the development of Pt catalysts. The large size of N-GTs provides a better platform than common CNTs (with diameters usually less than 30 nm), thus favorably anchoring the deposited metal nanoparticles. While the as-prepared N-GTs were found to show good intrinsic ORR activity and stability in acidic electrolytes, by modification with well-dispersed Pt nanoparticles, the unique Pt/N-GT hybrid materials were found to provide excellent performance that is superior to commercial Pt/C catalysts (Figure 3.12) [63]. The observed enhancements are likely due to the complementary ORR active sites on N-GTs, their highly graphitized structure formed during the high-temperature heat-treatment process, and the favorable interactions between the nitrogen dopant species and Pt nanoparticles.

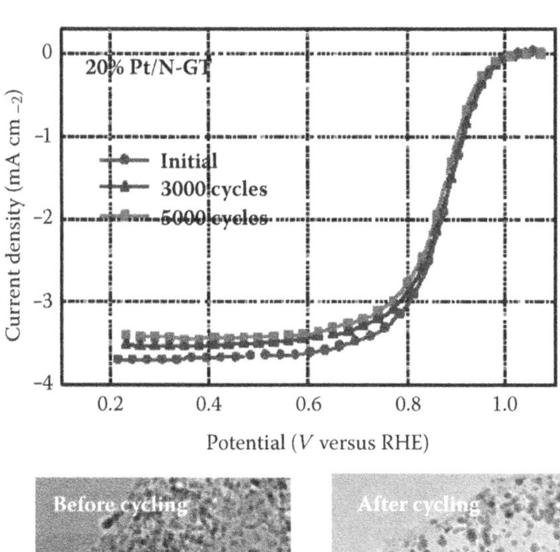

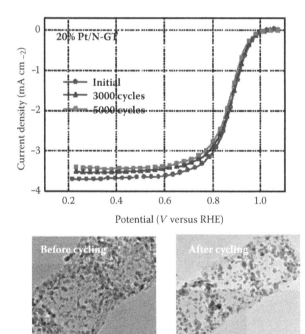

FIGURE 3.12 ORR steady-state RDE polarization curves recorded with a rotating speed of 900 rpm for Pt/N-GT (top). (Bottom) TEM images of Pt/N-GT before (left) and after (right) 5000 potential cycles. (Li, Q., H. Pan, D. Higgins, R. Cao, G. Zhang, H. Lv, K. Wu, J. Cho, and G. Wu: Metal–organic framework-derived bamboo-like nitrogen-doped graphene tubes as an active matrix for hybrid oxygen-reduction electrocatalysts. *Small*. 2014. doi: 10.1002/smll.201402069. Copyright Wiley-VCH Verlag GmbH & Co. KGaA. Reprinted with permission.)

This presents a novel strategy to develop new nanomaterials with improved ORR activity and reduced Pt contents for PEFC applications. Further development of such type of hybrid oxygen-reduction Pt catalysts requires continuously optimizing the integration of highly active NPMCs with well-dispersed Pt alloy nanoparticles.

3.4 ALKALINE-BASED ENERGY STORAGE AND CONVERSION TECHNOLOGIES

3.4.1 Anion-Exchange Membrane Fuel Cells

Although great effort has been devoted to exploring low-Pt and even nonprecious metal ORR catalysts for the PEFCs [15,18,19], the practical resolution regarding long-term durability and high activity in the strongly acidic nature of Nafion is still greatly limited [64]. Alternatively, the use of alkaline media in fuel cell applications presents many advantages, both in electrocatalytic activity and in materials stability, over acidic media [65–67].

The decreased extent of specific anion adsorption means that, in general, most electrocatalytic processes should become more favorable in alkaline solutions relative to those in acidic media [68]. While the electrocatalytic advantages of using alkaline media are significant, the improved materials stability afforded by the use of alkaline electrolytes may be even more important. A wide variety of materials have exhibited comparable corrosion resistance in alkaline media to that of precious metals, making great promise to utilize precious-metal-free cathode catalysts in fuel cells and metal–air batteries [69]. Additionally, water and ionic transport within the OH^--conducting electrolytes also benefit from a more favorable direction of the electro-osmotic drag, away from the cathode, thereby mitigating cathode flooding, a major issue for PEFCs and direct methanol fuel cells (DMFCs) [70]. This opposite direction of the electro-osmotic drag could also reduce the methanol crossover in alkaline-based DMFCs [71].

It is well known that alkaline fuel cells (AFCs) that use liquid electrolytes have been developed from 1960s to 1980s and were successfully applied in space programs [72]. Recently, AFCs, equipped with anion-exchange membranes that are able to overcome the deleterious problems associated with carbonate precipitates, have resulted in significant improvements in performance [12,73,74]. Although improving the conductivity and stability of anion-exchange polymer electrolytes is a top priority in AFC development, developing cost-effective cathode catalysts for ORR with sufficient activity and durability in alkaline media remains the major challenge. In turn, development of alkaline-based anion-exchange membrane fuel cells (AEMFCs) will provide a tremendous opportunity for use of NPMC as a replacement for expensive precious metals. To date, a wide variety of nonprecious catalytic materials have been explored as ORR cathode catalysts in alkaline media. They include nitrogen-doped carbon materials [75], macrocycles (e.g., porphyrins [24] and phthalocyanines [76]), and metal oxides [77] (e.g., manganese dioxide [78], perovskites [79], and spinels [80]). Among them, nitrogen-doped carbon materials with or without transition metals have been known as a promising class of catalyst for the ORR electrocatalysts in alkaline media. They have exhibited facile ORR kinetics with minimal peroxide generation, as well as their depolarized character such as insensitivity to methanol and halide ions [38,81,82].

As evidenced by the rotating disk electrode (RDE) testing in alkaline aqueous electrolyte, several M–N–C catalysts recently developed using the high-temperature approach show significantly higher electrocatalytic activity for the ORR than commercial Pt-based catalysts [28,38,83]. For example, a fully broken-in Fe–N–CNT/CNP composite catalyst demonstrates a record-breaking ORR onset potential of 1.16 V versus reversible hydrogen electrode (RHE) and half-wave RDE potential ($E_{½}$) of 0.95 V versus RHE, which are both at least 50 mV higher than those recorded with a reference Pt/C catalyst (Figure 3.13) [83]. As important, the same catalyst maintains excellent activity even after 10,000 potential cycles, indicating excellent electrochemical stability in aqueous electrolytes. AEMFC performance using this catalyst technology, however, has not been demonstrated.

Although these carbon-based catalysts have exhibited superior ORR activity and stability in aqueous alkaline media, few of them have been studied in the real electrode at the interface with an anion ionomer. It was found that the difference of ORR activity observed in the RDE and in fuel cell testing is very likely. Recently, a reverse discrepancy of activity was observed with a FePc/C catalyst in alkaline anion-exchange membrane fuel cells and in RDE, with fuel cell performance significantly inferior to that in RDE [84]. On the basis of the DFT calculation, it is deduced that OH adsorption on the catalyst is an important factor that mitigates the ORR activity and stability in alkaline media. Generally, OH adsorption has two negative effects on the kinetics of oxygen reduction: (i) these species block a part of the real surface area and (ii) change the Gibbs energy of adsorption of oxygen reduction intermediates [85]. It was pointed out that the electric field in fuel cell cathodes leads to more severe impact of the OH-binding on FePc/C catalysts, relative to the conditions in RDE tests [86]. Thus, studying the electrochemical properties, including anion adsorption/desorption behavior at the carbon catalysts, hydrophobicity/hydrophilicity at the interfaces, and charge-transfer rates at the catalyst/ionomer interface, is valuable to the development of efficient cathodes for AFCs.

Despite the challenge of studying the ORR catalysts in real AEMFCs, some research groups have studied such nano-carbon composite catalysts by using robust anion-exchange membranes, showing promising fuel cell performance [39,87]. For example, in Popov's group at the University of South Carolina, as shown in Figure 3.14 [87], an EDA-derived binary CoFe catalyst (EDA–CoFe–C) was prepared via a heat treatment at 900°C with dominant nitrogen-doped onion-like carbon structures. The EDA–CoFe–C cathode was successfully demonstrated in an AEMFC using an A201 membrane (Tokuyama Corporation, Japan), composed of a hydrocarbon main chain and quaternary ammonium groups as ion-exchange sites [87]. In a cell test, the OCVs were found to be 0.97 and 1.04 V for the EDA–CoFe–C and Pt/C catalysts, respectively. The corresponding maximum power densities were measured at 177 and 196 mW cm^{-2}. At high potentials, the performance of EDA–CoFe–C was only slightly lower than that of Pt/C. At intermediate potentials, both catalysts showed very similar activity. However, the lower performance of EDA–CoFe–C observed at low potentials may be attributed to high mass transfer and/or ionic resistance in the cathode, caused by a high catalyst loading and considerable thickness of the electrode. Adopting NPMC ORR catalysts to AEMFCs is challenging yet extremely valuable because of their great potential to reduce fuel cell costs

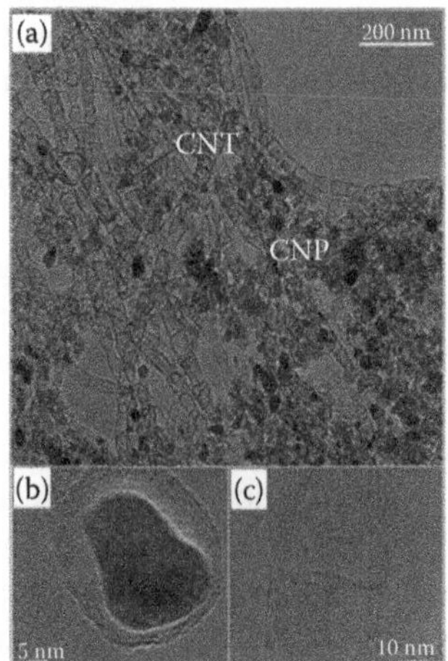

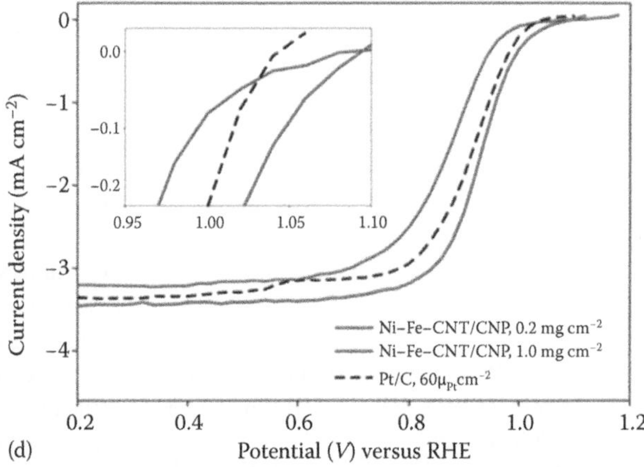

FIGURE 3.13 N-Fe–CNT/CNP catalyst: (a–c) morphology and (d) ORR polarization plots before a cycling durability test. (Reprinted with permission from Macmillan Publishers Ltd. *Nature Communications*, Chung, H.T., J.H. Won, and P. Zelenay. Active and stable carbon nanotube/nanoparticle composite electrocatalyst for oxygen reduction, 4, 1922, Copyright 2013.)

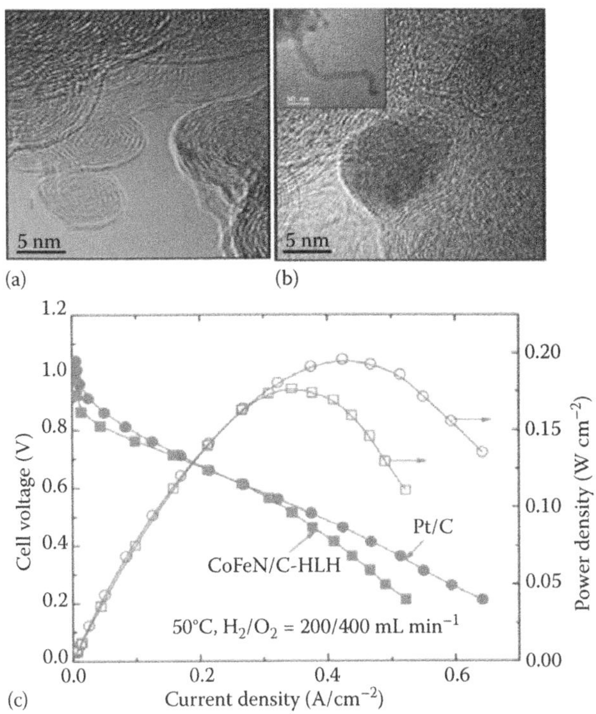

FIGURE 3.14 Morphologies of the USC-developed EDA–FeCo–C catalyst (a, b) and corresponding cathode performance in an AEMFC (c). (Reprinted from *Journal of Power Sources*, 196, Li, X.G., B.N. Popov, T. Kawahara, and H. Yanagi, Non-precious metal catalysts synthesized from precursors of carbon, nitrogen, and transition metal for oxygen reduction in alkaline fuel cells, 1717–1722, Copyright 2011, with permission from Elsevier.)

by replacing Pt catalysts without sacrificing performance. Therefore, we believe that investigating the ORR activities of NPMCs with qualified hydroxide conducting ionomvers and integrating them into high-performance cathodes are vitally important for AEMFC technologies. Further research goals are to fully optimize the compatibility of NPMCs with ionomer and membrane by systematically investigating the ORR kinetic activity, mass transfer phenomena, and structural stability of NPMCs.

3.4.2 Rechargeable Zinc–Air Batteries

Owing to their high energy and power density, high safety, and economic viability, zinc–air batteries are important energy-storage technology for stationary and transportation in the future. Usually, highly concentrated potassium hydroxide solution (e.g., 6.0 M) is used as the electrolyte in Zn–air batteries. However, the low efficiency of currently used ORR and OER catalysts still limits the development of Zn–air batteries. Robust bifunctional catalysts are needed for the oxygen cathode, aiming to minimize the high ORR and OER overpotentials, corresponding to battery discharge and charge processes.

Metal oxides, especially manganese oxides, are important ORR catalysts dominantly used in current Zn–air batteries owing to their exceptional stability [43,88–99]. However, their electrical conductivity during the electrocatalysis is not sufficient yet. Thus, an effective strategy is for combining manganese oxides with conductive nanocarbon materials, such as CNTs and graphene, which is proved to be an effective way to improve electrical conductivity and chemical stability. Recently, one example has been demonstrated via a facile solution-based route to anchor Mn_3O_4 on the surface of ionic liquid–modified reduced graphene oxide (rGO) [91]. The large surface area of the rGO couples with active oxides, potentially providing increasing stability of oxide particles along with the accessibility of active sites by reactants. Importantly, the catalytic activity of this type of hybrid catalyst was greatly dependent on the loading of oxides on graphene nanosheets. For example, compared with a higher Mn_3O_4 (52.5%) content, the hybrid with a lower Mn_3O_4 (19.2%) content exhibited better activity and higher four-electron selectivity during the ORR. The hybrid catalyst generated a maximum peak power density of 120 mW cm^{-2} in a primary Zn–air cell using 6.0 M KOH as an electrolyte.

Compared with primary Zn–air batteries only requiring ORR catalysts, significant effort has been focused on developing ORR/OER bifunctional catalysts for electrically rechargeable Zn–air batteries [77,100–104]. Recently, significant progress in the development of ORR/OER bifunctional catalyst has been made by Chen's group at University of Waterloo, as shown in Figure 3.15 [105]. A new class of core–corona structured catalyst (CCBC) with high activity and good durability was successfully demonstrated for rechargeable Zn–air batteries. The CCBC is composed of N-CNTs as a high ORR active component and lanthanum nickelate ($LaNiO_3$) as a high OER active component. The $LaNiO_3$ also acts as a support material for the synthesis of N-CNTs via a CVD process. A synergistic effect is very likely between the core material of $LaNO_3$ and the corona material of N-CNTs, which results in the excellent performance of rechargeable Zn–air batteries.

Very recently, a novel composite bifunctional catalyst consisting of CoO/N-CNT as an ORR catalyst and Ni–Fe-layered double hydroxide as an OER catalyst (NiFe-layered double hydroxide–LDH/CNT) [106] was used in a Zn–air battery. In a 6.0 M KOH electrolyte, the CoO/N-CNT catalyst showed higher ORR activity relative to state-of-the-art commercial Pt/C catalyst. Meanwhile, the NiFe LDH/CNT outperformed the well-known best benchmark catalyst of Ir/C. In a rechargeable Zn–air battery using a tri-electrode configuration, the bifunctional catalysts exhibited an unprecedented small charge–discharge voltage polarization of ~0.70 V at 20 mA cm^{-2}, demonstrating high reversibility and stability during deeply charge–discharge cycling tests up to 200 h (Figure 3.16) [106]. Thus, an optimal interaction between ORR and OER component in bifunctional cathodes can yield high energy efficiency for rechargeable Zn–air batteries. Further improvement of the performance of rechargeable Zn–air batteries will still rely on novel concepts and nanocomposite materials in catalyst design and synthesis.

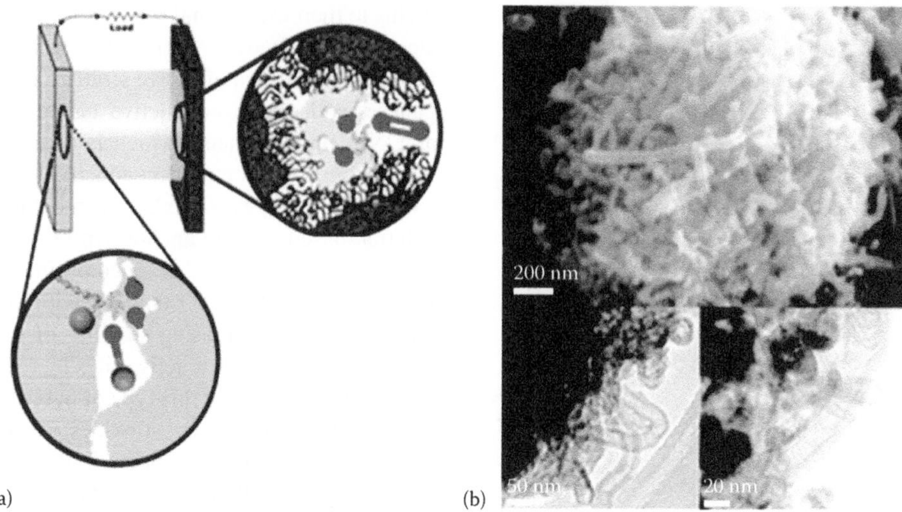

FIGURE 3.15 (a) Scheme of a Zn–air battery and the reactions taking places on the electrodes. The CCBC catalyst is applied onto the positive electrode, which catalyzes the ORR and OER reactions. (b) SEM and TEM images of the CCBC illustrating the N-CNT on the surface of the core particle. (Reprinted with permission from Chen, Z., A. Yu, D. Higgins, H. Li, H. Wang, and Z. Chen. Highly active and durable core–corona structured bifunctional catalyst for rechargeable metal–air battery application. *Nano Letters*, 12, 1946–1952. Copyright 2012 American Chemical Society.)

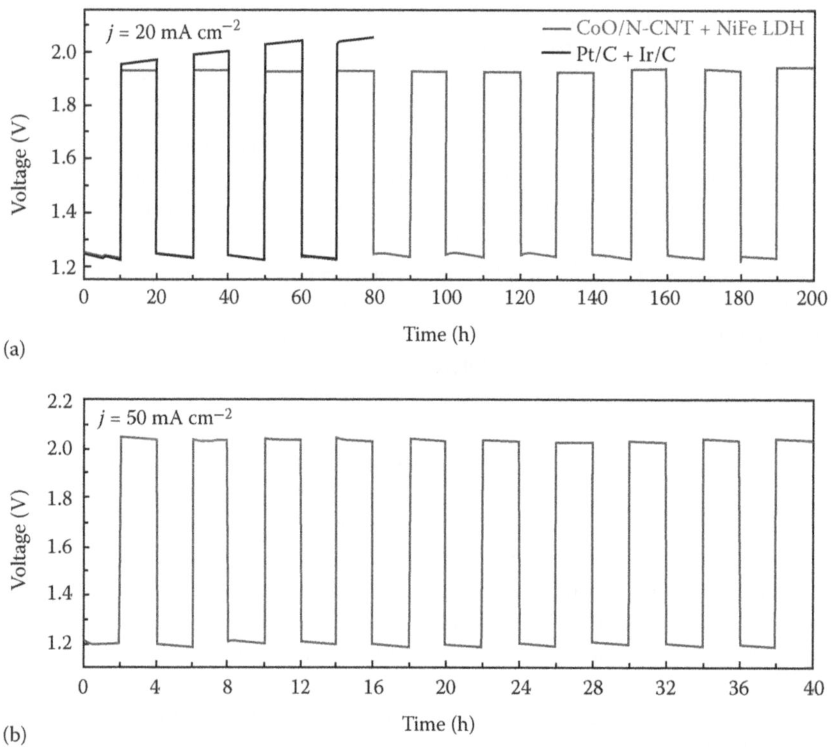

FIGURE 3.16 (a) Cycling performance of the tri-electrode Zn–air battery using CoO/N-CNT and NiFe LDH at 20 mAcm^{-2} and a 20-h cycle period compared with the tri-electrode battery using Pt/C and Ir/C. (b) The same CoO/N-CNT and NiFe LDH electrodes from (a) were used for a subsequent cycling experiment with a fresh Zn anode at 50 mA cm^{-2} and a 4-h cycle period. (From Li, Y.G., M. Gong, Y.Y. Liang, J. Feng, J.E. Kim, H.L. Wang, G.S. Hong, B. Zhang, and H.J. Dai, *Nature Communications*, 4, 2013.)

3.5 NEW OPPORTUNITIES FROM Li-O_2 ELECTROCHEMISTRY

Compared with other currently studied electrochemical energy storage devices, lithium–oxygen (Li–O_2) batteries have been viewed as the utmost energy storage technology to meet the transportation application in the future, owing to their highest theoretical energy density (5200 Wh kg^{-1}) [107–109]. The exceptionally improved energy density is due to an open architecture in porous air cathode, capable of continuously taking reactant O_2 directly from ambient atmosphere. As a result, the unique battery chemistry and electrode architecture provide a greatly increased specific energy density (theoretical value of 5200 Wh kg^{-1}) [108]. Generally, Li–O_2 batteries can be divided into four different categories based on the electrolyte used in batteries: nonaqueous, aqueous, hybrid, and all-solid-state batteries. Currently, owing to the insufficient ionic conductivity of solid-state electrolyte [110], liquid electrolytes including nonaqueous and aqueous systems are extensively investigated. Furthermore, the use of a nonaqueous media in a Li–O_2 battery proved to be more feasible than aqueous electrolyte to alleviate parasitic corrosion on Li metal at the anode along with providing higher energy density [5,7]. Thus, to date, the nonaqueous configuration has attracted the most attention worldwide compared with other electrolyte-based Li–O_2 batteries.

Differing from aqueous systems where H_2O or H_2O_2 is produced during the ORR associated with the discharging process, however, insoluble lithium–oxygen species (e.g., Li_2O_2) are formed in nonaqueous media and passivate the catalysts, eventually leading to the termination of the discharge process. In addition, the intermediates in Li–O_2 electrochemical reactions, for example, O^{2-}, O_2^{2-}, and LiO_2/LiO^{2-}, are very reactive to easily decompose most organic solvents [111–113]. Thus, this leads to the formation of discharging products of Li_2CO_3, LiOH, and lithium alkyl carbonates [114,115], instead of the desired product Li_2O_2. During the charge process, the catalysts can be exposed again, corresponding to OER, when an external potential is applied, $E > E_0$, accompanied by the decomposition of these solid-state lithium–oxygen compounds. Thus, an efficient OER catalyst to catalyze a fast decomposition is needed to recover active sites for the ORR in next discharge–charge cycle. However, obtained ORR and OER mechanisms in aqueous electrolyte are not effective in nonaqueous Li^+ electrolytes yet, wherein various reactants/products and reaction intermediates are often insoluble. The optimal catalyst properties required for efficiently catalyzing oxygen reactions in nonaqueous electrolytes are therefore different from those in aqueous media.

Like aqueous media, in nonaqueous Li^+–O_2 systems, large ORR and OER overpotentials are also observed on currently used carbon cathode catalysts, causing significant losses in the battery discharge voltage (2.5 V) and low overall charge–discharge efficiency [116]. Specifically, the voltage gap between the charge and discharge of an Li–O_2 battery is larger than 1.0 V [9], which results in a much lower voltage efficiency of <60%, relative to an Li-ion battery (>90%). An additional challenge for the cathode catalysts for nonaqueous Li–O_2 batteries is that the catalysts should be designed to generate ORR products that can be efficiently decomposed during the subsequent OER for the rechargeable Li^+–O_2 system. Thus, the development of bifunctional ORR/OER catalysts with high activity and durability is desperately needed. It was found that, from the electrocatalysis point of view, catalysts play a more significant role in the ORR than in the OER [117]. The discharge potentials were found to be greatly dependent on the cathode catalysts [28,35]. Compared with traditional carbon, no difference in catalytic activity for the OER is observed using the Au, Pt, or MnO_2 catalysts [117]. The OER during the charge process is greatly dependent on many noncatalysis-related factors, such as the diffusion rate of the insoluble species (Li_2O_2) and electron transport rate on Li_2O_2 [118].

Hence, it is generally believed that electrocatalysts and the resulting electrode structures are critical to improving the power density, cycling capability, and round-trip energy efficiency of Li–O_2 batteries [88,119]. Especially, lowering the overpotential during discharge and charge is of prime importance in order to avoid carbon corrosion and to diminish electrolyte oxidation. The fundamental principles of electrocatalyst design for Li–O_2 nonaqueous systems could be learned from aqueous oxygen electrochemistry, especially fuel cells and alkaline metal–air batteries. As shown in Figure 3.17 [120], recent efforts in the development of nanostructured carbon-based electrocatalysts for nonaqueous Li–O_2 batteries are highlighted, with special focus on ORR catalysts, in a recent perspective article [120]. Catalyst materials discussed include metal-free carbon catalysts, nanostructured transition metal–nitrogen–carbon composite catalysts, and transition metal/nanocarbon composite catalysts. Among various carbon-based catalyst formulations, nitrogen-doped nanocarbon with or without transition metals (e.g., Co, Fe, Mn, or Cu) have been recognized as the most promising ones for Li–O_2 batteries because of its reasonable balance among catalytic activity, durability, and cost [109,121].

Besides the activity, the stability of the cathode material is also a serious issue in the application of Li–O_2 batteries. Thus far, most of the published results are obtained with carbon or carbon-based cathodes. However, similar to carbon corrosion in fuel cells, carbon materials also undergo corrosion challenges at high operation potentials, especially for the charging process and high oxygen circumstances in Li–O_2 batteries. Recently, it has been demonstrated that carbon is unstable on charging above 3.5 V in the presence of Li_2O_2, undergoing oxidative decomposition to form Li_2CO_3 (Figure 3.18) [122]. Superoxide radicals generated at the cathode during discharge have also been proven to react with carbon that contains activated double bonds or aromatics to form epoxy groups and carbonates, which limits the rechargeable capability of Li–O_2 cells [123]. This stability issue still represents the largest technical barrier in developing practical Li–O_2 batteries.

Exploring stable electrolytes also remains a grand challenge for rechargeable Li–O_2 batteries. A large variety of electrolytes, including propylene carbonate [43], dimethoxyethane

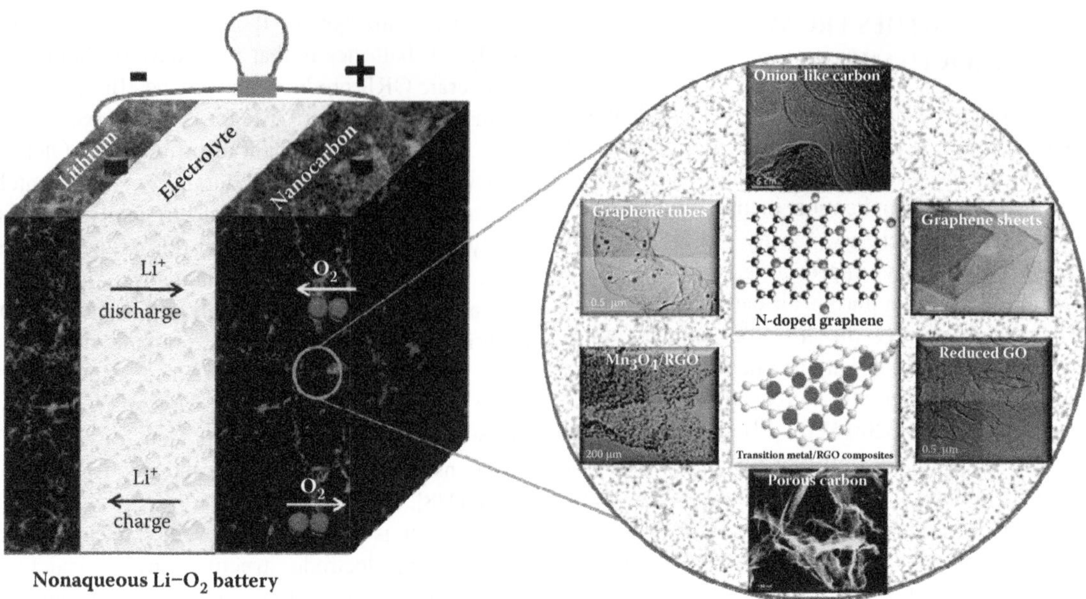

FIGURE 3.17 Nanocarbon composite catalysts for rechargeable Li–air battery cathodes. (From Li, Q., R. Cao, J. Cho, and G. Wu. Nanostructured carbon-based cathode catalysts for nonaqueous lithium–oxygen batteries. *Physical Chemistry Chemical Physics*, 16, 13568–13582, 2014. Reproduced by permission of The Royal Society of Chemistry.)

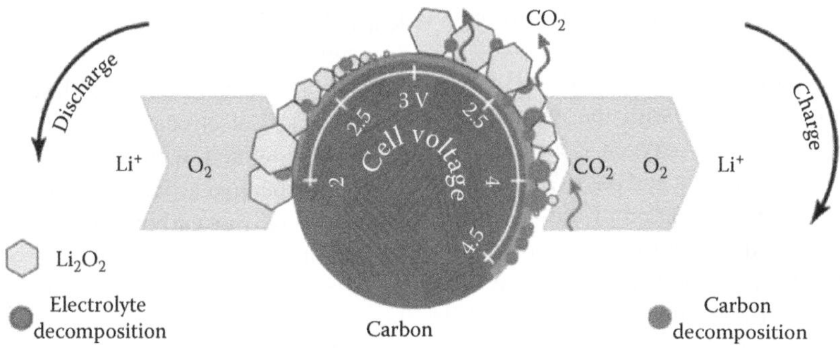

FIGURE 3.18 Illustration of the decomposition process of carbon electrodes during discharge and charge in nonaqueous $Li–O_2$ batteries. (Reprinted with permission from Thotiyl, M.M.O., S.A. Freunberger, Z.Q. Peng, and P.G. Bruce, The carbon electrode in nonaqueous $Li–O_2$ cells. *Journal of American Chemical Society*, 135, 494–500. Copyright 2013 American Chemical Society.)

[115], tetraethylene glycol dimethyl ether [35,124], dimethylformamide [125], dimethylsulfoxide [107,126], tetramethylene sulfone [127], etc., have been extensively studied. However, most of these electrolytes still suffer from rapid degradation with cycling. Recently, liquid electrolyte was replaced with a solid Li-ion conductor to circumvent the decomposition problem of liquid electrolytes, which may provide an alternative approach [128].

Overall, on the basis of our current understanding of the existing catalysts, we believe that the following research directions are important to the development of highly efficient cathode catalysts for $Li–O_2$ batteries: (i) design the catalysts with optimal distribution of macropores, mesopores, and micropores enabling efficient decomposition of discharging products; (ii) improve the catalyst stability by exploring new support materials and synthesis strategies; (iii) fundamentally understand the ORR/OER mechanisms in $Li–O_2$ batteries and their relationship with various catalyst active sites using both theoretical calculations (molecular/electronic level) and experimental methods.

3.6 SUMMARY AND PERSPECTIVE

Energy conversion and storage through electrochemical reactions are the most important energy technologies available today, including low-temperature fuel cells and metal–air batteries, and they greatly rely on electrocatalysis of oxygen reactions. The high reaction overpotentials require catalysts containing large amount of precious metals such as Pt and Ir to enhance the reaction activity and durability, leading to a grand challenge for the widespread application of these clean energy technologies. The development of cost-effective and robust electrocatalysts for these oxygen reactions has become an important research topic for electrochemical energy technologies. On the basis of our research effort in the last 8 years, we reviewed the current challenge and progress in exploring oxygen electrocatalysis for energy applications, with an emphasis on the development of nonprecious metal catalysts. Compared with other studied catalyst formulations, transition metal–involved nitrogen-doped carbon composite catalysts (M–N–C) have become the most promising alternatives to Pt for oxygen reduction in both acidic and alkaline electrolytes. Further performance improvement requires novel concepts to design catalysts with strong interaction among transition metal, nitrogen, and carbon during the processing of precursors (e.g., metal, nitrogen, carbon, and template), heat treating, and posttreatments. High surface area with mesoscale pore structures in catalysts would increase the density of active sites, generating high current density and facilitating mass transfer. Selection of advanced supporting materials or templates such as graphene and MOFs may provide new opportunities on boosting the catalyst activity. Thus far, the nature of the nonprecious active sites in such M–N–C catalysts still remains unclear. Whether the transition metals directly participate in active sites and how the metals coordinate with nitrogen and carbon are still a conundrum. As direct probing of the active sites on the top layers of catalyst surface is very challenging, theoretical calculation and simulation are powerful tools to provide insights into the origin of active sites. During the exploration of nonprecious metal catalysts, we proposed a new concept by properly integrating M–N–C catalysts with Pt nanoparticles, aiming to enhance catalytic activity, reduce Pt content, and improve the stability of Pt nanoparticles.

Compared with conventional acidic electrolyte for fuel cells, alkaline media provide a new opportunity for highly efficient conversion and storage of energy. Although the largest challenge is related to anion-exchange polymer electrolytes, cost-effective and active oxygen cathode catalysts are still required, with good compatibility with ionomer in electrodes, for maximum charge transfers. Unique Li–O_2 electrochemistry is able to yield the highest energy density, however, current batteries exhibit far more sufficient performance owing to several technical barriers, including catalyst development. Uses of liquid organic electrolyte greatly limit the battery performance, leading to poor cycling stability and serious safety issues. Lack of fundamental understanding of the ORR and OER in nonaqueous electrolytes cannot provide rational guidance for catalyst design and synthesis. Thus, mechanistic studies are very necessary for eventually realizing the high energy density offered by the exceptional Li–O_2 electrochemistry.

REFERENCES

1. Alamri, B.R. and A.R. Alamri. Technical review of energy storage technologies when integrated with intermittent renewable energy, in *Sustainable Power Generation and Supply. SUPERGEN '09*, 2009.
2. Rodriguez, G.D. A utility perspective of the role of energy storage in the smart grid, in *Power and Energy Society General Meeting*. IEEE, 2010.
3. Abraham, K.M. and Z. Jiang. A polymer electrolyte-based rechargeable lithium/oxygen battery. *Journal of the Electrochemical Society* 1996; 143: 1–5.
4. Van Mierlo, J. and G. Maggetto. Fuel cell or battery: Electric cars are the future. *Fuel Cells* 2007; 7: 165–173.
5. Girishkumar, G., B. McCloskey, A.C. Luntz, S. Swanson, and W. Wilcke. Lithium–air battery: Promise and challenges. *The Journal of Physical Chemistry Letters* 2010; 1: 2193–2203.
6. Katsounaros, I., S. Cherevko, A.R. Zeradjanin, and K.J. Mayrhofer. Oxygen electrochemistry as a cornerstone for sustainable energy conversion. *Angewandte Chemie International Edition* 2014; 53: 102–121.
7. Bruce, P.G., L.J. Hardwick, and K.M. Abraham. Lithium–air and lithium–sulfur batteries. *MRS Bulletin* 2011; 36: 506–512.
8. Lu, Y.C., H.A. Gasteiger, M.C. Parent, V. Chiloyan, and Y. Shao-Horn. The influence of catalysts on discharge and charge voltages of rechargeable Li–O_2 batteries. *Electrochemical and Solid State Letters* 2010; 13: A69–A72.
9. Shao, Y., S. Park, J. Xiao, J.-G. Zhang, Y. Wang, and J. Liu. Electrocatalysts for nonaqueous lithium–air batteries: Status, challenges, and perspective. *ACS Catalysis* 2012; 2: 844–857.
10. Black, R., J.-H. Lee, B. Adams, C.A. Mims, and L.F. Nazar. The role of catalysts and peroxide oxidation in lithium–oxygen batteries. *Angewandte Chemie International Edition* 2012; 52: 392–396.
11. Shively, D., J. Gardner, T. Haynes, and J. Ferguson. Energy storage methods for renewable energy integration and grid support, in *Energy 2030 Conference, 2008. ENERGY 2008*. IEEE, 2008.
12. Pan, J., C. Chen, L. Zhuang, and J.T. Lu. Designing advanced alkaline polymer electrolytes for fuel cell applications. *Accounts of Chemical Research* 2012; 45: 473–481.
13. Wu, G. and P. Zelenay. Nanostructured nonprecious metal catalysts for oxygen reduction reaction. *Accounts of Chemical Research* 2013; 46: 1878–1889.
14. Lai, L., J.R. Potts, D. Zhan, L. Wang, C.K. Poh, C. Tang, H. Gong, Z. Shen, J. Lin, and R.S. Ruoff. Exploration of the active center structure of nitrogen-doped graphene-based catalysts for oxygen reduction reaction. *Energy Environmental Science* 2012; 5: 7936–7942.
15. Jaouen, F., E. Proietti, M. Lefevre, R. Chenitz, J.P. Dodelet, G. Wu, H.T. Chung, C.M. Johnston, and P. Zelenay. Recent advances in non-precious metal catalysis for oxygen-reduction reaction in polymer electrolyte fuel cells. *Energy Environmental Science* 2011; 4: 114–130.
16. Chen, Z., D. Higgins, A. Yu, L. Zhang, and J. Zhang. A review on non-precious metal electrocatalysts for PEM fuel cells. *Energy Environmental Science* 2011; 4: 3167–3192.
17. Li, Q., R. Cao, J. Cho, and G. Wu. Nanocarbon electrocatalysts for oxygen reduction in alkaline media for advanced energy conversion and storage. *Advanced Energy Materials* 2014; 4: 1301415.
18. Lefevre, M., E. Proietti, F. Jaouen, and J.P. Dodelet. Iron-based catalysts with improved oxygen reduction activity in polymer electrolyte fuel cells. *Science* 2009; 324: 71–74.
19. Wu, G., K.L. More, C.M. Johnston, and P. Zelenay. High-performance electrocatalysts for oxygen reduction derived from polyaniline, iron, and cobalt. *Science* 2011; 332: 443–447.
20. Subramanian, N.P., X. Li, V. Nallathambi, S.P. Kumaraguru, H. Colon-Mercado, G. Wu, J.-W. Lee, and B.N. Popov. Nitrogen-modified carbon-based catalysts for oxygen reduction reaction in polymer electrolyte membrane fuel cells. *Journal of Power Sources* 2009; 188: 38–44.

21. Nallathambi, V., J.-W. Lee, S.P. Kumaraguru, G. Wu, and B.N. Popov. Development of high performance carbon composite catalyst for oxygen reduction reaction in proton exchange membrane fuel cells. *Journal of Power Sources* 2008; 183: 34–42.
22. Wu, G., C.M. Johnston, N.H. Mack, K. Artyushkova, M. Ferrandon, M. Nelson, J.S. Lezama-Pacheco, S.D. Conradson, K.L. More, D.J. Myers, and P. Zelenay. Synthesis–structure–performance correlation for polyaniline–Me–C non-precious metal cathode catalysts for oxygen reduction in fuel cells. *Journal of Materials Chemistry* 2011; 21: 11392–11405.
23. Wu, G., C. Dai, D. Wang, D. Li, and N. Li. Nitrogen-doped magnetic onion-like carbon as support for Pt particles in a hybrid cathode catalyst for fuel cells. *Journal of Materials Chemistry* 2010; 20: 3059–3068.
24. Gojković, S.L., S. Gupta, and R.F. Savinell. Heat-treated iron(III) tetramethoxyphenyl porphyrin chloride supported on high-area carbon as an electrocatalyst for oxygen reduction: Part II. Kinetics of oxygen reduction. *Journal of Electroanalytical Chemistry* 1999; 462: 63–72.
25. Wu, G., M. Nelson, S. Ma, H. Meng, G. Cui, and P.K. Shen. Synthesis of nitrogen-doped onion-like carbon and its use in carbon-based CoFe binary non-precious-metal catalysts for oxygen-reduction. *Carbon* 2011; 49: 3972–3982.
26. Niwa, H., K. Horiba, Y. Harada, M. Oshima, T. Ikeda, K. Terakura, J.-I. Ozaki, and S. Miyata. X-ray absorption analysis of nitrogen contribution to oxygen reduction reaction in carbon alloy cathode catalysts for polymer electrolyte fuel cells. *Journal of Power Sources* 2009; 187: 93–97.
27. Ferrandon, M., A.J. Kropf, D.J. Myers, K. Artyushkova, U. Kramm, P. Bogdanoff, G. Wu, C.M. Johnston, and P. Zelenay. Multitechnique characterization of a polyaniline–iron–carbon oxygen reduction catalyst. *The Journal of Physical Chemistry C* 2012; 116: 16001–16013.
28. Li, Q., P. Xu, W. Gao, S.G. Ma, G.Q. Zhang, R.G. Cao, J. Cho, H.L. Wang, and G. Wu. Graphene/graphene tube nanocomposites templated from cage-containing metal–organic frameworks for oxygen reduction in Li–O_2 batteries. *Advanced Materials* 2014; 26: 1378–1386.
29. Kou, R., Y. Shao, D. Wang, M.H. Engelhard, J.H. Kwak, J. Wang, V.V. Viswanathan, C. Wang, Y. Lin, and Y. Wang. Enhanced activity and stability of Pt catalysts on functionalized graphene sheets for electrocatalytic oxygen reduction. *Electrochemistry Communications* 2009; 11: 954–957.
30. Li, Q., G. Wu, D.A. Cullen, K.L. More, N.H. Mack, H.T. Chung, and P. Zelenay. Phosphate-tolerant oxygen reduction catalysts. *ACS Catalysis* 2014; 4: 3193–3200.
31. Zelenay, P., H. Chung, U. Martinz, and G. Wu. *DOE EERE Annual Merit Review Meeting Presentation*, Washington, DC, 2014.
32. Novoselov, K.S., A.K. Geim, S.V. Morozov, D. Jiang, Y. Zhang, S.V. Dubonos, I.V. Grigorieva, and A.A. Firsov. Electric field effect in atomically thin carbon films. *Science* 2004; 306: 666–669.
33. Geng, D., S. Yang, Y. Zhang, J. Yang, J. Liu, R. Li, T.-K. Sham, X. Sun, S. Ye, and S. Knights. Nitrogen doping effects on the structure of graphene. *Applied Surface Science* 2011; 257: 9193–9198.
34. Ismach, A., C. Druzgalski, S. Penwell, A. Schwartzberg, M. Zheng, A. Javey, J. Bokor, and Y. Zhang. Direct chemical vapor deposition of graphene on dielectric surfaces. *Nano Letters* 2010; 10: 1542–1548.
35. Wu, G., N.H. Mack, W. Gao, S.G. Ma, R.Q. Zhong, J.T. Han, J.K. Baldwin, and P. Zelenay. Nitrogen doped graphene-rich catalysts derived from heteroatom polymers for oxygen reduction in nonaqueous lithium–O_2 battery cathodes. *ACS Nano* 2012; 6: 9764–9776.
36. Wu, G., M.A. Nelson, N.H. Mack, S.G. Ma, P. Sekhar, F.H. Garzon, and P. Zelenay. Titanium dioxide-supported non-precious metal oxygen reduction electrocatalyst. *Chemical Communications* 2010; 46: 7489–7491.
37. Wu, G., K.L. More, P. Xu, H.L. Wang, M. Ferrandon, A.J. Kropf, D.J. Myers, S.G. Ma, C.M. Johnston, and P. Zelenay. A carbon-nanotube-supported graphene-rich non-precious metal oxygen reduction catalyst with enhanced performance durability. *Chemical Communications* 2013; 49: 3291–3293.
38. Wu, G., H.T. Chung, M. Nelson, K. Artyushkova, K.L. More, C.M. Johnston, and P. Zelenay. Graphene-enriched Co_9S_8–N–C non-precious metal catalyst for oxygen reduction in alkaline media. *ECS Transactions* 2011; 41: 1709–1717.
39. He, Q.G., Q. Li, S. Khene, X.M. Ren, F.E. Lopez-Suarez, D. Lozano-Castello, A. Bueno-Lopez, and G. Wu. High-loading cobalt oxide coupled with nitrogen-doped graphene for oxygen reduction in anion-exchange-membrane alkaline fuel cells. *The Journal of Physical Chemistry C* 2013; 117: 8697–8707.
40. Wang, L., X. Zhao, Y. Lu, M. Xu, D. Zhang, R.S. Ruoff, K.J. Stevenson, and J.B. Goodenough. $CoMn_2O_4$ spinel nanoparticles grown on graphene as bifunctional catalyst for lithium–air batteries. *Journal of the Electrochemical Society* 2011; 158: A1379–A1382.
41. Cheng, F., J. Shen, B. Peng, Y. Pan, Z. Tao, and J. Chen. Rapid room-temperature synthesis of nanocrystalline spinels as oxygen reduction and evolution electrocatalysts. *Nature Chemistry* 2011; 3: 79–84.
42. Du, G., X. Liu, Y. Zong, T.S.A. Hor, A. Yu, and Z. Liu. Co_3O_4 nanoparticle-modified MnO_2 nanotube bifunctional oxygen cathode catalysts for rechargeable zinc–air batteries. *Nanoscale* 2013; 5: 4657–4661.
43. Chen, Z., A.P. Yu, R. Ahmed, H.J. Wang, H. Li, and Z.W. Chen. Manganese dioxide nanotube and nitrogen-doped carbon nanotube based composite bifunctional catalyst for rechargeable zinc–air battery. *Electrochimica Acta* 2012; 69: 295–300.
44. Li, Q., P. Xu, B. Zhang, H. Tsai, J. Wang, H.-L. Wang, and G. Wu. One-step synthesis of Mn_3O_4/reduced graphene oxide nanocomposites for oxygen reduction in nonaqueous Li–O_2 batteries. *Chemical Communications* 2013; 49: 10838–10840.
45. Stamenkovic, V.R., B.S. Mun, M. Arenz, K.J.J. Mayrhofer, C.A. Lucas, G. Wang, P.N. Ross, and N.M. Markovic. Trends in electrocatalysis on extended and nanoscale Pt–bimetallic alloy surfaces. *Nature Materials* 2007; 6: 241–247.
46. Holby, E.F., G. Wu, P. Zelenay, and C.D. Taylor. Structure of Fe–N_x–C defects in oxygen reduction reaction catalysts from first-principles modeling. *The Journal of Physical Chemistry C* 2014; 118: 14388–14393.
47. Bing, Y.H., H.S. Liu, L. Zhang, D. Ghosh, and J.J. Zhang. Nanostructured Pt–alloy electrocatalysts for PEM fuel cell oxygen reduction reaction. *Chemical Society Reviews* 2010; 39: 2184–2202.
48. Meier, J.C., C. Galeano, I. Katsounaros, J. Witte, H.J. Bongard, A.A. Topalov, C. Baldizzone, S. Mezzavilla, F. Schüth, and K.J.J. Mayrhofer. Design criteria for stable Pt/C fuel cell catalysts. *Beilstein Journal of Nanotechnology* 2014; 5: 44–67.
49. Nakano, H., W.Z. Li, L.B. Xu, Z.W. Chen, M. Waje, S. Kuwabata, and Y.S. Yan. Carbon nanotube and carbon black supported platinum nanocomposites as oxygen reduction electrocatalysts for polymer electrolyte fuel cells. *Electrochemistry* 2007; 75: 705–708.
50. Li, Q., P. Xu, B. Zhang, G. Wu, H.T. Zhao, E.G. Fu, and H.L. Wang. Self-supported Pt nanoclusters via galvanic replacement from Cu_2O nanocubes as efficient electrocatalysts. *Nanoscale* 2013; 5: 7397–7402.

51. Wu, G., Y.-S. Chen, and B.-Q. Xu. Remarkable support effect of SWNTs in Pt catalyst for methanol electrooxidation. *Electrochemistry Communications* 2005; 7: 1237–1243.
52. Chen, Y.G., J.J. Wang, H. Liu, R.Y. Li, X.L. Sun, S.Y. Ye, and S. Knights. Enhanced stability of Pt electrocatalysts by nitrogen doping in CNTs for PEM fuel cells. *Electrochemistry Communications* 2009; 11: 2071–2076.
53. Higgins, D.C., J.Y. Choi, J. Wu, A. Lopez, and Z.W. Chen. Titanium nitride–carbon nanotube core–shell composites as effective electrocatalyst supports for low temperature fuel cells. *Journal of Materials Chemistry* 2012; 22: 3727–3732.
54. Higgins, D.C., D. Meza, and Z.W. Chen. Nitrogen-doped carbon nanotubes as platinum catalyst supports for oxygen reduction reaction in proton exchange membrane fuel cells. *The Journal of Physical Chemistry C* 2010; 114: 21982–21988.
55. Kundu, S., T.C. Nagaiah, X.X. Chen, W. Xia, M. Bron, W. Schuhmann, and M. Muhler. Synthesis of an improved hierarchical carbon-fiber composite as a catalyst support for platinum and its application in electrocatalysis. *Carbon* 2012; 50: 4534–4542.
56. Wu, G., D. Li, C. Dai, D. Wang, and N. Li. Well-dispersed high-loading Pt nanoparticles supported by shell–core nanostructured carbon for methanol electrooxidation. *Langmuir* 2008; 24: 3566–3575.
57. Zhu, C.Z. and S.J. Dong. Recent progress in graphene-based nanomaterials as advanced electrocatalysts towards oxygen reduction reaction. *Nanoscale* 2013; 5: 1753–1767.
58. Higgins, D., M.A. Hoque, M.H. Seo, R.Y. Wang, F. Hassan, J.Y. Choi, M. Pritzker, A.P. Yu, J.J. Zhang, and Z.W. Chen. Development and simulation of sulfur-doped graphene supported platinum with exemplary stability and activity towards oxygen reduction. *Advanced Functional Materials* 2014; 24: 4324–4324.
59. Wu, G., R. Swaidan, D. Li, and N. Li. Enhanced methanol electro-oxidation activity of PtRu catalysts supported on heteroatom-doped carbon. *Electrochimica Acta* 2008; 53: 7622–7629.
60. Cao, R., R. Thapa, H. Kim, X. Xu, M.G. Kim, Q. Li, N. Park, M.L. Liu, and J. Cho. Promotion of oxygen reduction by a bio-inspired tethered iron phthalocyanine carbon nanotube-based catalyst. *Nature Communications* 2013; 4: 2076.
61. Wu, G., Z. Chen, K. Artyushkova, F.H. Garzon, and P. Zelenay. Polyaniline-derived non-precious catalyst for the polymer electrolyte fuel cell cathode. *ECS Transactions* 2008; 16: 159–170.
62. Wu, G., K. Artyushkova, M. Ferrandon, A.J. Kropf, D. Myers, and P. Zelenay. Performance durability of polyaniline-derived non-precious cathode catalysts. *ECS Transactions* 2009; 25: 1299–1311.
63. Li, Q., H. Pan, D. Higgins, R. Cao, G. Zhang, H. Lv, K. Wu, J. Cho, and G. Wu. Metal–organic framework-derived bamboo-like nitrogen-doped graphene tubes as an active matrix for hybrid oxygen-reduction electrocatalysts. *Small* 2014: doi:10.1002/smll.201402069.
64. Gasteiger, H.A., S.S. Kocha, B. Sompalli, and F.T. Wagner. Activity benchmarks and requirements for Pt, Pt–alloy, and non-Pt oxygen reduction catalysts for PEMFCs. *Applied Catalysis B: Environment* 2005; 56: 9–35.
65. Lu, S.F., J. Pan, A.B. Huang, L. Zhuang, and J.T. Lu. Alkaline polymer electrolyte fuel cells completely free from noble metal catalysts. *Proceedings of the National Academy of Sciences of the United States of America* 2008; 105: 20611–20614.
66. Spendelow, J.S. and A. Wieckowski. Electrocatalysis of oxygen reduction and small alcohol oxidation in alkaline media. *Physical Chemistry Chemical Physics* 2007; 9: 2654–2675.
67. Wu, G., G. Cui, D. Li, P.-K. Shen, and N. Li. Carbon-supported $Co_{1.67}Te_2$ nanoparticles as electrocatalysts for oxygen reduction reaction in alkaline electrolyte. *Journal of Materials Chemistry* 2009; 19: 6581–6589.
68. Markovic, N.M. and P.N. Ross. Surface science studies of model fuel cell electrocatalysts. *Surface Scientific Reports* 2002; 45: 121–229.
69. Fernandez, J.L., D.A. Walsh, and A.J. Bard. Thermodynamic guidelines for the design of bimetallic catalysts for oxygen electroreduction and rapid screening by scanning electrochemical microscopy. M–Co (M: Pd, Ag, Au). *Journal of American Chemical Society* 2005; 127: 357–365.
70. Matsuoka, K., Y. Iriyama, T. Abe, M. Matsuoka, and Z. Ogumi. Alkaline direct alcohol fuel cells using an anion exchange membrane. *Journal of Power Sources* 2005; 150: 27–31.
71. Scott, K., E. Yu, G. Vlachogiannopoulos, M. Shivare, and N. Duteanu. Performance of a direct methanol alkaline membrane fuel cell. *Journal of Power Sources* 2008; 175: 452–457.
72. McLean, G.F., T. Niet, S. Prince-Richard, and N. Djilali. An assessment of alkaline fuel cell technology. *International Journal of Hydrogen Energy* 2002; 27: 507–526.
73. Varcoe, J.R. and R.C.T. Slade. Prospects for alkaline anion-exchange membranes in low temperature fuel cells. *Fuel Cells* 2005; 5: 187–200.
74. Varcoe, J.R., R.C.T. Slade, and E. Lam How Yee. An alkaline polymer electrochemical interface: A breakthrough in application of alkaline anion-exchange membranes in fuel cells. *Chemical Communications* 2006: 1428–1429.
75. Gong, K.P., F. Du, Z.H. Xia, M. Durstock, and L.M. Dai. Nitrogen-doped carbon nanotube arrays with high electrocatalytic activity for oxygen reduction. *Science* 2009; 323: 760–764.
76. Song, C., L. Zhang, and J. Zhang. Reversible one-electron electro-reduction of O_2 to produce a stable superoxide catalyzed by adsorbed Co(II) hexadecafluoro-phthalocyanine in aqueous alkaline solution. *Journal of Electroanalytical Chemistry* 2006; 587: 293–298.
77. Jörissen, L. Bifunctional oxygen/air electrodes. *Journal of Power Sources* 2006; 155: 23–32.
78. Mao, L., T. Sotomura, K. Nakatsu, N. Koshiba, D. Zhang, and T. Ohsaka. Electrochemical characterization of catalytic activities of manganese oxides to oxygen reduction in alkaline aqueous solution. *Journal of the Electrochemical Society* 2002; 149: A504–A507.
79. Hammouche, A., A. Kahoul, D.U. Sauer, and R.W. De Doncker. Influential factors on oxygen reduction at $La_{1-x}Ca_xCoO_3$ electrodes in alkaline electrolyte. *Journal of Power Sources* 2006; 153: 239–244.
80. Heller-Ling, N., M. Prestat, J.L. Gautier, J.F. Koenig, G. Poillerat, and P. Chartier. Oxygen electroreduction mechanism at thin $Ni_xCo_{3-x}O_4$ spinel films in a double channel electrode flow cell (DCEFC). *Electrochimica Acta* 1997; 42: 197–202.
81. Shao, Y.Y., S. Zhang, M.H. Engelhard, G.S. Li, G.C. Shao, Y. Wang, J. Liu, I.A. Aksay, and Y.H. Lin. Nitrogen-doped graphene and its electrochemical applications. *Journal of Materials Chemistry* 2010; 20: 7491–7496.
82. Higgins, D.C. and Z.W. Chen. Nitrogen doped carbon nanotube thin films as efficient oxygen reduction catalyst for alkaline anion exchange membrane fuel cell. *ECS Transactions* 2010; 28: 63–68.
83. Chung, H.T., J.H. Won, and P. Zelenay. Active and stable carbon nanotube/nanoparticle composite electrocatalyst for oxygen reduction. *Nature Communications* 2013; 4: 1922.

84. Guo, J.S., H. He, D. Chu, and R.R. Chen. OH-binding effects on metallophthalocyanine catalysts for O_2 reduction reaction in anion exchange membrane fuel cells. *Electrocatalysis* 2012; 3: 252–264.
85. Elezovic, N.R., B.M. Babic, L. Gajic-Krstajic, P. Ercius, V.R. Radmilovic, N.V. Krstajic, and L.M. Vracar. Pt supported on nano-tungsten carbide as a beneficial catalyst for the oxygen reduction reaction in alkaline solution. *Electrochimica Acta* 2012; 69: 239–246.
86. Chen, R.R., J.S. Guo, and A. Hsu. Non-Pt cathode electrocatalysts for anion-exchange-membrane fuel cells, in *Electrocatalysis in Fuel Cells, Lecture Notes in Energy*, 9, M. Shao, Editor. 2013, Springer-Verlag: London.
87. Li, X.G., B.N. Popov, T. Kawahara, and H. Yanagi, Non-precious metal catalysts synthesized from precursors of carbon, nitrogen, and transition metal for oxygen reduction in alkaline fuel cells. *Journal of Power Sources* 2011; 196: 1717–1722.
88. Débart, A., A.J. Paterson, J. Bao, and P.G. Bruce. Alpha-MnO_2 nanowires: A catalyst for the O_2 electrode in rechargeable lithium batteries. *Angewandte Chemie International Edition* 2008; 47: 4521–4524.
89. Lee, J.-S., G.S. Park, H.I. Lee, S.T. Kim, R. Cao, M. Liu, and J. Cho. Ketjen black carbon supported amorphous manganese oxides nanowires as highly efficient electrocatalyst for oxygen reduction reaction in alkaline solutions. *Nano Letters* 2011; 11: 5362–5366.
90. Crisostomo, V.M.B., J.K. Ngala, S. Alia, A. Dobley, C. Morein, C.-H. Chen, X. Shen, and S.L. Suib. New synthetic route, characterization, and electrocatalytic activity of nano-sized manganite. *Chemistry Materials* 2007; 19: 1832–1839.
91. Lee, J.-S., T. Lee, H.-K. Song, J. Cho, and B.-S. Kim. Ionic liquid modified graphene nanosheets anchoring manganese oxide nanoparticles as efficient electrocatalysts for Zn–air batteries. *Energy Environmental Science* 2011; 4: 4148–4154.
92. Huang, Y., Y. Lin, and W. Li. Controllable syntheses of α- and δ-MnO_2 as cathode catalysts for zinc–air battery. *Electrochimica Acta* 2013; 99: 161–165.
93. Kong, F. Synthesis of rod and beadlike Co_3O_4 and bi-functional properties as air/oxygen electrode materials. *Electrochimica Acta* 2012; 68: 198–201.
94. Zhang, G.Q. and X.G. Zhang. MnO/MCMB electrocatalyst for all solid-state alkaline zinc–air cells. *Electrochimica Acta* 2004; 49: 873–877.
95. Chervin, C.N., J.W. Long, N.L. Brandell, J.M. Wallace, N.W. Kucko, and D.R. Rolison. Redesigning air cathodes for metal–air batteries using MnO*x*-functionalized carbon nanofoam architectures. *Journal of Power Sources* 2012; 207: 191–198.
96. Wei, Z., W. Huang, S. Zhang, and J. Tan, Carbon-based air electrodes carrying MnO_2 in zinc–air batteries. *Journal of Power Sources* 2000; 91: 83–85.
97. Truong, T.T., Y. Liu, Y. Ren, L. Trahey, and Y. Sun. Morphological and crystalline evolution of nanostructured MnO_2 and its application in lithium–air batteries. *ACS Nano* 2012; 6: 8067–8077.
98. Cao, Y., Z. Wei, J. He, J. Zang, Q. Zhang, M. Zheng, and Q. Dong. α-MnO_2 nanorods grown *in situ* on graphene as catalysts for Li–O_2 batteries with excellent electrochemical performance. *Energy Environmental Science* 2012; 5: 9765–9768.
99. Jung, K.-N., J.-I. Lee, S. Yoon, S.-H. Yeon, W. Chang, K.-H. Shin, and J.-W. Lee. Manganese oxide/carbon composite nanofibers: Electrospinning preparation and application as a bi-functional cathode for rechargeable lithium–oxygen batteries. *Journal of Materials Chemistry* 2012; 22: 21845–21848.
100. Nikolova, V., P. Iliev, K. Petrov, T. Vitanov, E. Zhecheva, R. Stoyanova, I. Valov, and D. Stoychev. Electrocatalysts for bifunctional oxygen/air electrodes. *Journal of Power Sources* 2008; 185: 727–733.
101. Kannan, A.M., A.K. Shukla, and S. Sathyanarayana. Oxide-based bifunctional oxygen electrode for rechargeable metal/air batteries. *Journal of Power Sources* 1989; 25: 141–150.
102. Chakkaravarthy, C., A.K.A. Waheed, and H.V.K. Udupa. Zinc–air alkaline batteries—A review. *Journal of Power Sources* 1981; 6: 203–228.
103. Blurton, K.F. and A.F. Sammells. Metal/air batteries: Their status and potential—A review. *Journal of Power Sources* 1979; 4: 263–279.
104. Toussaint, G., P. Stevens, L. Akrour, R. Rouget, and F. Fourgeot. Development of a rechargeable zinc–air battery. *ECS Transactions* 2010; 28: 25–34.
105. Chen, Z., A. Yu, D. Higgins, H. Li, H. Wang, and Z. Chen. Highly active and durable core–corona structured bifunctional catalyst for rechargeable metal–air battery application. *Nano Letters* 2012; 12: 1946–1952.
106. Li, Y.G., M. Gong, Y.Y. Liang, J. Feng, J.E. Kim, H.L. Wang, G.S. Hong, B. Zhang, and H.J. Dai. Advanced zinc–air batteries based on high-performance hybrid electrocatalysts. *Nature Communications* 2013; 4, 1805.
107. Peng, Z.Q., S.A. Freunberger, Y.H. Chen, and P.G. Bruce. A reversible and higher-rate Li–O_2 battery. *Science* 2012; 337: 563–566.
108. Black, R., B. Adams, and L.F. Nazar. Non-aqueous and hybrid Li–O_2 batteries. *Advanced Energy Materials* 2012; 2: 801–815.
109. Cao, R., J.S. Lee, M.L. Liu, and J. Cho. Recent progress in non-precious catalysts for metal–air batteries. *Advanced Energy Materials* 2012; 2: 816–829.
110. Kitaura, H. and H. Zhou. Electrochemical performance of solid-state lithium–air batteries using carbon nanotube catalyst in the air electrode. *Advanced Energy Materials* 2012; 2: 889–894.
111. Wilshire, J. and D.T. Sawyer. Redox chemistry of dioxygen species. *Account of Chemical Research* 1979; 12: 105–110.
112. Hassoun, J., F. Croce, M. Armand, and B. Scrosati. Investigation of the O_2 electrochemistry in a polymer electrolyte solid-state cell. *Angewandte Chemie International Edition* 2011; 50: 2999–3002.
113. Islam, M.M., T. Imase, T. Okajima, M. Takahashi, Y. Niikura, N. Kawashima, Y. Nakamura, and T. Ohsaka. Stability of superoxide ion in imidazolium cation-based room-temperature ionic liquids. *The Journal of Physical Chemistry A* 2009; 113: 912–916.
114. Freunberger, S.A., Y.H. Chen, N.E. Drewett, L.J. Hardwick, F. Barde, and P.G. Bruce. The lithium–oxygen battery with ether-based electrolytes. *Angewandte Chemie International Edition* 2011; 50: 8609–8613.
115. McCloskey, B.D., D.S. Bethune, R.M. Shelby, G. Girishkumar, and A.C. Luntz. Solvents' critical role in nonaqueous lithium–oxygen battery electrochemistry. *The Journal of Physical Chemistry Letters* 2011; 2: 1161–1166.
116. Mitchell, R.R., B.M. Gallant, C.V. Thompson, and Y. Shao-Horn. All-carbon-nanofiber electrodes for high-energy rechargeable Li–O_2 batteries. *Energy Environmental Science* 2011; 4: 2952–2958.
117. McCloskey, B.D., R. Scheffler, A. Speidel, D.S. Bethune, R.M. Shelby, and A.C. Luntz. On the efficacy of electrocatalysis in nonaqueous Li–O_2 batteries. *Journal of American Chemical Society* 2011; 133: 18038–18041.

118. McCloskey, B.D., D.S. Bethune, R.M. Shelby, T. Mori, R. Scheffler, A. Speidel, M. Sherwood, and A.C. Luntz. Limitations in rechargeability of Li–O_2 batteries and possible origins. *The Journal of Physical Chemistry Letters* 2012; 3: 3043–3047.
119. Lu, Y.C., Z.C. Xu, H.A. Gasteiger, S. Chen, K. Hamad-Schifferli, and Y. Shao-Horn. Platinum–gold nanoparticles: A highly active bifunctional electrocatalyst for rechargeable lithium–air batteries. *Journal of American Chemical Society* 2010; 132: 12170–12171.
120. Li, Q., R. Cao, J. Cho, and G. Wu. Nanostructured carbon-based cathode catalysts for nonaqueous lithium–oxygen batteries. *Physical Chemistry Chemical Physics* 2014; 16: 13568–13582.
121. Shao, Y.Y., F. Ding, J. Xiao, J. Zhang, W. Xu, S. Park, J.G. Zhang, Y. Wang, and J. Liu, Making Li–air batteries rechargeable: Material challenges. *Advanced Functional Materials* 2013; 23: 987–1004.
122. Thotiyl, M.M.O., S.A. Freunberger, Z.Q. Peng, and P.G. Bruce, The carbon electrode in nonaqueous Li–O_2 cells. *Journal of American Chemical Society* 2013; 135: 494–500.
123. Itkis, D.M., D.A. Semenenko, E.Y. Kataev, A.I. Belova, V.S. Neudachina, A.P. Sirotina, M. Hävecker, D. Teschner, A. Knop-Gericke, P. Dudin, A. Barinov, E.A. Goodilin, Y. Shao-Horn, and L.V. Yashina. Reactivity of carbon in lithium–oxygen battery positive electrodes. *Nano Letters* 2013; 13: 4697–4701.
124. Trahey, L., N.K. Karan, M.K.Y. Chan, J. Lu, Y. Ren, J. Greeley, M. Balasubramanian, A.K. Burrell, L.A. Curtiss, and M.M. Thackeray. Synthesis, characterization, and structural modeling of high-capacity, dual functioning MnO_2 electrode/electrocatalysts for Li–O_2 cells. *Advanced Energy Materials* 2013; 3: 75–84.
125. Chen, Y., S.A. Freunberger, Z. Peng, F. Barde, and P.G. Bruce. Li–O_2 battery with a dimethylformamide electrolyte. *Journal of American Chemical Society* 2012; 134: 7952–7957.
126. Xu, D., Z.L. Wang, J.J. Xu, L.L. Zhang, and X.B. Zhang. Novel DMSO-based electrolyte for high performance rechargeable Li–O_2 batteries. *Chemical Communications* 2012; 48: 6948–6950.
127. Xu, D., Z.L. Wang, J.J. Xu, L.L. Zhang, L.M. Wang, and X.B. Zhang. A stable sulfone based electrolyte for high performance rechargeable Li–O_2 batteries. *Chemical Communications* 2012; 48: 11674–11676.
128. Zhang, T. and H.S. Zhou. A reversible long-life lithium–air battery in ambient air. *Nature Communications* 2013; 4.

Section II

Advanced Materials and Technologies for Lithium-Ion Rechargeable Batteries

4 Advanced Materials for Lithium-Ion Batteries

Zhaoping Liu, Yonggao Xia, Wei Wang, Zhen Wei, Junli Shi, Lan Xia, Laifen Qin, and Shaojie Han

CONTENTS

Recently, the rapid advancement of science and technology became the engine for great industrial and economic growth. Through this growth, the people of developed countries came to enjoy a comfortable and convenient way of life. Electric power is the key energy source in the current and future society, which supports a life of ease and the world's socioeconomic system [1]. As symbolized by the recent rapid growth in the use of compact portable telephones, computers, electric vehicles (EVs), hybrid EVs (HEVs) and backup of power systems, it is also evident that batteries, particularly secondary batteries, have become an essential part of our electricity-dependent day and age. There can be no doubt that society's consumption of electricity and dependence on it will increase in the future. As batteries provide a high added-value energy source, which makes life in a networked society even more convenient, there also can be no doubt that the use of batteries will likewise increase.

Generally, secondary batteries can be categorized into several types: lead–acid, Ni–Cd, Ni–MH, and Li-ion batteries (LIBs). Among these batteries, LIBs have attracted the most attention. It can be formed into a wide variety of shapes and sizes so as to efficiently fill available space in the devices. LIBs are lighter than other equivalent secondary batteries. The energy is stored in these batteries through the movement of lithium ions. The key advantage of using Li-ion chemistry is the high open circuit voltage that can be obtained in comparison to aqueous batteries (such as lead–acid, Ni–Cd, and Ni–MH). Moreover, LIBs do not suffer from the memory effect. They also have a low self-discharge rate of approximately 5% per month, compared with >30% per month in Ni–MH batteries and 10% per month in Ni–Cd batteries.

The basic structure and working principle of LIBs is illustrated in Figure 4.1. Generally speaking, one LIB consists of a cathode, an ion-conductive electrolyte, and an anode separated by a separator. During charging, lithium ions are extracted from the cathode and go into the electrolyte, while lithium ions in the electrolyte insert into the anode. The charge compensation is fulfilled through an external circuit.

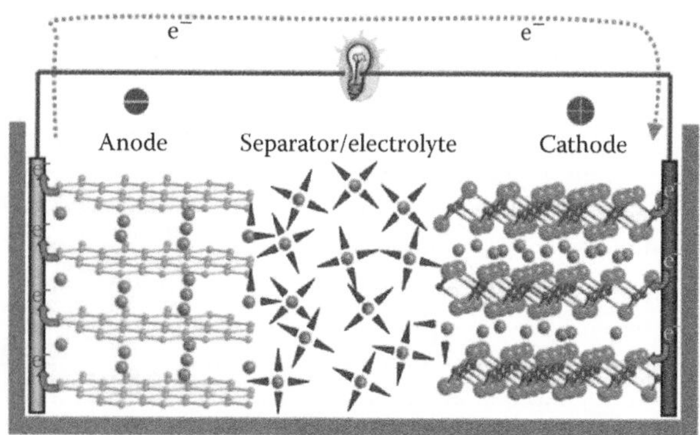

FIGURE 4.1 Schematic representation of a typical LIB. (From Meng, Y.S., and Dompablo, M.E.A., *Energy Environ. Sci.*, 2, 589, 2009.)

During discharge, lithium ions are de-intercalated from the anode and intercalated into the cathode.

High energy density (related to both average working voltage and reversible capacity) and long cycle life (related to the stability of structure and the electrode–electrolyte interface) are simultaneously required for a high performance of LIBs. Therefore, some aspects should be considered in the design of new electrode materials for advanced LIBs. In general, structural and chemical stabilities, safety, and the availability of redox couple at a suitable energy are the primary considerations. Ideally, an electrode material should simultaneously match such conditions listed as follows:

1. The cathode and the anode should have an open structure to permit reversible lithium insertion.
2. The stability of the electrode and the electrolyte is a requirement for a long cycle life. The insertion/extraction reaction has a topotactic character, and both insertion and extraction of guest lithium ions into and from the host compound should ideally not modify the host structure. On the other hand, the oxidation of electrolyte should be avoided for oxidation and reduction.
3. The equivalent weight of both electrodes must be low to assure specific capacity values of practical interest.
4. The mobility of Li^+ ions and electrons in both the positive electrode and negative electrode must be high to assure fast kinetics of the electrochemical process, and thus, fast charge and discharge rates.
5. The cost of battery and the potential influence on the environment should be always kept in mind in advanced battery design. It is one of the future challenges to develop cheaper and hazardless electrode materials with excellent battery performance.

4.1 CATHODE MATERIALS

4.1.1 Introduction

The cathode material is one of the four key components of LIBs (cathode, anode, separator, and electrolyte), which accounts for 30–40% manufacturing costs in a typical battery design; it also determines the safety, energy density, working voltage, cost, and cycling life of LIBs. Therefore, it is important to develop promising cathode materials for the current LIB technology.

Table 4.1 gives a comparison data among the existing cathode materials for LIBs. The working voltage of cathode materials strongly depends on the redox couple and the crystal structure of materials. Therefore, on the basis of the voltage region of lithium intercalation/de-intercalation in LIBs, cathode materials could be divided into the 3-V class, 4-V class, or 5-V class. $LiFePO_4$ could be classified as a 3-V class cathode material; it shows excellent safety attribute

TABLE 4.1
Comparison Data of the Existing Cathode Materials for LIBs

	Cathode	Average Voltage (V)	Practical Capacities (mAh/g)	Cycle Life (n)	Safety	Cost	Technology Maturity
Layer	$LiCoO_2$	3.7	140	1000	Acceptable	High	Mature
	$LiNi_{0.8}Co_{.15}Al_{0.05}O_2$ (NCA)	3.7	190	1000	Acceptable	High	Mature
	$LiNi_{1/3}Mn_{1/3}Co_{1/3}O_2$ (NMC)	3.6	160	1500	Acceptable	Acceptable	Mature
	Li_2MnO_3–$LiMO_2$ (Li-rich NMC)	3.5	300	–	–	Low	Basic research
Spinel	$LiMn_2O_4$	4.1	100	500	Good	Low	Mature
	$LiNi_{0.5}Mn_{1.5}O_4$	4.7	130	1000	Acceptable	Acceptable	Basic research
Olivine	$LiFePO_4$	3.4	160	2500	Excellent	Low	Mature
	$LiMnPO_4$	4.0	160	2500	Excellent	Low	Basic research

to its lower oxidative power and stability of a strong covalent bond in PO_4^{3-}; 4-V class cathode materials, $LiCoO_2$, $LiNi_{0.8}Co_{.15}Al_{0.05}O_2$ (NCA) and $LiNi_{1/3}Mn_{1/3}Co_{1/3}O_2$ (NMC), still mainly suffer from high cost and safety concerns; 5-V class cathode material, $LiNi_{0.5}Mn_{1.5}O_4$, has gained considerable attention by virtue of its high energy density, low cost, and acceptable thermal stability. In the future, the development of alternative high-voltage electrolytes will be necessary for their successful practical application.

$LiFePO_4$ (M = Fe and Mn) and Mn–spinel possess good safety characteristics, low cost, and environmental friendliness. These cathode materials are very suitable for large-scale LIB applications such as EVs, HEVs, and backup of power systems. For example, lightweight laminated structure manganese-based ($LiMn_2O_4$) batteries have been applied successfully in Nissan LEAF EV. It can drive 200 km on a single charge, and the total travel distance of an EV is about 100,000 km over a 10-year period without replacing batteries. This demonstrates that the capacity fading of manganese-based cathode at elevated temperatures has been greatly improved by metal doping and blending with NCA cathode materials.

4.1.2 Thermal Stability of Cathode Materials

Nowadays, LIBs have been extensively studied because of their high performance and potential. They have been widely utilized in the portable electronics industry and are being pursued as power sources for HEV and EV applications [2–4]. However, there are also potential safety problems in their use: the risk of explosion caused by either shorting or an external temperature increase. Generally, thermal runaway of LIBs occurs if the heat output exceeds the thermal diffusion [5,6]. The safety of LIBs is mainly related to the thermal stability of their constituent materials. The possible exothermic reactions are as follows: (i) the chemical reduction of the electrolyte by the anode; (ii) the thermal decomposition of the electrolyte; (iii) the oxidation of the electrolyte on the cathode; and (iv) the thermal decomposition of the cathode. These exothermic reactions in the battery can cause thermal runaway of the battery. The expression "thermal runaway" is usually used to describe the situation in which cells catch fire or explosion. Therefore, to prevent thermal runaway, the thermal stability of materials in battery has to be investigated.

The thermal stability of LIBs has been investigated by differential scanning calorimetry (DSC) or accelerating rate calorimetry (ARC) in order to improve the battery's safety during practical applications. The thermal behavior of lithium nickel oxide ($LiNiO_2$) [7–11], lithium manganese oxide ($LiMn_2O_4$) [12,13], lithium cobalt oxide ($LiCoO_2$) [14–16], and lithium phosphate ($LiFePO_4$) [4,17], as the cathode materials of the LIBs, has been investigated. The structure of these materials is unstable at the charged state (delithiation), and they can release oxygen at elevated temperatures. Among them, $LiCoO_2$ as a cathode material has been widely used in LIB. However, the thermal stability of charged Li_xCoO_2 is poor at elevated temperatures.

Baba et al. reported the thermal stability of $Li_{0.49}CoO_2$ by using DSC measurements (Figure 4.2) [5], and found two main exothermic peaks between 190°C and 400°C. The exothermic peak starting from 190°C can be attributed to a decomposition of solvent caused by the active surface of the delithiated $Li_{0.49}CoO_2$. The exothermic heat from 190°C to 230°C based on the cathode weight was 420 ± 120 J/g. Meanwhile, the exothermic heat of 1000 ± 250 J/g starting from 230°C mainly resulted from the oxidation reaction of the electrolyte. Roth et al.

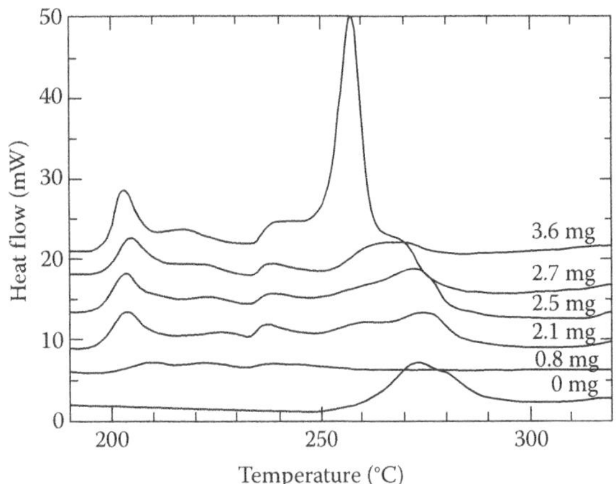

FIGURE 4.2 DSC profiles of chemically delithiated $Li_{0.49}CoO_2$ with electrolyte for various cathode weights in 3 μL electrolyte. (From Baba, Y., Okada, S., and Yamaki, J., *Solid State Ionics*, 148, 311, 2002.)

TABLE 4.2
DSC Results of Cathode Materials

Positive Materials	Onset Temperature (°C)	Maximum Temperature (°C)	Heat (J/g)
$LiNiO_2$	182	209	1300
$LiNi_{0.7}Co_{0.2}Ti_{0.05}Mg_{0.05}O_2$	175	220	1600
$LiNi_{0.8}Co_{0.2}O_2$	197	228	1600
$LiCoO_2$	181	256	1100
$LiMn_2O_4$	209	280	860
$LiFePO_4$	221	252	520
$LiNi_{3/8}Co_{1/4}Mn_{3/8}O_2$	270	297	290

Source: Roth, E., Doughty, D., and Franklin, J., *J. Power Sources*, 134, 222, 2004.

studied the reactions between the charged Li_xCoO_2, Li_xNiO_2, and $Li_xMn_2O_4$ with the electrolyte as shown in Table 4.2 [18], and found that the thermal stability increased as follows: Li_xNiO_2 < $LixCoO_2$ < $Li_xMn_2O_4$. Also, the thermal stability became much poorer as x values decreased. MacNeil et al. [17] also investigated several cathode materials (4.4 V versus Li/Li^+) with the 1 M $LiPF_6$ ethylene carbonate (EC) + diethyl carbonate [DEC] electrolyte. The thermal stability of these cathode materials with the electrolyte increased as follows:

$$LiNiO_2 < LiNi_{0.7}Co_{0.2}Ti_{0.05}Mg_{0.05}O_2 < LiNi_{0.8}Co_{0.2}O_2 < LiCoO_2 < LiMn_2O_4 < LiFePO_4 < LiNi_{3/8}Co_{1/4}Mn_{3/8}O_2$$

However, NMC Li-ion cells can withstand prolonged exposure to higher abusive elevated temperatures than can charge $LiCoO_2$ Li-ion cells. However, once charged, NMC begins to react with the electrolyte near 260–300°C and the reaction rate accelerates rapidly with temperature. On the other hand, substitutions of Al for Ni in $Li[Ni_{1-z}Al_z]O_2$ [19] and substitutions of Mg and Ti for Ni in $Li[Ni_{1-z}Mg_{z/2}Ti_{z/2}]O_2$ [20] have been shown to reduce the rate of exothermic reaction between the charged electrode and the electrolyte. Hence, the impact of Al substitution for Co in $Li[Ni_{1/3}Mn_{1/3}Co_{1-z}Al_z]O_2$ [21–24] on the exothermic reaction rate between the charged electrode material and electrolyte was also reported (Figure 4.3), and the Al-substitution played a very import role for the safety of MCN cells. Al-substituted NMC materials should play a key role in large-sized LIBs for EVs and HEVs where low cost, superior safety, and high energy density are all required.

Although the thermal stability of cathode materials has been extensively studied, some of the mechanism still has not been clarified thus far. For example, the reaction between cathode and electrolyte is not clear, which has to be examined deeply in the future. However, the exothermic reactions between the cathode and electrolyte and the self-decomposition of the cathode is a significant issue for the safety of LIBs. To improve the safety of the battery, materials with great thermal stability (such as $LiFePO_4$) has to be applied as cathode for LIB. On the basis of the above considerations, the thermal stability of cathode materials is very important to battery

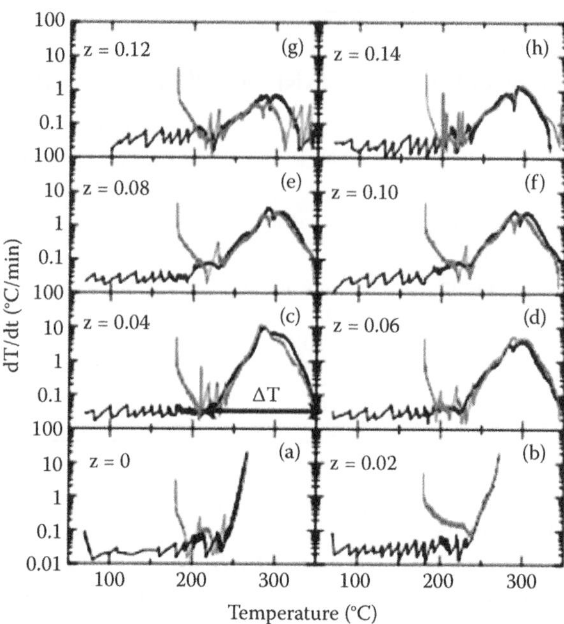

FIGURE 4.3 Self-heating rate versus temperature for delithiated $Li[Ni_{1/3}Mn_{1/3}Co_{1/3-z}Al_z]$ (4.3 V) reacting with 30 mg 1 M $LiPF_6$ EC/DEC with starting temperatures of 70°C (black lines) and 180°C (red lines). (a) z = 0, (b) z = 0.02, (c) z = 0.04, (d) z = 0.06, (e) z = 0.08, (f) z = 0.10, (g) z = 0.12, (h) z = 0.14. (From Zhou, F., Zhao, X., Lu, Z., Jiang, J., and Dahn, J., *Electrochem. Commun.*, 10, 1168, 2008.)

safety. The thermal stability is closely related to the type of material, the structure, and the surface of material. Therefore, (i) a stable material can be prepared by optimizing the synthesis condition or method; (ii) the thermal stability of material can be improved by using doping or coating methods. In a word, researchers have paid attention to the thermal stability of battery materials, and started to investigate it extensively, and tried to utilize a different approach to solve the safety problem of LIBs in their practical applications such as EVs and HEVs.

4.1.3 Layered Cathode Materials for LIBs

4.1.3.1 $LiNi_{1-x-y}Co_xMn_yO_2$

A layered cathode material for LIBs was first raised by Goodenough in 1980 [25], and the first commercial LIB was fabricated by Sony in 1990. $LiCoO_2$ possesses the advantages of high working voltage, high cycling stability, and good rate capability, and becomes the most successfully commercialized material. Other layered materials, especially $LiNi_{1-x-y}Co_xAl_yO_2$ and $LiNi_xCo_yMn_{1-x-y}O_2$, are very competitive for the next-generation battery for their high energy/power density.

The layered $LiMO_2$ materials have an α-$NaFeO_2$ crystal structure, as shown in Figure 4.4. For M = Co, Ni, it ascribes to hexagonal system (R-3m), while for M = Mn, the John–Taylor effect of Mn^{3+} causes crystal deformation, so $LiMnO_2$ ascribes to the monoclinic system (C/2m). Oxygen atoms form a cubic close-packed arrangement; M and lithium occupy the 3a and 3b octahedral positions. For $LiCoO_2$, lattice constant a = 0.28 nm, c = 1.41 nm. Its theoretical capacity is 274 mAh/g, and its actual capacity is controlled at 140 mAh/g

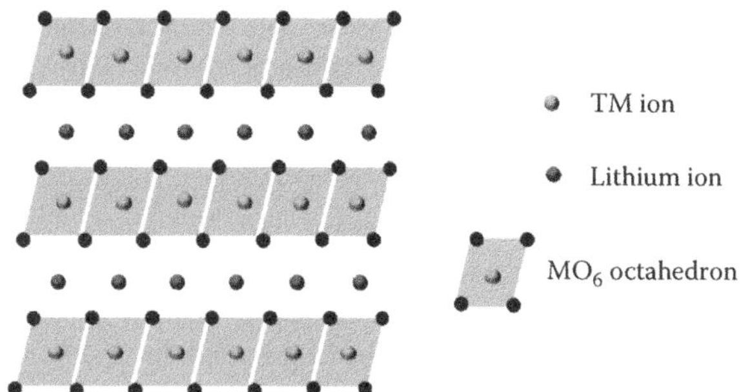

FIGURE 4.4 Schematic image of a layered structure of $LiMO_2$.

to achieve good cycling stability. However, $LiCoO_2$ confronts problems of low energy density and poor thermal stability. In $LiNiO_2$ crystal, its lattice constant $a = 0.29$ nm, $c = 1.42$ nm. The bond energy of Li–O bond is much lower than that of Ni–O bond, so Li–O bond has the characteristics of an ionic bond, and Li^+ can intercalate/extract between the Ni layer and the O layer [26]. Despite having high actual capacity (190–210 mAh/g) [27], $LiNiO_2$ is not used widely, which is because of its poor cycling stability and safety issue. For layered $LiMnO_2$, it tends to transform to the spinel phase during the charge–discharge process and is very unstable for cathode material use.

Layered $LiNi_{1-x-y}Co_xAl_yO_2$ is intensively researched for its potential application in EVs, owing to its superior capacity and rate capability [28,29]. Among them, $LiNi_{0.8}Co_{0.15}Al_{0.05}O_2$ (NCA) is the best one. By doping Co and Al, the Ni^{2+} content in the 3a position is reduced, which inhibits phase transformation from H2 (hexagonal) to H3 (hexagonal) and improves the reversible capacity, cycling performance, and thermal stability [30]. At present, there has been one commercial NCA product. The energy density of Panasonic's NCR18650A NCA battery can reach >250 Wh/kg (Figure 4.5), and it has been used to power of the TESA EV (Model S and Model X).

Ternary $LiNi_xCo_yMn_{1-x-y}O_2$ ($x > 0$, $y > 0$) material was raised by Liu et al. in 1999 [31]. Benefiting from a synergetic effect, the electrochemical performance of the ternary material is better than any single element. Co can reduce ion mixing, stabilize structure, and improve conductivity; Ni offers high capacity; and Mn sustains the layered structure, enhances safety, and cuts down the cost. $LiNi_{1/3}Co_{1/3}Mn_{1/3}O_2$ was studied the earliest and has already been commercialized, and its actual capacity reaches above 200 mAh/g in the voltage range of 3.0–4.5 V (versus Li^+/Li). $LiNi_{0.5}Co_{0.2}Mn_{0.3}O_2$ and $LiNi_{0.4}Co_{0.2}Mn_{0.4}O_2$ are also developed because of their high capacity and low cost.

FIGURE 4.5 Photograph of Panasonic's NCR18650A NCA battery.

Despite the layered cathode materials having competitive advantages to be the next-generation LIB candidates, their electrochemical performance still needs to be improved for practical applications, including issues of high voltage cycling performance, rate ability, and safety. Surface coating and bulk doping are two widely used methods to improve the performance, which are facile and effective. The material surface can be modified by coating other inert materials, such as oxides, fluorides, phosphates, or carbon [32–39]. These materials are chemically stable during the charge–discharge process, which protects active materials from electrolyte erosion or enhances the conductivity. Thereby, the electrochemical performance, especially high voltage cycling and rate capability, can be improved. The coating film should be very thin so that Li^+ and electron can get through easily. The cycling performance of $LiCoO_2$ is poor when the charge voltage is above 4.2 V due to its structure instability; however, after TiO_2 or ZrO_2 coating, the capacity remains above 140 mAh/g after 70 cycles [32,33]. Lee et al. reported that the NCA surface turns to zigzag shape after cycling, while the shape remains well after $Ni_2(PO_4)_3$ coating [40]. Sun et al. coated $LiNi_{1/3}Co_{1/3}Mn_{1/3}O_2$ by AlF_3 to improve cycling and rate capability, and meanwhile the interface impedance did not grow after long-term cycling [41].

On the other hand, doping other elements into layered cathode material can also greatly improve performance. Doping elements include cation Li, Al, In, Ni, Mn, Fe, and anion F, B, etc. [32,42–46]. The key point of doping is to ensure the right position and that the correct element is substituted. General element doping is suitable for any one of $LiCoO_2$, NCA, and ternary material, which can make the structure stable, and improve cycling and rate capability.

As mentioned above, the performance of layered materials can be significantly enhanced by surface modification and doping. Single materials such as $LiCoO_2$, NCA, and $LiNi_{1/3}Co_{1/3}Mn_{1/3}O_2$ may not be qualified, and two or more materials may be used together to solve the issue. Developing these materials undoubtedly brings a bright prospect for high-energy-density LIBs.

4.1.3.2 Lithium-Rich Cathode Material

Lithium-rich manganese-based layered material, denoted as $xLi_2MnO_3{\cdot}(1-x)LiMO_2$ (M is a transition metal such as Ni, Co, Mn; $0 \le x \le 1$), can deliver very high specific capacity and energy density. Because of the large Mn content, this material owns the advantages of low cost, high safety, and being environmentally benign. The Li-rich Mn-based material was discovered when developing Mn-based material as LIB cathode

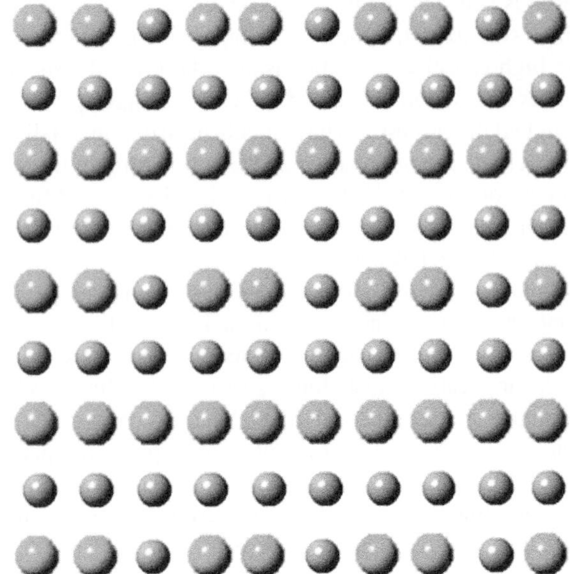

FIGURE 4.6 Schematic image of Li_2MnO_3 phase and $LiMO_2$ phase mixed in atomic scale.

material. In 1999, Thackeray et al. [46] used Li_2MnO_3 to stabilize $LiMO_2$ (M = Ni, Co, Mn) for the first time, and gained a solid solution material. Dahn et al. [47,48] did further research and found that a voltage plateau appeared at 4.5 V when elevating the charge cutoff voltage of the first cycle, which corresponded to electrochemical activation of Li_2MnO_3. A high capacity exceeding 200 mAh/g was achieved in this condition.

Lithium-rich material possesses a two-layered structure: the Li_2MnO_3 phase of the monoclinic system (C/2m) and the $LiMO_2$ of the hexagonal system (R-3m). The (001) crystal plane position of the Li_2MnO_3 phase exactly overlaps the (003) crystal plane position of the $LiMO_2$ phase, owning the same crystal plane distance of 0.47 nm; thus, these two phases can achieve mixed construction in atomic scale, as shown in Figure 4.6. The gray atomic layer is a lithium layer and the green atomic layer is a transition metal layer, while the Li_2MnO_3 layer is composed of both lithium and transition metal atomic layers.

During the charge process of the first cycle, a capacity below 4.4 V belongs to Li extraction from $LiMO_2$, while a capacity above 4.4 V derives from Li_2MnO_3 activation. At a high voltage of 4.55 V, lithium extracts from Li_2MnO_3, and meanwhile the valence of O^{2-} increases. In previous works [49–55], both O_2 and O^{1-} are observed, which confirms the activation of Li_2MnO_3, and this is the origin of the high capacity excess of 250 mAh/g. In the first charge process, two lithium ions extract from Li_2MnO_3, but only one lithium ion reinserts in the discharge process, accompanied by oxygen release. Such mechanism results in a low coulombic efficiency of the first cycle.

Moreover, lithium-rich material also confronts a unique problem of voltage decay. During the charge–discharge cycling, the discharge voltage curve decreases gradually. Such a problem makes it impossible to monitor the voltage–capacity situation, and leads to the uniformity and safety problems of LIBs. Moreover, according to electric energy, $W = U{\cdot}I{\cdot}t$, the output energy density decreases with voltage decay. Wang et al. [56] studied the structural change of $Li_{1.2}Ni_{0.2}Mn_{0.6}O_2$ material before and after electrochemical cycling. A layered–spinel mixed structure was observed after cycling, and its Li intercalation/extraction voltage is lower than that of the layered structure. It is accepted that the cause of voltage decay is as follows: the layered structure is unstable during the charge–discharge process, the transition-metal ion migrates to the lithium vacancy, and the layered phase converts to spinel. The formed layered–spinel combined structure owns a lower voltage plateau, leading to a gradual voltage decrease. The illustration in Figure 4.7 shows a layered to spinel phase transformation process.

As mentioned above, despite Li-rich Mn-based cathode material having a high specific capacity, there are still a few drawbacks for practical applications: low first-cycle coulombic efficiency, an irreversible capacity as high as 40–100 mAh/g, poor rate capability, 1 C capacity <200 mAh/g, poor cycling capability, high charging voltage causing electrolyte decomposition, and voltage decay during cycling. From the current literature, the performance is improved by approaches as follows:

1. Surface coating

   Generally, by coating inert materials such as oxides, fluorides, or phosphates on the surface of Li-rich materials, the reaction of O^{1-} with the electrolyte

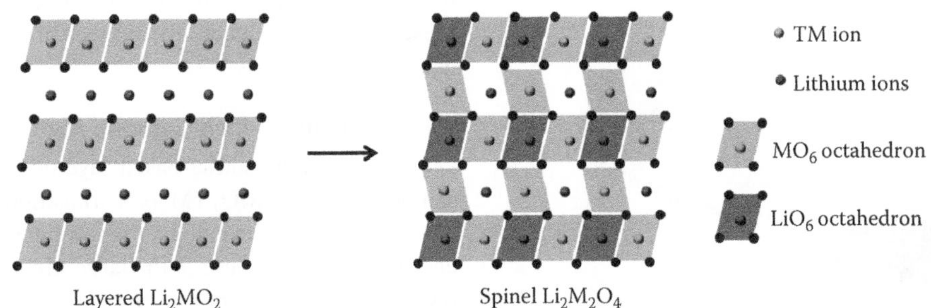

FIGURE 4.7 Illustration of metal ion transition, layered to spinel transformation of lithium-rich material.

can be restrained. Moreover, part of the transition-metal ions may diffuse into the crystal lattice of the base material, which can improve the structural stability of the material. Wu et al. [57] coated the $Li_{1.2}Mn_{0.54}Ni_{0.13}Co_{0.13}O_2$ material by 3% Al_2O_3. The first cycle discharge capacity increased from 253 to 285 mAh/g, and the irreversible capacity decreased from 75 to 41 mAh/g. Also, the cycling stability was improved, and the capacity retention of 50 cycles was enhanced to 94%, while the value was 90% for pristine material.

On the other hand, drawing on the experience of improving the rate capability of $LiFePO_4$ by carbon coating, researches also used carbon to coat the Li-rich material surface. Liu et al. [58] coated a graphite film at the surface of $Li_{1.15}Ni_{0.25}Mn_{0.6}O_2$ material through the vacuum sputtering method, and the first cycle discharge capacity, rate capability, and cycling stability were improved. The 5-C discharge capacity reached as high as 150 mAh/g. Wu et al. [59] prepared $Li_{1.2}Mn_{0.54}Ni_{0.13}Co_{0.13}O_2$ by the co-precipitation method, and then coated a carbon film at the surface. The morphology did not change after coating. The first cycle discharge capacity increased from 250 to 259 mAh/g at 18 mA/g, and the discharge capacity at 180 mA/g was 206.4 mAh/g, which was 8.9% higher than the value before carbon coating.

2. Combination with receptor-type cathode material

   To solve the problem of high irreversible capacity of Li-rich material, a new way of combining with receptor-type cathode material was developed, such as V_2O_5, $Li_4Mn_5O_{12}$, and LiV_3O_8 [60,61]. A large amount of Li ions irreversibly extracted from Li-rich material can reinsert into these materials. By tuning the ratio of Li-rich material and receptor-type material, the first cycle coulombic efficiency of composite material can reach as high as 100%. Taking $Li_{1.2}Mn_{0.54}Ni_{0.13}Co_{0.13}O_2$–$V_2O_5$ composite material as an example, when the V_2O_5 percentage is 25%, the initial discharge capacity reached 300 mAh/g. However, the cycling ability of V_2O_5, $Li_4Mn_5O_{12}$, and LiV_3O_8 is not good, so the cycling performance of the composite material is worse than that of pristine material.

3. Surface modification

   By treating Li-rich material in aqueous or gaseous phase solution of acid or oxidizing reagent, Li and O can be extracted from Li_2MnO_3 in advance, which is much beneficial for first cycle coulombic efficiency and rate capability. Xia et al. treated Li-rich material by $Na_2S_2O_8$ aqueous solution [62], and the surface structure is controlled and the electrochemical performance is enhanced. Li^+ in the Li_2MnO_3 region can be successfully extracted by $Na_2S_2O_8$ aqueous solution and the corresponding lattice oxygen is oxidized to formal O_2^{2-} species and retained in the Li_2MnO_3 region, other than being oxidized to O_2 and released from particle surface. After the treated materials are annealed at elevated temperatures, the generated formal O_2^{2-} species disappears, resulting in a spinel layer forming at the particle surface, as shown in Figure 4.8. Johnson et al. [63] and Kim et al. [64] initiated the acid treatment method. By immersing $0.5Li_2MnO_3{\cdot}0.5LiNi_{0.44}Co_{0.25}Mn_{0.31}O_2$ material in 0.1 M nitric acid, Li_2O could be preextracted. The initial coulombic efficiency was improved from 78% to 95%. However, the cycling performance became worse. Abouimrane et al. [65] raised a way to modify the valence state of transition metal ion in Li-rich material surface. By calcining the $Li_{1.12}Mn_{0.55}Ni_{0.145}Co_{0.1}O_2$ material in H_2(3.5%)/Ar gas, some Mn^{4+} ions were reduced to Mn^{3+}, forming a structure of one element with two valence states in an octahedron, which is beneficial for ion migration. The rate capability was improved by this method, exhibiting 196 mAh/g at 320 mA/g.

4. Preparation in special structure

   Materials with a special structure, such as nanowire, nanoplate, or core–shell structure, may have unique properties and show electrochemical performance beyond expectation. Wei et al. [66] reported that rate ability can be improved significantly by controlling the growing direction of nanomaterial. The 6-C discharge capacity can reach as high as 197 mAh/g. They ascribed the good rate performance to outstanding growing orientation. When Li ions insert/de-insert Li-rich material crystal, the channel with the lowest energy is perpendicular to the *c*-axis.

5. Tuning the cutoff voltage

   The electrochemical properties of Li-rich material relate closely to the charge–discharge condition. Thus, it is an important and effective way to enhance the material performance by tuning the cutoff voltage, including the upper limit and the lower limit. For example, in high upper limit voltage, more capacity can be released, but the cycling performance may become poor due to a severe charge–discharge condition. Ito et al. [67] studied the morphology change of $Li_{1.2}Ni_{0.17}Co_{0.07}Mn_{0.56}O_2$ material before and after electrochemical cycling. Small cracks appeared at the particle surface from the initial cycle, which may probably be caused by an oxygen loss process. They applied a method of charging the material in low cutoff voltages (2.0–4.5 V, 2.0–4.6 V, and 2.0–4.7 V) for two cycles, then mixing the charge–discharge voltage range at 2.0–4.8 V. By this method, the capacity retention of 50 cycles was enhanced from 68% to 90%.

In summary, the Li-rich Mn-based material shows good application prospects, and is expected as a key material for the next-generation high-capacity LIB. However, for large-scale applications, this material still needs a lot of research. For example, in the reaction mechanism aspect, during the first

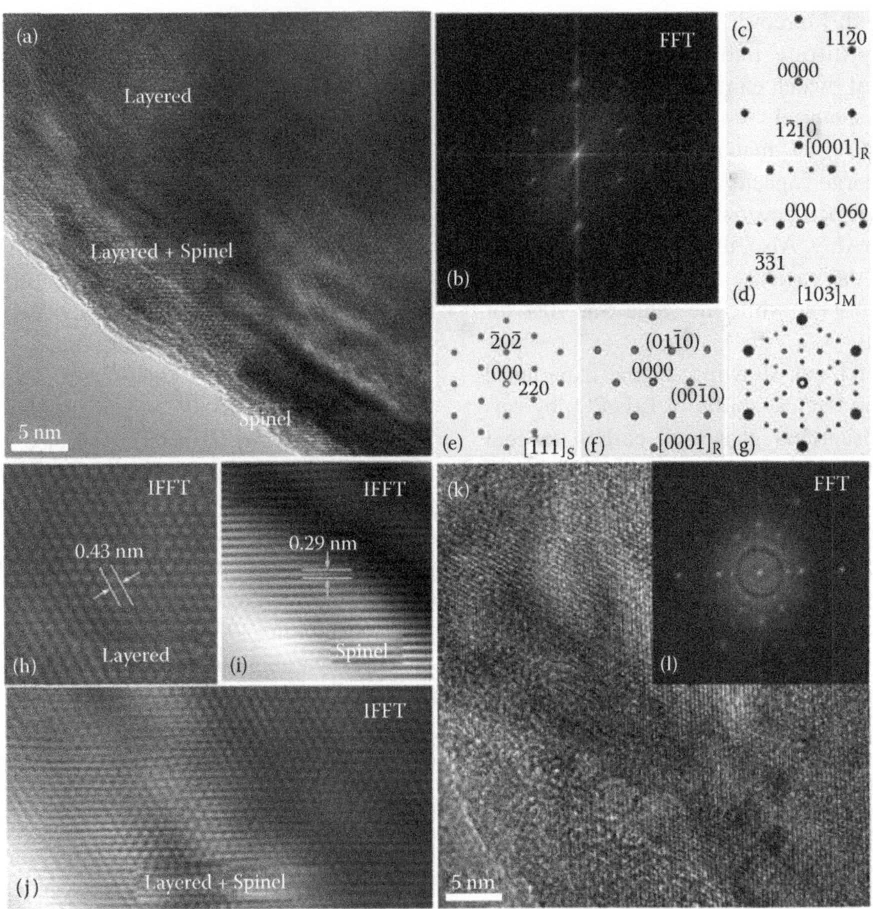

FIGURE 4.8 (a) High-resolution transmission electron microscopic (HRTEM) identification of annealed samples. (b) The fast flourier transformation (FFT) to Panel a. (c) Simulated selected area electron diffraction (SAED) pattern of rhombohedral phase along [0001] zone axis. (d) Simulated SAED pattern of monoclinic phase along (103) zone axis. (e) Simulated SAED pattern of cubic spinel phase along [−111] zone axis. (f) Simulated SAED pattern of forbidden {10–10} reflection along [0001] zone axis. (g) The superimposition of the above simulated SAED patterns. (h), (i), and (j) are inverse FFT (IFFT) images providing lattice fringes for layered and spinel phases. (l) The FFT to Panel k. (From Han, S.J., Qiu, B., Wei, Z., Xia, Y.G., and Liu, Z.P., *J. Power Sources*, 268, 683, 2014.)

charging process, even taking both the oxygen loss theory and the proton-exchange theory, there are many issues to be explained as to the 4.5-V platform, such as the whereabouts of oxygen, whether part of oxygen participates in the subsequent charge–discharge cycle, and whether O^{1-} or Mn^{5+} exist in the electrochemical process. The proton in the proton-exchange theory stems from the oxidation and decomposition of electrolyte; however, it has not been reported how much impact this process will make on the following cycle performance and safety concern. Does the proton-exchange phenomenon only exist in Li-rich material, or does it also exist in other layered materials? Moreover, do phase transformation and the voltage decay phenomenon only exist in Li-rich material, or do other layered materials also confront this issue?

Most current theoretical studies have focused on the first charging process, while there are not enough researches and expositions on the subsequent cycles, which is more important for application use. On the basis of the current theory, after lithium extraction in the initial charge process, the oxide framework ($MnO_2 \cdot MO_2$) is formed, and how does O^{1-}exist steadily in the structure? When charging other layered materials to high voltage to form deep Li intercalation/extraction, such as $LiMO_2$ (M = Co, Ni, Mn, $Ni_{0.5}Mn_{0.5}$, and $Ni_{1/3}Co_{1/3}Mn_{1/3}$, etc.), whether they have the same oxygen loss mechanism or not is unknown. Meanwhile, the effect of efforts to improve the rate performance of the material is not realistic, and the lack of research on low-temperature performance is obvious.

For these unresolved scientific issues, well-designed experimental schemes and accurate laboratory facilities are needed, as well as understanding the electrochemical mechanism thoroughly and developing appropriate chemical/physical methods based on the mechanism. To develop the Li-rich Mn-based material as the next-generation LIB cathode material, efforts from both the aspects of basic theory and practical applications are needed.

4.1.4 Olivine Cathode Materials for LIBs

4.1.4.1 $LiFePO_4$ Cathode Materials

In recent years, the phospho-olivines $LiMPO_4$ (where M = Fe, Mn, Co, Ni) are now recognized as attractive alternatives. In spite of the same olivine structure for all the samples, the potential is found to be, respectively, 3.4 [68], 4.0 [69], 4.8 [70], and 5.2 V [71]. The theoretical energy density in Wh/g of all phospho-olivines is even larger than the $LiCoO_2$ and $LiMn_2O_4$.

In these systems, the overwhelming advantage of the Fe-based compound $LiFePO_4$ has been a focus for its inexpensiveness and natural abundance, as well as less toxicity than Co, Ni, and Mn. This is the first time an Fe-based compound as a promising cathode material was studied and reported. It is well known that Li_xFeO_2 is rather unstable and the ratio of ionic radii is not suitable for the semiempirical tolerance criterion [72]. To overcome this problem, large polyanions (XO_4) (X = S, Mo, P, W, Si etc.) with a stable structure were studied by Goodenough's group [68,73–79]. The strong X–O covalence supports a relatively high voltage. For example, the open-circuit voltage versus lithium is 3.6 V for $LiFe_2(SO_4)_3$, 2.8 V for $Li_3Fe_2(PO_4)_3$ and Li_2FeSiO_4, etc. Moreover, each sample shows a large capacity achieved over 100 mAh/g. Among these Fe-based compounds, $LiFePO_4$ exhibits the highest energy in the current electrolyte window (≤4.3 versus Li /Li).

The $LiMPO_4$ (M = Fe, Mn, Co, Ni) compounds have the ordered olivine structure (shown in Figure 4.9) [80]. In fact, olivine is referred to one large family of compounds, M_2XO_4, which has the common generalized AB_2O_4 formula as the spinel, and consists of a hexagonal close-packed (hcp) oxygen array, in which X atoms are located in one-eighth of tetrahedral sites. On the other hand, M atoms are located in half of the octahedral sites (denoted M1 and M2). The MO_6 octahedrals are linked through shared corners in the *bc*-plane The LiO_6 octahedrals form edge-sharing chains along the *b*-axis. One MO_6 octahedron shares edges with two LiO_6 octahedral and a PO_6 tetrahedron. PO_4 groups share one edge with an MO_6 octahedron and two edges with the LiO_6 octahedral.

It is generally accepted that $LiCoO_2$ possesses a two-dimensional (2D) diffusion plane, while $LiMn_2O_4$ possesses 3D

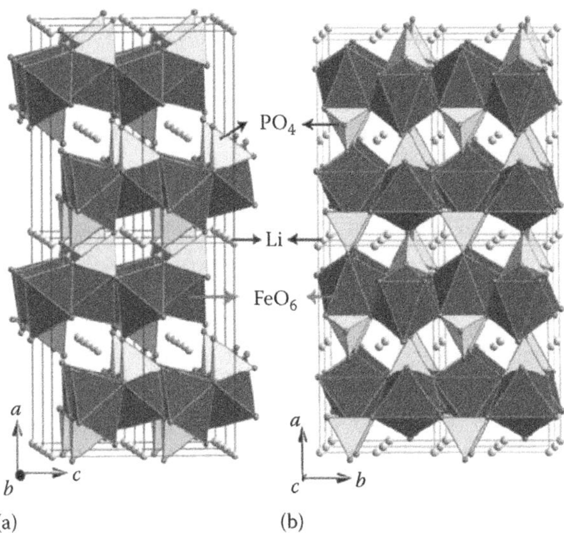

FIGURE 4.9 Crystal structure of $LiFePO_4$ cathode materials. Olivine $LiFePO_4$ viewed along the (a) *b*- and (b) *c*-axis. Tetrahedral, PO_4; octahedral, FeO_6. (From Gong, Z.L. and Yang, Y., *Energy Environ. Sci.*, 4, 3223, 2011.)

channels. For $LiFePO_4$, Goodenough et al. [68] suggested firstly that it would be analogous to dimension extraction/insertion in the $LiMO_2$ layer oxides with M = Co, Ni; this crystal structure provides two spatial diffusion tunnels, running parallel to the *c*-axis and *a*-axis, respectively. However, recent researches have shown that the diffusion in $LiFePO_4$ is on dimension because the other migration pathways have much high-energy barriers resulting very low possibility of lithium-ion migration by first-principles calculation [81]. The theoretical studies of the mean square displacements also have shown that lithium ions can move easily only diffusely along the *b*-direction.

The electronic conductivity of $LiFePO_4$ is lower than that of the lithium storage cathode recently used, such as $LiMn_2O_4$ (~10^{-4}), $LiCO_2$ (~10^{-3}), and $LiNiO_2$ (~10^{-2}) [82–84] by factor of more than 10^5, only exhibiting values of 10^{-9}–10^{-10} at room temperature [85]. Thackeray et al. [86] suggest that the low electronic conductivity of $LiFePO_4$ may be due to the strongly distorted continuous network of FeO_6. $LiMnPO_4$, with the same structure, also shows low electronic conductivity (10^{-10}–10^{-11}), which is even much worse than that of $LiFePO_4$.

As mentioned above, low electric conductivity and lithium ion diffusion coefficient have been the main problems when $LiFePO_4$ is studied as promising cathode materials, which leads to its lower discharge capacity, high polarization, and poor rate capability. Many attempts such as particle-size minimization, carbon coating, and foreign metal doping have been developed to improve the electrochemical performance of the $LiFePO_4$ cathode.

I. Particle size minimization

To reduce the diffusion distance of lithium ion within particles and to increase the contact area between active materials and electrolyte, particle-size minimization is an effective way to improve the rate capability of $LiFePO_4$. Until now, many synthesis methods have been developed to prepare the cathode material with nanoparticles, such as solid-state reactions [87], hydrothermal method [88], co-precipitation [89], sol–gel method [90], and so on. Here, two kinds of main synthetic methods are introduced.

1. The solid-state method is a simple and conventional method to synthesize $LiFePO_4$ cathode materials. This synthetic method has been widely adopted by many large materials manufacturers (BYD Co. Ltd., Aleees Co. Ltd., Valence Co. Ltd., etc.). The typical solid-state reaction route for $LiFePO_4/C$ cathode materials is shown in Figure 4.10. The mixtures of all starting materials are grinded extensively. Then, the slurry is fed into a spray-drying machine to produce a homogenous precursor. Finally, the precursor is sintered at 600–700°C under an N_2 gas flow. In this procedure, wet high-energy ball milling is required for nanoscale homogenizing of slurry. The primary particle size of the final $LiFeO_4$ product depends on the size of the precursor and the calcination temperature.

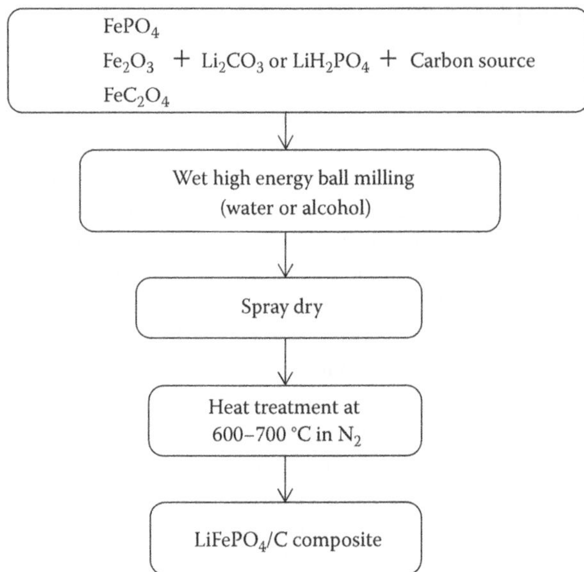

FIGURE 4.10 Typical solid-state reaction route for $LiFePO_4$/C cathode materials.

In general, the primary particle size of $LiFeO_4$ cathode materials can be controlled from 100 to 300 nm by the solid-state method.

2. The hydrothermal method is also a promising one for designing ideal $LiFePO_4$ nanomaterials because of its low temperature, short time, high yield, and scalable process. Compared with the conventional solid-state method, the hydrothermal method could obtain smaller nanoparticles of < 100 nm. Moreover, the morphology of the $LiFePO_4$ final product is easily controlled by adding various kinds of surfactants and adjusting the pH value of the reaction suspension. This synthetic method has been successfully industrialized practically by Osaka Cement Co. Ltd. Ou et al. [88] systematically investigated the effect of hydrothermal temperature (120°C, 140°C, 150°C, 160°C, and 175°C) on the crystallinity and morphology of $LiFePO_4$. Scanning electron microscopy (SEM) shows that the crystal morphology develops from a diamond plate to a polygonal plate and the thickness of the flakelet decreases from 130–150 nm to 80–90 nm with increasing the temperature (Figure 4.11a–d). Among these samples, the sample prepared at 160 (denoted as LFP-60) exhibits the largest discharge capacity of 161 mAh/g (Figure 4.11f). It is believed that, to prepare $LiFePO_4$ with good electrochemical performance and prevent the generation of thick particles in the early stage of the reaction, a fast heating apparatus should be used and the hydrothermal reaction temperature should be set above 160°C.

II. Carbon coating

Carbon coating has been generally introduced to prepare $LiFePO_4$/C composite with excellent electrochemical performance. The carbon layer not only can improve the surface electronic conductivity but also can prevent the growth of $LiFePO_4$ particles at high annealing temperatures [72]. In addition, carbon can act as a reducing agent to allow reduction of Fe^{3+} to Fe^{2+} during sintering; thus, cheap raw material (Fe_2O_3, $FePO_4$, etc.) can be used to prepare $LiFePO_4$ materials. Although the surface electronic conductivity increases with the increase of carbon content, a large amount of carbon will decrease the energy density of material significantly owing to its inactive and bulky character. The tap density and rate capability is a very important factor for practical application in the cathode materials of LIBs. Therefore, it is important for $LiFePO_4$ to carefully note the effect of carbon coating on capacity, rate capability, and tap density [91].

To minimize the amount of carbon, a uniform carbon layer and a high-electronic-conductivity carbon are necessary. A uniform carbon layer on $LiFePO_4$ can be prepared by *in situ* carbonization of an organic precursor. The electronic conductivity of the carbon coating layer strongly depends on the calcination temperature and the precursor. Carbon coatings synthesized at high temperatures (>700°C) have higher electronic conductivity than those synthesized at low temperatures (<600°C). However, the particle size of $LiFePO_4$ materials easily grows when samples are synthesized at high temperatures of >700, which could lead to decrease in the electrochemical performance. Furthermore, as Chen and Dahn have already reported [91], small particle size is a more important factor than carbon coating itself. Recently, graphene as a highly conductive carbon has been used to enhance the electronic conductivity of $LiFePO_4$ materials by many groups [92–94]. As shown in Figure 4.12a and b, the graphene sheets have a relatively regular stacking state in LFP/G, but glucose-derived carbon coating layers have no layered structure in LFP/C. XPS and Raman tests further prove that the thermally reduced graphene has a higher degree of graphitization than the glucose-derived carbon (Figure 4.12c,d). A higher degree of graphitization results in higher composite conductivities, and thus higher rate capabilities of composites (Figure 4.12e). Moreover, at 10 C charge rate and 20 C discharge rate, the LFP/C can deliver a discharge capacity of 90 mAh/g as shown in Figure 4.12f; after 1000 cycles, the capacity retention is about 70% of the initial capacity. In contrast, both LFP/G and LFP/(G + C) show much better cycling stability at the high rates [94].

III. Foreign metal doping

Chiang et al. [95] first reported on supervalence ion doping in the Li-site to enhance the intrinsic electronic conductivity of $LiFePO_4$ by a factor of ~10^8,

FIGURE 4.11 SEM images of $LiFePO_4$ synthesized by hydrothermal method at (a) 120°C, (b) 140°C, (c) 150°C, (d) 160°C, and (e) 175°C. (f) Charge and discharge curves of LFP/C samples measured at 0.1 C. (From Ou, X., Pan, L., Gu, H., Wu, Y., Lu, J., *J. Mater. Chem.*, 22, 9064, 2012.)

and this result became the focus of worldwide attention. However, controversial results were communicated by another group. Ravet et al. [96] suggested that the enhanced conductivity may come from the residual carbon from the supervalence metal organic precursor. Up to now, no general consensus on the effectiveness of supervalence ion doping has yet been reached.

Two-valence ion doping into Fe-site can also be an effective way of increasing the electronic conductivity. Wang et al. [97] reported that the electronic conductivities of Fe-site-doped samples are almost two orders of magnitude higher than that of undoped sample, reaching about 10^{-7} S/cm. Similar work was also reported by Ni et al. They showed that the Mg-doped $LiFePO_4$ samples display comparable conductivity of 10^{-4} S/cm to the undoped one [98]. As reported thus far in the literature, it seems that Li-site doping for the increase of electronic conductivity is more effective than Fe-site doping, but Li-site doping will block the 1D tunnel of lithium diffusion. Consequently, although foreign metal doping could enhance electronic conductivity to a certain degree, the ionic conduction may be affected adversely if the foreign metal is doped into the Li-site.

4.1.4.2 $LiMnPO_4$ Cathode Materials

Although $LiFePO_4$ has excellent cycling performance and was widely used in lithium ion batteries, the problem is that the potential of the material is only about 3.4 V versus Li/Li^+. Compared with $LiFePO_4$, $LiMnPO_4$ is of particular interest, of Mn^{3+}/Mn^{2+} of 4.1 V versus Li/Li+, which is compatible with conventional nonaqueous organic electrolytes of commercialized lithium-ion batteries. Therefore, $LiMnPO_4$ cathode material has become a hot research topic in recent years.

In 1997, Goodenough et al. first investigated and showed that hardly any lithium ion can be extracted from $LiMnPO_4$ [68]. Yamada et al. [99] suggested that $LiMnPO_4$ is an insulator with a 2-eV band gap, where $LiFePO_4$ is a

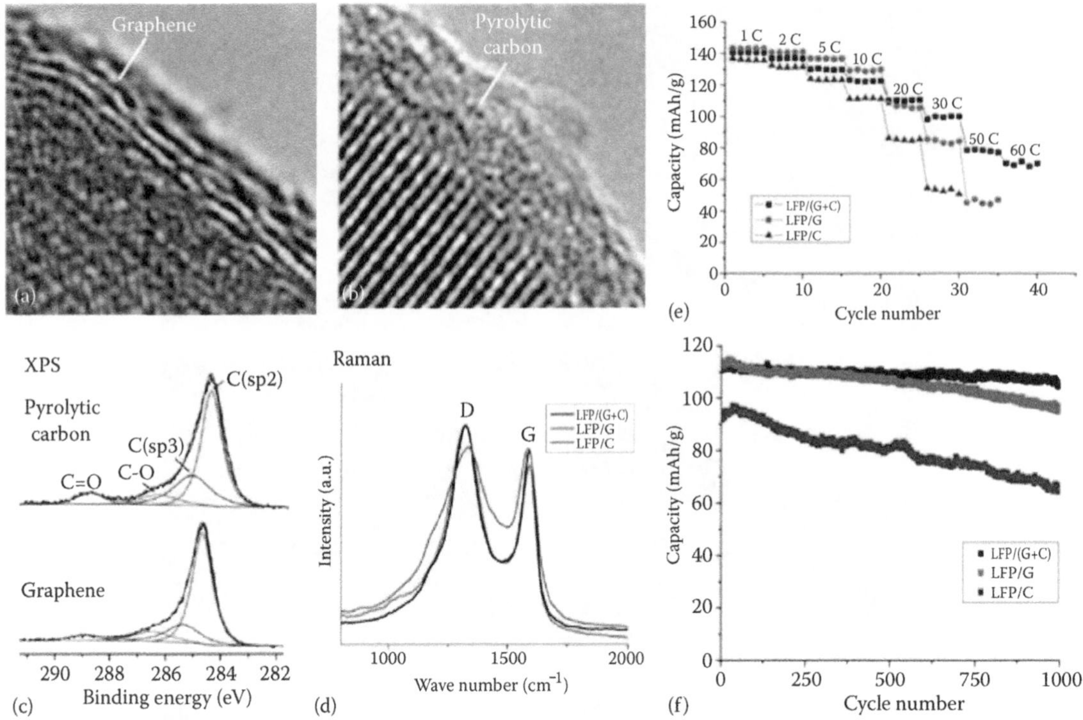

FIGURE 4.12 (a) HRTEM images focusing on the surface of individual LFP nanoparticles in LFP/G. (b) HRTEM images focusing on the surface of individual LFP nanoparticles in LFP/C. (c) Raman spectra of LFP/C, LFP/(G+C), and LFP/G. (d) XPS spectra of LFP/C, LFP/(G+C), and LFP/G. (e) Comparison of rate capability of LFP/C, LFP/(G+C), and LFP/G. (f) Comparative cycling performances of LFP/C, LFP/(G+C), and LFP/G operated under 10 C charging and 20 C discharging. (From Zhou, X., Wang, F., Zhu, Y., and Liu, Z., *J. Mater. Chem.*, 21, 3353, 2011.)

semiconductor with ca. 0.3-eV band gap by their first-principle calculations. In addition, the lithium-ion diffusivity of $LiMnPO_4$ is about 10^{-16}, also much worse than that of $LiFePO_4$ (10^{-14}). Both poor electronic conductivity and low lithium-ion diffusivity of $LiMnPO_4$ cathode materials lead to its very low discharge capacity and poor rate capability. How to use the appropriate modification methods to improve the electrochemical performance of $LiMnPO_4$ has been a research focus in the study field. At present, the performance of $LiMnPO_4$ can be improved by doping, reducing particle size and controlling the morphology, and inhibiting the dissolution of manganese.

Ion doping or substitution is an effective way to improve the electrochemical performance of the $LiMnPO_4$ cathode material. Doping other elements to form a solid solution can restrain the Jahn–Teller distortion, and can affect the grain size and internal transport properties of $LiMnPO_4$. Currently, the doping elements commonly used are Fe, Co, Ni, Cu, Zn, Mg, Ti, V, etc. [100–105]. Among these doping elements, Fe is the most effective element to enhance the performance of $LiMnPO_4$. $LiMnPO_4$ and $LiFePO_4$ have the same olive structure; Fe and Mn can form $LiFe_xMn_{1-x}PO_4$ solid solution with any ratio. As shown in Figure 4.13, Martha et al. prepared carbon-coated nanoparticles of $LiMn_{0.8}Fe_{0.2}PO_4$ by ball milling [100]. Its particle size is about 25–60 nm in diameter with a 5-nm-thick carbon film. C–$LiMn_{0.8}Fe_{0.2}PO_4$ shows the initial discharge capacity of 160–165 mAh/g (close to the theoretical value of 170 mAh/g) at low rates, and the discharge capacity is more than 90 mAh/g even at 10 C.

Another element that has been shown to improve the performance of $LiMnPO_4$ is vanadium. Vanadium can be doped into the host lattice of $LiMnPO_4$ or form composites. Xia et al. [104] studied the effect of V additive on the electrochemical performance of $LiMnPO_4$. A series of $(1-x)LiMnPO_4{\cdot}xLi_3V_2(PO_4)_3$ were synthesized through the solid-state method. The results show partial mutual doping between the LMP and LVP phases. The $x = 0.4$ sample displays the best electrochemical performance with the discharge specific capacity of 154 mAh/g at 0.05 C, and >95 mAh/g at 0.5 C after 50 cycles (Figure 4.14). The outstanding performance can be ascribed to the existence of the open 3D framework of the electronically conductive LVP phase and the doping of V into the crystal structure of LMP. Similar results were reported by Su et al. [105]. The rational cation substitution and the proper doping content are beneficial to improve the electrochemical performance of $LiMnPO_4$.

It is well known that a decrease in the particle size could shorten the Li-ion and electron transport length, and may ease the strain between the two-phase $LiMnPO_4$ and $MnPO_4$. Then, the electrode performance of $LiMnPO_4$ can be improved. Moreover, morphology control is important to enhance the electrochemical performance of $LiMnPO_4$, since the Li-ion diffusion in the olive structured $LiMnPO_4$ is 1D and restricted to the [010] direction. The oriented growth

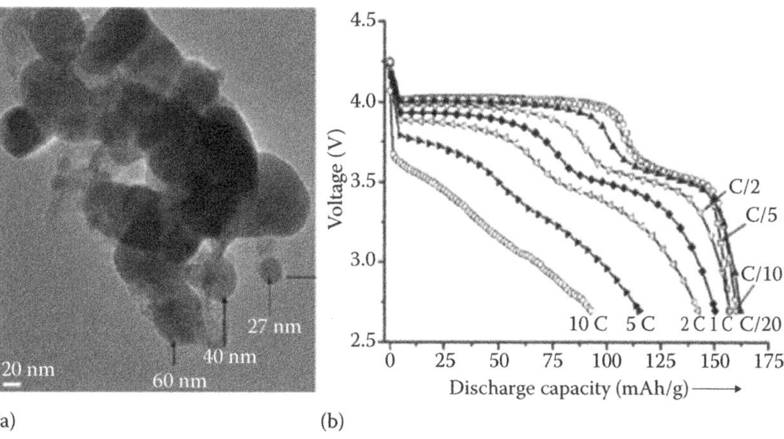

FIGURE 4.13 (a) TEM images and (b) discharge curves at various rates of C-$LiMn_{0.8}Fe_{0.2}PO_4$. (From Hu, L., Qiu, B., Xia, Y., Qin, Z., Qin, L., Zhou, X., and Liu, Z., *J. Power Sources*, 248, 264, 2014.)

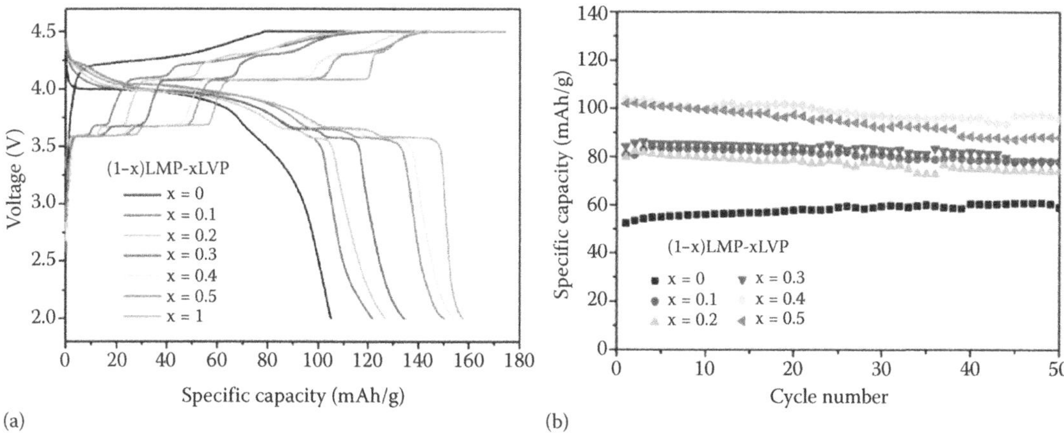

FIGURE 4.14 (a) First charge–discharge curves; (b) cycling performances of the samples. (From Su, L., Li, X., Ming, H., Adkins, J., Liu, M., Zhou, Q., and Zheng, J., *J. Solid State Electrochem.*, 18, 755, 2014.)

of LMP nanocrystals has gained much attention in recent years. Qin et al. [106] successfully synthesized morphology-controlled monodispersed $LiMnPO_4$ nanocrystals by a solvothermal method in a mixed solvent of water and PEG400 (Figure 4.15). By increasing the pH value of the reaction suspension, $LiMnPO_4$ nanoparticles from a nanorod to a thick nanoplate (~50 nm in thickness) and to a smaller thin nanoplate (20–30 nm in thickness) were observed. Electrochemical measurements confirm that the $LiMnPO_4$ thin nanoplates display the best charge–discharge performance, thick nanoplates the intermediate, and nanorods the worst, which can be mainly ascribed to the difference in their morphologies and particle sizes in three dimensions. $LiMnPO_4$ thin nanoplates, whose dimension along the *b*-direction is smaller than those of thick nanoplates and nanorods, hence have shorter diffusion channels for Li ions. Meanwhile, the smaller the particle size is in the *bc*-plane, the smaller the probability is of finding defects and the faster the boundary displacement is.

Similar to the $LiMn_2O_4$, $LiMnPO_4$ cathode materials also face the problem of Mn dissolution, especially at elevated temperatures. Consequently, reduction of Mn dissolution is a very important factor for its practical application. Surface treatment of $LiMnPO_4$ can reduce the manganese dissolution. Zaghib et al. [107] reported on $LiMnPO_4$ encapsulated with the active material $LiFePO_4$ film (Figure 4.16). In addition, the $LiFePO_4$ shell allows for the coating of the composite while it is difficult to deposit the carbon at the surface of $LiMnPO_4$. More important, the $LiFePO_4$ layer may avoid side reactions with the electrolyte, since the electrochemical properties do not degrade upon cycling. The results indicate that electrochemical properties of the carbon-coated $LiFePO_4$–$LiMnPO_4$ composite are improved in terms of power density, aging upon cycling, and capacity with respect to the carbon-coated $LiMn_{2/3}Fe_{1/3}PO_4$ solid solution with comparable Fe/Mn ratio. Moreover, isovalent co-doping can decrease Mn dissolution in the electrolyte. Balaya et al. [108] investigated the doping of Fe^{2+} and/or Mg^{2+} in an $LiMnPO_4$ cathode material to enhance its lithium storage performance. The tests show that co-doped $LiMn_{0.9}Fe_{0.05}Mg_{0.05}PO_4/C$ exhibits better storage performance than $LiMn_{0.9}Fe_{0.1}PO_4/C$ and $LiMn_{0.9}Mg_{0.1}PO_4/C$ with the capacity of 159 mAh/g at 0.1 C and 96% of its initial capacity at 1 C after 200 cycles. This is attributed to the suppressed Mn dissolution in the electrolyte compared with the other samples.

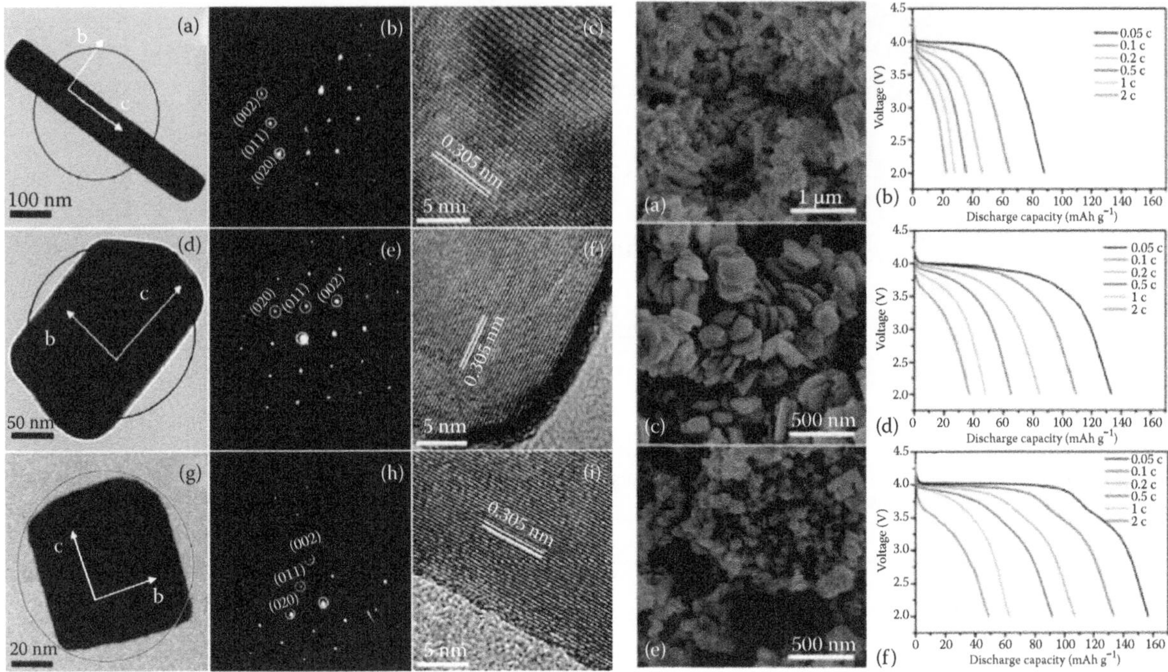

FIGURE 4.15 TEM images, SAED patterns, SEM images, and discharge curves of $LiMnPO_4$ nanorods (left: a–c; right: a and b), $LiMnPO_4$ thick nanoplates (left: d–f; right: c and d), and $LiMnPO_4$ thin nanoplates (left: g–i; right: e and f). (From Zaghib, K., Trudeaua, M., Guerfi, A., Trottier, J., Mauger, A., Veillette, R., and Julien, C.M., *J. Power Sources*, 204, 177, 2012.)

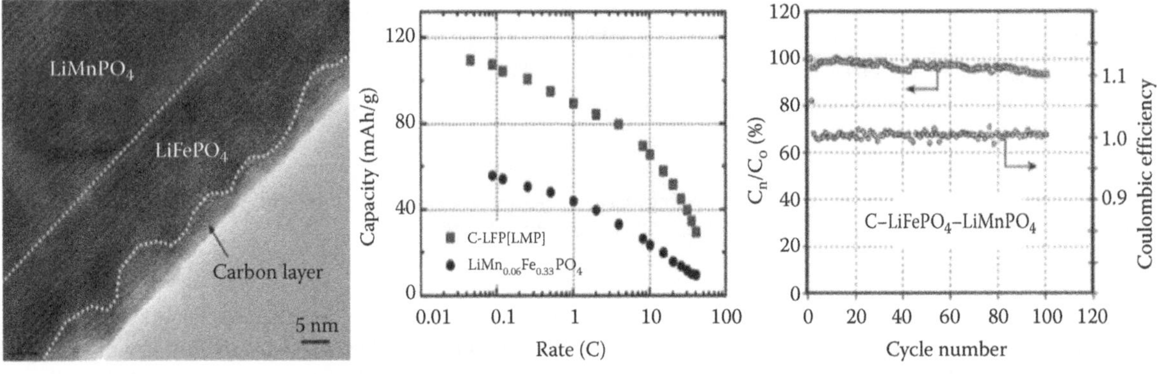

FIGURE 4.16 TEM image of C–$LiFePO_4$–$LiMnPO_4$ and the electrochemical performance of $LiFePO_4$–$LiMnPO_4$ and $LiMn_{2/3}Fe_{1/3}PO_4$. (From Gummow, R.J., Kock, A., and de Thackeray, M., *Solid State Ionics*, 69, 59, 1994.)

4.1.5 Spinel Cathode Materials for LIBs

4.1.5.1 $LiMn_2O_4$ Spinel

Sine spinel-type lithium manganese oxide is a material having 3D lithium-ion passages, and has the advantages of low cost, environmental friendliness, high safety, etc.; it is suitable for use in large-scale LIBs for EVs and energy storage devices for smart grid, solar, and wind stations. However, the problem of capacity fading limits further application of Mn–spinel, especially at elevated temperatures. Several reasons have been suggested to explain its degraded cycling performance, including Jahn–Teller distortion, oxygen deficiency, Mn dissolution into the electrolyte, etc. [109–111].

Oxygen deficiency is a key factor in controlling the cycle performance of Mn–spinel cathode for LIBs [112]. In the typical charge–discharge curve for oxygen-defect spinel, $LiMn_2O_{4-\delta}$ (as shown in Figure 4.17), a 3.2-V discharge plateau and large irreversible capacity usually appears.

The oxygen-defect spinel could be considered as a Schottoky-type oxygen defect (Figure 4.18) [112]. One oxygen vacancy gives rise to three Mn ions with the coordination number of 5 (MnO_5), which links with three MnO_6 octahedrons across O atom. Therefore, a 1 mol defect oxygen forms 12 Mn atoms (three sets of MnO_5–$(MnO_6)_3$), which contribute to the two equivalent plateaus at 3.2 and 4.5 V. These capacities are expressed as C3.2 V and C4.5 V, respectively. Assume that the molar ratio of $Mn^{3+}/(Mn^{3+} + Mn^{4+})$ is 0.5 and 1 mol Mn^{3+} in $LiMn_2O_4$ delivers 148 mAh/g capacity, we could obtain the following equation for the capacity of 12δ mol Mn in $LiMn_2O_{4-\delta}$: C3.2 V + C4.5 V = 148 × 12δ × 0.5.

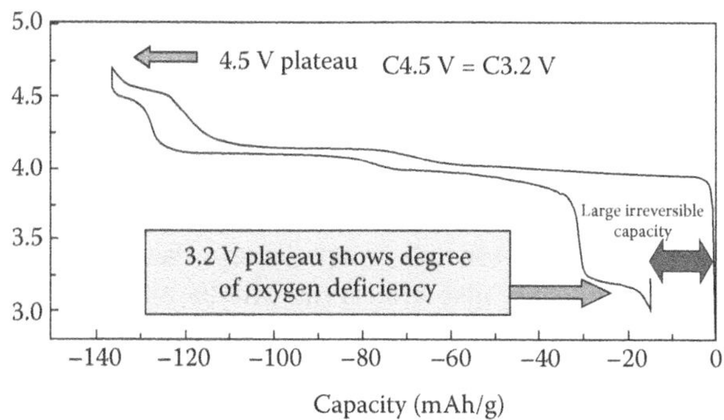

FIGURE 4.17 Typical charge–discharge curve for oxygen defect spinel, $LiMn2O_{4-\delta}$. (From Xia, Y., Zhang, Q., Wang, H., Nakamura, H., Noguchi, H., and Yoshio, M., *Electrochim. Acta*, 52, 4708, 2007.)

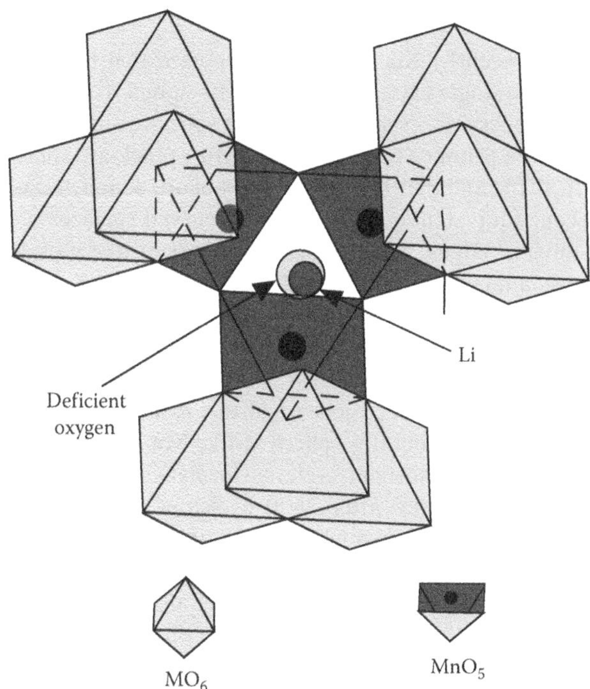

FIGURE 4.18 Structure of oxygen-deficient spinel. (From Xia, Y., Zhang, Q., Wang, H., Nakamura, H., Noguchi, H., and Yoshio, M., *Electrochim. Acta*, 52, 4708, 2007.)

Since C3.2 V equals to C4.5 V, C3.2 V could be expressed as Equation 4.1

$$C3.2\ V = 148 \times 12\delta \times 0.5 \times 0.5 = 444\delta \quad (4.1)$$

The relation between cycling performance and oxygen defect degree for Li–Mn–O spinel system can be seen in Figure 4.19 [112], where the capacity retention after the first 50 cycles along with various oxygen-defect degrees δ were described. All the points are distributed around a straight line. The extrapolation of this line to δ = 0 could give the crossover point of 100%. Therefore, it indicated that the most important factor for improving cycling performance is oxygen stoichiometry in spinel cathode material when the spinel cathode was cycled at room temperature.

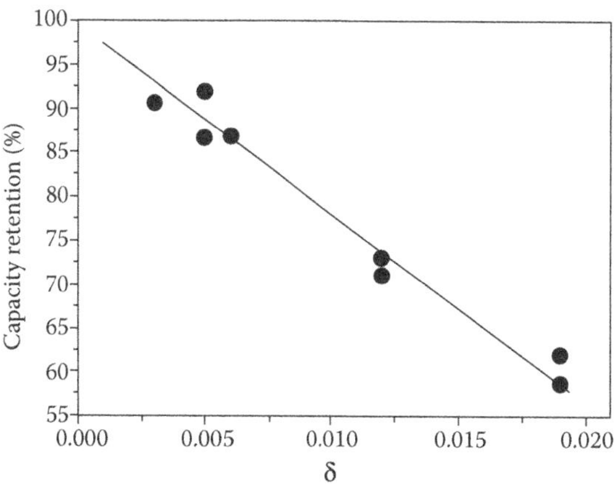

FIGURE 4.19 Dependence of capacity retention after first 50 cycles and oxygen deficiency degree (δ) at room temperature. (From Xia, Y., Zhang, Q.,Wang, H., Nakamura, H., Noguchi, H., and Yoshio, M., *Electrochim. Acta*, 52, 4708, 2007.)

Foreign metal doping plays a role in helping the formation of oxygen stoichiometric spinels. Xia et al. [113] have carefully investigated the effect of doping of various foreign metals on the initial discharge capacity of the 3.2 V plateau (C3.2 V) for $LiMn_{1.9}M_{0.1}O_{4-\delta}$ (M: Mg, Al, Fe, Cr, Co, Ni, Li, and Mn; δ: oxygen-defect content) spinel samples (Figure 4.20). Their results showed that Mg and Al doping could suppress 3.2-V capacities effectively even at the sintering temperature is as high as 900–1000°C. In addition, to select the suitable doping metal, the valance state of the doped metal ion also needs to be considered because the theoretical capacity of the 4-V region increases with increase in the oxidation number of the doped metal ion in the 16d site of $LiMn_2O_4$ at the same doping level in mole fraction.

Although foreign metal doping can improve the cycling stability of Mn–spinel cathode material at elevated temperatures, it still cannot suppress Mn dissolution from the spinel surface in the electrolyte. Moreover, Mn ions can gradually destroy the solid electrolyte interface (SEI) layer of the anode carbon, which leads to Mn ions in the electrolyte not only causing a capacity fading of the cathode spinel but also

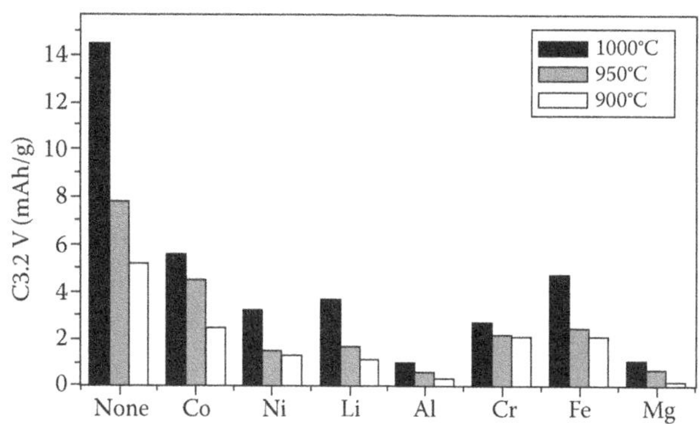

FIGURE 4.20 Initial discharge capacity of the 3.2 V plateau (C3.2 V) for various metal (M) doped spinel samples ($LiM_{0.1}Mn_{1.9}O_4$) prepared at 900–1000°C in air. (From Takahashi, M., Yoshida, T., Ichikawa, A., Kitoh, K., Katsukawa, H., Zhang, Q., and Yoshio, M., *Electrochim. Acta*, 51, 5508, 2006.)

deteriorates the performance of the anode carbon at elevated temperatures. The degradation of the cycle performance becomes more severe. Therefore, it is believed that reduction of Mn dissolution is a very important factor, especially in the full cell. To overcome the problem of Mn dissolution, mainly the following two approaches have been made:

1. High-temperature sintering: Spinel cathode materials with lower surface area are usually prepared at a high temperature (≥900°C). However, such spinels often present oxygen deficiency. For the sake of overcoming this problem, Takahashi et al. reported that annealing at 800°C after 1000°C calcinations can reduce the oxygen-defect content in the Mg-doped spinel, and they showed improved discharge capacity even after storage at 60°C [114]. Xia et al. proposed a two-step method to decrease Mn dissolution and improve cycling performance of spinel cathode materials at elevated temperatures [114]. Two heating processes are used. High-temperature heating in the initial step helps the development of crystal growth and leads to a decrease in specific surface area. The oxygen deficiency is repaired in the second step by heating at relatively a lower temperature (ca. 600–700°C) with excess Li salt.
2. Surface modification: Metal oxide (MgO, Al_2O_3, ZnO, etc.) [115–117] surface modification is an effective method for minimizing the Mn dissolution of spinel $LiMn_2O_4$. However, metal oxide coating material can be easily converted to a corresponding fluoride material by HF from the electrolyte during charge–discharge cycling. Therefore, a growing number of researchers believe that stable metal fluoride (AlF_3, FeF_3, LaF_3, etc.) [118–120] and phosphate compound ($FePO_4$ and $AlPO_4$) [121,122] layers result in less decomposition in the SEI layer and better cycling performance. In addition, coating other lithium compounds ($LixCoO_2$ and $LiNi_{1-x}CoO_2$) [123,124] on the surface of $LiMn_2O_4$ was also developed by some research groups. This kind of structure is also called a core–shell structure. Because $LiCoO_2$ and $LiNi_{1-x}CoO_2$ have a higher capacity than $LiMn_2O_4$, the capacity and cycling stability of the coated $LiMn_2O_4$ are greatly improved.

Recently, many research groups believe that the primary particle morphology also affects the Mn dissolution of Mn–spinel cathode material. Generally, the most common primary particle morphologies are octahedral (Figure 4.21a) and polyhedron (Figure 4.21b). Thackeray et al. supposed that the predicted ordering of surface energies are (111) > (110) > (001) by first-principles density functional theory simulations [125]. Koga et al. confirmed that the degree of corrosion of crystal plane of Mn–spinel was (111) > (110) > (001) [126], which is in good agreement with the calculation result reported by Thackeray.

More recently, Xia et al. [127] reported that they could control the single crystal particle of Mn–spinel from octahedron to spheroidal (Figure 4.22). The spherical-like spinels, namely the primary particles of the Mn–spinel, are spheroidal in shape and the (111) planes thereof are connected with adjacent equivalent crystal planes by curved surfaces without obvious edges. The spinel with spherical-like morphology has a much smaller (111) plane area compared with the single crystal octahedral or polyhedral Mn–spinel products. The decrease of the plane area could reduce the Mn dissolution during the cycling process, which led to better cycling performance. The electrochemical test results showed that 18650 batteries using this spherical-like Mn–spinel cathode material has an excellent cycle performance with a capacity retention of 80% after 2500 cycles at room temperature. Even at an elevated temperature of 60°C, the capacity retention still reached 80% after 400 cycles. Consequently, reducing the (111) planes of Mn–spinel also is a key factor for improving the cycling performance of oxygen stoichiometric Mn–spinel. Spherical-like spinels will provide better cycling performance at high-temperature and high-filling capability.

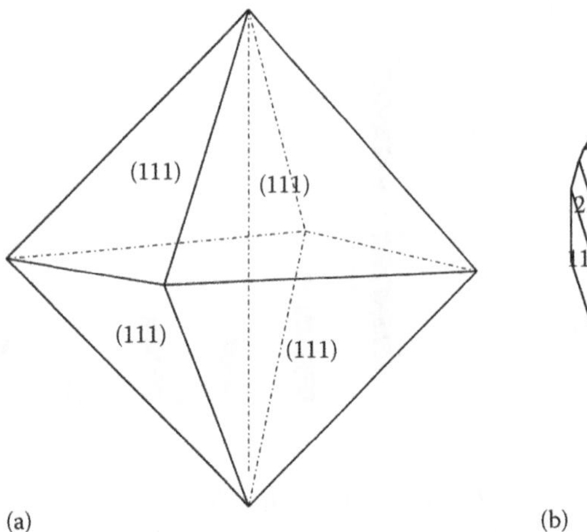

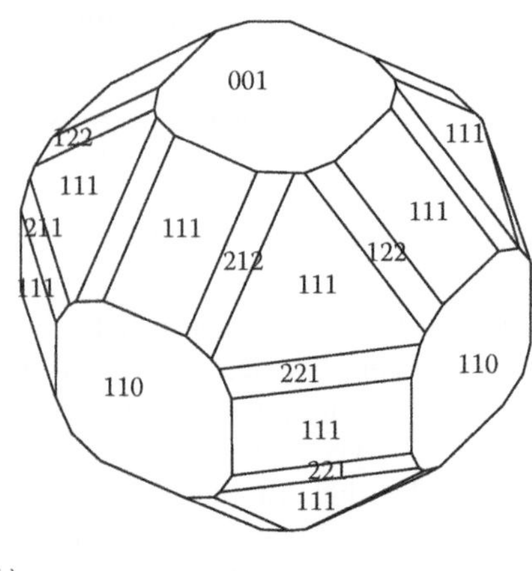

FIGURE 4.21 $LiMn_2O_4$ model structure of octahedron (a) and polyhedron (b). (From Lee, S., Hu, H., Xia, Y., Zhao, L., and Xiao, F., *Chin. Sci. Bull.* (Chinese version), 58, 3350, 2013.)

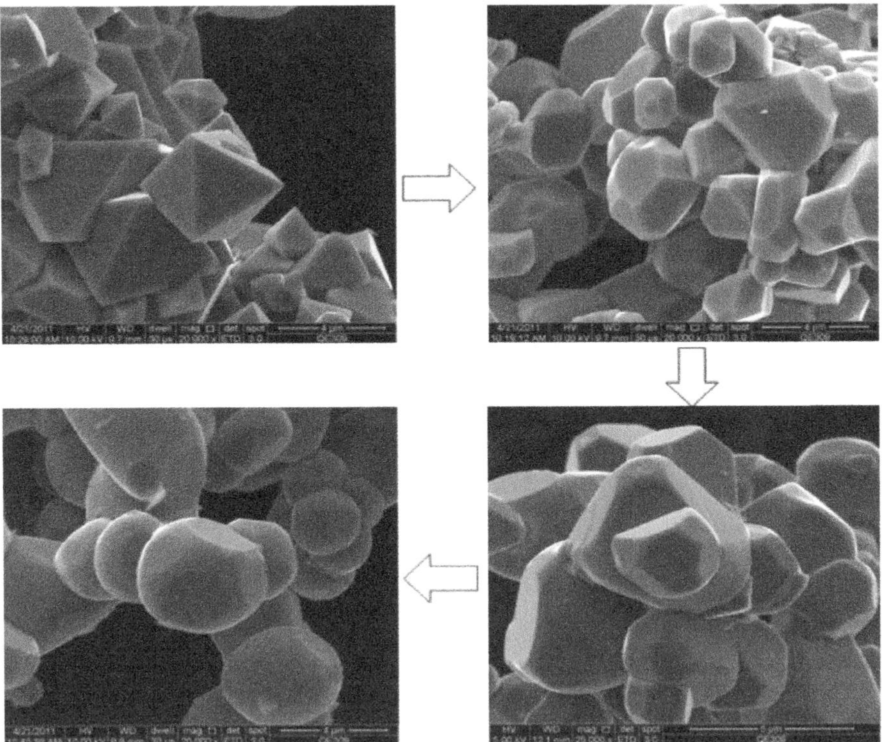

FIGURE 4.22 Morphology control of Mn–spinel. (From Lee, S., Hu, H., Xia, Y., Zhao, L., and Xiao, F., *Chin. Sci. Bull.* [Chinese version], 58, 3350, 2013.)

4.1.5.2 High-Voltage $LiNi_{0.5}Mn_{1.5}O_4$ Spinel

The spinel $LiNi_{0.5}Mn_{1.5}O_4$ is a kind of 5-V cathode material. Since it displays high energy density with an operating voltage at around 4.7 V and a capacity above 135 mAh/g, it has become a promising cathode candidate for next-generation LIBs.

By definition, the spinel cathode offers robust bulk electrochemical properties due to its crystallographic structure. The $LiNi_{0.5}Mn_{1.5}O_4$ spinel has the benefit of a robust cubic close-packed crystal structure made up of strong edge-shared MO_6 octahedral (M = Mn or Ni). Depending on different heating conditions, $LiNi_{0.5}Mn_{1.5}O_4$ has two possible structures: face-centered spinel (Fd3m) and primitive simple cubic crystal ($P4_332$) [128,129]. It is theoretically possible to obtain a disordered structure in which Ni^{2+} and Mn^{4+} are distributed randomly among the 16d octahedral sites, which would result in a space group of Fd3m (Figure 4.23a) [130]. In contrast, due to the 3:1 ratio of Mn/Ni, it is also theoretically possible to obtain a perfectly ordered structure in which all Ni^{2+} ions are fully coordinated by six Mn^{4+} nearest-neighbors in the lattice, which would result in a space group of $P4_332$ (Figure 4.23b). In the ordered phase, the larger Ni^{2+} ions (ionic radius 0.69 Å) occupy only the 4b sites that give more room than the 16d sites of the normal spinel structure. The cation distribution in the $P4_332$ symmetry is then Li on 8c, Ni on 4b, and Mn on 12d, and the O(1) and O(2) oxygen ions occupy the 24e and 8c Wyckoff positions, respectively. The net result is thus a significant optimization of space occupation leading to a reduced unit cell volume.

X-ray diffraction (XRD) is thought to be the simplest technique to distinguish between these two structures. Figure 4.23c show a comparison of the XRD patterns of disordered Fd3m and ordered $P4_332$ samples [131]. Although the patterns are rather similar in the characteristic spinel peaks present, there are two noteworthy features that differentiate the samples. In the disordered Fd3m pattern, the weak reflections at 37.5°, 43.7°, and 63.7° can be assigned to a nickel-rich $Ni_{1-x}Li_xO$ rock salt impurity phase. To obtain a disordered sample, the material is typically heated to 800–1000°C during the final sintering process [129,132,133]. At these high temperatures, the lattice oxygen sometimes will be released from the $LiNi_{0.5}Mn_{1.5}O_4$ framework to form oxygen vacancies. To compensate for the corresponding charge loss, the Mn^{4+} is reduced to Mn^{3+}. The presence of Mn^{3+} in the spinel lattice and the deviation of the Ni/Mn ratio from the ideal value of 1:3 discourage the ordering between Mn and Ni. Moreover, the larger ionic radius of Mn^{3+} (0.645 Å) compared with Mn^{4+} (0.530 Å) results in a larger cell volume. Interestingly, the disordered structure can transform to the ordered structure by being reannealed at a lower temperature of 700°C [128,129,134,135]. This lower-temperature annealing allows the reintegration of the rock salt phase into the spinel lattice and eliminates the Mn^{3+} content in the spinel phase. In the ordered $P4_332$ pattern, the small peaks at 15.4°, 57.5° indicate a superstructure caused by cation ordering, which decreases the degree of symmetry [136].

The charge–discharge profiles of the ordered and disordered samples are different due to the degree of cation

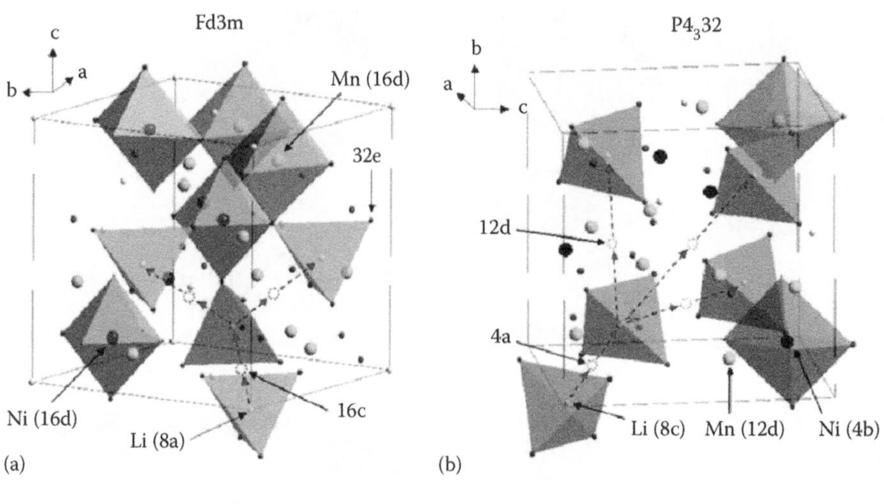

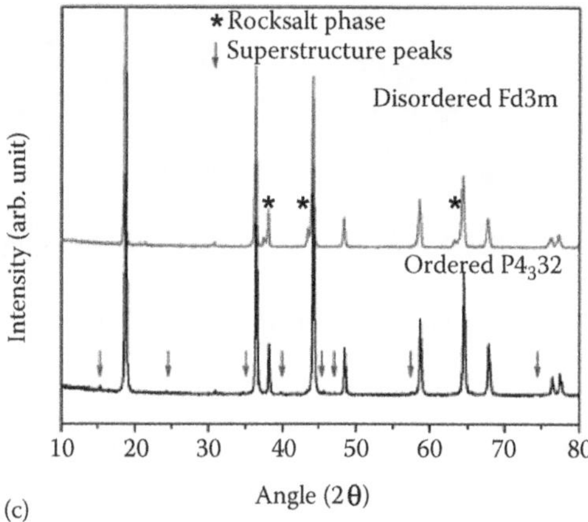

FIGURE 4.23 Crystal structures of $LiNi_{0.5}Mn_{1.5}O_4$with (a) Fd3m and (b) P4$_3$32 space groups. (From Yang, J.G., Han, X.P., Zhang, X.L., Cheng, F.Y., and Chen, J., *Nano Res.*, 6, 679, 2013.) (c) XRD patterns of $LiNi_{0.5}Mn_{1.5}O_4$ with Fd3m and P4$_3$32 space groups. (From Liu, D., Zhu, W., Trottier, J., Gagnon, C., Barray, F., Guerfi, A., Mauger, A., Groult, H., Julien, C.M., Goodenough, J.B., and Zaghib, K., *RSC Adv.*, 4, 154, 2014.)

ordering as well as the amount of impurity phase and Mn^{3+} ions present in the material. Figure 4.24 shows the representative charge and discharge curves of the ordered and disordered samples [137]. The long, flat plateau present at around 4.7 V represents the redox reaction of the $Ni^{2+/3+/4+}$ couple, and the activity near 4.0 V can be attributed to the $Mn^{3+/4+}$ couple [138]. For the disordered Fd3m sample, there is a clear separation between the $Ni^{2+/3+}$plateau at around 4.6 V and the $Ni^{3+/4+}$ plateau at around 4.8 V, besides the extra $Mn^{3+/4+}$ plateau at around 4.0 V. However, for the ordered P4$_3$32 sample, no obvious voltage plateau can be found at around 4.0 V, indicating that there is little or no Mn^{3+} in the spinel lattice. Meanwhile, the Ni plateau at around 4.7 V is continuous with no obvious break between the $Ni^{2+/3+}$ and $Ni^{3+/4+}$ reactions, supporting the premise that a two-phase mechanism dominates rather than a solid-solution regime [139].

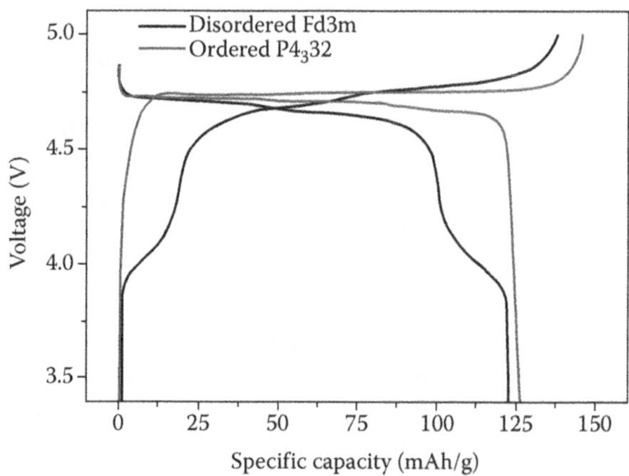

FIGURE 4.24 Typical electrochemical charge and discharge profiles of ordered and disordered $LiNi_{0.5}Mn_{1.5}O_4$ spinel cathodes. (From Manthiram, A., Chemelewski, K., and Lee, E.S., *Energ. Environ. Sci.*, 7, 1339, 2014.)

As mentioned above, the electrochemical performance of the ordered and disordered samples is different. Although the ordered samples deliver higher power, the samples with some cation disorder give more stable long-term performance and higher rate capability [140,141]. This has largely been attributed to the improved electrical conductivity contributed by both the increased probability of proximity between active Ni redox centers as well as the presence of hopping carriers introduced by the small amounts of Mn^{3+} present in a disordered sample [140]. Therefore, the face-centered spinel (Fd3m) is more promising to become a cathode material.

At present, the major obstacle restricting $LiNi_{0.5}Mn_{1.5}O_4$ for commercialization is the capacity fade resulting from an aggressive side reaction with the electrolyte that occurs during sustained high-voltage operation [137]. One of the most popular strategies to combat this issue has been the substitution of small amounts (2–8%) of other cations for nickel and/or manganese in the structure [142]. Thus far, the reported elements contain Al [143], Fe [144–146], Cu [147], Co [148,149], Ti [150–152], Cr [153,154], Mg [155], Zn [156], and Ru [157]. Among them, the metal ions with a valence of 3+ are the most popular. The substitution of 3+ valence ions for Ni induces a charge compensation from a corresponding amount of manganese, giving rise to mixed Mn^{3+} and Mn^{4+} ions in the lattice. Many of the dopant ions may not be electrochemically active in the operating voltage window, but the Mn^{3+} ions participate in the redox reaction, which serves two benefits. First, the activity of manganese helps offset the loss of capacity from replacement of the Ni by dopant ions. Additionally, the Mn^{3+} ions act as internal carriers, assisting the hopping conduction mechanism and enhancing the rate of charge transfer. Cation doping has also been attributed to a decrease in the capacity fade with cycling, as can be seen in Figure 4.25 [158].

Although the above methods have been identified to improve the electrochemical performance in half-cell configurations with the lithium–metal anode, significant capacity fade is still seen in full cell configurations with the graphite anode. The inferior performance in full cells could originate from melt-ion dissolution due to the high operating voltage. The dissolved metal ions migrate to the anode and subside on the graphite surface to form the SEI layer during cycling, resulting in gradual trapping of lithium ions and consequent rapid capacity fade [159]. Surface modifications are effective approaches to promote the stability of the high-voltage spinel cathode. The inert surface coatings on the surface of the active material and graphite anode can prevent side reactions with the electrolyte and dissolution of metal ions.

For the future, development of alternative electrolytes that have no or less reaction with the cathode surface and prevent metal-ion dissolution will be necessary for the successful utilization of the high-voltage spinel cathodes in next-generation practical cells. In addition, a great deal of effort should also be focused on realizing a robust anode surface that will not reduce the dissolved metal ions.

4.2 ANODE MATERIALS

Although Li-based batteries have been investigated for decades, Sony Co., Japan, brought the LIBs into real life since the 1990s, by using lithium cobalt oxide as the cathode and graphite as the anode. The most important advancement in this technique is replacing the Li metal anode by graphite, which significantly improved the safety and stability of LIBs. It explains how important the anode is in LIB applications.

During the following 20 years, the carbon-based anodes play a major role in LIBs, since the carbon exhibits a relatively higher capacity compared with the cathode materials and thus meets the battery requirements at that time. Recently, the high-speed development of portable electronic devices, electrical tools, and EVs rises with higher requirement of energy storage, especially the LIBs [160]. Various advanced cathode materials have been successfully developed, including lithium iron phosphate cathode ($LiFePO_4$), lithium manganese phosphate cathode ($LiMnPO_4$), and lithium-rich manganese-based oxides cathode, etc. [161]. It is known that

$$1/C_{total} = 1/C_{cathode} + 1/C_{anode} + 1/C_{constant}$$

where the C_{total}, $C_{cathode}$, and C_{anode} are the capacity of total cell, cathode, and anode, respectively. $C_{constant}$ is the contribution from other components in battery, including the electrolyte,

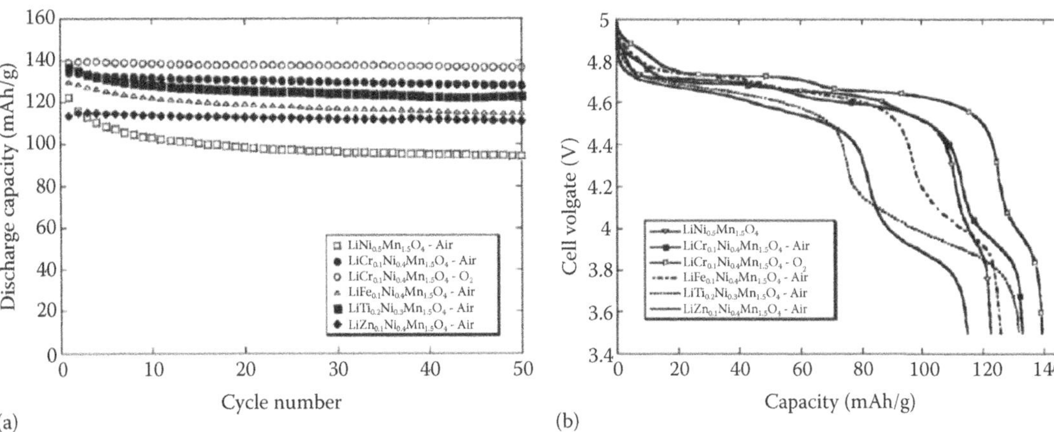

FIGURE 4.25 (a) Cyclic performance and (b) discharge profiles of cells made of $LiNi_{0.5-x}Mn_{1.5}M_xO_4$ (M = Mn, Cr, Fe, Ti, Zn) cathode materials synthesized by mechano-chemical process. (From Oh, S.H., Jeon, S.H., Cho, W.I., Kim, C.S. and Cho, B.W., *J. Alloys Comp.*, 452, 389, 2008.)

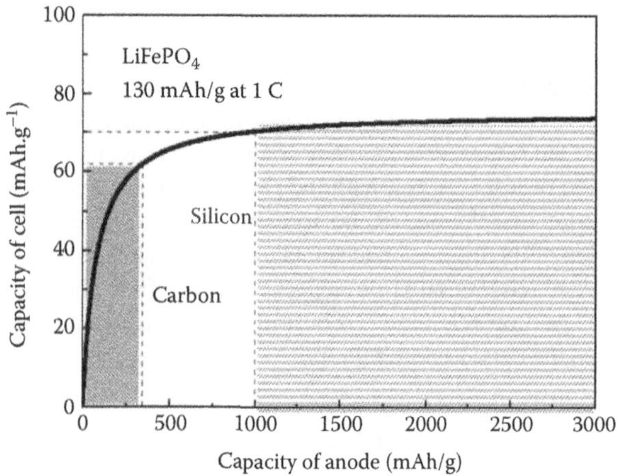

FIGURE 4.26 Total capacity of 18650 LIB as a function of anode capacity.

separator, current collectors, etc. To further improve battery performance, it is important to develop more outstanding anode materials compared with the traditional carbon-based anodes. As shown in Figure 4.26, when using $LiFePO_4$ as a cathode material with a capacity of 130 mAh/g, the capacity of the total cell can be promoted by more than 20% by improving the capacity of the anode from 350 to 1000 mAh/g.

Both the theoretical and experimental results have predicted the potential novel anode materials that can store/release energy through the alloying/de-alloying, intercalation/de-intercalation, and conversion reactions. It can be briefly separated into two categories: (i) Metal-based anodes, such as silicon (Si), tin (Sn), germanium (Ge), etc. They can form alloys with Li ions, and exhibit large gravimetric capacity. (ii) Metal oxide–based anodes, including transitional metal oxide, titanium oxide, and other metal oxides. Overall, the new anode materials suffer from the following problems: (a) large capacity accompanied with a large volume expansion that results in an unstable electrode structure and hence rapid capacity degradation; (b) poor electrical conductivity, especially for the metal oxide, resulting in low rate capability; and (c) the unstable SEI leads to low coulombic efficiency. Many efforts have been made for decades. Considering that a few papers have reviewed the previous progress, we are going to talk about recent representative developments of a new technology that benefits a lot from nanotechnology.

4.2.1 Advanced Metal-Based Anodes

Like the carbon-based anodes, various LIB anodes can store Li ions by alloying/de-alloying reaction, including Si, Sn, Ge, etc. There is no doubt that Si is the hottest material among these materials, which shows its bright future for practical applications. Thus, studies on Si-based anodes can represent the typical advantages and challenges of all similar anodes, which will be described in detail.

Si can form various alloys with Li ions, such as $Li_{22}Si_5$, $Li_{15}Si_5$, and Li_7Si_3, etc. Accordingly, it can exhibit the highest capacity of 4200 mAh/g, which is 10 times higher than that of commercially used carbon-based anodes [162,163]. Actually, the $Li_{22}Si_5$ alloy can only form at high temperatures. Normally, the capacity of Si can only reach 3579 mAh/g [164,165]. It shows a discharge plateau of around 0.05 V and a charge plateau of around 0.5 V, which is essentially good for improving the open-circuit voltage of the full battery [166]. In the meantime, Si is one of the most abundant elements on Earth, and thus, its cost is relatively low. Therefore, Si is the best candidate for the next-generation LIB anodes.

However, accompanying the insertion of many Li ions, Si anodes suffer from a large volume expansion of 400% [167]. Owing to the repeating expanding/shrinking during the discharge–charge cycles, the electrode becomes unstable, resulting in rapid capacity degradation. Up to now, most of the researches in this field are focused on how to accommodate the volume change during the electrochemical process. Before this, many efforts have been made and some literature reports have summarized the approaches to achieve better battery performance for Si-based anodes [168]. These were mainly focused on (i) Si-inactive material composites. For example, Ni–Si alloy has been widely investigated, and it was found that lower capacity but longer cycle life can be obtained because Ni is an inactive material that could buffer the volume expansion and also improve the electrical conductivity in the electrode. (ii) Si-active material composite. Mainly, carbon is used as the active material since it is lightweight and highly conductive, and it has a low volume expansion of 9%. The related information has been described in detail [168]. However, the improvement of battery performance is rather limited by using these methods. The intrinsic physics and the corresponding solution are still unknown.

4.2.1.1 Size Effect of Si-Based Anodes

During the insertion of Li ions into Si bulk materials, pulverization usually occurs in the Si bulk films and Si bulk particles. It leads to capacity degradation and low coulombic efficiency due to the formation of new SEI films. Besides that, the resulting fragments have a weak electrical connection to the electrode and hence are not able to run the electrochemical cycle again. This motivates reducing the size or dimension of Si based on three advantages in nanostructures [169]: (i) the volume expansion in nanostructures is lower than that of microstructures; (ii) there is much shorter diffusion length of electrons and Li ions in nanostructures that a higher rate capability can be expected; and (iii) nanostructures have much higher surface area and reaction activity.

Kumta et al. conducted a systematic study on the interfacial properties of 2D a-Si/Cu films [170]. The Si films on Cu substrates broke into small islands after starting the electrochemical cycling process. The thinner film with a thickness of 250 nm exhibited better cycle life than that of the thicker one with a thickness of 1 μm. More important, although the Si islands form at the initial electrochemical cycling process, the capacity started to decrease when the Si islands lost the attachment to the Cu substrates. This result strongly suggests that the connection between active materials and conducting network plays a critical role in capacity degradation.

As a further step, Cui et al. designed and fabricated 1D Si nanowire arrays as LIB anodes [171]. As shown in Figure 4.27a, it is proposed that an Si nanowire with a small diameter would not be broken, and the gaps between the nanowires could accommodate the volume expansion of Si nanowires. Meanwhile, each nanowire is directly connected to the substrate, which is efficient for electron transport. Finally, the Si nanowire arrays revealed an ultrahigh discharge capacity of around 4200 mAh/g in the first cycle, which is close to the theoretical capacity of Si, as presented in Figure 4.27b. The diameter of Si nanowires after Li insertion was about 1.6 times of that of the original ones. However, the high capacity only persisted for 10 cycles. The high performance discovered in Si nanowires attracted broad attention immediately in the field of energy storage. Many approaches have been successfully developed to fabricate various kinds of Si nanowires, including the chemical vapor deposition (CVD) method, metal-assisted etching method, and solution phase method [172–176]. Zhu et al. prepared free-standing Si nanowire arrays through the metal-assisted etching method [177]. With the help of carbon coating, the capacity reaches 3657 mAh/g without the use of a current collector and any binder materials. The solution-grown Si nanowires composite also exhibited a high initial capacity of 2500 mAh/g, but a low cycle life of fewer than 10 cycles [175]. After replacing the carbon black with multiwalled carbon nanotubes (CNTs), the cycle life significantly improved owing to the formation of a conducting network of CNTs. Yang et al. successfully built a free-standing Si/carbon paper by interweaving the Si nanowires through a simple precipitation process [178]. The resulting Si/carbon paper showed a high discharge capacity of around 2000–2500 mAh/g for 20 cycles. The cycle life can be further improved by reducing the size of nanowires, the addition of carbon coating, and the limitation of cycling voltage windows. The free-standing Si nanowires paper can also be fabricated by the free-catalyst thermal evaporation in a vertical high-frequency induction furnace [179].

After many electrochemical tests on the Si nanowire–based anode, the cycle life of these nanowire-based electrodes was found to be greatly limited to tens of cycles, although high capacity can be observed. To understand the failure mechanism, the structure evolution of Si nanowires during the electrochemical process has been investigated carefully. Huang et al. developed an advanced *in situ* transmission electron microscopy that can clearly observe the electrochemical behavior of single Si nanowires [165]. It was found that the Si nanowire started to bend as a helical wire after reacting with the Li ions. With the help of carbon coating, the charging rate can be improved by at least 10 times compared with the pristine Si nanowires. It provides direct proof that the improvement of electron transport is important for the rate capability. It was also found that the anisotropic expansion results in a remarkable dumbbell-shaped cross section of Si nanowire with <112> growth direction [180]. Further experiments indicated the nature of highly anisotropic expansion along <110> and <111>, and thus different morphologies can be formed on the nanowire with different orientations after electrochemical cycling [181]. On the other hand, Cui et al. investigated the structure evolution of Si nanowires and ZnO nanowires that showed the formation of nanopores in a stepwise manner [182]. However, on the basis of these studies, the failure mechanism of nanowire arrays is still unknown. As a possible proof, mechanical property studies on CNT@α-Si coaxial nanowires demonstrated that α-Si shell mainly broke at the anchoring point to the Si substrate, which suggested that the weakness on 1D nanowire was at the root [183]. The theoretical study of interfacial properties by the combined density functional theory and *ab initio* molecular dynamics calculations also indicated that the small decrease of adhesion strength was related to the Li segregation [184].

Although 1D Si nanowires exhibit rather impressive battery performance, the low cycle life and the complex fabrication process greatly hinder it from the practical applications. With a further reduced dimension, 0D nanoparticles are potentially competent for this purpose with advanced battery performance and also low cost since they can be produced by various facile methods. Usually, micro-sized particles would break into small pieces, thus leading to a significant capacity loss in 10 cycles [185]. Both the theoretical and experimental works have demonstrated that nanoparticles could possibly bear the repeated expanding/shrinking. Huang et al. systematically investigated the *in situ* structure evolution

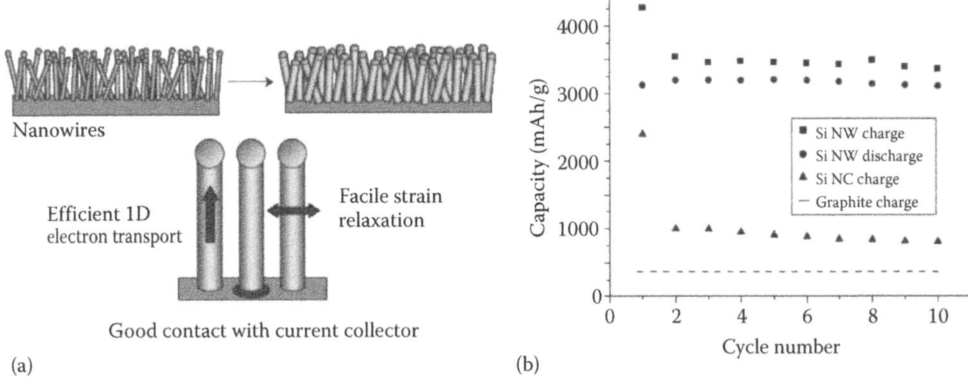

FIGURE 4.27 (a) Schematic of morphological change of Si nanowire during electrochemical cycling and (b) capacity of Si nanowires and nanocrystals. (From Maranchi, J.P., Hepp, A.F., Evans, A.G., Nuhfer, N.T., and Kumta, P.N., *J. Electrochem. Soc.*, 153, A1246, 2006.)

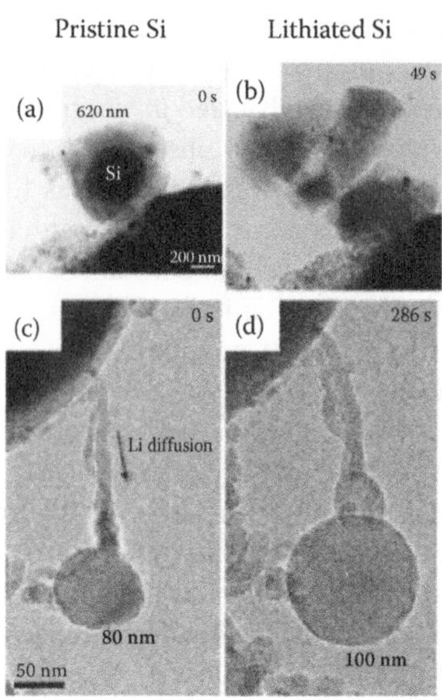

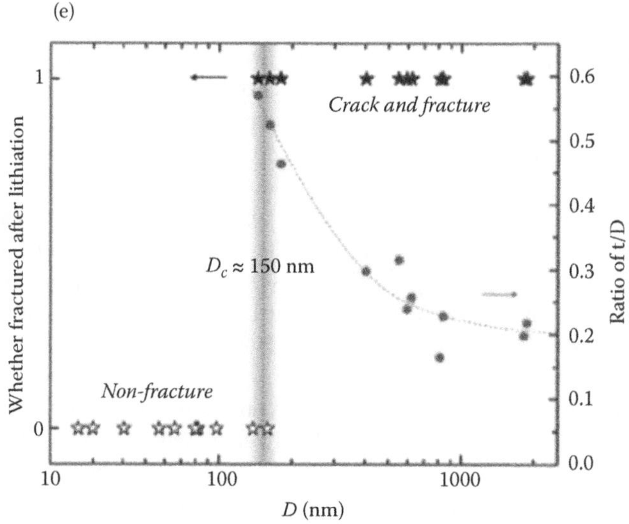

FIGURE 4.28 TEM images of (a) pristine Si nanoparticle with D = 620 nm and (b) corresponding lithiated Si nanoparticle. TEM images of (c) pristine Si nanoparticle with D = 80 nm and (d) corresponding lithiated Si nanoparticle. (e) Statistics showing the critical size around 150 nm. (From Ryu, J.H., Kim, J.W., Sung, Y.E., and Oh, S.M., *Electrochem. Solid-State Lett.* 7, A306, 2004.)

of Si nanoparticles in sizes of 80, 160, 620 and 940 nm. As shown in Figure 4.28a and b, the Si nanoparticles with a size of 620 nm broke into pieces after the lithiation. However, decreasing the size to 80 nm, the Si nanoparticle can survive after the lithiation, as demonstrated in Figure 4.28c and d. It was concluded in Figure 4.28e based on the experimental results that the shell of nanoparticles in size of 150–2000 nm could break during the electrochemical tests. The crack would be eliminated when the size of nanoparticles was less than the critical size of 150 nm [186]. Liu et al. fabricated Si nanoparticles on graphene sheets through the reduction of an SiO_2 precursor with magnesium [187]. The size of Si nanoparticles reached 10–30 nm. However, it revealed a stable capacity of >1000 mAh/g for 20 cycles, and then the capacity dramatically dropped. Cho et al. performed systematical studies on the size effect of Si nanoparticles on the electrochemical properties [188]. The best battery performance was obtained in the Si nanoparticles with a diameter of around 10 nm. With further carbon coating, the coulombic efficiency and capacity retention were significantly improved. Similar results can be realized by a flash heat treatment that generates SiO_2/C coating on Si nanoparticles. Thus, the battery performance can be significantly improved with a capacity of 1150 mAh/g for 500 cycles [189].

4.2.1.2 Si/Conducting Network Composite

Although it has been proven that the Si nanoparticles can survive during the electrochemical cycling, the electrode made with Si nanoparticles is still of low cycle life. As mentioned above, the study on the a-Si/Cu films indicated that the capacity degradation originated from the loss of electrical connection between active Si and conducting substrate. Many efforts have been made to enhance the electrical connection by establishing the Si-based composite on the conducting network. On the other hand, 3D conducting networks have been demonstrated to be important in energy storage [190]. Thus, high battery performance can be expected in an Si–conducting network composite.

Wang et al. coated the Si on the 2D $TiSi_2$ nanosheets [191]. The resulting heterostructures exhibited a high capacity of around 2000 mAh/g for 40 cycles at the operation potential of 0.03–3 V. With limiting the operation potential to 0.15–3 V, the capacity was reduced to about 1000 mAh/g but remained unchanged for 50 cycles. The Si nanoparticles graphene paper composites had a high initial capacity of 2000 mAh/g. If the Si is uniformly coated on the nanofiber network, the composite exhibited a high capacity of 2000 mAh/g for 30 cycles without capacity fade. With further replacing the nanofiber network by CNT films, the Si/CNT films can be free-standing, with the corresponding capacity reaching 2000–2400 mAh/g with a cycle life of 100 [192]. It was proposed that the CNTs in this structure can relax the large strain during electrochemical cycling. In another way, Si can also be coated on aligned CNTs. With double carbon coating, the capacity can be maintained at 1500 mAh/g for 45 cycles [193]. Cui et al. also coated Si on the CNT sponge, and an ultrahigh areal capacity of 40 mAh/cm^2 was achieved, which is around 10 times higher than that of commercially used LIBs [194]. A detailed study disclosed that the many nanopores formed after cycling are probably due to the large expansion mismatch at the interface between Si and CNT. This might lead to a rapid capacity decrease. Similar electrochemical properties were also observed in the Si-coated 3D Cu nanonetwork [195]. Cho et al. demonstrated advanced flexible graphite silicon hybrid electrodes using 3D current collector for flexible batteries [196]. Up to now, most of the examples of directly coating Si on conducting materials presented high capacity but low cycle life within 50 cycles. Yushin et al. reported a hierarchical bottom–up approach to Si coated on annealed carbon black, and another C CVD coating on the Si surface

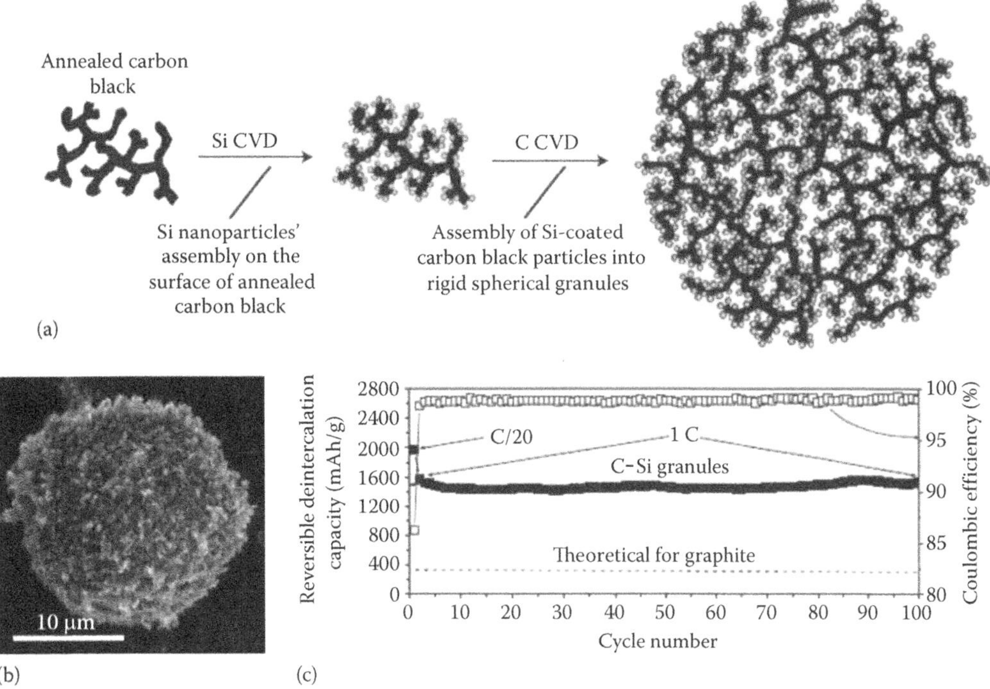

FIGURE 4.29 (a) Schematic of hierarchical assembly of Si/C nanocomposite. (b) SEM image of the as-prepared Si/C nanocomposite granule. (c) Battery performance of as-prepared Si/C nanocomposite electrode. (From Magasinski, A., Dixon, P., Hertzberg, B., Kvit, A., Ayala, J., and Yushin, G., *Nat. Mater.*, 9, 353, 2010.)

assembles the carbon black particles into rigid spherical granules, as shown in Figure 4.29a and b [197]. In this structure, amounts of pores embedded in granules can accommodate the volume expansion while the connected anneal carbon black could provide efficient electron transport. Thus, a high capacity of around 1600 mAh/g was obtained without any capacity degradation for over 100 cycles, as shown in Figure 4.29c. It also exhibited a high capacity of 1000 mAh/g at a high discharge–charge rate of 8 C, which was about 10 times higher than that of annealed carbon black.

4.2.1.3 Porous Si Composite

Pores or gaps between active Si materials have been demonstrated to be very good for volume expansion accommodation. A large amount of work has been done toward this purpose, basically in two ways: (i) porous Si as anode to generate the pores inside Si that can accommodate the volume expansion by themselves; (ii) Si embedded in a porous composite to generate the pores surrounding the Si nanoparticles in the matrix.

Cho et al. first reported a novel approach to fabricate 3D porous Si particles [198]. They formed the bulk particle in size of tens of micrometers, which had many pores with a size of 40 nm. Uniform carbon was naturally coated on the surface during the reaction. Benefiting from this unique structure, it presented a high capacity of 2800 mAh/g at 0.2 C with a cycle life of over 100 cycles. Similar micro-/nanostructured porous Si particles were obtained by Yu et al. [199]. With the help of Ag nanoparticles coated inside, the discharge capacity was as high as about 3000 mAh/g with capacity retention for more than 100 cycles. Ozin et al. and Cui et al. used the SiO_2 spheres as the scarifying templates, the interconnected inverse-opal hollow Si films can be fabricated by CVD Si coating, and then SiO_2 is removed by HF [200,201]. The larger pore, higher coulombic efficiency, and better capacity retention would lead to a relatively low volumetric capacity. By using carbonates as templates, various hollow Si can be fabricated after coating Si and then removing the template. The as-prepared hollow Si exhibited a discharge capacity of 960 mA/g while 71% of the capacity can be retained after 500 cycles [202]. The best record was achieved by Cui et al. that Si was coated inside the CNT with a double-walled structure, a significantly impressive cycle life of 6000 cycles was achieved, which was supposed to benefit from the hollow structure and also the stable SEI films [203].

To further improve the battery performance, nanopores or mesopores should be created inside the Si particles in two ways. One is to reduce the mesoporous SiO_2 by Mg since the mesostructured SiO_2 can be easily obtained through the soft-template method with the desired size and morphology [204]. Another way is to etch the highly doped Si, which can generate the mesopores in Si matrix. Zhou et al. reported that the porous doped Si nanowires can reach an ultrahigh cycle life of over 2000 cycles [205]. Besides that, the capacity can remain more than 1000 mAh/g at a high discharge rate of 4.5 C.

However, all the methods for preparing the porous Si mentioned above are rather complex and expensive. The samples were synthesized in very low quantity. Compared with the porous Si, Si nanoparticles can be much easier to fabricate; however, they exhibit unsatisfactory battery performance. As shown in Figure 4.30a and b, Cui et al. designed a novel

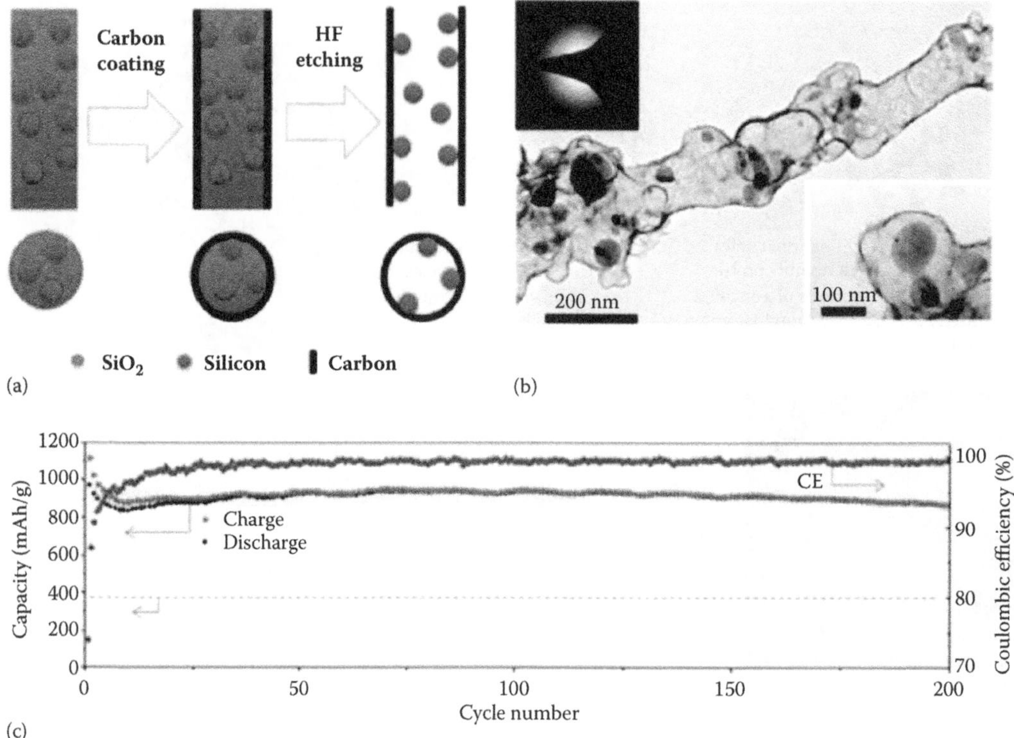

FIGURE 4.30 (a) Schematic of formation of Si/C with engineered empty space. (b) TEM image showing the empty space in Si/C nanotube. (c) Battery performance of as-prepared Si/C composite. (From Wu, H., Zheng, G., Liu, N., Carney, T.J., Yang, Y., and Cui, Y., *Nano Lett.*, 12, 904, 2012.)

Si/C composite by engineering an empty space between Si nanoparticles inside carbon channels [206]. Compared with the rapid capacity decrease for the pure Si nanoparticles or carbon-coated Si nanoparticles, the Si/C composite with empty space inside retained a capacity of 800 mAh/g for 200 cycles without obvious decrease, as presented in Figure 4.30c. A similar improvement was also achieved by Choi et al. [207]. They successfully fabricated core–shell fibers with Si nanoparticles embedded in the carbonized PAN channels. The empty space formed between Si nanoparticles can accommodate the volume expansion while keeping the good electrical connection between the Si nanoparticles and carbonized PAN channel. The capacity is around 800 mAh/g at 3 C without obvious capacity degradation for 300 cycles. Considering the possibility for large-scale production, Choi et al. developed a spray-drying method to fabricate the porous Si/C composite [208]. Nanopores can be embedded inside the matrix by removing the sacrificed SiO_2 nanoparticles. The discharge capacity can reach 1400 mAh/g with a cycle life of more than 150.

4.2.1.4 Other Approaches

The volume expansion would not only lead to the pulverization or the expansion of electrode but also the detachment of active materials from the conducting network due to expansion mismatch at the interface. For example, the volume expansion for the Si is about 400%, and that for the carbon is about 9%, and the inactive Cu substrate would not expand. To solve this problem, experiments had been majorly conducted in three ways as shown in Figure 4.31. (i) Increasing the

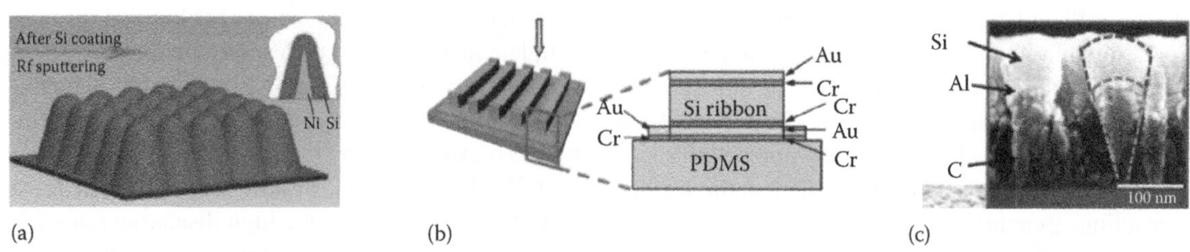

FIGURE 4.31 Schematic of approaches to buffer the volume expansion mismatch at interface: (a) increasing the roughness of current collector, (b) using flexible current collector substrate, and (c) building a strain-graded interface. (From Zhang, S., Du, Z., Lin, R., Jiang, T., Liu, G., Wu, X., and Weng, D., *Adv. Mater.*, 22, 5378, 2010; Son, S.B., Kim, S.C., Kang, C.S., Yersak, T.A., Kim, Y.C., Lee, C.G., Moon, S.H., Cho, J.S., Moon, J.T., Oh, K.H., Lee, S.H., *Adv. Energ. Mater.*, 2, 1226, 2012; Krishnan, R., Lu, T.M., and Koratkar, N., *Nano Lett.*, 11, 377, 2011.)

roughness of substrate: using the CNT arrays or Ni nanocones as the substrate, the high roughness can buffer the expansion mismatch because the deformation can gradually increase from the inactive substrate to the active Si [209]. (ii) Using the flexible conducting substrate: polydimethylsiloxane (PDMS) with metal coating was employed as the conducting substrate so that the soft polymer PDMS could expand/shrink as the active Si expand/shrink during the electrochemical cycling. Thus, the stability of the electrode can be maintained [210]. (iii) Establishing buffering interface: Koratkar et al. grew a thin layer of Al between the Si and C [211]. It formed a strain-graded interface because the volume expansion of Al is about 94%, which is in the middle between that of Si and C. The C–Al–Si composite exhibited a capacity of 350 mA/g at an ultra-high discharge rate of 60 C. The cycle life reached 100 cycles.

On the other hand, the stability of Si-based electrode can also be improved by using high-viscosity binder materials. It has been demonstrated that the heat-treated poly(vinylidene fluoride) (PVDF) shows a much better battery performance in Si electrode [212]. Cho et al. reported that the highly cross-linked polymeric binder exhibits a much longer cycle life than the one using individual poly(acrylic acid) (PAA) or carboxymethyl cellulose (CMC) [213]. With further development of the binder materials, the alginate, a high-modulus natural polysaccharide extracted from brown algae, could significantly improve the cycle life of Si-based electrodes [214]. A high capacity of about 2000 mAh/g was achieved for 100 cycles using the alginate binder, compared with <10 cycles using CMC binder and PVDF binder.

4.2.1.5 Other Metal-Based Anodes

Besides the Si anodes, there are also other metals available for storing the Li ions, such as Ge and Sn. Since they are rather similar to Si in that a high capacity of around 1000–2000 mAh/g can be obtained accompanied by a large volume expansion, the solutions are almost focused on several similar approaches. Thus, we would not describe the details of each one but quickly summarize the recent success of advanced Ge and Sn anodes.

Ge has a lower theoretical capacity of 1600 mAh/g compared with the 4200 mAh/g for Si, and hence has a much lower volume expansion and thus slightly better cycle life. The Ge nanowires directly grown on current collector can retain the capacity of ~900 mAh/g for over 1000 cycles [215]. Maier et al. fabricated the carbon-coated Ge nanoparticles with an average diameter of 62 nm [216]. The capacity in the first cycle reached 1200 mAh/g and gradually decreased to around 900 mAh/g. The cycle life could easily reach over 200 cycles. Mullins discovered that the determining factor was the selection of electrolyte [217]. By replacing the normally used EC with fluoroethylene carbonate (FEC) as the electrolyte, the capacity can be maintained at around 1200 mAh/g for 2500 cycles. Even at a high rate of 10 C, the capacity is also as high as 700 mAh/g. The advanced achievement in this Ge electrode is due to the formation of an FEC-derived SEI that could protect the electrode from oxidation. Similar to Si, the 3D porous architecture is also good for improving the battery performance of Ge. The 3D porous Ge nanoparticles exhibited a capacity of 1400 mAh/g without obvious degradation for 100 cycles [218]. Other works about Ge nanostructures with different morphologies also presented rather impressive battery performance [219].

Sn has a theoretical capacity of about 990 mAh/g with the formation of alloy $Li_{22}Sn_5$. A little different from Si and Ge, Sn^{2+} ions can be more easily reduced to Sn. Thus, Sn can be fabricated through rather many approaches with convenient structure control, such as electrochemical deposition and solvothermal methods, which were barely used for synthesizing Si and Ge. Sn or Ni–Sn nanowire arrays were also designed and fabricated as Li-ion anodes. By employing the 3D PAA template instead of conventional PAA template, Yang et al. successfully fabricated the 3D Ni–Sn nanowire networks [220]. The connection between the nanowires protected the nanowire arrays from agglomeration that much higher areal capacity and better cycle life were achieved. By coating Ni–Sn on the Cu nanowire arrays, a capacity of 500 mAh/g can be obtained with a cycle life of over 200 cycles [221]. There are many works focused on Sn composite with different structure optimization, such as RGO–Sn, CNT–Sn, graphene–Sn nanosheets, etc. [222,223]. Wan et al. also tried to engineer the empty space in the carbon sphere with Sn nanoparticle embedded. The initial capacity was higher than 1000 mAh/g but decreased rapidly in tens of cycles [224]. By embedding Sn in the porous carbon matrix, the ultrasmall Sn nanoparticles can have a stable capacity of ~750 mAh/g for 200 cycles [225].

4.2.2 Metal Oxide Anodes

4.2.2.1 Metal Oxide–Based Alloying/De-Alloying Reaction

This kind of metal oxide essentially undergoes the reaction to form the Li–M (M = Si, Ge, Sn etc.) alloy acting like the individual metal. For example, (i) $SnO_2 + 4\ Li^+ + 4\ e^- = Sn + 2Li_2O$, (ii) then $5Sn + 22Li = Li_{22}Sn_5$. Owing to the higher formula weight, the theoretical capacity would be a little lower than that of the pure metal. That is, about 782 mAh/g for SnO_2, 875 mAh/g for SnO, and about 2000 mAh/g for SiO_x. The electrochemical process on SnO_2 nanowire had been observed by Huang et al. that the SnO_2 nanowire elongated 60% with a diameter expansion of 45% [226]. Focusing on the volume expansion and low electrical conductivity problems, the structure manipulation, the composite, and the building of conducing network were widely investigated. Various shapes of SnO_2 have been fabricated for Li-ion anodes, including nanoparticles, nanorods, nanotubes, and hollow spheres [227–230]. The introduction of conducting coating had been proven to be good for improving the battery performance of SnO_x anodes. It was reported by Guo et al. that binding SnO_2 nanocrystals in graphene sheets by Sn–N bonding can significantly improve the battery performance [231]. A high capacity of around 1000 mAh/g can be maintained for 500 cycles in the SnO_2/G composite. They also found that the ultra-uniform SnO_x/carbon nanohybrids can retain the capacity of 600 mAh/g for 200 cycles [232].

Recently SiO_x-based anodes have attracted broad attention since SiO_2 is one of the cheapest and most abundant materials on Earth. Sohn et al. simply milled SiO_2 for 24 h, which resulted in samples that exhibited a capacity of about 800 mAh/g compared with almost zero for the pristine SiO_2 [233]. Auger electron microscopy analysis indicated that the O/Si ratio after 24 h milling was changed from 2 to 1.91, which might lead to the ability to store Li ions. The battery performance of SiO can also be enhanced by carbon coating and TiO_2 coating. The SiO nanotubes fabricated by thermal degradation of PDMS in PAA template exhibited a capacity of 1000 mAh/g without any capacity degradation for 100 cycles [234]. Similar to Si, the selection of binder is also very important for battery performance. It has been demonstrated by Konno et al. that the use of PAA in SiO anodes can significantly improve the cycle life compared with that of sodium carboxymethyl cellulose (CMCNa), poly(vinyl alchohol) (PVA), and PVDF [235]. Furthermore, the hierarchically designed SiO_x/SiO_y bilayer nanomembranes exhibited a stable capacity of 1200 mAh/g for 100 cycles [236].

4.2.2.2 Transitional Metal Oxide by Conversion Reaction

Transitional metal oxides would react with Li ions by the conversion reaction. Owing to the native low electrical conductivity, many efforts are made on building the electron transport path for promoting the battery performance of transitional metal oxide/conducting network composite. Tarascon et al. employed the Cu nanowire arrays as the current collector for direct and efficient electron transport. The rate capability can be significantly improved using this architecture [237]. Almost 80% of the initial capacity can be retained, even increasing the C-rate from C/32 to 8C. This work helps not only the Fe_3O_4 anodes but also paved a novel route to improve rate capability, which greatly influenced recent researches. Besides the sustained development of various morphologies, how to build the conducting network in the oxide-based electrode becomes extremely important. Dillon et al. used single-walled CNTs to interweave into Fe_3O_4 nanoparticles [238]. The resulting composite exhibited a high capacity of 1000 mAh/g with a cycle life of more than 100 cycles. It is worthwhile to mention that the rate capacity of this composite was able to compete with the that from coating Fe_3O_4 on Cu nanowire arrays. Fan et al. developed a facile approach to coat Fe_3O_4 on aligned CNT scaffolds, which presented a capacity of 800 mAh/g for over 100 cycles [239]. More recently, graphene, a novel 2D atomic crystal, was to be of great electrical, mechanical, and thermal properties, which are suggested to be suitable for LIB applications. Especially in oxide-based electrodes, many works treaded the transitional metal oxide nanostructures with graphene, which have demonstrated success [240–242].

4.2.2.3 Titanium Oxide–Based Anode

Titanium oxide–based anode (TiO_2 and $Li_4Ti_5O_{12}$, etc.) is one kind of novel anode material for power LIB applications because of its high safety and long cycle life benefiting from its unique structure. TiO_2 has different structures, such as rutile, anatase, brookite, TiO_2 (B) etc. [243]. However, all of them suffer from the low electrical conductivity due to the intrinsic properties of oxide. To overcome the disadvantages, recent work has been majorly focused on two ways. (i) Structure modification: various methods have been developed to tailor the morphology of nanostructured TiO_2 and $Li_4Ti_5O_{12}$, especially the mesoporous structure. (ii) Optimization of electrons transport: the electrical behavior can be improved by coating conducting materials, building the conducting network, or mixing with CNT or graphene.

Since TiO_2 is a rather popular material used in not only LIBs but also new energy field such as photovoltaics, photochemical catalysis, etc., many approaches have been developed to fabricate various TiO_2 structures with well-controlled sizes, shapes, and architectures. One of the TiO_2 structures, mesoporous TiO_2, has attracted excessive attention. First, benefiting from the sufficient choices of titanium-based metal–organic compound, mesoporous TiO_2 can be easily fabricated with precise control. Second, the mesostructures have been demonstrated to be good for significantly improving the rate capability in various battery materials. Brown et al. compared the rate performance of TiO_2 nanopowder and mesoporous TiO_2–B [244]. The capacity of mesoporous TiO_2–B reaches 250 mAh/g, compared with <200 mAh/g for TiO_2 nanopowder. More important, the capacity of mesoporous TiO_2 can remain 70% of the initial capacity as the C-rate increases from C/10 to 10 C, whereas that for nanopowder is <50%. The mesoporous TiO_2 prepared by different routes shows the similar advantages. It is also the same for the mesoporous $Li_4Ti_5O_{12}$ [245].

To further improve the battery performance using the same TiO_2 raw materials, the key is to build an unrestricted electron transport path. Inspired by the structure design of Fe_3O_4 on Cu nanowires, Cheah et al. coated TiO_2 on the Al nanorods, which exhibited 10 times higher areal capacity compared with the 2D microbattery [246]. The record was greatly improved by Yang et al. that over 100 times higher area capacity can be reached in 3D Ni/TiO_2 due to the elimination of nanowire agglomeration [247]. Considering the complexity and low gravimetric capacity of the overall electrode, carbon coating was widely used similarly to other battery electrodes. Chen et al. reported that the porous $Li_4Ti_5O_{12}$ with N-doped carbon coating can retain about 80% of the initial capacity as the C-rate increases from C/2 to 10 C, while that of the pristine $Li_4Ti_5O_{12}$ is <10% [248]. On the other hand, the battery performance of TiO_2 can also be greatly improved by using CNT or graphene to form the well-connected conducting network [249]. Liu at al. manipulated the growth of TiO_2 on the RGO by calcination of hydroxyltitanium oxalate (HTO), as shown in Figure 4.32a through g [250]. Impressive rate performance was observed as shown in Figure 4.32h. More than 50% of the initial capacity can be retained even at an ultrahigh discharge rate of 30 C for f-TiO_2/G. The as-built $LiMn_2O_4$-f-TiO_2/G full cell can retain 80% of the initial capacity for over 200 cycles as demonstrated in Figure 4.32i. A similar conclusion can also be obtained in other graphene-wrapped TiO_2 anodes [251].

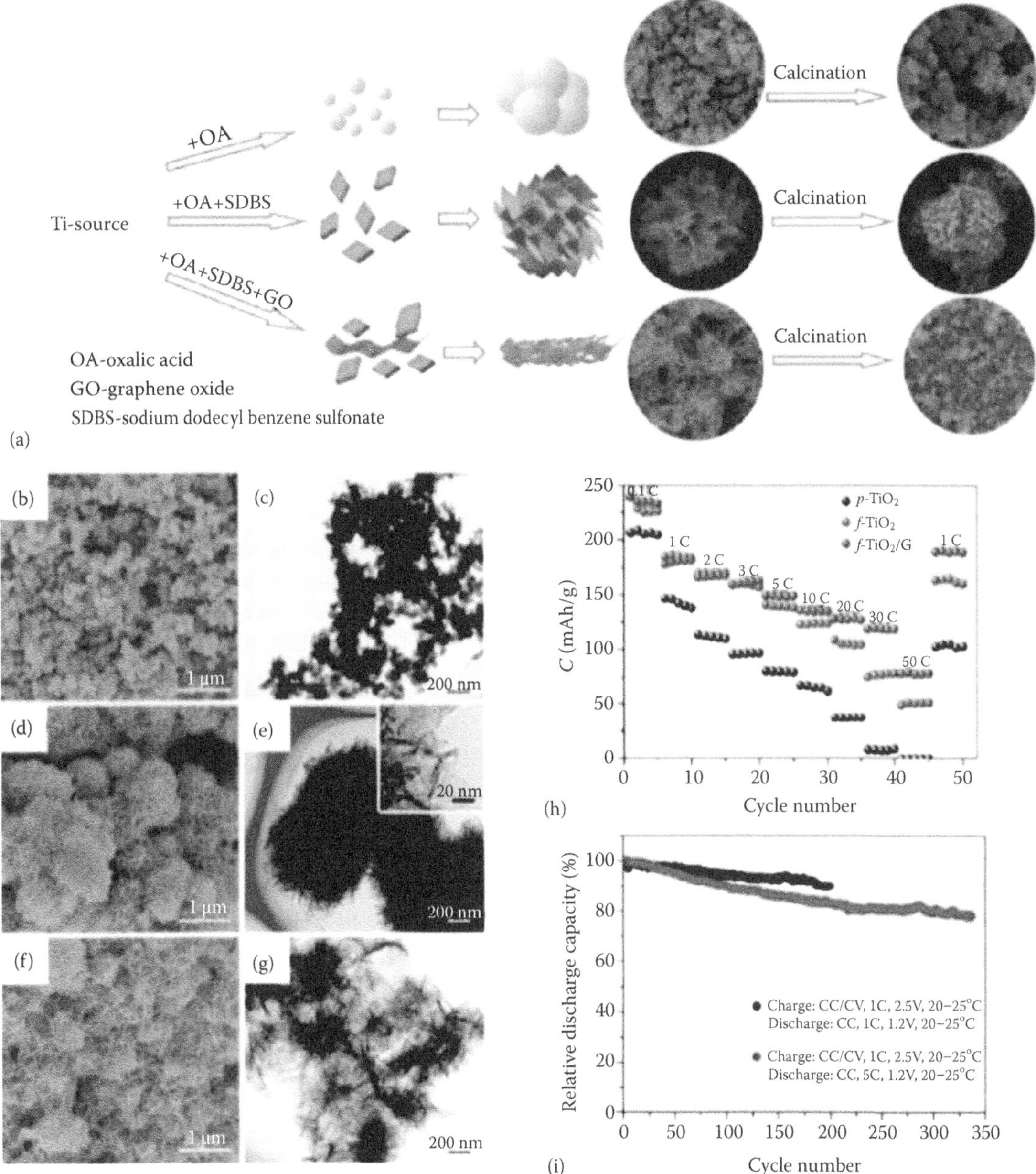

FIGURE 4.32 (a) Schematic of the preparation processes and growth mechanisms of TiO_2-based nanocomposites. SEM and TEM images of *p*-HTO (b, c), *f*-HTO (d, e), and *f*-HTP/GO (f, g). (h) Rate capability of *p*-TiO_2, *f*-TiO_2, and *f*-TiO_2/G. (i) Capacity of $LiMn_2O_4$-*f*-TiO_2/G full cell. (From Xin, X., Zhou, X., Wu, J., Yao, X., and Liu, Z., *ACS Nano*, 6, 11035, 2012.)

4.3 SEPARATOR

4.3.1 Introduction

Separators are usually porous membranes. They are another crucial component in LIBs that play an important role in retaining liquid electrolyte, preventing physical contact between the electrodes and conducting lithium ions. This means they should be good electronic insulators and excellent ion conductors [252].

4.3.2 Separator Requirements and Characterization

The properties of separators would largely influence the interfacial properties and internal resistance of LIBs, and ultimately affect the battery capacity, cycling performance, safety, etc. [253]. The main requirements on separator properties are as follows:

1. Pore structure

   The mobility of lithium ions between cathode and anode relies on the pore structure of the separators, since pores are the main location for the liquid electrolyte and lithium ions and they determine the transport route of lithium ions in the charging–discharging process of LIB. The pore structure could be described by the porosity, pore size, and pore distribution. Generally, the suitable porosity is deemed as 40–50% and the pore size is about 0.03–0.12 μm [254,255]. A too high porosity may cause internal short circuit and poor mechanical strength, while when the porosity is too low, the internal resistance of the battery would increase, yielding a poor C-rate capacity [256]. Changes of the pore size have similar effects. The pores are required to be uniformly distributed to ensure the uniformity of the electrode interface and current density of the batteries.

2. Thickness

   The thickness of the separators should be as small as possible, which could give rise to higher energy/power density. However, the premise is to ensure the mechanical strength of the separators. The thickness of the separator is commonly required to be smaller than 25 μm. For the high-power LIBs, e.g., batteries for (H)EVs, the thickness of the separators is usually 40 μm.

3. Air permeability

   The air permeability reflects the resistance of the separators, which could be represented by the Gurley value. The Gurley value (s/cm^3) is defined as the time for a settled amount of air to pass through a fixed area under a given pressure [257].

4. Tensile strength

   The tensile strength depends on the preparation method of the separators. Generally, high porosity corresponds to poor mechanical strength. For the separator with a thickness of 25 μm, the mechanical strength should be higher than 1000 kgf/cm^2 [258].

5. Puncture strength

   Owing to the surface of the electrodes being always rough, and the lithium dendrites would also form during the application of LIBs, separators always work under pressure. To effectively avoid battery shorting, the puncture strength of the separators must be not less than 300 g/mil [259].

6. Wettability

   For rapidly developing large-sized batteries, especially (H)EVs and grid storage systems, the superior wettability of the entire separator is strongly demanded [260]. The wettability is usually evaluated by using the liquid electrolyte contact angle (CA) and the retention ratio [261].

7. Dimensional stability

   The separator should be flat without wrinkles in use.

8. Thermal stability

   The thermal stability of separators is the most important factor for battery safety. For the commonly used LIB separators, the shrinkage is required to be less than 5% after being treated at 90°C for 1 h. Recently, the application of LIBs in (H)EVs has become wider, which requires much higher battery safety. High-performance separators with superior thermal stability are urgently required [262]. The separators are needed to keep dimensional stability when treated at 150°C [263].

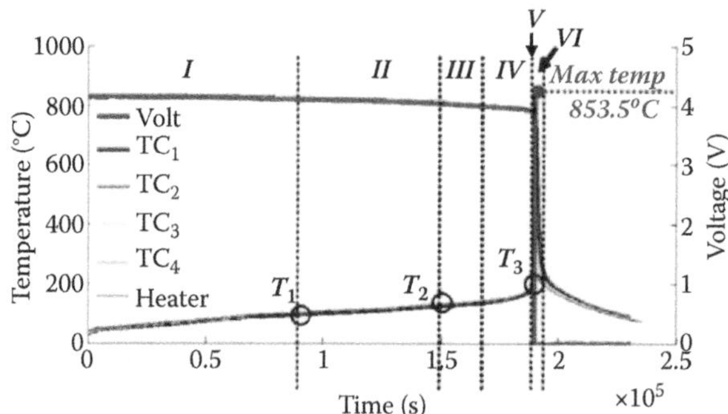

FIGURE 4.33 (*T, V*)–*t* curves of LIBs. (From Feng, X., Fang, M., He, X., Ouyang, M., Lu, L., Wang, H., and Zhang, M., *J. Power Sources*, 255, 294, 2014.)

9. Shutdown

   The shutdown characteristic of separators is deemed to be able to further enhance the safety of the LIBs. As shown in Figure 4.33, when battery short occurs, the temperature of the batteries would increase sharply, causing various side reactions and further inducing thermal runaway [264]. When the batteries are at a high temperature, separator shutdown could effectively prevent the thermal runaway by the internal pore closure [265]. However, the development of nonwoven separators with enhanced thermal resistance (melting temperature being >180°C) has now posed a challenge to the shutdown characteristic of separators. It is still a major point of contention whether this characteristic is indispensable for high-safety Li-ion batteries.

10. Chemical stability

   Since separators in LIBs are always in a long-term redox environment, they should own enough chemical stability during use. They also should not degrade and lose the mechanical strength.

4.3.3 Separator Classification

Recently, various kinds of separators have been developed. They could be divided into inorganic composite separator, polyolefin separator, nonwoven separator, polymer electrolyte, and ion-exchange membrane. The description, manufacture method, modification, and development of the separator are discussed below.

4.3.3.1 Inorganic Separator

Inorganic separators are usually prepared with inorganic particles, and sometimes a small amount of binder is used (this kind of separator is also called inorganic composite separator [259]). For example, Figure 4.34 gives the SEM images of an Al_2O_3 inorganic separator [266]. This kind of separator was prepared by a double-sintering process, in which the ethylene diamine tetraacetic acid (EDTA) was used as a pore-forming agent. The first sintering process at 1000°C is done to remove

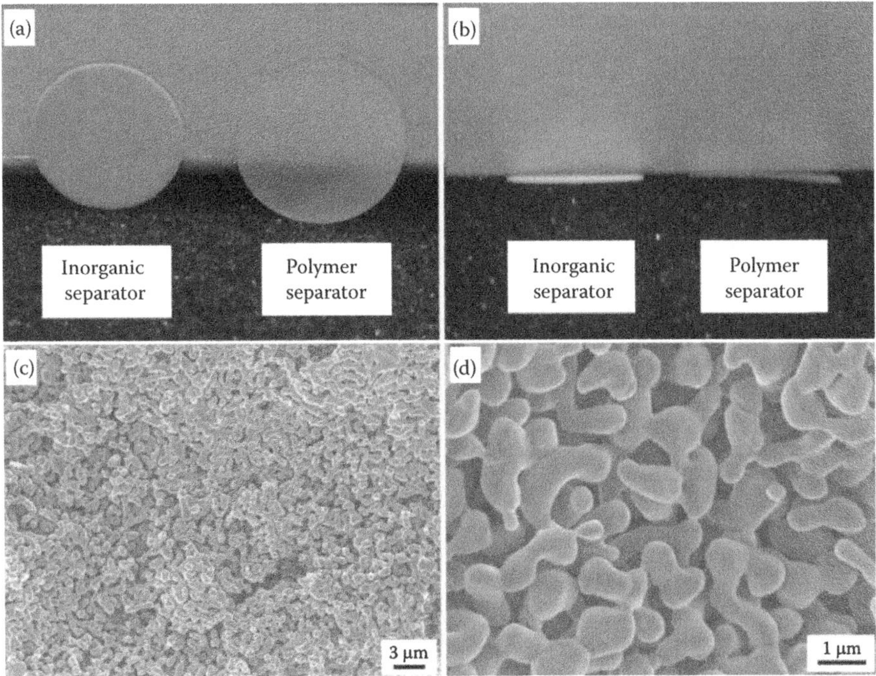

FIGURE 4.34 Photographs (a, b) and SEM images (c, d) of an Al_2O_3 inorganic separator and a polymer separator. (From Xiang, H., Chen, J., Li, Z., and Wang, H., *J. Power Sources*, 196, 8651, 2011.)

the EDTA for the formation of the pores. The second sintering process at 1500°C is done to enhance the mechanical strength of the separators. Zhang et al. [267] have reported a $CaCO_3$/polymer binder composite separator. The preparation process of the separator is that the solution containing the $CaCO_3$ and polymer binder is prepared and hot-rolled into a self-standing membrane with the evaporation of the solvent. Other inorganic particles used for the separators include MgO [268], γ-$LiAlO_2$ [17], TiO_2 [269], and so on. To obtain a well-dispersed inorganic composite separator, Carlson et al. [270] reported the application of the sol–gel method. In this process, the Boehmite sol is mixed with the polymer binder PVA, and then the homogenous solution is coated onto the polyester (PET) sheet. Finally, the sheet is removed by the water–isopropanol solution, giving rise to an inorganic composite separator.

Because of the existence of a large amount of inorganic particles, the separators usually show superior wettability with the nonaqueous liquid electrolyte and excellent thermal stability. However, the biggest drawback of this kind of separators lies in the poor mechanical strength, which limited their application in large-scale practical production process. To solve this problem and to effectively use the advantage of the inorganic separators, various kinds of composite separator using the organic porous separators as the mechanical support are developed [271,272], which would be introduced later.

4.3.3.2 Polyolefin Separator

Polyolefin separators are the most widely used ones not only in commercialized LIBs, but also in fundamental researches [273,274]. The advantages of the polyolefin separators are their low cost, excellent chemical resistance, good mechanical strength, etc. Therefore, this kind of separators occupies a leading position in liquid LIBs.

I. Preparation method

Currently, the commercialized polyolefin separators include polypropylene (PP) separator, polyethylene (PE) separator, and PP/PE/PP trilayer separator (Figure 4.35). The preparation methods for the polyolefin separators could be divided into the dry process [275] and the wet process [276]. The pore structure of separators prepared by the dry process (uniaxial stretching process) usually is slit-like, as shown in Figure 4.35d, while separators prepared by the wet process usually has a netty pore structure (Figure 4.35c).

The dry process represents the melting spinning–cold stretching method. In this process, the polymer (usually PP) is melted and extruded passing through a shaped die to form a precursor film. Then, the film is quenched and stretched for pore creation, giving rise to the porous separator. The principle of pore forming is shown in Figure 4.36. Pores are mainly formed by the fragmentation of the neighboring crystal lamellae during stretching [277]. To obtain the optimal pore structure, the polymer is required to have very high crystallinity (>80%). Thus, the dry process is usually used for the preparation of the porous PP separators. Meanwhile, a modification agent for blending purpose is not suitable for this kind of separators because the introduction of another substance would obviously decrease the polymer's crystallinity. The pore structure of the

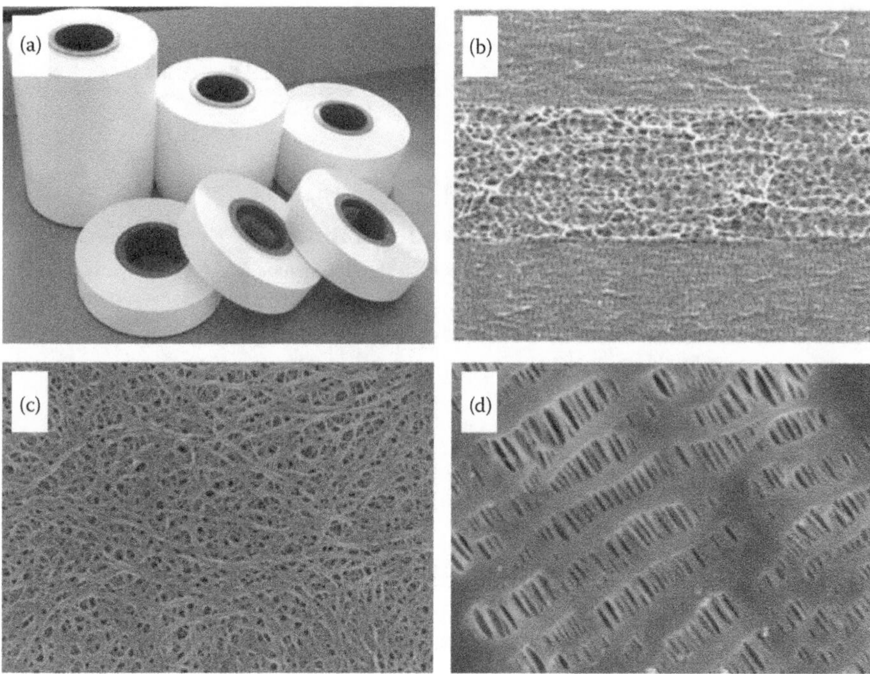

FIGURE 4.35 Photograph (a) and the SEM images of the PP/PE/PP trilayer separator (b), PE separator (c) and PP separator (d). (From Xiang, H., Chen, J., Li, Z., and Wang, H., *J. Power Sources*, 196, 8651, 2011.)

final separator is deemed to be determined by the properties of the precursor film, the annealing conditions, and the stretching and heat-setting conditions [259]. The representative products are the PP separators prepared by Celgard Company.

The wetting process also refers to the thermally induced phase separation (TIPS) process. In this process, the polymer is dissolved in the diluents to form a homogenous casting solution at high temperature (usually higher than the melting point of the polymer). The casting solution is extruded to form a liquid precursor film and then the film is pulled into the cooling bath to form the solid precursor film. The diluent is further removed by the extractant to obtain the porous membrane. Commonly, the uniaxial or biaxial stretching will be followed to achieve better pore structure. The apparatus for the TIPS process is shown in Figure 4.37 [278]. The advantages of the wetting process lie in the adjustable pore structure and the possibility that a second or additional component could be blended into the separator matrix for chemical modification purpose. A typical representative as the manufacturer for the separators prepared by the wetting process is Asahi Kasei Chemicals.

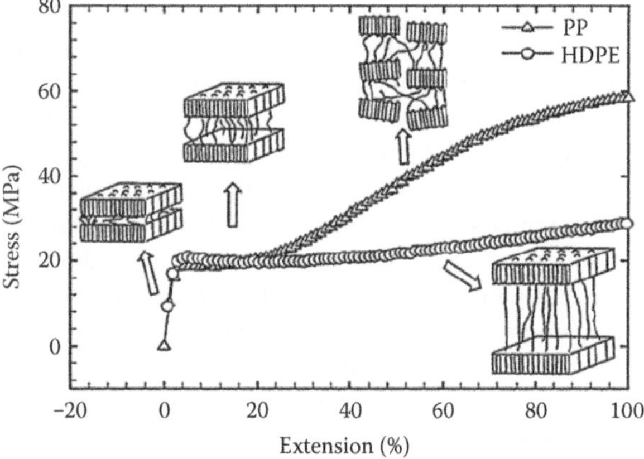

FIGURE 4.36 Sketch map of the pore-forming processes during the stretching process. (From Tabatabaei, S.H., Carreau, P.J., and Ajji, A., *J. Membr. Sci.*, 345, 148, 2009.)

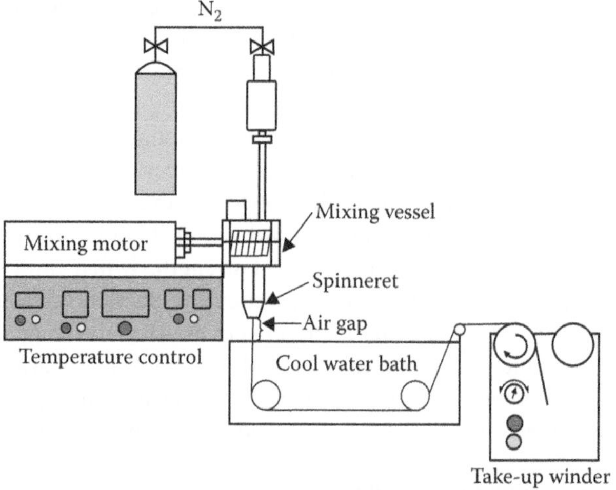

FIGURE 4.37 Schematic diagram of a batch-type extrusion apparatus for membrane preparation via a thermally induced phase separation process. (From Matsuyama, H., Okafuji, H., Maki, T., Teramoto, M., and Kubota, N., *J. Membr. Sci.*, 223, 119, 2003.)

II. Separator modification

1. Improving the wettability

The drawbacks of polyolefin separators are their hydrophobic property and lower surface energy, which cause poor compatibility with the liquid electrolyte. This property usually yields a very low electrolyte uptake for separators and a poor stability of liquid electrolyte trapped in the separators [279]. The poor wettability of the separator may also generate additional ionic conduction resistance in the interfacial layer of the separator and electrodes [280], which would deteriorate battery capacity and cycle lives. Furthermore, since achieving fast and uniform wetting for separators by liquid electrolyte is becoming an urgent requirement for high-power LIBs, the successful application of the polyolefin separators in the EVs and energy storage field is still a challenge [281]. To overcome these disadvantages, it is necessary to modify the surface properties of the polyolefin separator to increase the surface energy of the separators and to enhance the affinity of the separator with the polar liquid electrolyte. Various efforts have been done, including surface coating, chemical grafting, blending, and so on. Detailed introductions are as follows.

a. Surface coating

In comparison, surface coating is a convenient and effective way for separator modification, which is realized by depositing or coating the functional layer onto the membrane surface. The electrochemical performances of the modified separators are strongly dependent on the chemical composition, thickness, and morphologies of the coating layer. Jeong et al. [282] reported a modified PE separator coated with acrylonitrile (AN)–methyl methacrylate (MMA) copolymer. Introduction of the polar polymer onto the separator surface effectively enhanced the wettability of the separators. Meanwhile, the influences of the thickness of the coating layer on the electrochemical performances were also investigated. As shown in Figure 4.38, with the thickness slightly increasing, the capacity retention of the cells was effectively improved. The reason was considered to be the enhanced interfacial contact between the separator and the electrodes due to the adhesive gel formed form the coating layer.

To enhance the interaction between the coating layer and the separator substrate, Sohn et al. [283,284] investigated the performances of the electron beam irradiation–treated coated separators. This study found

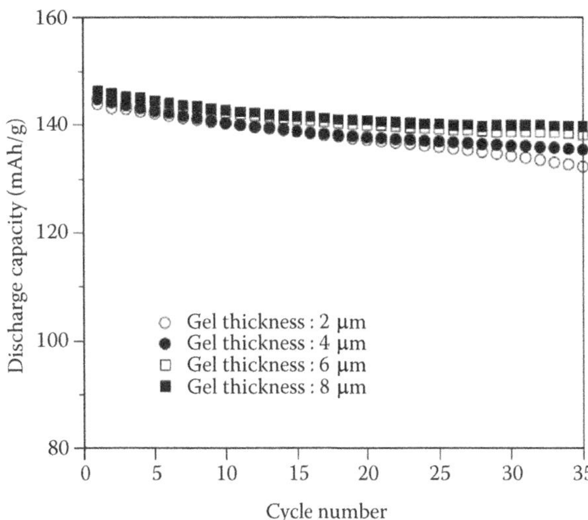

FIGURE 4.38 Dependence of the discharge capacity on the thickness of the coating layer (tested with the assembly of $LiCoO_2$/separator/lithium metal). (From Jeong, Y.-B. and Kim, D.-W., *Solid State Ionics*, 176, 47, 2005.)

that a porous structure was formed on the coating layer when the modified separators were treated under appropriate conditions, which is conducive to increasing the ion conductivity. Moreover, the irradiation-treated separators showed enhance thermal resistance due to the formation of the chemical bond between the coating layer and the separator. Kim et al. [285] also explored the gamma ray irradiation–treated modified separators. An obvious increase on the electrolyte uptake and the ion conductivity were achieved. The electrolyte uptake was increased from 120% to 140% and the ion conductivity from 3×10^{-4} to 4.5×10^{-4} S/cm.

A gel polymer electrolyte layer could be formed when the coating layer absorbs enough liquid electrolytes. Thus, a dense coating layer usually causes an obvious decrease in the ion conductivity, because the diffusion of lithium ions is more difficult. To solve this problem, Park et al. [286] developed a kind of close-packed poly(methylmethacrylate) (PMMA) nanoparticle array–coated PE separator, as shown in Figure 4.39. The well-connected interstitial voids between the PMMA nanoparticles allow the fast transport of lithium ions in the coating layer, yielding higher ion conductivity and improved discharge C-rate performance.

Inspired by the superior adhesive ability of polydopamine onto all kinds of substrates, the polydopamine coating layer was also used for separator modification [287,288]. Ryou et al. [289] first investigated the

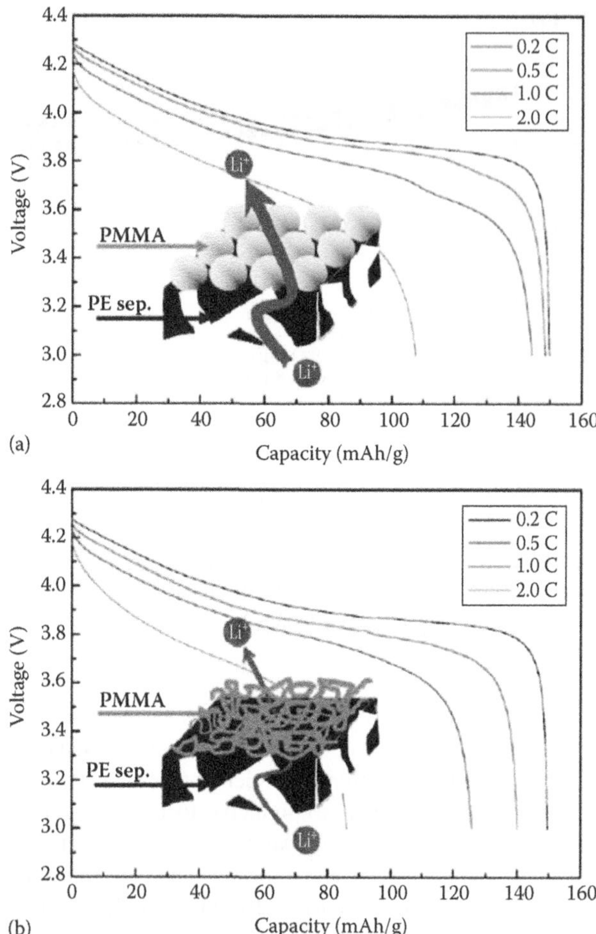

FIGURE 4.39 Comparison of the C-rate capacities of the PMMA nanoparticles (a) and the dense PMMA (b) coated PE separators (tested with the assembly of $LiCoO_2$/separator/lithium metal). (From Park, J.-H., Park, W., Kim, J.H., Ryoo, D., Kim, H.S., Jeong, Y.U., Kim, D.-W., and Lee, S.-Y., *J. Power Sources*, 196, 7035, 2011.)

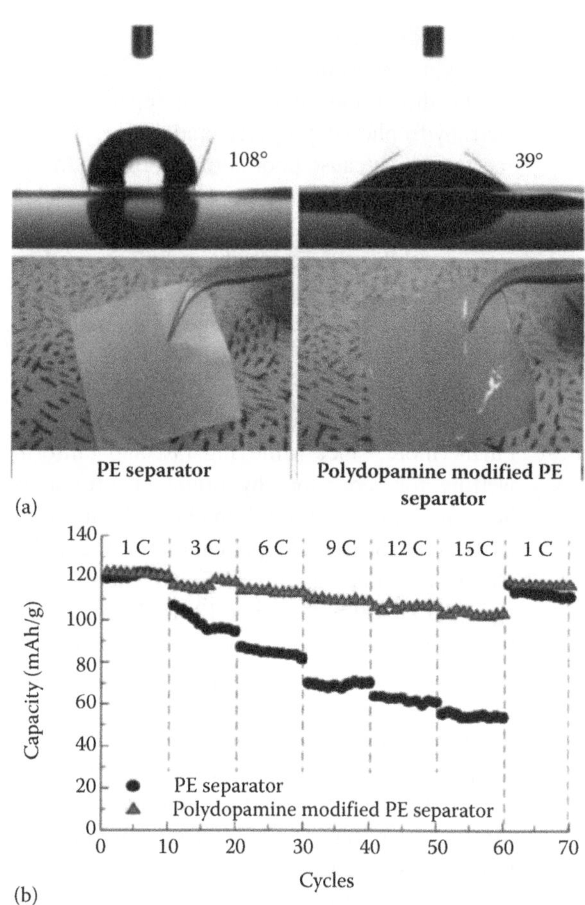

FIGURE 4.40 Contact angle (a) and the C-rate capacity (b) of the PE separator and the polydopamine-modified PE separator (tested with the assembly of $LiMn_2O_4$/separator/lithium metal). (From Ryou, M.-H., Lee, Y.M., Park, J.-K., and Choi, J.W., *Adv. Mater.*, 23, 3066, 2011.)

electrochemical performances of the polydopamine-treated PE separator. As shown in Figure 4.40, the contact angle of the modified separators obviously decreased, suggesting an enhanced interaction between the separators and the liquid electrolyte. A superior C-rate performance was also achieved. Meanwhile, Lee et al. [283] have proved that the introduction of the polydopamine could effectively enhance the thermal stability of the separators. However, the main drawback of this method is that it requires a long time (several hours) to realize the effective coating of polydopamine, which might hinder its application on a large scale.

b. Chemical grafting

The surface coating process usually yields an unstable modification layer. To improve the long-term durability of the modified separators, the chemical grafting method has achieved wide attention. Owing to the strong chemical inertness of polyolefin materials, irradiation grafting methods, such as plasma irradiation and electron beam irradiation, etc., are commonly used for separator modification [290–292]. The grafted polar monomers or polar segments could contribute to improved wettability and electrochemical performances. However, these modification techniques usually deteriorate the mechanical strength of the separators. To solve this problem, Fang et al. [293] prepared a kind of methoxypolyethylene glycol (MPEG) grafted PP separator with polydopamine as the transition interface (Figure 4.41). The advantages of this modification method are its low cost, low energy consumption, and high security. The introduction of the MPEG layer effectively enhanced the interfacial stability of the separators and the electrodes. Moreover, the cyclability of the modified separators was obviously improved. To further increase the ion conductivity of the

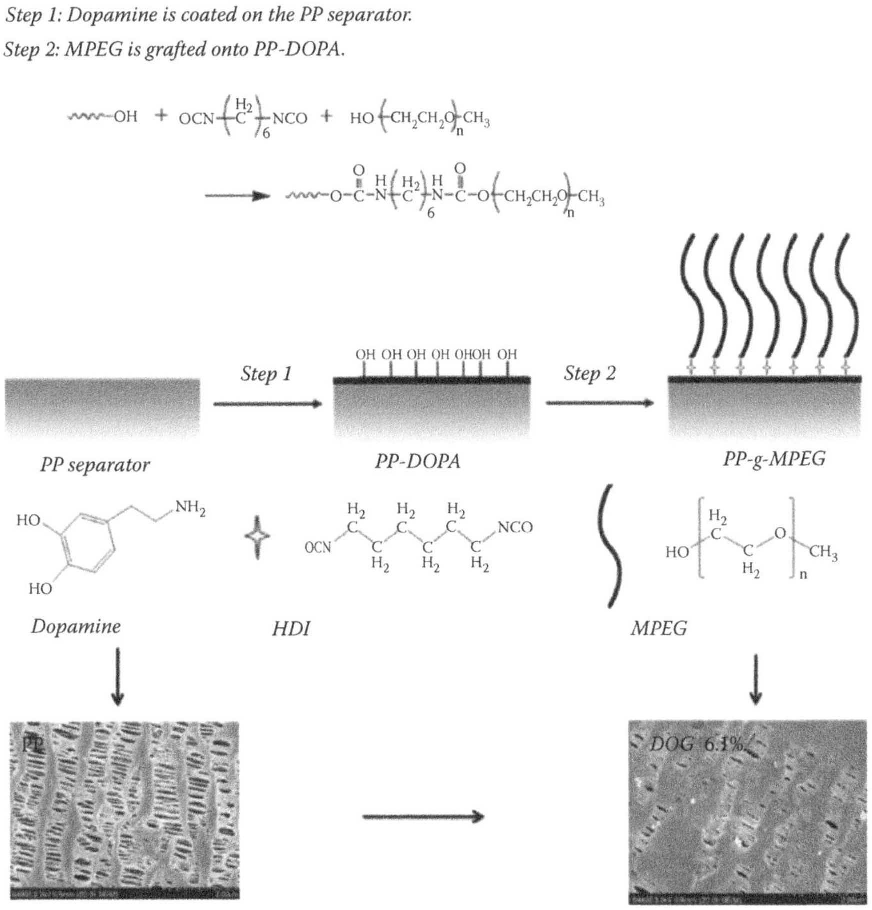

FIGURE 4.41 Flow diagram of the preparation process of the MPEG grafted PP separators (PP-g-MPEG). (From Fang, L.-F., Shi, J.-L., Zhu, B.-K., and Zhu, L.-P., *J. Membr. Sci.*, 448, 143, 2013.)

modified separators, Gwon et al. [294,295] treated the PMMA grafted PP separators with a nonsolvent-induced phase separation process to create pores on the modification layer. The porous structure helped the lithium ions to migrate fast through the separators during the cycling process, yielding better electrochemical performances.

c. Blending

The blending method is widely used for membrane modification, which is also a simple and effectively way [296]. By choosing the optimum blending reagent and preparation conditions, the blending method could not only improve the properties of the membrane surface effectively, but also change the properties of the whole membrane matrix [297].

To improve the performances of PE-based separators, Shi et al. [298] incorporated the polyether chains into PE separator by blending with poly(ethylene-*block*-ethylene glycol) (PE-*b*-PEG) into the PE matrix via the TIPS process. The ionic conductivity of liquid electrolyte–soaked blend separators was up to 1.24 × 10^{-3} S/cm at 25°C. The interface resistance between the separator and electrode was effectively reduced by the surface-enriched polyether chains. As shown in Figure 4.42, the contact angle of the blend separators with the liquid electrolyte decreased obviously with increasing PEG content (from M0-D to M3-D), suggesting an enhanced interaction between the separator and the liquid electrolyte. Better cycling performance was also obtained. The advantages of the blending method via the TIPS process lies in that the preparation and modification process could be achieved simultaneously. Meanwhile, the separator structure could be controlled more easily and higher porosity could be realized. This is a promising way for high-performance separator production in the large scale.

d. Gel incorporation

The gel polymer electrolyte owns better liquid electrolyte retention ability and performs at a higher security [299]. The main drawback lies in the poor mechanical strength. On the basis of this consideration,

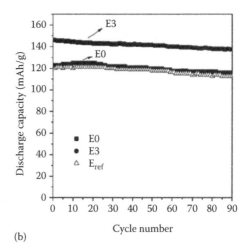

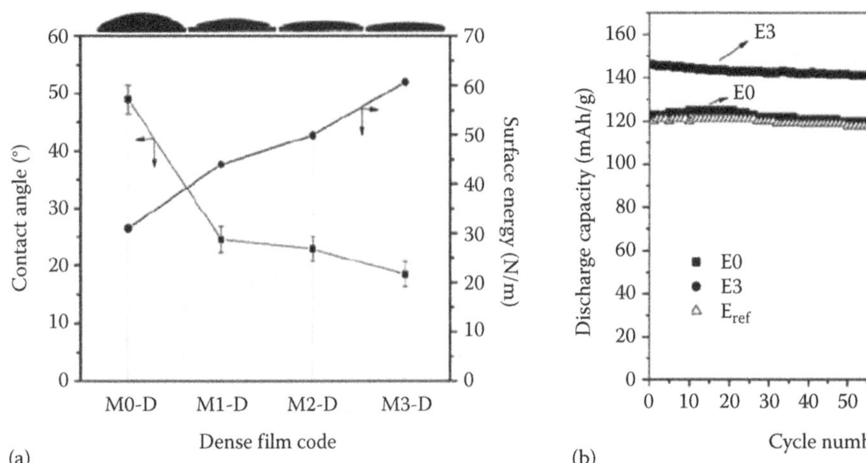

FIGURE 4.42 (a) Contact angle and surface energy of the blending separators. (b) Discharge capacity of the pure PE separator (E0), the blending separator (E3), and the commercialized PE separator (E_{ref}) with the cycle numbers (tested with the assembly of $LiFePO_4$/separator/graphite). (From Shi, J.-L., Fang, L.-F., Li, H., Liang, Z.-Y., Zhu, B.-K., and Zhu, L.-P., *J. Membr. Sci.*, 429, 355, 2013.)

Li et al. [300] first introduced the cross-linked poly(ethylene oxide) (PEO) gel polymer electrolyte into the pores of the PP separators. A kind of "active separator" was obtained, for which the PP separators provided enough thermal stability and mechanical strength and the gel part conferred improved wettability. This work provided a new concept for separator design. Wang et al. [301] reported similar work.

2. Improving the thermal stability

As discussed above, the thermal stability of the separators is very important for battery safety. However, the melting point of polyolefin materials is about 130–165°C. When the temperature is further increased, the polyolefin-based separators would tend to meltdown, causing short circuit and security incidents. It can be said that separator thermal resistance has been a key limiting factor for the application of polyolefin separators in high-power lithium batteries [302]. Thus, enhancing the thermal stability of polyolefin separators is also an important research direction. The reported modification methods include surface coating and chemical cross-linking.

a. Surface coating

Introducing various kinds of inorganic nanoparticles is the most commonly used method [303,304]. Jeong et al. [305] reported a kind of closely packed SiO_2 nanoparticle/poly(vinylidene fluoride-hexafluoropropylene) (PVDF-HFP) layer–coated PE separator (Figure 4.43a), and investigated the influences of particle size (40 and 530 nm) on the battery's electrochemical performances. Owing to the existence of the heat-resistant SiO_2 coating layer, the composite

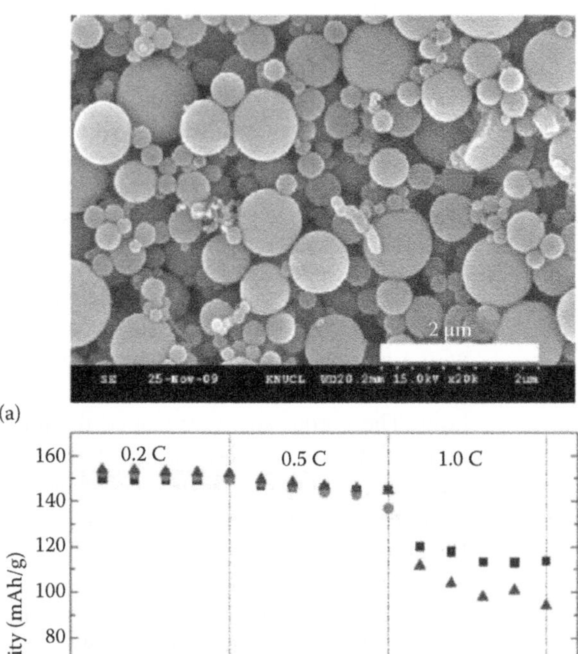

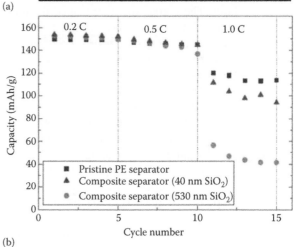

FIGURE 4.43 (a) SEM images of the closely packed SiO_2 nanoparticles/PVDF coated PE separators and (b) the C-rate performances of cells containing PE separators and the modified separators (tested with the assembly of $LiCoO_2$/separator/graphite). (From Jeong, H.-S. and Lee, S.-Y., *J. Power Sources*, 196, 6716, 2011.)

separators showed significant reduction in thermal shrinkage. Meanwhile, a comparison found that the small-sized (40 nm) nanoparticles in the coating layer contributed to a more facile ion transport and better C-rate performance (Figure 4.43b). A series of commercialized ceramic-coated separators have been widely promoted and used in LIBs. For example, Hitachi Maxell has produced a kind of plate-like ceramic coated separator that showed little dimensional change even after being treated at 180°C. The company used its accumulation in the technologies of tape manufacturing to form inorganic nanoparticles with uniform shape and size. The plate-like ceramic would not obviously increase the separator thickness. Therefore, even if the separators are very thin, they can still show excellent thermal stability.

Chung et al. [306] prepared surface-coated PE separators with high thermal resistance polymers (poly(ethylene glycol dimethacrylate). The shutdown temperature and the meltdown temperature were reported to be increased to 142°C and 155°C.

b. Chemical cross-linking

Previous reports [307,308] have pointed that high-energy irradiation could induce internal cross-link reactions for the PE material and improve the heat resistance of the PE material. On the basis of those studies, Kim et al. [309] treated the PE separators by gamma ray. The study found that the cross-link of polyethylene is strongly effective against the thermal shrinkage. The melt integrity temperature is increased from 146°C to 166°C. However, it should be noted that this method could only be applied to PE base separators, since the PP materials would tend to degrade when they are deposited at high-energy irradiation.

4.3.3.3 Nonwoven Separator

Nonwoven fabric is obtained by the random orientation and distribution of the fibers to form the network structure, and is sticky fixed by mechanical, thermal, and chemical methods. The used fibers could be classified into glass fibers, ceramic fibers, and synthetic fibers. The typical features of the nonwoven separators are the very high porosity (60–80%) and the 3D pore structure.

1. Preparation method

Formation of the nonwoven webs includes a wet process, e.g., the papermaking process [310], electrospinning technique [311,312], wet-laid method [313,314], and a dry process (a melt blowing process) [315,316].

Zhang et al. [317] prepared aramide-based nonwoven separators via the papermaking process, which is illustrated in Figure 4.44. The obtained separators had a porosity of 70% and showed better uptake ability for the liquid electrolyte than the commercialized PP separator, yielding better cycling performances. More notably, the nonwoven separators owned excellent thermal stability, which showed no obvious dimensional change when they are treated at 250°C for 0.5 h. The superior thermal resistance comes from the high melting temperature of the separator matrix (aramide).

Electrospinning is a particularly low-cost and versatile method for manufacturing nanofibers and nonwoven mets [318]. The apparatus for the electrospinning process is shown in Figure 4.45. Various kinds of materials can be used in the electrospinning process, such as PE, PP, PVDF, PVDF-HFP, poly(vinyl chloride) (PVC), and so on [319–322]. Recently, with the development of the

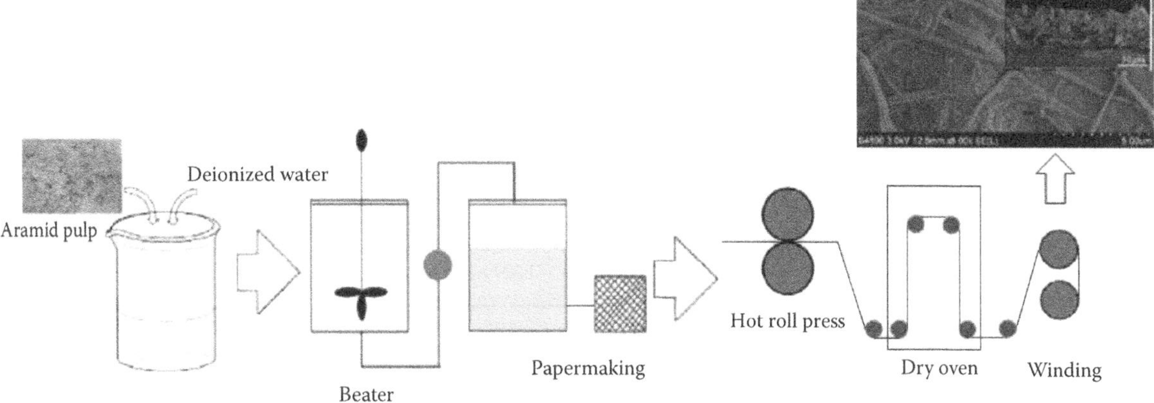

FIGURE 4.44 Schematic of a papermaking process for aramide nonwoven separators. (From Zhang, J., Kong, Q., Liu, Z., Pang, S., Yue, L., Yao, J., Wang, X., and Cui, G., *Solid State Ionics*, 245–246, 49, 2013.)

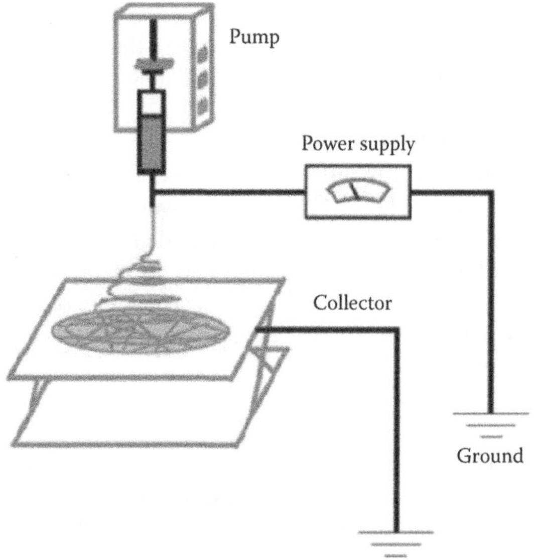

FIGURE 4.45 Schematic diagram of the electrospinning apparatus. (From Dong, Z., Kennedy, S.J., and Wu, Y., *J. Power Sources*, 196, 4886, 2011.)

more sophisticated electrospinning technology, nonwoven separators with heat-resistant engineering polymer, such as polyimide (PI) [323], PET [324], poly(phthalazinone ether sulfone ketone) [325], etc., as matrix have attracted lots of attention for high-power LIBs. The heat-resistant engineering polymers usually have a higher melting temperature (>250°C) than polyolefin materials (melting temperature: 130–165°C). Thus, when the LIBs are deposited at a high temperature (e.g., 180°C), this kind of nonwoven separators could still keep dimensional stability and effectively isolate the electrodes, further preserving battery short circuit and explosion and giving enhanced battery safety. Furthermore, the high porosity and large pore size of the nonwoven separators could yield higher C-rate capability of the LIBs (as shown in Figure 4.46) [326]. Thus, this kind of separator is now deemed to be a promising candidate for the separators used in high-power LIBs.

Because of the outstanding characteristics of the nonwoven separators, many well-known companies are now committed to the development and promotion of commercialized nonwoven separators. For example, DuPont Company has introduced polyamide nonwoven products in 2011. Mitsubishi Paper Mills Limited has also produced a series of PET-based nonwoven separators via the wet-laid process.

2. Separator modification

The shortcomings of the nonwoven separators lie in the excessively large pore size and broad pore size distribution. These features usually cause internal short circuits, self-discharge, overgrowth of the lithium dendrite, and electrolyte leakages, which would obviously deteriorate the battery performances [327]. To resolve the limitations of the nonwoven separators, Zhu et al. [328] prepared a kind of PVDF gel polymer electrolyte–doped nonwoven separator. The gel part could effectively prevent liquid electrolyte leakage and provide a more intimate contact with the electrodes to enhance the interfacial stability [77]. However, the fully populated pores are not conducive to the rapid diffusion of the lithium ion, which might further restrict the improvement of battery power performances. Lee et al. [329–331] developed various kinds of colloidal particle and inorganic nanoparticle composite nonwoven separators. The introduction of the particles helped decrease the pore size of the nonwoven substrate. Meanwhile, the well-connected interstitial voids between the particles could provide an easy access for liquid electrolyte penetration and lithium-ion transportation. Shi et al. [332,333] reported a PI matrix–enhanced

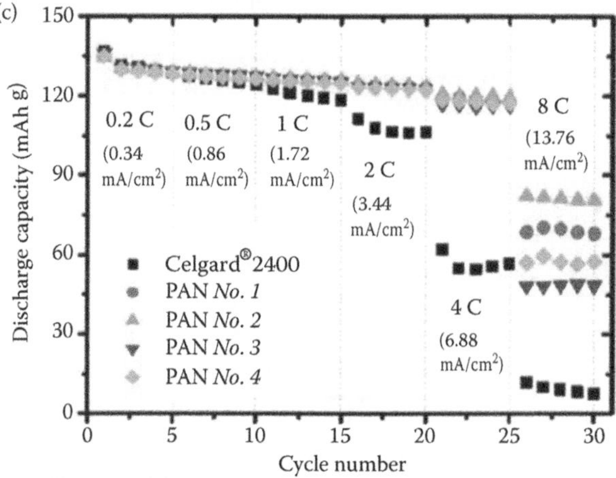

FIGURE 4.46 SEM images of the PAN nonwoven separators with (a) 380-nm nanofibers and (b) 250-nm nanofibers, and (c) C-rate capacities of the cells containing the nonwoven separators and the PP separator (tested with the assembly of $LiCoO_2$/separator/graphite). (From Cho, T.-H., Tanaka, M., Onishi, H., Kondo, Y., Nakamura, T., Yamazaki, H., Tanase, S., and Sakai, T., *J. Power Sources*, 181, 155, 2008.)

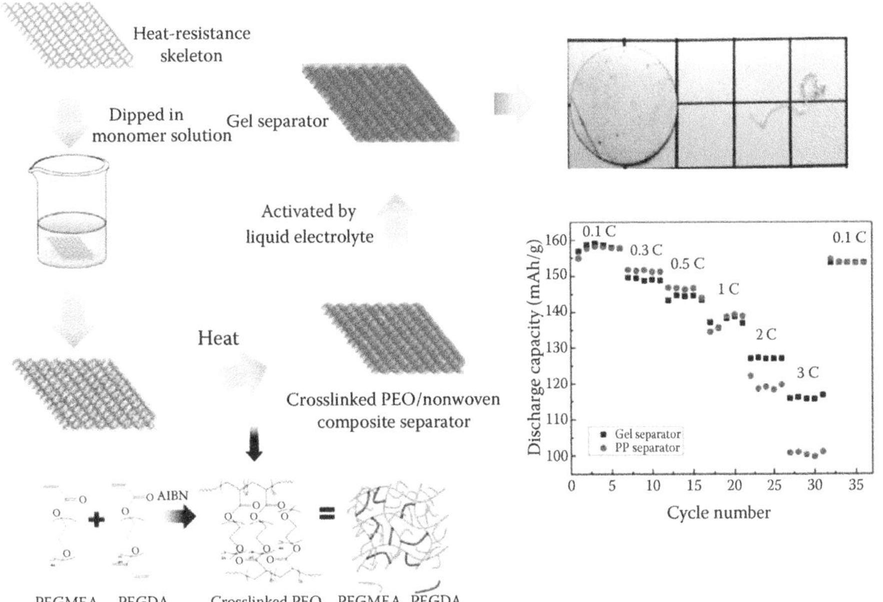

FIGURE 4.47 Schematic of the preparation process of the gel separator and the thermal resistance (180°C, 1 h), and C-rate capacities of the gel separator and the PP separator (tested with the assembly of $LiFePO_4$/separator/lithium). (From Shi, J.L., Hu, H., Xia, Y., Liu, Y., and Liu, Z., *J. Mater. Chem. A*, 2, 9134, 2014.)

cross-linked gel separator that was manufactured by introducing the cross-linked PEO gel by a simple dip-coating and heat-treatment method into the PI nonwoven separators (Figure 4.47). The cross-linked gel part gave separators enhanced affinity with liquid electrolyte and could solve the limitation of long-term durability of the composite separators in use. Moreover, the separators possessed better cyclability and rate capability than the traditional PP separator, implying a promising potential application in high-power, high-safety LIBs.

To overcome the problems of the pristine nonwoven separators and to further optimize the separator performances, the modified/composite nonwoven separators have also been launched in the separator market, e.g., the "Nanobase X" produced by Mitsubishi Paper Mills Limited, which is a ceramic-coated nonwoven separator, and the SYMMETRIX HPX and HPXF separators of Porous Power Technologies, which are composite separators with PVDF and nonwoven webs.

4.3.3.4 Polymer Electrolyte

Polymer electrolyte has attracted considerable attention in recent years because of its higher safety, in which the liquid electrolyte is fixed by the polymer chains or no liquid electrolyte exist; thus, the leakage of the electrolyte could be effectively hindered to avoid security incidents [334,335]. Based on the phase state, the polymer electrolyte could be divided into the solid polymer electrolyte and the gel polymer electrolyte.

The solid polymer electrolyte is defined as the system in which lithium salts are dissolved in the ion-solvating polymers (PEO, PPO, PMMA, etc.) [336]. The motion of the lithium ions is coupled to the free volume and local segmental mobility of the flexible polymer chains. The electrochemical performances strongly depend on the chain structure, molecular weight, crystallinity, and glass transition temperature (T_g) of the polymers. Generally, low crystallinity and low T_g correspond to high ion conductivity of the solid polymer electrolyte. Conversely, the ion conductivity is relatively low [337]. Now, studies on the solid polymer electrolyte are mainly focused on the improvement of ion conductivity, which is the main limitation for the large-scale application of the solid polymer electrolyte. Various efforts have been done, including blending [338], cross-linking [339], introducing inorganic nanoparticles [340], and so on. The objective of these studies is to decrease the crystallinity of the polymer matrix and enhance the mobility of the lithium ions.

The concept of gel polymer electrolyte was first proposed by Feuillade, which is prepared by introducing the organic solvent into the solid polymer electrolyte. In comparison with the solid polymer electrolyte, the transport of lithium ions is also dependent on the liquid electrolyte solvent, contributing relatively higher ion conductivity. Currently, the extensively investigated gel polymer electrolytes include the PEO- [341,342], PAN- [343,344], PVDF- [345,346], and PMMA- [347] based systems. Then, the PVDF-based gel polymer electrolytes have been commercialized, e.g., the SYMMETRIX HP series of Porous Power Technologies.

Poor mechanical strength and thermal stability are now the core issues of the gel polymer electrolytes. To overcome these problems, Kim et al. [348] introduced Al_2O_3 nanoparticles onto both sides of a PMMA gel polymer electrolyte, as shown in Figure 4.48. The dimensional stability and thermal stability of the gel polymer electrolyte were greatly improved. Jung et al. [349] reported a kind of PMMA/PVC composited gel polymer electrolyte prepared by the electrospinning method. The high

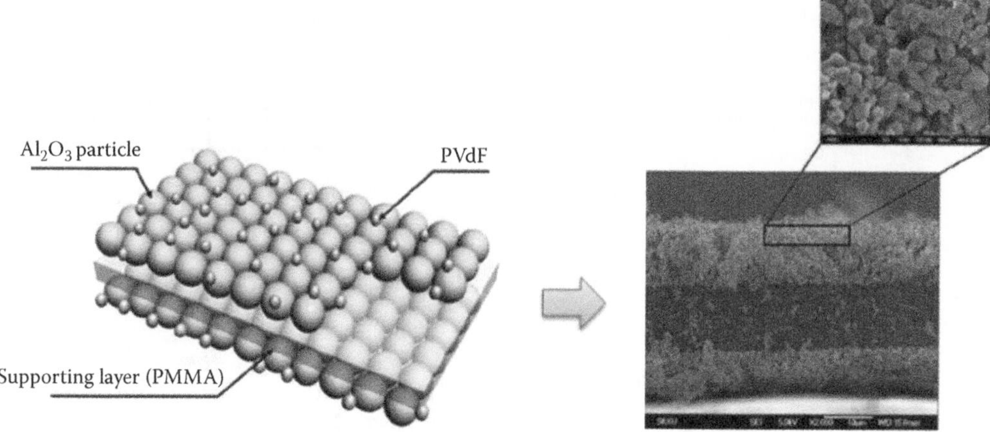

FIGURE 4.48 Schematic illustration and the SEM image of the trilayer gel polymer electrolyte. (From Kim, M., Han, G.Y., Yoon, K.J., and Park, J.H., *J. Power Sources*, 195, 8302, 2010.)

porosity and fully interconnected pore structure gave the gel polymer electrolyte a high ion conductivity of 7.8×10^{-3} S/cm.

4.3.3.5 Ion-Exchange Membrane

Ion-exchange membrane is an ionic group–containing polymer film. The commonly used ion-exchange membranes in LIBs are lithiated perfluorinated sulfonic polymer films [350]. They are swollen by the nonaqueous solvent to form polymer electrolyte. The unique feature of this kind of polymer electrolyte is the absence of the additional lithium salts in the electrolyte. The charged species are bound to the side chains of the polymers. That is to say that the swollen ion-exchange membranes act as both separator and electrolyte. In the charge–discharge process of LIBs, the only mobile species, i.e., lithium ions, migrate along the direction of the electric field, giving rise to a transference number of unity [351].

Work on the ion-exchange membrane is still in the basic research stage, e.g., the design and regulation of polymer chain structure, since the material cost is relatively high and the ion conductivity needs to be further improved. However, Liu et al. [352] have reported the application of the ion-exchange membrane in improving the high-temperature retention of $LiMn_2O_4$ cathode LIBs, as shown in Figure 4.49. Thus, these studies might provide some inspiration for the separator/polymer electrolyte designation in the LIBs containing manganese cathode materials, i.e., $LiMn_2O_4$, $LiMnPO_4$, etc.

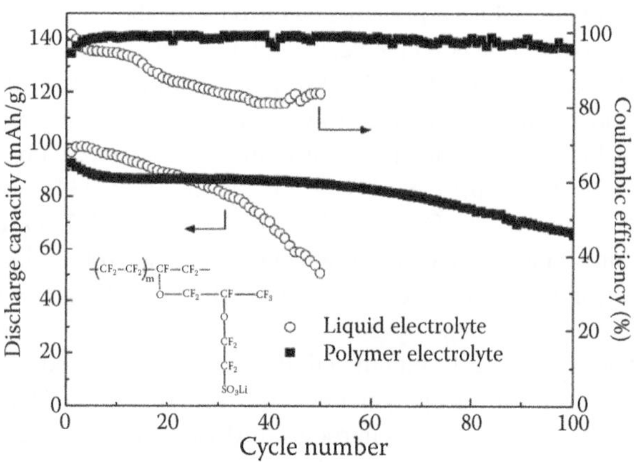

FIGURE 4.49 Chemical structure of lithiated perfluorinated sulfonic acid and cycling stability of the Li/ion-exchange membrane/$LiMn_2O_4$ battery. (From Liu, Y., Tan, L., and Li, L., *Chem. Commun.*, 48, 9858, 2012.)

In conclusion, all kinds of separators have their own advantages and disadvantages. The polyolefin separators are the most widely used commercial separators in today's market. However, this type of separators is deemed to be suitable for small cells (<3 Ah) [353], which are used in portable electronic products, e.g., mobile phones, laptop computers, digital cameras, and so on. For high-power LIBs used in (H) EVs and energy storage, higher safety standards and higher performance requirements necessitate separators to have better thermal stability, flame retardancy, and electrochemical performance, which are also the future directions of the separators.

4.4 NONAQUEOUS LIQUID ELECTROLYTES

4.4.1 Introduction

The nonaqueous electrolytes are an indispensable constituent in LIBs, and their role is to conduct electricity by means of the transportation of charge carried between the pair of electrodes. The nonaqueous electrolyte should be a poor electronic conductor but a good ionic conductor. Because the electrolyte is being sandwiched between positive and negative electrodes, in close interaction with both electrodes, the key to a safe and high-performance LIB lies in the identification of a suitable electrolyte [354–356]. Generally, the electrolyte refers to a solution comprising the lithium salts and solvents, and functional additives. In accordance with the basic requirements for LIBs, the ideal electrolyte generally should meet the following prerequisites [357]:

1. High chemical and thermal stability
2. High ionic conductivity (>10^{-3} S/cm)
3. A wide electrochemical window between its lowest unoccupied molecular orbital (LUMO) and its highest occupied molecular orbital (HOMO)
4. Remain inert to other cell components such as the separator, electrode substrates, and cell packaging materials
5. With safe, nontoxic, and economical components

During the past three decades, tremendous research efforts have been devoted to investigating the electrochemical performance of a wide variety of electrolytes, as outlined in Figure 4.50. In the early era of LIBs, alkyl carbonates (e.g., binary solvent mixtures such as EC and either dimethyl carbonate [DMC], ethyl methyl carbonate [EMC], or DEC) and the lithium salt, lithium hexafluorophosphate ($LiPF_6$), were used in commercialized LIBs. During the last decade, the introduction of high-voltage positive materials such as $LiNi_{0.5}Mn_{1.5}O_4$, $LiMPO_4$ (M = Mn, Co, V), Li_2MPO_4F (M = Ni, Co), and lithium-rich layer oxides has fostered the development of high-energy LIBs [358]. However, the requirement of high-voltage cathode materials is a great challenge for the conventional electrolytes, which tend to be oxidized at ~4.5 V. Although the oxidization potential of organic carbonate solvents is about 5 V, the transition-metal ions could catalyze the oxidation reaction and accelerate the decomposition of electrolytes at lower potentials, leading to rapid capacity fading [359–365]. Therefore, searching for matched electrolytes is essential to the realization of high-voltage LIBs. The present studies on high-voltage electrolytes [366] mainly focus on novel stable solvents and electrolyte additives.

At the same time, the safety issue continues to be the biggest obstacle retarding the commercial applications of high-rate and high-capacity LIBs in EVs and renewable power stations [367–369]. It is now well recognized that the use of highly flammable organic carbonate electrolytes in current lithium battery technology, which might cause serious hazards of firing and explosion under abusive conditions such as electrical overcharge, short-circuit, or high thermal impact, is a major cause for the hazardous behaviors of LIBs [355,370]. To overcome this problem, much effort has been focused on utilizing flame-retardant solvents as either an electrolyte additive or a co-solvent for lowering the combustibility of the nonaqueous electrolytes [371,372]. Therefore, searching for flame-retardant electrolytes agrees with improving the safety of LIBs.

In this section, we review recent advances in key electrolyte systems developed for currently prevailing lithium rechargeable batteries, with a focus on classic electrolytes, high-voltage electrolytes, and flame-retardant electrolytes.

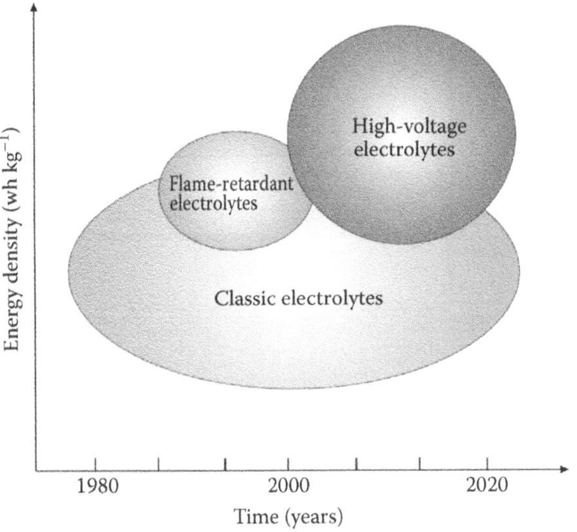

FIGURE 4.50 Comparison of the different electrolytes for rechargeable lithium batteries.

4.4.2 Classic Electrolytes

Most compositions of Li-ion electrolytes are based on solutions of one or more lithium salts in mixtures of two or more solvents, and single-solvent formulations are very rare, if there are any. The reason behind this mixed solvent formulation is that any individual compound, for example, high fluidity versus high dielectric constant, can hardly meet the diverse and often contradicting requirements of battery applications; therefore, solvents of very different physical and chemical natures are often used together to perform various functions simultaneously. A mixture of salts, on the other hand, is usually not used, because the anion choice is usually limited, and the performance advantages or improvements are not readily demonstrated. This section presents an overview of the various components of electrolyte solutions for lithium rechargeable batteries. More emphasis is paid toward solvents and lithium salts. The physicochemical and functional parameters relevant to electrochemical stabilities are also presented.

4.4.2.1 Solvents

Solvents are the main component of the electrolyte, which contain carbonates, ether, and carboxylic acid esters. The electrochemical performance of the electrolyte depends essentially on the various parameters of the solvents (i.e., viscosity, dielectric constant, melting point, boiling point, flash point, and the electrochemical windows) [357]. A good solvent is used to achieve low resistance, long life, and safety for the LIB. Here, recent progress in solvents applied in the currently prevailing electrolytes in LIBs is reviewed.

Carbonates were found to be the best and most suitable solvents for commercialized LIBs, as their properties are crucial for LIBs with high chemical and electrochemical stability, a wide electrochemical window, and a wide operating temperature range. These carbonates can be classified into cyclic and noncyclic groups according to their molecular structure. EC and propylene carbonate (PC) are two well-known examples of the cyclic carbonates, whereas the noncyclic carbonates include DMC, EMC, and DEC, as illustrated in Figure 4.51.

Among these carbonates, PC [373] has undoubtedly attracted the most attention throughout the entire history of

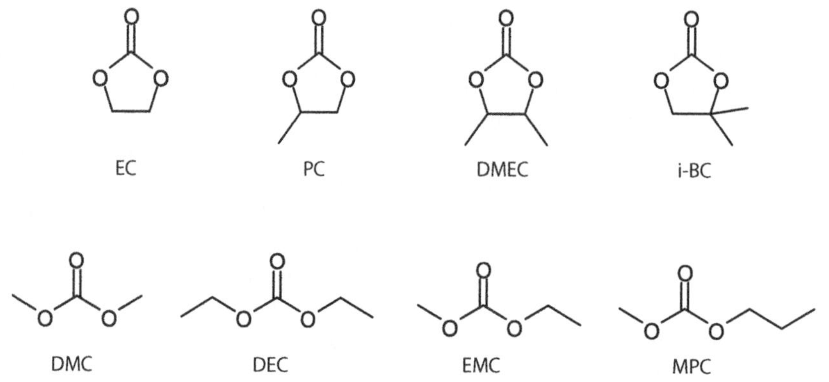

EC PC DMEC i-BC

DMC DEC EMC MPC

FIGURE 4.51 Chemical structures of carbonates.

lithium batteries, especially in the past decade. Until recently, the solvent mixture of PC and dimethoxyethane (DME) is still being used in lithium primary batteries. Because PC has a low melting point of −49.2°C, a high boiling point of 241.7°C, and a high flash point of 132°C, the PC-based electrolyte can enhance the low temperature and safety performance of the electrolyte. Unfortunately, although PC is able to substantially improve the low-temperature performance of the electrolyte, it is difficult for practical application because PC can decompose reductively on the graphite electrode at a potential of ~0.80 V in earlier work, and cannot form a stable SEI on graphite to prevent the further decomposition of PC solvent [374,375]. For example, it was found that PC could lead to the exfoliation of a lamellar structure graphite negative electrode, which makes the use of an SEI additive necessary [376–379]. Therefore, EC is used in almost all commercial compositions because it can provide the distinctive protective layer on the surface of graphite. Despite the seemingly minute difference in molecular structure between EC and PC, EC has a higher dielectric constant (89.78) than PC (64.92). EC was found to form an effective, compact, and stable SEI film on graphite anode, which prevents continuous electrolyte reduction, significantly improving the cycling performance of the batteries [380,381]. However, because of its high melting point (~36°C) and high viscosity, its single-solvent electrolyte solutions cannot be used as a low-temperature electrolyte. Therefore, the addition of low-viscosity, low-dielectric-constant, and low-boiling-point and low-flash-point solvents such as linear carbonates, e.g., DMC, DEC, EMC, and methyl propyl carbonate (MPC), is mandatory to enlarge the temperature range of use. Such binary mixture solutions based on cyclic carbonates (EC) and linear dialkyl carbonates are widely used as classic electrolytes for LIBs [382]. As a rule of thumb, a cyclic carbonate is desired for its high dielectric constant to reduce ion pairing or to increase the mobility of ions in the electric field; a linear carbonate is selected for its low viscosity for faster lithium-ion diffusion in the electrolyte. Though the addition of low-viscosity and low-boiling-point solvents, linear carbonates, is desired for their low viscosity for faster lithium-ion diffusion in the electrolyte to improve the low-temperature performance of the electrolyte, the flammability of the liquid electrolyte should be lowered greatly for the safety of LIBs [383]. Therefore, many studies have been focused on the development of the flame-retardant electrolytes, or even nonflammable electrolytes.

The liquid range of a nonaqueous electrolyte system is defined at the upper limit by the temperature at which one of its components begins to vaporize and at the lower limit by the temperature at which one of its components begins to crystallize. Surprisingly, despite the significance of the liquid range to practical applications, there are rather few studies dedicated to the thermophysical properties of the electrolytes. To try to delineate a temperature range for LIBs, Ding et al. [384,385] published a phase diagram (Figure 4.52) that provided a complete representation for all the binary combinations of some common carbonates: EC, PC, DMC, EMC, DEC, dimethyl ethylene carbonate, and isobutylene carbonate. The results showed that all the combinations formed simple eutectic systems, although the details of their phase diagrams

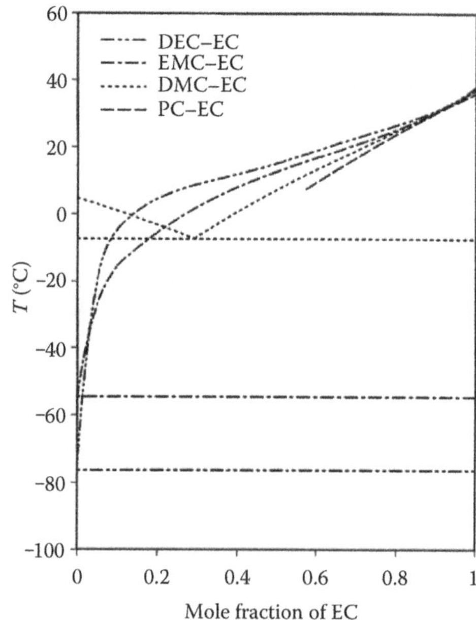

FIGURE 4.52 Liquid/solid phase diagrams of DEC–EC, EMC–EC, DMC–EC, and PC–EC combined with EC as the common component. (From Ding, M.S., Xu, K., Zhang, S., and Jow, T.R., *J. Electrochem. Soc.*, 148, A299, 2001.)

greatly varied, showing that DEC melted at −74.3°C instead of the currently accepted value of −43°C, and that DEC was less effective than either EMC or DMC in bringing down the temperatures of the liquids' binary mixtures with EC. These results followed the trend we had observed earlier in which an expansion toward the low temperature of the liquid region of a binary system was facilitated by two components having close melting points and similar molecular structures. Because EMC has a low freezing point (−55°C), Plichta et al. [382] used EMC as a co-solvent with EC and DMC in a ternary solution to increase the liquid range of the LIB electrolyte. Thus, Li/$LiCoO_2$ and graphite/$LiCoO_2$ button cells, using 1 M $LiPF_6$/EC–DMC–EMC (1:1:1, by volume) as the electrolyte, were found to be operable at temperatures down to −40°C. Moreover, carbonate-based electrolytes used in lithium-ion electrochemical cells were found to undergo ester-exchange reactions, where the use of DMC and DEC resulted in the *in situ* formation of EMC [386]. The reaction was found to be reversible and occurs during the first charge cycle of the $LiCoO_2$/petroleum coke lithium-ion system (Equation 4.2).

$$2\text{EMC} \Leftrightarrow \text{DMC} + \text{DEC} \tag{4.2}$$

In the 1980s, ethers were widely preferred by researchers as an alternative candidate because of their low viscosity and resultant high ionic conductivity. Similarly, these ethers can be classified, according to their molecular structure, into cyclic and noncyclic groups. The cyclic carbonates include tetrahydrofuran (THF), 2-methyltetrahydrofuran (2-MeTHF), 1,3-dioxolane (DOL), 4-methyl-1,3-dioxolane (4-MeDOL), and so on (Figure 4.53). THF- and DOL-based binary solvents with PC have been examined mainly as the electrolytes for lithium primary batteries. However, DOL cannot be used in LIBs, which could be attributed to the polymerization of the DOL at temperatures greater than 100°C and at voltages higher than 4 V [387–389]. Although 2-MeTHF shows a low boiling point (79°C), low flash point (−11°C), and low oxidation on the cathode surface, 2-MeTHF is found to form an effective protective film (SEI) on the lithium anode that sufficiently suppresses the formation of dendritic lithium. Sasaki et al. [390] reported that the addition of 2-MeTHF to the $LiPF_6$/EC-DMC electrolyte was helpful in suppressing the formation of lithium dendrite on the nickel electrode, and then improved the lithium cycling efficiency. Several linear ethers such as dimethoxymethane, DME, dimethoxypropane, and diethylene glycol dimethyl ether, have been studied as co-solvents to improve the low-temperature performance of the electrolyte. However, these compounds tend to show a high vapor pressure and low flash point, which would decrease the safety of application at high temperatures such as 60°C.

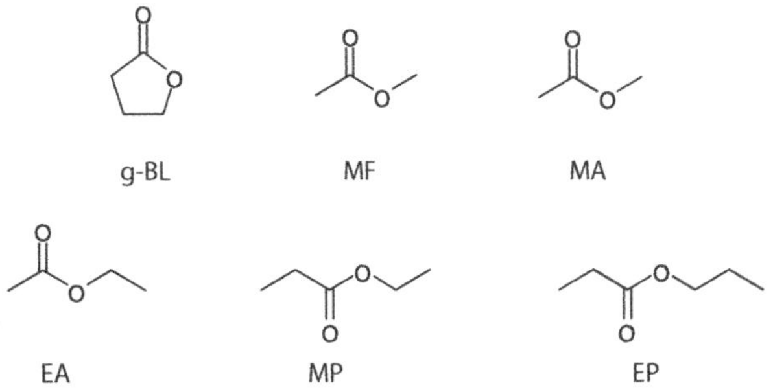

FIGURE 4.54 Chemical structures of carboxylic acid esters.

Carboxylic acid esters can also be classified into cyclic and noncyclic groups. γ-Butyrolactone (γ-BL) [391] is a well-known example of cyclic carboxylic acid esters, and the noncyclic carbonates, including formic acid methyl ester, methyl acetate, ethyl acetate, methyl propionate, and ethyl propionate (Figure 4.54). γ-BL has a high dielectric constant (39), which is lower than that of PC (64.92). γ-BL-based electrolytes have poor ionic conductivity compared with PC-based electrolytes. Because γ-BL is rather toxic and undergoes hydrolysis with only a trace amount of H_2O, it is impractical as an electrolyte solvent for industry purposes. Linear carboxylic esters with low melting point and low viscosity have been studied as co-solvents to enhance the low-temperature performance of the electrolyte. Lu et al. [392] found that linear fluorinated carboxylic esters are also suitable as co-solvents to improve the low-temperature performance and the nonflammability of the electrolyte, because of their higher flash points, although their melting and boiling points are close to the corresponding nonfluorinated esters. It could be concluded that the trifluoroacetate with a long carbon chain has a weak ability to dissociate $LiPF_6$ salt into free ions, and simultaneously decreases the mobility of solvated Li^+ in the modified electrolyte. Thus, as a low-temperature co-solvent, the carbon-chain length of the alcohol group in the trifluoroacetate structure should be selected to be as short as possible.

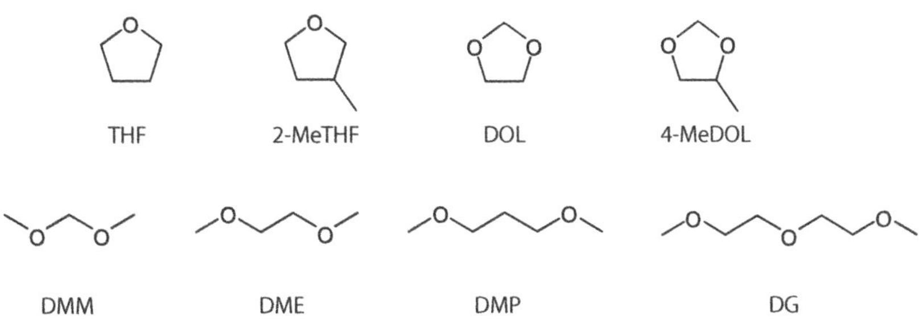

FIGURE 4.53 Chemical structures of ethers.

4.4.2.2 Lithium Salts

Lithium salts are used to conduct Li^+ ions in electrolyte solutions for lithium rechargeable batteries. The solution resistance, the surface film resistance, and the charge-transfer resistance were all found to depend on the electrolyte composition. In past studies, $LiClO_4$, $LiBF_4$, $LiAsF_6$, and $LiPF_6$ were used as Li-salt electrolytes. $LiClO_4$ is known to be a strong oxidizing agent that readily reacts with organic species at higher temperature or high current charging, possibly leading to explosion. For this reason, this salt cannot be used in practical batteries. However, it is one of the best candidates being used in electrochemical experiments because it generally exhibits very good electrochemical stability. $LiAsF_6$ has a restricted application in the electrolyte of LIBs because of its high costs and its highly toxic in nature relative to the use of As. Among the lithium salts, $LiPF_6$ has been extensively studied by many researchers, and although it has a serious moisture problem, it remains the only candidate to be widely used in commercial LIBs. To overcome the drawbacks of $LiPF_6$, lithium bis(trifluoromethanesulfonyl)imide ($Li[N(CF_3SO_2)_2]$ or LiTFSI) [393,394], and lithium tris(trifluoromethanesulfonyl) methide ($Li[C(CF_3SO_2)_3]$) or LiTFSM) [395,396] have been discussed to provide electrolytes having a high conductivity and a good thermal and chemical stability. Unfortunately, these lithium salts show severe corrosion behavior toward Al current collector. As recently reported, new types of lithium salts, such as lithium bis(oxalate)borate (LiBOB), dilithium-dodecafluorododecaborate ($Li_2B_{12}F_{12}$) [397], and lithium fluoroalkylphosphate (LiFAP) [398], are also discussed, from their synthesis to their compete electrochemical properties, such as ionic conductivity, electrochemical stability, cyclability, SEI film formation, and so forth, with the appropriate mechanisms [399].

After more than a decade of exploration, the nonaqueous electrolyte used in commercialized LIBs generally contains a certain amount of lithium salts, e.g., 1.0 M $LiPF_6$, in a solvent mixture of a cyclic carbonate, such as EC, and a linear carbonate, either DMC, EMC, or DEC. As a rule of thumb, a cyclic carbonate is desired for its high dielectric constant to reduce ion pairing or to increase the mobility of ions in the electric field; a linear carbonate is selected for its low viscosity for faster lithium-ion diffusion in the electrolyte.

4.4.3 High-Voltage Electrolytes

Higher energy and power densities are urgently needed for next-generation HEVs and EVs and smart grid devices. Increasing the cell voltage is an effective way to increase both densities since the delivered energy and power are proportional to the working voltage. In recent years, high-voltage cathode materials, such as $LiNi_{0.5}Mn_{1.5}O_4$, $LiCoPO_4$, $Li_3V_2(PO_4)_3$, Li_2CoPO_4F, and lithium-rich layer oxides, have been investigated extensively. Unfortunately, at a high working voltage, the cell may become unstable, as the conventional organic carbonate electrolyte is not stable around 4.5 V versus Li^+/Li, which tend to suffer from severe oxidative decomposition, resulting in the formation of a resistive and unstable surface film consisting of inorganic lithium salts and organic carbonates in the cathode, and correspondingly leads to a severe deterioration of the cycling performance. Although the oxidization potentials of organic carbonates are about 5 V, the transition-metal ions could catalyze the oxidation reaction and accelerate the decomposition of electrolytes at lower potentials, leading to rapid capacity fading. Therefore, searching for matched electrolytes is essential to the realization of high-voltage LIBs. In this section, we summarize the recent progress in high-voltage electrolytes, which mainly focus on novel stable solvents and functional additives.

4.4.3.1 Novel Stable Solvents

1. Dinitrile solvents

   From the work of Ue et al. on liquid electrolytes for electrical double-layer capacitors, it was demonstrated that aliphatic dinitrile solvents like glutaronitrile (GLN) (n = 3) and adiponitrile (AND) (n = 4) exhibit exceptional resistance to electrochemical oxidation to voltages reaching as high as ~5 V versus SCE (~8.3 versus Li^+/Li) [400,401], which is stable than most of the aprotic solvents used in LIBs including the more stable sulfone family. Besides the extra-anodic stability, AND and GLN show the best thermal (high boiling point and flash point) and physical (high dielectric constant and low viscosity) properties of all solvents in the dinitrile family. This makes it the aliphatic dinitrile solvent with the best thermal and physical properties. However, they are prone to easy reduction and poor compatibility with lithium or graphite anodes, which does not result in a compact, ionically conducting SEI, necessary for the functioning of the battery. Therefore, dinitrile solvent can be used with carbonate solvents (EC or EMC) as co-solvent and LiTFSI/LiBOB as salt in graphite anode, permitting the formation of a stable SEI at the surface of the MCMB graphitic electrode that protects the dinitrile solvent from undergoing reductive decomposition reactions [402,403]. Rather good capacity and cycling behavior was achieved with a 1 M LiTFSI, 0.1 M LiBOB, and EC/dinitrile (AND or GLN) (1:1) electrolyte composition in a cell incorporating graphite/$LiCoO_2$ electrodes [402]. The use of LiTFSI salt, however, limits the high-voltage application of this particular mixture because it corrodes aluminum at high voltage (>5 V). Thus, $LiPF_6$ and $LiBF_4$ are more suitable Li salt electrolytes than LiTFSI. Yaser et al. [404] reported that LiTFSI could be successfully substituted by $LiBF_4$ in electrolytes consisting of EC/DMC/nitrile (nitrile = AND, PMN, SUN, and SEN) with $LiBF_4$ and LiBOB as co-salts, and was cycled in commercial $LiCoO_2$ (4.2 V) and the high voltage material $LiNi_{0.5}Mn_{1.5}O_4$ (4.9 V) half cells for 50 cycles with good electrochemical performance. A sebaconitrile-based electrolyte, 1 M $LiBF_4$/EC+DMC+sebaconitrile (25:25:50 by vol%), exhibits excellent electrochemical stability above 6 V against Li^+/Li using a $LiFePO_4$ cathode. In addition, this oxidation-resistant electrolyte was able to prove that Li_2NiPO_4F has actually ca. 5.3 V redox potential against Li anode through cyclic voltammetry and charge–discharge cycle test [405]. The butyronitrile was also investigated as a co-solvent to the EC/EMC electrolyte to improve the high-temperature performance of $LiNi_{0.5}Mn_{1.5}O_4$/MCMB cells [406]. Very recently, Gmitter et al. [407] investigated the addition of vinylene carbonate (VC) and FEC additives to 1 M LiTFSI, 0.25 M $LiBF_4$ in single AND electrolyte solution in MCMB half cells and MCMB/$LiCoO_2$ full cells, and showed an evidence for the formation of a stable SEI and good capacity retention. Duncan et al. [408] successfully demonstrated the possibility of using dinitrile-based electrolyte solutions in high-voltage lithium batteries (Figure 4.55). It was shown that dinitriles with chain lengths of n = 4–8, used alone or mixed with a second solvent EC, could not

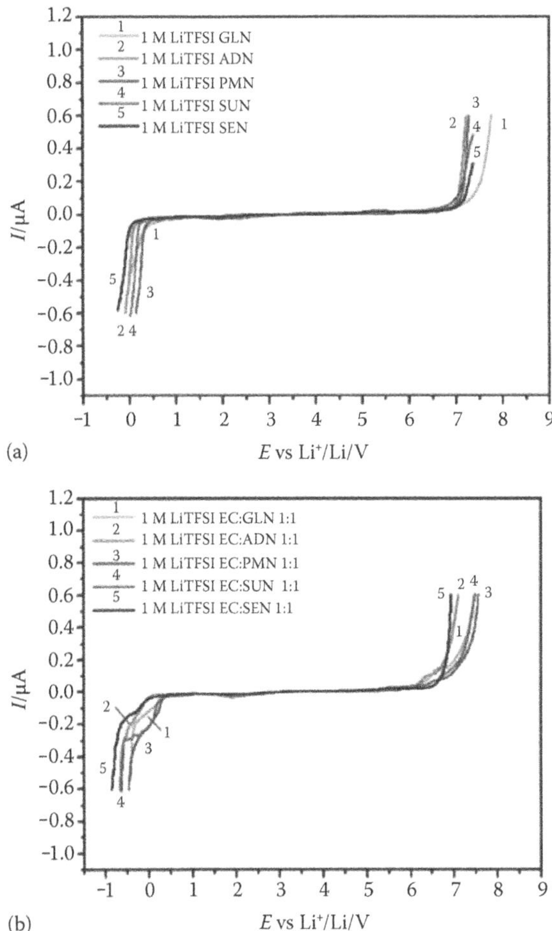

FIGURE 4.55 Linear sweep voltammetry scans of (a) 1 M LiTFSI dinitrile; (b) 1 M LiTFSI EC/dinitrile 1:1 (dinitrile = GLN, ADN, PMN, SUN, and SEN) electrolyte solutions using Pt working electrode and Ag wire reference electrode. Scan rate of 10 mV/s. (From Duncan, H., Salem, N., and Abu-Lebdeh, Y., *J. Electrochem. Soc.*, 160, A838, 2013.)

sustain Li/LMNO batteries (poor capacities with low coulombic efficiencies) except when DMC was used as a third solvent and $LiBF_4$ as salt and LiBOB as co-salt. This was investigated by the examination of cycled cells by XPS and is attributed to the ability of DMC to sacrificially protect from the oxidative decomposition of the $LiBF_4$ salt at high voltage.

2. Sulfones

Sulfones are well known for their high anodic stability, hence potential applications in high-voltage rechargeable LIBs. Electrolyte solutions based on molecular solvents with the sulfone functional group, $-SO_2-$, were reported by Angell to exhibit a rather uniform high stability against oxidative decomposition, which occurred at around 5.8 V versus Li^+/Li [409]. Shao et al. [410,411] computed oxidation potentials for a series of sulfone-based molecules functionalized with fluorine, cyano, ester, and carbonate groups by using a quantum chemistry method within a continuum solvation model. They found that the strong electron-withdrawing groups can greatly increase the oxidation potentials of sulfones [411]. Because dimethyl sulfone has a very high melting point (108.9°C), they cannot be used as a single solvent in LIBs. Adding unsymmetric aliphatic groups can lower the melting point. For example, ethyl methyl sulfone (EMS) and some unsymmetric sulfones with different functional group were reported with even lower melting points [412–414]. In contrast with the stability on the cathode surface, the ability to form a stable SEI film could not be found on graphite anodes by sulfones [412]. The fluorination of the alkyl groups in sulfones can overcome this shortcoming. It has also been reported that the introduction of additives such as VC [415–417], lithium bis(oxalate)borate (LiBOB) [418,419], p-toluenesulfonyl isocyanate [420], and hexamethylene diisocyanate [421] can be used as a film-forming additive in sulfone-based electrolytes to promote SEI film formation, giving a cycling performance equal to the conventional carbonate electrolytes. In addition, the co-solvents are also introduced to decrease the melting point and viscosity, and improve the wettability and conductivity [422–425]. Xing et al. [426], using molecular dynamic simulations, showed that, in a tetramethylene sulfone (TMS)/DMC/$LiPF_6$ electrolyte, DMC is located approximately 0.8 Å further from the positive electrode than in EC/DMC/$LiPF_6$, indicating that it might may be more difficult to oxidize DMC in TMS-based electrolytes, which is consistent with experimentally reported increases in the oxidative stability of the latter (Figure 4.56).

The excellent electrochemical stabilities of sulfone-based electrolytes have been confirmed by various applications. For example, EMS was found to be capable of supporting the complete delithiation process of a layered transition metal oxide cathode

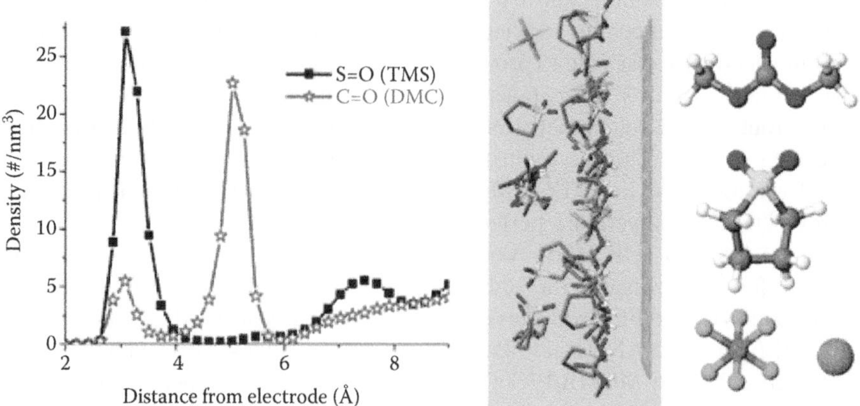

FIGURE 4.56 Density profiles of S=O and C=O for the TMS/DMC/$LiPF_6$ electrolytes near surfaces of the positive electrode using MD simulations. (From Xing, L., Vatamanu, J., Borodin, O., Smith, G.D., and Bedrov, D., *J. Phys. Chem. C*, 116, 23871, 2012.)

up to 5.4 V [427]. TMS and EMS were examined in the $Li_4Ti_5O_{12}/LiNi_{0.5}Mn_{1.5}O_4$ cell by Abouimrane et al. [428]. When TMS was blended with EMC, the $Li_4Ti_5O_{12}/LiNi_{0.5}Mn_{1.5}O_4$ cell delivered an initial capacity of 80 mAh/g and cycled fairly well for 1000 cycles under 2-C rates. Demeaux et al. [429] investigated the stability of sulfones such as TMS, EMS, methyl isopropyl sulfone (MIS), and ethyl isopropyl sulfone (EIS) toward reduction using LTO/Li and $Li_{4+x}Ti_5O_{12}/Li_4Ti_5O_{12}$ symmetric test cells to evaluate the behavior of alkylsufones at their interfaces during cycling. Because sulfones did not participate in the formation of surface layer, symmetric $Li_{4+x}Ti_5O_{12}/Li_4Ti_5O_{12}$ cells in sulfone-based electrolytes exhibited reasonable electrochemical performances as compared with the EC/DMC or EC/EMC standard mixtures.

3. Ionic liquid (ILs)

ILs that consist of cations and anions are also known to show nonvolatile, nonflammable, and higher anodic stability [430–436]. Especially, the piperidinium and pyrrolidinium salts show higher oxidization potentials of ~6 V versus Li^+/Li, which make ILs candidate electrolyte solvents for high-voltage LIBs [437,438]. However, the disadvantages of ILs, including higher viscosity and lower ionic conductivity than conventional organic solvents, lead to poor electrochemical performance of the cells with ILs as electrolytes. The addition of additives such as VC [439,440], and ethylene sulfite [441] as electrolyte additives, or the introduction of carbonates or other solvents as a co-solvent [442–444], can improve the compatibility of IL-based electrolyte toward lithium anodes and enhance the rate capacity and low-temperature performance. For example, Li et al. [445] reported that the safety and the high-voltage performance of Li-rich cathode material $Li[Li_{0.2}Mn_{0.54}Ni_{0.13}Co_{0.13}]O_2$ were improved using a mixture of organic carbonate solvents and PP_{13}TFSI ionic liquid. Xiang et al. [424] reported novel IL/sulfone mixed electrolytes based on *N*-butyl-methyl piperidiniumbis(trifluoromethylsulfonyl)imide (PP_{14}-TFSI) and TMS at high voltage and safe electrolytes for use in LIBs, as shown in Figure 4.57. The mixed electrolytes exhibited excellent anodic stability of above 5 V, high ionic conductivities up to 10^{-3} S/cm at room temperature, and good nonflammability properties. It was found that the 0.5 M LiDFOB/(60%) PP_{14}–TFSI/(40%)TMS mixed electrolyte exhibited better compatibility with the $Li_{1.2}Ni_{0.2}Mn_{0.6}O_2$ electrode than a commercial electrolyte (1 M $LiPF_6$ in EC/DMC, 1:1) because of the good electrochemical stability of the PP_{14}–TFSI/TMS mixed electrolyte. Mun et al. [446] studied the surface film on the high-voltage $LiNi_{0.5}Mn_{1.5}O_4$ positive electrode at elevated temperatures in two different electrolytes: $LiPF_6$/organic carbonate and LiTFSI/IL (propylme thylpyrrolidiniumbis(trifluoromethylsulfonyl)imide, PMPyr–TFSI). The inorganic fluorinated species were dominant in the film derived from the carbonate electrolyte, whereas organic carbon species from the IL. This organic-rich film deposited in the earlier period of cycling seems to effectively passivate the positive electrode presumably due to uniform coverage. As a result, the film does not grow with cycling and the electrode polarization is not enough to give a stable cycling behavior. In addition, ILs can be used as electrolyte additives for high-voltage LIB to stabilize carbonate-based electrolytes on the surface of 5-V class cathode materials [447].

4. Fluorinated solvents

Although dinitriles, sulfones, and ILs can provide high anodic stability, they suffer from their intrinsic high viscosity, low dielectric constant, and low conductivity. More important, they do not form an effective SEI film on carbonaceous anode materials. Therefore, the fluorinated carbonate solvents seem more appropriate for high-voltage LIBs in the current stage.

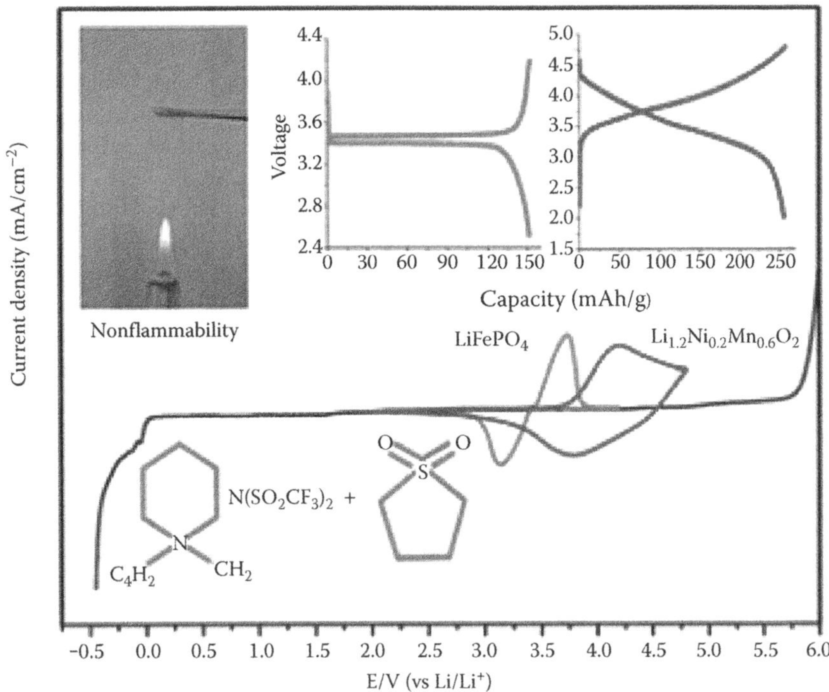

FIGURE 4.57 High-voltage and safe electrolytes based on ionic liquid and sulfone for LIBs. (From Xiang, J., Wu, F., Chen, R., Li, L., and Yu, H., *J. Power Source*, 233, 115, 2013.)

Fluorinated cyclic and linear carbonate compounds possess desirable physical properties imparted by the presence of the fluorine substituent, such as lower melting point, higher flash point, increased oxidation stability, and favorable SEI film forming characteristics on carbon. For example, FEC and 3,3,3-trifluoropropylene carbonate (TFPC) were widely studied as an SEI formation additive for LIBs using PC-based electrolyte [448,449], silicon anodes [450–453], and other anodes [454–460]. Smart et al. [461] reported a series of partially and fully fluorinated linear carbonates incorporated into multicomponent ternary and quaternary carbonate-based electrolytes and evaluated their charge–discharge behavior at low temperatures in Li–carbon and carbon–$LiNi_{0.8}Co_{0.2}O_2$ cells. However, much less attention was given to the investigation of the effect of fluorinated carbonates on the increased oxidation stability in high-voltage LIBs. The fluorinated solvent exhibits higher oxidation stability than classic solvents owing to the high electronegativity and low polarizability of the fluorine atom. In 2008, Kobayashi et al. [462] reported that the oxidation potential of FEC was higher than that of PC on platinum electrode and activated carbon electrode. Moreover, Sasaki et al. [463] also demonstrated that the oxidative decomposition potentials of the fluorinated MPC derivatives tend to increase with increasing numbers of fluorine atoms. The beneficial effect of the electrolytes containing these solvents has been extensively evaluated with high-voltage cathode materials. Fridman et al. [464] reported that excellent performance of the $LiNi_{0.5}Mn_{1.5}O_4$/Si cells was achieved by replacing the conventional 1 M $LiPF_6$ EC/DMC electrolyte solution with 1 M $LiPF_6$ FEC/DMC electrolyte. Amine's group [465–467] reported that the cycling performance of $LiNi_{0.5}Mn_{1.5}O_4$/Li, $LiNi_{0.5}Mn_{1.5}O_4$/$Li_4Ti_5O_{12}$, and $LiNi_{0.5}Mn_{1.5}O_4$/graphite cells was remarkably enhanced in the all-fluorinated electrolytes composed of fluorinated cyclic carbonate (F-AEC), fluorinated linear carbonate (F-EMC), and fluorinated ether (E-EPE) compared with the conventional electrolytes even at elevated temperatures (55°C). Recently, Markevich et al. [468] reported that an excellent cycling stability of $LiCoPO_4$/Li and $LiNi_{0.5}Mn_{1.5}O_4$/Si cells, and a reasonable cycling performance of $LiCoPO_4$/Si full cells were achieved by replacing the commonly used co-solvent EC by FEC in electrolyte solutions for high-voltage LIBs (Figure 4.58). FEC participates in the formation of protective surface films on these high-voltage cathodes, which are much more effective and are formed much faster than the surface films formed in EC-based electrolyte solutions. These data suggest that fluorinated carbonate may be a suitable electrolyte candidate for transition metal oxide cathodes at high voltage (5 V versus Li^+/Li).

4.4.3.2 Additives

Electrolyte additives are known to be an effective and economic approach to improve the stability of the cathode–electrolyte interface [358]. The additives for high-voltage LIBs are mainly based on the mechanism of assisting in the formation of stable interfacial films on the cathodes that decrease the decomposition of the electrolytes under high potentials. In the past decade, many organic and inorganic compounds have been identified as effective electrolyte additives, including LiBOB, LiDFOB, phosphides, sulfonate esters, carboxyl anhydrides, and monomers that can be electrochemically polymerized [469–478]. Among them, LiBOB has attracted interest as

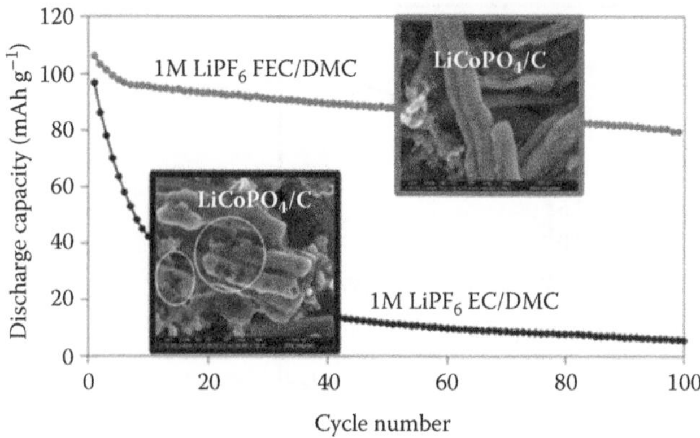

FIGURE 4.58 Curves of discharge capacity versus cycle number obtained upon galvanostatic cycling between 3.5 and 5.2 V at a rate of C/5 h for $LiCoPO_4$ electrodes in EC-based (below) and FEC-based (above) electrolyte solutions (30°C). (From Markevich, E., Salitra, G., Fridman, K., Sharabi, R., Gershinsky, G., Garsuch, A., Semrau, G., Schmidt, M.A., and Aurbach, D., *Langmuir*, 30, 7414, 2014.)

one of the promising additives. Several studies have reported that the LiBOB additive effectively improved the cycling performance of $LiNi_{0.5}Mn_{1.5}O_4$/Li and $LiNi_{0.42}Fe_{0.08}Mn_{1.5}O_4$ (LNFMO)/graphite cells [418,479–484], as shown in Figure 4.59. It has been speculated that such improvement may originate from the inhibition of HF or PF_5 generation or the formation of a protective electrode/electrolyte film on the cathode that suppresses oxidation of the electrolyte. Phosphides such as tri(hexafluoro-iso-propyl)phosphate [485] and tris(trimethylsilyl)phosphate [486–488] exhibited similar surface-film-forming characters and effectively enhanced the electrochemical performance of high-voltage cathodes. Recently, Zhu et al. [489] reported that a new family of polyfluoroalkyl-substituted ethylene carbonates was examined as electrolyte additives in $Li_{1.2}Ni_{0.15}Mn_{0.55}Co_{0.1}O_2$/graphite full cells. The results showed that PFO–EC improved capacity

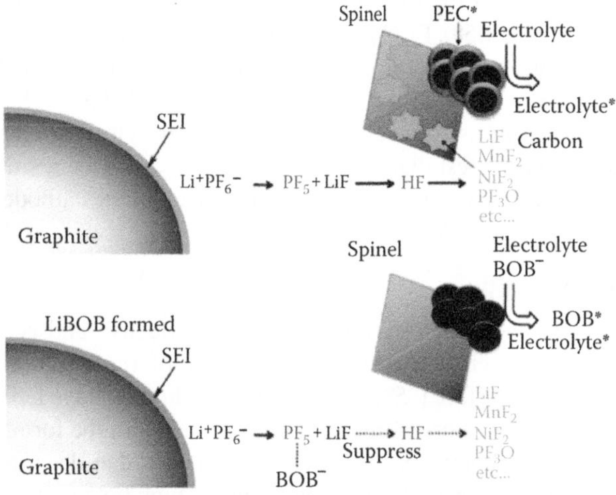

FIGURE 4.59 Proposed improvement mechanisms for the LiBOB additive on the cathode surface. (From Pieczonka, N.P.W., Yang, L., Balogh, M.P., Powell, B.R., Chemelewski, K., Manthiram, A., Krachkovskiy, S.A., Goward, G.R., Liu, M., and Kim, J.-H., *J. Phys. Chem. C*, 117, 22603, 2013.)

retention and lowered impedance increase in full cells cycled between 2.2 and 4.6 V. Improvements in cell performance from the PFO–EC additive can be attributed to synergistic effects of the performance on both electrodes. The sacrificial oxidation of PFO–EC at the cathode forms surface films that reduce cell impedance increase; PFO–EC also reacts at the anode to generate a more stable SEI. Additionally, 1,3-propanediolcyclic sulfate [490] and glutaric anhydride [491] formed a denser SEI on the graphite anode that suppressed the anodic deposition of manganese oxide species resulting from the "cross-talk" effect, so as to improve the performance of the high-voltage cathodes. This approach offers a new approach to the design of high-voltage LIBs with advanced electrolyte additives.

4.4.4 Flame-Retardant Electrolytes

4.4.4.1 Organic Phosphides

The most common strategy to enhance the original electrolyte's safety is introducing flame-retardant additives. Organic phosphides are widely investigated owing to their high flame-retarding effectiveness [492], good miscibility with organic carbonates, and low cost, as mentioned above. They mainly include alkyl phosphates, benzyl phosphates, alkyl phosphites, and cyclophosphazenes [493–501]. Trimethyl phosphate (TMP) is the earliest and most thoroughly studied flame retardant for LIBs. Despite the perceptible reduction of the flammability of the electrolyte, TMP is readily co-intercalated into the graphite anode, as well as decomposes on the graphite anode, causing severe capacity fading of the batteries [494,495]. The reductive stability can be improved by adopting triethyl phosphate (TEP) and tributyl phosphate (TBP) with increased carbon numbers in the alkyl groups, or TPP with phenyl replacing the alkyl groups. For instance, Hyung et al. [496] found that less than 5 wt.% of TPP present in the electrolyte significantly increased the onset reaction temperature of the fully charged graphite anode with the electrolyte from 160°C to 210°C. Also, TPP showed no negative effect on the cycling characteristics of the graphite–$LiNi_{0.8}Co_{0.2}O_2$ cell. Moreover, hexamethoxy-cyclotriphosphazene (HMPN) [493] was also found not only to be highly flame retardant but also extremely stable at low potentials. It was shown that only 1.5% HMPN in the electrolyte can significantly reduce the self-heat rate and noticeably improve the charge–discharge capacity of the Li–$LiNi_{0.8}Co_{0.2}O_2$ cell. Other than the P(V) phosphates, the phosphites with P(III) seem to have better compatibility with the low-potential carbon anodes, as they are able to facilitate SEI film formation and deactivate PF_5 [498].

4.4.4.2 Fluorinated Phosphates

Fluorination is a successful attempt to alleviate the trade-off of alkyl phosphates between flammability and cell performance [502–505]. In contrast to alkyl phosphates, the fluorinated phosphates have many advantages: First, both P and F are breakers for radical chain reactions; the combination of these two elements can advance the flame-retardant effectiveness with less usage. Next, fluorinated phosphates are chemically and electrochemically stable, and fluorine is favorable to stable SEI film formation, and thus can realize the electrochemical compatibility of alkyl phosphates with the carbon anodes. Third, F atoms can impair the intermolecular viscous force and reduce the moving resistance of molecules and ions, so that fluorinated phosphates have lower viscosity than their corresponding alkyl phosphates and show less impact on the conductivity of electrolyte. Xu et al. [502,503] evaluated a series of fluorinated alkyl phosphates including, bis(2,2,2-trifluoroethyl) methylphosphate and (2,2,2-trifluoroethyl)diethyl phosphate, and found that the presence of fluorine not only improved the reductive stability but also increased the flame-retarding ability. Tris(2,2,2-trifluoroethyl)phosphate (TFP) displayed the best comprehensive performance. The electrolyte became nonflammable with 20% TFP as co-solvent, which, however, had no any adverse impact on the long-term cell stability. Zhang et al. [506] reported that the addition of about 15% tris(2,2,2-trifluoroethyl) phosphite (TTFP) could make a 1 mol/kg $LiPF_6$/PC/EC/EMC (3:3:4, by weight) electrolyte nonflammable at the expense of 20% reduction in ionic conductivity. TTFP could effectively suppress PC decomposition and graphite exfoliation so as to increase the cycling efficiency of the graphite electrode. Recently, Zeng et al. [507] reported that bis(2,2,2-trifluoroethyl) methylphosphonate (TFMP) used as a flame retardant could reduce the flammability of the electrolyte with minimal impact on the electrochemical performance of the electrode. The results demonstrated that the addition of 20 vol% TFMP can make the electrolyte be hardly flammable, exhibiting highly efficient flame retardancy. Moreover, the 20 vol% TFMP-containing electrolyte also exhibited an insignificant influence on graphite anode with a high initial capacity of 346 mAh/g and excellent capacity retention of 93% over 100 cycles.

Despite some positive results, the organic phosphides, fluorinated phosphate fire-retardant additives, in all, realized thermal safety for LIBs at the cost of their electrochemical performance. Fox example, these compounds usually show higher viscosity than the routine carbonate solvents, which would no doubt compromise the capacity utilization, power, and low-temperature performance of Li-ion batteries. Still, the electrochemical compatibility with electrodes is another problem despite a small content in electrolyte. Solutions should be proceeding rational structural modification, in the search for effective SEI formation additives, etc. Moreover, despite greatly improved thermal stability, there is still a large amount of flammable carbonate compounds existing, making the electrolyte vulnerable to fire hazards. These liquid carbonate compounds could be oxidized or decomposed into flammable gases under overcharge or overheating. Thus, it is necessary to develop biphasic (liquid and gas phases) flame retardants that can not only scavenge gas radicals but also can carbonize the condensed carbonate to prevent both gaseous and liquid combustion.

4.4.4.3 Fluorinated Solvents

Fluorinated solvents having a high flash point, and even no flash point, can also be used as either an electrolyte additive or a co-solvent for lowering the combustibility of the traditional electrolytes. Fluorinated solvents reported thus far, including halogenated carbonates (such as Cl–EC [375,508] and TFPC [449]), and hydrofluoroethers (HFEs) [509–514]. Arai et al. [509–511] eliminated the electrolyte flash point by using a nonflammable solvent of methyl nonafluorobuyl ether (MFE) as the main component of the electrolyte solvent. The 1 M LiBETI ($LiN[SO_2C_2F_5]_2$)–MFE/EMC (80:20 vol%) is a representative of novel electrolytes having no FP that the authors proposed. The graphite/$LiCoO_2$ 18,650 cylindrical cell using this electrolyte did not show thermal runaway when a nail penetrated it even when the cell was overcharged. Naoi et al. [512,513] also used HFEs of 2-trifluomethyl-3-methoxyperfluoropentane (TMMP) and 2-(trifluoro-2-fluoro-3-difluoropropoxy)-3-difluoro-4-fluoro-5-trifluoropentane (TPTP) as co-solvents of EC + DEC electrolytes. Although a dosage of more than 50 vol% of TMMP is required to have good nonflammability, the TMMP dosage in the EC + DEC electrolytes using $LiPF_6$ is limited to less than 30 vol% owing to low miscibility of the TMMP. Recently, Wong et al. [514] prepared a unique nonflammable electrolyte composed of low molecular weight perfluoropolyethers and bis(trifluoromethane)sulfonamide lithium salt. As displayed in Figure 4.60, these electrolytes exhibited thermal stability beyond 200°C and a remarkably high transference number of at least 0.91 (more than double

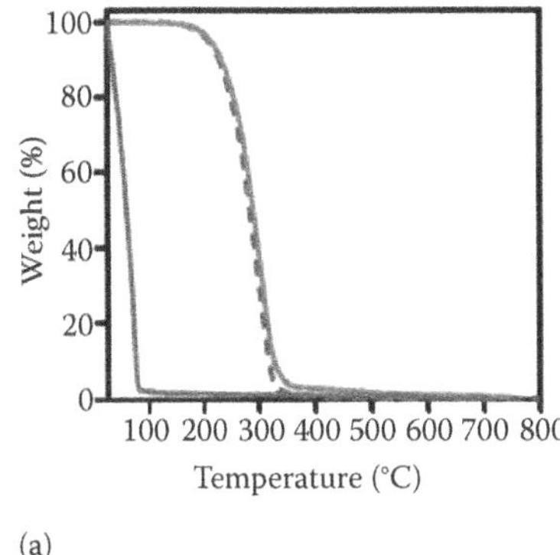

(a)

Electrolyte	T_d (5%)	Sustained burning time	Flash point
DMC	34°C	221 ± 7 s	18°C
$PFPE_{1000}$-diol	210°C	No burning observed	None <200°C
$PFPE_{1000}$–DMC	212°C	No burning observed	None <200°C

(b)

FIGURE 4.60 Thermal stability and flammability of PTPEs. (a) TG curves for thermal decompositions of DMC (solid line, left), $PFPE_{1000}$-diol (dash line), and $PFPE_{1000}$–DMC (solid line, right). (b) Corresponding decomposition temperature (5%), sustained burning characteristics, and flash points of these materials. (From Wong, D.H.C., Thelen, J.L., Fu, Y., Devaux, D., Pandya, A.A., Battaglia, V.S., Balsara, N.P., and DeSimone, J.M., *Proc. Natl. Acad. Sci.*, 111, 3327, 2014.)

that of conventional electrolytes). $Li/LiNi_{1/3}Co_{1/3}Mn_{1/3}O_2$cells made with this electrolyte showed good performance in galvanostatic cycling, confirming their potential as rechargeable lithium batteries with enhanced safety and longevity.

4.4.4.4 Bifunctional Flame-Retardant Additives

It should be noted that a single technology discussed above can only provide safety protection at a particular abusive condition. Recently, a few flame-retardant additives have been reported to be able to perform potential clamping safety protections. Yang's group found that a TMPP additive has both flame-retarding and voltage-clamping abilities [515]. With 10% TMPP content, the self-extinguishing time of 1 M $LiClO_4$/EC + DEC electrolyte dropped from the original 38 s/g to 22 s/g. TMPP also could be polymerized at 4.35 V (versus Li^+/Li) to form a conducting polymer to carry the overcharge current. Since this molecule has not much influence on the electrode performance, it is possible to use it for both overcharge protection and flame retardancy of LIBs. Also, the aforementioned shuttle additive TEDBPDP itself is a flame retardant due to its phosphate groups, it can elevate the ignition temperature of the electrolyte [516]. These experimental results throw a novel strategy to design safety additives with multiple protection functionalities.

4.5 CONCLUSIONS AND PERSPECTIVES

The state of the art of lithium-ion cells, high-voltage cathode materials, $LiNi_{0.5}Mn_{1.5}O_4$, $LiCoPO_4$, and so forth has been fully explored. However, this effort is still hampered by the limitation of the decomposition potential of the traditional electrolytes. Extending the stability of the electrolytes is desperately needed when using high-voltage cathode materials. At the same time, the safety issue continues to be a major obstacle that prevents the commercial applications of large-scale LIBs in EVs and energy storage devices for smart grid, solar, and wind stations. Reducing the combustibility of the electrolytes is a promising effort for improving the safety of LIBs. The latest research indicates that the fluorinated carbonate solvents, which exhibit improved oxidation stability due to the high electronegativity and low polarity of fluorine atoms, appear to be a suitable electrolyte candidates for $LiNi_{0.5}Mn_{1.5}O_4$ cathodes at high voltage (5 V versus Li^+/Li). Results from mixed fluorinated solution are rather impressive, e.g., F-AEC, F-EMC, and F-EPE; however, further studies are required to understand the mechanism. Additionally, the HFEs, which are partially fluorinated, have also attracted much attention as nonflammable electrolytes for LIBs owing to their high flash point. Thus, this type of fluorinated solvent offers a great prospective nonflammable electrolyte for the application in 5-V LIBs.

REFERENCES

1. Meng, Y.S., and Dompablo, M.E.A. First principles computational materials design for energy storage materials in lithium ion batteries. *Energy & Environmental Science* 2009; 2:589–609.
2. Doi, T., Zhou, M., Zhao, L., Okada, S., and Yamaki, J. Influence of irreversible reactions at non-graphitizable carbon electrodes on their thermal stability in Li-ion batteries. *Electrochemistry Communications* 2009; 11:1405–1408.
3. Zhou, M., Zhao, L., Doi, T., Okada, S., and Yamaki, J. Thermal stability of FeF_3 cathode for Li-ion batteries. *Journal of Power Sources* 2010; 195:4952–4956.
4. Zhou, M., Zhao, L., Okada, S., and Yamaki, J. Thermal characteristics of a FeF_3 cathode via conversion reaction in comparison with $LiFePO_4$. *Journal of Power Sources* 2011; 196:8110–8115.
5. Baba, Y., Okada, S., and Yamaki, J. Thermal stability of Li_xCoO_2 cathode for lithium ion battery. *Solid State Ionics* 2002; 148:311–316.
6. Tobishima, S., and Yamaki, J. A consideration of lithium cell safety. *Journal of Power Sources* 1999; 81–82:882–886.
7. Biensan, P., Simon, B., Peres, J., Guibert, A., Broussely, M., Bodet, J., and Perton, F. On safety of lithium-ion cells. *Journal of Power Sources* 1999; 81–82:906–912.
8. Zhang, Z., Fouchard, D., and Rea, J. Differential scanning calorimetry material studies: Implications for the safety of lithium-ion cells. *Journal of Power Sources* 1998; 70:16–20.
9. Cho, J., Jung, H., Park, Y., Kim, G., and Lim, H. Electrochemical properties and thermal stability of $Li_aNi_{1-x}Co_xO_2$ cathode materials. *Journal of the Electrochemical Society* 2000; 147:15–20.
10. Arai, H., Okada, S., Sakurai, Y., and Yamaki, J. Electrochemical and thermal behavior of $LiNi_{1-z}M_zO_2$ (M = Co, Mn, Ti). *Journal of the Electrochemical Society* 1997; 144:3117–3125.
11. Li, W., Currie, J., and Wolstenholme, J. Influence of morphology on the stability of $LiNiO_2$. *Journal of Power Sources* 1997; 68:565–569.
12. Dahn, J., Fuller, E., Obrovac, M., and Sacken, U. Thermal stability of Li_xCoO_2, Li_xNiO_2 and λ-MnO_2 and consequences for the safety of Li-ion cells. *Solid State Ionics* 1994; 69:265–270.
13. MacNeil, D., Hatchard, T., and Dahn, J. A comparison between the high temperature electrode/electrolyte reactions of Li_xCoO_2 and $Li_xMn_2O_4$. *Journal of the Electrochemical Society* 2001; 148:A663–A667.
14. Saito, Y., Kanari, K., Takano, K., and Masuda, T. A calorimetric study on a cylindrical type lithium secondary battery by using a twin-type heat conduction calorimeter. *Thermochimica Acta* 1997; 296:75–85.
15. MacNeil, D., Christensen, L., Landucci, J., Paulsen, J., and Dahn, J. An autocatalytic mechanism for the reaction of Li_xCoO_2 in electrolyte at elevated temperature. *Journal of the Electrochemical Society* 2000; 147:970–979.
16. MacNeil, D., and Dahn, J. The reaction of charged cathodes with nonaqueous solvents and electrolytes: I. $Li_{0.5}CoO_2$. *Journal of the Electrochemical Society* 2001; 148:A1205–A1210.
17. MacNeil, D., Lu, Z., and Chen, Z. A comparison of the electrode/electrolyte reaction at elevated temperatures for various Li-ion battery cathodes. *Journal of Power Sources* 2002; 108:8–14.
18. Roth, E., Doughty, D., and Franklin, J. DSC investigation of exothermic reactions occurring at elevated temperatures in lithium-ion anodes containing PVDF-based binders. *Journal of Power Sources* 2004; 134:222–234.
19. Ohzuku, T., Ueda, A., and Kouguchi, M. Synthesis and characterization of $LiAl_{1/4}Ni_{3/4}O_2$(R-3m) for lithium-ion (shuttlecock) batteries. *Journal of the Electrochemical Society* 1995; 142:4033–4039.

20. Gao, Y., Yakovleva, M., and Ebner, W. Novel $LiNi_{1-x}Ti_{x/2}Mg_{x/2}O_2$ compounds as cathode materials for safer lithium-ion batteries. *Electrochemical and Solid-State Letters* 1998; 1:117–119.
21. Zhou, F., Zhao, X., Lu, Z., Jiang, J., and Dahn, J. The effect of Al substitution on the reactivity of delithiated $LiNi_{1/3}Mn_{1/3}Co_{(1/3-z)}Al_zO_2$ with non-aqueous electrolyte. *Electrochemistry Communications* 2008; 10:1168–1171.
22. Zhou, F., Zhao, X., Jiang, J., and Dahn, J. Advantages of simultaneous substitution of Co in $LiNi_{1/3}Mn_{1/3}Co_{1/3}O_2$ by Ni and Al. *Electrochemical and Solid-State Letters* 2009; 12:A81–A83.
23. Zhou, F., Zhao, X., and Dahn, J. Synthesis, electrochemical properties, and thermal stability of Al-doped $LiNi_{1/3}Mn_{1/3}Co_{(1/3-z)}Al_zO_2$ positive electrode materials. *Journal of the Electrochemical Society* 2009; 156:A343–A347.
24. Zhou, F., Zhao, X., Lu, Z., Jiang, J., and Dahn, J. The effect of Al substitution on the reactivity of delithiated $LiNi_{(0.5-z)}Mn_{(0.5-z)}Al_{2z}O_2$ with nonaqueous electrolyte. *Electrochemical and Solid-State Letters* 2008; 10:A155–A157.
25. Mizushima, K., Jones, P.C., Wiseman, P.J., and Goodenough, J.B. LixCoO$_2$ ($0 < x < -1$): A new cathode material for batteries of high energy density. *Materials Research Bulletin* 1980; 15:783–789.
26. Benco, L., Barras, J.L., Atanasov, M., Claude, A.D., and Deiss, E. First-principles prediction of voltages of lithiated oxides for LIBs. *Solid State Ionics* 1998; 112:255–259.
27. Cha, C.S., and Yang, H.X. Recent advances in experimental methods applied to lithium battery researches. *Journal of Power Sources* 1993; 43–44:145–155.
28. Bang, H.J., Joachin, H., and Yang, H. Contribution of the structural changes of $LiNi_{0.8}Co_{0.15}Al_{0.05}O_2$ cathodes on the exothermic reactions in li-ion cells. *Journal of the Electrochemical Society* 2006; 153:A731–A737.
29. Yoon, W.S., Balasubramanian, M., and Yang, X.Q. Time-resolved XRD study on the thermal decomposition of $Li1_{-x}Ni_{0.8}Co_{0.15}Al_{0.05}O_2$ cathode materials for Li-Ion batteries. *Electrochemical and Solid-State Letters* 2005; 8:A83–A86.
30. Lee, K.K., Yoon, W.S., and Kim, K.B. Characterization of $LiNi_{0.85}Co_{0.10}M_{0.05}O_2$ (M = Al, Fe) as a cathode material for lithium secondary batteries. *Journal of Power Sources* 2001; 97:308–312.
31. Liu, Z.L., Yu, A.S., and Lee, J.Y. Synthesis and characterization of $LiNi_{1-x-y}Co_xMnyO_2$ as the cathode materials of secondary lithium batteries. *Journal of Power Sources* 1999; 81:416–419.
32. Wang, Z.X., and Huang, X.J. Performance improvement of surface-modified $LiCoO_2$ cathode materials: An infrared absorption and X-ray photoelectron spectroscopic investigation. *Journal of the Electrochemical Society* 2003; 150:A199–A208.
33. Cho, J., and Kim, Y.J. Zero-strain intercalation cathode for rechargeable Li-ion cell. *Angewandte Chemie International Edition* 2001; 40:3367.
34. Kannan, A.M., Rabenberg, L., and Marithiram, A. High capacity surface-modified $LiCoO_2$ cathodes for lithium-ion batteries. *Electrochemical and Solid-State Letters* 2003; 6:A16–A18.
35. Chen, Z.H., and Dahn, J.R. Methods to obtain excellent capacity retention in $LiCoO_2$ cycled to 4.5 V. *Electrochimica Acta* 2004; 49:1079–1090.
36. Kweon, H.J., and Park, D.G. Surface modification of $LiSr_{0.002}Ni_{0.9}Co_{0.1}O_2$ by overcoating with a magnesium oxide. *Electrochemical and Solid-State Letters* 2000; 3:128–130.
37. Myung, S.T., Izumi, K., Komaba, S., Sun, Y.K., Yashiro, H., and Kumagai, N. Role of alumina coating on Li–Ni–Co–Mn–O particles as positive electrode material for lithium-ion batteries. *Chemistry of Materials* 2005; 17:3695–3704.
38. Li, D.C., Yasuhiro, K., and Koichi, K. Preparation and electrochemical characteristics of $LiNi_{1/3}Mn_{1/3}Co_{1/3}O_2$ coated with metal oxides coating. *Journal of Power Sources* 2006; 160:1342–1348.
39. Cho, J., Tim, T.J., and Tim, J. Synthesis, thermal, and electrochemical properties of $AlPO_4$-Coated $LiNi_{0.8}Co_{0.1}Mn_{0.1}O_2$ cathode materials for a Li-ion cell. *Journal of the Electrochemical Society* 2004; 151:A1899–A1904.
40. Lee, D.J., Scrosati, B., and Sun, Y.K. $Ni_3(PO_4)_2$-coated $LiNi_{0.8}Co_{0.15}Al_{0.05}O_2$ lithium battery electrode with improved cycling performance at 55°C. *Journal of Power Sources* 2011; 196:7742–7746.
41. Sun, Y.K., Cho, S.W., and Lee, S.W. AlF_3-coating to improve high voltage cycling performance of $LiNi_{1/3}Co_{1/3}Mn_{1/3}O_2$ cathode materials for lithium secondary batteries. *Journal of the Electrochemical Society* 2007; 154:A168–A172.
42. Kim, J.M., Kumagai, N., and Cho, T.H. Synthesis, structure, and electrochemical characteristics of overlithiated $Li_{1+x}(Ni_zCo_{1-2z}Mn_z)_{1-x}O_2$ ($z = 0.1$–0.4 and $x = 0.0$–0.1) positive electrodes prepared by spray-drying method. *Journal of the Electrochemical Society* 2008; 155:A82–A89.
43. Kim, J., Kumagai, N., and Komaba, S. Improved electrochemical properties of $Li_{1+x}(Ni_{0.3}Co_{0.4}Mn_{0.3})O_{2-\delta}$ (x = 0, 0.03 and 0.06) with lithium excess composition prepared by a spray drying method. *Electrochimica Acta* 2006; 52:1483–1490.
44. Liu, D., Wang, Z., and Chen, L. Comparison of structure and electrochemistry of Al- and Fe-doped $LiNi_{1/3}Co_{1/3}Mn_{1/3}O_2$. *Electrochimica Acta* 2006; 51:4199–4203.
45. Kim, G.H., Myung, S.T., Kim, H.S., and Sun, Y.K. Synthesis of spherical $LiNi_{(1/3-z)}Co_{(1/3-z)}Mn_{(1/3-z)}Mg_zO_2$ as positive electrode material for lithium-ion battery. *Electrochimica Acta* 2006; 51:2447–2453.
46. Johnson, C.S., Korte, S.D., Vaughey, J.T., Thackeray, M.M., Bofinger, T.E., Shao-Horn, Y., and Hackney, S.A. Structural and electrochemical analysis of layered compounds from Li_2MnO_3. *Journal of Power Sources* 1999; 81–82:491–495.
47. Lu, Z., MacNeil, D.D., and Dahn, J.R. Layered cathode materials $LiNi_x\ Li_{(1/3-2x/3)}\ Mn_{(2/3-x/3)}O_2$ for lithium-ion batteries. *Electrochemical and Solid-State Letters* 2001; 4:A191–A194.
48. Lu, Z., and Dahn, J.R. Understanding the anomalous capacity of Li/Li $Ni_x\ Li_{(1/3-2x/3)}Mn_{(2/3-x/3)}\ O_2$ cells using *in situ* x-ray diffraction and electrochemical studies. *Journal of the Electrochemical Society* 2002; 149:A815–A822.
49. Croy, J.R., Kim, D.H., Balasubramanian, M., Gallagher, K., Kang, S.H., and Thackeray, M.M. Countering the voltage decay in high capacity xLi$_2$MnO$_3$·(1–x)LiMO$_2$ electrodes (M = Mn, Ni, Co) for Li^+-ion batteries. *Journal of the Electrochemical Society* 2012; 159:A781–A790.
50. Mohanty, D., Sefat, A.S., Li, J.L., Meisner, R.A., Rondinone, A.J., Payzant, E.A., Abraham, D.P., Wood, D.L., and Daniel, C. Correlating cation ordering and voltage fade in a lithium–manganese-rich lithium-ion battery cathode oxide: A joint magnetic susceptibility and TEM study. *Physical Chemistry Chemical Physics* 2013; 15:19496–19509.
51. Yu, H.J., Kim, H., Wang, Y.R., He, P., Asakura, D., Nakamuraa, Y., and Zhou, H.S. High-energy "composite" layered manganese-rich cathode materials via controlling Li_2MnO_3 phase activation for lithium-ion batteries. *Physical Chemistry Chemical Physics* 2012; 14:6584–6595.

52. Yu, X.Q., Lyu, Y.C., Gu, L., Wu, H.M., Bak, S.M., Zhou, Y.N., Amine, K., Ehrlich, S.N., Li, H., Nam, K.W., and Yang, X.Q. Understanding the rate capability of high-energy-density Li-rich layered $Li_{1.2}Ni_{0.15}Co_{0.1}Mn_{0.55}O_2$ cathode materials. *Advanced Energy Materials* 2013, 4:201300950 DOI:10.1002/aenm.201300950.
53. Qiu, S., Chen, Z.X., Pei, F., Wu, F.Y., Wu, Y., Ai, X.P., Yang, H.X., and Cao, Y.L. Synthesis of monoclinic $LiLi_{0.2}Mn_{0.54}Ni_{0.13}Co_{0.13}O_2$ nanoparticles by a layered-template route for high-performance Li-ion batteries. *European Journal of Inorganic Chemistry* 2013; 16:2887–2892.
54. Xiong, X., Wang, Z., Guo, H., Zhang, Q., and Li, X. Enhanced electrochemical properties of lithium-reactive V_2O_5 coated on the $LiNi_{0.8}Co_{0.1}Mn_{0.1}O_2$ cathode material for lithium ion batteries at 60°C. *Journal of Materials Chemistry A* 2013; 1:1284–1288.
55. Cheng, F., Xin, Y., Chen, J., Lu, L., Zhang, X., and Zhou, H. Monodisperse $Li_{1.2}Mn_{0.6}Ni_{0.2}O_2$ microspheres with enhanced lithium storage capability. *Journal of Materials Chemistry A* 2013; 1:5301–5308.
56. Gu, M., Belharouak, I., Zheng, J.M., Wu, H.M., Xiao, J., Genc, A., Amine, K., Thevuthasan, S., Baer, D.R., Zhang, J.G., Browning, N.D., Liu, J., and Wang, C.M. Formation of the spinel phase in the layered composite cathode used in Li-ion batteries. *ACS Nano* 2013; 7:760–767.
57. Wu, Y., and Manthiram, A. Effect of surface modifications on the layered solid solution cathodes $(1-z)$ $LiLi_{1/3}Mn_{2/3}O_2-(z)$ $LiMn_{0.5-y}Ni_{0.5-y}Co_{2y}O_2$. *Solid State Ionics* 2009; 180:50–56.
58. Liu, J., Wang, Q., and Reeja, J.B. Carbon-coated high capacity layered $LiLi_{0.2}Mn_{0.54}Ni_{0.13}Co_{0.13}O_2$ cathodes. *Electrochemistry Communications* 2010; 12:750–753.
59. Wu, X.B., Dong, Z.X., and Zheng, J.M. The Effect of Carbon Coating on the Electrochemical Performance of Li $[Li_{0.2}Mn_{0.54}Ni_{0.13}Co_{0.13}]$ O_2 Positive Electrode Material for Lithium Ion Battery. *Journal of Xiamen University* 2008; 47:224–227.
60. Gao, J., and Manthiram, A. High capacity $LiLi_{0.2}Mn_{0.54}$ $Ni_{0.13}Co_{0.13}O_2–V_2O_5$ composite cathodes with low irreversible capacity loss for lithium ion batteries. *Electrochemistry Communications* 2009; 11:84–86.
61. Gao, J., and Manthiram, A. Eliminating the irreversible capacity loss of high capacity layered $LiLi_{0.2}Mn_{0.54}Ni_{0.13}Co_{0.13}O_2$ cathode by blending with other lithium insertion hosts. *Journal of Power Sources* 2009; 191:644–647.
62. Han, S.J., Qiu, B., Wei, Z., Xia, Y.G., and Liu, Z.P. Surface structural conversion and electrochemical enhancement by heat treatment of chemical pre-delithiation processed lithium-rich layered cathode material. *Journal of Power Sources* 2014; 268:683–691.
63. Johnson, C.S., Kim, J.S., and Lefief, C. The significance of the Li_2MnO_3 component in "composite" $xLi_2MnO_3·(1-x)$ $LiMn_{0.5}Ni_{0.5}O_2$ electrodes. *Electrochemistry Communications* 2004; 6:1085–1091.
64. Kim, J.S., Johnson, C.S., and Vaughey, J.T. Pre-conditioned layered electrodes for lithium batteries. *Journal of Power Sources* 2006; 153:258–264.
65. Abouimrane, A., Compton, O.C., and Deng, H. Improved rate capability in a high-capacity layered cathode material via thermal reduction. *Electrochemical and Solid-State Letters* 2011; 14:A126–A129.
66. Wei, G.Z., Lu, X., and Ke, F.S. Crystal habit-tuned nanoplate material of $LiL_{i1/3-2x/3}Ni_xMn_{2/3-x/3}O_2$ for high-rate performance lithium-ion batteries. *Advanced Materials* 2010; 22:4364–4367.
67. Ito, A., Li, D., and Ohsawa, Y. A new approach to improve the high-voltage cyclic performance of Li-rich layered cathode material by electrochemical pre-treatment. *Journal of Power Sources* 2008; 183:344–346.
68. Padhi, A.K., Nanjundaswamy, K.S., and Goodenough, J.B. Phospho-olivines as positive-electrode materials for rechargeable lithium batteries. *Journal of the Electrochemical Society* 1997; 144:1188–1194.
69. Li, G.H., Azuma, H., and Tohda, M. $LiMnPO_4$ as the cathode for lithium batteries. *Electrochemical and Solid-State Letters* 2002; 5:A135–A137.
70. Okada, S., Sawa, S., Egashira, M., Yamaki, J., Tabuchi, M., Kageyama, H., Konishi, T., and Yoshino, A. Cathode properties of phospho-olivine $LiMPO_4$ for lithium secondary batteries. *Journal of Power Sources* 2001; 97–98:430–432.
71. Woffenstine, J., and Allen, J. Ni^{3+}/Ni^{2+} redox potential in $LiNiPO_4$. *Journal of Power Sources* 2005; 14:389–390.
72. Yamada, A., Chung, S.C., and Hinokuma, K. Optimized $LiFePO_4$ for lithium battery cathodes. *Journal of the Electrochemical Society* 2001; 148:A224–A229.
73. Manthiram, A., and Goodenough, J.B. Lithium insertion into $Fe_2(MO_4)_3$ frameworks: Comparison of M = W with M = Mo. *Journal of Solid-State Chemistry* 1987; 71:349–360.
74. Manthiram, A., and Goodenough, J.B. Lithium insertion into $Fe_2(SO_4)_3$ frameworks. *Journal of Power Sources* 1989; 26:403–408.
75. Padhi, A.K., Nanjundaswamy, K.S., Masquelier, C., Okada, S., and Goodenough, J.B. Effect of structure on the Fe^{3+}/Fe^{2+} redox couple in iron phosphates. *Journal of the Electrochemical Society* 1997; 144:1609–1613.
76. Goodenough, J.B. Mapping of Redox Energies. *Molecular Crystals and Liquid Crystals* 1998; 311:1.
77. Goodenough, J.B. Synopsis of the Lithium-Ion Battery Markets. in: *Lithium Ion Batteries*, Wakihara, M., and Yamamoto, O. Editors, Chapter 1, Kodansha, Tokyo, Japan, 1998.
78. Goodenough, J.B., and Manivannan, V. Cathodes for lithium-ion batteries: Some comparisons. *Denki Kagakuoyobi Kogyo Butsuri Kagaku* 1998; 66:1173.
79. Padhi, A.K., Nanjundaswamy, K.S., and Goodenough, J.B. Phospho-olivines as positive-electrode materials for rechargeable lithium batteries. *Journal of the Electrochemical Society* 1997; 144:1188–1194.
80. Gong, Z.L., and Yang, Y. Recent advances in the research of polyanion-type cathode materials for Li-ion batteries. *Energy and Environmental Science* 2011; 4:3223–3242.
81. Ouyang, C., Shi, S., Wang, Z., Huang, X., and Chen, L. First-principles study of Li ion diffusion in $LiFePO_4$. *Physical Review B* 2004; 69:104303.
82. Molenda, J., and Bak, A. Modification in the electronic structure of cobalt bronze $LixCoO_2$ and the resulting electrochemical properties. *Solid State Ionics* 1989; 36:53–58.
83. Numata, Y., and Tabuchi, T. Verwey-type transition and magnetic properties of the $LiMn_2O_4$ spinels. *Journal of Solid-State Chemistry* 1997; 131:138–143.
84. Kawai, H., Nagata, M., Kageyama, H., and Tukamoto, H. 5 V lithium cathodes based on spinel solid solutions $Li_2Col + XMn_{3-X}O_8$: $-1 \leq X \leq 1$. *Electrochimica Acta* 1999; 45:315–327.
85. Andersson, A.S., Kalska, B., Haggstrom, L., and Thomas, J.O. Lithium extraction/insertion in $LiFePO_4$: An X-ray diffraction and Mössbauer spectroscopy study. *Solid State Ionics* 2000; 130:41–52.
86. Thackeray, M.M. Lithium-ion batteries: An unexpected conductor. *Nature Materials* 2002; 1:81–82.

87. Xia, Y., Yoshio, M., and Noguchi, H. Improved electrochemical performance of $LiFePO_4$ by increasing its specific surface area. *Electrochimica Acta* 2006; 52:240–245.
88. Ou, X., Pan, L., Gu, H., Wu, Y., and Lu, J. Temperature-dependent crystallinity and morphology of $LiFePO_4$ prepared by hydrothermal synthesis. *Journal of Materials Chemistry* 2012; 22:9064–9068.
89. Liu, Y., and Cao, C. Enhanced electrochemical performance of nano-sized $LiFePO_4$/C synthesized by an ultrasonic-assisted co-precipitation method. *Electrochimica Acta* 2010; 55:4694–4699.
90. Sanchez, M.A.E., Brito, G.E.S., Fantini, M.C.A., Goya, G.F., and Matos, J.R. Synthesis and characterization of $LiFePO_4$ prepared by sol–gel technique. *Solid State Ionics* 2006; 177:497–500.
91. Chen, Z., and Dahn, J.R. Reducing carbon in $LiFePO_4$/C composite electrodes to maximize specific energy, volumetric energy, and tap density. *Journal of the Electrochemical Society* 2002; 149:A1184–A1189.
92. Tang, Y., Huang, F., Bi, H., Liu, Z., and Wan, D. Highly conductive three-dimensional graphene for enhancing the rate performance of $LiFePO_4$ cathode. *Journal of Power Sources* 2012; 203:130–134.
93. Fan, Q., Lei, L., Xu, X., Yin, G., and Sun, Y. Direct growth of $FePO_4$/graphene and $LiFePO_4$/graphene hybrids for high rate Li-ion batteries. *Journal of Power Sources* 2014; 257:65–69.
94. Zhou, X., Wang, F., Zhu, Y., and Liu, Z. Graphene modified $LiFePO_4$ cathode materials for high power lithium ion batteries. *Journal of Materials Chemistry* 2011; 21:3353–3358.
95. Chung, S.Y., Bloking, J.T., and Chiang, Y.M. Electronically conductive phospho-olivines as lithium storage electrodes. *Nature Materials* 2002; 1:123–128.
96. Ravet, N., Abouimrane, A., and Armand, M. From our readers: On the electronic conductivity of phospho-olivines as lithium storage electrodes. *Nature Materials* 2003; 2:702.
97. Wang, D., Li, H., Shi, S., Huang, X., and Chen, L. Improving the rate performance of $LiFePO_4$ by Fe-site doping. *Electrochimica Acta* 1999; 50:2955–2958.
98. Ni, J., Zhao, Y., Chen, J., Gao, L., and Lu, L. Site-dependent electrochemical performance of Mg doped $LiFePO_4$. *Electrochemistry Communications* 2014; 44:4–7.
99. Yamada, A., Hosoya, M., Chung, S.C., Kudo, Y., Hinokuma, K., Liu, K.Y., and Nishi, Y. Olivine-type cathodes: Achievements and problems. *Journal of Power Sources* 2003; 119:232–238.
100. Martha, S.K., Grinblat, J., Haik, O., Zinigrad, E., Drezen, T., Miners, J.H., Exnar, I., Kay, A., Markovsky, B., and Aurbach, D. $LiMn_{0.8}Fe_{0.2}PO_4$: An advanced cathode material for rechargeable lithium batteries. *Angewandte Chemie International Edition* 2009; 48:8559–8563.
101. Hu, L., Qiu, B., Xia, Y., Qin, Z., Qin, L., Zhou, X., and Liu, Z. Facile synthesis and electrochemical properties of two dimensional layered MoS_2/graphene composite for reversible lithium storage. *Journal of Power Sources* 2014; 248:264–268.
102. Ni, J., and Gao, L. Effect of copper doping on $LiMnPO_4$ prepared via hydrothermal route. *Journal of Power Sources* 2011; 196:6498–6501.
103. Wang, D., Ouyang, C., Drézen, T., Exnar, I., Kay, A., Kwon, N.H., Gouerec, P., Miners, J.H., Wang, M., and Grätzela, M. Improving the electrochemical activity of $LiMnPO_4$ via Mn-site substitution. *Journal of the Electrochemical Society* 2010; 157:A225–A229.
104. Qin, L., Xia, Y., Qiu, B., Cao, H., Liu, Y., and Liu, Z. Synthesis and electrochemical performances of (1–x) $LiMnPO_4{\cdot}xLi_3V_2(PO_4)_3$/C composite cathode materials for lithium ion batteries. *Journal of Power Sources* 2013; 239:144.
105. Su, L., Li, X., Ming, H., Adkins, J., Liu, M., Zhou, Q., and Zheng, J. Effect of vanadium doping on electrochemical performance of $LiMnPO_4$ for lithium-ion batteries. *Journal of Solid-State Electrochemistry* 2014; 18:755–762.
106. Qin, Z., Zhou, X., Xia, Y., Tang, C., and Liu, Z. Morphology controlled synthesis and modification of high-performance $LiMnPO_4$ cathode materials for Li-ion batteries. *Journal of Materials Chemistry* 2012; 22:21144–21153.
107. Zaghib, K., Trudeaua, M., Guerfi, A., Trottier, J., Mauger, A., Veillette, R., and Julien, C.M. New advanced cathode material: $LiMnPO_4$ encapsulated with $LiFePO_4$. *Journal of Power Sources* 2012; 204:177–181.
108. Ramar, V., and Balaya, P. Enhancing the electrochemical kinetics of high voltage olivine $LiMnPO_4$ by isovalent co-doping. *Physical Chemistry Chemical Physics* 2013; 15:17240–17249.
109. Gummow, R.J., Kock, A., and de Thackeray, M. Improved capacity retention in rechargeable 4 V lithium/lithium–manganese oxide (spinel) cells. *Solid State Ionics* 1994; 69:59–67.
110. Jang, D.H., Shin, Y.J., and Oh, S.M. Dissolution of spinel oxides and capacity losses in 4 V Li/$LixMn_2O_4$ Cells. *Journal of the Electrochemical Society* 1996; 143:2204–2211.
111. Deng, B., Nakamura, H., and Yoshio, M. Superior capacity retention of oxygen stoichiometric spinel $Li_{1+x}Mn_{2-x}O_{4+\delta}$ at elevated temperature. *Electrochemical and Solid-State Letters* 2005; 8:A171–A174.
112. Xia, Y., Wang, H., Zhang, Q., Nakamura, H., Noguchi, H., and Yoshio, M. Oxygen deficiency, a key factor in controlling the cycle performance of Mn-spinel cathode for lithium-ion batteries. *Journal of Power Sources* 2007; 166:485–491.
113. Xia, Y., Zhang, Q., Wang, H., Nakamura, H., Noguchi, H., and Yoshio, M. Improved cycling performance of oxygen-stoichiometric spinel $Li_{1+x}Al_yMn_{2-x-y}O_{4+\delta}$ at elevated temperature. *Electrochimica Acta* 2007; 52:4708–4714.
114. Takahashi, M., Yoshida, T., Ichikawa, A., Kitoh, K., Katsukawa, H., Zhang, Q., and Yoshio, M. Effect of oxygen deficiency reduction in Mg-doped Mn-spinel on its cell storage performance at high temperature. *Electrochimica Acta* 2006; 51:5508–5514.
115. Kannan, A.M., and Manthiram, A. Surface/chemically modified $LiMn_2O_4$ cathodes for lithium-ion batteries. *Electrochemical and Solid-State Letters* 2002; 5:A167–A169.
116. Lee, S.W., Kim, K.S., Moon, H.S., Kim, H.J., Cho, B.W., Cho, W.I., Ju, J.B., and Park, J.W. Electrochemical characteristics of Al_2O_3-coated lithium manganese spinel as a cathode material for a lithium secondary battery. *Journal of Power Sources* 2004; 126:150–155.
117. Sun, Y.K., Lee, Y.S., Yoshio, M., and Amine, K. Synthesis and electrochemical properties of ZnO-coated $LiNi_{0.5}Mn_{1.5}O_4$ spinel as 5 V cathode material for lithium secondary batteries. *Electrochemical and Solid-State Letters* 2002; 5:A99–A102.
118. Zhao, S., Bai, Y., Chang, Q., Yang, Y., and Zhang, W. Surface modification of spinel $LiMn_2O_4$ with FeF_3 for lithium ion batteries. *Electrochimica Acta* 2013; 108:727–735.
119. Lee, D.J., Lee, K.S., Myung, S.T., Yashiro, H., and Sun, Y.K. Improvement of electrochemical properties of $Li_{1.1}Al_{0.05}Mn_{1.85}O_4$ achieved by an AlF_3 coating. *Journal of Power Sources* 2011; 196:1353–1357.
120. Chen, Q., Wang, Y., Zhang, T., Yin, W., Yang, J., and Wang, X. Electrochemical performance of LaF_3-coated $LiMn_2O_4$ cathode materials for lithium ion batteries. *Electrochimica Acta* 2012; 83:65–72.
121. Qing, C., Bai, Y., Yang, J., and Zhang, W. Enhanced cycling stability of $LiMn_2O_4$ cathode by amorphous $FePO_4$ coating. *Electrochimica Acta* 2011; 56:6612–6618.

122. Liu, D., He, Z., and Liu, X. Increased cycling stability of $AlPO_4$-coated $LiMn_2O_4$ for lithium ion batteries. *Materials Letters* 2007; 61:4703–4706.
123. Cho, J., Kim, G.B., Lim, H.S., Kim, C.S., and Yoo, S.I. Improvement of structural stability of $LiMn_2O_4$ cathode material on 55°C cycling by sol–gel coating of $LiCoO_2$. *Electrochemical and Solid-State Letters* 1999; 2:607–609.
124. Park, S.C., Kim, Y.M., Han, S.C., Ahn, S., Ku, C.H., and Lee, J.Y. The elevated temperature performance of $LiMn_2O_4$ coated with $LiNi_{1-X}Co_XO_2$ (X = 0.2 and 1). *Journal of Power Sources* 2002; 107:42–47.
125. Benedek, R., and Thackeray, M.M. Simulation of the surface structure of lithium manganese oxide spine. *Physical Review B* 2011; 83:195439.1–195439.8.
126. Koga, K., Masukuni, H., Matsumiti, K., Uegami, M., Hiroaki, M., Kazutoshi, M. Positive electrode active material used for non-aqueous-electrolyte secondary battery, comprises lithium manganate particles containing lithium and manganese as main component and having cubic crystal spinel structure, WO2010084855 (2010).
127. Lee, S., Hu, H., Xia, Y., Zhao, L., and Xiao, F. Morphology-control preparation and electrochemical performance of Mn–spinel cathode materials. *Chinese Science Bulletin* (Chinese Version) 2013; 58:3350–3356.
128. Zhong, Q., Bonakdarpour, A., Zhang, M., Gao, Y., and Dahn, J.R. Synthesis and electrochemistry of $LiNi_xMn_{2-x}O_4$. *Journal of the Electrochemical Society* 1997; 144:205–213.
129. Song, J., Shin, D.W., Lu, Y., Amos, C.D., Manthiram, A., and Goodenough, J.B. Role of oxygen vacancies on the performance of $LiNi_{0.5-x}Mn_{1.5+x}O_4$ (x = 0, 0.05, and 0.08) spinel cathodes for lithium-ion batteries. *Chemistry of Materials* 2012; 24:3101–3109.
130. Yang, J.G., Han, X.P., Zhang, X.L., Cheng, F.Y., and Chen, J. Spinel $LiNi_{0.5}Mn_{1.5}O_4$ cathode for rechargeable lithium ion batteries: Nano vs micro, ordered phase ($P4_332$) vs disordered phase. *Nano Research* 2013; 6:679–687.
131. Liu, D., Zhu, W., Trottier, J., Gagnon, C., Barray, F., Guerfi, A., Mauger, A., Groult, H., Julien, C.M., Goodenough, J.B., and Zaghib, K. Spinel materials for high-voltage cathodes in Li-ion batteries. *RSC Advances* 2014; 4:154–167.
132. Hwang, B.J., Wu, Y.W., Venkateswarlu, M.M., Cheng, Y., and Santhanam, R. Influence of synthesis conditions on electrochemical properties of high-voltage $Li_{1.02}Ni_{0.5}Mn_{1.5}O_4$ spinel cathode material. *Journal of Power Sources* 2009; 193:828–833.
133. Lee, E.S., Nam, K.W., Hu, E., and Manthiram, A. Influence of cation ordering and lattice distortion on the charge–discharge behavior of $LiMn_{1.5}Ni_{0.5}O_4$ spinel between 5.0 and 2.0 V. *Chemistry of Materials* 2012; 24:3610–3620.
134. McCalla, E., and Dahn, J.R. The spinel and cubic rocksalt solid-solutions in the Li–Mn–Ni oxide pseudo-ternary system. *Solid State Ionics* 2013; 242:1–9.
135. McCalla, E., Rowe, A.W., Shunmugasundaram, R., and Dahn, J.R. Structural study of the Li–Mn–Ni oxide pseudoternary system of interest for positive electrodes of Li-Ion batteries. *Chemistry of Materials* 2013; 25:989–999.
136. Kim, J.H., Yoon, C.S., Myung, S.T., Prakash, J., and Sun, Y.K. Phase transitions in $Li_{1-\delta}Ni_{0.5}Mn_{1.5}O_4$ during cycling at 5 V. *Electrochemical and Solid-State Letters* 2004; 7:A216–A220.
137. Manthiram, A., Chemelewski, K., and Lee, E.S. A perspective on the high-voltage $LiMn_{1.5}Ni_{0.5}O_4$ spinel cathode for lithium-ion batteries. *Energy and Environmental Science* 2014; 7:1339–1350.
138. Terada, Y., Yasaka, K., Nishikawa, F., Konishi, T., Yoshio, M., and Nakai, I. *In situ* XAFS analysis of $Li(Mn, M)_2O_4$ (M = Cr, Co, Ni) 5 V cathode materials for lithium-ion secondary batteries. *Journal of Solid-State Chemistry* 2001; 156:286–291.
139. Kunduraci, M., and Amatucci, G.G. The effect of particle size and morphology on the rate capability of 4.7 V $LiMn_{1.5+\delta}Ni_{0.5-\delta}O_4$ spinel lithium-ion battery cathodes. *Electrochimica Acta* 2008; 53:4193–4199.
140. Kim, J.H., Myung, S.T., Yoon, C.S., Kang, S.G., and Sun, Y.K. Comparative study of $LiNi_{0.5}Mn_{1.5}O_{4-\delta}$ and $LiNi_{0.5}Mn_{1.5}O_4$ cathodes having two crystallographic structures: Fd3m and $P4_332$. *Chemistry of Materials* 2004; 16:906–914.
141. Wang, L.P., Li, H., Huang, X.J., and Baudrin, E. A comparative study of Fd-3m and $P4_332$ "$LiNi_{0.5}Mn_{1.5}O_4$". *Solid State Ionics* 2011; 193:32–38.
142. Oh, S.H., Jeon, S.H., Cho, W.I., Kim, C.S., and Cho, B.W. Synthesis and characterization of the metal-doped high-voltage spinel $LiNi_{0.5}Mn_{1.5}O_4$ by mechanochemical process. *Journal of Alloys and Compounds* 2008; 452:389–396.
143. Yang, S.H., and Middaugh, R.L. Redox reactions of cobalt, aluminum and titanium substituted lithium manganese spinel compounds in lithium cells. *Solid State Ionics* 2001; 139:13–25.
144. Leon, B., Lloris, J.M., Vicente, C.P., and Tirado, J.L. Structure and lithium extraction mechanism in $LiNi_{0.5}Mn_{1.5}O_4$ after double substitution with iron and titanium. *Electrochemical and Solid-State Letters* 2006; 9:A96–A100.
145. Alcantara, R., Jaraba, M., Lavela, P., Lloris, J.M., Vicente, C.P., and Tirado, J.L. Synergistic effects of double substitution in $LiNi_{0.5-y}Fe_yMn_{1.5}O_4$ spinel as 5 V cathode materials. *Journal of the Electrochemical Society* 2005; 152:A13–A18.
146. Liu, J., and Manthiram, A. Understanding the improved electrochemical performances of Fe-substituted 5 V spinel cathode $LiMn_{1.5}Ni_{0.5}O_4$. *Journal of Physical Chemistry C* 2009; 113:15073–15079.
147. Eli, Y.E., Vaughey, J.T., Thackeray, M.M., Mukerjee, S., Yang, X.Q., and McBreenc, J. $LiNi_xCu_{0.5-x}Mn_{1.5}O_4$ spinel electrodes, superior high-potential cathode materials for Li batteries: I. Electrochemical and structural studies. *Journal of the Electrochemical Society* 1999; 146:908–913.
148. Alcantara, R., Jaraba, M., Lavela, P., and Tirado, J.L. New $LiNi_yCo_{1-2y}Mn_{1+y}O_4$ spinel oxide solid solutions as 5 V electrode material for Li-ion batteries. *Journal of the Electrochemical Society* 2004; 151:A53–A58.
149. Ito, A., Li, D., Lee, Y., Kobayakawa, K., and Sato, Y. Influence of Co substitution for Ni and Mn on the structural and electrochemical characteristics of $LiNi_{0.5}Mn_{1.5}O_4$. *Journal of Power Sources* 2008; 185:1429–1433.
150. Kim, J.H., Myung, S.T., Yoon, C.S., Oh, I.H., and Sun, Y.K. Effect of Ti substitution for Mn on the structure of $LiNi_{0.5}Mn_{1.5-x}Ti_xO_4$ and their electrochemical properties as lithium insertion material. *Journal of the Electrochemical Society* 2004; 151:A1911–A1918.
151. Liu, G.Q., Yuan, W.S., Liu, G.Y., and Tian, Y.W. The electrochemical properties of $LiNi_{0.5}Mn_{1.2}Ti_{0.3}O_4$ compound. *Journal of Alloys and Compounds* 2009; 484:567–569.
152. Alcantara, R., Jaraba, M., Lavela, P., Tirado, J.L., Biensan, P., and Peres, J.P. Structural and electrochemical study of new $LiNi_{0.5}Ti_xMn_{1.5-x}O_4$ spinel oxides for 5 V cathode materials. *Chemistry of Materials* 2003; 15:2376–2382.
153. Liu, D., Lu, Y., and Goodenough, J.B. Rate properties and elevated-temperature performances of $LiNi_{0.5-x}Cr_{2x}Mn_{1.5-x}O_4$ ($0 \le 2x \le 0.8$) as 5 V cathode materials for lithium-ion batteries. *Journal of the Electrochemical Society* 2010; 157:A1269–A1273.

154. Aklalouch, M., Amarilla, J.M., Saadoune, I., and Rojo, J.M. $LiCr_{0.2}Ni_{0.4}Mn_{1.4}O_4$ spinels exhibiting huge rate capability at 25 and 55 °C: Analysis of the effect of the particle size. *Journal of Power Sources* 2011; 196:10222–10227.
155. Ooms, F.G.B., Kelder, E.M., Schoonman, J., Wagemaker, M., and Mulder, F.M. High-voltage $LiMg_{\delta}Ni_{0.5-\delta}Mn_{1.5}O_4$ spinels for Li-ion batteries. *Solid State Ionics* 2002; 143:143–153.
156. Xu, X.X., Yang, J., Wang, Y.Q., and Wang, J.L. $LiNi_{0.5}Mn_{1.5}O_{3.975}F_{0.05}$ as novel 5 V cathode material. *Journal of Power Sources* 2007; 174:1113–1116.
157. Du, G.D., Li, Y.N., Yang, J., and Wang, J.L. Fluorine-doped $LiNi_{0.5}Mn_{1.5}O_4$ for 5 V cathode materials of lithium-ion battery. *Materials Research Bulletin* 2008; 43:3607–3613.
158. Oh, S.H., Jeon, S.H., Cho, W.I., Kim, C.S., and Cho, B.W. Synthesis and characterization of the metal-doped high-voltage spinel $LiNi_{0.5}Mn_{1.5}O_4$ by mechanochemical process. *Journal of Alloys and Compounds* 2008; 452:389–396.
159. Kim, J.H., Pieczonka, N.P.W., Li, Z., Wu, Y., Harris, S., and Powell, B.R. Understanding the capacity fading mechanism in $LiNi_{0.5}Mn_{1.5}O_4$/graphite Li-ion batteries. *Electrochimica Acta* 2013; 90:556–562.
160. Han, E., Kang, H.M., Liu, C.C., Nealey, P.F., and Gopalan, P. Graphoepitaxial assembly of symmetric block copolymers on weakly preferential substrates. *Advanced Materials* 2010; 22:4325–4329.
161. Ellis, B.L., Lee, K.T., and Nazar, L.F. Positive electrode materials for Li-ion and Li-batteries. *Chemistry of Materials* 2010; 22:691–714.
162. Wen, C.J., and Huggins, R.A. Electrochemical investigation of the lithium–gallium system. *Journal of the Electrochemical Society* 1981; 128:1636–1641.
163. Wang, J., Raistrick, D., and Huggins, R.A. Behavior of some binary lithium alloys as negative electrodes in organic solvent-based electrolytes. *Journal of the Electrochemical Society* 1986; 133:457–460.
164. Obrovac, M.N., and Christensen, L. Structural changes in silicon anodes during lithium insertion/extraction. *Electrochemical and Solid-State Letters* 2004; 7:A93–A96.
165. Liu, X.H., Zhang, L.Q., Zhong, L., Liu, Y., Zheng, H., Wang, J.W., Cho, J.H., Dayeh, S.A., Picraux, S.T., Sullivan, J.P., Mao, S.X., Ye, Z.Z., and Huang, J.Y. Ultrafast electrochemical lithiation of individual Si nanowire anodes. *Nano Letters* 2011; 11:2251–2258.
166. Cui, L.F., Ruffo, R., Chan, C.K., Peng, H., and Cui, Y. Crystalline-amorphous core–shell silicon nanowires for high capacity and high current battery electrodes. *Nano Letters* 2009; 9:491–495.
167. Boukamp, B.A., Lesh, G.C., and Huggins, R.A. All-solid lithium electrodes with mixed-conductor matrix. *Journal of the Electrochemical Society* 1981; 128:725–729.
168. Kasavajjula, U., Wang, C.S., and Appleby, A.J. Nano- and bulk-silicon-based insertion anodes for lithium-ion secondary cells. *Journal of Power Sources* 2007; 163:1003–1039.
169. Bruce, P.G., Scrosati, B., and Tarascon, J.M. Nanomaterials for rechargeable lithium batteries. *Angewandte Chemie International Edition* 2008; 47:2930–2946.
170. Maranchi, J.P., Hepp, A.F., Evans, A.G., Nuhfer, N.T., and Kumta, P.N. Interfacial properties of the a-Si/Cu: Active–inactive thin-film anode system for lithium-ion batteries. *Journal of the Electrochemical Society* 2006; 153:A1246–A1253.
171. Chan, C.K., Peng, H.L., Liu, G., McIlwrath, K., Zhang, X.F., Huggins, R.A., and Cui, Y. High performance lithium battery anodes using silicon nanowires. *Nature Nanotechnology* 2008; 3:31–35.
172. Gudiksen, M.S., and Lieber, C.M. Diameter-selective synthesis of semiconductor nanowires. *Journal of the American Chemical Society* 2000; 122:8801–8802.
173. Yu, D.P., Bai, Z.G., Ding, Y., Hang, Q.L., Zhang, H.Z., Wang, J.J., Zou, Y.H., Qian, W., Xiong, G.C., Zhou, H.T., and Feng, S.Q. Nanoscale silicon wires synthesized using simple physical evaporation. *Applied Physics Letters* 1998; 72:3458–3460.
174. Mallet, J., Molinari, M., Martineau, F., Delavoie, F., Fricoteaux, P., and Troyon, M. Growth of silicon nanowires of controlled diameters by electrodeposition in ionic liquid at room temperature. *Nano Letters* 2008; 8:3468–3474.
175. Chan, C.K., Patel, R.N., O'Connell, M.J., Korgel, B.A., and Cui, Y. Solution-grown silicon nanowires for lithium-ion battery anodes. *ACS Nano* 2010; 4:1443–1450.
176. Huang, Z.P., Geyer, N., Werner, P., de Boor, J., and Gosele, U. Metal-assisted chemical etching of silicon: A review. In memory of Prof. Ulrich Gösele. *Advanced Materials* 2011; 23:285–308.
177. Huang, R., Fan, X., Shen, W.C., and Zhu, J. Carbon coated silicon nanowire arrays films for high performance lithium ion battery anodes. *Applied Physics Letters* 2009; 95:133119.
178. Wang, W., Tian, M., Wei, Y.J., Lee, S.H., Lee, Y.C., and Yang, R.G. Binder free three dimensional silicon/carbon nanowire networks for high performance lithium ion battery anodes. *Nano Energy* 2013; 2:943–950.
179. Pang, C.L., Cui, H., Yang, G.W., and Wang, C.X. Flexible transparent and free-standing silicon nanowires paper. *Nano Letters* 2013; 13:4708–4714.
180. Liu, X.H., Zheng, H., Zhong, L., Huang, S., Karki, K., Zhang, L.Q., Liu, Y., Kushima, A., Liang, W.T., Wang, J.W., Cho, J.H., Epstein, E., Dayeh, S.A., Picraux, S.T., Zhu, T., Li, J., Sullivan, J.P., Cumings, J., Wang, C., Mao, S.X., Ye, Z.Z., Zhang, S., and Huang, J.Y. Anisotropic swelling and fracture of silicon nanowires during lithiation. *Nano Letters* 2011; 11:3312–3318.
181. Yang, H., Huang, S., Huang, X., Fan, F., Liang, W., Liu, X.H., Chen, L.Q., Huang, J.Y., Li, J., Zhu, T., and Zhang, S. Orientation dependent interfacial mobility governs the anisotropic swelling in lithiated silicon nanowires. *Nano Letters* 2012; 12:1953–1958.
182. Choi, J.W., McDonough, J., Jeong, S., Yoo, J.S., Chan, C.K., and Cui, Y. Stepwise nanopore evolution in one dimensional nanostructures. *Nano Letters* 2010; 10:1409–1413.
183. Liao, H., Karki, K., Zhang, Y., Cumings, J., and Wang, Y. Interfacial mechanics of carbon nanotube/amorphous Si coaxial nanostructures. *Advanced Materials* 2011; 23:4318–4322.
184. Stouranara, M.E., Xiao, X.C., Qi, Y., Johari, P., Lu, P., Sheldon, B.W., Gao, H.J., and Shenoy, V.B. Li segregation induces structure and strength changes at the amorphous Si/Cu interface. *Nano Letters* 2013; 13:4759–4768.
185. Ryu, J.H., Kim, J.W., Sung, Y.E., and Oh, S.M. Failure modes of silicon powder negative electrode in lithium secondary batteries. *Electrochemical and Solid-State Letters* 2004; 7:A306–A309.
186. Liu, X.H., Zhong, L., Huang, S., Mao, S.X., Zhu, T., and Huang, J.Y. Size-dependent fracture of silicon nanoparticles during lithiation. *ACS Nano* 2012; 6:1522–1531.
187. Xin, X., Zhou, X., Wang, F., Yao, X., Xu, X., Zhu, Y., and Liu, Z. A 3D porous architecture of Si/graphene nanocomposite as high-performance anode materials for Li-ion batteries. *Journal of Materials Chemistry* 2012; 22:7724–7730.
188. Kim, H., Seo, M., Park, M.H., and Cho, J. A critical size of silicon nano anodes for lithium rechargeable batteries. *Angewandte Chemie International Edition* 2010; 49:2146–2149.

189. Hassan, F.M., Chabot, V., Elsayed, A.R., Xiao, X.C., and Chen, Z.W. Engineered Si electrode nanoarchitecture: A scalable postfabrication treatment for the production of next-generation Li-ion batteries. *Nano Letters* 2014; 14:277–283.
190. Zhang, H., Yu, X., and Braun, P.V. Three-dimensional bicontinuous ultrafast-charge and -discharge bulk battery electrodes. *Nature Nanotechnology* 2011; 6:277–281.
191. Zhou, S., Liu, X., and Wang, D. Si/TiSi$_2$ heteronanostructures as high capacity anode material for Li ion batteries. *Nano Letters* 2010; 10:860–863.
192. Cui, L.F., Yang, Y., Hsu, C.M., and Cui, Y. Carbon silicon core shell nanowires as high capacity electrode for lithium ion batteries. *Nano Letters* 2009; 9:3370–3374.
193. Fu, K., Yildiz, O., Bhanushali, H., Wang, Y.X., Stano, K., Xue, L.G., Zhang, X.W., and Bradford, P.D. Aligned carbon nanotube-silicon sheets: A novel nano-architecture for flexible lithium ion battery electrodes. *Advanced Materials* 2013; 25:5109–5114.
194. Hu, L., Wu, H., Gao, Y., Cao, A., Li, H., McDough, J., Xie, X., Zhou, M., and Cui, Y. Silicon carbon nanotube coaxial sponge as Li ion anodes with high areal capacity. *Advanced Energy Materials* 2011; 1:523–527.
195. Zhang, H., and Braun, P.V. Three-dimensional metal scaffold supported bicontinuous silicon battery anodes. *Nano Letters* 2012; 12:2778–2783.
196. Kim, S.W., Yun, J.H., Son, B.K., Lee, Y.G., Kim, K.M., Lee, Y.M., and Cho, K.Y. Graphite/silicon hybrid electrodes using a 3D current collector for flexible batteries. *Advanced Materials* 2014; 26:2977–2982.
197. Magasinski, A., Dixon, P., Hertzberg, B., Kvit, A., Ayala, J., and Yushin, G. High-performance lithium-ion anodes using a hierarchical bottom-up approach. *Nature Materials* 2010; 9:353–358.
198. Kim, H., Han, B., Choo, J., and Cho, J. Three dimensional porous silicon particles for use in high performance lithium secondary batteries. *Angewandte Chemie International Edition* 2008; 47:10151–10154.
199. Yu, Y., Gu, L., Zhu, C., Tsukimoto, S., van Aken, P.A., and Maier, J. Reversible storage of lithium in silver coated three dimensional macroporous silicon *Advanced Materials* 2010; 22:2247–2250.
200. Yao, Y., McDowell, M.T., Ryu, I., Wu, H., Liu, N.A., Hu, L.B., Nix, W.D., and Cui, Y. Interconnected silicon hollow nanospheres for lithium ion battery anodes with long cycle life. *Nano Letters* 2011; 11:2949–2954.
201. Esmanski, A., and Ozin, G.A. Silicon inverse opal based macroporous materials as negative electrodes for lithium ion batteries. *Advanced Functional Materials* 2009; 19:1999–2010.
202. Huang, X.K., Yang, J., Mao, S., Chang, J.B., Hallac, P.B., Fell, C.R., Metz, B., Jiang, J.W., Hurley, P.T., and Chen, J.H. Controllable synthesis of hollow Si anode for long-cycle-life lithium-ion batteries. *Advanced Materials* 2014; 26:4326–4332.
203. Wu, H., Chan, G., Choi, J.W., Ryu, I., Yao, Y., McDowell, M.T., Lee, S.W., Jackson, A., Yang, Y., Hu, L., and Cui, Y. Stable cycling of double walled silicon nanotube battery anodes through solid electrolyte interphase control. *Nature Nanotechnology* 2012; 7:310–315.
204. Feng, X.J., Yang, J., Gao, P.F., Wang, J.L., and Nuli, Y. Facile approach to an advanced nanoporous silicon/carbon composite anode material for lithium ion batteries. *RSC Advances* 2012; 2:5701–5706.
205. Ge, M.Y., Rong, J.P., Fang, X., and Zhou, C.W. Porous doped silicon nanowires for lithium ion battery anode with long cycle life. *Nano Letters*; 2012; 12:2318–2323.
206. Wu, H., Zheng, G., Liu, N., Carney, T.J., Yang, Y., and Cui, Y. Engineering empty space between Si nanoparticles for lithium-ion battery anodes. *Nano Letters* 2012; 12:904–909.
207. Hwang, T.H., Lee, Y.M., Kong, B.S., Seo, J.S., and Choi, J.W. Electrospun core–shell fibers for robust silicon nanoparticle-based lithium ion battery anodes. *Nano Letters* 2012; 12:802–807.
208. Jung, D.S., Hwang, T.H., Park, S.B., and Choi, J.W. Spray drying method for large-scale and high-performance silicon negative electrodes in Li-ion batteries. *Nano Letters* 2013; 13:2092–2097.
209. Zhang, S., Du, Z., Lin, R., Jiang, T., Liu, G., Wu, X., and Weng, D. Nickel nanocone-array supported silicon anode for high-performance lithium-ion batteries. *Advanced Materials* 2010; 22:5378–5382.
210. Son, S.B., Kim, S.C., Kang, C.S., Yersak, T.A., Kim, Y.C., Lee, C.G., Moon, S.H., Cho, J.S., Moon, J.T., Oh, K.H., and Lee, S.H. A highly reversible nano-Si anode enabled by mechanical confinement in an electrochemically activated $Li_xTi_4Ni_4Si_7$ matrix. *Advanced Energy Materials* 2012; 2:1226–1231.
211. Krishnan, R., Lu, T.M., and Koratkar, N. Functionally strain-graded nanoscoops for high power li-ion battery anodes. *Nano Letters* 2011; 11:377–384.
212. Li, J., Christensen, L., Obrovac, M.N., Hewitt, K.C., and Dahn, J.R. Effect of heat treatment on Si electrodes using polyvinylidene fluoride binder. *Journal of the Electrochemical Society* 2008; 155:A234–A238.
213. Koo, B., Kim, H., Cho, Y., Lee, K.T., Choi, N.S., and Cho, J. A highly cross linked polymeric binder for high performance silicon negative electrode in lithium ion batteries. *Angewandte Chemie International Edition* 2012; 51:8762–8767.
214. Kovalenko, I., Zdyrko, B., Magasinski, A., Hertzberg, B., Milicev, Z., Burtovyy, R., Luzinov, I., and Yushin, G. A major constituent of brown algae for use in high capacity Li ion batteries. *Science* 2011; 334:75–79.
215. Kennedy, T., Mullance, E., Geaney, H., Osiak, M., O'Dwyer, C., and Ryan, K.M. High-performance germanium nanowire-based lithium-ion battery anodes extending over 1000 cycles through *in situ* formation of a continuous porous network. *Nano Letters* 2014; 14:716–723.
216. Cui, G., Gu, L., Zhi, L., Kaskhedikar, N., van Aken, P.A., Müllen, K., and Maier, J. A germanium carbon nanocomposite material for lithium batteries. *Advanced Materials* 2008; 20:3079–3083.
217. Klavetter, K.C., Wood, S.M., Lin, Y.M., Snider, J.L., Davy, N.C., Chockla, A.M., Romanovicz, D.K., Korgel, B.A., Lee, J.W., Heller, A., and Mullins, C.B. A high rate germanium particle slurry cast Li ion anode with high coulombic efficiency and long cycle life. *Journal of Power Sources* 2013; 238:123–136.
218. Park, M.H., Kim, K., Kim, J., and Cho, J. Flexible dimensional control of high capacity Li ion battery anodes: From 0D hollow to 3D porous germanium nanoparticle assemblies. *Advanced Materials* 2010; 22:415–418.
219. Park, M.H., Cho, Y., Kim, K., Kim, J., Liu, M., and Cho, J. Germanium nanotubes prepared by using the Kirkendall effect as anodes for high rate lithium batteries. *Angewandte Chemie International Edition* 2011; 50:9647–9650.
220. Tian, M., Wang, W., Wei, Y., and Yang, R. Stable high areal capacity lithium ion battery anodes based on three dimensional Ni–Sn nanowire networks. *Journal of Power Sources* 2012; 211:46–51.
221. Hassoun, J., Panero, S., Simon, P., Taberna, P.L., and Scrosati, B. High rate long life Ni–Sn nanostructured electrodes for lithium ion batteries. *Advanced Materials* 2007; 19:1632–1635.

222. Luo, B., Wang, B., Li, X., Jia, Y., Liang, M., and Zhi, L. Graphene confined Sn nanosheets with enhanced lithium storage capability. *Advanced Materials* 2012; 24:3538–3543.
223. Wang, Y., Wu, M., Jiao, Z., and Lee, J.Y. Sn@CNT and Sn@C@CNT nanostructures for superior reversible lithium ion storage. *Chemistry of Materials* 2009; 21:3210–3215.
224. Zhang, W.M., Hu, J.S., Guo, Y.G., Zheng, S.F., Zhong, L.S., Song, W.G., and Wan, L.J. Tin-nanoparticles encapsulated in elastic hollow carbon spheres for high-performance anode material in lithium-ion batteries. *Advanced Materials* 2008; 20:1160–1165.
225. Zhu, Z.Q., Wang, S.W., Du, J., Jin, Q., Zhang, T., Cheng, F.G., and Chen, J. Ultrasmall Sn nanoparticles embedded in nitrogen-doped porous carbon as high-performance anode for lithium-ion batteries. *Nano Letters* 2014; 14:153–157.
226. Huang, J.Y., Zhong, L., Wang, C.M., Sullivan, J.P., Xu, W., Zhang, L.Q., Mao, S.X., Hudak, N.S., Liu, X.H., Subramanian, A., Fan, H.Y., Qi, L.A., Kushima, A., and Li, J. In Situ Observation of the Electrochemical Lithiation of a Single SnO_2 Nanowire Electrode. *Science* 2010; 330:1515–1520.
227. Liu, J., Li, W., and Manthiram, A. Dense core–shell structured SnO_2/C composites as high performance anodes for lithium ion batteries. *Chemical Communications* 2010; 46:1437–1439.
228. Paek, S.M., Yoo, E., and Honma, I. Enhanced cyclic performance and lithium storage capacity of SnO_2/graphene nanoporous electrodes with three dimensionally delaminated flexible structure. *Nano Letters* 2009; 9:72–75.
229. Wang, Y., Zeng, H.C., and Lee, J.Y. Highly reversible lithium storage in porous SnO_2 nanotubes with coaxially grown carbon nanotube overlayers. *Advanced Materials* 2006; 18:645.
230. Lou, X.W., Chen, J.S., Chen, P., and Archer, L.A. One-pot synthesis of carbon coated SnO_2 nanocolloids with improved reversible lithium storage properties. *Chemistry of Materials* 2009; 21:2868–2874.
231. Zhou, X., Wan, L.J., and Guo, Y.G. Binding SnO_2 nanocrystals in nitrogen doped graphene sheets as anode materials for lithium ion batteries. *Advanced Materials* 2013; 25:2152–2157.
232. Zhou, X.S., Dai, Z.H., Liu, S.H., Bao, J.C., and Guo, Y.G. Ultra-uniform Sn*Ox*/carbon nanohybrids toward advanced lithium-ion battery anodes. *Advanced Materials* 2014; 26:3943–3949.
233. Chang, W.S., Park, C.M., Kim, J.H., Kim, Y.U., Jeong, G., and Sohn, H.J. Quartz (SiO_2): A new energy storage anode materials for Li-ion batteries. *Energy & Environmental Science* 2012; 5:6895–6899.
234. Favors, Z., Wang, W., Bay, H.H., George, A., Ozkan, M., and Ozkan, C.S. Stable cycling of SiO_2 nanotubes as high performance anodes for lithium ion batteries. *Scientific Reports* 2014; 4:4605–4615.
235. Komaba, S., Shimomura, K., Yabuuchi, N., Ozeki, T., Yui, H., and Konno, K. Study on polymer binders for high capacity SiO negative electrode of Li ion batteries. *Journal of Physical Chemistry C* 2011; 115:13487–13495.
236. Zhang, L., Deng, J.W., Liu, L.F., Si, W.P., Oswald, S., Xi, L.X., Kundu, M., Ma, G.Z., Gemming, T., Baunack, S., Ding, F., Yan, C.L., and Schmidt, O.G. Hierarchically designed SiO_x/SiO_y bilayer nanomembranes as stable anodes for lithium ion batteries. *Advanced Materials* 2014; 26:4527–4532.
237. Taberna, P.L., Mitra, S., Poizot, P., Simon, P., and Tarascon, J.M. High rate capabilities Fe_3O_4 based Cu nano architecture electrodes for lithium ion battery applications. *Nature Materials* 2006; 5:567–573.
238. Ban, C., Wu, Z., Gillaspie, D.T., Chen, L., Yan, Y., Blackburn, J.L., and Dillon, A.C. Nanostructured Fe_3O_4/SWNT electrode: Binder free and high rate Li ion anode. *Advanced Materials* 2010; 22:E145.
239. Wu, Y., Wei, Y., Wang, J.P., Jiang, K.L., and Fan, S.S. Conformal Fe_3O_4 sheath on aligned carbon nanotube scaffolds as high-performance anodes for lithium ion batteries. *Nano Letters* 2013; 23:818–823.
240. Zhou, G., Wang, D.W., Li, F., Zhang, L., Li, N., Wu, Z.S., Wen, L., Lu, G.Q., and Cheng, H.M. Graphene wrapped Fe_3O_4 anode material with improved reversible capacity and cyclic stability for lithium ion batteries. *Chemistry of Materials* 2010; 22:5306–5313.
241. Su, J., Cao, M., Ren, L., and Hu, C. Fe_3O_4 graphene nanocomposites with improved lithium storage and magnetism properties. *Journal of Physical Chemistry C* 2011; 115:14469–14477.
242. Wu, Z.S., Ren, W., Wen, L., Gao, L., Zhao, J., Chen, Z., Zhou, G., Li, F., and Cheng, H.M., Graphene anchored with Co_3O_4 nanoparticles as anode of lithium ion batteries with enhanced reversible capacity and cyclic performance. *Nano Letters* 2010; 4:3187–3194.
243. Yang, Z., Choi, D., Kerisit, S., Rosso, K.M., Wang, D., Zhang, J., Graff, G., and Liu, J. Nanostructures and lithium electrochemical reactivity of lithium titanites and titanium oxides: A review. *Journal of Power Sources* 2009; 192:588–598.
244. Liu, H., Bi, Z., Sun, X.G., Unocic, R.R., Paranthaman, M.P., Dai, S., and Brown, G.M., Mesoporous TiO_2–B microspheres with superior rate performance for lithium ion batteries. *Advanced Materials* 2011; 23:3450–3454.
245. Kang, E., Jung, Y.S., Kim, G.H., Chun, J., Wiesner, U., Dillon, A.C., Kim, J.K., and Lee, J. Highly improved rate capability for a lithium ion battery nano-$Li_4Ti_5O_{12}$ negative electrode via carbon coated mesoporous uniform pores with a simple self-assembly method. *Advanced Functional Materials* 2011; 21:4349–4357.
246. Cheah, S.K., Perre, E., Rooth, M., Fondell, M., Harsta, A., Nyholm, L., Boman, M., Gustafsson, T., Lu, J., Simon, P., and Edstrom, K. Self-supported three-dimensional nanoelectrodes for microbattery applications. *Nano Letters* 2009; 9:3230–3233.
247. Wang, W., Tian, M., Abdulagatov, A., George, S.M., Lee, Y.C., and Yang, R. Three-dimensional Ni/TiO_2 nanowire network for high areal capacity lithium ion microbattery applications. *Nano Letters* 2012; 12:655–660.
248. Zhao, L., Hu, Y.S., Li, H., Wang, Z., and Chen, L. Porous $Li_4Ti_5O_{12}$ coated with N-doped carbon from ionic liquids for Li-ion batteries. *Advanced Materials* 2011; 23:1385–1388.
249. Shen, L., Yuan, C., Luo, H., Zhang, X., Xu, K., and Zhang, F. *In-situ* growth of $Li_4Ti_5O_{12}$ on multi-walled carbon nanotubes: Novel coaxial nanocables for high rate lithium ion batteries. *Journal of Materials Chemistry* 2011; 21:761.
250. Xin, X., Zhou, X., Wu, J., Yao, X., and Liu, Z. Scalable synthesis of TiO_2/graphene nanostructured composite with high-rate performance for lithium ion batteries. *ACS Nano* 2012; 6:11035–11043.
251. Wang, D., Wang, C., Choi, D., Li, J., Saraf, L.V., Yang, Z., Zhang, J., Nie, Z., Kou, R., Aksay, I.A., Liu, A.J., and Hu, D. Self-assembled TiO_2–graphene hybrid nanostructures for enhanced Li-Ion insertion. *ACS Nano* 2009; 3:907–914.
252. Arora, P., and Zhang, Z.M. Battery separators. *Chemical Reviews* 2004; 104:4419–4462.
253. Shi, J.L., Fang, L.F., Li, H., Zhang, H., Zhu, B.K., and Zhu, L.P. Improved thermal and electrochemical performances of PMMA modified PE separator skeleton prepared via dopamine-initiated ATRP for lithium ion batteries. *Journal of Membrane Science* 2013; 437:160–168.

254. Chen, J., Wang, S. Ding, L., Jiang, Y., and Wang, H. Performance of through-hole anodic aluminum oxide membrane as a separator for lithium-ion battery. *Journal of Membrane Science* 2014; 461:22–27.
255. Braham, K.M., and Alamgir, M. Polymer electrolytes reinforced by Celgard(r) membrane. *Journal of Electrochemical Society* 1995; 142:683–687.
256. Jeong, H.S., Hong, S.C., and Lee, S.Y. Effect of microporous structure on thermal shrinkage and electrochemical performance of Al_2O_3/poly(vinylidene fluoride-hexafluoropropylene) composite separators for lithium-ion batteries. *Journal of Membrane Science* 2010; 364:177–182.
257. Kim, M., and Park, J.H. Inorganic thin layer coated porous separator with high thermal stability for safety reinforced Li-ion battery. *Journal of Power Sources* 2012; 212:22–27.
258. Zhang, S.S. A review on the separators of liquid electrolyte Li-ion batteries. *Journal of Power Sources* 2007; 164:351–364.
259. Huang, X.S. Separator technologies for lithium ion batteries. *Journal of Solid State Electrochemistry* 2011; 15:649–662.
260. Kim, J.H., Kim, J.H., Choi, E.S., Yu, H.K., Kim, J.H., Wu, Q.S., Chun, J., Lee, S.Y., and Lee, S.Y. Colloidal silica nanoparticle-assisted structural control of cellulose nanofiber paper separators for lithium-ion batteries. *Journal of Power Sources* 2013; 242:533–540.
261. Li, H., Chen, Y.M., Ma, X.T., Shi, J.L., Zhu, B.K., and Zhu, L.P. Gel polymer electrolytes based on active PVDF separator for lithium ion battery. I: Preparation and property of PVDF/poly(dimethylsiloxane) blending membrane. *Journal of Membrane Science* 2011; 379:397–402.
262. Zhang, J., Liu, Z., Kong, Q., Zhang, C., Pang, S., Yue, L., Wang, X., Yao, J., and Cui, G. Renewable and superior thermal-resistant cellulose-based composite nonwoven as lithium-ion battery separator. *ACS Applied Materials & Interfaces* 2013; 5:128–134.
263. Miao, Y.-E., Zhu, G.-N., Hou, H., Xia, Y.-Y., and Liu, T. Electrospun polyimide nanofiber-based nonwoven separators for lithium-ion batteries. *Journal of Power Sources* 2013; 226:82–86.
264. Feng, X., Fang, M., He, X., Ouyang, M., Lu, L., Wang, H., and Zhang, M. Thermal runaway features of large format prismatic lithium ion battery using extended volume accelerating rate calorimetry. *Journal of Power Sources* 2014; 255:294–301.
265. Yoo, S.H., and Kim, C.K. Enhancement of the meltdown temperature of a lithium ion battery separator via a nanocomposite coating. *Industrial and Engineering Chemistry Research* 2009; 48:9936–9941.
266. Xiang, H., Chen, J., Li, Z., and Wang, H. An inorganic membrane as a separator for lithium-ion battery. *Journal of Power Sources* 2011; 196:8651–8655.
267. Zhang, S.S., Xu, K., and Jow, T.R. An inorganic composite membrane as the separator of Li-ion batteries. *Journal of Power Sources* 2005; 140:361–364.
268. Prosini, P.P., Villano, P., and Carewska, M. A novel intrinsically porous separator for self-standing lithium-ion batteries. *Electrochimica Acta* 2002; 48:227–233.
269. Kim, K.M., Park, N.G., Ryu, K.S., and Chang, S.H. Characteristics of PVdF-HFP/TiO_2 composite membrane electrolytes prepared by phase inversion and conventional casting methods. *Electrochimica Acta* 2006; 51:5636–5644.
270. Carlson, S.A., Ying, Q., Deng, Z., and Skotheim, T.A. Separator for electrochemical cells. U.S. Patent 6,306,545 (2001).
271. Choi, E.-S., and Lee, S.-Y. Particle size-dependent, tunable porous structure of a SiO_2/poly(vinylidene fluoride-hexafluoropropylene)-coated poly(ethylene terephthalate) nonwoven composite separator for a lithium-ion battery. *Journal of Materials Chemistry* 2011; 21:14747–14754.
272. Jeong, H.-S., Kim, D.-W., Jeong, Y.U., and Lee, S.-Y. Effect of phase inversion on microporous structure development of Al_2O_3/poly(vinylidene fluoride-hexafluoropropylene)-based ceramic composite separators for lithium-ion batteries. *Journal of Power Sources* 2010; 195:6116–6121.
273. Song, J., Ryou, M.-H., Son, B., Lee, J.-N., Lee, D.J., Lee, Y.M., Choi, J.W., and Park, J.-K. Co-polyimide-coated polyethylene separators for enhanced thermal stability of lithium ion batteries. *Electrochimica Acta* 2012; 85:524–530.
274. Song, K.W., and Kim, C.K. Coating with macroporous polyarylate via a nonsolvent induced phase separation process for enhancement of polyethylene separator thermal stability. *Journal of Membrane Science* 2010; 352:239–246.
275. Sadeghi, F., Ajji, A., and Carreau, P.J. Analysis of microporous membranes obtained from polypropylene films by stretching. *Journal of Membrane Science* 2007; 292:62–71.
276. Ihm, D.W., Noh, J.G., and Kim, J.Y. Effect of polymer blending and drawing conditions on properties of polyethylene separator prepared for Li-ion secondary battery. *Journal of Power Sources* 2002; 109:388–393.
277. Tabatabaei, S.H., Carreau, P.J., and Ajji, A. Microporous membranes obtained from PP/HDPE multilayer films by stretching. *Journal of Membrane Science* 2009; 345:148–159.
278. Matsuyama, H., Okafuji, H., Maki, T., Teramoto, M., and Kubota, N. Preparation of polyethylene hollow fiber membrane via thermally induced phase separation. *Journal of Membrane Science* 2003; 223:119–126.
279. Kim, J.Y., Lee, Y., and Lim, D.Y. Plasma-modified polyethylene membrane as a separator for lithium-ion polymer battery. *Electrochimica Acta* 2009; 54:3714–3719.
280. Lee, J.Y., Lee, Y.M., Bhattacharya, B., Nho, Y.-C., and Park, J.-K. Separator grafted with siloxane by electron beam irradiation for lithium secondary batteries. *Electrochimica Acta* 2009; 54:4312–4315.
281. Cho, J.-H., Park, J.-H., Kim, J.H., and Lee, S.-Y. Facile fabrication of nanoporous composite separator membranes for lithium-ion batteries: Poly(methyl methacrylate) colloidal particles-embedded nonwoven poly(ethylene terephthalate). *Journal of Materials Chemistry* 2011; 21:8192–8198.
282. Jeong, Y.-B., and Kim, D.-W. The role of an adhesive gel-forming polymer coated on separator for rechargeable lithium metal polymer cells. *Solid State Ionics* 2005; 176:47–51.
283. Sohn, J.-Y., Gwon, S.-J., Choi, J.-H., Shin, J., and Nho, Y.-C. Preparation of polymer-coated separators using an electron beam irradiation. *Nuclear Instruments and Methods in Physics Research Section B: Beam Interactions with Materials and Atoms* 2008; 266:4994–5000.
284. Sohn, J.-Y., Im, J.S., Gwon, S.-J., Choi, J.-H., Shin, J., and Nho, Y.-C. Preparation and characterization of a PVDF-HFP/PEGDMA-coated PE separator for lithium-ion polymer battery by electron beam irradiation. *Radiation Physics and Chemistry* 2009; 78:505–508.
285. Kim, K.J., Kim, J.-H., Park, M.-S., Kwon, H.K., Kim, H., and Kim, Y.-J. Enhancement of electrochemical and thermal properties of polyethylene separators coated with polyvinylidene fluoride-hexafluoropropylene co-polymer for Li-ion batteries. *Journal of Power Sources* 2012; 198:298–302.

286. Park, J.-H., Park, W., Kim, J.H., Ryoo, D., Kim, H.S., Jeong, Y.U., Kim, D.-W., and Lee, S.-Y. Close-packed poly(methyl methacrylate) nanoparticle arrays-coated polyethylene separators for high-power lithium-ion polymer batteries. *Journal of Power Sources* 2011; 196:7035–7038.
287. Cao, C., Tan, L., Liu, W., Ma, J., and Li, L. Polydopamine coated electrospunpoly(vinyldiene fluoride) nanofibrous membrane as separator for lithium-ion batteries. *Journal of Power Sources* 2014; 248:224–229.
288. Lee, Y., Ryou, M.-H., Seo, M., Choi, J.W., and Lee, Y.M. Effect of polydopamine surface coating on polyethylene separators as a function of their porosity for high-power Li-ion batteries. *Electrochimica Acta* 2013; 113:433–438.
289. Ryou, M.-H., Lee, Y.M., Park, J.-K., and Choi, J.W. Mussel-inspired polydopamine-treated polyethylene separators for high-power Li-ion batteries. *Advanced Materials* 2011; 23:3066–3070.
290. Gao, K., Hu, X., Yi, T., and Dai, C. PE-*g*-MMA polymer electrolyte membrane for lithium polymer battery. *Electrochimica Acta* 2006; 52:443–449.
291. Li, S., and Gao, K. The study on methyl methacrylate graft-copolymerized composite separator prepared by pre-irradiation method for Li-ion batteries. *Surface and Coatings Technology* 2010; 204:2822–2828.
292. Kim, J.Y., Lee, Y.B., and Lim, D.Y. Plasma-modified polyethylene membrane as a separator for lithium-ion polymer battery. *Electrochimica Acta* 2009; 54:3714–3719.
293. Fang, L.-F., Shi, J.-L., Zhu, B.-K., and Zhu, L.-P. Facile introduction of polyether chains onto polypropylene separators and its application in lithium ion batteries. *Journal of Membrane Science* 2013; 448:143–150.
294. Gwon, S.-J., Choi, J.-H., Sohn, J.-Y., Lim, Y.-M., Nho, Y.-C., and Ihm, Y.-E. Battery performance of PMMA-grafted PE separators prepared by pre-irradiation grafting technique. *Journal of Industrial and Engineering Chemistry* 2009; 15:748–751.
295. Gwon, S.-J., Choi, J.-H., Sohn, J.-Y., An, S.-J., Ihm, Y.-E., and Nho, Y.-C. Radiation grafting of methyl methacrylate onto polyethylene separators for lithium secondary batteries. *Nuclear Instruments and Methods in Physics Research Section B: Beam Interactions with Materials and Atoms* 2008; 266:3387–3391.
296. Yang, J.-M., Wang, N.-C., and Chiu, H.-C. Preparation and characterization of poly(vinyl alcohol)/sodium alginate blended membrane for alkaline solid polymer electrolytes membrane. *Journal of Membrane Science* 2014; 457:139–148.
297. Zhang, C., Bai, Y., Sun, Y., Gu, J., and Xu, Y. Preparation of hydrophilic HDPE porous membranes via thermally induced phase separation by blending of amphiphilic PE-*b*-PEG copolymer. *Journal of Membrane Science* 2010; 365:216–224.
298. Shi, J.-L., Fang, L.-F., Li, H., Liang, Z.-Y., Zhu, B.-K., and Zhu, L.-P. Enhanced performance of modified HDPE separators generated from surface enrichment of polyether chains for lithium ion secondary battery. *Journal of Membrane Science* 2013; 429:355–363.
299. Li, L., Wang, J., Yang, P., Guo, S., Wang, H., Yang, X., Ma, X., Yang, S., and Wu, B. Preparation and characterization of gel polymer electrolytes containing *N*-butyl-*N*-methylpyrrolidini umbis(trifluoromethanesulfonyl) imide ionic liquid for lithium ion batteries. *Electrochimica Acta* 2013; 88:147–156.
300. Li, H., Ma, X.-T., Shi, J.-L., Yao, Z.-K., Zhu, B.-K., and Zhu, L.-P. Preparation and properties of poly(ethylene oxide) gel filled polypropylene separators and their corresponding gel polymer electrolytes for Li-ion batteries. *Electrochimica Acta* 2011; 56:2641–2647.
301. Wang, S.-H., Hou, S.-S., Kuo, P.-L., and Teng, H. Poly(ethylene oxide)-co-poly(propylene oxide)-based gel electrolyte with high ionic conductivity and mechanical integrity for lithium-ion batteries. *ACS Applied Materials & Interfaces* 2013; 5:8477–8485.
302. Zhang, J., Liu, Z., Kong, Q., Zhang, C., Pang, S., Yue, L., Wang, X., Yao, J., and Cui, G. Renewable and superior thermal-resistant cellulose-based composite nonwoven as lithium-ion battery separator. *ACS Applied Materials & Interfaces* 2013; 5:128–134.
303. Zhang, Z., Lai, Y., Zhang, Z., Zhang, K., and Li, J. Al_2O_3-coated porous separator for enhanced electrochemical performance of lithium sulfur batteries. *Electrochimica Acta* 2014; 129:55–61.
304. Fu, D., Luan, B., Argue, S., Bureau, M.N., and Davidson, I.J. Nano SiO_2 particle formation and deposition on polypropylene separators for lithium-ion batteries. *Journal of Power Sources* 2012; 206:325–333.
305. Jeong, H.-S., and Lee, S.-Y. Closely packed SiO_2 nanoparticles/poly(vinylidene fluoride-hexafluoropropylene) layers-coated polyethylene separators for lithium-ion batteries. *Journal of Power Sources* 2011; 196:6716–6722.
306. Chung, Y.S., Yoo, S.H., and Kim, C.K. Enhancement of meltdown temperature of the polyethylene lithium-ion battery separator via surface coating with polymers having high thermal resistance. *Industrial & Engineering Chemistry Research* 2009; 48:4346–4351.
307. Sohn, J.-Y., Gwon, S.-J., Choi, J.-H., Shin, J., and Nho, Y.-C. Preparation of polymer-coated separators using an electron beam irradiation. *Nuclear Instruments and Methods in Physics Research Section B: Beam Interactions with Materials and Atoms* 2008; 266:4994–5000.
308. Croonenborghs, B., Smith, M.A., and Strain, P. X-ray versus gamma irradiation effects on polymers. *Radiation Physics and Chemistry* 2007; 76:1676–1678.
309. Kim, K.J., Kim, Y.H., Song, J.H., Jo, Y.N., Kim, J.-S., and Kim, Y.-J. Effect of gamma ray irradiation on thermal and electrochemical properties of polyethylene separator for Li ion batteries. *Journal of Power Sources* 2010; 195:6075–6080.
310. Chun, S.-J., Choi, E.-S., Lee, E.-H., Kim, J.H., Lee, S.-Y., and Lee, S.-Y. Eco-friendly cellulose nanofiber paper-derived separator membranes featuring tunable nanoporous network channels for lithium-ion batteries. *Journal of Materials Chemistry* 2012; 22:16618–16626.
311. Kim, Y.-J., Kim, H.-S., Doh, C.-H., Kim, S.H., and Lee, S.-M. Technological potential and issues of polyacrylonitrile based nanofiber non-woven separator for Li-ion rechargeable batteries. *Journal of Power Sources* 2013; 244:196–206.
312. Jiang, W., Liu, Z., Kong, Q., Yao, J., Zhang, C., Han, P., and Cui, G. A high temperature operating nanofibrous polyimide separator in Li-ion battery. *Solid State Ionics* 2013; 232:44–48.
313. Kritzer, P. Nonwoven support material for improved separators in Li-polymer batteries. *Journal of Power Sources* 2006; 161:1335–1340.
314. Takahashi, T., Terazono, S., Kamei, R., and Takiyama, E. Separator for electrochemical cells. U.S. Patent 5,525,409 (1996).
315. Fabbricante, A.S., Ward, G.F., and Fabbricante, T.J. Separator for electrochemical cells. U.S. Patent 6,114,017 (2000).
316. Zucker, J. Separator for electrochemical cells. U.S. Patent 6,692,868 (2004).
317. Zhang, J., Kong, Q., Liu, Z., Pang, S., Yue, L., Yao, J., Wang, X., and Cui, G. A highly safe and inflame retarding aramid lithium ion battery separator by a papermaking process. *Solid State Ionics* 2013; 245–246:49–55.
318. Dong, Z., Kennedy, S.J., and Wu, Y. Electrospinning materials for energy-related applications and devices. *Journal of Power Sources* 2011; 196:4886–4904.

319. Xin, Z., Yan, S., Ding, J., Yang, Z., Du, B., and Du, S. Surface modification of polypropylene nonwoven fabrics via covalent immobilization of nonionic sugar-based surfactants. *Applied Surface Science* 2014; 300:8–15.
320. Wang, C., Feng, R., and Yang, F. Enhancing the hydrophilic and antifouling properties of polypropylene nonwoven fabric membranes by the grafting of poly(N-vinyl-2-pyrrolidone) via the ATRP method. *Journal of Colloid and Interface Science* 2011; 357:273–279.
321. Zhang, F., Ma, X., Cao, C., Li, J., and Zhu, Y. Poly(vinylidene fluoride)/SiO_2 composite membranes prepared by electrospinning and their excellent properties for nonwoven separators for lithium-ion batteries. *Journal of Power Sources* 2014; 251:423–431.
322. Angulakshmi, N., and Stephan, A.M. Electrospun trilayer polymeric membranes as separator for lithium-ion batteries. *Electrochimica Acta* 2014; 127:167–172.
323. Wang, Q., Song, W.-L., Wang, L., Song, Y., Shi, Q., and Fan, L.-Z. Electrospun polyimide-based fiber membranes as polymer electrolytes for lithium-ion batteries. *Electrochimica Acta* 2014; 132:538–544.
324. Hao, J., Lei, G., Li, Z., Wu, L., Xiao, Q., and Wang, L. A novel polyethylene terephthalate nonwoven separator based on electrospinning technique for lithium ion battery. *Journal of Membrane Science* 2013; 428:11–16.
325. Lu, C., Qi, W., Li, L., Xu, J.L., Chen, P., Xu, R.Q., Han, L., and Yu, Q. Electrochemical performance and thermal property of electrospun PPESK/PVDF/PPESK composite separator for lithium-ion battery. *Journal of Applied Electrochemistry* 2013; 43:711–720.
326. Cho, T.-H., Tanaka, M., Onishi, H., Kondo, Y., Nakamura, T., Yamazaki, H., Tanase, S., and Sakai, T. Battery performances and thermal stability of polyacrylonitrilenano-fiber-based nonwoven separators for Li-ion battery. *Journal of Power Sources* 2008; 181:155–160.
327. Jeong, H.-S., Kim, J.H., and Lee, S.-Y. A novel poly(vinylidene fluoride-hexafluoropropylene)/poly(ethylene terephthalate) composite nonwoven separator with phase inversion-controlled microporous structure for a lithium-ion battery. *Journal of Materials Chemistry* 2010; 20:9180–9186.
328. Zhu, Y., Wang, F., Liu, L., Xiao, S., Chang, Z., and Wu, Y. Composite of a nonwoven fabric with poly(vinylidene fluoride) as a gel membrane of high safety for lithium ion battery. *Energy & Environmental Science* 2013; 6:618–624.
329. Lee, J.R., Won, J.H., Kim, J.H., Kim, K.J., and Lee, S.Y. Evaporation-induced self-assembled silica colloidal particle-assisted nanoporous structural evolution of poly(ethylene terephthalate) nonwoven composite separators for high-safety/high-rate lithium-ion batteries. *Journal of Power Sources* 2012; 216:42–47.
330. Jeong, H.S., Choi, E.S., Lee, S.Y., and Kim, J.H. Evaporation-induced, close-packed silica nanoparticle-embedded nonwoven composite separator membranes for high-voltage/high-rate lithium-ion batteries: Advantageous effect of highly percolated, electrolyte-philic microporous architecture. *Journal of Materials Chemistry* 2012; 415–416:513–519.
331. Jeong, H.S., Choi, E.S., and Lee, S.Y. Composition ratio-dependent structural evolution of SiO_2/Poly(vinylidene fluoride-hexafluoropropylene)-coated poly(ethylene terephthalate) nonwoven composite separators for lithium-ion batteries. *Electrochimica Acta* 2012; 86:317–322.
332. Shi, J.L., Hu, H., Xia, Y., Liu, Y., and Liu, Z. Polyimide matrix enhanced cross-linked gel separator with three-dimensional heat-resistance skeleton for high safety and high power lithium ion battery. *Journal of Materials Chemistry A* 2014; 2:9134–9141.
333. Shi, J.L., Shen, T., Hu, H.S., Xia, Y.G., and Liu, Z.P. Sandwich-like heat-resistance composite separators with tunable pore structure for high power high safety lithium ion batteries. *Journal of Power Sources* 2014; 271:134–142.
334. Shubha, N., Prasanth, R., Hoon, H.H., and Srinivasan, M. Dual phase polymer gel electrolyte based on non-woven poly(vinylidenefluoride-co- hexafluoropropylene)-layered clay nanocomposite fibrous membranes for lithium ion batteries. *Materials Research Bulletin* 2013; 48:526–537.
335. Liao, Y., Rao, M., Li, W., Tan, C., Yi, J., and Chen, L. Improvement in ionic conductivity of self-supported P(MMA-AN-VAc) gel electrolyte by fumed silica for lithium ion batteries. *Electrochimica Acta* 2009; 54:6396–6402.
336. Liao, C., Sun, X.-G., and Dai, S. Crosslinked gel polymer electrolytes based on polyethylene glycol methacrylate and ionic liquid for lithium ion battery applications. *Electrochimica Acta* 2013; 87:889–894.
337. Nguyen, C.A., Xiong, S., Ma, J., Lu, X., and Lee, P.S. Toward electrochromic device using solid electrolyte with polar polymer host. *Journal of Physical Chemistry B* 2009; 113:8006–8010.
338. Masuda, Y., Nakayama, M., and Wakihara, M., Fabrication of all solid-state lithium polymer secondary batteries using PEG-borate/aluminate ester as plasticizer for polymer electrolyte. *Solid State Ionics* 2007; 178:981–986.
339. Wetjen, M., Kim, G.T., Joost, M., Winter, M., and Passerini, S. Temperature dependence of electrochemical properties of cross-linked poly(ethylene oxide)-lithium bis(trifluoromethanesulfonyl)imide-*N*-butyl-*N*-methylpyrrolidiniumbis(trifluoromethanesulfonyl)imide solid polymer electrolytes for lithium batteries. *Electrochimica Acta* 2013; 87:779–787.
340. Prasanth, R., Shubha, N., Hng, H.H., and Srinivasan, M. Effect of nano-clay on ionic conductivity and electrochemical properties of poly(vinylidene fluoride) based nanocomposite porous polymer membranes and their application as polymer electrolyte in lithium ion batteries. *European Polymer Journal* 2013; 49:307–318.
341. Angulakshmi, N., Nahm, K.S., Nair, J.R., Gerbaldi, C., Bongiovanni, R., Penazzi, N., and Stephan, A.M. Cycling profile of $MgAl_2O_4$-incorporated composite electrolytes composed of PEO and $LiPF_6$ for lithium polymer batteries. *Electrochimica Acta* 2013; 90:179–185.
342. Tang, Z., Wang, J., Chen, Q., He, W., Shen, C., Mao, X.-X., and Zhang, J. A novel PEO-based composite polymer electrolyte with absorptive glass mat for Li-ion batteries. *Electrochimica Acta* 2007; 52:6638–6643.
343. Kurc, B. Gel electrolytes based on poly(acrylonitrile)/sulpholane with hybrid TiO_2/SiO_2 filler for advanced lithium polymer batteries. *Electrochimica Acta* 2014; 125:415–420.
344. Amaral, F.A., Dalmolin, C., Canobre, S.C., Bocchi, N., Rocha-Filho, R.C., and Biaggio, S.R. Electrochemical and physical properties of poly(acrylonitrile)/poly(vinyl acetate)-based gel electrolytes for lithium ion batteries. *Journal of Power Sources* 2007; 164:379–385.
345. Li, W., Xing, Y., Xing, X., Li, Y., Yang, G., and Xu, L. PVDF-based composite microporous gel polymer electrolytes containing a novel single ionic conductor $SiO_2(Li^+)$. *Electrochimica Acta* 2013; 112:183–190.
346. Ferrari, S., Quartarone, E., Mustarelli, P., Magistris, A., Fagnoni, M., Protti, S., Gerbaldi, C., and Spinella, A. Lithium ion conducting PVdF-HFP composite gel electrolytes based on *N*-methoxyethyl-*N*-methylpyrrolidiniumbis(trifluoromethanesulfonyl)-imide ionic liquid. *Journal of Power Sources* 2010; 195:559–566.

347. Xiao, Q., Wang, X., Li, W., Li, Z., Zhang, T., and Zhang, H. Macroporous polymer electrolytes based on PVDF/PEO-*b*-PMMA block copolymer blends for rechargeable lithium ion battery. *Journal of Membrane Science* 2009; 334:117–122.
348. Kim, M., Han, G.Y., Yoon, K.J., and Park, J.H. Preparation of a trilayer separator and its application to lithium-ion batteries. *Journal of Power Sources* 2010; 195:8302–8305.
349. Jung, H.R., and Lee, W.J. Electrochemical characteristics of electrospunpoly(methyl methacrylate)/polyvinyl chloride as gel polymer electrolytes for lithium ion battery. *Electrochimica Acta* 2011; 58:674–680.
350. Liu, Y., Cai, Z., Tan, L., and Li, L. Ion exchange membranes as electrolyte for high performance Li-ion batteries. *Energy & Environmental Science* 2012; 5:9007–9013.
351. Cai, Z., Liu, Y., Liu, S., Li, L., and Zhang, Y. High performance of lithium-ion polymer battery based on non-aqueous lithiated perfluorinated sulfonic ion-exchange membranes. *Energy & Environmental Science* 2012; 5:5690–5693.
352. Liu, Y., Tan, L., and Li, L. Ion exchange membranes as electrolyte to improve high temperature capacity retention of $LiMn_2O_4$ cathode lithium-ion batteries. *Chemical Communications* 2012; 48:9858–9860.
353. Orendorff, C.J., Lambert, T.N., Chavez, C.A., Bencomo, M., and Fenton, K.R. Polyester separators for lithium-ion cells: Improving thermal stability and abuse tolerance. *Advanced Energy Materials* 2013; 3:314–320.
354. Aurbach, D. Review of selected electrode–solution interactions which determine the performance of Li and Li ion batteries. *Journal of Power Sources* 2000; 89:206–218.
355. Balakrishnan, P.G., Ramesh, R., and Prem Kumar, T. Safety mechanisms in lithium-ion batteries. *Journal of Power Sources* 2006; 155:401–414.
356. Etacheri, V., Marom, R., Elazari, R., Salitra, G., and Aurbach, D. Challenges in the development of advanced Li-ion batteries: A review. *Energy & Environmental Science* 2011; 4:3243–3262.
357. Xu, K. Nonaqueous liquid electrolytes for lithium-based rechargeable batteries. *Chemical Reviews* 2004; 104:4303–4418.
358. Hu, M., Pang, X., and Zhou, Z. Recent progress in high-voltage lithium ion batteries. *Journal of Power Sources* 2013; 237:229–242.
359. Talyosef, Y., Markovsky, B., Salitra, G., Aurbach, D., Kim, H.J., and Choi, S. The study of $LiNi_{0.5}Mn_{1.5}O_4$ 5-V cathodes for Li-ion batteries. *Journal of Power Sources* 2005; 146:664–669.
360. Aurbach, D., Markovsky, B., Talyossef, Y., Salitra, G., Kim, H.-J., and Choi, S. Studies of cycling behavior, ageing, and interfacial reactions of $LiNi_{0.5}Mn_{1.5}O_4$ and carbon electrodes for lithium-ion 5-V cells. *Journal of Power Sources* 2006; 162:780–789.
361. Fang, X., Ding, N., Feng, X.Y., Lu, Y., and Chen, C.H. Study of $LiNi_{0.5}Mn_{1.5}O_4$ synthesized via a chloride-ammonia co-precipitation method: Electrochemical performance, diffusion coefficient and capacity loss mechanism. *Electrochimica Acta* 2009; 54:7471–7475.
362. Horino, T., Tamada, H., Kishimoto, A., Kaneko, J., Iriyama, Y., Tanaka, Y., and Fujinami, T. High voltage stability of interfacial reaction at the $LiMn_2O_4$ thin-film electrodes/liquid electrolytes with boroxine compounds. *Journal of the Electrochemical Society* 2010; 157: A677–A681.
363. Yang, L., Ravdel, B., and Lucht, B.L. Electrolyte reactions with the surface of high voltage $LiNi_{0.5}Mn_{1.5}O_4$ cathodes for lithium-ion batteries. *Electrochemical and Solid-State Letters* 2010; 13: A95–A97.
364. Kim, J.-H., Pieczonka, N.P.W., Li, Z., Wu, Y., Harris, S., and Powell, B.R. Understanding the capacity fading mechanism in $LiNi_{0.5}Mn_{1.5}O_4$/graphite Li-ion batteries. *Electrochimica Acta* 2013; 90:556–562.
365. Pieczonka, N.P.W., Liu, Z., Lu, P., Olson, K.L., Moote, J., Powell, B.R., and Kim, J.-H. Understanding transition-metal dissolution behavior in $LiNi_{0.5}Mn_{1.5}O_4$ high-voltage spinel for lithium ion batteries. *Journal of Physical Chemistry C* 2013; 117:15947–15957.
366. Xu, M., Dalavi, S., and Lucht, B.L. Electrolytes for lithium-ion batteries with high-voltage cathodes, Chapter 4, in: *Lithium Batteries: Advanced Technologies and Applications*, Scrosati, B., Abraham, K.M., Schalkwijk, W.V., and Hassoun, J. Editors, 2013, pp. 71–87.
367. Lee, K.T., Jeong, S., and Cho, J. Roles of surface chemistry on safety and electrochemistry in lithium ion batteries. *Accounts of Chemical Research* 2013; 46:1161–1170.
368. Bandhauer, T.M., Garimella, S., and Fuller, T.F. A critical review of thermal issues in lithium-ion batteries. *Journal of the Electrochemical Society* 2011; 158: R1–R25.
369. Choi, N.-S., Chen, Z., Freunberger, S.A., Ji, X., Sun, Y.-K., Amine, K., Yushin, G., Nazar, L.F., Cho, J., and Bruce, P.G. Challenges facing lithium batteries and electrical double-layer capacitors. *Angewandte Chemie International Edition* 2012; 51:9994–10024.
370. Wang, Q., Ping, P., Zhao, X., Chu, G., Sun, J., and Chen, C. Thermal runaway caused fire and explosion of lithium ion battery. *Journal of Power Sources* 2012; 208:210–224.
371. Zhang, S.S. A review on electrolyte additives for lithium-ion batteries. *Journal of Power Sources* 2006; 162:1379–1394.
372. Xia, L., Li, S., Ai, X., and Yang, H. Safety enhancing methods for Li-ion batteries progress in chemistry. *Progress in Chemistry* 2011; 23:328–335.
373. Aurbach, D., Daroux, M.L., Faguy, P.W., and Yeager, E. Identification of surface films formed on lithium in propylene carbonate solutions. *Journal of the Electrochemical Society* 1987; 134:1611–1620.
374. Dey, A.N., and Sullivan, B.P. The electrochemical decomposition of propylene carbonate on graphite. *Journal of the Electrochemical Society* 1970; 117:222–224.
375. Shu, Z.X., McMillan, R.S., Murray, J.J., and Davidson, I.J. Use of chloroethylene carbonate as an electrolyte solvent for a graphite anode in a lithium-ion battery. *Journal of the Electrochemical Society* 1996; 143:2230–2235.
376. Wang, B., Qu, Q.T., Xia, Q., Wu, Y.P., Li, X., Gan, C.L., and van Ree, T. Effects of 3,5-bis(trifluoromethyl)benzeneboronic acid as an additive on electrochemical performance of propylene carbonate-based electrolytes for lithium ion batteries. *Electrochimica Acta* 2008; 54:816–820.
377. Krämer, E., Schmitz, R., Niehoff, P., Passerini, S., and Winter, M. SEI-forming mechanism of 1-fluoropropane-2-one in lithium-ion batteries. *Electrochimica Acta* 2012; 81:161–165.
378. Li, B., Xu, M., Li, T., Li, W., and Hu, S. Prop-1-ene-1,3-sultone as SEI formation additive in propylene carbonate-based electrolyte for lithium ion batteries. *Electrochemistry Communications* 2012; 17:92–95.
379. Walkowiak, M., Waszak, D., Schroeder, G., and Gierczyk, B. Enhanced graphite passivation in Li-ion battery electrolytes containing disiloxane-type additive/co-solvent. *Journal of Solid State Electrochemistry* 2010; 14:2213–2218.
380. Pistoia, G. Nonaqueous batteries with $LiClO_4$–ethylene carbonate as electrolyte. *Journal of the Electrochemical Society* 1971; 118:153–158.

381. Fong, R., von Sacken, U., and Dahn, J.R. Studies of lithium intercalation into carbons using nonaqueous electrochemical cells. *Journal of the Electrochemical Society* 1990; 137:2009–2013.
382. Plichta, E.J., and Behl, W.K. A low-temperature electrolyte for lithium and lithium-ion batteries. *Journal of Power Sources* 2000; 88:192–196.
383. Ein-Eli, Y., McDevitt, S.F., and Laura, R. The superiority of asymmetric alkyl methyl carbonates. *Journal of the Electrochemical Society* 1998; 145: L1–L3.
384. Ding, M.S., Xu, K., and Jow, T.R. Liquid–solid phase diagrams of binary carbonates for lithium batteries. *Journal of the Electrochemical Society* 2000; 147:1688–1694.
385. Ding, M.S., Xu, K., Zhang, S., and Jow, T.R. Liquid/solid phase diagrams of binary carbonates for lithium batteries part II. *Journal of the Electrochemical Society* 2001; 148: A299–A304.
386. Takeuchi, E.S., Gan, H., Palazzo, M., Leising, R.A., and Davis, S.M. Anode passivation and electrolyte solvent disproportionation: Mechanism of ester exchange reaction in lithium-ion batteries. *Journal of the Electrochemical Society* 1997; 144:1944–1948.
387. Dan, P., Geronov, J., Luski, S., Megenitsky, E. and Aurbach, D. Nonaqueous safe secondary cell. U.S. Patent 5,506,068 (1996).
388. Mengeritsky, E., Dan, P., Weissman, I., Zaban, A., and Aurbach, D. Safety and performance of Tadiran TLR-7103 rechargeable batteries. *Journal of the Electrochemical Society* 1996; 143:2110–2116.
389. Aurbach, D., Zinigrad, E., Teller, H., and Dan, P. Factors which limit the cycle life of rechargeable lithium (metal) batteries. *Journal of the Electrochemical Society* 2000; 147:1274–1279.
390. Sasaki, Y., Hosoya, M., and Handa, M. Lithium cycling efficiency of ternary solvent electrolytes with ethylene carbonate–dimethyl carbonate mixture. *Journal of Power Sources* 1997; 68:492–496.
391. Fujimoto, M., Kida, Y., Nohma, T., Takahashi, M., Nishio, K., and Saito, T. Electrochemical behaviour of carbon electrodes in some electrolyte solutions. *Journal of Power Sources* 1996; 63:127–130.
392. Lu, W., Xie, K., Pan, Y., Chen, Z.-X., and Zheng, C.-M. Effects of carbon-chain length of trifluoroacetate co-solvents for lithium-ion battery electrolytes using at low temperature. *Journal of Fluorine Chemistry* 2013; 156:136–143.
393. Foropoulos, J., and DesMarteau, D.D. Synthesis, properties, and reactions of bis((trifluoromethyl)sulfonyl) imide, $(CF_3SO_2)_2NH$. *Inorganic Chemistry* 1984; 23:3720–3723.
394. PéTer, L., and Arai, J. Anodic dissolution of aluminium in organic electrolytes containing perfluoroalkylsulfonyl imides. *Journal of Applied Electrochemistry* 1999; 29:1053–1061.
395. Dominey, L.A., Koch, V.R., and Blakley, T.J. Thermally stable lithium salts for polymer electrolytes. *Electrochimica Acta* 1992; 37:1551–1554.
396. Ue, M., Murakami, A., and Nakamura, S. Anodic stability of several anions examined by *ab initio* molecular orbital and density functional theories. *Journal of the Electrochemical Society* 2002; 149: A1572–A1577.
397. Arai, J., Matsuo, A., Fujisaki, T., and Ozawa, K. A novel high temperature stable lithium salt ($Li_2B_{12}F_{12}$) for lithium ion batteries. *Journal of Power Sources* 2009; 193:851–854.
398. Gnanaraj, J.S., Zinigrad, E., Asraf, L., Sprecher, M., Gottlieb, H.E., Geissler, W., Schmidt, M., and Aurbach, D. On the use of $LiPF_3(CF_2CF_3)_3$ (LiFAP) solutions for Li-ion batteries. Electrochemical and thermal studies. *Electrochemistry Communications* 2003; 5:946–951.
399. Aravindan, V., Gnanaraj, J., Madhavi, S., and Liu, H.-K. Lithium-ion conducting electrolyte salts for lithium batteries. *Chemistry—A European Journal* 2011; 17:14326–14346.
400. Ue, M., Ida, K., and Mori, S. Electrochemical properties of organic liquid electrolytes based on quaternary onium salts for electrical double-layer capacitors. *Journal of the Electrochemical Society* 1994; 141:2989–2996.
401. Ue, M., Takeda, M., Takehara, M., and Mori, S. Electrochemical properties of quaternary ammonium salts for electrochemical capacitors. *Journal of the Electrochemical Society* 1997; 144:2684–2688.
402. Abu-Lebdeh, Y., and Davidson, I. High-voltage electrolytes based on adiponitrile for Li-ion batteries. *Journal of the Electrochemical Society* 2009; 156: A60–A65.
403. Abu-Lebdeh, Y., and Davidson, I. New electrolytes based on glutaronitrile for high energy/power Li-ion batteries. *Journal of Power Sources* 2009; 189:576–579.
404. Duncan, H., Salem, N., Abu-Lebdeh, Y., and Davidson, I. Dinitrile-based electrolytes for lithium ion batteries. Meeting Abstracts 2012; MA2012-01: 343.
405. Nagahama, M., Hasegawa, N., and Okada, S. High Voltage Performances of Li_2NiPO_4F cathode with dinitrile-based electrolytes. *Journal of the Electrochemical Society* 2010; 157: A748–A752.
406. Oh, B., Ofer, D., Rempel, J., Sriramulu, S., and Barnett, B. Novel Li-ion electrolytes for extended temperature and voltage conditions. Meeting Abstracts 2010; MA2010-02: 592.
407. Gmitter, A.J., Plitz, I., and Amatucci, G.G. High concentration dinitrile, 3-alkoxypropionitrile, and linear carbonate electrolytes enabled by vinylene and monofluoroethylene carbonate additives. *Journal of the Electrochemical Society* 2012; 159: A370–A379.
408. Duncan, H., Salem, N., and Abu-Lebdeh, Y. Electrolyte formulations based on dinitrile solvents for high voltage Li-ion batteries. *Journal of the Electrochemical Society* 2013; 160: A838–A848.
409. Xu, K., and Angell, C.A. High anodic stability of a new electrolyte solvent: Unsymmetric noncyclic aliphatic sulfone. *Journal of the Electrochemical Society* 1998; 145: L70–L72.
410. Shao, N., Sun, X.-G., Dai, S., and Jiang, D.-E. Electrochemical windows of sulfone-based electrolytes for high-voltage Li-ion batteries. *Journal of Physical Chemistry B* 2011; 115:12120–12125.
411. Shao, N., Sun, X.-G., Dai, S., and Jiang, D.-E. Oxidation potentials of functionalized sulfone solvents for high-voltage Li-ion batteries: A computational study. *Journal of Physical Chemistry B* 2012; 116:3235–3238.
412. Xu, K., and Angell, C.A. Sulfone-based electrolytes for lithium-ion batteries. *Journal of the Electrochemical Society* 2002; 149: A920–A926.
413. Sun, X.-G., and Angell, C.A. New sulfone electrolytes: Part II. Cyclo alkyl group containing sulfones. *Solid State Ionics* 2004; 175:257–260.
414. Sun, X.G., and Angell, C.A. New sulfone electrolytes for rechargeable lithium batteries: Part I. Oligoether-containing sulfones. *Electrochemistry Communications* 2005; 7:261–266.
415. Sun, X., and Angell, C.A. Ether sulfones with additives as electrolytes for applications in rechargeable lithium ion batteries. Meeting Abstracts 2008, MA2008-01: 162.
416. Watanabe, Y., Kinoshita, S.-I., Wada, S., Hoshino, K., Morimoto, H., and Tobishima, S.-I. Electrochemical properties and lithium ion solvation behavior of sulfone-ester mixed electrolytes for high-voltage rechargeable lithium cells. *Journal of Power Sources* 2008; 179:770–779.

417. Sun, X., and Angell, C.A. Doped sulfone electrolytes for high voltage Li-ion cell applications. *Electrochemistry Communications* 2009; 11:1418–1421.
418. Li, C., Zhao, Y., Zhang, H., Liu, J., Jing, J., Cui, X., and Li, S. Compatibility between $LiNi_{0.5}Mn_{1.5}O_4$ and electrolyte based upon lithium bis(oxalate)borate and sulfolane for high voltage lithium-ion batteries. *Electrochimica Acta* 2013; 104:134–139.
419. Mao, L., Li, B., Cui, X., Zhao, Y., Xu, X., Shi, X., Li, S., and Li, F. Electrochemical performance of electrolytes based upon lithium bis(oxalate)borate and sulfolane/alkyl sulfite mixtures for high temperature lithium-ion batteries. *Electrochimica Acta* 2012; 79:197–201.
420. Wu, F., Xiang, J., Li, L., Chen, J., Tan, G., and Chen, R. Study of the electrochemical characteristics of sulfonylisocyanate/sulfone binary electrolytes for use in lithium-ion batteries. *Journal of Power Sources* 2012; 202:322–331.
421. Wu, F., Zhu, Q., Li, L., Chen, R., and Chen, S. A diisocyanate/sulfone binary electrolyte based on lithium difluoro(oxalate) borate for lithium batteries. *Journal of Materials Chemistry A* 2013; 1:3659–3666.
422. Cui, X., Zhang, H., Li, S., Zhao, Y., Mao, L., Zhao, W., Li, Y., and Ye, X. Electrochemical performances of a novel high-voltage electrolyte based upon sulfolane and γ-butyrolactone. *Journal of Power Sources* 2013; 240:476–485.
423. Demeaux, J., De Vito, E., Le Digabel, M., Galiano, H., Claude-Montigny, B., and Lemordant, D. Dynamics of Li4Ti5O12/sulfone-based electrolyte interfaces in lithium-ion batteries. *Phys Chem Chem Phys* 2014; 16: 5201-5212.
424. Xiang, J., Wu, F., Chen, R., Li, L., and Yu, H. High voltage and safe electrolytes based on ionic liquid and sulfone for lithium-ion batteries. *Journal of Power Sources* 2013; 233:115–120.
425. Li, S., Zhao, W., Cui, X., Zhao, Y., Li, B., Zhang, H., Li, Y., Li, G., Ye, X., and Luo, Y. An improved method for synthesis of lithium difluoro(oxalato)borate and effects of sulfolane on the electrochemical performances of lithium-ion batteries. *Electrochimica Acta* 2013; 91:282–292.
426. Xing, L., Vatamanu, J., Borodin, O., Smith, G.D., and Bedrov, D. Electrode/electrolyte interface in sulfolane-based electrolytes for Li ion batteries: A molecular dynamics simulation study. *Journal of Physical Chemistry C* 2012; 116:23871–23881.
427. Lu, Z., and Dahn, J.R. Can all the lithium be removed from T2 $Li_{2/3}[Ni_{1/3}Mn_{2/3}]O_2$. *Journal of the Electrochemical Society* 2001; 148: A710–A715.
428. Abouimrane, A., Belharouak, I., and Amine, K. Sulfone-based electrolytes for high-voltage Li-ion batteries. *Electrochemistry Communications* 2009; 11:1073–1076.
429. Demeaux, J., De Vito, E., Le Digabel, M., Galiano, H., Claude-Montigny, B., and Lemordant, D. Dynamics of $Li_4Ti_5O_{12}$/sulfone-based electrolyte interfaces in lithium-ion batteries. *Physical Chemistry Chemical Physics* 2014; 16:5201–5212.
430. Sakaebe, H., and Matsumoto, H. *N*-methyl-*N*-propylpiperidini umbis(trifluoromethanesulfonyl)imide (PP13-TFSI)-novel electrolyte base for Li battery. *Electrochemistry Communications* 2003; 5:594–598.
431. Matsumoto, H., Sakaebe, H., and Tatsumi, K. Preparation of room temperature ionic liquids based on aliphatic onium cations and asymmetric amide anions and their electrochemical properties as a lithium battery electrolyte. *Journal of Power Sources* 2005; 146:45–50.
432. Galiński, M., Lewandowski, A., and Stępniak, I. Ionic liquids as electrolytes. *Electrochimica Acta* 2006; 51:5567–5580.
433. Han, H.-B., Liu, K., Feng, S.-W., Zhou, S.-S., Feng, W.-F., Nie, J., Li, H., Huang, X.-J., Matsumoto, H., Armand, M., and Zhou, Z.-B. Ionic liquid electrolytes based on multi-methoxyethyl substituted ammoniums and perfluorinated sulfonimides: Preparation, characterization, and properties. *Electrochimica Acta* 2010; 55:7134–7144.
434. Seki, S., Serizawa, N., Takei, K., Miyashiro, H., and Watanabe, M. Glyme-$LiN(SO_2F)_2$ complex electrolyte for lithium-ion secondary batteries. Meeting Abstracts 2011; MA2011-02: 1277.
435. Le, M.-L.-P., Alloin, F., Strobel, P., Leprêtre, J.-C., Cointeaux, L., and Valle, C. Electrolyte based on fluorinated cyclic quaternary ammonium ionic liquids. *Ionics* 2012; 18:817–827.
436. Dokko, K., Tachikawa, N., Yamauchi, K., Tsuchiya, M., Yamazaki, A., Takashima, E., Park, J.W., Ueno, K., Seki, S., Serizawa, N., and Watanabe, M. Solvate ionic liquid electrolyte for Li–S batteries. *Journal of the Electrochemical Society* 2013; 160: A1304–A1310.
439. Borgel, V., Markevich, E., Aurbach, D., Semrau, G., and Schmidt, M. On the application of ionic liquids for rechargeable Li batteries: High voltage systems. *Journal of Power Sources* 2009; 189:331–336.
438. Lewandowski, A., and Świderska-Mocek, A. Ionic liquids as electrolytes for Li-ion batteries—An overview of electrochemical studies. *Journal of Power Sources* 2009; 194:601–609.
439. Jin, J., Li, H.H., Wei, J.P., Bian, X.K., Zhou, Z., and Yan, J. Li/$LiFePO_4$ batteries with room temperature ionic liquid as electrolyte. *Electrochemistry Communications* 2009; 11: 1500–1503.
440. Srour, H., Rouault, H., and Santini, C. Imidazolium based ionic liquid electrolytes for Li-ion secondary batteries based on graphite and $LiFePO_4$. *Journal of the Electrochemical Society* 2013; 160: A66–A69.
441. Sano, H., Sakaebe, H., and Matsumoto, H. Effect of organic additives on electrochemical properties of Li anode in room temperature ionic liquid. *Journal of the Electrochemical Society* 2011; 158: A316–A321.
442. Xiang, H.F., Yin, B., Wang, H., Lin, H.W., Ge, X.W., Xie, S., and Chen, C.H. Improving electrochemical properties of room temperature ionic liquid (RTIL) based electrolyte for Li-ion batteries. *Electrochimica Acta* 2010; 55:5204–5209.
443. Kühnel, R.S., Böckenfeld, N., Passerini, S., Winter, M., and Balducci, A. Mixtures of ionic liquid and organic carbonate as electrolyte with improved safety and performance for rechargeable lithium batteries. *Electrochimica Acta* 2011; 56:4092–4099.
444. Kim, K., Cho, Y.-H., and Shin, H.-C. 1-Ethyl-1-methyl piper idiniumbis(trifluoromethanesulfonyl)imide as a co-solvent in Li-ion batteries. *Journal of Power Sources* 2013; 225:113–118.
445. Li, H., Pang, J., Yin, Y., Zhuang, W., Wang, H., Zhai, C., and Lu, S. Application of a nonflammable electrolyte containing PP13TFSI ionic liquid for lithium-ion batteries using the high capacity cathode material $Li[Li_{0.2}Mn_{0.54}Ni_{0.13}Co_{0.13}]O_2$. *RSC Advances* 2013; 3:13907–13914.
446. Mun, J., Yim, T., Park, K., Ryu, J.H., Kim, Y.G., and Oh, S.M. Surface film formation on $LiNi_{0.5}Mn_{1.5}O_4$ electrode in an ionic liquid solvent at elevated temperature. *Journal of the Electrochemical Society* 2011; 158: A453–A457.
447. Wang, Z., Cai, Y., Wang, Z., Chen, S., Lu, X., and Zhang, S. Vinyl-functionalized imidazolium ionic liquids as new electrolyte additives for high-voltage Li-ion batteries. *Journal of Solid State Electrochemistry* 2013; 17:2839–2848.
448. McMillan, R., Slegr, H., Shu, Z.X., and Wang, W. Fluoroethylene carbonate electrolyte and its use in lithium ion batteries with graphite anodes. *Journal of Power Sources* 1999; 81–82:20–26.

449. Arai, J., Katayama, H., and Akahoshi, H. Binary mixed solvent electrolytes containing trifluoropropylene carbonate for lithium secondary batteries. *Journal of the Electrochemical Society* 2002; 149: A217–A226.
450. Choi, N.-S., Yew, K.H., Lee, K.Y., Sung, M., Kim, H., and Kim, S.-S. Effect of fluoroethylene carbonate additive on interfacial properties of silicon thin-film electrode. *Journal of Power Sources* 2006; 161:1254–1259.
451. Nakai, H., Kubota, T., Kita, A., and Kawashima, A. Investigation of the solid electrolyte interphase formed by fluoroethylene carbonate on Si electrodes. *Journal of the Electrochemical Society* 2011; 158: A798–A801.
452. Etacheri, V., Haik, O., Goffer, Y., Roberts, G.A., Stefan, I.C., Fasching, R., and Aurbach, D. Effect of fluoroethylene carbonate (FEC) on the performance and surface chemistry of Si-nanowire Li-ion battery anodes. *Langmuir* 2012; 28:965–976.
453. Lin, Y.-M., Klavetter, K.C., Abel, P.R., Davy, N.C., Snider, J.L., Heller, A., and Mullins, C.B. High performance silicon nanoparticle anode in fluoroethylene carbonate-based electrolyte for Li-ion batteries. *Chemical Communications (Cambridge, UK)* 2012; 48:7268–7270.
454. Chockla, A.M., Klavetter, K.C., Mullins, C.B., and Korgel, B.A. Solution-grown germanium nanowire anodes for lithium-ion batteries. *ACS Applied Material Interfaces* 2012; 4:4658–4664.
455. Elazari, R., Salitra, G., Gershinsky, G., Garsuch, A., Panchenko, A., and Aurbach, D. Li ion cells comprising lithiated columnar silicon film anodes, TiS2 cathodes and fluoroethylene carbonate (FEC) as a critically important component. *Journal of the Electrochemical Society* 2012; 159: A1440–A1445.
456. Wilhelm, H.A., Marino, C., Darwiche, A., Monconduit, L., and Lestriez, B. Significant electrochemical performance improvement of TiSnSb as anode material for Li-ion batteries with composite electrode formulation and the use of VC and FEC electrolyte additives. *Electrochemistry Communications* 2012; 24:89–92.
457. Baggetto, L., Keum, J.K., Browning, J.F., and Veith, G.M. Germanium as negative electrode material for sodium-ion batteries. *Electrochemistry Communications* 2013; 34:41–44.
458. Darwiche, A., Sougrati, M.T., Fraisse, B., Stievano, L., and Monconduit, L. Facile synthesis and long cycle life of SnSb as negative electrode material for Na-ion batteries. *Electrochemistry Communications* 2013; 32:18–21.
459. Webb, S.A., Baggetto, L., Bridges, C.A., and Veith, G.M. The electrochemical reactions of pure indium with Li and Na: Anomalous electrolyte decomposition, benefits of FEC additive, phase transitions and electrode performance. *Journal of Power Sources* 2014; 248:1105–1117.
460. Yu, D.Y.W., Hoster, H.E., and Batabyal, S.K. Bulk antimony sulfide with excellent cycle stability as next-generation anode for lithium-ion batteries. *Science Reports* 2014; 4:1–6.
461. Smart, M.C., Ratnakumar, B.V., Ryan-Mowrey, V.S., Surampudi, S., Prakash, G.K.S., Hu, J., and Cheung, I. Improved performance of lithium-ion cells with the use of fluorinated carbonate-based electrolytes. *Journal of Power Sources* 2003; 119–121:359–367.
462. Kobayashi, D., Takehara, M., Nanbu, N., Ue, M., and Sasaki, Y. Physical and electrochemical properties of fluoroethy lene carbonate for electric double layer capacitors. Meeting Abstracts 2008; MA2008-02: 166.
463. Sasaki, Y., Satake, H., Tsukimori, N., Nanbu, N., Takehara, M., and Ue, M. Physical and electrolytic properties of partially fluorinated methyl propyl carbonate and its application to lithium batteries. *Electrochemistry* 2010; 78:467–470.
464. Fridman, K., Sharabi, R., Elazari, R., Gershinsky, G., Markevich, E., Salitra, G., Aurbach, D., Garsuch, A., and Lampert, J. A new advanced lithium ion battery: Combination of high performance amorphous columnar silicon thin film anode, 5 V $LiNi_{0.5}Mn_{1.5}O_4$ spinel cathode and fluoroethy lene carbonate-based electrolyte solution. *Electrochemistry Communications* 2013; 33:31–34.
465. Hu, L., Zhang, Z., and Amine, K. Fluorinated electrolytes for Li-ion battery: An FEC-based electrolyte for high voltage $LiNi_{0.5}Mn_{1.5}O_4$/graphite couple. *Electrochemistry Communications* 2013; 35:76–79.
466. Hu, L., Zhang, Z., and Amine, K. Electrochemical investigation of carbonate-based electrolytes for high voltage lithium-ion cells. *Journal of Power Sources* 2013; 236:175–180.
467. Zhang, Z., Hu, L., Wu, H., Weng, W., Koh, M., Redfern, P.C., Curtiss, L.A., and Amine, K. Fluorinated electrolytes for 5 V lithium-ion battery chemistry. *Energy & Environmental Science* 2013; 6:1806–1810.
468. Markevich, E., Salitra, G., Fridman, K., Sharabi, R., Gershinsky, G., Garsuch, A., Semrau, G., Schmidt, M.A., and Aurbach, D. Fluoroethylene carbonate as an important component in electrolyte solutions for high-voltage lithium batteries: Role of surface chemistry on the cathode. *Langmuir* 2014; 30:7414–7424.
469. von Cresce, A., and Xu, K. Electrolyte additive in support of 5 V Li ion chemistry. *Journal of the Electrochemical Society* 2011; 158: A337–A342.
470. Tan, S., Zhang, Z., Li, Y., Li, Y., Zheng, J., Zhou, Z., and Yang, Y. Tris(hexafluoro-iso-propyl)phosphate as an SEI-forming additive on improving the electrochemical performance of the $Li[Li_{0.2}Mn_{0.56}Ni_{0.16}Co_{0.08}]O_2$ cathode material. *Journal of the Electrochemical Society* 2012; 160: A285–A292.
471. Von Cresce, A., and Xu, K. Phosphate-based compounds as additives for 5-volt lithium-ion electrolytes. *ECS Transactions* 2012; 41:17–22.
472. Xing, L.Y., Hu, M., Tang, Q., Wei, J.P., Qin, X., and Zhou, Z. Improved cyclic performances of $LiCoPO_4$/C cathode materials for high-cell-potential lithium-ion batteries with thiophene as an electrolyte additive. *Electrochimica Acta* 2012; 59:172–178.
473. Alloin, F., Tall, O.E., Leprêtre, J.C., Cointeaux, L., Boutafa, L., Guindet, J., and Martin, J.F. Chemically controlled protective film based on biphenyl derivatives for high potential lithium battery. *Electrochimica Acta* 2013; 112:74–81.
474. Jin, Z., Gao, H., Kong, C., Zhan, H., and Li, Z. A novel phosphate-based flame retardant and film-forming electrolyte additive for lithium ion batteries. *ECS Electrochemistry Letters* 2013; 2: A66–A68.
475. Kang, K.S., Choi, S., Song, J.-H., Woo, S.-G., Jo, Y.N., Choi, J., Yim, T., Yu, J.-S., and Kim, Y.-J. Effect of additives on electrochemical performance of lithium nickel cobalt manganese oxide at high temperature. *Journal of Power Sources* 2014; 253:48–54.
476. Li, Z.D., Zhang, Y.C., Xiang, H.F., Ma, X.H., Yuan, Q.F., Wang, Q.S., and Chen, C.H. Trimethylphosphite as an electrolyte additive for high-voltage lithium-ion batteries using lithium-rich layered oxide cathode. *Journal of Power Sources* 2013; 240:471–475.
477. Ryou, M.-H., Lee, J.-N., Lee, D.J., Kim, W.-K., Choi, J.W., Park, J.-K., and Lee, Y.M. 2-(Triphenylphosphoranylidene) succinic anhydride as a new electrolyte additive to improve high temperature cycle performance of $LiMn_2O_4$/graphite Li-ion batteries. *Electrochimica Acta* 2013; 102:97–103.

478. Tarnopolskiy, V., Kalhoff, J., Nádherná, M., Bresser, D., Picard, L., Fabre, F., Rey, M., and Passerini, S. Beneficial influence of succinic anhydride as electrolyte additive on the self-discharge of 5 V $LiNi_{0.4}Mn_{1.6}O_4$ cathodes. *Journal of Power Sources* 2013; 236:39–46.
479. Hu, M., Wei, J., Xing, L., and Zhou, Z. Effect of lithium difluoro(oxalate)borate (LiDFOB) additive on the performance of high-voltage lithium-ion batteries. *Journal of Applied Electrochemistry* 2012; 42:291–296.
480. Ha, S.-Y., Han, J.-G., Song, Y.-M., Chun, M.-J., Han, S.-I., Shin, W.-C., and Choi, N.-S. Using a lithium bis(oxalato) borate additive to improve electrochemical performance of high-voltage spinel $LiNi_{0.5}Mn_{1.5}O_4$ cathodes at 60°C. *Electrochimica Acta* 2013; 104:170–177.
481. Li, S.R., Sinha, N.N., Chen, C.H., Xu, K., and Dahn, J.R. A consideration of electrolyte additives for $LiNi_{0.5}Mn_{1.5}O_4$/$Li_4Ti_5O_{12}$ Li-ion cells. *Journal of the Electrochemical Society* 2013; 160: A2014–A2020.
482. Pieczonka, N.P.W., Yang, L., Balogh, M.P., Powell, B.R., Chemelewski, K.R., Manthiram, A., Krachkovskiy, S.A., Goward, G.R., Liu, M., and Kim, J.-H. Impact of lithium bisoxalate borate electrolyte additive on the performance of high-voltage spinel/graphite Li-ion batteries. *Journal of Physical Chemistry C* 2013; 117:22603–22612.
483. Pieczonka, N.P.W., Yang, L., Balogh, M.P., Powell, B.R., Chemelewski, K., Manthiram, A., Krachkovskiy, S.A., Goward, G.R., Liu, M., and Kim, J.-H. Impact of lithium bis(oxalate)borate electrolyte additive on the performance of high-voltage spinel/graphite Li-ion batteries. *Journal of Physical Chemistry C* 2013; 117:22603–22612.
484. Dalavi, S., Xu, M., Knight, B., and Lucht, B.L. Effect of added LiBOB on high voltage ($LiNi_{0.5}Mn_{1.5}O_4$) spinel cathodes. *Electrochemical and Solid-State Letters* 2011; 15: A28–A31.
485. von Cresce, A., and Xu, K. Electrolyte additive in support of 5 V Li ion chemistry. *Journal of the Electrochemical Society* 2011; 158: A337–A342.
486. Rong, H., Xu, M., Xing, L., and Li, W. Enhanced cyclability of $LiNi_{0.5}Mn_{1.5}O_4$ cathode in carbonate based electrolyte with incorporation of tris(trimethylsilyl)phosphate (TMSP). *Journal of Power Sources* 2014; 261:148–155.
487. Yan, G., Li, X., Wang, Z., Guo, H., and Wang, C. Tris(trimethylsilyl)phosphate: A film-forming additive for high voltage cathode material in lithium-ion batteries. *Journal of Power Sources* 2014; 248:1306–1311.
488. Zhang, J., Wang, J., Yang, J., and NuLi, Y. Artificial interface deriving from sacrificial Tris(trimethylsilyl)phosphate additive for lithium rich cathode materials. *Electrochimica Acta* 2014; 117:99–104.
489. Zhu, Y., Casselman, M.D., Li, Y., Wei, A., and Abraham, D.P. Perfluoroalkyl-substituted ethylene carbonates: Novel electrolyte additives for high-voltage lithium-ion batteries. *Journal of Power Sources* 2014; 246:184–191.
490. Felix, F., Cheng, J.-H., Hy, S., Rick, J., Wang, F.M., and Hwang, B.-J. Mechanistic basis of enhanced capacity retention found with novel sulfate-based additive in high voltage Li-ion batteries. *Journal of Physical Chemistry C* 2013; 117:22619–22626.
491. Bouayad, H., Wang, Z., Dupré, N., Dedryvère, R., Foix, D., Franger, S., Martin, J.F., Boutafa, L., Patoux, S., Gonbeau, D., and Guyomard, D. Improvement of electrode/electrolyte interfaces in high-voltage spinel lithium-ion batteries by using glutaric anhydride as electrolyte additive. *Journal of Physical Chemistry C* 2014; 118:4634–4648.
492. Korobeinichev, O.P., Ilyin, S.B., Bolshova, T.A., Shvartsberg, V.M., and Chernov, A.A. The chemistry of the destruction of organophosphorus compounds in flames—III: The destruction of DMMP and TMP in a flame of hydrogen and oxygen. *Combustion and Flame* 2000; 121:593–609.
493. Wang, X., Yasukawa, E., and Kasuya, S. Nonflammable trimethyl phosphate solvent-containing electrolytes for lithium-ion batteries: II. The use of an amorphous carbon anode. *Journal of the Electrochemical Society* 2001; 148: A1066–A1071.
494. Wang, X., Yasukawa, E., and Kasuya, S. Nonflammable trimethyl phosphate solvent-containing electrolytes for lithium-ion batteries: I. Fundamental properties. *Journal of the Electrochemical Society* 2001; 148: A1058–A1065.
495. Xu, K., Ding, M.S., Zhang, S., Allen, J.L., and Jow, T.R. An attempt to formulate nonflammable lithium ion electrolytes with alkyl phosphates and phosphazenes. *Journal of the Electrochemical Society* 2002; 149: A622–A626.
496. Hyung, Y.E., Vissers, D.R., and Amine, K. Flame-retardant additives for lithium-ion batteries. *Journal of Power Sources* 2003; 119–121:383–387.
497. Ota, H., Kominato, A., Chun, W.-J., Yasukawa, E., and Kasuya, S. Effect of cyclic phosphate additive in non-flammable electrolyte. *Journal of Power Sources* 2003; 119–121:393–398.
498. Yao, X.L., Xie, S., Chen, C.H., Wang, Q.S., Sun, J.H., Li, Y.L., and Lu, S.X. Comparative study of trimethylphosphite and trimethyl phosphate as electrolyte additives in lithium ion batteries. *Journal of Power Sources* 2005; 144:170–175.
499. Mandal, B.K., Padhi, A.K., Shi, Z., Chakraborty, S., and Filler, R. Thermal runaway inhibitors for lithium battery electrolytes. *Journal of Power Sources* 2006; 161:1341–1345.
500. Mandal, B.K., Padhi, A.K., Shi, Z., Chakraborty, S., and Filler, R. New low temperature electrolytes with thermal runaway inhibition for lithium-ion rechargeable batteries. *Journal of Power Sources* 2006; 162:690–695.
501. Wang, X., Yamada, C., Naito, H., Segami, G., and Kibe, K. High-concentration trimethyl phosphate-based nonflammable electrolytes with improved charge–discharge performance of a graphite anode for lithium-ion cells. *Journal of the Electrochemical Society* 2006; 153: A135–A139.
502. Xu, K., Ding, M.S., Zhang, S., Allen, J.L., and Jow, T.R. Evaluation of fluorinated alkyl phosphates as flame retardants in electrolytes for Li-ion batteries: I. Physical and electrochemical properties. *Journal of the Electrochemical Society* 2003; 150: A161–A169.
503. Xu, K., Zhang, S., Allen, J.L., and Jow, T.R. Evaluation of fluorinated alkyl phosphates as flame retardants in electrolytes for Li-ion batteries: II. performance in cell. *Journal of the Electrochemical Society* 2003; 150: A170–A175.
504. Nam, T.-H., Shim, E.-G., Kim, J.-G., Kim, H.-S., and Moon, S.-I. Diphenyloctyl phosphate and tris(2,2,2-trifluoroethyl) phosphite as flame-retardant additives for Li-ion cell electrolytes at elevated temperature. *Journal of Power Sources* 2008; 180:561–567.
505. Özmen-Monkul, B., and Lerner, M.M. The first graphite intercalation compounds containing tris(pentafluoroethyl)trifluorophosphate. *Carbon* 2010; 48:3205–3210.
506. Zhang, S.S., Xu, K., and Jow, T.R. Tris(2,2,2-trifluoroethyl) phosphite as a co-solvent for nonflammable electrolytes in Li-ion batteries. *Journal of Power Sources* 2003; 113:166–172.
507. Zeng, Z., Jiang, X., Wu, B., Xiao, L., Ai, X., Yang, H., and Cao, Y. Bis(2,2,2-trifluoroethyl) methylphosphonate: An novel flame-retardant additive for safe lithium-ion battery. *Electrochimica Acta* 2014; 129:300–304.

508. Winter, M., and Novák, P. Chloroethylene carbonate, a solvent for lithium-ion cells, evolving CO_2 during reduction. *Journal of the Electrochemical Society* 1998; 145: L27–L30.
509. Arai, J. A novel non-flammable electrolyte containing methyl nonafluorobutyl ether for lithium secondary batteries. *Journal of Applied Electrochemistry* 2002; 32:1071–1079.
510. Arai, J. No-flash-point electrolytes applied to amorphous carbon/$Li_{1+x}Mn_2O_4$ cells for EV use. *Journal of Power Sources* 2003; 119–121:388–392.
511. Arai, J. Nonflammable methyl nonafluorobutyl ether for electrolyte used in lithium secondary batteries. *Journal of the Electrochemical Society* 2003; 150: A219–A228.
512. Naoi, K., Iwama, E., Ogihara, N., Nakamura, Y., Segawa, H., and Ino, Y. Nonflammable hydrofluoroether for lithium-ion batteries: Enhanced rate capability, cyclability, and low-temperature performance. *Journal of the Electrochemical Society* 2009; 156: A272–A276.
513. Naoi, K., Iwama, E., Honda, Y., and Shimodate, F. Discharge behavior and rate performances of lithium-ion batteries in nonflammable hydrofluoroethers(II). *Journal of the Electrochemical Society* 2010; 157: A190–A195.
514. Wong, D.H.C., Thelen, J.L., Fu, Y., Devaux, D., Pandya, A.A., Battaglia, V.S., Balsara, N.P., and DeSimone, J.M. Nonflammable perfluoropolyether-based electrolytes for lithium batteries. *Proceedings of the National Academy of Sciences of the United States of America* 2014; 111:3327–3331.
515. Feng, J.K., Cao, Y.L., Ai, X.P., and Yang, H.X. Tri-(4-methoxythphenyl) phosphate: A new electrolyte additive with both fire-retardancy and overcharge protection for Li-ion batteries. *Electrochimica Acta* 2008; 53:8265–8268.
516. Zhang, L., Zhang, Z., Wu, H., and Amine, K. Novel redox shuttle additive for high-voltage cathode materials. *Energy & Environmental Science* 2011; 4:2858–2862.

5 Advanced Technologies for Lithium-Ion Rechargeable Batteries

Qizheng Li and Xueliang Sun

CONTENTS

5.1 FABRICATION/MANUFACTURING OF RECHARGEABLE Li-ION BATTERIES

The lithium-ion battery (LIB) was invented by Sony in 1991 and has grown rapidly as the power source of choice for portable electronic devices, especially digital cameras, cell phones, and laptop computers, over the past two decades. LIBs are essential to life in the modern world. New applications such as grid energy storage and electric vehicles (EVs) require LIBs with larger capacity and longer calendar lives than those for portable electronics applications. As an important energy storage system, advanced LIB technology is attributed to the tremendous work of researchers and engineers worldwide. However, differing from theoretical research work in academic laboratories, the LIB industry pays more attention to an easier and more effective battery assembly procedure, the needs of their customers and, most important, a remarkable cost reduction. Generally speaking, three different kinds of LIBs are currently available on the market: cylindrical, prismatic, and pouch. All of them have mature manufacturing processes, especially cylindrical cells. A typical process outline of lithium-ion cell fabrication is given in Figure 5.1.

5.1.1 Cylindrical Batteries

Cylindrical batteries already have fully automated production lines, as the vast majority use the 18650 universal cell format. For example, a PaTech 130PPM fully automatic assembly line, made by PaTech Company in Korea, can make 130 pieces per minute. Hundreds of millions of 18650 batteries are produced every year. Cylindrical batteries offer many advantages, such as high production efficiency and low manufacturing cost, consistent performance, and no swelling during the service lifetime. Therefore, cylindrical batteries are suitable for battery pack applications, such as for use in laptops, power tools, and EVs. Figure 5.2 shows the internal construction of an 18650 cell [1].

5.1.2 Prismatic Batteries

Prismatic cells were the main power source for cell phones in the 2000s. Figure 5.3 shows the internal construction of a prismatic cell [2]. According to various demands from different customers, the cell size (length, width, thickness) can be changed by modifying the cell design and applying a different

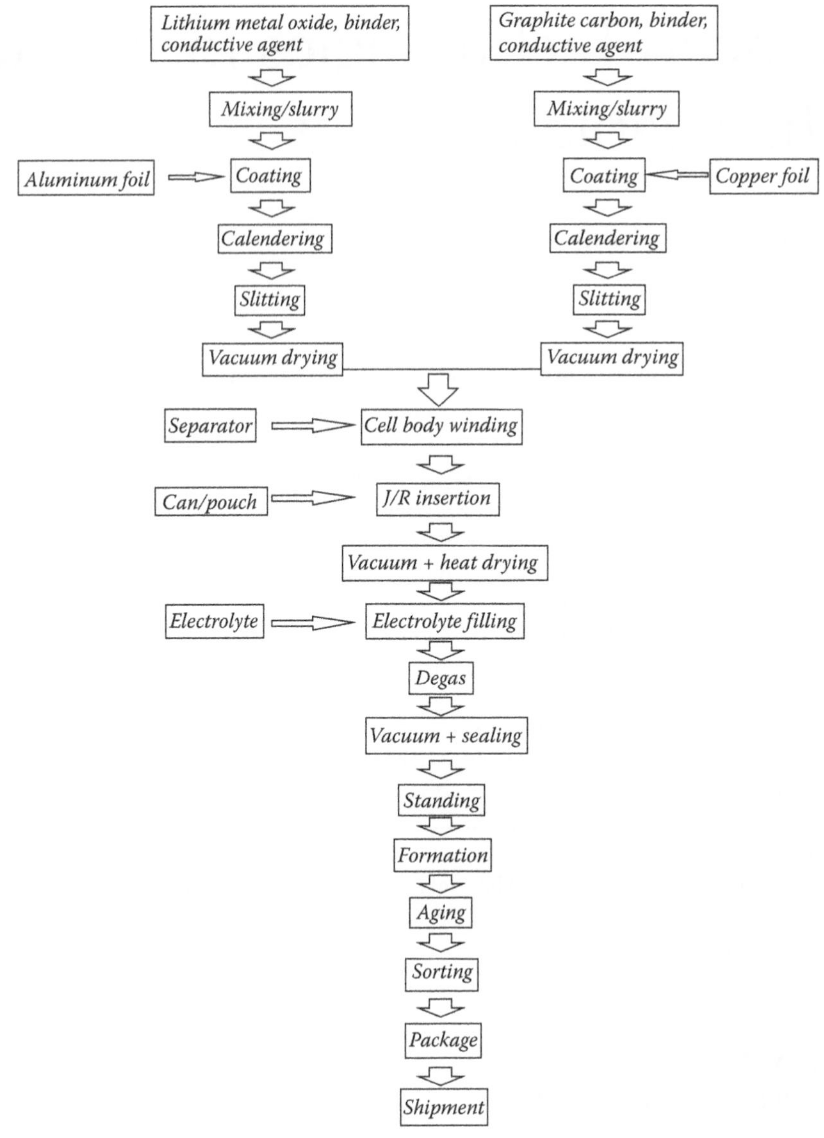

FIGURE 5.1 Process outline for Li-ion cell fabrication.

aluminum can. Compared with cylindrical cells, the biggest challenge for prismatic cells is swelling control. Generally after 500 cycles, the maximum cell swelling is 8–10%. In the last few years, the global prismatic battery demand has remained relatively constant around 1600 million per year, because more new products prefer to utilize pouch cells as their power sources for new devices. However, prismatic cells will continue to be used for a long time on account of their lower manufacturing cost.

5.1.3 Pouch Batteries

Pouch cells, a form of rechargeable lithium-ion batteries in pouch format, have been rapidly seizing market share in the last few years because of their excellent safety and diversity of cell shape. The pouch cell can be divided into two types: those using a polymer electrolyte and those with a liquid electrolyte. As for the pouch cell using a polymer electrolyte, the result is a "plastic" cell in which every component inside the pouch is tightly stuck together by a thermal polymerization reaction. Therefore, the cell can theoretically be very thin, it is easier to control the cell swelling, and there is no risk of electrolyte leakage. For the pouch cell using a liquid electrolyte, the electrochemical system is exactly the same as for prismatic and cylindrical cells. Compared with prismatic cells, cell swelling control is a more difficult problem because the pouch material is an aluminum laminate film that is much softer than the aluminum can. There are a few processes that are crucial to improve the cell swelling, such as cell body hot pressing, cold pressing, and baking treatments. Figure 5.4 shows the internal construction of a pouch cell. All pouch cells have excellent safety performance because of the aluminum laminate film, which acts as a buffer to prevent the cell from explosion when it is exposed to

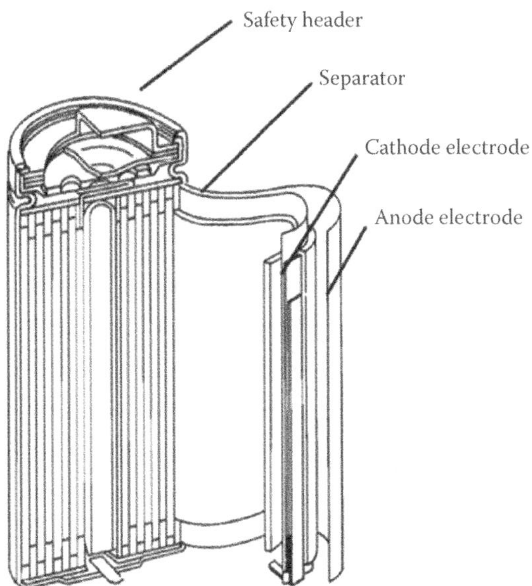

FIGURE 5.2 Expanded view of the internal construction of a cylindrical cell. (From Nishi, Y., Performance of the first lithium ion battery and its process technology. In *Lithium Ion Batteries: Fundamentals and Performance*, ed. M. Wakihara and O. Yamamoto. eds. Tokyo: Kodansha, Weinheim: Wiley-VCH, 1998.)

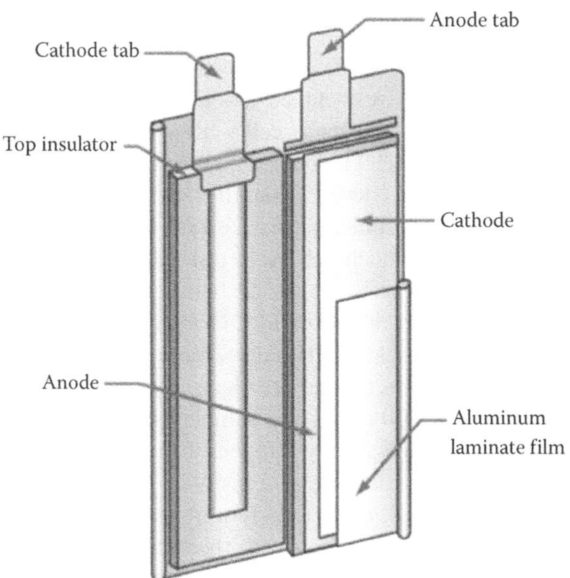

FIGURE 5.4 Expanded view of the internal construction of a pouch cell. (From VanZwol, J., *Electronic Design*, 43, 2011.)

extreme, harsh conditions, such as short circuiting and overcharging. On the basis of the reserved space for batteries in an electric device, pouch cells can be designed to have different shapes, even curved batteries that are applied in smart watches. However, this diversity of the cell model is the main cause of its low level of automatic production and long manufacturing cycle.

5.2 Li-ION BATTERY PERFORMANCE TESTING AND DIAGNOSIS

For each cell model, cell performance has to be confirmed in the pilot stage before mass production, to ensure that they match the customer's demands. At the same time, the customer will also conduct product audits according to the auditing instructor. The main tests can be categorized as follows: cycling performance, storage performance, and safety and reliability.

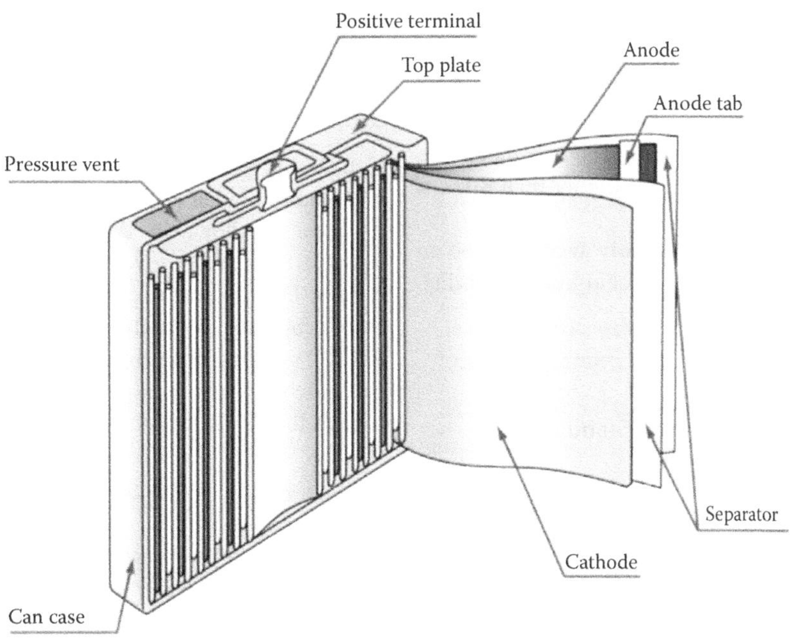

FIGURE 5.3 Expanded view of the internal construction of a prismatic cell. (From VanZwol, J., *Electronic Design*, 43, 2011.)

5.2.1 Cycling Performance

Cycling performance, one of the most important indicators of battery quality, gives detailed charge/discharge data at different temperatures during a cell's lifespan, which include high temperature (HiT), room temperature (RT), and low temperature (LowT) tests. Generally speaking, cycle tests have different rate requirements according to the type of cell. An example of a pouch cell cycle test standard is given in Table 5.1. Sample cells are cycled between 3.0 and 4.35 V with an applied current rate of 0.7 C charge/0.5 C discharge at a temperature of 25 ± 3°C. However, for prismatic and cylindrical cells, the current rates are changed to 1 C charge/1 C discharge. After 500 cycles, the samples pass the test if the remaining retention capacity is higher than 80% of the initial capacity.

High-temperature cycle tests require the sample cells to cycle between 3.0 and 4.35 V with the same current rates of the RT test, but at a higher temperature of 45 ± 3°C. After 400 cycles, the remaining retention capacity should be higher than 80% of the initial capacity. On the contrary, low-temperature cycle tests demand lower charging rates because of the difficulty involved in charging a cell at low temperatures. According Table 5.1, compared with RT and HiT cycle tests, the LowT cycle test charge current rate is 0.3 C, much lower than 0.7 C, and the discharge current rate is the same. After 300 cycles, the sample cells pass the test if the remaining retention capacity is higher than 80% of the initial capacity. HiT and LowT cycle tests can help cell engineers understand the practical cycling profile in extreme circumstances and prevent some predictable safety issues.

Interval testing is a cycling performance analysis method combined with high-temperature storage. For instance, Dell's interval test for pouch cells in laptops has a total of 80 cycles and the test is carried out at 40°C. Every cycle contain three steps: (i) the cell is galvanostatically charged at a rate of 0.8 C to the voltage of 4.35 V, followed by constant voltage charging to 0.05 C; (ii) the fully charged cell is stored for the next 24 h; (iii) the cell is discharged at 0.5 C with a cutoff voltage of 3 V. After 80 cycles, the cell meets the requirements if the retention capacity is higher than 80% of the initial capacity.

Rate capability testing is a commonly used method to understand a cell's charging and discharging capability.

TABLE 5.1
Example of a Pouch Cell Cycle Test Standard

	Condition	Ambient Temperature
Charge (CC–CV)	0.7 C 4.35 V, 0.05 C cutoff	25 ± 3°C, 45 ± 3°C
	0.3 C 4.35 V, 0.05 C cutoff	12 ± 3°C
Discharge (CC)	0.5 C 3.0 V cutoff	25 ± 3°C, 45 ± 3°C
	0.5 C 3.0 V cutoff	12 ± 3°C

Note: CC, constant current; CV, constant voltage.

TABLE 5.2
Example of a Pouch Cell Charge Rate Test Standard

	Condition	Temperature
Charge (CC–CV)	1.0 C to 4.35 V, 4.35 V to 0.05 C	45 ± 2°C
	1.0 C to 4.35 V, 4.35 V to 0.05 C	25 ± 2°C
	0.4 C to 4.35 V, 4.35 V to 0.05 C	20 ± 2°C
	0.3 C to 4.35 V, 4.35 V to 0.05 C	15 ± 2°C and 10 ± 2°C
Discharge (CC)	0.5 C to 3.0 V	25 ± 2°C

TABLE 5.3
Example of a Pouch Cell Discharge Rate Test Standard

	C-Rate	Temperature
Charge (CC–CV)	0.7 C 4.35 V, 0.05 C cutoff	25 ± 2°C
Discharge (CC)	0.1 C, 0.2 C, 0.5 C, 1 C, and 1.5 C	25 ± 2°C, 45 ± 2°C, 60 ± 2°C
	0.1 C, 0.2 C, and 0.5 C	10 ± 2°C, 0 ± 2°C, −20 ± 2°C

Applications in power tools and EVs especially require high rate capability. For the pouch cells used in smart phones, Tables 5.2 and 5.3 show examples of standard charge and discharge rate tests, respectively. According to Table 5.2, the charge rate test is carried out at three different rates at five different temperatures. Low-rate charging is needed when a cell is tested at a temperature lower than 25°C. Before performing the discharge rate tests, the cell is galvanostatically charged at a rate of 0.7 C to 4.35 V at 25°C, followed by constant voltage charging until the current reaches 0.05 C. Compared with the charge rate test, the discharge rate test has a wider temperature range, from −20°C to 60°C, and a higher discharge current.

5.2.2 Storage Performance

Cell storage performance indicates how cell parameters vary when it is stored at different temperatures for a certain period of time. As shown in Table 5.4, a typical pouch cell storage test standard is given, and the data are recorded every 3 days. Detailed data gathered from the test includes recoverable capacity, internal resistance (IR), open circuit voltage (OCV), swelling (for prismatic and pouch cell), and abnormal cell state from visual inspection. The storage test also directly reveals the cell safety during transportation.

5.2.3 Safety and Reliability

Rechargeable lithium-ion batteries are a fast-growing technology already widely used in daily life because of their light weight, high voltage, and high-energy density. In addition, new applications, such as EVs and grid energy systems, demand higher capacities by using battery packs. Therefore, it is extremely important that battery manufacturers evaluate

TABLE 5.4
Example of a Pouch Cell Storage Test Standard

Temperature	SOC	Storage Time	Samples
25 ± 3°C	25%, 50%, 75%, 100%	30 days, 60 days, 120 days	4 SOC * 3 storage * 10 samples = 120 cells
45 ± 3°C	25%, 50%, 75%, 100%	30 days, 60 days, 120 days	4 SOC * 3 storage * 10 samples = 120 cells
60 ± 3°C	50%, 100%	30 days	2 SOC * 1 storage * 10 samples = 20 cells
85 ± 3°C	100%	6 h	10 cells

the safety of such batteries before their equipment is marketed. The purpose of this is to reduce the potential of an accident occurring and minimize the damage caused if one does happen. Recently, several accidents involving lithium-ion batteries have been reported. In January 2013, the Boeing 787 Dreamliner, introduced in 2011, reported five accidents in 7 days, most of which involved their batteries and electrical systems. For safety considerations, the United States Federal Aviation Administration ordered a review into the design and manufacture of the Boeing 787. Also in 2013, the Tesla Model S EV, launched in June 2012, had four serious fire accidents, the cause of which was related to damage to the lithium-ion batteries sustained during car accidents. The safety margin was still not good enough as car sales reached 22,300 globally that year. These accidents indicate the potential risks of LIBs in some extreme cases. Furthermore, the fire caused by the car accident was not easy to put out because firing of a cell can lit up the whole battery pack, even causing an explosion. Eventually, this will lead to casualties and enormous property losses. An example of pouch cell safety test standard is given in Table 5.5, and both fresh cells and cycled cells have to pass this test.

Reliability testing is a safety evaluation for fresh cells based on the concept of not destroying the sample cell. However, the test should be stopped if the sample catches fire or smokes during the experiment. Generally speaking, the purpose of a reliability test is to make sure the battery will be absolutely secure in five circumstances, listed in Table 5.6. Actually, these tests, to a large extent, ensure the batteries' safety during transportation.

Many standard safety tests have been set for different applications, including international standards (such as UL1642, UL2054, UN38.3/DOT, IEEE1725/1625, IEC62133, ICAO/IATA, BATSO, etc.) and national standards (such as BG18287 in China, JIS C 8711:2000 in Japan, etc.). As an LIB manufacturer, the key point is to meet the battery specifications of various customers according to their different applications in smart phones, laptops, UPS, EVs, and so on.

TABLE 5.5
Example of a Pouch Cell Safety Test Standard

Test Items		Condition (20 ± 5°C Unless Temperature Is Specified)	Criteria
Fresh cell	UL1642 short-circuit test (1)	Max resistance 0.1 ohm. Test without TCO	NS, NF, NE, T (cell skin) <150°C
	UL1642 short-circuit test (2)	55 + 5°C, max resistance 0.1 ohm. Test without TCO	NS, NF, NE, T (cell skin) <150°C
	UL1642 forced-discharge test	Max resistance 0.1 ohm. Test without TCO	NS, NF, NE
	UL1642 abnormal charging test	4.40 V 3 C, test without TCO	NS, NF, NE
	UL1642 projectile test	Steel wire mesh with 20 opening per inch (25.4 mm). Wire diameter 0.017 in (0.43 mm)	No penetration of wire screen
	UL1642 heating test	5 + 2°C/min to 130 + 2°C, 1 h test without TCO	NS, NF, NE
	UL1642 crush test	13 kN (3000 lb) until 2500 psig (17.2 MPa) 1.25 in (32 mm) diameter piston. Both the wide and narrow sides of cells	NS, NF, NE
	UL1642 impact test	5/8 in (15.8 mm) diameter bar 20 pound (9.1 kg) weight drop from 24 in	NS, NF, NE
	Nail penetration test	Nailing speed 100 mm/s, fresh nail of stainless steel, 2.5 mm diameter, full charge at 4.35 V C/20, OCV > 4.315 V before test	NS, NF, NE
	Overvoltage protection limit test	1 C 4.6 V 3 h	NS, NF, NE
	Overcharge test	0.5 C 10 V 3 h	Reference only
Cycled cell	UL1642 short-circuit test (1)	Max resistance 0.1 ohm. Test without TCO	NS, NF, NE, T (cell skin) <150°C
	UL1642 heating test	5 + 2°C/min to 130 + 2°C, 1 h test without TCO	NS, NF, NE
	Nail penetration test	Nailing speed 100 mm/s, fresh nail of stainless steel, 2.5 mm diameter, full charge at 4.35 V C/20, OCV > 4.30 V before test	NS, NF, NE

Note: NE, no explosion; NF, no fire; NL, no leakage; NS, no smoke.

TABLE 5.6
Example of a Pouch Cell Reliability Test Standard

Test Items	Condition (20 ± 5°C Unless Temperature Is Specified)	Criteria
IEC62133 free fall	Fully charged, 1 m drop onto concrete	NF, NE
UL1642 low-pressure test (altitude simulation)	6 h at an absolute pressure of 11.6 kPa (1.68 psi)	NL, NS, NF, NE
UL1642 vibration test	Amplitude 0.8 mm, 1 Hz/min,10–55 Hz, 90–100 min	NL, NS, NF, NE
UL1642 temperature cycling test	−40–70°C 10 cycles, 4 h at each temperature, 2 h at 20°C, temperature transition by 30 min	NL, NS, NF, NE
UL1642 shock test	Three mutually perpendicular directions 3 ms, 75 g min average, 125–175 g peak acceleration	NL, NS, NF, NE

Note: NE, no explosion; NF, no fire; NL, no leakage; NS, no smoke.

5.3 DEGRADATION MECHANISMS AND MITIGATION STRATEGIES

Battery failure is an abnormal phenomenon in which the battery cannot be charged or discharged correctly in its service lifespan. Technologically, most of these "dead" batteries are caused by a complicated process involving internal and external factors. Internal factors mainly form in the manufacturing process. For example, active materials are not well distributed in the slurry; rolling pressure of the electrode is unstable in calendering; moisture control fails in the cell assembly stage; the anode electrode cannot cover the whole cathode electrode because of low precision of the winding machine; and time and temperature of formation are not well modified. As for external factors, the reasons usually have to do with maloperation or accidents when the battery is working in a device, such as using the wrong charger, extreme service environment, storing a fully charged battery, and dropping it. Therefore, battery failure mechanisms are divided into two main categories: internal failure mechanisms and external failure mechanisms.

5.3.1 Internal Failure Mechanisms

Electrode degradation is the main reason for the internal failure mechanism. A typical lithium-ion battery electrode consists of current collector, binder, conductive agent, and active materials, which are the most essential materials for a battery. More than 80% of battery performance is determined by the cathode and anode materials. However, a battery cannot work normally if any one of these materials is degraded.

Irreversible capacity losses can be observed in the first few charge/discharge cycles. One common concept is that a passivating film is formed on the cathode and anode by the reaction between electrolyte and active materials, universally known as the solid electrolyte interphase (SEI) [3–7]. This solid layer is electrically insulating yet provides significant ionic conductivity, and thus the interphase makes it possible for lithium ions to move between electrodes. In the cycling process of the cell, lithium-ion insertion/extraction causes a volume change for both electrodes, especially for the graphite anode (around 6%). Internal stress is produced during this reaction, and rupture occurs at the surface of graphite particles. Therefore, new graphite surfaces are exposed to the electrolyte, and consumption of Li and electrolyte occurs in the next cycle to form a new SEI. This is one of the reasons for battery capacity fading during cycling tests [8–10]. As for $LiCoO_2$ cathodes, the crystal structure will collapse when extra Li ions migrate into graphite, eventually leading to potential safety issues [11–14]. Meanwhile, growing SEI causes an increase in the battery's IR and polarization potential [15], leading to electrode degradation.

The function of the binder in an electrode is to glue the active materials together and tightly stick them onto the current collector. The best choices of binders in industrial production are polyvinylidene fluoride (PVdF) and carboxyl methylcellulose plus styrene butadiene rubber. These binders lose elasticity and strength after hundreds of volume changes during long-term cycling. Ultimately, the contact between active materials or between the active materials and foil will be reduced, resulting in low electrode conductivity and even separation between the active material and current collector. In this case, the battery will eventually fail due to increasing IR and capacity fading.

The current collector is also an important component of lithium-ion batteries and its stability also affects their performance. Usually in commercial batteries, aluminum foil and copper foil are used as the cathode and anode current collectors, respectively. Some surface treatments of current collector can greatly enhance the cohesive force between foils and binders [16,17]. However, the side effect is a corrosion reaction, such as pitting corrosion, which occurs on the surface of the current collector. Corrosion products, such as copper oxides and copper fluorides, can be found on the surface of Cu foil, which will directly increase the battery's interface resistance. Furthermore, Cu^{2+} can be deposited on the surface of the electrode during electrochemical oxidation/reduction processes. The formation of Cu dendrites can cause internal short circuiting by impaling the separator and connecting with the cathode. The current collector is surface pretreated for practical use [18], improving its adhesion to active materials and reducing its corrosion rate [19,20]. Another method for improving the anticorrosion performance of the foil is modifying the electrolyte formula with certain additives [21–23].

Other than the active materials, the electrolyte is an essential element of a battery's electrochemical reaction, providing a conductive medium for lithium ions to move between the electrodes. A battery electrolyte typically consists of lithium inorganic or organic salts, aprotic organic solvents, and functional additives. Part of the electrolyte will be degraded during each lithium-ion intercalation/deintercalation cycle in both electrodes. Moreover, degradation of the electrolyte generates gas and causes dead areas in the electrodes, leading to incomplete electrochemical reaction. This situation accelerates the capacity fading of a battery. The most effective method to solve this problem is applying functional additives, which can significantly improve the performance of SEI by inducing its forming and recombination [24,25].

The separator, another key component in a battery, prevents electrical short circuiting between the cathode and anode, as well as allows the rapid transport of ionic charge carriers [26]. Commercially available microporous polyolefin membranes, dominated by polypropylene (PP) and polyethylene (PE), are the most widely used materials for separators in the LIB industry because they can generally satisfy the requirements of low ionic resistance, good mechanical and thermal stability, and excellent chemical inertness. However, aging is a common issue for these materials, and thus it is inevitable to experience separator aging in long-term cycling processes. If the chemical stability is not good enough [27], the battery will suffer increasing electrode polarization and IR, and finally capacity fading. The separator shrinks when a battery is in a high-temperature storage test at 85°C due to poor thermal and mechanical stability in high temperature. Cell safety is a serious problem if the separator shrinks so much that the electrodes contact each other. A higher closing temperature for separator is crucial to prevent a battery from thermal runaway [28,29]. In fact, a composite separator is a better choice for industrial production. For instance, a PP/PE/PP composite separator produced by Celgard takes advantage of the PE material with a lower temperature of pore closing (around 130°C), while the PP material maintains a good mechanical performance up to 160°C.

5.3.2 External Failure Mechanisms

Battery failure is a continuous degradation process caused by internal failure mechanisms as well as external factors. Other than the cell's physical structure and electrochemical system, its method of use also plays a key role in a longer battery life. Misoperation, such as overcharging, overdischarging, and free fall, can cause potential hazards. For instance, cycling at elevated temperatures will cause the degradation of electrolyte via side reactions and thus generate gas. The internal pressure continuously increases until the current interrupt device works to prevent further damage in cylindrical batteries. A battery's thermal stability in high temperature is crucial for safety considerations, and is dependent on the thermal stability of the electrolyte and electrolyte/electrode. Typically, the thermal stability of the electrolyte/electrode chemical reaction determines the battery's heat resistance. Cells can be overcharged when the cell voltage is incorrectly detected by the charger, or when the wrong charger is used. Long-term overcharging will lead to irreversible capacity loss due to some lithium ions getting stuck in the graphite. Cell pack designs can provide external protection in some extreme situations, such as falling into water and external short circuiting. The practical use of lithium-ion cells is now possible with the help of protective electronic circuits and devices to compensate for their low thermal stability and poor tolerance to overcharging.

Consequently, guiding customers to correctly use lithium-ion batteries is essential, which consists of proper battery charging, correctly operating electronic devices, and storing lithium-ion batteries in a suitable environment. Differing from NiCd and NiMH rechargeable batteries, there is no memory effect for lithium-ion batteries. Therefore, long-term overcharging is not necessary for a new lithium-ion battery in the first few charge/discharge cycles. Actually, lithium-ion batteries can be charged whenever you want to in your daily life, regardless of its state of charge (SOC). On the other hand, some batteries can be damaged by repeated deep discharge, caused by collapse of graphite's layered structure. These batteries cannot be charged anymore and may even explode in some cases. Therefore, charging the battery before the electronic device shuts down is a good habit for proper maintenance. Generally speaking, depth of discharge is only needed once a month. The correct method of storage for LIBs is at 40–60% SOC (OCV ≈ 3.85 V) in a low-temperature (0–20°C) dry environment for months. All of the battery's capacity will be recovered after a few cycles at room temperature. Batteries should be fully discharged and then charged to 50% SOC every 6 months to store the battery for more than 1 year. For security reasons, cell engineers strongly suggest making certain that batteries are stored far away from any sources of heat.

5.4 ADVANCED DESIGNS OF RECHARGEABLE Li-ION BATTERIES

With the development of portable devices, smart grids, EVs, and hybrid EVs, better batteries are required to meet our growing energy demand. In practical applications, higher capacity is the most important standard as a "better battery," and this is why a 5% capacity increase every year is a common concept in the LIB market. According to the history of LIBs in the last two decades, this goal can be achieved from the following three aspects.

5.4.1 Improvement of Cell Electrochemical System

Taking high-energy-density pouch cells as an example, one of the simplest methods of improving the energy density of lithium-ion batteries is to increase the operating voltage of the cells by increasing the working potential of the $LiCoO_2$ electrode. Changing from 4.2 to 4.3 V, then from 4.3 to 4.35 V [30–34], for a specific cell model, the capacity has been increased around 5–7%. The main reason for this is because the $LiCoO_2$ cathode material will deliver extra Li^+ to the anode to support the full cell specific capacity elevation [30,31,34,35].

Nowadays, the 4.35 V electrochemical platform is already very mature and a 4.4 V cell system is being developed, which will hopefully replace the previous one within the next 3 years.

5.4.2 Modification of Cell Structure

In a cell platform, such as a 4.35 V battery system, cell capacity can be improved by modifying the cell design. Changing the cell design involves using new electrode materials that can provide better performance. Other than the electrochemical performance of the cathode and anode materials, other physical parameters (such as tap density, size, Brunauer–Emmett–Teller surface area, etc.) are crucial to initial evaluation and greatly influence the cell design. Tap density, one of the most important engineering parameters for an electrode material, reflects the weight of powder material per unit volume and the compacted density of the pressing step in electrode preparation. In other words, a higher value of tap density means more active material per unit volume in the electrode and more energy density in a cell model. Electrode material vendors take responsibility to improve the machinability and electrochemical performance of materials by optimizing the production process and surface treatments. For instance, surface-modified graphite (SMG; provided by Hitachi Chemical, Japan) is one kind of high-energy graphite anode. On the basis of the knowledge and experience of electrode materials, Hitachi, one of the biggest material suppliers in the world, developed SMG series products that are widely applied in LIBs. The compacted density improved from 1.6 g/cm^3 (in 2010) to 1.7 g/cm^3 (in 2013) for the same coating level, and the anode thickness after pressing was decreased by around 6%. Meanwhile, the compacted density of $LiCoO_2$ cathode material was modified from 3.8 to 4.1 mg/cm^3, and the cathode thickness was decreased by around 5%. Therefore, in a cell model, more than 10% of the volume is free for capacity improvement.

5.4.3 Optimization of Auxiliary Materials

Decreasing the ratio of binder and conductive agent in the electrodes results in more active material that can deliver higher capacity. Battery engineers have two choices: (i) Pretreat the electrode material with the conductive agent. For example, some vendors will supply composite $LiCoO_2$ powder with carbon nanotubes (CNTs, as conductive agent), in which the CNTs have been evenly distributed in the powder. The advantage of this material is that an extra conductive agent is not needed in the slurry process and the CNTs make up only 0.5% by weight of the electrode. Furthermore, the battery's IR is reduced significantly by using this electrode because well-distributed CNTs provide a better path for electron movement. (ii) Although binders are electrochemically inactive materials in battery electrodes, they can have an important influence on the electrode performance. New binder innovation has not only reduced the proportion of the binder, but also improved the adhesion strength to hold the active material together [36–41]. In addition, the binder may play many other roles in LIB electrodes, such as large electrolyte uptake and easy absorption of electrolyte [42]; stabilizing the active material's surface [43]; and improving rate capability and cycling behavior [44–50], environmental friendliness, and nontoxicity [51].

The current collector is the key component for electron conduction between the electrode materials and the external circuit in lithium-ion batteries, and the collector should be electrochemically inactive when it is in contact with the electrolyte (Figure 5.5). Ensuring sufficient mechanical strength and excellent ductility, a thinner foil is preferred for LIBs because it frees up volume for more active materials and other components. Aluminum is commonly considered to be the most suitable current collector material for the positive electrode in lithium-ion batteries on account of its excellent machinability, low density, good electrical and thermal conductivity, and low cost. Compared with 20-μm Al current collectors 4 years ago, 16-μm Al foil dominates the market now, and some thinner foils, 13 and 12 μm, are being evaluated by LIB manufacturers. According to the suppliers, once the new foils pass certification, they will be able to begin mass production soon. Meanwhile, copper foil, as the most widely utilized anode current collector in the LIB industry, changed from 10 μm (in 2010) to 8 μm (in 2013), and 7-μm foil will be available soon.

Developing a functional coating on the surface of the current collector is another option to improve its performance [52–55], in terms of electrical conductivity, anticorrosion, and chemical stability. As a new methodology, researchers propose a new strategy using a foam-type three-dimensional (3D) substrate instead of the foil-type current collector [56]. The 3D substrate has highly tunable structures, such as thickness,

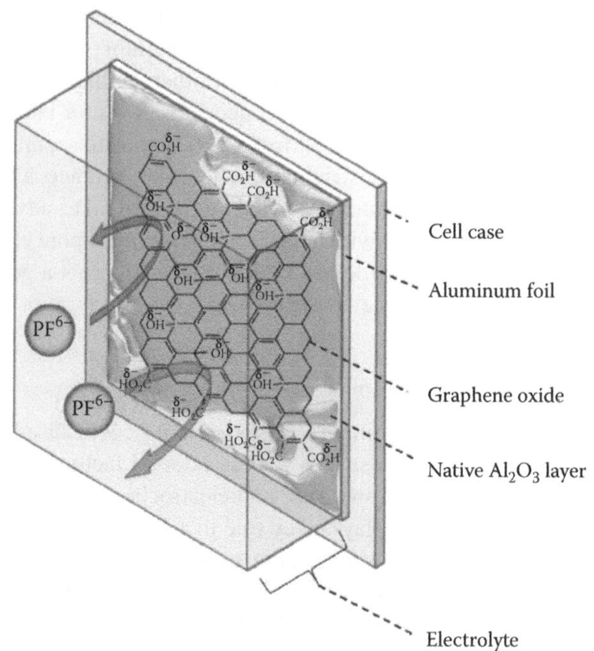

FIGURE 5.5 Schematic representation of the corrosion inhibition properties of graphene oxide when used on an aluminum current collector. The graphene oxide film blocks PF^{6-} ion flux by electrostatic repulsion, protects the native Al_2O_3 layer, and establishes direct contact with the Al surface through the defect sites of native Al_2O_3. (From Prabakar, S.J.R., Hwang, Y.H., Bae, E.G., Lee, D.K., and Pyo, M., *Carbon*, 52, 128, 2013.)

porosity, and pore size. By optimizing the structure of the 3D substrate and the annealing conditions, a great improvement in lithium-ion batteries using alloy-based negative electrodes will be realized. In addition, this methodology would also lead to the simplification of the manufacturing process and reduce the environmental load (Figure 5.6).

Among the various components in Li-ion batteries, separators play an essential role to prevent electrical short circuiting between the cathode and anode, as well as allow the rapid transport of ionic charge carriers (Figure 5.7). Commercially available microporous polyolefin membranes, predominantly PP and PE, are the most widely used materials for separators because they can generally satisfy the aforementioned requirements. Four years ago, 20–25 μm PE or PE/PP composite separators dominated the LIB industry. However, 16-μm separators are the most widely used now because of the requirement of improving the specific energy as well as the specific power of batteries. Some companies claim that the 12-μm separator is ready to mass production. Conventional separators also have some disadvantages, including insufficient wettability, low liquid electrolyte retention, and poor thermal resistance. Hence, some functional composite separators have been proposed to overcome these drawbacks. In the past few years, functional ceramic separators have been studied as candidates for improving thermal integrity and mechanical properties [57–64]. Advanced coating techniques have been developed to produce ceramic separators. The most common ceramic separators, including 9 μm polyolefin substrates and one-sided/two-sided 3 μm ceramic coatings, have already been applied in LIBs. Despite their excellent performance, ceramic separators have a few issues that battery engineers need to consider. For instance, typically, ceramic coating materials are nanostructured Al_2O_3 or SiO_2, which can very easily absorb water from the atmosphere during shipment or processing of the separators. Once the water combines with nano-Al_2O_3/SiO_2, it is very difficult to get rid of it, even when heated for a very long time. Therefore, moisture control is a key point during production, shipment, storage, and application. In the process of winding step, friction decreasing is another change in which the machine stress parameters have to be adjusted (Figure 5.8).

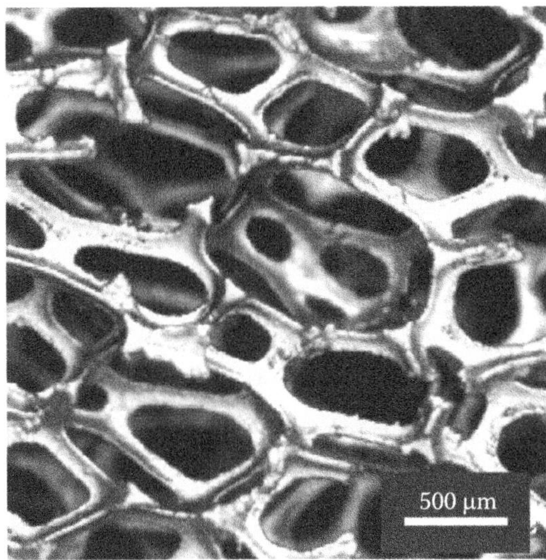

FIGURE 5.6 Optical microscope image of the foil's 3D structure. (From Hu, Y., Liu, J., Ding, R., Wang, K., Jiang, J., Ji, X., Li, Y., and Huang, X., *Thin Solid Films*, 518, 6876, 2010.)

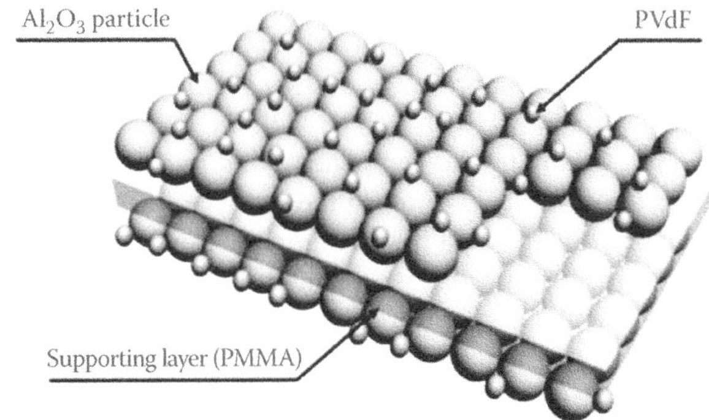

FIGURE 5.7 Schematic diagram of an organic/inorganic trilayer separator. (From Kim, M., Han, G.Y., Yoon, K.J., and Park, J.H., *Journal of Power Sources*, 195, 8302, 2010.)

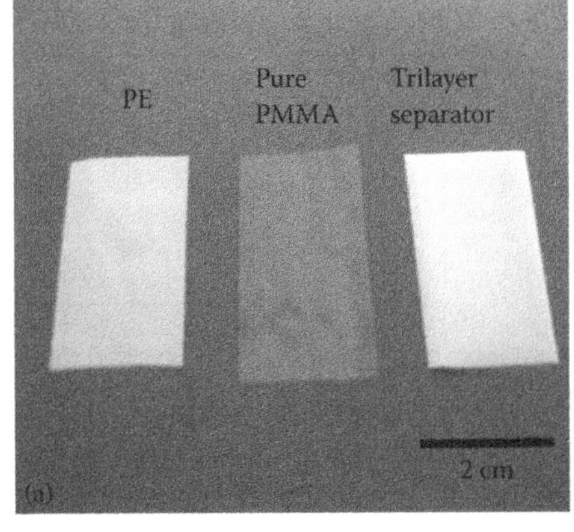

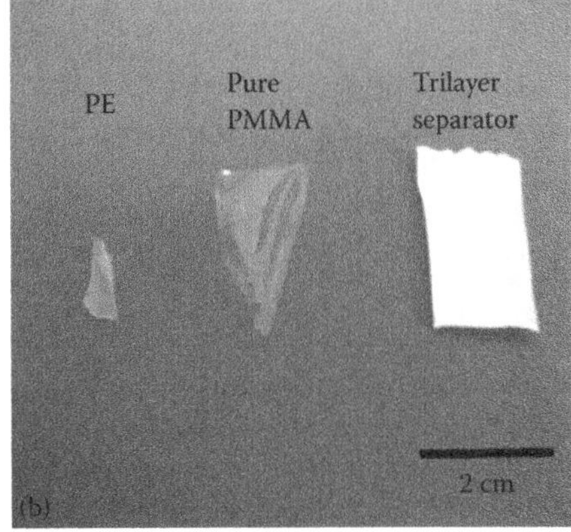

FIGURE 5.8 Photographs of PE, pure PMMA, and Trilayer separator (a) before and (b) after being stored at 150°C for 20 min. (From Kim, M., Han, G.Y., Yoon, K.J., and Park, J.H., *Journal of Power Sources*, 195, 8302, 2010.)

5.5 TECHNICAL CHALLENGES AND PERSPECTIVES

Although hundreds of millions of LIBs have been produced in the last few years, LIB technology is still developing and achieving higher performance and production efficiency. The challenges of the current lithium-ion battery industry lie in the following aspects.

5.5.1 Limits of Current Industrial Production Materials

There are a few kinds of cathode materials that are applied in commercial LIBs, including $LiCoO_2$, $LiMn_2O_4$, $LiFePO_4$, and $LiNi_xCo_yMn_{1-x-y}O_2$. Among them, $LiCoO_2$ is currently the most popular because of its facile fabrication process and well-balanced electrochemical performance in terms of practical application [65]. It has a specific capacity of 140 mAh/g between 3.0 and 4.2 V versus Li/Li^+. Reversible delithiation of Li_xCoO_2 is typically limited to $x = 0.5$, corresponding to ≈4.2 V. The conventional method to increase the energy density of commercial LIBs is raising the charge cutoff voltage. However, microstructural damage has been previously observed in $LiCoO_2$ composite powder cathodes after cycling to 4.35 V [66,67], as well as deterioration of cycling performance and thermal stability. To increase the capacity and energy density utilization of the cathode, efforts have been made to improve the cycling stability above 4.2 V by surface treatment with Al [68–77], Ti [78–80], or Mg [35,81–84]. Changing the components and ratio of the electrolyte is another way to prolong the service of $LiCoO_2$ [85–87]. Even for $LiCoO_2$ that has been modified for high voltage electrochemical platforms, increasing battery energy density is very difficult by simply raising the cutoff voltage [88]. Nowadays, some manufacturers are starting to use a composite cathode material of $LiCoO_2$ and $LiNi_{1/3}Co_{1/3}Mn_{1/3}O_2$ to meet the requirement of capacity improvement and cost reduction. For this application, the electrolyte formula has to been adjusted to prevent side reactions at high temperatures and voltages [89–108]. A possible candidate cathode for next-generation LIBs is a Li-rich cathode [109–115], such as $Li_{1.2}Ni_{0.15}Mn_{0.55}Co_{0.1}O_2$. However, to take advantage of this new class of materials, new electrolyte systems are needed because the charge voltages approach and exceed 4.5 V versus Li/Li^+. Most conventional carbonate-based organic solvents start oxidizing at voltages >4.5 V versus Li/Li^+. Furthermore, there is a long way to go for industrialization because the material synthesis process is not yet ready for mass production.

All anodes used in commercial LIBs are made of graphite [116–119], including natural graphite, artificial graphite, soft carbon, hard carbon, and meso-carbon microbeads. In high-energy-density LIBs (per weight and volume), the practical capacity achieved almost reaches the theoretical capacity for graphite materials: 372 mAh/g. After modification by the material vendor (such as Hitachi, Nippon Carbon, JFE Chemical, Shanshan, BTR, etc.), the tap density of graphite is greatly improved, which means the anode electrode can be pressed much thinner. This is a part of cell design to increase the volume energy density by reducing the physical thickness of the anode. However, graphite materials' tap density cannot be increased infinitely. The relatively low capacity of the electrodes (370 mAh/g for graphite) limits the total specific energy of the cell. It is time to introduce new anode materials for the next generation of commercial LIBs (Figure 5.9). Many researchers have worked to find alternative anode materials in the last 10 years. Other than graphite, graphene, tin alloys, and silicon alloys have been promising candidates to replace graphite. Silicon, in particular, has been studied extensively due to its high gravimetric capacity (about 10 times that of graphite) [120–125] and low toxicity. Si is also the second-most abundant element in the earth's crust, and the semiconductor industry has developed a large and mature infrastructure for processing Si. One of the main challenges preventing the implementation of silicon-based anodes is the ~300% volume change that occurs during lithiation/delithiation of silicon material [126,127]. These large volume changes can cause mechanical fracture and loss of electrical contact with the current collector, which results in capacity fade [128]. These shortcomings of Si anodes in LIBs have been ameliorated by utilizing nanostructured Si, as nanostructures are much more robust with respect to fracture than bulk material. Great progress in nanostructure synthesis has enabled researchers to probe the lithiation and delithiation behavior of anodes composed of various Si geometries, including pillars, spheres, and wires. A few Japanese companies (Hitachi, Shin-Etsu) have claimed that Si–O–C composite anode materials are ready for mass production, and they have already sent their silicon-based anode materials to many LIB manufacturers for initial evaluation in real batteries. By mixing 5% (wt.) Si–O–C with graphite, full cell capacity can be improved by 7%, reaching 600 Wh/L (Figure 5.10).

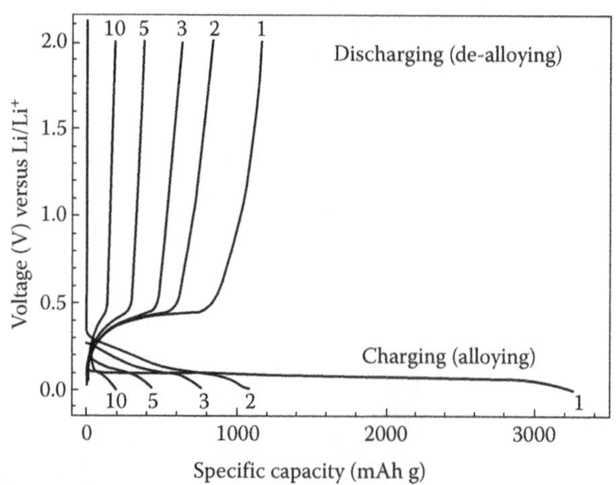

FIGURE 5.9 Galvanostatic charge/discharge profiles for a micro-Si (10-μm particle) anode. (From Wu, H., and Cui, Y., *Nano Today*, 7, 414, 2012.)

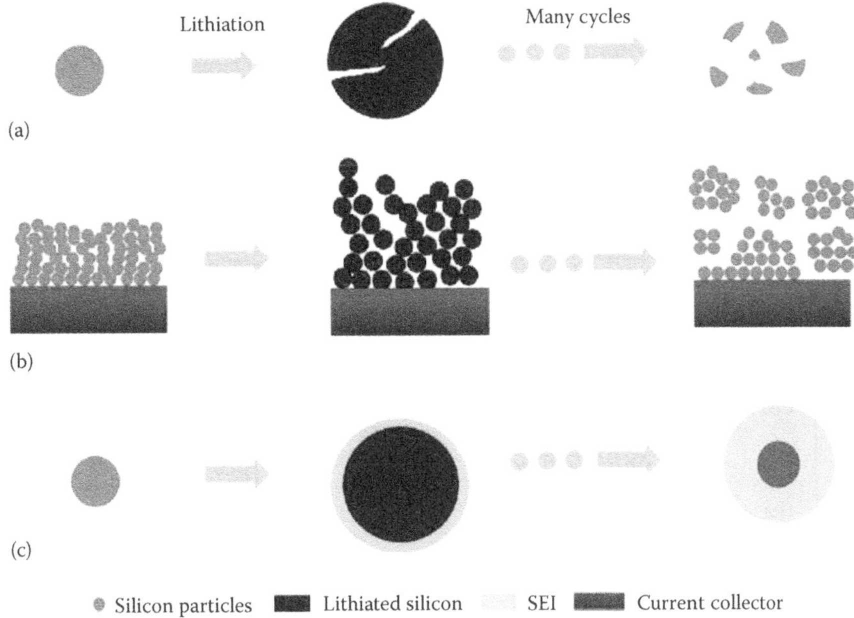

FIGURE 5.10 Si electrode failure mechanism: (a) material pulverization; (b) morphology and volume change of the entire electrode; (c) continuous SEI growth. (From Wu, H., and Cui, Y., *Nano Today*, 7, 414, 2012.)

The electrolyte greatly affects the electrochemical performance of LIBs, and the bottleneck effect of the electrolyte becomes especially problematic in developing high-energy-density and high-cutoff-voltage lithium-ion batteries. According to the current trend, new 4.4 V LIBs will replace the current 4.35 V batteries within the next few years. Therefore, many LIB manufacturers are working on new electrolyte formulas based on 4.4 V, and even exceeding 4.5 V, Li-ion chemistry [129–136]. Other challenges come from the introduction of new electrode materials, such as Li-rich-layered oxide cathodes [137–141] and Si alloy anodes [142–151]. Many researchers and world-class institutions are studying new additives, solvents, Li salts, and the combination of these components to build next-generation Li-ion battery systems.

5.5.2 Simplify the Assembly Process, Improve the Degree of Industrial Automation

The development of lithium-ion batteries in the automotive industry and smart grid will have tremendous consequences on the battery market in the near future. One of the key points of future wide-scale production is the development of new technology-based processes using a fully automated production line to reduce cost. Compared with prismatic and cylindrical batteries, the assembly process of pouch batteries is much longer. The main reason for this is that the process is much more complex. A typical pouch cell assembly process includes more than 30 steps, which makes it very difficult to control and the production yield is greatly reduced. The steps that can be easily simplified are after treatment, including standing, formation, clod pressing, degasing, and aging. Improving mass production of pouch cells is a big challenge because the parameters are constantly changing. In addition to their utilization in traditional electronic devices (such as smart phones, laptops, etc.), very small pouch cells have been produced for new mobile devices (such as Google Glass, electric model planes, etc.). For applications in smart watches, pouch cells can be designed as a curved cell hidden in the watchband. Battery manufacturers want simple adjustment of equipment, so that the production line can easily adapt to produce different sized LIBs. The improvement of automation will greatly reduce the influence of artificial factors, elevating the production efficiency and yield.

5.5.3 New Demands from Customers

With the limitation of current electrode materials, it is not easy to increase 5% capacity every year in the near future. The biggest challenge for future batteries is to find a way to prolong their longevity, especially for EVs and large-scale energy storage systems that harvest the electricity produced by wind, tide, and solar. Recently, some customers have started to pay close attention to other technologies. They believe that this is another way to promote the competitiveness of their products. According to the roadmap of customers, the baseline for super-long-lifespan batteries are more than 1000 cycles and, after that, the reversible capacity must still be higher than 75% of the initial capacity For pouch cells, the battery's volumetric energy density should be 530–560 Wh/L, which is exactly the commercial battery's standard in 2013. Although the capacity demand is not high, changing the cell design for more than 1000 cycles is not easy for current commercial electrode materials (lithium metal oxide and graphitic carbon). The design can be improved by following several steps:

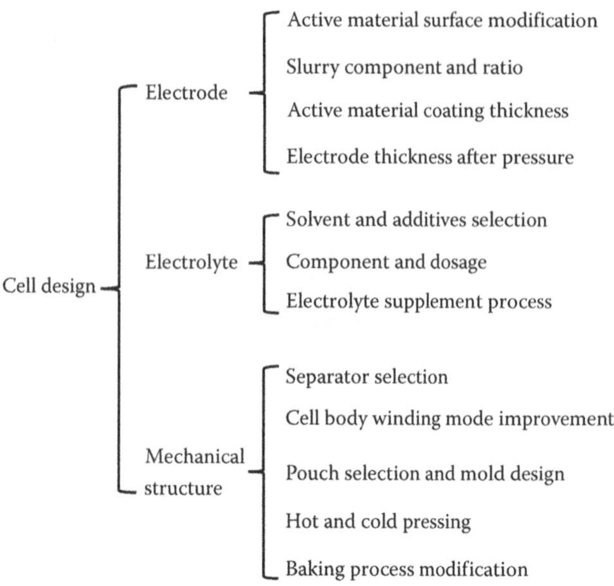

Every aspect mentioned above is for reducing the electrochemical polarization while the battery is cycling, and these will eventually extend the service time of the battery.

The lithium-ion battery industry needs to find a balance between performance and cost. On the premise of meeting the cell specifications of the customer, lower cost is preferred, while ensuring the stable and fast development of LIB mass production. From the viewpoint of an LIB manufacturer, the best way to do this is by intensively studying LIB electrochemical systems and selecting the most effective raw materials. Furthermore, enhancing the ability of automated mass production is essential by developing newly assembled machines for improving the production efficiency and yield.

REFERENCES

1. Nishi, Y., Performance of the first lithium ion battery and its process technology. In *Lithium Ion Batteries: Fundamentals and Performance*, eds. M. Wakihara and O. Yamamoto. Tokyo: Kodansha; Weinheim: Wiley-VCH, 1998.
2. VanZwol, J. Use Li-ion batteries in your next mobile computer. *Electronic Design* 2011; 43–46.
3. Aurbach, D., Zinigrad, E., Cohen, Y., and Hanan, T. A short review of failure mechanisms of lithium metal and lithiated graphite anodes in liquid electrolyte solutions. *Solid State Ionics* 2002; 148:405–416.
4. Chusid, O., Gofer, Y., Aurbach, D., Watanabe, M., Momma, T., and Osaka, T. Studies of the interface between lithium electrodes and polymeric electrolyte systems using *in situ* FTIR spectroscopy. *Journal of Power Sources* 2001; 97–98:632–636.
5. Yang, C.R., Wang, Y.Y., and Wan, C.C. Composition analysis of the passive film on the carbon electrode of a lithium-ion battery with an EC-based electrolyte. *Journal of Power Sources* 1998; 72:66–70.
6. Munichandraiah, N., Scanlon, L.G., and Marsh, R.A. Surface films of lithium: An overview of electrochemical studies. *Journal of Power Sources* 1998; 72:203–210.
7. Peled, E. The electrochemical behavior of alkali and alkaline earth metals in nonaqueous battery systems—The solid electrolyte interphase model. *Journal of the Electrochemical Society* 1979; 126:2047–2051.
8. Kim, Y., Veith, G.M., Nanda, J., Unocic, R.R., Chi, M., and Dudney, N.J. High voltage stability of $LiCoO_2$ particles with a nano-scale Lipon coating. *Electrochimica Acta* 2011; 56:6573–6580.
9. Yang, L., Markmaitree, T., and Lucht, B.L. Inorganic additives for passivation of high voltage cathode materials. *Journal of Power Sources* 2011; 196:2251–2254.
10. Yazami, R., Ozawa, Y., Gabrisch, H., and Fultz, B. Mechanism of electrochemical performance decay in $LiCoO_2$ aged at high voltage. *Electrochimica Acta* 2004; 50:385–390.
11. Menetrier, M., Saadoune, I., Levasseur, S., and Delmas, C. The insulator–metal transition upon lithium deintercalation from $LiCoO_2$: Electronic properties and 7Li NMR study. *Journal of Materials Chemistry* 1999; 9:1135–1140.
12. Wang, H., Jang, Y.-I., Huang, B., Sadoway, D.R., and Chiang, Y.-M. TEM study of electrochemical cycling-induced damage and disorder in $LiCoO_2$ cathodes for rechargeable lithium batteries. *Journal of the Electrochemical Society* 1999; 146:473–480.
13. Ohzuku, T., Ueda, A., Nagayama, M., Iwakoshi, Y., and Komori, H. Comparative study of $LiCoO_2$, $LiNi_{1/2}Co_{1/2}O_2$ and $LiNiO_2$ for 4 volt secondary lithium cells. *Electrochimica Acta* 1993; 38:1159–1167.
14. Reimers, J.N., and Dahn, J.R. Electrochemical and *in situ* x-ray diffraction studies of lithium intercalation in Li_xCoO_2. *Journal of the Electrochemical Society* 1992; 139:2091–2097.
15. Kim, J.S., and Park, Y.T. Characteristics of surface films formed at a mesocarbon microbead electrode in a Li-ion battery. *Journal of Power Sources* 2000; 91:172–176.
16. Kim, T.K., Li, X., and Wang, C. Temperature dependent capacity contribution of thermally treated anode current collectors in lithium ion batteries. *Applied Surface Science* 2013; 264:419–423.
17. Shu, J., Shui, M., Huang, F., Xu, D., Ren, Y., Hou, L., Cui, J., and Xu, J. Comparative study on surface behaviors of copper current collector in electrolyte for lithium-ion batteries. *Electrochimica Acta* 2011; 56:3006–3014.
18. NakaJima, T., Mori, M., Gupta, V., Ohzawa, Y., and Iwata, H. Effect of fluoride additives on the corrosion of aluminum for lithium ion batteries. *Solid State Sciences* 2002; 4(11–12):1385–1394.
19. Wu, H.C., Lee, E., Wu, N.L., and Jow, T.R. Effects of current collectors on power performance of $Li_4Ti_5O_{12}$ anode for Li-ion battery. *Journal of Power Sources* 2012; 197:301–304.
20. Kanamura, K. Anodic oxidation of nonaqueous electrolyte on cathode materials and current collectors for rechargeable lithium batteries. *Journal of Power Sources* 1999; 81–82:123–129.
21. Cho, E., Mun, J., Chae, O.B., Kwon, O.M., Kim, H.T., Ryu, J.H., Kim, Y.G., and Oh, S.M. Corrosion/passivation of aluminum current collector in bis(fluorosulfonyl)imide-based ionic liquid for lithium-ion batteries. *Electrochemistry Communications* 2012; 22:1–3.
22. Myung, S.T., Sasaki, Y., Sakurada, S., Sun, Y.K., and Yashiro, H. Electrochemical behavior of current collectors for lithium batteries in non-aqueous alkyl carbonate solution and surface analysis by ToF–SIMS. *Electrochimica Acta* 2009; 55:288–297.
23. Peng, C., Yang, L., Zhang, Z., Tachibana, K., Yang, Y., and Zhao, S. Investigation of the anodic behavior of Al current collector in room temperature ionic liquid electrolytes. *Electrochimica Acta* 2008; 53:4764–4772.

24. Rollins, H.W., Harrup, M.K., Dufek, E.J., Jamison, D.K., Sazhin, S.V., Gering, K.L., and Daubaras, D.L. Fluorinated phosphazene co-solvents for improved thermal and safety performance in lithium-ion battery electrolytes. *Journal of Power Sources* 2014; 263:66–74.
25. Shin, J.S., Han, C.H., Jung, U.H., Lee, S.I., Kim, H.J., and Kim, K. Effect of Li_2CO_3 additive on gas generation in lithium-ion batteries. *Journal of Power Sources* 2002; 109:47–52.
26. Zhang, S.S. A review on the separators of liquid electrolyte Li-ion batteries. *Journal of Power Sources* 2007; 164:351–364.
27. Aroura, P., and Zhang, Z. Battery separators. *Chemical Reviews* 2004; 104:4419–4462.
28. Roth, E.P., Doughty, D.H., and Pile, D.L. Effects of separator breakdown on abuse response of 18650 Li-ion cells. *Journal of Power Sources* 2007; 174:579–583.
29. Tobishima, S., and Yamaki, J. A consideration of lithium cell safety. *Journal of Power Sources* 1999; 81–82:882–886.
30. Hudaya, C., Park, J.H., Lee, J.K., and Choi, W. SnO_2-coated $LiCoO_2$ cathode material for high-voltage applications in lithium-ion batteries. *Solid State Ionics* 2014; 256:89–92.
31. Park, J.H., Cho, J.H., Kim, J.S., Shim, E.G., and Lee, S.Y. High-voltage cell performance and thermal stability of nanoarchitectured polyimide gel polymer electrolyte-coated $LiCoO_2$ cathode materials. *Electrochimica Acta* 2012; 86:346–351.
32. Han, G.B., Ryou, M.H., Choi, K.Y., Lee, Y.M., and Park, J.K. Effect of succinic anhydride as an electrolyte additive on electrochemical characteristics of silicon thin-film electrode. *Journal of Power Sources* 2010; 195:3709–3714.
33. Meng, X.L., Dou, S.M., and Wang, W.L. High power and high capacity cathode material $LiNi_{0.5}Mn_{0.5}O_2$ for advanced lithium-ion batteries. *Journal of Power Sources* 2008; 184:489–493.
34. Jang, Y.I., Dudney, N.J., Blom, D.A., and Allard, L.F. Electrochemical and electron microscopic characterization of thin-film $LiCoO_2$ cathodes under high-voltage cycling conditions. *Journal of Power Sources* 2003; 119–121:295–299.
35. Mladenov, M., Stoyanova, R., Zhecheva, E., and Vassilev, S. Effect of Mg doping and MgO-surface modification on the cycling stability of $LiCoO_2$ electrodes. *Electrochemistry Communications* 2001; 3:410–416.
36. Chen, D., Yi, R., Chen, S., Xu, T., Gorgin, M.L., and Wang, D. Facile synthesis of graphene-silicon nanocomposites with an advanced binder for high-performance lithium-ion battery anodes. *Solid State Ionics* 2014; 254:65–71.
37. Choi, J., Ryou, M.H., Son, B., Song, J., Park, J.K., Cho, K.Y., and Lee, Y.M. Improved high-temperature performance of lithium-ion batteries through use of a thermally stable co-polyimide-based cathode binder. *Journal of Power Sources* 2014; 252:138–143.
38. Ryou, M.H., Kim, J., Lee, I., Kim, S., Jeong, Y.K., Hong, S., Ryu, J.H., Kim, T.S., Park, J.K., Lee, H., and Choi, J.W. Mussel-inspired adhesive binders for high-performance silicon nanoparticle anodes in lithium-ion batteries. *Advanced Materials* 2013; 25:1571–1576.
39. Wu, Q., Ha, S., Prakash, J., Dees, D.W., and Lu, W. Investigations on high energy lithium-ion batteries with aqueous binder. *Electrochimica Acta* 2013; 114:1–6.
40. Xue, Z., Zhang, Z., and Amine, K. Cross-linkable urethane acrylate oligomers as binders for lithium-ion battery. *Electrochemistry Communications* 2013; 34:86–89.
41. Chen, Z., Christensen, L., and Dahn, J.R. Large-volume-change electrodes for Li-ion batteries of amorphous alloy particles held by elastomeric tethers. *Electrochemistry Communications* 2003; 5:919–923.
42. Fu, Z., Feng, H.L., Xiang, X.D., Rao, M.M., Wu, W., Luo, J.C., Chen, T.T., Hu, Q.P., Feng, A.B., and Li, W.S. A novel polymer composite as cathode binder of lithium ion batteries with improved rate capability and cyclic stability. *Journal of Power Sources* 2014; 261:170–174.
43. Zhao, H., Zhou, X., Park, S.J., Shi, F., Fu, Y., Ling, M., Yuca, N., Battaglia, V., and Liu, G. A polymerized vinylene carbonate anode binder enhances performance of lithium-ion batteries. *Journal of Power Sources* 2014; 263:288–295.
44. Hu, S., Li, Y., Yin, J., Wang, H., Yuan, X., and Li, Q. Effect of different binders on electrochemical properties of $LiFePO_4$/C cathode material in lithium ion batteries. *Chemical Engineering Journal* 2014; 237:497–502.
45. Chai, L., Qu, Q., Zhang, L., Shen, M., Zhang, L., and Zheng, H. Chitosan, a new and environmental benign electrode binder for use with graphite anode in lithium-ion batteries. *Electrochimica Acta* 2013; 105:378–383.
46. Xu, J., Chou, S.L., Gu, Q.F., Liu, H.K., and Dou, S.X. The effect of different binders on electrochemical properties of $LiNi_{1/3}Mn_{1/3}Co_{1/3}O_2$ cathode material in lithium ion batteries. *Journal of Power Sources* 2013; 225:172–178.
47. Mancini, M., Nobili, F., Tossici, R., and Marassi, R. Study of the electrochemical behavior at low temperatures of green anodes for Lithium ion batteries prepared with anatase TiO_2 and water soluble sodium carboxymethyl cellulose binder. *Electrochimica Acta* 2012; 85:566–571.
48. Chong, J., Xun, S., Zheng, H., Song, X., Liu, G., Ridgway, P., Wang, J.Q., and Battaglia, V.S. A comparative study of polyacrylic acid and poly(vinylidene difluoride) binders for spherical natural graphite/$LiFePO_4$ electrodes and cells. *Journal of Power Sources* 2011; 196:7707–7714.
49. Courtel, F.M., Niketic, S., Duguay, D., Abu-Lebdeh, Y., and Davidson, I.J. Water-soluble binders for MCMB carbon anodes for lithium-ion batteries. *Journal of Power Sources* 2011; 196:2128–2134.
50. Jarvis, C.R., Macklin, W.J., Macklin, A.J., Mattingley, N.J., and Kronfli, E. Use of grafted PVdF-based polymers in lithium batteries. *Journal of Power Sources* 2001; 97–98: 664–666.
51. Kim, G.T., Jeong, S.S., Joost, M., Rocca, E., Winter, M., Passerini, S., and Balducci, A. Use of natural binders and ionic liquid electrolytes for greener and safer lithium-ion batteries. *Journal of Power Sources* 2011; 196:2187–2194.
52. Doberdò, I., Löffler, N., Laszczynski, N., Cericola, D., Penazzi, N., Bodoardo, S., Kim, G.T., and Passerini, S. Enabling aqueous binders for lithium battery cathodes. Carbon coating of aluminum current collector. *Journal of Power Sources* 2014; 248:1000–1006.
53. Luo, F., Zhou, H., Vullum-Bruer, F., Tran, T.D., and Chen, D. Synthesis of carbon nanofibers@MnO_2 3D structures over copper foil as binder free anodes for lithium ion batteries. *Journal of Energy Chemistry* 2013; 22:78–86.
54. Prabakar, S.J.R., Hwang, Y.H., Bae, E.G., Lee, D.K., and Pyo, M. Graphene oxide as a corrosion inhibitor for the aluminum current collector in lithium ion batteries. *Carbon* 2013; 52:128–136.
55. Hu, Y., Liu, J., Ding, R., Wang, K., Jiang, J., Ji, X., Li, Y., and Huang, X. Copper nanowall array grow on bulk Fe–Co–Ni alloy substrate at room temperature as lithium-ion battery current collector. *Thin Solid Films* 2010; 518:6876–6882.
56. Yao, M., Okuno, K., Iwaki, T., Awazu, T., and Sakai, T. Long cycle-life $LiFePO_4$/Cu-Sn lithium ion battery using foam-type three-dimensional current collector. *Journal of Power Sources* 2010; 195:2077–2081.

57. Jeong, H.S., Hong, S.C., and Lee, S.Y. Effect of microporous structure on thermal shrinkage and electrochemical performance of Al_2O_3/poly(vinylidene fluoride-hexafluoropropylene) composite separators for lithium-ion batteries. *Journal of Membrane Science* 2010; 364:177–182.
58. Jeong, H.S., and Lee, S.Y. Closely packed SiO_2 nanoparticles/poly(vinylidene fluoride-hexafluoropropylene) layers-coated polyethylene separators for lithium-ion batteries. *Journal of Power Sources* 2011; 196:6716–6722.
59. Park, J.H., Cho, J.H., Park, W., Ryoo, D., Yoon, S.J., Kim, J.H., Jeong, Y.U., and Lee, S.Y. Close-packed SiO_2/poly(methyl methacrylate) binary nanoparticles-coated polyethylene separators for lithium-ion batteries. *Journal of Power Sources* 2010; 195:8306–8310.
60. Choi, E.S., and Lee, S.Y. Particle size-dependent, tunable porous structure of a SiO_2/poly(vinylidene fluoride-hexafluoropropylene)-coated poly(ethylene terephthalate) nonwoven composite separator for a lithium-ion battery. *Journal of Materials Chemistry* 2011; 21:14747–14754.
61. Fang, J., Kelarakis, A., Lin, Y.W., Kang, C.Y., Yang, M.H., Cheng, C.L., Wang, Y., Giannelis, E.P., and Tsai, L.D. Nanoparticle-coated separators for lithium-ion batteries with advanced electrochemical performance. *Physical Chemistry Chemical Physics* 2011; 13:14457–14461.
62. Kim, M., Han, G.Y., Yoon, K.J., and Park, J.H. Preparation of a trilayer separator and its application to lithium-ion batteries. *Journal of Power Sources* 2010; 195:8302–8305.
63. Cao, J.H., Zhu, B.K., and Xu, Y.Y. Structure and ionic conductivity of porous polymer electrolytes based on PVDF–HFP copolymer membranes. *Journal of Membrane Science* 2006; 281:446–453.
64. Tian, Z., He, X., Pu, W., Wan, C., and Jiang, C. Preparation of poly(acrylonitrile-butyl acrylate) gel electrolyte for lithium-ion batteries. *Electrochimica Acta* 2006; 52:688–693.
65. Li, C., Zhang, H.P., Fu, L.J., Liu, H., Wu, Y.P., Rahm, E., Holze, R., and Wu, H.Q. Cathode materials modified by surface coating for lithium ion batteries. *Electrochimica Acta* 2006; 51:3872–3883.
66. Chen, Z., and Dahn, J.R. Methods to obtain excellent capacity retention in $LiCoO_2$ cycled to 4.5 V. *Electrochimica Acta* 2004; 49:1079–1090.
67. Levasseur, S., Menetrier, M., and Delmas, C. On the $Li_xCo_{1-y}Mg_yO_2$ system upon deintercalation: Electrochemical, electronic properties and ^{7}Li MAS NMR studies. *Journal of Power Sources* 2002; 112:419–427.
68. Liu, J., and Manthiram, A. Improved electrochemical performance of the 5 V spinel cathode $LiMn_{1.5}Ni_{0.42}Zn_{0.08}O_4$ by surface modification. *Journal of the Electrochemical Society* 2009; 156:A66–A72.
69. Cho, J., Kim, B., Lee, J.G., Kim, Y.W., and Park, B. Annealing-temperature effect on various cutoff-voltage electrochemical performances in $AlPO_4$-nanoparticle-coated $LiCoO_2$. *Journal of the Electrochemical Society* 2005; 152: A32–A36.
70. Kweon, H.J., Park, J.J., Seo, J.W., Kim, G.B., Jung, B.H., and Lim, H.S. Effects of metal oxide coatings on the thermal stability and electrical performance of $LiCoCO_2$ in a Li-ion cell. *Journal of Power Sources* 2004; 126:156–162.
71. Oh, S., Lee, J.K., Byun, D., Cho, W.I., and Cho, B.W. Effect of Al_2O_3 coating on electrochemical performance of $LiCoO_2$ as cathode materials for secondary lithium batteries. *Journal of Power Sources* 2004; 132:249–255.
72. Lee, J.G., Kim, B., Cho, J., Kim, Y.W., and Park, B. Effect of $AlPO_4$-nanoparticle coating concentration on high-cutoff-voltage electrochemical performances in $LiCoO_2$. *Journal of the Electrochemical Society* 2004; 151:A801–A805.
73. Cho, J. Dependence of $AlPO_4$ coating thickness on overcharge behaviour of $LiCoO_2$ cathode material at 1 and 2 C rates. *Journal of Power Sources* 2004; 126:186–189.
74. Cho, J., Kim, H., and Park, B. Comparison of overcharge behaviour of $AlPO_4$-coated $LiCoO_2$ and $LiNi_{0.8}Co_{0.1}Mn_{0.1}O_2$ cathode materials in Li-ion cells. *Journal of the Electrochemical Society* 2004; 151:A1707–A1711.
75. Kim, B., Lee, J.G., Choi, M., Cho, J., and Park, B. Correlation between local strain and cycle-life performance of $AlPO_4$-coated $LiCoO_2$ cathodes. *Journal of Power Sources* 2004; 126:190–192.
76. Cho, J., Kim, Y.J., and Park, B. $LiCoO_2$ cathode material that does not show a phase transition from hexagonal to monoclinic phase. *Journal of the Electrochemical Society* 2001; 148:A1110–A1115.
77. Cho, J., Kim, Y.J., and Park, B. Novel $LiCoO_2$ cathode material with Al_2O_3 coating for a Li ion cell. *Chemistry of Materials* 2000; 12:3788–3791.
78. Kim, J., Noh, M., Cho, J., Kim, H., and Kim, K.B. Controlled nanoparticle metal phosphates (metal = Al, Fe, Ce, and Sr) coatings on $LiCoO_2$ cathode materials. *Journal of the Electrochemical Society* 2005; 152:A1142–A1148.
79. Fey, G.T.-K., Yang, H.Z., Kumar, T.P., Naik, S.P., Chiang, A.S.T., Lee, D.C., and Lin, J.R. A simple mechano-thermal coating process for improved lithium battery cathode materials. *Journal of Power Sources* 2004; 132:172–180.
80. Amatucci, G.G., Tarascon, J.M., and Klein, L.C. Cobalt dissolution in $LiCoO_2$-based non-aqueous rechargeable batteries. *Solid State Ionics* 1996; 83:167–173.
81. Shi, S., Ouyang, C., Lei, M., and Tang, W. Effect of Mg-doping on the structural and electronic properties of $LiCoO_2$: A first-principles investigation. *Journal of Power Sources* 2007; 171:908–912.
82. Iriyama, Y., Kurita, H., Yamada, I., Abe, T., and Ogumi, Z. Effects of surface modification by MgO on interfacial reactions of lithium cobalt oxide thin film electrode. *Journal of Power Sources* 2004; 137:111–116.
83. Liu, H., Wu, Y.P., Rahm, E., Holze, R., and Wu, H.Q. Cathode materials for lithium ion batteries prepared by sol–gel methods. *Journal of Solid State Electrochemistry* 2004; 8:450–466.
84. Levasseur, S., Ménétrier, M., and Delmas, C. On the dual effect of Mg doping in $LiCoO_2$ and $Li_{1+\delta}CoO_2$: Structural, electronic properties, and ^{7}Li MAS NMR studies. *Chemistry of Materials* 2002; 14:3584–3590.
85. Yang, J., Zhao, P., Shang, Y., Wang, L., He, X., Fang, M., and Wang, J. Improvement in high-voltage performance of lithium-ion batteries using bismaleimide as an electrolyte additive. *Electrochimica Acta* 2014; 121:264–269.
86. Zuo, X., Fan, C., Xiao, X., Liu, J., and Nan, J. High-voltage performance of $LiCoO_2$/graphite batteries with methylene methanedisulfonate as electrolyte additive. *Journal of Power Sources* 2012; 219:94–99.
87. Lee, J.N., Han, G.B., Ryou, M.H., Lee, D.J., Song, J. Choi, J.W., and Park, J.K. *N*-(triphenylphosphoranylidene) aniline as a novel electrolyte additive for high voltage $LiCoO_2$ operations in lithium ion batteries. *Electrochimica Acta* 2011; 56:5195–5200.
88. Brandt, A., and Balducci, A. Theoretical and practical energy limitations of organic and ionic liquid-based electrolytes for high voltage electrochemical double layer capacitors. *Journal of Power Sources* 2014; 250:343–351.
89. Xu, J., Hu, Y., Liu, T., and Wu, X. Improvement of cycle stability for high-voltage lithium-ion batteries by in-situ growth of SEI film on cathode. *Nano Energy* 2014; 5:67–73.

90. Hu, L., Amine, K., and Zhang, Z. Fluorinated electrolytes for 5-V Li-ion chemistry: Dramatic enhancement of $LiNi_{0.5}Mn_{1.5}O_4$/graphite cell performance by a lithium reservoir. *Electrochemistry Communications* 2014; 44:34–37.
91. Xue, L., Ueno, K., Lee, S.Y., and Angell, C.A. Enhanced performance of sulfone-based electrolytes at lithium ion battery electrodes, including the $LiNi_{0.5}Mn_{1.5}O_4$ high voltage cathode. *Journal of Power Sources* 2014; 262:123–128.
92. Demeaux, J., Lemordant, D., Galiano, H., Caillon-Caravanier, M., and Claude-Montigny, B. Impact of storage on the $LiNi_{0.4}Mn_{1.6}O_4$ high voltage spinel performances in alkylcarbonate-based electrolytes. *Electrochimica Acta* 2014; 116:271–277.
93. Lee, E.H., Park, J.H., Cho, J.H., Cho, S.J., Kim, D.W., Dan, H., Kang, Y., and Lee, S.Y. Direct ultraviolet-assisted conformal coating of nanometer-thick poly(tris(2-(acryloyloxy)ethyl) phosphate) gel polymer electrolytes on high-voltage $LiNi_{1/3}Co_{1/3}Mn_{1/3}O_2$ cathodes. *Journal of Power Sources* 2013; 244:389–394.
94. Cui, X., Zhang, H., Li, S., Zhao, Y., Mao, L., Zhao, W., Li, Y., and Ye, X. Electrochemical performances of a novel high-voltage electrolyte based upon sulfolane and γ-butyrolactone. *Journal of Power Sources* 2013; 240:476–485.
95. Arbizzani, C., Giorgio, F.D., Porcarelli, L., Mastragostino, M., Khomenko, V., Barsukov, V., Bresser, D., and Passerini, S. Use of non-conventional electrolyte salt and additives in high-voltage graphite/$LiNi_{0.4}Mn_{1.6}O_4$ batteries. *Journal of Power Sources* 2013; 238:17–20.
96. Hu, L., Zhang, Z., and Amine, K. Electrochemical investigation of carbonate-based electrolytes for high voltage lithium-ion cells. *Journal of Power Sources* 2013; 236:175–180.
97. Ha, S.Y., Han, J.G., Song, Y.M., Chun, M.J., Han, S.I., Shin, W.C., and Choi, N.S. Using a lithium bis(oxalate) borate additive to improve electrochemical performance of high-voltage spinel $LiNi_{0.5}Mn_{1.5}O_4$ cathodes at 60°C. *Electrochimica Acta* 2013; 104:170–177.
98. Hu, L., Zhang, Z., and Amine, K. Fluorinated electrolytes for Li-ion battery: An FEC-based electrolyte for high voltage $LiNi_{0.5}Mn_{1.5}O_4$/graphite couple. *Electrochemistry Communications* 2013; 35:76–79.
99. Li, C., Zhao, Y., Zhang, H., Liu, J., Jing, J., Cui, X., and Li, S. Compatibility between $LiNi_{0.5}Mn_{1.5}O_4$ and electrolyte based upon lithium bis(oxalate)borate and sulfolane for high voltage lithium-ion batteries. *Electrochimica Acta* 2013; 104:134–139.
100. Xu, M., Liu, Y., Li, B., Li, W., Li, X., and Hu, S. Tris (pentafluorophenyl) phosphine: An electrolyte additive for high voltage Li-ion batteries. *Electrochemistry Communications* 2012; 18:123–126.
101. Eom, J., Kim, G.M., and Cho, J. Storage characteristics of $LiNi_{0.8}Co_{0.1+x}Mn_{0.1-x}O_2$ (x = 0, 0.03, and 0.06) cathode materials for lithium batteries. *Journal of the Electrochemistry Society* 2008; 155:A239–A245.
102. Lee, K.-S., Myung, S.-T., Amine, K., Yashiro, H., and Sun, Y.-K. Structural and electrochemical properties of layered $Li[Ni_{1-2x}Co_xMn_x]O_2$ (x = 0.1–0.3) positive electrode materials for Li-ion batteries. *Journal of the Electrochemistry Society* 2007; 154:A971–A977.
103. Woo, S.U., Park, B.C., Yoon, C.S., Myung, S.T. Prakash, J., and Sun, Y.K. Improvement of electrochemical performances of $Li[Ni_{0.8}Co_{0.1}Mn_{0.1}]O_2$ cathode materials by fluorine substitution. *Journal of the Electrochemistry Society* 2007; 154:A649–A655.
104. Mijung, N., Lee, Y., and Cho, J. Water adsorption and storage characteristics of optimized $LiCoO_2$ and $LiNi_{1/3}Co_{1/3}Mn_{1/3}O_2$ composite cathode material for Li-ion cells. *Journal of the Electrochemistry Society* 2006; 153:A935–A940.
105. Itou, Y., and Ukyo, Y. Performance of $LiNiCoO_2$ materials for advanced lithium-ion batteries. *Journal of Power Sources* 2005; 146:39–44.
106. Liu, H.S., Zhang, Z.R., Gong, Z.L., and Yang, Y. Origin of deterioration for $LiNiO_2$ cathode material during storage in air. *Electrochemical and Solid-State Letters* 2004; 7:A190–A193.
107. Abraham, D.P., Twesten, R.D., Balasubramanian, M., Petrov, I., McBreen, J., and Amine, K. Surface changes on $LiNi_{0.8}Co_{0.2}O_2$ particles during testing of high-power lithium-ion cells. *Electrochemistry Communications* 2002; 4:620–625.
108. Arai, H., Okada, S., Sakurai, Y., and Yamaki, J. Thermal behaviour of $Li_{1-y}NiO_2$ and decomposition mechanism. *Solid State Ionics* 1998; 109:295–302.
109. Mohanty, D., Kalnaus, S., Meisner, R.A., Rhodes, K.J., Li, J., Payzant, E.A., Wood III, D.L., and Daniel, C. Structural transformation of a lithium-rich $Li_{1.2}Co_{0.1}Mn_{0.55}Ni_{0.15}O_2$ cathode during high voltage cycling resolved by in situ X-ray diffraction. *Journal of Power Sources* 2013; 229:239–248.
110. Oishi, M., Fujimoto, T., Takanashi, Y., Orikasa, Y., Kawamura, A., Ina, T., Yamashige, H., Takamatsu, D., Sato, K., Murayama, H., Tanida, H., Arai, H., Ishii, H., Yogi, C., Watanabe, I., Ohta, T., Mineshige, A., Uchimoto, Y., and Ogumi, Z. Charge compensation mechanisms in $Li_{1.16}Ni_{0.15}Co_{0.19}Mn_{0.50}O_2$ positive electrode material for Li-ion batteries analyzed by a combination of hard and soft X-ray absorption near edge structure. *Journal of Power Sources* 2013; 222:45–51.
111. M. Gu, M., Belharouak, I., Zheng, J., Wu, H., Xiao, J., Genc, A., Amine, K., Thevuthasan, S., Baer, D.R., Zhang, J.G., Browning, N.D., Liu, J., and Wang, C. Formation of the spinel phase in the layered composite cathode used in Li-ion batteries. *ACS Nano* 2013; 7:760–767.
112. Li, G.R., Feng, X., Ding, Y., Ye, S.H., and Gao, X.P. AlF_3-coated $Li(Li_{0.17}Ni_{0.25}Mn_{0.58})O_2$ as cathode material for Li-ion batteries. *Electrochimica Acta* 2012; 78:308–315.
113. Tang, Z., Wang, Z., Li, X., and Peng, W. Influence of lithium content on the electrochemical performance of $Li_{1+x}(Mn_{0.533}Ni_{0.233}Co_{0.233})_{1-x}O_2$ cathode materials. *Journal of Power Sources* 2012; 208:237–241.
114. Zhang, H.Z., Qiao, Q.Q., Li, G.R., Ye, S.H., and Gao, X.P. Surface nitridation of Li-rich layered $Li(Li_{0.17}Ni_{0.25}Mn_{0.58})O_2$ oxide as cathode material for lithium-ion battery. *Journal of Materials Chemistry* 2012; 22:13104–13109.
115. Zhao, Y., Zhao, C., Feng, H., Sun, Z., and Xia, D. Enhanced electrochemical performance of $Li[Li_{0.2}Ni_{0.2}Mn_{0.6}]O_2$ modified by manganese oxide coating for lithium-ion batteries. *Electrochemical and Solid-State Letters* 2011; 14:A1–A5.
116. Buiel, E., and J.R. Dahn, J.R. Li-insertion in hard carbon anode materials for Li-ion batteries. *Electrochimica Acta* 1999; 45:121–130.
117. Xing, W., Xue, J.S., and Dahn, J.R. Optimizing pyrolysis of sugar carbons for use as anode materials in lithium-ion batteries. *Journal of the Electrochemical Society* 1996; 143:3046–3052.
118. Dahn, J.R., Zheng, T., Liu, Y.H., and Xue, J.S. Mechanisms for lithium insertion in carbonaceous materials. *Science* 1995; 270:590–593.
119. Ohzuku, T., Iwakoshi, Y., and Sawai, K. Formation of lithium–graphite intercalation compounds in nonaqueous electrolytes and their application as a negative electrode for a lithium ion (shuttlecock) cell. *Journal of the Electrochemical Society* 1993; 140:2490–2498.
120. Kim, J.S., Choi, W., Cho, K.Y., Byun, D., Lim, J., and Lee, J.K. Effect of polyimide binder on electrochemical characteristics of surface-modified silicon anode for lithium ion batteries. *Journal of Power Sources* 2013; 244:521–526.

121. Ryu, I., Choi, J.W., Cui, Y., and Nix, W.D. Size-dependent fracture of Si nanowire battery anodes. *Journal of the Mechanics and Physics of Solids* 2011; 59:1717–1730.
122. Yamada, M., Ueda, A., Matsumoto, K., and Ohzuku, T. Silicon-based negative electrode for high-capacity lithium-ion batteries: "SiO"-carbon composite. *Journal of the Electrochemical Society* 2011; 158:A417–A421.
123. Lee, J.K., Smith, K.B., Hayner, C.M. and Kung, H.H. Silicon nanoparticles–graphene paper composites for Li ion battery anode. *Chemical Communications* 2010; 46:2025–2027.
124. Takamura, T., Ohara, S., Uehara, M., Suzuki, J., and Sekine, K. A vacuum deposited Si film having a Li extraction capacity over 2000 mAh/g with a long cycle life. *Journal of Power Sources* 2004; 129:96–100.
125. Ohara, S., Suzuki, J., Sekine, K., and Takamura, T. A thin film silicon anode for Li-ion batteries having a very large specific capacity and long cycle life. *Journal of Power Sources* 2004; 136:303–306.
126. Berla, L.A., Lee, S.W., Ryu, I., Cui, Y., and Nix, W.D. Robustness of amorphous silicon during the initial lithiation/delithiation cycle. *Journal of Power Sources* 2014; 258:253–259.
127. McDowell, M.T., Lee, S.W., Wang, C., and Cui, Y. The effect of metallic coatings and crystallinity on the volume expansion of silicon during electrochemical lithiation/delithiation. *Nano Energy* 2012; 1:401–410.
128. Wu, H., and Cui, Y. Designing nanostructured Si anode for high energy lithium ion batteries. *Nano Today* 2012; 7:414–429.
129. Aravindan, V., Cheah, Y.L., Ling, W.C., and Madhavi, S. Effect of LiBOB additive on the electrochemical performance of $LiCoPO_4$. *Journal of the Electrochemical Society* 2012; 159:A1435–A1439.
130. Shao, N., Sun, S.G., Dai, S., and Jiang, D.E. Electrochemical windows of sulfone-based electrolytes for high-voltage Li-ion batteries. *The Journal of Physical Chemistry B* 2011; 115:12120–12125.
131. Lee, Y.S., Lee, K.S., Sun, Y.K., Lee, Y.M., and Kim, D.W. Effect of an organic additive on the cycling performance and thermal stability of lithium-ion cells assembled with carbon anode and $LiNi_{1/3}Co_{1/3}Mn_{1/3}O_2$ cathode. *Journal of Power Sources* 2011; 196:6997–7001.
132. Kitagawa, T., Azuma, K., Koh, M., Yamauchi, A., Kagawa, M., Sakata, H., Miyawaki, H., Nakazono, A., Arima, H., Yamagata, M., and Ishikawa, M. Application of fluorine-containing solvents to $LiCoO_2$ cathode in high voltage operation. *Electrochemistry* 2010; 78:345–348.
133. Lee, K.S., Sun, Y.K., Noh, J., Song, K.S., and Kim, D.W. Improvement of high voltage cycling performance and thermal stability of lithium-ion cells by use of a thiophene additive. *Electrochemistry Communications* 2009; 11:1900–1903.
134. Sakaebe, H., and Matsumoto, H. *N*-methyl-*N*-propylpiperidinium bis(trifluoromethanesulfonyl)imide (PP13-TFSI)—Novel electrolyte base for Li battery. *Electrochemistry Communications* 2003; 5:594–598.
135. Xu, K., and Angell, C.A. Sulfone-based electrolytes for lithium-ion batteries. *Journal of the Electrochemical Society* 2002; 149:A920–A926.
136. Ue, M., Takeda, M., Takehara, M., and Mori, S. Electrochemical properties of quaternary ammonium salts for electrochemical capacitors. *Journal of the Electrochemical Society* 1997; 144:2684–2688.
137. Zhu, Y., Casselman, M.D., Li, Y., Wei, A., and Abraham, D.P. Perfluoroalkyl-substituted ethylene carbonates: Novel electrolyte additives for high-voltage lithium-ion batteries. *Journal of Power Sources* 2014; 246:184–191.
138. Zhu, Y., Li, Y., Bettge, M., and Abraham, D.P. Electrolyte additive combinations that enhance performance of high-capacity $Li_{1.2}Ni_{0.15}Mn_{0.55}Co_{0.1}O_2$-graphite cells. *Electrochimica Acta* 2013; 110:191–199.
139. Li, Z.D., Zhang, Y.C., Xiang, H.F., Ma, X.H., Yuan, Q.F., Wang, Q.S., and Chen, C.H. Trimethyl phosphate as an electrolyte additive for high-voltage lithium-ion batteries using lithium-rich layered oxide cathode. *Journal of Power Sources* 2013; 240:471–475.
140. Jeong, H.S., Choi, E.S., Lee, S.Y., and Kim, J.H. Evaporation-induced, close-packed silica nanoparticle-embedded nonwoven composite separator membranes for high-voltage/high-rate lithium-ion batteries: Advantageous effect of highly percolated, electrolyte-philic microporous architecture. *Journal of Membrane Science* 2012; 415–416:513–519.
141. Abouimrane, A., Odom, S.A., Tavassol, H., Schulmerich, M.V., Wu, H., Bhargava, R., Gewirth, A.A., Moore, J.S., and Amine, K. 3-Hexylthiophene as a stabilizing additive for high voltage cathodes in lithium-ion batteries. *Journal of the Electrochemical Society* 2013; 160:A268–A271.
142. Fridman, K., Sharabi, R., Elazari, R., Gershinsky, G., Markevich, E., Salitra, G., Aurbach, D., Garsuch, A., and Lampert, J. A new advanced lithium ion battery: Combination of high performance amorphous columnar silicon thin film anode, 5 V $LiNi_{0.5}Mn_{1.5}O_4$ spinel cathode and fluoroethylene carbonate-based electrolyte solution. *Electrochemistry Communications* 2013; 33:31–34.
143. Lee, S., and Oh, E.S. Performance enhancement of a lithium ion battery by incorporation of a graphene/polyvinylidene fluoride conductive adhesive layer between the current collector and the active material layer. *Journal of Power Sources* 2013; 244:721–725.
144. Nadimpalli, S.P.V., Sethuraman, V.A., Dalavi, S., Lucht, B., Chon, M.J., Shenoy, V.B., and Guduru, P.R. Quantifying capacity loss due to solid-electrolyte-interphase layer formation on silicon negative electrodes in lithium-ion batteries. *Journal of Power Sources* 2012; 215:145–151.
145. Ulldemolins, M., Cras, F.L., Pecquenard, B., Phan, V.P., Martin, L., and Martinez, H. Investigation on the part played by the solid electrolyte interphase on the electrochemical performances of the silicon electrode for lithium-ion batteries. *Journal of Power Sources* 2012; 206:245–252.
146. Han, G.B., Lee, J.N., Choi, J.W., and Park, J.K. Tris(pentafluorophenyl) borane as an electrolyte additive for high performance silicon thin film electrodes in lithium ion batteries. *Electrochimica Acta* 2011; 56:8997–9003.
147. Chan, C.K., Ruffo, R., Hong, S.S., and Cui, Y. Surface chemistry and morphology of the solid electrolyte interphase on silicon nanowire lithium-ion battery anodes. *Journal of Power Sources* 2009; 189:1132–1140.
148. Li, M.Q., Qu, M.Z., He, X.Y., and Yu, Z.L. Effects of electrolytes on the electrochemical performance of Si/graphite/disordered carbon composite anode for lithium-ion batteries. *Electrochimica Acta* 2009; 54:4506–4513.
149. Kobayashi, Y., Seki, S., Mita, Y., Ohno, Y., Miyashiro, H., Charest, P., Guerfi, A., and Zaghib, K. High reversible capacities of graphite and SiO/graphite with solvent-free solid polymer electrolyte for lithium-ion batteries. *Journal of Power Sources* 2008; 185:542–548.
150. Inose, T., Watanabe, D., Morimoto, H., and Tobishima, S. Influence of glyme-based nonaqueous electrolyte solutions on electrochemical properties of Si-based anodes for rechargeable lithium cells. *Journal of Power Sources* 2006; 162:1297–1303.
151. Tobishima, S., Morimoto, H., Aoki, M., Saito, Y., Inose, T., Fukumoto, T., and Kuryu, T. Glyme-based nonaqueous electrolytes for rechargeable lithium cells. *Electrochimica Acta* 2004; 49:979–987.

6 Application of Lithium-Ion Batteries in Vehicle Electrification

Qiangfeng Xiao, Bing Li, Fang Dai, Li Yang, and Mei Cai

CONTENTS

6.1 INTRODUCTION

In the 1980s, Professor John B. Goodenough at the University of Texas, Austin, developed crucial cathode materials for the rechargeable lithium-ion battery (LIB) [1]. However, the first commercial LIB was released by Sony and Asahi Kasei in 1991 [2]. In the most common configuration, an LIB is formed by a graphite anode, a lithium metal oxide (e.g., $LiCoO_2$) cathode, and a separator soaked with a liquid solution of a lithium salt. As compared with conventional lead–acid and nickel–metal hydride (Ni–MH) batteries, lithium-ion batteries are lighter, have higher operational voltage, and have a higher energy density—ranging from 100 to 265 Wh/kg. Such features make them the power sources of choice for portable electronic devices such as cell phones, laptop computers, digital cameras/video cameras, and portable audio/game players. In 2011, laptops and cellphones dominated battery markets and accounted for $9.7 billion in revenue [3]. Recently, clean energy and sustainable development have promoted the application of lithium-ion batteries in both stationary electrical energy storage, to improve grid reliability and utilization, and in transportation electrification, to reduce greenhouse gas emissions and decrease dependence on oil. The Boston Consulting Group has reported that, by 2020, the global market for advanced batteries for electric vehicles (EVs) is expected to reach US $25 billion [4], which is more than double the size of today's entire LIB market for consumer electronics.

In this chapter, the focus will be on the application of lithium-ion batteries in EVs, future market perspectives, and their great impact on energy consumption and emission reductions. The value chain of EV batteries comprises seven steps: raw material and component production, cell production, module production, assembly of modules into the battery pack including the battery management system and the thermal management system, integration of the pack into the vehicle, operation during the life of vehicle, and reuse and recycling [5]. The following sections will primarily relate to the first four battery pack manufacture steps, and will include the history of EVs, EV battery requirements, current battery technology, and a prospectus.

6.2 EV HISTORY

The history of the EVs can be tracked back parallel to the history of the internal combustion engine (ICE) vehicles. In the late 1800s, the invention of rechargeable lead–acid batteries triggered the birth of the first EV [6,7]. The Bersey Cab used by the London Electric Cab Company could drive 50 miles per charge. The early years of the 1900s witnessed the golden period of EVs, whose number was almost double that of ICE vehicles. However, Henry Ford's assembly line brought lightweight and low-priced ICE vehicles onto the market, and those vehicles quickly supplanted EVs because the batteries at that time had low energy density, long charging time, and poor durability. From 1930 to 1960, battery-powered EVs went into hibernation.

The energy crisis in the 1970s spawned resurgent interest in the development of high-density, low-cost batteries for EVs to help reduce dependence on foreign oil. In this period, the best available lead–acid EV technology yielded a range of less than 50 miles. For example, the Sebring-Vanguard CitiCar, introduced to the EV market in 1974, had a range of approximately 40 miles. General Motors (GM) undertook a program to develop Zn/NiOOH batteries for electric automobiles and pickup trucks. The Electrovette had an urban range of 60 miles, and could accelerate like gasoline-powered vehicles.

In the 1990s, California established the zero-emission vehicle (ZEV) mandate, requiring large manufacturers to sell ZEVs as share of all vehicles. GM initially mass produced the EV1 followed by the S-10 electric pickup, Chrysler EPIC minivan, Ford Ranger pickup, Honda EV Plus, Nissan Alta EV, and Toyota RAV4 [8]. Recently, the idea of vehicle electrification has spread worldwide. Various governments are funding research to develop advanced battery technology and providing financial incentives for consumer purchases. By 2014, GM, Ford, Audi, Fiat, BMW, Volkswagen, Honda, Toyota, Mitsubishi, Tesla, BYD, and others have launched various EVs into the market.

6.3 EV CLASSIFICATION AND GENERAL BATTERY REQUIREMENTS

The rechargeable battery is as essential to the operation of an EV as the ICE is to the operation of a traditional vehicle. Power density and energy density are the two fundamental characteristics. The former determines the amount of energy that can be delivered in a given period of time, affecting how fast a vehicle accelerates, while the latter determines how much total energy is available, affecting the range a vehicle can travel. As compared with lead–acid and nickel–hydride batteries, lithium-ion batteries are the most promising power source in view of their superior power and energy density. As shown in Figure 6.1, a 300-kg LIB with an energy density of 120 Wh/kg can drive a 250-km range, while a lead–acid or Ni–MH battery with the same weight can only offer 63 km or 130 km, respectively [9]. Besides power and energy density, battery performance is characterized by a number of other properties such as cell voltage, C-rate, state of charge (SOC), depth of discharge (DOD), charge efficiency, self-discharge rate, operation temperature range, cycle life, and other characteristics. Battery performance requirements are determined by the specific vehicle application. Depending on their propulsion power and energy storage systems, EVs can be generally categorized as hybrid EVs (HEVs), plug-in hybrid EVs (PHEVs), and pure battery EVs (BEVs) [10,11]. They are introduced in the following sections along with their specific battery requirements.

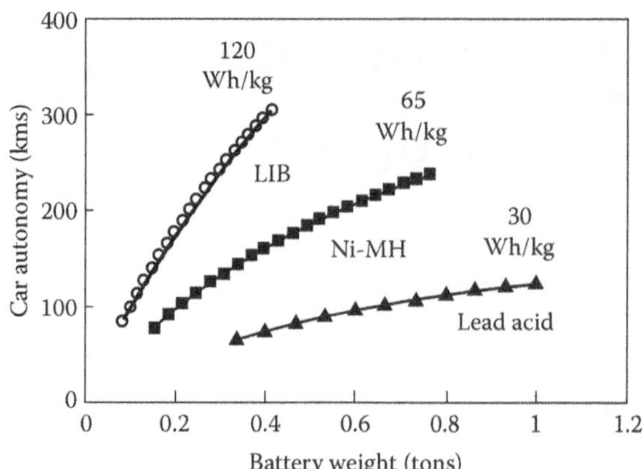

FIGURE 6.1 Car autonomy (vehicle range) as a function of battery weight and specific energy. Vehicle energy consumption: 135 Wh/(t km), vehicle weight: 800 kg. (From Broussely, M., Planchat, J.P., Rigobert, G., Virey, D., Sarre, G., *J. Power Sources* 68, 8, 1997.)

An HEV combines a conventional ICE propulsion system with an electric propulsion system. As shown in Figure 6.2, there are two basic drivetrain architectures. In the parallel hybrid, the ICE and the electric motor are both connected to the mechanical transmission and can simultaneously power the vehicle. In the series hybrid, the ICE charges the batteries or powers the motor directly via a generator and only an electrical motor powers the vehicle [12]. Depending on the degree of hybridization, an HEV can be further subclassified from micro- to full HEV. The battery in an HEV is only required to have a small amount of energy (1–2 kWh) but must deliver high power; these batteries have a high power/energy (P/E)

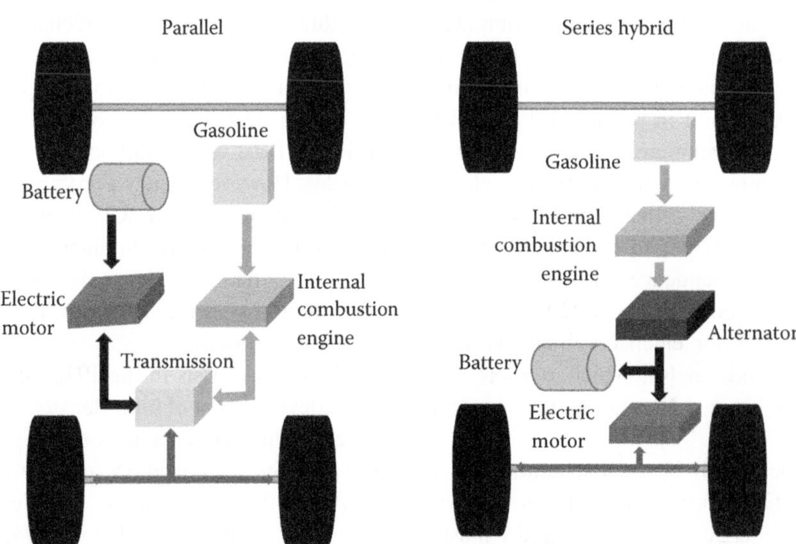

FIGURE 6.2 Basic HEV drivetrain architecture: series versus parallel design.

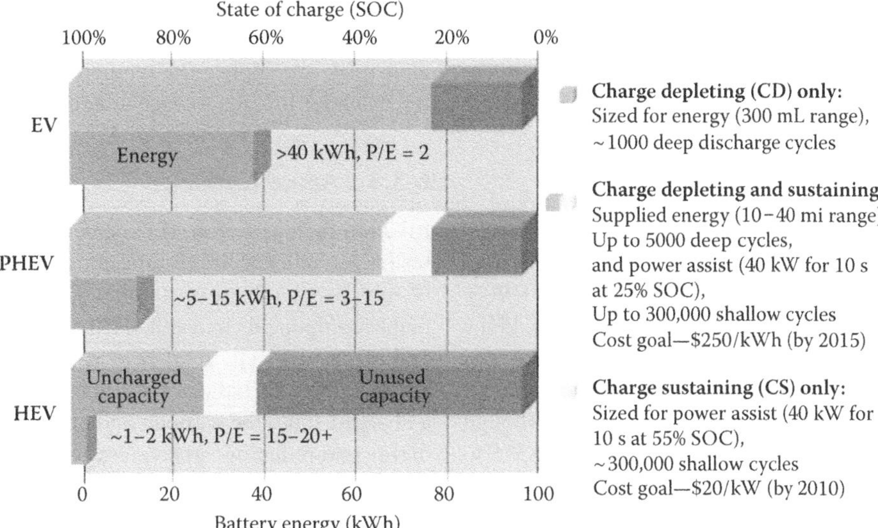

FIGURE 6.3 Battery performance requirement by vehicle application.

value ranging from 15 to 20 h^{-1}, as shown in Figure 6.3 [13]. In addition, the SOC of these batteries must be maintained at an intermediate level in order to frequently either deliver peak power to the drivetrain or accept high power from regenerative braking. Batteries for HEV are designed for a 300,000 shallow cycles over their lifetime.

PHEVs refer to hybrid vehicles with larger-capacity batteries than ordinary HEVs that can also be charged from the electric grid. PHEVs can be viewed as an intermediate technology between HEVs and BEVs. To reap the advantage of PHEVs, two basic modes are combined over the travelled distance, as shown in Figure 6.4 [14]. In the charge-depleting (CD) operating mode, the vehicle is powered purely or mostly by the battery that discharges from a fully charged state to a minimum level. In this mode, the operation can be an all-electric (AE) or blended operation with both battery and engine providing power. In AE mode, the vehicle behaves as a BEV. Once the battery is depleted, the vehicle switches to charge-sustaining (CS) mode. In CS mode, the vehicle is primary powered by ICE. As is the case in HEVs, the battery in CS mode provides power assist, and can be charged by either the ICE or regenerative braking. Owing to these two operation modes, PHEV batteries are required to have both energy and power performance, resulting in a medium P/E range of 3–15 h^{-1}. Generally, PHEV batteries will have a usable energy ranging from 5 to 15 kWh [13], depending on the distance that the vehicle can drive in the CD mode; these batteries can achieve about 5000 cycles over their lifetime.

BEVs use an electric motor powered by batteries as the vehicle's sole propulsion. Batteries for BEVs require very high energy because of the longer AE driving range, resulting in the lowest P/E ratio of any electrified vehicle. These batteries generally require a 1000-cycle durability. The battery size of EVs is larger than that for PHEVs or HEVs. For example, Tesla's Model S has a battery pack with energy 85 kWh, which contains 7104 LIB cells and provides a 265-mile range [15].

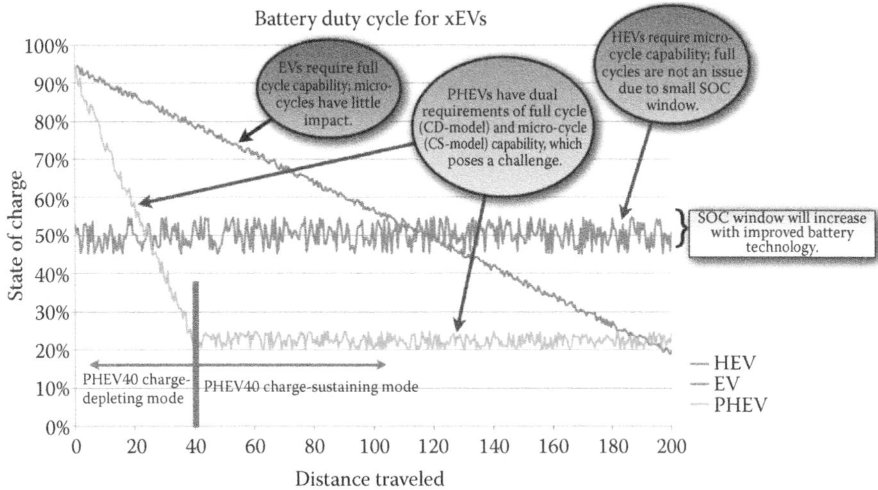

FIGURE 6.4 Illustration of discharge cycle for HEVs, PHEVs, and BEVs.

Besides the general power and energy ranges required from the battery by HEVs, PHEVs, and BEVs, other factors such as cost, life, and abuse tolerance also play important roles in the development of EVs.

6.3.1 Cost

A rough estimate of the battery cost at pack level is 75–85% for the cells and 15–25% for the battery assembly, thermal, and electrical management system. At the cell level, 80–90% is the material cost and 10–20% is the labor cost [16]. If the cost is further broken down, the cathode-active material accounts for ~49%, the electrolyte for ~23%, and the anode-active material for ~11% of the total cell cost [17]. Although data are confidential, batteries cost reasonably from $375 to $750 per kWh, meaning a 40-kWh battery can cost as much as $30,000. As LIBs are suitable power sources for both BEVs (requiring high energy) and HEVs (requiring high power), they have a potential cost evolution advantage in the future owing to accumulative production volumes.

6.3.2 Life

The battery is the most critical component of an EV. During the average lifetime of a car, the battery should be able to provide energy to the vehicle powertrain continuously. Inside the battery, many complex chemical reactions occur during charge or discharge, and even at rest. Therefore, both cycle life and calendar life should be considered to assess the battery lifespan [18]. The former refers to the number of discharge–charge cycles the battery can handle at a specific DOD (which is determined by the application) before it reaches its end of life [19]; that is, loses 20% of energy content according to the automotive industry. In reality, cycle life depends on the operation conditions, including the charging and discharging rates, DOD, and other conditions such as temperature. Calendar life is defined as the expected life span of the battery under storage or periodic mild cycling conditions. It can be strongly related to the temperature and SOC during storage. Generally, the higher the temperature and SOC, the shorter the calendar life due to accelerated side reactions.

6.3.3 Abuse Tolerance

Like the fuel tank of an IEC vehicle, the battery systems represent a challenge in certain abusive environments because of the energy release in failures [20]. There are many flammable components in a cell, such as the electrolyte, separator, and electrodes. The total heat generated from both cathode and anode is an indicator for abuse tolerance evaluation. In a charged state, the cathode and anode can be viewed as oxidizing and reducing agents, respectively, which might chemically react together or react with the electrolyte under certain circumstances (e.g., such as internal short circuits, external short circuits, or battery overcharge). Such reactions generate hotspots and trigger thermal runaway in the worst scenario. These abuse tolerance concerns play a role in determining both the choice of materials, and battery design.

In brief, battery performance requirements depend on the type of EV: HEV, PHEV, or BEV. Energy and power are the two primary factors and determine the car's autonomy. Derived from the energy density and specific energy, the battery volume and mass can be obtained. Allowing for the limited space in the car, volume is the prime consideration in pack design. Battery cost and life are key economic factors and play important roles in the decision to enter mass production. Abuse tolerance is also a big concern in the implementation of EVs. The Department of Energy (DOE) assessed all the necessary battery attributes for an EV to be competitive with current vehicles, and set the system-level battery performance goals shown in Table 6.1 [21]. Current battery systems fall short of the DOE goals principally in life and system price. Their detailed status will be discussed in the following sections.

TABLE 6.1
DOE Goals for System-Level Performance for HEV, PHEV, and EV

Performance Criteria	HEV-Min	HEV-Max	PHEV-Min	PHEV-Max	EV-Short Term
Specific power, W/kg	625	667	750	316	300
Specific energy, Wh/kg	7.5	8.3	57	97	150
Power density, W/L	782	889	1125	475	460
Energy density, Wh/L	9	11	90	145	230
Pack energy, kWh	0.3	0.5	3.4	11.6	40
Pack weight, kg	40	60	60	120	267
Pack volume, L	32	45	40	80	174
Life, years	15	15	15	15	10
Cycle life	300,000	300,000	5000	5000	1000
System price, $	500	800	1700	3400	6000

Note: The min and max categories refer to designs that meet the basic HRV or PHEV application but have a minimum or maximum amount of battery.

6.4 BATTERY CHEMISTRY

There are four principal LIB cathode materials used today for automotive applications, namely lithium nickel cobalt aluminum (NCA), lithium manganese spinel (LMO), lithium nickel manganese cobalt (NMC), and lithium iron phosphate (LFP). As for the anode, graphite-based carbons are the dominant active materials. When high power is needed, lithium titanate (LTO) is used as the anode material. On the basis of combinations of these categories of cathode and anode materials, various lithium-ion batteries can be fabricated. Each material has distinct advantages and disadvantages in terms of specific energy, specific power, abuse tolerance, cost, life span, and performance.

The layered lithium transition metal oxides represent the most successful family of positive electrode materials since the commercialization of LIBs. $LiCoO_2$ was the first commercialized compound and dominates consumer electronics; however, the high cost of these cells and concerns about abuse tolerance issues have led to the development of alternative materials that can offer lower cost, a longer life, and improved abuse tolerance. Substituted nickel oxides, such as $LiNi_{1-y-z}Co_yAl_zO_2$, or NCA, have been developed [22–24]. The partial substitution of Ni with Co is effective at reducing the Ni migration into the Li layer. Co also helps reduce oxygen loss at high SOC, and improves the abuse tolerance. The presence of electrochemically inert Al prevents the complete removal of all the lithium and minimizes structural collapse. On charging, Ni is oxidized first to Ni^{4+}, then the Co oxidizes to Co^{4+}. The high specific capacity (180–200 mAh/g) and good power capability make this material attractive for vehicular applications. Another well-known compound is $LiNi_{1-y-z}Mn_yCo_zO_2$ (NMC), which integrates the parent compounds' advantages [25–27]. Among the myriad stoichiometries possible, the symmetric compound $Li[Ni_{1/3}Co_{1/3}Mn_{1/3}]O_2$ is the best known. In this material Ni, Co, and Mn adopt the oxidation states of +2, +3, and +4, respectively. During removal of the first two-third of the total capacity, Ni^{2+} is oxidized to Ni^{4+}, followed by Co^{3+} to Co^{4+} during the extraction of the last one-third of the capacity at high potential. Mn is the inactive element and maintains its oxidation state during the electrochemical process. Li $[Ni_{1/3}Co_{1/3}Mn_{1/3}]$ O_2 exhibits high capacity (160–170 mAh/g), small volume change (1–2%), and good rate performance. As compared with $LiCoO_2$ and NCA, this material shows improved abuse tolerance properties at a high SOC.

The spinel cathode $LiMn_2O_4$ (LMO), originally proposed by Thackeray et al., has been extensively developed by Bellcore Labs [28–30]. This material exhibits a cubic structure (space group Fd-3m), where there are Li-ion tunnels intersecting three-dimensionally through the manganese oxide skeleton. In theory, LMO can be either charged or discharged initially and cycled over the range of $0 < x < 2$ in $Li_xMn_2O_4$. For $0 < x < 1$, the crystal structure is maintained as the cubic phase with a 4-V plateau. For $1 < x < 2$, a 3-V plateau occurs and the structure transfers into a tetragonal phase due to the cooperative Jahn–Teller effect of Mn^{3+} when it exceeds 50% occupancy. For practical applications, the discharge is limited to the 4-V plateau with the capacity of 100–120 mAh/g. As compared with other cathode materials, LMO is safe, abundant, environmentally benign, low cost, and has high power capability. Such properties make LMO an attractive cathode material for large-scale LIB applications. The shortcomings of LMO include the low capacity and Mn dissolution. Strategies including doping, surface coating, and utilization of nonacidic electrolyte (e.g., LiBOB) have been used to mitigate the Mn dissolution and capacity fading.

$LiFePO_4$ has attracted great interest due to its unique properties, which include (i) a reasonably high capacity (170 mAh/g), (ii) a flat 3.5 V versus Li voltage plateau generated by a two-phase electrochemical process, (iii) low reactivity with the electrolyte, (iv) high intrinsic abuse tolerance due to the strength of the P–O covalent bond, and (v) potentially low cost due to the abundant resources [31–33]. This material was first reported by Goodenough and coworkers; it is known as phosphor-olivine and adopts an orthorhombic structure (space group Pnma). The one-dimensional Li^+ diffusion channel of the olivine limits the mobility of Li ions. The resulting low ion diffusion rate and the intrinsic low conductivity have been a hurdle to its commercialization. In the early 2000s, these hurdles were significantly lowered by A123 Systems Inc., when they made the material in a nanoparticulate form, coating the surface with a carbon layer, and doping the material with various elements. As for the disadvantages of LFP, the low tap density of nanostructured LFP leads to a lower energy density, compared with other cathode materials, and the voltage is not as high as several electrode materials being developed today.

Graphite is the most common anode used in lithium-ion batteries. It can react with lithium to form LiC_6 corresponding to a theoretical specific capacity of 372 mAh/g and can pair with most of the current cathode materials [34–36]. The potential of graphitic anodes is about 100 mV versus lithium potential. During the lithiation–delithiation, the volume expansion/contraction is about 9%, which accounts for the good cyclability of graphite as the anode material. In addition, it is abundant and low cost. All the above properties make graphite the dominant anode. However, to solve the solid electrolyte interphase (SEI) formation and lithium plating issues associated with graphite electrodes, it is necessary to develop alternative anodes that operate at higher voltage with respect to lithium potential. For example, LTO is one of the attractive candidates due to its unique properties, including (i) a flat 1.5 V versus Li voltage plateau by a two-phase electrochemical process; (ii) zero strain during cycling, which leads to high cycling stability; (iii) no SEI formation; and (iv) high rate and very low temperature charge–discharge capability [37–39]. As for the disadvantages, its relatively low capacity and high operation voltage result in lower energy density. The characteristics of the major cathodes and anodes are summarized in Table 6.2.

TABLE 6.2
Characteristics of Current Electrode Materials Employed in EVs

Electrode	Potential versus Li/Li+ (V)	Specific Capacity (mAh/g)	Advantages	Disadvantages
			Cathodes	
$LiNi_{0.8}Co_{0.15}Al_{0.05}O_2$	3.8	180–200	High capacity and voltage, excellent rate performance	Abuse tolerance, cost and resource limitations of Ni and Co
$LiNi_{1/3}Mn_{1/3}Co_{1/3}O_2$	3.8	160–170	High capacity and voltage, moderate abuse tolerance	Cost and resource limitations of Ni and Co
$LiMn_2O_4$ variants	4.1	100–120	High voltage, moderate abuse tolerance, excellent rate performance, low cost and abundance of Mn	Poor cycle life due to Mn dissolution, low capacity
$LiFePO_4$	3.45	150–170	Excellent abuse tolerance, cycling, and rate capability, low cost and abundance of Fe	Low voltage, capacity, and energy density
			Anodes	
Graphite	0.1	372	Long cycle life, abundant, low cost	Relatively low energy density, plating and solid electrolyte interface formation
$Li_4Ti_5O_{12}$	1.5	175	Zero strain, good cycling, high efficiencies	High voltage, low capacity and energy density

6.5 FROM CELLS TO MODULES TO BATTERY PACKS

Just like small portable lithium-ion cells, EV cells are built from three primary layers, namely, cathode, separator, and anode. Depending on their packing structure, the cells can be classified into cylindrical, prismatic, or pouch configuration [40,41]. In the case of cylindrical or prismatic cells, the three primary layers are rolled and sealed in metal cans with the electrolyte. In the pouch cell case, these layers are enclosed in the laminate pouch with their edge heat-sealed. Each configuration has its advantages and disadvantages. Cylindrical cells are economical to manufacture and have good mechanical stability and high energy density, but have low packing efficiency. As compared with the cylindrical cells, prismatic cells have higher packing efficiency, slightly lower energy density, and are more expensive to manufacture. Pouch cells are light and cost-effective to manufacture and provide design freedom on dimensions. The pouch cells have become the first choice for high-capacity PHEV or EV packs. It is worth

TABLE 6.3
State-of-the-Art Lithium-Ion Cell Employed in EV

	Cell Maker	Chemistry (Anode/Cathode)	Capacity Ah	Configuration	Voltage (V)	Weight (kg)	Volume (L)	Energy Density (Wh/L)	Specific Energy (Wh/kg)	Used in Company	Used in Model
1	AESC	G/LMO-NCA	33	Pouch	3.75	0.80	0.40	309	155	Nissan	Leaf
2	LG Chem	G/LMO-NMC	15	Pouch	3.8	0.38	0.19	300	150	GM	Volt
3	Li-Tec	G/NMC	52	Pouch	3.65	1.25	0.60	316	152	Daimler	Smart
4	Li Energy Japan	G/LMO-NMC	50	Prismatic	3.7	1.70	0.85	218	109	Mitsubishi	i-MiEV
5	Samsung	G/NMC–LMO	64	Prismatic	3.7	1.80	0.97	243	132	Fiat	500
6	Lishen Tianjin	G/LFP	16	Prismatic	3.25	0.45	0.23	226	116	Coda	EV
7	Toshiba	LTO/NMC	20	Prismatic	2.3	0.52	0.23	200	89	Honda, Mitsubishi[a]	Fit, i-MiEV[a]
8	Panasonic	G/NCA	3.1	Cylindrical	3.6	0.045	0.018	630	248	Tesla	Model S

[a] Toshiba Super Charge ion Battery (SCiB) technology was introduced into the i-MiEV by Mitsubishi in 2011.

mentioning that for cylindrical and prismatic cells, the maximum cell temperature is in the core of the cell. In the case of pouch cells, the highest temperature is near the cell terminal tabs, which makes it easier to dissipate heat. As for disadvantages, pouch cells have a swelling issue during cycling. In Table 6.3, the key characteristics of cells currently used in EVs are shown [42]. At the cell level, the specific energy and energy density of current lithium-ion batteries usually ranges from 90 to 160 Wh/kg and 200 to 320 Wh/L, respectively. The Toshiba cells have the lowest specific energy and energy density in the table due to the low capacity and high voltage plateau of the LTO anode. Using an 18650 cylindrical design, Panasonic cells deliver the highest specific energy (248 Wh/kg) and energy density (630 Wh/L) in the table. Such differences arise from the design comprising the abuse tolerance, reliability, durability, P/E ratio, and cost for a particular application. In an EV, the cells must be connected in series and parallel to form modules, and those modules are joined to make battery packs that can meet the vehicle's specific energy and power requirements. Typically, the packs include four main components: (i) LIB cells, (ii) the mechanical structure and/or modules, (iii) a battery management system, and (iv) a thermal management system. Generally, the specific energy of a battery pack is only 60–65% of the cell's specific energy.

6.6 CURRENT BATTERY TECHNOLOGY

As lithium-ion batteries can fulfill the HEV battery performance requirements, we will focus on the applications of current lithium-ion batteries in PHEVs and BEVs with GM's Chevrolet Volt, Toyota's Prius, and Nissan's Leaf as examples.

The GM Chevrolet Volt was launched in 2010 and has become the bestselling PHEV in the world. As of June 2014, the global sales of Volt and its variant Ampera models have surpassed 77,000 units [42]. The 2011 Chevrolet Volt has a combination of a 1.4-L ICE and a 16-kWh (10.4 kWh usable) LIB pack [43–45]. The electric motor has a peak output of 111 kW (149 hp) delivering 273 lb·ft (370 N·m) of torque [43]. The battery system comprises 288 lithium-ion pouch cells, each with a 15 Ah capacity and a 3.8 V nominal voltage. Three Li-ion cells are connected in parallel to create cell groups of 45 Ah. A total of 96 cell groups are connected in series with a nominal system voltage is of 360 V [46]. The cell groups are integrated in nine modules assembled into three sections to form a "T"-shaped design. The pouch cells were developed and are manufactured by LG Chem with NMC-based materials as the cathode and graphite as the anode. The battery pack can be charged by either the engine or an outlet (14 h at 120 V and about 4 h at 240 V) [43]. The Volt operates as a pure battery EV with a range of 38 miles until the battery capacity drops to a predetermined threshold from full charge. From there, its ICE powers an electric generator to extend the vehicle's range as far as 379 miles [43].

The Toyota Prius Plug-in Hybrid was unveiled in September 2011 and is the second best-selling PHEV. As of June 2014, about 60,000 Prius PHVs have been sold worldwide [47]. The hybrid powertrain system consists of a gasoline engine and a pair of electric motors. The 1.8-L DOHC 16-valve VVT-i gasoline engine can deliver a power of 73 kW (98 hp) at 5200 rpm, and a torque of 105 lb·ft (142 N·m) at 4000 rpm. One motor has power of 60 kW (80 hp) and works to power the compact, lightweight transaxle, and the other one has power of 42 kW (56 hp) and works as the electric power source for battery regeneration and as a starter for the gasoline engine. The battery pack is 4.4 kWh LIB codeveloped with Panasonic [47–49]. The pack is installed in the rear of the vehicle with a nominal voltage of 209 V and weight of 80 kg. The LIB pack can be charged in 180 min at 120 V or in 90 min at 240 V. During the charge-depleting stage, the vehicle can drive 11 miles (18 km) in a blended operation mode. Afterward, the drive mode switches to the CS mode and operates similarly to a standard Prius. A total range of 540 miles (870 km) can be reached after both fully charged battery and gasoline tank are depleted [47].

Nissan's Leaf is a pure BEV released in the United States and Japan in December 2010. By mid-January 2014, global sales had reached up to 100,000 units, which accounted for a 45% market share of worldwide pure EVs [50]. The drivetrain system is a front-mounted synchronous electric motor that can provide a power of 80 kW (110 hp) and a torque of 280 N·m (210 ft·lb) to propel the front axle. The battery pack built by Automotive Energy Supply Corporation (AESC) is 24 kWh LIB, has nominal voltage of 345 V, and weighs 294 kg. The pack contains 48 modules, and each module has four laminated prismatic cells that use LMO–NCA as the cathode and graphite as the anode [42,50–52]. The specific energy of the cells is 155 Wh/kg. As the heaviest part, the pack is positioned below the seats and rear foot space to maintain the center of gravity as low as possible. Depending on driving style, load, traffic conditions, weather (i.e., wind and atmospheric density), and accessory use, the driving range can vary from 62 to 138 miles [50].

6.7 PROSPECTUS ON THE FUTURE OF EV BATTERY

Current lithium-ion batteries, through two decades of optimizing their intercalation chemistry, are approaching their energy density limits. They have enjoyed great commercial success in portable electronic devices and are now being employed for vehicle electrification. In the coming years, further optimization of existing chemistries and cell and pack designs will dominate the development of EV batteries, providing ongoing, incremental, energy density improvement. A big leap forward will be the realization of Si-based anode materials as well as high-voltage, high-capacity cathode materials for the next-generation battery. To further extend driving range, improve reliability, and reduce cost, new chemistries beyond LIBs (such as conversion reactions) will be required to make batteries with higher energy density (Figure 6.5) [53]. As an example, the Li–S battery is one of the most promising candidates for the development of a next-generation battery, in view

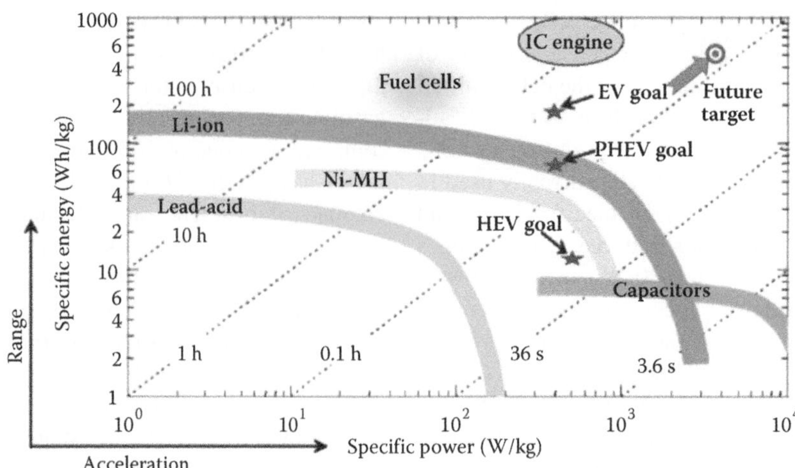

FIGURE 6.5 Ragone plot of current energy storage systems, and future targets for vehicle propulsion.

of its high capacity, low cost, and small environmental footprint. Although Li–S batteries are plagued by issues such as polysulfide shuttling and the low conductivity of S and Li_2S, great progress has been made in recent years owing to the advancement of nanotechnology [54–56]. On the basis of the current technology, a specific energy of 350 Wh/kg, which doubles that of LIB, has been demonstrated in large pouch cells. The authors believe that further improvement can be achieved by optimizing the designs of electrodes/cells, and utilizing newly developed novel materials. If the development of Li–S cells is successful, a low-cost all-EV with a range in excess of 300 miles per charge could be anticipated. As for the other high-energy batteries such as lithium–air battery, there are still tremendous challenges to be addressed before their applications in vehicle electrification could be realized [57,58].

REFERENCES

1. Mizushima, K., Jones, P.C., Wiseman, P.J., and Goodenough, J.B. 1980. Li_xCoO_2 ($0 < x < -1$): A new cathode material for batteries of high energy density. *Mater. Res. Bull.* 15(6): 783–799.
2. Imanishi, N., Takeda, Y., and Yamamoto, O. 1998. Development of the carbon anode in lithium ion batteries. In *Lithium Ion Batteries Fundamentals and Performance*, eds. Wakihara, O., and Yamamoto, M. Wiley-VCH, Tokyo.
3. Pillot, C. 2012. The worldwide battery market 2011-2025. Available at http://www.academia.edu/5823687/The_worldwide_battery_market_2011-2025_Director_AVICENNE_ENERGY (accessed June 16, 2014).
4. Xing, Y.J., Ma, E.W.M., Tsui, K.L., and Pecht, M. 2011. Battery management systems in electric and hybrid vehicles. *Energies* 4: 1840–1857.
5. Dinger, A., Martin, R., Mosquet, X., Rizoulis, D., Russo, M., and Sticher, G. 2010. Batteries for electric cars challenges, opportunities, and the outlook to 2020. Boston Consulting Group, Boston. Available at http://www.bcg.com/documents/file36615.pdf (accessed June 16, 2014).
6. Corrigan, D., and Masias, A. 2011. Batteries for electric and hybrid vehicles. In *Linden's Handbook of Batteries* (4th Ed.), ed. Reddy, T.B. McGraw Hill, New York.
7. Electric Auto Association. 2014. EV history. Available at http://www.electricauto.org/?page=evhistory (accessed June 16, 2014).
8. Idaho National Laboratory. 2014. The history of electric cars. Available at http://avt.inel.gov/pdf/fsev/history.pdf (accessed August 16, 2014).
9. Broussely, M., Planchat, J.P., Rigobert, G., Virey, D., and Sarre, G. 1997. Lithium-ion batteries for electric vehicles: Performance of 100 Ah cells. *J. Power Sources* 68: 8–12.
10. Howell, D. 2011. Annual progress report for energy storage R&D. U.S. Department of Energy, Washington, DC. Available at http://energy.gov/sites/prod/files/2014/03/f8/2010_energy_storage.pdf (accessed August 16, 2014).
11. Boulanger, A.G., Chu, A.C., Maxx, S., and Waltz, D.L. 2011. Vehicle electrification: Status and issues. *Proc. IEEE* 99(6): 1116–1138.
12. Ehsani, M., Gao, Y., and Miller, J.M. 2007. Hybrid electric vehicles: Architecture and motor drives. *Proc. IEEE* 95(4): 719–728.
13. DOE. 2007. Plug-in hybrid electric vehicle R&D plan. Available at http://www1.eere.energy.gov/vehiclesandfuels/pdfs/program/phev_rd_plan_02-28-07.pdf.
14. Anderson, D. 2008. *Status and Trends in the HEV/PHEV/EV Battery Industry*. Rocky Mountain Institute, Snowmass, CO, 2008.
15. Wikipedia. 2015. Tesla Model S. http://en.wikipedia.org/wiki/Tesla_Model_S (accessed January 16, 2015).
16. Broussely, M. 2010. Battery requirements for HEVs, PHEVs and EVs. In *Electric and Hybrid Vehicles*, ed. Pistoia, G. Elsevier, Oxford.
17. Gaines, L., and Cuenca, R. 2000. Costs of lithium-ion batteries for vehicles. Rep. ANL/ESD-42, Argonne Natl. Lab., Argonne, IL. Available at http://www.transportation.anl.gov/pdfs/TA/149.pdf (accessed June 16, 2014).
18. SAFT. 2014. Lithium ion battery life. Available at http://www.google.com/url?sa=t&rct=j&q=&esrc=s&frm=1&source=web&cd=6&ved=0CEkQFjAF&url=http%3A%2F%2Fwww.saftbatteries.com%2Fforce_download%2Fli_ion_battery_life__TechnicalSheet_en_0514_Protected.pdf&ei=IGL3U62tPNKfyAT7z4DICQ&usg=AFQjCNFYc_xZ_rYtc3em9se7VtFBxioelQ&sig2=v6zSVlP7Ff1al2sel4-YCA (accessed June 16, 2014).

19. Dhameja, S. 2002. Electric vehicle batteries. In *Electric Vehicle Battery System*, ed. Dhameja, S. Newnes, Boston.
20. Wen, J.W., Yu, Y., and Chen, C.H. 2012. A review on lithium-ion batteries safety issues: Existing problems and possible solutions. *Mater. Expr.* 2(3): 197–212.
21. Cairns, E.J., and Albertus, P. 2010. Batteries for electric and hybrid-electric vehicles. *Annu. Rev. Chem. Biomol. Eng.* 1: 299–320.
22. Albrecht, S., Kumpers, J., Kruft, M., Malcus, S., Vogler, C., Wahl, M., and Wohlfahrt-Mehrens, M. 2003. Electrochemical and thermal behavior of aluminum- and magnesium-doped spherical lithium nickel cobalt mixed oxides $Li_{1-x}(Ni_{1-y-z}Co_yM_z)O_2$ (M = Al, Mg). *J Power Sources* 119–121: 178–183.
23. Whittingham, M.S. 2004. Lithium batteries and cathode materials. *Chem. Rev.* 104: 4271–4301.
24. Chen, C.H., Liu, J., Stoll, M.E., Henriksen, G., Vissers, D.R., and Amine, K. 2004. Aluminum-doped lithium nickel cobalt oxide electrodes for high-power lithium-ion batteries. *J Power Sources* 128: 278–285.
25. Yoon, W.S., Grey, C.P., Balasubramanian, M., Yang, X.Q., Fischer, D.A., and McBreen, J. 2004. Combined NMR and XAS study on local environments and electronic structures of electrochemically Li-ion deintercalated $Li_{1-x}Co_{1/3}Ni_{1/3}Mn_{1/3}O_2$ electrode system. *Electrochem. Solid-State Lett.* 7(3): A53–A55.
26. Yoon, W.S., Balasubramanian, M., Chung, K.Y., Yang, X.Q., McBreen, J., Grey, C.P., and Fischer, D.A. 2005. Investigation of the charge compensation mechanism on the electrochemically Li-ion deintercalated $Li_{1-x}Co_{1/3}Ni_{1/3}Mn_{1/3}O_2$ electrode system by combination of soft and hard X-ray absorption spectroscopy. *J Am Chem Soc.* 127(49): 17479–17487.
27. Koyama, Y., Tanaka, I., Adachi, H., Makimura, Y., and Ohzuku, T. 2003. Crystal and electronic structures of superstructural $Li_{1-x}[Co_{1/3}Ni_{1/3}Mn_{1/3}]O_2$ ($0 \le x \le 1$). *J Power Sources* 119–121: 644–648.
28. Thackeray, M.M., Shao-Horn, Y., Kahaian, A.J., Kepler, K.D., Vaughey, J.T., and Hackney, S.A. 1998. Structural fatigue in spinel electrodes in high voltage (4 V) $Li/Li_xMn_2O_4$ cells. *Electrochem. Solid-State Lett.* 1(1): 7–9.
29. Ellis, B.L., Lee, K.T., and Nazar, L.F. 2010. Positive electrode materials for Li-ion and Li-batteries. *Chem. Mater.* 22: 691–714.
30. Tarascon, J.M., McKinnon, W.R., Coowar, F., Bowmer, T.N., Amatucci, G., and Guyomard, D. 1994. Synthesis conditions and oxygen stoichiometry effects on Li insertion into the spinel $LiMn_2O_4$. *J. Electrochem. Soc.* 141: 1421–1431.
31. Padhi, A.K., Nanjundaswamy, K.S., and Goodenough, J.B. 1997. Phospho-olivines as positive electrode materials for rechargeable lithium batteries. *J. Electrochem. Soc.* 144: 1188–1194.
32. Choi, N.S., Chen, Z.H., Freunberger, S.A. et al. 2012. Challenges facing lithium batteries and electrical double-layer capacitors. *Angew. Chem. Int. Ed.* 51: 9994–10024.
33. Chung, S.-Y., Bloking, J.T., and Chiang, Y.-M. 2002. Electronically conductive phospho-olivines as lithium storage electrodes. *Nat. Mater.* 1: 123–128.
34. Kamali, A.R., and Fray, D.J. 2010. Review on carbon and silicon based materials as anode materials for lithium ion batteries. *J. New Mat. Electrochem. Syst.* 13: 147–160.
35. Wu, Y.P., Rahm, E., and Holze, R. 2003. Carbon anode materials for lithium ion batteries. *J. Power Sources* 114: 228–236.
36. Arakawa, M., and Yamaki, J. 1995. Anodic oxidation of propylene carbonate and ethylene carbonate on graphite electrodes. *J. Power Sources* 54: 250–254.
37. Yi, T.F., Jiang, L.J., Shu, J., Yue, C.B., Zhu, R.S., and Qiao, H.B. 2010. Recent development and application of $Li_4Ti_5O_{12}$ as anode material of lithium ion battery. *J. Phys. Chem. Solids* 71: 1236–1242.
38. Ohzuku, T., Ueda, A., and Yamamoto, N. 1995. Zero-strain insertion material of Li $[Li_{1/3}Ti_{5/3}]$ O_4 for rechargeable lithium cells. *J. Electrochem. Soc.* 142: 1431–1435.
39. Ouyang, C.Y., Zhong, Z.Y., and Lei, M.S. 2007. *Ab initio* studies of structural and electronic properties of $Li_4Ti_5O_{12}$ spinel. *Electrochem. Commun.* 9: 1107–1112.
40. Pesaran, A.A., Kim, G.-H., and Keyser, M. May 13–16, 2009. Integration issues of cells into battery packs for plug-in and hybrid electric vehicles. In *EVS-24 International Battery, Hybrid and Fuel Cell Electric Vehicle Symposium*, Stavanger, Norway. Available at http://www.nrel.gov/vehiclesandfuels/energystorage/pdfs/45779.pdf (accessed June 16, 2014).
41. Guerrero, C.P.A., Li, J.S., Biller, S., and Xiao, G.X. August 21–24, 2010. Hybrid/electric vehicle battery manufacturing: The state-of-the-art. In *6th Annual IEEE Conference on Automation Science and Engineering*, Marriott Eaton Centre Hotel, Toronto, Ontario, Canada. Available at http://ieeexplore.ieee.org/stamp/stamp.jsp?tp=&arnumber=5584739 (accessed June 16, 2014).
42. Anderman, M. 2013. Assessing the future of hybrid and electric vehicles: The 2014 xEV industry insider report. Available at https://www.advancedautobat.com/industry-reports/2014-xEV-Industry-report/Executive-Summary-Selections.pdf (accessed June 16, 2014).
43. Wikipedia. 2014. Chevrolet Volt. Available at http://en.wikipedia.org/wiki/Chevrolet_Volt (accessed June 16, 2014).
44. Cobb, J. 2011. Chevrolet Volt battery animation 2011. Available at http://www.autoguide.com/auto-news/2011/03/gm-developing-a-buick-version-of-chevy-volt.html/2011-chevrolet-volt-battery-animation (accessed June 16, 2014).
45. Insideevs. 2013. Chevrolet Volt range expands to 38 miles, gets 98 MPGe. Available at http://insideevs.com/2013-chevrolet-volt-range-expands-to-38-miles-gets-98-mpge/ (accessed June 16, 2014).
46. Matthé, R., and Eberle, U. The Voltec system—Energy storage and electric propulsion. In *Lithium-Ion Batteries Advances and Applications*, ed. Pistoia, G., 151–176. Elsevier, Amsterdam.
47. Wikipedia. 2014. Toyota Prius Plug-in Hybrid. Available at http://en.wikipedia.org/wiki/Toyota_Prius_Plug-in_Hybrid (accessed June 16, 2014).
48. Taylor, A. 2012. Toyota Prius Plug-in Hybrid car review. Available at http://www.theregister.co.uk/Print/2012/10/29/review_toyota_prius_plug_in_hybrid_car/ (accessed June 16, 2014).
49. Thomas. 2013. Toyota—Erhöhung der Lithium-Ionen Batterieproduktion geplant, 2013. Available at http://www.greenmotorsblog.de/elektromobilitaet/toyota-%E2%80%93-erhohung-der-lithium-ionen-batterieproduktion-geplant/13213/ (accessed June 16, 2014).
50. Wikipedia. 2014. Nissan Leaf. Available at http://en.wikipedia.org/wiki/Nissan_Leaf (accessed June 16, 2014).
51. EV Sales. 2013. Available at http://ev-sales.blogspot.com/2013_12_01_archive.html (accessed June 16, 2014).
52. Yoney, D. 2010. Is the Nissan Leaf battery pack under-engineered? Available at http://green.autoblog.com/2010/01/25/is-the-nissan-leaf-battery-pack-under-engineered/ (accessed June 16, 2014).

53. Srinivasan, V. 2008. Batteries for vehicular applications. In *AIP Conference Proceedings Physics of Sustainable Energy*, Berkeley, CA, 1044: 283–296.
54. Ji, X., Lee, K.T., and Nazar, L.F. 2009. A highly ordered nanostructured carbon–sulphur cathode for lithium–sulphur batteries. *Nat. Mater.* 8: 500–506.
55. Hayashi, A., Ohtomo, T., Mizuno, F., Tadanaga, K., and Tatsumisago, M. 2004. Rechargeable lithium batteries, using sulfur-based cathode materials and $Li_2S–P_2S_5$ glass-ceramic electrolytes. *Electrochim. Acta* 50: 893–897.
56. Wang, H., Yang, Y., Liang, Y. et al. 2011. Graphene-wrapped sulfur particles as a rechargeable lithium-sulfur battery cathode material with high capacity and cycling stability. *Nano Lett.* 11(7): 2644–2647.
57. Abraham, K.M., and Jiang, Z. 1996. A polymer electrolyte-based rechargeable lithium/oxygen battery. *J. Electrochem. Soc.* 143: 1–5.
58. Rahman, M.A., Wang, X.J., and Wen, C. 2014. A review of high energy density lithium–air battery technology. *J. Appl. Electrochem.* 44: 5–22.

Section III

Advanced Materials and Technologies for Metal–Air Rechargeable Batteries

7 Advanced Materials for Zn–Air Rechargeable Batteries

Dong Un Lee, Jing Fu, and Zhongwei Chen

CONTENTS

7.1 INTRODUCTION

Rechargeable zinc–air batteries are an advanced electrochemical energy conversion and storage system under a class of metal–air battery technology. While researchers anticipate that the energy density of lithium-ion batteries may double with the use of high-voltage and high-capacity cathode materials combined with alloy anode materials, zinc–air battery technologies have also recently attracted immense attention as an alternative to lithium-ion batteries owing to their extremely high energy density. The specific energy density of zinc–air battery is approximately 5 and 25 times higher than those of lithium-ion and lead–acid batteries, respectively. In addition, safe operation of zinc–air batteries makes them highly suitable for a wide range of applications, particularly for electric and hybrid electric vehicles. Rechargeable zinc–air batteries can be recharged either electrically or mechanically depending on the type of application; however, all systems utilize oxygen in atmospheric air as the main source of fuel. This means that no fuel reservoir is required, which allows manufacturing of zinc–air batteries to be very compact and lightweight. These characteristics are ideal for electric vehicle applications, as the size and weight specifications of the battery pack can be easily met. Furthermore, oxygen drawn from atmospheric air is virtually unlimited in quantity, which makes zinc–air systems highly cost-competitive. Components used in zinc–air batteries are also relatively nontoxic, including the alkaline electrolyte compared with strongly acidic electrolytes currently used in lead–acid batteries. Other notable advantages of zinc–air batteries are the natural abundance of zinc and oxygen, environmental benignity, and nontoxic nature of metallic zinc compared with other metal anodes.

Typically, rechargeable zinc–air batteries can be categorized into two main types depending on the charging mechanism. The first is a mechanically rechargeable zinc–air battery where exhausted zinc electrode after discharge is replaced by a fresh zinc plate along with newly replenished electrolyte. The exhausted zinc and electrolyte can be easily regenerated to be used again. This system is advantageous as the battery can be instantaneously recharged; however, it may not be suitable for large-scale applications as recharging stations are required to replace large quantities of zinc electrodes and electrolyte. The second type is an electrically rechargeable zinc–air battery where current is applied to the battery in order to reverse the discharge reaction, thereby regenerating the zinc electrode and electrolyte. Unlike the mechanical system, this method of recharging the battery is ideal for applications such as electric vehicles, as the battery can be easily recharged by plugging into an electrical source. However, electrically recharging the battery is not only time consuming, but more important, faces a number of technical challenges that must be addressed for commercialization.

7.2 PRINCIPLE OF OPERATION

The main components of a (rechargeable) zinc–air battery (two-electrode system) are zinc and air electrodes, electrolyte, and separator, as shown in Figure 7.1. When the battery discharges, electrons released at the zinc anode travel through an external load to the air cathode, producing zinc ions. At the air cathode, air diffuses into the electrode from the atmosphere, and then the oxygen in the air reacts at the three-phase reaction site, which is the point of interaction of oxygen (gas), electrolyte (liquid), and electrocatalyst (solid).

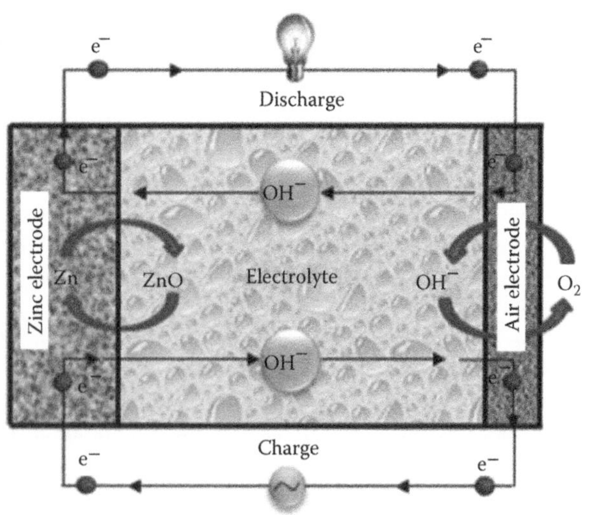

FIGURE 7.1 Schematic of a rechargeable zinc–air battery during charging and discharging processes. (Reprinted from *Appl. Energy*, 128, Pei, P.C., Wang, K.L. and Ma, Z., Technologies for extending zinc–air battery's cyclelife: A review, 315–324, Copyright 2014, with permission from Elsevier.)

At this point, oxygen is reduced to hydroxide ions via an oxygen reduction reaction (ORR) facilitated by the electrocatalyst in the presence of electrolyte. The generated hydroxide ions then migrate from the reaction site past a separator to the zinc anode, where they react with the zinc ions to form zincate ions, which then further decomposes into zinc oxide, thereby completing the cell reaction. During battery charging, the process is reversed via an oxygen evolution reaction (OER) at the three-phase reaction site, and zinc is plated back to the original zinc electrode [1].

7.3 OXYGEN ELECTROCHEMICAL REACTIONS

As briefly mentioned in the previous section, the two main electrochemical reactions that govern the operation of a rechargeable zinc–air battery are ORR and OER, which correspond to the battery's discharge and charge processes. These reactions must be catalyzed owing to the slow kinetics with the use of bifunctional electrocatalysts. The ORR at the air cathode proceeds in several steps: oxygen diffusion from the atmosphere to the catalyst surface, oxygen absorption on the catalyst surface, transfer of electrons from the anode to oxygen, weakening and breaking of the oxygen bond, and removal of the hydroxide ions from the catalyst surface to the electrolyte [2]. The OER involves the reverse process of ORR; however, it is very complicated and difficult to describe as the reaction involves a series of complex electrochemical reactions with multistep electron-transfer processes [2].

7.4 AIR ELECTRODE

Regardless of the type of rechargeable zinc–air battery, the system must have an efficient air electrode that allows diffusion of active oxygen species from the atmosphere. This air "breathing" electrode must be hydrophobic so that electrolyte (normally aqueous solution) inside the battery is not leaking to the outer environment. Typically, an air electrode consists of a gas diffusion layer and a catalytic active layer, prepared

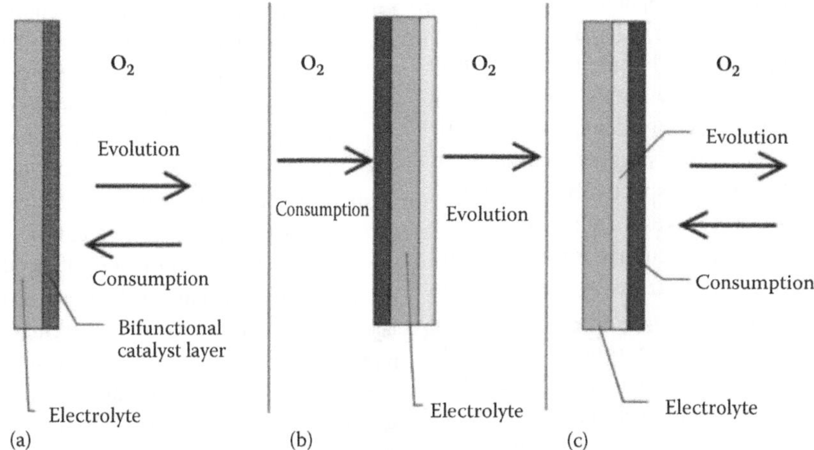

FIGURE 7.2 Schematic of typical microstructure of an air electrode: (a) single electronode, single catalytic layer, (b) seperate electrodes, single catalytic layer, and (c) single electrode, multiple catalytic layers. (Reprinted from *J. Power Sources*, 155, Jörissen, L., Bifunctional oxygen/air electrodes, 23–32, Copyright 2006, with permission from Elsevier.)

by laminating the two together with a metal mesh as a current collector, as shown in Figure 7.2 [3]. The gas diffusion layer is composed of carbon-based materials mixed with hydrophobic binding polymer such as polytetrafluoroethylene (PTFE) as a wet-proofing agent. The PTFE layer allows diffusion of air through the pores created by the carbon material, while preventing aqueous electrolyte from leaking to the air side. The catalyst layer typically consists of an active oxygen catalyst mixed with high-surface-area carbon black and binding polymer to enhance dispersion and active surface area.

In addition to the hydrophobic and porosity requirements of an air electrode, using oxygen in atmospheric air as the source of fuel requires the electrode to be catalytically active toward the oxygen reactions. Active catalysts facilitate the ORR and OER during the battery discharge and charge, respectively, to lower the large overpotentials associated with these reactions. Reducing overpotentials are critical for operating the battery at superior voltages, resulting in higher energy efficiencies.

7.4.1 Bifunctional Electrocatalyst

As mentioned, efficient bifunctional electrocatalysts are required to lower the overpotentials of both ORR and OER during the battery's discharge and charge processes, respectively. Precious metal-based catalysts such as platinum and palladium and their alloys are known to be highly effective for ORR; however, low OER activity and high materials costs have limited their use. Because zinc–air batteries typically utilize an alkaline solution as electrolyte instead of an acidic solution, expensive precious metal-based catalysts can be replaced with transition metal oxides such as spinel and perovskite oxides and nanostructured carbon supports that are known to be bifunctionally active [2,3]. Generally, the selection of an appropriate electrocatalyst involves a compromise between electrocatalytic activity, thermodynamic stability, corrosion resistance, fabrication and materials cost, and long-term stability. These properties can be tuned by changing the composition and geometry of the catalyst. The composition is directly related to the electronic properties of the catalyst, which determines the strength of surface-intermediate bonds, hence the electrocatalytic activity, while the geometry is related to the actual surface areas, and active site density, which can be controlled by the catalyst morphology. In the following sections, advanced eletrocatalysts based on transition metal oxides, and organic/inorganic hybrids are introduced as bifunctional electrode materials for rechargeable zinc–air batteries.

7.4.1.1 Transition Metal Oxide–Based Catalyst

In alkaline electrolytes, nonnoble transition metal oxides or mixed transition oxides can be effectively utilized to replace precious metal–based catalysts with similar battery performance. With a wide range of transition metals available, the composition, crystal structure, and morphology of the catalyst can be properly tailored to obtain a desired electronic structure resulting in good electrical properties and ion mobility for the electrochemical oxygen reactions. Spinel-type oxide catalysts are particularly interesting for efficient oxygen catalysis as they exhibit relatively high bifunctionality, excellent electrochemical durability, and low electrical resistance. For instance, spinel-type Co_3O_4 has been extensively investigated as active and durable transition metal oxide catalyst to minimize the cost and weight of electrically rechargeable zinc–air batteries [4]. In fact, spinel Co_3O_4 is composed of mixed valence states of the same cobalt cation (Co^{2+} and Co^{3+} in tetrahedral and octahedral sites, respectively), and this is helpful for the bifunctional activity as it provides donor–acceptor chemisorption sites for reversible oxygen adsorption. The low electrical resistance of spinel oxides is also due to an electron-hopping process between these mixed valencies, which have a relatively low activation barrier. Similarly, mixed-metal spinel oxides are also largely used for bifunctional ORR and OER catalysts. For example, two different phases of cobalt–manganese nanostructures, tetragonal and cubic, have been synthesized at room temperature by a reduction–recrystallization method demonstrating excellent bifunctional activities as shown in Figure 7.3 [5]. These catalysts can potentially replace the conventional platinum-based catalysts, which is extremely expensive and shows limited OER activity and electrochemical instability, to make rechargeable zinc–air battery more commercially viable.

7.4.1.2 Organic/Inorganic Hybrid Catalyst

Organic/inorganic hybrid catalysts are typically composed of metal or metal oxide–based catalysts aforementioned in Section 7.4.1.1, combined with carbon-based materials such as carbon nanotubes and graphene nanosheets. The carbon materials usually act as both catalyst support and active electrocatalyst. As a catalyst support, carbon-based materials that usually have relatively high surface area enable good dispersion of inorganic catalysts for increased availability of catalytically active sites as well as enhance the electrical properties of the hybrid catalyst especially when combined with transition metal oxides with intrinsically low electrical conductivity. As an active electrocatalyst, carbon-based materials doped with hetero atoms such as nitrogen, boron, and sulfur are relatively more active toward the ORR, complementing the relatively higher oxygen evolution activity of inorganic catalysts. This complementary nature of hybrid catalysts often results in a synergistic effect between the two components where the catalytic activity of the hybrid catalysts exceeds the activities of individual components.

In a hybrid catalyst, the interaction between the organic and inorganic components has an important effect on the catalyst's performance. Dai and coworkers have reported strongly coupled transition metal oxide–based catalysts with graphitic carbon species such as graphene and carbon nanotubes for advanced electrocatalysts [6,7]. Using CoO/carbon nanotube hybrid (ORR active) and Ni–Fe-layered double hydroxide (OER active) catalysts, they demonstrated rechargeable zinc–air battery performance exceeding precious metal platinum and iridium catalysts in terms of both catalytic activity and durability [6]. Another novel hybrid bifunctional catalyst based on core–corona structure (CCBC) has been introduced by Chen and coworkers consisting of lanthanum nickel oxide–based core covered by nitrogen-doped carbon

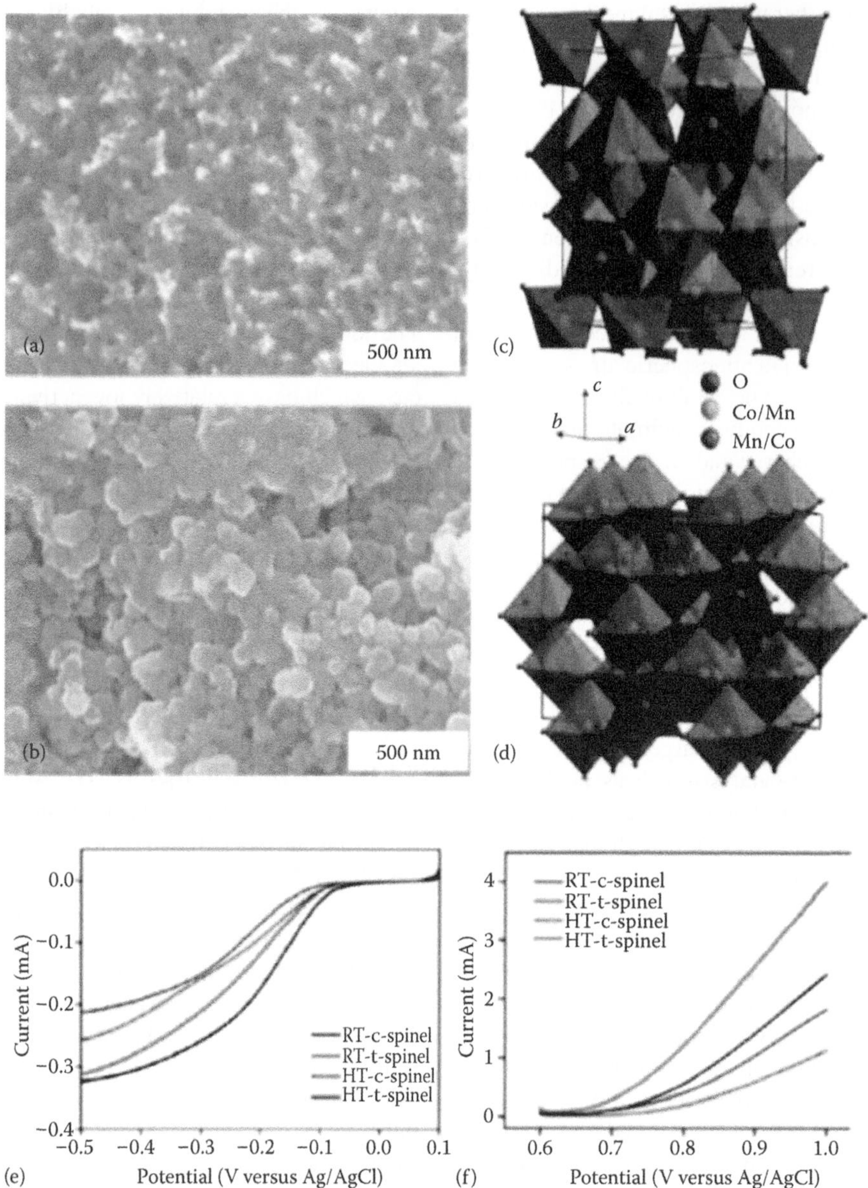

FIGURE 7.3 SEM images and crystal structures of cobalt–manganese spinel oxides: (a, c) tetragonal-phase (RT-t-spinel); (b, d) cubic-phase spinel (RT-c-spinel). Voltammograms of the (e) ORR and (f) OER recorded on these catalysts. (Reprinted by permission from Macmillan Publishers Ltd. *Nat. Chem.*, Cheng, F.Y., Shen, J.A., Peng, B., Pan, Y.D., Tao, Z.L., and Chen, J., 3, 79–84, copyright 2011.)

nanotubes (NCNT) as shown in Figure 7.4 [8]. The lanthanum nickel oxide–based core is shown to be highly OER active, while the nitrogen-doped graphene is active toward ORR. Upon rechargeable zinc–air battery testing, the hybrid CCBC catalyst demonstrates superior battery performance including durability up to 75 cycles without voltage fading, whereas both charge and discharges voltages of $LaNiO_3$ core tested itself show significant degradation. The improved electrochemical performance of CCBC is attributed to strongly connected NCNT to the core material, resulting in increased overall electrical conductivity and electrochemical durability from the graphitic walls of carbon nanotubes. CCBC is an exemplary hybrid catalyst that is both active and durable, and cost-effective for the next-generation rechargeable zinc–air battery system.

7.5 ELECTROLYTE

The ideal electrolytes, which provide the bridge for ion transportation between electrodes inside the cell, should be (i) chemically inert toward the electrode materials under open circuit condition, (ii) electrochemically stable during the operation over a wide range of current demands and temperatures, and (iii) have a high specific ionic conductivity and low coefficient of dynamic viscosity to ensure appropriate ion migration, as well as to minimize the concentration polarization of

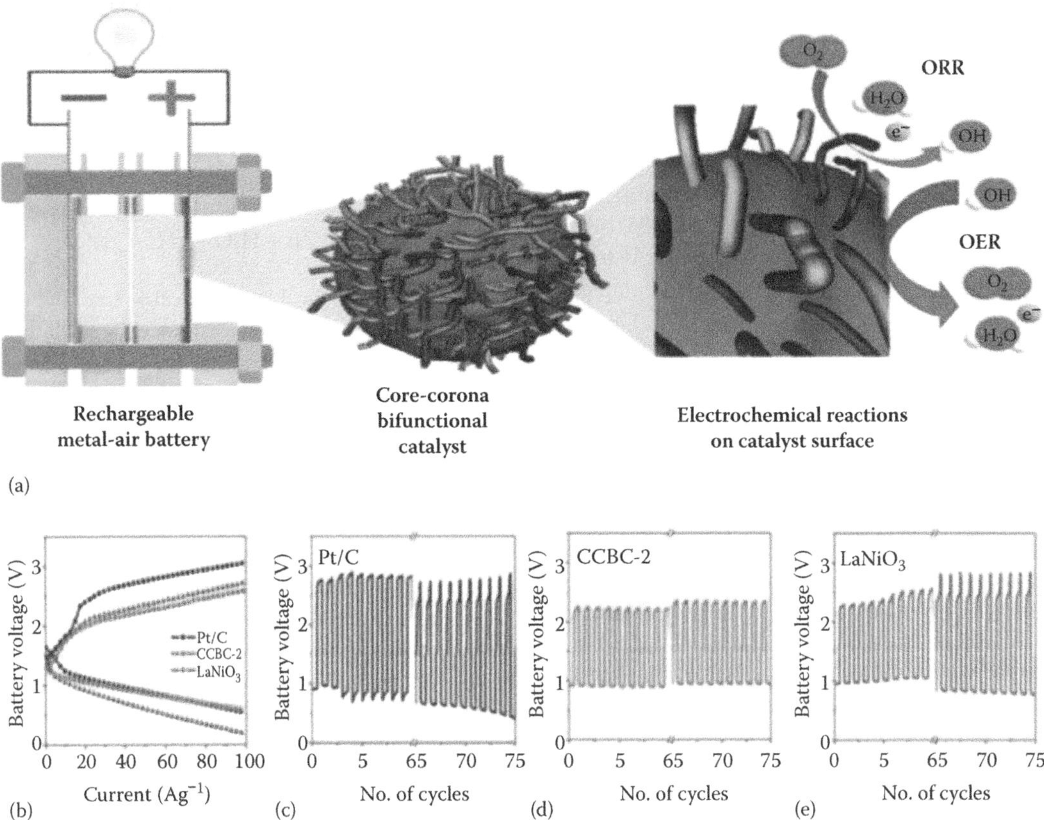

FIGURE 7.4 (a) Schematic of CCBC for rechargeable metal–air battery. (b) Charge–discharge polarization curves of Pt/C, CCBC-2, and $LaNiO_3$. Durability cycling of (c) Pt/C, (d) CCBC-2, and (e) $LaNiO_3$. (Reprinted with permission from Chen, Z., Yu, A.P., Higgins, D., Li, H., Wang, H.J., and Chen, Z.W., *Nano Lett.*, 12, 1946–1952. Copyright 2014 American Chemical Society.)

the electrodes. The electrolytes utilized in the conventional zinc–air batteries are typically aqueous solutions that have a high ionic conductivity. Nonaqueous electrolytes such as solid polyelectrolytes, gel electrolytes, and ionic liquids also offer great potentials for applications such as solid-state zinc–air batteries. Table 7.1 shows the typical ranges of specific conductivities for various types of electrolyte systems [9].

7.5.1 Aqueous Electrolyte

7.5.1.1 Alkaline Electrolyte

Commonly, alkaline solutions are used for zinc–air batteries because metallic zinc is relatively stable and catalyst corrosion is less severe compared with acidic electrolytes that readily react with the metal anode. Typical alkaline electrolytes used in zinc–air batteries are potassium hydroxide (KOH) and sodium hydroxide (NaOH). KOH is most widely used as it exhibits superior ionic conductivity of K^+ compared with Na^+ as well as higher oxygen diffusion coefficients, and lower viscosity [10]. To reduce the resistance of electrolytes and increase the ionic conductivity, high concentrations are normally utilized, such as 6.0 M KOH. The main drawback of using aqueous alkaline electrolytes is the formation of carbonate species. The alkaline electrolytes have high solubility for carbon dioxide (CO_2). The CO_2 from the atmosphere diffuses into the cell from air cathode and consumes the electrolytes by forming the carbonate crystals, which decrease the molarity of hydroxide ions leading to the passivation of zinc anode, and irreversibly impede air access by precipitating carbonate in the pores of air cathode. Moreover, the crystallization can produce a pathway for direct electrolyte leakage if this occurs. With the present large zinc–air battery systems, the problem of carbonation can be controlled by scrubbing the CO_2 from the atmosphere. To prevent this, purified air using a selective membrane only permeable to O_2 or pure oxygen is often utilized to feed the air cathode.

Moreover, electrodeposition of metallic zinc during charging frequently results in dendrite formation and shape change;

TABLE 7.1
Conductivity Ranges of Various Electrolytes at Ambient Temperature

Electrolyte System	Conductivity (S cm^{-1})
Aqueous electrolytes	0.1–0.55
Molten salts	~10^{-1}
Inorganic electrolytes	10^{-2}–10^{-1}
Ionic liquids	10^{-4}–10^{-2}
Polymer electrolytes	10^{-7}–10^{-3}

thereby, many additives have been developed for rechargeable zinc–air batteries to the KOH electrolyte to improve the morphology of zinc deposition and extend the cycle life of cells [11,12]. Furthermore, a separator is generally necessary to retard zinc dendrite growths, preventing internal short circuits. The current commercial separators applied in conventional zinc–air batteries, which are mostly based on polypropylene (i.e., Celgard 4560 and Celgard 5550), are not capable of ion selection. On the one hand, zinc is highly soluble in alkaline electrolytes as forming zincate complexes (i.e., $Zn(OH)_4^{2-}$) that are able to migrate throughout the entire electrolyte body and reach the cathode. This results in detrimental contamination of the air cathode if zinc ions are oxidized by oxygen and further zinc oxide begun to precipitate into the pores of the cathode. In some cases, the catalysts in the cathode are not stable enough and partially dissolve into catalyst compound ions over time under highly concentrated alkaline electrolytes. The impurities from the catalyst may migrate across the separator to leach into zinc metal to produce active metal sites that are capable of catalyzing hydrogen evolution.

7.5.1.2 Nonalkaline Electrolyte

It is noted that the problem of dendrite growths in zinc–air batteries can be addressed through proprietary electrolyte designs. To avoid dendrites as well as carbonates forming associated with the use of alkaline electrolytes in zinc–air batteries, some acidic aqueous electrolytes are evaluated. An organic acid electrolyte, methane sulfonic acid (MSA), was proposed for rechargeable zinc–air batteries, in which the anionic form of MSA^- formed a complex with a zinc ion (Zn^{2+}) during discharge, and wherein the acid in the electrolyte reduced dendrite formation during charge [13]. The reactions during discharging are as follows:

$$\text{Cathode: } O_2 + 2CH_3SO_3H + 2e^- \rightarrow H_2O + 2CH_3SO^{3-}$$

$$\text{Anode: } Zn \rightarrow Zn^{2+} + 2e^-$$

$$Zn^{2+} + 2CH_3SO^{3-} \rightarrow Zn(CH_3SO_3)_2$$

However, the major problem of using acid electrolytes is that zinc reacts vigorously accompanied by the rapid evolution of hydrogen gas. Moreover, it should be recognized that in acid media, the catalytic activity and durability of the electrodes are more critical compared with the case of alkaline electrolytes.

Researchers also look into neutral electrolytes as safer and more robust alternatives with respect to traditional alkaline electrolytes for rechargeable zinc–air batteries. Recently, a type of near-neutral electrolytes based on zinc chloride and ammonium chloride were reported for rechargeable zinc–air battery applications [14]. The prototyped cell tests proved that this type of chloride electrolyte system can sustain more than 1000 h and hundreds of discharge–charge cycles, under discharge–charge capacity ranging from 20 to 120 mAh. In addition, no zinc dendrite formation was observed after prolonged cycling tests.

7.5.2 Nonaqueous Electrolyte

The zinc metal is thermodynamically unstable with respect to aqueous solutions since the zinc redox potential is less than that of hydrogen (see Equation 7.1), and the high solubility of zinc ions and compounds (i.e., $Zn(OH)_2$, $Zn(OH)_4^{2-}$) in alkaline solution further accelerates zinc corrosion.

$$Zn + H_2O \rightarrow H_2 + ZnO \tag{7.1}$$

Moreover, the problem of water loss by evaporation of electrolytes results in performance degradation due to the open system of the air cathode in zinc–air batteries. On the one hand, the water loss causes the electrolytes to concentrate, and therefore promotes carbonation and increases the transfer resistance of hydroxyl ions. On the other hand, insufficient water in the cathode is not capable to support the destruction of peroxide and subsequent hydroxyl formation during ORR. Therefore, recent studies have been investigating the possibility of replacing the current caustic aqueous electrolytes (KOH), with the alternatives of nonaqueous electrolytes, such as gel electrolytes, solid polymer electrolytes, ionic liquids, and ceramic/glassy electrolytes. The nonaqueous electrolytes can offer merits, such as improved redox stability, ion selectivity, absence of water loss and leakage, and dendritic recrystallization.

7.5.2.1 Gel Electrolyte

One of the alternative ways to minimize water loss and to eliminate the possibility of leakage during the operation in zinc–air batteries is to gel the liquid electrolytes. The gel electrolytes incorporate a low-molecular-weight solvent medium (normally is water) via hydrogen bonding, to assist ion conducting and mass transfer processes. They are not typical polymer electrolytes but rely on certain types of polymer hosts as gelling agents or electrolyte absorbents to immobilize alkali (KOH) for ion transport. The polymer gelling agents must be alkali stable and have high capability of storing electrolytes. Cellulose or starch derivatives such as sodium carboxymethylcellulose and sago, vinyl polymers such as poly(vinyl alcohol) (PVA), hydroponics gel and agar, and poly(acrylic acid) have been reported as good gelling agents for applications in zinc–air batteries [15].

For example, Othman et al. investigated a hydroponics gel as a gelling agent by blending 12 wt.% KOH electrolyte for primary zinc–air batteries [16]. This gel electrolyte was found to be capable of sustaining discharge loads of 5, 50, and 100 mA with corresponding capacities of 229, 165, and 115 mAh. A further study from this group demonstrated that it was beneficial to the zinc–air batteries' discharge capabilities by applying a thin agar layer in gelled KOH on the intimate interface of the electrodes/electrolyte. Zhu and coworkers prepared a rubber-like gel electrolyte based on acrylate/KOH system reaching a high specific conductivity of 0.288 S cm^{-1} at room temperature [17]. This gel electrolyte exhibited almost the same chemical and electrochemical stability as aqueous alkaline solution, and showed good performance characteristics for application of zinc–air batteries.

7.5.2.2 Solid Polymer Electrolyte

A new family of zinc–air batteries by utilizing solid polymer electrolytes (membranes) has drawn great attention due to the advantages of potentially high-energy-density power sources. Because of the rigid structure of the membranes, those solid polymer electrolytes can also serve as the separators and are compliant to construct solid-state zinc–air batteries in which the membranes conform to the volume changes of the anode that typically occur during discharge and charge cycles. The solid polymer electrolytes must be designed to meet the requirements, including ionic conductivity, physical and chemical stability, electrical insulation, ion selectivity, and air permeability. Here, the solid alkaline polymer electrolytes are divided into two categories: heterogeneous membranes and homogeneous membranes.

The heterogeneous membranes mainly result from the blending of water-soluble polymer matrix and metal hydroxide salts (most of the time, KOH). This type of polymer complexes has the physical properties of the polymer host and the electrochemical and conductive properties of KOH. A high ionic conductivity of the polymer membranes is desired to ensure an efficient transfer of the hydroxyl ions from one electrode to the other. Concerning the polymers, poly(ethylene oxide) (PEO), PVA, and blends with KOH have been developed for the applications in solid-state zinc–air batteries [18,19]. For example, Yang demonstrated various types of alkaline polymer electrolytes based on PVA, such as PEO/PVA/glass fiber system and PVA/poly(acrylic) acid system. Those composite polymer electrolytes exhibited ionic conductivities in the range of 10^{-2}–10^{-1} S cm^{-1} at room temperature as well as good electrochemical stability, and exhibited high potential for application in the solid-sate primary zinc–air cell. However, since the polymer combinations are inherent to KOH, the limitations such as the loss of ionic conductivity due to leaking out of KOH and the carbonation may impede their applications in zinc–air batteries. Many efforts were made to hold the KOH in the polymer matrix, including dissolution of a "plasticizer" solvent such as ethylene or propylene carbonate and backbone modification.

Another strategy is to employ alkaline anion-exchange membranes (AEMs), a type of homogeneous membranes consisting exclusively of the cationic (i.e., ammonium) groups and the mobile counter ions (hydroxyl groups), as nonaqueous electrolyte candidates for zinc–air batteries. Those hydroxyl ion-conducting polymers have been developed widely in the range of electrochemical energy applications such as fuel cells. In zinc–air batteries, the AEM is joined directly and positioned between the porous catalyst layer of the cathode (usually the gas diffusion layer) and the zinc anode. Table 7.2 lists the major commercial alkaline AEMs [20]. The ionic conductivities strongly depend on the proportion of charge carrier groups and the water uptake of the membranes. Despite their homogeneous fixed ionic charge distribution over the entire polymer matrix, the inadequate ionic conductivity and insufficient long-term chemical stability, which are also noted in alkaline fuel cells, still limit their application to meet high-energy-density zinc–air batteries in the future.

7.5.2.3 Room Temperature Ionic Liquid Electrolyte

A problem with zinc–air batteries is the disparity in the reaction kinetics of the zinc anode and the air cathode. Since the oxidation of metallic zinc in aqueous alkaline solution is fast, the hydrogen evolution reaction (HER) (Equation 7.2) at the cathode competes with the ORR (Equation 7.3) during discharge, which decreases the columbic efficiency of discharge. This also promotes the rapid corrosion of anode material. The same issue of the unproductive HER during charge occurring at the anode may require additional overcharge before the cell is fully charged. Moreover, the difference in charging efficiency results in electrolytes drying out or even fast premature failure due to additional water consumption from the HER.

$$2H_2O + 2e^- \rightarrow H_2 + 2OH^- \quad E^0 = -0.828\ \text{V} \tag{7.2}$$

$$2OH^- \rightarrow 1/2O_2 + H_2O + 2e^- \quad E^0 = 0.401\ \text{V} \tag{7.3}$$

In addition, when the cell is discharged in the presence of a high alkaline electrolyte, the formation of passivation film of zinc oxide hinders the regeneration of zinc ions during the recharge process. Therefore, researchers have begun to focus on developing an electrolyte that is water–solvent free, non-alkaline, and can help improve the zinc morphology during a redox reaction. Room temperature ionic liquids (RTILs), which are molten salts generally composed of large organic cations and organic/inorganic anions, possess the many advantages of low volatility, nonflammability, high thermal stability, wide electrochemical window, and the capability of modifying metal electrodepositions. These important characteristics make RTILs attractive as potential electrolyte alternatives to avoid the issues noted as above for rechargeable zinc–air battery applications. For example, RTILs can prevent zinc dendrite formation at the anode, achieve a longer cycle life as they slow down the drying out of the electrolyte from water evaporation, suppress the self-discharge of zinc, and eliminate carbonation.

Some common ionic liquid ion families are divided into groups of cations and anions in Figure 7.5 [21]. RTILs with immidazolium and pyrrolidinium cations, together with [bis(trifluoromethanesulfonyl)imide] imide and dicyanamide anions, are among the most promising candidates for use in batteries.

The reversible electrochemistry of zinc has been studied in a wide range of RTILs. The RTILs as electrolytes generally comprise a type of soluble inorganic or organic zinc salt as zinc ion carriers and an ionic liquid as solvent to facilitate zinc ion transfer. One example is from halide of zinc-based ionic liquids. However, the disadvantage of such RTILs is their sensitivity to air and water, which limits their applications in zinc–air batteries under open air conditions. Liu studied the deposition of zinc from a variety of zinc salts that were soluble in air- and water-stable ionic liquids, namely 1-butyl-1methylpyrrolidinium trifluoromethylsulfonate ([$Py_{1,4}$] TfO) and 1-ethyl-3-methylimidazolium trifluoromethylsulfonate ([EMIm]TfO), and they exhibited a typical redox couple

TABLE 7.2
Properties and Nature of Commercial Anion-Exchange Membranes

Membrane and Manufacturer	Structure	IEC (meq g^{-1})	Thickness (mm)	IC (mS cm^{-1})	Resistance (Ω cm^{-2})
Tokuyama Co. Ltd., Japan [256]					
Neosepta AMX	PS/DVB	1.4–1.7	0.12–0.18		2.0–3.5
Neosepta ACM	PS/DVB	1.5	0.12		4.0–5.0
Neosepta AM-1	PS/DVB	1.8–2.2	0.12–0.16		1.3–2.0
Neosepta AM-3	PS/DVB	1.3–2.0	0.11–0.16		2.8–5.0
Neosepta AHA	PS/DVB	1.15–1.25	0.18–0.24		0.18–0.24
Neosepta ACS	PS/DVB	1.4–2.0	0.12–0.20		3.0–6.0
Neosepta AFN	PS/DVB	2.0–3.5	0.15–0.13		0.2–1.0
Neosepta AFX	PS/DVB	1.5–2.0	0.14–0.17		0.7–1.5
Neosepta ACH-45T		1.3–2.0			
A 201 (developing code A-006)		1.7	28	29	
A 901		1.7	10	11.4	
Ionics Inc., USA [257]					
103PZL 183	Heterogeneous membrane	1.2	0.60		4.9
AR103QDP		1.95–2.20	0.56–0.69		14.5
AS204SZRA		2.3–2.7	0.48–0.66		6.2–9.3
AR112-B		1.3–1.8	0.48–0.66		20–28
RAI Research Corp., USA [258]					
R-5030-L	LDPF (IPN)	0.9	0.24		4.0–7.0
R-1030	IPN-fluorinated	1.0	0.1		0.7–1.5
CSMCRI, Bhavnagar India [259]					
IPA	LDPE/HDPE (IPN)	0.8–0.9	0.16–0.0.18		2.0–4.0
HGA	Heterogeneous PVC	0.4–0.5	0.22–0.25		5.0–7.0
Tosoh Corporation Japan [260]					
Tosflex	Nafion				
Asahi Chemical Industry Co. Japan [261]					
Aciplex A-192			>0.15		1.8–2.1
Aciplex A-501-SB			0.14–0.13		2.0–3.0
Aciplex A201			0.22–0.24		3.6–4.2
Aciplex A221			0.17–0.19		1.4–1.7
Asahi Glass Co. Ltd., Japan [262]					
Selemion AMV	PS-b-EB-b-PS		0.11–0.15		1.5–3.0
Selemion ASV			0.11–0.15		2.3–3.5
Selemion DSV			0.13–0.17		
AMV	PS/butadiene	1.9	0.14		2.0–4.5
FuMA-Tech GmbH Germany					
FAS		1.1	0.10–0.12		2–4
FAB		0.8	0.09–0.11		2–4
FAN		0.3	0.09–0.11		2–4
FAA		1.1	0.08–0.10		2.4
FAD		1.3	0.08		1.2
MEGA a.s., Czech Republic [263]					
Ralex MH-PES		1.8	0.55 (Dry)		<8
Ralex AMH-SE		1.8	0.7 (Dry)		<13
PCA Polymerchemie Altmeier GmbH, Germany					
PC 100 D		1.2 quat.	0.08–0.1		5

(Continued)

TABLE 7.2 (CONTINUED)
Properties and Nature of Commercial Anion-Exchange Membranes

Membrane and Manufacturer	Structure	IEC (meq g^{-1})	Thickness (mm)	IC (mS cm^{-1})	Resistance (Ω cm^{-2})
PC 200 D		1.3 quat.	0.08–0.1		2
PC Acid 35		1.0 quat.	0.08–0.1		
PC Acid 70		1.1 quat.	0.08–0.1		
PC Acid 100		0.57 quat.	0.08–0.1		
Solvay S.A., Belgium [264]					
Morgane ADP		1.3–1.7	0.13–0.17		1.8–2.9
Morgane AW		1.0–2.0	0.13–0.17		0.9–2.5
Tianwei Membrane Co. Ltd., China [265]					
TWEDG		1.6–1.9	0.16–0.21		3–5
TWDDG		1.9–2.1	0.18–0.23		<3
TWAPB		1.4–1.6	0.16–0.21		5–8
TWANS		1.2–1.4	0.17–0.20		6–10
TWAHP		1.2–1.4	0.20–0.21		<2
TWAEDI		1.6–1.8	0.18–0.21		6–8
Shanghai Chemical Plant of China, China [266]					
PE3352	Heterogeneous PE	1.8–2.0	0.45		13.1
Institute of Plastic Materials. Moscow [238]					
MA-40	Heterogeneous	0.6	0.15		5.0

Source: Wang, Y.J., Qiao, J.L., Baker, R. and Zhang, J.J., *Chem. Soc. Rev.*, 42, 5768, 2013. Reproduced with permission of The Royal Society of Chemistry.

associated with deposition/stripping of zinc from the employed electrolytes. In particular to the application in zinc–air batteries, Mega investigated a family of chelating RTILs that are able to solubilize the zinc ions to avoid oxide/hydroxide formation in the presence of water and also release the metal ion easily during charge [22]. However, as compared with aqueous alkaline solutions or alkali-doped gelled types and polyelectrolytes, the issues of dissolution of metallic zinc, the catalytic activities with respect to the ORR and the OER, the lower ionic conductivity, and the wettability of electrodes in RTILs are challenges to facilitating their practical applications in zinc–air batteries. Thus far, the investigation of functionalized RTILs as zinc–air battery electrolytes is still in its preliminary stage, and more efforts are needed to meet the complex requirements for zinc–air battery applications.

For the purposes of enabling the desired electrochemical reactions, a trace amount of water is added in RTILs. A benefit to adding water to the ionic liquids is that the proton released from the water oxidation reaction transiently acidified the electrolyte, thereby promoting the solvation of the precipitation of zinc oxide. In addition, small quantities of water in ionic liquids may substantially decrease viscosity, resulting in an increase in conductivity.

7.6 ZINC ELECTRODE

For zinc–air batteries to be electrically rechargeable, not only a proper bifunctional catalyst is required at the cathode to catalyze both ORR and OER, but also the zinc electrode at the anode must be reversible. One of the critical challenges of developing electrically rechargeable zinc–air batteries is creating zinc electrode that is chemically/electrochemically stable and morphologically reversible upon extended cycling. One of the two main challenges of developing rechargeable zinc electrode is hydrogen evolution reaction due to zinc corrosion in the presence of hydroxide ions in the alkaline electrolyte. This reaction not only increases the gas pressure buildup in the cell but also reduces the zinc metal utilization, lowering the energy density of the battery. Previously, mercury was alloyed with zinc metal to increase the overpotential for hydrogen evolution; however, this alloy is no longer in use because of the health and environmental issues associated with the use of mercury. Hence, alloying zinc with other metals such as Hg, Pb, and Cd that exhibit high overpotential for hydrogen evolution has also been introduced; however, their use is limited because of their toxic nature. The second challenge is the shape change of the zinc electrode during charging, which leads to the formation of zinc dendrites upon the battery charge–discharge cycling process, which is detrimental to the lifetime of the battery. Furthermore, during battery discharge, zinc metal is oxidized to generate electrons and form zinc(II) ions, which in turn forms zincate ions, $Zn(OH)^{2-}_{4\,(aq)}$. This reaction proceeds until the solubility of the zincate ions reaches a saturation point in the hydroxide electrolyte. After exceeding this point, the zincate ions decompose to form a white solid product zinc oxide, ZnO, which is

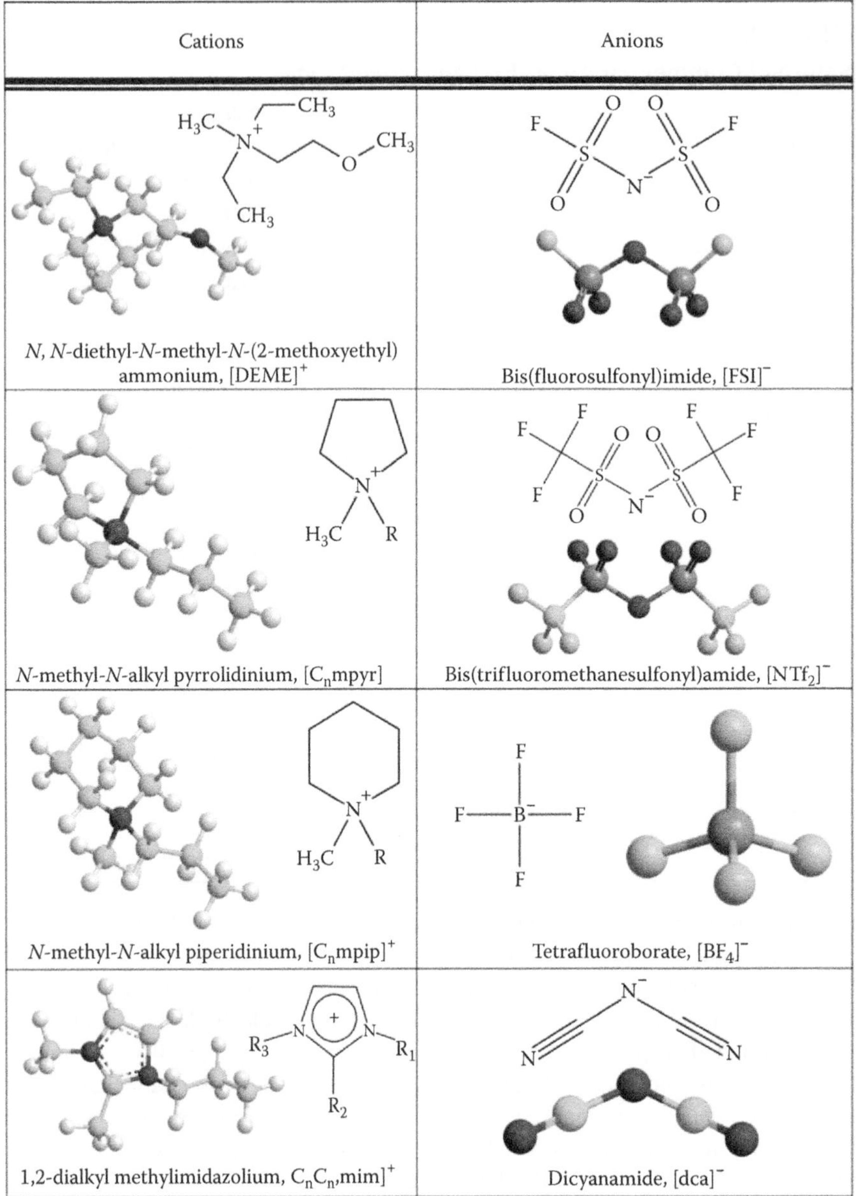

Cations
Anions
N, N-diethyl-N-methyl-N-(2-methoxyethyl) ammonium, [DEME]+
Bis(fluorosulfonyl)imide, [FSI]−
N-methyl-N-alkyl pyrrolidinium, [CnmpyR]
Bis(trifluoromethanesulfonyl)amide, [NTf2]−
N-methyl-N-alkyl piperidinium, [Cnmpip]+
Tetrafluoroborate, [BF4]−
1,2-dialkyl methylimidazolium, CnCn'mim]+
Dicyanamide, [dca]−

FIGURE 7.5 Common ionic liquid ion families appearing in energy applications. *R groups typically denote ethyl, propyl, and butyl groups. (From MacFarlane, D.R., Tachikawa, N., Forsyth, M., Pringle, J.M., Howlett, P.C., Elliott, G.D., Davis, J.H., Watanabe, M., Simon, P., and Angell, C.A., *Energy Environ. Sci.* 7, 232–250. Reproduced by permission of The Royal Society of Chemistry.)

FIGURE 7.6 Various morphologies of zinc anode: (a) zinc powder, (b) zinc dendrite, and (c) zinc fiber. (Reprinted from *J. Power Sources*, 163, Zhang, X.G., Fibrous zinc anodes for high power batteries, 591–597, Copyright 2006, with permission from Elsevier.)

an insulator, limiting the energy capacity and cycle life of the battery. Because supersaturation of the zincate ions in alkaline solution is time dependent, it is very difficult to make a rechargeable zinc–air battery. Therefore, understanding the chemistry and behavior of zinc anode in alkaline electrolyte is necessary in order to create a truly electrically rechargeable zinc–air battery. One practical method of improving the performance of the zinc–air battery is to increase the active surface area of the zinc electrode to enhance the zinc ion solubility and interaction with the electrolyte. For example, various morphologies of high-surface-area zinc electrodes have been reported in the literature, such as zinc flakes, ribbons, and fibers as shown in Figure 7.6 [23], for improved battery performance compared with that of planar zinc electrodes. However, the downside of increasing the zinc surface area is that it also promotes hydrogen evolution, which lowers the effective zinc utilization thereby limiting the energy density of the zinc–air battery.

7.7 CURRENT STATUS

7.7.1 Primary Zinc–Air Batteries

Currently, only primary zinc–air batteries are commercially available, manufactured by companies such as Duracell and Energizer. The primary applications of zinc–air batteries are in hearing aids and railway traffic lights and other applications where the longevity of battery life is important because of difficulties in replacing the battery. For example, zinc–air batteries used for hearing aids are sold in a button cell form with holes on the air electrode side sealed by a sticker to prevent it from discharging. Once the sticker is removed, the battery is activated and starts discharging as oxygen from the atmosphere diffuses into the battery. Inside the cell, the zinc anode is usually composed of a paste consisting of zinc powder, zinc oxide powder, and binding polymer. The zinc anode is separated from the air cathode by a separator that allows the movement of hydroxide ions during oxygen electrochemical reactions. The air electrode is composed of a gas diffusion layer deposited with an electrocatalyst such as manganese oxide to facilitate ORR, sandwiched with a layer of PTFE membrane on the air side to prevent the leakage of electrolyte. For rechargeable zinc–air battery systems, they are still largely under development in corporate and academic settings. Even though a number of companies such as EOS claim to have developed rechargeable zinc–air systems for commercial implementation, no large-scale production and usage have been realized.

7.7.2 Technical Challenges of Rechargeable Zinc–Air Batteries

Commercialization of electrically rechargeable zinc–air batteries have yet been truly realized because of a number of technical challenges that must be addressed to make them suitable for practical usage. The challenges are associated with each component of zinc–air battery. First, the electrocatalysts used in air electrodes must be bifunctionally active toward both the oxygen reduction and evolution reactions in order to reduce the overpotentials associated with these reactions. Also, the electrocatalysts must be sufficiently durable in alkaline conditions especially at high battery voltages experienced during charging in order to prolong the active lifetime of the battery. In addition to electrocatalysts, the air electrode architecture is important and designed in mind of the OER during charging of the battery. The oxygen gas evolved from the surface of the electrocatalysts must be able to escape the electrode easily in order to maintain good contact between electrolyte and active catalyst area for the OER to occur consistently throughout the duration of charging of the battery. Second, the concentration of aqueous electrolyte must remain consistent to maintain an optimum level of ionic conductivity during the electrochemical oxygen reactions. At the same time, the separator must be stable at the optimum electrolyte concentration and needs to remain highly permeable to hydroxide ions but impermeable to zinc ions even during an extended period of use. Most important, unlike in primary zinc–air batteries where the process of zinc plating is absent, the separator in a rechargeable zinc–air battery must be able to physically block any dendritic structures growing toward the air electrode during charging in order to prevent premature shorting of the battery. Lastly, zinc electrodes have a number of technical challenges, one of which is the shape change of the electrode's surface during charging where zinc ions are reduced back to metallic zinc. The zinc deposition typically occurs nonuniformly on the surface, leading to the growth of needle-like dendrites toward the direction of the air electrode. Uncontrolled growth of dendrites eventually penetrates the separator and makes contact with the air electrode, shorting the battery. In addition to nonuniform zinc reduction, hydrogen evolution must be prevented during charging, which involves a reaction with zinc that reduces the overall zinc utilization. This has a significant negative impact on the energy density of the zinc–air battery; hence, the zinc electrode must be made to remain inactive toward hydrogen evolution.

REFERENCES

1. Pei, P. C., Wang, K. L. and Ma, Z. Technologies for extending zinc–air battery's cycle life: A review. *Applied Energy* 2014; 128:315–324.
2. Cao, R., Lee, J. S., Liu, M. L. and Cho, J. Recent progress in non-precious catalysts for metal–air batteries. *Advanced Energy Materials* 2012; 2:816–829.
3. Muller, S., Striebel, K. and Haas, O. LA0.6CA0.4COO3—A stable and powerful catalyst for bifunctional air electrodes. *Electrochimica Acta* 1994; 39:1661–1668.
4. Lee, D. U., Choi, J. Y., Feng, K., Park, H. W. and Chen, Z. W. Advanced extremely durable 3D bifunctional air electrodes for rechargeable zinc–air batteries. *Advanced Energy Materials* 2014; 4:1301389.
5. Cheng, F. Y., Shen, J. A., Peng, B., Pan, Y. D., Tao, Z. L. and Chen, J. Rapid room-temperature synthesis of nanocrystalline spinels as oxygen reduction and evolution electrocatalysts. *Nature Chemistry* 2011; 3:79–84.

6. Li, Y. G., Gong, M., Liang, Y. Y., Feng, J., Kim, J. E., Wang, H. L., Hong, G. S., Zhang, B. and Dai, H. J. Advanced zinc–air batteries based on high-performance hybrid electrocatalysts. *Nature Communications* 2013; 4:1805.
7. Wang, H. L. and Dai, H. J. Strongly coupled inorganic–nano-carbon hybrid materials for energy storage. *Chemical Society Reviews* 2013; 42:3088–3113.
8. Chen, Z., Yu, A. P., Higgins, D., Li, H., Wang, H. J. and Chen, Z. W. Highly active and durable core–corona structured bifunctional catalyst for rechargeable metal–air battery application. *Nano Letters* 2012; 12:1946–1952.
9. Reddy, T. *Linden's Handbook of Batteries*, 4th Ed. (New York: McGraw-Hill, 2010).
10. See, D. M. and White, R. E. Temperature and concentration dependence of the specific conductivity of concentrated solutions of potassium hydroxide. *Journal of Chemical and Engineering Data* 1997; 42:1266–1268.
11. Lee, C. W., Sathiyanarayanan, K., Eom, S. W., Kim, H. S. and Yun, M. S. Novel electrochemical behavior of zinc anodes in zinc/air batteries in the presence of additives. *Journal of Power Sources* 2006; 159:1474–1477.
12. Ein-Eli, Y., Auinat, M. and Starosvetsky, D. Electrochemical and surface studies of zinc in alkaline solutions containing organic corrosion inhibitors. *Journal of Power Sources* 2003; 114:330–337.
13. Clarke, R. L. US Patent 7582385 B2. (Google Patents, 2009).
14. Goh, F. W. T., Liu, Z. L., Hor, T. S. A., Zhang, J., Ge, X. M., Zong, Y., Yu, A. S. and Khoo, W. A near-neutral chloride electrolyte for electrically rechargeable zinc–air batteries. *Journal of the Electrochemical Society* 2014; 161:A2080–A2086.
15. Masri, M., Nazeri, M. and Mohamad, A. Sago gel polymer electrolyte for zinc–air battery. *Advances in Science and Technology* 2011; 72:305–308.
16. Othman, R., Basirun, W. J., Yahaya, A. H. and Arof, A. K. Hydroponics gel as a new electrolyte gelling agent for alkaline zinc–air cells. *Journal of Power Sources* 2001; 103:34–41.
17. Zhu, X. M., Yang, H. X., Cao, Y. L. and Ai, X. P. Preparation and electrochemical characterization of the alkaline polymer gel electrolyte polymerized from acrylic acid and KOH solution. *Electrochimica Acta* 2004; 49:2533–2539.
18. Fauvarque, J. F., Guinot, S., Bouzir, N., Salmon, E. and Penneau, J. F. Alkaline poly(ethylene oxide) solid polymer electrolytes—Application to nickel secondary batteries. *Electrochimica Acta* 1995; 40:2449–2453.
19. Mohamad, A. A., Mohamed, N. S., Yahya, M. Z. A., Othman, R., Ramesh, S., Alias, Y. and Arof, A. K. Ionic conductivity studies of poly(vinyl alcohol) alkaline solid polymer electrolyte and its use in nickel–zinc cells. *Solid State Ionics* 2003; 156:171–177.
20. Wang, Y. J., Qiao, J. L., Baker, R. and Zhang, J. J. Alkaline polymer electrolyte membranes for fuel cell applications. *Chemical Society Reviews* 2013; 42:5768–5787.
21. MacFarlane, D. R., Tachikawa, N., Forsyth, M., Pringle, J. M., Howlett, P. C., Elliott, G. D., Davis, J. H., Watanabe, M., Simon, P. and Angell, C. A. Energy applications of ionic liquids. *Energy & Environmental Science* 2014; 7:232–250.
22. Kar, M., Winther-Jensen, B., Forsyth, M. and MacFarlane, D. R. Chelating ionic liquids for reversible zinc electrochemistry. *Physical Chemistry Chemical Physics* 2013; 15:7191–7197.
23. Zhang, X. G. Fibrous zinc anodes for high power batteries. *Journal of Power Sources* 2006; 163:591–597.

8 Advanced Materials for Li–Air Rechargeable Batteries

Hossein Yadegari and Xueliang Sun

CONTENTS

8.1 INTRODUCTION

The rechargeable lithium–air (Li–O_2) battery has been attracting a great amount of consideration in recent years since the first report on the feasibility of this system using a nonaqueous polymeric electrolyte membrane by Abraham and Jiang [1]. Li–O_2 batteries, which use a different chemistry from intercalating lithium-ion batteries (LIBs), have a phenomenal theoretical specific energy of 5200 Wh kg^{-1} (based on the mass of lithium metal and gained oxygen during discharge). This specific energy, which is high enough to compete with that of gasoline (13,000 Wh kg^{-1}), motivates the investigation of Li–O_2 batteries for potential applications in electrical vehicles (EVs). Besides, relatively low conversion efficiency of the present internal combustion engines (less than 13%) limits the practical specific energy of gasoline to values as low as 1700 Wh kg^{-1} [2]. Considering the 90% conversion efficiency of electrical engines of EVs, Li–O_2 battery system needs to deliver around 33% of its theoretical specific energy to have the same practical specific energy as gasoline, which is not inconceivable. Various challenges facing the development of Li–O_2 battery system and corresponding solutions introduced from the materials point of view are summarized in the present context.

8.2 FUNDAMENTALS OF Li–O_2 CELL

The high specific energy densities seen for metal–air batteries is related to the use of high-energy alkaline metals as negative, and oxygen, from ambient air, as positive electrode materials. There are four different configurations for an Li–O_2 battery system based on the type of used electrolyte [2], as is depicted in Figure 8.1.

1. Nonaqueous: The most investigated Li–O_2 battery is based on an aprotic nonaqueous solvent in which lithium metal is stable and forms a solid electrolyte interface (SEI). The discharge product, however, is insoluble lithium peroxide (Li_2O_2) that accumulates on the air electrode surface. In addition, Li_2O_2 is an

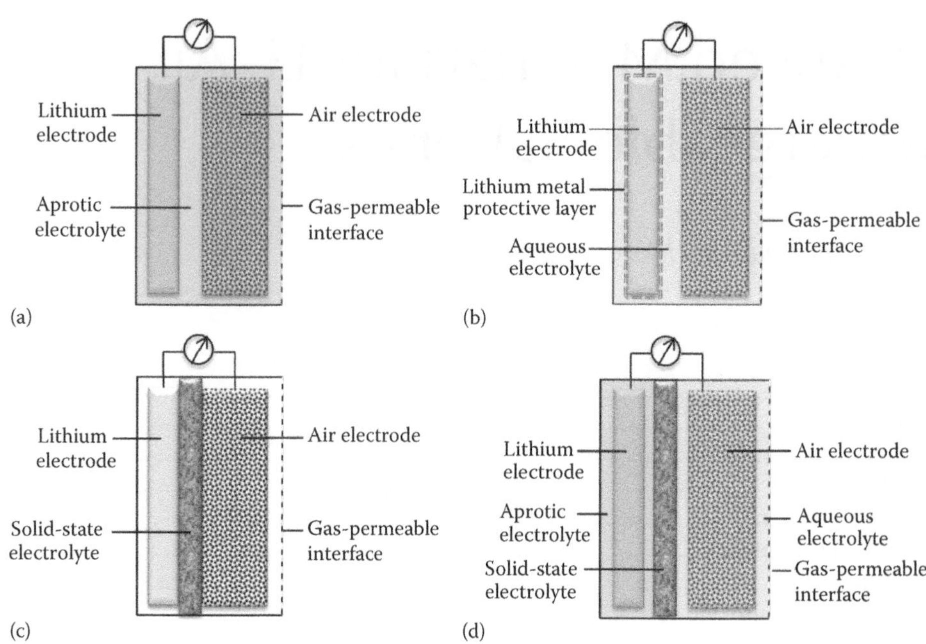

FIGURE 8.1 Four different configurations for Li–O_2 battery system based on the type of the used electrolyte: (a) aprotic nonaqueous electrolyte, (b) aqueous electrolyte, (c) solid-state electrolyte, and (d) dual aqueous/nonaqueous electrolyte.

electrical insulator and imposes a large overpotential to the charge reaction.

2. Aqueous: Uses a protic aqueous solvent in which lithium metal is not stable and must be protected by an interface. The advantage of this system is that the discharge products are soluble, so the system can be recharged more easily. The main challenge of this system, however, is protecting Li metal in an aqueous electrolyte using an efficient lithium conducting interfacial membrane.
3. Solid-state Li–O_2 battery: The liquid electrolyte is replaced by a more stable solid-state electrolyte to provide a safer battery system.
4. Dual aqueous/nonaqueous: Uses a dual aqueous and nonaqueous solvent system in which lithium metal is in an aprotic nonaqueous electrolyte and positive electrode is in an aqueous electrolyte separated by an interface membrane.

Most researches of Li–O_2 battery systems have been focused on aprotic systems in the past decade. During the discharge of this system, Li metal oxidizes and produces Li^+ ions and electrons. The produced electrons go through the external circuit to the positive electrode and reduce the dissolved oxygen to peroxide ions $\left(O_2^{2-}\right)$. At the same time, the Li^+ ions go through the electrolyte to the air electrode and combine with the peroxide ions to produce solid lithium peroxide (Li_2O_2) as the main discharge product [3]:

$$\text{Negative electrode:} \quad 2Li \rightarrow 2Li^+ + 2e^- \tag{8.1}$$

$$\text{Positive electrode:} \quad O_2 + 2e^- \rightarrow O_2^{2-} \tag{8.2}$$

$$\text{Overall:} \quad 2Li + O_2 \rightarrow Li_2O_2 \quad E^0 = 3.10\ \text{V} \tag{8.3}$$

The basic discharge reaction in an aqueous electrolyte can also be summarized as follows [2,4]:

$$\text{Negative electrode:} \quad \text{Aprotic} \quad 2Li \rightarrow 2Li^+ + 2e^- \tag{8.4}$$

$$\text{Positive electrode:} \quad \text{Basic} \quad 1/2O_2 + H_2O + 2e^- \rightarrow 2OH^- \tag{8.5}$$

$$\text{Acidic} \quad 1/2O_2 + 2H^+ + 2e^- \rightarrow H_2O \tag{8.6}$$

$$\text{Overall:} \quad \text{Basic} \quad 2Li + 1/2O_2 + H_2O \rightarrow 2Li^+ + 2OH^- \tag{8.7}$$

$$\text{Acidic} \quad 2Li + H_2O \rightarrow 2Li+ + H_2O \tag{8.8}$$

Various mechanisms for O_2 reduction in Li^+ nonaqueous electrolytes have been proposed [5]. However, the mechanism proposed by Bruce et al. [6,7] is more matched with practical observations. In the proposed mechanism by Bruce et al., the discharge reaction of Li–O_2 in nonaqueous solvents involves the formation of a superoxide $\left(O_2^-\right)$ intermediate that then binds to Li^+, forming a lithium superoxide (LiO_2) intermediate on the air electrode surface. LiO_2 is thermodynamically unstable and transforms to more stable Li_2O_2 via a disproportionation reaction:

$$O_2 + e^- \rightarrow O_2^- \tag{8.9}$$

$$O_2^- + Li^+ \rightarrow LiO_2 \tag{8.10}$$

$$2LiO_2 \rightarrow Li_2O_2 + O_2 \tag{8.11}$$

In the proposed charging mechanism, however, Li_2O_2 directly decomposes to oxygen and lithium without passing through the LiO_2 intermediate. More evidences in accordance with this mechanism have recently been revealed [8]. The kinetics of discharge reaction in an Li–O_2 cell was studied and a first-order disproportionation reaction from an oxygen-rich superoxide-like phase to an Li_2O_2 phase was observed. Interestingly, the oxygen-rich component showed a much smaller overpotential during charge (3.2–3.5 V) than the Li_2O_2 component (~4.2 V). The authors related the formation of the superoxide-like component to the porosity of the activated carbon that was used as the air electrode. The results also showed that the portion of oxygen-rich component in the discharge product increases with increase of the discharge current density, which has been related to the increased LiO_2 nucleation. A similar behavior is also reported by Nazar et al. [9]. Furthermore, a theoretical calculation by Ceder's research group suggests that the decomposition of oxygen-rich off-stoichiometric $Li_{2-x}O_2$ compound is kinetically more favored over the decomposition of Li_2O_2 [10].

8.3 CHALLENGES AND OBSTACLES

8.3.1 Negative Electrode

Lithium metal has the most negative electrochemical potential (−3.040 versus standard hydrogen electrode) and lowest molar mass (6.941 g mol^{-1}) among all the metals. These properties result in the highest theoretical specific energy (11,680 Wh kg^{-1}) and specific capacity (3860 mAh g^{-1}) when using lithium metal as a negative electrode material. As a result, it has become the electrode material of choice for high-energy battery systems like Li–O_2 and Li–S. Nevertheless, rechargeable batteries based on a lithium metal anode have not been commercialized yet. Safety concerns and low cycling performance are two of the main obstacles in this regard. It is well known that the lithium metal anodes form dendrite structures during consecutive discharge–charge cycles, as a result of sequential formation and cracking of the SEI layer in the presence of incompatible organic electrolytes [11–13]. Dendritic structures may ultimately penetrate into the separator and reach the positive electrode, creating a short circuit inside the battery that leads to smoke or even fire in the presence of volatile organic electrolytes. The lithium anode material is also lost because of corrosion and passivation of the dendrite structures during battery cycling [11]. In addition, continuous formation of the SEI layer on high-surface-area dendritic structures results in consumption of the electrolyte, compromising the cycle life of the battery [13]. Addressing the problems relating to the implementation of lithium metal as negative electrode materials is an active research topic [12,13]. Two main approaches are being followed as the potential solutions. The first involves mechanical suppression of the dendrite structure by using an interfacial or protective layer on lithium electrode. The layers, which are also called *ex situ* or artificial SEI, are composed of Li-ion conductive polymers, ceramics, or glass [12,13]. The second approach involves the *in situ* formation of a stable SEI layer using various organic solvents, lithium salts, and functional additives [13].

Another difficulty associated with the utilization of lithium metal in Li–O_2 cells is the contamination of the electrode by moisture and oxygen from the positive electrode. However, an appropriate protective layer, such as LISICON, would undoubtedly improve the cyclability and safety of lithium metal negative electrodes for applications in not only Li–O_2 batteries, but also other types of high-energy lithium batteries [8]. Replacing lithium metal with other negative electrodes materials like Si or Sn, which use different chemistry than the conversion reaction of the lithium electrode, is another option at the cost of a lower specific energy.

8.3.2 Electrolyte

The stability of the electrolyte is one of the most severe challenges facing the development of Li–O_2 batteries even on the research scale. Such a desired electrolyte should tolerate the highly oxidative environment of an Li–O_2 cell for a long cycling life and enable the cell to form and decompose Li_2O_2 reversibly. It is well recognized that the electrolyte not only affects the oxygen reduction reaction (ORR) and oxygen evolution reaction (OER) mechanisms, but also the chemical composition of discharge products and reversibility of the cell [14]. Dozens of electrolytes are proposed involving combinations of various aprotic solvents with different lithium salts. A good review regarding Li–O_2 electrolytes is carried out by Ein-Eli's research group [14]. The most popular electrolytes that are employed in Li–O_2 cells can be categorized as carbonates or ethers; however, Li–O_2 electrolytes are not limited to these two groups. The more popular electrolyte systems in Li–O_2 cells are briefly reviewed here.

8.3.2.1 Alkyl Carbonates

Carbonate solvents include different combinations of ethylene carbonate (EC), propylene carbonate (PC), and dimethyl carbonate. Carbonates are common solvents in rechargeable Li-ion batteries because of their high stability and low volatility. Thus, the research on Li–O_2 batteries was initiated using the same carbonate-based electrolytes [1]. However, later studies revealed that the carbonate-based solvents are unstable in an Li–O_2 environment [15,16]. The discharge reaction of Li–O_2 in nonaqueous solvents involves the formation of a superoxide $\left(O_2^-\right)$ intermediate. Addition of the superoxide anion radical into the ethereal carbon atom of the organic carbonates results in the ring opening of the carbonate and the formation of a peroxy anion (ROO^-) species that is even more reactive than the starting superoxide. Accordingly, carbonate-based electrolytes decompose to form high-molecular-weight products, including lithium alkyl carbonates, lithium carbonate (Li_2CO_3), $C_3H_6(OCO_2Li)_2$, HCO_2Li, and CH_3CO_2Li [14]. In addition, it is proven by McCloskey et al. by the use of atom labeling and differential electrochemical mass spectrometry (DEMS) analysis that CO_2 gas is evolved instead of O_2 in the charge cycle and also that the electrolyte decomposition was the primary source of CO_2 formation during charging [15].

Owing to these side reactions, carbonate-based electrolytes are not considered as a viable electrolyte for the Li–O_2 cell.

8.3.2.2 Ethers

Following the reports on instability of carbonate solvents, most research in this field switched to ether-based electrolytes, which are shown to be more stable than carbonates [17,18]. Ethers are more stable than carbonates toward nucleophilic attacks by superoxide intermediates, have relatively low vapor pressure, and show an extended oxidation potential window up to 4.5 V versus Li/Li^+ and hence might be considered good candidates for Li–O_2 electrolyte. Despite the higher stability found in ether-based electrolytes and Li_2O_2 being the main product at the first discharge cycle, it was found that the formed Li_2O_2 on the first discharge is accompanied by electrolyte decomposition, producing a mixture of lithium carbonate, lithium alkyl carbonates, polyethers/esters, CO_2, and H_2O [19]. Formation of carbonate-based parasitic side products with high interfacial resistance at the substrate/products and products/electrolyte interfaces is believed to impose a large charging overpotential to the Li–O_2 cell. Increasing charging overpotential to over 4 V leads to more electrolyte decomposition at consecutive cycles, which results in the continuing increase of the charging overpotential and fading the cell's performance. Formation of these carbonate-based parasitic side products must be prevented to have a truly rechargeable Li–O_2 cell. The mutual interactions of electrolyte/electrode should be considered in this context. Therefore, neither carbonate- nor ether-based electrolytes are truly stable in Li–O_2 batteries, and searching for a stable solvent remains a major challenge.

8.3.2.3 Dimethyl Sulfoxide

Dimethyl sulfoxide (DMSO)-based electrolytes have also been employed into Li–O_2 cells, demonstrating good stability especially when using noncarbon electrodes [20,21]. However, small amounts of side products from decomposition of DMSO solvent ($DMSO_2$, Li_2SO_3, and Li_2SO_4) were found when using a carbon electrode. DMSO-based electrolytes also interact with the Li metal anode requiring protection of the negative electrode surface from the electrolyte, which may also result in an increase of the internal resistance and a decrease of the specific energy of the cell. The relatively high vapor pressure of DMSO also decreases its viability as an electrolyte for Li–O_2 batteries.

8.3.2.4 Ionic Liquids

Ionic liquids (ILs) possess some key characteristics such as a wide potential window, negligible vapor pressure, and a high ionic conductivity and hydrophobicity making them attractive for application in Li–O_2 cells [22]. The application of ILs in Li–O_2, however, is limited mainly by their poor Li^+ solubility. The study of the ORR mechanism in ILs electrolytes by Abraham et al. showed that Li_2O_2 is formed from the chemical decomposition reaction of LiO_2 [23]. It should also be mentioned that imidazolium-based ILs are not stable enough at the air cathode and also are not stable against reduction at Li-deposition potentials and thus cannot be used alongside Li metal anodes. While electrolytes based on pyrrolidinium and piperidinium IL families are shown to be fairly stable against peroxide radical attack [14], more studies are needed to investigate and address the challenges regarding the use of ILs as an electrolyte for Li–O_2 cells.

8.3.2.5 Solid-State Electrolytes

Solid-state electrolytes offer unique characteristics as electrolytes for Li–O_2 battery systems, including safety, durability, wide operational temperature, and potential range, and also the ability to prevent lithium dendrite formation. The latter characteristic is especially remarkable in Li–O_2 cells, since they act as a barrier against diffusion of ambient gases and moisture toward the Li–anode. However, the low Li^+ ion conductivity of currently available solid-state electrolytes is the main drawback that prevents them from being widely employed in practical battery systems [14]. The application of solid-state electrolytes in Li–O_2 cells is reviewed by Sun [24]. NASICON-type lithium-ion solid conductors are among the most extensively studied solid-state electrolytes, since they are commercialized. Zhang et al. [25] reported an Li–O_2 cell employing a solution of CH_3COOH (HOAc)-H_2O-CH_3COOLi (LiOAc) as an electrolyte under ambient atmosphere and a NASICON-type $Li_{1.35}Al_{0.25}Ti_{1.75}S_{0.3}P_{2.7}O_{12}$ (LTAP) glass ceramic as a solid electrolyte. The Li–O_2 cell consisted of a lithium metal active layer, a lithium-ion conducting polymer buffer layer, a water-stable LTAP protective layer, the HOAc–H_2O–LiOAc electrolyte, and a carbon air cathode including superfine Pt particles (1–4 nm) as catalyst. Flat discharge and charge plateaus with a discharge capacity of 225 mAh g^{-1} were obtained using 56% HOAc at a current density of 0.5 mA cm^{-2}, 60°C under 3 atm of air. The Li–O_2 cell retained a discharge–charge capacity of 250 mAh g^{-1} within 15 cycles.

Zhou et al. [26] also reported an all-solid-state Li–O_2 cell using a lithium anode, an $Li_{1+x}Al_yGe_{2-y}(PO_4)_3$ inorganic solid electrolyte and an air electrode composed of carbon nanotubes (CNTs) and an inorganic solid electrolyte. The porous layer of the air electrode was composed of $Li_{1+x}Al_yGe_{2-y}(PO_4)_3$ (LAGP) and CNT particles (LAGP@CNT). The thickness of the air electrode was around 15 mm. The all-solid-state Li/CE/LAGP@CNT–air cell demonstrated good rate performance, allowing the cell to be discharged and charged at a high current density of 10 A g^{-1}. The authors related the observed good rate performance of the cell to the multiwalled CNTs, which act as both a nanosized catalyst and a continuous electron conduction path in the air electrode. Using the current density of 200 mA g^{-1}, the all-solid-state Li/CE/LAGP@CNT–air cell exhibited a specific discharge capacity of more than 1500 mAh g^{-1}; however, less than 20% of initial capacity was retained after 10 consecutive cycles in the voltage range between 2.0 and 5.0 V (versus Li/Li^+). In another report by the same research group, a solid Li^+ ion conductor was combined with a gel cathode to enhance the electrode/electrolyte interface in the Li–O_2 cell [27]. The solid Li^+ ion conductor in this study was $Li_{1.35}Ti_{1.75}Al_{0.25}P_{2.7}Si_{0.3}O_{12}$ (LTAP) and a mixture of single-wall CNT (SWCNT) and an imidazolium-based

IL was used as the gel cathode electrode. The resulting cell was tested under the ambient air and delivered a high specific discharge capacity of 56,800 mAh g^{-1} (based on SWCNT weight) at the first discharge cycle using a current density of 200 mA g^{-1}. The cell sustained 100 consecutive discharge and charge cycles with limited discharge capacity of 2000 mAh g^{-1}. In addition, spectroscopic studies demonstrated that Li_2O_2 is formed during the discharge of the cell and converted chemically to LiOH and Li_2CO_3 as a result of reaction with H_2O and CO_2 from the air.

Although solid-state electrolytes bring some promise to the Li–O_2 battery system, the associated challenges with these electrolytes such as lack of flexibility toward the large volume changes in discharge and charge cycles, low electrode/electrolyte interface, and low Li^+ ion conductivity should be addressed for the solid-state Li–O_2 battery system to become practical. In addition, most of the available solid-state electrolytes are unstable against lithium metal owing to the presence of high valence transition metals that can be reduced by lithium. Mechanical strength, ease of fabrication, and cost are other important parameters that should be considered.

8.3.2.6 Electrolyte Salts and Binders

In addition to solvent instability, binder and electrolyte salts are also shown to be attacked by superoxide ions and produce decomposition products in the discharged electrode [5]. $LiPF_6$, the most widely used lithium electrolyte salt, is shown to be decomposed during the discharge cycle [28]. Nazar et al. found lithium oxalate in discharge products of Li–O_2 cell using bis(oxalato)borate (LiBOB) in PC as the electrolyte, through *ex situ* x-ray diffraction (XRD) and infrared spectroscopic observations, which has been related to the decomposition of LiBOB as a result of reactions with superoxide radicals [29]. Furthermore, several studies have revealed additional evidence relating the instability of lithium salts and reactions with Li_2O_2 during the discharge cycle. Veith et al. used binder-free carbon foam as the air electrode in tetraethylene glycol dimethyl ether (tetraglyme) with various lithium salts as the electrolyte [30]. They detected that the Li_2O_2 discharge product was contaminated by decomposed halide species. It was concluded that in the absence of the binder, the decomposed halide species originate from the electrolyte anions of $LiClO_4$, $LiBF_4$, $LiPF_6$, and LiTFSI. In another study by Xu et al., the authors employed LiTFSI lithium salt in six different nonaqueous solvents as the electrolyte and polytetrafluoroethylene (PTFE) as the binder [31]. They also observed LiF in discharge products of the cell, which could be a result from the decomposition of the binder and/or lithium salt. Younesi et al. have also reported on the instability of poly(vinylidene) fluoride (PVDF) binder using x-ray photoelectron spectroscopy (XPS). They employed nonfluorinated lithium salt of $LiB(CN)_4$ in polyethylene glycol dimethyl ether and tetraglyme as the electrolyte, and observed a new peak in the F 1s spectrum after cell cycling, which suggests the decomposition of PVDF binder in the Li–O_2 environment [32,33]. Utilizing free-standing binder-free air electrode materials would eliminate the decomposition problem of the binder in Li–O_2 cell; however, truly stable and reliable electrolytes for this system are still lacking.

8.3.3 Positive Electrode

As mentioned before, the produced Li_2O_2 is insoluble in the nonaqueous electrolyte and hence accumulates on the air electrode surface. Such an air electrode should have a proper porous structure with an appropriate pore volume and pore size distribution, in addition to the general characteristics such as conductivity, chemical stability, high surface area, and low cost. The porous structure is responsible for oxygen diffusion into the positive electrode material, formation and storage of the discharge products, and also decomposition of the produced discharge products during the charge cycle [12]. Furthermore, the final performance of the metal–air system strongly depends on the efficiency of the air electrode. It is shown that the discharge capacity of metal–air cells is limited to the air electrode's capacity to store the discharge products [13,34]. The ability of the air electrode to accommodate the discharge products (Li_2O_2) determines the discharge capacity of the electrode and hence the whole battery. Two different processes control the air electrode's ability to accommodate the discharge product, as shown in Figure 8.2 [35]:

1. At relatively lower discharge current densities, a dense layer of discharge product covers the active surface of the electrode. The insulator nature of the deposited layer increases the charge transfer resistance and results in a voltage drop.
2. At relatively higher discharge current densities, transportation of electroactive species, i.e., lithium ions and dissolved oxygen molecules, limits the capacity of the cell. In fact, mass transport resistance has a more important role at higher current densities.

Furthermore, the solid Li_2O_2 must convert back to Li metal and oxygen during the charge cycle. However, since the Li_2O_2 is an electrical insulator, accumulated discharge products impose a large overpotential to the charge reaction [2]. This high charge overpotential increases the required energy for charging the system and hence decreases the energy efficiency of the battery. Various sources have been suggested for this high charge overpotential by different researchers. Bruce et al., for example, investigated ORR and OER in a nonaqueous Li^+ electrolyte and observed LiO_2 as the intermediate of the ORR [7]. The unstable LiO_2 intermediate thereafter disproportionates to the more stable Li_2O_2 oxide. On the charging cycle, however, they found that Li_2O_2 directly decomposes to produce oxygen without passing through an LiO_2 intermediate. As a result, the oxygen reduction and evolution reaction in nonaqueous electrolytes may not be considered as a classic reversible electrochemical reaction involving a reversible redox couple. On the other side, formation of parasitic side products rather than Li_2O_2 as a result of electrolyte decomposition as well as electronic ohmic voltage drop are among the

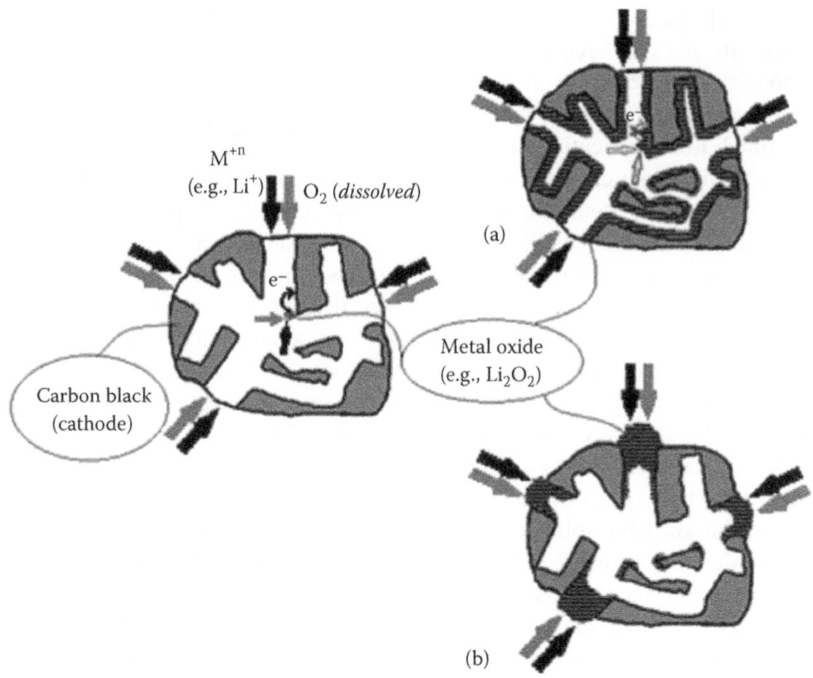

FIGURE 8.2 Two different limiting criteria for the capacity of the Li–O_2 air electrode: (a) charge transfer resistance at relatively lower discharge current densities and (b) mass transfer resistance at relatively higher current densities. (Adapted from Kraytsberg, A., Ein-Eli, Y., *Nano Energy*, 2, 468, 2013. With permission.)

other suggested possibilities for the observed high charging overpotential [35].

Along these lines, the major outcome of all these limitations is the poor cycling performance from most of the reported Li–O_2 batteries. Degradation of air electrode materials as a result of reactions with highly reactive superoxide and peroxide ions, decomposition of electrolyte components, and lithium metal devastation by contamination with oxygen and moisture are given as potential explanations for the observed limited cycling performances. However, most research on Li–O_2 batteries concentrate on the air electrode side, where the oxygen diffuses inside the positive electrode, oxygen reduction and evolution reactions take place, and discharge products are produced and decomposed. Research on the air electrode materials may be separated into two main streams [35]:

1. Increase of discharge capacity by improving the accommodation ability of the air electrode to load more discharge products
2. Decrease of charging overpotential by utilizing different electrocatalysts

Employing advanced nanomaterials as the air electrode of Li–O_2 cell to overcome the limitations and challenges facing Li–O_2 batteries development is the subject of the following section.

8.3.3.1 Carbon Nanotubes and Nanofibers

Carbon has been one of the materials of choice for application in energy storage and conversion field, since it is conductive, stable, lightweight, inexpensive, and also can be synthesized in different structures such as nanoparticles (NPs), nanotubes, nanorods, nanosheets, and porous structures. Different carbon structures have also been synthesized and employed as the air electrode of Li–O_2 batteries. Nanotube-based structures are among the most investigated carbon structures in this regard. One-dimensional (1D) carbon nanomaterials have been intensively investigated in various applications ranging from energy conversion and storage to medical and biosensing. The high electrical conductivity and extended surface area of CNT makes it an attractive choice as the air electrode of Li–O_2 batteries.

Freestanding SWNT and carbon nanofiber (CNF) mixed buckypaper was synthesized using a filtration method and applied as positive electrode of an Li–O_2 cell by Zhang et al. [36]. The cell showed a specific capacity of 2540 mAh g^{-1} at a current density of 0.1 mA cm^{-2}. In addition, it was observed that the discharge capacity is also related to the thickness of the air electrode and decreases with increase of the electrode thickness. Using electrochemical impedance spectroscopy (EIS), the authors found that the charge transfer resistance (R_{ct}) increased dramatically after the discharge reaction. It is believed that the increase in R_{ct} is mainly due to the oxygen deficiency inside the air electrode. On the basis of the EIS and scanning electron microscopy (SEM) results, blocking of the air electrode a result of nonuniform deposition of discharge product is proposed as the limiting factor for the discharge capacity. Shao-Horn's research group also synthesized hollow carbon fibers (~30 nm in diameter) on a ceramic porous substrate and applied it as the air electrode in an Li–O_2 cell [37]. The obtained air electrode showed a high specific capacity of 4720 mAh g^{-1} based on the mass of carbon using 0.1 M $LiClO_4$

in 1,2-dimethoxyethane (DME) as the electrolyte. This high specific capacity has been attributed to low carbon packing, high utilization of the carbon mass, and void volume for Li_2O_2 formation. However, normalized specific capacity based on the combined weight of carbon air electrode and total Li_2O_2 formed upon discharge gives the value of 944 mAh g^{-1} at the same current density, which is much lower than when normalized based on the mass of carbon (Figure 8.3). This is due to the mass of discharged product being calculated as four times higher than the carbon's mass, showing the considerable influence of the discharge product's mass on the specific capacity of the Li–O_2 cells. In addition, the Li–O_2 cell based on CNFs maintained little more than half of its original specific capacity after 10 cycles.

In another study, a freestanding hierarchically porous CNT (FHP-CNT) film with bimodal porosity was synthesized as a binder-free electrode for an Li–O_2 cell [38]. To synthesize the air electrode, a colloid suspension of CNTs and polystyrene (200 nm in diameter) was used as a template and was formed by vacuum filtration followed by annealing at 500°C under the nitrogen atmosphere to leave a porous CNT film. A porosity of 81% in comparison with bulk graphite was achieved using the described synthesis method. The FHP-CNT electrode showed a specific capacity of 4683 mAh g^{-1} at a current density of 50 mA g^{-1} and an electrode loading of 0.11 mg cm^{-2} in an ether-based electrolyte. The specific capacity was also decreased in this case to 937 mAh g^{-1} when considering the mass of the discharge product in the calculations. The Li–O_2 cell exhibited 14 reversible cycles at limited discharge depth of 1000 mAh g^{-1} based on the carbon electrode's mass. The bimodal nature of porous electrode is the advantage of the synthesized air electrode, providing both broad oxygen diffusion paths and large void volumes for accommodation of discharge products, as emphasized by the authors.

Aurbach's research group also fabricated activated carbon microfiber (ACM) by heat treating commercial carbon microfiber at 900°C under a carbon dioxide atmosphere for 10 h [39]. The ACM electrode exhibited a maximum specific capacity of 4116 mAh g^{-1} based on its carbon mass at a current density of 0.025 mA cm^{-2} using 1 M LiTFSI in triglyme as the electrolyte. SEM micrographs before and after discharge of the air electrode to 2.0 V show uniform distribution of discharge products on individual fibers without blocking the diffusion paths between the fibers (Figure 8.4a–d). The electrochemical response of ACM electrode was also compared with a carbon black composite electrode as a reference. A unique electrochemical behavior was observed for ACM electrode upon the second and subsequent discharge cycles (Figure 8.4f). Although a regular discharge plateau was observed at around 2.7 V in the first cycle, the second and subsequent discharge curves showed reduction processes that start at 4.2 V, sloping down continuously to reach to 2.7 V again. This behavior was not observed in the case of the carbon black composite electrode and hence is a characteristic of the ACM electrode.

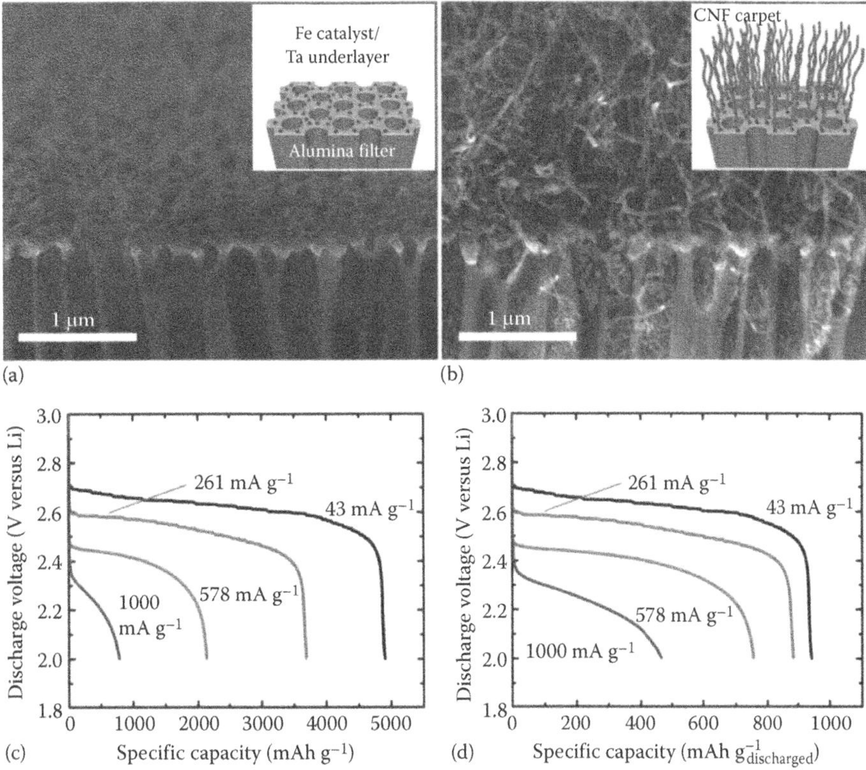

FIGURE 8.3 SEM micrographs of the porous anodized aluminum oxide (a) before and (b) after nanofiber growth. First discharge capacity of carbon nanofiber electrode at different current densities of 43, 261, 578, and 1000 mA g^{-1} based on the weight of (c) carbon and (d) carbon + Li_2O_2. (Adapted from Mitchell, R. R., Gallant, B. M., Thompson, C. V., Shao-Horn, Y., *Energy Environ. Sci.*, 4, 2952, 2011. With permission.)

FIGURE 8.4 SEM micrographs of ACM electrode before (a and b) and after (c and d) discharge to 2 V in Li–O_2 cell. Discharge and charge curves at different current densities (e) and cycling performance (f) of ACM electrode in triglyme–LiTFSI electrolyte. (Adapted from Etacheri, V., Sharon, D., Garsuch, A., Afri, M., Frimer, A. A., Aurbach, D., *J. Mater. Chem. A*, 1, 5021, 2013. With permission.)

Two potential reasons are discussed to explain the observed behavior. First, the large surface area of the ACM electrode promotes the decomposition reaction of electrolyte in the first charge cycle, and reduction of accumulated decomposition products causes the sloping region. The authors believe that the higher volume/mass ratio of the electrolyte in this case made it possible to observe the sloping region that has not been reported before. The possible catalytic effect of the high-surface-area carbon fibers to enhance both the decomposition process of the electrolyte solution and the further reduction of the solution reaction products during the discharge process is also considered as the second potential reason. Although no more analytical results were revealed to confirm the proposed reasons, it seems that the decomposition of electrolyte plays an important role. In addition, the better cycling performance of the ACM electrode (83% capacity retention after five cycles) compared with that of carbon black composite electrode (14% capacity retention after five cycles) is correlated to the electrode's macrostructure, including interwoven fibers and hierarchical porosity by meso- and micropores, which provides ideal diffusion paths and three-phase boundaries for ORRs.

8.3.3.2 Doped Carbon Nanotubes

Doping of carbon materials with heteroatoms such as nitrogen, sulfur, phosphorus, and bromine can cause some structural defects on the carbon and also increase the conductivity of the electrode material. Doping of carbon with nitrogen atoms has attracted much attention because conjugation between the nitrogen lone-pair electrons and graphene π-systems may create nanostructures with desired properties [40,41]. Three different N structures were identified to be present in nitrogen-doped CNT (N-CNT) samples: graphite-like, pyridine-like, and molecular N_2. The higher content of pyridinic N structure would lead to more metallic behavior of CNTs. In addition, the pyridinic N content is considered to be responsible for the wall roughness and interlinked morphologies [42].

The electrochemical behaviors of CNTs and N-CNTs were compared in an Li–O_2 cell by Sun's research group [43]. The nitrogen content of the N-CNT sample used in this work was determined to be 10.2% by the peak area ratio between N and C+N from XPS analysis. The N-CNT electrode material showed a slightly higher discharge voltage, lower charging overpotential, and around a 50% larger specific capacity in comparison with a pristine CNT electrode. The electrolyte was 1 M $LiPF_6$ in PC/EC in 1:1 weight ratio. The increase of electrical conductivity and also the stronger interaction between carbon and the product's molecules, as a result of doping negatively charged nitrogen atoms in the carbon structure, were given as the explanation for the improved electrochemical performance of the N-CNT electrode in the Li–O_2 cell. More recently, Mi et al. investigated the effects of nitrogen doping on the discharge performance of the CNT electrode in Li–O_2 cells [44]. CNTs and N-CNTs were synthesized using a floating catalyst chemical vapor deposition method, and the content of nitrogen in the N-CNT sample was determined to be 1.52% by XPS. They found that N-doping in CNTs promotes the discharge capacity by 30–40% when using both carbonate- and ether-based electrolytes in Li–O_2 cells. In fact, the doped N atoms improve the electron transmission between the O_2 molecules and CNTs, and also decrease the energy barrier of O_2 dissociation. It is also discussed that the N-doping produces more available sites for O_2 adsorption on the neighboring atoms by imposing a positive charge on them. The morphology of the discharge products was also compared on CNT and N-CNT air electrodes. A denser population of discharge products was observed on the N-CNT electrode. Furthermore, discharge products on the N-CNT electrode showed a more uniform film consisting of small particles, while the discharge products on the CNT electrode tended to aggregate into large clusters. The smaller discharge product particles can be related to the presence of more nucleation sites on the N-CNT surface and might result in lower polarization during the charge process. It is concluded that three main reasons are jointly contributing to improve the discharge performance of N-CNT electrode material: improved conductivity, more nucleation sites around the nitrogen dopants, and less agglomeration of discharge products.

8.3.3.3 Graphene

Graphene nanosheets (GNSs), a 2D carbon nanostructure, have attracted huge amounts of attention and application in energy-related research, as well as Li–O_2 battery investigations [45]. It can be expected that the unique characteristics of GNSs, such as a large surface area (2639 m^2 g^{-1}), high electrical conductivity (~64 mS cm^{-1}), and existence of surface defects, would enhance the carbon's efficiency for application as an air electrode in metal–air batteries [35].

Sun's research group was among the first research groups that investigated GNS as the air electrode in a nonaqueous Li–O_2 cell [46]. GNSs were synthesized via the oxidation of graphite powder using the modified Hummers method. SEM and TEM micrographs showed a thin, wrinkled structure for the synthesized GNS sample (Figure 8.5a,b). The air electrode was prepared by casting a mixture of the GNSs and a PVDF binder with a weight ratio of 9:1 on a separator sheet, where the active material loading was around 0.4 mg cm^{-2}. The GNS air electrode delivered a high specific capacity of 8705 mAh g^{-1} at a current density of 75 mA g^{-1} using 1 M $LiPF_6$ in PC/EC (1:1 weight ratio) as the electrolyte. In fact, GNSs provide a large surface area and proper void volume to accommodate solid discharge products of the Li–O_2 cell. Direct observations of the air electrode materials after discharge revealed that discharge products are deposited on both sides of the GNS with a higher density at the edge sites (Figure 8.5c,d). The higher density of discharge products on the edges of GNSs suggests that edge-plane carbon sites are more electrochemically active for the ORR. In addition, the GNS air electrode showed a slightly higher discharge plateau compared with the commercial carbon black control electrodes (Figure 8.5e). The higher activity of edge-plane carbons can also explain the higher discharge plateau of GNS electrode, since GNS electrode material possesses much more edge-plane carbon sites compared with graphitic carbon black samples. The chemical compositions of discharge products were determined to be Li_2CO_3 (major) and Li_2O_2 (minor) using XRD. The presence of lithium carbonate as the major discharge product is correlated to the reaction of the superoxide intermediate with the carbonate-based electrolyte. Either way, the columbic efficiency of the GNS electrode for charging was very low in this case.

GNSs were also employed to prepare a hierarchical porous structure for application as an air electrode of an Li–air cell [47]. GNSs in this work were prepared by thermal expansion and simultaneous reduction of graphite oxide. The C/O ratio of graphite oxide was increased from ~2 to ~15 after thermal reduction at 1050°C for 30 s, while longer residence times increased this ratio to values as high as 100 or even more as a result of CO_2 dismissing. The electrical conductivity of the prepared GNSs was also doubled with an increase of the C/O ratio from 15 to 100. The as-prepared GNS samples with C/O ratios of 14 and 100 were then dispersed in a microemulsion solution containing PTFE binder with the weight ratio of 75:25. The final air electrode was formed by casting and drying the dispersed microemulsion. The prepared air electrode using GNSs with C/O ratios of 14 and 100 showed discharge capacities of around 15,000 and 8000 mAh g^{-1}, respectively, using 1 M of LITFSI in triglyme as the electrolyte and an oxygen pressure of 2 atm. It should be mentioned that a prototype cell was constructed using the air electrode consisting of GNS with a C/O ratio of 14 and tested in ambient air with an

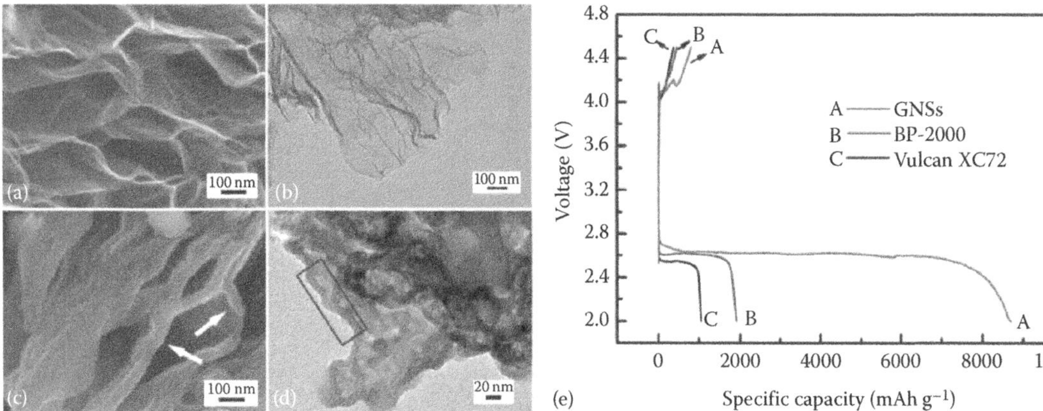

FIGURE 8.5 SEM (a, c) and TEM (b, d) micrographs of GNS electrode before (a, b) and after (c, d) discharge to 2 V in Li–O_2 cell. (e) Discharge and charge curves of Li–O_2 cell using a GNS electrode and two commercial carbon air electrode at a current density of 75 mAh g^{-1} using 1-M $LiPF_6$ in PC/EC as the electrolyte. (Adapted from Li, Y., Wang, J., Li, X., Geng, D., Li, R., Sun, X., *Chem. Commun.*, 47, 9438, 2011. With permission.)

oxygen partial pressure of 0.21. The prototype cell delivered a specific capacity of 5000 mAh g^{-1} on the first discharge cycle. The main discharge product was determined to be Li_2O_2 by means of XRD and selected area electron diffraction (SAED). In addition, density functional theory (DFT) calculations showed that the nucleation and growth of Li_2O_2 is preferred near functionalized lattice defect sites on graphene with functional groups, as a result of the relatively stronger interaction between deposited Li_2O_2 monomers at the defect sites. Two factors are emphasized by the authors to be responsible for the improved discharge performance of the graphene-based air electrode: first, the morphology of the graphene-based air electrode in which macro-sized tunnels facilitate continuous oxygen flow into the air electrode, and meso- and micropores provide three-phase regions for the ORR; second, a facilitated Li_2O_2 deposition mechanism on the graphene surface near functionalized defect sites.

To further improve the 3D structure of graphene-based air electrodes, Wang et al. synthesized a FHP carbon (FHPC) structure via the sol–gel method using graphene oxide (GO) gel on nickel foam as the base skeleton [48]. The authors discussed that the acidic nature of GO contributes to the slight etching of nickel foam, producing NiOOH by-products in the presence of the gel solution. The NiOOH by-product was then decomposed in the carbonization process at 800°C under N_2 atmosphere into NiO and reduced by the carbon to form Ni particles (Figure 8.6a–d). The conclusion was that Ni particles improve the electrical conductivity and adjust the porous structures. The prepared FHPC air electrode was applied in an Li–O_2 cell using 1 M of LiTFSI in DME at an oxygen pressure of 2 atm. The discharge capacity at a current density of 0.2 mA cm^{-2} (280 mA g^{-1}) reached 11,060 mAh g^{-1} (Figure 8.6e). However, the calculated specific capacity was based on the carbon weight, and considering the weight of nickel foam in the calculations may dramatically decrease the specific capacity. The discharge product deposited on the FHPC air electrode after the discharge cycle was examined by means of XRD, which was detected to be mainly Li_2O_2. The Li–O_2 cell showed 10 consecutive discharge and charge cycles with restriction of the discharge capacity to 2000 mAh g^{-1} at a current density of 5 mA cm^{-2}.

More recently, Zhang et al. performed an interesting study to investigate the effect of structural surface defects on the cycling efficiency of graphene materials in Li–O_2 cells [49]. The graphene electrode used in this study was prepared by using the electrochemical leavening process. Two pieces of highly ordered pyrolytic graphite paper were used as the working and counter electrodes in a three-electrode electrochemical cell containing an aqueous solution of Na_2SO_4 (1 M) as the electrolyte. A constant anodic current of 200 mA cm^{-2} was applied for 6 h to the graphite paper to insert electrolyte anions into the graphitic structure (Figure 8.7a–d). The resulting paper, which can be considered as a transient state between the pristine graphite paper and fully exfoliated graphene sheets, was annealed in an argon atmosphere at temperatures of 400°C (H400), 600°C (H600), and 800°C (H800) for 30 min to remove the formed defects during the electrochemical oxidation. Raman spectroscopy analysis of the as-prepared graphene foams revealed that the I_D/I_G ratio (which refers to the defect density in the samples) constantly decreases from 0.71 for the as-prepared graphene foam (before annealing) to 0.07 for the H800 sample. Lower values for the I_D/I_G ratio showed that the surface defects were removed in the annealing process. All the samples were also tested as the air electrode of an Li–O_2 cell using 1 M $LiCF_3SO_3$ in tetra(ethylene) glycol dimethyl ether (TEGDME) as the electrolyte (Figure 8.7e). Interestingly, with increase of the annealing temperature, the discharge voltage of cells increased and the charging overpotential decreases. The round-trip efficiencies (the specific discharging energy/specific charging energy) were calculated on the basis of the discharge and charge cycles, which showed a noticeable increase from 51% for the as-prepared graphene foam to 80% for the H800 sample. The increase in round-trip efficiency is correlated to the lower amounts of defects in annealed graphene foams. The authors discussed that the defect sites (which are oxygen-containing functional group

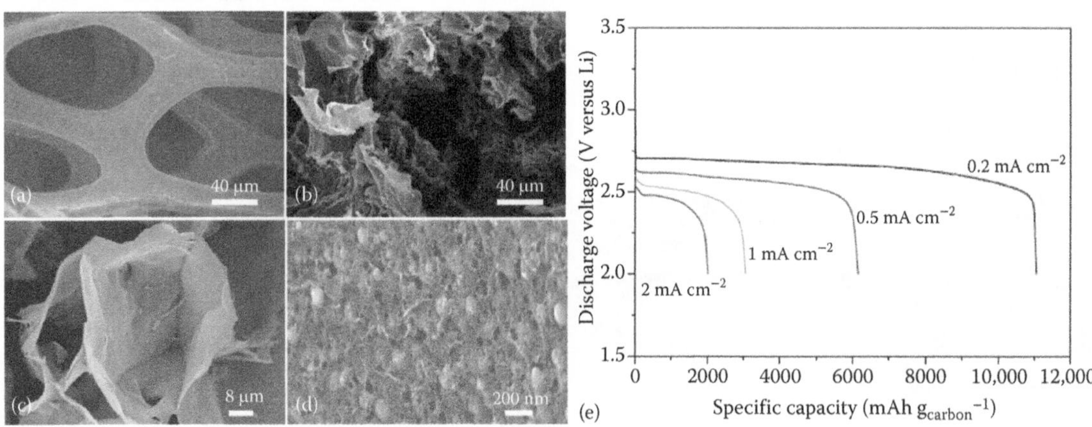

FIGURE 8.6 SEM micrographs of the pristine nickel foam (a) and the as-prepared FHPC electrode with different magnifications (b and c). (e) Discharge curves of FHPC electrode in Li–O_2 cell using a 1-M LITFSI in DME electrolyte at an oxygen pressure of 2 atm with different current densities. (Adapted from Wang, Z. L., Xu, D., Xu, J. J., Zhang, L. L., Zhang, X. B., *Adv. Funct. Mater.*, 22, 3699, 2012. With permission.)

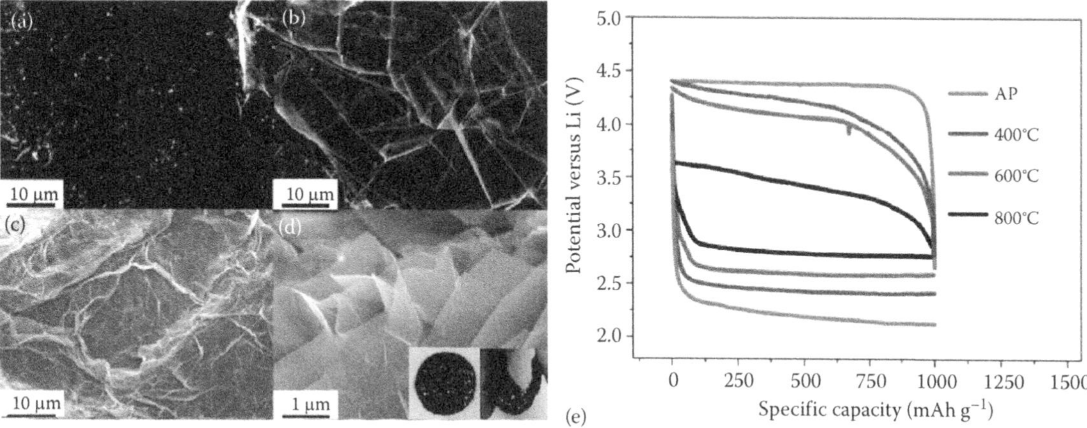

FIGURE 8.7 SEM micrographs of the graphite paper after electrochemical leavening for (a) 0 h, (b) 2 h, (c) 4 h, and (d) 6 h in 1 M Na_2SO_4 at 200 mA cm^{-2}. (e) Discharge and charge curves of graphene foam annealed at different temperatures as the air electrode of Li–O_2 cell using 1-M $LiCF_3SO_3$ in TEGDME as the electrolyte. (Adapted from Zhang, W., Zhu, J., Ang, H., Zeng, Y., Xiao, N., Gao, Y., Liu, W., Hng, H. H., Yan, Q., *Nanoscale*, 5, 9651, 2013. With permission.)

like C–O–C, C=O, and C–OH) slow down the charge transfer during the discharge and charge processes, suppressing the oxygen reduction and evolution reactions. The H800 electrode showed the best cycling performance among all the samples, exhibiting stable discharge voltages at 2.8 V and charge voltages below 3.7 V for 20 cycles. The authors concluded then that the carbon atoms connected to the functional groups of C–O–C, C=O, and C–OH are more active to oxidation and are converted to carbonates or other side products during the charge process. Accumulation of such side products in the Li_2O_2/electrolyte or Li_2O_2/carbon substrate interfaces may result in the loss of electrical contact and lead to the increase of the charge overpotential. These findings about the role of structural surface defects are in contrary with previous results. Although a part of the observed improved electrochemical behavior of the graphene-based foams after removal of surface defects can be correlated to the improved electrical conductivity, more comparative studies are needed to make a general conclusion.

8.3.3.4 Doped Graphene

As was mentioned earlier, doping of carbon materials with heteroatoms such as nitrogen, sulfur, and phosphorus can increase the electrocatalytic activity of carbon materials by forming structural defects on the carbon. In the case of nitrogen doping for instance, the spin density and charge distribution of carbon atoms will be influenced by the neighbor nitrogen dopants and induce an activation region on the graphene surface. These activated regions can directly contribute to catalytic reactions including the ORR [50]. Sun's research group synthesized and examined both N- and S-doped GNSs (N-GNS and S-GNS) as the air electrode of Li–O_2 cells [51,52]. A GNS sample was synthesized by oxidation of graphite powder using the modified Hummers method and a N-GNS sample was prepared by post-heating of the GNS under high purity ammonia mixed with Ar at 900°C for 5 min. To synthesize S-GNS, GNS was dispersed in a solution of *p*-toluenesulfonic acid in acetone and the slurry was stirred at room temperature until the solvent totally evaporated. The resulting product was dried at 100°C overnight and finally calcined at 900°C under Ar atmosphere for 1 h. XPS studies showed three different types of nitrogen in the N-GNS (pyridinic, pyrrolic, and graphitic) and the nitrogen content was also determined to be 2.8%. The presence of sulfur in GNS with a relative percentage of 1.9% was also confirmed by XPS analysis.

The air electrodes were prepared from synthesized electrode materials using a PVDF binder and tested in an Li–O_2 cell containing 1 M $LiPF_6$ in tetraglyme as the electrolyte. The N-GNS sample exhibited a 40% higher discharge capacity (11,660 mAh g^{-1}) in comparison with pristine GNS (8530 mAh g^{-1}) at a current density of 75 mA g^{-1}. Formation of Li_2O_2 as the discharge product was confirmed by analysis of both GNS and N-GNS electrodes after discharge. Furthermore, SEM micrographs of the electrode materials after discharge revealed that the discharge products have a smaller particle size on the N-GNS electrode when compared with those on the pristine GNS electrode. The presence of more active sites (defects and functional groups) as a result of homogeneously distributed nitrogen species on N-GNS surface is argued by the authors to provide more nucleation sites and result in deposition of discharge products with a smaller size. These observations are consistent with the results obtained by the comparison of CNTs and N-CNTs as the air electrodes of Li–O_2 cell [44]. The improved discharge performance of the N-GNS electrode has also been attributed to the presence of the surface defects introduced by nitrogen doping. In the case of sulfur doping, the S-GNS electrode exhibited a lower discharge capacity (4300 mAh g^{-1}) in comparison with the GNS electrode (8700 mAh g^{-1}). However, the S-GNSs showed a higher charge capacity (4100 mAh g^{-1} compared with 170 mAh g^{-1} for GNS electrode). To explain the lower discharge capacity for the S-GNS electrode, the authors cited first-principle calculations, revealing that the oxygen adsorption energy is not increased by doping sulfur into graphene. Therefore, the discharge capacity of S-GNS in the Li–O_2 cell could not be improved. Furthermore, SEM micrographs revealed that the discharge products have a different morphology on S-GNS and pristine GNS electrode materials. Some nanorod structures were observed in the case of S-GNS electrode, while the discharge products on GNS electrode had an amorphous shape. The authors concluded that S-GNS acts as a semiconductor and diminishes the attraction between the produced O_2^- intermediate and active electrode sites, therefore allowing the superoxide ions to diffuse away from electrode surface and make the nanorod structure. It is also mentioned that the short diffusion paths and higher void volumes resulting from randomly oriented nanorods are responsible for the decreased discharge capacity and enhanced charge performance in the case of the S-GNS electrode material. This study also highlights the role of Li_2O_2 morphology on the charge and discharge characteristics of Li–O_2 cells.

8.3.3.5 Hierarchical Carbon Materials

Hierarchical structures provide a high surface area, increased number of active sites, and bimodal porosity, which are required characteristics of air electrodes. Although nanomaterials offer short ion and electron diffusion paths and hence high rate capability, single-mode nanomaterials exhibit limited cycle life due to the isolation of nano-domains especially in thicker electrodes. However, hierarchical materials combine the advantages of nano- and micro-sized materials together to have highly efficient electrode materials with an increased rate capability and cycle life [53]. Particularly in Li–O_2 cells, hierarchical structures present both macroscale channels to transfer O_2 and Li^+ onto the reaction sites, and nano-scale structures to increase the surface area and accommodate more amounts of discharge products. The bimodal-pore concept is well recognized in drug and catalyst industries, but is a relatively new and the most promising material in battery technology, especially in terms of energy density and power density [54].

Guo et al. synthesized mesoporous/macroporous carbon sphere array (MMCSA) to be used as the air electrode for Li–O_2 cells [55]. The MMCSA sample was synthesized using a template-assisted method, presenting a Brunauer–Emmett–Teller (BET) surface area of 360 $m^2\ g^{-1}$ (Figure 8.8a). The as-prepared MMCSA sample then was mixed with carbon black (at different mass ratios from 0 to 80 wt.% of MMCSA) and a PVDF (20 wt.%) binder and mounted onto a carbon paper disk to form the air electrode of Li–O_2 cell working in TEGDME–LiTFSI electrolyte. The specific discharge capacity of the Li–O_2 cell at a current density of 50 mA g^{-1} exhibited a considerable increasing trend from 2100 to 7000 mAh g^{-1} using electrodes with MMCSA contents of 0 to 80 wt.%, respectively (Figure 8.8b). However, an increase of MMCSA content to 100 wt.% resulted in the decrease of the specific discharge capacity to 4100 mAh g^{-1}, which has been related to the poor mechanical strength of the final air electrode. The air electrode containing 30 wt.% MMCSA showed the best cycling performance (30 discharge and charge cycles restricted to the discharge capacity of 1000 mAh g^{-1}) among the samples.

8.3.3.6 Instability of Carbon in an Li–O_2 Environment

As mentioned before, the research on Li–air batteries started with the use of carbonate-based electrolytes. However, the application of carbonate-based electrolytes was halted upon revealing the instability of these electrolytes in an Li–O_2 environment. It was found that carbonate-based electrolytes involve a decomposition reaction to produce lithium carbonate and alkyl lithium carbonate species [16]. Most of the researchers switched to ether-based electrolytes for Li–O_2 battery system thereafter. However, lithium carbonate was still being detected as a minor by-product of Li–O_2 batteries with ether-based electrolytes [19]. McCloskey et al. used XPS and isotope labeling coupled with DEMS to study the discharge–charge reaction mechanism of Li–O_2 in an ether-based electrolyte [19]. DEMS measurements of galvanostatic discharge–charge experiments using 1 M LiTFSI in DME as the electrolyte and different carbon-based cathodes (Avcarb

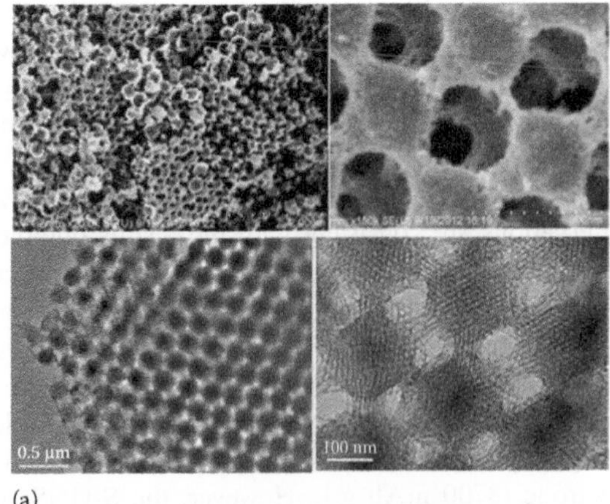

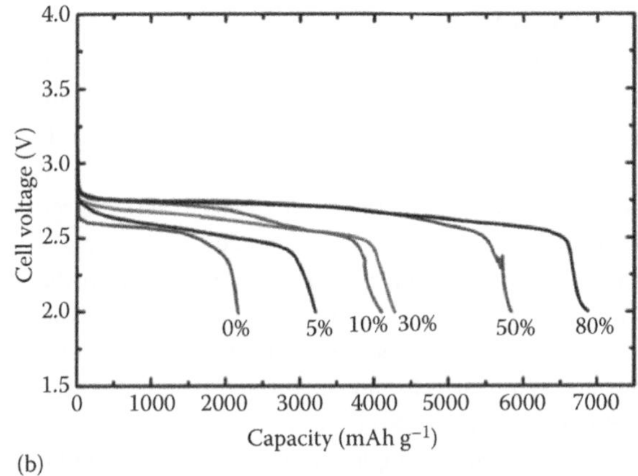

FIGURE 8.8 (a) SEM and TEM images of MMCSA sample. (b) Discharge curves of Li–O_2 cells at a current density of 50 mA g^{-1} with different wt.% of MMCSA in porous catalytic electrodes. (Adapted from Guo, Z., Zhou, D., Dong, X., Qiu, Z., Wang, Y., Xia, Y., *Adv. Mater.*, 25, 5668, 2013. With permission.)

P50 C paper, XC72, Super P, and Ketjen black [KB]) showed that lithium carbonate is always produced during the charge reaction. The authors then used an isotopic carbon electrode made of 99% ^{13}C to trace the source of this carbonate and measure the evolved gases during the charge reaction using DEMS. They found that both $^{13}CO_2$ and $^{12}CO_2$ evolve during the charge cycle, implying that solid carbonate by-products come both from reactions of Li_2O_2 with the carbon cathode and decomposition of the electrolyte. In addition, XPS analysis on a smooth glassy carbon cathode in a bulk electrolysis cell revealed that some carbonate is formed during discharge and is not removed unless charged to >4 V. On the basis of the DEMS and XPS results, the authors concluded that the carbonate is formed both by chemical reactions of Li_2O_2 with the carbon substrate at the Li_2O_2/carbon interface during discharge and by decomposition of the electrolyte at the Li_2O_2/electrolyte interface during discharge and charge (Figure 8.9). They also argued that during the charge cycle, the Li_2O_2 layer also becomes partially covered by carbonates (Li_2CO_3 and $LiRCO_3$) via the electrochemical reaction of Li_2O_2 with the electrolyte, which results in a continuous increase of the charging potential.

Shortly after publishing their results, Shao-Horn's research group used x-ray absorption near edge structure (XANES) to study the discharge–charge reaction mechanism of Li–O_2 in an ether-based electrolyte [18]. A binder-free vertically aligned CNT (VACNT) electrode in a similar (DME-based) electrolyte was applied in this study. Two modes of total electron yield (TEY) and bulk fluorescence yield (FY) were used to record XANES spectra, which enabled them to obtain analytical information either from surface or bulk of the discharge products, respectively. They found that Li_2CO_3 predominately forms at the surface of CNTs in discharged electrodes (Figure 8.10a,b). However, no evidences were found regarding the presence of carbonates at the Li_2O_2/electrolyte interface. In addition, it was observed that Li_2CO_3 accumulates continuously upon cycling (Figure 8.10c). The XANES results were also accompanied by TEM observation and SAED analysis. The authors related the accumulation of parasitic carbonate species to the changes in the charging voltage profile during cycling (Figure 8.10e). They concluded that Li_2CO_3 becomes increasingly difficult to oxidize at higher cycles and agglomerates on the electrode surface.

Bruce's research group also investigated the stability of carbon in the Li–O_2 cell [56]. The hydrophobicity/hydrophilicity of a ^{13}C carbon electrode was adjusted using acid treatment or Fenton's reagent, and the discharge–charge mechanism of the Li–O_2 cell was studied in a DMSO or tetraglyme electrolyte with the aid of DEMS and FTIR. Interestingly, the results showed that the dominant discharge product was Li_2O_2 in both electrolytes, and the main side reactions involved the electrolyte rather than carbon decomposition. It was also found that the side product was Li_2CO_3 in the case of carbon composed of a hydrophobic surface in contact with the DMSO electrolyte, while relatively similar amounts of lithium carboxylates were also formed in the case of tetraglyme or hydrophilic carbon. On subsequent charging up to ~4 V, Li_2O_2, Li_2CO_3, and lithium carboxylates were oxidized, while simultaneously additional Li_2CO_3 and likely lithium carboxylates were formed from electrolyte decomposition. Application of a ^{13}C carbon electrode also revealed that formation of parasitic species was largely related to the electrolyte decomposition during discharge in these electrolytes. On the charge cycle, however, continued electrolyte decomposition and $^{12}CO_2$

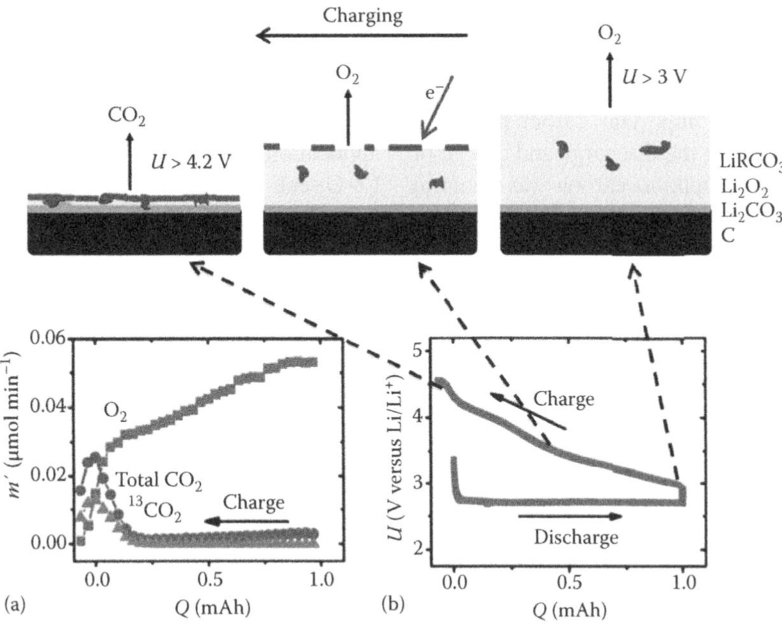

FIGURE 8.9 (a) Quantitative evolution rates (m') for O_2, $^{13}CO_2$, and $^{12}CO_2$ + $^{13}CO_2$ during charging measured by quantitative DEMS. (b) Cell voltage versus discharge capacity for Li–O_2 discharge and charge. Top panel shows a schematic related to the decomposition of discharge products and its correlation with charging voltage. (Adapted from McCloskey, B. D., Speidel, A., Scheffler, R., Miller, D. C., Viswanathan, V., Hummelshøj, J. S., Nørskov, J. K., Luntz, A. C., *J. Phys. Chem. Lett.*, 3, 997, 2012. With permission.)

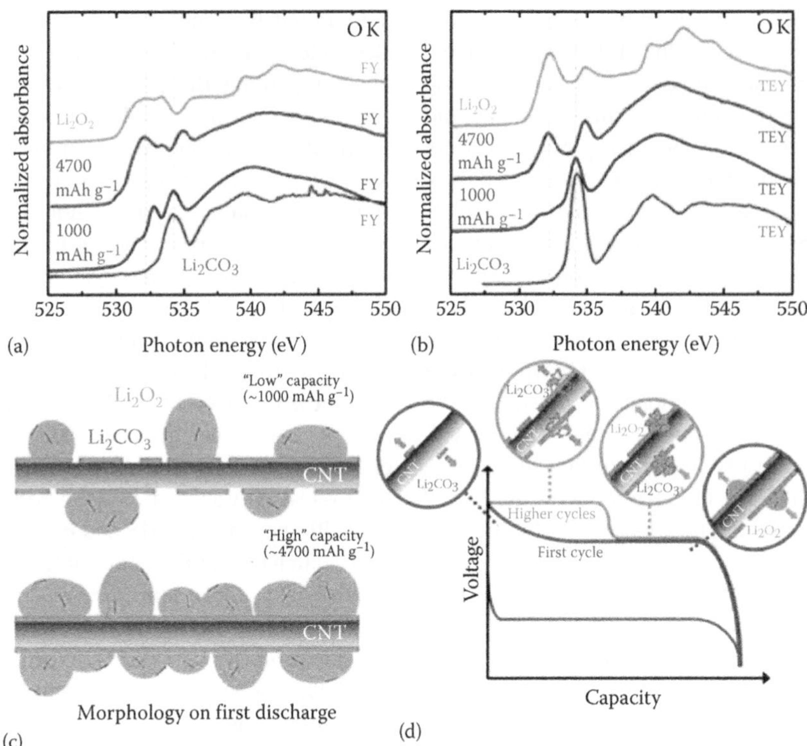

FIGURE 8.10 (a) O K-edge FY and (b) O K-edge TEY spectra of electrodes discharged to 1000 and 4700 mAh g^{-1} on the first discharge. (c) Schematic of discharge products formed at low and high capacity on VACNTs on the first discharge. (d) Schematic morphological evolution of the discharge product during the first and higher cycle numbers and its influence on the charging voltage. (Adapted from Gallant, B. M., Mitchell, R. R., Kwabi, D. G., Zhou, J., Zuin, L., Thompson C. V., Shao-Horn, Y., *J. Phys. Chem. C*, 116, 20800, 2012. With permission.)

evolution was observed at potentials below 3.5 V, whereas $^{13}CO_2$ related to carbon oxidation was only observed above 3.5 V. These results suggest that carbon is relatively stable below 3.5 V on discharge or charge (especially so for hydrophobic carbon), but is unstable when charging above 3.5 V in the presence of Li_2O_2. The authors concluded that direct chemical reaction with Li_2O_2 is not primarily responsible for carbon decomposition, but more important, carbon promotes electrolyte decomposition during the discharge and charge of an Li–O_2 cell. In addition, hydrophobic carbon was found to be more stable and less able to promote electrolyte decomposition than its hydrophilic counterpart. It is also emphasized that the stability of the electrode and the electrolyte cannot be considered in isolation, but also depends on the synergy between the electrolyte and the electrode.

Despite the small differences in the proposed mechanisms, all above-mentioned studies suggest that the carbon substrate is not quite stable in an Li–O_2 environment. The Li_2O_2 produced during the discharge reaction reacts with the carbon substrate and/or electrolyte to form Li_2CO_3 and lithium carboxylates at the carbon/Li_2O_2 or Li_2O_2/electrolyte interfaces, which subsequently increases the charging potential of the cell. McCloskey et al. also showed theoretically that a monolayer of carbonate at the Li_2O_2/carbon interface causes a ~10- to 100-fold decrease in the exchange current density due to the interfacial resistance to charge transport, which can be translated to the charging overpotential [19]. Since the oxidation of carbonate-based by-products on the charging cycle is incomplete, these species accumulate during cycling and result in electrode passivation and capacity fading. Replacing the carbon electrode material [18,19] and trying to decrease the charging potential to below 3.5 V [56] are proposed as potential solutions in order to make a stable cycling Li–O_2 cell. Reducing the charge overpotential, which is also beneficial to enhance the energy efficiency of the Li–O_2 cell, has been the subject of a considerable number of studies. Synthesis and application of electrocatalytic materials as the air electrode of Li–O_2 cell to reduce the charging overpotential of the cell is the subject of the following section.

8.3.4 Catalyst

The high charging overpotential of the Li–O_2 cell is considered one of the major challenges that limit the cyclability of the cell due to the decomposition of cell components, such as the carbon substrate and electrolyte, at high positive potentials. Thus, reducing the charge overpotential of the Li–O_2 cell has always been considered as one of the potential solutions to develop rechargeable Li–O_2 batteries. Many different catalytic electrode materials based on transition metals with various structures are illustrated to enhance the ORR and OER characteristics by either increasing the discharge voltage and capacity or decreasing the charge overpotential [57]. Although the electrocatalytic activity of these materials is well accepted, the mechanism of the catalytic activity is a point of controversy and is not truly understood [35,58]. On one side, formation of the very first layers of solid products during the discharge cycle is argued to reduce the catalytic activity very quickly [58]. This is supported by McCloskey et al.'s work, which used the DEMS method to study cells with various added cathode nanocatalysts (Au, Pt, MnO_2) in both carbonate- and ether-based electrolytes [59]. Although similar decreases in charging potentials were observed in carbonate-based electrolyte, analysis of evolved gases by means of DEMS revealed that CO_2 is predominantly evolved from all cells during charge, with only a small amount of O_2 evolution during the initial stages of cell charging (Figure 8.11a). The evolution of CO_2 during charge is correlated to the decomposition of carbonate species formed during the discharge cycle and hence the lowering of the cell potential during charging is supposed to be due to catalysis of CO_2 evolution. In an ether-based electrolyte, where the dominant discharge product is Li_2O_2, all catalysts exhibited similar discharge and charge voltages, while O_2 and CO_2 evolution had the largest and smallest values, respectively, in the cell using the carbon cathode (Figure 8.11b). Thus, the conclusion was that conventional oxygen evolution electrocatalysis has no effects on Li–O_2 electrochemistry but it does influence the decomposition of the electrolyte solvent. Similarly, the authors argued that the formation of insoluble and immobile species (Li_2O_2 and/or LiO_2 intermediate) on the cathode surface precludes transport of these species to active catalytic sites, making conventional OER electrocatalysis in Li–O_2 cells implausible.

On the other side, there are several reports on catalytic activity of such transition metals and metal oxides in an Li–O_2 cell with either carbonate- or ether-based electrolytes. It is believed that the ORR at the cathode of nonaqueous metal–air cells is composed of several consecutive electrochemical and nonelectrochemical stages, including initial charge transfer to dissolve oxygen, appearance of a low soluble oxygen–metal intermediate, charge transfer stage into the prepared oxygen–metal intermediate, and final agglomeration into globular metal oxide precipitate. Accordingly, it is suggested by Kraytsberg and Ein-Eli [35] that the catalyst may affect the choice of the specific reaction path and thus influences the appearance and structure of the specific oxygen–metal intermediates. The catalyst may alter the size, structure, density, and electronic conductivity of the metal oxide products, since various intermediates present different mobilities and diffusion rates. In such a way, it would be possible to explain the observed increase of the discharge capacity of Li–O_2 cells using an electrocatalyst as a positive electrode. The formation

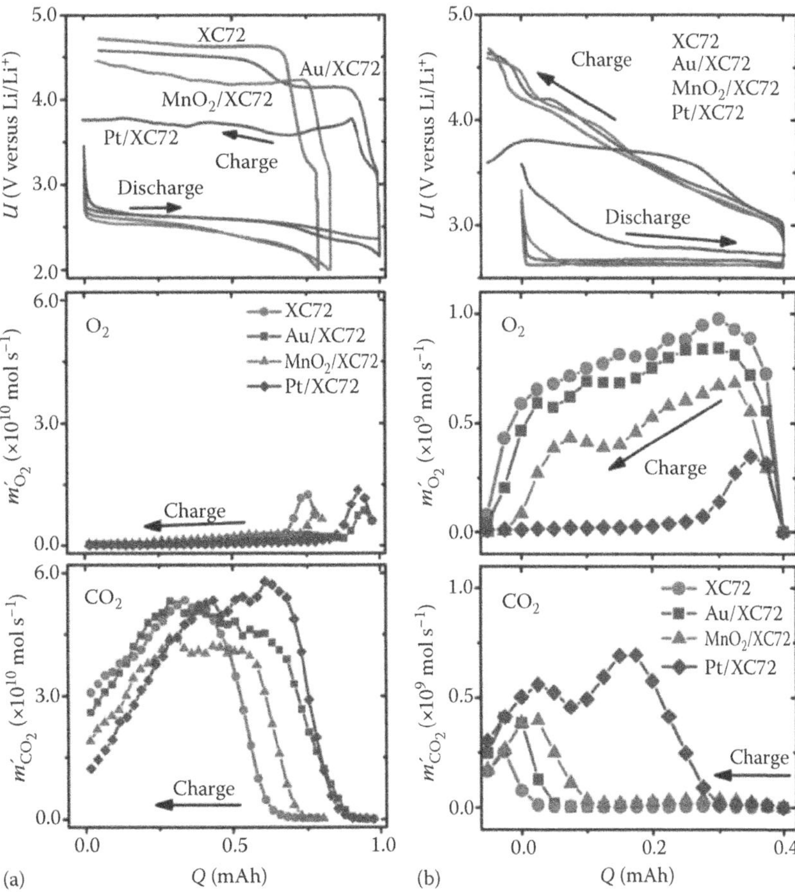

FIGURE 8.11 Discharge–charge curves (top panels) and corresponding O_2 (middle panels) and CO_2 (bottom panels) evolution during charging of cells using various cathode catalysts in (a) 1PC:1DME and (b) DME. m_i' is the molar generation of species "i", U is the cell output voltage, and Q is the cell charge. (Adapted from McCloskey, B. D., Scheffler, R., Speidel, A., Bethune, D. S., Shelby, R. M., Luntz, A. C., *J. Am. Chem. Soc.*, 133, 18038, 2011. With permission.)

of more conductive metal oxide products in the presence of a catalyst may also result in the increase of the discharge potential as well as the decrease of the charge overpotential in Li–O_2 cells [35]. Since a big portion of Li–O_2 studies are based on the catalytic effect, the synthesis and application of metal and metal oxide catalysts in Li–O_2 cells are also reviewed here, keeping in mind that catalysts may act as a two-edged knife to the catalysis of either the electrolyte or discharge products in the cell.

8.3.4.1 Precious Metals

Noble metals like Pt and Au have been abundantly employed as electrocatalysts to reduce the overpotential and enhance the reaction rates in hydrogen fuel cells. Accordingly, Shao-Horn et al. used PtAu NPs as a catalyst in Li–air positive electrodes [60]. PtAu composite was chosen since Au and Pt are the most active catalysts for the ORR and OER, respectively. PtAu NPs (<10 nm, 40 $m^2\ g^{-1}_{PtAu}$) in this work were synthesized by reducing $HAuCl_4$ and H_2PtCl_6 and then loaded onto Vulcan carbon to yield 40 wt.% PtAu/C. PtAu NPs showed significantly enhanced discharge and charge catalytic effects as the air electrode. The authors compared bare carbon, Pt/C, and Au/C electrode materials in an Li–air cell and found that the discharge voltages with PtAu/C are comparable to those with Au/C, while charging voltages with PtAu/C are comparable to those with Pt/C (Figure 8.12). The average charging potential of the PtAu/C electrode was 3.6 V, which is substantially lower (by 900 mV) than that of pure carbon. The round-trip efficiency of the PtAu/C cathode was calculated to be 73%, which is a large improvement when compared with 57% as calculated for the pure carbon cathode.

Later, Bruce et al. showed that truly reversible formation and decomposition of Li_2O_2 for extended number of cycles is plausible on a nanoporous gold (NPG) electrode [20]. They used an electrolyte composed of 0.1 M $LiClO_4$ in DMSO, since neither carbonate- nor ether-based electrolytes are quite stable in an Li–O_2 environment. To stabilize the Li anode electrode against the DMSO electrolyte, Li metal foil was kept in a 0.1 M $LiClO_4$–PC electrolyte for 3 days before being used as the anode in the Li–O_2 cell. The constructed cell exhibited a specific capacity of over 300 mAh g^{-1} and maintained more than 95% of its initial capacity after 100 consecutive discharge and charge cycles. The relatively low specific discharge and charge capacity is related to the heavy air electrode material (Au) and is equivalent to 3000 mAh g^{-1} for a similar carbon electrode. The authors employed several analytical techniques, including infrared, Raman, and DEMS to identify the chemical composition of the discharge products at different cycles. It was found that Li_2O_2 was reversibly formed and decomposed entirely during the cycles. In addition, a small amount (<1%) of side product composed of Li_2CO_3 and HCO_2Li was also detected. No S-containing compound was found during analysis, which means that the DMSO-based electrolyte is quite stable in the Li–O_2 cell. To confirm these results, a similar procedure was also repeated using a carbon electrode in the same electrolyte. Surprisingly, they found that 15% of the discharge products were composed of the Li_2CO_3 and HCO_2Li side products at the end of the discharge cycle. The charging overpotential of the cell was also significantly higher than that of the NPG cell. The results of this systematic study clearly highlight that the mutual effect of electrode/electrolyte interface should be considered to study the stability and cyclability of the Li–O_2 cell. It was also mentioned that the charging overpotential was reduced and the cell kinetic was considerably enhanced on the NPG electrode.

More recently, Lim et al. also reported on synthesis of hierarchical air electrode based on an aligned carbon structure and a Pt catalyst [61]. In this study, the air electrode was cross-woven by CVD-prepared CNT fibrils on a nickel mesh as a current collector and the Pt catalyst was added using a DC sputtering technique with the weight percentage of 66% (Figure 8.13a–d). The air electrode was then tested in an Li–O_2 cell using 1 M $LiPF_6$ in a TEGDME electrolyte. The authors discussed that in contrast to wet-chemical deposition methods,

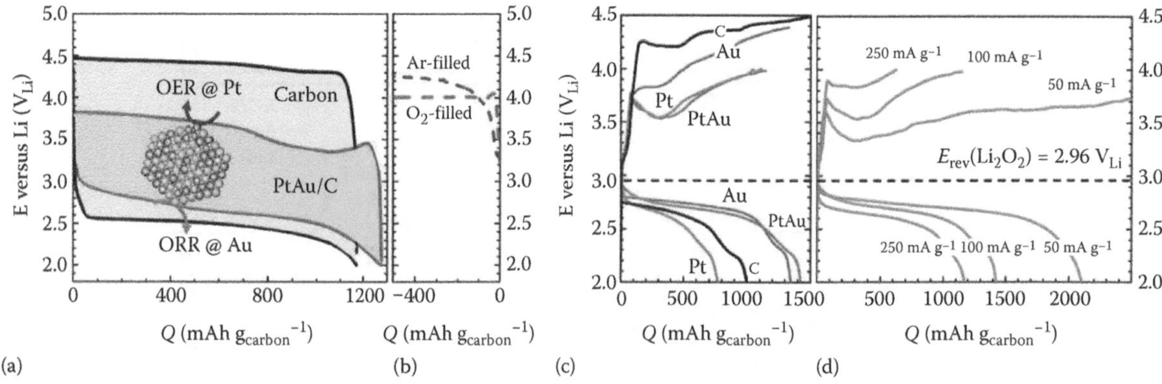

FIGURE 8.12 (a) Discharge–charge curves of Li–O_2 cell using carbon (85 mA g^{-1}) and PtAu/C (100 mA g^{-1}) positive electrodes in the third cycle. (b) Background measurement during charging at 100 mA g^{-1} of Ar- and O_2-filled cells for PtAu/C electrode. (c) First discharge–charge curves of Li–O_2 cell using carbon (85 mA g^{-1}) and Au/C, Pt/C, and PtAu/C (100 mA g^{-1}) electrodes. (d) First discharge–charge curves of Li–O_2 cell using PtAu/C electrode at current densities of 50, 100, and 250 mA g^{-1}. (Adapted from Lu, Y. C., Xu, Z., Gasteiger, H. A., Chen, S., Hamad-Schifferli, K., Shao-Horn, Y., *J. Am. Chem. Soc.*, 132, 12170, 2010. With permission.)

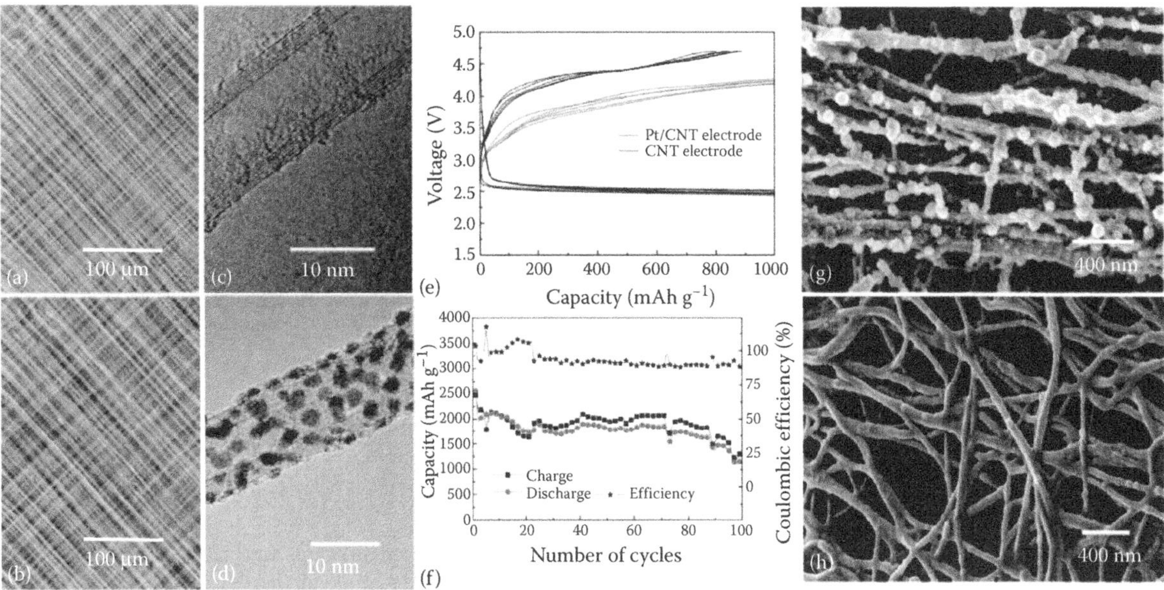

FIGURE 8.13 SEM (a, b) and TEM (c, d) images of CNT (a, c) and Pt/CNT (b, d) electrodes. (e) Discharge and charge curves of the CNT and Pt/CNT electrodes at a current density of 2 A g^{-1}. (f) Cyclability of the $Li–O_2$ cell with a Pt/CNT electrode at a current density of 2 A g^{-1}. SEM images of the CNT (g) and Pt/CNT (h) electrodes after full discharge to 2.0 V. (Adapted from Lim, H. D., Song, H., Gwon, H., Park, K. Y., Kim, J., Bae, Y. et al., *Energy Environ. Sci.*, 6, 3570, 2013. With permission.)

the physical sputtering technique enabled them to control the structure and loading weight of the catalyst. The Pt-loaded CNT fibril (Pt/CNT) showed reduced charging overpotentials by about 500 mV at a current density of 2 A g^{-1} and a limited discharge depth of 1000 mAh g^{-1} (Figure 8.13e). It has also been mentioned that the decrease in the charging overpotential is only seen at a low depth of discharge, and no significant catalytic effect was observed at deep discharge (2500 mAh g^{-1}). Passivation of catalytic sites by an insulating layer of solid discharge products was suggested as a potential reason. More than 130 consecutive cycles were achieved using the Pt/CNT electrode at a limited discharge depth of 1000 mAh g^{-1} compared with less than 70 cycles for the CNT electrode under the same conditions. In addition, the Pt/CNT electrode also exhibited 100 cycles at a full discharge depth with around 50% capacity retention (Figure 8.13f), even though the reduction of the overpotential at a full discharge depth was not considerable. To explain this discrepancy, the morphology of discharge products formed on the CNT and Pt/CNT electrodes were compared (Figure 8.13g,h). It was illustrated that a uniform layer of discharge products is formed in the presence of the Pt, while larger particles were formed on the bare CNT surface. This is another interesting example of altering the structure and morphology of the discharge products using an electrocatalyst, which was discussed earlier. Another possible scenario for enhanced cyclic performance of the Pt/CNT electrode compared with the bare CNT electrode might be related to the role of Pt to prevent the formation of the Li_2CO_3 layer at the Li_2O_2/carbon interface. However, such a comparison in terms of chemical composition of discharge products on CNT and Pt/CNT electrodes is lacking here and should be addressed in future studies in order to truly understand the catalyst's role in the $Li–O_2$ cell.

Palladium (Pd) is also among the metals that have been predicted to exhibit high catalytic activity toward Li^+–ORR and –OER [62]. Xu et al. synthesized freestanding honeycomb-like Pd-modified hollow spherical carbon deposited onto carbon paper and applied it as the air electrode of $Li–O_2$ cell [63]. To prepare the air electrode, functionalized silica spheres were deposited onto a piece of carbon paper using an electrophoretic technique. Then, the sample was loaded with Pd NPs and covered with a carbon shell. The silica template was washed out with hydrofluoric acid to leave a hollow spherical carbon (with Pd/P-HSC or without Pd/HSC) structure with macropores of 500-nm diameter and 9-nm wall thickness (Figure 8.14). As-prepared material was tested as the air electrode of an $Li–O_2$ cell using an electrolyte composed of lithium triflate ($LiCF_3SO_3$) in TEGDME. The P-HSC electrode delivered a specific discharge capacity of more than 12,000 mAh g^{-1} at a current density of 500 mA g^{-1} and maintained 5900 mAh g^{-1} at 1500 mA g^{-1}. Using a restricted specific discharge capacity of 3000 mAh g^{-1}, the P-HSC electrode showed a lower charging overpotential by 300 mV when compared with a Super P carbon (SP) control electrode at a current density of 300 mA g^{-1} (Figure 8.14b). The HSC electrode also exhibited a lower charging overpotential by 190 mV. The electrocatalytic activity of the P-HSC electrode has been attributed to the several factors, including a high active surface area (127 m^2 g^{-1}) and the number of surface defects, which provide more reaction sites for the ORR and OER; the hierarchically porous structure and Pd NPs modification, which facilitate the continuous oxygen and lithium ion flow into and out of the cathode; and the well-tailored deposition behavior and morphology of the discharge product that benefits the ORRs and OERs. In addition, the P-SHC electrode tolerated 205 consecutive

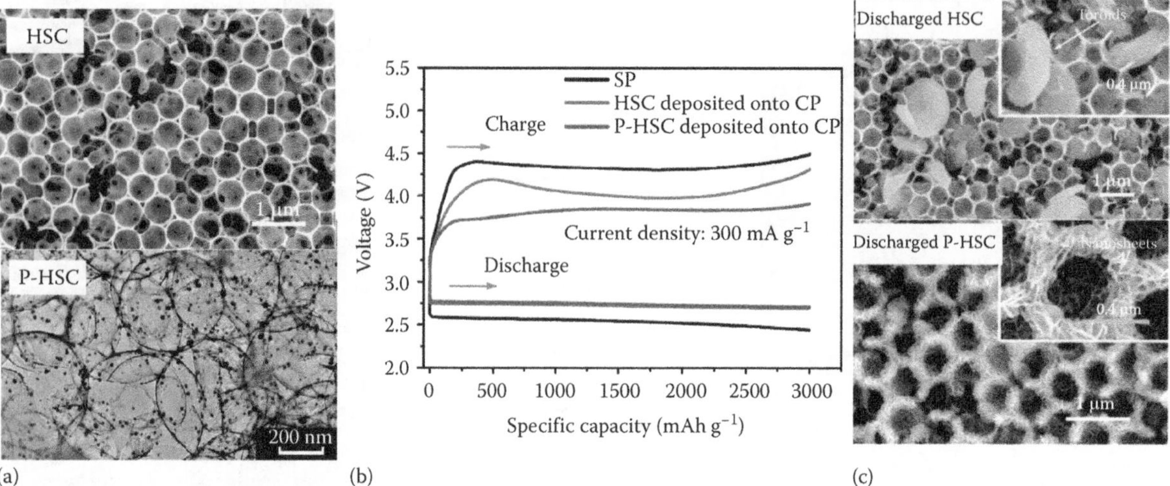

FIGURE 8.14 (a) SEM and TEM images of HSC and P-HSC, respectively. (b) First limited discharge and charge curves of Li–O_2 cells at a current density of 300 mA g^{-1}. (c) SEM images of HSC and P-HSC electrodes discharged to 3000 mAh g^{-1}. (Adapted from Xu, J. J., Wang, Z. L., Xu, D., Zhang, L. L., Zhang, X. B., *Nat. Commun.*, 4, 2438, 2013. doi:10.1038/ncomms3438. With permission.)

discharge and charge cycles at a current density of 300 mA g^{-1} using a limited discharge depth of 1000 mAh g^{-1}. The reversible cycle number decreased to 61 and 33 with an increase of the discharge depth to 2000 and 5000 mAh g^{-1}, respectively. Checking the morphology of discharge products also revealed interesting details about the electrocatalytic activity of the Pd NPs. While large toroidal discharge products were formed on the HSC and SP electrodes, unique nanosheets with <10 nm thicknesses uniformly grew onto the wall of the P-HSC electrode (Figure 8.14c). The chemical composition of the discharge product was identified using XRD to be Li_2O_2 in all cases. This is an interesting example of the role of the catalyst in altering the structure of the discharge product and its influence on the charging overpotential of the Li–O_2 cell.

Another Pd-based catalyst was fabricated by Amine's research group and was used as the air electrode of an Li–O_2 cell [64]. Super P Li (SPL) conductive carbon black (62 m^2 g^{-1}) was used as the catalyst support material in this study. Al_2O_3 and Pd NPs were loaded using three cycles of atomic layer deposition (ALD) techniques. It is noticed that the deposited Al_2O_3 layer on the carbon surface covers the surface defects containing carbonyl, alcohol, and ether groups, since these sites are reactive toward the trimethylaluminium precursor (ALD precursor for Al_2O_3 deposition). The as-prepared catalyst (Pd/SPL) was then mixed with a binder (80:20 wt.%) to make the air electrode and was tested using a TEGDME-based electrolyte. Electrochemical charge and discharge tests restricted to a specific capacity of 1000 mAh g^{-1} revealed that the Al_2O_3 layer increased the charging overpotential of the cell by 0.2 V, probably due to the insolating nature of the aluminum oxide. Nevertheless, Pd/SPL electrode exhibited considerably lower charging potential around 3.2 V (charge overpotential of 0.2 V). Direct observations of the discharge products of Li–O_2 indicated that the Li_2O_2 formed on the Pd/SPL electrode is composed of smaller toroid-shaped particles than the discharge products in the absence of Pd NPs. DFT calculation showed that partially oxidized Pd NP forms bonds to the defect-free graphene surface, which provides a good interface for electron transfer. A higher charge transfer rate may explain the increased discharge potential of 0.2 V on the Pd/SPL electrode. However, the electronic properties of the lithium peroxide discharge product produced in the presence of the Pd NPs are shown to lower the charging overpotential of the cell. It was suggested that the Pd NPs can serve as the nucleation sites for growth of the discharge products resulting in the nanocrystalline nature of the Li_2O_2 formed on the Pd/SPL electrode. Small 2–10 nm grains and grain boundaries can provide a mechanism for good electronic transport during the charge cycle. The presence of the Li_2O_3 layer as an inhibitive barrier toward the decomposition of the electrolyte is also mentioned as another potential cause of the increased discharge potential. Despite a relatively low charging overpotential, the Li–O_2 cell using a Pd/SPL cathode electrode did not show better cycling performance than 10 cycles with a restricted discharge capacity of 500 mAh g^{-1}. Degradation of the Li anode due to an oxygen crossover effect, and poisoning of the Pd catalyst by contaminants or passivation and decomposition of the TEGDME-based electrolyte have been proposed as potential reasons for the poor cyclability of the cell.

Ruthenium NPs were also examined by Sun et al. as the electrocatalyst for an Li–O_2 cell [65]. Ruthenium nanocrystals supported on carbon black in this study were synthesized by using a soft template method followed by a heat treatment step. The tri-block copolymer F127 was adsorbed on the surface of carbon black via hydrophobic interactions, and then $RuCl_3$ was added to the mixture in which Ru ions were absorbed on the long chain of the tri-block copolymer through weak coordination bonds between alkylene oxide segments and metal ions. The final catalyst (Ru-CB) powder was obtained by heat treatment of the prepared gel in a reducing atmosphere, containing 34 wt.% Ru (determined by thermal gravimetric analysis). The catalyst powder was mixed with PVDF (90:10 wt.%)

and then coated on a glass fiber separator. The air electrode was tested in the Li–O_2 cell using a DMSO-based electrolyte. The Ru-CB electrode exhibited a specific discharge capacity of 9800 mAh g^{-1} at a current density of 200 mA g^{-1}, which showed a 57% increase over the CB electrode. In addition, the charging overpotential of the cell was significantly lower (370 mV) on the Ru-CB electrode, resulting in 85% of the discharge capacity being recovered during a charge cycle below 3.5 V in the first cycle, which corresponds to a high energy efficiency of 83.37%. The Ru-CB electrode showed 150 and 40 stable consecutive cycles at a current density of 200 mA g^{-1} and using limited discharge capacities of 1000 and 4000 mAh g^{-1}, respectively. The ability of Ru toward oxygen adsorption is proposed as a potential reason for the high activity of Ru-CB electrode toward the ORR and OER in Li–O_2 cell.

Jian et al. covered CNTs with a layer of ruthenium oxide to prepare a core–shell structure of CNT@RuO_2 and used the composite material as the catalyst for an Li–O_2 cell [66]. The core–shell structure was synthesized using a sol–gel method in which ruthenium was precipitated in the presence of CNTs and followed by annealing of the resultant precipitate at an elevated temperature. The air electrodes were fabricated by casting CNTs or CNT@RuO_2 with PTFE as the binder (90:10 wt.%) and coating the paste onto a Ti mesh. The air electrodes were then tested in an Li–O_2 cell using a TEGDME-based electrolyte. The CNT@RuO_2 electrode delivered a capacity of 1130 mAh g^{-1} (based on the CNT@RuO_2 weight) at a current density of 100 mA g^{-1} and maintained a capacity of 790 mAh g^{-1} at 500 mA g^{-1}. The Li–O_2 cell using a CNT@RuO_2 air electrode exhibited a considerably reduced charging overpotential by 0.98 V, corresponding to a round-trip efficiency of about 79%. The CNT@RuO_2 air electrode showed 20 and 100 stable discharge and charge cycles using limited discharge capacities of 500 and 300 mAh g^{-1}, respectively. XRD and SAED revealed that the discharge products were mainly composed of Li_2O_2. However, trace amounts of Li_2CO_3 were also detected via FTIR of the discharged electrodes, which have been related to the decomposition of the electrolyte. It was also discussed that based on the intensity of Li_2CO_3 peaks in the FTIR spectra, the amount of carbonate side products on CNT@RuO_2 electrode were much lower than that of on the CNT electrode, indicating the inhibitive role of the RuO_2 layer regarding the formation of Li_2CO_3.

8.3.4.2 Transition Metal Oxides

Bruce et al. were among the first researchers who considered the electrocatalytic activity of transition metal oxides in Li–O_2 cells [67–69]. They examined a number of potential catalyst materials, which have shown good oxygen electrocatalytic activity in aqueous media, including Co_3O_4, Fe_2O_3, CuO, and $CoFe_2O_4$ in the Li–O_2 cell [67,68]. The air electrodes in all cases were fabricated by casting a mixture of Super P carbon, the appropriate catalyst, and Kynar2801 in a molar ratio of 95:2.5:2.5 and tested using 1 M $LiPF_6$ in PC as the electrolyte. Among all the examined catalysts, Fe_2O_3 showed the highest initial specific capacity (2700 mAh g^{-1}), while Fe_3O_4, CuO, and $CoFe_2O_4$ exhibited the best capacity retention. Meanwhile, Co_3O_4 gave a compromise between the specific capacity (2000 mAh g^{-1}) and the cycling retention. However, the mechanism of capacity enhancement or retention was not discussed.

The same research group also synthesized and examined various MnO_x catalysts, including α-MnO_2 in bulk and nanowire form; β-MnO_2 in bulk and nanowire form; and γ-MnO_2, λ-MnO_2, Mn_2O_3, and Mn_3O_4 in Li–O_2 cell [69]. The air electrodes were prepared by casting a mixture of Super P carbon, the appropriate catalyst, and Kynar2801 in a molar ratio 95:2.5:2.5 and tested using 1 M $LiPF_6$ in PC as the electrolyte. It was found that α-MnO_2 nanowires exhibit the highest specific capacity (initial discharge capacity of 3000 mAh g^{-1} based on carbon weight) with the better capacity retention of the examined manganese oxides (Figure 8.15). The chemical

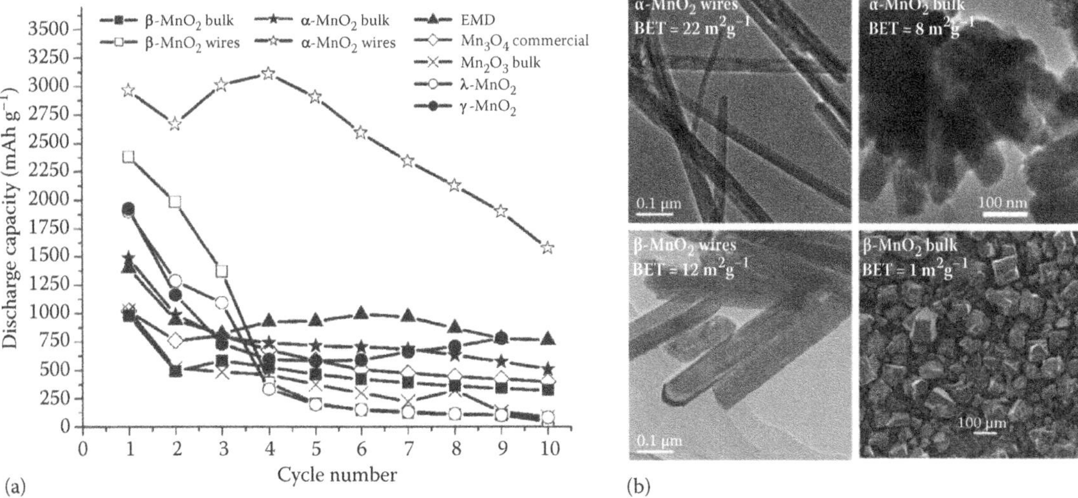

FIGURE 8.15 (a) Variation of discharge capacity with cycle number for porous electrodes containing manganese oxides as catalysts. (b) SEM and TEM images of α-MnO_2 and β-MnO_2 in bulk and nanowire forms. (Adapted from Débart, A., Paterson, A. J., Bao, J., Bruce, P. G., *Angew. Chem. Int. Ed.*, 47, 4521, 2008. With permission.)

composition of the discharge product was also determined by means of Raman spectroscopy to be Li_2O_2. Higher initial specific capacity was also observed for β-MnO_2 in nanowire form compared with bulk form, leading to the conclusion that the higher surface area of the nanowire catalysts is responsible for the enhanced specific capacity. To explain the higher catalytic activity of the α-MnO_2 nanowires, the authors argued that the tunneling structure of α polymorph accommodates the Li^+ and O^{2-} ions in close vicinity, resulting in a subsequent incorporation of the ions.

Several studies were also carried out to synthesize and apply various transition metal oxide/carbon composites in the Li–O_2 cell [70–75]. However, these early studies were performed in carbonate-based electrolytes, which are not stable in an Li–O_2 environment and undergo the decomposition reaction in the presence of the highly oxidative superoxide intermediate [16]. Therefore, more studies are required to confirm the results of these studies with a more stable electrolyte, since the reported catalytic activity might be due to the decomposition of the electrolyte solvent. In the present context, we focus mainly on the results that are obtained in more stable ether-based electrolytes.

More recently, Zahoor et al. compared α- and γ-MnO_2 materials with different structures as the positive electrode of an Li–O_2 cell in an ether-based electrolyte [76]. The sea urchin–shaped α-MnO_2 and flower-like γ-MnO_2 NPs were synthesized using a hydrothermal method. As-prepared materials were mixed with conductive KB carbon (1:2) in isopropyl alcohol and then pressed onto a Ni mesh to prepare the air electrode. The air electrodes were tested using a 1-M LiTFSI, TEGDME electrolyte. The urchin α-MnO_2 showed a specific capacity of 6125 mAh g^{-1} at a current density of 0.1 mA cm^{-2} during discharge and charge cycles between 2.0 and 4.3 V. On the other hand, the control carbon and flower γ-MnO_2 electrodes showed their first specific discharge capacities to be slightly higher than 2000 and 3674 mAh g^{-1}, respectively. It is discussed that α-MnO_2 contains more defects and OH^- groups that are beneficial to surface adsorption of O_2 and dissociation of O–O bonds. Using a limited discharge depth of 0.8 mAh, α- and γ-MnO_2 nanomaterials exhibited 35 and 20 consecutive discharge and charge cycles, respectively.

To further investigate the role of defects on the behavior of electrocatalysts in the Li–O_2 cell, Nazar et al. studied the effect of defects and vacancies in nanofibrous manganese oxides ($Na_{0.44}MnO_2$) [77]. $Na_{0.44}MnO_2$ nanowires were synthesized by a hydrothermal method followed by acid leaching in nitric acid (concentrated HNO_3, room temperature), and dehydrated by heat treatment to induce controllable defect formation. It was revealed on the basis of the XRD data and detection of Mn^{2+} in the acidic solution that acid leaching of $Na_{0.44}MnO_2$ also involves a redox-driven extraction of Mn^{2+} formed from the disproportionation of Mn^{3+} to Mn^{4+} and Mn^{2+}, along with the probable formation of oxygen vacancies as well as sodium/proton ion-exchange mechanisms. TEM investigation also indicated a partial amorphization of the surface and the presence of a damaged layer caused by partial extraction of Na^+ and Mn^{2+} ions (Figure 8.16a). In addition, the BET surface area of the $Na_{0.44}MnO_2$ nanowires increased from 15.7 to 49.2 m^2 g^{-1} after acid leaching. The electrochemical performances of the pristine and acid-leached $Na_{0.44}MnO_2$ nanowires as the catalysts (cathode composition: KB carbon, catalyst, and binder with 50:20:120 weight ratio) were examined in an Li–O_2 cell using a 1-M $LiPF_6$, TEGDME electrolyte (Figure 8.16b). The acid-leached sample exhibited a discharge capacity of 11,000 mAh g^{-1}, which is almost double that of pristine $Na_{0.44}MnO_2$ (5700 mAh g^{-1}). It should also be noticed that the specific discharge values are calculated on the basis of the mass of the carbon support. The dominant discharge product was proven to be Li_2O_2 by XRD, and the acid-leached sample showed a slightly lower charging overpotential when compared with pristine and carbon black electrodes. More investigation about the morphology of discharge products revealed that Li_2O_2 toroid deposition occurred only on the outer edge of the pristine $Na_{0.44}MnO_2$ and carbon electrodes, while the central electrode area was bare. In the case of the acid-leached catalyst, however, the entire electrode

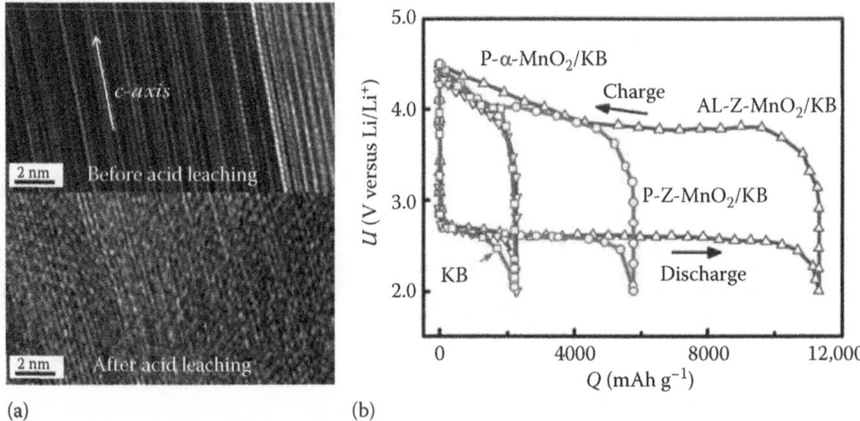

FIGURE 8.16 (a) HRTEM images of the pristine and acid-leached $Na_{0.44}MnO_2$ nanowires. (b) First discharge and charge curves of pristine (P-Z-MnO_2/KB) and acid-leached $Na_{0.44}MnO_2$ (AL-Z-MnO_2/KB), α-MnO_2 (P-α-MnO_2/KB), and carbon (KB) electrodes. (Adapted from Lee, J. H., Black, R., Popov, G., Pomerantseva, E., Nan, F., Botton, G. A., Nazar, L. F., *Energy Environ. Sci.*, 5, 9558, 2012. With permission.)

surface was being uniformly covered with Li_2O_2 toroids. This behavior might be related to the increased number of defect sites that can be served as nucleation sites for Li_2O_2 formation.

Amine's research group studied the mechanism of the capacity fading with cycling on a porous carbon-supported MnO_2 nanorods electrode in an Li–O_2 cell [78]. The electrode materials were fabricated using a chemical precipitation method in the presence of oxidized high surface area Super P Li (SPL) carbon with different ratios (MnO_2 versus SPL carbon: 10 and 50 wt.%). The cathode was prepared by casting a mixture of the as-prepared MnO_2/SPL carbon and Kynar 2801 binder in a weight ratio of 50:50 onto an aluminum grid, and it was tested in an electrolyte of 1-M $LiCF_3SO_3$ in TEGDME. Both samples showed a similar discharge capacity of around 1400 mAh g^{-1} (based on carbon + catalyst weight) at a current density of 100 mA g^{-1}. Although the same capacity was recovered during the charge cycle at 4.0 V, the capacity dropped to 800 mAh g^{-1} in the second discharge. Using a limited discharge depth of 500 mAh g^{-1}, over 50 consecutive cycles were achieved. The discharge products of the cell were also examined at the first and tenth cycles using XPS. Li 1s XPS results indicated that Li_2O_2 forms and decomposes completely during discharge and charge of the Li–O_2 cell. The authors argued that no change of the relative intensity of the O–C=O peak is an indication of the TEGDME-based electrolyte's stability upon cycling. In contrast, the C 1s spectra from the carbon electrode containing the same catalyst cycled in a PC-based electrolyte showed carbonate type components, suggesting that the PC electrolyte decomposed during cell's discharge and charge. The dominant discharge product of the cell was discovered to be Li_2CO_3 after only five discharge and charge cycles. This study highlights the possibility of the catalytic decomposition of carbonate-based electrolytes in Li–O_2 cell.

Thapa et al. also used Pd metal to replace the carbon binder in an MnO_2-based positive electrode of the Li–O_2 cell [79]. The air electrode was formed by casting a mixture of electrolytic manganese oxide, Pd metal powder, and PTFE with a molar ratio of 70:20:10. The mixture was then pressed onto a stainless-steel mesh. Replacing lightweight carbon with heavy Pd metal dramatically decreased the specific capacity of the prepared positive electrode (158 mAh g^{-1}). However, the charging overpotential of the cell was also considerably decreased to 0.4 V (89% energy efficiency). In addition, the charging overpotential of the cell was further decreased to 0.3 V with increase of the Pd/MnO_2 ratio. The discharge products of the cell were also analyzed by means of Raman spectroscopy to be composed of Li_2O_2 and Li_2O. No Li_2CO_3 was detected as the discharge side product, even though a carbonate-based electrolyte was used. Therefore, the authors concluded that the Pd-based positive electrode prevents the formation of Li_2CO_3 during the discharge cycle and hence decrease the charging overpotential. The authors then added acetylene black to the air electrode in order to increase the active surface area of MnO_2. The specific capacity of the electrode materials increased to 257 mAh g^{-1} by adding acetylene black to the air electrode, and the cell showed 20 stable consecutive cycles. Moreover, this study further highlights the role of the catalyst on choosing a specific reaction path to alter physicochemical properties of the discharge products in an Li–O_2 cell, which was briefly discussed before. Moreover, DFT calculations demonstrated the intrinsic activity of Au, Ag, Pt, Pd, Ir, and Ru, on the close-packed and step edge surfaces with respect to the adsorption energy of O on the metals, for application as catalyst for Li^+–ORR [62]. It was found that these six metals form a volcano-like trend, with Pd and Pt at the top, for Li^+–ORR. In addition, it is predicted that the catalyst particles with more edges and defects would be more active than single-crystal close-packed surfaces.

Nazar at al. synthesized a metallic mesoporous pyrochlore using the surfactant templating method as a catalyst for the Li–O_2 cell [80]. The synthesis route was based on the cooperative assembly of an initially amorphous lead/ruthenium hydroxide precursor with a cationic organic surfactant. Crystallization of the wall structure was effected by kinetic control of the reaction processes using an oxidizing agent (sodium hypochlorite) to gradually increase the oxidation state of the transition metal in the template hydroxide precursor. The transformation of the hydroxo bridges to oxo bridges results in condensation, to give a wall structure comprising a nanocrystalline metal oxide. The organic framework that supports the mesostructure during crystallization was then removed by ethanol extraction. Using the pyrochlore/carbon catalyst, a discharge capacity of 7200 mAh g^{-1} was obtained. On the second and third cycles, the reversible capacity increased to reach a maximum of 10,400 mAh g^{-1}. In contrast, the reversible capacity of the control carbon electrode dropped to less than 1000 mAh g^{-1} by the third cycle. This high electrochemical performance is attributed to a combination of the presence of additional binding sites on the high-surface-area pyrochlore, higher electronic conductivity (leading to better electronic transport), and a correspondingly lowered activation energy for oxygen reduction. It is also mentioned that better mass transport for intermediate species such as LiO_2 may also play an important role. The effective OER properties of the MP pyrochlore during charging result in better peroxide stripping during charging (i.e., at a lower potential than for carbon) and reactivation of the sites. Thus, the reversible capacity is ascribed to the good catalytic properties for the oxidation of Li_2O_2 that is formed on discharge (OER). Another possible reason for the increase of the capacity is that the Ru or Pb oxidation state changes slightly after the first cycle to result in a more active catalyst. It was also found from diffraction analysis of the discharged electrode that only Li_2O_2 was produced, with no Li_2O or LiOH. Importantly, the peroxide exhibits a shorter coherence length than that formed with carbon alone (13 nm versus 18 nm). This is based on Scherrer analysis of the average of the XRD peaks, indicating that the catalyst determines the morphology of the product. The improvement of the electrochemical behavior has been attributed to the tailored properties of MP pyrochlore, which are vital to catalytic reactions, including a high fraction of exposed oxygen vacancies, porosity that enables good diffusion to the active sites, and a nanoscale conductive network with metallic conductivity. In addition, the role of the pyrochlore as a catalyst for mass transport in the electrochemical reaction (which differs from an electrocatalyst) is also mentioned as a potential role.

Carbides and nitrides are two classes of compounds that may possess the required characteristics of a desired air electrode, like high electric conductivity, high chemical and electrochemical stability, low density, nontoxicity, and low cost. Titanium carbide (TiC) was examined by Bruce's research group to be used as the electrocatalyst in the air electrode of an Li–O_2 cell [21]. Commercial TiC nanopowder was mixed with a PTFE (95:5 wt.%) binder in isopropanol, and the prepared slurry was then coated onto a stainless-steel mesh current collector. The as-prepared air electrode was tested in two electrolytes: 0.5 M $LiClO_4$ in DMSO and 0.5 M $LiPF_6$ in TEGDME. Partially charged $Li_{1-x}FePO_4$ was employed as the negative electrode to avoid possible contamination from reactions during cycling between an Li metal anode and DMSO or TEGDME. The cell exhibited a specific capacity of 350 mAh g^{-1} based on the TiC weight at a current density of 1 mA cm^{-2} in DMSO and maintained more than 98% of the initial specific capacity after 100 consecutive cycles (Figure 8.17a). Switching the electrolyte from DMSO to TEGDME did not change the specific capacity of the electrode, but increased the charging overpotential of the cell upon the cycling for 25 cycles. The authors also used several spectroscopic techniques, including FTIR, NMR, and XRD to identify the chemical composition of the discharge products of the cell. The results revealed that Li_2O_2 was the major (> 98%) product of the cell before and after cycling the cell in both of the DMSO and TEGDME electrolytes. To further investigate the side products of the cell, an analytical technique based on the selective reaction of decomposition products present at the cathode by an inorganic acid or Fenton's reagent were performed in which treating the cathode with acid releases CO_2 from the inorganic carbonate, whereas Fenton's reagent releases CO_2 from organics, such as carboxylates. The produced CO_2 from either inorganic acid or Fenton's reagent was then measured by mass spectrometry, and the measured values were correlated to the carbonates or carboxylates side products. The dominant side products were confirmed to be composed of inorganic carbonate (Li_2CO_3) in both DMSO and TEGDME electrolytes; however, the quantity of decomposition products for TEGDME was more than twice that for DMSO, indicating higher stability of DMSO in the Li–O_2 cell (Figure 8.17b). Comparing the amount of Li_2CO_3 formed on the TiC electrode in both electrolytes revealed that there is a significant increase in the proportion of Li_2CO_3 during cycling for TEGDME, which has been related to the increasing decomposition of electrolyte with cycle number or more likely accumulation of Li_2CO_3 due to its incomplete oxidation during charging. To further confirm the stability of TiC, the results of carbonate analysis on TiC electrode in DMSO electrolyte was also compared with those for carbon electrode in the same electrolyte (Figure 8.17c). The comparison indicated that 16% of the discharge products on the carbon electrode is composed of Li_2CO_3 after only five discharge and charge cycles, which shows a 40 times increase over the TiC electrode (0.4%). These results, once again, highlights the mutual effect of electrode/electrolyte in Li–O_2 cell

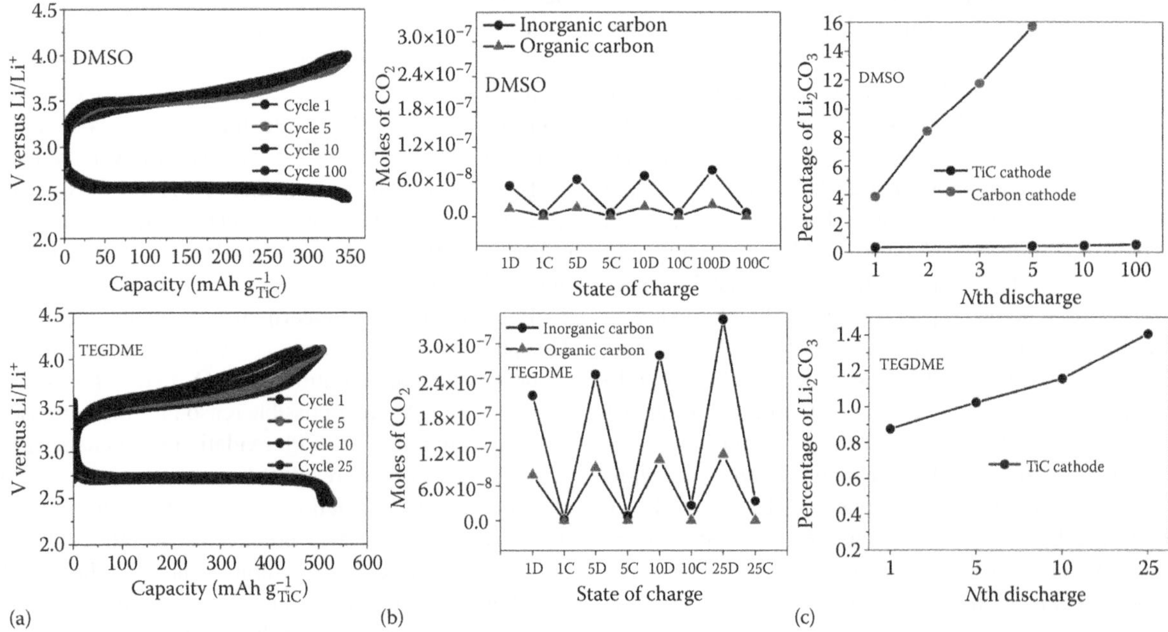

FIGURE 8.17 (a) Discharge and charge curves of the TiC electrode; (b) moles of CO_2 evolved from the TiC cathodes, removed from the cells at the states of charge indicated, and then reacted with an inorganic acid and Fenton's reagent; (c) proportion of Li_2CO_3 accumulated on the TiC cathode at the end of a discharge cycle expressed as a percentage of Li_2O_2 in the electrode; in DMSO and TEGDME electrolytes. (Adapted from Ottakam Thotiyl, M. M., Freunberger, S. A., Peng, Z., Chen, Y., Liu, Z., Bruce, P. G., *Nat. Mater.*, 12, 1050, 2013. With permission.)

behavior. Furthermore, XPS study of the TiC electrode surface demonstrated a significant proportion of TiO_2 and TiOC species, and also the proportion of TiO_2 increased on the first discharge to become the dominant species on the surface. On the basis of the XPS observation, it is suggested that the stability of the TiO_2-rich surface layer (with some TiOC) present on the TiC particles, combined with the electronic conductivity of TiC, is responsible for the stable and reversible Li_2O_2 formation/decomposition in the Li–O_2 cell.

Iron–nitrogen–carbon (Fe/N/C) catalysts have been synthesized and studied extensively as low-cost alternatives to Pt for ORR in both acidic and alkaline fuel cells, and recently by Shui et al. as the air electrode of an Li–O_2 cell [81]. The Fe/N/C composite in this study was prepared by a thermolysis method from a transition metal ligated by nitrogen-containing organic compounds over a high-surface-area carbon support. During the thermolysis process, the organic components in the organometallic precursor decompose, and fractions of nitrogen and carbon integrate into the carbonaceous support with dispersed iron. As-prepared catalyst was then mixed with PVDF (80:20 wt.%) and painted onto a circular piece of carbon paper. The control α-MnO_2/carbon (40 wt.% of α-MnO_2) was also prepared in a similar way. The electrodes were tested using 1 M $LiCF_3SO_3$ in TEGDME as the electrolyte. The Fe/N/C composite demonstrated a slightly higher discharge potential and significantly lower charge potential in comparison with α-MnO_2. Li–O_2 cells with Fe/N/C and α-MnO_2 electrodes were cycled over a controlled discharge capacity, and gas chromatography analysis was performed to determine the composition of the gas released at the end of the charge cycles. They found that only oxygen was released from the cell with the Fe/N/C catalyst, whereas both O_2 and CO_2 were found in the gas from the cell with α-MnO_2 as the catalyst (Figure 16b). The enhanced cycling performance led to achieve 50 consecutive charge–discharge cycles under a controlled discharge capacity of 500 mAh g^{-1}.

8.3.4.3 Soluble Catalysts

The concept of a soluble catalyst is well known in electrocatalysis as the redox mediator. A redox mediator is an electroactive molecule that is also stable in both the reduced and oxidized forms. The redox mediator is usually being applied to enhance the kinetics of a sluggish electrochemical reaction. In an anodic electrochemical process, for example, the oxidation of the redox mediator takes place at the electrode surface in the solution and produces an oxidized form of the mediator. Thereafter, the oxidized form of the mediator reacts chemically with the target species, producing the initial reduced form of the redox mediator and oxidation products of the target. An efficient redox mediator should have a less positive redox potential compared with the target species (to be oxidized easier); however, its redox potential should still be close enough to the redox potential of the target species to enable oxidization of the target. In addition, the redox mediator should have stable reduced and oxidized forms in the chemical environment of the solution. Bruce et al. demonstrated the applicability of the redox mediator concept in Li–O_2 battery system for the first time to enhance the charging characteristics of the cell [82]. They examined three different redox mediators, tetrathiafulvalene (TTF), ferrocene (FC), and *N*, *N*, *N′*,*N′*-tetramethyl-*p*-phenylenediamine (TMPD), in an Li–O_2 cell composed of a NPG air electrode [20], partially charged $Li_{1-x}FePO_4$ negative electrode [21], and 1 M of $LiClO_4$ in DMSO as the electrolyte (see also Section 8.3.4.1). To determine the oxidizing capability of the redox mediators toward Li_2O_2, certain amounts of the oxidized forms of the mediators were reacted with excess of Li_2O_2 and the evolved oxygen was monitored using mass spectroscopy. All the three mediators reacted with Li_2O_2 and released the oxygen gas; however, only in the case of TTF was the amount of evolved oxygen that expected for the quantity of the mediator in the solution according to a ratio of electrons to O_2 of 2 e^-/O_2. The TTF mediator was then tested in the Li–O_2 cell at different current densities. Addition of the TTF redox mediator (10 mM TTF) had a great effect on the charging overpotential of the cell, especially at higher current densities (Figure 8.18a,b). While the charging overpotential of the cell increased with the increase of current density in the absence of the redox mediator, the cells containing TTF were oxidized at a constant potential throughout the charging process and showed only a very modest increase in charging potential with an increase in the current density from 0.078 to 1 mA cm^{-2}. Using a restricted discharge capacity of 300 mAh g^{-1} (based on the weight of gold), the cells containing the TTF mediator exhibited 100 consecutive cycles (Figure 8.18c,d). FTIR, Raman, and DEMS techniques were employed to confirm the formation and decomposition of Li_2O_2 during the cycling of Li–O_2 cells containing the TTF redox mediator. DEMS results showed that the ratio 2 e^-/O_2 (the ratio expected for oxidation of Li_2O_2) increased slightly to 2.14 e^-/O_2 after 100 cycles, indicating a small degree of decomposition of the TTF mediator. It has been argued that the solid insulating Li_2O_2 discharge products, which are not in close contact with the surface of the air electrode, impose a charging overpotential to the cell in the absence of the redox mediator due to the hindered charge transport during oxidation. This is in contrast to the TTF redox mediator, which undergoes oxidation at the electrode surface and then oxidizes the Li_2O_2 particles that are not directly connected to the electrode surface. Although addition of the redox mediator into the electrolyte showed significant enhancements toward the rate capability and charging performance of the Li–O_2 cell with a restricted discharge depth, the effect of the discharge depth on the performance of such redox mediator has not been discussed. With the increase of discharge depth, the surface of the air electrode would be covered with a thicker layer of insulating Li_2O_2, which, in turn, might affect the oxidation of the redox mediator at the electrode surface as well.

More recently, Lim et al. combined the idea of a soluble catalyst with a woven porous CNT fibril air electrode to enhance the cyclability of the Li–O_2 cell [83]. The synthesis of CNT fibril air electrode and its application in Li–O_2 cell has been discussed earlier in this text (see also Figure 8.13). Lithium iodide, LiI, was employed as the redox mediator in

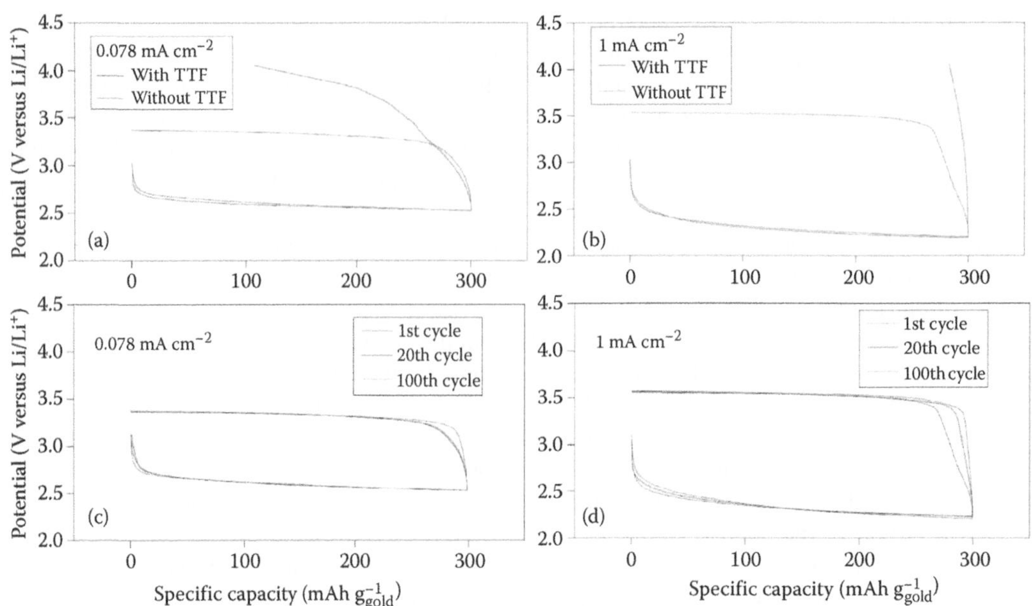

FIGURE 8.18 Discharge and charge curves of Li–O_2 cell without and with TTF redox mediator (10 mM) at current densities of 0.078 mA cm^{-2} (a) and 1 mA cm^{-2} (b); cycling stability of Li–O_2 cells containing TTF redox mediator at current densities of 0.078 mA cm^{-2} (c) and 1 mA cm^{-2} (d). (Adapted from Chen, Y., Freunberger, S. A., Peng, Z., Fontaine, O., Bruce, P. G., *Nat. Chem.*, 5, 489, 2013. With permission.)

this work. During the charge cycle, iodine (I^-) ions oxidize at the electrode surface to produce I_3^- or I_2. This is followed by the chemical reaction of I_3^- or I_2 with the Li_2O_2 discharge products of the cell and produces the initial I^-, Li^+, and O_2. It is emphasized that the hierarchically aligned porous air electrode with micro- and nanopores can prohibit clogging of the air electrode by the discharge products and lead to a thin and uniform formation of Li_2O_2 on the CNT surface, maintaining the rapid pathway for both the soluble catalyst and the reaction products. The discharge and charge characteristics of CNT fibril air electrodes with and without the LiI redox mediator and also with a solid Pt catalyst are compared in Figure 8.19a. While a CNT fibril air electrode without any catalyst exhibited a relatively high overpotential (>1.5 V) during the charge cycle, a lower overpotential of around 1 V was observed in the presence of a solid Pt catalyst. However, an even lower overpotential of around 0.7 V was resulted with a soluble LiI catalyst. In addition, the applicability of the soluble LiI catalyst was tested in two different electrolyte salts ($LiPF_6$ or LiTFSI), indicating that the LiI catalytic behavior is independent of the electrolyte type. Using a CNT fibril air electrode and LiI soluble catalyst, the Li–O_2 cell exhibited 900 consecutive restricted discharge and charge cycles (Figure 8.19b). Furthermore, the Li–O_2 cell maintained the discharge capacity of 3000 mAh g^{-1}

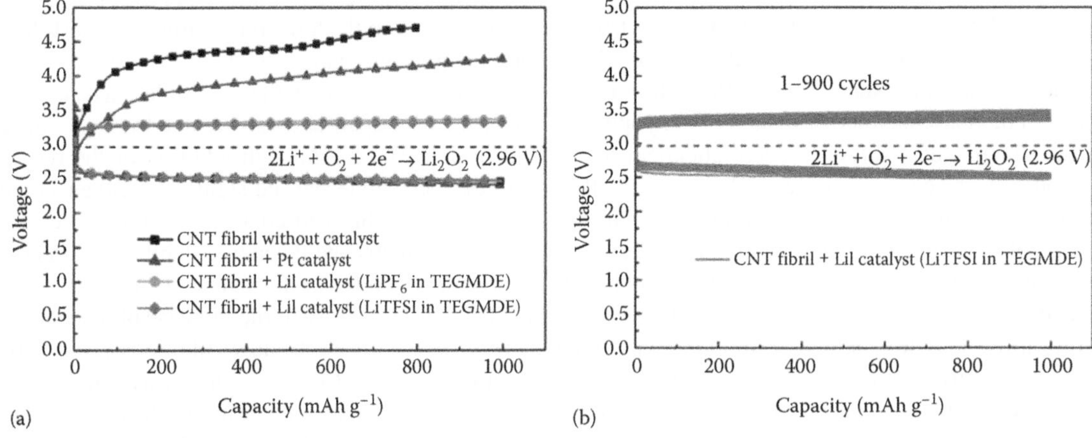

FIGURE 8.19 (a) Discharge and charge profiles of the CNT fibril electrodes with and without a solid Pt catalyst or soluble LiI catalyst at a discharge depth of 1000 mAh g^{-1} and a current density of 2000 mA g^{-1}. (b) Consecutive discharge and charge profiles of the Li–O_2 cell with a CNT fibril air electrode and a soluble LiI catalyst. (Adapted from Lim, H. D., Song, H., Kim, J., Gwon, H., Bae, Y., Park K. Y. et al., *Angew. Chem. Int. Ed.*, 53, 3926, 2014. With permission.)

for 300 cycles. This is the highest reported cycle number for an Li–O_2 cell up to the time of writing this text. The authors believe that the activity of the soluble redox mediator is synergistically enhanced through combination with the CNT fibril air electrode, since the Li–O_2 cells with a conventional carbon electrode (KB and binder) showed a higher charging potential (>3.5 V) with limited cyclability in the presence of the soluble LiI catalyst compared with the cell using the CNT fibril air electrode. These results confirm that the architecture of the air electrode may significantly influence the performance of a soluble redox mediator and also the whole Li–O_2 cell. Moreover, suppression of the side reactions, such as decomposition of the electrolyte and oxidation of the carbon cathode, as a result of the decreased charging potential to below 3.5 V, has also been mentioned as another contributing factor to the enhancement of the cyclability of the cell.

Iron phthalocyanine (FePc) has also been employed as the soluble catalyst in an Li–O_2 cell [84]. Using cyclic voltammetry, Sun et al. found that in the absence of oxygen, FePc exhibits two redox transitions related to Fe^{III}/Fe^{II} and Fe^{II}/Fe^{I} in a DMSO–LiTFSI electrolyte. In the presence of oxygen, however, the FePc coordinates with oxygen and the Fe^{II}/Fe^{I} transition shifts to more positive potentials. This redox transition is described as the result of both Fe^{II}/Fe^{I} reactions and ORR: $(FePc–O_2)^0/(FePc–O_2)^-$. The $(FePc–O_2)^-$ can be further reduced to $(FePc–O_2)^{2-}$ and form FePc–Li_2O_2 in the presence of Li^+ ions. For FePc to act as a catalyst, Li_2O_2 should dissociate from the FePc and nucleate in the electrolyte. The FePc coordination complex can also be oxidized around 3.65 V. It has been mentioned that the FePc may serve as a shuttle to transfer the charge from the electrode surface to the lithium peroxide in the OER process. Addition of the FePc to Li–O_2 cells using a carbon fibers air electrode containing TEGDME and DMSO electrolytes demonstrated that the FePc soluble catalyst slightly increases the discharge capacity of the cell. It was also observed that the morphology of the discharge products of the cell is also different in the absence and presence of the FePc. Without the FePc, the products formed during discharge only covers the surface of the carbon fibers, while with the FePc in solution, the products partially fills the space between the carbon fibers. The authors argued that the FePc acts not only as a redox mediator, but also as a molecular shuttle of O_2^- species between the surface of the air electrode and the insulator Li_2O_2 discharge products. To enhance the cyclability of the cell, the carbon fiber air electrode was exchanged with a graphene sponge and the cell was tested in a DMSO electrolyte. Using a restricted discharge capacity of 1000 mAh g^{-1}, the cell failed at the 21st cycle without the FePc. In the presence of the FePc, however, the discharge curves exhibited a flat plateau at 2.69 V fading only to 2.67 V after 130 cycles.

Research on soluble catalysts for Li–O_2 cells has only just begun and it is expected that the performance of the Li–O_2 cells will be greatly enhanced by combination of the soluble catalysts with other solid catalysts using efficient architectures of the air electrode.

8.4 CONCLUSION AND OUTLOOK

The Li–O_2 battery system has a promising potential to be applied in EVs and increase the practical travel range by a factor of 3. However, current problems associated with the Li–O_2 system should be properly addressed before this system can be used practically. The main challenge at the present time is the poor cyclability of the Li–O_2 cell. This problem is partly due to the metal anode, but the main problem in this regard is the accumulation of the discharge product on the air electrode surface. The Li_2O_2 discharge product is an insulator resulting in accumulation of this insulator material diminishing the further electrochemical reaction of the air electrode, hence limiting the cell capacity during the discharge cycle. Deposited insulator materials also increase the charge overpotential and cause instability problem for the electrolyte and electrode materials. In addition, it is recognized that carbon-based materials are unstable in the presence of Li_2O_2 discharge products, especially at potentials above 3.5 V. Li_2O_2 discharge products react with carbon substrates and/or electrolytes to form Li_2CO_3 and lithium carboxylate side products at the carbon/Li_2O_2 or Li_2O_2/electrolyte interfaces, which subsequently increases the charging potential of the cell. Since the oxidation of carbonate-based side products on the charge cycle is incomplete, these species accumulate at the air electrode surface during cycling and result in electrode passivation and even more increase of the charging overpotential. Increased charging overpotential causes more electrolyte and air electrode decomposition reactions in turn, and leads to the capacity fading. Design of the new air electrode materials with an appropriate pore structure and a high surface area that can store a large amount of solid discharge product as a uniform thin film, may decrease the charge overpotential and enhance the cyclability of the metal–air cell. A bimodal porous structure with a 3D framework may take the advantage of both high surface area substrates for accommodation of solid discharge products and also macroporous channels for transferring the oxygen and metal ions. On the other side, different catalyst systems based on noble and transition metals are also being examined for employment in the Li–O_2 cell in order to improve the kinetics of the cell. More recently, application of soluble catalysts in Li–O_2 cell has brought new promise to achieve a truly rechargeable Li–O_2 battery system.

Solid-state electrolytes are also considered as a potential solution to the instability of the conventional organic electrolytes in an Li–O_2 environment. In terms of safety, elimination of flammable organic liquid electrolytes from the system results in safer operation of the battery systems, especially at elevated temperatures. Another notable advantage of solid-state electrolytes is the ability to maintain the ionic conductivity below the freezing point of liquid electrolytes, which is beneficial for operation of the battery cell at low temperatures. However, the ionic conductivity of most solid-state electrolytes is still too low at ambient temperatures and needs improvement. The combination of all the above-mentioned ideas may help in attaining a rechargeable high energy Li–O_2 battery system in the close future.

REFERENCES

1. Abraham, K., Jiang, Z. 1996. A polymer electrolyte-based rechargeable lithium/oxygen battery. *J. Electrochem. Soc.* 143:1–5.
2. Girishkumar, G., McCloskey, B., Luntz, A. C., Swanson, S., Wilcke, W. 2010. Lithium–air battery: Promise and challenges. *J. Phys. Chem. Lett.* 1:2193–2203.
3. Ogasawara, T., Débart, A., Holzapfel, M., Novak, P., Bruce, P. G. 2006. Rechargeable Li_2O_2 electrode for lithium batteries. *J. Am. Chem. Soc.* 128:1390–1393.
4. Wang, Y., Zhou, H. 2010. A lithium–air battery with a potential to continuously reduce O_2 from air for delivering energy. *J. Power Sources* 195:358–361.
5. Lu, Y. C., Gallant, B. M., Kwabi, D. J., Harding, J. R., Mitchell, R. R., Whittingham, M. S., Shao-Horn, Y. 2013. Lithium–oxygen batteries: Bridging mechanistic understanding and battery performance. *Energy Environ. Sci.* 6:750–768.
6. Freunberger, S. A., Chen, Y., Drewett, N. E., Hardwick, L. J., Barde, F., Bruce, P. G. 2011. The lithium–oxygen battery with ether-based electrolytes. *Angew. Chem., Int. Ed.* 50:8609–8613.
7. Peng, Z., Freunberger, S. A., Hardwick, L. J., Chen, Y., Giordani, V., Bardé, F., Novák, P., Graham, D., Tarascon, J. M., Bruce, P. G. 2011. Oxygen reactions in a non-aqueous Li^+ electrolyte. *Angew. Chem., Int. Ed.* 123:6475–6479.
8. Zhai, D., Wang, H. H., Yang, J., Chun Lau, K., Li, K., Amine, K., Curtiss, L. A. 2013. Disproportionation in Li–O_2 batteries based on a large surface area carbon cathode. *J. Am. Chem. Soc.* 135:15364–15372.
9. Adams, B. D., Radtke, C., Black, R., Trudeau, M. L., Zaghib, K., Nazar, L. F. 2013. Current density dependence of peroxide formation in the Li–O_2 battery and its effect on charge. *Energy Environ. Sci.* 6:1772–1778.
10. Kang, S. Y., Mo, Y., Ong, S. P., Ceder, G. 2013. A facile mechanism for recharging Li_2O_2 in Li–O_2 batteries. *Chem. Mater.* 25:3328–3336.
11. Aurbach, D., Cohen, Y. 1996. The application of atomic force microscopy for the study of Li deposition processes. *J. Electrochem. Soc.* 143:3525–3532.
12. Kraytsberg, A., Ein-Eli, Y. 2011. Review on Li–air batteries—Opportunities, limitations and perspective. *J. Power Sources* 196:886–893.
13. Xu, W., Wang, J., Ding, F., Chen, X., Nasybulin, E., Zhang, Y., Zhang, J. G. 2014. Lithium metal anodes for rechargeable batteries. *Energy Environ. Sci.* 7:513–537.
14. Balaish, M., Kraytsberg, A., Ein-Eli, Y. 2014. A critical review on lithium–air battery electrolytes. *Phys. Chem. Chem. Phys.* 16:2801–2822.
15. McCloskey, B. D., Bethune, D. S., Shelby, R. M., Girishkumar, G., Luntz, A. C. 2011. Solvents' critical role in nonaqueous lithium–oxygen battery electrochemistry. *J. Phys. Chem. Lett.* 2:1161–1166.
16. Freunberger, S. A., Chen, Y. H., Peng, Z. Q., Griffin, J. M., Hardwick, L. J., Barde, F., Novak, P., Bruce, P. G. 2011. Reactions in the rechargeable lithium–O_2 battery with alkyl carbonate electrolytes. *J. Am. Chem. Soc.* 133:8040–8047.
17. Laoire, C. Ó., Mukerjee, S., Plichta, E. J., Hendrickson, M. A., Abraham, K. M. 2011. Rechargeable lithium/TEGDME–$LiPF_6/O_2$ battery. *J. Electrochem. Soc.* 158:A302–A308.
18. Gallant, B. M., Mitchell, R. R., Kwabi, D. G., Zhou, J., Zuin, L., Thompson, C. V., Shao-Horn, Y. 2012. Chemical and morphological changes of Li–O_2 battery electrodes upon cycling. *J. Phys. Chem. C* 116:20800–20805.
19. McCloskey, B. D., Speidel, A., Scheffler, R., Miller, D. C., Viswanathan, V., Hummelshøj, J. S., Nørskov, J. K., Luntz, A. C. 2012. Twin problems of interfacial carbonate formation in nonaqueous Li–O_2 batteries. *J. Phys. Chem. Lett.* 3:997–1001.
20. Peng, Z., Freunberger, S. A., Chen, Y., Bruce, P. G. 2012. A reversible and higher-rate Li–O_2 battery. *Science* 337:563–566.
21. Ottakam Thotiyl, M. M., Freunberger, S. A., Peng, Z., Chen, Y., Liu, Z., Bruce, P. G. 2013. A stable cathode for the aprotic Li–O_2 battery. *Nat. Mater.* 12:1050–1056.
22. Kuboki, T., Okuyama, T., Ohsaki, T., Takami, N. 2005. Lithium–air batteries using hydrophobic room temperature ionic liquid electrolyte. *J. Power Sources* 146:766–769.
23. Allen, C. J., Hwang, J., Kautz, R., Mukerjee, S., Plichta, E. J., Hendrickson, M. A., Abraham, K. M. 2012. Oxygen reduction reactions in ionic liquids and the formulation of a general ORR mechanism for Li–air batteries. *J. Phys. Chem. C* 116:20755–20764.
24. Sun, Y. 2013. Lithium ion conducting membranes for lithium–air batteries. *Nano Energy* 2:801–816.
25. Zhang, T., Imanishi, N., Shimonishi, Y., Hirano, A., Takeda, Y., Yamamoto, O., Sammes, N. 2010. A novel high energy density rechargeable lithium/air battery. *Chem. Commun.* 46:1661–1663.
26. Kitaura, H., Zhou, H. 2012. Electrochemical performance and reaction mechanism of all-solid-state lithium–air batteries composed of lithium, $Li_{1+x}Al_yGe_{2-y}(PO_4)_3$ solid electrolyte and carbon nanotube air electrode. *Energy Environ. Sci.* 5:9077–9084.
27. Zhang, T., Zhou, H. 2013. A reversible long-life lithium–air battery in ambient air. *Nat. Commun.* 4:1817. DOI:10.1038/ncomms2855.
28. Shao, Y., Ding, F., Xiao, J., Zhang, J., Xu, W., Park, S., Zhang, J. G., Wang, Y., Liu, J. 2013. Making Li–air batteries rechargeable: Material challenges. *Adv. Funct. Mater.* 23:987–1004.
29. Oh, S. H., Yim, T., Pomerantseva, E., Nazar, L. F. 2011. Decomposition reaction of lithium bis(oxalato)borate in the rechargeable lithium–oxygen cell. *Electrochem. Solid State Lett.* 14:A185–A188.
30. Veith, G. M., Nanda, J., Delmau, L. H., Dudney, N. J. 2012. Influence of lithium salts on the discharge chemistry of Li–air cells. *J. Phys. Chem. Lett.* 3:1242–1247.
31. Xu, W., Hu, J. Z., Engelhard, M. H., Towne, S. A., Hardy, J. S., Xiao, J., et al. 2012. The stability of organic solvents and carbon electrode in nonaqueous Li–O_2 batteries. *J. Power Sources* 215:240–247.
32. Younesi, R., Hahlin, M., Treskow, M., Scheers, J., Johansson, P., Edstrom, K. 2012. Ether based electrolyte, $LiB(CN)_4$ salt and binder degradation in the Li–O_2 battery studied by hard X-ray photoelectron spectroscopy (HAXPES). *J. Phys. Chem. C* 116:18597–18604.
33. Younesi, R., Hahlin, M., Bjorefors, F., Johansson, P., Edstrom, K., Li–O_2 battery degradation by lithium peroxide (Li_2O_2): A model study. *Chem. Mater.* 25:77–84.
34. Bruce, P. G., Freunberger, S. A., Hardwick, L. J., Tarascon, J. M. 2012. Li–O_2 and Li–S batteries with high energy storage. *Nat. Mater.* 11:19–29.
35. Kraytsberg, A., Ein-Eli, Y. 2013. The impact of nano-scaled materials on advanced metal–air battery systems. *Nano Energy* 2:468–480.
36. Zhang, G. Q., Zheng, J. P., Liang, R., Zhang, C., Wang, B., Hendrickson, M., Plichta, E. J. 2010. Lithium–air batteries using SWNT/CNF buckypapers as air electrodes. *J. Electrochem. Soc.* 157:A953–A956.
37. Mitchell, R. R., Gallant, B. M., Thompson, C. V., Shao-Horn, Y. 2011. All-carbon-nano fiber electrodes for high-energy rechargeable Li–O_2 batteries. *Energy Environ. Sci.* 4:2952–2958.

38. Liu, S., Wang, Z., Yu, C., Zhao, Z., Fan, X., Ling, Z., Qiu, J. 2013. Free-standing, hierarchically porous carbon nanotube film as a binder-free electrode for high-energy Li–O_2 batteries. *J. Mater. Chem. A* 1:12033–12037.
39. Etacheri, V., Sharon, D., Garsuch, A., Afri, M., Frimer, A. A., Aurbach, D. 2013. Hierarchical activated carbon microfiber (ACM) electrodes for rechargeable Li–O_2 batteries. *J. Mater. Chem. A* 1:5021–5030.
40. Gong, K., Du, F., Xia, Z., Durstock, M., Dai, L. 2009. Nitrogen-doped carbon nanotube arrays with high electrocatalytic activity for oxygen reduction. *Science* 323:760–764.
41. Kichambare, P., Kumar, J., Rodrigues, S., Kumar, B. 2011. Electrochemical performance of highly mesoporous nitrogen doped carbon cathode in lithium–oxygen batteries. *J. Power Sources* 196:3310–3316.
42. Choi, H. C., Park, J., Kim, B. 2005. Distribution and structure of N atoms in multiwalled carbon nanotubes using variable-energy X-ray photoelectron spectroscopy. *J. Phys. Chem. B* 109:4333–4340.
43. Li, Y., Wang, J., Li, X., Liu, J., Geng, D., Yang, J., Li, R., Sun, X. 2011. Nitrogen-doped carbon nanotubes as cathode for lithium–air batteries. *Electrochem. Commun.* 13:668–672.
44. Mi, R., Liu, H., Wang, H., Wong, K. W., Mei, J., Chen, Y., Lau, W. M., Yan, H. 2014. Effects of nitrogen-doped carbon nanotubes on the discharge performance of Li–air batteries. *Carbon* 67:744–752.
45. Kim, H., Lim, H. D., Kima, J., Kang, K. 2014. Graphene for advanced Li/S and Li/air batteries. *J. Mater. Chem. A* 2:33–47.
46. Li, Y., Wang, J., Li, X., Geng, D., Li, R., Sun, X. 2011. Superior energy capacity of graphene nanosheets for a nonaqueous lithium-oxygen battery. *Chem. Commun.* 47:9438–9440.
47. Xiao, J., Mei, D., Li, X., Xu, W., Wang, D., Graff, G. L., et al. 2011. Hierarchically porous graphene as a lithium–air battery electrode. *Nano Lett.* 11:5071–5078.
48. Wang, Z. L., Xu, D., Xu, J. J., Zhang, L. L., Zhang, X. B. 2012. Graphene oxide gel-derived, free-standing, hierarchically porous carbon for high-capacity and high-rate rechargeable Li–O_2 batteries. *Adv. Funct. Mater.* 22:3699–3705.
49. Zhang, W., Zhu, J., Ang, H., Zeng, Y., Xiao, N., Gao, Y., Liu, W., Hng, H. H., Yan, Q. 2013. Binder-free graphene foams for O_2 electrodes of Li–O_2 batteries. *Nanoscale* 5:9651–9658.
50. Wang, H., Maiyalagan, T., Wang, X. 2012. Review on recent progress in nitrogen-doped graphene: Synthesis, characterization, and its potential applications. *ACS Catal.* 2:781–794.
51. Li, Y., Wang, J., Li, X., Geng, D., Banis, M. N., Li, R., Sun, X. 2012. Nitrogen-doped graphene nanosheets as cathode materials with excellent electrocatalytic activity for high capacity lithium–oxygen batteries. *Electrochem. Commun.* 18:12–15.
52. Li, Y., Wang, J., Li, X., Geng, D., Banis, M. N., Tang, Y., Wang, D., Li, R., Sham, T. K., Sun, X. 2012. Discharge product morphology and increased charge performance of lithium–oxygen batteries with graphene nanosheet electrodes: The effect of sulphur doping. *J. Mater. Chem.* 22:20170–20174.
53. Xiao, J., Zheng, J., Li, X., Shao, Y., Zhang, J. G. 2013. Hierarchically structured materials for lithium batteries. *Nanotechnology* 24:424004.
54. Williford, R. E., Zhang, J. G. 2009. Air electrode design for sustained high power operation of Li/air batteries. *J. Power Sources* 194:1164–1170.
55. Guo, Z., Zhou, D., Dong, X., Qiu, Z., Wang, Y., Xia, Y. 2013. Ordered hierarchical mesoporous/macroporous carbon: A high-performance catalyst for rechargeable Li–O_2 batteries. *Adv. Mater.* 25:5668–5672.
56. Ottakam Thotiyl, M. M., Freunberger, S. A., Peng, Z., Bruce, P. G. 2012. The carbon electrode in nonaqueous Li–O_2 cells. *J. Am. Chem. Soc.* 135:494–500.
57. Wang, J., Li, Y., Sun, X. 2013. Challenges and opportunities of nanostructured materials for aprotic rechargeable lithium–air batteries. *Nano Energy* 2:443–467.
58. Christensen, J., Albertus, P., Sanchez-Carrera, R. S., Lohmann, T., Kozinsky, B., Liedtke, R., Ahmed, J., Kojic, A. 2012. A critical review of Li/air batteries. *J. Electrochem. Soc.* 159:R1–R30.
59. McCloskey, B. D., Scheffler, R., Speidel, A., Bethune, D. S., Shelby, R. M., Luntz, A. C. 2011. On the efficacy of electrocatalysis in nonaqueous Li–O_2 batteries. *J. Am. Chem. Soc.* 133:18038–18041.
60. Lu, Y. C., Xu, Z., Gasteiger, H. A., Chen, S., Hamad-Schifferli, K., Shao-Horn, Y. 2010. Platinum–gold nanoparticles: A highly active bifunctional electrocatalyst for rechargeable lithium–air batteries. *J. Am. Chem. Soc.* 132:12170–12171.
61. Lim, H. D., Song, H., Gwon, H., Park, K. Y., Kim, J., Bae, Y., et al. 2013. A new catalyst-embedded hierarchical air electrode for high-performance Li–O_2 batteries. *Energy Environ. Sci.* 6:3570–3575.
62. Dathar, G. K. P., Shelton, W. A., Xu, Y. 2012. Trends in the catalytic activity of transition metals for the oxygen reduction reaction by lithium. *J. Phys. Chem. Lett.* 3:891–895.
63. Xu, J. J., Wang, Z. L., Xu, D., Zhang, L. L., Zhang, X. B. 2013. Tailoring deposition and morphology of discharge products towards high-rate and long-life lithium–oxygen batteries. *Nat. Commun.* 4:2438. DOI:10.1038/ncomms3438.
64. Lu, J., Lei, Y., Lau, K. C., Luo, X., Du, P., Wen, J. et al. 2013. A nanostructured cathode architecture for low charge overpotential in lithium–oxygen batteries. *Nat. Commun.* 4:2383. DOI:10.1038/ncomms3383.
65. Sun, B., Munroe, P., Wang, G. 2013. Ruthenium nanocrystals as cathode catalysts for lithium–oxygen batteries with a superior performance. *Sci. Rep.* 3:2247. DOI:10.1038/srep02247.
66. Jian, Z., Liu, P., Li, F., He, P., Guo, X., Chen, M., Zhou, H. 2014. Core–shell-structured CNT@RuO_2 composite as a high-performance cathode catalyst for rechargeable Li–O_2 batteries. *Angew. Chem. Int. Ed.* 53:442–446.
67. Débart, A., Bao, J., Armstrong, G., Bruce, P. G. 2007. An O_2 cathode for rechargeable lithium batteries: The effect of a catalyst. *J. Power Sources* 174:1177–1182.
68. Débart, A., Bao, J., Armstrong, G., Bruce, P. G. 2007. Effect of catalyst on the performance of rechargeable lithium/air batteries. *ECS Trans.* 3:225–232.
69. Débart, A., Paterson, A. J., Bao, J., Bruce, P. G. 2008. α-MnO_2 nanowires: A catalyst for the O_2 electrode in rechargeable lithium batteries. *Angew. Chem. Int. Ed.* 47:4521–4524.
70. Li, J. X., Wang, N., Zhao, Y., Ding, Y. H., Guan, L. H. 2011. MnO_2 nanoflakes coated on multi-walled carbon nanotubes for rechargeable lithium–air batteries. *Electrochem. Commun.* 13:698–700.
71. Cheng, H., Scott, K. 2010. Carbon-supported manganese oxide nanocatalysts for rechargeable lithium–air batteries. *J. Power Sources* 195:1370–1374.
72. Ida, S., Thapa, A. K., Hidaka, Y., Okamoto, Y., Matsuka, M., Hagiwara, H., Ishihara, T. 2012. Manganese oxide with a card-house-like structure reassembled from nanosheets for rechargeable Li–air battery. *J. Power Sources* 203:159–164.
73. Jin, L., Xu, L. P., Morein, C., Chen, C. H., Lai, M., Dharmarathna, S., Dobley, A., Suib, S. L. 2010. Titanium containing γ-MnO_2 (TM) hollow spheres: One-step synthesis and catalytic activities in Li/air batteries and oxidative chemical reactions. *Adv. Funct. Mater.* 20:3373–3382.

74. Thapa, A. K., Ishihara, T. 2011. Mesoporous α-MnO_2/Pd catalyst air electrode for rechargeable lithium–air battery. *J. Power Sources* 196:7016–7020.
75. Zhang, G. Q., Zheng, J. P., Liang, R., Zhang, C., Wang, B., Au, M., Hendrickson, M., Plichta, E. J. 2011. α-MnO_2/carbon nanotube/carbon nanofiber composite catalytic air electrodes for rechargeable lithium–air batteries. *J. Electrochem. Soc.* 158:A822–A827.
76. Zahoor, A., Jang, H. S., Jeong, J. S., Christy, M., Hwang, Y. J., Nahm, K. S. 2014. Comparative study of nanostructured α and γ MnO_2 for lithium oxygen battery application. *RSC Adv.* 4:8973–8977.
77. Lee, J. H., Black, R., Popov, G., Pomerantseva, E., Nan, F., Botton, G. A., Nazar, L. F. 2012. The role of vacancies and defects in $Na_{0.44}MnO_2$ nanowire catalysts for lithium–oxygen batteries. *Energy Environ. Sci.* 5:9558–9565.
78. Qin, Y., Lu, J., Du, P., Chen, Z., Ren, Y., Wu, T., Miller, J. T., Wen, J., Miller, D. J., Zhanga, Z., Amine, K. 2013. *In situ* fabrication of porous-carbon-supported α-MnO_2 nanorods at room temperature: Application for rechargeable Li–O_2 batteries. *Energy Environ. Sci.* 6:519–531.
79. Thapa, A. K., Saimen, K., Ishihara, T. 2010. Pd/MnO_2 air electrode catalyst for rechargeable lithium/air battery. *Electrochem. Solid-State Lett.* 13:A165–A167.
80. Oh, S. H., Black, R., Pomerantseva, E., Lee, J. H., Nazar, L. F. 2012. Synthesis of a metallic mesoporous pyrochlore as a catalyst for lithium–O_2 batteries. *Nat. Chem.* 4:1004–1010.
81. Shui, J. L., Karan, N. K., Balasubramanian, M., Li, S. Y., Liu, D. J. 2012. Fe/N/C composite in Li–O_2 battery: Studies of catalytic structure and activity toward oxygen evolution reaction. *J. Am. Chem. Soc.* 134:16654–16661.
82. Chen, Y., Freunberger, S. A., Peng, Z., Fontaine, O., Bruce, P. G. 2013. Charging a Li–O_2 battery using a redox mediator. *Nat. Chem.* 5:489–494.
83. Lim, H. D., Song, H., Kim, J., Gwon, H., Bae, Y., Park, K. Y. et al. 2014. Superior rechargeability and efficiency of lithium–oxygen batteries: Hierarchical air electrode architecture combined with a soluble catalyst. *Angew. Chem. Int. Ed.* 53:3926–3931.
84. Sun, D., Shen, Y., Zhang, W., Yu, L., Yi, Z., Yin, W., Wang, D., Huang, Y., Wang, J., Wang, D., Goodenough, J. B. 2014. A solution-phase bifunctional catalyst for lithium–oxygen batteries. *J. Am. Chem. Soc.* 136:8941–8946.

9 Advanced Materials for Na–Air Batteries

Yongliang Li, Hossein Yadegari, Jianhong Liu, and Xueliang Sun

CONTENTS

9.1 INTRODUCTION

As the amount of fossil fuels decreases worldwide together with the issue of global warming, it becomes necessary to develop clean and renewable energy storage and conversion systems for transportation. Lithium-ion batteries (LIBs) are promising alternatives to the internal combustion engine for hybrid electric vehicles, plug-in hybrid electric vehicles, and pure electric vehicles. However, the energy density of current LIBs is limited by the intercalation chemistry occurring within the electrode materials, which will not meet the rapid growth of energy demand for future electric vehicle applications. Lithium–air batteries have been intensively studied during the last few years, and they have shown extremely high discharge capacities [1–3]. They consist of a lithium metal anode, a liquid electrolyte comprising lithium salts dissolved in organic solvents, a microporous polymer separator, and a porous cathode electrode. The theoretical energy density of lithium–air battery is about 3600 Wh kg^{-1}, which is eight times that of LIBs. Moreover, the batteries could be charged when electrocatalysts that facilitate the oxygen evolution reaction (OER) are applied in cathodes.

However, there are several critical issues with lithium–air batteries should be addressed before their practical applications. First, the discharge product, Li_2O_2, is very reactive to the electrolyte, leading to an irreversible capacity and the consumption of the electrolyte during battery cycling. Second, Li_2O_2 is an electrical-insulating material and as the discharge process continues, more and more product deposits on the surface of the electrode and eventually the discharge process is terminated because of the passivation effect. Third, on the charge process, dendrites will form on the lithium anode, which may cause serious safety problems. Moreover, the supply of lithium resource would probably run out in the foreseeable future, which is insufficient to satisfy the increasing demand of batteries.

Sodium (Na), an element with wide availability and low cost, is a very promising material for meeting the large-scale grid energy storage needs [4]. In addition, this material is the fourth most abundant element on Earth, having the potential to be an alternative source to lithium [5]. Recently, sodium has been investigated to replace lithium in air battery systems [6–8]. The Na–air battery is one of the environmentally friendly electrochemical power sources having a high energy density. Although the specific capacity of the Na–air battery (1600 Wh kg^{-1}) is much lower than that of the lithium–air battery, it is considerably higher than that of LIBs. Similar to lithium–air battery, the Na–air battery is composed of an Na metal anode, Na-salt-containing electrolyte, and a porous cathode, while the cathode active material, oxygen, is obtained from the environment during the discharge process (Figure 9.1).

Replacing lithium with sodium also has other advantages, such as the reaction mechanism for sodium is very similar to that of lithium in the positive electrode, making it possible to replace the anode materials without changing the configuration of the system. The working potential of sodium battery is lower than that of lithium battery; if rechargeable sodium batteries could be developed, low voltages are suitable for the charge process and therefore the electrolyte would become more stable in these systems and, also, the energy efficiency will be increased.

In this chapter, we review the advanced materials for rechargeable Na–air batteries, including electrolyte, cathode, and anode materials. In addition, the development of current collector materials is also reviewed. Finally, the perspective of material development direction is discussed.

9.2 BASIC ELECTROCHEMISTRY OF Na–AIR BATTERIES

The electrochemical reactions for Na–air batteries are listed below:

$$Na^+ + O_2 + e^- \rightarrow NaO_2 \quad E = 2.26\ \text{V} \tag{9.1}$$

$$2Na^+ + O_2 + 2e^- \rightarrow Na_2O_2 \quad E = 2.33\ \text{V} \tag{9.2}$$

$$4Na^+ + O_2 + 4e^- \rightarrow 2Na_2O \quad E = 1.95\ \text{V} \tag{9.3}$$

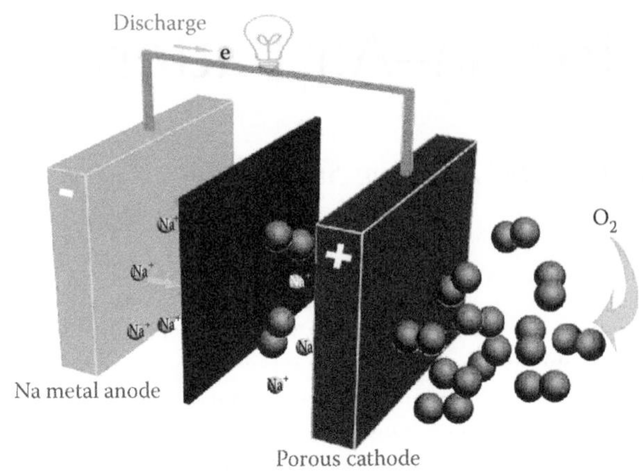

FIGURE 9.1 Schematic diagram of Na–air battery. (Reproduced from Li, Y. et al. *Chem. Commun.*, 49, 11731, 2013. With permission from Royal Society of Chemistry.)

As shown from Equations 9.1 through 9.3, the redox potentials of Na–air batteries are different, resulting in different discharge products. Hartmann et al. demonstrated that sodium superoxide (NaO_2) is the main discharge product of the Na–air battery [8]. They suggested that the formation of Na_2O_2 (E = 2.33 V) is thermodynamically favored than that of NaO_2 (E = 2.26 V); however, from the kinetic point of view, the formation of NaO_2 only requires one electron transfer, which is relatively easier. Moreover, NaO_2 is considered to be an electronic conductor, which may show different properties for the charge process.

The charge occurs at a voltage ~2.3 V at low current densities, which is close to the potential for decomposition of NaO_2 (Figure 9.2a). The finding clearly indicates the formation and decomposition of NaO_2 is reversible. In contrast, the polarization for lithium–air battery is significant during the charge process (Figure 9.2b). It is believed that for the oxygen-rich NaO_2 phase, the metal–oxygen bonding in the lattice is weaker than that of Li_2O_2 and NaO_2 can decompose reversely to form Na^+ ions and oxygen during charging. Several groups have reported different electrochemical properties of Na–air batteries; however, the detailed electrochemical reaction

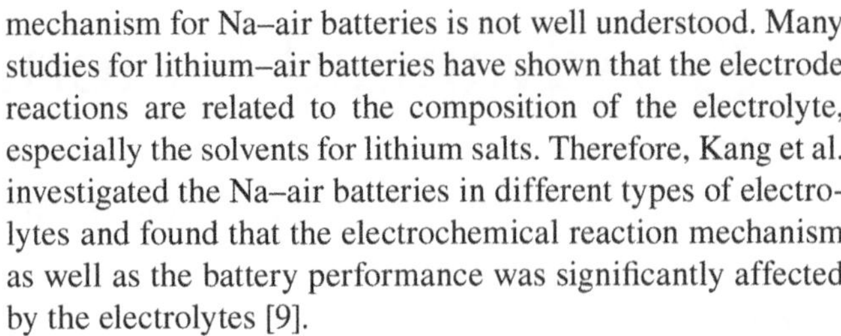

mechanism for Na–air batteries is not well understood. Many studies for lithium–air batteries have shown that the electrode reactions are related to the composition of the electrolyte, especially the solvents for lithium salts. Therefore, Kang et al. investigated the Na–air batteries in different types of electrolytes and found that the electrochemical reaction mechanism as well as the battery performance was significantly affected by the electrolytes [9].

9.3 ELECTROLYTE MATERIALS

The electrolyte applied in Na–air battery systems can be divided into two categories based on solvents: carbonate-based and non-carbonate-based (usually ethers) electrolytes. Kang et al. reported that sodium carbonate formed and decomposed during cycling in the carbonate-based battery, while hydrated sodium oxides and sodium hydroxide were the main reaction products in the non-carbonate-based electrolyte. This phenomenon is similar to that for lithium–air batteries; however, the hydrated phases are formed due to the strongly hygroscopic nature of sodium oxides when contacted with water, which is from the decomposition of the electrolyte.

Carbonates are employed as solvents of electrolytes for LIBs owing to their high oxidation potentials and relatively low viscosity, and have been initially used in lithium–air battery systems. However, it is found that carbonates decompose during discharge, leading to side reactions that affect the cycle performance of batteries. A similar behavior was observed in the Na–air battery (Figure 9.3). The proposed reaction mechanisms for Na–air batteries during discharge and charge processes in the carbonate-based electrolyte are shown in Figure 9.4. The initial degradation of propylene carbonate (PC) during discharge is identical to that for lithium–air batteries with the same electrolyte [10].

During discharge, (i) O_2 is reduced by reacting with the electron to form O_2^-, which is highly reactive; (ii) O_2^- attacks the ethereal C atoms of the CH_2 group in PC by S_N2 substitution; (iii) ring opening of PC to form peroxo alkyl carbonate due to the imbalance in the bonding between the two O atoms; (iv) H_2O, CO_2, and some by-products are formed because of

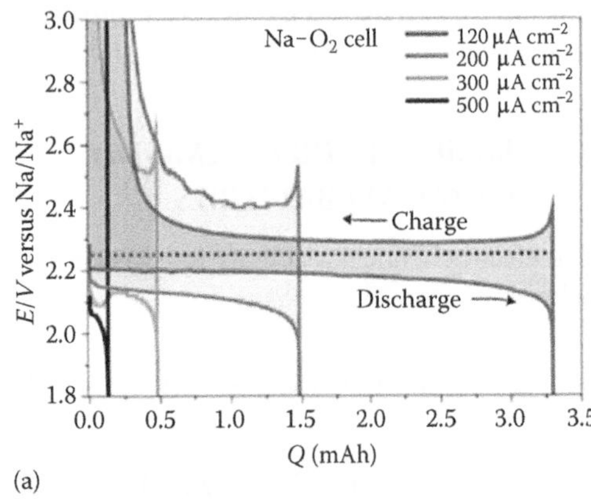

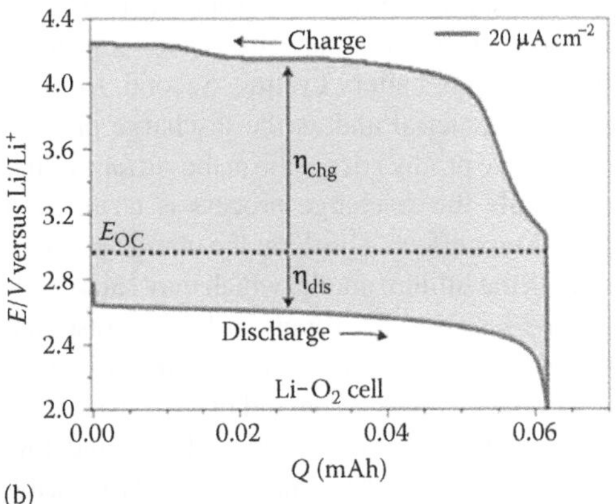

FIGURE 9.2 Electrochemical characterization of (a) Na–O_2 and (b) Li–O_2 cells. (Reproduced from Hartmann, P. et al. *Nat. Mater.*, 12, 228, 2013. With permission from Nature.)

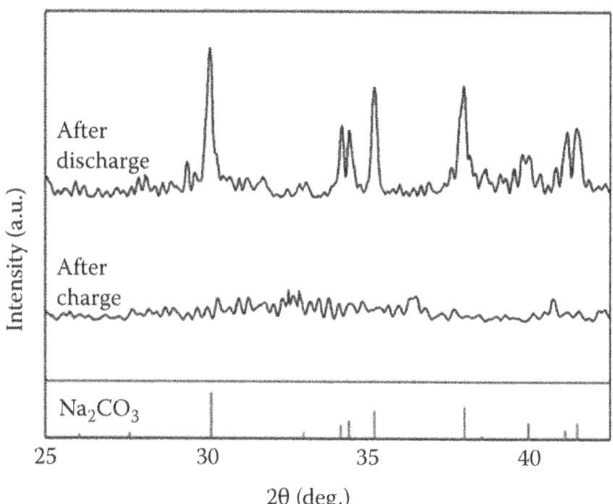

FIGURE 9.3 X-ray diffraction measurements of Na–air battery electrodes after discharge and charge. (Reproduced from Kim, J. et al. *Phys. Chem. Chem. Phys.*, 15, 3623, 2013. With permission from Royal Society of Chemistry.)

the decomposition of peroxo radical species in the presence of O_2; (v) CO_2 reacts with the O_2^- to form an intermediate $C_2O_6^{2-}$ species; and (vi) the intermediate product reacts with Na^+ ions to form Na_2CO_3 and O_2. While for charge, (i) Na_2CO_3 decomposes to form the peroxo radical species; (ii) $Na_2C_2O_6$ forms due to dimerization of the radical species; (iii) Na^+ and e^- extraction; (iv) $Na_2C_2O_6$ decomposes to form $NaCO_4$ and CO_2; and (v) Na^+, O_2, and CO_2 form by decomposition of $NaCO_4$. Moreover, (vi) the formation of O_2^- would promote the decomposition of PC to H_2O and CO_2.

Ethers are also studied for lithium–air batteries because of their attributes, for example, capability of operating with a lithium anode, high oxidation potential, inexpensiveness, and low volatility of high molecular weight samples. Although ethers are more stable than carbonates during discharge and charge processes for lithium–air batteries, they are not suitable for long-term operation due to the decomposition. The reaction mechanisms during discharge and charge in tetraethylene glycol dimethyl ether (TEGDME) electrolyte for Na–air batteries are proposed by Kang et al. As shown in Figure 9.5, the O_2^- will also attack the electrolyte to form an unstable peroxo ether radical, which decomposes to evolve H_2O, CO_2, and some by-products. In contrast to the PC electrolyte, sodium oxides and sodium hydroxides are the main discharge products. While for lithium–air batteries, lithium peroxide is the discharge product and neither the hydrated form of Li_2O_2 nor LiOH was found. This is because Na_2O_2 is highly hygroscopic compared with Li_2O_2, which readily absorbs water molecules that were produced in the oxidative decomposition reaction. Subsequently, the charge mechanism is more complex than that of lithium–air battery with the same electrolyte because of the involvement of water in Na–air battery.

As discussed above, the reaction mechanisms of Na–air batteries are different and depend on the type of electrolyte. However, both carbonates and ethers will decompose during the operation of an Na–air battery; therefore, a suitable alternative is required for developing this new battery system [11–13].

9.4 CATHODE MATERIALS

The discharge product of Na–air battery is insoluble in the electrolyte but deposits on the surface of electrodes, which is similar to lithium–air battery. It is believed that porous electrodes with optimum porosity and effective catalytic site

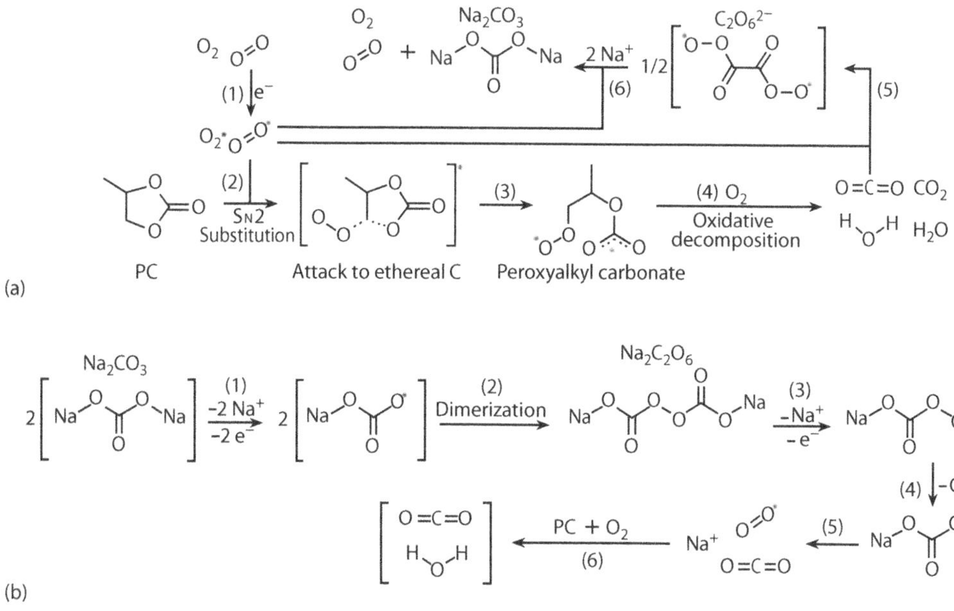

FIGURE 9.4 Schematic diagram of the proposed reaction mechanisms for the (a) discharge and (b) charge processes with PC electrolyte. (Reproduced from Kim, J. et al. *Phys. Chem. Chem. Phys.*, 15, 3623, 2013. With permission from Royal Society of Chemistry.)

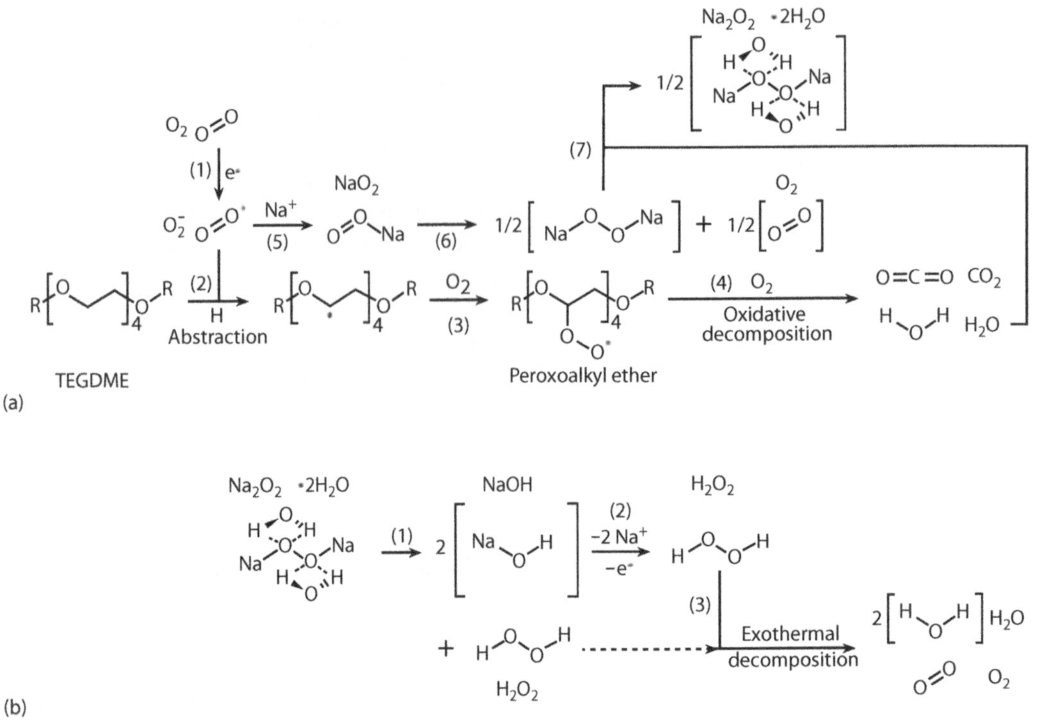

FIGURE 9.5 Schematic diagram of the proposed reaction mechanisms for the (a) discharge and (b) charge processes with TEGDME electrolyte. (Reproduced from Kim, J. et al. *Phys. Chem. Chem. Phys.*, 15, 3623, 2013. With permission from Royal Society of Chemistry.)

distribution are desirable. Several different nanostructured air electrodes for batteries were investigated.

Carbon powder, such as Super P carbon, was studied as a cathode material, and the discharge capacity was 1500 mAh g^{-1} [14]; however, there are few reports on a systematic study of the parameters of carbon powder materials (particle size, surface area, pore size, conductivity, etc.) for the Na–air battery. Fu et al. synthesized diamond-like carbon (DLC) material, and the DLC thin film electrode delivered a discharge capacity of ~1900 mAh g^{-1} [15].

One-dimensional (1D) materials have been studied in Na–air battery systems. Hartmann et al. employed carbon fibers as a gas diffusion layer for the battery, and demonstrated that the overpotential of the electrode was significantly lower than that of lithium–air battery and the battery reactions are reversible at a current density of 0.2 mA cm^{-2} [8]. Jian et al. directly used a carbon nanotube (CNT) paper without binder as electrode and a large discharge capacity of 7530 mAh g^{-1} was achieved, which is due to not only the improved electronic conductivity but also the large void spaces inside the electrode resulting from the interpenetrating CNT network (Figure 9.6) [16]. In addition, the binder-free CNT paper can be thinner and more porous, which is important for accommodating of the discharge products, leading to large specific capacity.

Graphene nanosheet (GNS) cathode could deliver an extremely high discharge capacity and was an ideal 3D and

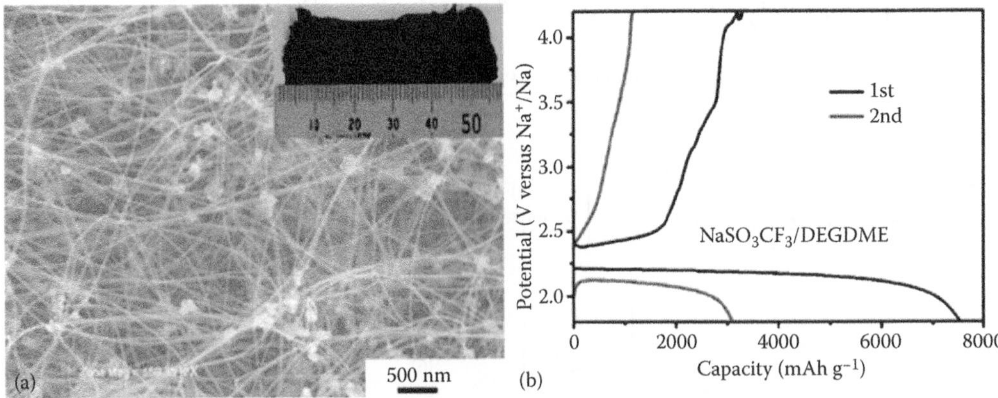

FIGURE 9.6 (a) Scanning electron microscopy (SEM) image of CNT paper electrode and (b) first two discharge/charge cycles of Na–air battery with a cutoff voltage of 1.8–4.2 V. (Reproduced from Jian, Z. et al. *J. Power Sources.*, 251, 466, 2014. With permission from Elsevier.)

three-phase electrochemical air electrode due to its unique morphology and structure in Na–air battery. Liu et al. reported that a high discharge capacity of ~9300 mAh g^{-1} was obtained with a low overpotential [17]. Moreover, GNS was demonstrated as an efficient catalyst for both the oxygen reduction reaction (ORR) and OER in the rechargeable Na–air battery, which is due to the large surface area and high electron mobility. These features of GNSs could provide more active sites for the reactions and enhance the electron transportation during the discharge/charge process.

Recently, Li et al. employed nitrogen-doped GNSs (N-GNSs) as electrode materials for the Na–air battery and found that the discharge capacity was two times greater than that of pristine counterpart, and have superior electrocatalytic activity (Figure 9.7) [18]. The enhanced performance of N-GNSs is attributed to the active sites introduced by nitrogen doping. Interestingly, the size of the discharge product on N-GNSs is much smaller than those on GNSs, which is similar for lithium–air battery, and this is also due to the defective sites introduced by nitrogen doping.

A carbon- and binder-free electrode was synthesized by Liu et al. They deposited $NiCo_2O_4$ nanosheets on Ni foam by a solvothermal method and the as-prepared material was directly applied as electrode for Na–air battery (Figure 9.8)

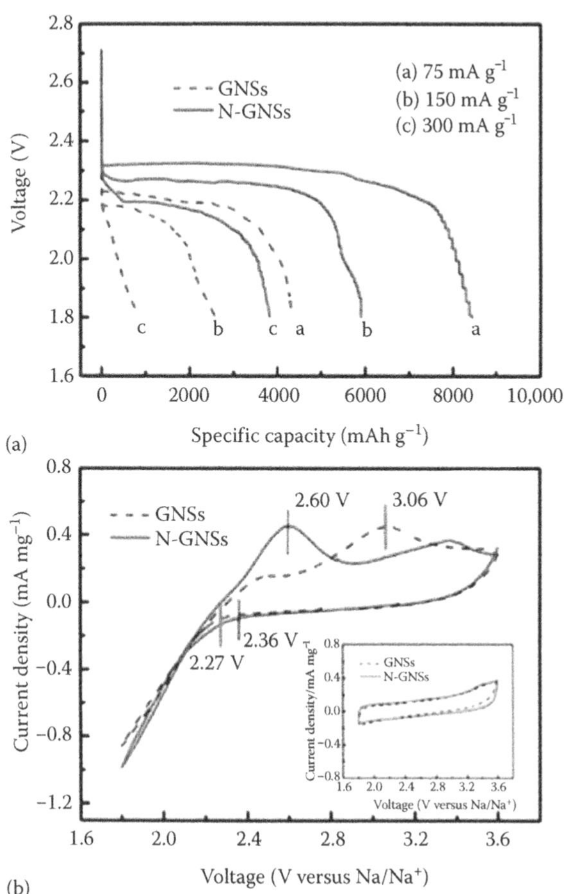

FIGURE 9.7 (a) Voltage profiles and (b) CV curves of GNS and N-GNS electrodes. (Reproduced from Li, Y. et al. *Chem. Commun.*, 49, 11731, 2013. With permission from Royal Society of Chemistry.)

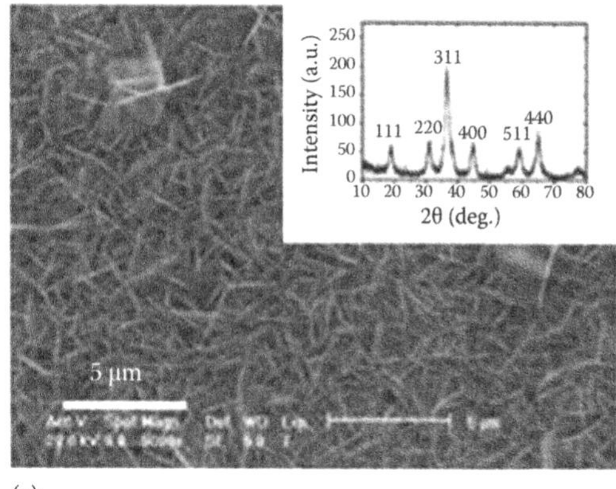

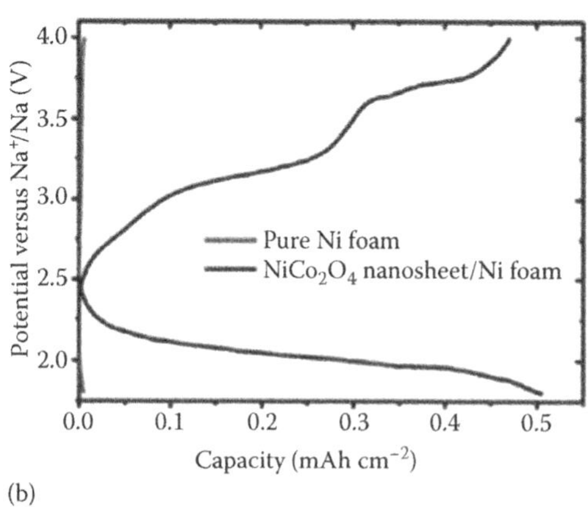

FIGURE 9.8 (a) SEM image of $NiCo_2O_4$ nanosheet electrode and (b) voltage profiles of pure Ni foam and $NiCo_2O_4$ nanosheet/Ni foam electrodes. (Reproduced from Liu, W. et al. *Electrochem. Commun.*, 45, 87, 2014. With permission from Elsevier.)

[19]. $NiCo_2O_4$ nanosheets/Ni showed a high electrocatalytic activity for ORR and OER, and the battery exhibited an initial discharge capacity of ~1800 mAh g^{-1} with a low polarization of 0.96 V at 20 mA g^{-1}.

9.5 ANODE MATERIALS

The Na anode also presents a challenge in the practical application of the battery since the metal is reactive to moisture, which may be introduced from the environment during the operation. In addition, the decomposition of the electrolyte will also produce water, which should be considered. Another issue is dendrite formation during charging, which may also cause a safety problem. Peled et al. proposed a new concept of operating the Na–air battery at a temperature above the melting point of Na [7,20]. It is expected that the surface tension of the liquid metal could prevent the formation of dendrites. Moreover, the higher operating temperature can also accelerate the electrode kinetics and reduce electrolyte resistance, and thus increase the power of the battery. However, liquid sodium is well known for its highly corrosive characteristic that hinders its practical applications. Therefore, it is important to develop protection layers on sodium metal, such as a sodium contacting conductor or a polymer-based membrane.

9.6 CURRENT COLLECTOR MATERIALS

Several materials have been applied as current collector materials for Na–air batteries, such as nickel [15], titanium [16], stainless steel [18], and aluminum [20]. It is noted that as reported in lithium–air battery system, nickel will promote the decomposition of the electrolyte during the charge process if the potential is greater than 3.5 V; therefore, the stability of the current collector should be considered.

9.7 CONCLUSION AND PERSPECTIVE

Na–air batteries have received much attention owing to their potential as an energy system for future electric vehicles. However, before their practical application, there are still many challenges and obstacles for the materials that need to be overcome. First, the development of an electrolyte with high stability, high oxygen solubility/diffusivity, high ion conductivity, and low volatility is urgent in this area since this is essential for the battery system. Consequently, the reaction mechanisms of the electrode in the novel electrolyte should be studied in detail to understand the chemistry. Second, design and construction of a cathode with improved porosity, conductivity, and catalytic activity is critical for battery capacity and cyclability. Finally, the safety issue resulting from the Na anode should also be considered, and a possible approach is to develop a protection layer that prevents the formation of a dendrite structure. Research on Na–air battery is at the early stages, and the understanding of this novel electrochemical energy system is still largely incomplete. Numerous systematic and detailed studies, especially in materials research, should be pursued in the future.

REFERENCES

1. Abraham K. and Jiang, Z. 1996. A polymer electrolyte-based rechargeable lithium/oxygen battery. *J. Electrochem. Soc.* 143:1.
2. Débart A., Paterson A., Bao J., and Bruce P. 2008. α-MnO_2 nanowires: A catalyst for the O_2 electrode in rechargeable lithium batteries. *Angew. Chem. Int. Ed.* 47:4521–24.
3. Girishkumar G., McCloskey B., Luntz A., Swanson S., and Wilcke W. 2010. Lithium–air battery: Promise and challenges. *J. Phys. Chem. Lett.* 1:2193–203.
4. Lu X., Xia G., Lemmon J., and Yang Z. 2010. Advanced materials for sodium-beta alumina batteries: Status, challenges, and perspectives. *J. Power Sources* 195:2431–42.
5. Kim S., Seo D., Ma X., Ceder G., and Kang K. 2012. Electrode materials for rechargeable sodium-ion batteries: Potential alternatives to current lithium-ion batteries. *Adv. Energy Mater.* 2:710–21.
6. Ha S., Kim J., Choi A., Kim Y., and Lee K. 2014. Sodium–metal halide and sodium–air batteries. *ChemPhysChem* 15:1971–82.
7. Peled E., Golodnitsky D., Hadar R., Mazor H., Goor M., and Burstein L. 2013. Challenges and obstacles in the development of sodium–air batteries. *J. Power Sources* 244:771–6.
8. Hartmann P., Bender C., Vracar M., Dürr A., Garsuch A., Janek J., and Adelhelm P. 2013. A rechargeable room-temperature sodium superoxide (NaO_2) battery. *Nat. Mater.* 12:228–32.
9. Kim J., Lim H., Gwon H., and Kang K. 2013. Sodium–oxygen batteries with alkyl-carbonate and ether based electrolytes. *Phys. Chem. Chem. Phys.* 15:3623–29.
10. Bruce P., Freunberger S., Hardwick L., and Tarascon J. 2012. Li–O_2 and Li–S batteries with high energy storage. *Nat. Mater.* 11:19–29.
11. Lee B., Seo D., Lim H., Park I., Park K., and Kim J. 2014. First-principles study of the reaction mechanism in sodium–oxygen batteries. *Chem. Mater.* 26:1048–55.
12. Hartmann P., Grübl D., Sommer H., Janek J., Bessler W., and Adelhelm P. 2014. Pressure dynamics in metal–oxygen (metal–air) batteries: A case study on sodium superoxide cells. *J. Phys. Chem. C* 118:1461–71.
13. Kang S., Mo Y., Ong S., and Ceder G. 2014. Nanoscale stabilization of sodium oxides: Implications for Na–O_2 batteries. *Nano Lett.* 14:1016–20.
14. Das S., Xu S., and Archer L. 2013. Carbon dioxide assist for non-aqueous sodium–oxygen batteries. *Electrochem. Commun.* 27:59–62.
15. Sun Q., Yang Y., and Fu Z. 2012. Electrochemical properties of room temperature sodium–air batteries with non-aqueous electrolyte. *Electrochem. Commun.* 16:22–5.
16. Jian Z., Chen Y., Li F., Zhang T., Liu C., and Zhou H. 2014. High capacity Na–O_2 batteries with CNT paper as binder-free air cathode. *J. Power Sources* 251:466–9.
17. Liu W., Sun Q., Yang Y., Xie J., and Fu Z. 2013. An enhanced electrochemical performance of a sodium–air battery with graphene nanosheets as air electrode catalysts. *Chem. Commun.* 49:1951–53.
18. Li Y., Yadegari H., Li X., Banis M., Li R., and Sun X. 2013. Superior catalytic activity of nitrogen-doped graphene cathodes for high energy capacity sodium–air batteries. *Chem. Commun.* 49:11731–33.
19. Liu W., Yin W., Ding F., Sang L., and Fu Z. 2014. $NiCo_2O_4$ nanosheets supported on Ni foam for rechargeable nonaqueous sodium–air batteries. *Electrochem. Commun.* 45:87–90.
20. Peled E., Golodnitsky D., Mazor H., Goor M., and Avshalomov S. 2011. Parameter analysis of a practical lithium- and sodium-air electric vehicle battery. *J. Power Sources* 196:6835–40.

Section IV

Advanced Materials and Technologies for Lead–Acid Rechargeable Batteries

10 Advanced Technologies for Lead–Acid Rechargeable Batteries

Jon L. Anderson

CONTENTS

10.1 OVERVIEW

Lead–acid batteries and lead–acid technology celebrated its sesquicentennial anniversary in 2010. In the 150 subsequent years, since Gaston Plante presented his set of nine electrochemical cells to the French Academy of Sciences, the advances in lead–acid technology have been such that in 2012, lead–acid batteries accounted for ~$32 billion dollars of a ~$50 billion dollar annual global battery market for rechargeable batteries [1].

Recent developments in lead–acid batteries have been, as most advances are, market driven. Changing application requirements, competing technologies, new markets, and emerging markets have led to the current trends and innovations in lead–acid technology.

Chief among recent advances are the numerous variations on "lead–carbon" technology. The primary market drivers for the advances and proliferation in lead–carbon technology are the emergence of applications with a requirement for partial state-of-charge operation (PSoC), high-rate partial state-of-charge operation (HRPSoC), and greater dynamic charge acceptance (DCA), such as stop–start technology in automobiles and many of the emerging renewable energy and grid-scale energy storage systems (ESSs). The expanded duty of the battery, in automobiles, and the emphasis on using the battery to increase the overall system efficiency, in both automotive and stationary systems, has prompted battery manufacturers to explore new ways of improving performance.

In addition to new application requirements, the drive for higher performance, longer service life, and lower costs (both initial and overall) in existing applications for standby power systems, motive power systems, and automotive batteries has led to other advances in lead–acid batteries, such as "pure lead" technologies for VRLA batteries and other material advances. These technical advances for lead–acid batteries continue to allow the technology to maintain its strong market position despite the advances in other technologies, such as lithium-ion batteries.

The key benefit and advantage of lead–acid technology in these applications remains the cost. Safety and reliability are additional factors leading to the continued large-scale deployment and use of the lead–acid batteries. The mature infrastructure, large manufacturing base, low material costs, and high rate of recycle (estimated to be from >90% up to 99.9% in some areas [2]) all play a role in the continued marketability of the lead–acid battery.

10.2 Pb–ACID BATTERY BASICS (FIGURE 10.1)

As can be expected, with a 150-year-old technology, the proliferation of subsets of the technology, designs, and formats of lead–acid batteries is quite extensive. The traditional breakdown of modern lead–acid technology has typically been a bimodal exercise, the cells are either vented lead–acid (VLA or "flooded") or valve regulated lead–acid (VRLA) designs. In both cases, the system is described by Equation 10.1, with the main technical difference between the two technologies being the operation of an internal oxygen recombination cycle (Equations 10.3 through 10.5) versus the continuous breakdown, release, and, in some cases, replenishment of the electrolyte (Equation 10.2) [3].

$$PbO_2 + Pb + 2H_2SO_4 \underset{\text{Discharge}}{\overset{\text{Charge}}{\leftrightarrows}} 2PbSO_4 + 2H_2O(l) \quad (10.1)$$

(VLA)

$$H_2O \rightarrow H_2 + \frac{1}{2}O_2 \quad (10.2)$$

(VRLA reaction at the positive electrode)

$$2H_2O \rightarrow O_2 + 4H^+ + 4e^- \quad (10.3)$$

(VRLA reactions at the negative electrode)

$$Pb + \frac{1}{2}O_2 \rightarrow PbO \quad (10.4)$$

$$PbO + H_2SO_4 \rightarrow 2PbSO_4 + H_2O \quad (10.5)$$

$$PbSO_4 + 2H^+ + 2e^- \rightarrow Pb + H_2SO_4 \quad (10.6)$$

A further differentiation of technologies within both VLA and VRLA includes positive plate technologies, such as tubular, prismatic, or Plante, and for VRLA batteries the electrolyte immobilization methods, gelled versus absorbent glass mat (AGM). As if that was not enough, the design variation within each of the above categories can be further split by market and application requirements. As an example, a 12-V VRLA AGM monobloc for telecommunications, uninterruptable power systems (UPS), and deep cycle may look identical and have entirely different designs and performance characteristics, making them more or less well suited for specific application conditions. Regardless of the differences, the

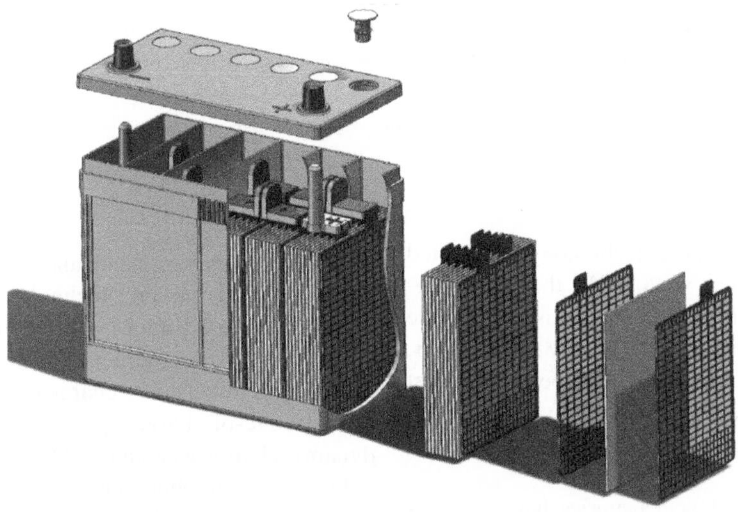

FIGURE 10.1 Exploded view of a 12-V VRLA AGM lead–acid battery. (Courtesy of C&D Technologies Inc.)

TABLE 10.1
Typical Lead–Acid Battery Performance

Battery/Description	Application	Power Density (W/kg)	Specific Energy (Wh/kg)	Energy Density (W/L)	Cycle Life 80% DoD	Cycle Life 50% DoD	Typical Calendar Life (25°C) (years)
12-V VRLA (AGM)	UPS/Telecomm	100–320	30–35	100–266	200–300	300–400	5–10
12-V VRLA (AGM)	Deep cycle/PV	100–216	25–35	65–70	500–700	1000–1800	5–10
12-V VRLA (AGM)	SLI	200–215	30–35	100–266	200–300	300–400	3–7
12-V VRLA (gel)	Telecomm/utility	30–45	33–36	60–70	200–400	200–800	5–7
2-V VRLA (AGM)	UPS/telecom	35–45	30–35	100–110	600–1200	800–1800	10–20
2-V VRLA (AGM)	PV/motive	15–20	30–35	40–50	1200–1800	2400–3000	7–20

Source: Various manufacturers' published data.

defining reactions of the cells' operation are still the previously illustrated equations.

In addition to the variants described previously, other emerging technologies such as bipolar batteries, carbon foam electrodes, asymmetrical supercapacitors, and integrated carbon electrodes have been added to the growing list of lead–acid technologies that are either fully commercialized or in the process of being brought to commercialization in the coming years. Unlike the aforementioned batteries, some of these emerging technologies are at least partially moving away from the standard lead–acid reactions and providing new ways of defining lead–acid technology. As previously stated, the majority of the recent advances have been highlighted and studied in VRLA batteries; however, almost all of them have also been applied to VLA batteries.

In advance of the introduction of the emerging technologies for lead–acid batteries, a summary of current battery capability is required. Table 10.1 summarizes the current state of the art for lead–acid batteries. All of the stated battery performances are based on existing commercially available products.

10.3 Pb CARBON BATTERIES

For the majority of lead–acid battery applications, from standby power systems to SLI (starting lighting and ignition), the primary failure mode is the degradation of the positive electrode, usually due to either positive grid corrosion or "softening" of positive active material (PAM), resulting in loss of the active material's cohesive integrity. The cause of these failure modes is a function of the charge required to bring the lead–acid cell back to full state of charge (SoC), for the positive grid corrosion, and the repeated morphology changes that occur during the charge/discharge reaction of the positive electrode for softening.

However, in many emerging applications for ESSs, the additional requirements of high cycle life, possibly at temperature and at PSoC, the primary failure mode is shifting to the negative electrode.

Current state-of-the-art lead–acid negative electrodes utilize a number of additives to improve the performance and longevity of the cell [3]. Lignosulfonates are added to maintain the high surface area of the negative active material (NAM) to improve utilization; barium sulfate is added to provide nucleation sites for the reaction product, $PbSO_4$, preventing large crystals from forming. Large crystals with a limited surface area are difficult to convert back to Pb on charge. Finally, carbon black is added to increase the conductivity of the plate to improve charge acceptance. While other additives can and are used within the industry, these three constitute the overwhelming majority of additives and are commonly referred to as the "expander" package.

In applications, as presented later in this chapter, typical electrode technology would provide very good initial performance but degrade very rapidly as the system continues to operate. The reasons and mechanisms for this are very well understood. In a PSoC operation, the negative electrode is held at various SoCs with a percentage of the active material converted to $PbSO_4$. This $PbSO_4$ can recrystallize over time and converts to what is commonly referred to as "hard sulfates" [3], which in reality are simply very large sulfate crystals with low surface area when compared with initial crystals formed during normal discharge. In turn, these crystals become sites for preferential crystal growth. The resulting sulfate crystals are difficult to convert back into Pb on recharge, and a steady decrease in available capacity results over time as more and more $PbSO_4$ is formed. In addition to suffering from sulfation in a PSoC operation, the negative electrode also limits the charge acceptance of the cell. In a pulse charge operation, the majority of the current is converted to hydrogen evolution. This leads to dry-out of the cell and also reduces the capacity over time.

The emerging solution to the above problem is the development and proliferation of the lead–carbon battery. Lead–carbon batteries and lead–carbon technology is a generic term for a number of variations on the use of carbon materials in combination with lead–acid technologies.

A number of variations on the concept of a lead–carbon battery are being utilized within the industry (Figure 10.2):

1. Integrated lead–carbon electrodes ranging from carbon enhanced active materials to mostly carbon formulations of the NAM

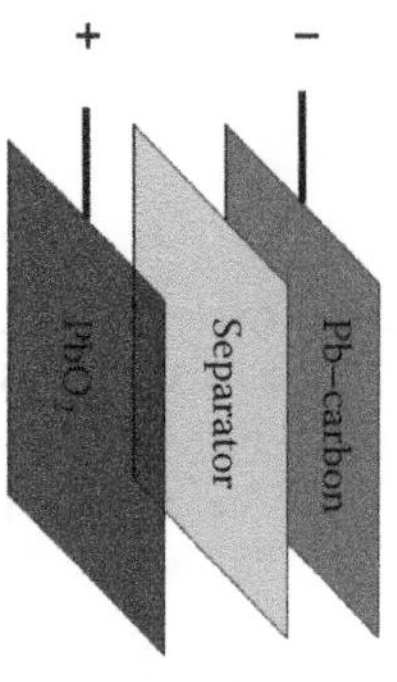

1. Pb–carbon negative electrode

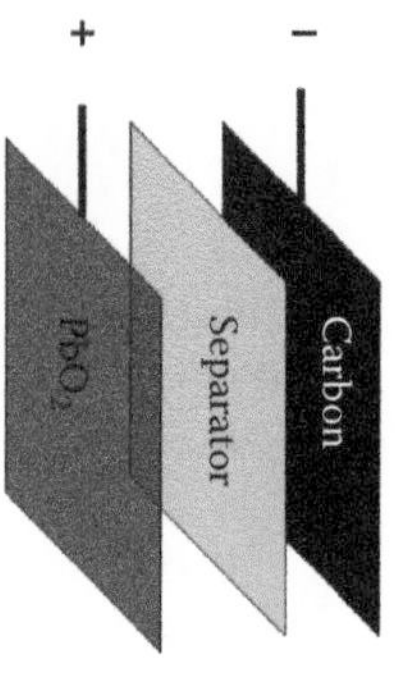

2. Carbon negative electrode

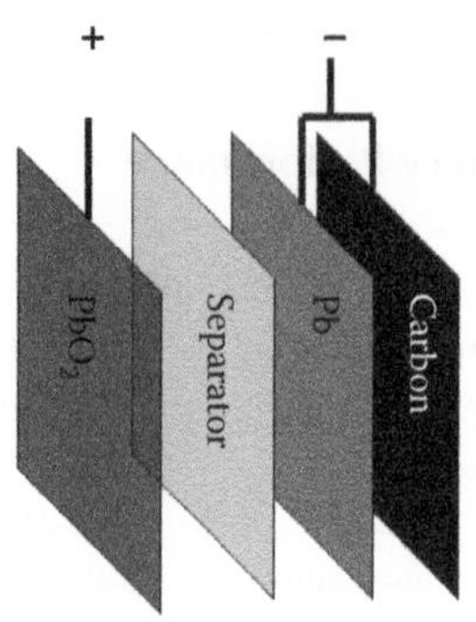

3. Pb and carbon negative electrodes in parallel

FIGURE 10.2 Graphical depiction of the various lead–carbon battery systems. From left to right: (1) integrated lead carbon, (2) asymmetrical supercapacitor, (3) dual negative electrode system (UltraBattery®). (From J. Anderson, Grid scale energy storage, in: *Proceedings of the 2011 Energy Storage—2nd Renewable Power Generation & Energy Storage System Industry Outlook*, September 9, Beijing, China, 2011.)

2. True asymmetrical supercapacitors utilizing full carbon electrodes (carbon on a conductive substrate) and not a Pb/C system for the NAM
3. Dual-electrode systems with Pb electrodes and carbon electrodes used in parallel with the PbO_2 electrode [4]
4. Lead-impregnated carbon foam electrodes and carbon current collectors instead of lead and lead alloys, sometimes in conjunction with carbon-enhanced active materials

10.3.1 Carbon as an Active Material and in the Active Material (Figure 10.2, Ref. [5])

As previously stated, carbon, in a number of forms but most commonly carbon black, has been used as an additive in the NAM of lead–acid batteries since nearly the introduction of the pasted negative plate, and the actual influence and function has probably been debated and investigated ever since that first addition [3]. The often-quoted benefits of carbon in the NAM range from improving formation efficiency and reducing end-of-charge voltage to providing a practical visual indicator during manufacturing that allows for clear differentiation from the positive electrodes before formation.

Among the most common forms of carbon investigated and used in lead–carbon batteries, are carbon blacks, graphite and activated carbons. Among these forms of carbon, almost every possible variation seems to have been trialed and reported on at some point in the history of lead–acid technology. Since many better detailed references exist for the types of carbons discussed in this section, only a brief overview of each carbon type will be presented here with a primary focus on the description and properties of importance in lead–acid battery technology.

Table 10.2 illustrates the range of properties among the commonly investigated carbon materials. While all of the materials have properties that fall within the range of the others, depending on the properties required by the application, it should be apparent that some materials would provide better and even different results than others.

10.3.1.1 Carbon Black

As stated prior, the earliest forms of carbon used in the NAM were carbon blacks, "lamp blacks" or "furnace blacks," and even today the majority of lead–acid batteries continue to use anywhere from 0.1% to 0.2%, by oxide weight, of carbon black in the NAM. The various grades used can range from fairly common industrial grades to highly specialized "battery

TABLE 10.2
Generalized Properties for Three Carbon Types: Graphite, Carbon Black, and Activated Carbon

Carbon	Surface Area (m^2/g)	Capacitance (F/g)	Conductivity (Percolation Dry $PbSO_4$) (wt.%)	Pore Volume (cm^3/g)
Graphite (G)	1–20	1–5	2–3	0–0.1
Carbon black (CB)	50–1700	5–100	2–4	0.1–0.3
Activated carbon (AC)	500–2000	50–200	5–7	0.5–1.3

Source: P.S. Walmet. *Evaluation of Lead/Carbon Devices for Utility Applications* (Sandia Report SAND2009-5537). Albuquerque, NM: Sandia National Laboratories, 2009.

grade" materials, with the primary difference between the two being the impurity levels of the materials.

Carbon black is virtually pure elemental carbon in the form of colloidal particles that are produced by incomplete combustion or thermal decomposition of gaseous or liquid hydrocarbons under controlled conditions. Its physical appearance is that of a black, finely divided pellet or powder. Current worldwide production is about 18 billion pounds per year (8.1 million metric tons). Approximately 90% of carbon black is used in rubber applications, 9% as a pigment, and the remaining 1% as an essential ingredient in hundreds of diverse applications, including lead–acid batteries [6].

Among the material properties of carbon blacks, the primary concern to the battery manufacturer is the impurity content of the material in terms of soluble organics, and metallic impurities such as iron and manganese. The impurity requirements indirectly dictate the raw materials and in some cases the processes used in the manufacture of the materials. In addition to impurities, the physical properties of surface area, usually expressed by Brunauer–Emmett–Teller (BET) or iodine absorption, and oil absorption number (OAN) are the other most commonly specified properties for carbon blacks used in advanced lead–acid batteries. More recently, the properties of percolation threshold with regard to $PbSO_4$, specific capacitance, and wettability or water absorption have been added to the list of specified properties of interest to the battery industry.

Figure 10.3 demonstrates the aggregate and particle nature of carbon blacks. Individual particles are on the order of nanometers and aggregates range from 4 to 6 μm. Carbon blacks are typically pelletized for shipment and handling (PSD d50 for the pellets range from 100 to 450 μm); the existing body of work indicates that the pellets are readily broken apart in the grinding action that occurs in mixing, as large particles are not evident after processing. Nonpelletized forms of carbon black, called "fluffy," have been trialed by the author, and the required water addition to maintain consistent processing was three to five times the amount required for pelletized versions of the same materials, limiting both the amount of addition and the processability of the final formulation. Therefore, all experimental references to carbon blacks throughout this work are in the pelletized versions, unless otherwise stated.

As Table 10.3 illustrates, the properties of surface area and structure are the defining properties for carbon blacks. It is these properties that have been investigated in the performance of the negative electrode in lead–acid batteries. For practical purposes (commercially available and economically viable), the range of surface area for carbon blacks used in batteries is from <100 to ~1500 m²/g. The majority of the carbons being used in common lead–acid batteries are typically at the low end of the range (~150 m²/g), while lead–carbon battery materials range from 250 to 1500 m²/g, depending on the intended purpose of the battery. As the cost of the material scales proportionally to the surface area, it is easy to understand why the differentiation has occurred.

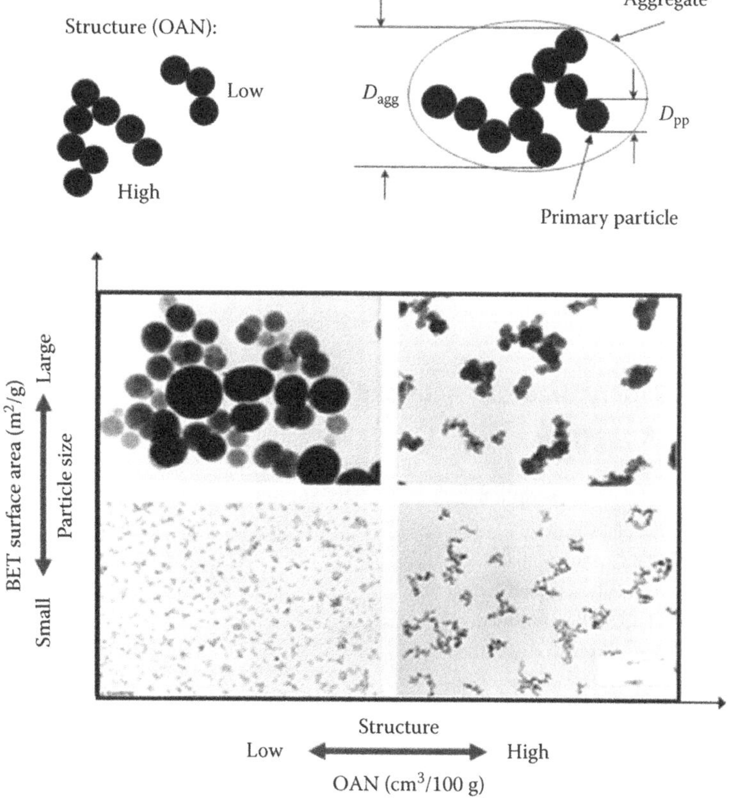

FIGURE 10.3 Structure and relative relationship between structure and surface area for carbon blacks. (Courtesy of Cabot Corporation.)

TABLE 10.3
Various Grades of Carbon Black: Properties and Characteristics

Material/Manufacturer/Brand	Description	BET Surface Area (ASTM D6556) (m^2/g)	OAN (ASTM D2414) (mL/100 g)	Impurities Fe (ppm)	Particle Size d50 (PSD) (μm)
Sid Richardson N220	Low S.A. low structure	100–500	100–150	<10	~4
Cabot PBX™09	Low S.A. low structure	200–300	100–150	<10	~4
Cabot PBX™51	High S.A. low structure	1200–1500	100–200	<10	~4
Experimental battery Grade CB	Med S.A. high structure	500–1000	250–300	<20	~4
Cabot black Pearls®2000	High S.A. high structure	>1500	200–350	<30	~4

Source: Materials manufacturers' specifications and analysis.

10.3.1.2 Activated Carbon

The second form of carbon heavily investigated and utilized in lead–carbon systems is activated carbons. Activated carbons are physically different from the previously discussed carbon blacks; they have a much larger particle size and an inherent pore structure, which is a function of the initial raw materials and processing conditions. Activated carbons are used for a number of types of applications: filtration, purification, and as catalyst beds. In addition, activated carbons are used in electrochemical double-layer capacitors (DLCs). It is this last application that provides most of the materials for lead–acid batteries. Among the primary physical properties of activated carbons of interest for battery applications are particle size; surface area; impurity content; pore size; the ratio between micro-, meso-, and macroporosity; and total pore volume.

The physical structure of activated carbons resembles a porous mass with IUPAC porosity definitions (Table 10.4):

- Macropores (>50 nm diameter)
- Mesopores (2–50 nm diameter)
- Micropores (<2 nm diameter)

Historically, activated carbons are not typically utilized in the NAM of lead–acid batteries.

10.3.1.3 Graphite

Graphite, in a number of forms, has also been investigated as a possible material for use in the NAM of lead–acid batteries. Graphite, as a material, has numerous forms and properties that vary between forms. Similar to carbon blacks and activated carbons, the typical properties of interest include the particle size, particle shape, and level of impurities. It can be physically similar to activated carbons; however, unlike activated carbons, it does not have an inherent pore structure and typically has a limited internal surface area. Therefore, particle size and shape plays a greater role in determining the surface area of the material. The layered structures of the graphene planes make the edges of the material more reactive than the plane surface [7]. Table 10.5 lists the properties of various graphite materials.

What should be apparent from the previous sections is both the diversity of properties within each type of carbon material and the very disparate properties between the different materials. These two facts, coupled with the further variables of processing conditions, negative material formulations, and electrode design, can partially explain both the number of proposed mechanisms for the operation of carbon within the negative plates of lead–acid batteries, as well as the lack of a unified demonstration of that proposed mechanism.

TABLE 10.4
Typical Properties of Common Battery Grades of Activated Carbons

Manufacturer/Material/Grade	Description	BET Surface Area (ASTM D6556) (m^2/g)	% Porosity (Micro, Meso, Macro)	Particle Size (μm)	Pore Volume (cm^3/g)
Engineered experimental (high surface area, microporous material)	Specialty activated carbon for DLCs/Pb–acid batteries	1721	90% Micro	6.5	0.76
Engineered experimental (high surface area, micro-/mesoporous material)	Specialty activated carbon for DLCs/Pb–acid batteries	1886	60/40% Micro/meso	4	1.48
Engineered experimental (medium surface area, mesoporous material)	Specialty activated carbon for DLCs/Pb–acid batteries	810	80% Meso	6.5	0.91
Norit® DLC Supra	Specialty steam activated carbon for DLCs	2090	60/40% Micro/meso	D_{50} = 5–10 D_{90} = 13–20	1.2
Norit® Azo AC	Steam activated carbon	700	–	75–150	–

Source: Manufacturers' specified properties and analysis.

TABLE 10.5
Typical Properties of Common Battery Grades of Graphite

Material/Grade	Description	BET Surface Area (ASTM D6556) (m^2/g)	Particle Size (μm)	Pore Volume (cm^3/g)	Impurities Fe (ppm)
Superior Graphite APH 2939	Flake graphite	8	11.4	–	43
Superior Graphite ABG 1010	Expanded graphite	22	–	0.12	15
TIMREX® MX15	Synthetic graphite	9.5	9	–	40

Source: P.S. Walmet. *Evaluation of Lead/Carbon Devices for Utility Applications* (Sandia Report SAND2009-5537). Albuquerque, NM: Sandia National Laboratories, 2009.

10.3.2 Integrated Carbon in the Negative Electrode

The predominant form of the lead–carbon battery remains the integrated lead–carbon design with negative electrodes ranging from slightly more than a standard lead–acid battery up to 5% by oxide weight carbon formulations, with 0.5% to 1.5% being the norm for commercially available products.

Using existing technology for the manufacturing of the negative electrode, the addition of moderate (0.5–5%) levels of carbon is possible, as has been demonstrated by Pavlov et al. [8]. At these levels, depending on the properties of the additive, the interspersed particles are not in sufficient quantities to affect the interconnected structure of the electrode. However, as the loading increases to higher levels of carbon (5–15%), the volumetric loading in the system reaches more than 35%. At this level, the quantity is sufficient to occupy enough of the system volume to prevent the interconnections of the skeletal structure of the NAM, resulting in insufficient cohesive strength of the active material and a condition known as shedding of the active material. As the name implies, shedding is simply the disintegration of the active material as it is utilized, due to the volumetric changes of the electrode, during the charge/discharge process. The volumetric change disrupts the cohesive bonds of the material, and it becomes separated from the current collector and is no longer available for the reaction.

Pavlov et al. [8] has demonstrated that the addition of electrochemically activated carbon (EAC) to the NAM can increase PSoC operation in high-rate applications by an order of magnitude, from 1000 microcycles to more than 10,000 cycles. Mosely et al. [9] summarizes some of the possible functions of the carbon additives in HRPSoC cycling:

1. Capacitive contribution—The added carbon acts as an asymmetric supercapacitor, storing charge at high rates in the electric double layer and spontaneously discharging, converting $PbSO_4$ into Pb.
2. Electronic conductivity—The carbon particles maintain the conductivity of the active material in the presence of increased amounts of $PbSO_4$, which are electrically insulating. These conductive pathways then facilitate charging despite the increased resistance of the plates.
3. Restriction of crystal growth—Carbon prevents the progressive growth of $PbSO_4$ crystals, maintaining surface area and improving charging characteristics [10].
4. Hydrogen overpotential and impurities—Certain forms of carbon contains elements that can suppress the evolution of hydrogen at the negative electrode, improving charge efficiency.

In addition to the four functions listed here, numerous works have investigated and proposed additional functions of carbon in the NAM, for example, the influence on the specific surface area of the NAM [11].

The specific capacitance of carbon materials has been proposed as a possible benefit to HRPSoC operation [9] and DCA. Capacitance is proportional to surface area in carbon blacks, similar to activated carbons, with measured specific capacitance values in F/g being nearly equivalent for materials with similar BET surface areas. Figure 10.4 shows the measured capacitance of various carbon blacks and activated carbons under equivalent conditions.

Assuming that the carbon additive in the electrode functions similarly to a nonfaradaic electrode of an asymmetric supercapacitor device, such as one based on the $PbO_2/PbSO_4$ system, then it is possible to determine the relative contribution of the capacitance of the carbon to the discharge or charge capacity.

Pell and Conway [12] calculated the effective equivalent weight for carbon in an asymmetric aqueous DLC. Using a 200 g/mol e^- equivalent for a 400 F/g activated carbon with 2000 m^2/g, the relative contribution to the discharge capacity of a typical HRPSoC cycle of ~5% depth of discharge (DoD) (76 s at 3 C) [13], would be a mere 0.02%, at a 1% loading by weight of the NAM. Furthermore, this number is a maximum as it assumes complete availability of the surface area and specific capacitance within the active material. For the materials described above, it would be even less of a contribution, as the material surface area, specific capacitance, and final available surface are significantly less than the example presented above.

The question then becomes, is a relatively minor contribution of the capacitance of the carbon (<0.02%) significant to the charge acceptance of the electrode and the overall performance in an HRPSoC application? Nikolov et al. [14]

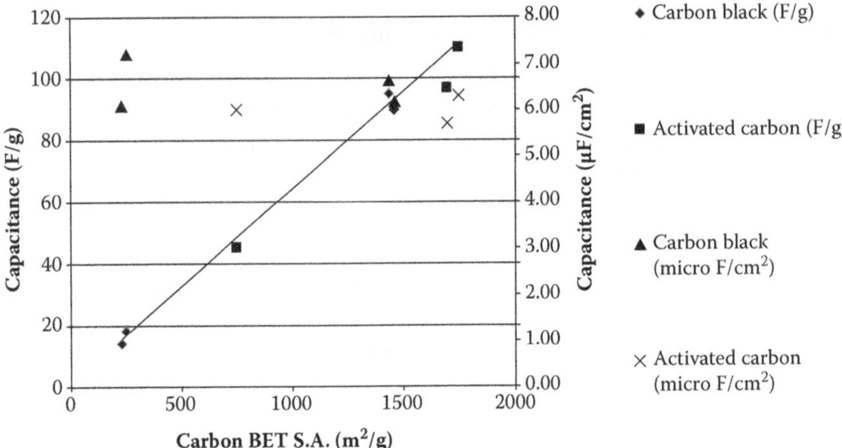

FIGURE 10.4 Capacitance versus surface area (S.A.) of various carbon blacks and activated carbons. (Based on P. Atanassova, N. Hardman, A.D. Pasquier, and M. Oljaca, Role of carbon blacks for improved cycleability and charge acceptance in advanced lead acid batteries, In: *13th ELBC Conference*, September 28, Paris, France, 2012. By permission of Cabot Corporation.)

demonstrate that the charge acceptance improvement with carbon additives only occurs over the first 10 s of charge when the battery is at 90% SoC. In addition, Pavlov et al. found that the HRPSoC cycle performance for both activated carbon and carbon black was highly dependent on the DoD. The cycle life at 0.5% DoD is an order of magnitude higher than at 3.0% DoD, which is interesting as the contribution of the capacitance of the carbon additive would only account for ~50% of the charge or discharge capacity at 0.5% DoD at the highest carbon loading tested [15].

While the capacitance increase of the NAM, due to addition of carbon, can provide some improvement of the initial charge acceptance and HRPSoC performance of the negative electrode, capacitance cannot be the only function of carbon in the NAM. The performance improvement is not directly proportional to the added capacitance that the carbon provides. Furthermore, even lower loadings of carbons, with surface areas and specific capacitance values at the low end of the spectrum of properties, still provide clear improvement of DCA and HRPSoC operation [8,12,13,15], further disputing the role of capacitance on the increased performance.

Similar to capacitance, the impact of the electrical conductivity of the carbon on the NAM is also of interest, specifically which materials and material properties, in conjunction with the conductivity, affect the overall conductivity of the NAM. Of further interest is, how those properties influence the conductivity and charge behavior of the NAM, as a function of the SoC, and how they influence the PSoC behavior and high-rate charge acceptance.

As stated previously, imparting increased conductivity to the NAM is an attributed benefit of carbon blacks used in the expander package of lead–acid cells. Conductive carbons are commonly used as electrical conductivity–enhancing additives in a number of applications; in mixtures of conducting and insulating materials, the conductivity of the mixture is well explained by the percolation theory. Furthermore, the electrical properties mainly depend on the dispersion, particle size, and aggregate structure [16].

Applying the above principles, dry powder studies with $PbSO_4$ show the expected correlation between the particle size, surface area, and the weight percent required to achieve the percolation threshold for carbon blacks. Comparing graphite and activated carbons to carbon blacks, we see a similar influence of some forms of graphite in the NAM [13], while activated carbons require much higher loadings to achieve the same results under similar test conditions.

Figure 10.5 shows the dry powder resistivity of the various forms of carbon listed in Table 10.6 by weight percent in $PbSO_4$. The figure further illustrates both the impact of

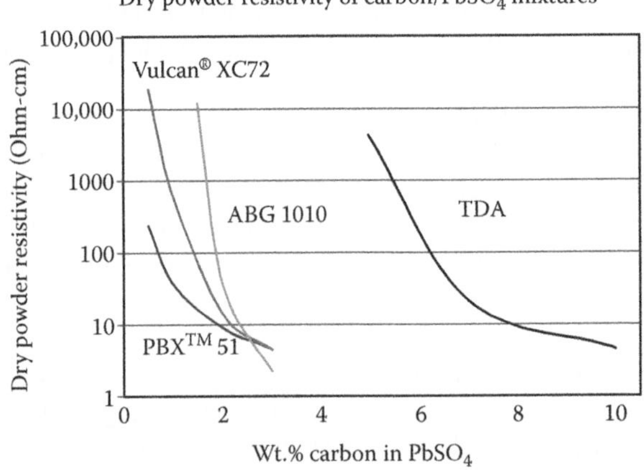

FIGURE 10.5 Dry powder resistivity versus weight percent loading for carbon blacks, expanded graphite, and activated carbon. (Based on P. Atanassova, N. Hardman, A.D. Pasquier, and M. Oljaca, Role of carbon blacks for improved cycleability and charge acceptance in advanced lead acid batteries, In: *13th ELBC Conference*, September 28, Paris, France, 2012. Courtesy of Cabot Corporation.)

TABLE 10.6
Carbon Properties for Conductivity

Material	Carbon Type	BET S.A. (m^2/g)	Particle Size (μm)[a]	Dry Powder Resistivity at 1 wt.% in $PbSO_4$ (Ohm-cm)	Percolation Threshold (wt.%)
Cabot PBX™ 51	High-surface-area CB	1435	3.8/162	20	2%
Cabot Vulcan® XC72	Low-surface-area CB	240	6.1/430	200	3%
ABG1010	Flake graphite	22	Oct-50	>100,000	3%
TDA	Activated carbon	~1800	–	>100,000	8%

Source: P. Atanassova, N. Hardman, A.D. Pasquier, M. Oljaca. Role of carbon blacks for improved cycleability and charge acceptance in advanced lead acid batteries. In: *13th ELBC Conference*, September 28, Paris, France, 2012. By permission of Cabot Corporation.

[a] d50 (PSD) for carbon blacks (CB) with sonication and without sonication.

material properties within a group of carbons, as well as the difference between the various forms of carbon on the targeted properties in the active material, at levels below the percolation threshold. The carbon blacks show an increase in conductivity influence with increasing surface area and decreasing particle size. Referring back to the aggregate nature of carbon blacks and the relationship between the structure, surface area, and particle size, the higher-surface-area materials should be able to form continuous networks at lower levels than those materials with smaller surface areas.

Graphite shows a similar percolation threshold to carbon blacks; however, the influence on conductivity shows a much tighter range required to influence the conductivity at levels below the percolation threshold. One explanation of this is both the larger particle size of the material and the discrete particulate nature of the material, which would be less able to fill all interstitial sites, limiting the ability of generating the required continuous network to the same extent as the agglomerated carbon black materials. The activated carbon supports these conclusions in that it is both a discrete particulate material and has the largest particle size resulting in the lowest influence on the conductivity of the powder mixture at equal loadings and requiring nearly three to four times the amount by weight to achieve the same conductivity.

Figure 10.6 demonstrates the influence of the aforementioned properties of carbon blacks in the NAM of a 12-V VRLA battery. The carbon materials and loadings from Table 10.7 were added to the NAM, and 12-V VRLA deep

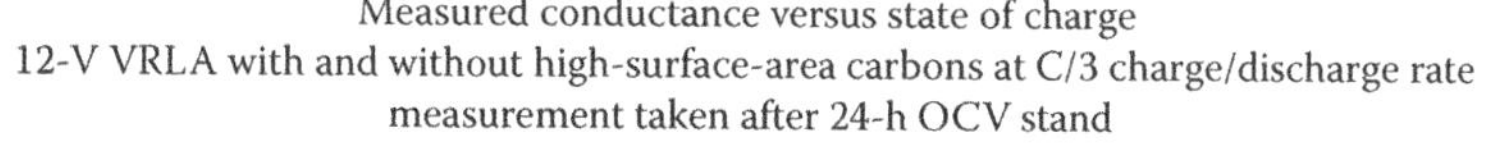

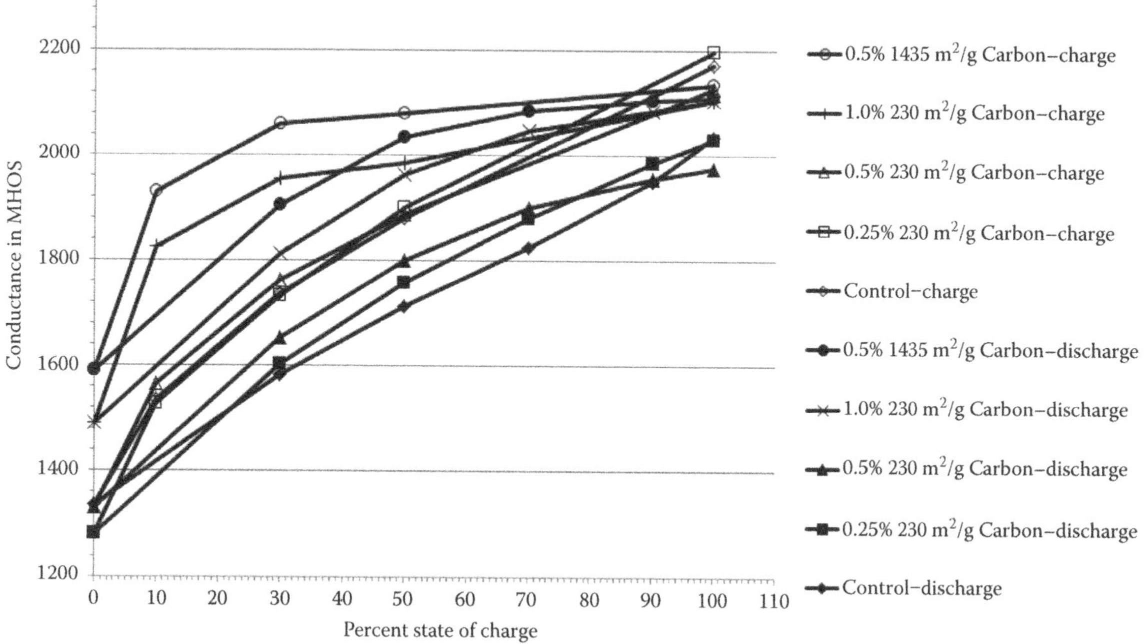

FIGURE 10.6 Measured conductivity versus state of charge in VRLA batteries with various loadings of carbon black. (Courtesy of C&D Technologies Inc.)

TABLE 10.7
Carbon Properties for Conductivity in 12 VRLA Batteries

Material	Carbon Type	Wt.% Loading in the NAM (by Oxide wt.)	BET S.A. (m^2/g)	Particle Size (μm)[a]	Estimated Relative Conductivity
PBX™ 51	High-surface-area CB	0.5	1435	3.8/162	10
PBX™ 09	Low-surface-area CB	1	240	4.2/273	7.5
PBX™ 09	Low-surface-area CB	0.5	240	4.2/273	1
PBX™ 09	Low-surface-area CB	0.25	240	4.2/273	0.1
Control	Low-surface-area CB	0.16	140	–	<0.1

Source: C&D Technologies Inc.
[a] d50 (PSD) for carbon blacks (CB) with sonication and without sonication.

cycle batteries were assembled; other than the carbon additives, all other parameters of the batteries were maintained. After initial conditioning, the units were discharged at a progressively increasing DoD from 100% SoC to 0% SoC at the C/3 rate, and the conductance of the batteries and open circuit voltage (OCV) were measured at specific SoCs (100%, 90%, 70%, 50%, 30%, and 0%); measurements were taken following 24-h OCV rest after each step. Following the final measurement at 0% SoC, the batteries were charged at the same C/3 rate (with a 2.4 VPC limit) and the same procedures were followed to take the same measurements at 10%, 30%, 50%, 70%, 90%, and 100% SoC.

The results for the measured conductivity values follow the trends as predicted by the dry powder measurements, mainly 0.5% 1435 m^2/g > 1.0% 240 m^2/g > 0.5% 240 m^2/g > 0.25% 240 m^2/g ≈ control, with the highest-surface-area material providing the greatest increase in conductance even at 50% lower loading than the lower-surface-area material. Furthermore, the lower-surface-area material provided decreasing improvement with decreasing loading in the NAM, such that at the lowest loading the result are similar to the control. On the basis of the results and correlation to the dry powder study, it seems unlikely that similar results could be achieved with either graphite materials or activated carbons, or at least for the materials tested, as the loading requirements to achieve the same levels of conductivity would most likely affect the cells in other ways, such as active material cohesion [8], which would confound the results.

Reviewing the impact that the carbon addition in the NAM has on the measured conductance of the batteries, a 10–25% increase, it appears technically feasible that the increased conductivity of the NAM with higher loadings of carbons could contribute to both the PSoC and HRPSoC operation of the battery over the long term, particularly in light of the fact that the greatest measured impact occurs over the SoC range that a battery would normally be required to operate in during a PSoC application.

However, the results for activated carbon and graphite would suggest that the influence on conductivity alone is not the overriding benefit of carbon in the NAM, although studies on expanded graphite have shown the highest HRPSoC cycling performance at loadings as high as 2% [17], which supports the previous conclusions. Additional studies have shown the same HRPSoC cycle life with equal weight percent loadings of activated carbons, which would be significantly less than the amount required to achieve similar changes in conductance behavior [17].

It is also worth noting that the lowest loadings of carbon black did not have any appreciable influence on the measured conductance, at any SoC. The fact that this is the typical loading used in most commercially manufactured batteries on the market today would suggest that the benefits beyond a visual indicator, during the early stages of formation of the electrodes, or for deep discharge recovery, are most likely minimal.

Probably the greatest measurable impact that carbon additives can have on the NAM is the increase in the surface area of the material. As Table 10.2 indicates, the surface area of the carbons used in lead–carbon batteries range from 1 to 2000 m^2/g, with the majority of the materials being from 200 to 1500 m^2/g. By comparison, the average BET surface for the NAM is ~0.5 m^2/g. Figure 10.7 illustrates the impact to surface area that carbon additions can have on the dry unformed (DUF) and formed NAM. With only 1 wt.% addition to the NAM, the final electrode surface area is increased by up to an order of magnitude [18].

Furthermore, the increased surface area can translate directly to the performance of the negative electrode, for better or worse, depending on the carbons selected. In the lead–acid cell, the overvoltage is a function of the current density, i, which is current over the electrode surface area and expressed as A/m^2. The surface area can be either the geometric area or the internal surface of the porous electrode [3]. At small overvoltages, the expression for current density is explained by Equation 10.7. At large overvoltages, the relationship between the polarization of the electrode and the surface area can be simplified to Equation 10.8, which upon taking the logarithm provides the Tafel equation (Equation 10.9) [3]. Therefore, the current increases exponentially with potential, or conversely reducing current density by increasing the surface area reduces the overvoltage of the electrode for the same current.

$$i = i_o \frac{F}{RT}(E - E_o) \tag{10.7}$$

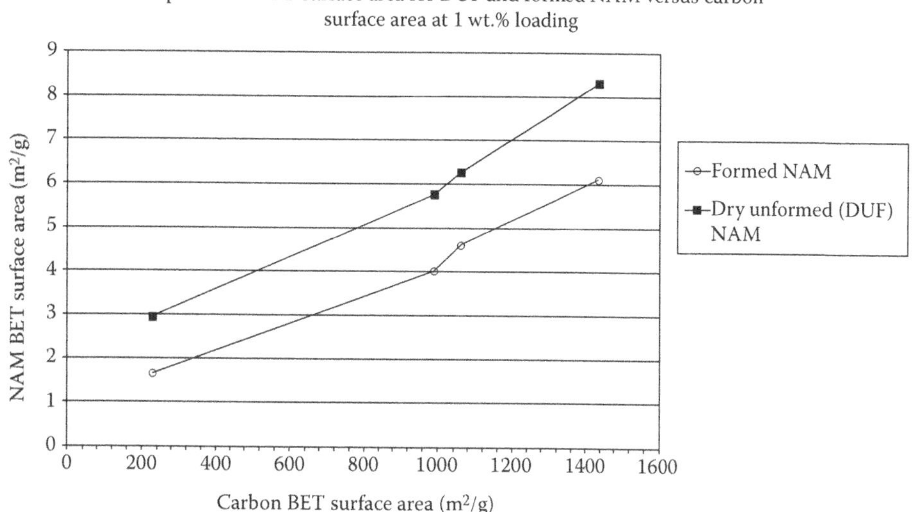

FIGURE 10.7 Measured NAM BET surface area versus carbon surface area for dry unformed and formed negative electrodes. (Based on J. Anderson, S. Mraz, and D. Boyer. Enabling renewable energy transmission—Advanced lead carbon energy storage system for transmission utilization improvement. In: *Proceedings of EESAT11*, October 25, San Diego, CA, 2011.)

$$i = i_o \exp\left[\frac{\alpha F}{RT}(E - E_o)\right] \tag{10.8}$$

$$E - E_o = a - b \log i \tag{10.9}$$

Tafel experiments on lead–carbon cells with 1 wt.% loadings of carbon blacks are listed in Table 10.3. Figure 10.8 illustrates the reduction in negative polarization as a function of NAM surface area, which can be attributed to the carbon additions. For equal weight percent loadings, the greatest reduction in the negative polarization occurs with the highest-surface-area carbon (NAM BET S.A. 8 m²/g), while the lowest surface area carbon shows least impact to polarization, although it still shows some influence compared with the control. Consequently, the positive polarization is correspondingly increased in the cells with carbon negative electrodes.

To further demonstrate the influence of the carbon surface area and consequently the NAM surface on the Tafel behavior of the negative electrode, cells were assembled with negative electrodes with equivalent surface area loadings of the carbon blacks used in the previous experiment. The NAM formulations included 1% 230 m²/g, 0.25% 1060 m²/g, 0.17% 1435 m²/g of carbon black, respectively, with a targeted NAM BET surface area of 3.25–3.5 m²/g.

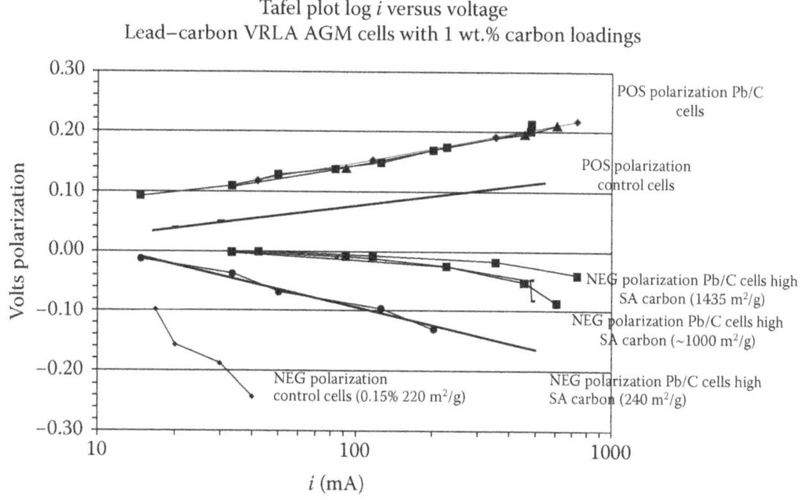

FIGURE 10.8 Tafel plot log *i* versus polarization voltage for 1 wt.% carbon additives. (From J. Anderson, S. Mraz, and D. Boyer. Enabling renewable energy transmission—Advanced lead carbon energy storage system for transmission utilization improvement. In: *Proceedings of EESAT11*, October 25, San Diego, CA, 2011.)

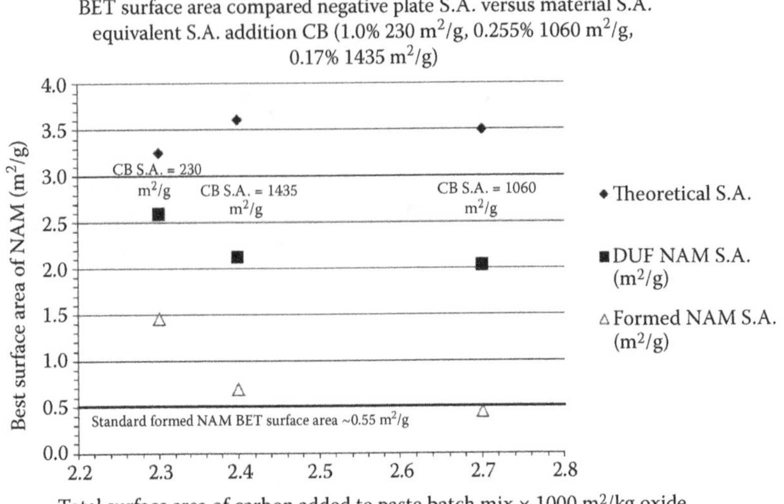

FIGURE 10.9 BET surface area study of negative plates with equivalent surface area loadings of carbon blacks. Theoretical surface area, DUF, and formed NAM values of surface area are shown. (Courtesy of C&D Technologies Inc.)

The expectation was the Tafel behavior of the cells would be similar. The results were very different than expected for the simple fact that the lower weight percent loadings resulted in almost no appreciable increase in the surface area of the formed NAM material. The surface area target and results are depicted in Figure 10.9. As the figure clearly shows, the final measured BET surface area, for the high surface area carbon, at the lowered loadings in the NAM, resulted in surface areas resembling the NAM without any carbon additions. Nevertheless, the results inadvertently support the conclusion that without a measurable change in surface area of the electrode, the Tafel behavior is unchanged by the addition of carbon. In addition, the result would suggest that there is a lower limit for the addition of carbon to the NAM, or at least a lower limit for carbon blacks.

The fact that there is no discernible impact on the surface area of the formed NAM at 0.25% and 0.17% loadings and the DUF surface area is lower than the theoretical values for all of the samples suggests that a certain portion of any carbon addition is lost both during the paste mixing and plate processing, and during the electrolyte filling and electrode formation process. It is surmised that the loss during paste processing is due to encapsulation of the carbon into the crystal structure of DUF NAM. The loss during formation could have a number of causes, including encapsulation into the structure of the NAM, which is proposed by Pavlov [17], or migration into the electrolyte during the filling and soaking process, which would lead to possible oxidation at the positive electrode during the formation.

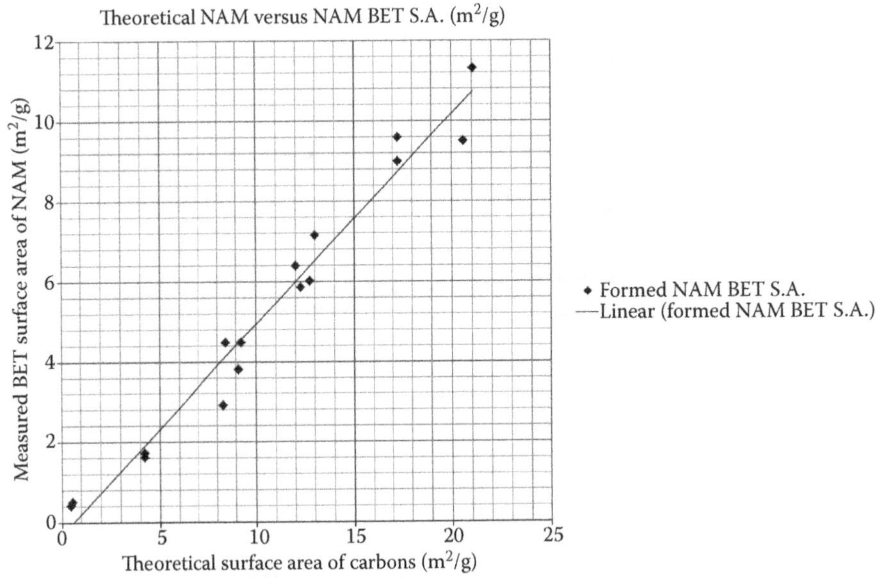

FIGURE 10.10 Influence of the surface area of activated carbons on the BET surface of the NAM. (Courtesy of C&D Technologies.)

Similar to carbon blacks, activated carbons added to the NAM provide the same increase to the measured BET surface area of the electrode (Figure 10.10). However, depending on the material, the relative impact to the polarization of the electrode is significantly less than with carbon blacks. This would suggest that the difference in the method of integration, of the different forms of carbon, into the structure of the active material [19] influences the relative impact the material has on the performance of the NAM. To further understand the influence of the BET surface area on the current acceptance, cells were assembled with 0% to 0.5% of the first three activated carbons listed in Table 10.4, according to a 16-factor fractional factorial Plackett–Burman design of experiments. The overall influence to the surface area of the NAM was similar to the results with carbon blacks as shown in Figure 10.8; however, the charge acceptance of the cells showed an inverse relationship to surface area, which is the exact opposite effect as with carbon blacks.

At this point, the specific properties of activated carbons that influence the effectiveness of the materials integration into the active material are not completely understood. However, the following section provides some insight into the possible influence with the other components of the expander package.

10.3.2.1 Processing and the Interrelationships between Carbon and Other Expander Components

Translating the above properties of carbons into the active material and determining their specific influences on the performance of the electrode first requires integration of the material into the active mass. The phrase "typical active material formulation" is almost an oxymoron within the battery industry, however; all mix formulas include PbO, H_2SO_4, and H_2O, plus any additional additives including the previously mentioned expander package and fibers.

Without getting into the nuances and chemistry of the paste mixing and paste formulations, which can be as varied as the battery designs themselves, the primary issue that needs to be overcome, with the use of higher loadings of carbon additives, is the hydrophilic nature of the materials. As an example, the typical weight percent of H_2O in the paste is 14–18%, depending on the process and process conditions. As Figure 10.11 demonstrates, the amount of H_2O by weight, to achieve the rather qualitative property of "wetted," takes two to four times the weight of the carbon addition. This translates to 5% to 25% increase in required water addition to the paste to achieve the same available moisture for the NAM mix.

One point of interest is that the wettability of the carbon blacks is a function of the structure as measured by the OAN, and is not dependent on the surface area of the materials. However, activated carbons show a greater correlation between the surface area of the material and the water absorption, presumably because for activated carbons, surface area is a function of the pore structure and volume. The one known caveat to the relationship between OAN and the wettability is related to the form of the carbon blacks; as previously stated, the unpelletized material has a significantly greater wettability ratio, about 14:1, which is four times the ratio for the equivalent pelletized material.

In addition to the changes required, in the formulation of the NAM, to incorporate the carbons into the paste mix, the interactions between the other components within the NAM need to be considered. For example, the addition of the lignosulfonate into the carbon before mixing with water can greatly reduce the water required to wet the materials due to the absorption on the surface of the carbon. It has been proposed that the absorption of the lignosulfonate on the carbon surface impedes its function in the electrode proportionally to the surface area of the carbon. Furthermore, the absorption of the lignin is partly responsible for the improved performance in HRPSoC [19]. What has been clearly demonstrated

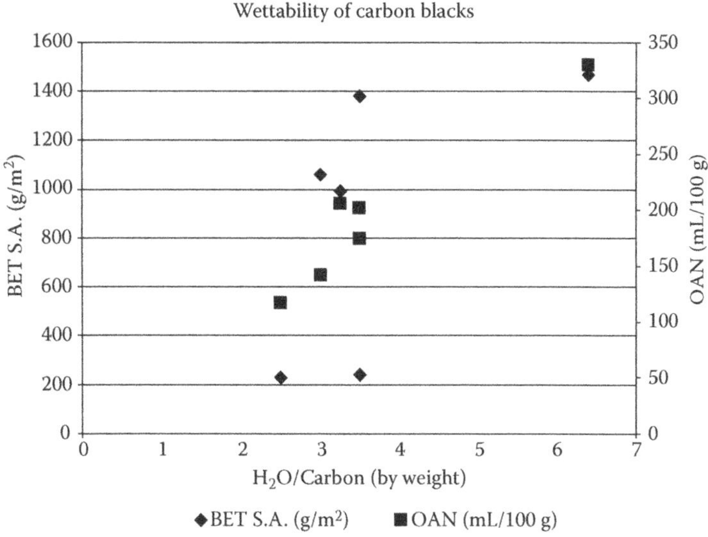

FIGURE 10.11 Wettability: ratio of water to carbon by weight required to prewet material for addition to the negative paste mix. (Courtesy of C&D Technologies Inc.)

is that the interaction between the expander components and the carbon can have significant consequences to the high-rate discharge and charge performance of the electrode, and can influence the final pore structure of the NAM [11]. It may also explain the almost contradictory results reported on the performance of the same materials in different NAM formulations.

10.3.2.2 Performance Influence of Lead–Carbon Technology

From a performance standpoint, the influence of carbon on the NAM can be broken down into the following areas of impact: specific capacity, PSoC, HRPSoC, DCA, and charge behavior.

In terms of specific capacity, the effect of carbon additions to NAM materials appears to be minimal, with possibly a slight decrease in the active material utilization at lower discharge rates and with loadings up to 1%, in either experimental cells or full-scale battery tests. Table 10.8 outlines the properties of the carbons and the electrodes tested along with the relative performance of the NAM in Ah/kg.

Analysis of the data does not show any significant correlation between the carbon properties and the utilization of the active material at the tested rate of 0.65 mA/cm^2. However, Atanassova et al. found that the decrease in the average pore size of the NAM, with the addition of higher-surface-area carbon blacks, led to lower cold-temperature high-rate performance [11]. This could be a function of the decrease of the "penetration depth," as defined by Bode, as the decrease in pore diameter increases the effective resistance and current density [3]. The amount of carbon addition does impact the relative utilization from the first to second discharge, which may be an indication that the influence on the conductivity of the active mass improves the formation efficiency of the electrode.

The performance of the lead–acid battery for PSoC, HRPSoC, and DCA is significant when compared with traditional batteries. As previously stated, a demonstrated order of magnitude increase in the HRPSoC, with the addition of carbon into the active mass, is achievable [15] and a two to three times increase in DCA is possible with proper selection, loading, and integration of carbon in the NAM [11]. Furthermore, the suppression of sulfation in PSoC operation can result in a significant increase in the cycle performance of lead–acid batteries at moderate to low discharge rates as demonstrated in Figure 10.12.

The function of the carbon in the NAM can also positively influence the performance of multicell high-voltage strings and large stationary battery installations with multiple strings in parallel, similar to those discussed later in this chapter. In higher-voltage systems, the tendency for the cells to diverge in voltage and SoC as they age is a well-understood phenomena [2]. It is usually caused by small initial variations in cell behavior, primarily during charging, that leads to pronounced performance degradation. The mechanism is related to sulfation of the negative electrode, the increased cell resistance due to internal positive grid corrosion, and loss of water from the electrolyte. This phenomenon is so common that most battery manufacturers provide specific instructions on how and when to equalize the cells, in an effort to mitigate the influence of these effects. Furthermore, the impact is greatest in cycle applications that tend to result in chronic undercharging and continuous PSoC, such as renewable ESSs. The reduction of negative polarization by the addition of carbon in the NAM has the effect of providing increased POS polarization, under typical constant voltage charge scenarios, and increased charge currents resulting in additional equalization of the cells during operation.

10.3.3 Asymmetrical Supercapacitors

Supercapacitors or electrochemical capacitors (ECs) are a promising form of energy storage, when compared with the conventional capacitor; however, at 5–10 Wh/kg, they are still significantly lower than traditional electrochemical storage devices. Electrochemical DLCs (ECDLCs) utilizing carbon as the active material are typically symmetric capacitors in that they use the same materials in the positive and negative electrodes of the energy storage device [12]. These devices

TABLE 10.8
Influence of Carbon Properties on the NAM

Wt.% Carbon Black	OAN (mL/100 g)	BET S.A. Carbon (m^2/g)	BET S.A. DUF NAM (m^2/g)	Pore Volume (cm^3/g)	Average Pore Size (nm)	1st Discharge (Ah/kg)	2nd Discharge (Ah/kg)
0.15	Control	220	1.8	0.008	18.0	126.6	144.3
0.25	192	240	1.60	0.007	17.0	135.5	148.4
0.50	330	1470	3.60	0.011	13.0	119.3	129.5
1.00	165	1435	7.5	0.016	9.0	125.4	123.9
1.00	174	240	2.50	0.009	15.0	124.1	124.5
1.00	117	230	2.60	0.013	19.0	120.6	132.9
1.00	202	1380	7.90	0.020	10.0	122.6	128.9
1.00	206	990	4.70	0.015	13.0	146.6	143.4
1.00	117	230	2.5	0.011	18	139.5	142.2

Source: C&D Technologies Inc.

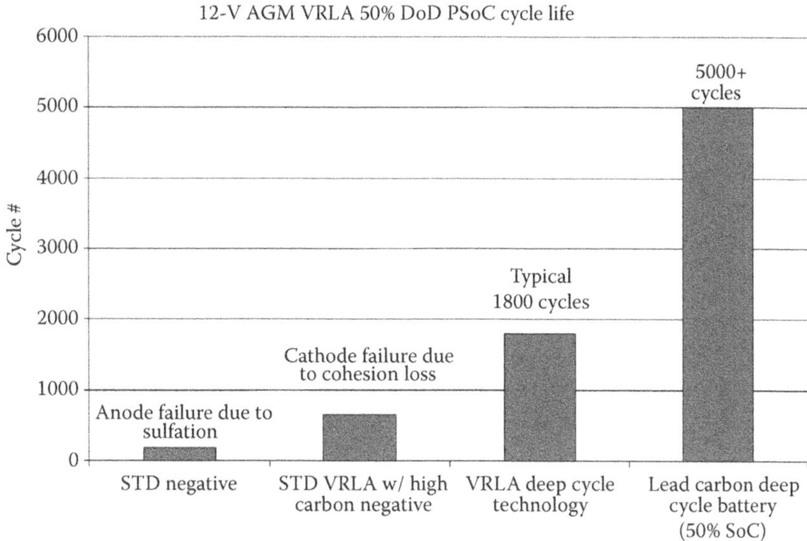

FIGURE 10.12 Relative PSoC cycle performance of various battery technologies and the typical failure modes. (From J. Anderson, S. Mraz, and D. Boyer. Enabling renewable energy transmission—Advanced lead carbon energy storage system for transmission utilization improvement. In: *Proceedings of EESAT11*, October 25, San Diego, CA, 2011.)

store energy at the electrolyte/carbon interface through the reversible ion absorption onto the carbon surface, charging the double-layer capacitance [20,21]. The capacitance of the device can be described according to Equation 10.10, the energy density according to Equation 10.10, and the power according to Equation 10.12 [20]:

$$C = \varepsilon A/d(l) \tag{10.10}$$

$$ED = \frac{1}{2}CV^2 \tag{10.11}$$

$$P = V^2/4R \tag{10.12}$$

where ε is the electrolyte dielectric constant, *A* is the surface area of the carbon, and *d* is the distance between the center of the ion and the carbon surface in Equation 10.10; *V* is the potential difference across the device for Equation 10.11; and *R* is the resistance of the device. Combining the equations, we see that the energy density and power of the device is function of the electrolyte, surface area of the carbon, and operating voltage.

In an aqueous electrolyte, the typical double-layer capacitance of carbons is in the range of 5–20 μF/cm², with the carbons previously reviewed in conjunction with Pb–carbon batteries (Figure 10.4) having capacitance values at the low end of the range (Table 10.9).

The concept of the hybrid–asymmetrical supercapacitor attempts to take advantage of the power performance benefits of a carbon ECDLC electrode with the storage capacity of the faradaic battery electrode. In the hybrid–asymmetrical supercapacitor, as it relates to advanced lead–acid batteries, the reaction at the positive electrode is described by the equation for the lead dioxide electrode (Equation 10.13) in the lead–acid battery.

$$PbSO_4 + 2H_2O \leftrightarrows PbO_2 + SO_4^{2-} + 4H^+ + 2e^- \tag{10.13}$$

TABLE 10.9
Capacitance of Carbons Utilized in Lead–Carbon Batteries

Material	Material Type	Capacitance (F/g)	Surface Area (m²/g)	Capacitance (μF/cm²)
PBX™ 51	Carbon black	95	1450	6.6
PBX™ 09	Carbon black	18	250	7.2
Vulcan® XC72	Carbon black	15	220	6.8
TDA 1	Activated carbon	45	750	6.0
TDA 2	Activated carbon	110	1800	6.1

Source: P. Atanassova, N. Hardman, A.D. Pasquier, and M. Oljaca. Role of carbon blacks for improved cycleability and charge acceptance in advanced lead acid batteries. In: *13th ELBC Conference*, September 28, Paris, France, 2012. With permission of Cabot Corporation.

TABLE 10.10
Comparison of Estimated Asymmetric Supercapacitor Performance to STD VRLA

	Deep Cycle VRLA [22]	High-Power VRLA [22]	Asymmetric Supercapacitor[a] [23]
Weight kg (BCI group 24)	25	26.5	24[a]
Ah capacity	75 at C/20 rate	80 at C/20 rate	50
Wh/kg	35.9 at C/20 rate	36.2 at C/20 rate	25
Wh/L[b]	83.1	88.9	55.5
Wh (C/20 rate)	898	960	600[a]

Source: C&D Technologies Inc. and Axion Power.
[a] Calculated values from available data (25 Wh/kg and 50 Ah capacity—assumes avg. 2.0 VPC during discharge).
[b] Volume of BCI group 24, ~10.8 L.

The negative electrode utilizes a high-surface-area carbon DLC electrode. One such system based on the $PbO_2/PbSO_4$ positive electrode and a carbon DLC for the negative electrode is the Axion Power *PbC*® battery. The proposed equation for the negative electrode reaction is as follows [22]:

$$nC_{6^{x-}}(H^+)_x \leftrightarrows nC_{6^{(x-2)-}}(H^+)_{x-2} + 2H^+ + 2e^- \text{ (discharged)} \quad (10.14)$$

The *PbC* battery has a claimed energy density of 25 Wh/kg and 50 Ah capacity in a BCI group 24 battery size [24]. Table 10.10 compares the calculated capacity of an asymmetric supercapacitor with similar capacity and comparison to both a 12-V AGM VRLA battery for deep cycle applications and a high-power battery, both BCI group 24 size.

While the results show the expected lower specific energy of the asymmetric device, the expectation is that the power capability, both discharge capability and DCA, should be improved over a standard VRLA battery. Because of the lack of a Pb negative electrode, and more specifically the sulfation reaction at the negative electrode, PSoC cycling should not be an issue. As the positive half-cell still relies on the $PbO_2/PbSO_4$ reaction, the overall cycle life will continue to be a function of the design and formulation of the positive electrode.

10.3.4 UltraBattery®

The next variation in the lead–carbon technology is the UltraBattery, which is similar to the asymmetric supercapacitor in that it relies on a carbon electrode. It is dissimilar in that it also relies on a typical $Pb/PbSO_4$ negative electrode as well. In this way, it might be more similar and better explained by the equations describing the integrated lead–carbon batteries, albeit to a greater extent due to the greater capacitance that is possible with addition of a carbon electrode in the system.

The UltraBattery was developed by Australia's Commonwealth Scientific and Industrial Research Organization (CSIRO). Commercialized in 2005, the UltraBattery technology has been licensed to the Furukawa Battery Company Ltd (Japan) and East Penn Manufacturing (United States) for various applications throughout the United States, Japan, Thailand, Mexico, Canada, and, most recently, China [25].

Similar to the genesis of other variations of the lead–carbon battery, the intended target applications were ones where standard lead–acid batteries performed very poorly, primarily hybrid electric vehicle (HEV) applications and renewable energy storage [25]. Similar to the previously discussed benefits of carbon addition to the NAM, the UltraBattery technology shows increased DCA, and very good PSoC and HRPSoC cycle life [26,27]. The carbon electrode is designed to minimize hydrogen evolution, through additives, and the demonstrated gas evolution is comparable to a Pb negative electrode under the same conditions [28].

In reference to the influence of the capacitance of parallel carbon electrode, the specific capacity of the carbon electrode is presented as ~10× that of the lead–acid electrodes or ~0.12 Ah/g versus 0.015 Ah/g for the battery electrode (both measured at 1 A/g current density) [28]. As with the integrated lead–carbon systems, the relative benefit of the carbon electrode should be a function of the specific capacity contribution to the charge and discharge reaction. For the UltraBattery design, this will be dependent on the physical design of the electrode versus the amount of the addition as in an integrated carbon electrode.

It should be noted that Lam and Louey [28] do demonstrate an almost four times improvement over an integrated carbon electrode design, referenced as containing carbon black and graphite, on a simple cycle test at 40°C. The results are similar to the differentiation in HRPSoC cycle that Pavlov et al. was able to obtain in similar experiments with various carbon loadings [15].

Among the variants on lead–carbon batteries, the UltraBattery probably has the most extensive and widely published third-party-test data set (Table 10.11). Original tests done at Sandia National Labs demonstrated the benefit of the technology in PSoC cycles, which showed results comparable to $LiFePO_4$ cells, while the advanced VRLA batteries

TABLE 10.11
Comparison of Performance of the Stationary Furukawa UltraBattery versus the Performance of a Deep Cycle VRLA Cell

	Deep Cycle 2-V VRLA [22]	UltraBattery (UB1000) [29]
Cell weight (kg)	71.2	75
Ah capacity (C/10 rate)	1020	1000
Cell volume (L)	20.1	26.4
Wh/kg (C/10 rate)	28.7	26.7
Wh/L (C/10 rate)	101	75.8
Wh (C/20 rate)	2004	2000

Source: C&D Technologies and J. Furukawa, M. Miura, H. Yoshida, W. Tezuka, L.T. Lam. Performance of UltraBatteries produced from different grid thicknesses. *Proceedings from the 12th ELBC Conference*, September 21–24, Istanbul, Turkey, 2010.

Note: Watt hour values calculated from available data, assuming avg. 2.0 VPC on discharge.

on test provided only 20% of the cycle life [26]. In addition, extensive tests done under the Advanced Lead Acid Battery Consortium (ALABC) programs involving a hybrid Honda Insight test vehicle outfitted with the UltraBattery has demonstrated 150K actual miles [27].

With an impressive number of demonstration projects in progress and the renewal of tests under more recent standardized cycle regimes, the demonstrated capability of the UltraBattery should translate to greater commercialization and adoption.

10.3.5 Lead Foam and Carbon Foam Electrodes

Beyond the three types of carbons described in previous sections, there exist many additional carbon materials that may be of interest to the lead–acid battery researcher; however, as many are either cost-prohibitive, too early in the development cycle for commercialization, or simply have not been researched to a great extent, they have been mostly excluded here. Nevertheless, one material that is of interest in the development of advanced lead–acid batteries and has not been covered in previous sections is carbon foams.

Non-Pb current collectors have been and continue to be utilized within the battery industry. For example, products still exist that utilize an expanded Cu mesh, electroplated with Pb to prevent corrosion, as the negative current collector in some special purpose battery designs. The intent is to increase the overall conductivity of the cell and improve the high-rate performance. In this design, the top frame and burning lug are cast onto the plated substrate and then the entire "grid" is manufactured according to more conventional processes.

There have been a number of variants of this technology applied to lead–acid batteries over the years. In previous incarnations, it was referred to as composite grids or sometimes as lead foam electrodes. More recently, it utilizes carbon foams, which can either be considered another variant on the lead–carbon battery or it can be viewed as a novel electrode design, which may also provide some benefits similar to lead–carbon technologies, by nature of the materials utilized in the construction of the electrode. The technology has been trialed as both a replacement for the negative electrode, where bare carbon foam can be utilized, and for the positive electrode, which typically necessitates the addition of a stable barrier coating to protect the underlying carbon from oxidation. In most instances, metallic Pb is utilized as the barrier film; this method, although unnecessary, has also been utilized for the negative electrodes.

The primary feature of these electrodes is the use of a porous sheet of carbon foam, in combination with a "lug" in the standard lead–acid battery grid/plate configuration, which would allow for the use of industry standard manufacturing equipment. The last requirement is an overriding fact in most lead–acid battery development. The manufacturing base

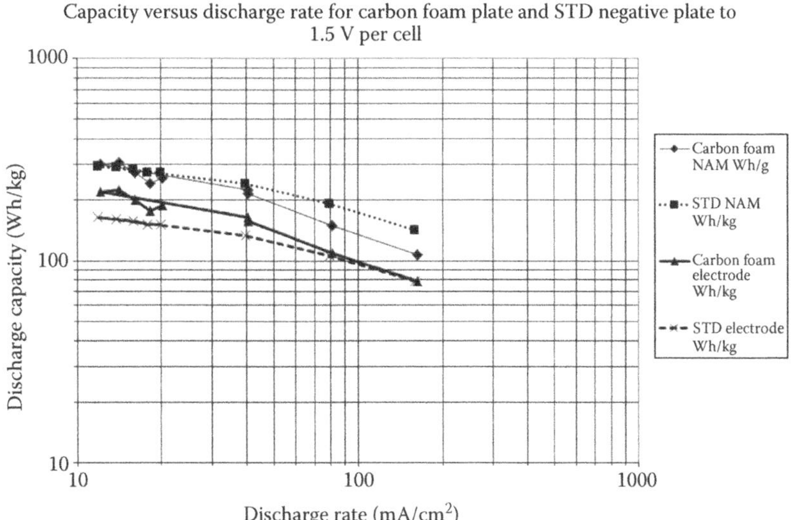

FIGURE 10.13 Specific capacity of carbon foam and STD VRLA negative electrodes. (Courtesy of C&D Technologies Inc.)

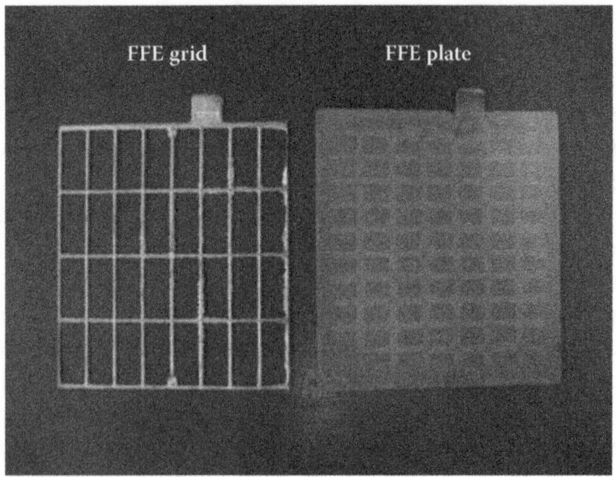

FIGURE 10.14 Firefly Energy® grid and plate—lightweight grid and pasted carbon foam electrode. The embed grid provides the additional current path for the plate lug, which is bonded using normal battery assembly processes.

for lead–acid technology is fairly standardized and firmly entrenched. For better or worse, any deviation from standard manufacturing methods obviously requires a much greater commitment of capital and consequently leads to greater difficulty for widespread adoption within the industry. This is perhaps the reason why the adoption and proliferation of integrated carbon electrodes in the marketplace has been much broader than for some of the other lead–carbon technologies.

Figure 10.13 compares specific capacity (Wh/kg of the electrode) and the active material utilization (Wh/kg of the NAM only) of a carbon foam electrode and an electrode using a standard cast grid. The NAM material utilization efficiency of a carbon foam electrode is similar to the cast electrode at lower discharge rates, while at higher rates the utilization is ~25% lower. This is most likely attributed to the higher resistance of the carbon foam structure. However, on an electrode basis, the carbon foam electrode outperforms the standard negative electrode at the lower rates due to the lower contribution weight of the carbon foam versus the cast grid, which is roughly 50% lower weight than the standard electrode.

A recent incarnation of the carbon foam electrode is the Firefly® battery, which was commercialized in 2011 but not widely adopted. The Firefly technology utilizes a unique composite current collector in place of a typical lead alloy grid.

The negative electrode uses a reticulated carbon foam sheet with a minimal lead alloy "grid" structure sintered into the structure to provide a current path and a standard lead alloy lug for bonding to the internal bus work the normal cast on strap (COS) manufacturing process (Figure 10.14). The stated benefits of the electrode design were first and foremost an inherent pore structure and surface area for the NAM, as well as a carbon surface for properties similar to lead–carbon batteries. The electrode was slurry pasted in a proprietary process to permeate the pores of the raw carbon electrode with lead oxide.

For comparison, Table 10.12 compares the performance of a group 29 VRLA AGM carbon foam battery with commercial AGM batteries (deep cycle and high power) of an equivalent BCI group size. Comparing performance values, the carbon foam technology provides only slightly improved overall battery weight despite the lower weight electrodes. This is a typical design trade-off where an increase in active material is utilized to increase overall performance (capacity and cycle life) while negating the weight benefit of the technology. The data do clearly show the high power limitation of the carbon foam in the CCA, which is comparable to the lower-performing deep cycle battery.

The carbon foam grid provides a 10% increase in the specific energy and allows an increase in active materials, which can provide superior cycle life. However, the deciding factor for a technology can be the economics. The one primary drawback of the carbon foam electrode is the cost to manufacture, which, compared to a standard grid design, is enough to offset the technical improvements it provides.

TABLE 10.12
Comparison of a BCI Group 29 Carbon Foam Electrode Performance versus STD VRLA AGM Technologies

	Deep Cycle AGM VRLA [22]	High Power AGM VRLA [22]	Group 29 Carbon Foam Negative
Weight kg (BCI group 31)	31	34.4	33
Rated Ah capacity (C/10)	92	98	102
Wh/kg	36	36	40
Wh/L[a]	87	98	100
Wh (C/20 rate)	1104	1248	1350
CCA (−18°C)	600	860	592
25-A reserve capacity	180	220	218

Source: C&D Technologies.
[a] Volume of BCI group size battery 29/31, ~12.74 L.

10.4 BIPOLAR LEAD–ACID BATTERIES

It is generally believed that bipolar batteries as a concept dates back to the 1920s, where they were developed for use in high-energy experiments and patents for bipolar lead–acid batteries dating from the 1970s up to the present, with new design variations and technical advances being developed every year. The question is why does a technical concept that manufacturers have tried repeatedly to develop, without uniform success, continue to evoke interest for commercialization? The answer lies in the very apparent benefits of successfully developing a bipolar design.

The concept of the bipolar battery is very straightforward; the term bipolar refers to the electrode design that would provide both positive and negative polarity. The design as illustrated in Figure 10.15 shows how the cells are configured with the PAMs and NAMs sharing a current collector, similar to the two sides of a coin, thus making a bipolar electrode. The bipolar electrode design typically utilizes a frame that isolates the edges of the electrode electrically and provides the sealing surface for the assembly of the cell, to isolate it ionically, and for the final assembly of the battery stack.

The bipolar electrodes are arranged with a porous separator between the opposing electrode surfaces, similar to a typical lead–acid cell, and each pair of the opposing "face" of the electrodes becomes a single 2-V cell, with the addition of the electrolyte. The cells are stacked together in series configurations for the required battery voltage, with each stack being sealed to the corresponding neighboring electrode pair, isolating each cell ionically from its neighbor as well as from the external environment.

This arrangement is vastly different from a typical lead–acid battery design that utilizes individual electrodes connected to a common internal bus, typically in parallel plate configurations, and relies on the individual cell partitions to isolate the cells from ionic transfer while using the internal bus work to provide electric conductivity and connect the cells in a series or parallel configuration, as depicted in Figure 10.1.

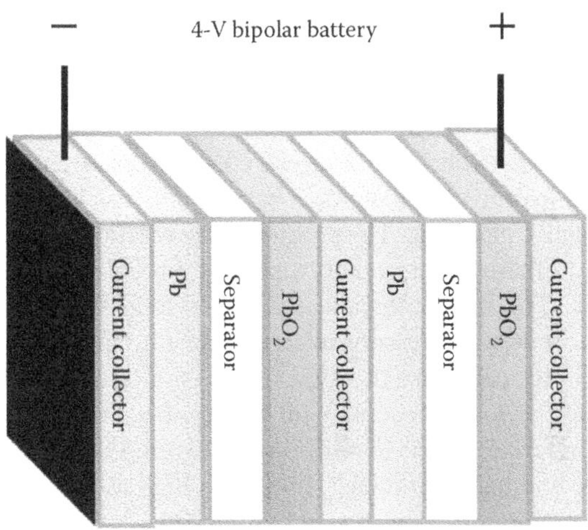

FIGURE 10.15 Schematic diagram: 4-V bipolar battery configuration with a shared current collector.

However, other than the design of the electrodes and battery assembly process, the bipolar lead–acid battery functions according to the same principles as all other lead–acid batteries. The same additives, active material formulations, and processes can be utilized, although the unique manufacturing requirements for the electrodes have led to some very unique and highly proprietary processes in manufacturing. Historically, the assembly of the battery on a commercial scale has been the Achilles heel of the bipolar battery design.

The bipolar design offers improvements in energy density, active material utilization, and power density when compared with a typical lead–acid battery design. The design benefit comes first from the shared current collectors, which reduces overall component weight increasing the gravimetric energy density. In addition, the lack of need for additional internal cell-to-cell connections further reduces component weights. As these components are effectively 98% lead and account for up to ~25% of the weight of the battery, their reduction or elimination can be significant.

Furthermore, with proper design of the terminal cells, the current density across the current collector should be very uniform, allowing for extremely efficient usage of the active materials. By comparison, the internal connections in a standard lead–acid battery, with a single connection point of the electrodes to the bus, generate inherently higher current densities at the lug, which leads to nonuniform usage of the active materials, particularly at higher discharge rates. In an application with repeated high-rate charge/discharge cycles, the active material is preferentially utilized at locations within the electrodes with the highest current densities. This results in premature failure of the electrode and is particularly common for the positive electrode and the PAMs.

Finally, the electrode stack design provides for an efficient configuration of high-power systems by allowing for a relatively simple increase in battery voltage while keeping amperage low, without adding significantly to the inactive components such as grids and bus work. Additionally, the previously stated benefit of uniform current density is of further benefit in high-power systems as the area of the electrode becomes the cross-sectional area of the current carrying member; therefore, a 100-A load on a single 15 cm × 15 cm electrode will result in 0.44 A/cm^2 current density, at a nominal 200 W of power. By comparison, a typical 100-Ah cell composed of seven positive and eight negative (15 cm × 15 cm) electrodes result in the same power but a ~40 A/cm^2 current density, and associated resistance, at the lug where the cross-sectional area is 0.2 cm^2. To counteract the impact of current density, typical lead–acid designs increase the current carrying component, which improves high-rate performance but reduces gravimetric energy density.

To achieve the bipolar design, the requirements for the current collector, in terms of design and material selection, are probably the most critical component. To be considered as a current collector, the material should have the following properties:

1. The material must be electrically conductive but prevent ionic transfer between the electrodes.

TABLE 10.13
Comparison of Estimated Bipolar Battery Performance to STD VRLA

	STD VRLA [22]	Bipolar (Calculated)	Ebonex Bipolar [30]
Weight (U1 size)	12	9.5	–
W/kg	180–220	290–300	250–320
Wh/kg	30–40	45–50	40–60
Wh/L	60–75	80–85	100–120
W/L	600–750	700–900	700–1000
Wh (4 C rate)	225	250	–

Source: C&D Technologies Inc.; K. Ellis, A. Hill, J. Hill, A. Lyons, and T. Partington, *Journal of Power Sources*, 136, 366, 2004.

2. It must be resistant to sulfuric acid and chemically and electrochemically stable at the potentials of the lead–acid system.
3. It should provide some affinity for both Pb and PbO_2 to provide a stable interface between active materials and current collector.
4. It should provide the basis for the assembly of the battery stack and sealing of the system both from internal ion transfer as well as for external seals.
5. Other considerations: low cost, light weight, nontoxicity, etc.

Owing to the varied requirements, a number of solutions for bipolar electrodes have been tried unsuccessfully in the past. For example, the use of Pb and Pb alloys would appear to be an easy solution as they are utilized in batteries today and meet requirements 1 through 3; however, the complications of encapsulating the lead current collector in a material suitable for constructing the plate stacks is not as simple as proposed. Furthermore, a lead sheet also poses problems for the adhesion of the active material due to the volumetric transformation that occurs during change and discharge.

It is for the reasons presented above that most bipolar solutions today rely on materials other than Pb or Pb alone as the current collector. One example of a current system is used by Atraverda, a United Kingdom–based company, which has developed a conductive ceramic Ebonex®. Ebonex is used in conjunction with a Pb substrate, and a polymer frame, as the material for the current collector. It is based on a high-purity titanium suboxide, Ti_4O_7, which has conductivity comparable to carbon (up to 300 S/cm) but with superior oxidation resistance, allowing it to be used at the positive electrode [30].

Table 10.13 presents the relative benefits of a bipolar design versus and typical VRLA battery; it also includes the projected performance of the Atraverda battery using Ebonex.

With the performance benefits as projected in Table 10.13 and assuming that the difficulties related the manufacture of the bipolar design do not add excessive cost to the battery, it is easy to understand why the development of the bipolar lead–acid technology will continue into the foreseeable future.

10.5 ADVANCED POSITIVE GRID DESIGNS

The overall goal in the development effort in batteries has been to provide improved product reliability, a reduction or elimination of battery maintenance, increased service life, and improved performance, i.e., energy density, reduced cost, and improved safety. Development of lead–acid cells and batteries in general and VRLA cells and batteries in particular have evolved along these lines.

Before maintenance-free technology, batteries used a Pb/Sb alloy for the positive grid. Conventional antimonial positive grid alloys cannot be used in maintenance-free cells. It is widely known that antimony shortens the life of such cells due to increased rates of gassing at both electrodes. Measurements have shown that antimony present in the NAMs or PAMs increases the rates of oxygen and hydrogen evolution. Antimony becomes present in both electrodes via corrosion of the positive grid [17].

As previously illustrated, the working principle of the VRLA cell is based on an internal oxygen recombination cycle. This is distinctly different than the working principle of maintenance-free flooded battery or the so-called sealed flooded lead–acid batteries. In maintenance-free flooded lead–acid batteries, the lack of maintenance is achieved primarily by designing in a reservoir of excess free liquid electrolyte (i.e., flooded) that is depleted over the life of the battery. The design of the cell is done in such a way that the reservoir of electrolyte is sufficient to maintain the performance of the cell over the life of the battery.

In a VRLA cell, oxygen produced at the positive electrode through hydrolysis during the charging process is reacted at the negative electrode. The product of the reaction is then converted to the initial reactants and H_2O through the charging process. The recombination is a closed-loop system by the use of a one-way pressure release valve to maintain the gas inside of the cell; thus, this type of cell is referred to as valve-regulated or sealed recombinant battery. Because the recombination cycle water loss through charging is minimized and recombination efficiency can be as high as 99% in some cases, the conservation of water in the cell allows the battery to operate without maintenance (watering). This is primarily achieved in one of two ways, both of which use the principle of immobilizing the electrolyte and developing gas paths through the separator medium. The gas paths are possible due to the cells operating in an unsaturated (<100%) state, the rate of gas transfer in liquid electrolyte is extremely low, and the recombination efficiency of such a cell is essentially zero for practical purposes. To achieve useful recombination efficiencies, the rate of gas transfer needs to be sufficiently high to maintain the electrolyte of the cell.

In an AGM cell, as the name implies, the bulk of the electrolyte is adsorbed and uniformly distributed in the separator material. Provided the material is not completely saturated, the development of gas can pass through unsaturated areas of the material to react at the opposing electrode [17]. The lack of water addition and lower state of saturation has immense implications for the design of the cell components. Although the process of recombination is essential to the long-term operation of the cell, if the rate of gas generation and subsequently the recombination rate increase significantly, the entire system can become unstable. The processes of recombination are exothermic and the increase in energy of the system can lead to an autocatalytic process known as thermal runaway.

If the energy of the recombination process is not dissipated, the temperature of the cell increases, which leads to an increase in the reaction kinetics of the processes in the cell and increases in the energy of the system. In addition, the recombination reactions at the negative electrode result in a dual effect to accelerate the entire process; first, the polarization of the negative electrode is reduced, resulting in an increase at the positive electrode that increases the rate of electrolysis and oxygen generation and, second, the effective discharging and charging of the negative results in an increase of the charging current and adds additional energy to the system. The increase in current and polarization also has the effect of accelerating the corrosion of the positive grid [17].

Water loss in a VRLA cell comes primarily from three sources: uncombined gas loss through the valve, loss through the container surfaces (which is minimal), and loss through corrosion of the components. The main source of corrosion occurs at the positive grid. Positive grid corrosion results in the irreversible loss of water in the cell. This water loss is critical for VRLA cells because of the increase in susceptibility to the thermal runaway of the element as the saturation passes below a critical value, the increase in the concentration of the electrolyte, and the loss of physical contact between the electrolyte and the active material due to dry-out of the separator. In addition to water loss, positive grid corrosion reduces the life of the battery. Particularly at high discharge rates, the propagation of intergranular corrosion increases the electrical resistance of the grid, reducing discharge capacity. As an example, the service life of a battery at a long discharge rate (10 h) may be 10 years, while the same service life at a discharge rate of 15 min may only be half of that due to the increase in resistance and loss of cohesive strength of the positive electrode due to corrosion.

Grids produced by direct casting methods suffer primarily from intergranular corrosion as the failure mode. The cast grain structure of Ca alloys results in inordinately high concentrations of Ca, the primary strengthening component of the alloy, segregating at the grain boundaries. The resulting boundary is prone to defects and corrosion due to the high oxidation rate of the alloyed elements. Significant work has been done to reduce both the onset of intergranular corrosion and its propagation following initiation. To this point, specific positive grid alloys have been developed to minimize the rate of intergranular corrosion.

The use of a rolled strip of grid has many advantages over a cast microstructure. The rolling of the strip or grid results in a fully recrystallized microstructure that significantly changes the properties of the material (Figure 10.16). The microstructure of the material not only increases the yield strength but also reduces the surface energy of the grain boundary, greatly reducing the susceptibility to intergranular corrosion. However, the use of a rolled grid for the positive electrode has severe limitations due to issues with the development of the interface between the grid and active material of the electrode. The strip can also be expanded to form a grid structure. The expanded positive strip has a number of disadvantages. First, the process imparts significant cold worked and most of

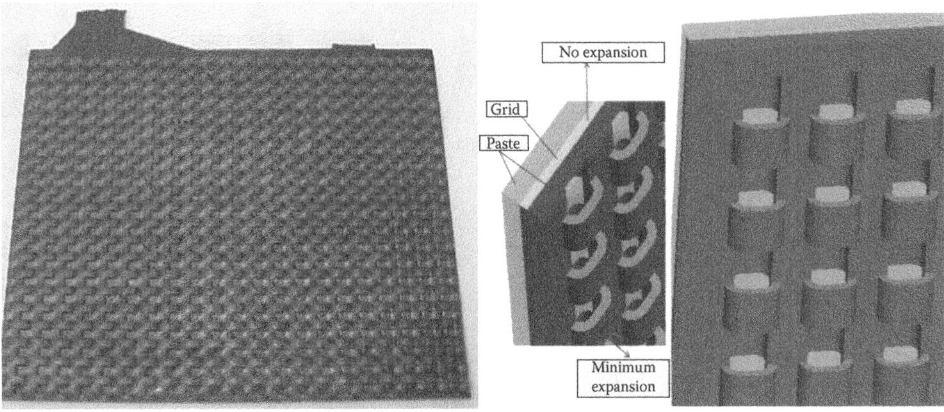

FIGURE 10.16 (Left and right) Rolled and cold forged grid and three-dimensional model of the grid structure. (Courtesy of C&D Technologies Inc.)

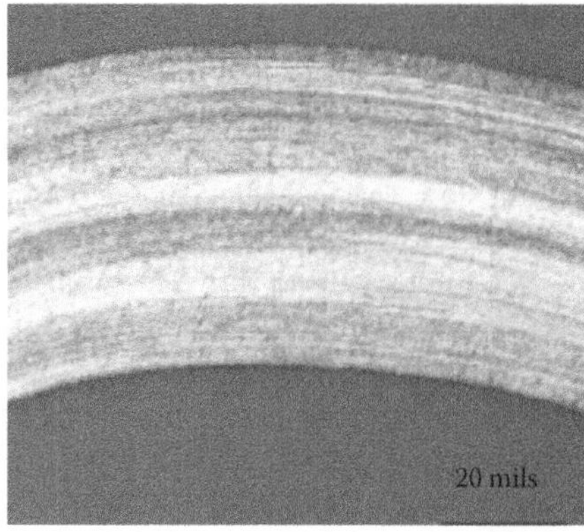

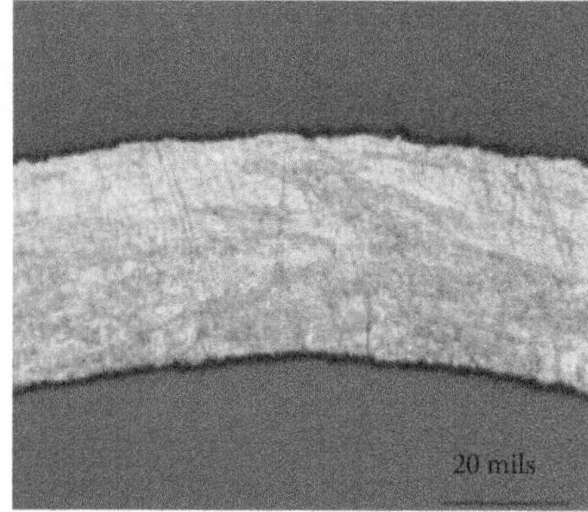

FIGURE 10.17 Corrosion study of rolled microstructure grid. (2.5 VPC at 0 and 100 days). (Courtesy of C&D Technologies Inc.)

the corrosion resistance of the boundary is lost due to defects introduced during expansion. The cold work generally results in retaining dislocations in nonrecrystallized deformed grains and microcracking in the areas of the most severe deformation. Second, the grid resulting from the expansion process is inherently prone to further expansion by design. This is an issue in operation as the reaction products of the discharge process in a lead–acid battery have a much greater volumetric density than the reactants; this, in turn, produces stress on the grid structure and leads to loss of adhesion between the active material and the grid.

The punched grid or cold forged electrode process avoids the issues above, the latter by utilizing a unique grid design and corrugation process that does not change the microstructure of the rolled base metal and does not introduce significant numbers of defects, such as microcracks, which will be prone to corrosion in later applications. The micrographs in Figure 10.17 show the relative corrosion rate of the corrugated positive grid following an extended overcharge test, of ~2400 Ah at 2.5 VPC. This is equivalent to ~16 years of service life in a standby application.

As the images clearly show, the extended overcharge, approximately equivalent to the entire service life of the battery, results in a cumulative base metal corrosion of ~7 mils, and the final microstructure does not show the initiation or propagation of intergranular corrosion. By contrast, a typical cast grid clearly shows indications of intergranular corrosion within a much shorter time frame (Figure 10.18).

The elimination of intergranular corrosion will maintain grid integrity over the life of the cell, improving or maintaining high-rate discharge performance and application life at high temperatures. In addition, the grid design has an inherently lower resistance, due to the minimal separation of the current collector and the active material, increasing both the discharge and charging performance of the positive electrode at higher discharge rates.

The net effect of this technology is the elimination of the primary failure mode in VRLA batteries, intergranular corrosion, and an extremely low resistance and highly efficient electrode design. These result in the following performance improvements.

1. Increased operating life at high temperatures
2. Increased high-rate performance
3. Increased operating life at high charge and discharge rates
4. Improved active material utilization efficiency (increased cycle life)
5. Improved battery reliability

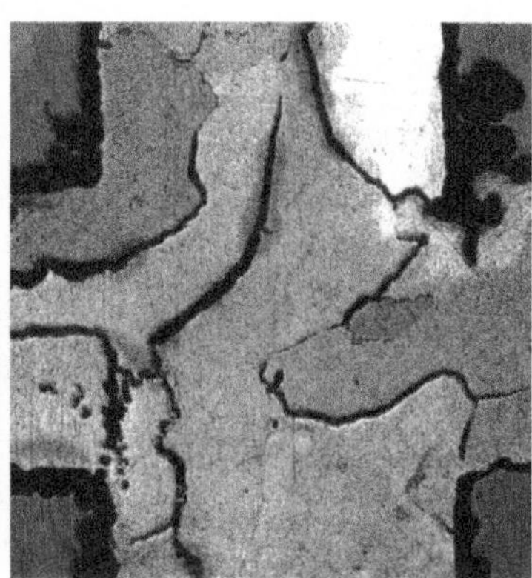

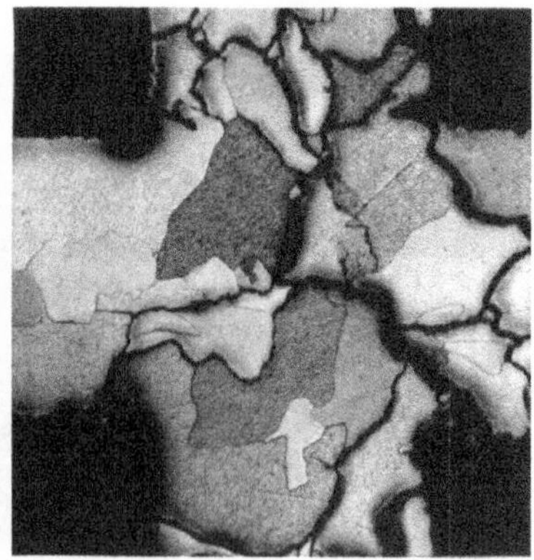

FIGURE 10.18 As-cast positive battery grid (PbCaSn) alloy on an accelerated float test at 60°C and 2.27 VPC, showing intergranular corrosion progression from 0 to 360 days. (Courtesy of C&D Technologies Inc.)

10.6 PURE Pb ACTIVE MATERIALS

In addition to the development of specialty alloys and processes for the manufacturing of the positive grids, the influence of the PAM and NAM structures and composition can have a significant impact on the function and overpotential of the battery electrodes, and significantly influence the long-term performance of the battery. Similar to how the addition of carbons can shift the polarization of the electrodes for cycle service, maintaining the polarization balance to favor the negative electrode can provide the same dramatic increase in performance for noncyclic, standby power, and high-temperature applications.

While ultra-high-purity components can benefit a sealed flooded system, such as an SLI battery, it is of even greater benefit in a VRLA system. Furthermore, design of the active material's surface area and pore structure to balance the polarization between the positive and negative electrodes can significantly reduce the charge currents and promote lower oxygen evolution at the positive electrode, and concurrently maintain the polarization at the negative electrode.

In this scenario, the VRLA battery electrodes are polarized similar to a flooded design with similar float currents. The overall impact to the battery performance is dramatic with lower float currents; there is reduced positive grid corrosion and water consumption. This can reduce the cell}s resistance increase and provide high discharge capability for much longer periods of service [31].

10.7 EMERGING APPLICATIONS

As stated in the introduction of this chapter, the drive for the majority of the developments in lead–acid technology is new and emerging applications. The common requirement for the majority of the emerging applications for energy storage is the ability of the system to operate in a PSoC condition, on a continuous or semicontinuous basis. There are a number of application-specific reasons for the requirement, of which the following are the most prominent.

1. Specific time of use requirements—In applications, such as demand response, peak shaving, etc., the battery has specific windows during which it can operate.
2. Improved or increased system efficiency—For lead–acid batteries, the efficiency of the charge of the battery is greatest at <90% SoC.
3. Bidirectional operation—Using the charge behavior of energy storage as the primary function or ancillary function of the system.
4. Increased service life—Similar to lithium-ion batteries, preventing overcharge can lead to increased service life, provided the battery is designed to operate in PSoC.

While the applications discussed in the following sections are not limited to advance lead–acid batteries, an attempt has been made to focus on the requirements, limitations, and system considerations as they apply specifically to lead–acid batteries.

10.7.1 Renewable Energy

The use of an ESS in conjunction with renewable energy generation systems (REGS) has numerous benefits and advantages over a stand-alone generation site. In addition, depending on the generation capacity, grid interconnection, and system load profile, the use of an ESS may be necessary to maintain the stability of the whole system. There are three primary scenarios where the use of REGS requires the use of an ESS: remote sites without a grid interconnect, grid interconnected sites with a high percentage of REGS, and microgrids that may or may not include both renewable and conventional energy generation.

10.7.1.1 Off-Grid

Renewable energy storage has been a common application for lead–acid batteries. The most basic and common application of energy storage with renewables are "off-grid" situations, primarily remote sites where the location makes providing AC power cost-prohibitive. In many instances, these sites included remote outposts and installations, as well as uninhabited systems such as cathodic protection of pipelines and remote relay stations. In this type of application, the ESS is absolutely necessary if continuous power to the site is required; in these applications, the ESS system has three primary functions.

1. Stores excess energy generated by the system during times of production
2. Provides the backup power during periods without generation, for example, at night with a photovoltaic (PV) system
3. Mitigates variation in the system either due to fluctuations in generating capacity, such as intermittent cloud cover on a PV array, or spikes in the system load profile.

To achieve each of these functions, the battery must be sized appropriately to maintain the load in the event of long periods without generation, in some cases for up to a week; be capable of daily charge/discharge cycles over the life of the system; and have a fast response time to changes in the load profiles or generation rates.

The current solution for the vast majority of these applications is either a flooded or VRLA lead–acid battery. The lead–acid chemistry, depending on the design, performs well in these applications. It has a relatively high cycle life at low depths of discharge (5–20%) and low discharge rates (C/20–C/100). The lead–acid chemistry is also extremely tolerant of both occasional overcharge and overdischarge, and has a millisecond response time to load fluctuations. In addition, the Pb–acid battery provides the most economical chemistry commercially available.

10.7.1.2 Grid Connected

In addition to the remote ESS applications previously described, the recent high penetration of REGS has demonstrated the need for energy storage in grid-connected systems. In the past, with only minor penetration of REGS with grid interconnects, the use of ESS was relegated to niche applications and demonstrations. The system cost versus the benefit of being able to generate off-peak and deliver the stored energy at peak times did not make financial sense.

However, with a high percentage of REGS online, the need for stored energy becomes more necessary due to the intermittent nature of the generation. The entire grid interconnect system in the United States was designed primarily with constant power generation in mind. As REGS make up more and more of the generating capacity, the inherent peaks and lulls in generation associated with it threaten to make the entire grid more and more unstable.

Previous solutions for grid stabilization relied on gas turbines to remain online at idle to respond to both ramping requirements throughout the day as the load profile changes from off-peak to peak usage, as well as to respond to minor spikes in the load requirements that occur continuously throughout the day. The use of gas turbines was only partly suited to this form of regulation; however, they have the fastest response time of most generation systems and the ability to regulate in 4-min increments was sufficient. The regulation does come at a price, however; a gas turbine runs at the highest efficiency at full generating capacity. In a ramping mode, it is extremely inefficient in terms of fuel consumption and generation capacity. This results in the emission of more greenhouse gases and greater fuel consumption with only minimal power generation. In addition, as the analog system response is relatively slow to a digital load, the amount of regulation capacity needed to mitigate small spikes in the demand can be on the order of 10 times the actual load because the generation must overshoot the load to respond to it in a timely manner (Figure 10.19).

With the addition of a large amount of generating capacity coming from REGS, this situation is only aggravated as the base load is no longer constant and the generation now adds additional instability to the system. The intermittency of REGS is an issue both in terms of the variation in output and in the rate of change at which it varies, both increasing and decreasing very rapidly. The variable output is also in both directions, generation both lulls and spikes at a very rapid rate. In this situation, the traditional generation is a poor solution to the problem. As it is slow in response, it requires even more capacity to be kept online and it cannot absorb excess power generation. For this application, the better answer is an ESS that has both a millisecond response time and the ability to mitigate both increases in load by discharging, as well as increases in generation output by being charged [32].

The use of an ESS also makes more financial sense in these applications, as the battery response is fast enough to closely follow the load profile requiring only sufficient capacity to match variation, freeing up the 10× generation capacity to be used at higher efficiencies when required by the overall load profile. This use of the ESS can also be used for ramping, to mitigate the need to operate turbines at partial capacity, as the ESS can maintain the load until it becomes most advantageous to bring more generation capacity online. Overall, this allows the system to operate at its highest efficiency, which not only reduces greenhouse gas emissions but also reduces fuel consumption and lowers the overall generation costs. In fact, these benefits and mode of operation are exactly the reason that energy storage is utilized in microgrids, the third scenario where energy storage is commonly utilized.

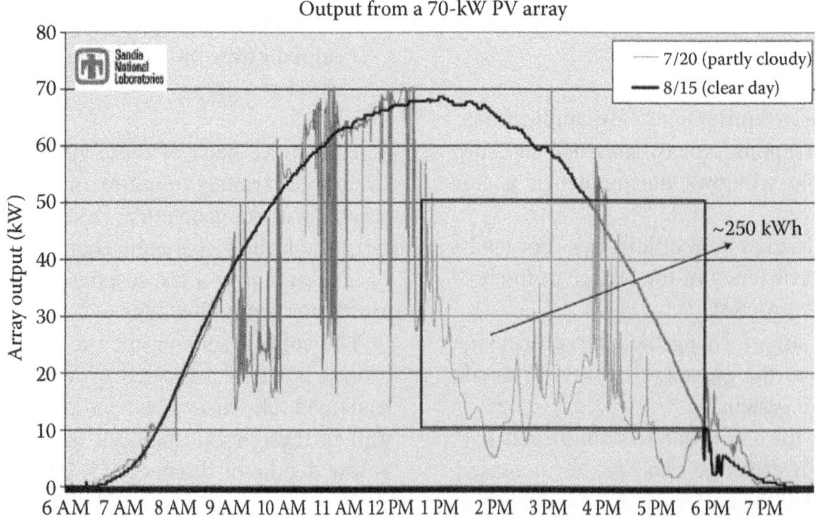

FIGURE 10.19 PV output Sandia National Labs (clear versus a cloudy day). (From J. Eyer and G. Corey. *Energy Storage for the Electricity Grid: Benefits and Market Potential Assessment Guide* (Sandia Report SAND2010-0815). Albuquerque, NM: Sandia National Laboratories, 2010.)

10.7.1.3 Microgrids

The third scenario that can benefit from utilizing an ESS is a microgrid with both renewable and conventional power generation. The situation is similar to the system with a high penetration of REGS only to a much greater extent as the size of the system makes it inherently more sensitive to variability, either in the load or generation capacity. A microgrid by nature is simply a smaller version of the entire grid interconnect system; in this situation, the benefits of an ESS are similar to the previous scenario, only the need for stabilization is much greater. In addition, the use of an ESS can be employed in peak shaving. Without an ESS, the generation capacity needs to exceed the highest load on the system, independent of the longevity or frequency with which the peak occurs. This translates to either having a large amount of available generating capacity for what could be either infrequent occurrences or short duration peaks, or a system capable of triaging load during times it exceeds capacity. Using an ESS in these situations not only mitigates the cost of increasing the generation capacity, but it also provides the aforementioned benefits, making the entire system more stable and hopefully more cost-effective.

10.7.2 Grid-Scale Energy Storage

The traditional role of lead–acid batteries in stationary applications has been primarily to provide backup power and, depending on location, power conditioning. In a typical application, the actual use (discharge) of the battery is fairly infrequent, and it remains on float charge for the majority of its service life. However, the use of energy storage in large grid-scale systems is more similar to some cycling applications with repeated charge/discharge operation. In these applications, the traditional standby technologies perform poorly compared with other ESSs. Even lead–acid batteries designed for cycling applications do not perform as well as alternative technologies without sizing the systems to such an extent or reducing the expected service life to such a level that the primary advantage, cost, almost reaches parity with other solutions.

With the development of lead–carbon technology for commercial products, many of the performance limitations of traditional lead–acid systems have been reduced or eliminated. The ability of lead–carbon batteries to operate in a PSoC and the stabilizing effect the technology has on the electrodes in cycling have increased the application possibilities for these systems, without increasing cost.

ESSs are particularly well suited for increasing the overall system efficiency of the electric grid. Increasing the renewable energy generation and transmission, reducing congestion at transmission bottlenecks, mitigating demand, and providing system stabilization are all possible with ESSs.

10.7.2.1 Congestion Relief, Contingency Mitigation Model

Congestion on the electric grid can be as large a problem as limited available generation and have as large an impact on the price of electricity. This is a common problem within and around urban areas and around certain geographic features, such as shorelines. The traditional causes of congestion occur due to the rapid expansion of load from increased urbanization and industrialization, versus the relatively expensive and slow process of increasing transmission capacity. More recent examples include the expansion of renewable generation being located in areas without enough available transmission capacity. In addition to the lack of capacity, the full utilization of

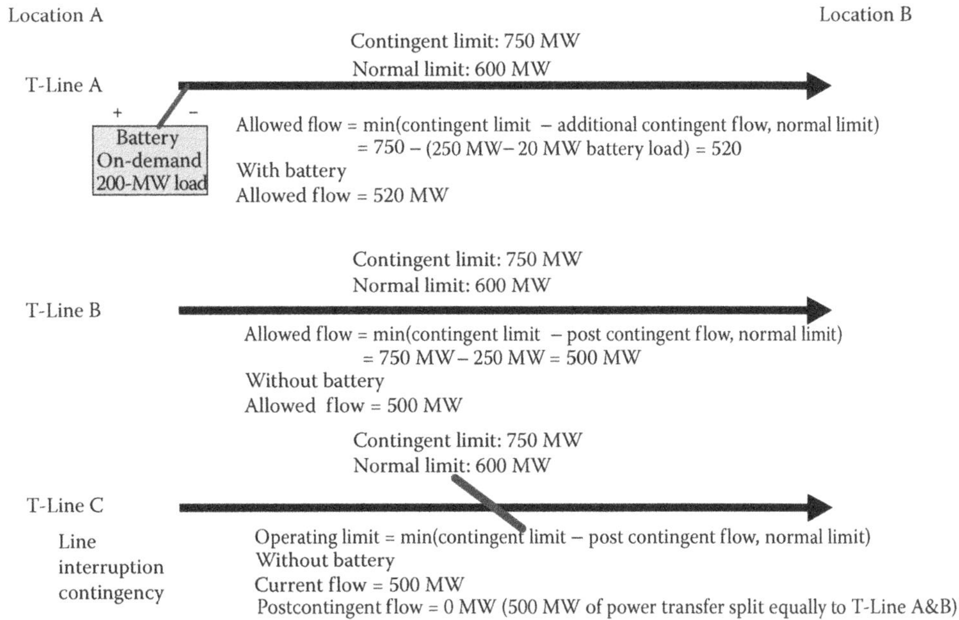

FIGURE 10.20 Functional diagram: contingency relief model for increasing transmission capacity on a capacity-constrained line using and energy storage system. (From J. Anderson, S. Mraz, and D. Boyer. enabling renewable energy transmission—Advanced lead carbon energy storage system for transmission utilization improvement. In: *Proceedings of EESAT11*, October 25, San Diego, CA, 2011.)

the available transmission is limited by the need to maintain enough excess capacity to mitigate a contingency event.

A contingency event is the loss of some transmission capacity, and contingency mitigation then requires the remaining transmission lines to immediately support the load that the lost transmission line was carrying to protect the entire system. As Figure 10.20 demonstrates, the three lines can each carry 750 MW, but are limited to only 500 MW each; this allows two lines to carry the load of all three lines in the event one line goes down. The benefit of this operation is system stability; one line going down does not cause an overload in the other two. However, the overall system utilization is only 66%.

The use of an ESS for contingency mitigation is an efficient method of increasing transmission capacity on existing lines. Placing the storage system up the line from a transmission bottleneck allows the line transmission capacity to be increased by an amount equal to the size of the storage system by providing load on demand. In the event of a contingency, the battery is charged for the amount of time necessary to mitigate the transmission contingency. The benefit is a continual increase in the line transmission capacity as the contingent limit is increased by the storage unit. The increased transmission capacity in transmission-constrained areas has the value of reducing the line congestion, which, in turn, will reduce the cost of all the power flowing on that line. In addition to the increased transmission capacity and reduced cost of energy in congested areas, the use of an ESS may also allow for the deferral of transmission upgrades.

The functional diagram in Figure 10.20 outlines the use of a 20 MW/25 MWh storage unit in this application. As the figure illustrates, during a line interruption contingency, on C, lines A and B must carry the entire load of all three lines. Without the ESS, this limits the transmission capacity of all three lines to 500 MW each, 100 MW less than their normal transmission limit. However, with the ESS, the line capacity can increase by 20 MW continually as the battery is available to accept the load, up to 20 MW, during a contingency event.

To operate in this application, the battery system must have the following performance requirements:

1. The system must be maintained in a PSoC to accept charge on demand.
2. The response time of the ESS to the contingency must occur in real time.
3. The system needs to provide at least 30 min of charge capacity as specified by the ISO for system adjustment to the contingency.

10.7.2.2 Demand Response

Demand response is the rapidly growing market for point-in-time load shedding, on demand, in response to peak demand for electricity. In a demand response scenario, rather than bringing additional generation on-line, "smart assets" are turned off or curtailed during peak demand events to lower the overall load on the system. As an example, smart monitors are commonly placed on HVAC systems in the northeast and elsewhere in the United States, allowing the local utility to dial back demand when required. In some markets, (New York Independent System Operator [NYISO]) the requirement for demand response from building operators has created a market for ESSs to operate behind the meter and seemingly curtail load, by discharging, while the building continues to operate at the normal load. As these events are usually daily occurrences centered around peak usage and peak cost times, the benefits are both financial, as the energy discharged is typically from a period of lower energy costs, and one of convenience for the building occupants, as the elevators and HVAC systems are not "curtailed" during the morning or evening rush.

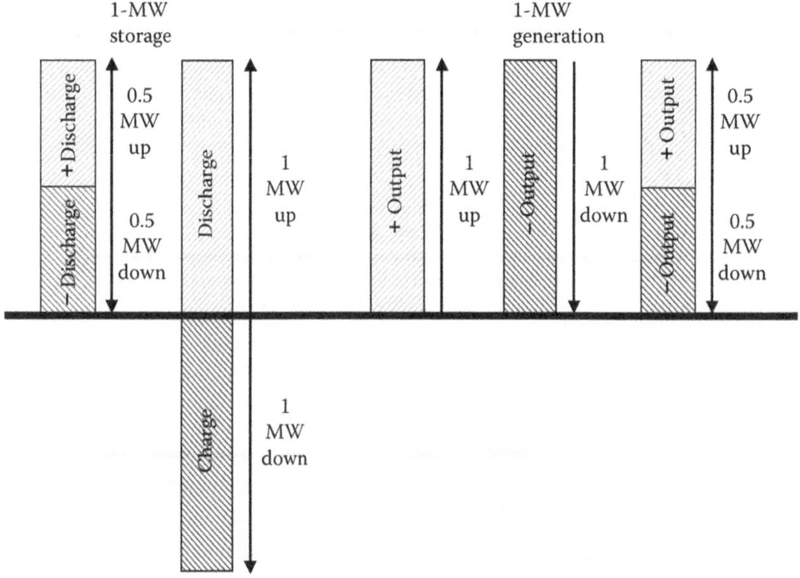

FIGURE 10.21 Regulation model for energy storage versus conventional generation. (From J. Eyer and G. Corey. *Energy Storage for the Electricity Grid: Benefits and Market Potential Assessment Guide* (Sandia Report SAND2010-0815). Albuquerque, NM: Sandia National Laboratories, 2010.)

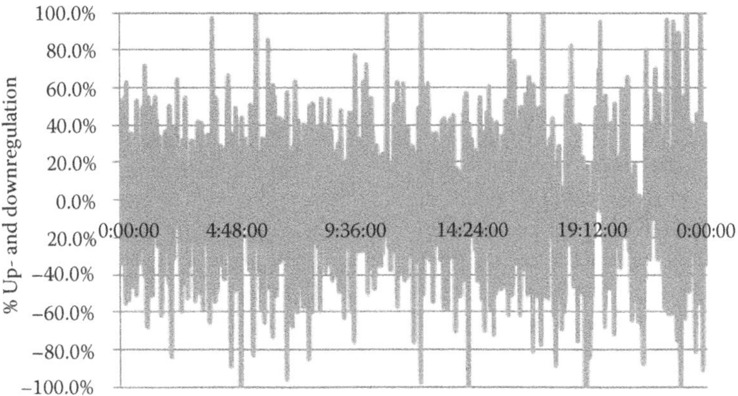

FIGURE 10.22 PJM fast regulation signal. (From Pennsylvania, Jersey, Maryland Independent System Operator.)

10.7.2.3 Ancillary Services

The ancillary services market is really the market for regulation services used by the ISO to stabilize the grid against changes in load, transmission, distribution, and generation, and is based on a number of different time scales [33]. For energy storage, the market of interest is the fast regulation market, which is similar to smoothing of renewable energy fluctuations.

Figure 10.21 depicts the relative advantage of using energy storage for regulation versus conventional generation resources [32]. Generation can provide 1 unit of upgeneration (increased output) or 1 unit of downgeneration (reduced output) for each unit bid into the market (in the case depicted, the unit is 1 MW), whereas a 1-MW energy storage device can provide 1 unit (1 MW) of upregulation by discharging and 1 MW of downregulation by charging with a net impact of 2 MW to the market.

This advantage, along with the speed at which energy storage can respond to changing requirements (in the millisecond range) versus minutes for traditional generation assets, is the reason that a separate fast regulation market has been developed in a number of regions in the United States. One of those regions, PJM ISO (Pennsylvania, Jersey, Maryland Independent System Operator), has a specific signal for fast regulation, which is depicted in Figure 10.22; the signal for the PJM market changes every 4 s.

Overall, the regulation market is a net zero energy application where the targeted total energy output increase for up regulation is equivalent to the energy absorption or decrease for down regulation. With the dynamic cycling required for the regulation market, it is easy to understand why lead–carbon batteries show a promising solution for these applications.

10.7.3 Stationary Power Application System Requirements

Table 10.14 summarizes the general power, energy, cycle, and calendar life requirements for the applications presented [34].

TABLE 10.14
Functions, Requirements, and Technologies for Stationary Power Applications

Application	System Function	Voltage (VDC)	Power (MW)	Energy (MWh)	% DoD	Cycle/Day	System Life (years)
Renewables off-grid	Stabilization and time shifting	48–1000	0.001–5	0.01–12	5–20	1	2–10
Renewables grid tied	Stabilization	480–1000	0.1–5	1–25	1–5	100 s	10–20
	Time shifting	480–1000	0.1–5	1–5	20–80	1	
Microgrids	Stabilization	48–480	0.1–5	0.001	0.1–1	1000 s	5–15
	Time shifting		0.1–4	0.5–20	30–60	1	
	Peak shaving		0.1–4	0.01–0.4	5–10	5–10	
Demand response	Time shifting	240–480	0.1–1	0.001	20–60	1	5–10
Regulation	Stabilization	480–1000	2–4	48	0.1–5	21,600	3–7

Source: J. Eyer and G. Corey. *Energy Storage for the Electricity Grid: Benefits and Market Potential Assessment Guide* (Sandia Report SAND2010-0815). Albuquerque, NM: Sandia National Laboratories, 2010; T. Hund, N. Clark, and W. Baca. DOE energy storage systems research program. Annual Peer Review. September 29–30, Washington, DC, 2008; J. Anderson, S. Mraz, and D. Boyer. Enabling Renewable Energy Transmission—Advanced lead carbon energy storage system for transmission utilization improvement. In: *Proceedings of EESAT11*, October 25, San Diego, CA, 2011; R. Lasseter and M. Erickson. Integration of battery-based energy storage element in the CERTS microgrid. Final Report. Madison, WI: US Department of Energy. October 27, 2009; B.J. Kirby. Frequency regulation basics and trends. US Department of Energy, 2005.

To meet the requirements for the various applications outlined previously, the battery needs to have a number of performance characteristics. The ideal system needs to have the following performance criteria:

1. High power—The system needs to have sufficient high-power capability to mitigate rapid changes in either the generation of the REGS or the load profile. The ability to provide a high-power battery will not only help with stabilization, but it will also reduce the required size of the ESS. The high-power capability needs to be for both charge and discharge.
2. Energy—The system should be able to not only stabilize intermittent spikes but also to provide continued backup power for ramping, peak load shaving, or load shifting.
3. PSoC operation—To adequately stabilize a REGS, the ESS system needs to accept energy as well as provide it. To achieve this, the system must be maintained in a PSoC to absorb excess generation capacity, which it cannot do at a full SoC.
4. Overcharge/overdischarge insensitive—As the system requirements will be constantly changing depending on local conditions, the most robust solution must be able to accept overcharge, and be resistant to overdischarge.
5. High cycle life—The operation of the system will be primarily a cycling application with numerous low DoD cycles for stabilization, as well as daily discharge cycles for load shifting and continuous power operations with REGS.
6. Long calendar life—In addition to high cycle life, the system needs to have a long operating life on both constant voltage charge and at open circuit, as in many cases the system usage is seasonal.
7. Maintenance-free operation—As previously stated, many applications are remote, unoccupied locations where periodic maintenance is either difficult or impossible to provide.
8. Predictability—The system should be capable of providing both state-of-health (SoH) and SoC information for prognostic capability and for integration in an active management operation when interconnected with a Smart Grid system.
9. Economical—As with any ESS, the cost of the system needs to be comparative to the benefits it achieves. An extremely expensive solution will be regulated to niche applications where other solutions will not work.

A system with these performance criteria would be well suited for the majority of the energy storage requirements, especially those applications in conjunction with REGS.

10.7.4 HEVs/Plug-In HEVs (PHEVs)

Similar to renewable energy generation, with politically and environmentally motivated regulation requirements and incentives generating the initial momentum, HEVs using stop–start technology, regenerative braking, and launch assist have become the new standard in vehicle offerings. As viewed from the battery manufacturer, the critical component and greatest opportunity for most of these features is the battery's ability to withstand the new demands placed on it.

As discussed throughout this section, the requirements for the battery, in terms of HRPSoC and DCA, for these new functions are well understood and the new lead–carbon technologies are playing a crucial role in the economic viability of these systems. Table 10.15 outlines the features related to each type of hybrid vehicle and the power requirements, efficiency improvements, and current technologies utilized in those applications.

10.7.5 HEV System Requirements

Most of the requirements for emerging stationary power applications can be translated to the requirements for HEV battery applications. Quantifying those requirements, Table 10.16 illustrates typical values for HEV application requirements for the battery.

While it would take significant improvement to the performance of lead–carbon batteries for them to be utilized in a parallel PHEV, their use today in microhybrids is providing valuable field experience for both the battery and automotive systems designers. This experience in parallel with continued development and improvement of both the vehicle

TABLE 10.15
Functions, Requirements, and Technologies for Common Hybrid Electric Vehicle Types

Hybrid Type	Electric Function	Power (kW)	V	% CO_2 Reduction	Flooded Pb–Acid SLI	EFB	Pb–Carbon VRLA	Ni–MH	Li Ion
Regular	SLI	2–4	12	0	×	×	×		
Micro	Stop/start regen braking	2–4	12	3–8		×	×		×
Mild	Launch and mild power assist	2–4	48	8–12			×	×	×
Moderate	Launch and mild power assist	5–15	100–150	12–20			×	×	×
Strong	Limited electric drive	25–60	150–350	20–35				×	×
Parallel PHEV	Extended electric drive	40–100	150–600	>35					×

Source: B. Monahov. Advanced lead battery consortium update. In: *BCI Committee Meetings*, October 9, Chicago, 2014.

TABLE 10.16
Performance Requirements for Common Hybrid Electric Vehicle Types

Hybrid Type	Electric Function	Power (kW)	System Voltage (VDC)	DCA (A/Ah)	DoD	No. of Cycles
Regular	SLI	2–4	12	0.25	<5%	<1000
Micro	Stop/start regen braking	2–4	12	Up to 1	5–10	1000
Mild	Launch and mild power assist	2–4	48	>1	Up to 30%	2000
Moderate	Launch and mild power assist	5–15	100–150	~2	40%	4000
Strong	Limited electric drive	25–60	150–350	~6	40%	6000
Parallel PHEV	Extended electric drive	40–100	150–600	10	50–60	10,000

Source: B. Monahov. Advanced lead battery consortium update. In: *BCI Committee Meetings*, October 9, Chicago, 2014.

TABLE 10.17
Battery Average Weight (kg/100 Ah) and Components Relative Weight Percent

	Weight (kg/100 Ah)	Grid (wt.%)	PAM/NAM (wt.%)	Strap/Terminals (wt.%)	Plastic (wt.%)	Separator (wt.%)	Electrolyte (wt.%)
12-V SLI	25	20–25%	35–40%	4–5%	3–4%	3–4%	25–30%
12-V VRLA (AGM)	31	25–30%	35–40%	3–5%	5–6%	2–3%	20–25%
2-V VRLA (AGM)	5.76	20–25%	40–45%	3–4%	3–4%	2–3%	20–25%

Source: D.A.J. Rand, P.T. Moseley, J. Garche, and C.D. Parker. *Valve-Regulated Lead–Acid Batteries*. Netherlands: Elsevier, 2004. Print; D. Pavlov. *Lead Acid Batteries Science and Technology*. Netherlands: Elsevier, 2011. Print; C&D Technologies Inc.

components and the battery is paving the way for the introduction of advanced lead–carbon batteries into the more the rigorous mild and moderate HEV applications.

10.8 ECONOMICS OF LEAD–ACID AND ENERGY STORAGE SYSTEMS

Economics will always play a role in the selection of an energy storage technology; for most ESSs, the majority of the cell cost is materials, and lead–acid is no different, with 65–70%, by weight, of the material in a 12-V VRLA AGM battery being Pb and Pb alloys. Breaking this down in terms of kg/100 Ah for a typical 12-V VRLA battery, the result is ~20–22 kg of Pb (Table 10.17).

Lead as a commodity is traded on a number of exchanges and over the past decades has ranged in price from around \$0.6/kg to over \$3.50/kg, with more recent values being around \$2/kg. It should be noted that in recent years, because the majority of Pb used in the manufacture of Pb acid batteries is from recycled batteries, coupled with the volatility of

TABLE 10.18
Economics of Energy Storage Comparison—Cycle Applications

	Deep Cycle 12-V VRLA (48-V Series)	12-V PbC VRLA (48-V Series)	Lithium Ion (NCM)	Sodium Nickel Chloride
Description	48 V 140 Ah	48 V 140 Ah	48 V 100–200 Ah	48 V 200–220 Ah
Battery price	\$1000–\$1250	\$1250–\$1500	\$2000–\$6000	\$9600–\$13,000
Battery price (\$/Wh)	\$0.16–\$0.20	\$0.20–\$0.22	\$0.5–\$0.7	\$0.95–\$1.20
Actual DoD used	50%	50%	80%	80%
Cycle life at DoD	1800	2600	~4500	2500–3000
Service life cost per cycle (\$/kWh)	\$0.17–\$0.22	\$0.15–\$0.18	\$0.2–\$0.22	\$0.49–\$0.53
Relative volume	2.2×	2.2×	1	1.5×
Relative weight	4.2×	4.2×	1	1.2×

Source: Various manufacturers' system pricing and published performance.

TABLE 10.19
Economics of Energy Storage—Standby Applications

	Lithium Ion (NCM)	Sodium Nickel Chloride	2-V VRLA	12-V VRLA
Description	48 V 100–200 Ah	48 V 200–220 Ah	48 V 1700 Ah	48 V 210 Ah
Battery price	$2000–$6000	$9600–$13,000	$15,000–$18,000	$1000–$1250
Battery price ($/Wh)	$0.5–$0.7	$0.95–$1.20	$0.19–$0.23	$0.10–$0.13
Float charge % SoC	90%	100%	100%	100%
Calendar life	15 years	15 years	20 years	10 years
Relative volume	1	1.5×	5.4×	2×
Relative weight	1	1.2×	5.1×	4×

Source: Various manufacturers' system pricing and published performance.

the commodities markets, the Pb industry and by extension the Pb–acid battery industry has been moving away from the commodity pricing for Pb. Instead, the index for the cost of the reclaimed batteries has been used as the basis for pricing; therefore, the cost of a new battery is based on the recycle value of the old one.

Furthermore, with a mature manufacturing base and supply infrastructure, an order of magnitude cost reduction in the manufacture of the battery does not seem likely, while incremental improvements are continuous but of less significant impact. Therefore, the greatest cost reduction can come from increasing the performance of the battery requiring smaller units, fewer batteries, or fewer replacements over the life of the system.

By comparison, lithium-ion technology continues to come down in cost both as a function of economies of scale with increased manufacturing capacity responding to demand, as well as to efficiency improvements within the manufacturing base. Tables 10.18 and 10.19 attempt to normalize the performance of various storage technologies for the application requirements of real-world systems, in an effort to provide a direct economic comparison for a cyclic stationary application and a standby power application.

As the tables show, the economic advantage of the lead–acid chemistry is significant for standby power applications, which explains the >95% market share. For cyclic applications, as previously stated, the reduction in cost for lithium ion along with the higher cycle performance at greater DoD has significantly narrowed the cost gap between lead–acid and lithium. However, with the introduction of lead–carbon batteries, the performance gap has been narrowed as well, placing some Pb carbon systems at parity in terms of the cost-to-performance ratio.

10.9 CONCLUSION

Even though more than 150 years have passed since the introduction of the first cells utilizing lead–acid technology, ever-changing market demands, new application requirements, and challenges from new technologies continue to push the development and further improvement of the lead–acid battery.

The introduction of lead–carbon technology to the negative electrode has come a long way to resolving the issues associated with PSoC and HRPSoC operation, such as sulfation, opening up new applications to lead–acid batteries and providing greater reliability in existing ones. The global trend toward improving efficiency and reducing the impact to the environment will require new and better ways to utilize available technologies, while stimulating the development of new technologies and advancements. Examples of these applications include the use of energy storage to facilitate a greater reliance on renewable energy and the continued further electrification of vehicles, requiring more and more from the battery.

While lead–acid batteries may not be ideally suited to every energy storage application, the high reliability, safety, low cost, and mature infrastructure for the total life cycle management of the battery will continue to make it an attractive solution for many of them.

REFERENCES

1. C. Pillot. The rechargeable battery market and main trends 2012–2025. Paper from the 30th International Battery Seminar and Exhibit, March 11, 2013.
2. D.A.J. Rand, P.T. Moseley, J. Garche, and C.D. Parker. *Valve-Regulated Lead–Acid Batteries*. Netherlands: Elsevier, 2004.
3. B. Hans. *Lead–Acid Batteries*. New York: Wiley, 1977.
4. L.T. Lam, R. Louey, N.P. Haigh, O.V. Lim, D.G. Vella, C.G. Phyland, L.H. Vu, J. Furukawa, T. Takada, D. Monma, and T. Kano. VRLA UltraBattery for high-rate partial-state-of-charge operation. *Journal of Power Sources* 174.1 (2007): 16–29.
5. J. Anderson. Grid scale energy storage. In: *Proceedings of the 2011 Energy Storage—2nd Renewable Power Generation & Energy Storage System Industry Outlook*, Beijing, China, September 9, 2011.
6. The International Carbon Black Association (ICBA). Home. January 27, 2015. Available at http://www.carbon-black.org.
7. J. Settelein, S. Hartmann, F. Güthlein, M. Gelbke, R. Wagner, V. Trapp, and G. Sextl. Study of the interaction between lead and carbon by electrochemical deposition. *Proceedings of the 9th International Conference on Lead Acid Batteries, LABAT 2014*, Albena, Bulgaria, June 10–13, 2014.

8. D. Pavlov, T. Rogachev, P. Nikolov, and G. Petkova. Mechanism of action of electrochemically active carbons on the processes that take place at the negative plates of lead–acid batteries. In: *Proc. of 7th International Conference on Lead–Acid Batteries, Varna, Bulgaria*, IEES, Varna, 2008.
9. P.T. Mosley. Consequences of including carbon in the negative plates of valve-regulated lead acid batteries exposed to high-rate partial-state-of-charge operation. *Journal of Power Sources* 191.1 (2009): 134–138.
10. M. Calabek, K. Micka, and P. Braca. Significance of carbon additive in negative lead–acid battery electrodes. *Journal of Power Sources* 158 (2006): 864–867.
11. P. Atanassova, N. Hardman, A.D. Pasquier, and M. Oljaca. Role of carbon blacks for improved cycleability and charge acceptance in advanced lead acid batteries. In: *13th ELBC Conference*, Paris, France, September 28, 2012.
12. W.G. Pell and B.E. Conway. Peculiarities and requirements of asymmetric capacitor devices based on combination of capacitor and battery-type electrodes. *Journal of Power Sources* 136 (2004): 334–345.
13. P.S. Walmet. *Evaluation of Lead/Carbon Devices for Utility Applications* (SANDIA REPORT SAND2009-5537). Albuquerque, NM: Sandia National Laboratories, 2009.
14. P. Nikolov, M. Matrakova, D. Pavlov, P. Attanossova, A. Du Pasquier, and M. Oljaca. Study of the interaction between lead and carbon by electrochemical deposition. *Proceedings of the 9th International Conference on Lead Acid Batteries, LABAT 2014, Carbon Additives for Improving Charge Acceptance of Flooded Lead–Acid Batteries*, Albena, Bulgaria, June 10–13, 2014.
15. D. Pavlov and P. Nikolov. Capacitive carbon and electrochemical lead–acid batteries and elementary processes on cycling. *Journal of Power Sources* 242 (2013): 380–399.
16. X. Jing, W. Zhao, and L. Lan. The effect of particle size on electric conducting percolation threshold in polymer/conducting particle composites. *Journal of Materials Science Letters* 19 (2000): 377–379.
17. D. Pavlov. *Lead Acid Batteries Science and Technology*. Netherlands: Elsevier, 2011.
18. J. Anderson, S. Mraz, and D. Boyer. Enabling renewable energy transmission—Advanced lead carbon energy storage system for transmission utilization improvement. In: *Proceedings of EESAT11*, San Diego, CA, October 25, 2011.
19. D. Pavlov, P. Nikolov, and T. Rogachev. Influence of expander components on the processes at the negative plates of lead–acid cells on high-rate partial-state-of-charge cycling. Part II. Effect of carbon additives on the processes of charge and discharge of negative plates. *Journal of Power Sources* 195 (2010): 4444–4457.
20. A. Ghosh and Y.H. Lee. Carbon-based electrochemical capacitors. *ChemSusChem* 5 (2012): 480–499.
21. O. Barbieri, M. Hahn, A. Herzog, and R. Kötz. Capacitance limits of high surface area activated carbons for double layer capacitors. *Carbon* 43 (2005): 1303.
22. Power you can depend on. C&D Technologies. Home. January 27, 2014. Available at http://www.cdtechno.com.
23. Axion PbC® technology. September 1, 2014. Available at http://www.axionpower.com.
24. E. Buiel, E. Dickinson, A.O. Stoermer, and S. Schaeck. Dynamic charge acceptance of lead acid batteries in micro hybrid board net. In: *Proceedings 12th ELBC Conference*, Istanbul, Turkey, September 21–24, 2010.
25. Ultrabattery® technology. UltraBattery: No ordinary battery. CSIRO. January 1, 2014. Available at http://www.csiro.au/Outcomes/Energy/Storing-renewable-energy/Ultra-Battery.
26. T. Hund, N. Clark, and W. Baca. DOE energy storage systems research program. In: *Annual Peer Review*, Washington, DC, September 29–30, 2008.
27. B. Monahov. Advanced lead battery consortium update. In: *BCI Committee Meetings*, Chicago, IL, October 9, 2014.
28. L.T. Lam and R. Louey. Development of ultra-battery for hybrid-electric vehicle applications. *Journal of Power Sources* 158 (2006): 1140–1148.
29. J. Furukawa, M. Miura, H. Yoshida, W. Tezuka, and L.T. Lam. Performance of UltraBatteries produced from different grid thicknesses. In: *Proceedings from the 12th ELBC Conference*, Istanbul, Turkey, September 21–24, 2010.
30. K. Ellis, A. Hill, J. Hill, A. Lyons, and T. Partington. The performance of Ebonex® electrodes in bipolar lead–acid batteries. *Journal of Power Sources* 136 (2004): 366–371.
31. R. Malley and A. Williamson. Large format VRLA products for uncontrolled temperature environment. In: *Proceedings of BattCon*, Orlando, FL, May 8, 2008.
32. J. Eyer and G. Corey. *Energy Storage for the Electricity Grid: Benefits and Market Potential Assessment Guide* (Sandia Report SAND2010-0815). Albuquerque, NM: Sandia National Laboratories, 2010.
33. B.J. Kirby. *Frequency Regulation Basics and Trends*. Madison, WI: United States Department of Energy, 2005.
34. R. Lasseter and M. Erickson. *Integration of Battery-based Energy Storage Element in the CERTS Microgrid*. Final Report. Madison, WI: US Department of Energy, October 27, 2009.

11 Lead–Acid Automotive Battery

Anson Lee and Jun Gou

CONTENTS

11.1 INTRODUCTION

Since the birth of the science of electrochemistry in the late 18th century, credited to Professors Luigi Galvani and Alessandro Volta at the University of Bologna and Pavia University, respectively [1], the lead–acid battery has been the mainstay of electrical power source for most commercial usage. This was true for both stationery and mobile applications.

For the automobile industry, the lead–acid battery has been the most reliable and the most cost-effective electrical power source since the early 1900s. Back in those days, there was virtually no electrical equipment on an automobile. The engines had to be started by a hand crank and the lighting equipment was either "oil lamps" or "gas lamps" [2]. It was the dawn of the automobile industry.

As electrical equipment such as the starter motor, ignition system, and electrical lighting were developed in the 1910s and 1920s, the lead–acid battery became an indispensable component of an automobile. The voltage chosen in that era was 6 V. The 6-V electrical system was the ideal voltage for lamps, and it was more than sufficient to drive the starter motor and the ignition system of engines in that time period.

By the 1930s, all automobiles had a 6-V lead–acid battery on board. It provided the electrical power for the starter motor, ignition system, and lighting. It also made possible several safety or convenience-type systems such as electric windshield wipers and electric horns.

11.2 LEAD–ACID CHEMISTRY

11.2.1 Electrochemistry

The lead–acid battery is a secondary battery that utilizes electrochemical reactions to convert chemical energy into electrical energy and vice versa. This secondary battery acts as energy storage when extra electrical energy is available, and it acts as a power source when a need for electrical energy in the system arises. Lead–acid batteries, like other electrochemical devices such as fuel cells and lithium-ion batteries, work based on electrochemical principles. Figure 11.1 is a sketch of an electrochemical cell displaying the operating principle.

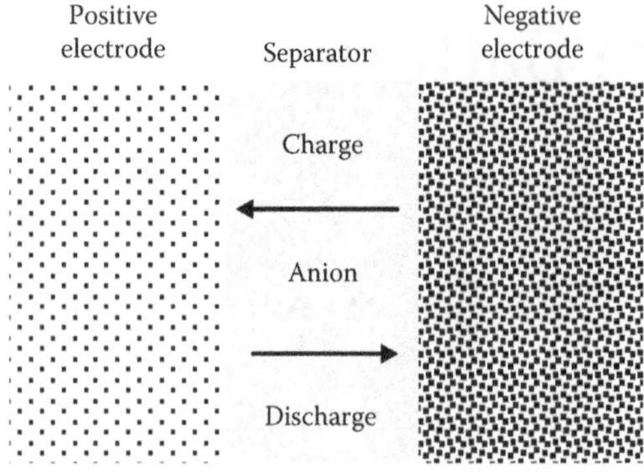

FIGURE 11.1 An electrochemical cell's typical structure: porous electrodes and electrically insulated but ion-permeable separator.

The cell comprises two pairs of electrodes with different potentials, for example, lead–acid batteries with $PbO_2/PbSO_4$ (1.685 V versus standard hydrogen electrode [SHE]) and $Pb/PbSO_4$ (−0.356 V versus SHE), and hydrogen fuel cells with H^+/H_2 (0 V versus SHE) and O_2/H_2O (1.229 V versus SHE). Internally, the two electrodes are immersed in the electrolyte, either liquid or solid, in which the ions are transferred from one side to the other when the electrodes are connected by an external circuit.

A lead–acid cell comprises Pb^{2+}/PbO_2 as the positive electrode and Pb/Pb^{2+} as the negative electrode. Sulfuric acid with a concentration of 4–5 M acts as the electrolyte. During a discharge process, the lead dioxide is reduced to Pb^{2+} at the positive side and lead is oxidized to Pb^{2+} at the negative side. As the reaction continues, more and more Pb^{2+} and SO_4^{2-} are produced in the electrolyte. When the solution of $PbSO_4$ becomes saturated, it precipitates as solid lead sulfate on both electrodes. During a charge process, the solid lead sulfate first dissolves into Pb^{2+} and SO_4^{2-} at both electrodes, then the Pb^{2+} is oxidized/reduced and PbO_2/Pb are formed on the reaction site of positive/negative electrodes. Equations 11.1 and 11.2 summarize the reactions.

$$PbO_2 + 3H^+ + HSO_4^- + 2e^- = PbSO_4 + 2H_2O \quad (11.1)$$

$$Pb + HSO_4^- = PbSO_4 + H^+ + 2e^- \quad (11.2)$$

Figure 11.2 shows the comparisons between the modeling results and the test data of a 96-Ah battery during discharge and charge processes. Figure 11.3 demonstrates the acid concentration evolution inside a cell of the battery. The discharge and charge processes are conducted at $I = 20$ A. It is indicated from Figure 11.3a that acid concentration decreases faster in the positive electrode than in the negative electrode; thus, H^+ diffuses from the negative electrode to the positive electrode. The steep drop at the interface between the positive electrode and the separator is due to the intense electrochemical reaction at the interface. Figure 11.3 shows the acid concentration recovery during the charge process. It can be seen that the acid concentration in the positive electrode increases gradually due to the H^+ production, and the diffusion is now from the positive side to the negative side.

The electrode potentials are governed by the Nerst equation (Equation 11.3) below. From this equation, it can be seen that during charging, the H^+/HSO_4^- concentration increases at both electrodes; thus, the potential of the positive electrode will become more positive and that of the negative will become more negative, which causes the cell voltage to rise. During discharge, the concentration of H^+/HSO_4^- decreases, which

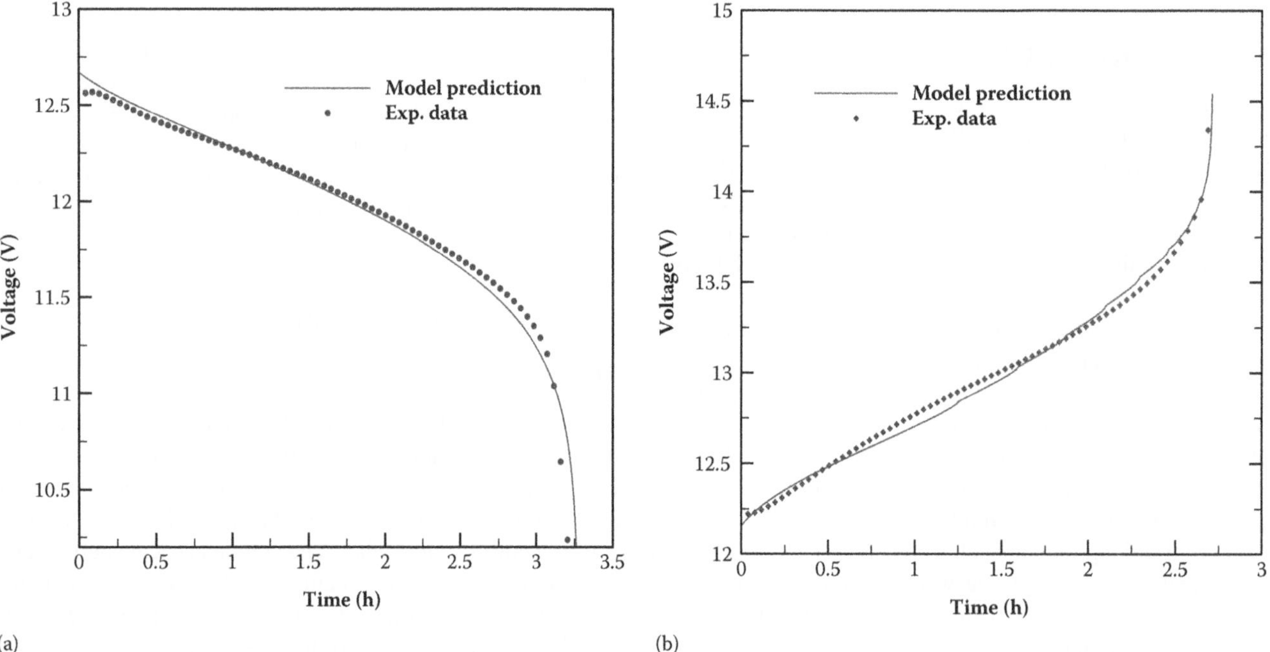

FIGURE 11.2 Battery discharge (a) and charge (b) curve comparisons between modeling results and experimental data.

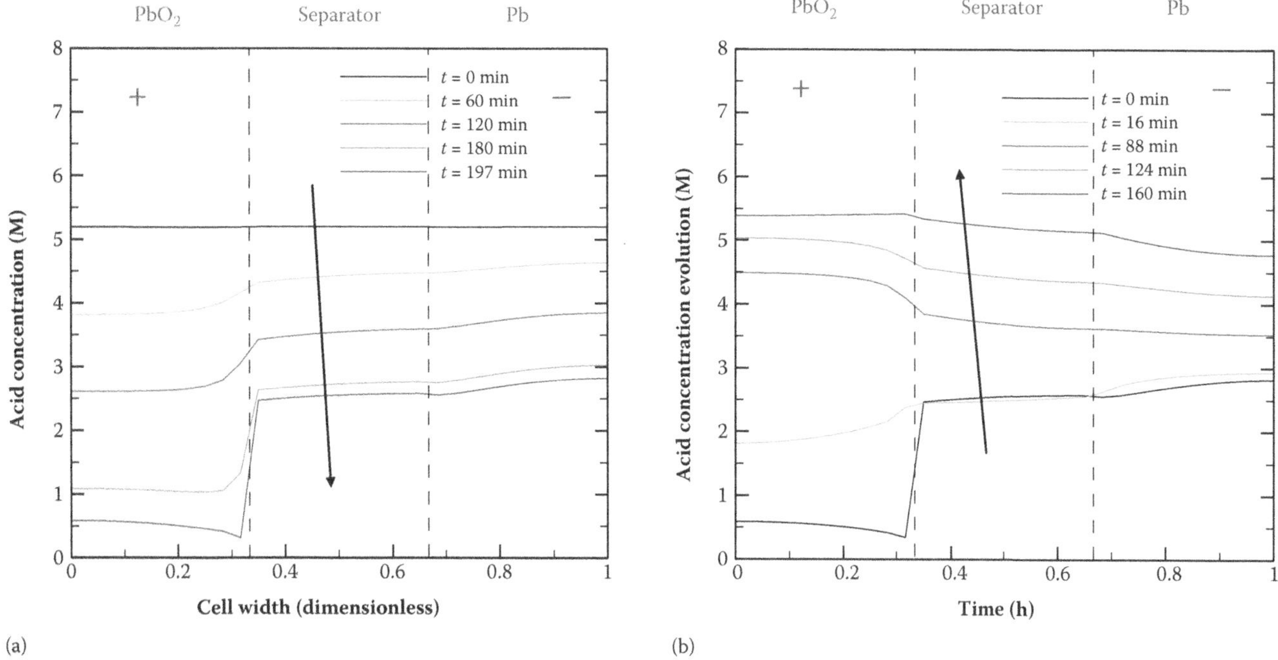

FIGURE 11.3 Acid concentration evolution during discharge (a) and charge (b) processes.

results in the positive electrode potential being less positive and the negative electrode potential being less negative; thus, the cell voltage drops.

$$E^+ = 1.685 - \frac{2.303RT}{nF}\log\left(\frac{1}{[H^+]^3\left[HSO_4^-\right]}\right)$$
$$E^- = -0.365 - \frac{2.303RT}{nF}\log\left(\frac{\left[HSO_4^-\right]}{[H^+]}\right) \quad (11.3)$$

11.2.2 Mechanical Structure and Bulk Materials

A lead–acid battery usually comprises six chambers that are connected in series, and each chamber has a voltage of 2.1 V. Cells in each chamber are connected in parallel. The number of paralleled cells scales up and down the battery capacity. The positive and negative plates of cells are connected by two individual connectors to collect current. The electrode is structured with a lead–alloy grid on which the active materials, lead dioxide paste and lead paste, are applied. The ingredients and the preparation process of electrode pastes are critical to battery performance and degradation, and they are usually proprietary to the industrial manufacturers. Here, we can discuss a general concept on the mechanical structure of lead–acid batteries.

11.2.3 Paste

The preparation of the active material paste of both electrodes usually starts with the mixing of lead oxide and water, and then with sulfuric acid. Adding water into lead oxide powder removes air from the porous structure to form a soft paste with plasticity. The second step is to add acid into the mixture to react with the lead oxide. By controlling the mixing temperature and ratio between lead oxide, water, and acid, lead sulfate with different characteristics can be formed and can be used for different applications. For example, when lead oxide (PbO_2) reacts with sulfuric acid, the stable product with a temperature below 160°F will probably be tribasic lead sulfate (TRB), while when temperature increases above 160°F, the stable product is tetrabasic lead sulfate (TTB). TRB is preferred for engine-start applications; however, TTB is more suitable for cycling applications [3].

11.2.4 Grid

The plate grid in a lead–acid battery is the skeleton of the plate. The most desirable feature for grids is corrosion resistance, especially for the positive plate. This is because the positive plate always bears higher potential during the charging process, which can easily cause grid corrosion if the additives of grid alloy are not properly done. Usually, the lead alloy grid is made with lead–antimony to improve the fluidity of metal and the tensile strength of the grid. However, such alloy can cause hydrogen gassing and could dry out the battery. To make the battery maintenance free, metals such as calcium, tin, and silver are used to replace antimony. Moreover, experimental cycling tests showed that the lack of antimony led to premature failures as antimony actually improves the adhesion between the active materials and the grids [4]. Therefore, the addition of antimony sulfate in the active material paste can be used.

11.3 VOLTAGE-LEVEL DEVELOPMENT

11.3.1 6-V, 12-V, and 24-V Systems Development

As mentioned earlier, the first automotive voltage level was chosen to be 6 V. The 6-V electrical system was the ideal voltage for electric lamps as the lamp filaments needed mechanical integrity. The electric starter and magneto devices for the ignition system were both developed based on a 6-V system. As electric windshield wiper and electric horns came to the automobile, the 6-V lead–acid battery became standard equipment on every automobile.

As the demand for engine power increased, new engine development resulted in higher displacements and higher compression ratios for the internal combustion engine. These newer engines required higher cranking and starting power and much higher breakdown voltages for the ignition system's spark plugs. The 6-V system was unable to provide these types of power, and a new 12-V system was developed out of necessity. The 12-V lead–acid battery soon became the industry norm. By the late 1950s, almost all automobiles have switched from the 6-V system to the 12-V system, and this 12-V system continues to be the standard for light-duty vehicles, i.e., passenger cars and light trucks for the next six decades including the present decade.

In a typical automobile today, excluding "hybrid" vehicles, regardless of the manufacturer, one will find a single 12-V lead–acid battery as the main source of electrical power, along with a 12-V alternator that charges the battery as well as supplies power to the various electrical and electronic devices on the vehicle during engine operation.

The battery acts as the chemical energy storage device, and the alternator generates the electrical energy being captured and stored by the battery through the electrochemical process described in the previous section. The battery must be able to provide the electrical current at low temperatures (as low as −40°C) to the electric starter to crank the engine, as well as to supply the electricity to all the electrical equipment on an automobile. The number of electrical and electronic equipment has been growing exponentially in the last three decades, and it is taxing the 12-V system to its capability limit.

Figure 11.4 depicts a typical automotive 12-V electrical power system with a 12-V lead–acid battery, a 12-V alternator (G), a 12-V starter (S), and the connection to the various 12-V electrical loads, represented by the box on the right-hand side of the diagram.

For the heavy-duty vehicles, which are invariably powered by large diesel engines requiring even higher power for cranking, starting, and ignition, not to mention all the ancillary

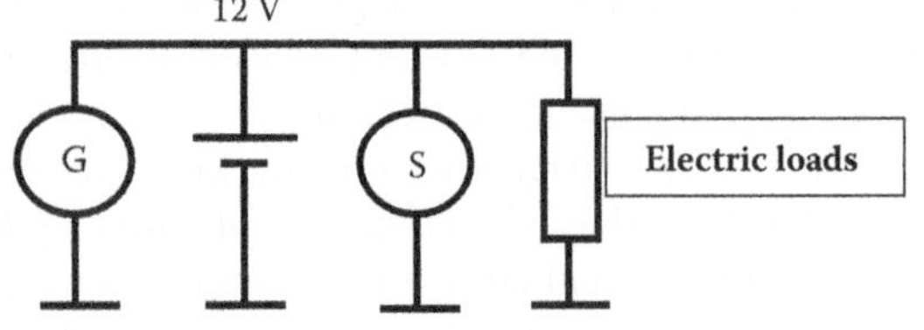

FIGURE 11.4 Simplified diagram of a 12-V automotive electrical system.

equipment, a 24-V system has been developed and many of today's heavy-duty trucks are powered by 24-V lead–acid batteries. Much of the aviation industry as well as many military vehicles has also been operating with 24-V systems owing to their much higher power requirements than passenger cars and light-duty trucks. The lead–acid battery is still the dominant electrochemistry being used for ground equipment; however, airborne equipment is usually "nickel cadmium" based due to its weight and performance advantages.

As can be seen from history, a common theme that has emerged is "high power requirements" need "high-voltage systems." This theme will set the stage for things to come in the coming years.

11.3.2 42-V Systems Development

In the 1990s, a number of luxury automobile manufacturers spearheaded by Mercedes-Benz initiated and formed a research group at the Massachusetts Institute of Technology (MIT) called the "MIT/Industry Consortium on Advanced Automotive Electrical/Electronic Components and Systems" [5]. The objective of the consortium was to study the future needs for automotive electrical power and the various avenues to satisfy this future need. The membership included more than 30 of all the major global automobile manufacturers, battery manufacturers, electronics manufacturers, wiring harness manufacturers, and academia. Meetings were held quarterly and the venues rotated between North America, Europe, and Asia depending on the location of the host member company. A high-voltage system called "42-V PowerNet" evolved from this intense research effort and a "42-V" standard was formed and specifications written around the year 2000. This voltage was to be the next voltage level for all future automobiles.

The 42-V PowerNet standard was developed with the premise that many mechanical subsystems would be replaced with electrical subsystems in the future, for reasons of control efficiency, packaging efficiency, and most important, fuel economy and overall system efficiencies. As the world grappled with economic issues based on oil and pollution issues due to automobile emissions, government regulations became more and more stringent. Subsystems such as steering, brakes, cooling, and even engine valves that are traditionally mechanical driven, by belts, gears, and shafts, which use engine power and are all considered "parasitic" power drains from the engine, could be replaced by electrical motors and power electronics that rely on batteries and need not be turned on when not needed. It was called "power on demand," and this design approach will reduce fuel consumption and vehicle emissions. An industry consensus was reached at the MIT consortium that the future electrical power requirement for a typical automobile would be three or four times the capability of the 12-V system. The list below contains some of the potential vehicle subsystems that could be converted to the 42-V PowerNet architecture [6]:

1. Power steering
2. Radiator cooling fans
3. Water pump

4. Heated windshield
5. Rear defroster
6. Heated seats and steering wheel
7. Heated catalytic converter
8. Cabin heaters
9. Active engine mounts
10. Electric air conditioner
11. Electromagnetic engine valves
12. Engine stop–start operation

Also considered was the safety limit of direct current (DC) voltage, which was determined to be 60-V DC. The 42-V PowerNet system was designed based on a "36-V lead–acid absorbent glass mat (AGM) battery," basically equivalent to three 12-V batteries connected in series. The "42-V" name was coined based on the charging voltage of the 36-V lead–acid AGM battery.

The emergence of the 42-V PowerNet did not mean the end of the 12-V power system, as many electrical loads on the vehicle will continue to operate at 12 V, such as lighting, small motors for windows, seats, door lock, etc., so there would be a "dual voltage system" having both 12 V and 42 V on board the same vehicle. Figure 11.5 is an illustration of one of the possible implementation of the dual-voltage system.

This implementation shows a dual battery system consisting of a 36-V lead–acid AGM battery (battery 2) and a 12-V lead–acid AGM battery (battery 1), connected by a "bidirectional DC to DC converter." The 12-V battery feeds the low-power vehicle loads, such as lighting, electronic sensors and small motors, etc., and the 36-V battery feeds the high-power loads such as electric steering system, electric cooling fan, and electric heaters. The 9-V minimum shown in this figure is the same 9-V minimum that has been used for conventional 12-V systems. The 55-V maximum shown is the specification of the 42-V system's maximum operating voltage. The 42-V PowerNet was a costly venture, and there were several manufacturers that proceeded with the development into production. Examples are the "Toyota Crown" and the "Saturn Vue" in the early and mid 2000s. Most manufacturers refrained from the expenditure for various business and technical reasons.

The worldwide recession in the late 2000s stopped most automobile companies from continuing the 42-V PowerNet development and the MIT consortium's membership dwindled. When hybrid electric vehicles (HEVs), which operate at hundreds of volts, were introduced and popularized by Toyota, all 42-V activities came to a screeching halt.

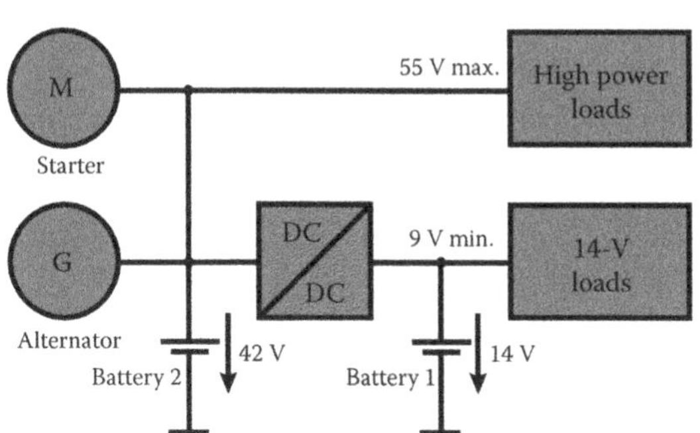

FIGURE 11.5 Example of a dual-voltage (14-V/42-V) system.

11.3.3 48-V Systems Development

In the 2010s, the high voltage discussions resurfaced primarily among the luxury automobile manufacturers in Germany. Five German automobile companies (Daimler-Benz, BMW, Audi, Porsche, and Volkswagen) convened and reaffirmed their common belief that a higher voltage for future automobiles is inevitable, for exactly the same reasons and rationales behind the development of the 42-V PowerNet at the MIT consortium. Through the VDA (German Auto Industry Association), they created a new automotive electrical power system standard at 48 V [7]. This voltage level is four times the 12-V level, but it is still below the 60-V DC safety threshold. The battery being considered for the 48-V system is a lithium-based battery, which has been under development for over a decade for plug-in HEVs (PHEVs), pure electric vehicles (EVs), and HEVs, all of which operate at several hundred volts. The advantage of lithium is its light weight, and therefore its power density and energy density. However, there are disadvantages and technical hurdles for lithium batteries to overcome, such as thermal runaway, cooling, battery control, and cell balancing issues.

Figure 11.6 depicts one implementation of the dual-voltage system consisting both the 12-V and the 48-V subsystems linked by a "bidirectional DC-to-DC converter," very similar to the dual-voltage (12 V/42 V) architecture described earlier.

This system implementation assumes an electric machine (M) operating at 48 V and as both a starter and a generator, supplying the high power loads, and feeding a bidirectional DC-to-DC converter to manage the 12-V bus and to charge the 12-V battery. The 12-V battery would most likely be a lead–acid battery, and the 48-V battery could be a lithium-based, a nickel-based, or a lead-based battery.

Different manufacturers would approach the design of their systems differently based on economics, battery availability, and the technical competence of their staff. The degree of 48 V conversion in an automobile will mostly depend on the vehicle class, target market, feature content, and price level. Similar to 42-V systems, the 48-V system will add cost and complexity to a vehicle. However, with the pressure of government regulations on fuel economy and emissions, as well as the customers' demand for performance, drivability, comfort, and convenience, there is little chance that the 48-V system architecture would follow the same fate as the 42-V system two decades ago.

The 48-V system will most likely be implemented on luxury vehicles initially, with the high-voltage system operating

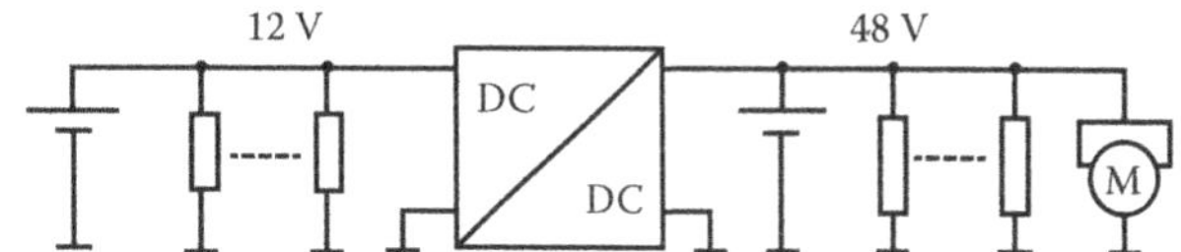

FIGURE 11.6 Example of a dual-voltage (12-V/48-V) system.

most of the power on demand features, such as engine stop–start, electric power steering, and electric cooling fans. Once the supply base developed 48-V products and the economies of scale established, it will probably migrate into the entire automotive industry.

11.4 AGM BATTERY

The AGM battery emerges in the early 1980s as a solution for lead-based batteries to reduce weight and improve reliability on military vehicles [8]. For military vehicle applications, the specific energy and power requirements are very demanding, much more so than passenger vehicles and commercial applications. The AGM battery provides more robustness and less maintenance.

AGM batteries are designed and developed differently from flooded batteries in that an absorbent glass mat separator is used as the major acid storage rather than an acid reservoir in a flooded battery. The AGM separator holds and stabilizes the acid in a porous structure, which makes the battery spill-proof and easier to transport. It also gives much more packaging flexibility as the vertical battery orientation is no longer required due to its spill-proof nature. The AGM battery reduces redundant acid storage that reduces the battery weight. It is referred to as the acid-starved or acid-limited lead–acid battery.

AGM batteries are often referred to as VRLA (valve regulated lead–acid) batteries. The "valve-regulated" design of AGM batteries resolves the issue of "dry-out" during overcharge in flooded batteries, where hydrogen and oxygen gases generated from electrolysis during overcharge are free to escape and the acid concentration will increase after the batteries overcharged repeatedly. The flooded batteries are maintained by adding distilled water to the reservoir when the water level is low. In AGM batteries, the separator is partially saturated (usually around 95%), leaving room for the gases generated on one side to travel to the other side. Oxygen generated at the positive side can diffuse through the separator to the negative side and become converted back to water, as in Figure 11.7. This is referred as the gassing cycle inside AGM batteries, chemically described by Equations 11.4 and 11.6 [9]. Hydrogen, once generated as described in Equations 11.5 and 11.7, does not reconvert and is allowed to vent through the valve in an AGM battery.

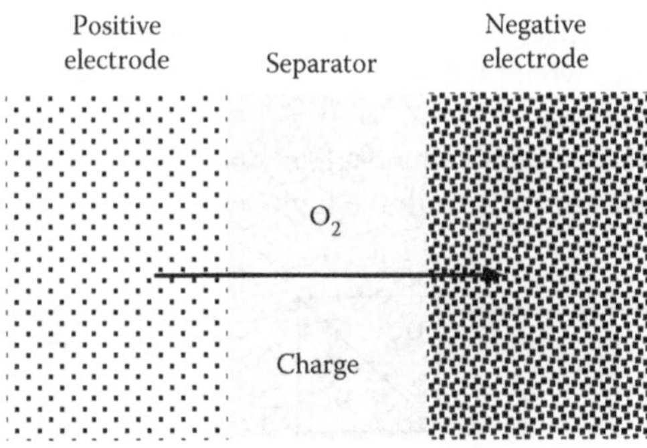

FIGURE 11.7 Oxygen transport during gassing cycle.

$$2H_2O \rightarrow O_2 + 4H^+ + 4e^- \quad (11.4)$$

$$H_2 \rightarrow 2H^+ + 2e^- \quad (11.5)$$

$$O_2 + 4H^+ + 4e^- \rightarrow 2H_2O \quad (11.6)$$

$$2H^+ + 2e^- \rightarrow H_2 \quad (11.7)$$

Another favorable feature of AGM batteries is that the separator stabilizes the electrolyte, which eliminates the possibility of acid stratification that occurs in flooded acid batteries. In flooded batteries, the acid in the reservoir stratifies in the vertical direction due to gravity, and the acid concentration in the bottom will be higher than in the top region [10]. The acid stratification results in different electrode utilization and more lead sulfate formation in the bottom that is difficult to recover. Lead sulfate formation is one of the major causes of failure of flooded batteries. In AGM batteries, acid is stabilized so the concentration is almost uniform in all directions, and the electrodes can be equally utilized without any lead sulfate "hard zone" to be recovered. This is one of the reasons why AGM batteries have longer lives than flooded batteries. Today, the lead AGM battery is the preferred battery technology for most European luxury cars owing to its superior performance characteristics in power, energy, and durability. It is viewed as an advanced battery technology strongly enhancing the images of many European luxury vehicle brands such as Mercedes, BMW, and Audi, etc. More in-depth discussions of AGM battery applications will be covered in Section 11.6.

11.5 LEAD–ACID BATTERY MATHEMATICAL MODEL

To model lead–acid batteries, there are usually two approaches: electrical equivalent circuit model and fundamental electrochemical model. The first approach is usually employed for control-oriented system simulation or on-board diagnosis purposes, which require computational efficiency, while the second is usually for battery-design purposes or thermal management simulation. In this section, only the electrochemical modeling development is reviewed, as it is more fundamental and material science based, and requires more effort to understand and achieve.

Shepherd [11] derived an equation describing the discharge process, which gave the cell voltage during discharge as a function of discharge time, current density, and other parameters. Schumacher-Grohn [12] followed the Shepherd equation and studied the charge process but using a different set of parameters. This work can be considered as one of the earliest theoretical approaches to predict the performance of a battery. In the Shepherd model, a set of coefficients were introduced for different terms composed of the voltage losses, which made the adjustment of model to experiments convenient. However, it was originally derived assuming all the model parameters were constant, which is far from reality. In Ref. [12], a completely separate set of parameters was used for charging and discharging. This leads to more freedom to adapt the model but also resulted in discontinuities when changing between charging and discharging. Moreover, the assumption employed that the structure is the same between charging and discharging also ignored the internal evolutions of geometric parameters.

Mathematical models [13–24] have been developed in one or two dimensions for flooded or VRLA batteries to understand the fundamentals inside the cells and capture the voltage behavior under different operating conditions. These models were developed physically from the basic conservation laws of mass, momentum, species, charge, and energy, and solved numerically such that the internal phenomena can be demonstrated and the voltage can be predicted. This is one of the most important merits of mathematical models over the Shepherd equation.

Newman and Tiedemann [25] first developed a porous electrode theory and applied it to model the discharge behavior of a lead–acid battery. Sunu [13] included the nonuniformity of acid concentration on the basis of the porous cell model, which was ignored by Newman and Tiedemann. Gu et al. [14] extended the models to predict cell behavior during discharge, rest, and charge cycles. These models were solved in one dimension across the thickness of a cell, and the acid stratification phenomenon along the height of the cell in a flooded battery was not accounted for.

Bernardi et al. [16] predicted the acid concentration and current density distribution along the cell height through a two-dimensional model. However, they failed to study the effect of stratification on performance.

Alavyoon et al. [15] studied the phenomenon of free convection and acid stratification in the reservoir and the corresponding effect on performance. This work, however, assumed a uniform current density across both electrodes, which actually decoupled the electrochemical kinetics and mass transport.

Gu et al. [19] coupled the electrochemical and transport processes and developed a comprehensive cell model for flooded battery on the basis of the porous electrode theory. This model was solved in a two-dimensional space and was capable of predicting the transients of acid stratification and their effect on performance during the charge and discharge processes. The internal information about the evolutions of porosity, and the state of charge during charge/discharge could also be demonstrated.

All of the aforementioned modeling work focused on the flooded lead–acid battery. Gu et al.'s model [19] has accomplished the fundamental modeling work of coupled electrochemical kinetics as well as mass and heat transfer physics. It can be used as a good performance predictor for a VRLA battery as well as a flooded battery when the internal gassing cycle, which is a unique feature of VRLA battery, does not occur. The internal gassing cycle only occurs after the charging voltage reaches a certain level before it consumes the charging current, so it is necessary to overcharge the VRLA battery to ensure a full charge.

Atlung and Zachau-Christiansen [17] first identified the incomplete charge of the negative plate due to the internal current of gassing and the resulting sulfation as a key failure mechanism of the VRLA battery. They also quantified the effect of the gassing cycle on recharge through a simple model, which is related to the saturation level in the AGM separator.

With the understanding of the major failure mechanisms in Atlung's model, Nelson et al. [26] studied the effect of charging algorithms on the cycle life of VRLA batteries, and concluded that high-finishing charging current could compensate for the insufficient charge caused by the gassing process.

Bernardi and Carpenter [18] developed a model of oxygen evolution and recombination in a lead–acid battery. However, infinitely rapid transport of oxygen across the separator was assumed so the transport process of oxygen through the separator and the effect of saturation in the separator on performance were ignored.

Huang and Nguyen [20,21] developed a two-dimensional transient un-isothermal model to account for the internal gassing cycle in VRLA batteries under an overcharge or a float charge condition. However, the capillary flow during the gassing cycle was neglected and it only focused on the thermal effects.

Newman and Tiedemann [22] proposed a lumped-parameter model to clarify the oxygen evolution and recombination mechanisms. Uniform current and concentration distribution were assumed in this model, so the effects of transport properties and separator design on the internal gas cycle were ignored.

Gu et al. [24] extended the model presented in Ref. [19] to include all gassing processes and accounted for the effects of the internal cycle on the recharge process. Owing to the partially saturated separator in a VRLA battery, the momentum equation was reduced to Darcy's law and the two-phase capillary flow was solved through the porous regions. As a result, the effects of separator saturation, transport properties, venting, and capillary flow on the internal gassing cycle and the charging process were all investigated. This model included primary and secondary reactions, plus all the phenomena affecting the performance behavior, such as acid stratification and the gassing process. In terms of the aging mechanism, this model was actually capable of simulating the water loss phenomenon due to gassing and the ensuing increased internal gassing cycle, and consequently incomplete charge influencing the cycle life. However, there was no other aging mechanism included.

After a variety of aging mechanism investigations in Refs. [17,27–30], aging models became a possible solution for predicting batteries' cycle life under different working cycles. Cherif et al. [31] proposed an empirical aging model based on the Shepherd equation. This work introduced the transient form of coefficients in the Shepherd equation, which were obtained by parameter identification in experiments. This model was one of the methods to estimate battery life. However, it bears the shortcomings of Shepherd's model, which is a simple estimation of cell voltage with incomplete physics in mass and heat transport. Cherif's model also failed to relate key physical parameters to the coefficients identified such as corrosion and sulfation, which are the major degradation mechanisms.

Shen et al. [32] presented a similar aging model, and the approach proposed was basically the same as Cherif's. Shen investigated the effect of different discharging currents on voltage behavior during the cycle life and showed the corresponding life curve. Similar to Cherif's work, the physics and the corresponding internal evolutions of parameters demonstrated were quite limited.

Schiffer et al. [33] proposed an empirical lifetime model that took into account the degradation caused by corrosion, acid stratification, gassing, and the sulfation mechanisms. This model employed a modified Shepherd's equation as its performance predictor, which included four terms with different physical meanings. This modified form introduced the effects of depth of discharge on the open-circuit potential and the impact of state of charge on the overpotential into Shepherd's equation. The performance predictor used in this model also included an aging model using a weighted ampere-hour (Ah) throughput approach. This model adequately showed the relationship between battery aging and the corresponding model parameters as well as the evolution of the intermediate parameters. However, because it was developed based on the Shepherd model, it was inherently lacking in the working principles of the lead–acid battery. It is insufficient for handling some basic phenomena, such as the coupling between electrochemical and transport processes. Furthermore, the set of coefficients of the modified Shepherd's equation has to be identified experimentally for each battery type, which makes the performance predictor heavily reliant on experiments.

Using the concept of a corrosion layer with low conductivity that grows during the lifetime of a battery in Ref. [33], Kim et al. [34] recently proposed to predict the corrosion layer growth and active mass loss by solving the Butler–Volmer equation of the corrosion reactions. This was the first time that the corrosion layer growth was theoretically predicted during the cycle life. However, in this work, corrosion alone was considered as a single aging mechanism to predict the variation of performance during cycle life, and it is insufficient to work as a full aging model.

To summarize, numerous fundamental performance models of lead–acid batteries have been developed over the years, and they are probably mature enough for some battery design and simulation work. However, there are many aging mechanism models that remain underdeveloped. The battery failure process is quite complicated and very sensitive to real-world cycling profiles and usage history. The modeling of the various aging processes still needs much more effort and research to make it more accurate and more useful in real-world applications.

11.6 ENVIRONMENTAL/"GREENHOUSE GAS" IMPACT

11.6.1 Impact of Government Regulations

As fuel economy and emissions regulations become more and more stringent, all automobile companies are developing technologies to comply with new regulations and to reduce fuel consumption and vehicle emissions. One of the most obvious and popular methods is the "stop–start" technology, which automatically turns the engine off during traffic stops. Upon the release of the brake pedal or other indications of traffic movement, the engine automatically starts again and the vehicle moves on. This is a standard feature on every HEV and it saves fuel during traffic stops, thereby producing a modest improvement on the fuel economy of the vehicle. A typical gain is about 3–4% during a US Federal Test Procedure drive cycle.

An HEV operates at hundreds of volts and its electric motor can start the engine quickly, providing almost a transparent stop–start operation to the driver. However, the cost of the high-voltage system is driving many carmakers to try to implement the stop–start feature at lower voltages such as 12, 42, or 48 V. Naturally, the lowest cost implementation will be the conventional 12-V system implementation. Of course, the stop–start cranking torque requirement of the engine is the determining factor of what voltage level one needs to use.

The MIT consortium did much research on the stop–start feature, and a consensus was reached that a 12-V implementation is possible for small engines (<2.0 L in displacement). Engines with large displacements (>2.0 L) should use 42 V or higher to ensure a seamless stop–start operation. This consensus was adhered to by some carmakers but not by the low-cost manufacturers. Many lower-cost producers are developing 12-V stop–start systems based on 12-V lead–acid batteries. This stop–start technology is driving the expansion of the usage of 12-V lead–acid batteries, both the "flooded" type and the "AGM" type.

11.6.2 Stop–Start Feature

One should consider the sequence of events during a stop–start operation to fully understand the impact of stop–start could have on the vehicle battery.

A typical driver would normally have the following vehicle systems operating when he/she comes to a stop at a traffic sign or traffic light: powertrain controller, body controller, chassis controller, climate controller, safety controller with all five subsystems' sensors and actuators, various lighting equipment, instrument cluster, radio and telematics systems, and various luxury features such as heated seats, heated steering wheel and rear defroster, etc.

When the engine stops, the alternator also stops and it does not produce any electrical output, so all the electrical loads mentioned above will be completely reliant on the vehicle battery. Depending on how much time is spent in the "engine off" mode, the battery discharge could be substantial and a deeper discharge will be detrimental to the cycle life of the lead–acid battery. In most cases, the stop event is short (seconds, not minutes) and the electrical loads are not very high during this "stop" period so the discharge is less than 1% or 2% of the capacity of the battery. This should not harm the lead–acid battery. However, in cases where the "stop" duration is long (minutes) and the vehicle electrical loads are high, the battery discharge could be as high as 5% of the capacity of the battery; this is considered a "deep" discharge, and a repetitive cycle of "deep" discharge will shorten the cycle life of a lead–acid battery. This is one of the reasons why many automobile manufacturers have switched from the flooded lead–acid battery technology to the AGM battery technology for their 12-V vehicles equipped with the stop–start feature, as the AGM battery was designed to handle the "deeper" cycling much better than the traditional flooded battery.

Today, the AGM battery is the preferred battery technology for most European luxury vehicles owing to its superior performance characteristics in power, energy, and durability. It is viewed as an advanced battery technology matching the images of many European luxury vehicle brands such as Mercedes, BMW, and Audi.

AGM batteries have lower internal resistance, as the ion migration resistance is lower since there is no acid reservoir. This lower electric resistance allows AGM batteries to deliver high current in a short pulse and to meet the high power requirement for starting events in stop–start applications. The lower resistance also means higher charge acceptance during the charging process. In stop–start applications, the battery experiences a high-rate discharge followed by a high-rate charge after every stop–start event. Therefore, the battery has to have high charge acceptance to recover the charge that has been released during the stop and the cranking event. AGM batteries can fulfill this duty cycle admirably.

Another advantage of AGM batteries is the deep discharge capability. As described earlier, a deep discharge could occur during stop–start events and AGM batteries are more robust than flooded batteries in terms of deep discharge. They can be deeply discharged down to very low state of charge (such as 40%) without losing the capability of restarting the vehicle, although control strategies for most vehicles with stop–start features would not allow batteries to discharge to such low levels.

The spill-proof and tight assembling feature also lead to a longer cycle life for AGM batteries in a hostile automotive environment on stop–start vehicles. The glass mat tightly holds the acid supply and all the electrode plates are compacted to make AGM batteries volume-efficient with improved longevity.

11.6.3 Voltage Fluctuation

Another major issue with the "12-V stop–start" system is the voltage fluctuation caused by the constant cranking of the engine. During engine cranking, the high current being delivered from the battery to the starter is in the hundreds of amperes. Large engines require as much as a thousand amperes for cranking, and this large current draw creates a voltage "dip" in the 12-V power system. This voltage dip could be as low as 7 or 8 V depending on the state of charge and the state of health of the battery (see Figure 11.8).

It is well known that the automotive 12-V system is designed to operate between 9 and 16 V, a voltage dip of less than 9 V is outside the normal operating window of the 12-V system, and many electronic modules or subsystems on the vehicle will react to this voltage dip as a system fault. Some systems will go into a default mode, some will shut itself down, and some will go through a rebooting and resetting process. All these reactions create a major disturbance and discontinuity of system operation to the driver and are completely unacceptable. To minimize or eliminate the voltage dip and to maintain the stability of the 12-V electrical system, many methods have been used. These methods include the use of DC-to-DC converters, capacitors, and a second supplemental battery. The most effective method is the use of a second supplemental battery.

The fundamental idea is to use any of these devices (DC-to-DC converter, capacitor or second battery) to power the voltage-sensitive loads during the cranking event, so the battery voltage on the cranking circuit would not affect the voltage of the power system feeding these sensitive loads.

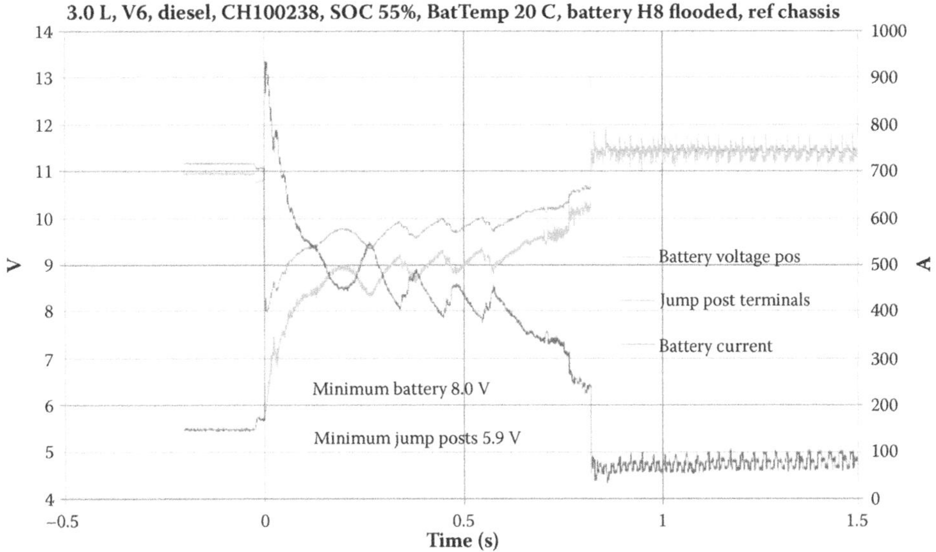

FIGURE 11.8 Typical battery voltage (blue) and battery current (red) during an engine crank event. (Courtesy of Chrysler Group LLC.)

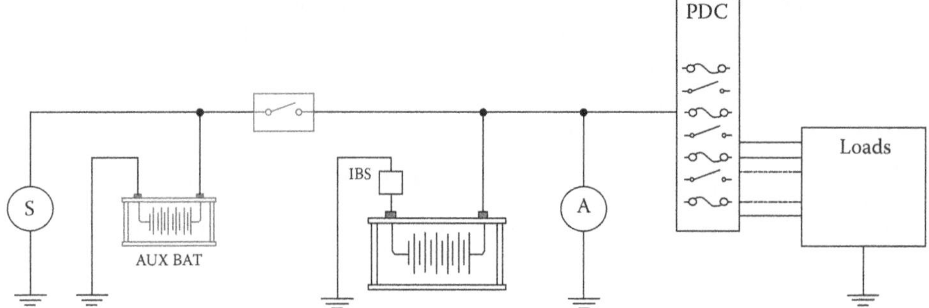

FIGURE 11.9 Dual 12-V battery system for stop–start.

All these designs work to a certain extent; however, the most effective system design is a "dual-battery" system design with two 12-V lead–acid or AGM batteries with their associated controls. Figure 11.9 depicts one of the configurations of a dual battery system design.

Figure 11.9 shows an auxiliary battery (AUX BAT) connected to the starter (S), and the vehicle loads connected to the alternator and the main battery with an IBS (intelligent battery sensor). The "relay" that connects the two batteries is closed during normal vehicle operation, so both batteries are powering the vehicle. During a stop–start, when the engine is ready to crank, the relay opens and separates the auxiliary battery from the main battery. This allows the auxiliary battery voltage to dip without affecting the voltage of the main battery. After the cranking and starting, the relay closes again and both batteries operate as one. The alternator charges both batteries as the relay is closed during engine operations. This is the fundamental design of a very effective and simple dual 12-V battery system for stop–start. Of course, the sizes and the capacities of the two batteries are dependent on each vehicle's requirements for engine cranking, vehicle electrical loads and features, as well as ignition-off draw design requirements.

The emergence of the low-cost 12-V stop–start systems in many vehicles has increased and expanded the usage of lead–acid batteries especially with the design of "dual 12-V battery systems." This will most likely remain to be the principal fuel-savings technology of many automakers for years to come.

11.7 IMPROVED LEAD–ACID BATTERIES WITH CARBON ADDITIVES

Much effort has been expended over the last two decades to improve the cycle life of VRLA batteries under high-rate partial state-of-charge cycling conditions. In this section, the new development of carbon additives in battery electrodes will be discussed.

Shiomi et al. [35] reported that increasing the amount of carbon in the negative electrodes of VRLA batteries reduces the $PbSO_4$ accumulation and extends the life of the battery, based on the cycle life test of the operational pattern of HEVs. Saez et al. [36] and Boden et al. [37] investigated several different carbon additive candidates, and carbon black was shown to work most effectively [36]. The mechanism by which carbon additives can help extend cycle life was explained in the work done by Moseley [38] and Bullock [39]. First, the carbon particles constitute an electric conductive network that improves the conductivity of negative electrodes. Second, carbon particles provide the nucleus on which the lead sulfate can land so that the lead sulfate particles will be smaller and easier to recover. Third, the carbon additives in the electrode may form electric double layers that support the high-rate charge/discharge events and relieve high-rate reactions for the lead electrode.

Although carbon additives can extend the cycle life of VRLA batteries to some degree, the improvement is very limited. A new PbO_2–carbon (PbC) hybrid design of the battery was brought into market in recent years. This new battery replaces the negative lead electrode with an AC (activated carbon) electrode, so the sulfation problem in the negative electrode is completely eliminated. It is claimed that the necessary components of this new battery are designed around the standard lead–acid batteries and assembled with the same practices as conventional lead–acid batteries. Therefore, the cost is similar to that of conventional lead–acid batteries. It is also claimed that PbC batteries can achieve more than 1500 cycles with 100% depth of discharge, as opposed to the 200–300 cycles for conventional batteries, all because it completely removes the failure mechanism of lead sulfation. However, the AC electrode adopted in the new hybrid battery has a lower overpotential of hydrogen evolution especially in acidic electrolytes, so hydrogen gassing would be the most challenging issue for the hybrid PbC battery. Because the negative electrode is purely an AC capacitor electrode, the PbC batteries inherit partially the energy density characteristics of supercapacitors. The available energy density is only about 20–30 Wh/kg, lower than conventional VRLA batteries. Also, the AC electrode has a higher self-discharge rate similar to supercapacitors.

The "UltraBattery" [40–43] uses a half-carbon half-lead hybrid negative plate to try to overcome the drawbacks mentioned above. Since a high-rate charge or discharge generates a high overpotential at the negative electrode, the severe gassing process occurring at the negative electrode reduces the charging efficiency and leaves the electrode insufficiently charged. "UltraBatteries" are designed to effectively reduce the high currents applied on the negative lead electrodes because the half-carbon capacitor plates are connected in parallel with the lead plates and they share the high input currents. During high-rate partial state of charge cycling operations, part of the high load current is absorbed and released by the carbon capacitor electrode during every discharge or charge pulse, which effectively protects the lead electrode from sulfation and extends the cycle life of the UltraBattery.

It was reported [40,41] that the cycle life of UltraBatteries is much longer than that of conventional batteries. The initial performance characteristics of prototype UltraBatteries, such as capacity, power, cold cranking, and self-discharge have been evaluated based on the US FreedomCAR Battery Test Manual (DOE/ID-11069, October 2003) [41]. Results show that the UltraBatteries meet or exceed respective targets of power, available energy, cold cranking, and self-discharge set for both minimum and maximum power-assist HEVs [41]. Three different cycling profiles have been tested, and the cycling performance was compared between UltraBatteries and Ni–MH batteries. The results showed that the cycle life of UltraBatteries was at least four times greater than that of lead–acid batteries and comparable to that of Ni–MH batteries.

As for the manufacturing of UltraBatteries, it is not trivial to install a carbon capacitor electrode in a lead–acid battery because the capacitor electrode has to accommodate the acidic environment and the working potential window. To develop an effective PbC battery, the carbon material has to be properly selected and the electrode has to be carefully designed. The objective of the following sections is to provide a literature review on studies and approaches conducted to achieve a large capacity and to inhibit the gassing phenomena for carbon-based capacitor electrodes. The first section describes the capacitance performance of two carbon candidates that have been mostly employed in supercapacitors: AC and carbon nanotubes (CNTs). The second section summarizes the methods to suppress hydrogen evolution reaction (HER) and to promote oxygen reduction reaction (ORR) at the carbon electrodes.

11.7.1 Capacity of Carbon-Based Capacitor Electrode

Carbon with a superspecific area, such as AC or CNTs, is widely considered as a candidate electrode material for energy storage devices with capacitor electrodes. The capacity of such an electrode, if the working potential window is assumed to be fixed, will depend on the capacitance of the fabricated carbon electrode. Therefore, specific capacitance is one of the most important parameters to be considered when carbon materials are selected for capacitor electrodes.

11.7.1.1 Activated Carbon

Usually, AC is prepared from carbon-rich organic precursor by a thermal method (dry distillation) to form carbonized organic precursors, which can be activated to increase pore volume either thermally or chemically [44], described briefly below:

1. Thermal method: treatment at 700–1000°C in the presence of oxidizing gases such as steam, CO_2, steam/CO_2 mixtures, or air.
2. Chemical method: treatment at 500–800°C in the presence of dehydrating substances such as $ZnCl_2$, H_3PO_4, or KOH, which are leached out afterward.

AC material samples produced by steam originated from coconut shells, phenolic resin, or coal exhibit a specific surface area of 1860–2120 m^2/g depending on the activation time [45]. Wang et al. [46] achieved 160 F/g and 1180 m^2/g by using petroleum coke, KOH, and vapor etching plus $FeCl_3$ catalysis. Wu et al. [47] prepared AC by placing firewood in a sealed oven at a temperature of 500°C in an oxygen-deficient environment. Specific capacitance up to 120 F/g was obtained. Qiao and Mochida [48] used petroleum coke for porous carbon preparation by KOH activation, which led to a specific surface area of 400–2900 m^2/g; however, the specific capacitance was as low as 44 F/g. In Ref. [49], various coal and pitch-derived carbonaceous materials were activated for 5 h at a temperature of 800°C. A specific surface area in the range of 1900–3200 m^2/g and specific capacitance of 200–320 F/g was obtained. Ref. [50] shows the effect of drying of carbon electrode on specific capacitance. The reasons could be the reduced weight and the improved accessible carbon surface occupied by moisture. The AC nanofiber web in Ref. [51] manifested a specific surface area in the range of 500–1220 m^2/g. The specific capacitance was found to be in the range of 35–202 F/g depending on the steam activation temperature. AC produced by leaching metals from TiC and ZrC [52,53], with highly tailorable specific capacitance and surface area at different synthesized temperature, was reported. Specific capacitance and specific surface area can be adjusted in the range of 80–180 F/g and 800–2000 m^2/g, respectively. To summarize, the available data to date for specific surface area and specific capacitance mostly fall in the range of 400–3000 m^2/g and 30–300 F/g.

11.7.1.2 Carbon Nanotubes

Much research and development have focused on CNTs as supercapacitor electrodes. For multiwalled CNTs (MWCNTs), specific capacitance in the range of 4–135 F/g was cited in Refs. [54,55]. For single-walled CNTs (SWCNT), a maximum specific capacitance of 180 F/g in KOH electrolyte was reported in Ref. [55].

The enhancement of specific capacitance of CNTs has been achieved by the addition of conducting polypyrrole [56,57] (170 F/g) and electrochemical oxidation in Refs. [58,59] (335 F/g). Also, the wettability of CNTs has a large impact on specific capacitance of CNTs. In Ref. [60], activated CNTs were obtained by using KOH as the activating agent, resulting in an increase in the specific surface area and specific capacitance. Honda et al. [61] reported that vertically aligned MWCNTs can provide a discharge capacity of 10–15 F/g even at an extremely high current density of 200 A/g, much higher than that of AC and very useful for high-power applications.

Addition of CNTs to AC instead of acetylene black or graphite powder for electrodes resulted in even higher specific capacitance [62,63]. Ref. [63] proved this by showing an increase of specific capacitance from 130 to 180 F/g. Fernandez et al. [64] studied the impact of the amount of SWCNT on specific capacitance and hydrogen evolution. Its results indicated that although SWCNT increased specific capacitance, it also increased the hydrogen reaction exchange current density and therefore promoted hydrogen gassing.

11.7.2 Gassing Evolution on Carbon Electrode

Carbon electrodes in energy-storage devices with aqueous electrolyte can easily cause gassing evolution if the working potential window is inappropriate. For electrochemical supercapacitors, the allowed working potential windows of electrodes, within which no gassing or negligible gas production can be observed, are usually characterized first to determine the achievable cell voltage. For PbC batteries, however, the potential window of negative carbon electrode is limited because the potentials of a positive electrode $PbO_2/PbSO_4$ and the voltage of the entire cell, usually around 1.7 V versus SHE (H_2SO_4, 5 M) and 2 V, have to be accommodated. Therefore, suitable material candidates and proper preparation methods for the carbon electrodes in PbC batteries need to be thoroughly studied, to suppress hydrogen evolution and to promote oxygen reduction in the specified potential window.

The gassing rate is affected by two properties of an electrode: surface area and surface activity. In terms of surface area, one must lower the specific area without sacrificing too much capacitance. The surface activity is usually referred to as kinetic parameters, such as the exchange current density of HER or the overpotential at the occurrence of significant hydrogen evolution in the negative electrodes.

Ref. [65] compared pyrolytic graphite with two orientations (pyrolytic edge and pyrolytic face) and vitreous carbon in promoting HER during charge when used as negative electrodes. Exchange current densities (i_0) of HER on these three electrodes were measured to indicate activities of producing hydrogen gas.

A higher exchange current density (i_0) means that the material would produce more hydrogen gas during charge. The result for these i_0 measurements was that "pyrolytic edge" (i_0 = 8.8E−7A/cm^2) was approximately the same as "vitreous carbon" (i_0 = 6.8E−7A/cm^2) and both of them were higher than "pyrolytic face" (i_0 = 5.8E−8A/cm^2) in HER. Also, a deactivation treatment process preceded each test for comparison. The electrodes were cathodically charged with hydrogen in a molar solution of perchloric acid at I = 0.2A. The measurement showed i_0 was reduced after the treatment. For example, i_0 dropped from 8.8E−7 to 4.7E−8 A/cm^2 for pyrolytic edge graphite.

The results showed that the surface activities of electrode materials in promoting HER were affected by their microstructure and could be reduced by some electrochemical deactivation processes. These results provided a magnitude of HER i_0 for carbon electrodes. The i_0 of HER for all carbon electrodes in acid electrolyte was suggested as 6.25E−8 A/cm^2 in Ref. [66].

The effects of adding antimony and phosphoric acid into the sulfuric acid electrolyte on hydrogen evolution reaction kinetics in lead–acid batteries were studied in Ref. [67]. Antimony (Sb) was confirmed to have an effect on promoting hydrogen evolution, and the addition of phosphoric acid turned out to have a suppressing effect on hydrogen evolution. Two sulfides, K_2S and FeS, were employed as additives into the electrolyte and electrode, respectively, to suppress hydrogen evolution in Fe/C air batteries in Ref. [68]. The results showed that both K_2S and FeS significantly suppressed hydrogen evolution in the Fe/C electrode during cathodic charging.

11.7.3 Lead–Carbon Summary

To make PbC batteries competitive with conventional lead–acid batteries, a higher specific energy must be achieved. The specific capacitance of the carbon capacitor electrodes largely determine the specific energy of the supercapacitors as well as the PbC batteries where the electrodes are used. AC is selected in most hybrid batteries because of its low cost and large specific capacitance up to 413 F/g, although the number is usually much lower in practice. Activated CNTs can reach up to 180 F/g (electrode) as reported. They have higher conductivities, which is beneficial for making electrodes with high specific power. Both AC and CNTs have superspecific areas that promote HERs during charging, and they must be properly prepared to suppress these reactions to ensure viability.

11.8 OTHER BATTERY CHEMISTRIES FOR 12-V STOP–START

While there is much work being done to improve the lead-based battery chemistry for stop–start, there are several other battery chemistries under consideration by various automobile companies for the 12-V stop–start system. The most notable chemistries are lithium-ion, nickel–metal hydride, and nickel–zinc.

Both lithium-ion batteries and nickel–metal hydride (Ni–MH) batteries are basically offshoots of development of high-voltage batteries for HEVs or pure EVs, so there is no need to discuss them here. However, the nickel–zinc battery deserves mentioning and some attention.

The nickel–zinc battery has been in existence since the 1900s; it was studied and improved periodically in the past century every few decades—1930s, 1960s, 1980s, and again in the 2010s [69]. The main drawback has been its limited cycle life capability. A more recent design from a company in California has been tested by several automotive companies using a 12-V stop–start test cycle with favorable results. An example of the 12-V stop–start test cycle can be seen in Figure 11.10 [70].

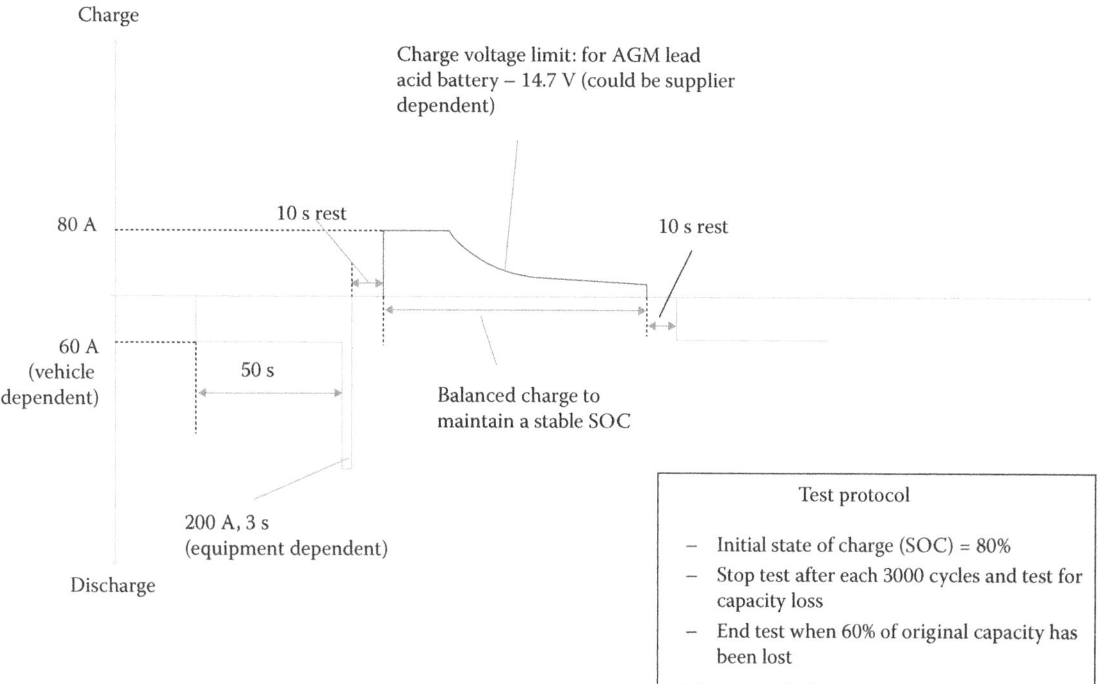

FIGURE 11.10 12-V stop–start test cycle profile. (From United States Council for Automotive Research (USCAR)—Electrical Power Systems Working Group (EPSWG), 12V stop start battery test cycle profile, May 2013.)

Various batteries could be subjected to the test cycle above and the results compared to determine which chemistry is best for the 12-V stop–start system. The latest nickel–zinc battery has performed very well in this test. This evaluation work is still ongoing in many automotive companies for the next-generation design, while almost all 12-V stop–start systems in production today have selected the lead AGM battery as their main battery of choice.

11.9 CONCLUSIONS

The lead–acid battery has served the automotive industry well for close to a century and will continue to do so although there are new power and energy requirements for many new vehicle features. The maturity, cost-effectiveness, and massive supply and recycling infrastructure of the lead–acid battery industry is unmatched by any other battery technologies attempting to enter the automotive market.

With the advent of HEVs, EVs, and PHEVs, there has been major development in the lighter-weight electrochemistries such as Ni–MH and lithium ion. However, these are high-voltage (hundreds of volts) systems where power and energy density is much more important than the conventional 12-V systems. There have been numerous market studies conducted by various entities trying to estimate the size of the market for these high-voltage systems. The conclusions have been consistent that it will be a "niche" market that could range between 10% and 25% of the global automotive market in 20 years. This means that the 12-V system will continue to be dominant for the majority of vehicles in the next two decades.

The newly standardized 48-V system could become more widespread; however, it is unlikely that it could overtake the conventional 12-V systems in popularity. It is highly likely that the 48-V architecture would only be used on high-end luxury vehicles. The 48-V batteries under development are mostly lithium based, but there will still be a 12-V system on board all 48-V vehicles, which means that a 12-V battery will still be utilized.

The "nickel–zinc" battery is readying itself again to compete with the lead–acid battery in the 12-V automotive market; however, much of its success depends on its cost-effectiveness and its performance. The jury is still out on this electrochemistry. If successful, it could play an important role in some automobile manufacturers' product lines.

The lead–acid battery industry has embarked on much research and development in the last two decades to improve the lead-based battery. This effort has resulted in the lead AGM battery, the "enhanced flooded" lead–acid battery, and the lead–carbon battery. These products will serve the majority of the automotive world in the next several decades.

REFERENCES

1. M. Barak, A. Wheaton. *Electrochemical Power Sources—Primary & Secondary Batteries*, Chapter 1.1.5, Institution of Electrical Engineers, Exeter, England, 1980.
2. Model T, 1908. Available at: www.history.com/topics/model-t.
3. R.V. Biagetti, M.C. Weeks, Tetrabasic lead sulfate as a paste material for positive plates. *Bell Syst. Tech. J.*, 49 (1970), 1305–1319.
4. R. Janakiraman, P.G. Balakrishnan, M. Devasahayam, S. Palanichamy, Lead–calcium alloy for lead-acid battery—Effect of additive on cycling. *Electrochemistry*, 4, 563–564, 1988.
5. J. Kassakian, T. Keim, *Proceedings—MIT/Industry Consortium on Advanced Automotive Electrical/Electronic Components and Systems—Program Review Meeting*, MIT School of Engineering Laboratory for Electromagnetic and Electronic Systems, Stuttgart, Germany, September 23–24, 1999.
6. J. Kassakian, T. Keim, *Proceedings—MIT/Industry Consortium on Advanced Automotive Electrical/Electronic Components and Systems—Program Review Meeting*, MIT School of Engineering Laboratory for Electromagnetic and Electronic Systems, Greenville, SC, January 25–26, 2000.
7. W. Diem, *Wards Auto World Mag.*, February 2012 issue, p. 27.
8. BU-201a: Absorbent Glass Mat (AGM), 2011. Available at: http://batteryuniversity.com/learn/article/absorbent_glass_mat_agm.
9. W.B. Gu, G.Q. Wang, C.Y. Wang, Modeling of overcharge process of VRLA batteries. *J. Power Sources* 108 (2002), 174–184.
10. W.B. Gu, C.Y. Wang, Numerical modeling of coupled electrochemcial and transport process in lead–acid batteries. *J. Electrochem. Soc.* 144(6) (1997), 2053–2061.
11. C.M. Shepherd, Design of primary and secondary cells—An equation describing battery discharge. *J. Electrochem. Soc.* 112 (1965), 657–664.
12. J. Schumacher-Grohn, Digitale Simulation regenerativer elektrischer Energieversorgungssysteme, PhD thesis, University of Oldensburg, 1991. Available at http://www.insel.eu.
13. W.G. Sunu, Mathematical model for design of battery, in *Electrochemical Cell Design*, R.E. White, editor, p. 357, Plenum Press, New York, 1984.
14. H. Gu, T.V. Nguyen, A mathematical model of a lead–acid cell. *J. Electrochem. Soc.* 134 (1987), 2953–2960.
15. F. Alavyoon, A. Eklund, F.H. Bark, R.I. Karlsson, D. Simonsson, Theoretical and experimental studies of free convection and stratification of electrolyte in lead–acid cell during recharge. *Electrochim. Acta* 36 (1991), 2153–2164.
16. D.M. Bernardi, H. Gu, A.Y. Schoene, Two-dimensional mathematical model of a lead–acid cell. *J. Electrochem. Soc.* 140 (1993), 2250–2258.
17. S. Atlung, B. Zachau-Christiansen, Failure mode of the negative plate in recombinant lead–acid batteries. *J. Power Sources* 52 (1994), 201–209.
18. D.M. Bernardi, M.K. Carpenter, A mathematical model of the oxygen-recombination lead–acid cell. *J. Electrochem. Soc.* 142 (1995), 2631–2642.
19. W.B. Gu, C.Y. Wang, B.Y. Liaw, Numerical modeling of coupled electrochemical and transport processes in lead–acid batteries. *J. Electrochem. Soc.* 144 (1997), 2053–2061.
20. H. Gu, T.V. Nguyen, A two-dimensional transient thermal model for valve-regulated lead–acid batteries under overcharge. *J. Electrochem. Soc.* 144 (1997), 2062–2068.
21. H. Gu, T.V. Nguyen, A transient nonisothermal model for valve-regulated lead–acid batteries under float. *J. Electrochem. Soc.* 144 (1997), 2420–2426.
22. J. Newman, W. Tiedemann, Simulation of recombinant lead–acid batteries. *J. Electrochem. Soc.* 144 (1997), 3081–3091.
23. W.B. Gu, C.Y. Wang, B.Y. Liaw, The use of computer simulation in the evaluation of electric vehicle batteries. *J. Power Sources* 75 (1998), 151–161.
24. W.B. Gu, G.Q. Wang, C.Y. Wang, Modeling the overcharge process of VRLA batteries. *J. Power Sources* 108 (2002), 174–184.
25. J. Newman, W. Tiedemann, Porous-electrode theory with battery applications. *AIChE J.* 21 (1975), 25–41.
26. R.F. Nelson, E.D. Sexton, J.B. Olson, M. Keyser, A. Pesaran, Search for an optimized cyclic charging algorithm for valve-regulated lead–acid batteries. *J. Power Sources* 88 (2000), 44–52.
27. L.T. Lam, N.P. Haigh, C.G. Phyland, A.J. Urban, Failure mode of valve-regulated lead–acid batteries under high-rate partial-state-of-charge operation. *J. Power Sources* 133 (2004), 126–134.
28. H.A. Catherino, F.F. Feres, F. Trinidad, Sulfation in lead–acid batteries. *J. Power Sources* 129 (2004), 113–120.
29. M.L. Soria, F. Trinidad, J.M. Lacadena, A. Sanchez, J. Valenciano, Advanced valve-regulated lead–acid batteries for hybrid vehicle applications. *J. Power Sources* 168 (2007), 12–21.
30. R.J. Ball, R. Kurian, R. Evans, R. Stevens, Failure mechanisms in valve regulated lead/acid batteries for cyclic applications. *J. Power Sources* 109 (2002), 189–202.
31. A. Cherif, M. Jraidi, A. Dhouib, A battery ageing model used in stand alone PV systems. *J. Power Sources* 112 (2002), 49–53.
32. W.X. Shen, C.C. Chan, E.W.C. Lo, K.T. Chau, Estimation of battery available capacity under variable discharge currents. *J. Power Sources* 103 (2002), 180–187.
33. J. Schiffer, D.U. Sauer, H. Bindner, T. Cronin, P. Lundsager, R. Kaiser, Model prediction for ranking lead–acid batteries according to expected lifetime in renewable energy systems and autonomous power-supply systems. *J. Power Sources* 168 (2007), 66–78.
34. U.S. Kim, C.B. Shin, S.M. Chung, S.T. Kim, B.W. Cho, Modeling of the capacity loss of a 12 V automotive lead–acid battery due to ageing and comparison with measurement data. *J. Power Sources* 190 (2009), 184–188.
35. M. Shiomi, T. Funato, K. Nakamura, K. Takahashi, M. Tsubota, Effects of carbon in negative plates on cycle-life performance of valve-regulated lead/ acid batteries. *J. Power Sources* 64 (1997), 147–152.
36. F. Saez, B. Martinez, D. Marin, P. Spinelli, F. Trinidada, The influence of different negative expanders on the performance of VRLA single cells. *J. Power Sources* 95 (2001), 74–190.
37. D.P. Boden, D.V. Loosemore, M.A. Spence, T.D. Wojcinski, Optimization studies of carbon additives to negative active material for the purpose of extending the life of VRLA batteries in high-rate. *J. Power Sources* 195 (2010), 4470–4493.
38. P.T. Moseley, Consequences of including carbon in the negative plates of valve-regulated lead–acid batteries exposed to high-rate partial-state-of-charge operation. *J. Power Sources* 191 (2009), 134–138.
39. K.R. Bullock, Carbon reactions and effects on valve-regulated lead–acid (VRLA) battery cycle life in high-rate, partial state-of-charge cycling. *J. Power Sources* 195 (2010), 4513–4519.
40. L.T. Lam, R. Louey, Development of ultra-battery for hybrid-electric vehicle applications. *J. Power Sources* 158 (2006), 1140–1148.

41. L.T. Lam, R. Louey, N.P. Haigh, O.V. Lim, VRLA Ultrabattery for high-rate partial-state-of-charge operation. *J. Power Sources* 174 (2007), 16–29.
42. A. Coopera, J. Furakawa, L. Lam, M. Kellawayd, The UltraBattery—A new battery design for a new beginning in hybrid electric vehicle energy storage. *J. Power Sources* 188 (2009), 642–649.
43. J. Furukawa, T. Takada, D. Monma, L.T. Lam, Further demonstration of the VRLA-type UltraBattery under medium-HEV duty and development of the flooded-type UltraBattery for micro-HEV applications. *J. Power Sources* 195 (2010), 1241–1245.
44. R. Strobel, J. Garche, P.T. Moseley, L. Jörissen, Hydrogen storage by carbon materials. *J. Power Sources* 159 (2006), 781–801.
45. M. Nakamura, M. Nakanishi, K. Yamamoto, Influence of physical properties of activated carbons on characteristics of electric double-layer capacitors, *J. Power Sources* 60(2) (1996), 225.
46. X.F. Wang, D. Wang, J. Liang, Performance of electric double layer capacitors using active carbons prepared from petroleum coke by KOH and vapor re-etching, *J. Mater. Sci. Technol.* 19(3) (2003), 265.
47. F.C. Wu, R.L. Tseng, C.C. Hu, C.C. Wang, Physical and electrochemical characterization of activated carbons prepared from firwoods for supercapacitors, *J. Power Sources* 138 (2004), 351.
48. W. Qiao, S.H. Yoon, I. Mochida, KOH activation of needle coke to develop activated carbons for high-performance EDLC, *Energy Fuels* 20(4) (2006), 1680.
49. K. Kierzek, E. Frackowiak, G. Lota, G. Gryglewicz, J. Machnikowski, Electrochemical capacitors based on highly porous carbons prepared by KOH activation, *J. Electrochim. Acta* 49(4) (2004), 515.
50. G.J.W. Radford, J. Cox, R.G.A. Wills, F.C. Walsh, Electrochemical characterisation of activated carbon particles used in redox flow battery electrodes. *J. Power Sources* 185 (2008), 1499–1504.
51. C. Kim, Electrochemical characterization of electrospun activated carbon nanofibres as an electrode in supercapacitors, *J. Power Sources* 142(1/2) (2005), 382.
52. J. Chmiola, G. Yushin, R.K. Dash, E.N. Hoffman, J.E. Fischer, M.W. Barsoum, Y. Gogotsi, Double-layer capacitance of carbide derived carbons in sulfuric acid. *Electrochem. Solid State Lett.* 8(7) (2005), A357–A360.
53. J. Chmiola, G. Yushin, R. Dash, Y. Gogotsi, Effect of pore size and surface area of carbide derived carbons on specific capacitance. *J. Power Sources* 158 (2006), 765.
54. E. Frackowiak, K. Metenier, V. Bertagna, F. Beguin, Supercapacitor electrodes from multiwalled carbon nanotubes, *Appl. Phys. Lett.* 77(15) (2000), 2421.
55. E. Frackowiak, K. Jurewicz, S. Delpeux, F. Beguin, Nanotubular materials for supercapacitors, *J. Power Sources* 97 (2001), 822.
56. E. Frackowiak, K. Jurewicz, K. Szoztak, S. Delpeux, F. Beguin, Nanotubular materials as electrodes for supercapacitors, *Fuel Process. Technol.* 77 (2002), 213.
57. E. Frackowiak, F. Beguin, Electrochemical storage of energy in carbon nanotubes and nanostructured carbons, *Carbon* 40(10) (2002), 1775.
58. J.S. Ye, X. Liu, H.F. Cui, W.D. Zhang, F.S. Sheu, T.M. Lim, Electrochemical oxidation of multi-walled carbon nanotubes and its application to electrochemical double layer capacitors, *Electrochem. Commun.* 7(3) (2005), 249.
59. C.G. Liu, H.T. Fang, F. Li, M. Liu, H.M. Cheng, Single-walled carbon nanotubes modified by electrochemical treatment for application in electrochemical capacitors, *J. Power Sources* 160 (2006), 758.
60. Q. Jiang, X.Y. Lu, Y. Zhao, X.M. Ren, L.J. Song, Effects of the Activating Agent Dosage on the Electrochemical Capacitance of Activated Carbon Nanotubes, *Mater* 21(5) (2006), 1253.
61. Y. Honda, T. Haramato, M. Takeshige, H. Shiozaki, T. Kitamura, M. Ishikawa, Aligned MWCNT sheet electrodes prepared by transfer methodology providing high-power capacitor performance, *Electrochem. Solid State Lett.* 10(4) (2007), A106.
62. Y. Show, K. Imaizumi, Decrease in equivalent series resistance of electric double-layer capacitor by addition of carbon nanotube into the activated carbon electrode, *Diamond Relat. Mater.* 15(11/12) (2006), 2086.
63. C. Portet, P.L. Taberna, P. Simon, E. Flahaut, Influence of carbon nanotubes addition on carbon–carbon supercapacitor performances in organic electrolyte, *J. Power Sources* 139(1/2) (2005), 371.
64. P.S. Fernandez, E.B. Castro, S.G. Real, M.E. Martins, Electrochemical behaviour of single walled carbon nanotubes—Hydrogen storage and hydrogen evolution reaction. *Int. J. Hydrogen Energy* 34 (2009), 8115–8126.
65. M.P.J. Brennan, O.R. Brown, Carbon electrodes: Part 1. Hydrogen evolution in acidic solution. *J. Appl. Electrochem.* 2 (1972), 43–49.
66. B. Pillay, J. Newman, The influence of side reactions on the performance of electrochemical double-layer capacitors. *J. Electrochem. Soc.* 143(6) (1996), 1806–1814.
67. S. Venugopalan, Kinetics of hydrogen-evolution reaction on lead and lead-alloy electrodes in sulfuric acid electrolyte with phosphoric acid and antimony additives. *J. Power Sources* 48 (1994), 371–384.
68. B.T. Hang, T. Watanabe, M. Egashira, I. Watanabe, S. Okada, J. Yamaki, The effect of additives on the electrochemical properties of Fe/C composite for Fe/air battery anode. *J. Power Sources* 155 (2006), 461–469.
69. U. Falk, A. Wheaton, *Electrochemical Power Sources—Primary & Secondary Batteries*, Chapter 5.6, Institution of Electrical Engineers, Exeter, England, 1980.
70. United States Council for Automotive Research (USCAR)—Electrical Power Systems Working Group (EPSWG), 12V stop start battery test cycle profile, May 2013.

Section V

Advanced Materials and Technologies for Fuel Cells

12 Advanced Materials for High-Temperature Solid Oxide Fuel Cells (SOFCs)

Kuan-Zong Fung

CONTENTS

12.1 INTRODUCTION

Among several types of fuel cells, the solid oxide fuel cell (SOFC) has shown advantages, such as high conversion efficiency, no need for noble metal catalysts, use of hydrocarbon fuel, no liquid in the fuel cell, etc. In recent years, a total of 10 MW Bloom Energy servers (100 kW per server) based on SOFC technology have been installed in California and North Carolina in the United States. The power generation of a 700-W unit of "Ene-Farm Type S," which realized a power generation efficiency of 46.5% (net AC, LHV) as a commercially available residential-use fuel cell system, was developed jointly by Osaka Gas, Aisin Seiki, Kyocera, and Toyota Motor. Although the cost of an SOFC unit is still high, the future looks promising for SOFCs.

The major components of an SOFC are the solid electrolyte, the anode, and the cathode (see Figures 12.1 and 12.2). In addition, an interconnect is also needed when cells are arranged in series as a stack. Since each component is operated under a different environment, it must provide a different function and meet special requirements. For example, a solid electrolyte and an interconnect need to provide ionic conduction and electronic conduction, respectively. Both of them are separated by the fuel and the oxidant. Thus, they need to have adequate stability against oxidizing and reducing atmospheres. On the other hand, cathode and anode need to have ionic–electronic conduction in addition to adequate stability in either an oxidizing or a reducing environment.

Fundamentally, the conductivity and stability of functional materials is strongly affected by their crystal structure and/or atomic arrangements. Therefore, in this chapter, some of the materials used in SOFCs are categorized on the basis of their unique crystal structures.

12.2 SOLID ELECTROLYTES

Solid electrolytes, namely solid-state ionic conductor, are materials that exhibit high ionic conduction (as high as 1 S cm^{-1}) with negligible electronic conduction. The mobile ions involved in the electrochemical reaction of SOFCs are either oxygen ions (O^{2-}) and/or protons (H^+). Since ionic conduction in an ionic conductor is highly dependent on its defect concentration and crystals structure, oxygen-ion conductors used for SOFC electrolytes commonly have similar crystal structures, mainly fluorite- and perovskite-type structures.

12.2.1 Oxygen-Ion Conductors with Fluorite-Type Structure

The cubic fluorite structure consists of relatively large oxygen ions that form 8-fold coordination. Every other cube is formed by the simple cubic packing of the oxygen ions having a cation at its center. On the basis of Pauling's rules, the radius ratio R_{cation}/R_{anion} needs to be greater than 0.73. Oxides with CaF_2 structure such as ZrO_2, CeO_2, and Bi_2O_3 have been extensively studied for the application.

12.2.1.1 Doped Zirconia or Zirconium Oxide

ZrO_2 with proper dopant or stabilizer is the most common electrolyte used for SOFC applications. However, ZrO_2 in its pure form cannot be used as a proper electrolyte because of its structural instability and low ionic conductivity. At temperatures from a melting point of 2680°C to 2370°C, undoped ZrO_2 shows a cubic (c) structure. With further cooling down to 2370°C, the cubic phase will change to a tetragonal (t) form with slight distortion. As the temperature reaches 1170°C and below, the tetragonal ZrO_2 exhibits a martensitic transformation into a monoclinic (m) form. The tetragonal–monoclinic transformation

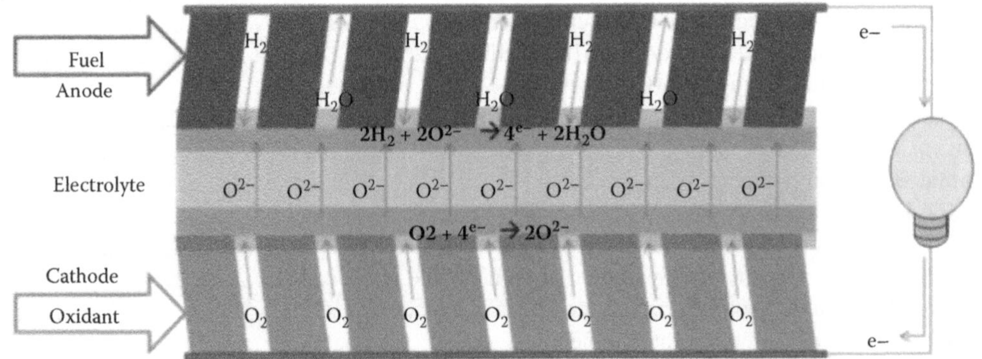

FIGURE 12.1 Schematic diagram of the operating concept of an SOFC.

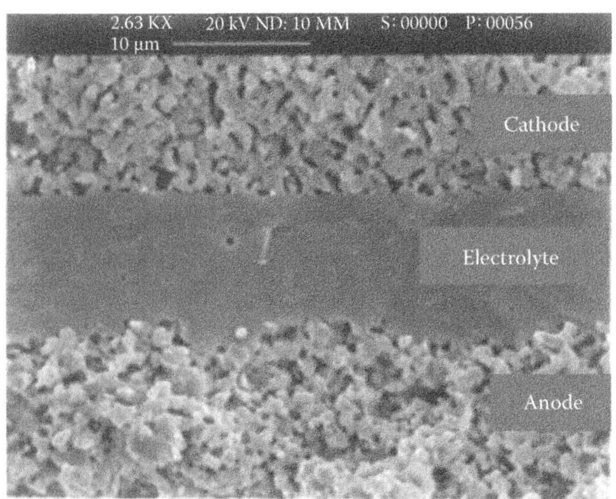

FIGURE 12.2 SOFC consists of two porous electrode layers and a dense electrolyte layer.

is accompanied with a large volume expansion (~5%) that may result in a catastrophic failure. Such a transformation may be rationalized by the undesired radius ratio of R_{Zr+4}/R_O of well below 0.732. Thus, destabilization of the fluorite-structure ZrO_2 is expected when the temperature is below 1170°C.

With proper addition of larger cations with lower valence, such as Y^{3+} and Ca^{2+}, not only the radius ratio, R_{cation}/R_{anion}, is greater than 0.73, but also the positive oxygen vacancies are also created due to compensation of YZr or CaZr. Consequently, the cubic phase of doped ZrO_2 may be stabilized to room temperature. In addition, the doped ZrO_2 shows enhanced conduction of oxygen ions. Thus, 8 mol% Y_2O_3–stabilized ZrO_2 (YSZ) is the most widely used electrolyte in SOFCs because YSZ exhibits adequate oxygen-ion conductivity (~0.02 S cm^{-1} at 800°C) as well as required stability in both oxidizing and reducing atmospheres. Other stabilizing oxides commonly used are CaO, MgO, and Sc_2O_3. All these oxides exhibit a relatively high solubility in ZrO_2 and form the fluorite structure. Among them, 10 mol% Sc_2O_3–stabilized ZrO_2 (10ScSZ) shows the highest ion conductivity of ~0.07 S cm^{-1} at 800°C [1]. Owing to its excellent chemical stability against reduction and adequate ionic conductivity at high temperatures, YSZ is still the most reliable and widely used electrolyte for SOFC applications.

12.2.1.2 Ceria or Cerium Oxide

Cerium oxide is another important electrolyte material that crystallizes in a fluorite-type structure. Unlike undoped ZrO_2, undoped CeO_2 shows a stable fluorite structure for a wide temperature range.

The main reason for CeO_2 to form a stable CaF_2 structure is because the ionic radius of Ce^{4+} is 0.97 Å [2], that is, more favorable than that of Zr^{4+} for 8-fold coordination of CaF_2-type structure. In undoped CeO_2, very limited intrinsic oxygen defects suppress its oxygen-ion conductivity in the range of 10^{-3}–10^{-4} S cm^{-1} at 800°C [3].

Similar to Y_2O_3 in ZrO_2, the addition of trivalent oxides (M_2O_3, M = Gd, Sm, or Y) into CeO_2 will induce the formation of positively charged oxygen vacancies due to the charge compensation mechanism through the following defect reaction:

$$M_2O_3 \xrightarrow[CeO_2]{} 2M'_{Ce} + 3O_O^x + V_O''$$

As a result, the addition of proper dopant(s) will significantly enhance the ionic conductivity of CeO_2.

For example, Gd_2O_3-doped CeO_2 and Sm_2O_3-doped CeO_2 (known as GDC and SDC) exhibit conductivity five times higher than that of YSZ at 800°C.

Although doped ceria shows much higher conductivities than YSZ, they tend to exhibit partially electronic conduction at low oxygen partial pressure or at temperature greater than 650°C. It is believed that the n-type semiconducting behavior of ceria is due to the presence of trivalent cerium ions (Ce^{3+}) that are charge compensated by the removal of oxygen ions from the anion sublattice under low $p(O_2)$. The reduction of Ce^{4+} to the Ce^{3+} at low $p(O_2)$ may be illustrated by the following equations:

$$O_O^x \rightarrow V_O^{\cdot\cdot} + 2e' + \frac{1}{2}O_2$$

$$2Ce_{Ce}^x + 2e' \rightarrow 2Ce'_{Ce}$$

Thus, the removal of oxygen ion will result in the reduction of two cerium ions from Ce^{4+} (shown as Ce_{Ce}^x) to Ce^{3+} (shown as Ce'_{Ce}).

Therefore, when ceria is directly used as the solid electrolyte, a lower OCV (open circuit voltage) than the theoretical value is obtained, as seen in Figure 12.3.

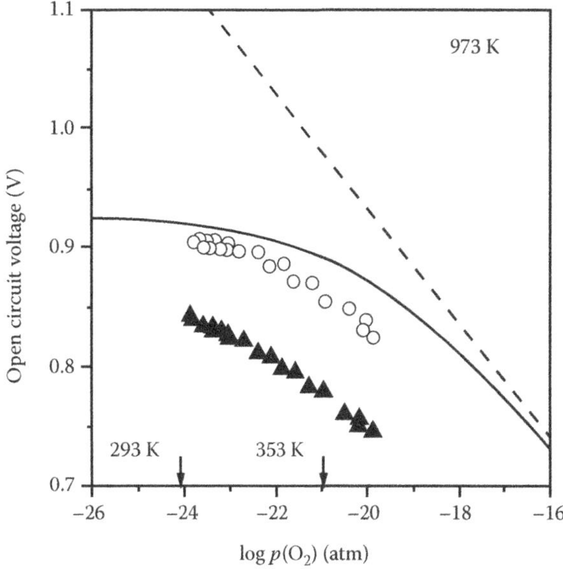

FIGURE 12.3 Open circuit voltage of cells A (○) and B (▲) as a function of $p(O_2)$ at 973 K. Electrolyte thickness: 1.193 mm (cell A) and 1.217 mm (cell B). Solid lines: theoretical OCV curves calculated by Matsui et al. Dotted lines: OCV curves calculated using Nernst equation. $p(O_2)$ values in pure H_2 humidified at 293 and 353 K are marked by arrows in the figure. (From T. Matsui, T. Kosaka, M. Inaba, A. Mineshige, and Z. Ogumi, *Solid State Ionics*, 176, 663, 2005.)

12.2.1.3 Bismuth Oxide

High-temperature bismuth oxide with cubic fluorite structure shows the highest oxygen-ion conductivity (~1 S cm^{-1} at 800°C) among many oxides. Bismuth oxide is known to exhibit two crystallographic polymorphs: the high-temperature cubic phase (δ) stable above ~730°C and the low-temperature monoclinic polymorph (α). In addition, two metastable phases have also been reported: tetragonal β-Bi_2O_3 and body-centered cubic γ-Bi_2O_3 [4]. For it to crystallize in the CaF_2-type structure (δ-Bi_2O_3), it is imperative that 25% of the oxygen sites are vacant. The high ionic conductivity of Bi_2O_3 in its cubic form (referred to as the δ-phase) has been attributed to the presence of such a large concentration of oxygen vacancies [4]. However, the δ to α phase transformation does not only result in significant volume change (reduction) but also show drastic conductivity decrease owing to more compact atomic arrangement.

Thus, the phase transformation of bismuth oxide needs to be avoided to prevent volume and conductivity changes. A cation with high valence or small radius can be used to stabilize δ-phase Bi_2O_3. Substituting the smaller cations (e.g., Y^{3+}, Dy^{3+}) in bismuth oxide results in a reduced lattice that inhibits diffusion of ions. Consequently, the high-temperature phase can be kinetically stabilized to lower temperature, although cubic to rhombohedral transformation may occur after annealing at ~600°C. Additionally, in some cases, high valence elements (e.g., Nb^{5+}, W^{6+}) as dopants for stabilizing the cubic structure are used, which results in the loss of oxygen vacancy and enhanced stability of the structure.

To further improve the conductivity of bismuth oxide, the content of dopant in fluorite structure should be minimized. However, degradation in phase stability may occur. Several doubly doped Bi_2O_3 electrolyte systems have been synthesized with the optimal dopant concentration. For example, Meng et al. showed that, with two dopants Y_2O_3 and $Pr_2O_{11/3}$, the cubic phase of bismuth oxide was stabilized with higher Bi content (88 mol%) than that of the single-dopant system (75 mol%) [5]. The reason for using two stabilizing dopants in Bi_2O_3 is to increase the entropies of the system. As the doping cations, vacancies and oxygen ions tend to arrange in a more disordered state, and the high-temperature δ phase is able to stabilize to lower temperature. More important, the doubly doped Bi_2O_3 electrolyte with higher Bi content shows higher conductivity than singly doped ones. Dy and W codoped Bi_2O_3 has shown conductivity as high as 0.569 S cm^{-1} at 700°C from Wachsman et al. [6]. Sr and Nb codoped Bi_2O_3 also shows a similar result with a conductivity of 0.3 S cm^{-1} at 700°C. The conductivity of stabilized bismuth oxide compared with different electrolyte materials is shown in Figure 12.4 [1,7].

12.2.1.3.1 *Stability against Reduction of Bismuth Oxide*

With such high conductivity, Bi_2O_3-based solid electrolytes are potential candidates for application in SOFCs. However, instability in a reducing atmosphere is the principal drawback of using them for SOFC electrolytes. After reduction in H_2 at ~400°C, the oxide electrolyte becomes decomposed because of the precipitation of metallic bismuth. The formation of Bi metal in spherical shape is due to its low melting point of 271°C (Figure 12.5).

For the enhancement of thermodynamic stability in a reducing atmosphere, it is essential that the activity of Bi_2O_3 is reduced [8]. The doping of high-valent cations such as Nb and W in Bi_2O_3 was found to effectively reduce the concentration of oxygen vacancy as well as enhance the stability of the

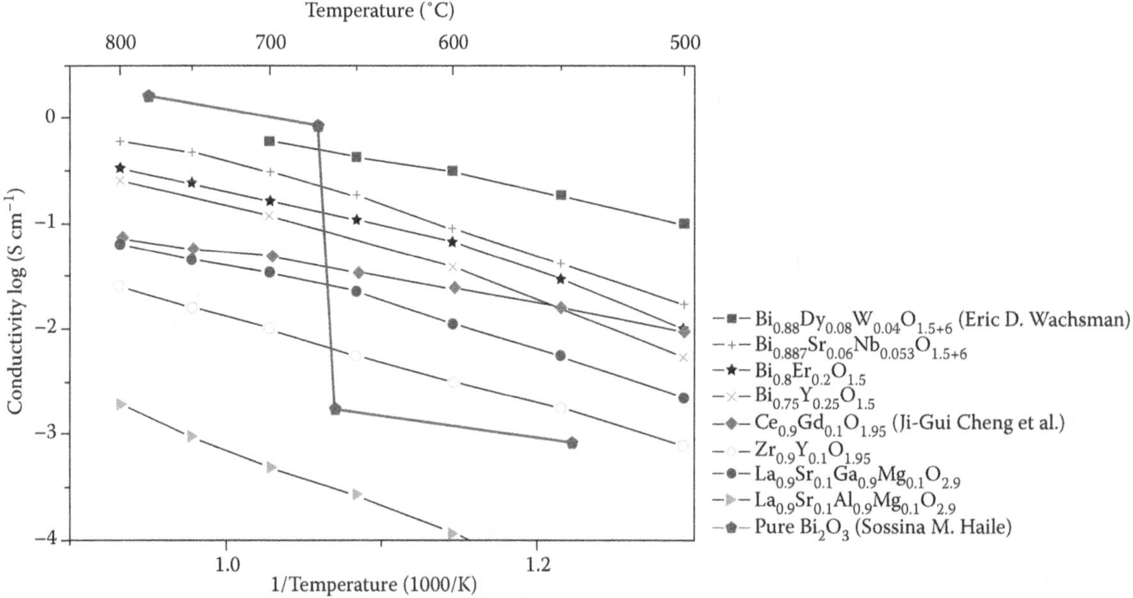

FIGURE 12.4 Conductivities of various types of oxygen conductors. (From S.M. Haile, *Acta Mater.*, 51, 5981, 2003; D.W. Jung, K.L. Duncan, and E.D. Wachsman, *Acta Mater.*, 58, 355, 2010; J.G. Cheng, S.W. Zha, J. Huang, X.Q. Liu, and G.Y. Meng, *Mater. Chem. Phys.*, 78, 791, 2003.)

FIGURE 12.5 SEM image of Bi_2O_3-based solid electrolytes after annealing in H_2 at ~400°C.

δ-phase. It is believed that W- or Nb-doped Bi_2O_3 also provides better stability against reduction in fuel environments.

12.2.1.3.2 Bilayered Electrolyte

As mentioned above, despite the high conductivity of doped Bi_2O_3 systems, they have as yet not been incorporated into SOFCs because of their low decomposition potential.

SOFC electrolytes with a bilayer design have been proposed and investigated to use highly conductive oxides with limited thermodynamic stability against low oxygen partial pressure [9]. In this design, a dense layer of a more stable electrolyte, such as doped CeO_2 or doped ZrO_2, is used as the electrolyte exposed to the fuel side. On the other hand, the less stable but highly conducting doped Bi_2O_3 is used as the electrolyte exposed to the air side. As a result, doped Bi_2O_3 is prevented from reduction to metallic Bi and allows the high flux of oxygen ions to be transported.

The results of Wachsman et al. demonstrate a functionally graded $Er_{0.2}Bi_{0.8}O_{1.5}$ (ESB)/$Sm_{0.2}Ce_{0.8}O_{1.9}$ (SDC) bilayer structure (Figure 12.6). The $p(O_2)$ at the ESB/SDC interface can be controlled to avoid ESB reduction by varying the thickness ratio of SDC and ESB layers [10]. In addition, Ce^{3+} and Ce^{4+} may coexist in cerium oxide electrolytes due to low $p(O_2)$ at the fuel side, resulting in mixed ionic and electronic conduction. The mixed conduction may reduce the activation polarization at the anode/ceria interface. Furthermore, electronic conduction can be blocked by the ESB layer and result in an acceptable OCV [10,11]. When the bilayer electrolyte is used for SOFC to operate at intermediate temperature, its low resistance contribution is beneficial to obtain high-performance cells with better reliability as shown in Figure 12.7.

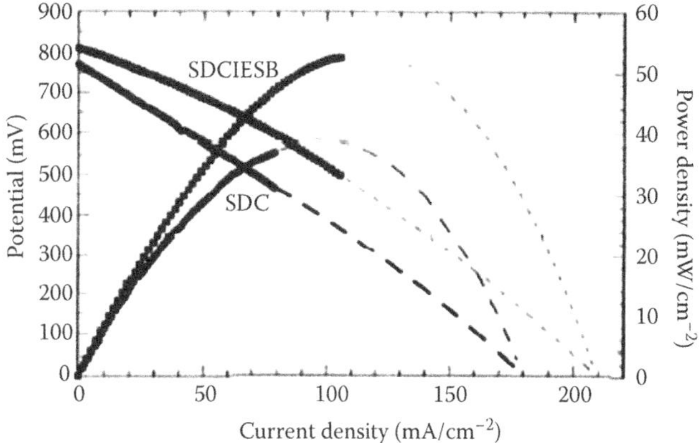

FIGURE 12.7 Comparison of cell performance of SDC and ESB/SDC electrolytes at 800°C. Cells are 0.8 mm thick with Pt–H_2/H_2O anode and Au–O_2 cathode. (From J.Y. Park and E.D. Wachsman, *Ionics*, 12, 15, 2006.)

12.2.2 Perovskite Structure

Another electrolyte system that received much attention and shows high conductivity and adequate stability is the perovskite-type oxide system. Perovskite oxides are usually formulated as ABO_3. "A" represents the large cation with ionic radius close to that of oxygen ions, such as La^{+3}, Sr^{+2}, etc. "B" represents the small cation with ionic radius that fits the octahedral sites that are surrounded by six close-packed oxygen ions. To enhance the ionic conductivity of perovskite oxides, aliovalent cations may be incorporated into both A-site and B-site cation sublattices. Consequently, oxygen vacancies are generated to compensate the charge of substituting ions.

Several oxides with perovskite structures have been studied as potential oxide-ion conductors [12,13]. Among them, $LaGaO_3$-based oxides have been investigated for their use as electrolyte materials for intermediate-temperature (<800°C) SOFCs [14].

$La_{0.8}Sr_{0.2}Ga_{0.8}Mg_{0.2}O_{2.8}$ (LSGM) is the most widely used perovskite electrolyte due to its high oxide-ion conductivity, 0.17 S cm^{-1} at 800°C, which is about 8.5 times as high as that of conventional YSZ electrolyte at the same temperature, over a wide range of oxygen partial pressure [15]. The enhanced

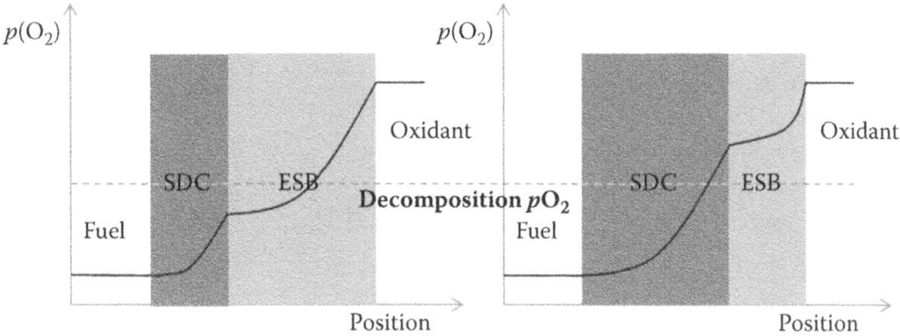

FIGURE 12.6 Conceptual representation of a bilayer electrolyte showing the effect of relative thickness on interfacial oxygen partial pressure, $p(O_2)$. (From J.Y. Park and E.D. Wachsman, *Ionics*, 12, 15, 2006.)

conductivity of LSGM can be explained by the following defect equation:

$$2MgO + La_2O_3 \xrightarrow[2LaGaO_3]{} 2Mg'_{Ga} + 2La^x_{La} + 5O^x_O + V^{\cdot\cdot}_O$$

$$2SrO + Ga_2O_3 \xrightarrow[2LaGaO_3]{} 2Sr'_{La} + 2Ga^x_{Ga} + 5O^x_O + V^{\cdot\cdot}_O$$

These reactions suggest that the substitution of Mg^{2+} for Ga^{3+} and Sr^{2+} for La^{3+} can increase the concentration of oxygen vacancies and enhance the ionic conductivity. However, the high cost of gallium compounds and their low mechanical strength are the main obstacles for using doped $LaGaO_3$ in SOFC application. Therefore, the replacement of Ga with an inexpensive element, such as Al, is highly desirable. It was found that the conductivity of $LaAlO_3$ was significantly affected by the addition of aliovalent cations. Rare-earth aluminates such as $(Ln_{1-x}M_x)AlO_{3-\delta}$ or $Ln(Al_{1-y}M_y)O_{3-\delta}$ (M: Mg^{2+}, Ca^{2+}, Sr^{2+}) show higher conductivity than that of $LnAlO_3$. Among these aluminates, Ca- and Ga-doped $NdAlO_3$ systems were found to exhibit an improved conductivity of about 0.0398 S cm^{-1} at 950°C although this value is still lower than that of YSZ based on the work of Ishihara [16]. The conductivity of $LnAlO_3$ may be further enhanced by the combined effect of charge compensation and lattice expansion. For $LaAlO_3$, the addition of Y ions to the Al site cation sublattice effectively expanded the lattice (~6%) and reduced the activation energy for the migration of oxygen ions. With the expansion of the perovskite lattice, not only was the activation energy for the migration reduced but also the dissolution of large Ba cations was enhanced. Furthermore, the incorporation of Ba into the A-site cation sublattice favors the creation of oxygen vacancies according to the following defect reaction:

$$2BaO + Al_2O_3 \xrightarrow[LaAlO_3]{} 2Ba'_{La} + 2Al^x_{Al} + 5O^x_O + V^{\cdot\cdot}_O$$

In this reaction, the effectively negative charge of Ba^{+2} on the trivalent cation site is compensated by the formation of positively charged oxygen vacancies. In other words, oxygen vacancies are created by the substitution of Ba for La. Thus, the concentration of oxygen vacancies increases as the concentration of the Ba ion dopant increases. Although the substitution of isovalent Y ions for the Al site did not create oxygen vacancies, the perovskite lattice was effectively expanded due to the addition of the Y ion. As expected, the activation energy for oxygen migration was effectively suppressed. With the combined effects of charge compensation and lattice expansion, the activation energy for oxygen conduction was significantly reduced from 125.28 kJ mol^{-1} (for undoped $LaAlO_3$) to 78.75 kJ mol^{-1} (for $La_{0.9}Ba_{0.1}Al_{0.9}Y_{0.1}O_3$). Consequently, the ionic conductivity of $La_{0.9}Ba_{0.1}Al_{0.9}Y_{0.1}O_3$ was substantially enhanced [17].

12.2.3 Proton Conductors

Because hydrogen has high reactivity and can be derived from various types of fuels such as hydrocarbons, alcohols, or coal, hydrogen is also very common fuel for operation of SOFCs. In addition to oxygen-ion conductors, proton (hydrogen-ion) conductors are also being used as solid electrolytes of SOFCs. Figure 12.8 shows the reactions in a proton-conductor SOFC.

Perovskite oxides that show proton conduction have been extensively studied since the 1980s. Recently, phosphates such as $LaPO_4$, LaP_3O_9, and $La_7P_3O_{18}$ (lanthanum oxophosphate), showed interesting proton conductivities (3 × 10^{-4} S cm^{-1} at 700°C) especially in humid atmospheres. At low temperatures ranging from 120°C to 300°C, inorganic solid acid compounds such as $CsHSO_4$ and CsH_2PO_4 exhibit high proton conductivity as high as 2.2 × 10^{-2} S cm^{-1} at 240°C. In comparison with high-temperature polymer electrolyte membrane fuel cell (HT-PEMFC), solid–acid electrolytes have demonstrated great potential for stable power generation based on H_2/O_2 with impressive proton transport and thermal stability at temperatures ranging between 230°C and 260°C [18].

In this section, proton conductors based on oxides with perovskite-type structures and used at high temperatures will be reviewed.

12.2.3.1 Proton Conductors with Perovskite Structure

In 1983, the proton conduction in doped strontium and barium cerates in H_2-containing atmospheres was first observed and reported by Iwahara et al. [19]. In the past 30 years, the proton conduction mechanism has been well established on the basis of extensive studies on doped cerates and zirconate perovskites. The presence of oxygen vacancies is crucial for the

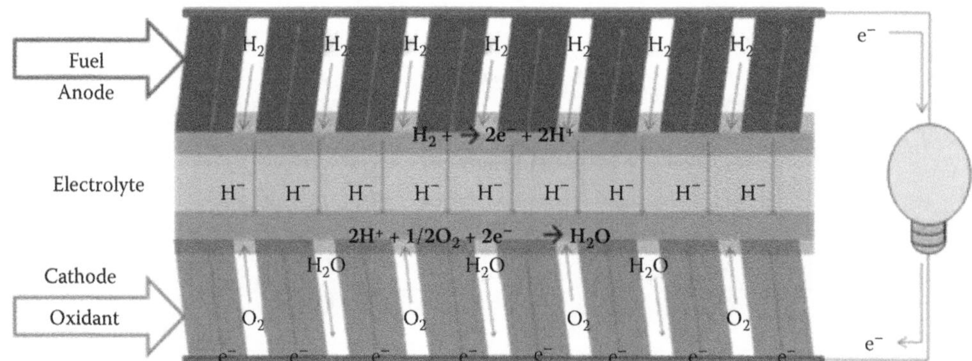

FIGURE 12.8 Schematic operation diagram of an SOFC using a proton conducting electrolyte.

formation of protonic defects [20]. In humidified atmosphere, water tends to absorb on the surface of proton conducting oxide that contains oxygen vacancies. Water molecule then dissociates into a hydroxide ion and a proton. The hydroxide ion fills into oxygen vacancy. At the same time, the proton forms a covalent bond with lattice oxygen [21].

Thus, the formation of protonic defects may be expressed in a defect reaction based on the Kröger–Vink notation:

$$H_2O_{(g)} + V_O^{\cdot\cdot} + O_O^x \rightarrow 2OH_H^{\cdot}$$

Therefore, to enhance proton conductivity in an ABO_3 perovskite, it is necessary to dope the B cation site with suitable trivalent metal ions, M, such as Y, Nd, Sm, Yb, In, Eu, Gd, etc. The partial substitution of the B cation with a trivalent dopant results in the creation of positively charged oxygen vacancies to compensate the negatively charged doping cation. Such substitution may be represented by the following reaction in Kröger–Vink notation:

$$2AO + M_2O_3 \xrightarrow[ABO_3]{} 2A_A^x + 2M_B' + V_O^{\cdot\cdot} + 5O_O^x$$

With enhanced concentration of oxygen vacancies, the proton conductivity of perovskite oxides increase accordingly. For example, the conductivity of 10% Y-doped $BaCeO_3$ is about 1.1×10^{-2} S cm^{-1} which is about three orders of magnitude higher than that of undoped $BaCeO_3$ at 600°C [22,23]. However, the electrical properties of perovskite proton conductors are strongly dependent on their porosities and grain sizes that may vary with processing parameters such as sintering temperature and duration.

Despite its high proton conductivity, the main problems in the use of doped $BaCeO_3$ are (i) poor chemical stability and (ii) induced electronic conduction under reducing atmosphere. For instance, doped $BaCeO_3$ electrolytes tend to react with carbon dioxide and form $BaCO_3$ and CeO_2 when SOFC uses hydrocarbon fuel at temperatures below 800°C. On the contrary, Y-doped $BaZrO_3$ showing a lower conductivity of 1.1×10^{-2} S cm^{-1} at 600°C indeed has better chemical stability than that of doped $BaCeO_3$ [22].

Since perovskite cerates and zirconates tend to form a solid solution easily, it is feasible to develop mixed cerates/zirconates protonic electrolyte with high conductivity and adequate chemical stability. Protonic conductors will then become promising electrolytes for intermediate-temperature SOFC applications.

12.3 ANODE OF SOFCs

The SOFCs, based on an oxide-ion conducting electrolyte, have several advantages over other types of fuel cells, including relatively inexpensive materials, relatively low sensitivity to impurities in the fuel, and very high efficiency. Direct use of available hydrocarbon fuels without first reforming them to hydrogen will greatly decrease the complexity and cost of the fuel cell system. Despite the excellent electrocatalytic properties of the commonly used Ni–YSZ cermet materials in SOFCs for operation in H_2 fuel, they will suffer from carbon deposition and sulfur poisoning when hydrocarbon fuels are used [24–29]. Therefore, there is still a great challenge in the development of new anode materials in SOFCs.

12.3.1 Requirements of Anode

A number of requirements should be satisfied for an anode material [24,27,30–34] to be considered useful in anode-supported planar SOFC applications, as follows:

1. Sufficient electronic conductivity and interconnected porosity for fuel gas transportation in a high-temperature reducing environment, effective anodic electrochemical reactions, etc.
2. Sufficient electrochemical activity for fuel oxidation and redox reactions, chemical stability, and thermal compatibility with other cell components
3. Extended triple-phase boundary (TPB) and sufficient tolerance toward a certain level of impurity/contaminants present in fuel gas
4. Sufficient catalytic activity and, thus, low polarization for electrochemical fuel oxidation
5. Low material cost and simple application technology due to the high operating temperature of SOFC

12.3.2 Development of Anode Materials

12.3.2.1 Conventional Ni–YSZ Cermet Anode Materials

The Ni in the cermet anode provides electronic conductivity and catalytic activity, both for direct oxidation and for steam reforming of methane. The YSZ provides a thermal expansion match with the YSZ electrolyte, and also ionic conductivity to extend the reaction zone in the anode, in addition to its function as a structural support for the anode that prevents Ni sintering [24–27]. Because NiO and YSZ do not form solid solutions, even at high temperatures, this green body can be sintered to form a NiO–YSZ composite and then reduced to form a porous Ni–YSZ cermet [35]. The amount of Ni is typically at least 30 vol% to achieve the percolation threshold for electronic conductivity [36]. This material fulfills most requirements of the anode. The disadvantages of this material are its poor redox stability, low tolerance to sulfur [37], carbon deposition when using hydrocarbon fuels, and the tendency of nickel agglomeration after prolonged operation. Especially, the low tolerance for carbon deposition makes this material inappropriate for operation with available hydrocarbon fuels [24–29,38]. Since nickel is an excellent catalyst for both steam reforming and hydrogen cracking, carbon deposition occurs rapidly when hydrocarbon was used as the fuel, unless excess steam is present to ensure steam reforming. The composition of the anode, particle sizes of the powders, and the manufacturing method are crucial to achieving high electronic conductivity, adequate ionic conductivity, and high activity for electrochemical, reforming, and shift reactions.

It can be evaluated from the previous two sections that Ni–YSZ cermet is currently a preferred anode since Ni and YSZ are essentially immiscible and nonreactive over a very wide temperature range. The subsequent development of a very fine microstructure can be maintained during service for a relatively long period of time. However, the resulting high interfacial surface area is susceptible to degradation with lower life cycle [39]. Some weaknesses of Ni–YSZ anode are listed below [40]:

1. The inability to oxidize commercial hydrocarbon fuels presents a problem for such Ni-based cermet anode. Modified cermet containing ceria, copper, and trace amounts of gold for the high catalyst action of Ni has been reported, which is more tolerant to direct hydrocarbon reactions [41–43].
2. Intolerance to repeated oxidation is another weakness of Ni-based cermet anode, which generates Ni coarsening. Therefore, any subsequent reoxidation cannot be accommodated in the available porosity of the cermet, and disrupts the mechanical integrity.
3. Performance of fuel cells with Ni–YSZ anode undergoes serious degradation upon subjecting to commercial natural gas or other hydrocarbon fuels. These gases contain traces of impurities/additives that act as catalyst poisons [38,44–46]. Therefore, commercialization of SOFC with natural gas/hydrocarbon fuels demands a desulfurizer unit that again adds to the maintenance cost.

12.3.2.2 Other Cermet Anode Materials

Instead of modifying Ni–YSZ cermet anodes, investigations were also performed on the use of electronic or mixed conducting oxides to make alternative metal oxide cermets such as CeO_2, Ru–Ni–GDC ($Ce_{0.9}Gd_{0.1}O_{1.95}$), and $Ce_{0.9}Gd_{0.1}O_{1.95}$.

Ceramics based on CeO_2 exhibit mixed ionic and electronic conductivity in a reducing atmosphere due to reduction of Ce^{4+} to Ce^{3+}. In addition, it is believed that the excellent catalytic activity of CeO_2-based materials stemmed from the oxygen-vacancy formation and migration associated with reversible CeO_2–Ce_2O_3 transition [47,48]. It has been reported that ceria-based ion conductors [49] have a high resistance to carbon deposition, which permits the direct supply of dry hydrocarbon fuels to the anode. The more effective method, however, is the addition of Ni, Co, and some noble metals, such as Pt, Rh, Pd, and Ru, which are beneficial to the reforming reactions of hydrocarbon, due to their functions of breaking the C–H bond more easily, especially in the case of Ru [50]. Hibino et al. [51,52] studied a thin ceria-based electrolyte film SOFC with an Ru-Ni-GDC ($Ce_{0.9}Gd_{0.1}O_{1.95}$) anode that was directly operated on hydrocarbons, including methane, ethane, and propane, at 600°C. They claimed that the role of the Ru catalyst in the anode reaction was to promote the reforming reactions of the unreacted hydrocarbons by the produced steam and CO_2, which avoided interference from steam and CO_2 in the gas-phase diffusion of the fuels. The resulting peak power density reached 750 mW cm^{-2} with dry methane, which was comparable to the peak power density of 769 mW cm^{-2} with wet (2.9 vol% H_2O) hydrogen.

Ramirez-Cabrera et al. [53] studied gadolinium-doped ceria, $Ce_{0.9}Gd_{0.1}O_{1.95}$ (CGO), as an anode material at 900°C in 5% CH_4 with steam/methane ratios between 0 and 5.5, as shown in Figure 12.9. The results revealed that this material is resistant to carbon deposition; the reaction rate was controlled by slow methane adsorption. Marina and Mogensen [49] studied the same catalyst for methane oxidation. It was also demonstrated that ceria has a low activity for methane oxidation but a high resistance to carbon deposition.

12.3.3 Conducting Oxides

In the context of direct utilization of available hydrocarbon fuels in SOFCs, the materials with mixed ionic and electronic conducting oxide may be desirable. The mixed conductivity

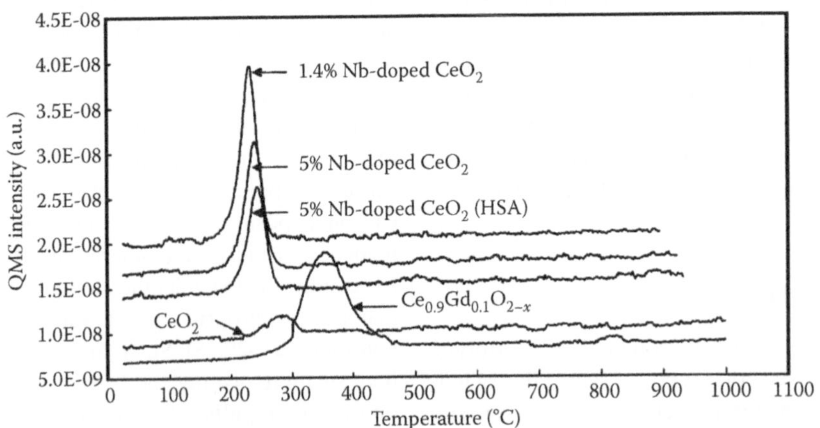

FIGURE 12.9 Comparison of CO_2 signals for different ceria specimens during TPO after isothermal reaction with 5% CH_4/95% Ar at 900°C. The baselines have been shifted for clarity. (From E. Ramirez-Cabrera, A. Atkinson, and D. Chadwick, *Solid State Ionics*, 136, 825, 2000.)

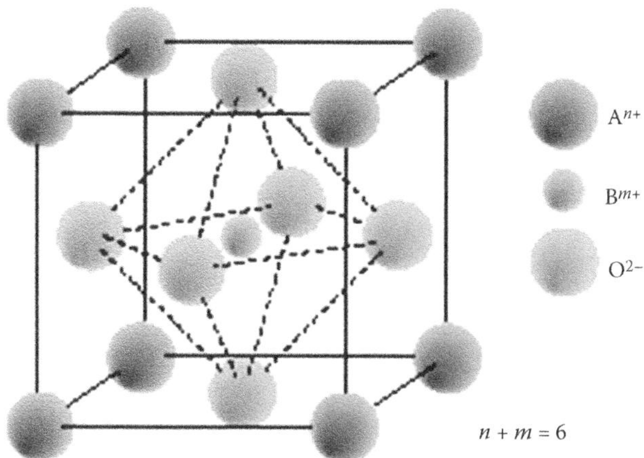

FIGURE 12.10 Unit cell of the ABO_3 perovskite structure. (From B.A. Boukamp, *Nat. Mater.*, 2, 294, 2003.)

extends the active zones where electrochemical reactions can occur by allowing O^{2-} ions to any position of interfaces between the anode and gas phase.

12.3.3.1 Perovskite Anode Materials

The perovskite-type oxide has the general formula ABO_3, in which A and B are cations with a total charge of +6. The lower valence A cations (such as La, Sr, Ca, and Pb) are large and reside on the larger spaces in the 12-fold oxygen coordinated holes; the B cations (such as Ti, Cr, Ni, Fe, Co, and Zr) occupy the much smaller octahedral holes (6-fold coordination). Full or partial substitution of the A or B cations with cations of different valence is possible. When the overall valence of the A-site and B-site cations ($n + m$) adds up to <6, the missing charge is made up by introducing vacancies at the oxygen lattice sites [54]. Figure 12.10 shows the typical structure of the cubic perovskite ABO_3. Recently, many new compounds with a perovskite structure have been offered as alternative anode materials in SOFCs.

Tao et al. [55] reported an oxygen-deficient perovskite, $La_{0.75}Sr_{0.25}Cr_{0.5}Mn_{0.5}O_3$, with comparable electrochemical performance to that of Ni–YSZ cermets and with good catalytic activity for the electro-oxidation of CH_4 at high temperatures. The electrode polarization resistance approaches 0.2 Ω cm² at 900°C in 97% H_2/3% H_2O, as shown in Figure 12.11. Very good performance is achieved for methane oxidation without using excess steam. The anode is stable in both fuel and air conditions, and shows stable electrode performance in methane. Both redox stability and operation in low steam hydrocarbons have been demonstrated, overcoming two of the major limitations of the Ni–YSZ cermet anodes. However, $La_{0.75}Sr_{0.25}Cr_{0.5}Mn_{0.5}O_3$ has a low electronic conductivity in the reducing anodic atmosphere and is not stable to sulfur impurities in the fuel [56].

Goodenough et al. [57] reported identification of the double perovskite $Sr_2Mg_{1-x}Mn_xMoO_{6-\delta}$ that can be used with natural gas as the fuel under operating temperatures of 650°C < T < 1000°C with long-term stability and tolerance to sulfur. This material has the characteristic of oxygen deficiency and is stable in a reducing atmosphere. The mixed valence Mo(VI)/Mo(V) subarray provides electronic conductivity with a large enough work function to accept electrons from a hydrocarbon. As a mixed ionic–electronic conductor (MIEC) that can accept electrons while losing oxygen, it also promises to be catalytically active to the oxidation of H_2 and hydrocarbons. The ability to lose oxygen while accepting electrons is realized because Mo(VI) and Mo(V) form molybdyl ions, which makes them stable in less than 6-fold oxygen coordination, and both Mg^{2+} and Mn^{2+} are stable in 4-fold as well as 6-fold

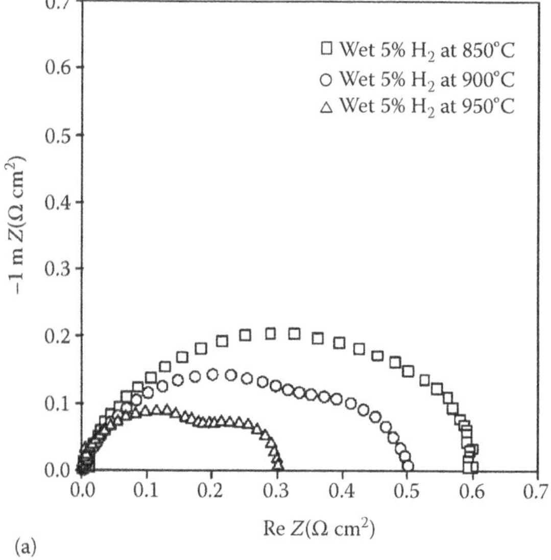

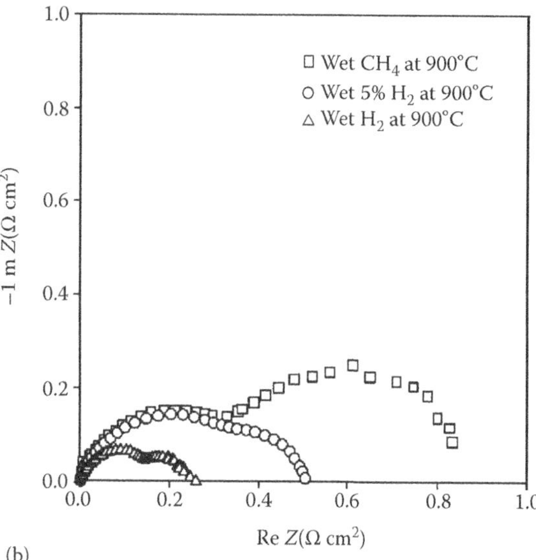

FIGURE 12.11 Electrode impedance of an optimized $La_{0.75}Sr_{0.25}Cr_{0.5}Mn_{0.5}O_3$ anode (a) in wet 5% H_2 (5% H_2, 3% H_2O, 92% Ar) at 850, 900, and 950°C; (b) at 900°C, in different humidified (3% H_2O) fuel gas compositions at 900°C. The electrode dispersions have been obtained in a three-electrode setup. The electrolyte contribution has been subtracted from the overall impedance. (From S. Tao and J.T. Irvine, *Nat. Mater.*, 2, 320, 2003.)

oxygen coordination [58]. The maximum power density reached 438 mW cm^{-2} for $Sr_2MgMoO_{6-\delta}$ at 800°C with dry CH_4 as the fuel.

Vernoux et al. [59] studied the catalytic and electrochemical characteristics of $La_{0.8}Sr_{0.2}Cr_{0.97}V_{0.03}O_3$ (LSCV). They found that this material exhibited a low activity toward CH_4 steam reforming at 800°C, although it has been shown to be stable over time and does not suffer from carbon deposition. When Ru as a steam reforming catalyst was added to this electrode to implement the gradual internal reforming (GIR) process, the LSCV–YSZ composite was a promising anode material.

Ruiz-Morales et al. [60] demonstrated symmetrical fuel cells (SFCs) using simultaneously the same material, $La_{0.75}Sr_{0.25}Cr_{0.5}Mn_{0.5}O_{3-\delta}$ (LSCM), at the anode and cathode sides. Owing to its enhanced electrochemical properties in both reducing and oxidizing conditions, LSCM-based SFCs offered promising performances, 0.5 and 0.3 W cm^{-2} at 950°C using H_2 and CH_4 as fuels, respectively.

Sauvet [61] studied $La_{1-x}Sr_xCr_{1-y}Ni_yO_{3-\delta}$ compositions as a novel anode. The author found that these materials have some catalytic activity, although less than that of Ni composites. For methane steam reforming, no carbon deposition was observed when the ratio of steam to methane is ≤1.

Sin et al. [62] investigated $La_{0.6}Sr_{0.4}Fe_{0.8}Co_{0.2}O_3$–$Ce_{0.8}Gd_{0.2}O_{1.9}$ (LSCFO–CGO) composite material as an anode for the direct electrochemical oxidation of methane in intermediate temperature ceria electrolyte–supported SOFCs. A maximum power density of 0.17 W cm^{-2} was obtained at 800°C. The anode did not show any structure degradation after the electrochemical testing, and no carbon deposits were detected. After high-temperature treatment in a dry methane stream in a packed-bed reactor, this material showed significant chemical and structural modifications. It is derived that the continuous supply of mobile oxygen anions from the electrolyte to the anode, promoted by the mixed conductivity of CGO at 800°C, stabilizes the perovskite structure near the surface under SOFC operation and open circuit conditions.

Vernoux et al. [63] investigated the performances of La(Sr)Cr(Ru,Mn)$O_{3-\delta}$ defective perovskite materials as an anode material. By inserting ruthenium in the B-site, the catalytic activity of this material for the steam reforming of methane has been improved. The catalytic performances are stable, and no carbon deposit was detected, even after 300 h of operation in a reactive mixture containing more methane than steam. It also has shown that the insertion of manganese in the B-site accelerates the electrochemical oxidation of hydrogen compared with the strontium-doped lanthanum chromite. It indicated that these perovskite materials are potential candidates as SOFC anode materials for implementing the GIR of methane.

12.3.3.2 Pyrochlore Anode Materials

The pyrochlore-type oxides, $A_2B_2O_7$, can be derived from fluorite, by removing one-eighth of the oxygens, ordering the two cations, and ordering the oxygen vacancies. The structure of pyrochlore-type oxides is shown in Figure 12.12. The

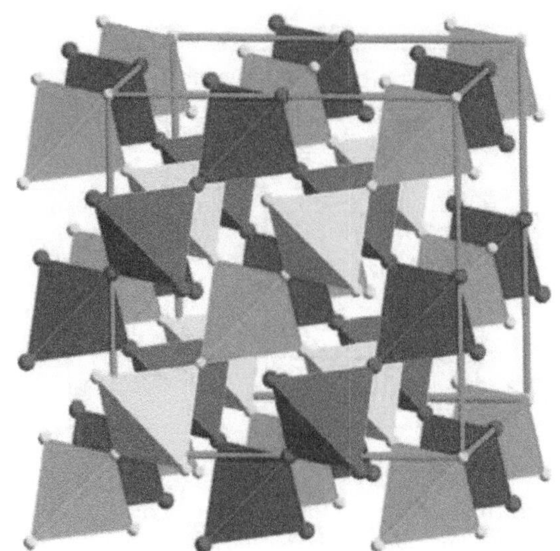

FIGURE 12.12 Pyrochlore, $A_2B_2O_7$ structure.

pyrochlore structure oxides, $Gd_2Ti_2O_7$ (GT), were also considered for use in SOFCs [64–66]. By replacing some of the Gd^{3+} with Ca^{2+}, oxygen vacancies in the A_2O network are created, significantly increasing the ionic conductivity. At 1000°C, the ionic conductivity of $(Gd_{0.98}Ca_{0.02})_2Ti_2O_7$ is about 10^{-2} S cm^{-1}, which is comparable to YSZ [67,68]. Mo-doped GT has been shown to have a very high mixed ionic and electronic conductivity under reducing conditions, making it suitable as an anode material [69]. It was found that the electrical conductivity of $Gd_2(Ti_{1-x}Mo_x)_2O_7$ is about 70 and 25 S cm^{-1} for x = 0.7 and 0.5, respectively, at $p(O_2)$ around 10^{-20} atm. However, this pyrochlore solid solution is only stable at a certain $p(O_2)$ range at high temperature. Mixed ionic–electronic conductivity may be obtained for zirconates containing cations that may change valences with changing oxygen partial pressure. Mogensen et al. [70] investigated three different pyrochlore systems: (i) $Gd_2Ti_2O_7$ modified by Mo, (ii) $Pr_2Zr_2O_{7\pm\delta}$ modified by the multivalent cations Mn and Ce on the Zr site, and (iii) $Pr_2Sn_2O_{7\pm\delta}$ modified by In on the Sn site. It is indicated that the pyrochlore solid solution $Gd_2TiMoO_{7\pm\delta}$ may be a possible candidate as an anode material in an SOFC based on a YSZ electrolyte. By optimizing the composition, these materials may be achieved an extension of the $p(O_2)$ range with stable pyrochlore structure and an increase in the total electrical conductivity.

12.3.3.3 Tungsten Bronze Anode Materials

Oxides with the general formula $A_2BM_5O_{15}$ (with M = Nb, Ta, Mo, W, and A or B = Ba, Na, etc.) show a tetragonal tungsten bronze structure or an orthorhombic tungsten bronze structure [71]. The structure of tungsten bronze–type compounds is shown in Figure 12.13 [72]. These oxides can be described by a framework of MO_6 octahedra sharing summits, delimiting tunnels of pentagonal, square, and triangular sections.

Slater et al. [73,74] investigated materials with the composition $(Ba/Sr/Ca/La)_{0.6}M_xNb_{1-x}O_{3-\delta}$ (M = Mg, Ni, Mn, Cr, Fe, In, Ti, Sn) used as anode materials for SOFCs. They found that the compounds (with M = Cr, Mn, Fe, Ni, Sn) are not

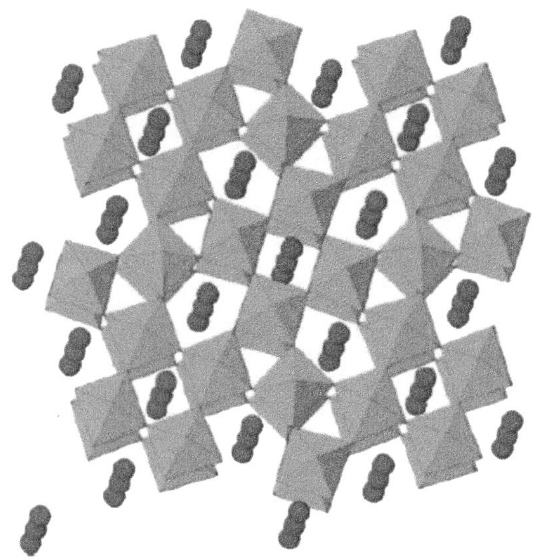

FIGURE 12.13 Tungsten bronze structure. (From S. Tao and J.T. Irvine, *Chem. Rec.*, 4, 83, 2004.)

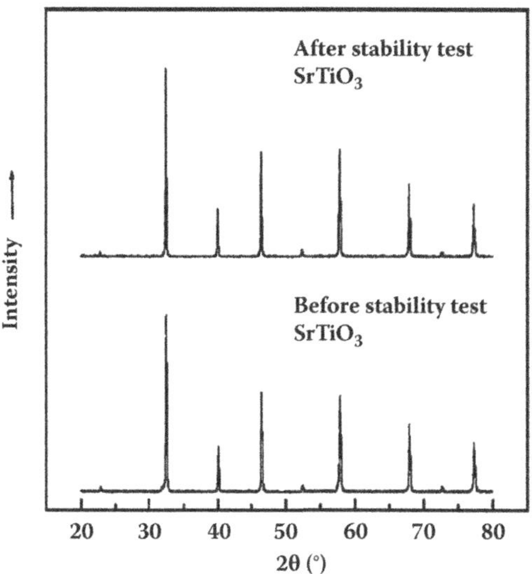

FIGURE 12.14 XRD patterns for the $SrTiO_3$ sample before and after exposure to 10% H_2S/3% H_2O/87% H_2 at 950°C for 5 days. (From Z. Cheng, S. Zha, and M. Liu, *J. Electrochem. Soc.*, 153, A1302, 2006.)

suitable for anode materials, either because of poor oxygen-exchange kinetics, possibly due to low oxide-ion conductivity or because of partial decomposition on prolonged heat treatment at 1000°C in reducing atmospheres. However, the compounds (with M = Mg or In) showed rather good conductivity characteristics and were stable under the prolonged reduction treatment, which are potential anodes for SOFCs.

Kaiser et al. [75] studied a wide range of different materials with tungsten bronze structures and the formula $(Ba/Sr/Ca/La)_{0.6}M_xNb_{1-x}O_{3-\delta}$(M = Mg, Ni, Mn, Cr, Fe, In, Ti, Sn) used as the anode in SOFCs. They found that $Sr_{0.2}Ba_{0.4}Ti_{0.2}Nb_{0.8}O_3$ exhibits the highest electronic conductivity of about 10 S cm^{-1} at $p(O_2)$ = 10–20 atm and at 930°C.

12.3.4 Sulfur-Tolerant Anode Materials

Matsusaki and Yasuda [38] studied the effect of process variables on the degree of H_2S poisoning effect for Ni-based anodes by cell impedance analysis. They found that SOFCs that utilize Ni–YSZ cermet anodes are susceptible to poisoning by sulfur contents as low as 2 ppm H_2S at 1273 K. While the performance loss is reversible at H_2S concentrations less than 15 ppm, the poisoning effect became more significant at lower cell operating temperatures and the poisoning time was relatively constant at low H_2S concentrations (<100 ppm). Their work also concluded that the relaxation times (for H_2S poisoning) decreased with increasing operating temperature. Therefore, it is highly desirable to develop anode materials that are not deactivated by sulfur-containing fuel gases. The desired sulfur-tolerant anode materials must be electronically conductive, chemically and thermally stable, and catalytically active for the oxidation of H_2S, H_2, and CO. Wang et al. [76] prepared $Y_{0.9}Ca_{0.1}FeO_3$ (CYF)-based sulfide material by exposing CYF to a gas mixture of 96% H_2 and 4% H_2S at 900°C, and they further obtained lithiated sulfide. They studied the performances of the lithiated sulfide material as the anode in H_2S-containing gases. They found that the addition of lithium into the sulfides significantly increases the electrical conductivity but does not lead to a significant change in phase structure. Its conductivity was about 0.02 S cm^{-1} at 650°C. However, the catalytic activity and long-term stability of this material as electrodes are not studied.

Recently, Zha et al. [77] reported a pyrochlore-based anode material, $Gd_2Ti_{1.4}Mo_{0.6}O_7$, that showed remarkable tolerance to sulfur-containing fuels. In a fuel gas mixture of 10% H_2S and 90% H_2, the anode/electrolyte interfacial resistance was only 0.2 Ω cm^2, demonstrating a peak power density of 342 mW cm^{-2}. The fuel cell was operated under these conditions for 6 days without any observable degradation, suggesting that this material have potential to make SOFCs powered by readily available sulfur-containing fuels.

Cheng et al. [78] analyzed the stability of various candidate materials for tolerance to sulfur in an SOFC anode atmosphere using the principles of thermodynamics. Their results help rule out a large number of candidates, such as most transition metal carbides, borides, nitrides, and silicides. Estimation of the thermochemical data for oxides with perovskite structure provides valuable predictions about the stability of those materials, as shown in Figure 12.14.

12.4 CATHODE

The cathode serves as the catalyst layer for oxygen reduction reaction, partially owing to the high activation energy and the slow kinetics of the cathode reaction whose operation temperature is 800–1000°C. A high operation temperature gives rise to serious problems, such as the cost of fabrication and maintenance, and the stability of cell components, accordingly setting rigorous requirements for cathode materials.

In recent years, great efforts have been devoted to the development of low- or intermediate-temperature SOFCs

(IT-SOFCs) operating at 500–800°C. Lowering the operating temperature can suppress the degradation of components and extend the range of acceptable material selection; this also serves to improve cell durability and reduce the system cost. However, reducing the operating temperature decreases the electrode kinetics and results in large interfacial polarization resistances. This effect is most pronounced for the oxygen reduction reaction (ORR) at the cathode. To lower the polarization resistance of the cathode, a favorable electronic and ionic conductivity as well as a high catalytic activity for oxygen reduction must be maintained.

1. *Overview of the Cathode Reaction Mechanism*

In SOFCs, the cathode functions as the site for the electrochemical reduction of oxygen. Figure 12.15 shows the principle of an SOFC. The porous electrodes (anode: fuel, cathode: oxidant) are separated by a gastight electrolyte. With the aid of electrons, the oxygen in the oxidant (e.g., air, O_2) will be reduced to oxygen ions O^{2-} at the cathode side. Starting from the cathode (the air electrode), molecular O_2 is first reduced to oxygen anions, using electrons external from the cell, in a half-cell reaction that can be written as follows [79,80]:

$$\frac{1}{2}O_2(\text{gas}) + 2e^-(\text{electrode}) = 2O^{2-}(\text{electrolyte}) \quad (12.1)$$

Oxygen transport is an important phenomenon for SOFCs because it is related to the cathode overpotential at the cathode/electrolyte interfaces. Figure 12.16 shows an oxygen reduction of SOFC around the O_2/cathode/electrolyte interface. The oxygen reduction can be divided into several elemental steps: (i) oxygen dissociative adsorption on the cathode surface; (ii) surface diffusion of oxygen on the cathode; (iii) incorporation of oxygen into the electrolyte via the TPB; (iv) oxide-ion diffusion in the bulk of the cathode; and (v) oxide-ion transfer from the cathode to the electrolyte [81].

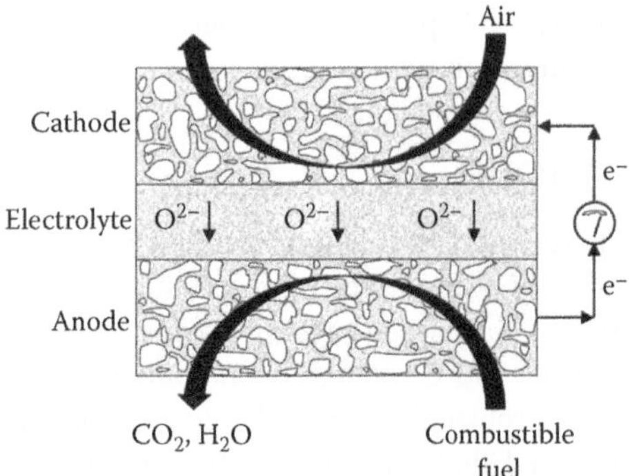

FIGURE 12.15 Diagram showing the operating principles for an SOFC.

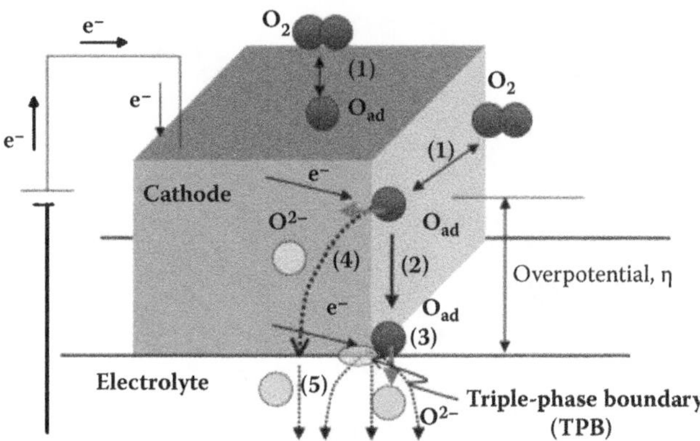

FIGURE 12.16 Schematic representation of oxygen transport around the O_2/cathode/electrolyte interface: (1) oxygen dissociation adsorption on cathode surface; (2) surface diffusion of adsorption oxygen; (3) incorporation of adsorption oxygen via TPB; (4) bulk diffusion of oxygen through cathode; (5) oxide-ion transfer at the cathode/electrolyte interface. (From T. Horita, K. Yamaji, N. Sakai, Y. Xiong, T. Kato, H. Yokokawa et al., *J. Power Sources*, 106, 224, 2002.)

2. *Triple-Phase Boundaries at Cathodes [82]*

A TPB is, generically, a point in two dimensions or a curve in three dimensions where three phases (here, gas, solid electrode, and electrolyte) come together simultaneously. TPBs are frequently cited in the fuel cell literature (cf. citations below), and they play a central role in a number of electrochemical systems. Figure 12.3 provides a simplified illustration of the TPB concept for an SOFC cathode. This multistep reaction involves all three phases and several processes, including diffusion, adsorption, electronic conduction, charge transfer reaction, and ionic conduction. In most instances, this entire reaction occurs almost exclusively in the immediate vicinity of TPBs via one of several possible reaction pathways [83].

a. The electrode surface pathway. The electrode surface pathway (Figure 12.17a) considers adsorption of oxygen species on the cathode electrode surface and their migration along the electrode surface toward the TPB (where the electrolyte, electrode, and gas phase meet), followed by complete ionization and ionic transfer into the electrolyte.
b. The electrode bulk pathway. The electrode bulk pathway (Figure 12.17b), like the electrode surface pathway, begins by considering adsorption of oxygen species on the cathode surface. However, in this case, the oxygen species are locally dissociated, ionized, and incorporated into the solid electrode. Oxide-ion transport then proceeds through the electrode bulk, followed by ionic transfer into the electrolyte.
c. The electrolyte surface pathway. The electrolyte surface pathway (Figure 12.17c) considers adsorption of oxygen species on the electrolyte surface and their migration along the electrolyte

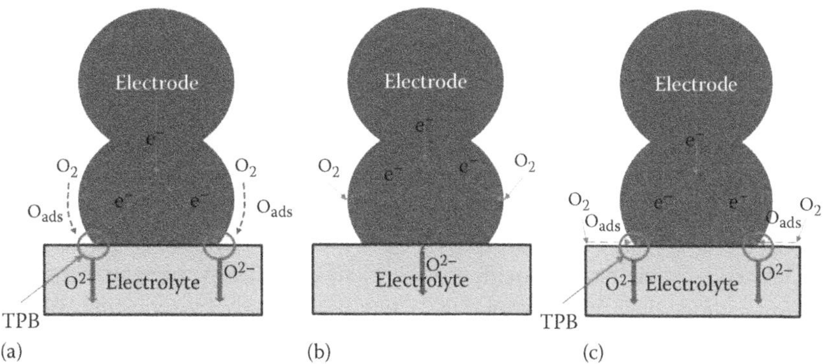

FIGURE 12.17 Possible reaction pathways in SOFC cathodes. (a) Path on electrode surface, (b) path through electrode bulk, and (c) path on electrolyte surface.

surface toward the TPB. Unless partial ionization (via electron donation) can be accomplished by the electrolyte, this pathway is geometrically similar to the electrode surface pathway. Frequently, this pathway is ignored in favor of the electrode surface pathway because of the smaller total surface area and extremely low electronic conductivity of most relevant electrolyte surfaces, and we will ignore it here.

It is important to understand TPB behavior in SOFC cathodes because the density and properties of the TPBs often determine the overall cathode performance. Because real SOFC cathodes are composite porous structures that exhibit relatively complex geometry/morphology, however, TPB properties have proven exceptionally difficult to determine experimentally.

3. *Cathode Requirements*

Oxygen transport is an important phenomenon for SOFCs because it is related to the cathode overpotential (η) at the cathode/electrolyte interfaces. To minimize the cathode overpotential, the cathode must have [84–86]

a. High electronic conductivity (preferably >100 S cm^{-1} under an oxidizing atmosphere)
b. A matched thermal expansion coefficient (TEC) and chemical compatibility with the electrolyte and interconnect materials
c. Adequate porosity to allow gaseous oxygen to readily diffuse through the cathode to the cathode/electrolyte interface
d. Stability under an oxidizing atmosphere during fabrication and operation
e. High catalytic activity for the ORR
f. Low cost

12.4.1 Category of the Cathode Material

The active cathode region needs to conduct both electrons and oxygen ions. Therefore, to produce high-charge transfer rates in SOFCs, the active cathode near the electrode/electrolyte interface should have a high mixed electronic and oxygen-ion conductivity.

Cathode materials are chosen to fulfill these requirements, because, among other characteristics, they are efficient electronic conductors at high temperature and stable under oxidizing conditions. Noble metal catalysts such as Au, Ag, and Pt fit these requirements, yet are too costly to be considered practical for commercial deployment. As an alternative, ceramic oxides with a perovskite (ABO_3) structure have proven to be a suitable and less expensive replacement. Three oxide structure types have been studied as cathode materials: perovskite, K_2NiF_4, and ordered double perovskites. The majority of perovskite-type oxides currently considered as cathodes are based on either the $La_{1-x}Sr_xCoO_3$ or $La_{1-x}Sr_xMnO_3$ materials. Therefore, it is sensible to deal with these two classes of materials individually and examine the performance of the related materials [87–90].

12.4.1.1 Perovskite-Type Structure Cathode Materials

Perovskite materials have been widely used as cathode materials in SOFCs [87,91]. To better design and optimize the cathode materials, it is necessary to first understand the fundamentals of the perovskite structure. A perovskite-type oxide has the general formula ABO_3, in which A and B are cations with a total charge of +6. The lower valence A cations (such as La, Sr, Ca, and Pb) are larger and coordinated to 12 oxygen anions, while the B cations (such as Ti, Cr, Ni, Fe, Co, and Zr) occupy the much smaller space and are coordinated to six oxygen anions. Full or partial substitution of A or B cations with cations of different valence is possible. When the overall valence of the A-site and B-site cations ($n + m$) adds up to less than six, introducing vacancies at the oxygen lattice sites makes up for the missing charge [54,92].

For most of the perovskite materials used as cathodes in SOFCs, the A-site cation is a mixture of rare and alkaline earths (such as La and Sr, Ca, or Ba), while the B-site cation is a reducible transition metal such as Mn, Fe, Co, or Ni (or a mixture thereof). Therefore, in most cases, a redox catalytic mechanism is usually provided by B-site cations [93–95]. The octahedral symmetry around the transition metal often promotes a metallic or semiconducting band structure at high temperatures, leading to high electronic conduction. With a rational choice of A- and

B-site cations, a large and stable number of oxygen-ion vacancies can be introduced at SOFC operating conditions, thus facilitating significant bulk ionic oxygen transport [96].

For $A_{1-x}A'_xBO_3$ perovskite-type oxides resulting in sub-stoichiometric oxygen-to-metal regions, an oxygen vacancy scenario results [97,98], e.g., in $Ln^{3+}_{1-x}Sr^{2+}_xM^{3+}O_{3-\delta}$; the oxygen vacancies arise either from substitution of A^{3+} by A^{2+} or from the partial reduction of B^{3+} or B^{4+} to B^{2+} and B^{3+}, respectively. Oxygen mobility through vacancies is the basis of oxide-ion conductivity. In general, most cathode materials rely on doping of both the A- and B-sites to improve electrical conductivity and electrocatalytic performance [99,100].

12.4.1.2 $Ln_{1-x}A_xM_{1-y}Mn_yO_{3-\delta}$, Where Ln = La, Nd, Pr, etc., A = Ca, Sr, and M = Transition Metal

The most popular material used for cathodes is strontium-doped lanthanum manganite of the composition $La_{1-x}Sr_xMnO_3$ (LSM) for fuel cells operating at 800–1000°C. This material belongs to the manganite family ($LaMnO_3$) of perovskites, where La is partially replaced by Sr [101–103].

The LSM does not increase the oxygen vacancy concentration, a common phenomenon in most of the other perovskite cathode materials studied, but rather oxidizes the manganese ion according to Equation 12.2.

$$Mn^x_{Mn} + SrO \xleftrightarrow{LaMnO_3} Sr'_{La} + Mn^{\cdot}_{Mn} + O^x_O \quad (12.2)$$

This reaction effectively increases the electron–hole concentration and improves the electrical conductivity. The electronic conductivity of LSM increases approximately linearly with increasing Sr concentration up to a maximum of around 50 mol% [104]. At a high temperature, $LaMnO_3$ undergoes a solid-phase reaction with YSZ to form $La_2Zr_2O_7$ (LZ) at the electrode electrolyte interface [105]. A little amount of Sr substitution decreases the reactivity of LSM compound with YSZ.

However, $SrZrO_3$ (SZ) forms when the Sr concentration is above about 30 mol% [106]. Therefore, an Sr content of 30 mol% is considered as optimal against the formation of unwanted electronically insulating phases. Incorporating a slight A-site deficiency in the materials can further decrease unwanted reactivity.

The relative amounts of LZ and SZ depend on the La/Sr ratio in the LSM [107] and increase with increasing reaction time and sintering temperature. The conductivity of these zirconates is two to three orders of magnitude lower than that of YSZ, and therefore must strictly be avoided to prevent high ohmic loss. In addition, it was found that nonstoichiometry at the La site in $LaMnO_3$ plays an important role in the reaction. Diffusion of Mn into YSZ leads to increases in La activity at the interface and promotes the interfacial reaction [108].

12.4.1.3 Other $Ln_{1-x}A_xM_{1-y}Mn_yO_{3-\delta}$

Generally, the electrical properties of manganite-based perovskite compounds are not sufficient for operation at temperatures below 800°C. The absence of oxygen vacancies in LSM restricts the reduction of oxygen to the TPB regions. This limitation is the primary reason why LSM does not have acceptable performance at lower temperatures. Usually, two approaches have been taken to improve the performance of LSM cathodes so that they may be used at lower temperatures. The first is to add a second ionically conducting phase to LSM. The second approach has been to replace La with other rare-earth elements or dope LSM with a cation (such as Co, Fe, or Ni) that promotes the formation of oxygen vacancies when strontium is doped on the A-site [109,110].

The cathodic overpotentials of manganite cathodes can be modified greatly by replacing different rare-earth cations at the A-site [111,112]. Ishihara et al. systematically studied the cathodic overpotentials of $Ln_{0.6}Sr_{0.4}MnO_{3-\delta}$ (Ln = La, Pr, Nd, Sm, Gd, Yb, or Y). The Sr-doped $PrMnO_3$ cathode showed low overpotential values even at decreased operating temperatures [113]. Sr-doped $PrMnO_3$ also exhibits a compatible TEC with YSZ. Consequently, they concluded that doped $PrMnO_3$ is a promising cathode material for IT-SOFCs. In addition, the formation of undesired pyrochlore phases $Ln_2Zr_2O_7$ between LSM cathode and YSZ electrolyte can be suppressed for those having smaller lanthanoids, especially the $Pr_{1-x}Sr_xMnO_3$ and $Sm_{1-x}Sr_xMnO_3$ system [114].

More Sr substitution and an increase of the atomic number of lanthanides are beneficial to adjust the TEC of $Ln_{1-x}Sr_x$-MnO_3 to match with those of the electrolytes. Kostogloudis and Ftikos found that $Nd_{1-x}Sr_xMnO_{3\pm\delta}$ (x = 0.4, 0.5) with a TEC of 12.3 × 10^{-6} K^{-1} demonstrated thermal and chemical compatibility with GDC electrolytes [115]. $Pr_{0.5}Sr_{0.5}MnO_3$ showed a high electrical conductivity of 226 S cm^{-1} at 500°C [116,117], which is expected to increase with temperature, indicating a potentially promising cathode material for SOFCs. In terms of chemical compatibility and electrical conductivity, however, Ca-doped ones have proven to be more promising. $Pr_{0.7}Ca_{0.3}MnO_3$ showed chemical stability and was thermally compatible (11.9 × 10^{-6} K^{-1}) with YSZ electrolyte materials [118]. Its conductivity is close to that of the previously mentioned Sr-doped $PrMnO_3$ with the same composition.

As for the doping at the B-site, Sc-doped manganite, $La_{0.8}Sr_{0.2}Mn_{1-x}ScxO_{3-\delta}$ (LSMS), is a potential cathode material for IT-SOFCs [119]. Nonstoichiometric defects are introduced into the perovskite lattice of LSMS samples by Sc substitution, which leads to an increased oxygen-ion mobility in the samples containing Sc. However, high-level doping of Sc (>10 mol%) results in the segregation of Sc_2O_3 secondary phases at elevated temperatures. The cells with LSMS cathodes exhibit higher performance, especially at lower temperatures, which can be ascribed to the increased oxygen vacancies in the LSMS. For this cathode material, the high cost of scandium may be a concern for applications. By doping Co in the B-site, the area-specific resistance (ASR) of $Sr_{0.8}Ce_{0.2}Mn_{0.8}Co_{0.2}O_{3-\delta}$ is 0.10 Ω cm^2 at 750°C, which is about 20 times lower than that of $Sr_{0.8}Ce_{0.2}MnO_{3-\delta}$ [120,121].

12.4.1.4 $La_{1-x}Sr_xCoO_{3-\delta}$

The use of lanthanum cobaltate materials as possible cathodes has been widely investigated in recent years [122]. Initially, of most interest were the $La_{1-x}Sr_xCoO_{3-\delta}$ (LSC) perovskites that were found to possess considerable oxide-ion conductivity and sufficient electronic conductivity to make them of interest as MIECs. They exhibit a rather high electronic density of states near the Fermi level (E_f) [123]. Petrov et al. proposed a defect model associated with the defect structure of $La_{1-x}Sr_xCoO_{3-\delta}$ in which strontium ions are assumed to occupy the regular La lattice sites Sr'_{La}, leading to predominantly electron holes [124]. To maintain electrical neutrality, the substitution of Sr ions must be compensated for by the formation of equivalent positive charges, comprising $Co^{\cdot}_{Co}$ and oxygen vacancies [$V^{\cdot\cdot}_O$].

The choice of electrolyte is seen to be of importance as highlighted by Gödickemeier et al. [125], where LSC is noted as the choice material for a ceria-based electrolyte although it reacts with zirconia and so is of little use with YSZ electrolytes. This work indicates the superior performance of LSC when tested with $Ce_{0.8}Sm_{0.2}O_{1.9}$ (CSO) and $Ce_{0.8}Gd_{0.2}O_{1.9}$ (CGO) electrolytes when compared with $La_{1-x}Sr_xMnO_3$. While of little use with YSZ electrolytes, LSC continues to receive attention because of this attractive performance with ceria-based electrolytes. It has also been found to be particularly useful with reference to the fundamental understanding of the kinetics of SOFC electrodes.

Of further interest is a study of the metal–insulator transition in LSC that investigated the effect of Sr content, temperature, and $p(O_2)$ on this transition. At room temperature, the metal–insulator transition was determined to occur at x ~0.25 and so to investigate the effect of $p(O_2)$ on this transition, a composition of $x = 0.3$ was studied. The authors established that at a $p(O_2)$ of 1×10^{-3} atm, LSC became insulating, whereas at a $p(O_2)$ of 2×10^{-3} atm it remained metallic over the entire temperature range. This is obviously of significance as the oxygen partial pressure at the SOFC electrode may well be subject to small variations and therefore drastic performance deterioration may be observed. A recent study [126] has also focused on the likely effect of structure on the magnitude of ionic conductivity in LSC. A variety of compositions from $0 < x < 0.8$ were studied, and evidence was presented for the existence of oxygen vacancy ordering in these materials where $x > 0.5$. At this composition, LSC adopts a cubic structure, whereas at $x < 0.5$ it is rhombohedral. When the composition reached $x = 0.7$, the presence of superstructure ordering was confirmed by high-resolution transmission electron microscopy and selected area electron diffraction measurements. It was proposed that in compositions where superstructure ordering was observed, the ionic conductivity would be lower than in disordered regions. The basis for this conclusion was an analysis of the atomic diffusion path lengths of the proposed superstructure that indicated that certain direct jump distances, such as (001), were energetically unfavorable. This was also true of the identical $x = 0.5$ and $y = 0.5$ cross sections. Evidently, this has great implications for the selection of LSC composition for the optimization of conductivity performance in a cathode material. Doping Cu into the Co site can further enhance the ionic conductivity and catalytic activity of $(La,Sr)CoO_{3-\delta}$, although the electrical conductivity of $(La,Sr)(Co,Cu)O_{3-\delta}$ was lower [127].

12.4.1.5 $La_{1-x}Sr_xFeO_{3-\delta}$

Lanthanum ferrite ($LaFeO_3$) is expected to be more stable than cobaltite perovskites because the Fe^{3+} ion has a stable electronic configuration $3d^5$. Sr-doped $LaFeO_3$ (LSF) cathodes have shown promising performance with respect to the power density and stability at 750°C [128]. In iron-based cathodes, reactivity with YSZ electrolyte is significantly reduced. In addition, TECs of the ferrite–perovskite are relatively close to those of the YSZ and CGO electrolyte.

The addition of Sr to $LaFeO_3$ creates Sr'_{Fe}, thus causing a charge unbalance. The charge neutrality is then maintained by the formation of Fe^{4+} ions or oxygen vacancies [$V^{\cdot\cdot}_O$]. The highest theoretical conductivity expected by Sr addition would be at $x = 0.5$, creating a maximum Fe^{4+}/Fe^{3+} ratio of 1:1 [129,130]. At high temperatures, LSF will lose oxygen to form oxygen vacancies at the cost of decreasing hole concentration. This reaction can be represented in Kröger–Vink notation:

$$2Fe^{\cdot}_{Fe} + O^x_O \rightarrow V^{\cdot\cdot}_O + \frac{1}{2}O_2 + 2Fe^x_{Fe}$$

where $Fe^{\cdot}_{Fe}$ represents Fe ions in the 4+ valence state (similar to holes, $2Fe^x_{Fe}$), dotes Fe cations in 3+ valence state, and $V^{\cdot\cdot}_O$ represents the oxygen vacancy. Under relatively high oxygen partial pressures (e.g., air or O_2), holes are the predominant charge carriers while at lower oxygen partial pressures, the defects of Sr'_{La} are compensated by oxygen vacancies. Because of the much higher mobility of the electrons or holes than that of oxygen ions, the total conductivity in ferrites is dominated by holes rather than oxygen ions. Therefore, the total conductivity in ferrites is dominated by a hole-conduction mechanism [131]. By incorporating La deficiency in $La_{0.8}Sr_{0.2}FeO_3$, Ralph et al. [132] found that the ASR of this cathode was significantly reduced and reached 0.1 Ω cm^2 at 800°C. The TEC of the $La_{0.75}Sr_{0.25}FeO_3$ cathode closely matches with those of CGO and YSZ. This cathode has demonstrated no degradation over a 500-h operation, indicating a promising material for lower-temperature SOFCs.

12.4.1.6 $La_{1-x}Sr_xFe_{1-y}Co_yO_{3-\delta}$

To overcome the technological problems associated with the LSC materials operating at elevated temperatures, doping strategies have been developed, and while A-site substitutions have already been discussed, it is more often the B-site substitutions that are of greater significance. One of the most promising systems to date has been produced by the doping of the B-site, with Fe for Co.

$La_{1-x}Sr_xFe_{1-y}Co_yO_3$ shows good electrical conductivity, a high oxygen surface exchange coefficient, and a good oxygen self-diffusion coefficient between 600°C and 800°C. The oxygen self-diffusion coefficient of LSCF is 2.6×10^{-9} cm^2 s^{-1} at 500°C, which is superior in performance to that of $La_{0.8}Sr_{0.2}MnO_3$ that has an oxygen self-diffusion coefficient of 10^{-12} cm^2 s^{-1} at 1000°C [133]. Although the electrical conductivity was found to decrease slightly as the Sr content decreased, the TECs were found to be the lowest for the composition with the highest A-site deficiency, that is, 13.8×10^{-6} K^{-1} for $La_{0.6}Sr_{0.4}Co_{0.2}Fe_{0.8}O_{3-\delta}$ at 700°C. This TEC value matches commonly used electrolytes [134].

The La/Sr ratio [135] was also found to have a marked effect on the performance of these materials with a peak conductivity of ~350 S cm^{-1} found at the $x = 0.4$ composition, which compares with ~160 S cm^{-1} for the $x = 0.2$ sample. It was also found that the peak conductivity occurred at 550°C, which, when combined with the results of the earlier study [136], suggest a composition of $La_{0.6}Sr_{0.4}Co_{0.8}Fe_{0.2}O_3$ as a candidate cathode material.

Variations of the Fe/Co ratio have been studied [137] in the $La_{0.84}Sr_{0.16}Co_{1-x}Fe_xO_3$ composition, and the effect on performance when used as a cathode on CGO followed. The optimum composition was reported to be at Co/Fe 0.7:0.3 with a conductivity of 643 S cm^{-1} at 800°C, which steadily increased over the 200–800°C temperature range. These results are in conflict with the early results of Tai et al. [135,136], as discussed earlier. However, Maguire et al. [137] have only tested a limited number of materials, but those that were tested formed a study of the material in contact with a CGO electrolyte. The CGO diffusion barrier layer is used to prevent the formation of low conductive compounds without negatively affecting the electrochemical performance.

Further investigation [138,139] of A-site-deficient LSCF has confirmed the performance advantage of increasing A-site deficiency. Although the electrical conductivity was found to decrease slightly as the Sr content decreased, the TECs were found to be lowest for the compositions with greatest A-site deficiency, that is, 13.8×10^{-6} °C^{-1} for $La_{0.6}Sr_{0.4}Co_{0.2}Fe_{0.8}O_{3-\delta}$ at 700°C. This value is relatively close to that of the common electrolytes and again indicates the value of LSCF as a cathode material.

12.4.2 K_2NiF_4-Type Structure Cathode Materials

The K_2NiF_4-type structure is usually formulated as A_2BO_4, which is described as a stacking of perovskite ABO_3 layers alternating with rock salt AO layers along the c-direction as shown in Figure 12.18 [140,141].

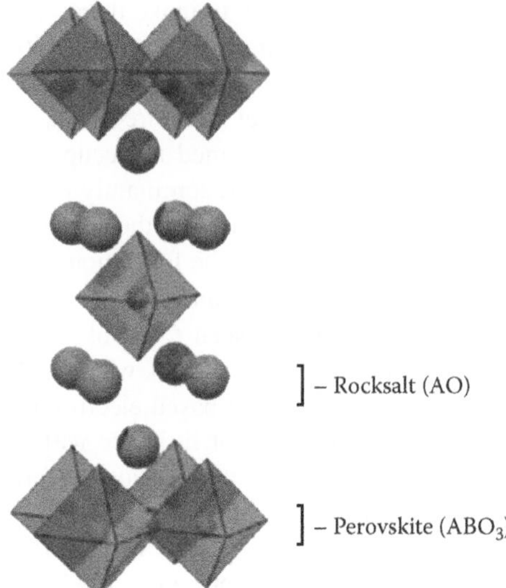

FIGURE 12.18 K_2NiF_4-type structure showing alternating AO and ABO_3 layers. (From E.N. Naumovich, M.V. Patrakeev, V.V. Kharton, A.A. Yaremchenko, D.I. Lopinovich, and F.M.B. Marques, *Solid State Sci.*, 7, 1353, 2005.)

Oxides with the perovskite-related K_2NiF_4 structure, for example, Ln_2NiO_{4+x} (Ln = La, Pr, Nd), are of interest for SOFC cathodes because of the high diffusivity of the interstitial oxygen ions [142]. The structure of Ln_2NiO_{4+x} can be described as a succession of $LaNiO_3$ perovskite layers alternating with LaO rock salt layers. The oxygen excess in La_2NiO_{4+x} is associated with the incorporation of interstitial oxygen anions into the rock salt layers where they are tetrahedrally coordinated by La^{3+} cations. In La_2NiO_{4+x} at ambient temperature, x can be as high as 0.18 and in Pr_2NiO_{4+x}, the maximum value of x is 0.22. At higher temperature, the diffusivity increases, although this is offset by the decrease in the concentration of oxygen interstitials. The advantages of lanthanum nickel oxide include, in addition to high oxygen mobility, a relatively low lattice expansion induced by variations in temperature and oxygen partial pressure [143]. As with the perovskite structure oxides, the Ln_2NiO_{4+x} compounds can incorporate chemical substitutions on both the La and Ni sites to give a wide range of different transport properties. Several systems have been investigated either by doping at the La position with alkaline earths or by forming solid solutions of different rare earths [144]. Measurements of the surface exchange and self-diffusion coefficient were made for $La_{1-y}Sr_yNiO_{4+x}$, $y = 0.0, 0.1$, by isotope exchange and depth profiling in the temperature range 640–842°C [145]. La_2NiO_{4+x} was found to have an oxygen diffusivity higher than that of $La_{0.6}Sr_{0.4}Co_{0.2}Fe_{0.8}O_{3-\delta}$ (LSCF) and 1 order of magnitude lower than the very good perovskite oxide mixed conductor $La_{0.3}Sr_{0.7}CoO_{3-\delta}$ (LSCO). The effects of strontium content on the electrical properties of $La_{2-x}Sr_xNiO_{4+x}$ have been reported [146].

Doping at the La site with alkaline earths (Sr, Ba, Ca) and other rare earths (Nd or Pr), or doping at the Ni site with other transition metals (typically Cu or Co) may lead to dramatic changes of the structural and physical properties. $La_2Ni_{1+x}Cu_xO_4$ ($0 < x < 1$) has been evaluated as a possible cathode material for SOFCs [147]. The TECs are in the range of $10.8–13.0 \times 10^{-6}$ K^{-1}. Lower ASR values are obtained with a LSGM electrolyte compared with those with 8YSZ as an electrolyte material. For the $La_2Ni_{0.6}Cu_{0.4}O_4$ composition, the total conductivity reached 87 S cm^{-1} at 580°C.

Although La_2NiO_{4+x} and related compounds have most of the intrinsic properties needed for a good cathode, the performance reported to date in SOFCs falls below expectations, therefore requiring further research.

12.4.3 Ordered Double Perovskites

In the mixed conducting oxides represented by the general formula $AA'Co_2O_{5+x}$ (A = Re, Y and A′ = Ba, Sr) [148], the combination of rare earths and Ba or Sr on the A-site leads to A-site ordering that is associated with rapid oxygen-ion transport. These compounds are related to the "112"-type structure

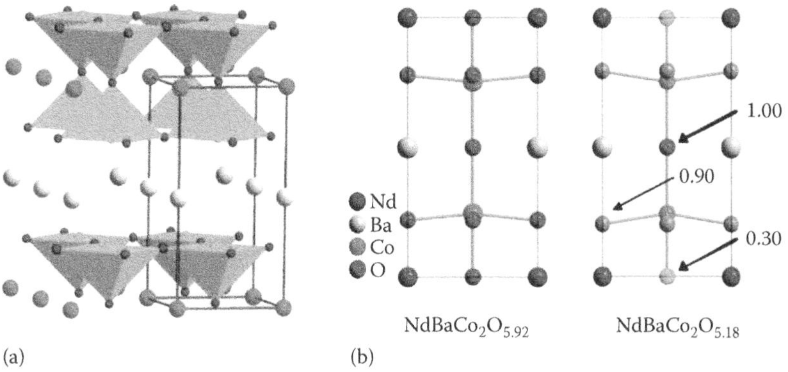

FIGURE 12.19 (a) Structure of $PrBaCo_2O_5$, Ba, and Pr cations and oxygen anions, shown as purple, gray, and red spheres, respectively. The cobalt ion coordination is represented by the blue pyramids. (b) Comparison of the site occupancies of two $NdBaCo_2O_{5+x}$ compositions.

first reported for $YBaFeCuO_5$ [149] and consist of double layers of square pyramidally coordinated cobalt cations. The Ba^{2+} cations are located in the Co double layers, and layers of lanthanide cations separate these layers. The structures of representative examples, $PrBaCo_2O_{5+x}$ and $NdBaCo_2O_{5+x}$ (x = 0.92, 0.18), of this class of compounds are shown in Figure 12.19 [150,151].

A considerable literature exists on the structural chemistry and low-temperature properties of the 112 phases, including the cobalt compounds. In contrast, little was known concerning their high-temperature properties until Taskin et al. [152] measured high rates of oxygen uptake in $GdBaB_2O_{5+x}$ (B = Mn, Co), together with clear indications that the vacancy layers are important to this process. Independently, measurements on $PrBaCo_2O_{5+x}$ were made on both thin films [153] and bulk materials [154], and showed high oxygen diffusivity and rapid surface exchange kinetics.

In other studies [88,155,156], the electrode performance of $GdBaCo_2O_{5+x}$ has been investigated at temperatures below 700°C by AC impedance spectroscopy. One potential problem with barium-containing systems is susceptibility to carbonate formation. Tarancon et al. studied the stability of $GdBaCo_2O_{5+x}$ and found very good stability in atmospheres of CO_2 (500 ppm to 100%) at temperatures up to 700°C [89].

The double perovskites are clearly an interesting class of cathode materials; however, at this stage, only limited data on a small number of compositions are available. More performance data under polarization conditions in complete cells are required, although some initial data for cathodes of $NdBaCo_2O_{5+x}$ and $PrBaCo_2O_{5+x}$ on a samarium-doped ceria electrolyte have been reported [157,158].

12.4.4 Surface Modification

In the past 15 years, extensive efforts have been devoted to developing new catalyst materials with varied infiltration approaches to enhance the performance of the SOFC cathode, especially for the catalytic activity improvement [159,160]. However, the long-term stability of the infiltration cathode is equally important and cannot be overlooked to evaluate whether a catalyst and a relevant infiltration process are effective.

Surface modification through infiltration is an effective approach to enhance cathode functionality while retaining the advantageous qualities of each constituent material. The main advantages of surface modification of electrodes through infiltration are 3-fold. First, it is an effective approach to enhancing electrocatalytic activity while improving the stability of the electrode at low cost through the application of a thin-film coating of a proper catalyst with high stability. Second, it promises to utilize a wide variety of active materials that cannot be otherwise used in a conventional electrode fabrication process because of high reactivity with the other components of the SOFC or high cost, since the infiltration process separates the formation temperature of the catalytic active phase from the high sintering temperature required to guarantee a good binding interface between electrolyte and electrode, and it requires much less material compared with the standard electrode fabrication process. Lastly, it allows establishment of nanosized and/or nanostructured electrodes. Given that nanomaterials have helped improve Li-ion batteries and electrochemical capacitors, it is not surprising that nanoarchitectured electrodes could outperform existing SOFC systems by creating higher surface areas of active electrocatalytic materials.

12.4.4.1 Mixed Ionic–Electronic Conductor (MIEC)

The length of the TPB correlates well with the interfacial resistances to electrochemical oxidation of fuels at the anode [161–163] and reduction of oxygen at the cathode. Therefore, an extension of the TPB and an increase of the number of active reaction sites become necessary to improve electrode performances. It can be achieved for fabrication electrode materials with a higher mixed ionic–electronic conductivity and optimizing the microstructure of electrodes.

A ubiquitous example of an advantageous material structure is embodied in the conventional composite cathode (Figure 12.20a), which consists of an electrolyte and an MIEC possessing high electronic conductivity. The percolated system exhibits superior performance compared with the MIEC alone [164], especially for those with poor ionic conductivity

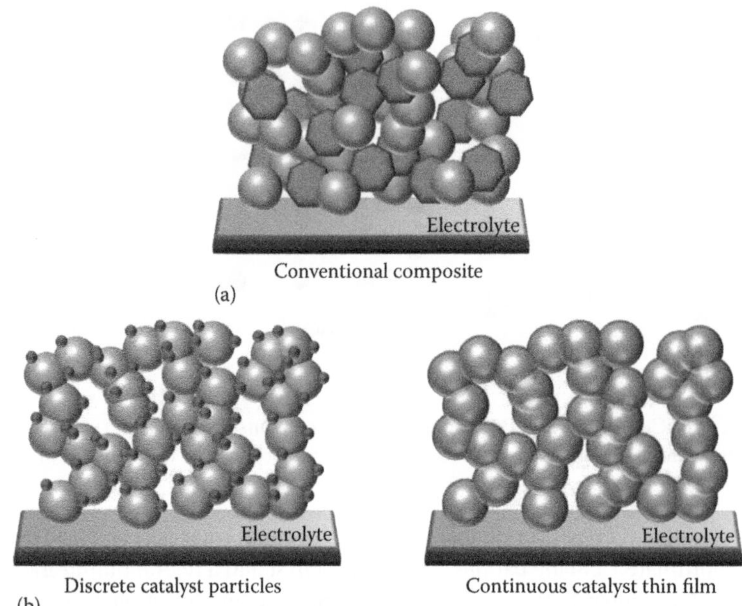

FIGURE 12.20 Schematic of composite cathode consisting of (a) a random mixture with a MIEC and an electrolyte material, and (b) a backbone and a catalyst coating (discrete particles or continuous thin films).

such as LSM. While such a random composite of two materials through a conventional mechanical mixing may provide optimization on the basis of bulk material properties, a superior approach is to engineer a cathode with targeted surface and bulk properties from different materials, e.g., use a superior MIEC as the backbone and an active coating catalyst [50,165], as schematically illustrated in Figure 12.20b. Ideally, the porous backbone provides excellent ambipolar conductivity, while the infiltrated surface coating possesses high stability and excellent catalytic activity toward the ORR.

Infiltration/impregnation processes have been widely used in fabrication of SOFC components in recent years, especially in laboratory-scale research [166–168]. In general, an infiltration/impregnation process involves the steps illuminated in Figure 12.8: (i) preparation of porous electrode backbone (Figure 12.21a). (The backbone is fired at high temperatures to ensure excellent bonding with the electrolyte, excellent connectivity for effective conduction of electron and oxygen ion, and good structural stability of the cathode under operating conditions.); (ii) infiltration of the electroactive

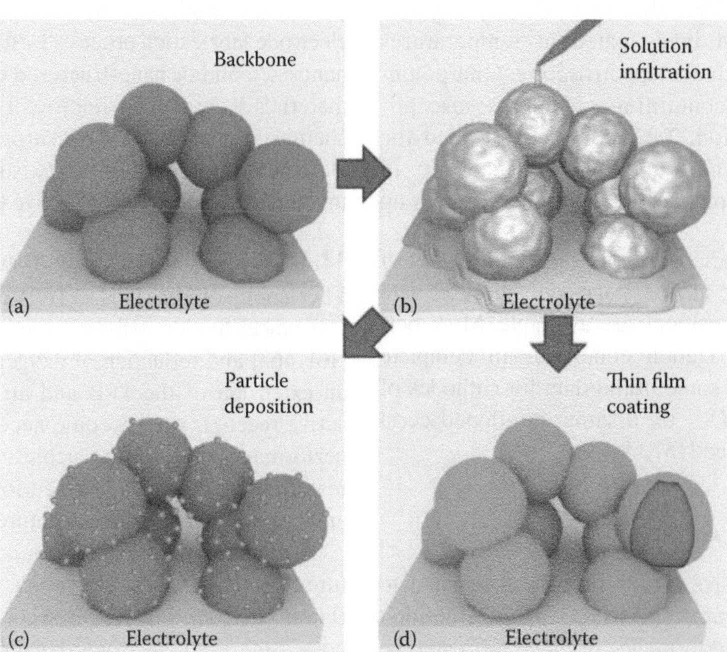

FIGURE 12.21 Schematic of a typical infiltration process: (a) an as-fired electrode backbone; (b) a process of solution drops entering into the electrode backbone; two typical morphologies of infiltrated electrode after thermal treatment: (c) particle deposition and (d) thin film coating.

surface phase (catalyst). A liquid solution or sol, containing the stoichiometric metal salt precursors with certain favorable surfactants and complex agents, is introduced into a presintered backbone (Figure 12.21b). After subsequent thermal treatment, two typical morphologies can be observed: discrete particles, as shown in Figure 12.21c, or a continuous and conformal thin film, as shown in Figure 12.21d. The catalyst/electrode coating introduced by infiltration can be fired at temperatures much lower than that needed for the backbone to form the desired phases. Furthermore, the infiltration solution may be engineered to allow control of the resultant backbone/infiltrate morphology. Morphological control of the infiltrate has been shown to influence electrode performance [159].

12.4.4.2 Noble Particle Deposition on MIEC Cathode

As the operating temperature is reduced, electrode polarization losses increase. The cathode thus becomes the limiting component for further progress. One way to overcome this limitation may be adding a certain amount of noble metal, such as palladium, silver, or platinum, to the cathode. The addition of noble metal phases to an active cathode layer is mainly used to enhance the ORR. The melting point of silver is 961°C. Metallic silver is a potential component for the cathode in SOFCs operated at less than 800°C because of its good catalytic activity, high electrical conductivity, and relatively low cost. A composite cathode consisting of high ionic conductivity copper-substituted bismuth vanadates ($Bi_2V_{0.9}Cu_{0.1}O_{5.35}$) and silver was used for low-temperature SOFCs by Xia et al. [169]. This composite cathode exhibits a remarkable catalytic activity for oxygen reduction at 500–550°C and greatly reduces the cathode–electrolyte interfacial polarization resistances down to about 0.53 Ω cm^2 at 500°C. The power density of an anode-supported cell consisting of a NiO–GDC anode, a GDC electrolyte, and a Ag–$Bi_2V_{0.9}$–$Cu_{0.1}O_{5.35}$ composite cathode reaches 0.231 W cm^{-2} at 500°C. Recently, Sakito et al. [170] observed that the infiltration of ~18 wt.% Ag fine particles into LSCF results in ~50% enhancement of power density. The Ni–GDC/GDC/LSCF–Ag cell shows a maximum power density of 0.415 W cm^{-2} at 530°C, which is very attractive for IT-SOFCs. Although silver-based cathodes show good electrochemical properties, it should be pointed out here that oxygen is easily dissolved into silver and forms an Ag/Ag_2O solid solution. In addition, even far below its melting point, silver is relatively mobile. Therefore, these concerns should be addressed before a long-term application of silver-based cathodes in SOFCs.

Pd is another noble metal widely used to improve the cathode performance in SOFCs. Sahibzada et al. [171] investigated the effect of the addition of Pd to an LSCF cathode and found the overall cell resistances decreased by 15.5% at 650°C and 40% at 550°C, indicating a promotion effect of Pd. However, this enhancement is not sustainable and is followed by a rapid degradation of power density in less than 24 h compared with reference cells fabricated without Pd additions. The alloying of Pd particles with Pt that has presumably migrated from the cathode current collector may be responsible for the observed degradation phenomenon, which is confirmed by scanning electron microscopy (SEM)/energy dispersive x-ray analysis (EDX) examinations. This is also indirectly confirmed by Haanappel's results [172] that the addition of Pt to cathode materials does not positively affect the electrochemical performance of cells. They found that neither infiltration of the cathode with Pd solution nor mixing with Pd black results in a positive effect. The catalytic effect was only observed with Pd loading on activated carbon. The well-dispersed Pd particles on an activated carbon support may be the reason for the increased catalytic effect. Usually, the particle size, distribution, and crystallinity of noble metals in the cathode play an important role on their catalytic activity. Therefore, a similar composition of materials prepared by different methods may show different results. As for the application of Pd on activated carbon, the stability and lifetime under SOFC operating conditions are potential concerns. It should be pointed out here that the addition of noble metals, in particular Pt, does not seem a good solution to improve the performances of cells since one target to commercialization of SOFC is to decrease price.

12.4.4.3 Thin Film Coating on MIEC Cathode

Another technical challenge associated with the LSCF cathode is the material's susceptibility to changes in the surface chemistry, structure, and morphology over time under high-temperature annealing and/or operating conditions. Such changes have been associated with strontium segregation [173,174]. Application of a continuous coating is predicted to suppress the segregation, and dense continuous LSM films have been successfully fabricated with the desired structure, composition, morphology, and thickness by two kinds of solution infiltration processes: nonaqueous [175] and water based [176]. In the former approach, 2-methoxyethanol and acetic acid were used to replace water as solvents, and strontium acetate and manganese acetate as metal organic precursors to replace nitrate precursors. In the latter one, only ethanol was added to improve the wettability, and proper amounts of polyvinyl pyrrolidone and glycine were employed as a surfactant and a complexing agent, respectively. This approach was straightforward, economically effective, and less chemically hazardous than alternatives. With application of the LSM infiltration, continuous and dense LSM films on LSCF substrates were fabricated [177,178]. The LSM-infiltrated LSCF cathode exhibited improved performance within a period of durability testing. The cell with LSM-infiltrated LSCF cathode initially had lower performance than the cell with blank LSCF cathode. However, the cell displayed a time-dependent activation that led to a considerable increase in performance in the first 200 h of operation. This increase for the infiltrated cell was drastically different from the rapid decrease in performance of the blank LSCF cell in the same time period. Given that the fabrication process of the cells was identical, the performance difference was attributed to the cathode.

12.4.5 Summary

Since cathode polarization still contributes considerably to energy loss in SOFC operation, one of the critical challenges facing the development of a new-generation SOFC is to create a highly active and stable cathode. In terms of the composition, oxygen nonstoichiometry and a presence of defects have a great effect on the ionic and electronic transport properties of cathode materials. Sr doping at the A-site or Sc substitution at the B-site in perovskite materials enhances the ionic conductivity, while Fe and/or Co doping on the B-site increases the electronic conductivity. However, Co doping also has great influence on the TEC of cathode materials. Alternatively, doping with a cation that contains multiple valence states, such as Pr or Sm, tends to improve the electronic conductivity. Although the electrochemical performance of some alternative cathode materials is significantly higher than that of conventional LSM cathodes, problems persist in the areas of chemical stability and thermal expansion match with other cell components. Cathode performance depends on the intrinsic properties of its components and on the microstructures. Significant progress has been made in performance enhancement in electrocatalytic activity and durability of the state-of-the-art cathodes through surface modification. Certain challenges of solution infiltration in SOFC cathode still exist: (i) morphological stability issues of nanoparticle-coated cathodes at SOFC operating temperatures, owing to sintering; (ii) additional cost associated with the infiltration process; and (iii) detrimental effects on the performance when unfavorable reactions occur between the infiltrate and the electrode backbone. However, the infiltration process has been demonstrated to be an effective approach for the modification of the electrode surface for better performance and stability.

12.5 INTERCONNECT

Irrespective of planar or tubular cell configuration, the role of interconnect is literally 2-fold; it provides electrical connection between the anode of one cell to the cathode of the neighboring one. It also acts as a physical barrier to protect the air electrode material from the reducing environment of the fuel on the fuel electrode side, and it equally prevents the fuel electrode material from contacting with the oxidizing atmosphere of the oxidant electrode side. Figure 12.22 shows the schematic of planar design for an SOFC. The criteria for the interconnect materials are the most stringent of all cell components. In particular, the chemical potential gradient stemming from considerable oxygen partial pressure differences between the oxidant and fuel places severe constraints on the choice of material for the interconnect. To perform their intended functions, interconnects need to fulfill the following conditions [179]:

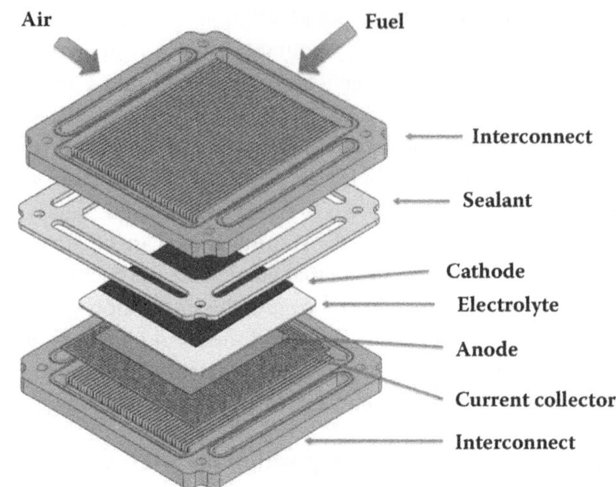

FIGURE 12.22 Planar design of an SOFC.

1. Under the SOFC operating environments, the interconnect must exhibit excellent electrical conductivity with preferably nearly 100% electronic conduction. This implies that not only the electronic transference number should be high, but also the absolute magnitude of the electrical conductivity should be reasonably large. In an ideal situation, the ohmic loss due to the introduction of an interconnect is noticeably small so that the power density of a stack does not show profound drop as compared with that of an individual cell. A value of 1 S cm^{-1} is a well-accepted minimum electrical conductivity for the usefulness of interconnects in the SOFC community [180]. This value is almost one to two orders of magnitude larger than the electrical conductivity of yttria-stabilized zirconia (YSZ), which is typically 0.1 S cm^{-1} at 1000°C and 0.02 S cm^{-1} at 800°C. YSZ is by far the most widely used solid electrolyte due to its excellent stability in both reducing and oxidizing atmospheres [36].
2. The interconnect should have adequate stability in terms of dimension, microstructure, chemistry, and phase at an operating temperature of around 800°C in both reducing and oxidizing atmospheres, since they are exposed to oxygen on one side and fuel on the other. The oxygen partial pressure on the cathode side ranges roughly from 10^{-4} to $10^{-0.7}$ atm, whereas on the anode side, it varies from 10^{-18} to 10^{-8} atm. Therefore, a significant oxygen partial pressure gradient builds up on two sides of the interconnect. Any dimensional change in the presence of oxidizing and reducing atmospheres is likely to yield mechanical stress that may be high enough to bring about cracking or warping to the sealing, thus drastically deteriorating the overall stack performance. The microstructure of interconnect should be relatively immune to chemical potential gradient so that no conceivable variation of electrical conductivity develops during its expected lifetime service. From the standpoint of environment conservation, the interconnect should preferably not contain any volatile species that subsequently react with possible contaminant gases such as H_2S or CO_2 in fuel. Phase transformation in the interconnect during operation inevitably accompanies drastic changes in its pertinent properties, and thus should be avoided [181].
3. The interconnect should display exceptionally low permeability for oxygen and hydrogen to minimize

the direct combination of oxidant and fuel during cell operation. It is clear from the Nernst equation that a small decrease in the partial pressure of oxygen and hydrogen, resulting from leakage of interconnects, may lead to a noticeable decline in the OCV [6–10]. This will significantly reduce the electrical efficiency of the cell. Cell efficiency is defined as the ratio between operating voltage and ideal voltage that is constant at the given condition (for instance, the ideal voltage is 1.227 V at 25°C for the reaction of pure hydrogen and oxygen gases). It is apparent from the following equation that a lower OCV leads to a lower operating voltage:

$$E^0 = E - IR_i - (\eta_a + \eta_c)$$

where E is OCV dependent on the temperature and type of fuel used, IR_i is the ohmic loss, I is the cell current, R_i is the internal resistance of the cell, and η_a and η_C are the anode and cathode polarization, respectively. For ceramic interconnects, the gas imperviousness can only be achieved via full densification during fabrication. Moreover, the high density of interconnects should be maintained throughout their service time.

4. The TEC of the interconnect should be comparable to those of electrodes and electrolyte between ambient and operating temperatures, so that the thermal stress developed during stack startup and shutdown could be minimized. Since the YSZ electrolyte shows a TEC around 10.5×10^{-6} °C^{-1}, the TEC of the interconnect is preferably close to this value. This constraint is more rigorous for ceramic interconnects than metallic ones, as metallic materials are generally more compliant and flexible to somehow accommodate stresses than their ceramic counterpart [36].
5. No reaction or interdiffusion between the interconnect and its adjacent components, specifically the anode and cathode, is allowed to occur under operation conditions. Reasonably good chemical compatibility is of extreme importance and constitutes a challenging task for SOFC stack development given its notably higher service temperature than other fuel cells. Any unexpected intermediate layers present at boundaries between the interconnect and its neighbors tend to not only substantially increase ohmic loss, but also markedly promote polarization losses. As a consequence, the stability of the interconnect with respect to its contacting materials is crucial to ensuring a sustained desirable stack performance [182].
6. The interconnect should possess fairly good thermal conductivity; 5 Wm^{-1} K^{-1} is considered to be the low limit. Especially in the case of planar stack configure ration, excellent thermal conductivity is highly desired and its advantage can be fully demonstrated. An interconnect with high thermal conductivity allows the heat generated at the cathode to be conducted to the anode where the endothermic fuel-reforming reaction takes place, which greatly facilitates the replacement of external reforming with internal reforming. The immediate benefit will be a measurable cost decrease of the stack as a whole [39].
7. Excellent oxidation, sulfidation, and carburization resistances are required attributes for the interconnect to qualify for application in SOFC-like environments. This does not pose any concern as long as perovskite-type ceramic oxides are utilized since the lattice diffusion coefficients of sulfur and carbon are negligibly small. The issue does arise as metallic interconnects are considered because formation of metal oxide scales in SOFC gases is a thermodynamically favored inevitable process [183]. The presence of sulfur and carbon-bearing gas species might exacerbate the issue [184]. Comparatively, resistance to CO_2 and CO gas attack is of more significance since the sulfur-containing gases can be largely eliminated by prior desulfurization treatment.
8. The interconnect should be easy to fabricate, which plays a key role in promoting mass production. The costs of raw materials and the manufacture processes for the interconnect are also supposed to be as low as possible so that they will not present hurdles to commercialization. Cost reduction of the interconnect is particularly attractive for anode-supported planar SOFC since it is the bulkiest part of all components [39].
9. The interconnect should also show adequate high-temperature strength and creep resistance. This requirement is of special relevance to the planar SOFC where the interconnect serves as a structural support.

12.5.1 Ceramic-Based Interconnect

Over the past several decades, efforts on the interconnect development have been primarily focused on the family of complex ceramic oxides with perovskite structure. Only a few such oxide systems can satisfy the rigorous requirements for the interconnect materials in SOFC. Lanthanum chromite ($LaCrO_3$) is currently the most common candidate material since it exhibits relatively high electronic conductivity in both fuel and oxidant atmospheres, moderate stability in the fuel cell environments, as well as fairly good compatibility with other cell components in terms of phase, microstructure, and thermal expansion. An earlier study [183] on $LaCrO_3$ indicates that it is a p-type conductor and becomes nonstoichiometric through the formation of cation vacancies. The negatively charged cation vacancies are electrically compensated by the concomitant appearance of positively charged electron holes. Electrical conduction in the undoped $LaCrO_3$ occurs by the small polaron mechanism via transport of electron holes. The p-type nonstoichiometric reaction is given by Equation 12.3

$$\frac{3}{2}O_2 \leftrightarrow V_{La}''' + V_{Cr}''' + 3O_O^x + 6h^n \quad (12.3)$$

where V_{La}''' and V_{Cr}''' refer to the La and Cr vacancy, respectively; O_O^x is the oxygen site; and h^n is the electron hole. To improve the electrical conductivity and TEC, $LaCrO_3$ is often doped at lanthanum or chromium, or both sites of the perovskite structure, for practical applications. Because of ionic radius similarity, strontium and calcium tend to replace La ions, whereas magnesium, iron, nickel, copper, and cobalt tend to take over Cr ions. As a matter of fact, in the tubular SOFC configuration, doped $LaCrO_3$ is still the most preferred and extensively employed interconnect.

The most daunting challenge for $LaCrO_3$ or doped $LaCrO_3$, up to now, is its extremely inferior sintering behavior in air due to easy volatilization of Cr(VI) species. It is almost impossible to sinter $LaCrO_3$ to full density in oxidizing atmospheres. The poor sinterability of $LaCrO_3$ has been ascribed primarily to the development of a thin layer of Cr_2O_3 at the interparticle neck at the initial stages of firing [185]. It is also noted that chromium evaporation, which would give rise to a chromium-deficient composition, retards sintering. To ensure full densification, several measures can be considered, although the added manufacturing complexity and cost are inevitably introduced. The proposed approaches include (i) developing a new approach to synthesize highly reactive $LaCrO_3$ powders. This holds the potential of lowering the sintering temperature by 100°C or 200°C. (ii) Carrying out sintering in reducing atmosphere followed by oxidation treatment to restore its high electrical conductivity. (iii) Adjusting the composition deliberately to assist in the formation of nonstoichiometric perovskite structure [186,187]. The sinterability of $LaCrO_3$ has been found to vary significantly with minor alteration in the material stoichiometry, although the mechanisms involved are complex and remain to be identified. It has also been reported that doping, which might increase the vacancy concentration or suppress the chromium volatilization, is effective in improving the density. An inherent disadvantage of the method is the formation of liquid phase that adversely influences the electrical conductivity and stability. (iv) Improving the sintering densities via conventional liquid-phase sintering process. This can be accomplished through the introduction of a liquid-phase sintering aid to enable the full densification of doped $LaCrO_3$ in oxidizing atmospheres to be achieved. For instance, Armstrong et al. added vanadium into the chromium site to form $Sr_3(VO_4)_2$, a proven effective sintering aid for acceptor-doped $LaCrO_3$. Unfortunately, the long-term stability problems associated with the liquid phase sintering may arise. (v) Firing $LaCrO_3$ between Cr_2O_3 plates (sandwich configuration) coupled with fast heating and cooling schedule improves the densification process [188,189].

12.5.2 Metallic Interconnect

Metallic interconnect was originally developed to get around the difficulties encountered in the design of planar SOFC, in particular the electrolyte-supported planar SOFC. One of the motivating forces behind the revolution in interconnect technology stems from the advent of an anode-supported planar SOFC design that has progressed substantially over the past several years. The concept of an anode-supported SOFC, in principle, enables the doped $LaCrO_3$ interconnect to be replaced by a metallic one due to the reduction of operating temperature below 800°C. The use of a 10–20-μm-thick electrolyte in the anode-supported version leads to a marked decrease of electrolyte ohmic polarization in comparison with the electrolyte-supported configuration where the electrolyte thickness is typically around 150 μm. A higher power density and easier fabrication are the two main features of an anode-supported planar configuration [190]. Recent works by Virkar et al. [191–193] have shown that, with the optimization of electrode structures to minimize activation and concentration polarization, a maximum power density in excess of 1.8 W cm^{-2} can be realized for an individual cell operating at 800°C. The replacement of electrochemical vapor deposition with tape casting [194–196] and plasma spraying [197–201] for the preparation of thin film electrolyte represents a major success in reducing the fabrication complexity and cost. It should be pointed out that the interconnect in the anode-supported design is one of the bulkiest items in the SOFC, and thus the total cost of the interconnect accounts for the largest raw material cost of the four main components. Apart from the roles of electrical connection between various cells and separation of fuel and oxidant gases, the bipolar plate (interconnect) also acts as a mechanical support for ceramic parts, constructional connection to the external inlets and outlets, as well as channel for distributing the gas in co-, cross-, and/or counterflow configurations. Obviously, $LaCrO_3$ ceramics cannot fulfill these conditions because it is costly to fabricate large and complex interconnects and the price of raw ceramics is high. Therefore, high cost and inferior workability of the ceramics are major restrictions in the production of large components.

By contrast, there are many advantages to preparing the interconnect from a metallic material, including higher electronic and thermal conductivity than $LaCrO_3$, low cost, easy manufacture, and good workability. The ohmic losses in the metallic bipolar plate are small enough to be neglected. The excellent heat conduction allows the heat generated at the air electrode to be easily transported to the fuel electrode where the endothermal reforming reaction occurs (internal reforming) and eliminates the use of cooling fluid that is usually excess air at the cathode electrode. Besides, a metallic "window-frame" design allows the use of relatively small areas of electrolyte/electrode structures that are in contact with the large metallic plate [202].

12.5.2.1 Problems for Metallic Materials as Interconnect

The application of ferritic stainless steel still poses many challenges even at reduced temperatures [203], which includes (i) unacceptably high oxidation rate, (ii) occurrence of buckling and spallation of the oxide scale when subjected to thermal cycling, and (iii) volatilization of high-valence Cr species in the form of CrO_3 or $Cr(OH)_2$.

12.5.2.2 Excessive Growth and Spallation of Oxide Scale

Evaluation of oxidation behavior indicates that chromia scales on ferritic stainless steels can grow to micrometers or even tens of micrometers after thousands of hours of exposure in the SOFC environment in the intermediate temperature range. Even without considering spalling and cracking, which are likely to occur, this scale growth would lead to an ASR increase and accompanying degradation in stack performance that are likely to be unacceptable. Use of optimized contact materials, such as a highly conductive oxide paste that is applied between the interconnect and cathode, can decrease the overall interfacial resistance [204]. However, as the thickness of the scale keeps increasing, spallation will occur due to the increased thermal stress between coatings and substrate, or, in a severe case, loss of hermeticity of the interconnect layer. Figure 12.23 shows the SUS 430 scale spallation after 1900 h of oxidation [205].

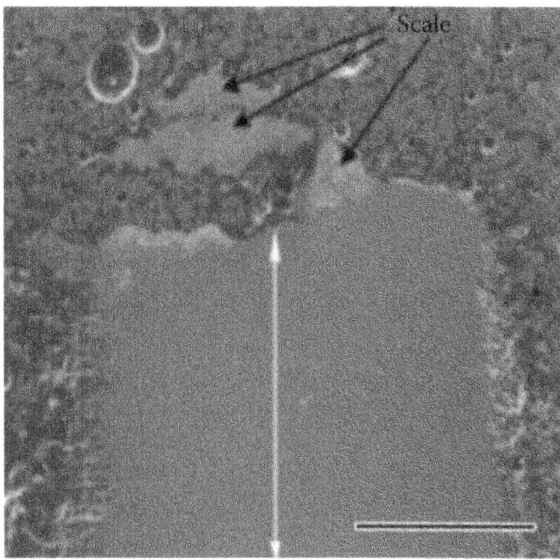

FIGURE 12.23 SEM cross-section micrograph of uncoated UNS430 stainless-steel rods after 1900 h oxidation in air at 800°C. (From X.H. Deng, P. Wei, M.R. Bateni, and A. Petric, *J. Power Sources*, 160, 1225, 2006.)

12.5.2.3 Chromium Poisoning

The vaporization of chromium from the metallic interconnect can lead to severe degradation of the electrical properties of SOFC. This has been observed by different groups at the cathode side of SOFC using Cr_2O_3-forming alloys as interconnects with Y_2O_3-doped ZrO_2 as the solid electrolyte and LSM as the cathode [206,207]. It is based on the vaporization of Cr_2O_3 from interconnect surface as $CrO_3(g)$ or $CrO_2(OH)_2(g)$ as major gaseous species with chromium in the 6+ oxidation state. The equilibrium gas pressures for CrO_xH_y species as a function of oxygen partial pressure at 800°C is displayed in Figure 12.24. The vapor pressure is becoming higher at higher oxygen partial pressures. Figure 12.25 presents the temperature dependence of vapor pressure of different volatile chromium species. It is clear that $CrO_2(OH)_2(g)$ shows the highest vapor pressure in the SOFC operation temperature range of 800–1000°C [208].

$$CrO_3(g) + H_2O(g) = CrO_2(OH)_2(g) \quad (12.4)$$

Chromium-containing vapor species formed from the interconnect material will electrochemically or chemically be reduced at the TPB. The resulting deposition can block the active electrode surface and degrade cell performance. The vapor pressures are higher in air, so such poisoning is most likely to occur at the cathode. This degradation can be represented by a decrease in cell voltage or an increase (i.e., more negative) in cell overvoltage [208,209]. The schematic diagram of the Cr poisoning process is displayed in Figure 12.26 [210].

12.5.3 Contact Resistance of Metallic Interconnect

Since a layer of protective oxide scale inevitably develops on a metal surface in an oxidizing environment at elevated temperatures, the ultimate commercialization of metallic interconnects depends, to a large extent, on the success of manipulating the increase in the contact resistance associated with the formation of the oxide layer. The large contact resistance of metallic interconnects has constituted a major concern due to the drastic drop of the electrical efficiency and premature failure of the stack over the projected service lifetime (40,000 h). With respect to this particular application, requirements for the oxide scale are as follows: (i) the growth rate of oxide in both cathodic (air) and anodic (fuel gas plus water vapor) atmospheres should be as sluggish as possible; (ii) the oxide scale should be strongly adherent to the alloy substrate to withstand thermal cycling without any spallation or delamination; (iii) the oxide scale should be homogeneous in terms of both microstructure and thickness, and be free of porosity or anomalous grain growth; and (iv) the oxide scale should be a good electronic conductor at the operating temperature.

The contact resistance of an oxidized metallic interconnect is usually characterized by the ASR, which is the product of electrical resistivity of the studied layer and its thickness. The acceptable ASR level for the metallic interconnect during service is generally considered to be below 0.1 Ω cm^2. For alloys that have been exposed to air at the operating temperature of SOFC, e.g., 800°C, for a period of time, oxide scale forms on both sides of the sample. The ASR of such an oxidized alloy can be expressed as

$$ASR = \tau_s l_s + 2\tau_o l_o \quad (12.5)$$

where τ_s and l_s are the resistivity and thickness of alloy substrate, respectively, and τ_o and l_o are the resistivity and thickness of oxide scale, respectively. In comparison with the resistivity of the oxide, the resistivity of the metallic substrate is so small that the contribution of the first term in the above equation can be neglected. As such, the ASR of an oxidized metallic interconnect is overwhelmingly dominated by that of the oxide layer on both surfaces so that

$$ASR = 2\tau_o l_o \quad (12.6)$$

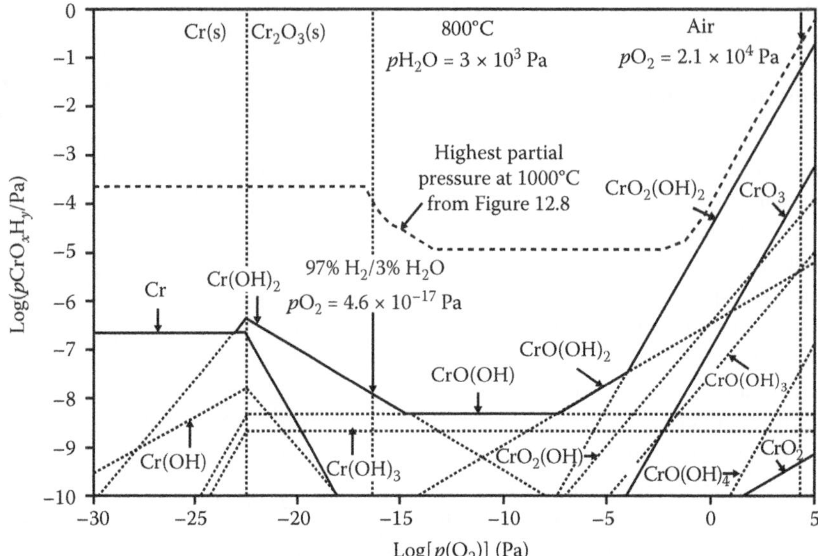

FIGURE 12.24 Equilibrium vapor pressures of chromium–oxygen–hydrogen gas species at 800°C with a water vapor pressure of 3 kPa using thermodynamic data from Ebbinghaus. (From J.W. Fergus, *Mater. Sci. Eng. A*, 397, 271, 2005.)

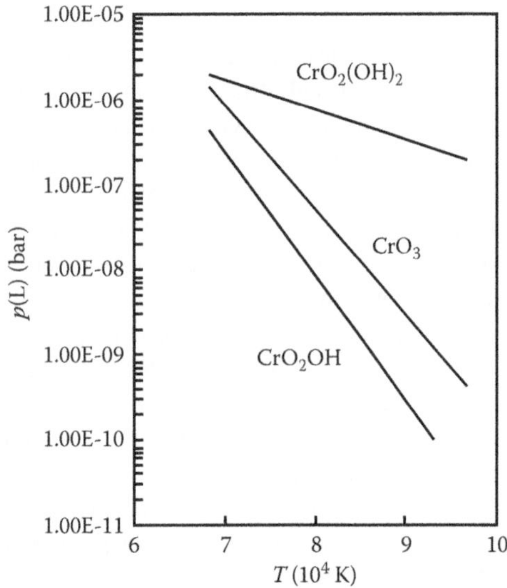

FIGURE 12.25 Vapor pressures of different volatile chromium species as a function of temperature. (From J.W. Fergus, *Mater. Sci. Eng. A*, 397, 271, 2005.)

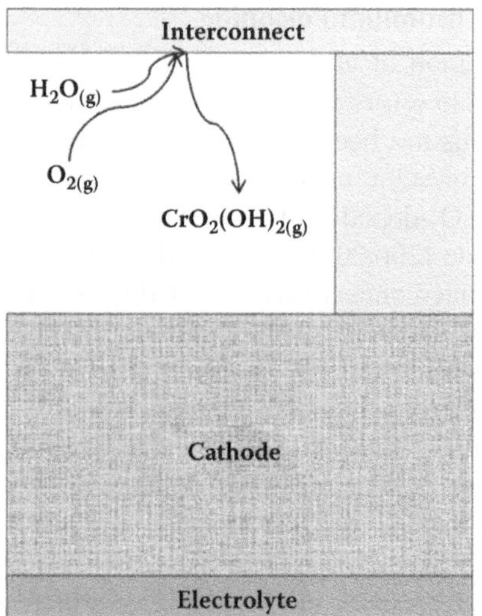

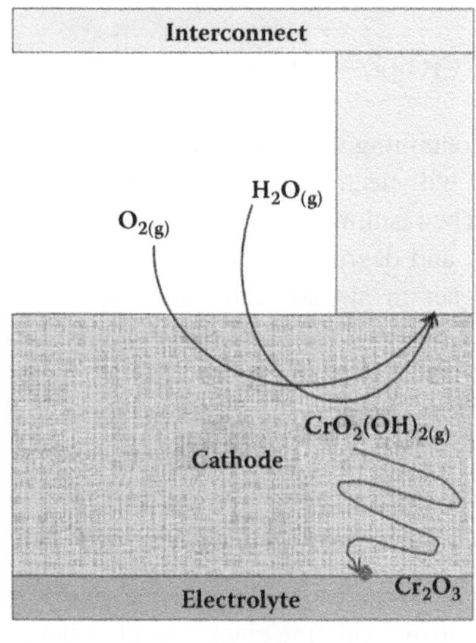

FIGURE 12.26 Schematic diagram of the Cr poisoning process.

It is clear that the occurrence of cracking, either within the oxide layer or along the scale/substrate interface, and the presence of porosity leads to a significant increase in the ASR of an oxide scale. Thereby, to facilitate the following derivation, a metallic interconnect is assumed to be free of such flaws and its resistivity can be represented by Equation 12.6.

The ASR of a metallic interconnect is very much influenced by the growth kinetics of the oxide layer upon its long-term exposure to oxidizing atmospheres. Essentially, the oxidation process involves the chemical reaction of a metal with gaseous oxygen in the atmosphere to yield a layer of a protective and thermodynamically stable oxide through either the inward diffusion of oxygen, or the outward diffusion of alloying elements, or both. The thickness of the oxide scale as a function of time (t) at a constant temperature is well documented to obey the general form of [211]

$$l_o^n = K_p t \quad (12.7)$$

where n is an exponent reflecting the oxidation mechanism, and K_p is the growth rate constant that depends on the absolute temperature T and the activation energy for the diffusion E_{ox} of the rate-limiting species, and can be empirically represented by the following expression:

$$K_p = K_o \exp\left(\frac{-E_{ox}}{kT}\right) \quad (12.8)$$

where K_o is a preexponent constant and k is the Boltzmann constant. It has been established that, in most cases, the growth of an oxide scale at steady-state oxidation follows a parabolic behavior, which means that the exponent n in Equation 12.7 equals 2 and K_p is often termed as the parabolic rate constant.

Likewise, the conduction of a metal oxide is a thermally activated process that involves the movement of small polarons via the transport of holes (or vacancies). The electrical resistivity, which is the inverse of conductivity, can be given as

$$\tau_o = \frac{1}{\sigma} = \frac{T}{\sigma_o \exp(-E_{co}/kT)} = \frac{T}{\sigma_o}\exp\left(\frac{E_{co}}{kT}\right) \quad (12.9)$$

where τ_o is a preexponent constant and E_{co} is the activation energy barrier for the conduction process. Therefore, Equation 12.6 can be expressed as

$$\text{ASR} = 2\frac{\sqrt[n]{K_p t}}{\sigma_o} T \exp\left(\frac{-(1/n)E_{ox} + E_{co}}{kT}\right) \quad (12.10)$$

In a situation where the oxide scale thickens in a parabolic fashion

$$\text{ASR} = 2\frac{\sqrt{K_p t}}{\sigma_o} T \exp\left(\frac{-(1/2)E_{ox} + E_{co}}{kT}\right) \quad (12.11)$$

By plotting ASR/T versus $1/T$ on a logarithmic scale, the activation energy term encompassing the contributions of both oxidation and conduction can be attained. It can be readily seen that the ASR changes in a parabolic manner with time at a fixed temperature. In principle, metals that develop slow-growing oxide scale with high electronic conductivity are highly preferred for interconnects. However, these two attributes are mutually exclusive for most metal oxides [53]. Any approach that is designed to suppress the oxidation kinetics and promote the electronic conductivity of the oxide scale in the SOFC operating environment is likely to lower the ASR value of the interconnect. The approaches that have been demonstrated to be effective in this aspect include (i) doping the oxide with an oxygen reactive element that alters the overall oxidation mechanism, (ii) doping the oxide with heterovalent metal ions that increase the hole concentration, and (iii) providing a complex oxide coating of adequate thickness that is impermeable to oxygen gas.

Before examining and comparing ASRs of various metallic interconnects under current development, and evaluating their application prospects to the planar anode-supported SOFC, it is necessary to describe the measurement approach that is well accepted and widely employed in many studies, although some minor details may vary. As shown schematically in Figure 12.27, basically, the contact resistance of the oxide scale is measured through a two-point, four-wire probe approach. A contact current in the order of 10 mA to several hundred milliamperes is introduced, and the voltage across the sample is measured with a multimeter. Different currents are employed initially to establish the fact the system exhibits ohmic behavior. Platinum gauzes are placed onto the test sample using platinum paste, and the measurement wires are spot welded to the gauzes. To ensure good contact between the gauze and the surface, an adequate load is applied in the through-thickness direction. In some cases, the platinum paste and gauze are replaced by platinum tab secured through a spring load of appropriate magnitude.

12.5.4 Materials for Metallic Interconnects

12.5.4.1 Chromium-Based Alloys

At temperatures as high as 900–1000°C, chromium-based oxide dispersion strengthened (ODS) alloys were specially used to replace $LaCrO_3$. A representative alloy is Ducrolloy (Cr–5Fe–1Y_2O_3) that was specifically designed by Plansee Company to match the TECs of other SOFC components. Some other chromium-based ODS alloys include Cr–5Fe–1.3La_2O_3, Cr–5Fe–0.5CeO_2, Cr–5Fe–0.3Ti–0.5Y_2O_3, and others.

The reason to choose chromium-based (chromia formation) alloys is because chromia has high conductivity compared with other oxides [212]. However, owing to its high chromium content, the chromium poisoning for cathode and excessive chromia growth are inevitable. The mechanism of chromia growth and chromium poisoning at the cathode side is shown in Figure 12.28 [213]. An excessively grown chromia layer will cause spallation after thermal cycles. Furthermore, ODS alloys are more difficult and costly to fabricate. Since melting can affect the dispersion of the oxides, these techniques are powder metallurgy based and designed to produce near-net-shape components. For example, a powder metallurgy technique for directly producing sheets of interconnects has been reported; however, the cost of the alloy was still very high.

12.5.4.2 Fe–Cr-Based Alloys

To get a continuous chromia layer, the substrate alloy should have enough Cr content. The critical Cr content has been summarized in the literature [214], and the critical minimum Cr

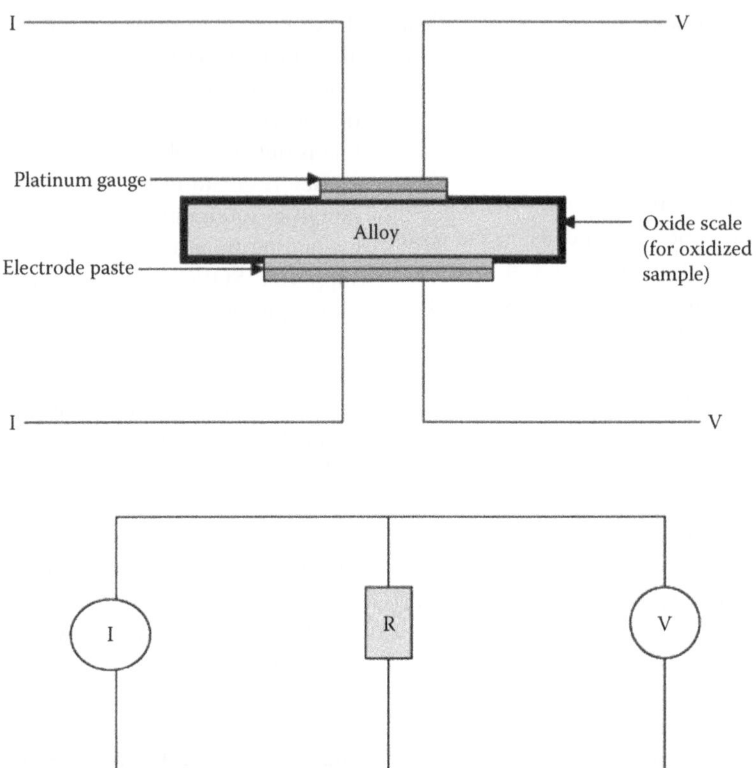

FIGURE 12.27 Schematic drawing for the measurement of area specific resistance across the sample with scale.

content is approximately 20–25% in order to ensure the formation of a protective, continuous Cr_2O_3 scale. Note that low Cr content (5% and 10%) has also been used in interconnects, but the oxidation resistance was reduced significantly when lowering the Cr content [215–217]. Low-Cr steels (<5% Cr) consist of nearly pure Fe oxide accompanied by internal oxide precipitates of Cr_2O_3 and/or $FeCr_2O_4$ spinels. With increasing Cr content, the scales become richer in spinel and chromia, which is accompanied by a decrease of the scale growth rate. The nominal composition of Fe–Cr-based alloys is shown in Table 12.1 [218].

Stainless steels are usually divided into four groups: (i) ferritic steels, (ii) austenitic steels, (iii) martensitic steels, and (vi) precipitation hardening steels [214]. Among them, usually the ferritic stainless steels are the most promising candidates for SOFC interconnect applications because of their body-centered cubic structure that makes the coefficient of thermal expansion (CTE) rather close to that of other SOFC materials. In addition, the processing methods of this type of alloys are rather simple [218–221]. However, the effect of substrate impurities, Si and Al, on the performance of the interconnect could not be neglected, especially silicon, which could form a continuous layer between substrate and scale. The Pacific Northwest National Laboratory (PNNL) has systematically investigated the $(Mn,Co)_3O_4$-coated SUS430 alloy (~0.5 wt.% Si), showing that ASR will increase sharply at 4000 h due to the formation of continuous silica layer. Accordingly, SUS 441, with the addition of Nb and Ti based on SUS430, has also been tested. The results showed that the ASR is rather

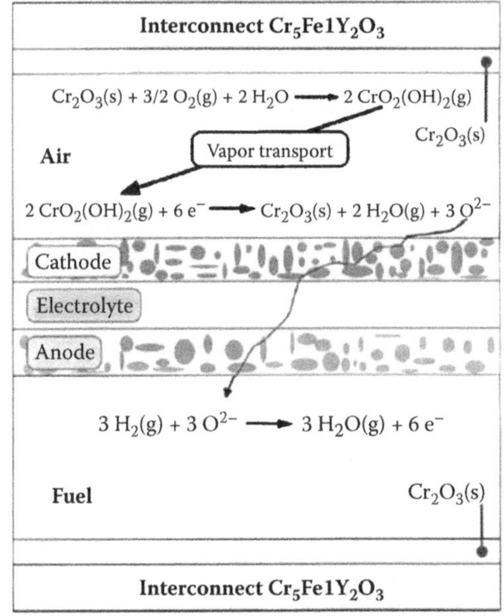

FIGURE 12.28 Cr transport at the cathode side of an SOFC. (From K. Hilpert, D. Das, M. Miller, D. H. Peck, and R. Weiss, *J. Electrochem. Soc.*, 143, 3642, 1996.)

low even for bare metal, because Nb tied up Si to prevent the formation of an SiO_2 layer at the scale/metal interface.

During the SOFC stack operation, the interconnect will face a reducing environment on the anode side and an oxidizing one on the cathode side. Therefore, exposure tests in air, to an $H_2O + H_2$ gas mixture simulating the anode side

TABLE 12.1
Nominal Composition of Fe-Based Alloys

	Concentration (wt.%)										
Alloy	**Fe**	**Cr**	**Mn**	**Mo**	**W**	**Si**	**Al**	**Ti**	**Y**	**Zr**	**La**
Fe–10Cr	Bal.	10	<0.02		<0.1						
1.4724	Bal.	13					1				
SUS 430	Bal.	16–17	0.2–1.0			0.4–1.0	<0.2				
Fe–17Cr–0.2Y	Bal.	17							0.2		
1.4016	Bal.	17									
Ferrotherm (1.4742)	Bal.	17–18	0.3–0.7		0.8–0.9	0.9–1.0					
Fe–18Cr–9W	Bal.	18			9						
Fe–20Cr–7W	Bal.	20			7	0.3	0.6			0.3	
Fe–20Cr	Bal.	20	<0.02		<0.1					0.2	0.04
AL 453	Bal.	22	0.3			0.3	0.6	0.02			0.1
1.4763(446)	Bal.	24–26	0.7–1.5	<0.05		0.4–1		<0.05			
FeCrMn(LaTi)	Bal.	16–25									
Fe–Cr–Mn	Bal.	16–25									
Fe–25CrDIN 50049	Bal.	25	0.3		0.7		0.01				
Fe–25Cr–0.1Y–2.5Ti	Bal.	25						2.5	0.1		
Fe–25Cr–0.2Y–1.6Mn	Bal.	25	1.6						0.2		
Fe–25Cr–0.4La	Bal.	25									0.4
Fe–25Cr–0.3Zr	Bal.	25								0.3	
Fe26CrTiY	Bal.	26	0.1	<0.02		<0.05	<0.05	0.3	0.4		
Fe26CrTiNbY	Bal.	26	Composition not provided but presumably the same as Fe26CrTiY with Nb								
Fe26CrMoNbY	Bal.	26	0.1	2	<0.05	<0.05	0.3	0.3			
E-Brite	Bal.	26–27	<0.1	1	0.03–0.2	<0.05	<0.05	<0.01			
Al29–4C	Bal.	27	0.3	4	0.3						
Fe–30Cr	Bal.	30	<0.02		<0.1						

service gas, and to air/H_2 gas dual atmosphere simulating real operation conditions, have been carried out in various reports. In the H_2 + H_2O gas mixture, the chromia morphology is slightly modified and the adhesion of the scale is improved. In dual atmosphere, the scales formed in air contained iron-rich spinel or Fe_2O_3 nodules, which were not present in the alloys exposed to air on both sides. This suggests that the mobility of iron is accelerated by hydrogen at the anode side [222,223].

12.5.4.3 Ni–Cr-Based Alloys

Compared with Fe–Cr-based alloys, Ni–Cr-based alloys always demonstrate better oxidation resistance and satisfactory-scale electrical conductivity. To get a continuous chromia layer, only 15% Cr was needed to establish reasonable resistance to hot corrosion, which is lower than Fe–Cr-based alloys, and the optimum content was 18%–19%. Also, Ni-based alloys are mechanically stronger. The nominal compositions of Ni-based alloy are shown in Table 12.2.

Most Ni–Cr-based alloys exhibited excellent oxidation resistance in moist hydrogen, growing a thin scale that was dominated with Cr_2O_3 and $(Mn,Cr,Ni)_3O_4$ spinels or Cr_2O_3 [224]; therefore, it could be used as clad metal or plated layer in the anode side [225,226]. In air oxidation, high Cr-containing alloys, such as Haynes 230 and Hastelloy S, formed a thin scale mainly composed of Cr_2O_3 and $(Mn,Cr,Ni)_3O_4$ spinel during high-temperature exposure; on the other hand, low-Cr-containing alloys, such as Haynes 242, developed a thick double-layer scale consisting of an NiO outside layer above a chromia-rich substrate, raising concern over its oxidation resistance for the interconnect applications [224].

The most significant problem to Ni–Cr-based alloys is the potential CTE mismatch to cell components. To take full advantages of the Ni-based alloys, novel designs of interconnects or stacks are necessary. The CTEs of Ni–Cr-based alloys containing W, Mo, Al, and Ti were calculated using Equation 12.12, a formulation derived for the CTE of Fe-free Ni alloys from room temperature to 700°C:

$$\mathrm{CTE} = 13.9 + 7.3 \times 10^{-2}[\mathrm{Cr}] - 8.0 \times 10^{-2}[\mathrm{W}] - 8.2 \times 10^{-2}[\mathrm{Mo}] - 1.8 \times 10^{-2}[\mathrm{Al}] - 1.6 \times 10^{-1}[\mathrm{Ti}] \quad (12.12)$$

The bracketed terms in this equation represent the concentration (in weight percent, wt.%) of the specific alloying element. According to Equation 12.12, most of the Ni-based alloys have higher CTE than that of other fuel cell components.

Jablonski and Alman from the National Energy Technology Laboratory (NETL) have developed a new series of Ni–Cr-based alloys containing W, Mo, Al, Ti, etc., that potentially can be used as an SOFC interconnect. After the test in moist

TABLE 12.2
Nominal Composition of Ni-Based Alloys for SOFC Interconnect Applications

	Concentration (wt.%)									
Alloy	**Ni**	**Cr**	**Fe**	**Co**	**Mn**	**Mo**	**Nb**	**Ti**	**Si**	**Al**
Inconel 600	Bal.	14–16	6–9		0.4–1			0.2–0.4	0.2–0.5	0.2
ASL528	Bal.	16	7.1		0.3			0.3	0.2	
Haynes R-41 (Rene 14)	Bal.	19	5	11	0.1	10		3.1	0.5	1.5
Inconel 718	Bal.	22	18	1	0.4	19				
Haynes 230	Bal.	22–26	3	5	0.5–0.7	1–2			0.3	
Hastelloy X	Bal.	24	19	1.5	1.0	5.3				
Ineonel 625	Bal.	25	5.4	1.0	0.6	5.7				
Nicrofer 6025HT	Bal.	25	9.5		0.1	0.5			0.5	0.15
Hastelloy G-30	Bal.	30	1.5	5	1.5	5.5	1.5	1.8	1	

air at room temperature and at elevated temperatures for more than 1000 h, it was found that the properties of alloy J5 were comparable to that of the commercial alloy Haynes 230. The nominal composition of J5 is Cr—12.5%, Ti—1%, Al—0.1%, Mo—22.5%, Mn—0.5%, Y—0.1%, and Ni—balance [227].

12.5.5 Modification Coating

Recently, various coatings have been developed for metallic SOFC interconnects mainly to mitigate excessive chromia growth and chromium poisoning. To qualify a coating material as viable, it should possess the following characteristics: (i) the diffusion coefficients of Cr and O in the coating should be as small as possible so that the transport of chromium and oxygen can be effectively hindered; (ii) it should be chemically compatible and stable with respect to substrate, electrodes, seal materials, and contact pastes; (iii) it should be thermodynamically stable in both oxidizing and reducing atmospheres over the applied temperature range; (iv) it must have low ohmic resistance to maximize electrical efficiency; (v) the TEC should be well matched to the substrate so that the coating is resistant to spallation during thermal cycling.

12.5.5.1 Nitride Coatings

Nitride coatings have been widely used in tool coatings owing to their superior wear resistance. On the other hand, this type of coating could supply an alternative coating for SOFC interconnect applications owing to their low resistance and high temperature stability.

Vacuum deposition, especially physical vapor deposition (PVD), has been widely used to prepare components and protective nitride coatings for SOFC interconnects, owing to the versatility of this technique as well as the ability to control composition and morphology [228]. Gannon et al. applied multilayer coatings consisting of repeated sections of CrN and a CrAlN superlattice [229]. The results showed that thinner bilayer (~1.1 nm) CrN/AlN superlattice coatings are more favorable for the SOFC interconnect application than a thicker 4.5-nm bilayer. Later, he deposited Cr–Al–O–N [230]. The introduction of oxide into nitride coatings reduces the Fe and Cr migration from the substrate. Both of these results suggest that PVD is an effective method to fabricate high-quality coatings for metallic interconnects.

TiAlN with 30% and 50% Al [231] and SmCoN [232] coatings have been investigated by our group. Both of them revealed that nitride coatings could remain stable at 700°C, which is helpful to inhibit Cr and Fe migration from substrate. In addition, a rather low ASR has been obtained in the short-term test. However, these methods have limitations, such as high capital cost and low deposition rate. Most importantly, nitride is rather unstable at temperatures higher than 600°C.

12.5.5.2 Perovskite Coatings

The structure of perovskite may be represented by the general formula ABO_3, where A is a lanthanide (such as La in most cases but sometimes Ce, Pr, or Nd) and B is a transition metal (such as Co, Mn, Fe, Cr, Cu, or V). Alkaline-earth metals such as Sr, Ca, or Ba can be substituted for sites A and a transition metal for site B [233]. For instance, the replacement of part of La^{3+} ions by Sr^{2+} can highly increase the electronic conductivity of perovskite in $LaCrO_3$, $LaMnO_3$, and $La(Co,Fe)O_3$.

$LaCrO_3$, a traditional interconnect material, has been deposited by Orlovskaya et al. [234] using r.f. sputtering. It was found that a two-step phase transformation occurred from the as-deposited amorphous phase. The first step is from the x-ray amorphous state to a major intermediate phase: monoclinic $LaCrO_4$ monazite. A finite amount of La_2CrO_6 phase was also formed. The second step is the transformation of $LaCrO_4$ into the $LaCrO_3$ perovskite phase. During this transition, only nanoporosity appears. These distinctive nanostructures have excellent potential for use as an SOFC interconnect coating. Another option is to coat these alloys with an oxide layer that subsequently reacts with chromia to form a chromite (or chromate). The reactive formation of an $LaCrO_3$ coating on SS-444 alloy via templated growth from thermally grown Cr_2O_3 and sputtered La_2O_3 layers has been demonstrated by Zhu et al. [235]. The results revealed that coated samples show much lower electrical resistance compared with the uncoated samples after a similar thermal exposure for 100 h at 850°C. However, this approach suffers the obvious drawback of difficulty in controlling the composition of the reaction products. Therefore, coating with perovskite directly on the surfaces of the metallic interconnects is preferred.

LSM has been widely used as an SOFC cathode material. Owing to its high electrical conductivity and thermal compatibility, and its stability in oxidizing environment, it has also been investigated as an interconnect coating. The presence of a LSM coating is crucial in maintaining the low level of contact resistance at elevated temperatures over extended periods of time. Slurry LSM coating was investigated on SUS430 substrate [236]. After sintering at 1200°C for 2 h in N_2, followed by heat treatment at 1000°C for 3 h in air, stable LSM-coated SUS 430 showed a low ASR and maintained an almost constant value for 2600 h. Additionally, interactions between LSM coatings and RA446 were explored [237]. In the early 500 h, LSM could react with Cr diffused from substrate to form $LaCrO_3$, $(Cr,Mn)_3O_4$, and Cr_2O_3. The latter two phases have nonnegligible Cr evaporation rates.

The reactions and the complicated sintering process are originated from the not fully dense coating and the inherent porosity of perovskite, which could supply a path for Cr migration. It is necessary to deposit fully dense coatings to inhibit Cr diffusion and increase conductivity. Recently, perovskite coatings of LSM and LSCF were deposited by aerosol deposition, which is based on the impact adhesion of fine particles, and nearly 95% dense coatings could be obtained. After 100 h of oxidation at 800°C, the ASR of LSM and Sr- and Fe-doped $LaCoO_3$ LSCF-coated alloys was 20.6 and 11.7 mΩ cm^2, respectively. Most importantly, no chromium was spotted in coatings by EDX line scan [238].

12.5.5.3 Spinel Coatings

Compared with perovskite, some spinels show better performance in preventing oxygen inward diffusion and Cr outward diffusion. However, Cr evaporation is still a problem for Cr-containing spinels. $MnCr_2O_4$ (Figure 12.29) has been found as a product of the reaction between chromium oxide and the Mn diffusion from Crofer 22 substrate; however, the Cr evaporation is nonnegligible. $CoCr_2O_4$ coatings cause no cell performance degradation in the 1000-h test, and a long-term test is necessary to prove the effectiveness of the coatings.

Among all Cr-free spinels, $(Mn,Co)_3O_4$ spinel is considered the most promising candidate for SOFC interconnect coatings. Previous work on Plansee Ducrolloy (Cr–5% Fe–1% Y_2O_3) indicated that an $(Mn,Co)_3O_4$ spinel layer could reduce chromium migration significantly, and predicted that the ASR at 10,000 h is as low as 0.024 mΩ cm^2 when using LSM and LSC contact paste [239]. Chen et al. reported $MnCo_2O_4$ coatings on the ferritic stainless-steel AISI430 via slurry coating followed by mechanical compaction and air heating [240]. Recently, Yang et al. at PNNL did a systematic study on $(Mn,Co)_3O_4$ spinel from conductivity, microstructure, CTE, and long-term ASR measurement of $(Mn,Co)_3O_4$ on ferritic stainless steel [241].

12.5.6 Summary

For SOFCs operating in a range of 650–800°C, cost-effective high-temperature oxidation-resistant alloys, such as ferritic stainless steels, are promising candidates for construction of interconnects. The suitability of an alloy for this particular application is dependent on its chemical composition, as well as the interconnect structure and stack design. Traditional alloys face challenges in meeting all of the requirements for the application. In recent years, considerable progress has been made in developing suitable high-temperature oxidation-resistant alloys for the interconnect application, and the degradation behavior of alloys under conditions that simulate SOFC operating conditions is better understood. However, challenges remain in terms of Cr volatility and long-term surface and electrical stability at SOFC operating temperatures that are allowed by the current technologies. As an alternative approach to bulk alloy development, surface modification of metallic interconnects via application of protective layers has proved to be viable for improvement of their stability and mitigation of adverse interactions with cell and stack components. In particular, protective layers fabricated from non-Cr-containing conductive oxides, e.g., $La_{0.8}Sr_{0.2}MnO_3$ perovskites and/or $(Mn,Co)_3O_4$ spinels, on the alloy such as Crofer22 APU or SS430 are among the most promising candidates. Further progress in terms of materials understanding and optimization will be necessary to achieve satisfactory cost-effectiveness and long-term interconnect performance.

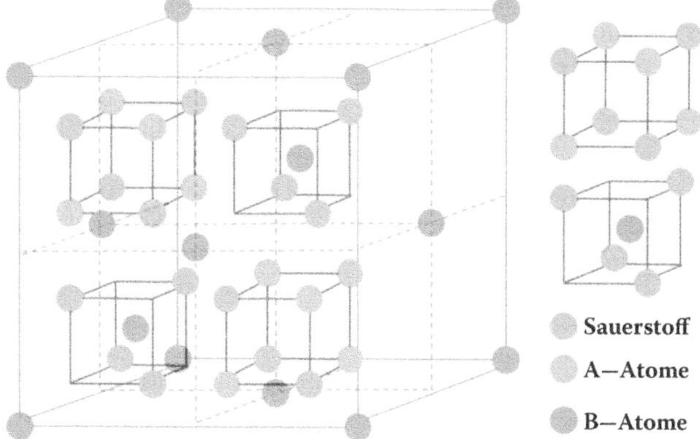

FIGURE 12.29 Unit cell of spinel structure.

REFERENCES

1. S. M. Haile, "Fuel cell materials and components," *Acta Materialia,* vol. 51, pp. 5981–6000, 2003.
2. D. Marrocchelli, S. R. Bishop, H. L. Tuller, and B. Yildiz, "Understanding chemical expansion in non-stoichiometric oxides: Ceria and zirconia case studies," *Advanced Functional Materials,* vol. 22, pp. 1958–1965, 2012.
3. H. Yahiro, Y. Eguchi, K. Eguchi, and H. Arai, "Oxygen ion conductivity of the ceria samarium oxide system with fluorite structure," *Journal of Applied Electrochemistry,* vol. 18, pp. 527–531, 1988.
4. K. Z. Fung, H. D. Baek, and A. V. Virkar, "Thermodynamic and kinetic considerations for Bi_2O_3-based electrolytes," *Solid State Ionics,* vol. 52, pp. 199–211, 1992.
5. G. Y. Meng, C. S. Chen, X. Han, P. H. Yang, and D. K. Peng, "Conductivity of Bi_2O_3-based oxide ion conductors with double stabilizers," *Solid State Ionics,* vol. 28, pp. 533–538, 1988.

6. D. W. Jung, K. L. Duncan, and E. D. Wachsman, "Effect of total dopant concentration and dopant ratio on conductivity of (DyO1.5)(*x*)–(WO3)(*y*)–(BiO1.5)1–*x*–*y*," *Acta Materialia,* vol. 58, pp. 355–363, 2010.
7. J. G. Cheng, S. W. Zha, J. Huang, X. Q. Liu, and G. Y. Meng, "Sintering behavior and electrical conductivity of Ce0.9Gd0.1O1.95 powder prepared by the gel-casting process," *Materials Chemistry and Physics,* vol. 78, pp. 791–795, 2003.
8. H. D. Baek and A. V. Virkar, "Thermodynamic investigations of Bi_2O_3–Mo (M = Ca, Sr, and Ba) systems using galvanic cells," *Journal of the Electrochemical Society,* vol. 139, pp. 3174–3182, 1992.
9. A. V. Virkar, "Theoretical analysis of solid oxide fuel cells with two-layer, composite electrolytes—Electrolyte stability," *Journal of the Electrochemical Society,* vol. 138, pp. 1481–1487, 1991.
10. J. Y. Park and E. D. Wachsman, "Stable and high conductivity ceria/bismuth oxide bilayer electrolytes for lower temperature solid oxide fuel cells," *Ionics,* vol. 12, pp. 15–20, 2006.
11. J. S. Ahn, M. A. Camaratta, D. Pergolesi, K. T. Lee, H. Yoon, B. W. Lee et al., "Development of high performance ceria/bismuth oxide bilayered electrolyte SOFCs for lower temperature operation," *Journal of the Electrochemical Society,* vol. 157, pp. B376–B382, 2010.
12. T. Ishihara, N. Jirathiwathanakul, and H. Zhong, "Intermediate temperature solid oxide electrolysis cell using $LaGaO_3$ based perovskite electrolyte," *Energy & Environmental Science,* vol. 3, pp. 665–672, 2010.
13. Z. H. Bi, B. L. Yi, Z. W. Wang, Y. L. Dong, H. J. Wu, Y. C. She et al., "A high-performance anode-supported SOFC with LDC–LSGM bilayer electrolytes," *Electrochemical and Solid State Letters,* vol. 7, pp. A105–A107, 2004.
14. T. Y. Chen and K. Z. Fung, "Comparison of dissolution behavior and ionic conduction between Sr and/or Mg doped $LaGaO_3$ and $LaAlO_{(3)}$," *Journal of Power Sources,* vol. 132, pp. 1–10, 2004.
15. T. Y. Chen and K. Z. Fung, "Synthesis of and densification of oxygen-conducting $La_{0.8}Sr_{0.2}Ga_{0.8}Mg_{0.2}O_{2.8}$ nanopowder prepared from a low temperature hydrothermal urea precipitation process," *Journal of the European Ceramic Society,* vol. 28, pp. 803–810, 2008.
16. T. Ishihara, H. Matsuda, and Y. Takita, "Oxide ion conductivity in doped $NdAlO_3$ perovskite-type oxides," *Journal of the Electrochemical Society,* vol. 141, pp. 3444–3449, 1994.
17. K. Z. Fung and T. Y. Chen, "Cathode-supported SOFC using a highly conductive lanthanum aluminate-based electrolyte," *Solid State Ionics,* vol. 188, pp. 64–68, 2011.
18. S. M. Haile, C. R. I. Chisholm, K. Sasaki, D. A. Boysen, and T. Uda, "Solid acid proton conductors: From laboratory curiosities to fuel cell electrolytes," *Faraday Discussions,* vol. 134, pp. 17–39, 2007.
19. H. Iwahara, H. Uchida, and S. Tanaka, "High-temperature type proton conductor based on $SrCeO_3$ and its application to solid electrolyte fuel cells," *Solid State Ionics,* vols. 9–10, pp. 1021–1025, 1983.
20. N. Bonanos, K. S. Knight, and B. Ellis, "Perovskite solid electrolytes—Structure, transport properties and fuel cell applications," *Solid State Ionics,* vol. 79, pp. 161–170, 1995.
21. K. D. Kreuer, "Proton-conducting oxides," *Annual Review of Materials Research,* vol. 33, pp. 333–359, 2003.
22. K. Katahira, Y. Kohchi, T. Shimura, and H. Iwahara, "Protonic conduction in Zr-substituted $BaCeO_3$," *Solid State Ionics,* vol. 138, pp. 91–98, 2000.
23. A. V. Kuzmin, V. P. Gorelov, B. T. Melekh, M. Glerup, and F. W. Poulsen, "Phase transitions in undoped $BaCeO_3$," *Solid State Ionics,* vol. 162, pp. 13–22, 2003.
24. S. C. Singhal and K. Kendall, *High Temperature Solid Oxide Fuel Cells: Fundamentals, Design, and Applications.* New York: Elsevier Inc., pp. P149–P153, 2003.
25. A. Atkinson, S. Barnett, R. J. Gorte, J. T. S. Irvine, A. J. Mcevoy, M. Mogensen et al., "Advanced anodes for high-temperature fuel cells," *Nature Materials,* vol. 3, pp. 17–27, 2004.
26. S. McIntosh and R. J. Gorte, "Direct hydrocarbon solid oxide fuel cells," *Chemical Reviews,* vol. 104, pp. 4845–4865, 2004.
27. S. P. Jiang and S. H. Chan, "A review of anode materials development in solid oxide fuel cells," *Journal of Materials Science,* vol. 39, pp. 4405–4439, 2004.
28. M. Mogensen and K. Kammer, "Conversion of hydrocarbons in solid oxide fuel cells," *Annual Review of Materials Research,* vol. 33, pp. 321–331, 2003.
29. J. Fergus, "Oxide anode materials for solid oxide fuel cells," *Solid State Ionics,* vol. 177, pp. 1529–1541, 2006.
30. N. Q. Minh and T. Takahashi, *Science and Technology of Ceramic Fuel Cells.* New York: Elsevier Inc., pp. P147–P148, 1995.
31. P. Holtappels, U. Vogt, and T. Graule, "Ceramic materials for advanced solid oxide fuel cells," *Advanced Engineering Materials,* vol. 7, pp. 292–302, 2005.
32. D. J. Brett, A. Atkinson, N. P. Brandon, and S. J. Skinner, "Intermediate temperature solid oxide fuel cells," *Chemical Society Reviews,* vol. 37, pp. 1568–1578, 2008.
33. R. N. Basu, *Recent Trends in Fuel Cell Science and Technology.* New York: Springer, pp. 286–331, 2007.
34. S. C. Singhal, "Advances in solid oxide fuel cell technology," *Solid State Ionics,* vol. 135, pp. 305–313, 2000.
35. N. Q. Minh, "Ceramic fuel cells," *Journal of the American Ceramic Society,* vol. 76, pp. 563–588, 1993.
36. D. W. Dees, T. D. Claar, T. E. Easler, D. C. Fee, and F. C. Mrazek, "Conductivity of porous Ni/ZrO_2–Y_2O_3 cermets," *Journal of the Electrochemical Society,* vol. 134, pp. 2141–2146, 1987.
37. Y. Matsuzaki and I. Yasuda, "The poisoning effect of sulfur-containing impurity gas on a SOFC anode: Part I. Dependence on temperature, time, and impurity concentration," *Solid State Ionics,* vol. 132, pp. 261–269, 2000.
38. W. Z. Zhu and S. C. Deevi, "Development of interconnect materials for solid oxide fuel cells," *Materials Science and Engineering A—Structural Materials Properties Microstructure and Processing,* vol. 348, pp. 227–243, 2003.
39. T. Matsui, T. Kosaka, M. Inaba, A. Mineshige, and Z. Ogumi, "Effects of mixed conduction on the open-circuit voltage of intermediate-temperature SOFCs based on Sm-doped ceria electrolytes," *Solid State Ionics,* vol. 176, pp. 663–668, 2005.
40. K. Ravindranathan, A. J. McEvoy, and B. El-Roustom, "Cermet anodes fully desensitized to sulfur," in *Proceedings of 8th European SOFC Forum in Lucerne, Switzerland,* 2008.
41. H. Kim, J. M. Vohs, and R. J. Gorte, "Direct oxidation of sulfur-containing fuels in a solid oxide fuel cell," *Chemical Communications,* pp. 2334–2335, 2001.
42. F. Besenbacher, I. Chorkendorff, B. S. Clausen, B. Hammer, A. M. Molenbroek, J. K. Norskov et al., "Design of a surface alloy catalyst for steam reforming," *Science,* vol. 279, pp. 1913–1915, 1998.
43. N. C. Triantafyllopoulos and S. G. Neophytides, "Dissociative adsorption of CH4 on NiAu/YSZ: The nature of adsorbed carbonaceous species and the inhibition of graphitic C formation," *Journal of Catalysis,* vol. 239, pp. 187–199, 2006.

44. S. Zha, Z. Cheng, and M. Liu, "Sulfur poisoning and regeneration of Ni-based anodes in solid oxide fuel cells," *Journal of the Electrochemical Society,* vol. 154, p. B201, 2007.
45. K. Sasaki, K. Susuki, A. Iyoshi, M. Uchimura, N. Imamura, H. Kusaba et al., "H_2S poisoning of solid oxide fuel cells," *Journal of the Electrochemical Society,* vol. 153, p. A2023, 2006.
46. K. Sasaki, K. Susuki, A. Iyoshi, S. Adachi, M. Uchimura, N. Imamura, Y. Shiratori, H. Kusaba, and Y. Teraoka, "Fuel impurity tolerance of solid oxide fuel cells," in *Proceedings of 7th European SOFC Forum in Lucerne, Switzerland*, 2006.
47. C. W. Sun, J. Sun, G. L. Xiao, H. R. Zhang, X. P. Qiu, H. Li et al., "Mesoscale organization of nearly monodisperse flower-like ceria microspheres," *Journal of Physical Chemistry B,* vol. 110, pp. 13445–13452, 2006.
48. N. Skorodumova, S. Simak, B. Lundqvist, I. Abrikosov, and B. Johansson, "Quantum origin of the oxygen storage capability of ceria," *Physical Review Letters,* vol. 89, p. 166601, 2002.
49. O. A. Marina and M. Mogensen, "High-temperature conversion of methane on a composite gadolinia-doped ceria-gold electrode," *Applied Catalysis A—General,* vol. 189, pp. 117–126, 1999.
50. M. J. Saeki, H. Uchida, and M. Watanabe, "Noble-metal catalysts highly-dispersed on Sm-doped ceria for the application to internal reforming solid oxide fuel cells operated at medium temperature," *Catalysis Letters,* vol. 26, pp. 149–157, 1994.
51. T. Hibino, A. Hashimoto, M. Yano, M. Suzuki, and M. Sano, "Ru-catalyzed anode materials for direct hydrocarbon SOFCs," *Electrochimica Acta,* vol. 48, pp. 2531–2537, 2003.
52. T. Hibino, A. Hashimoto, K. Asano, M. Yano, M. Suzuki, and M. Sano, "An intermediate-temperature solid oxide fuel cell providing higher performance with hydrocarbons than with hydrogen," *Electrochemical and Solid State Letters,* vol. 5, pp. A242–A244, 2002.
53. E. Ramirez-Cabrera, A. Atkinson, and D. Chadwick, "The influence of point defects on the resistance of ceria to carbon deposition in hydrocarbon catalysis," *Solid State Ionics,* vol. 136, pp. 825–831, 2000.
54. B. A. Boukamp, "The amazing perovskite anode," *Nature Materials,* vol. 2, pp. 294–296, 2003.
55. S. Tao and J. T. Irvine, "A redox-stable efficient anode for solid-oxide fuel cells," *Nature Materials,* vol. 2, pp. 320–323, 2003.
56. S. Zha, P. Tsang, Z. Cheng, and M. Liu, "Electrical properties and sulfur tolerance of $La_{0.75}Sr_{0.25}Cr_{1-x}Mn_xO_3$ under anodic conditions," *Journal of Solid State Chemistry,* vol. 178, pp. 1844–1850, 2005.
57. Y. H. Huang, R. I. Dass, Z. L. Xing, and J. B. Goodenough, "Double perovskites as anode materials for solid-oxide fuel cells," *Science,* vol. 312, pp. 254–257, 2006.
58. Y.-H. Huang, R. I. Dass, J. C. Denyszyn, and J. B. Goodenough, "Synthesis and characterization of $Sr_2MgMoO_{6-\delta}$," *Journal of the Electrochemical Society,* vol. 153, p. A1266, 2006.
59. P. Vernoux, M. Guillodo, J. Fouletier, and A. Hammou, "Alternative anode material for gradual methane reforming in solid oxide fuel cells," *Solid State Ionics,* vol. 135, pp. 425–431, 2000.
60. J. C. Ruiz-Morales, J. Canales-Vázquez, J. Peña-Martínez, D. M. López, and P. Núñez, "On the simultaneous use of $La_{0.75}Sr_{0.25}Cr_{0.5}Mn_{0.5}O_{3-\delta}$ as both anode and cathode material with improved microstructure in solid oxide fuel cells," *Electrochimica Acta,* vol. 52, pp. 278–284, 2006.
61. A. Sauvet, "Catalytic activity for steam methane reforming and physical characterisation of $La_{1-x}Sr_xCr_{1-y}Ni_yO_{3-\delta}$," *Solid State Ionics,* vol. 167, pp. 1–8, 2004.
62. A. Sin, E. Kopnin, Y. Dubitsky, A. Zaopo, A. S. Aricò, L. R. Gullo et al., "Stabilisation of composite LSFCO–CGO based anodes for methane oxidation in solid oxide fuel cells," *Journal of Power Sources,* vol. 145, pp. 68–73, 2005.
63. P. Vernoux, E. Djurado, and M. Guillodo, "Catalytic and electrochemical properties of doped lanthanum chromites as new anode materials for solid oxide fuel cells," *Journal of the American Ceramic Society,* vol. 84, pp. 2289–2295, 2001.
64. H. L. Tuller, "Oxygen-ion conduction and structural disorder in conductive oxides," *Journal of Physics and Chemistry of Solids,* vol. 55, pp. 1393–1404, 1994.
65. B. J. Wuensch, K. W. Eberman, C. Heremans, E. M. Ku, P. Onnerud, E. M. E. Yeo et al., "Connection between oxygen-ion conductivity of pyrochlore fuel cell materials and structural change with composition and temperature," *Solid State Ionics,* vol. 129, pp. 111–133, 2000.
66. M. Pirzada, R. W. Grimes, L. Minervini, J. F. Maguire, and K. E. Sickafus, "Oxygen migration in A(2)B(2)O(7) pyrochlores," *Solid State Ionics,* vol. 140, pp. 201–208, 2001.
67. S. Kramer, M. Spears, and H. L. Tuller, "Conduction in titanate pyrochlores—Role of dopants," *Solid State Ionics,* vol. 72, pp. 59–66, 1994.
68. S. A. Kramers and H. L. Tuller, "A novel titanate-based oxygen-ion conductor—$Gd_2Ti_2O_7$," *Solid State Ionics,* vol. 82, pp. 15–23, 1995.
69. O. Porat, C. Heremans, and H. L. Tuller, "Stability and mixed ionic electronic conduction in Gd-2(Ti1–*x*Mo*x*)(2)O-7 under anodic conditions," *Solid State Ionics,* vol. 94, pp. 75–83, 1997.
70. P. Holtappels, F. W. Poulsen, and M. Mogensen, "Electrical conductivities and chemical stabilities of mixed conducting pyrochlores for SOFC applications," *Solid State Ionics,* vol. 135, pp. 675–679, 2000.
71. M. Tournoux, M. Ganne, and Y. Piffard, "Htb-like 6-membered rings of octahedra in some new oxides—Structural aspects and related properties," *Journal of Solid State Chemistry,* vol. 96, pp. 141–153, 1992.
72. S. Tao and J. T. Irvine, "Discovery and characterization of novel oxide anodes for solid oxide fuel cells," *Chemical Record,* vol. 4, pp. 83–95, 2004.
73. P. R. Slater and J. T. S. Irvine, "Synthesis and electrical characterisation of the tetragonal tungsten bronze type phases, (Ba/Sr/Ca/La)(0) 6M*x*Nb1–*x*O_3–delta (M = Mg, Ni, Mn, Cr, Fe, In, Sn): Evaluation as potential anode materials for solid oxide fuel cells," *Solid State Ionics,* vol. 124, pp. 61–72, 1999.
74. P. R. Slater and J. T. S. Irvine, "Niobium based tetragonal tungsten bronzes as potential anodes for solid oxide fuel cells: Synthesis and electrical characterisation," *Solid State Ionics,* vol. 120, pp. 125–134, 1999.
75. A. Kaiser, J. L. Bradley, P. R. Slater, and J. T. S. Irvine, "Tetragonal tungsten bronze type phases (Sr1–*x*Ba*x*)(0.6) Ti0.2Nb0.8O3–delta: Material characterisation and performance as SOFC anodes," *Solid State Ionics,* vol. 135, pp. 519–524, 2000.
76. S. Z. Wang, M. L. Liu, and J. Winnick, "Stabilities and electrical conductivities of electrode materials for use in H_2S-containing gases," *Journal of Solid State Electrochemistry,* vol. 5, pp. 188–195, 2001.
77. S. W. Zha, Z. Cheng, and M. L. Liu, "A sulfur-tolerant anode material for SOFCsGd2Ti1.4Mo0.6O7," *Electrochemical and Solid State Letters,* vol. 8, pp. A406–A408, 2005.
78. Z. Cheng, S. Zha, and M. Liu, "Stability of materials as candidates for sulfur-resistant anodes of solid oxide fuel cells," *Journal of the Electrochemical Society,* vol. 153, p. A1302, 2006.

79. R. J. Gorte and J. M. Vohs, "Novel SOFC anodes for the direct electrochemical oxidation of hydrocarbons," *Journal of Catalysis,* vol. 216, pp. 477–486, 2003.
80. N. Q. Minh, "Ceramic fuel cells," *Journal of the American Ceramic Society,* vol. 76, pp. 563–588, 1993.
81. T. Horita, K. Yamaji, N. Sakai, Y. Xiong, T. Kato, H. Yokokawa et al., "Imaging of oxygen transport at SOFC cathode/electrolyte interfaces by a novel technique," *Journal of Power Sources,* vol. 106, pp. 224–230, 2002.
82. J. D. Fehribach and R. O'Hayre, "Triple phase boundaries in solid-oxide cathodes," *SIAM Journal on Applied Mathematics,* vol. 70, pp. 510–530, 2009.
83. J. Fleig, "Solid oxide fuel cell cathodes: Polarization mechanisms and modeling of the electrochemical performance," *Annual Review of Materials Research,* vol. 33, pp. 361–382, 2003.
84. C. Sun, R. Hui, and J. Roller, "Cathode materials for solid oxide fuel cells: A review," *Journal of Solid State Electrochemistry,* vol. 14, pp. 1125–1144, 2010.
85. P. J. Gellings and H. J. M. Bouwmeester, *The CRC Handbook of Solid State Electrochemistry.* Boca Raton, FL: CRC Press, 1997.
86. S. C. Singhal and K. Kendall, *High-Temperature Solid Oxide Fuel Cells: Fundamentals, Design, and Applications.* New York: Elsevier Advanced Technology, 2003.
87. S. J. Skinner, "Recent advances in perovskite-type materials for solid oxide fuel cell cathodes," *International Journal of Inorganic Materials,* vol. 3, pp. 113–121, 2001.
88. A. M. Chang, S. J. Skinner, and J. A. Kilner, "Electrical properties of GdBaCo2O5+*x* for ITSOFC applications," *Solid State Ionics,* vol. 177, pp. 2009–2011, 2006.
89. A. Tarancon, S. J. Skinner, R. J. Chater, F. Hernandez-Ramirez, and J. A. Kilner, "Layered perovskites as promising cathodes for intermediate temperature solid oxide fuel cells," *Journal of Materials Chemistry,* vol. 17, pp. 3175–3181, 2007.
90. Y. An, S. J. Skinner, and D. W. McComb, "Template-assisted fabrication of macroporous thin films for solid oxide fuel cells," *Journal of Materials Chemistry,* vol. 20, pp. 248–254, 2010.
91. Z. X. Xie, H. L. Zhao, Z. H. Du, T. Chen, N. Chen, X. T. Liu et al., "Effects of Co doping on the electrochemical performance of double perovskite oxide Sr2MgMoO6–delta as an anode material for solid oxide fuel cells," *Journal of Physical Chemistry C,* vol. 116, pp. 9734–9743, 2012.
92. T. Yu, X. B. Mao, and G. L. Ma, "Performance of cobalt-free perovskite La0.6Sr0.4Fe1–*x*Nb*x*O3–delta cathode materials for proton-conducting IT-SOFC," *Journal of Alloys and Compounds,* vol. 608, pp. 30–34, 2014.
93. X. J. Chen, Q. L. Liu, S. H. Chan, N. P. Brandon, and K. A. Khor, "High performance cathode-supported SOFC with perovskite anode operating in weakly humidified hydrogen and methane," *Electrochemistry Communications,* vol. 9, pp. 767–772, 2007.
94. H. Kishimoto, N. Sakai, T. Horita, K. Yamaji, M. E. Brito, and H. Yokokawa, "Cation transport behavior in SOFC cathode materials of La0.8Sr0.2CoO3 and La0.8Sr0.2FeO3 with perovskite structure," *Solid State Ionics,* vol. 178, pp. 1317–1325, 2007.
95. S. Tanasescu, N. D. Totir, and D. I. Marchidan, "Thermodynamic properties of some perovskite type oxides used as SOFC cathode materials," *Solid State Ionics,* vol. 119, pp. 311–315, 1999.
96. S. B. Adler, "Factors governing oxygen reduction in solid oxide fuel cell cathodes," *Chemical Reviews,* vol. 104, pp. 4791–4843, 2004.
97. A. Evans, C. Benel, A. J. Darbandi, H. Hahn, J. Martynczuk, L. J. Gauckler et al., "Integration of spin-coated nanoparticulate-based La0.6Sr0.4CoO3–delta cathodes into micro-solid oxide fuel cell membranes," *Fuel Cells,* vol. 13, pp. 441–444, 2013.
98. M. G. Bellino, J. G. Sacanell, D. G. Lamas, A. G. Leyva, and N. E. W. de Reca, "High-performance solid-oxide fuel cell cathodes based on cobaltite nanotubes," *Journal of the American Chemical Society,* vol. 129, pp. 3066–3067, 2007.
99. D. Han, Y. D. Liu, S. R. Wang, and Z. L. Zhan, "Co-infiltrating Pr0.6Sr0.4FeO3–Ce1–*x*Pr*x*O2 (x = 0.1, 0.3, 0.5, 0.7, 0.9) mixed oxides into the La0.9Sr0.1Ga0.8Mg0.2O3 skeleton for use as low temperature solid oxide fuel cell cathodes," *Electrochimica Acta,* vol. 143, pp. 168–174, 2014.
100. M. Shah and S. A. Barnett, "Solid oxide fuel cell cathodes by infiltration of La0.6Sr0.4Co0.2Fe0.8O3–(delta) into Gd-doped ceria," *Solid State Ionics,* vol. 179, pp. 2059–2064, 2008.
101. C. N. Chervin, S. M. Kauzlarich, and Q. Pham, "Synthesis of La1–*x*Sr*x*MnO3 nanopowders and fabrication of solid oxide fuel cell composite cathodes: Particle size effects on microstructure," *Abstracts of Papers of the American Chemical Society,* vol. 225, p. U50, 2003.
102. A. T. Duong and D. R. Mumm, "Microstructural optimization by tailoring particle sizes for LSM–YSZ solid oxide fuel cell composite cathodes," *Journal of the Electrochemical Society,* vol. 159, pp. B40–B53, 2012.
103. T. Horita, Y. P. Xiong, M. Yoshinaga, H. Kishimoto, K. Yamaji, M. E. Brito et al., "Determination of chromium concentration in solid oxide fuel cell cathodes: (La,Sr)MnO_3 and (La,Sr)FeO_3," *Electrochemical and Solid State Letters,* vol. 12, pp. B146–B149, 2009.
104. Q. M. Nguyen and T. Takahashi, *Science and Technology of Ceramic Fuel Cells.* Amsterdam: Elsevier Science, 1995.
105. O. Yamamoto, Y. Takeda, R. Kanno, and M. Noda, "Perovskite-type oxides as oxygen electrodes for high-temperature oxide fuel cells," *Solid State Ionics,* vol. 22, pp. 241–246, 1987.
106. C. Clausen, C. Bagger, J. B. Bildesorensen, and A. Horsewell, "Microstructural and microchemical characterization of the interface between $La_{0.85}Sr_{0.15}MnO_3$ and Y_2O_3-stabilized ZrO_2," *Solid State Ionics,* vol. 70, pp. 59–64, 1994.
107. G. Stochniol, E. Syskakis, and A. Naoumidis, "Chemical compatibility between strontium-doped lanthanum manganite and yttria-stabilized zirconia," *Journal of the American Ceramic Society,* vol. 78, pp. 929–932, 1995.
108. F. L. Chen and M. L. Liu, "Preparation of yttria-stabilized zirconia (YSZ) films on La(0.85)Sr(0.15)MnO(3) (LSM) and LSM–YSZ substrates using an electrophoretic deposition (EPD) process," *Journal of the European Ceramic Society,* vol. 21, pp. 127–134, 2001.
109. C. W. Sun, R. Hui, and J. Roller, "Cathode materials for solid oxide fuel cells: A review," *Journal of Solid State Electrochemistry,* vol. 14, pp. 1125–1144, 2010.
110. S. Y. Istomin and E. V. Antipov, "Cathode materials based on perovskite-like transition metal oxides for intermediate temperature solid oxide fuel cells," *Russian Chemical Reviews,* vol. 82, pp. 686–700, 2013.
111. M. Chen, B. H. Moon, S. H. Kim, B. H. Kim, Q. Xu, and B. G. Ahn, "Characterization of La0.6Sr0.4Co0.2Fe0.8O3–delta + La2NiO4+delta composite cathode materials for solid oxide fuel cells," *Fuel Cells,* vol. 12, pp. 86–96, 2012.
112. K. T. Lee and A. Manthiram, "Comparison of Ln(0.6)Sr(0.4) CoO(3–delta) (Ln = La, Pr, Nd, Sm, and Gd) as cathode materials for intermediate temperature solid oxide fuel cells," *Journal of the Electrochemical Society,* vol. 153, pp. A794–A798, 2006.

113. T. Ishihara, T. Kudo, H. Matsuda, and Y. Takita, "Doped perovskite oxide, $PrMnO_3$, as a new cathode for solid-oxide fuel cells that decreases the operating temperature," *Journal of the American Ceramic Society,* vol. 77, pp. 1682–1684, 1994.
114. Y. Sakaki, Y. Takeda, A. Kato, N. Imanishi, O. Yamamoto, M. Hattori et al., "Ln1–*x*Sr*x*MnO3 (Ln = Pr, Nd, Sm and Gd) as the cathode material for solid oxide fuel cells," *Solid State Ionics,* vol. 118, pp. 187–194, 1999.
115. G. C. Kostogloudis and C. Ftikos, "Characterization of Nd1–*x*Sr*x*MnO3+/–delta SOFC cathode materials," *Journal of the European Ceramic Society,* vol. 19, pp. 497–505, 1999.
116. G. C. Kostogloudis, N. Vasilakos, and C. Ftikos, "Preparation and characterization of Pr1–xSrxMnO3+/–delta is (x = 0, 0.15, 0.3, 0.4, 0.5) as a potential SOFC cathode material operating at intermediate temperatures (500–700 degrees C)," *Journal of the European Ceramic Society,* vol. 17, pp. 1513–1521, 1997.
117. G. C. Kostogloudis and C. Ftikos, "Oxygen nonstoichiometry in Pr1–*x*Sr*x*Co0.2B0.8O3–delta (B = Mn, Fe, x = 0.2, 0.4) perovskite oxides," *Journal of the European Ceramic Society,* vol. 27, pp. 273–277, 2007.
118. M. Rini, Y. Zhu, S. Wall, R. I. Tobey, H. Ehrke, T. Garl et al., "Transient electronic structure of the photoinduced phase of Pr0.7Ca0.3MnO3 probed with soft x-ray pulses," *Physical Review B,* vol. 80, p.155113, 2009.
119. X. Yue, A. Yan, M. Zhang, L. Liu, Y. Dong, and M. Cheng, "Investigation on scandium-doped manganate La0.8Sr0.2Mn1–*x*Sc*x*O3–δ cathode for intermediate temperature solid oxide fuel cells," *Journal of Power Sources,* vol. 185, pp. 691–697, 2008.
120. N. Jaiswal, D. Kumar, S. Upadhyay, and O. Parkash, "Effect of Mg and Sr co-doping on the electrical properties of ceria-based electrolyte materials for intermediate temperature solid oxide fuel cells," *Journal of Alloys and Compounds,* vol. 577, pp. 456–462, 2013.
121. H. T. Gu, H. Chen, L. Gao, Y. F. Zheng, X. F. Zhu, and L. C. Guo, "Effect of Co doping on the properties of Sr(0.8)Ce(0.2) MnO(3–delta) cathode for intermediate-temperature solid-oxide fuel cells," *International Journal of Hydrogen Energy,* vol. 33, pp. 4681–4688, 2008.
122. P. Moller, R. Kanarbik, I. Kivi, G. Nurk, and E. Lust, "Influence of microstructure on the electrochemical behavior of LSC cathodes for intermediate temperature SOFC," *Journal of the Electrochemical Society,* vol. 160, pp. F1245–F1253, 2013.
123. I. Kojima, H. Adachi, and I. Yasumori, "Electronic structures of the $LaCoO_3$, $LaFeO_3$, $LaAlO_3$ perovskite oxides related to their catalysis," *Surface Science,* vol. 130, pp. 50–62, 1983.
124. A. N. Petrov, O. F. Kononchuk, A. V. Andreev, V. A. Cherepanov, and P. Kofstad, "Crystal structure, electrical and magnetic properties of La1–*x*Sr*x*CoO3–*y*," *Solid State Ionics,* vol. 80, pp. 189–199, 1995.
125. M. Gödickemeier, K. Sasaki, L. J. Gauckler, and I. Riess, "Perovskite cathodes for solid oxide fuel cells based on ceria electrolytes," *Solid State Ionics,* vol. 86–88, Pt. 2, pp. 691–701, 1996.
126. R. H. E. van Doorn and A. J. Burggraaf, "Structural aspects of the ionic conductivity of La1–*x*Sr*x*CoO3–delta," *Solid State Ionics,* vol. 128, pp. 65–78, 2000.
127. K. Yasumoto, Y. Inagaki, M. Shiono, and M. Dokiya, "An (La,Sr)(Co,Cu)O3–delta cathode for reduced temperature SOFCs," *Solid State Ionics,* vol. 148, pp. 545–549, 2002.
128. F. Bidrawn, S. Lee, J. M. Vohs, and R. J. Gorte, "The effect of Ca, Sr, and Ba doping on the ionic conductivity and cathode performance of LaFeO(3)," *Journal of the Electrochemical Society,* vol. 155, pp. B660–B665, 2008.
129. L. M. Liu, K. N. Sun, X. K. Li, M. Zhang, Y. B. Liu, N. Q. Zhang et al., "A novel doped CeO_2–$LaFeO_3$ composite oxide as both anode and cathode for solid oxide fuel cells," *International Journal of Hydrogen Energy,* vol. 37, pp. 12574–12579, 2012.
130. H. G. Shi, G. M. Yang, Z. K. Liu, G. R. Zhang, R. Ran, Z. P. Shao et al., "High performance tubular solid oxide fuel cells with BSCF cathode," *International Journal of Hydrogen Energy,* vol. 37, pp. 13022–13029, 2012.
131. E. V. Bongio, H. Black, F. C. Raszewski, D. Edwards, C. J. McConville, and V. R. W. Amarakoon, "Microstructural and high-temperature electrical characterization of La1–*x*Sr*x*-FeO3–delta," *Journal of Electroceramics,* vol. 14, pp. 193–198, 2005.
132. J. M. Ralph, C. Rossignol, and R. Kumar, "Cathode materials for reduced-temperature SOFCs," *Journal of the Electrochemical Society,* vol. 150, pp. A1518–A1522, 2003.
133. V. Dusastre and J. A. Kilner, "Optimisation of composite cathodes for intermediate temperature SOFC applications," *Solid State Ionics,* vol. 126, pp. 163–174, 1999.
134. G. C. Kostogloudis and C. Ftikos, "Properties of A-site-deficient La0.6Sr0.4Co0.2Fe0.8O3-delta-based perovskite oxides," *Solid State Ionics,* vol. 126, pp. 143–151, 1999.
135. L. W. Tai, M. M. Nasrallah, H. U. Anderson, D. M. Sparlin, and S. R. Sehlin, "Structure and electrical–properties of La1–*x*Sr*x*Co1–*y*Fe*y*O3. 1. The system La0.8Sr0.2Co1–*y*Fe*y*O3," *Solid State Ionics,* vol. 76, pp. 259–271, 1995.
136. L. W. Tai, M. M. Nasrallah, H. U. Anderson, D. M. Sparlin, and S. R. Sehlin, "Structure and electrical properties of La1–*x*Sr*x*Co1–*y*Fe*y*O3. 2. The system La1–*x*Sr*x*Co0.2Fe0.8O3," *Solid State Ionics,* vol. 76, pp. 273–283, 1995.
137. E. Maguire, B. Gharbage, F. M. B. Marques, and J. A. Labrincha, "Cathode materials for intermediate temperature SOFCs," *Solid State Ionics,* vol. 127, pp. 329–335, 2000.
138. G. C. Kostogloudis, G. Tsiniarakis, and C. Ftikos, "Chemical reactivity of perovskite oxide SOFC cathodes and yttria stabilized zirconia," *Solid State Ionics,* vol. 135, pp. 529–535, 2000.
139. G. C. Kostogloudis, C. Ftikos, A. Ahmad-Khanlou, A. Naoumidis, and D. Stover, "Chemical compatibility of alternative perovskite oxide SOFC cathodes with doped lanthanum gallate solid electrolyte," *Solid State Ionics,* vol. 134, pp. 127–138, 2000.
140. E. N. Naumovich, M. V. Patrakeev, V. V. Kharton, A. A. Yaremchenko, D. I. Lopinovich, and F. M. B. Marques, "Oxygen nonstoichiornetry in La2Ni(M)O4+delta (M = Cu, Co) under oxidizing conditions," *Solid State Sciences,* vol. 7, pp. 1353–1362, 2005.
141. M. V. Patrakeev, E. N. Naumovich, V. V. Kharton, A. A. Yaremchenko, E. V. Tsipis, P. Nunez et al., "Oxygen nonstoichiometry and electron-hole transport in La2Ni0.9Co0.1O4+delta," *Solid State Ionics,* vol. 176, pp. 179–188, 2005.
142. V. V. Kharton, A. A. Yaremchenko, A. L. Shaula, M. V. Patrakeev, E. N. Naumovich, D. I. Loginovich et al., "Transport properties and stability of Ni-containing mixed conductors with perovskite- and K_2NiF_4-type structure," *Journal of Solid State Chemistry,* vol. 177, pp. 26–37, 2004.
143. A. A. Yaremchenko, V. V. Kharton, M. V. Patrakeev, and J. R. Frade, "p-Type electronic conductivity, oxygen permeability and stability of La2Ni0.9Co0.1O4+delta," *Journal of Materials Chemistry,* vol. 13, pp. 1136–1144, 2003.

144. J. Wan, J. B. Goodenough, and J. H. Zhu, "Nd(2–*x*)La(*x*) NiO(4+delta), a mixed ionic/electronic conductor with interstitial oxygen, as a cathode material," *Solid State Ionics,* vol. 178, pp. 281–286, 2007.
145. S. J. Skinner and J. A. Kilner, "Oxygen diffusion and surface exchange in La2–*x*Sr*x*NiO4+delta," *Solid State Ionics,* vol. 135, pp. 709–712, 2000.
146. A. Aguadero, M. J. Escudero, M. Perez, J. A. Alonso, V. Pomjakushin, and L. Daza, "Effect of Sr content on the crystal structure and electrical properties of the system La2–*x*Sr*x*NiO4+delta (0 <= *x* <= 1)," *Dalton Transactions,* vol. 36, pp. 4377–4383, 2006.
147. A. Aguadero, J. A. Alonso, M. J. Escudero, and L. Daza, "Evaluation of the La2Ni1–*x*Cu*x*O4+delta system as SOFC cathode material with 8YSZ and LSGM as electrolytes," *Solid State Ionics,* vol. 179, pp. 393–400, 2008.
148. A. Maignan, C. Martin, D. Pelloquin, N. Nguyen, and B. Raveau, "Structural and magnetic studies of ordered oxygen-deficient perovskites LnBaCo(2)O(5+delta), closely related to the '112' structure," *Journal of Solid State Chemistry,* vol. 142, pp. 247–260, 1999.
149. L. Barbey, N. Nguyen, A. Ducouret, V. Caignaert, J. M. Greneche, and B. Raveau, "Magnetic behavior of the 112-type substituted cuprate YBaCoCu1–*x*Fe*x*O5," *Journal of Solid State Chemistry,* vol. 115, pp. 514–520, 1995.
150. V. Pralong, V. Caignaert, S. Hebert, C. Marinescu, B. Raveau, and A. Maignan, "Electrochemical oxidation and reduction of the La0.2Sr0.8CoO3–delta phases: Control of itinerant ferromagnetism and magnetoresistance," *Solid State Ionics,* vol. 177, pp. 815–820, 2006.
151. V. Pralong, V. Caignaert, S. Hebert, A. Maignan, and B. Raveau, "Soft chemistry synthesis and characterizations of fully oxidized and reduced NdBaCo2O5+delta phases delta=0, 1," *Solid State Ionics,* vol. 177, pp. 1879–1881, 2006.
152. A. A. Taskin, A. N. Lavrov, and Y. Ando, "Achieving fast oxygen diffusion in perovskites by cation ordering," *Applied Physics Letters,* vol. 86, p. 091910, 2005.
153. Z. Yuan, J. Liu, C. L. Chen, C. H. Wang, X. G. Luo, X. H. Chen et al., "Epitaxial behavior and transport properties of PrBaCo2O5 thin films on (001) SrTiO3," *Applied Physics Letters,* vol. 90, p. 212111, 2007.
154. G. Kim, S. Wang, A. J. Jacobson, L. Reimus, P. Brodersen, and C. A. Mims, "Rapid oxygen ion diffusion and surface exchange kinetics in PrBaCo2O5+*x* with a perovskite related structure and ordered A cations," *Journal of Materials Chemistry,* vol. 17, pp. 2500–2505, 2007.
155. A. Tarancon, J. Pena-Martinez, D. Marrero-Lopez, A. Morata, J. C. Ruiz-Morales, and P. Nunez, "Stability, chemical compatibility and electrochemical performance of GdBaCo2O5+*x* layered perovskite as a cathode for intermediate temperature solid oxide fuel cells," *Solid State Ionics,* vol. 179, pp. 2372–2378, 2008.
156. A. Tarancon, A. Morata, G. Dezanneau, S. J. Skinner, J. A. Kilner, S. Estrade et al., "GdBaCO2O5+*x* layered perovskite temperature solid oxide fuel as an intermediate cell cathode," *Journal of Power Sources,* vol. 174, pp. 255–263, 2007.
157. H. T. Gu, H. Chen, L. Gao, Y. F. Zheng, X. F. Zhu, and L. C. Guo, "Oxygen reduction mechanism of NdBaCo2O5+delta cathode for intermediate-temperature solid oxide fuel cells under cathodic polarization," *International Journal of Hydrogen Energy,* vol. 34, pp. 2416–2420, 2009.
158. D. Chen, R. Ran, K. Zhang, J. Wang, and Z. Shao, "Intermediate-temperature electrochemical performance of a polycrystalline PrBaCo2O5+δ cathode on samarium-doped ceria electrolyte," *Journal of Power Sources,* vol. 188, pp. 96–105, 2009.
159. D. Ding, X. X. Li, S. Y. Lai, K. Gerdes, and M. L. Liu, "Enhancing SOFC cathode performance by surface modification through infiltration," *Energy & Environmental Science,* vol. 7, pp. 552–575, 2014.
160. J. Kim, S. Sengodan, G. Kwon, D. Ding, J. Shin, M. L. Liu et al., "Triple-conducting layered perovskites as cathode materials for proton-conducting solid oxide fuel cells," *ChemSusChem,* vol. 7, pp. 2811–2815, 2014.
161. C. W. Sun, Z. Xie, C. R. Xia, H. Li, and L. Q. Chen, "Investigations of mesoporous CeO_2–Ru as a reforming catalyst layer for solid oxide fuel cells," *Electrochemistry Communications,* vol. 8, pp. 833–838, 2006.
162. Z. V. Marinkovic, L. Mancic, J. F. Cribier, S. Ohara, T. Fukui, and O. Milosevic, "Nature of structural changes in LSM–YSZ nanocomposite material during thermal treatments," *Materials Science and Engineering A—Structural Materials Properties Microstructure and Processing,* vol. 375, pp. 615–619, 2004.
163. T. Fukui, K. Murata, S. Ohara, H. Abe, M. Naito, and K. Nogi, "Morphology control of Ni–YSZ cermet anode for lower temperature operation of SOFCs," *Journal of Power Sources,* vol. 125, pp. 17–21, 2004.
164. B. C. H. Steele, K. M. Hori, and S. Uchino, "Kinetic parameters influencing the performance of IT-SOFC composite electrodes," *Solid State Ionics,* vol. 135, pp. 445–450, 2000.
165. M. Watanabe, H. Uchida, M. Shibata, N. Mochizuki, and K. Amikura, "High-performance catalyzed reaction layer for medium-temperature operating solid oxide fuel cells," *Journal of the Electrochemical Society,* vol. 141, pp. 342–346, 1994.
166. S. P. Jiang, "A review of wet impregnation—An alternative method for the fabrication of high performance and nano-structured electrodes of solid oxide fuel cells," *Materials Science and Engineering A—Structural Materials Properties Microstructure and Processing,* vol. 418, pp. 199–210, 2006.
167. Z. Y. Jiang, C. R. Xia, and F. L. Chen, "Nano-structured composite cathodes for intermediate-temperature solid oxide fuel cells via an infiltration/impregnation technique," *Electrochimica Acta,* vol. 55, pp. 3595–3605, 2010.
168. T. Z. Sholklapper, C. P. Jacobson, S. J. Visco, and L. C. De Jonghe, "Synthesis of dispersed and contiguous nanoparticles in solid oxide fuel cell electrodes," *Fuel Cells,* vol. 8, pp. 303–312, 2008.
169. C. R. Xia and M. L. Liu, "Novel cathodes for low-temperature solid oxide fuel cells," *Advanced Materials,* vol. 14, pp. 521–522, 2002.
170. Y. Sakito, A. Hirano, N. Imanishi, Y. Takeda, O. Yamamoto, and Y. Liu, "Silver infiltrated La0.6Sr0.4Co0.2Fe0.8O3 cathodes for intermediate temperature solid oxide fuel cells," *Journal of Power Sources,* vol. 182, pp. 476–481, 2008.
171. M. Sahibzada, S. J. Benson, R. A. Rudkin, and J. A. Kilner, "Pd-promoted La0.6Sr0.4Co0.2Fe0.8O3 cathodes," *Solid State Ionics,* vol. 113, pp. 285–290, 1998.
172. V. A. C. Haanappel, D. Rutenbeck, A. Mai, S. Uhlenbruck, D. Sebold, H. Wesemeyer et al., "The influence of noble-metal-containing cathodes on the electrochemical performance of anode-supported SOFCs," *Journal of Power Sources,* vol. 130, pp. 119–128, 2004.
173. S. P. Simner, M. D. Anderson, M. H. Engelhard, and J. W. Stevenson, "Degradation mechanisms of La–Sr–Co–Fe–O3SOFC cathodes," *Electrochemical and Solid State Letters,* vol. 9, pp. A478–A481, 2006.

174. J. S. Hardy, J. W. Templeton, D. J. Edwards, Z. G. Lu, and J. W. Stevenson, "Lattice expansion of LSCF-6428 cathodes measured by *in situ* XRD during SOFC operation," *Journal of Power Sources,* vol. 198, pp. 76–82, 2012.
175. J. J. Choi, W. T. Qin, M. F. Liu, and M. L. Liu, "Preparation and characterization of (La0.8Sr0.2)(0.95)MnO3–delta (LSM) thin films and LSM/LSCF interface for solid oxide fuel cells," *Journal of the American Ceramic Society,* vol. 94, pp. 3340–3345, 2011.
176. D. Ding, M. F. Liu, Z. B. Liu, X. X. Li, K. Blinn, X. B. Zhu et al., "Efficient electro-catalysts for enhancing surface activity and stability of SOFC cathodes," *Advanced Energy Materials,* vol. 3, pp. 1149–1154, 2013.
177. M. L. Liu, M. E. Lynch, K. Blinn, F. M. Alamgir, and Y. Choi, "Rational SOFC material design: New advances and tools," *Materials Today,* vol. 14, pp. 534–546, 2011.
178. M. E. Lynch, L. Yang, W. T. Qin, J. J. Choi, M. F. Liu, K. Blinn et al., "Enhancement of La0.6Sr0.4Co0.2Fe0.8O3–delta durability and surface electrocatalytic activity by La0.85Sr0.15MnO3+/–delta investigated using a new test electrode platform," *Energy & Environmental Science,* vol. 4, pp. 2249–2258, 2011.
179. W. Z. Zhu and S. C. Deevi, "A review on the status of anode materials for solid oxide fuel cells," *Materials Science and Engineering A—Structural Materials Properties Microstructure and Processing,* vol. 362, pp. 228–239, 2003.
180. R. H. Till, W. W. Schertz, P. A. Nelson, and American Institute of Chemical Engineers, "Prospects for electric vehicles," *Proceedings of the 25th Intersociety Energy Conversion Engineering Conference: IECEC-90, August 12–17, 1990, Reno, Nevada.* New York: American Institute of Chemical Engineers, 1990.
181. S. C. Singhal, "Science and technology of solid-oxide fuel cells," *MRS Bulletin,* vol. 25, pp. 16–21, 2000.
182. S. P. S. Badwal, "Stability of solid oxide fuel cell components," *Solid State Ionics,* vol. 143, pp. 39–46, 2001.
183. W. Z. Zhu and S. C. Deevi, "Development of interconnect materials for solid oxide fuel cells," *Materials Science and Engineering A—Structural Materials Properties Microstructure and Processing,* vol. 348, pp. 227–243, 2003.
184. K. Natesan, "Corrosion performance of iron aluminides in mixed-oxidant environments," *Materials Science and Engineering A—Structural Materials Properties Microstructure and Processing,* vol. 258, pp. 126–134, 1998.
185. H. Yokokawa, N. Sakai, T. Kawada, and M. Dokiya, "Thermodynamic analysis of reaction profiles between $LaMo_3$ (M = Ni, Co, Mn) and ZrO_2," *Journal of the Electrochemical Society,* vol. 138, pp. 2719–2727, 1991.
186. N. Sakai, T. Kawada, H. Yokokawa, M. Dokiya, and T. Iwata, "Sinterability and electrical conductivity of calcium-doped lanthanum chromites," *Journal of Materials Science,* vol. 25, pp. 4531–4534, 1990.
187. M. Mori, Y. Hiei, and N. M. Sammes, "Sintering behavior and mechanism of Sr-doped lanthanum chromites with A site excess composition in air," *Solid State Ionics,* vol. 123, pp. 103–111, 1999.
188. L. W. Tai and P. A. Lessing, "Plasma spraying of porous electrodes for a planar solid oxide fuel cell," *Journal of the American Ceramic Society,* vol. 74, pp. 501–504, 1991.
189. L. W. Tai and P. A. Lessing, "Tape casting and sintering of strontium-doped lanthanum chromite for a planar solid oxide fuel-cell bipolar plate," *Journal of the American Ceramic Society,* vol. 74, pp. 155–160, 1991.
190. A. Manithiram and American Ceramic Society, "Developments in solid oxide fuel cells and lithium ion batteries: Alternative methods of sealing planar solid oxide fuel cells," in *Proceedings of the 106th Annual Meeting of the American Ceramic Society, Indianapolis, Indiana, USA (2004).* Westerville, OH: American Ceramic Society, 2005.
191. A. V. Virkar, J. Chen, C. W. Tanner, and J. W. Kim, "The role of electrode microstructure on activation and concentration polarizations in solid oxide fuel cells," *Solid State Ionics,* vol. 131, pp. 189–198, 2000.
192. J. W. Kim, A. V. Virkar, K. Z. Fung, K. Mehta, and S. C. Singhal, "Polarization effects in intermediate temperature, anode-supported solid oxide fuel cells," *Journal of the Electrochemical Society,* vol. 146, pp. 69–78, 1999.
193. C. W. Tanner, K. Z. Fung, and A. V. Virkar, "The effect of porous composite electrode structure on solid oxide fuel cell performance. 1. Theoretical analysis," *Journal of the Electrochemical Society,* vol. 144, pp. 21–30, 1997.
194. W. Z. Zhu and M. Yan, "Perspectives on the metallic interconnects for solid oxide fuel cells," *Journal of Zhejiang University of Science,* vol. 5, pp. 1471–1503, 2004.
195. D. M. Reed, H. U. Anderson, and W. Huebner, "Characterization of solid oxide fuel cells by use of an internal Pt voltage probe," *Journal of the Electrochemical Society,* vol. 143, pp. 1558–1561, 1996.
196. D. Simwonis, A. Naoumidis, F. J. Dias, J. Linke, and A. Moropoulou, "Material characterization in support of the development of an anode substrate for solid oxide fuel cells," *Journal of Materials Research,* vol. 12, pp. 1508–1518, 1997.
197. K. Barthel, S. Rambert, and S. Siegmann, "Microstructure and polarization resistance of thermally sprayed composite cathodes for solid oxide fuel cell use," *Journal of Thermal Spray Technology,* vol. 9, pp. 343–347, 2000.
198. H. C. Chen, J. Heberlein, and R. Henne, "Integrated fabrication process for solid oxide fuel cells in a triple torch plasma reactor," *Journal of Thermal Spray Technology,* vol. 9, pp. 348–353, 2000.
199. K. Okumura, Y. Aihara, S. Ito, and S. Kawasaki, "Development of thermal spraying-sintering technology for solid oxide fuel cells," *Journal of Thermal Spray Technology,* vol. 9, pp. 354–359, 2000.
200. S. Takenoiri, N. Kadokawa, and K. Koseki, "Development of metallic substrate supported planar solid oxide fuel cells fabricated by atmospheric plasma spraying," *Journal of Thermal Spray Technology,* vol. 9, pp. 360–363, 2000.
201. H. Tsukuda, A. Notomi, and N. Hisatome, "Application of plasma spraying to tubular-type solid oxide fuel cells production," *Journal of Thermal Spray Technology,* vol. 9, pp. 364–368, 2000.
202. E. Ivers-Tiffée, W. Wersing, M. Schießl, and H. Greiner, "Ceramic and metallic components for a planar SOFC," *Berichte der Bunsengesellschaft für physikalische Chemie,* vol. 94, pp. 978–981, 1990.
203. W. Z. Zhu and S. C. Deevi, "Opportunity of metallic interconnects for solid oxide fuel cells: A status on contact resistance," *Materials Research Bulletin,* vol. 38, pp. 957–972, 2003.
204. P. Jian, L. Jian, H. Bing, and G. Y. Xie, "Oxidation kinetics and phase evolution of a Fe–16Cr alloy in simulated SOFC cathode atmosphere," *Journal of Power Sources,* vol. 158, pp. 354–360, 2006.
205. X. H. Deng, P. Wei, M. R. Bateni, and A. Petric, "Cobalt plating of high temperature stainless steel interconnects," *Journal of Power Sources,* vol. 160, pp. 1225–1229, 2006.

206. H. Yokokawa, T. Horita, N. Sakai, K. Yamaji, M. E. Brito, Y. P. Xiong et al., "Thermodynamic considerations on Cr poisoning in SOFC cathodes," *Solid State Ionics,* vol. 177, pp. 3193–3198, 2006.
207. Y. Matsuzaki and I. Yasuda, "Electrochemical properties of a SOFC cathode in contact with a chromium-containing alloy separator," *Solid State Ionics,* vol. 132, pp. 271–278, 2000.
208. J. W. Fergus, "Metallic interconnects for solid oxide fuel cells," *Materials Science and Engineering A—Structural Materials Properties Microstructure and Processing,* vol. 397, pp. 271–283, 2005.
209. J. W. Fergus, "Effect of cathode and electrolyte transport properties on chromium poisoning in solid oxide fuel cells," *International Journal of Hydrogen Energy,* vol. 32, pp. 3664–3671, 2007.
210. J. W. Wu and X. B. Liu, "Recent development of SOFC metallic interconnect," *Journal of Materials Science & Technology,* vol. 26, pp. 293–305, 2010.
211. J. L. Gonzalez-Carrasco, P. Perez, P. Adeva, and J. Chao, "Oxidation behaviour of an ODS NiAl-based intermetallic alloy," *Intermetallics,* vol. 7, pp. 69–78, 1999.
212. J. W. Fergus, "Lanthanum chromite-based materials for solid oxide fuel cell interconnects," *Solid State Ionics,* vol. 171, pp. 1–15, 2004.
213. K. Hilpert, D. Das, M. Miller, D. H. Peck, and R. Weiss, "Chromium vapor species over solid oxide fuel cell interconnect materials and their potential for degradation processes," *Journal of the Electrochemical Society,* vol. 143, pp. 3642–3647, 1996.
214. Z. G. Yang, K. S. Weil, D. M. Paxton, and J. W. Stevenson, "Selection and evaluation of heat-resistant alloys for SOFC interconnect applications," *Journal of the Electrochemical Society,* vol. 150, pp. A1188–A1201, 2003.
215. S. J. Geng and J. H. Zhu, "Promising alloys for intermediate-temperature solid oxide fuel cell interconnect application," *Journal of Power Sources,* vol. 160, pp. 1009–1016, 2006.
216. S. J. Geng, J. H. Zhu, and Z. G. Lu, "Evaluation of Haynes 242 alloy as SOFC interconnect material," *Solid State Ionics,* vol. 177, pp. 559–568, 2006.
217. S. J. Geng, J. H. Zhu, M. P. Brady, H. U. Anderson, X. D. Zhou, and Z. G. Yang, "A low-Cr metallic interconnect for intermediate-temperature solid oxide fuel cells," *Journal of Power Sources,* vol. 172, pp. 775–781, 2007.
218. P. Piccardo, P. Gannon, S. Chevalier, M. Viviani, A. Barbucci, G. Caboche et al., "ASR evaluation of different kinds of coatings on a ferritic stainless steel as SOFC interconnects," *Surface & Coatings Technology,* vol. 202, pp. 1221–1225, 2007.
219. W. Qu, L. Jian, D. G. Ivey, and J. M. Hill, "Yttrium, cobalt and yttrium/cobalt oxide coatings on ferritic stainless steels for SOFC interconnects," *Journal of Power Sources,* vol. 157, pp. 335–350, 2006.
220. W. Qu, L. Jian, J. M. Hill, and D. G. Ivey, "Electrical and microstructural characterization of spinel phases as potential coatings for SOFC metallic interconnects," *Journal of Power Sources,* vol. 153, pp. 114–124, 2006.
221. I. Antepara, I. Villarreal, L. M. Rodriguez-Martinez, N. Lecanda, U. Castro, and A. Laresgoiti, "Evaluation of ferritic steels for use as interconnects and porous metal supports in IT-SOFCs," *Journal of Power Sources,* vol. 151, pp. 103–107, 2005.
222. Z. G. Yang, J. S. Hardy, M. S. Walker, G. G. Xia, S. P. Simner, and J. W. Stevenson, "Structure and conductivity of thermally grown scales on ferritic Fe–Cr–Mn steel for SOFC interconnect applications," *Journal of the Electrochemical Society,* vol. 151, pp. A1825–A1831, 2004.
223. Z. G. Yang, M. S. Walker, P. Singh, J. W. Stevenson, and T. Norby, "Oxidation behavior of ferritic stainless steels under SOFC interconnect exposure conditions," *Journal of the Electrochemical Society,* vol. 151, pp. B669–B678, 2004.
224. Z. G. Yang, G. G. Xia, and J. W. Stevenson, "Evaluation of Ni–Cr-base alloys for SOFC interconnect applications," *Journal of Power Sources,* vol. 160, pp. 1104–1110, 2006.
225. L. C. Chen, Z. G. Yang, B. Jha, G. G. Xia, and J. W. Stevenson, "Clad metals, roll bonding and their applications for SOFC interconnects," *Journal of Power Sources,* vol. 152, pp. 40–45, 2005.
226. K. A. Nielsen, A. R. Dinesen, L. Korcakova, L. Mikkelsen, P. V. Hendriksen, and F. W. Poulsen, "Testing of Ni-plated ferritic steel interconnect in SOFC stacks," *Fuel Cells,* vol. 6, pp. 100–106, 2006.
227. P. D. Jablonski and D. E. Alman, "Oxidation resistance and mechanical properties of experimental low coefficient of thermal expansion (CTE) Ni-base alloys," *International Journal of Hydrogen Energy,* vol. 32, pp. 3705–3712, 2007.
228. L. R. Pederson, P. Singh, and X. D. Zhou, "Application of vacuum deposition methods to solid oxide fuel cells," *Vacuum,* vol. 80, pp. 1066–1083, 2006.
229. P. E. Gannon, C. T. Tripp, A. K. Knospe, C. V. Ramana, M. Deibert, R. J. Smith et al., "High-temperature oxidation resistance and surface electrical conductivity of stainless steels with filtered arc Cr–Al–N multilayer and/or superlattice coatings," *Surface & Coatings Technology,* vol. 188, pp. 55–61, 2004.
230. A. Kayani, R. J. Smith, S. Teintze, M. Kopczyk, P. E. Gannon, M. C. Deibert et al., "Oxidation studies of CrAlON nanolayered coatings on steel plates," *Surface & Coatings Technology,* vol. 201, pp. 1685–1694, 2006.
231. X. B. Liu, C. Johnson, C. M. Li, J. Xu, and C. Cross, "Developing TiAlN coatings for intermediate temperature solid oxide fuel cell interconnect applications," *International Journal of Hydrogen Energy,* vol. 33, pp. 189–196, 2008.
232. J. W. Wu, C. M. Li, C. Johnson, and X. B. Liu, "Evaluation of SmCo and SmCoN magnetron sputtering coatings for SOFC interconnect applications," *Journal of Power Sources,* vol. 175, pp. 833–840, 2008.
233. F. Gaillard, J. P. Joly, A. Boreave, P. Vernoux, and J. P. Deloume, "Intermittent temperature-programmed desorption study of perovskites used for catalytic purposes," *Applied Surface Science,* vol. 253, pp. 5876–5881, 2007.
234. N. Orlovskaya, A. Nicholls, S. Yarmolenko, J. Sankar, C. Johnson, and R. Gemmen, "Microstructural characterization of La-Cr-O thin film deposited by RF magnetron sputtering on the stainless steel interconnect materials for SOFC application," *Fuel Cell Technologies: State and Perspectives NATO Science Series,* vol. 202, pp 355–371, 2005.
235. J. H. Zhu, Y. Zhang, A. Basu, Z. G. Lu, M. Paranthaman, D. F. Lee et al., "LaCrO3-based coatings on ferritic stainless steel for solid oxide fuel cell interconnect applications," *Surface & Coatings Technology,* vol. 177, pp. 65–72, 2004.
236. J. H. Kim, R. H. Song, and S. H. Hyun, "Effect of slurry-coated LaSrMnO(3) on the electrical property of Fe-Cr alloy for metallic interconnect of SOFC," *Solid State Ionics,* vol. 174, pp. 185–191, 2004.
237. Y. D. Zhen, S. P. Jiang, S. Zhang, and V. Tan, "Interaction between metallic interconnect and constituent oxides of (La, Sr)MnO3 coating of solid oxide fuel cells," *Journal of the European Ceramic Society,* vol. 26, pp. 3253–3264, 2006.

238. J. J. Choi, J. H. Lee, D. S. Park, B. D. Hahn, W. H. Yoon, and H. T. Lin, "Oxidation resistance coating of LSM and LSCF on SOFC metallic interconnects by the aerosol deposition process," *Journal of the American Ceramic Society,* vol. 90, pp. 1926–1929, 2007.
239. Y. Larring and T. Norby, "Spinel and perovskite functional layers between Plansee metallic interconnect (Cr–5 wt% Fe–1 wt% Y_2O_3) and ceramic (La0.85Sr0.15)(0.91) MnO3 cathode materials for solid oxide fuel cells," *Journal of the Electrochemical Society,* vol. 147, pp. 3251–3256, 2000.
240. X. Chen, P. Y. Hou, C. P. Jacobson, S. J. Visco, and L. C. De Jonghe, "Protective coating on stainless steel interconnect for SOFCs: Oxidation kinetics and electrical properties," *Solid State Ionics,* vol. 176, pp. 425–433, 2005.
241. Z. G. Yang, G. G. Xia, S. P. Simner, and J. W. Stevenson, "Thermal growth and performance of manganese cobaltite spinel protection layers on ferritic stainless steel SOFC interconnects," *Journal of the Electrochemical Society,* vol. 152, pp. A1896–A1901, 2005.

13 Advanced Technologies for High-Temperature Solid Oxide Fuel Cells

Zhe Lü, Bo Wei, and Zhihong Wang

CONTENTS

13.1 FABRICATION/MANUFACTURING OF SOLID OXIDE FUEL CELLS

13.1.1 Introduction

Solid oxide fuel cell (SOFC) is an all-solid electrochemical device that can convert chemical energy into electric energy via electrochemical oxidation of fuels. As shown in Figure 13.1, SOFC is composed of a solid oxide electrolyte (fast ionic conductor) and attached anode and cathode on its two sides to form a sandwich structure named positive electrode–electrolyte–negative electrode (PEN) assembly. On the anode, fuel is electrochemically oxidized and provides electrons to the external circuit. While on the cathode, oxygen gas is adsorbed and electrochemically reduced to oxygen ion by gaining the electrons transported from the external circuit. In an electrolyte with oxygen-ion conduction, oxygen ions diffuse through the electrolyte from cathode to anode. In a typical SOFC, both the electrolyte and cathode are oxide ceramic materials, although the anode is a porous cermet (such as Ni–YSZ) under operation condition after reduction, the as-prepared anode usually is in oxide ceramic status (e.g., NiO–YSZ composite).

One of the key tasks of SOFC development or application is to fabricate/manufacture qualified PEN with high performance and reliability. For this purpose, many techniques have been developed and employed. For example, a conventional casting process can be used to fabricate support components with certain geometric shape, and can also be employed to prepare thin or thick layers (called film or membrane) on the supports for SOFC.

13.1.2 Basic SOFC Configurations

Before presenting the fabrication/manufacturing techniques for SOFC, a brief introduction about configuration concepts is needed. Generally, SOFCs are grouped into four main

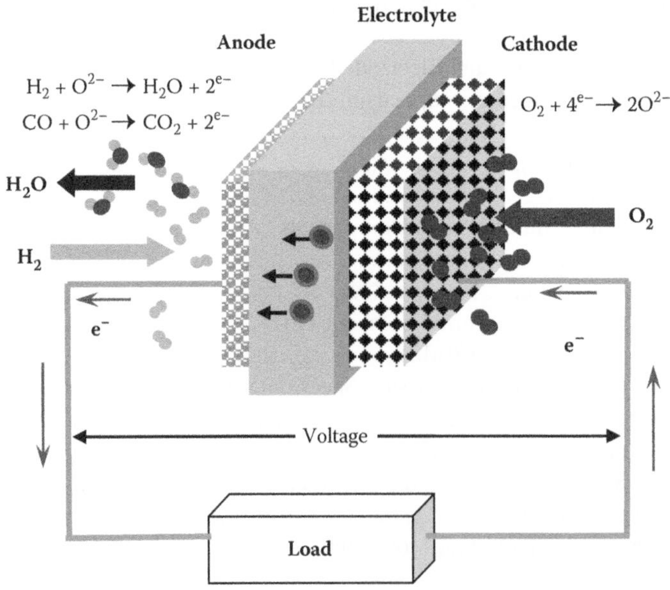

FIGURE 13.1 Schematic diagram of SOFC.

designs, i.e., tubular design, planar design, segmented-cell-in-series design, and monolithic design, according to the electrolyte component geometry [1,2]. Some further developed types (e.g., flat tubular SOFC) are usually related to the basic planar and tubular types, especially when the shape of the electrolyte component is considered. A sketch concerning fabrication still needs to be presented here, and a detailed review about these designs and other novel SOFC designs will be presented in Section 13.4.

The other types of fuel cells, such as alkaline fuel cells (AFCs), phosphoric acid fuel cells (PAFCs), and melt carbonate fuel cells (MCFCs), as well as proton-exchanged membrane fuel cells (PEMFCs), usually adopt a planar configuration. Although SOFCs are all in the solid state and can self-support, and more complex geometries may be selected, the planar geometry is still preferred for easier fabrication and convenient application.

A planar geometry can extend in two dimensions (2D). In planar-type SOFC, both ionic and electronic currents mainly migrate through the electrolyte and electrodes as well as the dipolar plate (interconnect) in the shortest direction perpendicular to the plate; even considering current collecting by the ribs of the interconnect, the transverse flow pathway is still short enough, thus lower ohmic resistance and larger current density and power density can be expected.

The primary issue of planar SOFC is the need of gastight sealing all around and the apertures for gas flow. At the same time, the sandwich-structure ceramic component with the 2D-like configuration leads to it being brittle when scaling up to a higher width–thickness ratio. The thermal mismatch between electrode/electrolyte can induce large stress during the heating–cooling cycle. Despite these shortcomings, the planar design is still a good choice for both experimental studies and some technical development.

Tubular SOFC is based on a long tube that supports all the components of a single cell or even a whole stack (segments in series; see Section 13.4.1). One gas (either fuel or oxidant) flows inside along the tube and is delivered to electrode on the way, and the other gas flows/feeds outside. The tube can be an inert porous ceramic that supports the electrode and electrolyte layers, or a porous electrode that bears a dense electrolyte film coated with another porous electrode. Compared with the planar SOFC, tubular SOFC reduces the requirement of gastight sealing, so it has achieved great success and demonstrated practicability for the application in power plants by Westinghouse at the end of 20th century [3]. The tubular configuration possesses many advantages: it has higher mechanical strength than the planar configuration and it is capable of solving problems related to cracking, thermocycling, quick start-up time, and sealing [4–6]. However, the large ohmic resistance caused by the long current path on the surface is the primary drawback of tubular SOFC. To overcome this drawback, flat tubular SOFC and microtubular SOFC have been developed in recent years.

Different kinds of fabrication techniques adapt various geometric characteristics, scales, and sizes. The ceramic techniques with good practical prospect are mainly introduced in this section. These techniques were developed to fabricate the components of SOFC, including supports with certain thickness and some thin coating layers adhered on the supports. These coatings and supports possess suitable microstructures (porous or dense) for their roles in SOFCs.

13.1.3 Fabrication/Manufacturing Techniques of SOFC Supports

The SOFC support has a regular shape with precise size and thickness so as to possess enough mechanical strength and assembling convenience. Besides the porous metal support that can be processed using general metal-shaping techniques, the ceramic supports for SOFC are fabricated via ceramic techniques, namely, using powder as raw material, followed by shaping and sintering.

13.1.3.1 Dry Pressing

Dry pressing (DP) is a simple and economic technique to prepare planar ceramic disc in laboratories. Powder is loaded in the chamber of a steel die, then pressed to a certain pressure using a press machine (as Figure 13.2a shown). Before pressing, the powder may be mixed with binder for pelleting to obtain a more homogeneous and dense body. The pressed pellets are fired in furnace to achieve suitable physical performances. It is usually employed to manufacture dense disc sample with thickness >0.1 mm for the purpose of basic electrical conductivity measurement and electrochemical performance evaluation of electrode materials.

In fact, besides the common usage in preparing supports for SOFC or electrolyte discs, DP is also used to prepare thin membrane on support simultaneously (see Figure 13.2b). Xia and Liu [7] have developed a DP method to fabricate anode-supported $Gd_{0.1}Ce_{0.9}O_{1.95}$ (GDC) electrolyte membrane for intermediate-temperature SOFC (as shown in Figure 13.3). The prepared the cell with a 20-μm-thick GDC membrane

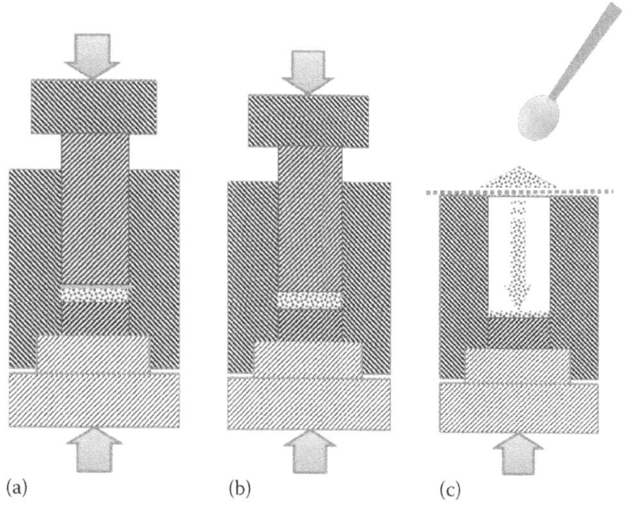

FIGURE 13.2 Schematic diagram of dry pressing methods: (a) dry pressing (DP), (b) bilayer DP or buffer DP, and (c) modified DP using filter.

FIGURE 13.3 Cross-section view of GDC membrane (a) on porous Ni–GDC anode substrate, (b) prepared via the dry pressing method. (Reproduced from *Solid State Ionics*, 144, Xia C.R., Liu M.L., Low-temperature SOFCs based on $Gd_{0.1}Ce_{0.9}O_{1.95}$ fabricated by dry pressing, 249–255, Copyright 2001, with permission from Elsevier.)

on NiO + 35 wt.% GDC composite anode and $Sm_{0.5}Sr_{0.5}CoO_3$ + 10% GDC composite cathode provided maximum power densities of 145 and 400 mW cm^{-2} at 500°C and 600°C, respectively. Xin et al. [8] developed a modified DP method to prepare NiO–YSZ anode–supported YSZ membrane by using nanocrystalline YSZ powder and introducing a screen to filter the electrolyte powders so as to uniformly distribute powder onto the anode substrate to form a thin YSZ layer (see Figure 13.2c). The single cell with 8 μm-thick YSZ membrane generated a power density of 0.79 W cm^{-2}.

Although DP is cost-effective and flexible to fabricate electrolyte pellets or anode–electrolyte bilayers, it still has the obvious disadvantages of being time consuming and a discontinuous operation that makes it less competitive for industrial production.

13.1.3.2 Tape Casting

Tape casting (TC) is a widely used technique in the ceramic industries to fabricate 2D thin and dense (or porous) films and plates (Figure 13.4). During the TP process, the slurry consisting of the ceramic powder, solvent, dispersants, binders, and plasticizers is cast onto a stationary or moving surface of a substrate tape [9]. Alcohols, ketones, or hydrocarbons are the generally used organic solvents for the purpose of obtaining highly concentrated suspension with suitable rheological properties and drying performance. Aqueous TP based on the slurry using water as solvent also attracted more attention for the environmental and health aspects [9]. The thickness of sintered support prepared via the TP technique can be controlled by setting the blade gap; the thickness range is generally from 100 to 800 μm.

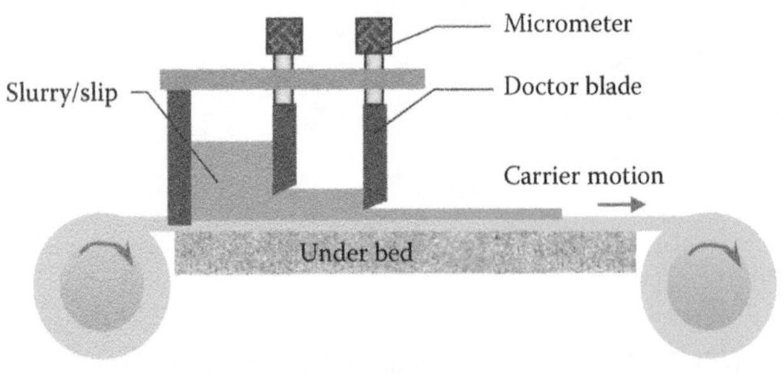

FIGURE 13.4 Schematic diagram of tape casting.

The application of TC in the SOFC field is to fabricate the mechanical bearing support for planar designs, i.e., dense electrolyte plates for electrolyte-supported SOFC [10,11] or porous Ni–YSZ substrates for anode-supported SOFC [12]. The anode-supported electrolyte bilayer can be manufactured by TC of the anode and electrolyte, respectively, and then hot lamination of the anode/electrolyte assembly followed by co-sintering [13,14] (Table 13.1). Chen et al. [15] combined TP with isostatic pressing to prepare YSZ slices for the application in electrolyte-supported SOFC. The isostatic pressing process improved the green tape density by 8–11% and the as-sintered membrane by 5–10%; the electrical performance was also improved.

13.1.3.3 Extrusion

Extrusion has been used in the ceramic industries to fabricate tubes or other shape green bodies with an elongation in one direction continuously [2,16]. In extrusion, a cohesive plastic material like clay is pushed to pass through the orifice of a rigid die of the desired cross section by a mechanical driving force, to form a green ceramic body (Figure 13.5a). The extruded tube is cut into segments with desired lengths by wire cutting, then debindered and sintered. This method is usually used to fabricate supports including electrolyte tubes or substrates for anode- or cathode-supported SOFC. For the electrolyte-supported one, the tubular green bodies are usually sintered directly to form a dense ceramic [17]; during the sintering process, the tubes are placed horizontally on a V-shaped holder (Figure 13.5b) or hung vertically to obtain straight tubes [18,19]. On the other hand, in preparing the support of membrane SOFC, the functional layer and electrolyte membrane are coated using other coating

TABLE 13.1
Concentrations of Slurry for Tape Casting

	Ceramic Powder	Solvent	Dispersant	Binder	Plasticizers
Nonaqueous tape casting [10]	YSZ, 55.40%	MEK/EtOH, 32.13%	Phosphate, 1.11%	PVB 6.26	PEG: 3.00%, PHT: 2.10%
Aqueous tape casting [14]	NiO: YSZ = 1:1	Distilled water	PAA	Acrylic latex emulsion	PEG

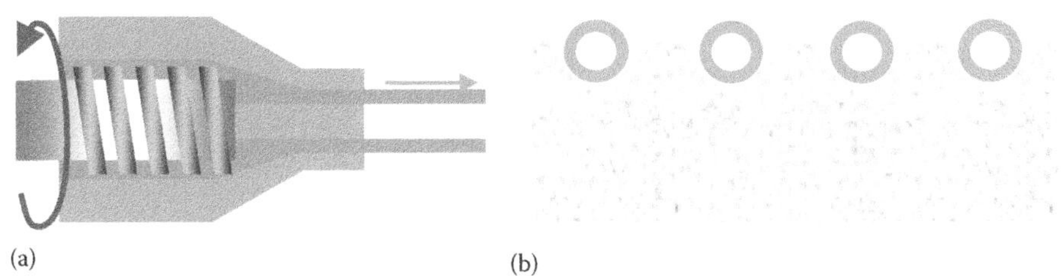

FIGURE 13.5 Schematic diagram of extrusion. (a) Extrusion and (b) sintering within a V-shape holder.

techniques (as described in Section 13.1.4) on the extruded tubular substrate, then a final co-sintering is allowed to proceed at a high temperature. Extrusion can also be combined with a phase-inversion method to fabricate hollow-fiber SOFCs [20].

13.1.3.4 Slip Casting

Slip casting is a traditional ceramic process with a long history. A plaster mold with a tube- or cone-shaped cavity is prepared by pouring a mixture of water and plaster powder into a container with a column, and following drying before the slip-casting process [21]. The stable suspension for slip casting is prepared by ball milling ceramic powders in solvent with dispersant and other additives. As Figure 13.6 shows, the suspension is continuously poured into the porous plaster mold, and by absorbing water in suspension, a green layer with suitable thickness is adhered on the cavity wall in the plaster mold. With the drying process, the green layer loses water, shrinks, and separates from the plaster wall; thus, it is taken out of the plaster mold and sintered to obtain a ceramic tube. Slip casting is a rather simple and cost-efficient technique for the fabrication of tubular supports; both dense YSZ electrolyte and porous NiO–YSZ anode can be fabricated.

Sui and Liu [21] used the slip-casting method to prepare cone-shaped YSZ electrolyte cylinders, then built up a three-cell stack that produced an open circuit voltage (OCV) of 2.9 V and 900 mW output at 800°C. He et al. [22] modified the slip-casting method by introducing a vacuum to prepare long thin-wall YSZ tubes with a thickness of only 0.2 mm. Ding and Liu prepared an NiO–YSZ substrate via the slip-casting method and deposited YSZ layer on the substrate, then co-sintered to form an anode–electrolyte bilayer [23]. Besides the fabrication of single-layer components, the slip-casting method was also employed to prepare YSZ/NiO–YSZ two-layer ceramic tube for SOFC by Okuyama and Nomura [24].

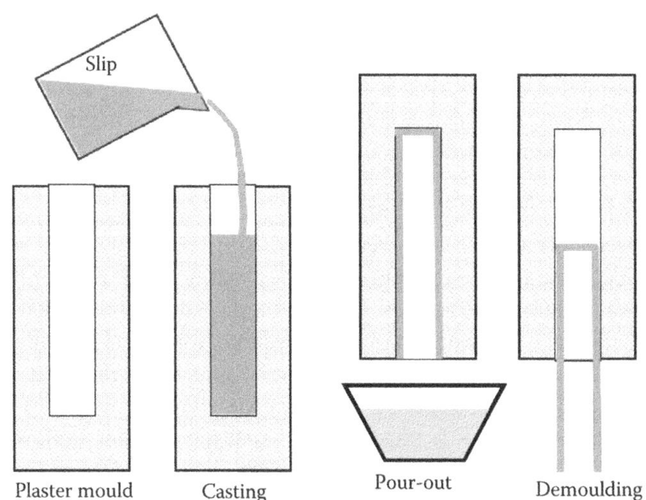

FIGURE 13.6 Schematic diagram of slip casting.

13.1.3.5 Gel Casting

Gel casting is a process used to prepare near-net-shape bodies in a mold using a suspension containing ceramic powder, premixed monomer, and cross-linking solution [25–28] (Figure 13.7). Typical compositions of mixture for gel casting are listed in Table 13.2. The primary advantages of gel casting are low cost, short forming time, good product homogeneity, high green capacity, and low-cost machining [27]. Cross-linking polymerization occurs under heating or catalysis condition to form a 3D network structure; after the slurry is solidified, ceramic objects shaped by the mold are obtained. This novel ceramic technique has been employed to fabricate electrode-supported substrates for SOFCs [28,29]. Furthermore, gel casting was also used as a synthesis method to prepare materials for SOFC [30].

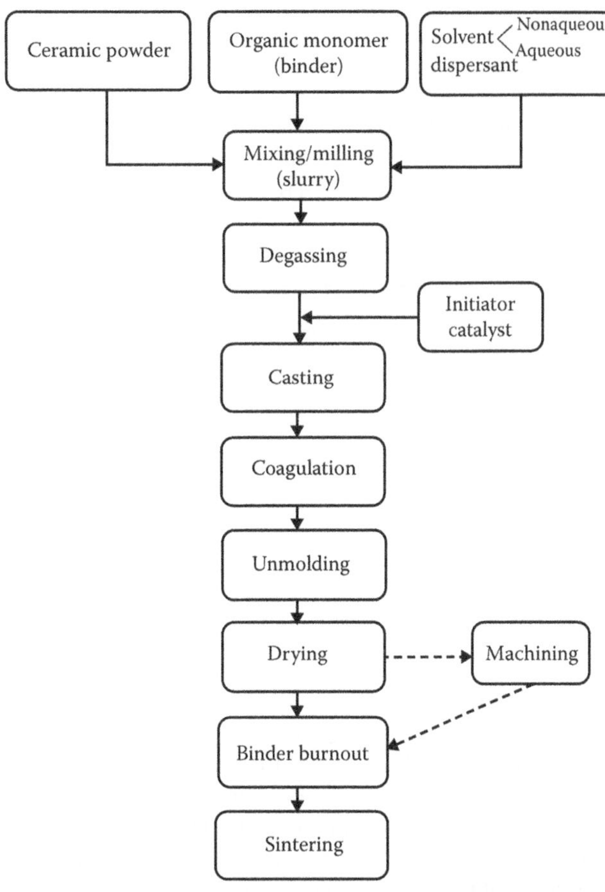

FIGURE 13.7 Detailed flowchart of gel-casting process. (Reproduced from *J. Eur. Ceram. Soc.*, 31, Yang J.L., Yu J.L., Huang Y., Recent developments in gelcasting of ceramics, 2569–2591, Copyright 2011, with permission from Elsevier.)

TABLE 13.2
Compositions of Mixture for Gel Casting

Content	Function	Weight Ratio
$(ZrO_2)_{0.92}(Y_2O_3)_{0.08}$ (8YSZ)	Ceramic powder	1.00
Methacrylamide (MAM)	Monomer	0.23
N,N'-methylene-bis-acrylamide (MBAM)	Cross-linker	0.01
Water	Solvent	0.61
Poly-acrylic ammonia (PAA) solution	Dispersant	0.02

Source: Huang W.L., Zhu Q., Xie Z., *J. Power Sources*, 162, 464, 2006.

13.1.4 Fabrication Techniques of Coatings for SOFC

13.1.4.1 Spin Coating

Spin coating is a common technique that has been used in the semiconductor and other industrial fields for a long time. The fluid (sol or slurry) with suitable viscosity is used as the film precursor for spin coating. During the spinning of substrate, the sol or slurry spreads on the flat substrate to form a smooth coating driven by centrifugal force, as shown in Figure 13.8.

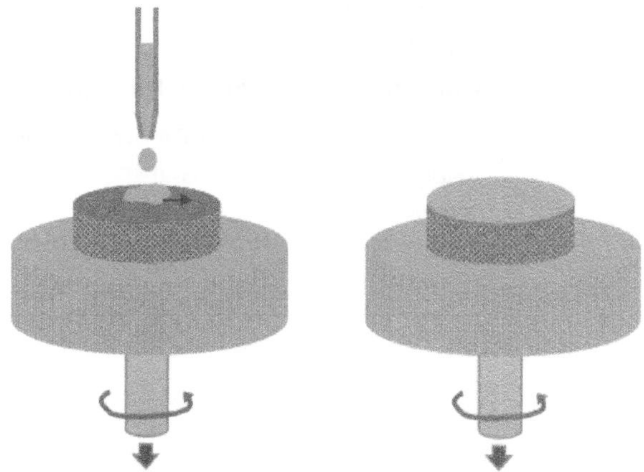

FIGURE 13.8 Schematic diagram of spin coating.

This technique was first employed to fabricate sol–gel thin films. Fang et al. [31] prepared $Ce_{1-x}Gd_xO_{2-x/2}$ thin films on different substrates. Kim et al. [32] prepared YSZ thin films on YDC substrates.

Spin coating based on a sol precursor with massive solvent needs a long time for drying and has a rather slow heating rate (e.g., ~0.5°C min^{-1}) to form a very thin film with high quality. In addition, to get an electrolyte membrane with several microns for SOFC, the sol spin coating, drying, and heat-treatment processes must be repeated for many times; thus, it is a time-consuming technology. Chen and Ai et al. [33,34] developed a slurry spin-coating method to prepare electrolyte membranes on porous anode supports by applying a slurry containing oxide powder and organic binder as precursors; thus, the thickness of each layer is several microns and more. As rapid heating rate is available, much less time is needed. On the basis of this pioneer investigation, Wang et al. [35] optimized the technique parameters of slurry spin coating and obtained a maximum power density of 2.005 W cm^{-2} using Ni–YSZ-supported 12-μm-thick YSZ membrane and Sm-doped CeO_2 (SDC)-infiltrated LSM as cathode. A spin-smoothing method was also developed by Wang et al. [36]; a preapplied thicker slurry coating was smoothed by the subsequent high-revolution-spinning process, thus a dense membrane with a thickness of 8 μm can be obtained in a single time processing, and a single cell based on this membrane produced a power density of 1.42 W cm^{-2} at 800°C.

13.1.4.2 Screen Printing

Screen printing is a universal industrial technology that is widely used in many fields. Screen printing use a paste/ink with suitable viscosity (few Pa s) and a special time-dependent rheological behavior [37,38]. The paste/ink for screen printing consists of ceramic powder, binder, solvent, and other additives. As shown in Figure 13.9, during the screen-printing process, the paste/ink is pressed by a squeegee in order to flow through the mesh of a screen and coat on the substrate surface. After removing the screen from the substrate, the paste fills the remaining channels of the screen and the cavities; the paste must stop flowing to obtain the accurate dimension

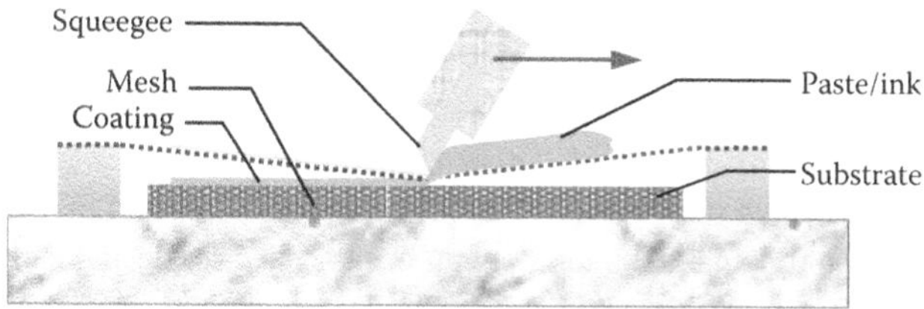

FIGURE 13.9 Schematic diagram of screen printing.

of the layer. Screen-printed layers after sintering have thicknesses ranging from 5 to 100 μm [16].

Screen-printing technology is usually used to prepare electrolyte membranes on anode substrates. Zhang et al. [39] prepared a gastight 12-μm-thick SDC film on NiO–SDC substrate using this method. The prepared single cell with a $Ba_{0.5}Sr_{0.5}Co_{0.8}Fe_{0.2}O_{3-\delta}$ cathode generated a maximum power density of 1280 mW cm^{-2}. Ge et al. [40] prepared a 31-μm-thick YSZ film on NiO–YSZ substrate (Figure 13.10). The maximum power densities of a single cell using this YSZ film and the LSM–YSZ composite cathode were 0.158, 0.318, 0.584, 0.964, and 1.30 W cm^{-2} at 650°C, 700°C, 750°C, 800°C, and 850°C, respectively.

Besides the preparation of electrolyte membranes on anode or cathode supports, the screen-printing technique has also been applied in the preparation of anode, cathode layers, and interconnector coatings on flat components of both planar SOFCs [41] and flat tubular ones. For example, Kim et al. fabricated the whole repeatable unit including the anode, electrolyte, cathode, and interconnector on a flat tubular support to form a segmented-in-series SOFC unit bundle [42].

13.1.4.3 Electrophoretic Deposition

The electrophoretic deposition (EPD) technique has a wide application range in processing ceramic materials and coatings. EPD has high versatility for use with different materials and their combinations, and possesses many advantages, including short processing time, simple equipment, little restriction of the shape of substrate, and no requirement of binder burnout [43]. As an inexpensive colloidal-processing method, during the EPD process, charged colloid particles in a stable suspension are driven by electrostatic force and deposited on a conductive substrate that works as an electrode in EPD, as shown in Figure 13.11.

The related suspension parameters influencing EPD include the size and zeta potential of particles, dielectric constant of

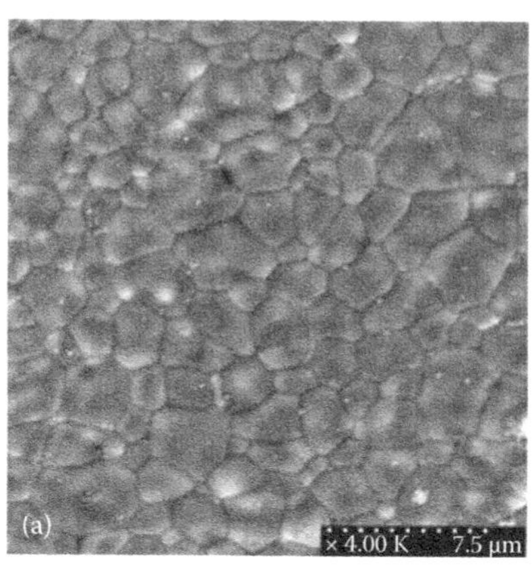

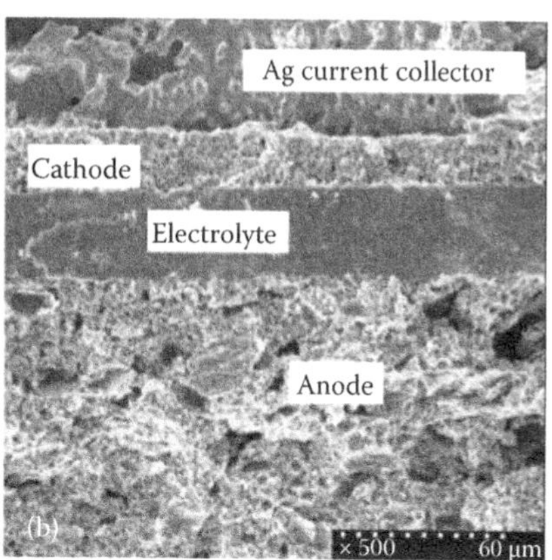

FIGURE 13.10 SEM images of YSZ membrane supported by Ni–YSZ substrate prepared via screen printing: (a) surface of YSZ film and (b) cross-sectional view of the cell. (Reproduced from *J. Power Sources*, 159, Ge X.D., Huang X.Q., Zhang Y.H., Lü Z., Xu J.H., Chen K.F., Dong D.W., Liu Z.G., Miao J.P., Su W.H., Screen-printed thin YSZ films used as electrolytes for solid oxide fuel cells, 1048–1050, Copyright 2006, with permission from Elsevier.)

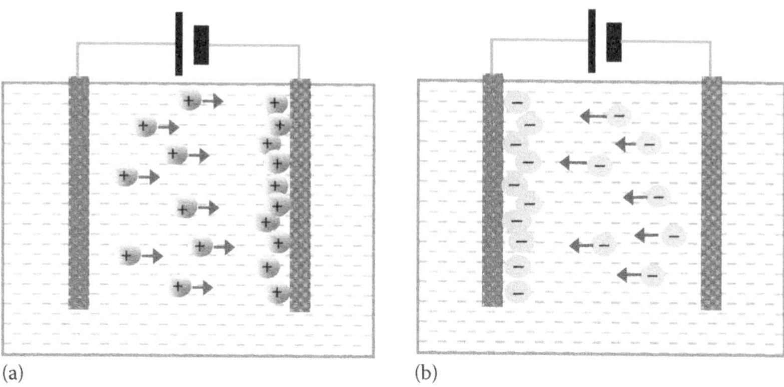

FIGURE 13.11 Schematic diagram of electrophoretic deposition: (a) cathode deposition and (b) anode deposition.

liquid, conductivity, and viscosity and the stability of suspension. At the same time, the deposition time, applied voltage, and concentration of solid in suspension can determine the deposition rate and thickness of coating, while the conductivity of the substrate is a critical parameter to the quality of the green film by EPD [43].

This method can be used to prepare thin and dense electrolyte films on anode substrates for SOFC applications. Majhi et al. [44] prepared YSZ film on NiO–YSZ substrate and the cell obtained an OCV of 1.03 V and a maximum power density of 624 mW cm^{-2} at 800°C. Besides the usual EPD process, Jia et al. employed this technique to modify the open pores in gravity-deposited YSZ films [45]; by the EPD process, the maximum power density was improved significantly. The EPD method has also been used to fabricate functionally graded NiO–YSZ composite films [46]. An additional advantage of EPD technology is the ability to fabricate film on the substrate with different shapes. For example, Caproni et al. deposited YSZ coating on graphite rods in an array matrix to obtain closed-end tubes, and showed the industrial scaling-up potential of EPD technology [47].

13.1.4.4 Dip Coating

Dip coating is a simple and economic ceramic process that can be used to prepare a thin coating layer on a planar or tubular substrate. Electrolyte thin film or electrode layers, even a complete sandwich structure with anode, electrolyte, and cathode, can be prepared using this technique. During the dip-coating process, the substrate is dipped into a suspension containing the powder, then pulled up from the suspension at a suitable speed (Figure 13.12). The thickness of the coating layer is mainly determined by the viscosity, fluid density and surface tension of the suspension as well as the pulling speed [16]. By repeating the dip-coating process, the thickness of coating can be increased in multiples and the defect in the coating may be mended by the subsequent coating. The coating is dried and then sintered at a high temperature to achieve the desired thickness and quality. It is easy to coat a layer on a porous substrate than on a dense one because the capillary force in open pores can help the adhesion [16].

Kim et al. prepared a dense YSZ film on an extrusion flat-tube anode substrate via a slurry dip-coating method [48]; the cell produced a power density of 225 mW cm^{-2} at 0.6 V. Panthi and Tsutsumi developed a new multistep dip-coating method [49], in which the NiO–YSZ anode layer was deposited first on a carbon rod and presintered, then the YSZ electrolyte and cathode layers were deposited and sintered subsequently. The single cell generated an OCV of 1.0 V and a maximum power density of 472 mW cm^{-2} at 850°C. Zhou et al. developed a vacuum dip-coating method to fabricate a dense YSZ film on a prefired NiO–YSZ anode substrate [50]. Using an infiltrated $Sm_{0.5}Sr_{0.5}CoO_3$ nanoparticle cathode in the screen-printed porous ScSZ layer, the cell reached a maximum power density of 0.85 W cm^{-2} with an area-specific resistance of 0.914 Ω cm^2 at 750°C.

13.1.4.5 Wet Powder Spraying

Although wet powder spraying (WPS) is another ceramic processing technology based on suspension, it is rather different from other techniques based on suspension as mentioned above. In those methods, the substrate must contact the mass suspension, while during the WPS process, the stable suspension is sprayed onto the substrate by using an air-spray gun with a continual supply of compressed air. The sprayed coating is dried and sintered as other ceramic coating methods. This method is widely used to prepare large, uniform thin films on different substrates in industrial scale for its simple and cost-efficient advantages. WPS is rather different from screen printing or slurry spin coating for its application to spraying on nonuniform or nonplanar geometries [16].

Wang et al. [51] used the WPS method to prepare a $Gd_{0.1}Ce_{0.9}O_{2-\delta}$ (CGO) interlayer with a thickness of 2–3 μm on a YSZ electrolyte to prevent the solid reactions between YSZ electrolyte and an $La_{0.6}Sr_{0.4}Co_{0.2}Fe_{0.8}O_{3-\delta}$ (LSCF) cathode. The properties of the CGO interlayer depend on the powder presintering temperature, the concentration of the dispersant, and the sintering temperature. The prepared dense CGO interlayer blocks Sr diffusion effectively. Guan et al. [52] developed a spray-modified pressing method, and sprayed a suspension containing $BaZr_{0.1}Ce_{0.7}Y_{0.2}O_{3-\delta}$ (BZCY) and 4 wt.% ZnO additive powders on prepressed NiO–BZCY anode support to form a bilayer.

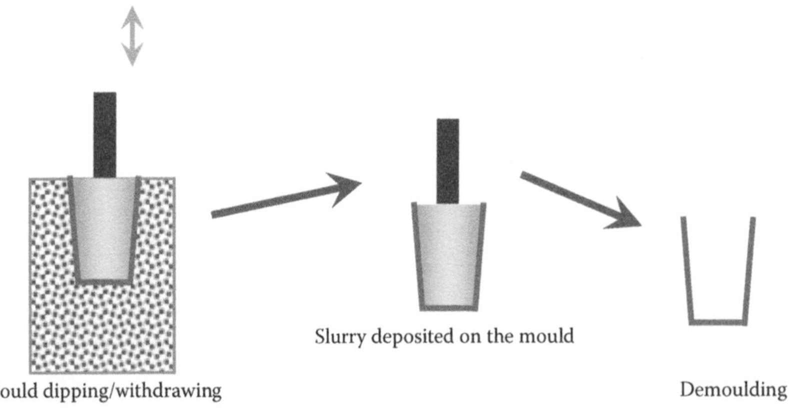

FIGURE 13.12 Schematic diagram of dip coating.

13.1.4.6 Roll Coating

Mücke et al. [53] developed a roll-coating method to prepare layers on a substrate prepared via TP. During the roll-coating process, the dip roll rotates in a liquor trough containing the suspension to transfer the suspension from the dip roll to a rubber-coated applicator roll. By moving the substrate under the applicator roll, a homogeneous coating is applied on the sample surface. The coating thickness can be controlled by adjusting the distance between the dip roll and the applicator roll, while the distance between the applicator roll and the transport roll is determined by the substrate thickness. During coating, only a linear load occurs between the two rolls, so there is no failure even with coating on thin and brittle presintered substrates; thus, it is considered to be better than the screen-printing process that results in a more inhomogeneous load distribution (Figure 13.13). This technique is considered to be a more promising process for the industrial production of SOFCs, owing to its high degree of automation [53].

13.1.4.7 Plasma Spraying

Plasma spraying is a promising technique for the deposition of different coatings in SOFCs owing to its advantages such as high deposition rate, cost-effectiveness, and flexibility for automatic production [54–56]. The substrates to be deposited can be porous ceramic, cermet, or metal. During plasma spraying, coatings are prepared by injecting the stock powders into an energetic plasma flow. When molten or softened droplets in plasma arrive at the substrate, they are cooled down quickly and solidified to form flat lamellae; subsequently, the coating layer is developed successfully [16]. Plasma spraying possesses the characteristics of fast deposition rate and easy masking for deposition of patterned structures, and is a process used to fabricate electrolytes for SOFCs. However, the porosity of an air plasma–sprayed electrolyte coating is high, resulting in a high gas permeability that causes a low OCV of SOFC and reduces the electrical conductivity of the electrolyte layer and output power [57]. This method has been used to prepare the various functional layers of SOFCs, including the anode [58], electrolyte [59], cathode [60,61], and interconnect [62]; it is also a key modification technique used to prepare oxide coating on an alloy interconnect in order to decrease the contacting resistance and prevent volatile chromium poisoning in the cathode [63]. As a well-established technology, plasma spraying has good application potential in the fabrication of SOFC components at temperatures below 700°C [59], especially when the low density is improved by a combination of high-velocity oxy-fuel and suspension feedstock [64].

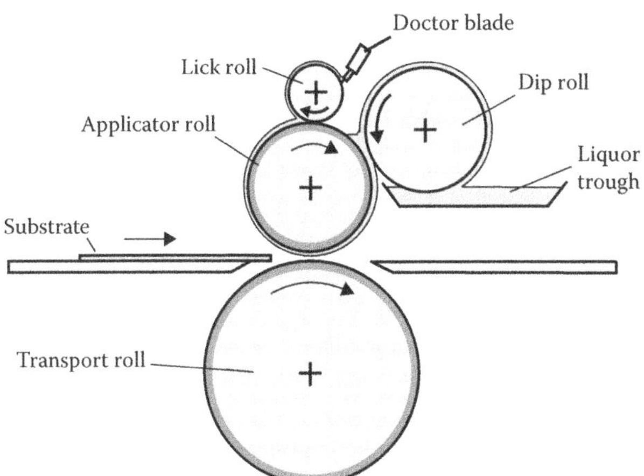

FIGURE 13.13 Schematic of the roll-coating process. (Reproduced from *J. Power Sources*, 196, Mücke R., Büchler O., Bram M., Leonide A., Ivers-Tifféeb E., Buchkremer H.P., Preparation of functional layers for anode-supported solid oxide fuel cells by the reverse roll coating process, 9528–9535, Copyright 2011, with permission from Elsevier.)

13.1.4.8 Electrochemical Vapor Deposition

Electrochemical vapor deposition (EVD) is a variant of the chemical vapor deposition (CVD) technique that is tailored for the deposition of dense films onto porous substrates [65]. It was first developed by Isenberg at Westinghouse in 1970s to fabricate dense ion or electron-conducting oxide films (e.g., electrolyte and interconnect materials) onto porous substrates for SOFC applications [66]. The formation of a dense layer in the EVD process needs two steps.

First, the open pores in the porous substrates is closed by the direct reaction of metal source reactants (e.g., $MeCl_x$; Me is the cation species) with an oxygen source reactant such as water vapor or a solid oxide such as NiO. The reaction equation for the process is as follows:

$$MeCl_x(g) + \frac{x}{2}H_2O(g) \rightarrow MeO_{x/2}(s) + xHCl(g) \quad (13.1)$$

The reactants are usually delivered to the opposite side of the porous substrate and diffuse into the substrate pores. The reaction sites are the pores that are deposited with the oxide products via the CVD process until the open pores are eventually closed so that any further direct contact and reaction of the reactants are prevented. Since the concentration and reaction rate of metal chloride are highest at the metal chloride side, pore closure always occurs at the metal chloride side of the porous substrate.

Second, after the CVD process is terminated, the growth mechanism will be driven by the electrochemical deposition process. In this process, the oxide films grow over the closed pores by solid-state diffusion of oxygen ions driven by the large oxygen activity gradient across the deposited film, since oxygen ions can be produced by the reduction of H_2O at the water vapor side, while oxygen ions are utilized at the metal chloride side to form oxide film. The reaction equations are as follows:

$$H_2O(g) + 2e^- \rightarrow H_2(g) + O^{2-} \quad (H_2O \text{ side}) \quad (13.2)$$

$$MeCl_x + \frac{x}{2}O^{2-} \rightarrow MeO_{x/2} + \frac{x}{2}Cl_2 + xe^- \quad (\text{chloride side}) \quad (13.3)$$

Although EVD and polarized EVD can obtain fully dense electrolyte layers, they still suffer from some drawbacks such as the high temperature required (~1200°C) and expensive equipment and materials, so their application is mainly limited to the fabrication of electrolyte thin films for tubular SOFCs, and they tend to be replaced by other techniques.

13.1.4.9 Pulsed Laser Deposition

Laser deposition is an important technique for depositing thin films of almost any material [67]. Pulsed laser deposition (PLD) is a widely used deposition technique to prepare a variety of oxide films on different substrates. During the PLD process, a pulsed laser beam is irradiated on the oxide target in a vacuum chamber and the target is evaporated to form a plasma plume over the target; thus, oxide film is deposited on the substrate in the plume. Many process variables, such as the wavelength and power density of the laser, background gas and its pressure, target composition, substrate–target distance, substrate temperature, substrate bias, and gas–surface interactions, can influence the deposition process and the performance of the deposited film [67]. PLD can produce dense films with good control of stoichiometry and grain microstructure [68]. As a typical thin film technique, the dense electrolyte film produced via PLD is near or less than 1 micron. The fuel cell based on PLD film can present high power density under an intermediate-temperature range for the rather low ohmic resistance in thin electrolyte layers. For example, Yan et al. reported a power density of 3270 mW cm^{-2} generated in a single cell based on an $La(Sr)Ga(Mg)O_3$ thin film prepared using the PLD method at 973 K [69].

A thin doped ceria interlayer is often used between zirconia and cathodes to prevent interfacial reactions [70]. However, it is difficult to obtain a fully dense ceria interlayer using ceramic methods, as a relative low sintering temperature (<1300°C) should be used to alleviate the formation of less conductive CeO_2–ZrO_2 solid solution. Lu et al. fabricated a dense SDC interlayer by PLD, and the fuel cell demonstrated improved performance with significantly reduced ohmic resistance [71].

13.1.4.10 Sputtering

Sputtering is a widely used physical vapor deposition (PVD) method to prepare alloy and compound films. In an evacuated chamber, the sputtering gas forms a self-sustained glow discharge, the positive ions in plasma are accelerated by high negative voltage to strike the target surface, and the target is sputtered by the momentum transferring process to move and the sputtering gas become deposited on the substrate or other surfaces [67]. For DC sputtering, a metal target is often used to prepare a metal film with inert sputtering gas. An oxide film can also be prepared by using a metal or alloy target and sputtering gas containing oxygen; thus, the method is named reactive DC magnetron sputtering. However, the surface of the metal target may be oxidized by oxygen, and it causes much lower sputtering yield than the original metal surface.

Using radiofrequency (RF) excitation, a nonconducting target or a metal target with an oxidized surface is also available. Although RF sputtering has been used for YSZ film deposition in the production of SOFCs [72], the deposition rate is low because of the low sputtering yield of YSZ (e.g., ~0.25 μm h^{-1}). On the other hand, reactive magnetron sputtering has a much higher sputtering yield (e.g., 5 μm h^{-1}) and deposition rate, and the magnetron sputtering provides ion irradiation of the film during deposition to form films with a high density at low temperature [67].

Bias-assisted sputtering can produce films with better quality than the normal sputtering since the bias voltage changes the potential distribution in the plasma; the deposited material can be sputtered off by argon ion with higher energy, thus the film morphology from porous to dense can be controlled by bias. Using bias-assist magnetron sputtering, a gastight electrolyte layer was deposited on anode substrates without any antecedent surface treatment or coating since the morphology of layer is changed from a columnar to a denser structure [73].

13.1.4.11 Electron-Beam PVD

Electron-beam PVD (EB-PVD) is a reliable technique to fabricate the large-scale production of thin films for industrial purposes [74]. During the EB-PVD process, the target material is evaporated using an electron beam heating device and deposited on the substrate. This technique can produce a reproducible film with an exact target composition, and it has a rather high deposition rate at low temperatures, e.g., a typical value is 1 μm·min^{-1}, which is much higher than that of magnetron sputtering [67]. The deposited films prepared via EB-PVD are crystallized without further annealing, which means both time and energy can be saved during the preparation. Jung et al. [74] prepared uniform large-scale YSZ thin films with a thickness of ~8 μm on an NiO–YSZ substrate with a 9-μm-thick functional layer; the maximum power density at 800°C was 0.76 W cm^{-2}. Besides electrolyte films, EB-PVD also can be used to prepare anode layers. Meng et al. prepared a gradient Ni–YSZ layer on an iron sheet [75], by altering the electron gun currents for the targets of YSZ and NiO; the element distributions along the direction across the coating varied gradually.

13.2 TESTING AND DIAGNOSIS OF SOFCS

The testing and diagnosis of SOFCs are very important, not only for the fundamental research and/or exploration of SOFC materials and techniques but also for the development of SOFC stacks and systems. These testing and diagnosis play essential roles in understanding and determining the reasons behind the inevitable performance degradation or even complete failure of cells and stacks. In this section, some primary testing and diagnosis techniques as well as some other characterization methods for SOFC materials and components are presented.

13.2.1 Single Cell and Stacks for Testing

13.2.1.1 Button Single Cell for Research Purpose

During the research and development (R&D) of SOFCs, the practicability of the material or the fabrication technique is usually evaluated in single cells under SOFC operation conditions. The most common configuration for the evaluation is planar-type fuel cells consisting of cathode, electrolyte, and anode (also named positive electrode, electrolyte, and negative electrode). The sandwich structure formed via layer by layer can be also called "coin cell" or "button cell" owing to its shape and size. Among the three layers of PEN, the part that plays the supporting role is usually prepared via the DP method, while the other two layers can be prepared using other ceramic coating techniques, such as screen printing [39] or slurry spin coating [33], and then they are sintered together.

To test the output performance of a fuel cell precisely, the effect of both the lead resistance and contact resistance on the connected nodes must be eliminated. Thus, during the testing of a fuel cell, a four-wire measurement technique similar to a low resistance measurement is employed, as Figure 13.14 shows. There are four leads connecting the fuel cell to testing instruments (such as electronic load, fuel cell testing system, or electrochemical station), i.e., two leads (separated current lead and potential lead) are connected to the anode and the other two are connected to the cathode. The potential measurement with special potential leads avoids the influence of voltage drops on the current leads and nodes when a huge current passes through them.

If the oxidant for the fuel cell is stationary air, the single cell is sealed on one end of a ceramic tube using sealant glass or even silver paste [76]. Fuel gas is transferred to the anode (fuel electrode) side using a pipe, while the cathode is exposed to open atmosphere.

Most laboratory SOFC studies are conducted on small single cells or short stacks. A testing process has a typical OCV of ~1.1 V (for a single cell) and a maximum current of usually <1 A, so the best instrument choice is a universal electrochemical station or a potentiostat that has a voltage and current measuring and controlling function, even providing an electrochemical impedance spectrum (EIS) function. Most electrochemical stations on the market can fully satisfy these demands. During the operation of a button cell, the fuel gas (such as hydrogen, methane, and propane) is supplied using a high-pressure gas cylinder, and the flow rate is controlled and measured using a mass flow controller. Of course, needle valve and glass flowmeter can be combined together to form the most simple and economic system. Hydrogen is a popular fuel gas for most basic investigations or demonstrations. The hydrogen consumed by a laboratory SOFC can also be prepared via water electrolysis using a hydrogen generator. Using a hydrogen generator may increase the equipment cost; however, the advantages are also obvious since the adjusting and controlling become more precise, and, at the same time, heavy and dangerous high-pressure hydrogen cylinders are not needed. The SOFC is usually operated or tested at a temperature range of 500°C to 1000°C.

13.2.1.2 SOFC Stack

An SOFC stack is composed of several or dozens of stack repeating units (SRUs) via a series or shunt connection. A stack in series can provide higher output voltage and power. Of course, the testing of stack needs more reactant supply, large operating space, and higher voltage and current range of the testing system. Since each single cell or SRU in a stack may be operated under different conditions, e.g., different fuel gas composition and local temperature, the voltage of each SRU may be different from others despite discharging under the same current; thus, the voltage monitoring of each SRU is usually needed, namely the testing system should have a multichannel voltage measurement function to measure the voltage of each cell during a constant current discharge process.

13.2.2 Electrochemical Performance Testing and Diagnosis of SOFCs

The testing of SOFCs is an electrochemical measuring process. The primary parameters of SOFCs, including OCV, maximum power density, internal resistance, degradation rate, etc., are usually measured with an electrochemical station, potentiostat, or electrical load.

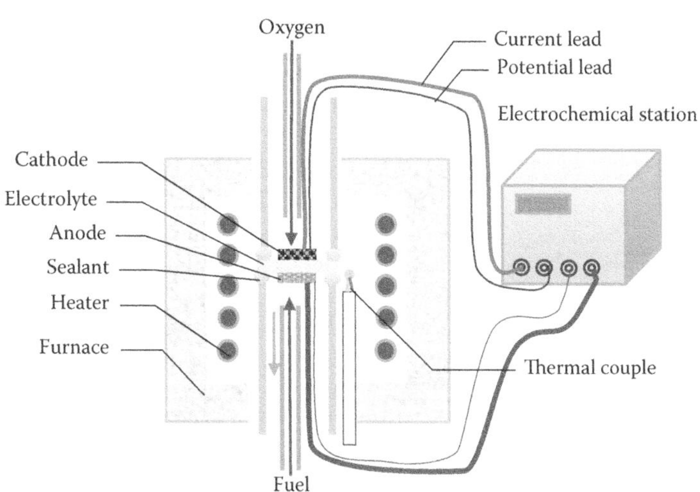

FIGURE 13.14 Schematic diagram of single cell test.

13.2.2.1 Open Circuit Voltage (OCV)

The OCV is the terminal voltage measured under open circuit condition, namely, in the absence of external circuit passage. It is concerned with the theoretical Nernst potential (V_{th}) and ion transport number (t_{ion}) of the electrolyte, and the OCV of an SOFC using a mixed conducting electrolyte is defined by Wagner's equation [75]:

$$OCV = t_{ion} \times \frac{RT}{4F} \ln\left(\frac{P'_{O_2}}{P''_{O_2}}\right) = t_{ion} \times V_{th} \tag{13.4}$$

The OCV may be lower than the theoretical one for the reasons of low t_{ion} of electrolyte with electron conduction [77] or fuel leakage [7] causing more complex reactions on the two electrodes.

13.2.2.2 Discharge Performance and Electrode Polarization

The terminal voltage and discharge current can be measured and recorded simultaneously when external load or the polarization current is changed; thus, a current–voltage (*I–V*) curve can be plotted. In addition, the power corresponding to current can be calculated to provide the relationship between power and current. For a single cell considered the effective area, power density and current density are usually calculated and adopted instead of power and current, in order to compare each other beyond the limit of cell size. The *I–V* curve of an SOFC is not a straight line in general (Figure 13.15), because only ohmic loss can cause a linear *I–V* response; the coexisting activation polarization causes curving up in the small current density region, while concentration polarization causes curving-down in the large current density region. To measure the electrochemical performance of a certain electrode (defined it as working electrode, WE), the reference electrode (RE) is introduced to the SOFC, and the voltage between the WE and the RE is measured to calculate the overpotential by compensating for the effect of electrolyte resistance.

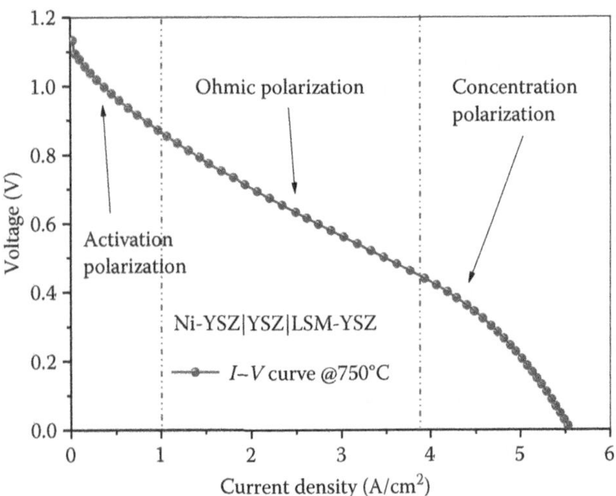

FIGURE 13.15 Typical output *I–V* curve of a single SOFC.

13.2.2.3 Electrochemical Impedance Spectrum (EIS)

The terminal voltage of an SOFC reduces with increasing current output, according to Ohm's law of the whole circuit; there is an internal resistance in SOFC. As discussed before, a nonlinear *I–V* curve means a nonconstant internal resistance since it consists of the contributions of different physical and chemical processes. To distinguish the contributions, the EIS technique is usually employed. AC impedance is a more general type of resistance. The modulus and phase of complex impedance depend on the frequency [78]. EIS measurement is used to measure the responses of an entire SOFC or only a single electrode (WE) to a small amplitude AC voltage with different frequencies applied on the cell or between the WE and the RE.

The EIS has many applications in the research and development of SOFCs. The EIS results can be presented as a Nyquist plot (ImZ~ReZ, Z'–Z'') and a Bode plot (|Z|~log*f* and θ~log*f*) [79] (Figure 13.6). On a Nyquist plot, an EIS of a whole SOFC or a certain electrode is usually composed of several depressed arcs one by one. Each arc means an electrochemical process with a specified characteristic frequency (time constant) and shows a peak on the Bode plot (Z''~log*f* or θ~log*f*). Generally, the analysis of EIS mainly aims to attribute these arcs to the respective related processes.

Measurement of the EIS and finding the relationship of each arc variation to alternating operation parameters can help achieve this goal. For example, an oxygen reduction reaction on the cathode was separated into several steps [80], then the rate equations were presented and the relationships of the resistance related to each step to oxygen partial pressure P_{O2} were derived. By measuring and analyzing the EIS of the cathode under different P_{O2}, the polarization resistance (R_p) and resistance of each step (R_i) were obtained, according to the relation $1/R_i \propto (P_{O2})^m$, where the *m* value can be calculated and used to attribute the processes and confirm the rate-determining step. A typical relationship between the *m* value and a related process is summarized in Table 13.3.

13.2.2.4 Stability Testing

To evaluate the stability and reliability of an SOFC, the single cell or stack may be operated for a long time (e.g., over thousands of hours) or a short time (e.g., about several hours or tens of hours), the output current (or voltage) is kept constant and the terminal voltage (or current) is recorded continuously, and the degradation rate can be calculated based on these data. The constant-current or constant-voltage discharge process may be interrupted to measure the EIS or *I–V* curve during the long-time testing.

13.2.2.5 Diagnosis of SOFC

During a long-term operation, an SOFC may suffer gradual performance degradation or different kinds of faults. Therefore, diagnosis is very important for the R&D of SOFC. Bareli et al. have classified the faults of SOFC and the detection techniques in their review paper [81]. As they described, the increase in ohmic resistance and the degradation of

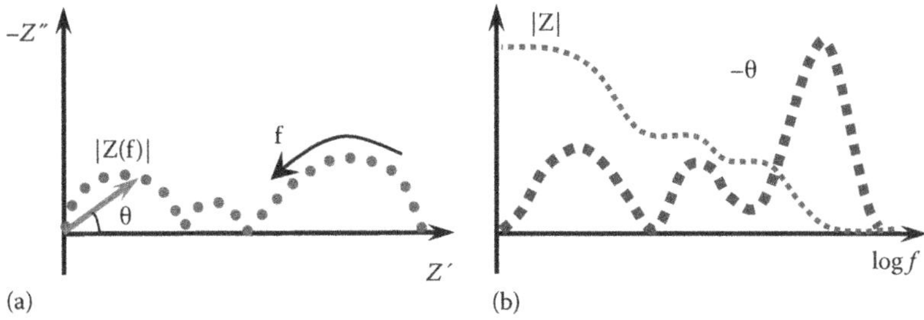

FIGURE 13.16 Schematic diagram of EIS: (a) Nyquist plot and (b) Bode plot.

TABLE 13.3
Processes during Oxygen Reduction on LSM Cathode

Step	Process	*m*
0	Diffusion $O_2 \rightarrow O_2$ (pore)	1
1	$O_{2(g)} \rightarrow 2O_{ad}$	1/2
2	$O_{ad} + e^- \rightarrow O^-_{ad}$	3/8
3	$O^-_{ad} \rightarrow O^-_{TPB}$	1/4
4	$O^-_{TPB} + e^- \rightarrow O^{2-}_{TPB}$	1/8
5	$O^{2-}_{TPB} + V_O^{\infty} \rightarrow O_o$	0

Source: Kim J.D., Kim G.D., Moon J.W., Park Y., Lee W.H., Kobayashi K., Nagai M., Kim C.E., *Solid State Ionics*, 143, 379, 2001.

electrode microstructure are the main two defaults in SOFC stacks.

Electrode delamination is a common reason that causes a considerable increase of internal resistance of single cell or stack. Both ohmic resistance and polarization resistance will be influenced by this phenomenon; an increase of ohmic resistance occurs proportionally to the delamination area, beyond that a reduction of the electro-active area [82]. The EIS technique is used to detect the change of ohmic resistance as well as polarization resistance during the SOFC operation [78], while imaging techniques, such as scanning electron microscopy [83] and x-ray computer tomography technique (see Section 13.3), can be employed to observe and evaluate the delamination. The active area above or below the delamination may be almost completely lost since the large aspect ratio (e.g., thin electrolyte) of SOFC causes the in-plane conduction of ionic species to become negligible [82]; its implication on the change in impedance spectrum is a simultaneous change on both series and polarization resistances by approximately the same proportion.

Other reasons resulting in ohmic resistance increase are the local shortage of fuel, anode reoxidation or oxide layer growth, and rib detachment [81]. Seal fault may cause local fuel shortage, anode reoxidation, and the growth of an oxide layer with lower electrical conductivity between the interconnect plate and electrodes. Rib detachment causes conduction interruption through the detached ribs, and more current flows through the other intact ribs and may lead to local overheating and resulting in the aggravation of oxidation [81].

The electrode microstructure degradation during long-term operation can be attributed to many reasons, such as sintering of the electrode during operation, poisoning of electrodes by the contaminants (Cr, B, S, C, Si, etc.), and so on [81]. The diagnosis related to these mechanisms usually needs further characterizations and discussion, as described in Section 13.3.

13.2.2.6 Fault Tree Analysis

A fault of an SOFC may be caused by several different kinds of lower-level events, while each event is usually caused by more basic events. To clarify the logical relation among these events and reasons, a suitable analysis approach is needed. Fault tree analysis (FTA) is such an approach. As Steiner et al. introduced in their review paper [84], FTA is a deductive tool that is used to link an undesired state to a combination of several lower-level events via a "top–down" approach. The fault tree representation appears to be a good way to clear out the connections and interactions between the different operating parameters responsible of a given fault [84].

Although the electrochemical measurements combined with other basic testing techniques around a running SOFC can provide the key performance data for the evaluation and diagnosis, thorough understanding of the experimental phenomena during an SOFC operation and the validation of related presumptions as well as the degradation mechanisms still need further characteristics of the basic materials, components, and tested samples using alternative nonelectrochemical techniques.

13.3 CHARACTERISTIC TECHNIQUES OF SOFC MATERIALS AND COMPONENTS

Many characteristic methods can be used during the R&D of SOFC materials and components, including the observation and measurement of the material's microstructure, surface morphology, spectrum analysis, as well as measurement of physical and/or chemical properties, and so on. The characterizations can provide abundant experimental evidences to clarify the role of materials and components in SOFC, in order to understand the mechanism of observed phenomena and degradation or failure.

13.3.1 X-Ray Diffraction

X-ray diffraction (XRD) is the most common characteristic technique that can reveal the lattice structure via x-ray coherent scattering by electrons in the atoms aligned in crystal. The XRD method can be used to determine phases of materials, to calculate lattice parameters, or even obtain more information by refinement [85,86]. By exploiting some effects of imperfections in the crystal system, such as amorphous structure, defect, and finite grain size, more information can be obtained, such as grain size, microstress, etc. Under the investigation of SOFC materials and components, as well as in some mechanisms, XRD is widely used as the preferred phase analysis technique. The XRD pattern must be presented for each novel material to determine if it is single phase or not. The evaluation of chemical compatibility between two adjacent materials is also determined by the XRD patterns. New peaks of impurity substances may appear in the XRD patterns of the interface or in a fired mixture; however, some element atoms may diffuse through the interface during the high-temperature process and only cause some XRD peak shift [87].

13.3.2 Scanning Electron Microscope

A scanning electron microscope (SEM) is an instrument that can obtain morphological information by detecting the signals produced by the interaction between the focused electron beam with energy of several keV to tens of keV and sample surface, including secondary electron (SE), back-scattering electron (BSE), x-ray excited by electrons, or absorbed electron current on sample. The SEM not only has higher magnification times and resolution but also larger depth of focus as well as better image quality than that of the optical microscope. In the R&D of SOFC, SEM is used to observe both the surface and cross-sectional morphologies of SOFC components (electrode, electrolyte layer, interconnect, or sealant) to evaluate the microstructures, such as dense electrolyte and interconnect, or porous electrolyte. In addition, the thickness of layers, particle size, and pore size in the component, as well as the toughness of surface or interface, can be determined by the analysis of SEM image.

In fact, the detecting modes of SEM have their special advantages. For example, the SE mode can offer distinct images with higher resolution than other modes; on the other hand, using the BSE mode, the regions containing different elements show different brightnesses, thus making the element distribution easy to distinguish. By analyzing the intensity of the excited characteristic x-ray with different energy, the distribution of elements on the sample surface can be obtained quantitatively.

13.3.3 Focused Ion Beam Scanning Electron Microscope (FIB-SEM)

Conventional SEM can only provide 2D images, which limited the observation of 3D space in SOFC components and the connection between different materials [88]. A novel SEM combined with focused ion beam (FIB) milling named dual-beam FIB-SEM has been developed and used to obtain 3D information. The sample can be milled to form a trench by the FIB on the sample surface, while the SEM is positioned on a tilt direction. After a thin section (50 nm) is removed from an exposed surface by the FIB, SEM imaging of the fresh surface is acquired, namely "cut-and-see" operation [89], and the process is repeated to yield a series of consecutive SEM images. Energy dispersive x-ray spectroscopy analysis is employed to correlate brightness with two solid phases at the beginning or the end of observation. Using this technique, high-quality volumetric data including both morphology and element distribution have been obtained, and 3D microstructure reconstruction of Ni–YSZ SOFC electrode has been accomplished first [88,90], as shown in Figure 13.17; then, other composite electrodes, such as LSM–YSZ cathode [91], were also investigated using this method, and both triple-phase boundary (TPB) length and tortuosity factors have be obtained.

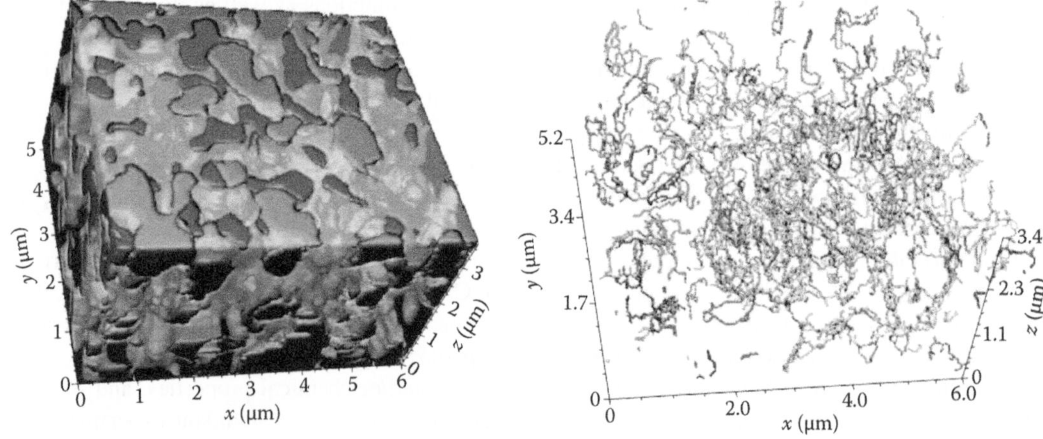

FIGURE 13.17 3D anode reconstruction (a) and 3D map of three phase boundaries in the anode (b). (Reproduced with permission from Macmillan Publishers Ltd., *Nat. Mater.*, Wilson J.R., Kobsiriphat W., Mendoza R., Chen H.-Y., Hiller J.M., Miller D.J., Thornton K., Voorhees P.W., Adler S.B., Barnett S.A., Three-dimensional reconstruction of a solid-oxide fuel-cell anode, 5, 541–544, Copyright 2006.)

13.3.4 Secondary Ion Mass Spectrometry (SIMS)

Using an ion beam with high energy bombarding on the surface of a sample, the surface atoms are sputtered off and become charged ions, named secondary ions. The analysis of ion charge/mass ratio using a mass analyzer can provide the information of surface compositions. SIMS is a sensitive analysis technique, e.g., for some elements, the detection limit can reach the parts per million order. SIMS can reveal the chemical composition of the real-surface and near-surface atomic layer, so it can provide more information than other simple analysis techniques of elements. Thus, it can play a large role in the measurement of tracer diffusion coefficients and surface-exchange coefficients of materials [92], and the investigation of the degradation mechanism of SOFC, since low concentration level impurities in SOFC stack components can be measured by the SIMS technique [93].

13.3.5 Inductively Coupled Plasma Atomic Emission Spectrometry

Inductively coupled plasma atomic emission spectrometry (ICP-AES), is based on the inductively coupled plasma torch method used to analyze the spectra of atom emission. It has a high accuracy and precision, low detection limit, quick measurement time, wide linear range, and simultaneous determination of multielements, etc. It has been used to determine the elements' composition in SOFC components [94] or even to detect a small quantity of contaminant in the electrode [95].

13.3.6 X-ray Fluorescence (XRF)

When x-ray, γ-ray, or electron beam with high enough energy irradiates on a sample, the electrons located on the inner shell and low energy level of atoms will be ionized, then x-ray fluorescence is emitted during the following de-excitation process. Since the energy of x-ray fluorescence is determined by the atomic energy level structure, each element can emit its particular x-ray fluorescence. Thus, the elements as well as their contents in a sample can be analyzed using the XRF technique. This technique is used to confirm the presence of a certain element [96] and analyze the composition [97] of elements in SOFC materials and components.

13.3.7 Dilatometer Technique

A typical SOFC has an anode–electrolyte–cathode sandwich structure that contact each other after the high-temperature sintering process. Since there may be rather large differences between the contact materials, the expansion is different during heating up, and thermal stress is generated on the interface. In addition, the difference of sintering shrinkage between two contact SOFC components (e.g., anode and electrolyte) can also cause stress or even deformation during the co-sintering process, and too large thermal stress may cause fracture or delamination. The thermal expansion coefficient (TEC) is a basic and key parameter of each SOFC material that needs to be measured during the development. During the cell and stack fabrication, the sintering shrinkage curve needs to be measured to draft an optimum sintering program to increase the yield. A dilatometer consists a programmed temperature furnace to heat up a bar sample that is clamped in sample holder by a push rod. When the sample expands or shrinks during heating up, the small displacement between the push rod and sample holder is measured via the displacement sensor of a linear variable displacement transducer; the thermal expansion curve or sintering shrinkage curve can be obtained after correction by compensating the thermal expansion of the sample holder and push rod.

The applications of a dilatometer in SOFC R&D are the measurements of TEC of SOFC materials [98], sintering performance [99], and the soft point of glass sealing materials [100]. Recently, chemical expansion induced by the change of nonstoichiometry has shown to be of importance in SOFCs. In addition to high-temperature XRD, a dilatometer is widely used for the measurement of isothermal dimensional change [101,102].

13.3.8 X-ray Photoelectron Spectroscopy

X-ray photoelectron spectroscopy (XPS) has the functions of elemental analysis for essentially the entire periodic table. XPS is a surface-sensitive technique since only the electrons from a depth of about 2–5 nm [103] can escape out under the excitation of soft x-ray photons with approximately keV-order energy. Besides the difference related to the distinct elemental atoms, the binding energies (B.E., equals $h\nu - K_e - \Phi$; $h\nu$—energy of x-rays, K_e—kinetic energy of photoelectrons, Φ—working function) in different chemical forms often give distinct results in the XPS, named chemical shift. The XPS technique can be used to investigate surface composition [104] and to identify the valence state of the atoms [105,106] in the sample and some species adsorbed on the sample surface [107]; at a certain depth, atoms may also be analyzed after sputtering with Ar ion [108]. Simner et al. have analyzed an anode-supported fuel cell after 500-h tests at 750°C/0.7 V, and XPS analysis revealed that Sr enrichment at the cathode–electrolyte and cathode–current collector interfaces was partially responsible for the high degradation [109]. Recently, synchrotron-based XPS coupling with electrochemical studies provides a capability for surface chemical measurements in the presence of gases at significant pressures (several Torr); the so-called ambient-pressure XPS was applied to *in situ* characterization of SOFC materials [110,111]. Huber et al. have studied the electrochemical activation and surface segregation of the $La_{0.75}Sr_{0.25}Cr_{0.5}Mn_{0.5}O_{3-\delta}$ electrode. The *in situ* XPS spectra clearly showed the Sr incorporation under cathodic polarization and the resegregation under anodic polarization [111], which agrees with the activation/deactivation mechanisms for the LSM electrodes proposed by Wang and Jiang [112].

13.3.9 X-ray Absorption Spectroscopy

X-ray absorption spectroscopy (XAS) is a technique based on modern synchrotron radiation facilities. A strong x-ray beam generated by a synchrotron system is split by varying degrees of monochromaticity, and the intensity is monitored by the first ion chamber; then, the beam passes through the sample and reference in turn and measures the intensities via other ion chambers. Thus, the absorbance at a given frequency can be calculated as defined by the Beer–Lambert law [113]. The absorbance shows a sharp increase with some fluctuation when x-ray photon energy equals the ionization energy of the atom, named an absorption edge. According to a shift away from the absorption edge, the fine structures of x-ray absorption can be classified by x-ray absorption near edge structure (XANES) and extended x-ray absorption fine structure (EXAFS). The basic information revealed by the absorption edge is the elemental constitution of a sample and the valence state of the atom; a higher valence state means a higher edge energy [114]. The x-ray absorption spectra of transition metals in the near edge region (XANES) are related to the electron transition from a lower level to an empty level, containing valuable chemical and structural information [115], while the position of the pre-edge peak strongly depends on the binding energy of the absorbing atom and consequently the effective charge, and they are usually used to investigate the interactions and electron states. EXAFS is more sensitive to the adjacent structure around the atom absorbing x-ray and can be used to probe the structure and radius of the different coordination shells around the central atom [114].

13.3.10 Other Special Characteristic Techniques for SOFC

A real SOFC stack is a relatively complex system. The degradation and failure of a stack during long-time operation and multiple thermal or redox cycles are related to chemical processes between contacted materials, fractures of sealant or cells, and depravation of interfaces. Thus, besides the usual *in situ* electrochemical testing and other general *ex situ* analysis methods, some special characterizing techniques are still needed.

Malzbender et al. have summarized some advanced nonelectrochemical techniques used at Forschungszentrum Jülich to characterize SOFC in their review paper [116]. Using these techniques, heat flow imaging, pore and defect in the anode and sealant, residual stress, and contact situation in the stack are addressed.

13.3.11 Thermography

The thermographic technique was used to obtain a visible thermal profile. Although *in situ* localization of such effects has not been possible up to now, the heat development of a short circuit resulting from the higher current can be localized and visualized by thermography. Although it is difficult to observe the thermography of an operating stack *in situ*, the stack can be removed from the stack testing setup and a thermography can be obtained by applying an external current at room temperature [117].

By the comparison of four different temperature measurements, including thermocouple, laser-guided thermometer, measurement of ohmic resistance via electrochemical impedance spectroscopy, and thermographic camera, a thermography method using high-temperature wind tunnel is considered to be the best one for determining the temperature gradient [118]. The thermocouple is the most common one used to measure and control temperature at one point precisely; however, it needs good contact with the component to be monitored and the uneven temperature distribution requires too many thermocouples. The EIS method can provide the practical core temperature of the cell under discharge condition, but it is only an average value for the operating cell or stack when there is a temperature difference at different regions of the cell. Although a laser-guided thermometer can realize noncontact temperature measurement point-by-point theoretically, it needs an open path to the operating stack and consumes more time to obtain a temperature distribution image. The thermographic method has higher space resolution and speed than the other three methods; however, the main drawbacks of this method are the requirements for an open stack setup (such as a high-temperature wind tunnel) and the emissivity effect of different materials on surfaces [118].

13.3.12 X-ray Computer Tomography Technique (CT)

The density differences in materials and components can be detected via the CT technique. Using this technique, the spatial distribution, size, and geometry of microstructural features with lower or higher density, can be elaborated in a nondestructive way; furthermore, it can be used to measure pores in the cell materials, to evaluate the contact situation in stacks, and to characterize the defect population in brittle planar SOFCs [116]. The weak spots in anode, including Ni cluster or pores, can be distinguished by the CT method for the different brightness on the images. The thermal mismatch may cause curvature of a planar cell with multilayer structure or residual stresses. The delamination caused by curvature effects can damage the physical contacts of geometrically stable planar cells and cause degradation or failure of the cells. Using the CT method, the contact situation in a post-operated SOFC stacks can be visualized nondestructively.

13.3.13 Raman Spectroscopy

A monochromatic laser light interacts with molecular vibrations or phonons in a solid sample, resulting in the energy of the laser photons being shifted up or down—this technique is named Raman spectroscopy. The shift in energy or wavelength gives information about the vibrational modes in the molecular or solid sample. Raman spectroscopy has some applications in the R&D of SOFC. The phase of zirconia containing ceramics can be analyzed by Raman spectroscopy [119]. In addition, results from electrolytes of long-term-tested real stacks have revealed the location dependence of the cubic to the tetragonal phase transformation, which was generally observed from Ni–cermet anodes and the electrolytes at the interface with the anodes. On the other hand, the electrolyte away from the anode side mainly remained cubic. Therefore, Ni dissolution into the cubic phase was considered to accelerate the conductivity degradation and the tetragonal phase formation. This finding is also important for the prediction of transformation speed and conductivity evolution of the electrolyte [120]. Peak shift analyses offer a novel approach to measure stress of zirconia-containing layers in SOFCs, e.g., the residual stress determination of YSZ thermal barrier coatings [119]. Micro-Raman spectroscopy allows measurements in very small areas (micrometer range) for the assessment of the residual stresses in the cross section and interlayers of coatings [116]. Using micro-Raman spectroscopy, the compounds produced during the poisoning process in the SOFC anode can be detected *in situ* or *ex situ* [121–123].

13.4 ADVANCED DESIGNS FOR SOFCS

13.4.1 Traditional Double-Chamber SOFC Stacks

At present, four stack configurations have been proposed and fabricated for traditional double-chamber SOFCs: planar design, tubular/microtubular design, segmented-cell-in-series design, and monolithic design [1,124–128]. The designs differ in the extent of dissipative losses within the cells, in the manner of sealing between fuel and oxidant channels, and in making cell-to-cell electrical connections in a stack of cells.

13.4.1.1 Planar Design

A single cell is usually composed of the anode, cathode, and electrolyte "sandwich." For the planar configuration, the components of the single cell are stacked together with interconnecting plates between them that connect the anode of one cell to the cathode of the next cell in the stack [124,126,129–133]. To produce significant amounts of power, practical SOFC stacks can be mainly divided into two types according to the assembled form: typical planar flat plate and radial planar configurations, as illustrated in Figure 13.18a,b [126,128]. For the typical planar flat plate, fuel flows into one face of the fuel cell stack and out the opposite face. Similarly, the oxidant flows into one face and out the opposite face [2]. However, for the radial planar configurations, fuel and oxidant flows do not flow in an open passage along an electrode surface, but diffuse through the porous electrode microstructure from the center to the periphery of the disk.

It can be seen that the flat-plate design permits several different ways of ducting gases in and out of the fuel cell [2]. In the most common flow geometry, the interconnect, having ribs on both sides, forms gas channels and serves as a bipolar gas separator contacting the anode and cathode of adjoining cells. Furthermore, fuel and oxidant flows in SOFC stacks can be arranged to be cross-flow, co-flow, or counterflow [1,134]. The selection of a particular flow configuration has significant effects on gas pressure, temperature, and current distributions within the stack. Various designs of the flow configurations on distributor plates have been proposed to increase the uniformity of gas pressure and distribution in fuel cell stacks to obtain optimum performances, such as serpentine, parallel serpentine, discontinuous, and interdigitated-type channel configurations [135–137].

Fabrication and stack assembly appear to be simpler for the flat-plate design as compared with the other designs because the electrolyte is commonly made by traditional TP. Furthermore, the electrodes could be fabricated on the sintered electrolyte by a variety of fabrication options (see Section 13.1), such as the slurry method, screen printing, and plasma spraying. Besides the simpler fabrication and low cost, the planar SOFC cell design offers other advantages over the tubular cell design. For example, the electrolyte can be made very thin in anode-supported cell designs to allow it to operate in the intermediate temperature region. The current path in the planar design is short, leading to lower ohmic loss and high power density [138]. On the basis of these advantages, planar SOFC systems have been developed and ranges from portable devices, small power systems, to distributed generation (DG) power plants.

The Solid State Energy Conversion Alliance (SECA), supported by the US Department of Energy, is attempting to increase the planar cell/stack power densities and reliability while simultaneously driving down operating temperature

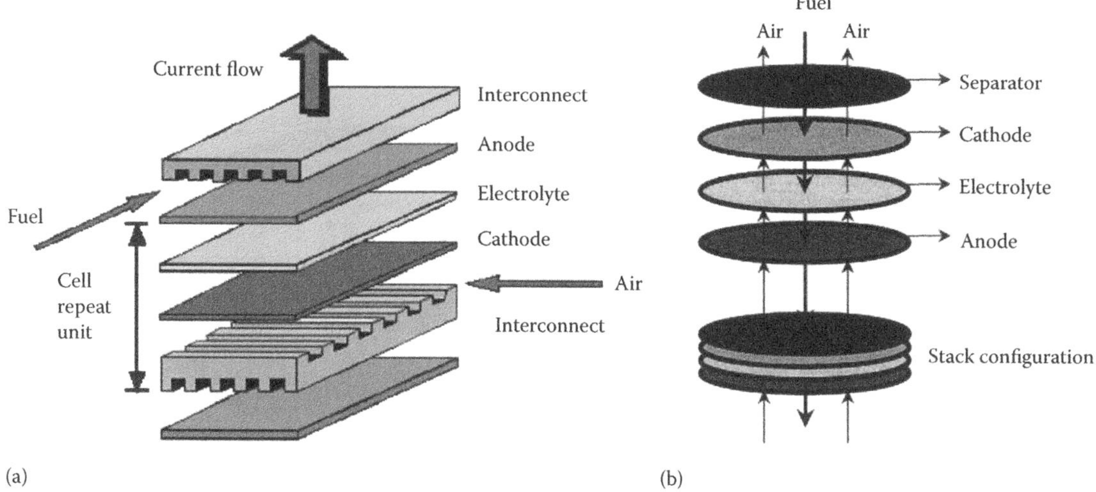

FIGURE 13.18 (a) Typical planar flat-plate and (b) radial planar configurations for planar cell stacks.

and costs. SECA's synergistic collaboration has driven projected SOFC stack costs from over \$1500 kW^{-1} in 2000 to around \$175 kW^{-1} in 2010 [139]. Bloom Energy (USA) is another commercial leader in terms of deployed planar SOFC units. They proposed that the 100-kW SOFC power generation "Bloom Box" can run on natural gas, and on renewable fuels like biogas, which has been proved to available for residential power generation and could be used for distributed power generation [140]. Furthermore, planar SOFCs are already beginning to make an appearance in transportation solutions as evidenced by Delphi's SOFC stacks as auxiliary power units for heavy duty trucks [138]. Research Center Jülich (Germany) has operated 20-kW planar SOFC stacks successfully. The consortium of Topsoe Fuel Cell A/S and Risø National Laboratory has identified planar SOFC technology as one of the most promising technologies for future power generation in commercial and industrial combined heat and power (CHP) applications in a power range of 0.2–5 MW [141]. They have tested several 50- or 75-cell stacks in the 1 kWe power range successfully on methane-rich reformate gas at a fuel utilization up to 92% [142,143]. In addition, some corporations and organizations, including Versa Power Systems (Canada), Ceres Power (Britain), Ceramic Fuel Cell Limited (Australia), Netherlands Energy Research Centre (Holland), and Sulzer Hexis (Switzerland), have also done their best to promote the commercial application of the planar SOFC system in CHP and DG, and some significant progress have be made.

However, lowering the operating temperatures and improving the reliability are still two main challenges for both planar cell and stack levels. Currently, continuous research is under way overcome the aforementioned challenges to put SOFCs on the path toward near-term commercial viability in a number of stationary power applications.

13.4.1.2 Tubular/Microtubular Design

Use of tubular geometry with the one-end closed feature realizes the seal-less design [127]. This is the most marked advantage of tubular geometry over its planar counterpart, where gas seals are necessary to separate air from fuel along the perimeters of cathode–electrolyte–anode and cathode–interconnect–anode [144]. In this design, the cell components are configured as thin layers on a closed-one-end tubular support (Figure 13.19a). For cell operation, oxidant is introduced to the cell through an injector tube near the closed end, where it traverses and exits from the open annulus between the support tube and the oxidant injector tube. Fuel flows on the outside of the support tube. At the open end of the cell, unreacted fuel is combusted using the oxidant stream exiting the cell [1,145]. Individual cells are bundled in series and parallel electrical connection to form a basic power-generating building block (Figure 13.19b).

Generally, the support tube, cathode, anode, and electrolyte in a tubular SOFC are formed by manufacturing extrusion, depositing, dipping, and the EVD process [1]. Planar and tubular designs are currently the two main assembly ways for SOFC stacks. Similar to planar SOFC stacks, various stacks based on the seal-less tubular design have been fabricated and tested to facilitate its practical application. In 1997, the consortium of Dutch and Danish utilities carried out a 100-kW SOFC power based on the tubular-type cells. The SOFC power consisted of 1152 air electrode–supported tubular type cells (2.2-cm diameter and 150-cm active length) and fabricated by Siemens Westinghouse Power (SWH) [125]. Subsequently, SWH proposed the first highest-efficiency, 200-kWe class pressurized SOFC/gas turbine (PSOFC/GT) hybrid system based on cathode-supported tubular SOFC technology [127]. In addition, in Japan, Hitachi has developed a module and a heat recovery system. The objectives of the 20-kW class cogeneration system are to achieve 40% of the electric power generating efficiency (HHV, AC) and 80% of the total energy efficiency (HHV) [146].

For tubular cells, power density depends on the inverse of cell diameter; the narrower, the better the performance. This observation led to the invention of microtubular SOFCs (MT-SOFCs) [147]. These tubes are on the scale of

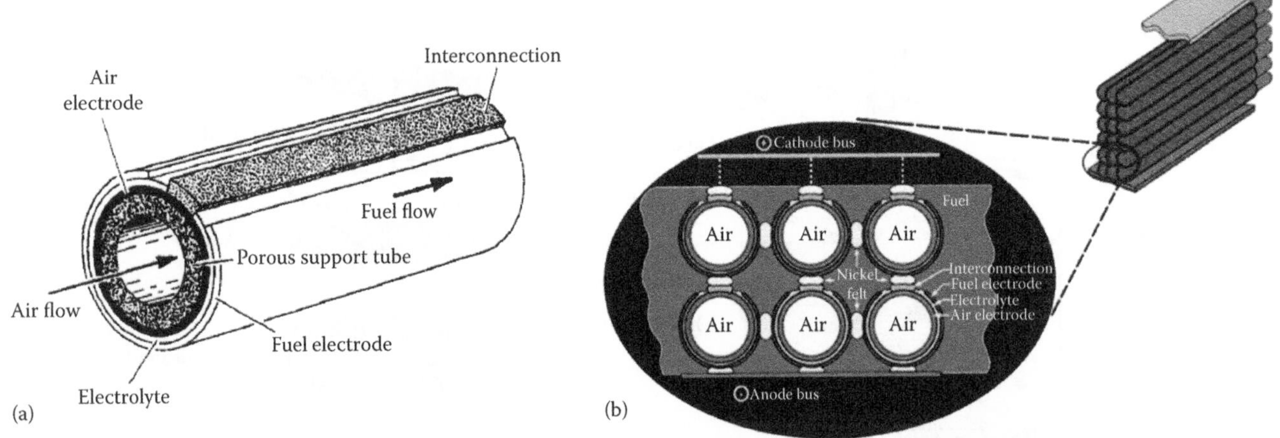

FIGURE 13.19 Schematic of (a) single cathode-supported tubular SOFC and (b) tubular SOFC bundles developed by Siemens/Westinghouse. (Reproduced from *J. Power Sources*, 237, Huang K., Singhal S.C., Cathode-supported tubular solid oxide fuel cell technology: A critical review, 84–97, Copyright 2013, with permission from Elsevier.)

millimeters, unlike the tens of centimeters for their predecessors. MT-SOFCs have the advantage of having short start-up times (on the order of a few seconds for a single cell), being resistant to thermal degradation on cycling, and having high power densities. The designs of MT-SOFC have also been investigated with anode and electrolyte-supported cells, as shown in Figure 13.20 [147,148]. In both designs, the support tube is longer than the active cell length. The first tube segment provides a gas inlet tube and the outlet section can be used as a combustor tube, where the fuel (hydrogen, short-chain hydrocarbons, etc.) and oxidant (oxygen or air) combust. Cathode-supported designs have also been investigated [149], but are not common in the literature because of the high polarization resistance of the cathode tube, weakening performance [150].

There are only a few design concepts described for MT-SOFC stacks. The most standard design is an array of MT-SOFCs with hydrogen flowing inside each cell in the array and air flowing around the outside of the cells. The lack of stack designs and optimization of the arrays in the literature is probably for two reasons. The first is that the problems associated with single cells have not been totally eradicated by most groups. Thus, they focus their efforts on single cells. The second reason is that groups associated with industrial partners will be guarding their stack designs carefully to keep their innovations hidden from their competitors [151]. However, an interesting cube-shaped MT-SOFC bundle was proposed by Funahashi et al. [152]. The reactor components were successfully assembled into a cube shaped MT-SOFC bundle (3 cm × 3 cm × 3 cm) with 36 tubular cells each having a diameter of 2.0 mm. In this type of MT-SOFC stack design, the cathode consists of blocks of matrices with several grooves that hold the MT-SOFCs. The cathode matrices are used as the current collectors, the guide for cell arrangement, and the oxidant flow path.

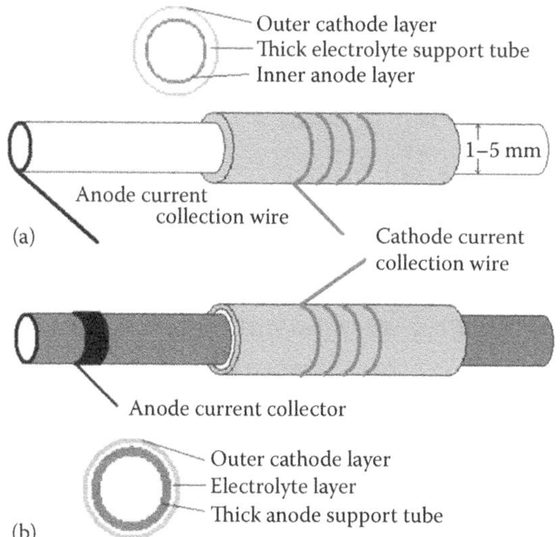

FIGURE 13.20 Basic microtubular fuel cell designs. (a) Electrolyte-supported and (b) anode-supported. (Reproduced from *J. Power Sources*, 196, Howe K.S., Thompson G.J., Kendall K., Micro-tubular solid oxide fuel cells and stacks, 1677–1686, Copyright 2011, with permission from Elsevier.)

13.4.1.3 Flattened Ribbed Tubular Design

To reduce the physical size and cost of SOFC generators, an alternate geometry cell has been investigated. Such alternate geometry cells combine all of the advantages of the tubular SOFCs, such as not requiring high-temperature seals while providing higher power per unit length and higher volumetric power density [153]. The important feature of this new design was the shortened current pathway by flattening the circular cross section into a rectangular shape in conjunction with multiple ribs connecting the two flattened surfaces; the formed cavities were still used as the channels for air delivery. These cells were known as high power density or HPD cells [127]. HPD5 with five air channels was among the first of several modifications to the cell design to enhance performance [154]. In an effort to increase the active area per cell and, consequently, the power per cell, a Delta 9 HPD cell with nine triangular air channels was developed. The Delta 9 HPD cell was extended to a wider Delta 13 cell, leading to further increase in power per cell, which ultimately evolved into a Delta 8 cell with eight air channels due to module compatibility and manufacturability considerations [127,154]. The progression of Siemens/Westinghouse's HPD cell geometries driven by the demands for high power density and power output is illustrated in Figure 13.21.

13.4.1.4 Segmented Cell-in-Series Design

The segmented cell-in-series design consists of segmented cells connected in electrical and gas flow series. The cells are either arranged as a thin-banded structure on a porous support tube (banded cell configuration) (Figure 13.22a) or fitted one into the other to form a tubular self-supporting structure (bell-and-spigot configuration) (Figure 13.22b) [1]. The interconnect provides sealing (and electrical contact) between the anode of one cell and the cathode of the next cell. In this design, the fuel flows from one cell to the next inside the tubular stack of cells, and the oxidant flows outside. In the banded cell configuration where the support tube (made either of ZrO_2 or Al_2O_3) is used, cells can be made with component thickness on the order of 100–250 μm. In the bell-and-spigot configuration, individual cells form into short cylinders of about 1.5 cm in diameter. The cells are about 0.3 mm thick to provide structural support [1]. The segmented-cell-in-series design offers the advantage of improved efficiency. However, gas manifolding and current collection within the stack are issues remaining to be solved [127]. In 1970, Westinghouse proposed a 100-kW coal burning fuel cell power plant using this SOFC concept, as the performance and lifetime of cells connected in series did not seem to be satisfactory at that time [125].

13.4.1.5 Monolithic Design

The monolithic design consists of thin cell components formed into a compact corrugated structure of either gas co-flow or cross-flow configurations [1], as Figure 13.23 shows.

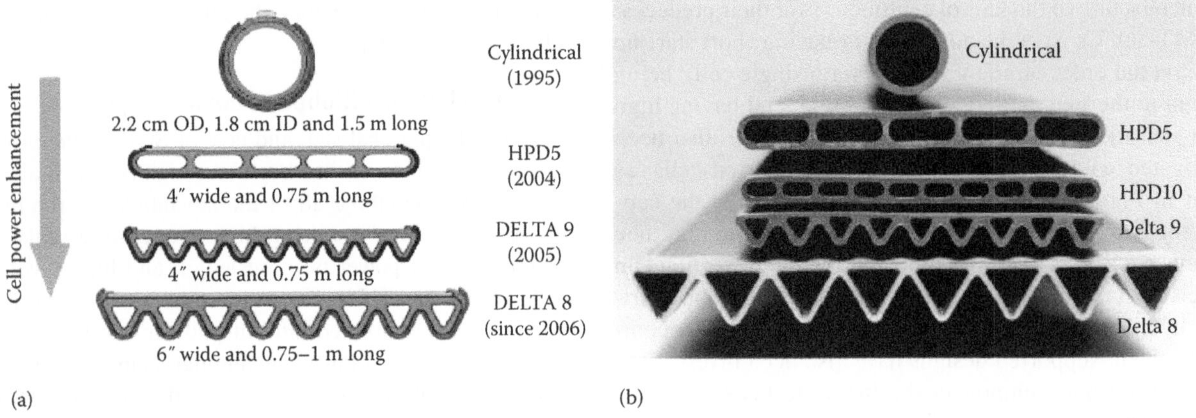

FIGURE 13.21 (a) Schematic and (b) actual image of HPD and Delta cells developed in Siemens/Westinghouse from the 1990s to 2000s. (Courtesy of Siemens/Westinghouse; reproduced from *J. Power Sources*, 237, Huang K., Singhal S.C., Cathode-supported tubular solid oxide fuel cell technology: A critical review, 84–97, Copyright 2013, with permission from Elsevier.)

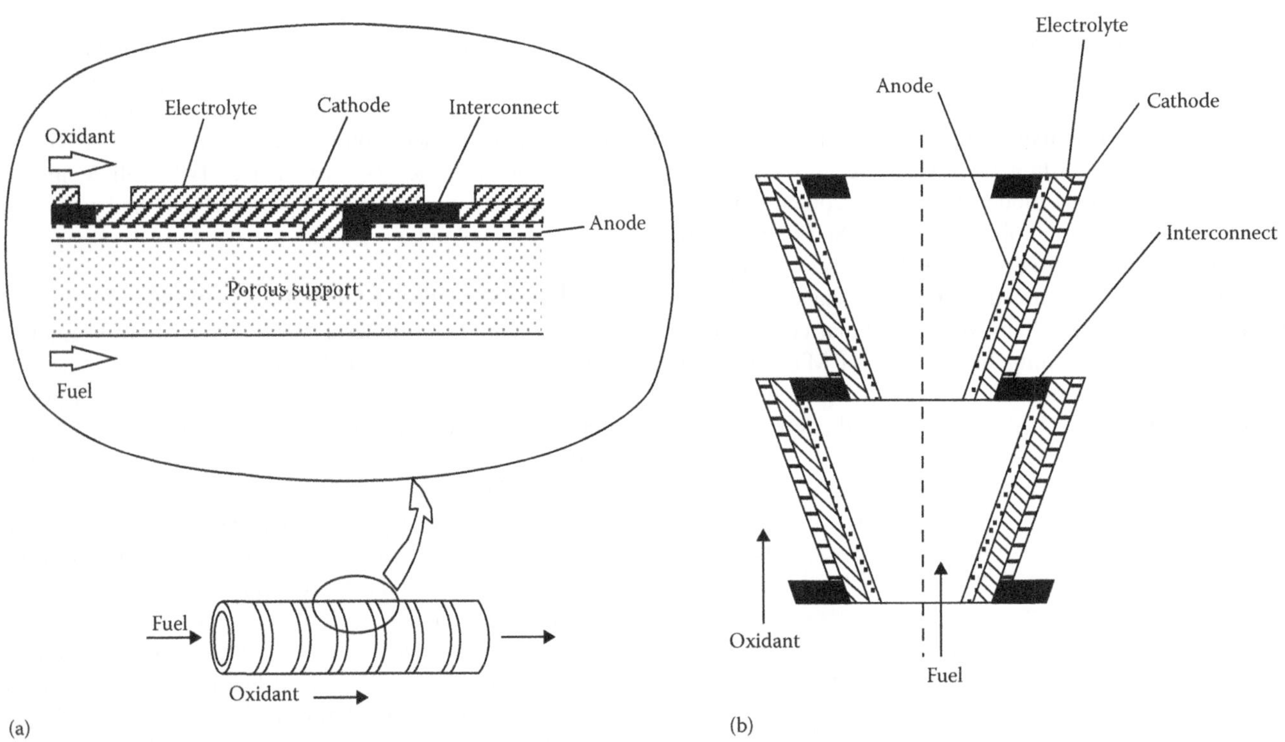

FIGURE 13.22 Segmented-cell-in-series design of solid oxide fuel cell: (a) banded cell configuration and (b) bell-and-spigot configuration. (From Minh N.Q.: Ceramic fuel cells. *J. Am. Ceram. Soc.* 1993. 76. 563–588. Copyright Wiley-VCH Verlag GmbH & Co. KGaA. Reproduced with permission.)

In the cross-flow version, the fuel cell consists of alternating flat layers of anode/electrolyte/cathode laminate and anode/interconnect/cathode laminate, separated by corrugated anode and cathode layers. The anode and cathode corrugations are oriented at right angles to each other. The major differences between the two versions are power density and gas manifolding. The cross-flow version results in a reduction in power density when compared with that of the co-flow. On the other hand, the cross-flow offers a simpler means of ducting gases into and out of the fuel cell ceramic structure. The key features of the monolithic design are small cell size and high power density. The power density for the monolithic SOFC is calculated to be about 8 or 4 kW kg^{-1}. The feasibility of the monolithic design has been demonstrated only on a laboratory scale because of the difficulty of development of suitable materials and fabrication processes [125]. A modified design of the monolithic configuration has been developed by Chubu Electric Power Co. and Mitsubishi Heavy Industry, and was

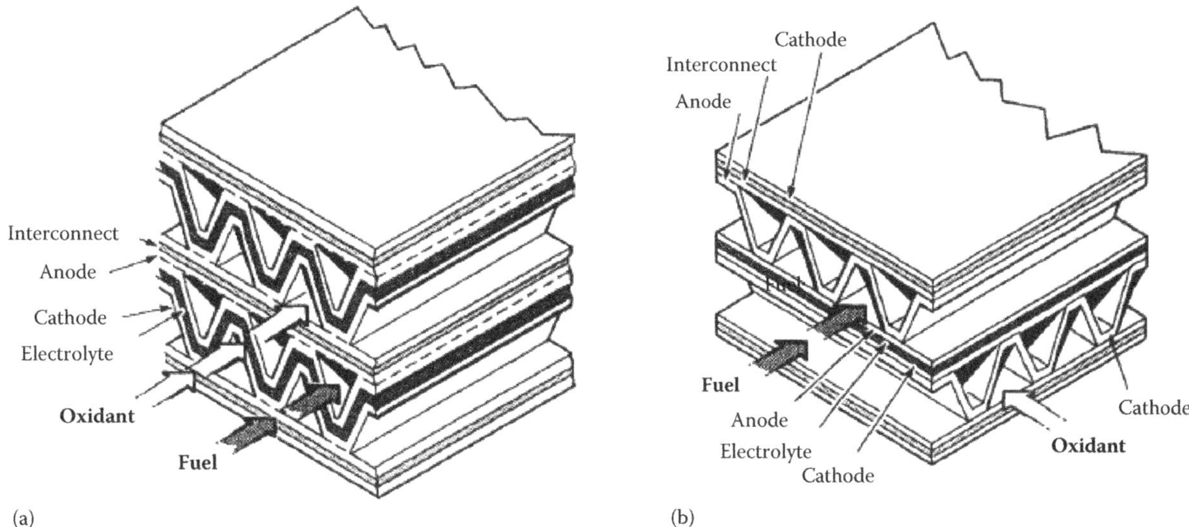

FIGURE 13.23 Monolithic SOFC design: (a) co-flow and (b) cross-flow configurations. (From Minh N.Q.: Ceramic fuel cells. *J. Am. Ceram. Soc.* 1993. 76. 563–588. Copyright Wiley-VCH Verlag GmbH & Co. KGaA. Reproduced with permission.)

named the Mono Block Layer Built (MOLB). A 5-kW scale MOLB-type SOFC was built, and successful results were obtained.

13.4.2 Novel Designs for SOFCs

13.4.2.1 Symmetric SOFC

As mentioned above, a standard SOFC usually consists of an anode, a cathode, and a gastight electrolyte. The gases reach each electrode compartment separately, and the anode was exposed to an oxidizing atmosphere while cathode was exposed to a reducing atmosphere. Obviously, the anode and cathode should have high stability in reducing and oxidizing conditions, respectively. Furthermore, they should have good chemical and TECs compatible with those of the other cell components. Thus, a symmetric SOFC using the same materials as anode and cathode has been proposed [155,156]. There are numerous further benefits derived from this concept. An SOFC with a symmetric structure should moderate compatibility problems between components, considering that both the anode–electrolyte and the electrolyte–cathode interfaces would be similar. Furthermore, such a system also simplifies notably the production of fuel cells since the electrodes may be fired using the same thermal process, which significantly reduces cost, and so is an important practical concept. More important, the possible coke formation and/or sulfur poisoning on the surface of the anode can be potentially eliminated, and thereby the anode is regenerated by operating the anode as the cathode in turn, where carbon or sulfur species absorbed on the electrode could be burnt out by the oxidant. Simultaneously, the durability of the cathode could also be improved in the low O_2 partial pressure in the TPB region of the cathode side [155–163].

However, to achieve reasonably high power densities, it becomes a key issue to develop an appropriate symmetric electrode material that is structurally stable and electrically conductive over a wide range of oxygen partial pressure, and have high catalytic activities for both oxygen reduction and hydrogen oxidation reactions. The perovskite $La_{0.75}Sr_{0.25}Cr_{0.5}Mn_{0.5}O_{3-\sigma}$ (LSCM) was shown to be an effective, redox-stable electrode that can be used for both cathode and anode SOFC operations, to provide a symmetric fuel cell system with good performance characteristics [155,156]. Zhu et al. further found that the performance of symmetric SOFC based on the LSCM anode could be improved with additional impregnation of a small amount of SDC or Ni [163]. Subsequently, researchers found that the symmetric electrode materials could also be obtained by substituting the common perovskite oxide cathodes with Cr, Ti, Mo, and Sc, which helped increase their chemical stability in reducing atmospheres, including that of $La_{0.3}Sr_{0.7}Fe_{1-x}Cr_xO_{3-\delta}$ [161], $Pr_{0.7}Ca_{0.3}Cr_{1-y}Mn_yO_{3-\delta}$ [164], $LaSr_2Fe_2CrO_{9-\delta}$ [157], $(La,Sr)(TiFe)O_{3-\delta}$ [165], $La_{0.8}Sr_{0.2}Sc_{0.8}Mn_{0.2}O_{3-\delta}$ [166], $La_{0.6}Sr_{0.4}Fe_{0.9}Sc_{0.1}O_{3-\delta}$ [167], and $Sr_2Fe_{1.5}Mo_{0.5}O_{6-\delta}$ [160].

13.4.2.2 Single-Chamber SOFCs

Single-chamber SOFC (SC-SOFC) is an interesting SOFC evolution. An SC-SOFC, in which both the anode and the cathode are exposed to unique fuel-oxidant mixtures, was proposed in 1961 by Eyraud et al. [168] This configuration, characterized by an eliminating sealing, greatly improves the fuel cell integrity over the dual-chamber SOFC and allows for good thermal and mechanical shock resistance [169–175]. SC-SOFCs can be mainly divided into two types in accordance with the configuration [169,176], as shown in Figure 13.24. The anode and cathode are located on the opposite sides of the electrolyte for the A-type SC-SOFC (Figure13.24a), while the electrodes are located on the same side of the electrolyte for the B-type SC-SOFC (Figure 13.24b).

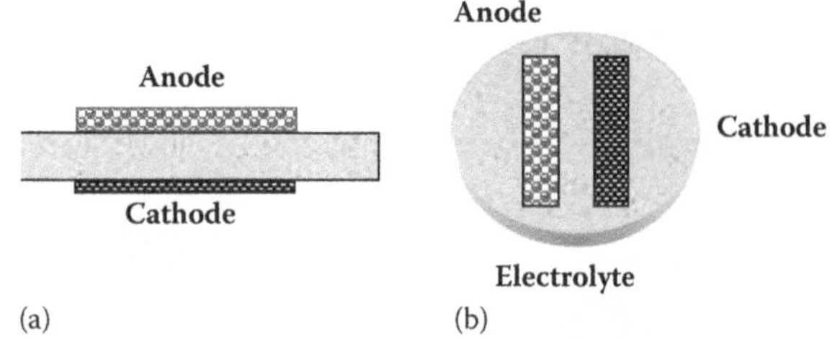

FIGURE 13.24 Schematic illustration of (a) A-type and (b) B-type SC-SOFCs.

For the A-type SC-SOFC configuration, the research is mainly focused on anode-supported, thin-film electrolyte fuel cells owing to their comparatively higher output performances that those of electrolyte-supported fuel cells [172,175,177–184]. Recently, significant progress has been made and many promising structures have been proposed in SC-SOFCs. Using the heat evolving from the partial oxidation of the fuel, Shao et al. demonstrated a thermally self-sustaining micro-SOFC stack with an anode-facing-anode configuration, as shown in Figure 13.25 [172].

The cell temperature of SC-SOFC stacks can be maintained at a sufficiently high level to allow power generation without an external heating source and exhibited a high power output and rapid start-up. Wei et al. [181] have proposed a star-shaped SC-SOFC microstack, which powered a USB fan stably. The advantage of this symmetric stack design consists of uniform gas distribution over the single cells and possible scale-up, as shown in Figure 13.26a. To improve the space utilization, Liu et al. have also fabricated an anode-facing cathode microstack with a novel cell array design [177], as shown in Figure 13.26b. To supply a relatively independent fuel and oxidant distribution over the respective electrodes and additionally extend the resident time of the gas flow over the cell, two gas channels with small vents were used as gas channels for a nonsealed SOFC microstack in a box-like stainless-steel chamber [184–186], as shown in Figure 13.26c. This novel stack design provides uniform reaction environment for the scaled-up stacks and avoids a previous mixing and the direct reaction between fuel and oxygen during the mixed gas flowing through the gas chamber. Such kind of stack has been assembled into a nonsealed SOFC module that produced an OCV of 6.4 V and a maximum power output of 8.18 W [184].

The B-type SC-SOFC is also called coplanar SC-SOFC, which can be easily connected in parallel and series to fabricate simple and compact SC-SOFC stacks to meet the power demand of portable devices. Recently, many promising structures related to B-type SC-SOFC have been proposed. To increase the effective electrode area, anode and cathode strips can be arranged in a comb-like electrode patterns [175,187–189], as shown in Figure 13.27a. Wang et al. provide a groove configuration by adding electrolyte wall fabricated between the anode and the cathode to reduce the conductive distance of oxygen-ion transport and form a gas separator between the electrodes, leading to a remarkable reduction in ohmic resistance, an elevation of open-circuit voltage, and, ultimately, an

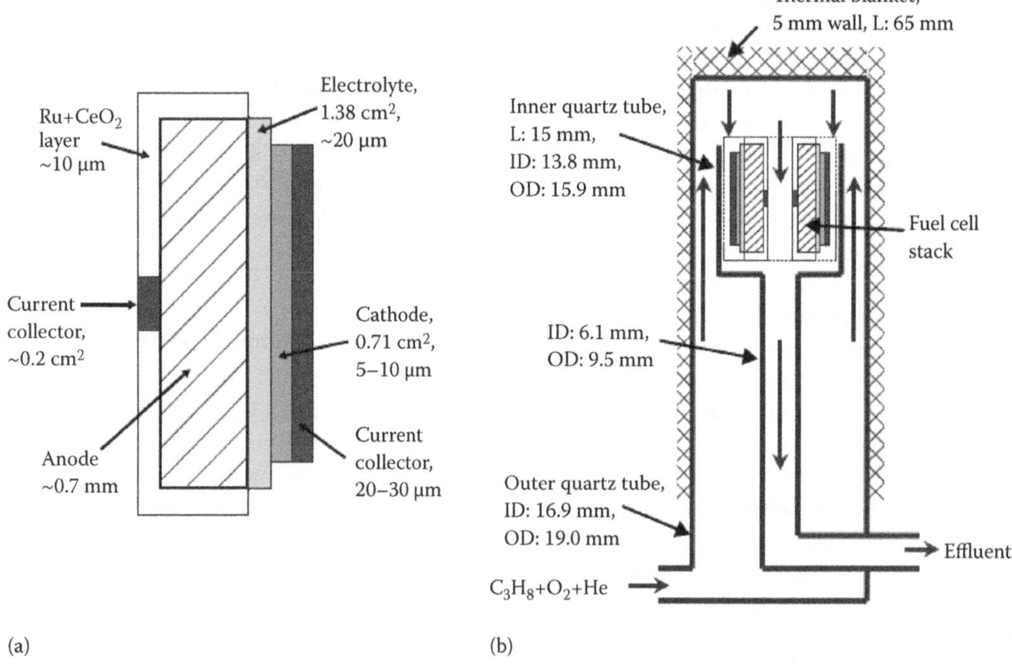

FIGURE 13.25 Schematic diagrams of the cell configuration (a) and reactor system (b) used in the thermally self-sustaining micro-SOFCs. (Reproduced with permission from Macmillan Publishers Ltd., *Nature*, Shao Z.P., Haile S.M., Ahn J., Ronney P.D., Zhan Z.L., Barnett S.A., A thermally self-sustained micro solid-oxide fuel-cell stack with high power density, 435, 795–798, Copyright 2005. From Nature Publishing Group.)

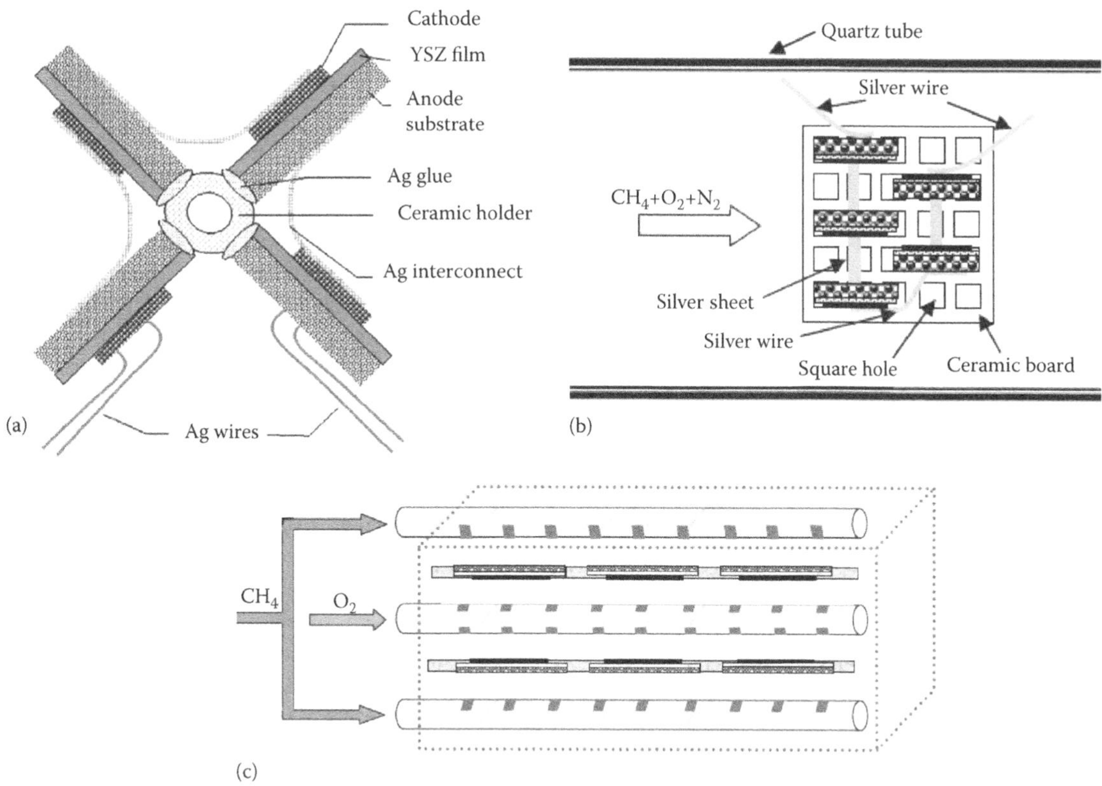

FIGURE 13.26 Schematic illustration of the microstacks with (a) star shape. (Reproduced from *Electrochem. Commun.*, 11, Wei B., Lü Z., Huang X.Q., Liu M.L., Jia D.C., Su W.H., A novel design of single-chamber SOFC micro-stack operated in methane-oxygen mixture, 347–350, Copyright 2009, with permission from Elsevier.) (b) Cell array. (From Liu M.L., Lü Z., Wei B., Huang X.Q., Chen K.F., Su W.H.: A novel cell-array design for single chamber SOFC microstack. *Fuel Cells*. 2009. 9. 717–721. Copyright Wiley-VCH Verlag GmbH & Co. KGaA. Reproduced with permission.) (c) Two gas channels. (From Tian Y., Lü Z., Wei B., Zhu X., Li W., Wang Z., Pan W., Su W.: Evaluation of a non-sealed solid oxide fuel cell stack with cells embedded in plane configuration. *Fuel Cells*. 2012. 12. 523–529. Copyright Wiley-VCH Verlag GmbH & Co. KGaA. Reproduced with permission.)

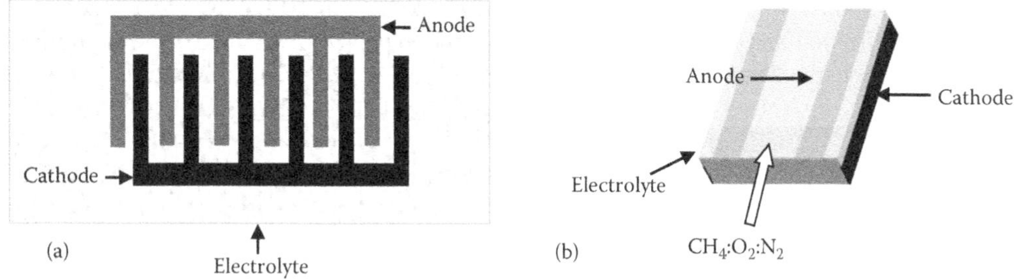

FIGURE 13.27 SC-SOFC with coplanar interdigitated electrodes (a) and right-angular configuration (b).

improved performance [173]. Subsequently, they further proposed a simpler right-angular configuration by arranging two electrodes symmetrically located on the two adjacent planes. This configuration can also make full use of the edge area of the electrolyte substrate and shorten the conductive channel of oxygen ion. An SC-SOFC stack with more compact configuration based on the right-angular configurations are also fabricated and operated successfully as shown in Figure 13.27b [190].

However, the low efficiency of the SC-SOFCs still imposes a severe technical hurdle to their practical use. To overcome the problem, on the one hand, SC-SOFCs have been proposed to generate electrical power from automotive exhaust gases. A 12-cell SC-SOFC stack delivered OCVs of 5–8 V and a power output of 1 W, making use of the sufficient heat and hydrocarbon fuels/oxygen in the exhaust gas of a motorcycle [183,191]. On the other hand, Shao et al. reported that an SC-SOFC integrated with a $GdNi/Al_2O_3$ partial oxidation catalyst in the same gas chamber was applied for the facile co-generation of electricity and synthesis gas from methane with zero waste gas emission [192]. The obtained results provide a novel option that promises the co-generation of electric power and synthesis gas from methane by SC-SOFC technology for the full utilization of fuel. However, it should be noted that the SC-SOFCs presented to date do not function as expected because of the selectivity of the anode and the limited selectivity of the cathode on a gas mixture of fuel and air [193]. Future research should, first of all, concentrate on the possible mechanisms and materials with suitable selectivity.

13.4.2.3 Direct-Flame SOFC

Direct-flame SOFC (DF-SOFC) is another type of sealant-free SOFC with an even simpler cell configuration that operates without any gas chamber [194,195]. The operation principle of a DF-SOFC is somewhat similar to the SC-SOFC, which is based on the combination of a combustion flame with a SOFC in a simple, "no-chamber" setup [195–197]. In this configuration, the anode is directly exposed to a fuel-rich flame while the cathode is open to ambient air. The flame serves not only as a fuel-flexible reformer through partial oxidation but also as a heat source to sustain the fuel cell operation illustrated in Figure 13.28 [198–200]. There are a number of advantages in DF-SOFC compared with DC-SOFC and SC-SOFC. (i) More simple setup: the DF-SOFC can be operated in a no-chamber setup, so that the anode is simply held into the exhaust gases of a combustion flame, and the cathode breathes ambient air. There is no need for external heat to initiate the fuel cell. (ii) Rapid start-up: the flame heat release brings the fuel cell rapidly to its operation temperature. (iii) Fuel versatility: since the intermediate species produced by the flame are similar for a wide variety of hydrocarbon fuels, the direct flame fuel cell is highly flexible in liquids such as ethanol, butanol, and kerosene and even solids such as wood and wax fuels [199,200].

However, the development of DF-SOFC is still in its infancy, and a number of questions regarding the DF-SOFC operation still remain [199,200]. For example, the current DF-SOFC has a low fuel utilization and performance owing to uncontrollable energy waste in the flame to sustain the fuel

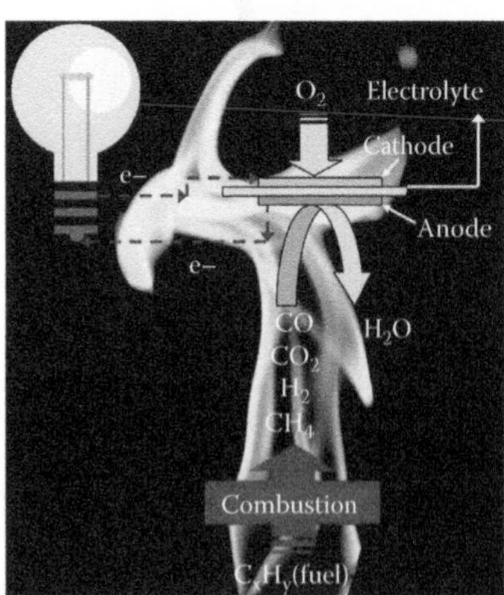

FIGURE 13.28 Schematic of direct-flame fuel cell. (Reproduced from *Prog. Energ. Combust.*, 37, Walther D.C., Ahn J., Advances and challenges in the development of power-generation systems at small scales, 583–610, Copyright 2011, with permission from Elsevier.)

cell temperature. Furthermore, the thermal shock resistance of DF-SOFC is poor. Rapid start-up and uneven temperature distribution of fuel cell generated by flame could induce significant thermal stress to the SOFC, especially for the dense electrolyte layer. It is well known that a dense electrolyte layer is easy to crack when it experiences a high heating or cooling rate. In addition, carbon coking is still an issue when the hydrocarbon fuels are employed. To solve these issues, many researchers have studied the feasibility of various fuels, operating parameters, and materials/configuration designs for DF-SOFC. Kronemayer et al. have reported that the flame operating conditions equivalence ratio, gas inflow velocity, and distance between burner and fuel cell have a strong influence on DF-SOFC performance. The influence is rather complex caused by the coupled flame chemistry, electrochemistry, and mass and heat transport [196]. The choice of fuel has a minor influence on DF-SOFC performance, while the operating temperature of DF-SOFC plays an important role in cell performance [195,196]. Developing anode-supported DF-SOFC instead of the electrolyte-supported DF-SOFC could improve thermal shock resistance and performance [200,201]. In addition, an Ru/CeO_2 catalytic layer has been used to suppress coke formation when the DF-SOFC operates on ethanol flame [201]. Zhu et al. reported that a micro-DF-SOFC stack based on infiltrated LSCM composite anode was successfully operated in a liquefied petroleum gas (LPG) flame during cooking [198]. Compared with the cell performance based on the conventional Ni–YSZ anode in the same conditions, the obtained results suggested that the perovskite oxide anode has good tolerance to carbon buildup [198,202], and the DF-SOFC with a new configuration and design has a potential for combined heat and power generation for many applications.

To sum up, SOFCs provide a highly efficient, low-pollution power generation technology, and systems based on SOFCs are ideal power generation systems—reliable, clean, quiet, environmentally friendly, and fuel conserving. SOFCs with four traditional configurations (planar, tubular, segmented cell-in-series, and monolithic) have been developed over the past years. It should be noted that each configuration has its typical characteristics, e.g., the planar configuration is simpler to fabricate and has a low cost, the tubular configuration is seal less, and the monolithic configuration has a small cell size and high power density. Among all designs of SOFCs, planar and tubular designs are presently two common configurations for a variety of electric power generation applications with high system efficiencies, which have been on the verge of commercial viability for years. However, high operating temperature is still the main issue for all kinds of SOFCs and the resulting materials, cost limitations, and operating complexities. In the future, more efforts are needed to reduce the operating temperate of SOFCs without sacrificing their performance.

Configuration design is very important for SOFCs because a reasonable design could lead to higher power densities with lower weight, size, and cost of fuel cell systems. Thus, symmetric and single-chamber, and direct-flame SOFCs have been proposed and developed. Such novel designs have their advantages: simplifying the production and/or operating process greatly improves the fuel cell integrity over the dual-chamber SOFC and allows for good thermal and mechanical shock resistance. However, it is evident that these designs have not fully matured and several technological challenges remain to be solved before these types of fuel cell can find practical applications. At present, besides further lowing the operating temperature, developing more suitable materials for complicated and rigorous operating conditions for these novel designs is also the priority.

13.5 TECHNICAL CHALLENGES AND PERSPECTIVES

Research and development of SOFCs have made great progress in the past decades. At present, the demo running and practical applications of distributed power supply system and combined heat power system have demonstrated the advantages of SOFCs and its good application prospect to the world. Facing the larger-scale and multipurpose promotion of SOFC commercial opportunities for the future, SOFC technologies still face some challenges. Herein, several challenges are discussed:

1. On the fabrication/manufacturing of components, parts, and modules of SOFCs, we will still face the selection and development of preparation techniques. The technologies that posses the ability to prepare components or parts with given shapes and large-enough sizes, will be preferred. However, these techniques should also have high production efficiency, so they should be convenient to realize mechanization and automation. At the same time, the rate of finished products as well as their yield are also very important; only in this way can it be possible to reduce the production cost of the components, parts, and modules of SOFCs.
2. On the quality control of components before assembling and *in situ* diagnosis of SOFCs, more novel techniques should be researched and developed. Besides the regular fabrication process and rigid quality controlling, a precognition to the performance and situation of components before assembling can undoubtedly improve the credit of the final performance and reliability of SOFC. Thus, some innovative techniques to play such a role are needed, for example, to establish the correspondence between the nondestructive examination results during the fabrication process and operating performance. An *in situ* diagnosis technique in an operating SOFC system with larger power is also needed to prevent the failure caused by improper operating condition and with some detectable forebodings. Proper management and timely fault treatment can avoid complete failure and may provide a possibility to extend the lifetime of SOFC systems.
3. On long-term stability, although the factors that may cause performance degradation and even failure have been investigated extensively and many strategies have been proposed, most presented approaches were just about elimination, delay, or reduction of the degradation. If a quick degradation mechanism is unavoidable, e.g., sulfur poisoning of anode when operating with some cheap fuel, the recovery or regeneration of a degenerated component in system should be considered.
4. On the operation with various and cheaper fuels, besides usual and preferred gas fuel such as hydrogen, methane, coal gas, or LPG, many solid and liquid fuels are also (at least on concept) available for SOFC. However, the operation of SOFC using cheaper fuels may suffer anode poisoning by the contaminants in fuels, and carbon-containing fuel may cause carbon deposition on anode. Thus, techniques to eliminate contaminants in fuel are needed. The technique to supply solid fuels continuously is important for the realization of direct carbon SOFC.

REFERENCES

1. Minh N.Q., Ceramic fuel cells. *J. Am. Ceram. Soc.*, 1993, 76: 563–588.
2. Menzler N.H., Tietz F., Uhlenbruck S., Buchkremer H.P., Stöver D., Materials and manufacturing technologies for solid oxide fuel cells. *J. Mater. Sci.*, 2010, 45: 3109–3135.
3. Hassmann K., SOFC Power plants, the Siemens–Westinghouse approach. *Fuel Cells*, 2001, 1: 78–84.
4. Sammes N.M., Du Y., Bove R., Design and fabrication of a 100 W anode supported micro-tubular SOFC stack. *J. Power Sources*, 2005, 145: 428–434.
5. Kim H.K., Park Y.M., Mohsen O.R., Fabrication and evaluation of the thin NiFe supported solid oxide fuel cell by co-firing method. *Energy*, 2010, 35: 5385–5390.
6. Lawlor V., Griesser S., Buchinger G., Olabi A.G., Cordiner S., Meissner D., Review of the micro-tubular solid oxide fuel cell Part I. Stack design issues and research activities. *J. Power Sources*, 2009, 193: 387–399.
7. Xia C.R., Liu M.L., Low-temperature SOFCs based on $Gd_{0.1}Ce_{0.9}O_{1.95}$ fabricated by dry pressing. *Solid State Ionics*, 2001, 144: 249–255.
8. Xin X.S., Lü Z., Zhu Q.S., Huang X.Q., Su W.H., Fabrication of dense YSZ electrolyte membranes by a modified dry-pressing using nanocrystalline powders. *J. Mater. Chem.*, 2007, 17: 1627–1630.
9. Hotza D., Greil P., Review: Aqueous tape casting of ceramic powders. *Mater. Sci. Eng.*, 1995, A202: 206–217.
10. He L.X., Wen T.L., Lü Z.Y., Effects of ZrO_2 powder characteristics and organic additives on preparation of tape casting film. *J. Inorg. Mater.*, 1998, 13: 117–121.
11. Suciu C., Tikkanen H., Wærnhus I., Goga F., Dorolti E., Water-based tape-casting of SOFC composite 3YSZ/8YSZ electrolytes and ionic conductivity of their pellets. *Ceram. Int.*, 2012, 38: 357–365.
12. Simwonis D., Thu Èlen H., Dias F.J., Naoumidis A., Sto Èver D., Properties of Ni/YSZ porous cermets for SOFC anode substrates prepared by tape casting and coat-mix® process. *J. Mater. Process. Technol.*, 1999, 92–93: 107–111.

13. Torres-Garibay C., Kovar D., Manthiram A., $Ln_{0.6}Sr_{0.4}Co_{1-y}Fe_yO_{3-\delta}$ (Ln = La and Nd; y = 0 and 0.5) cathodes with thin yttria-stabilized zirconia electrolytes for intermediate temperature solid oxide fuel cells. *J. Power Sources*, 2009, 187: 480–486.
14. Wang C.C., Luo L.H., Wu Y.F., Hou B.X., Sun L.L., A novel multilayer aqueous tape casting method for anode-supported planar solid oxide fuel cell. *Mater. Lett.*, 2011, 65: 2251–2253.
15. Chen M., Wen T.L., Huang Z., Wang P.C., Tu H.Y., Lü Z.Y., Study on isostatic pressing YSZ membrane fabricated by tap casting. *J. Inorg. Mater.*, 1999, 14: 745–750.
16. Prakash B.S., Kumar S.S., Aruna S.T., Properties and development of Ni/YSZ as an anode material in solid oxide fuel cell: A review. *Renew. Sustain. Energy Rev.*, 2014, 36: 149–179.
17. Hatchwell C., Sammes N.M., Brown I.W.M., Fabrication and properties of $Ce_{0.8}Gd_{0.2}O_{1.9}$ electrolyte-based tubular solid oxide fuel cells. *Solid State Ionics*, 1999, 126: 201–208.
18. Du Y.H., Sammes N.M., Tompsett G.A., Optimisation parameters for the extrusion of thin YSZ tubes for SOFC electrolytes. *J. Eur. Ceram. Soc.*, 2000, 20: 959–965.
19. Sammes N.M., Du Y.H., Fabrication and characterization of tubular solid oxide fuel cells. *Int. J. Appl. Ceram. Technol.*, 2007, 4: 89–102.
20. Droushiotis N., Doraswami U., Othman M.H.D., Li K., Kelsall G.H., Co-extrusion/phase inversion/co-sintering for fabrication of hollow fiber solid oxide fuel cells. *ECS Trans.*, 2009, 25: 665–672.
21. Sui J., Liu J., An electrolyte-supported SOFC stack fabricated by slip casting technique. *ECS Trans.*, 2007, 7: 633–637.
22. He T.M., Lü Z., Huang Y.L., Guan P.F., Liu J., Su W.H., Characterization of YSZ electrolyte membrane tubes prepared by a vacuum casting method. *J. Alloys Compd.*, 2002, 337: 231–236.
23. Ding J., Liu J., Fabrication and electrochemical performance of anode-supported solid oxide fuel cells by a single-step cosintering process. *J. Am. Ceram. Soc.*, 2008, 91: 3303–3307.
24. Okuyama R., Nomura E., Fabrication of YSZ/NiO-YSZ two-layer ceramics for tubular SOFC by multilayer slurry casting process. *J. Ceram Soc. Jpn.*, 1993, 101: 405–409.
25. Omatete O.O., Janney M.A., Strehlow R.A., Gelcasting—A new ceramic forming process. *Am. Ceram. Soc. Bull.*, 1991, 70: 1641–1649.
26. Young A.C., Omatete O.O., Janney M.A., Menchofer P.A., Gelcasting of alumina. *J. Am. Ceram. Soc.*, 1998, 81: 581–591.
27. Yang J.L., Yu J.L., Huang Y., Recent developments in gelcasting of ceramics. *J. Eur. Ceram. Soc.*, 2011, 31: 2569–2591.
28. Huang W.L., Zhu Q., Xie Z., Gel-cast anode substrates for solid oxide fuel cells. *J. Power Sources*, 2006, 162: 464–468.
29. Zhang L., Jiang S.P., Wang W., Zhang Y.J., NiO/YSZ, anode-supported, thin-electrolyte, solid oxide fuel cells fabricated by gel casting. *J. Power Sources*, 2007, 170: 55–60.
30. Zhang L., Zhang Y.J., Zhen Y.D., Jiang S.P., Lanthanum strontium manganite powders synthesized by gel-casting for solid oxide fuel cell cathode materials. *J. Am. Ceram. Soc.*, 2007, 90: 1406–1411.
31. Fang Q., Zhang J.Y., Preparation of $Ce_{1-x}Gd_xO_{2-x/2}$ thin films by UV assisted sol–gel method. *Surface Coatings Technol.*, 2002, 151–152: 100–104.
32. Kim S.G., Yoon S.P., Nam S.W., Hyun S.H., Hong S.A., Fabrication and characterization of a YSZ/YDC composite electrolyte by a sol–gel coating method. *J. Power Sources*, 2002, 110: 222–228.
33. Chen K.F., Lü Z., Ai N., Huang X.Q., Zhang Y.H., Xin X.S., Zhu R.B., Su W.H., Development of yttria-stabilized zirconia thin films via slurry spin coating for intermediate-to-low temperature solid oxide fuel cells. *J. Power Sources*, 2006, 160: 436–438.
34. Ai N., Lü Z., Chen K.F., Huang X.Q., Wei B., Zhang Y.H., Li S.Y., Xin X.S., Sha X.Q., Su W.H., Low temperature solid oxide fuel cells based on Sm0.2Ce0.8O1.9 films fabricated by slurry spin coating. *J. Power Sources*, 2006, 159: 637–640.
35. Wang J.M., Lü Z., Chen K.F., Huang X.Q., Ai N., Hu J.Y., Zhang Y.H., Su W.H., Study of slurry spin coating technique parameters for the fabrication of anode-supported YSZ Films for SOFCs. *J. Power Sources*, 2007, 164: 17–23.
36. Wang J.M., Lü Z., Huang X.Q., Chen K.F., Ai N., Hu J.Y., Su W.H., YSZ films fabricated by a spin smoothing technique and its application in solid oxide fuel cell. *J. Power Sources*, 2007, 163: 957–959.
37. Somalu M.R., Brandon N.P., Rheological studies of nickel/scandia-stabilized-zirconia screen printing inks for solid oxide fuel cell anode fabrication. *J. Am. Ceram. Soc.*, 2012, 95: 1220–1228.
38. Somalu M.R., Yufit V., Brandon N.P., The effect of solids loading on the screen-printing and properties of nickel/scandia-stabilized-zirconia anodes for solid oxide fuel cells. *Int. J. Hydrogen Energy*, 2013, 38: 9500–9510.
39. Zhang Y.H., Huang X.Q., Lü Z., Liu Z.G., Ge X.D., Xu J.H., Xin X.S., Sha X.Q., Su S.H., A screen-printed $Ce_{0.8}Sm_{0.2}O_{1.9}$ film solid oxide fuel cell with a $Ba_{0.5}Sr_{0.5}Co_{0.8}Fe_{0.2}O_{3-\delta}$ cathode. *J. Power Sources*, 2006, 160: 1217–1220.
40. Ge X.D., Huang X.Q., Zhang Y.H., Lü Z., Xu J.H., Chen K.F., Dong D.W., Liu Z.G., Miao J.P., Su W.H., Screen-printed thin YSZ films used as electrolytes for solid oxide fuel cells. *J. Power Sources*, 2006, 159: 1048–1050.
41. Torres-Garibay C., Kovar D., Manthiram A., $Ln_{0.6}Sr_{0.4}Co_{1-y}Fe_yO_{3-\delta}$ (Ln = La and Nd; y = 0 and 0.5) cathodes with thin yttria-stabilized zirconia electrolytes for intermediate temperature solid oxide fuel cells. *J. Power Sources*, 2009, 187: 480–486.
42. Kim D.W., Yun U.J., Lee J.W., Lim T.H., Lee S.B., Park S.J., Song R.H., Kim G., Fabrication and operating characteristics of a flat tubular segmented-in-series solid oxide fuel cell unit bundle. *Energy*, 2014, 72: 215–221.
43. Besra L., Liu M.L., A review on fundamentals and applications of electrophoretic deposition (EPD). *Prog. Mater. Sci.*, 2007, 52: 1–61.
44. Majhi S.M., Behura S.K., Bhattacharjee S., Singh B.P., Chongdar T.K., Gokhale N.M., Besra L., Anode supported solid oxide fuel cells (SOFC) by electrophoretic deposition. *Int. J. Hydrogen Energy*, 2011, 36: 14930–14935.
45. Jia L., Lü Z., Huang X.Q., Liu Z.G., Zhi Z., Sha X.Q., Li G.Q., Su W.H., Preparation of YSZ film by gravity-electrophoretic deposition and its application in SOFC. *Ceram. Int.*, 2007, 33: 631–635.
46. Zarabian M., Yar A., Vafaeenezhad S., Sani F.M.A., Simchi A., Electrophoretic deposition of functionally-graded NiO–YSZ composite films. *J. Eur. Ceram. Soc.*, 2013, 33: 1815–1823.
47. Caproni E., Gouvêa D., Muccillo R., Yttria-stabilized zirconia closed end tubes prepared by electrophoretic deposition. *Ceram. Int.*, 2011, 37: 273–277.
48. Kim J.H., Song R.H., Song K.S., Hyun S.H., Shin D.R., Yokokawa H., Fabrication and characteristics of anode-supported flat-tube solid oxide fuel cell. *J. Power Sources*, 2003, 122: 138–143.
49. Panthi D., Tsutsumi A., A novel multistep dip-coating method for the fabrication of anode-supported microtubular solid oxide fuel cells. *J. Solid State Electrochem.*, 2014, 18: 1899–1905.

50. Zhou Z.Y., Han D., Wu H., Wang S.R., Fabrication of planar-type SOFC single cells by a novel vacuum dip-coating method and co-firing/infiltration techniques. *Int. J. Hydrogen Energy*, 2014, 39: 2274–2278.
51. Wang D.F., Wang J.X., He C.R., Tao Y.K., Xu C., Wang W.G., Preparation of a $Gd_{0.1}Ce_{0.9}O_{2-\delta}$ interlayer for intermediate-temperature solid oxide fuel cells by spray coating. *J. Alloys Compd.*, 2010, 505: 118–124.
52. Guan B., Lü Z., Wang G., Wei B., Li W., Huang X., A performance study of solid oxide fuel cells with $BaZr_{0.1}Ce_{0.7}Y_{0.2}O_{3-\delta}$ electrolyte developed by spray-modified pressing method. *Fuel Cells*, 2012, 12: 141–145.
53. Mücke R., Büchler O., Bram M., Leonide A., Ivers-Tifféeb E., Buchkremer H.P., Preparation of functional layers for anode-supported solid oxide fuel cells by the reverse roll coating process. *J. Power Sources*, 2011, 196: 9528–9535.
54. Fauchais P., Montavon G., Lima R.S., Marple B.R., Engineering a new class of thermal spray nano-based microstructures from agglomerated nanostructured particles, suspensions and solutions: An invited review. *J. Phys. D. Appl. Phys.*, 2011, 44: 093001.
55. Henne R., Solid oxide fuel cells: A challenge for plasma deposition processes. *J. Therm. Spray Technol.*, 2007; 16: 381–403.
56. Hui R., Wang Z., Kesler O., Rose L., Jankovic J., Yick S., Thermal plasma spraying for SOFCs: Applications, potential advantages, and challenges. *J. Power Sources*, 2007, 170: 308–323.
57. Li C.J., Li C.X., Ning X.J., Performance of YSZ electrolyte layer deposited by atmospheric plasma spraying for cermet-supported tubular SOFC. *Vacuum*, 2004, 73: 699–703.
58. Yang Y.C., Chang T.H., Wu Y.C., Wang S.F., Porous Ni/8YSZ anode of SOFC fabricated by the plasma sprayed method. *Int. J. Hydrogen Energy*, 2012, 37: 13746–13754.
59. Hui R., Berghaus J.O., Decès-Petit C., Qu Q., Yick S., Legoux J.G., Moreau C., High performance metal-supported solid oxide fuel cells fabricated by thermal spray. *J. Power Sources*, 2009, 191: 371–376.
60. Harris J., Qureshi M., Kesler O., Deposition of composite LSCF–SDC and SSC–SDC cathodes by axial-injection plasma spraying. *J. Therm. Spray Technol.*, 2012, 21: 461–468.
61. Wang X.M., Li C.X., Li C.J., Yang G.J., Microstructure and polarization of $La_{0.8}Sr_{0.2}MnO_3$ cathode deposited by alcohol solution precursor plasma spraying. *Int. J. Hydrogen Energy*, 2012, 37: 12879–12885.
62. Purohit R.D., Nair S.R., Prakash D., Sinha P.K., Sharma B.P., Sreekumar K.P., Ananthapadmanabhan P.V., Das A.K., Gantayet L.M., Development of Ca-doped $LaCrO_3$ feed material and its plasma coating for SOFC applications. 23rd National Symposium on Plasma Science & Technology (PLASMA-2008). *J. Phys. Conf. Ser.*, 2010, 208: 012125.
63. Wu W., Guan W.B., Wang G.L., Liu W., Zhang Q.S., Chen T., Wang W.G., Evaluation of $Ni_{80}Cr_{20}/(La_{0.75}Sr_{0.25})_{0.95}MnO_3$ dual layer coating on SUS 430 stainless steel used as metallic interconnect for solid oxide fuel cells. *Int. J. Hydrogen Energy*, 2014, 39: 996–1004.
64. Killinger A., Kuhn M., Gadow R., High-velocity suspension flame spraying (HVSFS), a new approach for spraying nanoparticles with hypersonic speed. *Surf. Coat. Technol.*, 2006, 201: 1922–1929.
65. Choy K.L., Chemical vapor deposition of coatings. *Prog. Mater. Sci.*, 2003, 48: 57–170.
66. Isenberg A.O., In: *Proceedings of the Symposium on Electrode Materials and Processes for Energy Conversion and Storage*, McIntyre J.D.E., Srinivasan S., Will F.G., eds. Pennington, NJ: Electrochemical Society, 1977, p. 572.
67. Will J., Mitterdorfer A., Kleinlogel C., Perednis D., Gauckler L.J., Fabrication of thin electrolytes for second-generation solid oxide fuel cells. *Solid State Ionics*, 2000, 131: 79–96.
68. Coccia L.G., Tyrrell G.C., Kilner J.A., Waller D., Chater R.J., Boyd I.W., Pulsed laser deposition of novel materials for thin film solid oxide fuel cell applications: $Ce_{0.9}Gd_{0.1}O_{1.95}$, $La_{0.7}Sr_{0.3}CoO_3$ and $La_{0.7}Sr_{0.3}Co_{0.2}Fe_{0.8}O_3$. *Appl. Surf. Sci.*, 1996, 96–98: 795–801.
69. Yan J., Enoki M., Matsumoto H., Ishihara T., An intermediate temperature solid oxide fuel cell using a $La(Sr)Ga(Mg)O_3$ thin film prepared by pulsed laser deposition as electrolyte. *Electrochemistry*, 2005, 73: 945–950.
70. Mai A., Haanappel V.A.C., Uhlenbruck S., Tietz F., Stover D., Ferrite-based perovskites as cathode materials for anode-supported solid oxide fuel cells. Part I. Variation of composition. *Solid State Ionics*, 2005, 176: 1341–1350.
71. Lu Z., Zhou X.-D., Fisher D., Templeton J., Stevenson J., Wu N., Ignatiev A., Enhanced performance of an anode-supported YSZ thin electrolyte fuel cell with a laser-deposited $Sm_{0.2}Ce_{0.8}O_{1.9}$ interlayer. *Electrochem. Commun.*, 2010, 12: 179–182.
72. Wang L.S., Barnett S.A., Deposition, structure, and properties of cermet thin films composed of Ag and Y-stabilized zirconia. *J. Electrochem. Soc.*, 1992, 139: 1134–1140.
73. Uhlenbruck N.S., Sebold D., Haanappel V.A.C., Buchkremer H.P., Stöver D., Dense yttria-stabilised zirconia electrolyte layers for SOFC by reactive magnetron sputtering. *J. Power Sources*, 2012, 205: 157–163.
74. Jung H.Y., Hong K.S., Kim H., Park J.K., Son J.W., Kim J., Lee H.W., Lee J.H., Characterization of thin-film YSZ deposited via EB-PVD technique in anode-supported SOFCs. *J. Electrochem. Soc.*, 2006, 153: A961–A966.
75. Meng B., Sun Y., He X.D., Li M.W., Graded Ni–YSZ anode coatings for solid oxide fuel cell prepared by EB-PVD. *Mater. Sci. Technol.*, 2008, 24: 997–1001.
76. Liu J., Lü Z., Barnett S.A., Ji Y., Su W., Single solid oxide fuel cell testing using silver paste for sealing and current collection. In: *Proceedings Electrochemical Society (9th International Symposium on Solid Oxide Fuel Cells, SOFC IX)*, 2005, PV 2005–07: pp. 1976–1980.
77. Miyashita T., Current–voltage relationship considering electrode degradation using Sm-doped ceria electrolytes in SOFCs. *ECS Trans.*, 2011, 35: 583–592.
78. Huang Q.A., Hui R., Wang B.W., Zhang J.J., A review of AC impedance modeling and validation in SOFC diagnosis. *Electrochim. Acta*, 2007, 52: 8144–8164.
79. Barsoukov E., Macdonald J.R., *Impedance Spectroscopy Theory, Experiment, and Applications* (2nd Edition). New York: John Wiley & Sons, 2005.
80. Kim J.D., Kim G.D., Moon J.W., Park Y., Lee W.H., Kobayashi K., Nagai M., Kim C.E., Characterization of LSM–YSZ composite electrode by ac impedance spectroscopy. *Solid State Ionics*, 2001, 143: 379–389.
81. Barelli L., Barluzzi E., Bidini G., Diagnosis methodology and technique for solid oxide fuel cells: A review. *Int. J. Hydrogen Energy*, 2013, 38: 5060–5074.
82. Gazzarri J.I., Kesler O., Electrochemical AC impedance model of a solid oxide fuel cell and its application to diagnosis of multiple degradation modes. *J. Power Sources*, 2007, 167: 100–110.

83. Park K., Yu S., Bae J., Kim H., Fast performance degradation of SOFC caused by cathode delamination in long-term testing. *Int. J. Hydrogen Energy*, 2010, 35: 2670–8677.
84. Steiner N.Y., Hissel D., Moçotéguy P., Candusso D., Marra D., Pianese C., Sorrentino M., Application of fault tree analysis to fuel cell diagnosis. *Fuel Cells*, 2012, 12: 302–309.
85. Azad A.K., Irvine J.T.S., Synthesis, chemical stability and proton conductivity of the perovksites $Ba(Ce,Zr)_{1-x}Sc_xO_{3-\delta}$, *Solid State Ionics*, 2007, 178: 635–640.
86. Rojas A.R., Esparza-Ponce H.E., Fuentes L., Lopez-Ortiz A., Keer A., Reyes-Gasga J., *In situ* X-ray Rietveld analysis of Ni–YSZ solid oxide fuel cell anodes during NiO reduction in H_2. *J. Phys. D Appl. Phys.*, 2005, 38: 2276–2282.
87. Li S.Y., Lü Z., Wei B., Huang X.Q., Miao J.P., Cao G., Zhu R.B., Su W.H., A study of $(Ba_{0.5}Sr_{0.5})_{1-x}Sm_xCo_{0.8}Fe_{0.2}O_{3-\delta}$ as a cathode material for IT-SOFCs. *J. Alloys Compd.*, 2006, 426: 408–414.
88. Wilson J.R., Kobsiriphat W., Mendoza R., Chen H.-Y., Hiller J.M., Miller D.J., Thornton K., Voorhees P.W., Adler S.B., Barnett S.A., Three-dimensional reconstruction of a solid-oxide fuel-cell anode. *Nat. Mater.*, 2006, 5: 541–544.
89. Iwai H., Shikazono N., Matsui T., Teshima H., Kishimoto M., Kishida R., Hayashi D., Matsuzaki K., Kanno D., Saito M., Muroyama H., Eguchi K., Kasagi N., Yoshida H., Quantification of SOFC anode microstructure based on dual beam FIB-SEM technique. *J. Power Sources*, 2010, 195: 955–961.
90. Cronin J.S., Wilson J.R., Barnett S.A., Impact of pore microstructure evolution on polarization resistance of Ni–Yttria-stabilized zirconia fuel cell anodes. *J. Power Sources*, 2011, 196: 2640–2643.
91. Wilson J.R., Duong A.T., Gameiro M., Chen H.Y., Thornton K., Mumm D.R., Barnett S.A., Quantitative three-dimensional microstructure of a solid oxide fuel cell cathode. *Electrochem. Commun.*, 2009, 11: 1052–1056.
92. Yeh T.C., Routbort J.L., Mason T.O., Oxygen transport and surface exchange properties of $Sr_{0.5}Sm_{0.5}CoO_{3-\delta}$. *Solid State Ionics*, 2013, 232: 138–143.
93. Horita T., Kishimotoa H., Yamaji K., Britoa M.E., Yokokawa H., Degradation and durability of SOFC materials by the impurities. *ECS Trans.*, 2011, 30: 115–122.
94. Ecija A., Vidal K., Larrañaga A., Martínez-Amesti A., Ortega-San-Martín L., Arriortua M.I., Characterization of $Ln_{0.5}M_{0.5}FeO_{3-\delta}$ (Ln = La, Nd, Sm; M = Ba, Sr) perovskites as SOFC cathodes. *Solid State Ionics*, 2011, 201: 35–41.
95. Chen K.F., Hyodo J., Zhao L., Ai N., Ishihara T., Jiang S.P., Boron poisoning of (La, Sr)(Co, Fe)O_3 cathodes of solid oxide fuel cells. *ECS Trans.*, 2013, 57: 1821–1830.
96. Liu L.M., Sun K.N., Zhou X.D., Li X.K., Zhang M., Zhang N.Q., Sulfur tolerance improvement of Ni–YSZ anode by alkaline earth metal oxide BaO for solid oxide fuel cells. *Electrochem. Commun.*, 2012, 19: 63–66.
97. Rismanchian A., Chuang S.S.C., Electroless deposited Cu on the Ni/YSZ anode for the direct CH_4–SOFC. *ECS Trans.*, 2013, 57: 1429–1436.
98. Wei B., Lü Z., Li S.Y., Liu Y.Q., Liu K.Y., Su W.H., Thermal and electrical properties of new cathode material $Ba_{0.5}Sr_{0.5}Co_{0.8}Fe_{0.2}O_{3-\delta}$ for solid oxide fuel cells. *Electrochem. Solid-State Lett.*, 2005, 8: A428–A431.
99. Hu J.Y., Lü Z., Chen K.F., Huang X.Q., Ai N., Du X.B., Fu C.W., Wang J.M., Su W.H., Effect of composite pore-former on the fabrication and performance of anode-supported membranes for SOFCs. *J. Membrane Sci.*, 2008, 318: 445–451.
100. Wang R.F., Lü Z., Liu C.Q., Zhu R.B., Huang X.Q., Wei B., Ai N., Su W.H., Characteristics of a SiO_2–B_2O_3–Al_2O_3–$BaCO_3$–PbO_2–ZnO glass– ceramic sealant for SOFCs. *J. Alloys Compd.*, 2007, 432: 189–193.
101. Adler S.B., Chemical expansivity of electrochemical ceramics. *J. Am. Ceram. Soc.*, 2001, 84: 2117–2119.
102. Bishop S.R., Marrocchelli D., Chatzichristodoulou C., Perry N.H., Mogensen M.B., Tuller H.L., Wachsman E.D., Chemical expansion: Implications for electrochemical energy storage and conversion devices. *Annu. Rev. Mater. Res.*, 2014, 44: 205–239.
103. Lambert J.B., McLaughlin C.D., Shawl C.E., Xue L., X-ray photoelectron spectroscopy and archaeology. *Anal. Chem. News Feat.*, 1999, 1: 614–620.
104. Wu Q.H., Liu M.L., Jaegermann W., X-ray photoelectron spectroscopy of $La_{0.5}Sr_{0.5}MnO_3$. *Mater. Lett.*, 2005, 59: 1480–1483.
105. Liu X.M., Su W.H., Lü Z., Study on synthesis of $Pr_{1-x}Ca_xCrO_3$ and their electrical properties. *Mater. Chem. Phys.*, 2003, 82: 327–330.
106. Liu Z.G., Zheng Z.R., Huang X.Q., Lü Z., He T.M., Dong D.W., Sui Y., Miao J.P., Su W.H., The Pr^{4+} ions in Mg doped $PrGaO_3$ perovskites. *J. Alloys Compd.*, 2004, 363: 60–62.
107. Xu C.C., Zondlo J.W., Gong M.Y., Elizalde-Blancas F., Liu X.B., Celik I.B., Tolerance tests of H_2S-laden biogas fuel on solid oxide fuel cells. *J. Power Sources*, 2010, 195: 4583–4592.
108. Su R., Lü Z., Jiang S.P., Shen Y.B., Su W.H., Chen K.F., Ag decorated (Ba,Sr)(Co,Fe)O_3 cathodes for solid oxide fuel cells prepared by electroless silver deposition. *Int. J. Hydrogen Energy*, 2013, 38: 2413–2420.
109. Simner S.P., Anderson M.D., Engelhard M.H., Stevenson J.W., Degradation mechanisms of $La_{1-x}Sr_xCo_{1-y}Fe_yO_3$ SOFC cathodes. *Electrochem. Solid-State Lett.*, 2006, 9: A478–A481.
110. DeCaluwe S.C., Jackson G.S., Farrow R.L., McDaniel A.H., Gabaly F.E., McCarty K.F., Nie S., Linne M.A., Bluhmc H., Newbergc J.T., Liu Z., Hussain Z., *In situ* XPS for evaluating ceria oxidation states in SOFC anodes. *ECS Trans.*, 2009, 16 (51): 253–263.
111. Huber A.K., Falk M., Rohnke M., Luerssen B., Gregoratti L., Amati M., Janek J., *In situ* study of electrochemical activation and surface segregation of the SOFC electrode material $La_{0.75}Sr_{0.25}Cr_{0.5}Mn_{0.5}O_{3-\delta}$. *Phys. Chem. Chem. Phys.*, 2012, 14: 751–758.
112. Wang W., Jiang S.P., A mechanistic study on the activation process of (La, Sr)MnO_3 electrodes of solid oxide fuel cells. *Solid State Ionics*, 2006, 177: 1361–1369.
113. Acrivos J.V., Nguyen L., Norman T., Lin C.T., Liang W.Y., Honig J.M., Somasundaran P., Chemical analysis by X-ray spectroscopy near phase transitions in the solid state. *Microchem. J.*, 2002, 71: 117–131.
114. Haas O., Vogt U.F., Soltmann C., Braun A., Yoon W.S., Yang X.Q., Graule T., The Fe K-edge X-ray absorption characteristics of $La_{1-x}Sr_xFeO_{3-\delta}$ prepared by solid state reaction. *Mater. Res. Bull.*, 2009, 44: 1397–1404.
115. Blennow P., Hagen A., Hansen K.K., Wallenberg L.R., Mogensen M., Defect and electrical transport properties of Nb-doped $SrTiO_3$. *Solid State Ionics*, 2008, 179: 2047–2058.
116. Malzbender J., Steinbrech R.W., Singheiser L., A review of advanced techniques for characterising SOFC behaviour. *Fuel Cells*, 2009, 9: 785–793.
117. Menzler N.H., Batfalsky P., Blum L., Bram M., Gross S.M., Haanappel V.A.C., Malzbender J., Shemet V., Steinbrech R.W., Vinke I., Studies of material interaction after long term stack operation. In: *Proc. 7th Eur. SOFC Forum*, 2006, p. B104.

118. Lawlor V., Zauner G., Hochenauer C., Mariani A., Griesser S., Carton J.G., Klein K., Kuehn S., Olabi A.G., Cordiner S., Meissner D., Buchinger G., The use of a high temperature wind tunnel for MT-SOFC testing—Part I: Detailed experimental temperature measurement of an MT-SOFC using an avant-garde high temperature wind tunnel and various measurement techniques. *J. Fuel Cell Sci. Technol.*, 2010, 7: 061016.
119. Nomura K., Mizutani Y., Kawai M., Nakamura Y., Yamamoto O., Aging and Raman scattering study of scandia and yttria doped zirconia. *Solid State Ionics*, 2000, 132: 235–239.
120. Kishimoto H., Shimonosono T., Yamaji K., Brito M.E., Horita T., Yokokawa H., Phase transformation of stabilized zirconia on SOFC stacks. *ECS Trans.*, 2011, 35: 1171–1176.
121. Cheng Z., Liu M.L., Characterization of sulfur poisoning of Ni-YSZ anodes for solid oxide fuel cells using in situ Raman micro spectroscopy. *Solid State Ionics* 2007, 178:925–935.
122. Cheng Z., Abernathy H., Liu M.L., Raman spectroscopy of nickel sulfide Ni_3S_2. *J. Phys. Chem. C*, 2007, 111: 17997–18000.
123. Blinn K.S., Abernathy H., Li X.X., Liu M.F., Bottomley L.A., Liu M.L., Raman spectroscopic monitoring of carbon deposition on hydrocarbon-fed solid oxide fuel cell anodes. *Energy Environ. Sci.*, 2012, 5: 7913–7917.
124. Badwal S.P.S., Foger K., Solid oxide electrolyte fuel cell review. *Ceram. Int.*, 1996, 22: 257–265.
125. Yamamoto O., Solid oxide fuel cells: Fundamental aspects and prospects. *Electrochim. Acta*, 2000, 45: 2423–2435.
126. Kakac S., Pramuanjaroenkij A., Zhou X.Y., A review of numerical modeling of solid oxide fuel cells. *Int. J. Hydrogen Energy*, 2007, 32: 761–786.
127. Huang K., Singhal S.C., Cathode-supported tubular solid oxide fuel cell technology: A critical review. *J. Power Sources*, 2013, 237: 84–97.
128. Stambouli A.B., Traversa E., Solid oxide fuel cells (SOFCs): A review of an environmentally clean and efficient source of energy. *Renew. Sustain. Energy Rev.*, 2002, 6: 433–455.
129. Wen T.L., Wang D., Chen M., Tu H., Lü Z., Zhang Z., Nie H., Huang W., Material research for planar SOFC stack. *Solid State Ionics*, 2002, 148: 513–519.
130. Lee H.W., Kim S.M., Kim H., Jung H.Y., Jung H.G., Lee J.H., Song H., Kim H.R., Son J.W., Advanced planar SOFC stack with improved thermo-mechanical reliability and electrochemical performance. *Solid State Ionics*, 2008, 179: 1454–1458.
131. Wen T.L., Wang D., Tu H.Y., Chen M., Lu Z., Zhang Z., Nie H., Huang W., Research on planar SOFC stack. *Solid State Ionics*, 2002, 152: 399–404.
132. Guan W.B., Jin L., Wu W., Zheng Y.F., Wang G.L., Wang W.G., Effect and mechanism of Cr deposition in cathode current collecting layer on cell performance inside stack for planar solid oxide fuel cells. *J. Power Sources*, 2014, 245: 119–128.
133. Singhal S.C., Solid oxide fuel cells for stationary, mobile, and military applications. *Solid State Ionics*, 2002, 152: 405–410.
134. Minh N.Q., Solid oxide fuel cell technology—Features and applications. *Solid State Ionics*, 2004, 174: 271–277.
135. Maharudrayya S., Jayanti S., Deshpande A.P., Pressure losses in laminar flow through serpentine channels in fuel cell stacks. *J. Power Sources*, 2004, 138: 1–13.
136. Maharudrayya S., Jayanti S., Deshpande A.P., Flow distribution and pressure drop in parallel-channel configurations of planar fuel cells. *J. Power Sources*, 2005, 144: 94–106.
137. Maharudrayya S., Jayanti S., Deshpande A.P., Pressure drop and flow distribution in multiple parallel-channel configurations used in proton-exchange membrane fuel cell stacks. *J. Power Sources*, 2006, 157: 358–367.
138. Tanim T., Bayless D.J., Trembly J.P., Modeling of a 5 kW(e) tubular solid oxide fuel cell based system operating on desulfurized JP-8 fuel for auxiliary and mobile power applications. *J. Power Sources*, 2013, 221: 387–396.
139. Wachsman E.D., Marlowe C.A., Lee K.T., Role of solid oxide fuel cells in a balanced energy strategy. *Energy Environ. Sci.*, 2012, 5: 5498–5509.
140. Choudhury A., Chandra H., Arora A., Application of solid oxide fuel cell technology for power generation—A review. *Renew. Sustain. Energy Rev.*, 2013, 20: 430–442.
141. Fontell E., Kivisaari T., Christiansen N., Hansen J.B., Palsson J., Conceptual study of a 250 kW planar SOFC system for CHP application. *J. Power Sources*, 2004, 131: 49–56.
142. Christiansen N., Hansen J.B., Holm-Larsen H., Linderoth S., Larsen P.H., Hendriksen P.V., Hagen A., Solid oxide fuel cell development at Topsoe Fuel Cell A/S and Riso National Laboratory. *Solid Oxide Fuel Cells 10 (SOFC-X)*, Pts 1 and 2, 2007, 7: 31–38.
143. Astrom K., Fontell E., Virtanen S., Reliability analysis and initial requirements for FC systems and stacks. *J. Power Sources*, 2007, 171: 46–54.
144. George R.A., Bessette N.F., Reducing the manufacturing cost of tubular SOFC technology. *J. Power Sources*, 1998, 71: 131–137.
145. Singhal S.C., Advances in solid oxide fuel cell technology. *Solid State Ionics*, 2000, 135: 305–313.
146. Hosoi K., Ito M., Fukae M., Status of national project for SOFC development in Japan. *ECS Trans.*, 2011, 35: 11–18.
147. Howe K.S., Thompson G.J., Kendall K., Micro-tubular solid oxide fuel cells and stacks. *J. Power Sources*, 2011, 196: 1677–1686.
148. Tan X.Y., Yin W.N., Meng B., Meng X.X., Yang N.T., Ma Z.F., Preparation of electrolyte membranes for micro tubular solid oxide fuel cells. *Sci. China Ser. B*, 2008, 51: 808–812.
149. Liu Y., Hashimoto S.I., Nishino H., Takei K., Mori M., Suzuki T., Funahashi Y., Fabrication and characterization of micro-tubular cathode-supported SOFC for intermediate temperature operation. *J. Power Sources*, 2007, 174: 95–102.
150. Liu M.F., Dong D.H., Zhao F., Gao J.F., Ding D., Liu X.Q., Meng G.Y., High-performance cathode-supported SOFCs prepared by a single-step co-firing process. *J. Power Sources*, 2008, 182: 585–588.
151. Lawlor V., Griesser S., Buchinger G., Olabi A.G., Cordiner S., Meissner D., Review of the micro-tubular solid oxide fuel cell Part I. Stack design issues and research activities. *J. Power Sources*, 2009, 193: 387–399.
152. Funahashi Y., Shimamori T., Suzuki T., Fujishiro Y., Awano M., Fabrication and characterization of components for cube shaped micro tubular SOFC bundle. *J. Power Sources*, 2007, 163: 731–736.
153. Singhal S.C., Advances in solid oxide fuel cell technology. *Solid State Ionics*, 2000, 135: 305–313.
154. Iyengar A.K.S., Desai N.A., Vora S.D., Shockling L.A., Numerical investigation of a delta high power density cell and comparison with a flattened tubular high power density cell. *J. Fuel Cell Sci. Technol.*, 2010, 7: 061002.
155. Bastidas D.M., Tao S.W., Irvine J.T.S., A symmetrical solid oxide fuel cell demonstrating redox stable perovskite electrodes. *J. Mater. Chem.*, 2006, 16: 1603–1605.
156. Ruiz-Morales J.C., Canales-Vazquez J., Pena-Martinez J., Marrero-Lopez D., Nunez P., On the simultaneous use of $La_{0.75}Sr_{0.25}Cr_{0.5}Mn_{0.5}O_{3-\delta}$ as both anode and cathode material with improved microstructure in solid oxide fuel cells. *Electrochim. Acta*, 2006, 52: 278–284.

157. Zhou Q., Yuan C., Han D., Luo T., Li J.L., Zhan Z.L., Evaluation of $LaSr_2Fe_2CrO_{9-\delta}$ as a potential electrode for symmetrical solid oxide fuel cells. *Electrochim. Acta*, 2014, 133: 453–458.
158. Martinez-Coronado R., Aguadero A., Perez-Coll D., Troncoso L., Alonso J.A., Fernandez-Diaz M.T., Characterization of $La_{0.5}Sr_{0.5}Co_{0.5}Ti_{0.5}O_{3-\delta}$ as symmetrical electrode material for intermediate-temperature solid-oxide fuel cells. *Int. J. Hydrogen Energy*, 2012, 37: 18310–18318.
159. Zhang P., Guan G.Q., Khaerudini D.S., Hao X.G., Han M.F., Kasai Y., Sasagawa K., Abudula A., Properties of A-site nonstoichiometry $(Pr_{0.4})_xSr_{0.6}Co_{0.2}Fe_{0.7}Nb_{0.1}O_{3-\sigma}$ $(0.9 \le x \le 1.1)$ as symmetrical electrode material for solid oxide fuel cells. *J. Power Sources*, 2014, 248: 163–171.
160. Liu Q.A., Dong X.H., Xiao G.L., Zhao F., Chen F.L., A novel electrode material for symmetrical SOFCs. *Adv. Mater.*, 2010, 22: 5478–5482.
161. Chen M., Paulson S., Thangadurai V., Birss V., Sr-rich chromium ferrites as symmetrical solid oxide fuel cell electrodes. *J. Power Sources*, 2013, 236: 68–79.
162. Zhang P., Guan G.Q., Khaerudini D.S., Hao X.G., Xue C.F., Han M.F., Kasai Y., Abudula A., Evaluation of performances of solid oxide fuel cells with symmetrical electrode material. *J. Power Sources*, 2014, 266: 241–249.
163. Zhu X.B., Lü Z., Wei B., Huang X.Q., Zhang Y.H., Su W.H., A symmetrical solid oxide fuel cell prepared by dry-pressing and impregnating methods. *J. Power Sources*, 2011, 196: 729–733.
164. El-Himri A., Marrero-Lopez D., Ruiz-Morales J.C., Pena-Martinez J., Nunez P., Structural and electrochemical characterisation of $Pr_{0.7}Ca_{0.3}Cr_{1-y}Mn_yO_{3-\delta}$ as symmetrical solid oxide fuel cell electrodes. *J. Power Sources*, 2009, 188: 230–237.
165. Canales-Vazquez J., Ruiz-Morales J.C., Marrero-Lopez D., Pena-Martinez J., Nunez P., Gomez-Romero P., Fe-substituted $(La,Sr)TiO_3$ as potential electrodes for symmetrical fuel cells (SFCs). *J. Power Sources*, 2007, 171: 552–557.
166. Zheng Y., Zhang C., Ran R., Cai R., Shao Z.P., Farrusseng D., A new symmetric solid-oxide fuel cell with $La_{0.8}Sr_{0.2}Sc_{0.2}Mn_{0.8}O_{3-\delta}$ perovskite oxide as both the anode and cathode. *Acta Mater.*, 2009, 57: 1165–1175.
167. Liu X.J., Han D., Zhou Y.C., Meng X., Wu H., Li J.L., Zeng F.R., Zhan Z.L., Sc-substituted $La_{0.6}Sr_{0.4}FeO_{3-\delta}$ mixed conducting oxides as promising electrodes for symmetrical solid oxide fuel cells. *J. Power Sources*, 2014, 246: 457–463.
168. Eyraud C., Lenoir J., Gery M., Fuel cells utilizing the electrochemical properties of absorbates. *Compt. Rend.*, 1961, 252: 1599–1603.
169. Hibino T., Wang S.Q., Kakimoto S., Sano M., One-chamber solid oxide fuel cell constructed from a YSZ electrolyte with a Ni anode and LSM cathode. *Solid State Ionics*, 2000, 127: 89–98.
170. Hibino T., Iwahara H., Simplification of solid oxide fuel cell system using partial oxidation of methane. *Chem. Lett.*, 1993, 22: 1131–1134.
171. Shao Z.P., Haile S.M., A high-performance cathode for the next generation of solid-oxide fuel cells. *Nature*, 2004, 431: 170–173.
172. Shao Z.P., Haile S.M., Ahn J., Ronney P.D., Zhan Z.L., Barnett S.A., A thermally self-sustained micro solid-oxide fuel-cell stack with high power density. *Nature*, 2005, 435: 795–798.
173. Wang Z.H., Lü Z., Wei B., Huang X.Q., Chen K.F., Liu M.L., Pan W.P., Su W.H., A configuration for improving the performance of coplanar single-chamber solid oxide fuel cell. *Electrochem. Solid State Lett.*, 2010, 13: B14–B16.
174. Yano M., Tomita A., Sano M., Hibino T., Recent advances in single-chamber solid oxide fuel cells: A review. *Solid State Ionics*, 2007, 177: 3351–3359.
175. Kuhn M., Napporn T.W., Single-chamber solid oxide fuel cell technology—From its origins to today's state of the art. *Energies*, 2010, 3: 57–134.
176. Hibino T., Hashimoto A., Inoue T., Tokuno J., Yoshida S., Sano M., Single-chamber solid oxide fuel cells at intermediate temperatures with various hydrocarbon–air mixtures. *J. Electrochem. Soc.*, 2000, 147: 2888–2892.
177. Liu M.L., Lü Z., Wei B., Huang X.Q., Chen K.F., Su W.H., A novel cell-array design for single chamber SOFC microstack. *Fuel Cells*, 2009, 9: 717–721.
178. Liu M.L., Lü Z., Wei B., Zhu R.B., Huang X.Q., Chen K.F., Ai G., Su W.H., Anode-supported micro-SOFC stacks operated under single-chamber conditions. *J. Electrochem. Soc.*, 2007, 154: B588–B592.
179. Wei B., Lü Z., Huang X.Q., Liu M.L., Chen K.F., Su W.H., Enhanced performance of a single-chamber solid oxide fuel cell with an SDC-impregnated cathode. *J. Power Sources*, 2007, 167: 58–63.
180. Liu M.L., Lü Z., Wei B., Huang X.Q., Zhang Y.H., Su W.H., Effects of the single chamber SOFC stack configuration on the performance of the single cells. *Solid State Ionics*, 2010, 181: 939–942.
181. Wei B., Lü Z., Huang X.Q., Liu M.L., Jia D.C., Su W.H., A novel design of single-chamber SOFC micro-stack operated in methane-oxygen mixture. *Electrochem. Commun.*, 2009, 11: 347–350.
182. Liu M.L., Lü Z., Wei B., Huang X.Q., Zhang Y.H., Su W.H., Performance of an annular solid-oxide fuel cell micro-stack array operating in single-chamber conditions. *J. Power Sources*, 2010, 195: 4247–4251.
183. Yano M., Nagaoa M., Okamoto K., Inoa A.T., Uchiyarna Y., Uchiyama N., Hibino T., A single-chamber SOFC stack operating in engine exhaust. *Electrochem. Solid State Lett.*, 2008, 11: B29–B33.
184. Tian Y., Lü Z., Wei B., Zhu X., Li W., Wang Z., Pan W., Su W., Evaluation of a non-sealed solid oxide fuel cell stack with cells embedded in plane configuration. *Fuel Cells*, 2012, 12: 523–529.
185. Tian Y., Lü Z., Zhang Y., Wei B., Liu M., Huang X., Su W., Study of a single-chamber solid oxide fuel cell microstack with V-shaped congener-electrode-facing configuration. *Fuel Cells*, 2012, 12: 4–10.
186. Tian Y.T., Lü Z., Liu M.L., Zhu X.B., Wei B., Zhang Y.H., Huang X.Q., Su W.H., Effect of gas supply method on the performance of the single-chamber SOFC micro-stack and the single cells. *J. Solid State Electrochem.*, 2013, 17: 269–275.
187. Ahn S.J., Kim Y.B., Moon J., Lee J.H., Kim J., Co-planar type single chamber solid oxide fuel cell with micro-patterned electrodes. *J. Electroceram.*, 2006, 17: 689–693.
188. Kuhn M., Napporn T., Meunier M., Therriault D., Vengallatore S., Direct-write microfabrication of single-chamber solid oxide fuel cells with interdigitated electrodes. In: *Solid-State Ionics-2006*, vol. 972, Armstrong T.R., Masquelier C., Sadaoka Y., eds. Warrendale, PA: Materials Research Society; 2007, pp. 211–216.
189. Fleig J., Tuller H.L., Maier J., Electrodes and electrolytes in micro-SOFCs: A discussion of geometrical constraints. *Solid State Ionics*, 2004, 174: 261–270.
190. Wang Z.H., Lü Z., Wei B., Huang X.Q., Chen K.F., Pan W.P., Su W.H., A right-angular configuration for the single-chamber solid oxide fuel cell. *Int. J. Hydrogen Energy*, 2011, 36: 3147–3152.

191. Nagao M., Yano M., Okamoto K., Tomita A., Uchiyama Y., Uchiyama N., Hibino T., A single-chamber SOFC stack: Energy recovery from engine exhaust. *Fuel Cells*, 2008, 8: 322–329.
192. Shao Z.P., Zhang C.M., Wang W., Su C., Zhou W., Zhu Z.H., Park H.J., Kwak C., Electric power and synthesis gas co-generation from methane with zero waste gas emission. *Angew. Chem. Int. Edn.*, 2011, 50: 1792–1797.
193. Riess I., On the single chamber solid oxide fuel cells. *J. Power Sources*, 2008, 175: 325–337.
194. Sun L.L., Hao Y., Zhang C.M., Ran R., Shao Z.P., Coking-free direct-methanol-flame fuel cell with traditional nickel-cermet anode. *Int. J. Hydrogen Energy*, 2010, 35: 7971–7981.
195. Horiuchi M., Suganuma S., Watanabe M., Electrochemical power generation directly from combustion flame of gases, liquids, and solids. *J. Electrochem. Soc.*, 2004, 151: A1402–A1405.
196. Kronemayer H., Barzan D., Horiuchi M., Suganuma S., Tokutake Y., Schulz C., Bessler W.G., A direct-flame solid oxide fuel cell (DFFC) operated on methane, propane, and butane. *J. Power Sources*, 2007, 166: 120–126.
197. Huang T.J., Huang M.C., Temperature effect on electrochemical promotion of bulk lattice–oxygen extraction for hydrogen oxidation over SOFC anodes. *Int. J. Hydrogen Energy*, 2009, 34: 2731–2738.
198. Zhu X.B., Wei B., Lü Z., Yang L., Huang X.Q., Zhang Y.H., Liu M.L., A direct flame solid oxide fuel cell for potential combined heat and power generation. *Int. J. Hydrogen Energy*, 2012, 37: 8621–8629.
199. Walther D.C., Ahn J., Advances and challenges in the development of power-generation systems at small scales. *Prog. Energy Combust.*, 2011, 37: 583–610.
200. Wang K., Zeng P.Y., Ahn J.M., High performance direct flame fuel cell using a propane flame. *Proc. Combust. Inst.*, 2011, 33: 3431–3437.
201. Wang K., Ran R., Hao Y., Shao Z.P., Jin W.Q., Xu N.P., A high-performance no-chamber fuel cell operated on ethanol flame. *J. Power Sources*, 2008, 177: 33–39.
202. Zhu X.B., Lü Z., Wei B., Huang X.Q., Wang Z.H., Su W.H., Direct flame SOFCs with $La_{0.75}Sr_{0.25}Cr_{0.5}Mn_{0.5}O_{3-\delta}$/Ni coimpregnated yttria-stabilized zirconia anodes operated on liquefied petroleum gas flame. *J. Electrochem. Soc.*, 2010, 157: B1838–B1843.

14 Theoretical Modeling for Fundamental Understanding of High-Temperature Solid Oxide Fuel Cells

Zijing Lin

CONTENTS

14.1 INTRODUCTION

Fossil fuel is currently the predominant energy source. In all likelihood, fossil fuel will remain the dominant energy source for the remainder of this century. Consuming the depleting energy source in an economic and environmentally friendly way is of tremendous importance for a sustainable society. Solid oxide fuel cell (SOFC) is a device that converts chemical energy into electricity by means of an electrochemical (EC) process and offers great potential for high efficiency and environmentally benign electric energy generation. The high operating temperature of SOFC is advantageous for fuel flexibility and high efficiency as CO can be directly used as fuel. It also offers the possibility of adding a bottoming thermal engine cycle and/or heat cogeneration for even higher efficiency. Consequently, SOFC is attracting increasing attention in recent years [1].

The basic working principle of SOFC is illustrated in Figure 14.1. Although the design and operation of SOFC appears simple, the illustration is in fact rather deceptive. There are numerous physical processes involved, and the details of the physical processes in SOFC can be rather complicated. Moreover, many of the phenomena dominating the SOFC performance are strongly coupled, competing, and often not well understood [2–8]. As a result, despite intensive research activities and significant progress, current SOFCs remain an immature technology. For broad market acceptance, major improvement should be made on (i) lowering the cost of manufacture and operation, (ii) increasing service lifetime to match that of power generation techniques in common use, and (iii) increasing efficiency to approach the theoretical potential. This development relies on improving material choices, component and stack configuration designs, and optimizing operational conditions. As physical prototyping is expensive and time consuming, its ability to examine the effects of material parameters and operating conditions on SOFC performance is severely limited. Theoretical modeling

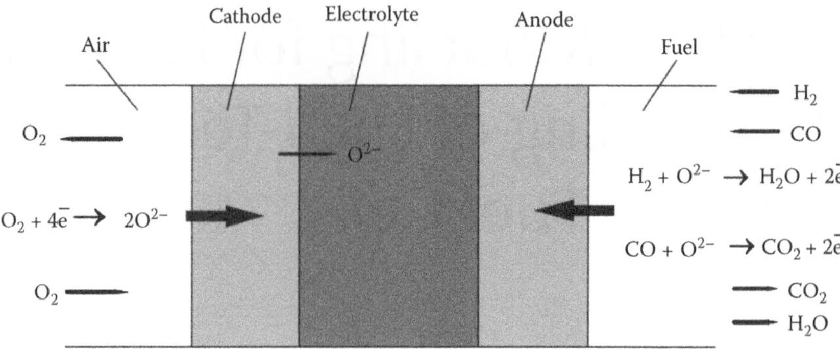

FIGURE 14.1 Basic working principle of SOFC.

based on the laws of physics and behaviors of materials for performance prediction is invaluable for the understanding and development of SOFC technology [9–14].

Theoretical modeling of SOFCs when compared with experimental results may improve the understanding of the phenomena observed. Validated models may be used to examine various parameters such as material properties, geometries, temperatures, fuels, etc., and determine their associated performance characteristics. Theoretical modeling is therefore an economic way to examine the effect of varying designs and operating parameters on the resulting electrical power, cell temperature, gas pressure and stress distributions, and fuel conversion efficiency, etc. Moreover, it can be used to answer questions such as how much the material properties need to be improved for the desired power generation, what geometric design parameters should be used to improve flow uniformity, or what the flow rates must be to avoid excessive temperature gradient and/or pressure drop. The capability of theoretical modeling to test the significance of various design features and the effectiveness of material property improvements, and select optimum feasible operating conditions offers the potential to direct the technology development. Therefore, theoretical modeling is a critical aspect of the SOFC technology development process.

There are numerous theoretical modeling works on SOFCs. Different models may emphasize different aspects of SOFC processes [9–17]. For example, SOFC may be modeled as an EC device, as a heat and mass exchanger, or as a chemical reactor. Modeling may examine steady-state performance or transient characteristics at component, cell, or stack level with continuum-level models. Alternatively, modeling may focus on one or a few related material properties using micro-, meso-, macro-, or multiscale models. The models may have different sophistication about the details of physical processes, and may be classified as single-field or multiphysics models. Alternatively, the models may be classified as zero-dimensional (0D), 1D, 2D, or 3D models. In short, there are a number of different ways to characterize the available theoretical models. For the convenience of referencing, the theoretical models are classified here broadly as continuum models and atomistic models. The continuum models are subdivided as 0D, 1D, 2D, and 3D models. The continuum models may deal with a single physical property or multiphysics processes. The model may also couple physical properties at different scales. Moreover, the models may focus on the steady-state performance or transient-state behaviors.

The majority of the existing theoretical modeling belongs to the category of continuum steady-state models. However, transient-state modeling and atomistic modeling are also critically important for the understanding of SOFCs and possess capabilities that are absent in the continuum steady-state models. For example, transient-state modeling is necessary for the understanding of the dynamic characteristics of the cell, such as start-up, cooling down, and responses to changes in the external load and other operating conditions. The dynamic behaviors of SOFCs need to be understood to design the required power plant with effective control. The importance of transient-state modeling has long been recognized, and the theoretical studies have been continuously performed [18–25]. Likewise, atomistic modeling is indispensible for revealing the mechanism for methane steam reforming reaction, carbon deposition processes, catalytic activity at TPB, degradation of materials due to microstructural and compositional changes, and conduction processes in the electrolyte and electrodes, etc. With the growing interest in the understanding of the detailed mechanisms of physical processes, research activities of atomistic modeling have been intensified [26–29] and are expected to increase rapidly in the near future. Nevertheless, to have a focus within a limited space, the transient-state modeling and atomistic modeling are omitted here. That is, the chapter will focus on describing the details about the theoretical foundation for the continuum steady-state models including the basic governing laws of physics and modeling examples. The chapter is not intended as a review of the literature, and the interested readers are referred to Refs. [10–17]. Instead, it is hoped that the chapter gives a self-contained description that is beneficial to the nonexperts to learn about the continuum steady-state modeling and the insights it is capable of revealing. Hopefully, the chapter is also helpful for students to gain the knowledge necessary for performing their modeling researches at the continuum steady-state level. Consistent with this goal, an attempt is made to discuss the modeling examples in a pedagogical way. To provide readers with up-to-date information, examples are chosen to emphasize the theoretical modeling conducted in recent years.

This chapter is organized as follows: Section 14.2 describes the basic theoretical principles for the continuum models, together with a brief overview of the numerical approaches. A simplification of the natural 3D system into a 0D model and modeling examples for the understanding of SOFCs are presented in Section 14.3. Analogous to Section 14.3, Sections 14.4 through 14.6 provide descriptions of the theoretical modeling at the levels of the 1D, 2D, and 3D models, respectively. The summary and concluding remarks are given in Section 14.7.

14.2 GENERAL THEORETICAL PRINCIPLES

To operate an SOFC, fuel and oxidant are required. Fuel is supplied through some gas channels and fed into a porous anode where fuel species such as H_2 and CO are oxidized via EC reactions. For hydrocarbon fuels that have not been fully reformed, fuel reformation reactions may take place inside the anode. The oxidant (usually oxygen in air) is fed into a porous cathode where it is reduced. Oxygen ions are conducted by a dense solid electrolyte layer that is sandwiched between the two electrodes. The cathode–electrolyte–anode block is often referred as a PEN (positive–electrolyte–negative) structure or MEA (membrane–electrode assembly). The electronic current accompanying oxidant reduction and fuel oxidation is drawn to external load in the form of electrical power. In other words, a working SOFC is involved with fuel and air transports, EC/chemical reactions, electronic and ionic current conductions, as well as heat productions associated with the reactions and mass and current transports. Naturally, the theoretical modeling of SOFCs focuses on the above physical processes using models established with the corresponding governing laws of physics.

14.2.1 Electrochemistry Model

The EC process is the heart of SOFC, and the relationship between the current and voltage (*I–V* relation) is the most important property of SOFCs. This is simply because the EC process generates the electrical power needed for SOFC applications. Understanding the complex interactions between various EC phenomena is critically important so that optimal choices of cell components, stack configurations, and operating conditions can be made, and the performance and efficiency of the SOFC can be improved. Moreover, the EC and chemical processes generate heat, causing temperature gradient and thermal stress that may affect the structural reliability and durability of the SOFC. Therefore, the voltage and current distribution in the PEN structure is one of the most important design issues of SOFCs and is coupled with the temperature distribution from the thermal and flow models and also with the EC reactions at the electrodes.

14.2.1.1 *I–V* Relation Model

The overall EC performance of an SOFC may be conveniently described by a relation between the operating cell voltage, V_{cell}, and the output current density, I. The voltage (potential) loss at a specified current with respect to the ideal thermodynamic performance, Nernst potential or open circuit voltage (OCV), is called overpotential or polarization. V_{cell} is lower than the OCV, E_{OCV}, owing to various polarizations such as activation, ohmic and concentration polarizations, and may be expressed as [9,30,31]

$$V_{cell} = E_{OCV} - \eta_{ohm} - \eta_{con} - \eta_{act} \tag{14.1}$$

where η_{ohm} is the ohmic overpotential due to the resistance to charge transport, η_{act} is the activation overpotential for the EC reactions, and η_{con} is the concentration polarization due to the resistance encountered by gas transport in porous electrodes. E_{OCV} may be calculated as

$$E_{OCV} = -\frac{G^0}{2F} + \frac{RT}{2F}\ln\frac{p_{H_2}^0\left(p_{O_2}^0\right)^{0.5}}{p_{H_2O}^0} \tag{14.2}$$

where ΔG^0 is the free energy change of the reaction $H_2 + 0.5O_2 \Leftrightarrow H_2O$ at the working temperature T, and when the partial pressures of all the reacting gas species are 1 atm [9]. The E_{OCV} for CO oxidation is the same as in Equation 14.2 if CO, CO_2, H_2, and H_2O are at equilibrium, which is often a good approximation due to the common finding that the water gas shift reaction, $CO + H_2O \rightleftharpoons H_2 + CO_2$, is sufficiently fast.

The output voltage of an SOFC is necessarily smaller than E_{OCV}, which is a little over 1 V for the normal SOFC operating conditions. For practical usage, repeating unit cells should be stacked together with interconnect plates providing electrical contact between adjacent anodes and cathodes to increase the output voltage and power. Structural models of a repeating unit cell and a stack of planar SOFCs (pSOFCs), as well as a pSOFC stack with frame and flow manifold, are shown in Figure 14.2.

η_{con} is the sum of the concentration polarizations of the anode, η_{con}^a, and cathode, η_{con}^c, which may be calculated as

$$\eta_{con}^a = \frac{RT}{2F}\ln\left(\frac{p_{H_2}^0 p_{H_2O}}{p_{H_2} p_{H_2O}^0}\right) \tag{14.3}$$

$$\eta_{con}^c = \frac{RT}{2F}\ln\left(\frac{p_{O_2}^0}{p_{O_2}}\right)^{0.5} \tag{14.4}$$

where $p_{H_2}\left(p_{H_2O}\right)$ and $p_{H_2}^0\left(p_{H_2O}^0\right)$ denote the partial pressure of H_2 (H_2O) at the anode three-phase boundary (TPB), where H_2 is oxidized, and at the anode–fuel channel interface, respectively. $p_{O_2}\left(p_{O_2}^0\right)$ is the partial pressure of O_2 at the cathode TPB (cathode–air channel interface).

η_{ohm} may be decomposed as the electronic and ionic ohmic overpotentials of the anode, cathode, and electrolyte and the overpotential due to contact ohmic resistance at the interface of cell components:

$$\eta_{ohm} = \eta_{ohm}^{a,e} + \eta_{ohm}^{a,i} + \eta_{ohm}^{c,e} + \eta_{ohm}^{c,i} + \eta_{ohm}^{ele,i} + \eta_{ASR} \tag{14.5}$$

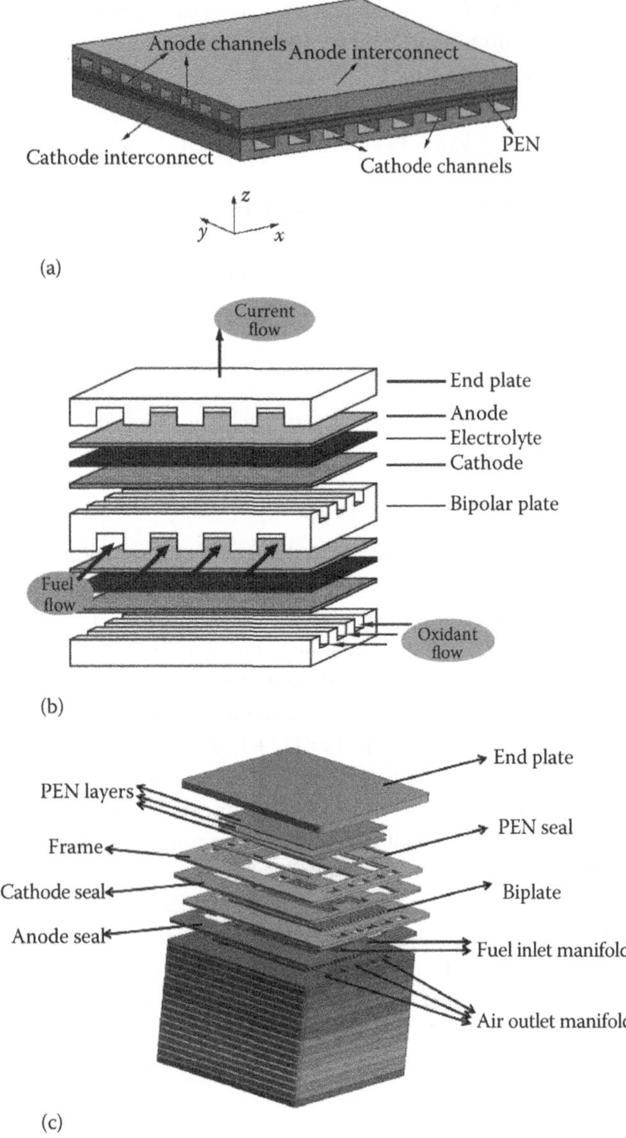

FIGURE 14.2 Schematic illustrations of (a) a unit cell, (b) a basic stack, and (c) a complete stack of pSOFCs.

Here $\eta_{ohm}^{a,e}$ $(\eta_{ohm}^{a,i})$ and $\eta_{ohm}^{c,e}$ $(\eta_{ohm}^{c,i})$ are, respectively, the electronic (ionic) ohmic overpotentials of the anode and the cathode. $\eta_{ohm}^{ele,i}$ is the ionic ohmic overpotential of the electrolyte. η_{ASR} denotes the overpotential due to the contact ohmic resistance due to adhesion problem or formation of highly resistive interfacial phase such as $La_2Zr_2O_7$ and $SrZrO_3$ [31]:

$$\eta_{ASR} = i_{el} \cdot R_{ASR} \tag{14.6}$$

where i_{el} is the electronic current density at the interface of cell components and R_{ASR} is the area-specific contact resistance. In addition, the ohmic polarization of the interconnect plate should be added to η_{ohm} when evaluating the stack performance.

In general, the dependences of polarizations on current density and temperature are nonlinear. Moreover, all the polarization terms are in principle location dependent, and depend on the current and temperature distributions and the mass transport.

14.2.1.2 Governing Equations for Charge Transport

Charge transports in the PEN structure are determined by electron conservation, ion conservation, and Ohm's law, and coupled with the EC reactions [32]:

$$\nabla \cdot \vec{i}_{el} = \nabla \cdot \left(-\sigma_{el}^{eff} \nabla \varphi_{el}\right) = \begin{cases} j_{TPB}^{c} \lambda_{TPB,eff}^{c} & \text{In cathode} \\ -j_{TPB}^{a} \lambda_{TPB,eff}^{a} & \text{In anode} \end{cases} \tag{14.7}$$

$$\nabla \cdot \vec{i}_{io} = \nabla \cdot \left(-\sigma_{io}^{eff} \nabla \varphi_{io}\right) = \begin{cases} -j_{TPB}^{c} \lambda_{TPB,eff}^{c} & \text{In cathode} \\ 0 & \text{In electrolyte} \\ j_{TPB}^{a} \lambda_{TPB,eff}^{a} & \text{In anode} \end{cases} \tag{14.8}$$

where $\vec{i}_{el}$ $(\vec{i}_{io})$ is the vector of electronic (ionic) current density, σ_{el}^{eff} (σ_{io}^{eff}) is the effective electronic (ionic) conductivity, and φ_{el} (φ_{io}) is the local electronic (ionic) potential. $\lambda_{TPB,eff}^{a(c)}$ is the anode (cathode) effective volume-specific TPB length. $j_{TPB}^{a(c)}$ is the anode (cathode) TPB length-specific transfer current density and is defined as positive when the current flows from the ionic phase to the electronic phase. j_{TPB} is produced by EC reaction and requires sacrificing some voltage to overcome the potential barrier of the EC reaction. The voltage sacrificed is called the local activation polarization, η_{act}. j_{TPB} is a nonlinear function of η_{act}, often expressed with the phenomenological Butler–Volmer (B–V) equation [32]:

$$j_{TPB} = j_0 \left[\exp\left(\frac{n_e \alpha_f F}{RT}\eta_{act}\right) - \exp\left(-\frac{n_e \alpha_r F}{RT}\eta_{act}\right)\right] \tag{14.9}$$

where α_f (α_r) is the forward (reverse) reaction symmetric factor and may have different values for anode and cathode. $\alpha_f + \alpha_r = 1$ when there is only one rate-determining step for the reaction. n_e is the number of electrons transferred per reaction, and is 2 for H_2 oxidation and 4 for O_2 reduction. j_0 is the local exchange transfer current per unit TPB length, which is often estimated as [33]

$$j_0^c = j_{0,ref}^c \exp\left(-\frac{E_{O_2}}{R}\left(\frac{1}{T} - \frac{1}{T_{ref}}\right)\right)\left(\frac{p_{O_2}}{p_{O_2}^0}\right)^{0.25} \tag{14.10}$$

$$j_0^a = j_{0,ref}^a \exp\left(-\frac{E_{H_2}}{R}\left(\frac{1}{T} - \frac{1}{T_{ref}}\right)\right)\left(\frac{p_{H_2} p_{H_2O}}{p_{H_2}^0 p_{H_2O}^0}\right) \tag{14.11}$$

where T is the absolute temperature and R is the universal gas constant. $j^{a}_{0,ref}$ $\left(j^{c}_{0,ref}\right)$ is the anode (cathode) exchange transfer current density at the reference temperature, T_{ref}, and is treated as an adjustable parameter for fitting the experimental result. E_{H_2} $\left(E_{O_2}\right)$ is the activation energy for the anode (cathode) EC reaction.

The anode and cathode activation polarizations, η^{a}_{act} and η^{c}_{act}, are determined by

$$\eta^{a}_{act} = \varphi_{el} - \varphi_{io} - \eta^{a}_{con} \tag{14.12}$$

$$\eta^{c}_{act} = \varphi_{io} - \varphi_{el} - \eta^{c}_{con} \tag{14.13}$$

The ohmic overpotentials in Equation 14.5, $\eta^{a,e}_{ohm}$, $\eta^{c,e}_{ohm}$, $\eta^{a,i}_{ohm}$, $\eta^{c,i}_{ohm}$, and $\eta^{ele,i}_{ohm}$, may be determined by the electronic (ionic) potential differences along the electronic (ionic) current flux paths. The difference in the electronic potentials at the two ending points of the cell (at the electrode surfaces here, or at the interconnect plates in general) corresponds to the measured cell output voltage [32].

14.2.2 Thermal Fluid Model

Thermal fluid modeling is a critically important part of SOFC modeling, as temperature and flow distributions strongly affect chemical and EC reactions. Temperature also affects material properties and thermal stress that may lead to mechanical failure. Flow and temperature are examined with the conservation laws in fluid mechanics [34–36].

14.2.2.1 Governing Equations for Mass Transport

The transports of gas species in porous electrodes and gas channels are governed by the mass balance equation:

$$\nabla \cdot \vec{N}_{\alpha} = R_{\alpha} \tag{14.14}$$

where $\vec{N}_{\alpha}$ and R_{α} are the molar flux and reaction rate of species α, respectively.

In the gas channels, $\vec{N}_{\alpha}$ may be determined with Fick's law [37]:

$$\vec{N}_{\alpha} = -D_{\alpha}\nabla C_{\alpha} + C_{\alpha}\vec{u} \tag{14.15}$$

where C_{α} is the molar concentration of species α and $\vec{u}$ is the vector of convection velocity. D_{α} is the diffusion coefficient of species α. R_{α} is zero in the oxidant channel. For hydrogen fuel, R_{α} is also zero as there is no reaction in the fuel channel. For hydrocarbon fuel, usually only the water gas shift reaction is involved in the fuel channel and R_{α} may be estimated by assuming that the contents of CO, CO_2, H_2, and H_2O are at equilibrium. In the porous cathode, R_{α} is zero for N_2 and

$$R_{O_2} = -j^{c}_{TPB}\lambda^{c}_{TPB,eff}/4F \tag{14.16}$$

For H_2 fuel in the anode, $R_{H_2} = -R_{H_2O} = -j_{TPB}\lambda_{TPB,eff}/2F$.

N_{α} may be described with Fick's law, the Stefan–Maxwell model, or the dusty-gas model [38,39]. Owing to the small pore sizes in the SOFC electrodes, only the dusty-gas model is considered to be reliable [37–39]. The dusty-gas model in molar units has the form [39]

$$\frac{N_i}{D^{eff}_{iK}} + \sum_{j=1}^{n}\frac{x_j N_i - x_i N_j}{D^{eff}_{ij}} = -\frac{1}{RT}\left(p\nabla x_i + x_i\nabla p + x_i\nabla p\frac{kp}{D^{eff}_{iK}\mu}\right) \tag{14.17}$$

where $x_i = (c_i/c_{tot})$ is the molar fraction of species i, c_i the molar concentration of species i, $c_{tot} = \left(\sum_j c_j\right)$ the total molar concentration of mixtures, p the total gas pressure, k the permeability coefficient, μ the viscosity coefficient, D^{eff}_{iK} the effective Knudsen diffusion coefficients of species i and D^{eff}_{ij} the effective binary diffusion coefficient.

It is generally more convenient to use an approximation of Equation 14.17 that has been shown to be very accurate [40]

$$\vec{N}_i = -c_{total}D^{eff}_{i}\nabla x_i - c_i\frac{\bar{k}_i}{\eta}\nabla p \tag{14.18}$$

where D^{eff}_{α} and $\bar{k}_i$ are, respectively, the effective diffusion coefficient and permeability coefficient of species α.

14.2.2.2 Governing Equations for Momentum Transport

The momentum transport in the gas channel may be described by the Navier–Stokes equation:

$$\nabla \cdot \left(\mu\left(\nabla\vec{u} + (\nabla\vec{u})^{T}\right)\right) - \nabla p = \rho(\vec{u}\cdot\nabla)\vec{u} \tag{14.19}$$

$$\nabla \cdot \vec{u} = 0 \tag{14.20}$$

Here ρ is the gas density.

The momentum transport in the porous electrodes may be modeled by the Brinkman equation that accounts for the viscous transport in the momentum balance and treats both the pressure and the flow velocity vector as independent variables [36]:

$$\frac{\mu}{B_0}\vec{u} = -\nabla p + \nabla\cdot\left(\frac{\mu}{\phi_g}\left(\nabla\vec{u} + (\nabla\vec{u})^{T}\right)\right) - \nabla\cdot\left(\frac{2\mu}{3\phi_g}\nabla\cdot\vec{u}I\right) \tag{14.21}$$

where B_0 is permittivity and ϕ_g is the electrode porosity.

14.2.2.3 Governing Equations for Heat Transport

The temperature field is governed by the energy balance. In the gas channel, it may be expressed as

$$\nabla\cdot(-k_f\nabla T_f)+\nabla\cdot(\rho_f c_p T_f\vec{u})=Q_f \tag{14.22}$$

Here k_f is the thermal conductivity of the fluid, c_p the specific heat, ρ_f the fluid density, and T_f the temperature. Q_f is the heat source, including (water gas shift) reaction heat, viscous and nonviscous volumetric heat generation, work by gravity, as well as turbulent flow kinetic energy. In SOFCs, it is often reasonable to neglect Q_f and assume incompressible flow, resulting in

$$\nabla\cdot(-k_f\nabla T_f)+\rho_f c_p\vec{u}\cdot\nabla T_f=0 \tag{14.23}$$

For the solid structure of the SOFC, the energy conservation equation may be written as

$$\nabla\cdot(-k^{\text{eff}}\nabla T)+\rho C_p\vec{u}\cdot\nabla T=Q \tag{14.24}$$

where Q is the heat source that includes the heat of methane steam reforming reaction, Q_{MSR}, water gas shift reaction, Q_{WGS} (and possibly other chemical reactions), the ohmic heat source, Q_{ohm}, activation heat source, Q_{act}, and the entropy heat source, Q_{entr}. The heat sources may be calculated as

$$Q_{ohm}=\frac{i_{el}^2}{\sigma_{el}}+\frac{i_{io}^2}{\sigma_{io}} \tag{14.25}$$

$$Q_{act}=i_{el}\eta_{act} \tag{14.26}$$

$$Q_{entr}=i_{el}\left(\frac{-T\Delta S}{2F}\right) \tag{14.27}$$

$$Q_{MSR}=-\Delta H_{MSR}r_{MSR} \tag{14.28}$$

$$Q_{WGS}=-\Delta H_{WGS}r_{WGS} \tag{14.29}$$

Here ΔS, ΔH_{MSR}, and ΔH_{WGS} are the entropy change of hydrogen combustion and the enthalpy changes of methane steam reforming and water gas shift reaction, respectively. r_{MSR} (r_{WGS}) is the reaction rate of methane steam reforming (water gas shift reaction).

k_{eff} is the effective thermal conductivity of the porous electrode material and may be evaluated as

$$k_{eff}=\phi_g k_f+(1-\phi_g)k_s \tag{14.30}$$

where k_s is the thermal conductivity of the solid material.

Heat transfer between cell components must be accounted for, often as boundary conditions for the governing equations. The heat transport due to the convective heat transfer may be described with Newton's law of cooling through the boundary condition:

$$k_{eff}\nabla T_s\cdot\vec{n}=h(T_f-T_s) \tag{14.31}$$

where $\vec{n}$ is the unit vector normal to the boundary, h is the heat transfer coefficient, and T_s is the temperature of the solid. The radiative heat transfer between the solid surface and the flow (or stack environment) may be calculated as

$$Q_{rad}=\varepsilon_{rad}\left(T_f^4-T_s^4\right) \tag{14.32}$$

where ε_{rad} is the surface emissivity of the solid material. The radiative heat transfer term should be added to the right hand side of Equation 14.31 for a consideration of both radiative and convective heat transfers.

14.2.3 Internal Methane Steam Reforming Reaction

Hydrocarbon may, in principle, be used as fuel for SOFC. The existing studies focus on methane and syngas; therefore, only reactions involving CH_4 and CO are considered here. CH_4 can be converted to H_2 and CO inside the anode by catalytic steam reforming

$$CH_4+H_2O\rightarrow 3H_2+CO \tag{14.33}$$

The details about the steam reforming process are rather complicated, and the current understanding is far from completed. On the basis of the experimental measurements and theoretical considerations, numerous kinetic expressions have been proposed for the steam reforming reaction. In general, a kinetic measurement of methane reforming rate may be fitted to an Arrhenius power law expression [2]:

$$r_{CH_4}=kP_{CH_4}^{\alpha}P_{H_2O}^{\beta}P_{H_2}^{\gamma}P_{CO_2}^{\delta}P_{CO}^{\varepsilon}\exp\left(\frac{-E_a}{RT}\right) \tag{14.34}$$

γ, δ, and ε are often found to be close to zero. k, α, β, and E_a show dependence on several factors such fuel composition and pressure and catalytic support, etc. Examples of the measured methane steam rate include [41]

$$r_{CH_4,Ach}=10^4\,\text{mol m}^{-1}\text{s}^{-1}\,\text{bar}^{-2}\,P_{CH_4}\exp\left(\frac{-82\text{ kJ mol}^{-1}}{RT}\right) \tag{14.35}$$

$$r_{CH_4,AhF}=8542\,\text{mol m}^{-1}\,\text{s}^{-1}\text{bar}^{-2}P_{CH_4}^{0.85}P_{H_2O}^{-0.35}\exp\left(\frac{-95\text{ kJ mol}^{-1}}{RT}\right) \tag{14.36}$$

$$r_{CH_4,Lei}=30.8\text{e}10\,\text{mol m}^{-2}\,\text{s}^{-1}\,\text{bar}^{-2}\,P_{CH_4}P_{H_2O}\exp\left(\frac{-205\text{ kJ mol}^{-1}}{RT}\right) \tag{14.37}$$

These results are believed to be correct for their corresponding experimental conditions [41]. For theoretical modeling, one should choose an expression that best corresponds to the targeted modeling condition.

H_2 and CO may convert into each other through the water–gas shift (WGS) reaction:

$$CO + H_2O \leftrightarrow H_2 + CO_2 \tag{14.38}$$

The shift reaction is often sufficiently fast, and the local mixture of H_2, CO, H_2O, and CO_2 is near equilibrium. An equilibrium limited rate expression may be used for the WGS reaction:

$$r_{WGS} = k_{WGS}\left(p_{CO}p_{H_2O} - \frac{p_{H_2}p_{CO_2}}{K_{eq,WGS}}\right) \tag{14.39}$$

where $K_{eq,WGS}$ is the temperature-dependent equilibrium constant for the WGS reaction and k_{WGS} is a sufficiently large constant that ensures the near-equilibrium condition. The overall rates for the steam reforming and WGS reactions may then be written as [9]

$$r_{H_2} = 3r_{CH_4} + r_{WGS} \tag{14.40}$$

$$r_{H_2O} = -r_{CH_4} - r_{WGS} \tag{14.41}$$

$$r_{CO} = r_{CH_4} - r_{WGS} \tag{14.42}$$

$$r_{CO_2} = r_{WGS} \tag{14.43}$$

These rate expressions should be combined with the EC reaction rates to determine the compositions of the fuel stream.

Both hydrogen and carbon monoxide may react with oxygen ions at the anodic TPB; however, the EC process is dominated by H_2 oxidation in most cases. As the contents of H_2, CO, H_2O, and CO_2 are near equilibrium due to the shift reaction, it is more convenient to assume that only H_2 oxidation is involved:

$$r_{H_2}^{EC} = -r_{H_2O}^{EC} = -j_{TPB}^{a}\lambda_{TPB,eff}^{a}/2F \tag{14.44}$$

$$r_{CO}^{EC} = r_{CO_2}^{EC} = 0 \tag{14.45}$$

The overall reaction rates, $r_\alpha + r_\alpha^{EC}$, are then used to determine the compositional change of the fuel stream.

14.2.4 Thermomechanical Modeling

SOFCs are manufactured by processing at temperatures that are higher than the working temperature. SOFCs may be considered to be stress free at the elevated processing temperature. Stresses are developed due to mismatch in the thermal expansion coefficients (CTEs) of the cell components. Stresses may be introduced during the assembly and sealing process of the stack. Stresses are also affected by the temperature distribution at the operating condition or in the processes of start-up or cooling down. Avoiding thermal–mechanical failure is critically important to the SOFC applications.

The thermomechanical modeling of SOFCs depends a lot on the mechanical characteristics of SOFC materials, and a variety of simplifications may have to be made. Here we illustrate the principle of a simple solid mechanics model and assume that all the SOFC materials undergo elastic deformation when subject to thermo–mechanical loads.

14.2.4.1 Stress–Strain Relation

In the solid mechanics model, assuming free deformations of materials, the stress–strain constitutive relation is written as

$$\sigma = D\xi_{el} + \sigma_0 \tag{14.46}$$

where σ is the stress tensor, D the elasticity matrix, ξ_{el} the elastic strain, and σ_0 the initial stress or the residual stress.

For an isotropic material with axial symmetry, D is expressed as [42]

$$D = \frac{E}{(1+\upsilon)(1-2\upsilon)}\begin{bmatrix} 1-\upsilon & \upsilon & \upsilon & 0 \\ \upsilon & 1-\upsilon & \upsilon & 0 \\ \upsilon & \upsilon & 1-\upsilon & 0 \\ 0 & 0 & 0 & \frac{1-2\upsilon}{2} \end{bmatrix} \tag{14.47}$$

Here E is the Young's modulus. υ is the Poisson's ratio defined by

$$\upsilon = \frac{E}{2G} - 1 \tag{14.48}$$

where G is the shear modulus of the material.

The total strain, ξ, consists of the elastic strain, ξ_{el}, the thermal strain, ξ_{th}, and the initial strain, ξ_0,

$$\xi = \xi_{el} + \xi_{th} + \xi_0 \tag{14.49}$$

The thermal strain is calculated as

$$\xi_{th} = \alpha(T - T_f) \tag{14.50}$$

where α is the CTE of the material, T is the temperature, and T_f is the stress-free temperature of the material. The initial strain and initial stress are related as

$$\sigma_0 = D\xi_0 \tag{14.51}$$

On the basis of the above the relationships, Equation 14.46 may be rewritten as

$$\sigma = D(\xi - \xi_{th} - \xi_0) + \sigma_0 = D[\xi - \alpha(T - T_f)] \tag{14.52}$$

It is usually assumed that the gravitational forces have a negligible contribution to the overall stress distribution, but may be added if necessary.

The solid mechanics model may be solved once the temperature field, mechanical properties of materials, stress-free temperature, and the mechanical constrain (load) are given. The temperature field may be obtained, e.g., through the thermofluid model described above.

14.2.4.2 Mechanical Properties of Composite Electrode

For dense composite materials, one way to compute their mechanical properties is to use the so-called composite sphere method that gives [43]

$$K_{com} = K_2 + \frac{\psi_1}{1/(K_1 - K_2) + 3\psi_2/(3K_2 + 4G_2)} \tag{14.53}$$

$$G_{com} = G_2 + \frac{\psi_1}{1/(G_1 - G_2) + 6\psi_2(3K_2 + 2G_2)/\left(5G_2(3K_2 + 4G_2)\right)} \tag{14.54}$$

Here K_{com} (G_{com}) is the bulk (shear) modulus of the dense composite material. K_2 (K_1) and G_2 (G_1) are the bulk and shear modulus of the matrix (impurity) in the composite, respectively.

The effective Young's modulus, E_{eff}, and shear modulus, G_{eff}, for a porous material may be calculated as [44]

$$E_{eff} = E_0 \frac{(1 - \phi_g)^2}{1 + (2 - 3\upsilon_0)\phi_g} \tag{14.55}$$

$$G_{eff} = G_0 \frac{(1 - \phi_g)^2}{1 + (11 - 19\upsilon_0)/(4 - 4\upsilon_0)\phi_g} \tag{14.56}$$

where the subscript "0" refers to the property of the dense material. According to these relationships, the effective Poisson ratio, υ_{eff}, for the porous material becomes

$$\upsilon_{eff} = \frac{1}{4} \frac{4\upsilon_0 + 3\phi_g - 7\upsilon_0\phi_g}{1 + 2\phi_g - 3\upsilon_0\phi_g} \tag{14.57}$$

The effective CTE of a composite material is often calculated as the weighted average of its constituents as [45]

$$\alpha_{com} = \frac{\alpha_1\psi_1K_1 + \alpha_2\psi_2K_2}{\psi_1K_1 + \psi_2K_2} \tag{14.58}$$

The effect of porosity on CTE is often neglected as it is shown to be insignificant [46]. Notice that the mechanical stress is sensitive to the CTEs, and the experimental results should be used if possible [33].

14.2.4.3 Failure Probability Analysis

The electrolyte and porous electrodes of SOFCs are ceramic or ceramic-like brittle materials. Other components such as frame, seal, and interconnect also have limited mechanical strengths and are prone to mechanical failure. A brittle material such as ceramic is known to exhibit a statistical strength distribution [47], and its risk of mechanical failure is analyzed in terms of failure probability. The failure probability, P_f, of a brittle material under a tensile stress, σ_{ten}, may be calculated using the Weibull method [48,49]:

$$P_f = 1 - \exp\left(-\int_V \left(\frac{\sigma_{ten}}{\sigma_0}\right)^m \frac{dV}{V_0}\right) \tag{14.59}$$

where σ_0 is the Weibull strength of the material at the reference volume, V_0, m the Weibull modulus, and V the volume of the material. When subjected to a multiaxial stress, the total survival probability of the material is a product of its survival probabilities for the three principal stresses, σ_i (i = 1, 2, 3):

$$P_f^m = 1 - \prod_{i=1}^{3} \exp\left(-\int_V \left(\frac{\sigma_i}{\sigma_0}\right)^m \frac{dV}{V_0}\right) \tag{14.60}$$

The three principal stresses of the stress tensor are determined by changing the basis so that the shear stress components are null. Normally, the principal stresses are ordered as $\sigma_1 > \sigma_2 > \sigma_3$. The stress is tensile when $\sigma_i > 0$, and compressive when $\sigma_i < 0$. The Weibull analysis applies only for tensile stresses, and $\sigma_i = 0$ should be used in evaluating Equation 14.60 when $\sigma_i < 0$ is encountered in the stress field.

When encountering a compressive stress, the mechanical failure or survival of the material is determined simply by whether the stress is above or below its compressive strength.

14.2.5 General Numerical Modeling Approaches

A theoretical modeling analysis usually involves the steps of formulation, solution, validation, and prediction. Formulation and solution are the essential part of a theoretical modeling. According to the purpose and the need of the study, a mathematical model based on the pertinent governing physical principle and some reasonable approximation is formulated. The mathematical model may be solved analytically or numerically. The analytical solution is the most desirable, but may be difficult or impossible to obtain in practice. In most cases, the model has to be solved approximately using some appropriate numerical technique. As there are approximations in the model and in the solution process, the results should be validated preferably by experimental data, or by comparison with other relevant analytical and numerical results. Validation by experimental data may be also meant to provide explanations to the experimental observations. Comparisons with the analytical or numerical results of previous models that are more specific than the current model are helpful for claiming the status of the new model as a more general and versatile model than the existing ones. The prediction step can significantly increase the usefulness of the model by providing information such as the effects of parameter changes and the optimal working conditions.

Sections 14.2.1 through 14.2.4 have provided a general description of the overall governing laws of physics for SOFCs. For most theoretical modeling works, the studies are concerned with a specific or subset of the physical problems. Specialized models or simplifications and assumptions may be made in accordance with the objective of the study, as will be illustrated below. Here we summarize briefly the general numerical approach for the solution of the coupled EC and thermal fluid model.

The partial differential equations (PDEs) for the transport of charge, mass, momentum, and energy usually need to be solved by a numerical method. The most common methods used are finite difference method (FDM), finite element method (FEM), and finite volume method (FVM). In these methods, the spatial (and temporal when dealing with transient state) domain is divided into numerous sections, a procedure called discretization or meshing. The generation of the mesh is dependent on the domain geometry and the details of knowledge required. The PDEs are then transformed into algebraic equations on the generated mesh in accordance with the numerical method employed. A computer code capable of solving the system of equations efficiently should then be developed. The computer code may be in-house developed, or a commercial software. The in-house developed software is often more flexible for incorporating new physical models, while the commercial software is often more versatile for having more modeling capabilities. The commercial software is mostly based on FEM, e.g., COMSOL® and MARC®, or FVM, e.g., FLUENT® and STAR-CD®. The FVM-based software seems to be numerically more efficient for thermofluid modeling, while the FEM based software may be advantageous for structural mechanics analysis.

The solutions of PDEs are critically dependent on their boundary conditions. The boundary conditions are dependent on a number of factors such as the cell geometry, stack configuration, and operating parameters. To be concrete, Tables 14.1 and 14.2 show, respectively, the examples of governing

TABLE 14.1
Example of Governing Equations Typically Applied in Modeling the SOFC Shown in Figure 14.3

Equations	Gas Chamber	Anode	Electrolyte	Cathode
Mass transport and conservation of momentum	Equations 14.14 and 14.15; Equations 14.18 through 14.20	Equation 14.14; Equation 14.18		Equation 14.14; Equation 14.18
Conservation of energy	Equation 14.24	Equation 14.24	Equation 14.24	Equation 14.24
Energy source		Equations 14.25 through 14.29	Equation 14.25	Equations 14.25 through 14.27
Charge transport		Equations 14.7 and 14.8	Equation 14.8	Equations 14.7 and 14.8
Electrochemical reaction (B–V equation)		Equation 14.11		Equation 14.10
Chemical reaction		Equation 14.33; Equation 14.38		
Reaction rate		Equation 14.34; Equations 14.39 through 14.45		Equation 14.16

TABLE 14.2
Typical Boundary Conditions Applied for Modeling the SOFC Shown in Figure 14.3

Boundary	Electric Field	Gas Flow	Mass Transport	Heat Transport
Fuel inlet		Laminar flow	CH_4, H_2, and H_2O concentration	Temperature (T_0)
Fuel outlet		Pressure (P_0)	Convective flux	Convective flux
Anode surface	Electric grounding	Normal stress normal flow ($c_{Total}RT$)	Continuous	Continuous
Electrolyte–anode interface	Electric insulation		Insulation	Continuous
Electrolyte–cathode interface	Electric insulation		Insulation	Continuous
Exposed electrolyte	Electric insulation			Convective cooling (T_0)
Cathode surface	Current density (i_{out})		O_2 and N_2 concentration	Convective cooling (T_0)
All others	Electric insulation	No slip	Insulation	Adiabatic

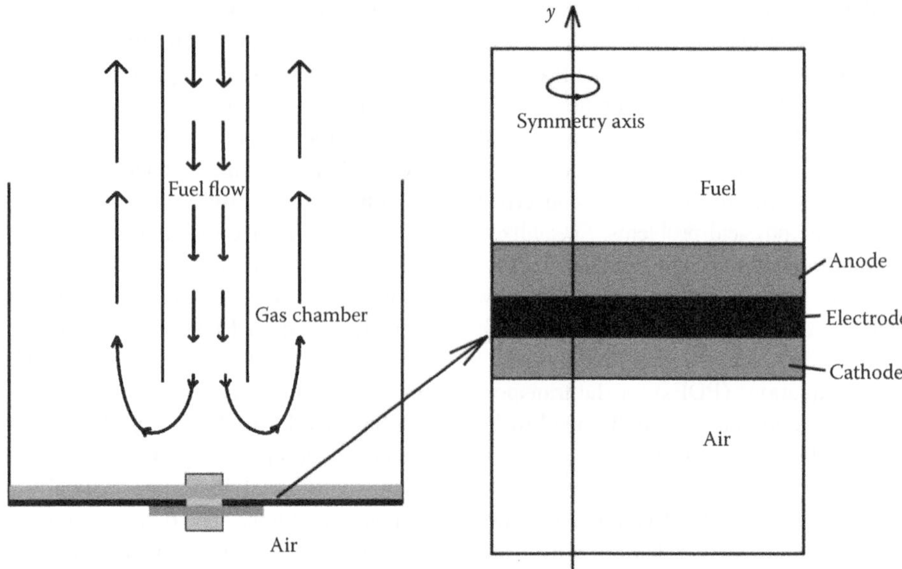

FIGURE 14.3 Model setup for a button cell operated in laboratory.

equations and boundary conditions for a special operating pSOFC illustrated in Figure 14.3. The cell is circular in shape and uses moisturized methane as the fuel. No reaction is assumed in the gas chamber.

14.3 0D MODELING

0D modeling is the simplest theoretical modeling approach. 0D modeling may be applied to the material component, cell and stack, as well as system of SOFCs. For example, at the level of material component such as the porous electrode, models of material properties may be built by treating the spatial distributions of the material composition and microstructure as statistically uniform. At the cell and stack levels, the *I–V* relations or other interested properties may be expressed in a mathematical form with lumped parameters for material properties and operating conditions. At the system level, the input–output characteristics may be parameterized for efficient computation so that the system may be controlled effectively to meet the varying external requirement. Therefore, 0D modeling can be a powerful tool for the fundamental understanding of SOFCs. The following gives some examples of 0D modeling at the material component and cell levels.

14.3.1 Lumped-Parameter Model for *I–V* Relation

Assuming that the electrolyte is a pure ionic conductor, the ionic current at the electrolyte–electrode interface equals the output current. Moreover, considering the reverse activation term is relatively small in comparison with the forward activation term, the Butler–Volmer equation may be approximated as

$$\eta_{\mathrm{act}}^{\mathrm{c(a)}} = \frac{RT}{n_{\mathrm{e}}^{\mathrm{c(a)}}\alpha_{\mathrm{f}}^{\mathrm{c(a)}}F}\sinh^{-1}\left(\frac{i}{\lambda_{\mathrm{TPB,eff}}^{\mathrm{c(a)}}j_0^{\mathrm{c(a)}}}\right) \quad (14.61)$$

Equation 14.61 reduces to the usual Tafel form, $\eta_{\mathrm{act}} = b\ln(i/i_0)$, for $i \gg i_0$, while eliminating the logarithmic divergence of the Tafel equation at $i = 0$.

The mass transport in porous electrode is mainly driven by concentration gradient. The concentration gradient of H_2 is maximal when p_{H_2} at the electrolyte–anode interface is zero, corresponding to a maximal amount of H_2 that may be transported to the TPB region for current generation. That is, there is a concentration limited current, i_{H_2}. On the basis of the analysis of the transport theory, i_{H_2} is expected to be proportional to $p_{H_2}^0$, the effective diffusion coefficient of H_2 and the inverse of the electrode thickness [9,30].

It is reasonable to think that p_{H_2} at the TPB varies linearly with the operating current [30], $p_{H_2}/p_{H_2}^0 = 1 - i/i_{H_2}$; the concentration polarization due to H_2 transport (Equation 14.3) becomes

$$\eta_{\mathrm{con}}^{\mathrm{H_2}} = -\frac{RT}{2F}\ln(1 - i/i_{\mathrm{H_2}}) \quad (14.62)$$

Assuming $p_{\mathrm{tot}}^0 = p_{\mathrm{tot}}^{\mathrm{TPB}}$, then $p_{\mathrm{H_2O}}^{\mathrm{TPB}} = p_{\mathrm{tot}}^0 - p_{\mathrm{H_2}}^{\mathrm{TPB}} = p_{\mathrm{H_2O}}^0 + \frac{ip_{\mathrm{H_2}}^0}{i_{\mathrm{H_2}}}$, the concentration polarization due to H_2O transport (Equation 14.3) is

$$\eta_{\mathrm{con}}^{\mathrm{H_2O}} = \frac{RT}{2F}\ln\left(1 + \frac{ip_{\mathrm{H_2}}^0}{i_{\mathrm{H_2}}p_{\mathrm{H_2O}}^0}\right) \quad (14.63)$$

Similar analysis for the cathode yields the concentration polarization for O_2 transport as

$$\eta_{\mathrm{con}}^{\mathrm{O_2}} = -\frac{RT}{4F}\ln(1 - i/i_{\mathrm{O_2}}) \quad (14.64)$$

Overall, the I–V relation is expressed as

$$\begin{aligned} V_{\text{cell}} = E_{\text{OCV}} - i\left(\frac{l_{\text{elt}}}{\sigma_{\text{i,elt}}} + \frac{l_{\text{c}}}{\sigma_{\text{e,c}}} + \frac{l_{\text{a}}}{\sigma_{\text{e,a}}} + \eta_{\text{ASR}}\right) \\ - \frac{RT}{4F\alpha_{\text{f}}^{\text{c}}}\sinh^{-1}\left(\frac{i}{j_{0\text{c}}\lambda_{\text{TPB,c}}^{\text{eff}}}\right) \\ - \frac{RT}{2F\alpha_{\text{f}}^{\text{a}}}\sinh^{-1}\left(\frac{i}{j_{0\text{a}}\lambda_{\text{TPB,a}}^{\text{eff}}}\right) \\ + \frac{RT}{2F}\ln\left(1 - \frac{i}{i_{\text{H}_2}}\right) - \frac{RT}{2F} \\ \ln\left(1 + \frac{iP_{\text{H}_2}^0}{i_{\text{H}_2}P_{\text{H}_2\text{O}}^0}\right) + \frac{RT}{4F}\ln\left(1 - \frac{i}{i_{\text{O}_2}}\right) \end{aligned} \quad (14.65)$$

where l_m is the layer thickness of component m, with the subscript m = elt, c, and a denotes the electrolyte, cathode, and anode, respectively. Equation 14.65 can be used to fit the experimental data, and the fit is often found to be highly satisfactory [9,30]. Notice, however, the number of fitting parameters should be kept at the minimal to be meaningful. Through the fitting, important performance parameters may then be determined for the tested cell.

Insights of fundamental importance may be gained by analyzing Equation 14.65. For example, the ionic conductivity of electrolyte is often much smaller than the electronic conductivities of the electrodes; it is helpful for reducing the ohmic polarization to use a thin electrolyte instead of a thin electrode. Therefore, an electrode-supported cell (with a thin electrolyte) is expected to have a higher performance than an electrolyte-supported cell (with a thick electrolyte) operating at the same condition. For an acceptable level of ohmic loss, electrode-supported cells can be operated at a lower temperature than that of electrolyte-supported cells, allowing for more choices of materials for SOFCs. The ohmic loss consideration also leads to intensive research activities of searching for new electrolyte materials with appropriate conductivities at the reduced operating temperature. Notice, however, the contact resistance, η_{ASR}, can be substantial, as observed experimentally [50]. Without adequate fabrication technique to reduce η_{ASR}, the benefit of reducing the electrolyte resistance may be substantially limited. This is one of the reasons that some cells with more conducting electrolytes show very limited performance improvements, or even reduced performance.

The activation polarization, η_{act}, is often a major component of the overall polarization. Materials with higher transfer current density, $j_0^{\text{a,c}}$, may be used to reduce η_{act}. Alternatively, a simple way to reduce η_{act} is to increase the TPB length, $\lambda_{\text{TPB,eff}}$ is small if TPB is found only at the electrolyte–electrode interface. $\lambda_{\text{TPB,eff}}$ can be significantly increased by mixing the electrolyte particles and the electronic conducting particles. This is an important role played by the composite electrode [51].

The concentration polarization, η_{con}, is often relatively small in comparison with the activation and ohmic polarizations. However, η_{con} can be substantial if the pore size is too small or the pore connectivity is low. As the diffusion coefficient of H_2 is about four times larger than that for air, and $P_{\text{H}_2}^0 > P_{\text{O}_2}^0$ in normal cases, the anode limiting current $\left(i_{\text{H}_2}\right)$ is more than a factor of 4 larger than the cathode limiting current $\left(i_{\text{O}_2}\right)$, with all other factors being equal. Hence, the I–V performance of an anode-supported design is expected to be superior to that of a cathode-supported design.

The pressure dependences of E_{OCV} shown in Equation 14.2 indicate that pressurized fuel may not effectively improve the cell performance as both $P_{\text{H}_2}^0$ and $P_{\text{H}_2\text{O}}^0$ are increased proportionally. However, the cell performance can be notably improved by pressurizing the air or replacing air with pure oxygen as the oxidant.

The I–V model may also be used for other important analysis, such as the overall heat generation, $i\left(\frac{-\Delta H}{2F} - V_{\text{cell}}\right)$, and the dependence of electrical efficiency on the output voltage and operating temperature, $V_{\text{cell}} / \frac{-\Delta H}{2F}$. Here, ΔH is the enthalpy change of the reaction $H_2 + 0.5O_2 \Leftrightarrow H_2O$ at the working condition.

14.3.2 Effective Electrode Property Models

The I–V model, Equation 14.65, requires a set of parameters for material properties, geometries (layer thicknesses), and operating conditions to predict the fuel cell performance. In turn, it may be used to examine the effect of changing parameters such as material properties on the cell performance. However, it cannot predict the effect of microstructure on the material properties and cell operation. To examine the influences of the electrode microstructure and cell design on the cell performance, models to predict crucial properties based on microstructural characteristics are required. A popular approach to achieve the goal is to apply a coordination number–based percolation model that relates the effective electrode properties to the microstructure parameters [32,51–54].

Figure 14.4 illustrates the microstructure of a typical composite SOFC electrode. The porous electrode is a binary system with a random mixture of electronic conducting electrode particles (denoted as ed) and ionic conducting electrolyte particles (denoted as el). If the relevant particles form a percolating network, the corresponding electrical conductivity is nonzero. The scaling behavior of the conductivity in the vicinity of percolation threshold is known by the percolation theory. However, the conductivity for the particle network that is substantially above the percolation threshold is not known exactly. On the basis of numerical simulations of hard-sphere packing systems, one recommended expression to calculate the electrical conductivity is [51]

$$\sigma_k^{\text{eff}} = \sigma_k^0[(1-\phi_{\text{g}})\psi_k P_k]^{\gamma} \quad (14.66)$$

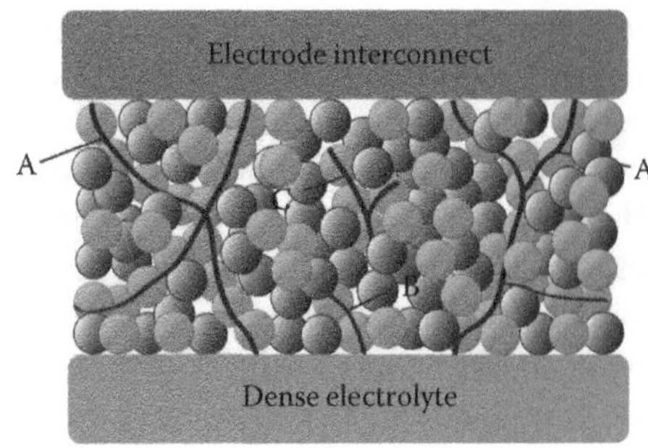

FIGURE 14.4 Illustration of a composite SOFC electrode as a random pack of electrode particles (light gray) and electrolyte particles (gray). The electrolyte particles that form the percolating network, that are not percolating but connected to the electrolyte for ion transport, and that are not percolating and isolated are denoted with A, B, and C, respectively.

where σ_k^0 is the electric conductivity of k-material in a dense solid, Ψ_k the volume fraction of the k-particles in the solid electrode structure. γ is the Bruggeman factor depending on the morphology of the fabricated electrode and may be in the range of 1.5–5 [55]. P_k is the probability of k-particles belonging to the percolated network and may be calculated as [54]

$$P_k = \left(1 - \left(\frac{3.764 - Z_{k,k}}{2}\right)^{2.5}\right)^{0.4} \quad (14.67)$$

where $Z_{k,k}$ is the average number of k-particles in contact with a k-particle. The average number of ℓ-particles in contact with a k-particle can be estimated as [51]

$$Z_{k,\ell} = 0.5\left(1 + r_k^2/r_\ell^2\right)\bar{Z}\frac{\psi_\ell/r_\ell}{\psi_k/r_k + \psi_l/r_l} \quad (14.68)$$

where $\bar{Z}$ is the overall average coordination number of all particles and is often set to 6 for a random packing of spheres [51], although a value of 6.7 seems more realistic [55]. $r_{\mathrm{ed(el)}}$ is the average radius of the ed (el) particle.

The TPB length for a pair of contacting el and ed particles is the contact perimeter between the particles. Overall, considering that the TPB is effective only when the contacting particles are a portion of the percolated network (particles belong to the A clusters in Figure 14.4), the effective TPB length per unit volume can be estimated as

$$\lambda_{\mathrm{TPB,eff}}^V = 2\pi \min(r_{\mathrm{ed}}, r_{\mathrm{el}}) \sin\theta\left(n_{\mathrm{ed}}^V Z_{\mathrm{ed,el}}\right) P_{\mathrm{el}} P_{\mathrm{ed}} \quad (14.69)$$

where $n_k^V = 3(1-\phi_g)\psi_k/4\pi r_k^3$ is the number of k-particles per unit volume in the composite electrode. θ is the angle of particle contact, which is normally not measured experimentally and treated as a fitting parameter. As fitting Equation 14.69 to the experimental data often yields a value of around 15^0 [54], θ is usually assigned with a value of 15^0. Notice that Equation 14.69 is the result of ignoring the contribution of particles belonging to the B clusters. The approximation is acceptable as the electrode is normally made to have high percolation probability.

A key parameter for describing the gas flow in a porous electrode is the so-called hydraulic radius, or the average pore radius, r_g. r_g may be estimated as [51]

$$r_g = \frac{2}{3(1-\phi_g)}\left(\frac{1}{\psi_{\mathrm{ed}}/r_{\mathrm{ed}} + \psi_{\mathrm{el}}/r_{\mathrm{el}}}\right) \quad (14.70)$$

The above expressions, Equations 14.66 through 14.70, do not appear simple, but they are in fact trivial to evaluate once the microstructure parameters of the electrode are specified. These expressions are very useful for understanding the effect of microstructure on the material properties and cell performance and may be used for designing high performing cells.

For example, the Ni–YSZ combination remains the state-of-the-art material choice for SOFC anode. However, the CTE of Ni is substantially larger than the YSZ electrolyte and LSM–YSZ cathode. The CTE mismatch of Ni–YSZ anode and YSZ electrolyte is detrimental to the mechanical stability of the cell. The higher the Ni content is, the higher the CTE mismatch will be. Fortunately, the electronic conductivity of Ni is very high, and a Ni content moderately higher than the percolation threshold (about 29% when the sizes of Ni and YSZ particles are similar) can provide sufficiently high conductivity for the anode. Therefore, the Ni volume fraction in the anode solid phase should be around 35%. The CTE mismatch between LSM and YSZ is small, and the conductivity of LSM is only moderate. Therefore, to maintain an adequate electronic conductivity of cathode and to maximize the cathode TPB length, the volume fraction of LSM and YSZ in the cathode should be similar.

As can be inferred from Equation 14.69, the effective TPB length increases with the reduced particle size as $\sim r_k^{-2}$. Reducing the particle size is beneficial for lowering the activation polarization. However, the concentration polarization may increase due to the reduced pore radius with the decreased particle size (Equation 14.70). As the EC reaction occurs mainly in the region very close to the electrode–electrolyte interface, to minimize the concentration polarization effect, a thin layer of composite electrode with small particles may be used there. This results in the current design of PEN structure typically consisting of five distinct layers: a relatively thick anode support layer with a relatively large Ni and YSZ particles, a porous anode interlayer with fine Ni and YSZ particles, a dense YSZ electrolyte layer, a porous cathode thin layer with fine LSM and YSZ particles, and a porous cathode current collecting layer with relative large LSM particles. A schematic illustration of this kind of PEN structure is shown in Figure 14.5. Notice that the cathode does not have the CTE mismatch problem encountered in the anode; the cathode current collector consists of LSM only in order to maximize the conductivity for the current collector.

The design of thin composite interlayer has some limitation. For example, the Ni volume fraction cannot be much reduced, and the CTE mismatch remains a serious concern. To significantly reduce the required Ni loading, a new design

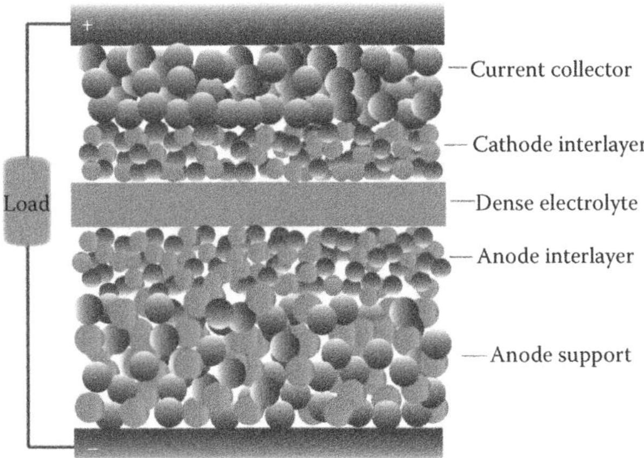

FIGURE 14.5 Schematic of the microstructures in a PEN with functional graded electrodes. Gray particles denote electrode particles (Ni or LSM) and light gray particles denote electrolyte particles (YSZ).

of nanoparticle infiltrated composite electrode is proposed and actively studied [56–60]. In such a design, nanosized Ni particles are coated on the surface of micron-sized core YSZ particle, as illustrated in Figure 14.6. As long as the nano-Ni particles on the surface of YSZ backbone form a percolating network, the electrode is electronic conducting. Owing to the small size of nano-Ni particles, the percolating threshold of Ni volume fraction can be reduced to be below 10% or less [61]. Applying the above-mentioned coordination-based percolation theory to the structural model of Figure 14.6, analytical models for the nano-Ni percolation threshold, both the electronic and ionic conductivities, the effective TPB length, as well as the hydraulic diameter of the electrode are derived [62–64]. The theoretical results for the size-dependent percolation threshold, temperature-dependent electronic conductivity, and the hydraulic diameter agree very well with the available experiments. The temperature dependent *I*–*V* relation thus obtained theoretically also agrees with the experimental data very well. These good agreements indicate that the model also provides good description of the ionic conductivity and TPB length owing to the strong and different temperature dependences of the ionic conductivity and the activation polarization. The validated model is further used to determine the optimal nanoparticle loading as 10–15 vol%. With this significantly reduced nano-Ni loading, high cell performance is assured, while the CTE mismatch is kept well within the acceptable level.

A major problem with the Ni-based anode is the coarsening of Ni particles due to the high working temperature of SOFC, affecting the long-term operational stability of the cell. Assuming the coarsening of Ni particles is surface diffusion driven and considering the statistical nature of the Ni-particle network confined by the YSZ backbone, a simple analytical expression for the Ni coarsening is derived using the concept of coordination number in the percolation theory [65]. The analytical model includes only one adjustable parameter representative of the initial morphology of the anode. The model is found to agree well with the available experiments on cells of different microstructure compositions and operated under different temperatures with low water vapor pressure. Thus, the surface diffusion process is validated as the dominant mechanism for the coarsening of Ni particles when the steam partial pressure is low. The model can be used to examine the long-term performance characteristics of SOFCs with different anode microstructure designs. As will be discussed below in 2D modeling, a conventional electrode design with a good initial performance may be accompanied with a relatively high degradation rate. However, by properly tuning the electrode microstructure, it is possible to obtain a cell with high performance for both short and long terms. Therefore, the model may be used to assist the design of durable high-performing cells.

Although the electrolyte is made as dense as possible, it is not a single crystal. It is fabricated from sintering the electrolyte material of some particle sizes. To optimize the particle size for the highest possible ionic conductivity of the most

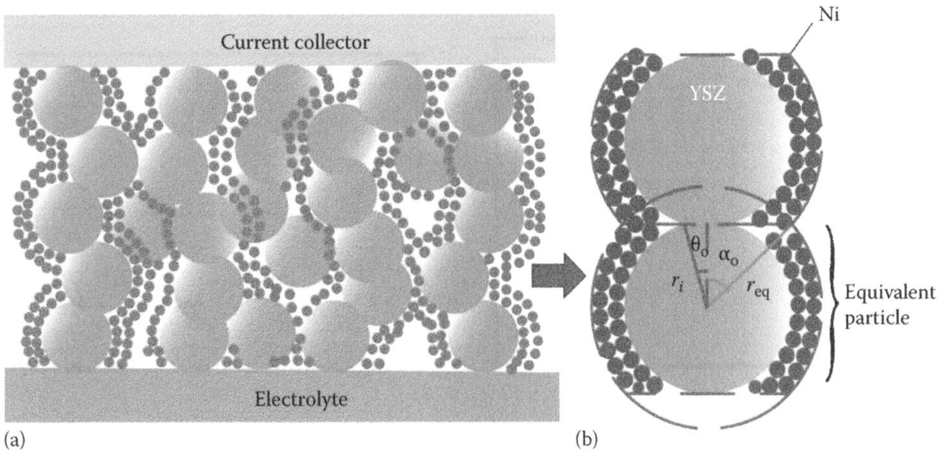

FIGURE 14.6 Sketch of (a) nanocomposite electrode and (b) model for the equivalent conduction element, a superparticle consisting of a core YSZ particle coated with nanosized Ni particles.

popular electrolyte, doped ZrO_2 and CeO_2, the grain size dependence of the ionic conductivity has been investigated intensively. Considering the discrete nature of lattice distortion at the grain boundary, the existing model for the curvature effect on defect formation energy is generalized and an analytical model for the grain size dependence of Schottky barrier is obtained [66]. The new model for the Schottky barrier is found to be in good agreement with a significant number of experimental data [66–68]. The model is then used to examine the grain size dependence of ionic conductivities of doped ZrO_2 and CeO_2. As seen in Figure 14.7, the total conductivity decreases with grain size at high temperature, but may show a maximum at about 200 nm at low temperature, providing a coherent explanation for numerous seemingly contradictory experimental observations. Moreover, it is concluded that the attempt to improve the ionic conductivity of doped ZrO_2 and CeO_2 by reducing the grain size is futile for the realistic working temperature of SOFCs. Therefore, the theoretical prediction may be helpful for avoiding further unnecessary and excessive experimental exploration on this subject.

14.4 1D MODELING

The 1D model can be used to reveal the distribution of physical quantities of interest along an appropriate chosen direction. Choosing the channel direction for a pSOFC in co-flow configuration, the 1D model may be used to examine the distribution of gas species, current density, and temperature along the channel direction. Alternatively, choosing the direction to be perpendicular to the cell plate, the 1D model can be used to examine the distribution of physical quantities within the PEN structure.

Considering a co-flow planar cell with a gas channel length of L and height of h, the cell uses moisturized hydrogen as the fuel. The fuel is at temperature T and is supplied with a velocity of v_0 at the channel entrance ($x = 0$). If the current density at the location x ($0 \le x \le L$) is $i(x)$, the H_2 consumption rate in a length element dx is $i\mathrm{d}x/2F$. The H_2 mass balance yields

$$vh\mathrm{d}C_{\mathrm{H}_2} = -\frac{i\mathrm{d}x}{2F} \tag{14.71}$$

where C_{H_2} denotes the molar concentration of H_2. Solving Equation 14.71 gives

$$C_{\mathrm{H}_2}(x) = C^0_{\mathrm{H}_2} - \int_0^x \frac{i(x)}{2hFv(x)}\mathrm{d}x \approx C^0_{H_2} - \frac{\int_0^x i(x)\,\mathrm{d}x}{2hFv_0} \tag{14.72}$$

Here $C^0_{\mathrm{H}_2} = p^0_{\mathrm{H}_2}/RT$ is the molar concentration of H_2 at $x = 0$. The last step in Equation 14.72 is exact for an isothermal cell. Similarly, one gets

$$C_{\mathrm{H_2O}}(x) \approx C^0_{\mathrm{H_2O}} - \int_0^x i(x)\,\mathrm{d}x/2hFv_0 \tag{14.73}$$

$$C_{\mathrm{O}_2}(x) \approx C^0_{\mathrm{O}_2} - \int_0^x i(x)\,\mathrm{d}x/4hFv_0 \tag{14.74}$$

The changes in the molar concentrations or partial pressures of gas species affect the local E_{OCV} as well as other quantities such as the concentration polarization in the I–V relation (Equation 14.65), causing nonuniform current density distribution for a given V_{cell}. Solving the coupled equations will produce the information about the distribution of current density and species concentration along the channel direction. The equations may be further coupled with the thermal fluid model to provide the temperature distribution.

The fuel utilization ratio, $\varepsilon_{\mathrm{fuel}}$, is an important parameter in determining the electrical conversion efficiency of a cell, $\varepsilon_{\mathrm{cell}} \sim \frac{\Delta G}{\Delta H}\frac{V_{\mathrm{cell}}}{E^0_{\mathrm{OCV}}}\varepsilon_{\mathrm{fuel}}$, and a large value of $\varepsilon_{\mathrm{fuel}}$ is therefore highly desirable. However, an excessively high fuel utilization ratio may cause fuel depletion and low current density near the channel exit region, resulting in a reduce average current

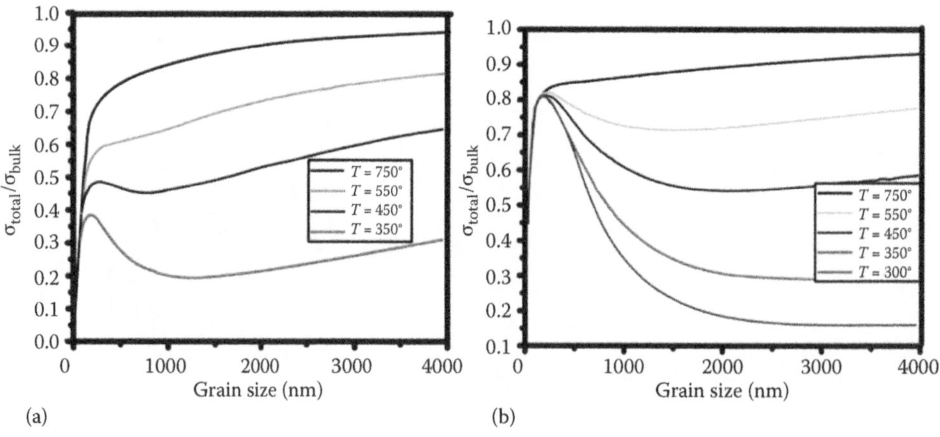

FIGURE 14.7 Ratio of total conductivity to ideal bulk conductivity as a function of grain size and temperature: (a) doped ZrO_2; (b) doped CeO_2.

density, $\bar{i} = \int_0^x i(x)\,\mathrm{d}x/\mathrm{L}$, and a low electrical power density of the cell, $\bar{i}V_{cell}$. A sample calculation on the current distributions of a model isothermal cell for different fuel utilization ratios has been performed. The results show that the average output current densities for ε_{fuel} of 50%, 70%, 80%, 90%, and 95% are 0.657, 0.588, 0.554, 0.502, and 0.466 A/cm^2, respectively. The reduction of $\bar{i}$ is 10% from ε_{fuel} = 50% to ε_{fuel} = 70%, but is 29% to ε_{fuel} = 95%. Therefore, a balanced approach should be adopted when choosing the cell operating conditions for practical applications.

The 1D model may be used to investigate the distributions of physical quantities such as the ionic and electronic currents (i_{io}, i_{el}), electrical potential, and gas species inside the PEN structure. For a cell with a five-layer PEN structure design shown schematically in Figure 14.5, a representative result for the distributions of i_{io} and i_{el} inside the PEN structure is shown in Figure 14.8. The sum of i_{io} and i_{el} is constant across the PEN structure and equals to the output current. i_{io} in the electrodes is concentrated in the area near the electrolyte. The closer the area to the electrolyte is, the larger i_{io} is. This is easily understandable: transporting i_{io} to a region far away from the electrolyte demands high ohmic loss, as the ionic conductivity is very low and orders of magnitude smaller than the electronic conductivity, diminishing the voltage loss available for activating the oxygen ion reduction reaction.

The numerical insight about the ionic current distribution revealed by 1D modeling can be used in designing the interlayer thickness for optimal cell performance. By using the relationship between the effective material properties and the microstructural parameters described above (Section 14.3.2), a comprehensive parametric study has been performed to find the optimal cathode interlayer thickness. Interestingly, the extensive parametric study concludes that a cathode interlayer thickness of 10–20 micron is optimal for any realistic material choice and microstructure design [32]. This finding is a very simple and useful guide for the experimental design of the cathode interlayer.

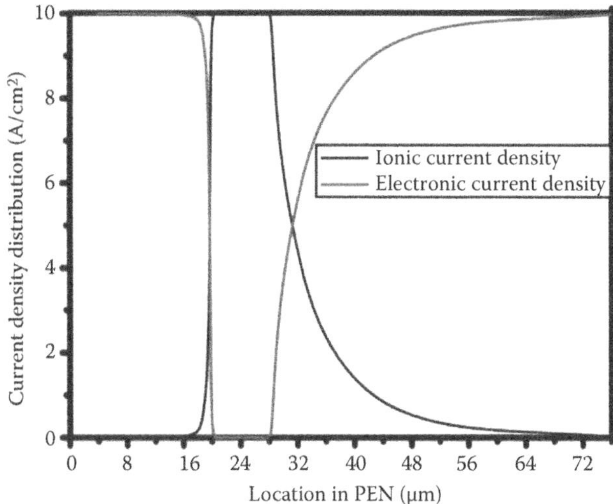

FIGURE 14.8 Typical electronic (gray line) and ionic (black line) current density distributions inside a PEN structure (the electrolyte is located at $20 \le x \le 28$ μm).

The information obtained from the 1D model may have other implications as well. When using hydrocarbon as fuel, it is widely observed that the operating current has a significant role of suppressing the carbon deposition propensity on the anode surface. Two mechanisms have been proposed for the effect of operating current on carbon deposition. One believes that the carbon deposition propensity is reduced due to the increased local concentration of water associated with the current generation. The other believes that the carbon is cleaned by combining with oxygen ion transported by the operating current. The two mechanisms are still under debate [2]. From the distribution of i_{io} provided by the 1D *I–V* model, it is clear that the first mechanism is preferable as the transport of steam is relatively easy, while the second mechanism is highly unlikely. Naturally, more studies are required. However, the 1D *I–V* model will be an essential element in the future study of this topic.

14.5 2D MODELING

In 2D modeling, a suitably chosen 2D cross section is used to represent a fuel cell or fuel cell stack. The choice of the 2D section varies with the problem of interest. For a pSOFC shown in Figure 14.2a, there are three natural choices for the 2D section. Choosing the cell plate as the 2D section, Figure 14.9a, the model may be used to examine the distributions of physical quantities such as current density, temperature, and gas species over the cell plate for flow configurations of co-, counter-, and cross-flow designs. However, this approach has to use lump-parameter models for the material properties and cell *I–V* characteristics and cannot provide detailed information inside the PEN structure. The 2D cross section of Figure 14.9b is reasonable for a co-flow or counterflow SOFC. The cross section includes basically all the components of a unit cell and may be extended to study the performance characteristics of an SOFC stack by repeating the 2D section in series. However, it is difficult for this 2D cross section to take into account the effect of the interconnect rib, the part of the interconnect that is in contact with the PEN structure and necessary for the current collection, on the cell operation. Another possible model choice for a co-flow or counterflow SOFC is shown in Figure 14.9c. Although less popular than those of Figure 14.9a and b, the model of Figure 14.9c can examine the important effects of current collection through the interconnect ribs that are absent in the former two models. Since the variations of physical quantities within a distance in the order of the cell thickness, ~1 mm, along the flow channel are rather small, the model can be an accurate description of the local conditions. Collecting a set of representative conditions, the model can provide a good description about the behavior of the whole cell plate. As the effect of current collection on the overall cell performance is very significant [31,69–71], the model of Figure 14.9c is in fact capable of providing a more

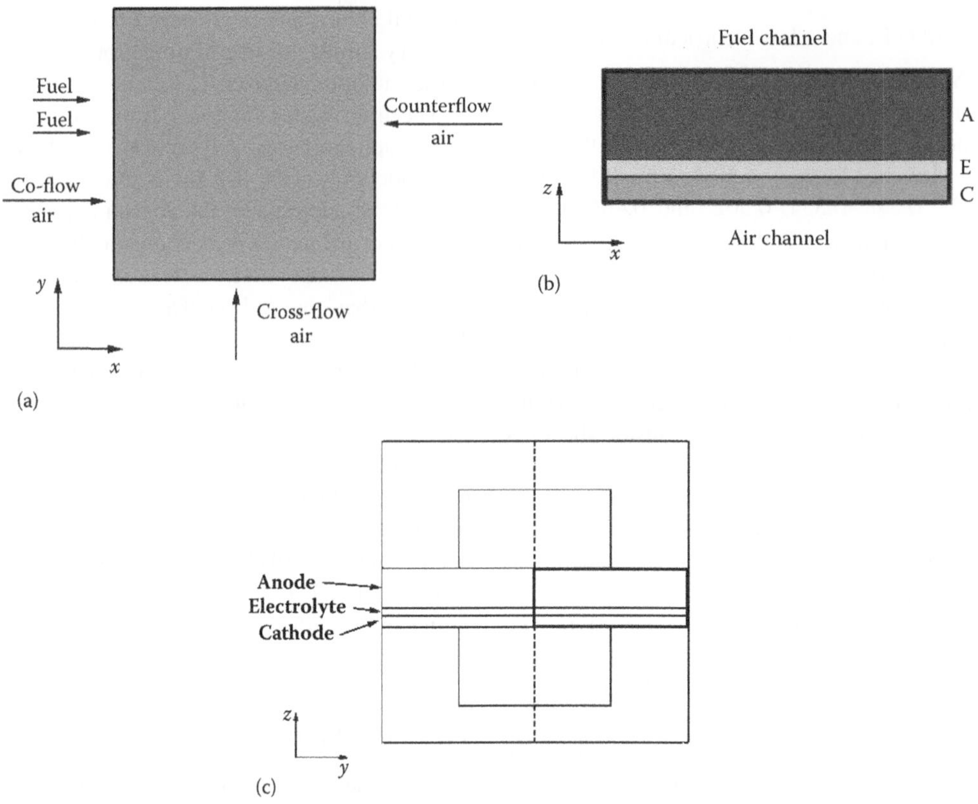

FIGURE 14.9 Choices of 2D cross section of a pSOFC unit cell for 2D modeling: (a) 2D section parallel to the cell plate; (b) 2D section perpendicular to the cell plate and parallel to the gas channels and (c) 2D section perpendicular to both the cell plate and the gas channels.

realistic description of the SOFC performance than that of Figure 14.9a and b.

As an example, Figure 14.10 shows some representative distributions of current density and fuel species at the anode–electrolyte interface and oxygen at the cathode–electrolyte interface based on the model of Figure 14.9c. As seen in Figure 14.10, the oxygen concentration at the cathode TPB region under the interconnect rib is low and the current density for under the rib region is much lower than that under the gas channel region. The rib interferes with the gas transport. The narrower the rib is, the more uniform the oxygen distribution is. The under-rib region is approximately a dead zone for current generation. However, the rib cannot be too narrow as the ohmic loss would be too high due to its inherent contact

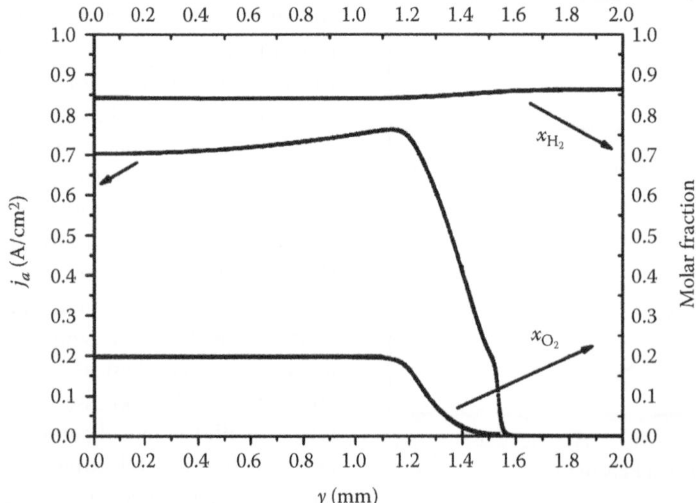

FIGURE 14.10 Distributions of current density, H_2, and O_2 molar fractions at the electrode–electrolyte interfaces ($0 \le x \le 1.2$ mm: the region under the gas channel, 1.2 mm $\le x \le 2$ mm: the region under the rib).

as well as bulk ohmic resistances. The trade-off between the gas transport and current collection represents an important design issue as the overall cell output is significantly affected [31,72]. The rib width for optimal cell output is found to be dependent on the contact resistance. However, the effects of other material and operating parameters such as the hydrogen concentration, layer thickness, porosity, and electrical conductivity of the cathode on the optimal rib width are negligible. The optimal rib width, d_{rib}^{opt}, can be written as

$$d_{rib}^{opt} = A + B \times d_{pitch} \tag{14.75}$$

where the pitch width, d_{pitch}, is the sum of the widths of channel and rib pair and is a design parameter dependent on the manufacturing technique and other considerations. The parameters A and B depend only on the contact resistance determined by fabrication technique. The specific values of A and B for any plausible contact resistance have been tabulated. The simple relationship, Equation 14.75, obtained by the 2D modeling is later confirmed by 3D modeling [73]. Equation 14.75 is a very useful guide that is easy to use for the broad SOFC engineering community.

In addition to providing rib design guidance, other important insights are also revealed by 2D modeling. For a typical cell of practical uses, the modeling shows that the average current density is only about half or less of that for an ideal button cell, satisfactorily explaining the widely observed puzzling phenomena. Moreover, the 2D model also provides guidance for the cathode design. For anode-supported pSOFCs, the conventional wisdom is to make the cathode a thin layer of about 50 microns in order to have a low cathode concentration polarization. However, this design for a practical cell means that the current generated under the channel region has to pass a narrow cross section in order to be collected by the interconnect rib, resulting in a large ohmic polarization. The thin cathode design is also detrimental for the oxygen to be transported to the TPB region under the rib. A cathode thickness of around 200 micron is near optimal and may significantly improve the cell performance for practical cathode material and microstructure designs [31,69].

To potentially combine the advantage of tubular SOFCs (tSOFCs) for thermal cycling endurance and the advantage of anode-supported pSOFC for high power density, a so-called microtubular SOFC (mtSOFC) design has been experimented. A geometric model for an anode-supported mtSOFC is shown in Figure 14.11a. There is an axial symmetry in the mtSOFC geometry. In this case, the dimensionality of the system is reduced and may be modeled completely by a 2D model, as sketched in Figure 14.11b. A comprehensive fully coupled thermal fluid electrochemistry model has been developed to analyze the EC performance of mtSOFC. On the basis of the material, geometric, and operational parameters designated in the experiment, the theoretical and experimental I–V relations are in very good agreement, illustrating the validity of the theoretical model [33,74]. The theoretical model has then been used to analyze the performance characteristics of mtSOFC, yielding useful information for its design. For example, changing the current collection from one side, as in the normal experimental practice, to both sides of the anode is found to significantly increase the cell output. As predicted by the numerical modeling results, the EC performance of the newly designed mtSOFC may be comparable to that of the pSOFC. In addition, a thermal mechanical model based on the principle described in Section 14.2.4 is also developed for the analysis of the mechanical behavior of mtSOFC. The thermal–mechanical analysis was performed using the typical PEN material set of LSM–YSZ–Ni and the temperature profile derived from the EC model [33]. It is found that the EC performance may be improved by increasing the Ni content or reducing the Ni particle size. However, the increased Ni content may substantially reduce the mechanical stability, as shown in Figure 14.12. Considering the mechanical safety using the commonly adopted criterion of a failure

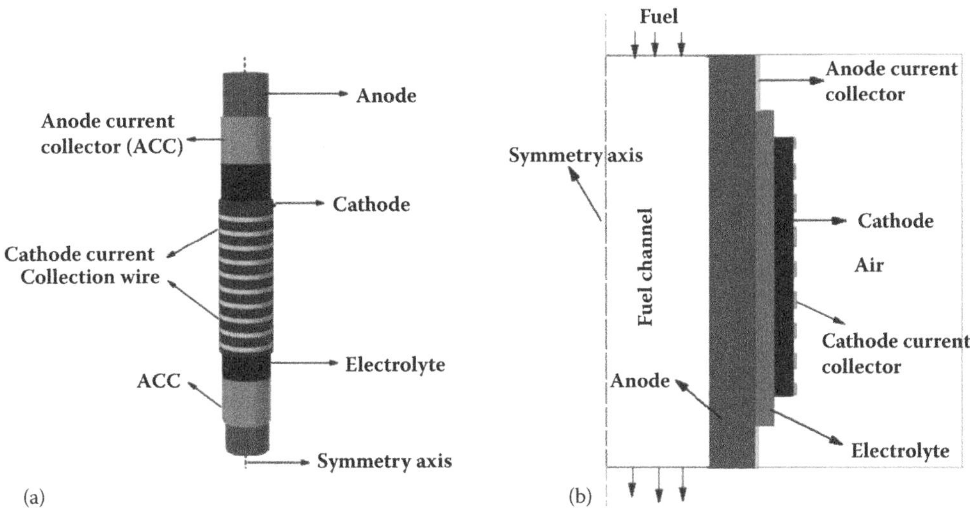

FIGURE 14.11 Model structure of mtSOFC: (a) 3D structure and (b) 2D computational modeling domain.

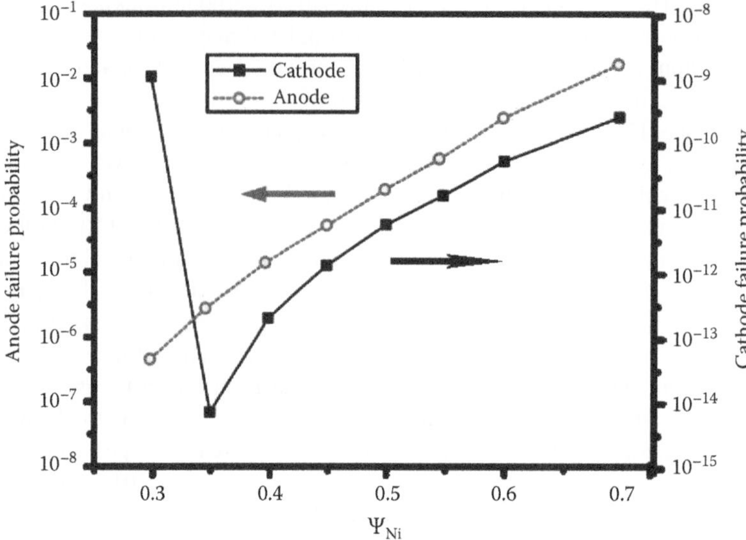

FIGURE 14.12 Dependence of the mechanical failure probabilities of the electrodes on the Ni volume fraction, Ψ_{Ni}, in the anode.

probability below 10^{-5} and the percolation threshold in conventional anode design, the suitable Ni content is determined to be in the range of 30–42 vol%. The result is in agreement with the conventional experimental wisdom, confirming the role of structural mechanical analysis in the design of SOFCs. Moreover, numerical results also show that the mechanical stability is not much affected by the LSM content. The LSM content should be determined by achieving high EC performance, and a range of 40–60 vol% is generally recommended.

As Ni particles tend to coarsen under the working temperature of SOFCs, the long-term performance stability of the Ni–anode is a serious concern. The long-term performance characteristics of SOFC has been analyzed with a theoretical model of Ni coarsening validated by experiments [65], and the above-described relationships between the effective properties and the electrode microstructure and composition. Figure 14.13 shows two representative results. For a conventional anode microstructure design with comparable sizes for Ni and YSZ particles, Figure 14.13a shows that a composition design with higher initial performance may degrade quickly with operational time and becomes inferior to a cell with lower initial performance. However, both higher initial performance and more stable long-term endurance may be obtained if the initial size of Ni particles is chosen to be about 30% smaller than that of YSZ particles, as shown in Figure 14.13b. Therefore, theoretical modeling can play an important role in guiding the design of high-performing and durable SOFCs.

There are rather a few experimental studies on the performance of SOFCs using dry methane as fuel. A few representative experimental setups may be described by Figure 14.3. Multiphysics 2D modeling has been used to examine the kinetics of methane steam reforming reaction under the condition of low steam content [75,76]. It is found that none of

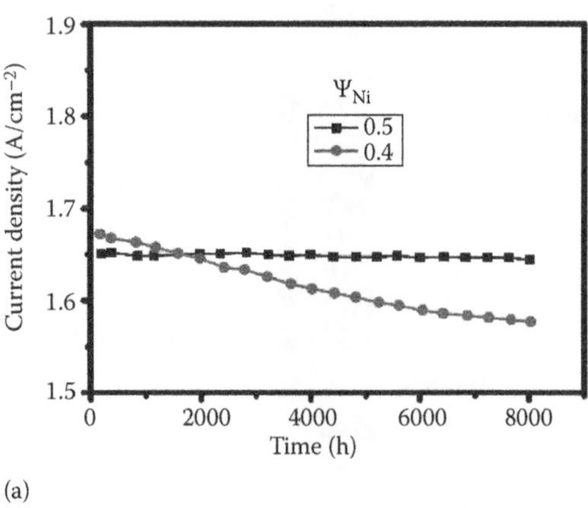

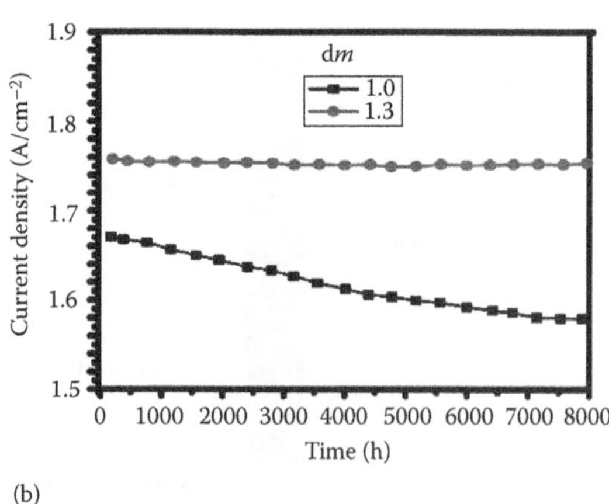

FIGURE 14.13 Effect of Ni coarsening on SOFC performances for (a) two volume fractions of Ni with the same YSZ and Ni particle size and (b) two size ratios of YSZ and Ni particles, d*m*, with the same volume fraction of Ni at 0.4.

the reforming kinetics expressions described in Section 14.2.3 may properly account for the experimental results about the dependence of OCV on the operating temperature [75]. The finding may be explained with the fact that the available kinetic expressions are deduced from the experiments conducted with much higher steam content and the reforming kinetics is strongly dependent on the steam content. The modeling results point out the need for more experimental study on the methane steam reforming kinetics if operating on dry methane fuel is a desired technology.

14.6 3D MODELING

A functional SOFC unit for practical application is an SOFC stack. A detailed knowledge about various factors affecting the behaviors of an SOFC stack is necessary for advancing the SOFC technology for market acceptance. Simplified 2D or lower-dimensional models are often incompetent for this purpose, and 3D modeling is required to gain the needed information. For example, it is important to know how to design the stack manifold so that each cell unit in the stack is supplied with an equal amount of fuel. It is also important to know if there is any hotspot or weak point that is prone to mechanical failure. In addition, it is helpful to know how the flow configuration designs of co-, counter-, and cross-flow affect the stack performance and temperature profile, etc. Adequate understanding of these critical design issues may be gained only through 3D modeling. In fact, 3D modeling may be desirable even at the cell component level. For example, a random packing of spherical particles and the associated percolation theory may be an oversimplified ideal model that deviates substantially from the true morphology of a porous electrode. A 3D microstructure reconstruction model is needed to provide an accurate description of physical parameters such as the particle size distribution, pore sizes, connectivity, effective TPB length, electrical conductivity, gas diffusivity, etc. By providing realistic property parameters, the 3D modeling enables the thermo–electrochemistry model to describe the cell behavior more accurately as well as assists the design of electrode microstructure for improved cell and stack performance. Clearly, 3D modeling is indispensable for the detailed knowledge of SOFC behaviors.

Insights critically important for stack designs have been obtained by 3D modeling [9,73,77–80]. For example, the cells in a stack are connected in series, and the total amount of current generation is the same for every cell. Consequently, the EC performance of an SOFC stack is constrained by the cell receiving the least amount of fuel/air supply. Similarly, the overall fuel/air utilization of the stack is limited by the least performing cell. Therefore, achieving flow supply uniformity is an important consideration for the stack flow manifold design. Only the manifold design with satisfactory flow uniformity is meaningful for practical applications. In such cases, all cells are operating under very similar conditions, if the boundary effect for a couple of bottom and top cells can be properly minimized. Consequently, a high flow uniformity achieved for a stack without current generation should also correspond to high flow uniformity for a working stack operated at the same temperature. In other words, the flow uniformity design may be determined by the numerical modeling of the flow field along.

CFD simulations have been employed to examine the effects of geometric design parameters along the flow path on the flow uniformity of pSOFC stacks [79]. The examined parameters include the channel height and length, the height of a unit cell, the inlet and outlet manifold widths, as well as the number of cells in the stack. The CFD results show that the flow uniformity generally decreases with the increased number of cells in the stack. While it is easy to obtain high flow uniformity for a short stack, say, a stack with fewer than 15 fuel cells, the flow uniformity can be very low for a naively designed long stack, say, a stack with more than 30 cells. The effect of the height of a repeating unit cell on the flow uniformity is very limited; however, the flow uniformity increases with the increased cell channel length or the reduced channel height. Increasing the cell length is not only beneficial for increasing the volumetric power density of the stack, but it is also helpful for improving the flow uniformity, and is highly recommended if allowed by the manufacturing technique. Reducing the cell channel height, however, has an undesirable effect of increased pressure drop, requiring an increased power demand of the gas pump and a reduced overall energy efficiency of the stack.

The ratio of the outlet manifold width, w_{out}, to the inlet manifold width, w_{in}, is found to be a key design parameter for the flow uniformity. Figure 14.14 illustrates the effects of w_{in} and w_{out} and the ratio, $\alpha = w_{out}/w_{in}$, on the flow uniformity of a 40-cell stack. The normalized mass flow rate for cell i is defined as $n_i = m_i / \sum_{j=1}^{N} m_j$, where m_i is the mass flow rate received by the ith cell and N is the number of cells in the stack. As seen in Figure 14.14, the flow uniformity for the conventional design ($\alpha = 1$) indicated in open literature is unsatisfactory, while the flow uniformity is much higher for $\alpha = 1.7$. The flow uniformity, $U_{flow} = \min(n_i)$, for $(w_{in}, \alpha) = (10\text{ mm}, 1)$, $(10\text{ mm}, 1.7)$, and $(17\text{ mm}, 1)$, is 0.35, 0.97, and 0.70, respectively. Moreover, it is noted that the pressure drop across a cell channel decreases substantially with the increased flow uniformity, as shown in Figure 14.14, another factor beneficial to the overall stack energy efficiency. It should also be noted that the flow uniformity may be improved by increasing w_{in} for the conventional design of $\alpha = 1$, but the improvement to $U_{flow} \geq 0.9$ may require an impractically large w_{in}. Clearly, choosing a proper α is an easy way to achieve high flow uniformity and low pressure drop. The optimal value of α is found to have a simple linear relation with the number of cells in the stack [79], with all others being equal. The optimal α is further found to be equally applicable to the manifold designs with two or three flow inlets, representing a robust result for guiding the experimental design.

Applications of 3D modeling with coupled electrochemistry and thermofluid models have been demonstrated [9,73,77]. To limit the demand of computational resource to

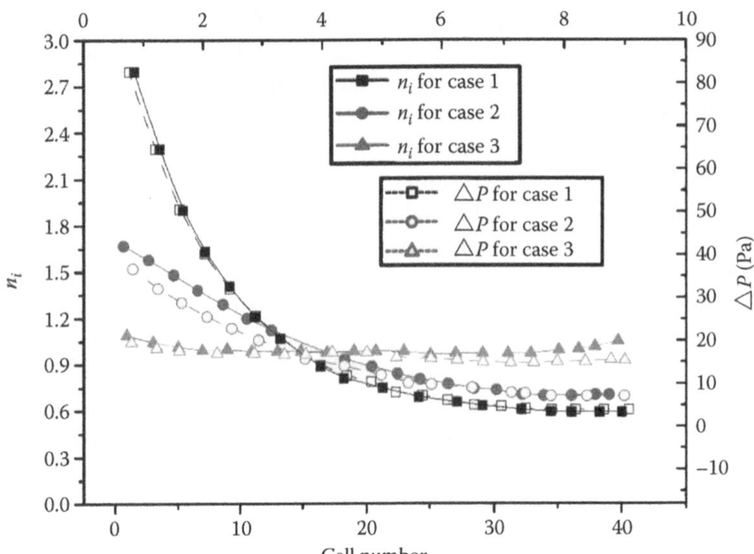

FIGURE 14.14 Distributions of the normalized air mass-flow rate, n_i, and the pressure drops across fuel cell gas channels, ΔP, in a 40-cell stack with three geometric designs. (w_{in}, α) = (10 mm, 1), (10 mm, 1.7), and (17 mm, 1) correspond to cases 1, 2, and 3, respectively.

an acceptable level, the electrochemistry model is usually simplified into some parameterized form, while the distributions of model quantities over the cell plate are supplied through the coupled thermofluid model. Moreover, the PEN element is often treated as a single material unit with appropriate chosen equivalent material properties [9,77]. Alternatively, a detailed electrochemistry model may be used, but the coupled electrochemistry and thermofluid model is applied to a representative portion of a cell [73]. Multiphysics 3D modeling has been used to examine and compare the performances of cells with co-, counter-, and cross-flow designs, and the performances of different cells in a stack. Detailed information about the distributions of temperature, current density, electrical potential, flow field, and gas species' concentration are obtained. The temperature profile is then used to calculate the mechanical stress distribution. Figure 14.15 shows some sample results for the stack temperature profile and mechanical stress distribution for a stack with a cross-flow configuration. The local temperature near some air channel exit region can be high, and

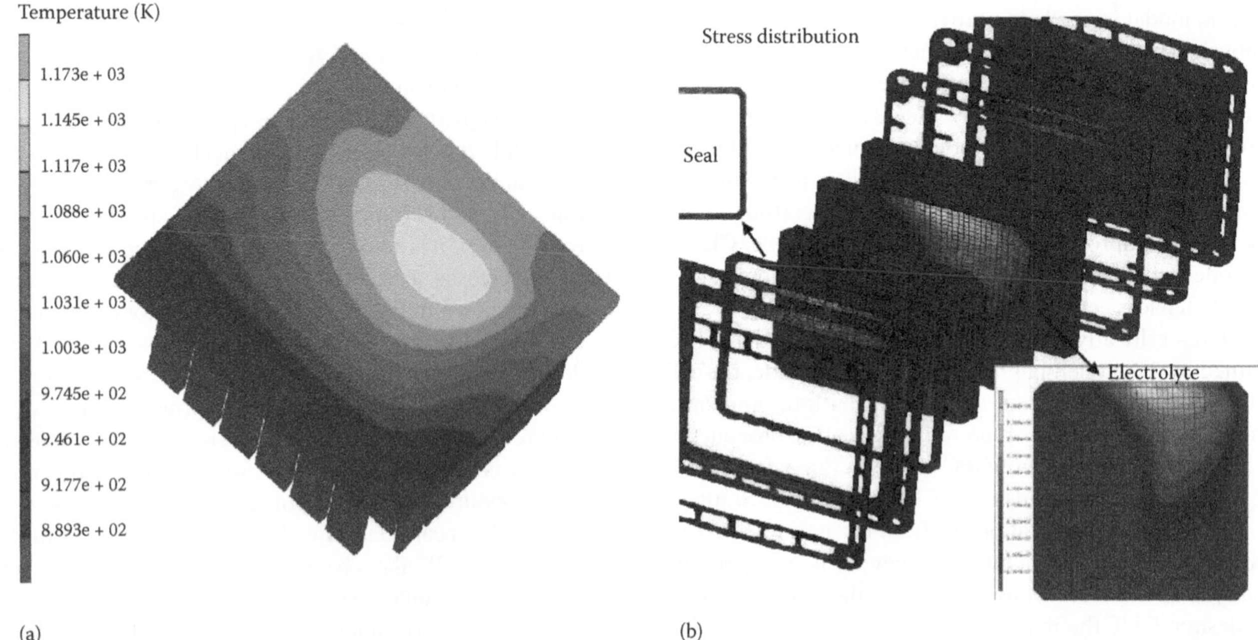

FIGURE 14.15 Sample temperature and stress results of a stack with a cross-flow design: (a) stack temperature profile and (b) mechanical stress distribution.

the thermal mechanical stress at that region is correspondingly high. The mechanical strengths of materials at that region are vital for the survival and durability of the SOFC stack. In particular, the mechanical strength of sealing material around the hotspot area should be properly enhanced. Moreover, modeling results show that the co-flow and counterflow configurations are often favorable for higher current output and better temperature uniformity, although the cross-flow design is preferred from the engineering point of view perceived initially. To realize the benefit of co-flow and counterflow configurations, manifold designs have been explored. A manifold design indicated in Figure 14.2c has thus been proposed. With this type of manifold design, it appears that the co-flow and counterflow configurations are equally easy to manufacture as the counterpart of cross-flow configuration. It is hoped that the engineering community will examine the manifold design so that the concern of mechanical failure associated with the thermal stress may be alleviated.

3D modeling may also be used to guide the design of some auxiliary components of an SOFC system. A coupled 3D thermofluid–thermomechanical model has been developed for the design and process optimization of a plate type air preheater [78]. Numerical results for the temperature profile have been validated by designated experimental measurements. The model is capable of identifying the component locations that are susceptible to high thermal stress. Consequently, the model should be able to reduce the prototyping cost and product development time in the design of efficient air preheater.

14.7 SUMMARY AND CONCLUDING REMARKS

We have described the fundamental principles for the continuum level of modeling of the steady states of SOFCs. The modeling approaches are categorized according to the dimensionalities of the models. Modeling analysis are presented to demonstrate the role of theoretical modeling played in the understanding of SOFCs that may also be used for directing improved designs of high performance SOFCs.

3D multiphysics modeling can provide important information about the interplay of various factors affecting the behaviors of SOFCs that are difficult to grasp with a lower-dimensional modeling. Owing to the inherent complexity of SOFC processes and the high computing power requirement, truly 3D multiphysics modeling is rare and at an immature stage. Great efforts are required to develop improved numerical algorithms so that 3D multiphysics modeling may be performed routinely. With the hard work of researchers in the field and the rapidly growing computer power, frequent occurrences of full-scale 3D multiphysics modeling may be expectable in the coming decade.

However, a full-scale 3D modeling will never be fast enough for many important applications, e.g., as a module for dynamic response control. Nevertheless, in addition to dedicated case studies, 3D modeling can be extremely useful for constructing improved lower-dimensional but much faster models by providing informed choice of reasonable approximations. A similar observation also applies to other nonzero-dimensional models so that simpler but reasonably accurate models may be developed for the broad use of the SOFC community. That is, developing accurate lower-dimensional models based on the knowledge provided from higher-dimensional models should also be a future research direction for theoretical modelers. This assessment is made for both steady-state and transient-state modeling.

As mentioned above, atomistic modeling is important for the in-depth understanding of SOFCs. Atomistic modeling results are also helpful for guiding the choice of property parameters used in the continuum models. Hence, more atomistic modeling and more coordination between atomistic modeling and continuum modeling are needed.

Many insights revealed by theoretical modeling are very helpful for the experimentalists. Conversely, theoreticians also need the help from the experimentalists by providing property parameters through dedicated measurements. Some parameters needed by theoretical model have been measured, but not published in the open literature. Others may require great effort on the part of experimentalists. This is true especially for those regarding long-term behaviors of materials. However, using reliable model parameters is essential for the success of the model predictions. Enhanced collaboration between theoretician and experimentalists is required for rapidly advancing the development of SOFC technology.

REFERENCES

1. Choudhury, A., Chandra, H., & Arora, A. 2013. Application of solid oxide fuel cell technology for power generation—A review. *Renewable and Sustainable Energy Reviews*, 20, 430–442.
2. Mogensen, D., Grunwaldt, J. D., Hendriksen, P. V., Dam-Johansen, K., & Nielsen, J. U. 2011. Internal steam reforming in solid oxide fuel cells: Status and opportunities of kinetic studies and their impact on modelling. *Journal of Power Sources*, 196, 25–38.
3. Jiang, S. P., & Chen, X. 2014. Chromium deposition and poisoning of cathodes of solid oxide fuel cells—A review. *International Journal of Hydrogen Energy*, 39, 505–531.
4. Shri Prakash, B., Senthil Kumar, S., & Aruna, S. T. 2014. Properties and development of Ni/YSZ as an anode material in solid oxide fuel cell: A review. *Renewable and Sustainable Energy Reviews*, 36, 149–179.
5. Shaigan, N., Qu, W., Ivey, D. G., & Chen, W. 2010. A review of recent progress in coatings, surface modifications and alloy developments for solid oxide fuel cell ferritic stainless steel interconnects. *Journal of Power Sources*, 195, 1529–1542.
6. Tucker, M. C. 2010. Progress in metal-supported solid oxide fuel cells: A review. *Journal of Power Sources*, 195, 4570–4582.
7. Liu, M., & Aravind, P. V. 2014. The fate of tars under solid oxide fuel cell conditions: A review. *Applied Thermal Engineering*, 70, 687–693.
8. Deleebeeck, L., & Hansen, K. K. 2014. Hybrid direct carbon fuel cells and their reaction mechanisms—A review. *Journal of Solid State Electrochemistry*, 18, 861–882.

9. Khaleel, M. A., Lin, Z., Singh, P., Surdoval, W., & Collin, D. 2004. A finite element analysis modeling tool for solid oxide fuel cell development: Coupled electrochemistry, thermal and flow analysis in MARC. *Journal of Power Sources*, 130, 136–148.
10. Wang, K., Hissel, D., Péra, M. C., Steiner, N., Marra, D., Sorrentino, M., Pianese, C., Monteverde, M., Cardone, P., & Saarinen, J. 2011. A review on solid oxide fuel cell models. *International Journal of Hydrogen Energy*, 36, 7212–7228.
11. Andersson, M., Yuan, J., & Sundén, B. 2010. Review on modeling development for multiscale chemical reactions coupled transport phenomena in solid oxide fuel cells. *Applied Energy*, 87, 1461–1476.
12. Hajimolana, S. A., Hussain, M. A., Daud, W. M., Soroush, M., & Shamiri, A. 2011. Mathematical modeling of solid oxide fuel cells: A review. *Renewable and Sustainable Energy Reviews*, 15, 1893–1917.
13. Grew, K. N., & Chiu, W. K. 2012. A review of modeling and simulation techniques across the length scales for the solid oxide fuel cell. *Journal of Power Sources*, 199, 1–13.
14. Huang, B., Qi, Y., & Murshed, M. 2011. Solid oxide fuel cell: Perspective of dynamic modeling and control. *Journal of Process Control*, 21, 1426–1437.
15. Kakac, S., Pramuanjaroenkij, A., & Zhou, X. Y. 2007. A review of numerical modeling of solid oxide fuel cells. *International Journal of Hydrogen Energy*, 32, 761–786.
16. Bavarian, M., Soroush, M., Kevrekidis, I. G., & Benziger, J. B. 2010. Mathematical modeling, steady-state and dynamic behavior, and control of fuel cells: A review. *Industrial & Engineering Chemistry Research*, 49, 7922–7950.
17. Colpan, C. O., Dincer, I., & Hamdullahpur, F. 2008. A review on macro-level modeling of planar solid oxide fuel cells. *International Journal of Energy Research*, 32, 336–355.
18. Achenbach, E. 1994. Three-dimensional and time-dependent simulation of a planar solid oxide fuel cell stack. *Journal of Power Sources*, 49, 333–348.
19. Padulles, J., Ault, G. W., & McDonald, J. R. 2000. An integrated SOFC plant dynamic model for power systems simulation. *Journal of Power Sources*, 86, 495–500.
20. Ota, T., Koyama, M., Wen, C. J., Yamada, K., & Takahashi, H. 2003. Object-based modeling of SOFC system: Dynamic behavior of micro-tube SOFC. *Journal of Power Sources*, 118, 430–439.
21. Sedghisigarchi, K., & Feliachi, A. 2004. Dynamic and transient analysis of power distribution systems with fuel cells—Part I: Fuel-cell dynamic model. *IEEE Transactions on Energy Conversion*, 19, 423–428.
22. Xue, X., Tang, J., Sammes, N., & Du, Y. 2005. Dynamic modeling of single tubular SOFC combining heat/mass transfer and EC reaction effects. *Journal of Power Sources*, 142, 211–222.
23. Jiang, W., Fang, R., Khan, J. A., & Dougal, R. A. 2006. Parameter setting and analysis of a dynamic tubular SOFC model. *Journal of Power Sources*, 162, 316–326.
24. Qi, Y., Huang, B., & Luo, J. 2008. 1-D dynamic modeling of SOFC with analytical solution for reacting gas-flow problem. *AIChE Journal*, 54, 1537–1553.
25. Gebregergis, A., Pillay, P., Bhattacharyya, D., & Rengaswemy, R. 2009. Solid oxide fuel cell modeling. *IEEE Transactions on Industrial Electronics*, 56, 139–148.
26. Frayret, C., Villesuzanne, A., Pouchard, M., & Matar, S. 2005. Density functional theory calculations on microscopic aspects of oxygen diffusion in ceria-based materials. *International Journal of Quantum Chemistry*, 101, 826–839.
27. Jones, G., Jakobsen, J. G., Shim, S. S., Kleis, J., Andersson, M. P., Rossmeisl, J., Abild-Pedersen, F., Bligaard, T., Helveg, S., Hinnemann, B., Rostrup-Nielsen, J. R., Chorkendorff, I., Sehested, J., & Nørskov, J. K. 2008. First principles calculations and experimental insight into methane steam reforming over transition metal catalysts. *Journal of Catalysis*, 259, 147–160.
28. Vogler, M., Bieberle-Hütter, A., Gauckler, L., Warnatz, J., & Bessler, W. G. 2009. Modelling study of surface reactions, diffusion, and spillover at a Ni/YSZ patterned anode. *Journal of the Electrochemical Society*, 156, B663–B672.
29. Blaylock, D. W., Ogura, T., Green, W. H., & Beran, G. J. 2009. Computational investigation of thermochemistry and kinetics of steam methane reforming on Ni (111) under realistic conditions. *The Journal of Physical Chemistry C*, 113, 4898–4908.
30. Kim, J. W., Virkar, A. V., Fung, K. Z., Mehta, K., & Singhal, S. C. 1999. Polarization effects in intermediate temperature, anode-supported solid oxide fuel cells. *Journal of the Electrochemical Society*, 146, 69–78.
31. Liu, S., Song, C., & Lin, Z. 2008. The effects of the interconnect rib contact resistance on the performance of planar solid oxide fuel cell stack and the rib design optimization. *Journal of Power Sources*, 183, 214–225.
32. Chen, D., Bi, W., Kong, W., & Lin, Z. 2010. Combined micro-scale and macro-scale modeling of the composite electrode of a solid oxide fuel cell. *Journal of Power Sources*, 195, 6598–6610.
33. Li, J., Kong, W., & Lin, Z. 2013. Theoretical studies on the electrochemical and mechanical properties and microstructure optimization of micro-tubular solid oxide fuel cells. *Journal of Power Sources*, 232, 106–122.
34. Cussler, E. L. 1984. *Diffusion: Mass Transfer in Fluid Systems*. New York: Cambridge University Press.
35. Bird, R. B., Stewart, W. E., & Lightfoot, E. N. 1960. *Transport Phenomena*. Madison, USA: John Wiley & Sons.
36. Incropera, F. P., & Dewitt, D. P. 2007. *Fundamentals of Heat and Mass Transfer*, 572–575. Singapore: Wiley & Sons.
37. Krishna, R., & Wesselingh, J. A. 1997. The Maxwell–Stefan approach to mass transfer. *Chemical Engineering Science*, 52, 861–911.
38. Bird, R. B., Stewart, W. E., & Lightfoot, E. N. (Eds.). 2002. *Transport Phenomenon*, 2nd ed. New York: John Wiley & Sons.
39. Mason, E. A., & Malinauskas, A. P. 1983. *Transport in Porous Media: The Dusty Gas Model*. New York: Elsevier.
40. Kong, W., Zhu, H., Fei, Z., & Lin, Z. 2012. A modified dusty gas model in the form of a Fick's model for the prediction of multicomponent mass transport in a solid oxide fuel cell anode. *Journal of Power Sources*, 206, 171–178.
41. Nagel, F. P., Schildhauer, T. J., Biollaz, S., & Stucki, S. 2008. Charge, mass and heat transfer interactions in solid oxide fuel cells operated with different fuel gases—A sensitivity analysis. *Journal of Power Sources*, 184, 129–142.
42. Serincan, M. F., Pasaogullari, U., & Sammes, N. M. 2010. Thermal stresses in an operating micro-tubular solid oxide fuel cell. *Journal of Power Sources*, 195, 4905–4914.
43. Hashin, Z., & Shtrikman, S. 1963. A variational approach to the theory of the elastic behaviour of multiphase materials. *Journal of the Mechanics and Physics of Solids*, 11, 127–140.
44. Hull, D., & Clyne, T. W. 1996. *An Introduction to Composite Materials*. New York: Cambridge University Press.
45. Vaßen, R., Czech, N., Mallener, W., Stamm, W., & Stöver, D. 2001. Influence of impurity content and porosity of plasma-sprayed yttria-stabilized zirconia layers on the sintering behaviour. *Surface and Coatings Technology*, 141, 135–140.

46. Green, D. J. 1998. *An Introduction to the Mechanical Properties of Ceramics*. New York: Cambridge University Press.
47. Laurencin, J., Delette, G., Lefebvre-Joud, F., & Dupeux, M. 2008. A numerical tool to estimate SOFC mechanical degradation: Case of the planar cell configuration. *Journal of the European Ceramic Society*, 28, 1857–1869.
48. Weibull, W. 1939. A statistical theory of the strength of metals. *Proceedings of the Royal Swedish Institute of Engineering Research*, 151, 1–45.
49. Cui, D., & Cheng, M. 2009. Thermal stress modeling of anode supported micro-tubular solid oxide fuel cell. *Journal of Power Sources*, 192, 400–407.
50. Zhao, F., & Virkar, A. V. 2005. Dependence of polarization in anode-supported solid oxide fuel cells on various cell parameters. *Journal of Power Sources*, 141, 79–95.
51. Chen, D., Lin, Z., Zhu, H., & Kee, R. J. 2009. Percolation theory to predict effective properties of solid oxide fuel-cell composite electrodes. *Journal of Power Sources*, 191, 240–252.
52. Bouvard, D., & Lange, F. F. 1991. Relation between percolation and particle coordination in binary powder mixtures. *Acta Metallurgica et Materialia*, 39, 3083–3090.
53. Suzuki, M., & Oshima, T. 1985. Comparison between the computer-simulated results and the model for estimating the co-ordination number in a three-component random mixture of spheres. *Powder Technology*, 43, 19–25.
54. Costamagna, P., Costa, P., & Antonucci, V. 1998. Micro-modelling of solid oxide fuel cell electrodes. *Electrochimica Acta*, 43, 375–394.
55. Sanyal, J., Goldin, G. M., Zhu, H., & Kee, R. J. 2010. A particle-based model for predicting the effective conductivities of composite electrodes. *Journal of Power Sources*, 195, 6671–6679.
56. Park, S., Craciun, R., Vohs, J. M., & Gorte, R. J. 1999. Direct oxidation of hydrocarbons in a solid oxide fuel cell: I. Methane oxidation. *Journal of the Electrochemical Society*, 146, 3603–3605.
57. Vohs, J. M., & Gorte, R. J. 2009. High-performance SOFC cathodes prepared by infiltration. *Advanced Materials*, 21, 943–956.
58. Gorte, R. J., & Vohs, J. M. 2009. Nanostructured anodes for solid oxide fuel cells. *Current Opinion in Colloid & Interface Science*, 14, 236–244.
59. Jiang, S. P. 2012. Nanoscale and nano-structured electrodes of solid oxide fuel cells by infiltration: Advances and challenges. *International Journal of Hydrogen Energy*, 37, 449–470.
60. Wilson, J. R., Kobsiriphat, W., Mendoza, R., Chen, H. Y., Hiller, J. M., Miller, D. J., & Barnett, S. A. 2006. Three-dimensional reconstruction of a solid-oxide fuel-cell anode. *Nature Materials*, 5, 541–544.
61. Zhan, Z., Bierschenk, D. M., Cronin, J. S., & Barnett, S. A. 2011. A reduced temperature solid oxide fuel cell with nanostructured anodes. *Energy & Environmental Science*, 4, 3951–3954.
62. Chen, M., Liu, T., & Lin, Z. 2013. Theory for the electrical conductivity of nano-particle-infiltrated composite electrode of solid oxide fuel cell. *ECS Electrochemistry Letters*, 2, F82–F84.
63. Chen, M., Song, C., & Lin, Z. 2014. Property models and theoretical analysis of novel solid oxide fuel cell with triplet nano-composite electrode. *International Journal of Hydrogen Energy*, 39, 13763–13769.
64. Chen, M., & Lin, Z. 2014. Theoretical models for effective electrical and electrochemical properties of nano-particle infiltrated electrode of solid oxide fuel cell. *International Journal of Hydrogen Energy*, 39, 15982–15988.
65. Gao, S., Li, J., & Lin, Z. 2014. Theoretical model for surface diffusion driven Ni-particle agglomeration in anode of solid oxide fuel cell. *Journal of Power Sources*, 255, 144–150.
66. Wang, B., & Lin, Z. 2014. A Schottky barrier based model for the grain size effect on oxygen ion conductivity of acceptor-doped ZrO_2 and CeO_2. *International Journal of Hydrogen Energy*, 39, 14334–14341.
67. Guo, X., & Zhang, Z. 2003. Grain size dependent grain boundary defect structure: Case of doped zirconia. *Acta Materialia*, 51, 2539–2547.
68. Chen, X. J., Khor, K. A., Chan, S. H., & Yu, L. G. 2002. Influence of microstructure on the ionic conductivity of yttria-stabilized zirconia electrolyte. *Materials Science and Engineering A*, 335, 246–252.
69. Jeon, D. H., Nam, J. H., & Kim, C. J. 2006. Micro-structural optimization of anode-supported solid oxide fuel cells by a comprehensive micro-scale model. *Journal of the Electrochemical Society*, 153, A406–A417.
70. Chung, B. W., Chervin, C. N., Haslam, J. J., Pham, A. Q., & Glass, R. S. 2005. Development and characterization of a high performance thin-film planar SOFC stack. *Journal of the Electrochemical Society*, 152, A265–A269.
71. Jung, H. Y., Choi, S. H., Kim, H., Son, J. W., Kim, J., Lee, H. W., & Lee, J. H. 2006. Fabrication and performance evaluation of 3-cell SOFC stack based on planar 10 cm × 10 cm anode-supported cells. *Journal of Power Sources*, 159, 478–483.
72. Kong, W., Li, J., Liu, S., & Lin, Z. 2012. The influence of interconnect ribs on the performance of planar solid oxide fuel cell and formulae for optimal rib sizes. *Journal of Power Sources*, 204, 106–115.
73. Liu, S., Kong, W., & Lin, Z. 2009. Three-dimensional modeling of planar solid oxide fuel cells and the rib design optimization. *Journal of Power Sources*, 194, 854–863.
74. Li, J., & Lin, Z. 2012. Effects of electrode composition on the electrochemical performance and mechanical property of micro-tubular solid oxide fuel cell. *International Journal of Hydrogen Energy*, 37, 12925–12940.
75. Liu, J., & Barnett, S. A. 2003. Operation of anode-supported solid oxide fuel cells on methane and natural gas. *Solid State Ionics*, 158, 11–16.
76. Lin, Y., Zhan, Z., & Barnett, S. A. 2006. Improving the stability of direct-methane solid oxide fuel cells using anode barrier layers. *Journal of Power Sources*, 158, 1313–1316.
77. Recknagle, K. P., Williford, R. E., Chick, L. A., Rector, D. R., & Khaleel, M. A. 2003. Three-dimensional thermo-fluid electrochemical modeling of planar SOFC stacks. *Journal of Power Sources*, 113, 109–114.
78. Peksen, M., Peters, R., Blum, L., & Stolten, D. 2011. 3D coupled CFD/FEM modelling and experimental validation of a planar type air pre-heater used in SOFC technology. *International Journal of Hydrogen Energy*, 36, 6851–6861.
79. Bi, W., Chen, D., & Lin, Z. 2009. A key geometric parameter for the flow uniformity in planar solid oxide fuel cell stacks. *International Journal of Hydrogen Energy*, 34, 3873–3884.
80. Bi, W., Li, J., & Lin, Z. 2010. Flow uniformity optimization for large size planar solid oxide fuel cells with U-type parallel channel designs. *Journal of Power Sources*, 195, 3207–3214.

15 Advanced Membrane Materials for Polymer Electrolyte Membrane Fuel Cells

David Aili, Jens Oluf Jensen, and Qingfeng Li

CONTENTS

15.1 INTRODUCTION

15.1.1 Polymer Electrolyte Membranes for Fuel Cells

Polymer electrolyte membrane (PEM)-based fuel cells are emerging as attractive energy conversion systems suitable for use in a wide range of applications from portable to stationary and automotive systems. After the pioneering work by Bacon on fuel cells with alkaline electrolyte shortly after World War II [1], acidic polymer membranes were first developed as electrolytes by Grubb and Niedrach [2] at General Electric in the earlier 1960s. The first membrane types were based on poly(phenolformaldehyde sulfonic acid) resins; however, the membranes were brittle and cracked in the dry state and were susceptible to hydrolysis [3]. The first membrane that exhibited sufficiently promising characteristics was made of poly(styrenesulfonic acid) (PSSA) and showed a lifetime of about 200 h at 60°C [4,5]. By cross-linking the PSSA into an inert fluorocarbon matrix for mechanical reinforcement, the sulfonated polymer membranes exhibited acceptable strength in both the wet and the dry state. The PSSA-based composite membranes were used in several Gemini space missions during the 1960s [6]. This type of composite membrane, however, suffered from lifetime limitations due to oxidation of the active phase during operation. With the poly(perfluorosulfonic acid) (PFSA) ionomer membranes, which found their first applications in the chloralkali process [7,8], breakthroughs were achieved in the PEM fuel cell development. Fuel cells based on this membrane type powered the Biosatellite mission in later 1960s and led to the renaissance of the polymer fuel cells from the late 1980s [8,9].

PFSA-based membranes exhibit good proton conductivity, chemical and mechanical stability, as well as high tear resistance and low gas permeability under the normal fuel cell operating conditions. However, a number of technical challenges are associated with this particular membrane type when used in fuel cells. Primarily, these include the high cost of the ionomer and especially the noble metal–based electrode materials. In addition, the narrow range of operational temperatures makes the thermal and water management critical. High purity of the fuel is required because of the risk of poisoning of the platinum electrocatalysts, especially by carbon monoxide, which is a by-product of hydrogen production via reforming of carbon-containing fuels.

Researchers worldwide have reported success in exploring new concepts for improving the properties of proton conducting membranes for fuel cells, which will be reviewed and summarized in this chapter.

15.1.2 Basic Requirements for Polymer Electrolyte Membranes

The type of electrolytes in fuel cells determines the operating conditions of the system as well as the electrochemistry at the electrodes. It is also after the type of electrolytes that the fuel cell types are named, which indicates that the electrolyte is the key material that determines the configuration and operational features of a fuel cell. The different requirements on the electrolyte are more or less linked to each other, and they are in many ways rather extensive. For the PEMs, the most important basic requirements are listed in Table 15.1.

15.1.3 Types of Polymer Electrolyte Membranes and Scope of Chapter

The sulfonated fluoropolymer membranes of the PFSA type have emerged as the benchmark materials for PEM fuel cells, showing performance that can be hardly matched by any alternative cation-exchange polymer membrane. Mainly motivated by the potential cost reductions, much research effort has been devoted to the development of alternative sulfonated materials based on partially fluorinated polymers or completely aromatic systems.

The proton transport in the membranes based on the sulfonated polymers is water mediated [10,11]. Careful control of the water balance in the cell is thus required, and it also limits the operating temperature to about 80°C. However, several critical challenges for the PEM fuel cell technology are directly associated with this low operating temperature, and it has been recognized that increased operating temperatures offer potential advantages from the kinetic and engineering points of view [12]. This has triggered the development of membranes in which the proton transport is not water mediated. The use of phosphoric acid–imbibed polymers, especially in the polybenzimidazole (PBI) family, has been demonstrated as one of the most promising approaches, and with this concept the operational temperature range of 120–200°C can be achieved. Much research effort has been devoted to the high-temperature PEM fuel cell technology during the last two decades, and it has now been developed to a stage where it is available on a commercial basis.

This chapter is devoted to a review of status and technical characteristics of five types of PEMs, starting with acidic polymer electrolyte fuel cell membranes in which the proton transport is water mediated, including PFSA, sulfonated partially fluorinated membranes, and aromatic hydrocarbon–based membranes. This is followed by an update on the acid–base or acid-doped electrolyte systems for operation

TABLE 15.1
Summary of the Fundamental Technological Requirements on the Polymer Electrolyte Membrane for Fuel Cell Applications

Requirement	Motivation
High proton conductivity	High proton conductivity is required to minimize the ohmic voltage losses, especially at high current loads.
Low electronic conductivity	The membrane has to be electronically insulating to avoid short circuiting of the cell.
Low reactant permeability	Reactant crossover results in poor fuel utilization and voltage losses due to mixed electrode potentials.
Mechanical strength and flexibility	Thermal cycling and humidity variations induce stresses in contact points between different cell components, which ultimately might lead to physical membrane failure.
Stability	The membrane has to be thermally stable and tolerant to oxidative and reductive environments, potential gradients, and extreme pH.
Compatibility	The membrane must be chemically compatible with the electrode materials and the fuel.
Availability and cost effectiveness	A necessity for mass production.

at temperatures in the 120–200°C range. Recent progress in alkaline membranes is then briefly discussed.

15.2 POLY(PERFLUOROSULFONIC ACID) MEMBRANES AND RELATED TECHNOLOGIES

15.2.1 Poly(perfluorosulfonic Acid) Membranes

15.2.1.1 Structure and Synthesis

Following the pioneering work on membranes based on sulfonated polystyrene by Grubb and Niedrach [2], the development of fuel cells with acidic polymer-based electrolytes was intensified in the late 1960s when the first PFSA membranes were launched by DuPont under the trade name Nafion® [13]. The PFSA membranes comprise a group of cation-exchange materials with densities normally around 2 g cm^{-3} and consist of a perfluorinated backbone with perfluorinated side chains containing terminal sulfonic acid groups, as shown in Figure 15.1.

The PFSAs are copolymers synthesized from tetrafluoroethylene (TFE) and a perfluorinated monomer containing the sulfonic acid functionality. The structure of the latter monomer can vary and determines the length of the sulfonic acid terminated side chains, and the ratio between the two co-monomers determine the concentration of functional groups (ion-exchange capacity [IEC] or equivalent weight, treated later) of the resulting polymer. The synthesis of PFSA-type materials requires highly dedicated process equipment and due to safety reasons, the use of TFE, which easily forms explosive peroxides in contact with air, is strongly regulated by the authorities in many countries. Most of the research and development on PFSA materials is thus carried out in industrial laboratories, and the literature describing the fundamental polymer chemistry is rather scarce. However, the synthetic approach of the co-monomer bearing the sulfonic acid functionality, which is subsequently copolymerized with TFE to give Nafion, is schematically illustrated in Figure 15.2.

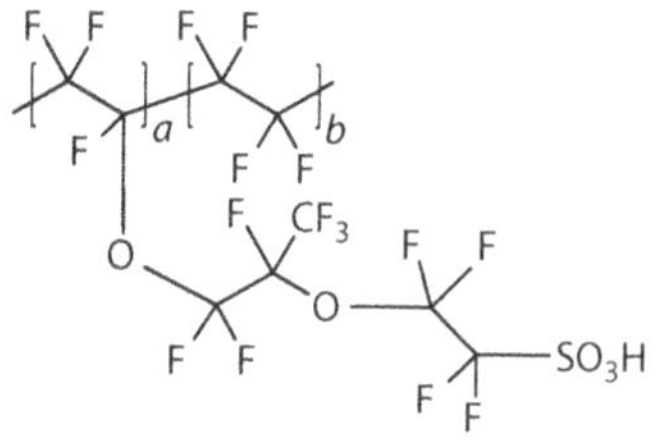

FIGURE 15.1 Chemical structure of Nafion.

15.2.1.2 Water Uptake and Morphology

When PFSA membranes are soaked in water or when subjected to an atmosphere with high relative humidity, a considerable water uptake occurs. During this process, the hydrophobic perfluorinated polymer backbones, on one hand, aggregate in regions with varying degrees of crystallinity, providing the structural integrity to the membrane. The hydrophilic acidic groups, on the other hand, form water-filled percolating ionic domains with extreme local acidity, providing pathways for proton conduction with water molecules as charge carriers, which will be discussed in more detail in Section 15.2.1.3.

There are many ways of representing the water content of PFSA membranes. The most straightforward is to report the water uptake in percent on the dry membrane basis or as the total water content of the wet membrane. However, it is often more convenient to normalize the water content as the degree of hydration, λ, defined as the number of water molecules per sulfonic acid group. Sometimes, the water content is reported as a volume fraction of the total membrane volume, calculated on the basis of the densities of the particular PFSA material and water. The water uptake of a PFSA membrane is strongly dependent on the thermal history and the IEC of the membrane, defined as mole acid equivalents per gram of polymer (eq. g^{-1}), or the equivalent weight (EW), defined as the average molecular weight of the polymer per mole sulfonic acid moieties (g $eq.^{-1}$). Typically for a Nafion 117 membrane, which has an equivalent weight of 1100 g $eq.^{-1}$ and a thickness of 175 μm (7 mil, 7 hundreds of an inch) in the dry state, a degree of hydration of about 19 is obtained after treatment in water or aqueous dilute sulfuric acid at 80°C for 2 h [14,15]. It corresponds to a water uptake of 31 wt.%, a total water content of 24 wt.%, or a water volume fraction of about 0.39.

Numerous studies on the nanomorphology of PFSA membranes have been carried out throughout the years after the famous cluster–network model was proposed based on x-ray diffraction studies by Gierke et al. in the early 1980s [16]. The parallel cylinder [17] and fibrillar bundle [18] models were later proposed, which have been further investigated by simulations [19] and nuclear magnetic resonance (NMR) studies [20]. However, a recent study points out the relatively large uncertainties in the experimentally determined water contents. On the basis of extensive small-angle x-ray (SAXS) studies on membranes with carefully controlled water contents, it is suggested that the water domains in PFSA membranes are locally flat and narrow [15]. High-resolution atomic force microscopy is another technique that can be employed

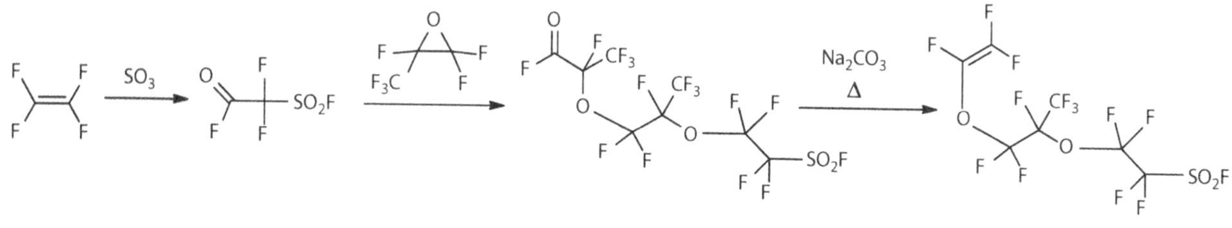

FIGURE 15.2 Schematic illustration of the synthesis of the co-monomer bearing the sulfonic acid precursor functionality in Nafion. (From Banerjee, S. and Curtin, D.E., *J. Fluor. Chem.*, 125, 1211, 2004.)

FIGURE 15.3 Two-dimensional schematic illustration of the percolation in a hydrated PFSA membrane. The gray strands represent the hydrophobic polymer backbones, while the orange and green circles represent water molecules and sulfonic acid groups, respectively.

to image the surface morphology of PFSA membranes on the nanoscale [18,21]. A schematic illustration of the nanostructure of a hydrated PFSA membrane, aiming at visualizing the hydrophobic/hydrophilic separation and the percolating network of water-filled domains, is shown in Figure 15.3.

15.2.1.3 Proton Conductivity

In PFSA membranes as well in alternative sulfonated polymers, the proton transport is mediated by the water molecules within the membrane. The water molecules solvate the protons originating from the sulfonic acid groups to form hydrated complexes such as hydronium cations (H_3O^+), Zundel cations ($H_5O_2^+$), or Eigen cations ($H_9O_4^+$) that are mobile and transport the protons according to the vehicle mechanism [22]. The process involves a parasitic transport of water molecules [23] of about 1–3 water molecules per proton during fuel cell operation and depending on the total water content of the membrane [24,25]. This process is accompanied by a continuous diffusive counterflow of water in the opposite direction. Water has a rather high self-diffusion coefficient of 10^{-7}–10^{-5} cm^2 s^{-1} in Nafion depending on the water content [26,27], which is essential for the maintenance of the water balance throughout the membrane cross-section during fuel cell operation.

Under optimal conditions at ambient pressure, when the degree of hydration is higher than, say 15, and at temperatures around 80°C, the conductivity of PFSA membranes can reach values above 0.1 S cm^{-1} [27]. The proton conductivity of PFSA materials at a certain degree of hydration generally increases with increasing IEC or with decreasing EW. However, increasing the IEC or decreasing the EW results in enhanced water uptake, extensive volume swelling, and decreased mechanical strength, which may ultimately lead to a complete dissolution of the membrane in water. Thus, a compromise between high IEC and good mechanical properties in the hydrated state has to be established.

15.2.1.4 Permeability and Solubility

The gas permeability of hydrogen and oxygen in dry Nafion has been determined to be 4.1×10^{-17} and 2.8×10^{-17} mol cm^{-1} s^{-1} Pa^{-1}, respectively, at 25°C [28]. Parthasarathy et al. [29] determined the diffusion and solubility coefficients for oxygen in Nafion 117 by an electrochemical method and reported values of 9.95×10^{-7} cm^2 s^{-1} and 9.34×10^{-6} mol cm^{-3} at 30°C, respectively, and 8.70×10^{-6} cm^2 s^{-1} and 4.43×10^{-6} mol cm^{-3} at 80°C, respectively. Extensive swelling of the membranes occurs in light alcohols, and the crossover rate of methanol is as high as about 10^{-8} mol cm^{-1} s^{-1} [30,31], i.e., several hundred times higher than that of hydrogen. The permeation of hydrogen and oxygen in Nafion 117 membranes corresponds to an equivalent current loss of less than 1 mA cm^{-2}, while the high methanol crossover corresponds to a leakage current density of 50–100 mA cm^{-2} in direct methanol fuel cells (DMFCs) [32]. This results in not only waste of fuel but also considerably lowered energy efficiency and cell performance due to the mixed electrode potential. Under fuel cell operating conditions, especially at higher temperatures, the methanol crossover rate will be even higher [33,34].

15.2.1.5 Mechanical Properties

Like other physicochemical properties of Nafion and other polymers in the PFSA family, the mechanical properties are highly dependent on the water content and temperature as well as the thermal history of the membrane. In addition, for PFSA membranes prepared by melt extrusion, the mechanical properties are generally anisotropic, i.e., depending on if they are measured perpendicular or parallel to the extrusion direction. For recast PFSA membranes, the mechanical properties can be very different and annealing under pressure is preferably done in order to develop the crystalline regions that provide most of the mechanical strength [35–37]. This should preferably be done when the membrane is in the neutral salt form, e.g., Na^+ or K^+, to avoid darkening due to the partial decomposition or anhydride formation of sulfonic acid.

Dynamic mechanical analysis (DMA) is a tool that has been extensively utilized for systematic studies of the viscoelastic properties of, e.g., Nafion 117 at different temperatures and water contents [27,38–41]. In general, water exhibits a strong plasticizing effect on Nafion membranes, but at elevated temperatures close to the glass transition temperature of the ionic regions it has been found to stiffen the structure by stabilizing the network of hydrophilic clusters [38]. In the lower temperature range and under atmosphere with a relative humidity of 85%, the elastic modulus of Nafion 117 decreases from 124 MPa at 30°C to 62 MPa at 90°C [38]. The two principal relaxations observed in the DMA data, often called α and β, have for a long time been speculated to be associated with the glass transitions of the ionic domains and the matrix, respectively [42]; however, the assignments were later reversed [43]. To explain the molecular origin of the thermal transitions and mechanical relaxations of Nafion, Page et al. [39] and Osborn et al. [41] carried out a comprehensive study on

Nafion 117 using a combination of techniques, including differential scanning calorimetry, DMA, SAXS, and solid-state ^{19}F-NMR. The β-relaxation centered at −20°C for H^+-Nafion 117 is now assigned as the glass transition temperature (T_g) of this material [41]. The more pronounced α-relaxation observed at temperatures around 100°C is, however, more important for the practical use of the membranes in fuel cells since a dramatic softening occurs due to the increased molecular mobility [41].

15.2.1.6 Thermal and Oxidative Stability

Even though PFSA membranes start to soften at temperatures in the 100–150°C range, the onset temperature of major decomposition in air is close to 300°C for H^+-Nafion and 400–450°C for Nafion 117 in the Li^+, Na^+, K^+, Rb^+, and Cs^+ salt forms [44], depending on the heating rate. Mass spectrometric analyses of the purge gas revealed that the decomposition products were initially sulfur dioxide, carbonyl fluoride, and thionyl fluoride, and in the higher temperature range hydrofluoric acid and C_xF_y and $C_xF_yO_z$ fragments were identified [45].

The radical oxidative resistance of PFSA membranes and other fuel cell membrane materials is often evaluated in an accelerated degradation test often called the Fenton test. During this test, the materials of interest are submerged in an aqueous solution of hydrogen peroxide, typically 3–30 wt.%, containing 3–30 ppm Fe^{2+} in order to catalyze the generation of hydroxyl and perhydroxyl radicals. Sampling is done after certain periods of time and the relative radical oxidative resistance is evaluated by, e.g., recording weight losses, molecular weight changes, or mechanical deterioration. At 68°C and in a Fenton solution containing 3 wt.% hydrogen peroxide and 4 ppm Fe^{2+}, which is normally replaced every 20–24 h, the H^+-Nafion membranes typically show a remaining mass of 95% after 200 h [46–48]. Fuel cell idling is also well known to severely enhance the membrane degradation. This is evidenced by increasing amounts of membrane decomposition products, e.g., sulfates and fluorides, in the effluent water [49–51] as well as gradually decreased open circuit voltage [50,52] indicating increasing gas crossover through the membrane.

15.2.2 Improved Poly(perfluorosulfonic Acid) Membranes

15.2.2.1 Short Side-Chain Poly(perfluorosulfonic Acid)

A wide range of different PFSA materials prepared by different synthetic approaches are today available from various suppliers, including Solvay Specialty Polymers (Aquivion®, formerly Hyflon® Ion), 3M™, Asahi Glass (Flemion®), Asahi Kasei (Aciplex®), and FuMA-Tech (fumion® F), as summarized in Table 15.2. The short side-chain PFSA membranes were intensively developed by Dow Chemical in the early 1990s [53]. However, owing to the challenging synthesis of the sulfonic acid–containing co-monomer, the development activities were stopped and the Dow PFSA membranes were withdrawn from the market [8]. A considerably more simple production process of the key co-monomer was later developed and patented by Solvay Solexis, and the short side-chain PFSA membrane based on this co-monomer was initially marketed under the trade name Hyflon Ion, which a few years later was renamed to Aquivion [54]. The short side-chain PFSA materials have been found to outperform the conventional Nafion-type membranes in fuel cells in terms of power density, especially at temperatures above 80°C [27,53–56]. The better performance is mainly due to a favorable combination of a high concentration of sulfonic acid groups and a high degree of crystallinity within the hydrophobic regions, which give the membrane better viscoelastic properties and thus better dimensional stability than Nafion [27,57]. Importantly, the α-relaxation associated with a dramatic softening of the membrane, which is observed at temperatures slightly above 100°C for Nafion-type membranes, is shifted to 160°C for the short side-chain PFSA membranes [54]. In addition, the short side-chain PFSA membranes show better hydration characteristics than Nafion-type membranes at elevated temperatures, which allows for operational temperatures up to 120°C [55]. Typically, the water uptake for a short side-chain membrane with an EW of 858 g $eq.^{-1}$ is higher than for Nafion-type membranes and at a degree of hydration λ of, say 15, the proton conductivity of the short side-chain membrane is 0.11 S cm^{-1} at room temperature compared with 0.06 S cm^{-1} for a Nafion 117 membrane [27].

TABLE 15.2
Commercial PFSA Membrane Materials and the Structure of the Corresponding Sulfonic Acid–Containing Monomers

Trade Name	EW Range (g $eq.^{-1}$)	Co-Monomer Structure	References
Nafion®	785–1500	CF_2=$CFOCF_2CF(CF_3)OCF_2CF_2SO_2F$	[7,8,58]
Flemion®	1000		
fumion® F	900–1400		
Aciplex®	1000–12000	CF_2=$CFOCF_2CF(CF_3)OCF_2CF_2CF_2SO_2F$	[8]
Dow	840–1150	CF_2=$CFOCF_2CF_2SO_2F$	[8,54,55]
Hyflon® Ion	850–1350		
Aquivion®	790–830		
3M™	700–1000	CF_2=$CFOCF_2CF_2CF_2CF_2SO_2F$	[59,60]

15.2.2.2 Poly(perfluorosulfonic Acid) Composite Membranes

For a fuel cell with an electrolyte of a certain conductivity, the area-specific resistance increases with increasing distance between the electrodes. The ohmic loss of a PEM fuel cell can thus be lowered by reducing the thickness of the membrane. By means of reinforcement, the thickness of PFSA membranes has been successfully reduced down to 5–30 μm with good conducting and mechanical properties. Owing to the effective back-diffusion of water from the cathode to the anode side through such thin membranes, water management and therefore the average conductivity is improved. One challenge for developing thinner membranes is the reduced mechanical strength, especially under swelling. Composite PFSA membranes with reinforcement, either by porous poly(tetrafluoroethylene) (PTFE) from Gore Fuel Cell Technologies or by laser-drilled polysulfone or polyimide from Giner Electrochemical Systems, show promising characteristics such as reduced swelling at high IECs [61]. Impregnation of other inert substrates, such as porous polypropylene [62,63], polyethylene [63], or PTFE [63,64], and co-electrospinning with polyphenylsulfone [65] are other examples of how the mechanical reinforcement of PFSA ionomer membranes has been tackled.

Aiming at, e.g., reduced methanol permeability, improved hydration and mechanical characteristics, nanocomposite systems based on PFSA ionomers have been extensively studied. On one hand, a large number of inorganic hygroscopic oxides have been considered, including silica (SiO_2) [66,67], titanium oxide (TiO_2) [68], zirconia (ZrO_2) [69], and tin oxide (SnO_2) [70]. On the other hand, much work has also been conducted on the development of PFSA electrolyte composite systems with hygroscopic inorganic proton conductors such as zirconium hydrogen phosphates [71–73], phophotungstic acid [74], or silicotungstic acid [75] as secondary components. Generally, a nanodispersed secondary inorganic phase enhances the elastic modulus of the membrane, especially at temperatures close to the α-relaxation, which, in combination with the improved hydration characteristics, could allow for a slightly elevated operating temperature. Even though there is no generally accepted understanding, it has been reported that such composite systems in some cases show improved conductivity at lower water contents in case that the secondary component has added to the total IEC of the membrane [11]. However, as reviewed by Alberti and Casciola [76], the incorporation of inorganic particles within the polymer matrix gives a dramatically reduced methanol permeability that is strongly beneficial for DMFC applications.

Replacing water with a thermally stable proton solvent that is considerably less volatile than water is another way to maintain the proton conductivity at temperatures above the boiling point of water. Phosphoric acid–imbibed Nafion was demonstrated by Savinell et al. [77] to give a system that could maintain proton conductivity above 10^{-2} S cm^{-1} under dry conditions at temperatures up to 180°C. However, in fuel cells based on this membrane type, the anode malfunctions because of the developing phosphoric acid imbalance [78]. Alternatively, different ionic liquids have been considered as dopants, and a Nafion 117 membrane that had been imbibed with 1-butyl-3-methyl imidazolium triflate has demonstrated proton conductivity of 0.06–0.11 S cm^{-1} at 150–180°C [79].

15.3 PARTIALLY FLUORINATED IONOMERS

15.3.1 Poly(α,β,β-Trifluorostyrene) and Copolymers

A series of alternative sulfonated membranes have been developed and extensively tested by Ballard Power Systems. The first generation of membranes, called BAM1G, was based on poly(phenylquinoxaline) with different degrees of sulfonation. Later, the second generation of advanced membranes (BAM2G) based on two distinct material types was developed. The first consisted of a series of sulfonated poly(2,6-diphenyl-1,4-phenylene oxide)s, whereas the second series consisted of sulfonated poly(arylene ether sulfone)s. However, the durability of the two first generations of membranes under fuel cell operating conditions was insufficient [80]. To improve the longevity, a new family of sulfonated membranes based on α,β,β-trifluorostyrene monomers was developed according to the early methodologies proposed by Prober [81] and named BAM3G [80]. The equivalent weight of the polymers ranged from 375 to 920 g $eq.^{-1}$, which resulted in high water uptake and good proton conductivity, and the perfluorinated structure ensured significant durability improvements compared with the previous generations of membranes in the series.

15.3.2 Radiation Grafted Membranes

Radiation grafting can be used to introduce acidic functionalities on activated sites in preformed polymer films, and has been extensively studied for the preparation of PEMs for fuel cells. The membrane preparation involves γ- or electron beam irradiation of the polymer base film to give activated sites through the generation of radicals. The irradiation can be carried out in the presence of the monomer or *in vacuo* or under inert atmosphere in order to trap the radicals so that they can react with the monomer in a subsequent process step. Alternatively, the base polymer film can be irradiated under air or oxygen-enriched atmosphere to generate hydroperoxides, from which the polymerization can be initiated by thermal treatment. The most commonly used monomers are styrene and divinylbenzene, on which the sulfonic acid groups can be readily introduced in a second sulfonation step.

The technique has been implemented on a large number of different base materials such as PTFE, poly(tetrafluoroethylene-*co*-hexafluoropropylene) (FEP), poly(ethylene-*alt*-tetrafluoroethylene) (ETFE), poly(vinylidene fluoride), and poly(tetrafluoroethylene-*co*-perfluorovinylether), as reviewed by Gubler et al. [82]. The properties of the membranes and the degree of sulfonation can be tuned by a number of parameters like irradiation dose, thickness of the base polymer film, the type of solvent, the composition of the solvent/monomer solution, and the subsequent sulfonation conditions.

The process is scalable and is preferably carried out on a roll-to-roll process based on low-cost starting materials. Fuel cell membrane electrode assemblies (MEAs) based on poly(styrene sulfonic acid) grafted FEP showed performance comparable to that of MEAs based on Nafion 112 and a service life in H_2/O_2 fuel cells of more than 200 h at 60°C and 500 mA cm^{-2}, after optimizing the membrane–electrode interfaces [83].

15.4 NONFLUORINATED SULFONATED POLYMERS

15.4.1 Sulfonated Poly(Arylene Ether)s

The development of cation-exchange materials based on alternative hydrocarbon polymers functionalized with pendant acidic groups is mainly motivated by a considerable potential cost reduction in combination with a reduced environmental impact due to the absence of fluorine [84,85]. In addition, they may also be more suitable for certain applications than the PFSA membranes since they, in many ways, have different properties [86].

The superior chemical stability of perfluorinated materials is connected to the high bond strength of the C–F bonds (about 485 kJ mol^{-1}), and owing to its hydrophobic nature the backbone structure repels nucleophiles, further protecting it from hydrolysis. Many aromatic structures are also known to exhibit excellent chemical stability; for comparison, the bond strength of C–H in a benzene ring is around 435 kJ mol^{-1}, whereas the C–H bond strength in aliphatic polymers is considerably lower (typically about 350 kJ mol^{-1}).

Much research effort has been devoted to the development of sulfonic acid functionalized systems, especially homopolymers and different copolymers based on poly(arylene ether)s owing to their availability and structure versatility, processability, and excellent chemical and thermal stability. Among different sulfonated poly(arylene ether) derivatives, the two most thoroughly studied polymers are based on poly(ether sulfone) (PSU) [84] and poly(ether ether ketone) (PEEK) [84,87], as shown in Figure 15.4. Both base polymers are commercially available, e.g., under the trade name Udel® (Solvay Specialty Polymers) or Ultrason® (BASF) and KetaSpire® (Solvay Specialty Polymers) or Victrex® PEEK (Victrex), respectively. Commercially, sulfonated poly(arylene ether ketone)s are available in the form of membranes and as dispersions from, e.g., FuMA-Tech (Fumion P, K, and E).

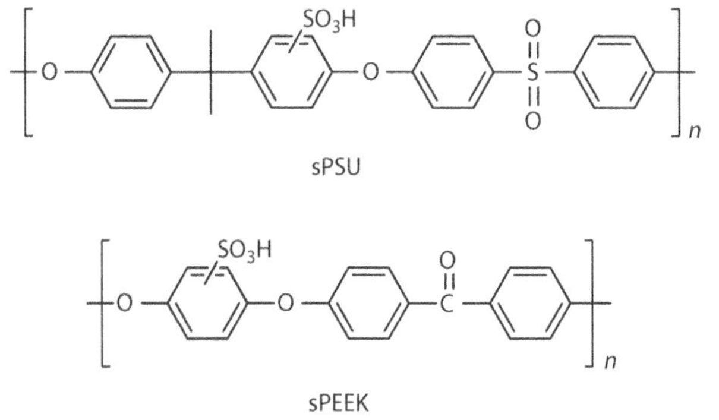

sPSU

sPEEK

FIGURE 15.4 Schematic illustration of sulfonated PSU (sPSU) and PEEK (sPEEK).

The most common way to introduce sulfonic acid functionalities on the polymer backbones is through postsulfonation by electrophilic aromatic sulfonation using sulfuric acid, oleum, chlorosulfonic acid, or sulfur trioxide complexes as reagents [87–90]. Different structure derivatives of the poly(arylene ether)s will thus require different sulfonation conditions when the electron density throughout the polymer backbone is altered. To avoid side reactions and partial decomposition of the polymer backbone, the electrophilic aromatic sulfonation conditions need to be delicately optimized. The degree of sulfonation can be somewhat controlled by adjusting the reaction conditions, and selectivity in terms of sulfonation pattern can be obtained by altering the electron density of the polymer backbone. However, the methodology does not give the precise control that is required for the optimization of the hyprophilic/hydrophobic properties, and thus the nanostructure and percolation, during hydration. Alternatively, sulfonated poly(arylene ether)s can be obtained by carrying out the polymerization from sulfonated monomers, allowing for precise control of the degree of sulfonation and the sulfonation patterns. This also makes it possible to introduce the sulfonic acid moieties on electron-deficient positions in the aromatic backbone, which generally cannot be accessed through the electrophilic aromatic sulfonation route [91,92].

Like for all the PEM systems described in the previous sections, the proton transport in polymer electrolytes based on sulfonated aromatic polymers is water mediated, which implies that careful control of the water content of the membrane is required. The hydrophobicity of the polymer backbone and the acidity of the sulfonic acid groups in the nonperfluorinated materials are considerably lower than that of the PFSA polymers. One of the greatest challenges of this approach is thus to obtain a good hydrophobic/hydrophilic nanostructure in which protons can be effectively conducted [86,93]. A current trend in this connection is to design and synthesize different structure analogues with different sulfonation patterns and compositions of hydrophilic and hydrophobic blocks [94]. Such membranes with high IECs of 2–5 meq. g^{-1} have demonstrated proton conductivities reaching well above 0.01 S cm^{-1}. Design and synthesis of microblock copolymers, which are statistical copolymers with short blocks, is an approach that is currently receiving much attention, aiming at further improvement of the morphological features of the sulfonated poly(arylene ether)s. This concept is, however, challenging since it requires macromonomers to be synthesized and isolated with sufficiently high purity for further polymerization [95].

A limited number of reports on polymers with protogenic moieties other than sulfonic acid, e.g., phosphinic acid [96] or phosphonic acid functionalized poly(arylene ether)s [97], are available in the literature; however, general synthetic methodologies are lacking. Moreover, the weaker acidity of the phosphinic- or phosphonic acid groups compared with that of sulfonic acid groups, may result in lower conductivity of the membrane.

15.4.2 Sulfonated Polyimides, Polybenzimidazoles, and Other Aromatic Polymers

The polyimides comprise a family of aromatic polymers with good film-forming properties, as well as excellent thermal, chemical, and mechanical stability [84,85]. Sulfonated polyimides, especially the six-membered ring derivatives obtained from 1,4,5,8-naphthalenetetracarboxylic dianhydride, have been thoroughly studied as PEMs for fuel cells. Conductivities higher than or comparable to that of PFSA membranes are commonly reported for such materials, and their low methanol permeability make them of particular interest for DMFC applications. Fuel cell performance comparable to that of cells with Nafion has been reported [98]. However, the susceptibility toward hydrolysis of the imide rings gives durability problems under humid conditions at temperatures above 70°C, even though they show much better stability than their five-membered ring counterparts. On the basis of a low molecular weight model system, this was investigated in detail by Genies et al. [99] using NMR and infrared (IR) spectroscopy. They also pioneered the field through the synthesis and characterization of a large number of different sulfonated polyimide random and segmented block copolymers with different IECs, as exemplified in Figure 15.5 [100].

Recent studies have shown that the hydrolytic stability of sulfonated polyimides can be further improved through electronic effects, by employing 4,4′-binaphthyl-1,1′,8,8′-tetracarboxylic acid anhydride, which gives a polyimide enhanced electron density at the carbonyl positions and thus reduced electrophilicity [101].

PBIs are another family of aromatic engineering plastics that can be sulfonated to give materials the water-mediated proton conductivity. Most work has been done on poly[2,2′-*m*-(phenylene)-5,5′-bibenzimidazole] (*m*PBI), which is a commercially available polymer and, in the form of spun fibers, mainly used for heat-resistant fabrics and, in the form of membranes, used for a large variety of separation processes [102]. One way to sulfonate *m*PBI is through electrophilic aromatic sulfonation by first doping the cast membrane with sulfuric acid or oleum followed by heat treatment at 450–600°C for a couple of minutes to give a zwitterionic polymer, also commonly referred to as stabilized PBI, as shown in Figure 15.6 [87,103–107]. Another way to introduce the sulfonic acid groups on the phenyl moieties is by using a sulfonated monomer in the condensation polymerization reaction [107–110]. By controlling the stoichiometry of the co-monomers, this methodology gives precise control of the sulfonation pattern

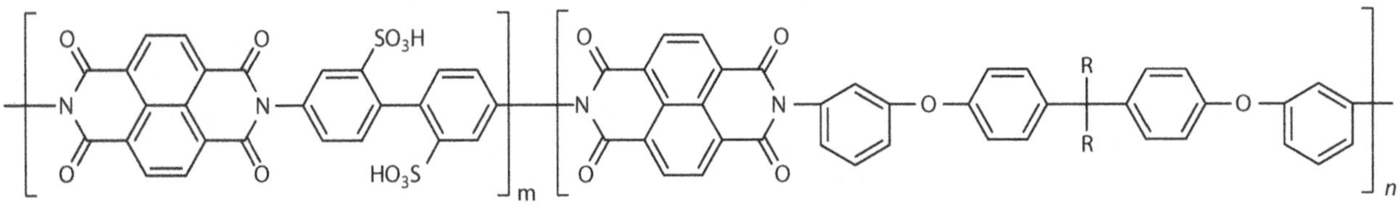

R	M	*n*	IEC (meq. g^{-1})
$-CH_3$	5	20	0.63
$-CH_3$	5	11.6	0.96
$-CH_3$	5	7.5	1.30
$-CH_3$	5	5.0	1.64
$-CF_3$	5	20	0.56
$-CF_3$	5	11.6	0.86
$-CF_3$	5	7.5	1.17
$-CF_3$	5	5.0	1.51
$-CF_3$	5	3.3	1.86

FIGURE 15.5 Examples of sulfonated polyimide copolymers synthesized and characterized by Genies et al. (From Genies, C., Mercier, R., Sillion, B., Cornet, N., Gebel, G., and Pineri, M., *Polymer*, 42, 359, 2001.)

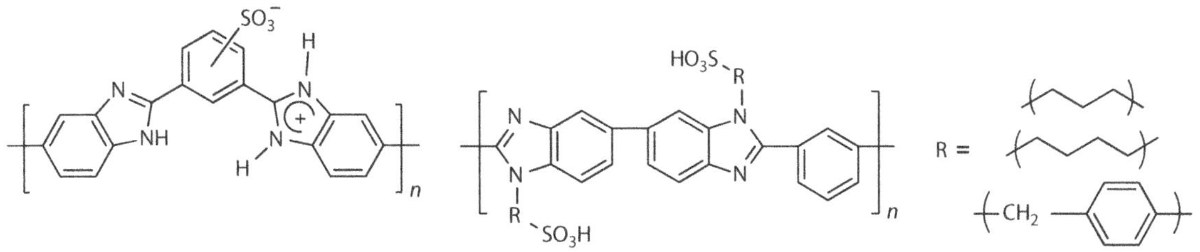

FIGURE 15.6 Schematic illustration of sulfonated PBI prepared by electrophilic aromatic sulfonation (left) and of *N*-sulfonated PBI (right).

and the degree of sulfonation. Alternatively, PBI can be sulfonated through N-functionalization by treating the polymer with a strong nonnucleophilic base such as an alkali metal hydride, and subsequently letting it react with a sulfonated alkyl or aryl halide through a substitution reaction. Different reagents have been investigated for this grafting reaction, including (4-bromomethyl)benzenesulfonate [111,112] or sultones [111,113], as schematically illustrated in Figure 15.6. In general, the proton conductivity of the sulfonated PBIs is relatively low compared with other sulfonated polymers owing to their amphoteric nature.

The polyphosphazenes are based on an inorganic backbone with alternating nitrogen and phosphorous atoms and are highly versatile in terms of varying the structure of the side chains and their functionalities. Sulfonation of polyphosphazenes with aromatic side chains can be done through electrophilic aromatic sulfonation using SO_3 in dichloroethane to give materials with IECs up to 2.0 meq. g^{-1} without excessive swelling in water [114]. As reviewed by Rozière and Jones [115], some work has also been done on sulfonated poly(phenyl quinoxaline)s, poly(phenylene oxide)s, and poly(phenylene sulfide)s.

15.5 ACID-DOPED MEMBRANES

15.5.1 Acid-Doped Polymers

As briefly mentioned in Section 15.2.1.7, many of the technological challenges of PEM fuel cells based on sulfonated membranes are associated with the low operating temperature of about 80°C, which is required by the water-mediated proton transport mechanism. It has been recognized that an elevated operating temperature brings advantages from kinetic and engineering points of view [12]. For example, increasing the operating temperature to 150–200°C will improve the CO tolerance by about three orders of magnitude, which implies that the cell can be fed directly with a hydrocarbon (or methanol) reformate without further CO cleanup. This has triggered the development of polymer-based membrane materials in which the proton transport is not water mediated. As briefly mentioned in Section 15.2.2.2, an early approach was to replace water as the proton solvent in PFSA membranes with phosphoric acid [77], which has the highest intrinsic proton conductivity of any known material [116]. In combination with its thermal stability and low vapor pressure, the proton conductivity of phosphoric acid remains high even under dry conditions at temperatures up to 200°C, which makes it of particular interest for this purpose. Polymers bearing sites that in relation to the doping acid are basic, such as ethers, hydroxyls, imines, amides or imides, interact with the acid dopant to an extent determined by the particular relative acidity and basicity. In other words, the basic polymers act as solvents promoting the dissociation of the acid to some extent. A number of basic polymers have been investigated for preparing acid–base electrolytes, such as poly(ethylene oxide), poly(vinyl alcohol), poly(acryl amide), and poly(ethylene imine).

15.5.2 Phosphoric Acid–Doped Polybenzimidazoles

15.5.2.1 Synthesis, Processing, and Membrane Preparation

Among the acid-doped polymers, the phosphoric acid and PBI (*m*PBI) adduct was first proposed as electrolyte for fuel cells operating at temperatures up to 200°C in 1995 [117]. Originally, it was considered particularly interesting as electrolyte material in DMFCs because of its low methanol crossover [117,118] originating from electro-osmotic drag [119] as well as from diffusion [120]; however, it seems to suffer from durability problems when operated with a direct feed of methanol as fuel [121,122].

The first synthesis methodology of PBIs was published in the early 1960s, and it was based on melt condensation polymerization from aromatic tetraamines and aromatic dicarboxylic acid derivatives [123]. This process gave relatively poorly defined PBIs due to pronounced side reactions, and the obtained polymers were of low molecular weight, as reviewed by Neuse [124]. A solution to this was to prepare low molecular weight prepolymers, which were further polymerized in a subsequent step by further heat treatment. Better control of the process, reduced side reactions such as branching or cross-linking, and a more well-defined polymer (lower polydispersity index) was obtained when the polymerization was carried out by homogeneous solution polymerization using polyphosphoric acid (PPA) as the polycondensation solvent, as first described by Iwakura et al. [125] a few years later. As schematically illustrated in Figure 15.7, this is typically done by reacting equimolar amounts of highly purified 3,3′-diaminobenzidine (DAB) and isophthalic acid in PPA at 170–200°C. High linear molecular weights of up to 150 kDa or even higher, as determined from viscosity measurements, can be obtained but it has to be balanced against the processability.

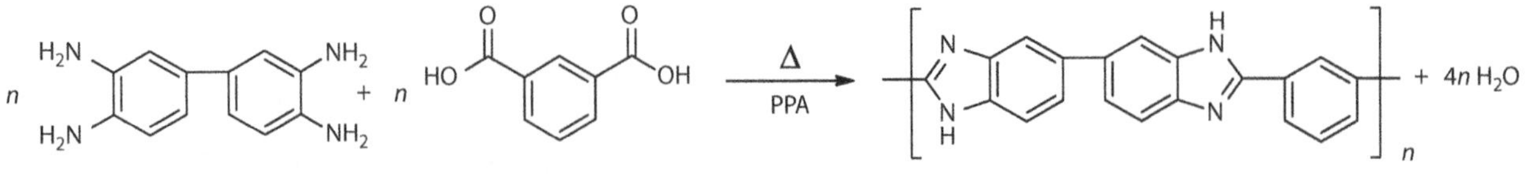

FIGURE 15.7 Schematic illustration of the homogeneous solution polymerization of *m*PBI.

The obtained *m*PBI, which has a density of about 1.3 g cm^{-3} and a glass transition temperature of 425–436°C, can subsequently be isolated through precipitation in water. After extensive washing with a dilute aqueous base and drying, it can be dissolved in some highly polar aprotic solvents, e.g., *N,N*-dimethylacetamide (DMAc), *N,N*-dimethylformamide, *N*-methyl-2-pyrrolidone, or dimethylsulfoxide. It is also soluble in concentrated sulfuric acid or alkali metal hydroxide or alkoxide solutions in light alcohols. Traditionally, DMAc has been the preferred organic solvent for *m*PBI since the viscosity of the obtained solution is suitable for dry spinning of *m*PBI fibers [102]. A small amount of LiCl is sometimes added to the solution as a stabilizer to prevent the polymer from phasing out during long-term storage [126]. With an average to high molecular weight, *m*PBI shows excellent film-forming properties, and membrane casting from solution can be done by slow evaporation of the solvent on a suitable substrate to give films as thin as a few micrometers. The membrane can subsequently be peeled off in water, and further treatment with hot water is required to completely remove the solvent residuals and eventual stabilizer. The highly hydrophilic *m*PBI membranes, which show water uptake of about 15–19 wt.% at room temperature [127], can be readily doped with acids by equilibration in aqueous solutions of, e.g., H_2SO_4, HNO_3, $HClO_4$, HCl, or H_3PO_4, to give highly proton conducting systems [128]. For the reasons discussed in the previous section, H_3PO_4 is of special interest for fuel cell applications at elevated temperatures.

In analogy to the degree of hydration for sulfonated polymer membranes, the acid content is often normalized as the acid doping level (ADL) defined as the number of phosphoric acid molecules per polymer repeat unit. The ADL is mainly dependent on the concentration of the bulk acid and the doping temperature. By equilibrating an *m*PBI membrane in 75 or 85 wt.% H_3PO_4 at room temperature for about 50 h, an ADL of about 5–6 or 10–11 is normally obtained, respectively.

An alternative way of preparing phosphoric acid–doped PBI membranes is to tape cast the hot crude PPA polymer solution on a substrate followed by hydrolysis of the PPA under an atmosphere with carefully controlled humidity. The hydrolysis of PPA yields orthophosphoric acid, which is a poorer solvent for PBI and induces a sol–gel transition to give a phosphoric acid–doped PBI membrane via a reduced number of processing steps [129,130]. The ADL is controlled by the temperature and humidity. With this process, the obtained membranes at ADLs as high as 30–40 display sufficient mechanical strength for processing and MEA fabrication. In this process, terephthalic acid is often used as co-monomer instead of isophtahlic acid, which gives poly[2,2′-*p*-(phenylene)-5,5′-bibenzimidazole] (*p*PBI). This structural isomer gives mechanically stronger membranes than *m*PBI. However, its versatility is limited by its poor solubility in organic solvents. It has also been shown that *m*PBI dissolves in a mixture of phosphoric acid and trifluoroacetic acid (TFA), and membranes can be cast from this solution by evaporating the TFA to give a phosphoric acid–doped membrane in one single-process step [131].

15.5.2.2 Membrane Characterization

When heated gradually from room temperature with a heating rate of 10°C min^{-1}, the major onset of thermal decomposition of PBI in air is around 500–550°C [132]; however, significant oxidation seems to occur when treated under air atmosphere at 350°C for 16 h [133]. The Fenton test, as described in Section 15.2.1.6, is commonly used for estimating the relative chemical stability of different PBI derivatives. Depending on the molecular weight, membranes of *m*PBI show a residual mass of 75–82% after 200 h in the Fenton solution. For Nafion 115, 95% remains under similar conditions [48].

Vibrational- and NMR spectroscopy are tools that are commonly used for structural identity confirmation of polymers and have also been employed for studying the acid doping chemistry of PBIs. For example, Bouchet and Siebert [134] and Li et al. [127] recorded the IR and Raman spectra, respectively, of a set of PBI membranes with different ADLs, whereas Hughes et al. [135] investigated the interaction between PBI and phosphoric acid by ^1H and ^{31}P magic-angle spinning NMR spectroscopy. All three techniques confirmed that ionic species were present in the membranes originating from the acid–base reaction between the polymer and the dopant. On a fundamental level, the doping chemistry has been described by Ma et al. [136], who estimated the equilibrium constant for the protonation of *m*PBI to 1.17×10^3. He et al. [137] calculated the complexation constants by the Scatchard method to 12.7 L mol^{-1} for ADLs up to 2 and 0.19 L mol^{-1} for the excess acid.

The proton conductivity of PBI with different ADLs and under different conditions has been extensively studied. It depends not only on ADL, temperature, and humidity, but also on the membrane fabrication procedure due to morphological effects. At a certain ADL, the proton conductivity is generally higher for membranes prepared by direct casting than for membranes prepared by postdoping [131,138]. On a slightly larger scale, the development of nanoporous membranes has proven promising for increasing the acid content of the membrane and thus the proton conductivity [139]. Like in phosphoric acid solutions, the conductivity of phosphoric acid–doped PBI follows the Arrhenius law [134], which suggests that the protons are predominately transported by the Grotthuss-type hopping mechanism, as also confirmed by measurements of the diffusion coefficients of ^1H and ^{31}P using solid-state NMR techniques [11,140]. This also implies that the electro-osmotic water drag coefficient is close to zero in accordance with experimental studies [119,141]. At an ADL of 2, the conductivity of PBI is in the range of 10^{-7} S cm^{-1} at room temperature [134]. When the membrane is doped with excess acid, the conductivity increases and reaches 0.04–0.07 S cm^{-1} at an ADL of 4–6 and at temperatures close to 200°C [136,142]. In the same temperature range and by further increasing the ADL to 10–11 by postdoping or to 32 by direct casting, the proton conductivity without active humidification increases to about 0.15 and 0.25 S cm^{-1}, respectively [48,129]. At any given temperature, the conductivity of a PBI membrane with a certain ADL increases with increasing partial pressure of water in the atmosphere [142].

Doping with phosphoric acid results in extensive volume swelling. For *m*PBI membranes with an ADL of 10–11, it typically ranges between 150% and 250% [143], which in turn gives a considerable increase of the hydrogen and oxygen permeability [28,144]. It also means that the intermolecular van der Waals forces that are responsible for the superior mechanical strength of pristine PBI are strongly reduced, meaning that the phosphoric acid acts as a plasticizer. This can, however, be compensated for by increasing the linear molecular weight of the polymer. For example, at room temperature and at an ADL of around 11, the tensile strength of an *m*PBI membrane increases from 5 to 30 MPa when the molecular weight is increased from 30 to 95 kDa [48].

Morphologically, the pristine as well as the acid-doped PBI membrane is practically amorphous, although heat treatment at 150°C for 1 week under dry conditions results in the development of regions with some degree of crystallinity, which is evidenced on a macroscopic scale by a considerable increase of the elastic modulus as well as the tensile strength [133].

15.5.2.3 Structural Variations, Alternative Polymers, and Composites

Fuel cells equipped with membranes based on *m*PBI or *p*PBI show significant durability and low cell voltage decay rates under mild operating conditions, i.e., at intermediate temperatures of about 150°C and continuous current loads of typically 200 mA cm^{-2} [133,145,146]. However, the *m*PBI was originally investigated for fuel cell use since it was the only commercially available polymer in this family of materials. The *p*PBI derivative could be synthesized from commercially available monomers. The most widely used membrane materials have thus not been optimized for fuel cell use in terms of stability, acid doping behavior, and performance. Much research effort has thus been devoted to the design, synthesis, and characterization of new backbone chemistries for this purpose, especially during the last decade. A selection of different PBI derivatives, among others, is listed in Table 15.3 [47,108,147–161].

The least complex structure of the PBIs is the 2,5-polybenzimidazole (AB-PBI). It is synthesized from 3,4-diaminobenzoic acid, which is a cheap monomer with lower toxicity than DAB used for *m*PBI and *p*PBI synthesis. The AB-PBI, like Py-PBI, has a higher concentration of basic moieties than *m*PBI, which gives a higher affinity for acids. An increased concentration of basic moieties can also be obtained by grafting heterocyclic moieties to, e.g., a PBI [162] or polysulfone backbone [163]. The properties of AB-PBI can be further improved by changing the sequence of the repeat units [148], and the regiochemistry of Py-PBI has been shown to greatly affect the characteristics of the polymer [149]. The dihydroxy *p*PBI has demonstrated the highest proton conductivity of any phosphoric acid–doped PBI derivatives, which may be explained by favorable interactions between the acid and the polymer [152,153].

As already discussed in the previous section, one way to reduce the dramatic softening of the membrane when doped with phosphoric acid is to increase the linear molecular weight of the polymer [48]. Another effective way to reduce the softening is to generate network structures through cross-linking. Thus, higher ADLs can be reached without compromising the mechanical strength. It also generally improves the oxidative resistance in the Fenton test. Many different cross-linking agents have been investigated, and most of them are bifunctional electrophiles such as α,α′-dibromo-*p*-xylene [164], dichloromethyl phosphinic acid [165], divinyl sulfone [132], chloromethyl polysulfone (CMPSU) [47], poly(vinylbenzyl chloride) [159], or 4,4′-diglycidyl(3,3′,5,5′-tetramethylbiphenyl) epoxy resin [166]. Alternatively, a cross-linked structure can be obtained by preparing polymer blends or interpenetrating networks with an acidic polymer as the secondary component. This cross-linking is ionic in nature rather than covalent, and many different binary acid–base polymer blend membranes have been investigated. With PBI derivatives as the basic components, the studied blends include, e.g., sulfonated poly(arylene ether sulfone)s [157,167–173], sulfonated poly(arylene ether ketone)s [167,168,174,175], sulfonated polyphosphazene [176], or PFSA [30,46,177,178]. Poly(vinyl phosphonic acid), which is a polymer with a high concentration of acidic groups, has been effectively immobilized in the PBI matrix to give an electrolyte system with good conductivity without further doping with phosphoric acid, and with excellent fuel cell durability [179,180]. Phosphoric acid–doped membranes based on alternative aromatic polymers such as polyimides [181] or pyridine-containing aromatic polyethers [182] are under active development and show promising characteristics and fuel cell performance.

Like for fuel cell membranes based on sulfonated polymers, another approach to tailor the characteristics of the phosphoric acid–doped PBI membranes in terms of proton conductivity, acid and water retention, dimensional stability, and mechanical properties, is to introduce a third component through the preparation of organic–inorganic composite systems. Promising results have been obtained by introducing hygroscopic inorganic materials such as titanium dioxide [183], modified laponite clay [184], or silica [185,186]. An alternative approach that also has proven successful is to use inorganic solid acids as additives, including zirconium phosphates ($Zr(HPO_4)_2 \cdot nH_2O$) [142], phosphotungstic acid ($H_3PW_{12}O_{40} \cdot nH_2O$, PWA) [142,187], silicotungstic acid ($H_4SiW_{12}O_{40} \cdot nH_2O$) [142,188,189], titanium oxysulfate

TABLE 15.3
Selected Structure Derivatives of PBI

Structure	Remarks	References
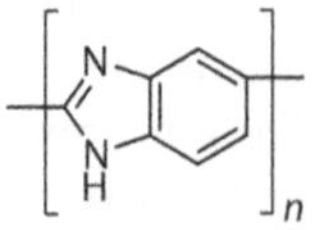 AB-PBI	Enhanced H_3PO_4 affinity	[108,147,148]
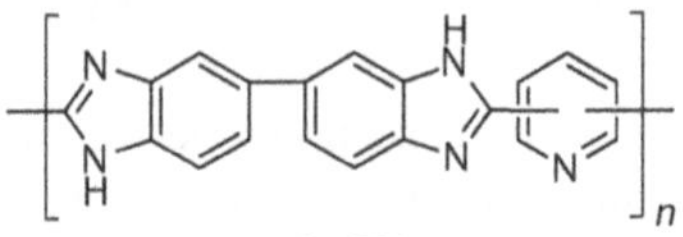 Py-PBI	Enhanced H_3PO_4 affinity	[149–151]
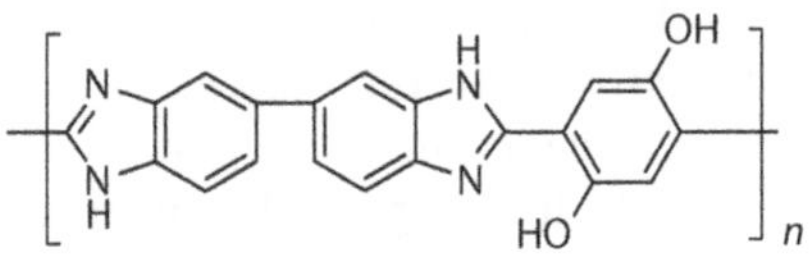 2OH-PBI	0.4 S cm^{-1} at 160°C, ADL 29	[152,153]
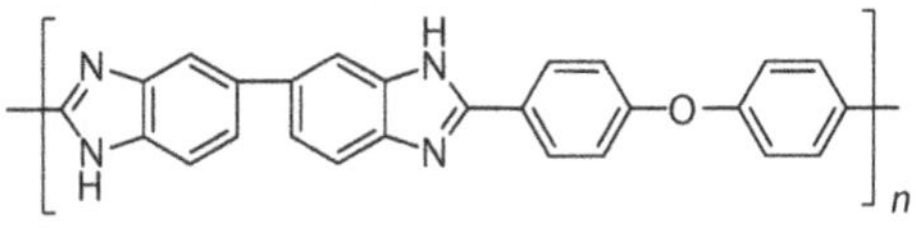 O-PBI	0.07 S cm^{-1} at 180°C, ADL 6	[154–156]
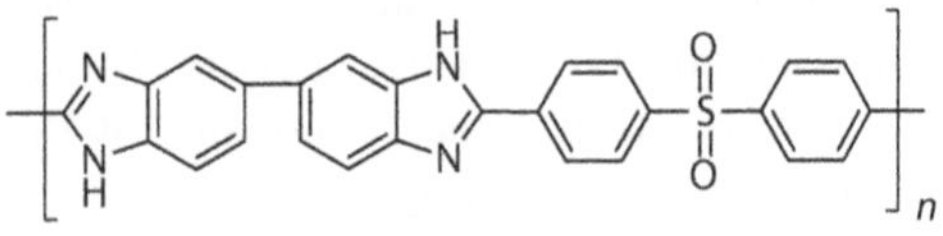 SO_2-PBI	Better oxidative stability and fuel cell performance than *m*PBI	[157–159]
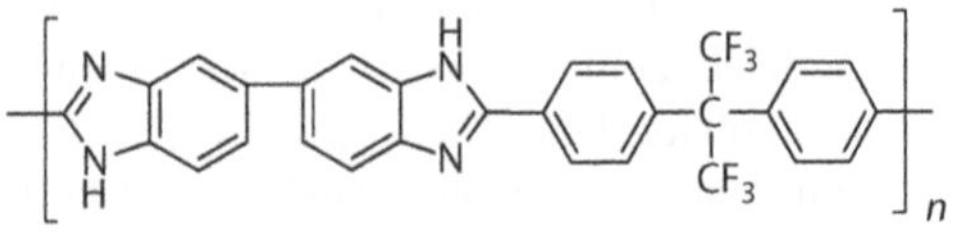 F6-PBI	Better oxidative stability and fuel cell performance than *m*PBI	[47,157,160,161]

($TiOSO_4$) [190], boron phosphate (BPO_4) [191], sulfonated silica [192], or sulfonated oligosilsequioxane [143]. In this way, the total acid concentration in the membrane can be increased, often without sacrificing the mechanical stability, leading to higher proton conductivity and thus enhanced fuel cell performance.

15.6 POLYMER MEMBRANES FOR ALKALINE FUEL CELLS

The fuel cells based on acidic membranes, and especially the PFSA-based systems, in general, show excellent performance in terms of power density, but depend on noble metals as catalyst materials. Much research effort is thus currently devoted to the development of nonnoble metal catalysts for acidic PEM fuel cells [193]. From an electrocatalyst point of view, alkaline media is less challenging and the electrodes can be made of abundant and relatively cheap materials [194]. This has resulted in a dramatically increased interest in developing polymers for alkaline fuel cell electrolytes and catalyst binders during the last few years, as recently reviewed [195–197].

Among the different anion-exchange groups that can be tethered to a polymer backbone, quaternary ammonium moieties are the most extensively studied. Quaternary ammonium functionalized polymers have been known since the early 1960s, and in the membrane form they are widely used as anion conductors in different electrodialysis processes. For fuel cell applications, the membranes need to be in the hydroxide form, in which they generally show stability limitations. The principal degradation mechanism is directly connected to the pronounced nucleophilicity of the hydroxide ion. The use of a polymer backbone with good alkali resistance, such as members in the polysulfone family of materials, limits the rate of chain scission. However, the quaternary ammonium groups are prone to attack by the hydroxide ions, causing degradation and rearrangement reactions [198]. In addition, the tethering of an electron-deficient moiety, such as a tertiary ammonium group, enhances the rate of hydrolysis of the polymer backbone [199]. Much effort is thus devoted to the development of polymers with anion-exchange moieties that are stable at high pH.

Quaternary ammonium groups can be readily introduced onto aromatic polymers such as poly(arylene ether sulfone)s through chloromethylation followed by amination with a tertiary alkyl or aryl amine to give the chloride salt of the ionomer, which subsequently can be converted into the hydroxide form by treatment with a dilute aqueous hydroxide salt solution. In analogy to the cation-exchange materials, the hydroxide ion conductivity mechanism in anion-exchange membranes is water mediated, and thus, strongly dependent on the water content of the membrane. However, little information is available about the morphology of anion-exchange membranes on the microscopic level and about the structure–conductivity relationship [200].

15.7 FUEL CELL APPLICATIONS

For fuel cells, the conductivity of the membrane has to be higher than about 5×10^{-2} S cm^{-1} to get reasonable *in situ* area-specific resistances of 0.1–0.3 Ω cm^2 with practical membrane thicknesses of 50–150 μm. As previously discussed, the proton transport through the membranes based on sulfonated polymers is water mediated, which limits the fuel cell operating temperature to about 80°C unless the system is pressurized to keep the membrane hydrated. Representative polarization curves for some different types of PEM fuel cells are shown in Figure 15.8 [201–205]. One is achieved with H_2/O_2 (3:5 bar) at 75°C with a PFSA membrane (Dow, 50 μm, EW 800 q $eq.^{-1}$) using the nanostructured catalysts from 3M with platinum loadings of 0.05 and 0.2 mg cm^{-2} on the anode and cathode, respectively [206]. This result, most likely the best PEM fuel cell performance thus far, showed the maximum current density of 5.7 A cm^{-2} at a cell voltage of 0.27 V. The peak power density was 1.88 W cm^{-2} at 0.48 V. Another set of data was obtained using Nafion 112 membrane operating at 80°C under ambient pressures of H_2 and air, more practically relevant conditions [207]. The typical performance is a current density of 1.4 A cm^{-2} at 0.54 V, which was achieved at cell temperature of 80°C where the anode fuel was humidified at 100°C and the cathodic air was humidified at 75°C.

Since the proton conductivity of PFSA membranes is water mediated, the fuel cells are typically operated at temperatures slightly below the boiling point of water at ambient pressure, which implies that a dual-phase water system will be difficult to avoid. When the humidification is too high, water condenses and the sophisticated nanostructures of the gas diffusion electrodes are flooded, causing severe mass transfer limitations.

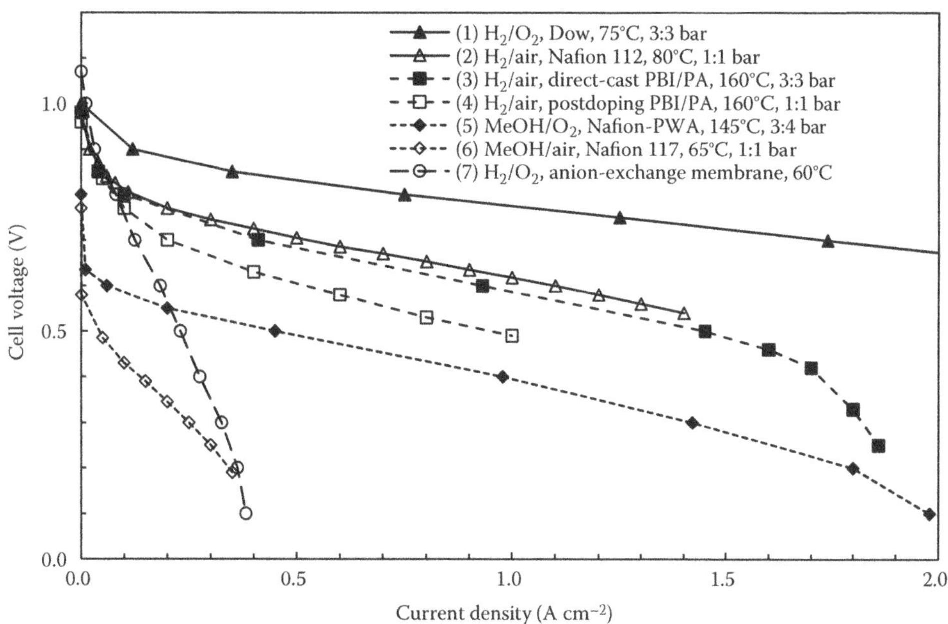

FIGURE 15.8 Polarization curves of selected PEM fuel cells based on different types of membranes. (1) H_2/O_2 (3:3 bar) cell using the Dow membrane (50 μm, EW 800 g $eq.^{-1}$), anode/cathode catalyst loadings 0.05:0.2 mg Pt cm^{-2}, 75°C (From Debe, M.K., Novel catalysts, catalyst supports and catalyst coated membrane methods, in: Vielsstich, W., Lamm, A., Gasteiger, H.A. (Eds.) *Handbook of Fuel Cells*, vol. 3, John Wiley & Sons, Chichester, pp. 576–589, 2003.); (2) H_2/air (1:1 bar) cell using the Nafion 112 membrane, 80°C (From Song, Y., Fenton, J.M., Kunz, H.R., Bonville, L.J., and Williams, M.V., *J. Electrochem. Soc.*, 152, A539, 2005.); (3) H_2/air (3:3 bar) cell using a PBI/PA membrane at 160°C (From Staudt, R., Development of polybenzimidazole-based high temperature membrane and electrode assemblies for stationary applications, in: *FY 2006 Annual Progress Report*, DOE Energy Program, pp. 718–721, 2006.); (4) H_2/air (1:1 bar) cell using a PBI/PA membrane at 160°C; (5) 2 mol L^{-1} MeOH/O_2 (3:4 bar) using SiO_2/PWA modified recast Nafion membrane 145°C (From Staiti, P., Arico, A.S., Baglio, V., Lufrano, F., Passalacqua, E., and Antonucci, V., *Solid State Ionics*, 145, 101, 2001.); (6) 1 mol L^{-1} MeOH/O_2 using Nafion 115 membrane, anode 2.5 mg cm^{-2} Pt/Ru, and cathode 2.5 mg cm^{-2} Pt, 65°C (From Bennett, B., Koraishy, B.M., and Meyers, J.P., *J. Power Sources*, 218, 268, 2012.); (7) H_2/O_2 (1:1 bar) cell using quaternary ammonium functionalized polysulfone (QAPS-OH) membrane (IEC 1.08 meq. g^{-1}), 60°C, 4 mg Pt cm^{-2}. (From Pan, J., Lu, S.F., Li, Y., Huang, A.B., Zhuang, L., and Lu, J.T., *Adv. Funct. Mater.*, 20, 312, 2010.)

Careful management of the water balance is therefore one of the key issues for the system design and operation.

Another challenge that has to be faced, which is related to the electrocatalysts materials, is the poor tolerance to fuel impurities, e.g., CO in the hydrogen stream, which is known to be highly temperature dependent [207]. At the typical operational temperature of a PFSA membrane electrolyte fuel cell, a CO content as low as 20 ppm in the fuel stream will result in a significant loss in cell performance [208].

Direct use of methanol as a fuel for fuel cells is desirable since methanol is a liquid under standard conditions, which is strongly advantageous when the storage and the distribution infrastructure of the fuel are considered. A DMFC polarization curve is shown in Figure 15.8, where the used membrane is Nafion 115 and the cell was fed with 1 mol L^{-1} CH_3OH (MeOH) and O_2 at 65°C [205]. The slow kinetics for electrochemical oxidation of methanol is clearly indicated and expected to improve by elevating the fuel cell temperatures. Composite membranes of PFSA with hygroscopic inorganic components improve the water retention as well as the mechanical stability at temperatures around or higher than the softening temperature of the pristine membrane, allowing for operation at temperatures above 100°C. A set of data by Staiti et al. [204] is selected to illustrate this type of membranes. The used membrane was recast Nafion modified with silica supported PWA tested in DMFCs at up to 145°C under pressures of 4 bar for O_2 and 3 bar for 2 mol L^{-1} MeOH vapor.

On the other hand, fuel cells based on phosphoric acid–doped membranes can operate under unhumidified conditions at temperatures up to at least 180°C. Figure 15.8 also shows the high-pressure (3:3 bar) H_2/air cell performance using direct cast PBI membranes [203] and ambient pressure (1:1 bar) H_2/air cell performance using postdoped PBI as electrolyte with data from the authors' group. Furthermore, efforts on developing anion conducting membranes are illustrated, where the used membrane was quaternary ammonium functionalized polysulfone with an IEC 1.08 meq. g^{-1} [206]. The cell was operated with hydrogen and oxygen at 60°C with high-loading (4 mg cm^{-2}) platinum catalysts on both the anode and cathode.

Generally speaking, owing to the high capital cost of the polymer-based fuel cell technology, the long-term durability is the most critical factor and lifetimes of many thousands of hours are normally required [209]. Different degradation mechanisms connected to different cell components are known. The polymer-based electrolytes suffer during fuel cell operation from degradation due to the harsh conditions with aggressive radicals, voltage gradients, and temperature and humidity fluctuations. This gradually damages the functionality of the membranes and may ultimately lead to cracks and pinholes. For the acid-doped membranes, acid loss can be fatal since it gradually increases the ohmic resistance in the cell. Depending on the operating conditions and how the end of life is defined, lifetimes ranging from a few thousands of hours up to 60,000 h have been reported for PFSA-based systems. Similarly, lifetimes up to 20,000 h have been reported for high-temperature PEM fuel cells based on phosphoric acid–doped membranes. Alkaline fuel cells based on anion-exchange membranes are still in an early development phase, and only a limited number of lifetime studies are available in the literature. Lifetimes exceeding hundreds of hours are rare. In general, and irrespectively of the type of membrane electrolyte, the longest lifetimes have been reported for fuel cells operating at a continuous and relatively low current load and at temperatures well below the maximum rated temperatures.

15.8 CONCLUDING REMARKS AND PERSPECTIVES

Fuel cells with polymer-based electrolytes represent an attractive approach to obtaining compact fuel cell systems with high power density for applications in portable, automotive, and stationary power systems. PFSA membranes have been the most prominent materials of the field, showing superior proton conductivity and chemical stability. With sulfonic acid moieties as protogenic functional groups, the proton transport within this type of membranes is water mediated, which limits the operating temperature to about 80°C. Much effort is made to develop proton conducting materials through different polymers and approaches to improve their properties. Among others, sulfonated polymers based on partially fluorinated or aromatic polymers are also under active development as cheaper alternatives to the PFSA materials; however, a viable and inexpensive substitute to the PFSA is yet to be developed. Alternative approaches based on modified PFSA membranes by means of, e.g., PTFE fibril reinforcements and inorganic–organic composites, have been successfully explored to enhance the mechanical and thermal stability and ultimately the long-term durability. Adoption of such membranes in practical stacks or power systems has still not been fully demonstrated. For fuel cells operating in the 100–200°C range, phosphoric acid–imbibed membranes based on aromatic heterocyclic polymers, especially the PBIs and poly(arylene ether)s, are of particular interest. For power units with liquid phase operation, e.g., DMFCs, membrane challenges include highly selective conduction of protons or proton carriers with low permeation rate of the fuel. Anion-exchange membranes as electrolytes in alkaline fuel cells are particularly favored with a view of a power system that is potentially noble metal free. This concept is still in an early development phase and facing challenges with respect to long-term durability. Alkaline fuel cells with liquid electrolyte were the first to be used in real practical applications; however, they have been considered less attractive for most applications because the electrolyte reacts with CO_2 from the ambient air passing over the cathode (carbonization). However, for the reverse process, water electrolysis, where hydrogen and oxygen are produced and not consumed, the liquid alkaline system has been the standard for decades. Alkaline membrane systems may similarly have a bright future in electrolyzers with a strong potential for reduced internal resistance through a zero gap configuration.

Research efforts for all membrane electrolytes are devoted to further improvement of the performance and durability and to reduction of the cost, which is expected to dramatically improve the viability and competitiveness of the technology.

ACKNOWLEDGMENTS

Funding of this work is acknowledged from the Danish Council for Strategic Research (4M Centre); the Danish Council for Independent Research, Technology and Production Science (11-117035/FTP); and the Danish ForskEL program.

ABBREVIATIONS

2OH-PBI	Dihydroxy PBI
AB-PBI	2,5-Polybenzimidazole
ADL	Acid doping level
DAB	3,3′-Diaminobenzidine
DMA	Dynamic mechanical analysis
DMAc	*N,N*-dimethylacetamide
DMFC	Direct methanol fuel cell
ETFE	Poly(ethylene-*alt*-tetrafluoroethylene)
EW	Equivalent weight
F6-PBI	Hexafluoropropylidene-containing PBI
FEP	Poly(tetrafluoroethylene-*co*-hexafluoropropylene)
Infrared	IR
MEA	Membrane electrode assembly
***m*PBI**	Poly[2,2′-*m*-(phenylene)-5,5′-bibenzimidazole]
NMR	Nuclear magnetic resonance
O-PBI	PBI containing ether linkages
PA	Phosphoric acid
PBI	Polybenzimidazole
PEEK	Poly(ether ether ketone)
PEM	Polymer electrolyte membrane (or proton-exchange membrane)
PEMFC	PEM fuel cell
PFSA	Poly(perfluorosulfonic acid)
PPA	Polyphosphoric acid
***p*PBI**	Poly[2,2′-*p*-(phenylene)-5,5′-bibenzimidazole]
PPO	Poly(phenylene oxide)
PPS	Poly(phenylene sulfide)
PSSA	Poly(styrenesulfonic acid)
PSU	Poly(ether sulfone)
PTFE	Poly(tetrafluoroethylene)
PWA	Phosphotungstic acid ($H_3PW_{12}O_{40}{\cdot}nH_2O$)
Py-PBI	Pyridine-containing PBI
SAXS	Small angle x-ray scattering
SO_2-PBI	Sulfone-containing PBI
TFA	Trifluoroacetic acid
TFE	Tetrafluoroethylene
T_g	Glass transition temperature

REFERENCES

1. Bacon, F.T. Fuel cells, past, present and future. *Electrochimica Acta* 1969; 14:569–585.
2. Grubb, W.T. and Niedrach, L.W. Batteries with solid ion-exchange membrane electrolytes. *Journal of the Electrochemical Society* 1960; 107:131–135.
3. Appleby, A.J. and Foulkes, F.R. *Fuel Cell Handbook*, Van Nostrand Reinhold, New York, 1989, pp. 285–286.
4. Wiley, R.H. and Venkatac, T.K. Rates of sulfonation of polystyrenes crosslinked with pure *p*-, 2:1/*m*:*p*-, and commercial divinylbenzenes. *Journal of Polymer Science Part A* 1965; 3:1063–1067.
5. Wiley, R.H. and Venkatac, T.K. Sulfonation of polystyrene crosslinked with pure *m*-divinylbenzene. *Journal of Polymer Science* 1966; 4:1892–1894.
6. Appleby, A.J. and Yeager, E.B. Solid polymer electrolyte fuel cells (SPEFCs). *Energy* 1986; 11:137–152.
7. Banerjee, S. and Curtin, D.E. Nafion® perfluorinated membranes in fuel cells. *Journal of Fluorine Chemistry* 2004; 125:1211–1216.
8. Rajendran, R.G. Polymer electrolyte membrane technology for fuel cells. *MRS Bulletin* 2005; 30:587–590.
9. Prater, K. The renaissance of the solid polymer fuel cell. *Journal of Power Sources* 1990; 29:239–250.
10. Hickner, M.A. Water-mediated transport in ion-containing polymers. *Journal of Polymer Science Part B: Polymer Physics* 2012; 50:9–20.
11. Kreuer, K.-D. Ion conducting membranes for fuel cells and other electrochemical devices. *Chemistry of Materials* 2014; 26:361–380.
12. Li, Q., He, R.H., Jensen, J.O. and Bjerrum, N.J. Approaches and recent development of polymer electrolyte membranes for fuel cells operating above 100°C. *Chemistry of Materials* 2003; 15:4896–4915.
13. Mauritz, K.A. and Moore, R.B. State of understanding of Nafion. *Chemical Reviews* 2004; 104:4535–4585.
14. Zawodzinski, T.A., Derouin, C., Radzinski, S., Sherman, R.J., Smith, V.T., Springer, T.E. and Gottesfeld, S. Water-uptake by and transport trough Nafion® 117 membranes. *Journal of the Electrochemical Society* 1993; 140:1041–1047.
15. Kreuer, K.-D. and Portale, G. A critical revision of the nano-morphology of proton conducting ionomers and polyelectrolytes for fuel cell applications. *Advanced Functional Materials* 2013; 23:5390–5397.
16. Gierke, T.D., Munn, G.E. and Wilson, F.C. The morphology in Nafion perfluorinated membrane products, as determined by wide-angle and small-angle x-ray studies. *Journal of Polymer Science Part B: Polymer Physics* 1981; 19:1687–1704.
17. Rubatat, L., Rollet, A.L., Gebel, G. and Diat, O. Evidence of elongated polymeric aggregates in Nafion. *Macromolecules* 2002; 35:4050–4055.
18. Rubatat, L., Gebel, G. and Diat, O. Fibrillar structure of Nafion: Matching Fourier and real space studies of corresponding films and solutions. *Macromolecules* 2004; 37: 7772–7783.
19. Schmidt-Rohr, K. and Chen, Q. Parallel cylindrical water nanochannels in Nafion fuel-cell membranes. *Nature Materials* 2008; 7:75–83.
20. Li, J., Park, J.K., Moore, R.B. and Madsen, L.A. Linear coupling of alignment with transport in a polymer electrolyte membrane. *Nature Materials* 2011; 10:507–511.
21. McLean, R.S., Doyle, M. and Sauer, B.B. High-resolution imaging of ionic domains and crystal morphology in ionomers using AFM techniques. *Macromolecules* 2000; 33: 6541–6550.
22. Kreuer, K.-D., Paddison, S.J., Spohr, E. and Schuster, M. Transport in proton conductors for fuel-cell applications: Simulations, elementary reactions, and phenomenology. *Chemical Reviews* 2004; 104:4637–4678.
23. Kreuer, K.-D., Rabenau, A. and Weppner, W. Vehicle mechanism, a new model for the interpretation of the conductivity

of fast proton conductors. *Angewandte Chemie International Edition* 1982; 21:208–209.
24. Zawodzinski, T.A., Springer, T.E., Davey, J., Jestel, R., Lopez, C., Valerio, J. and Gottesfeld, S. A comparative-study of water-uptake by and transport through ionomeric fuel-cell membranes. *Journal of the Electrochemical Society* 1993; 140:1981–1985.
25. Xie, G. and Okada, T. Water transport behaviour in Nafion-117 membranes. *Journal of the Electrochemical Society* 1995; 142:3057–3062.
26. Zawodzinski, T.A., Neeman, M., Sillerud, L.O. and Gottesfeld, S. Determination of water diffusion coefficients in perfluorosulfonate ionomeric membranes. *The Journal of Physical Chemistry* 1991; 95:6040–6044.
27. Kreuer, K.-D., Schuster, M., Obliers, B., Diat, O., Traub, U., Fuchs, A., Klock, U., Paddison, S.J. and Maier, J. Short-side-chain proton conducting perfluorosulfonic acid ionomers: Why they perform better in PEM fuel cells. *Journal of Power Sources* 2008; 178:499–509.
28. He, R.H., Li, Q., Bach, A., Jensen, J.O. and Bjerrum, N.J. Physicochemical properties of phosphoric acid doped polybenzimidazole membranes for fuel cells. *Journal of Membrane Science* 2006; 277:38–45.
29. Parthasarathy, A., Srinivasan, S., Appleby, A.J. and Martin, C.R. Temperature-dependence of the electrode-kinetics of oxygen reduction at the platinum Nafion® interface—A microelectrode investigation. *Journal of the Electrochemical Society* 1992; 139:2530–2537.
30. Wycisk, R., Chisholm, J., Lee, J., Lin, J. and Pintauro, P.N. Direct methanol fuel cell membranes from Nafion–polybenzimidazole blends. *Journal of Power Sources* 2006; 163:9–17.
31. Casciola, M., Bagnasco, G., Donnadio, A., Micoli, L., Pica, M., Sganappa, M. and Turco, M. Conductivity and methanol permeability of Nafion–zirconium phosphate composite membranes containing high aspect ratio filler particles. *Fuel Cells* 2009; 9:394–400.
32. Heinzel, A. and Barragan, V.M. A review of the state-of-the-art of the methanol crossover in direct methanol fuel cells. *Journal of Power Sources* 1999; 84:70–74.
33. Ise, M., Kreuer, K.-D. and Maier, J. Electroosmotic drag in polymer electrolyte membranes: An electrophoretic NMR study. *Solid State Ionics* 1999; 125:213–223.
34. Ren, X.M. and Gottesfeld, S. Electro-osmotic drag of water in poly(perfluorosulfonic acid) membranes. *Journal of the Electrochemical Society* 2001; 148:A87–A93.
35. Werner, S., Jörissen, L. and Heider, U. Conductivity and mechanical properties of recast Nafion films. *Ionics* 1996; 2:19–23.
36. Li, J., Yang, X., Tang, H. and Pan, M. Durable and high performance Nafion membrane prepared through high-temperature annealing methodology. *Journal of Membrane Science* 2010; 361:38–42.
37. Alberti, G., Narducci, R., Di Vona, M.L. and Giancola, S. Annealing of Nafion 1100 in the presence of an annealing agent: A powerful method for increasing ionomer working temperature in PEMFCs. *Fuel Cells* 2013; 13:42–47.
38. Bauer, F., Denneler, S. and Willert-Porada, M. Influence of temperature and humidity on the mechanical properties of Nafion® 117 polymer electrolyte membrane. *Journal of Polymer Science Part B: Polymer Physics* 2005; 43:786–795.
39. Page, K.A., Cable, K.M. and Moore, R.B. Molecular origins of the thermal transitions and dynamic mechanical relaxations in perfluorosulfonate ionomers. *Macromolecules* 2005; 38:6472–6484.
40. Casciola, M., Alberti, G., Sganappa, M. and Narducci, R. On the decay of Nafion proton conductivity at high temperature and relative humidity. *Journal of Power Sources* 2006; 162:141–145.
41. Osborn, S.J., Hassan, M.K., Divoux, G.M., Rhoades, D.W., Mauritz, K.A. and Moore, R.B. Glass transition temperature of perfluorosulfonic acid ionomers. *Macromolecules* 2007; 40:3886–3890.
42. Yeo, S.C. and Eisenberg, A. Physical–properties and supermolecular structure of perfluorinated ion-containing (Nafion) polymers. *Journal of Applied Polymer Science* 1977; 21: 875–898.
43. Kyu, T., Hashiyama, M. and Eisenberg, A. Dynamic mechanical studies of partially ionized and neutralized Nafion polymers. *Canadian Journal of Chemistry* 1983; 61:680–687.
44. Lage, L.G., Delgado, P.G. and Kawano, Y. Thermal stability and decomposition of Nafion® membranes with different cations. *Journal of Thermal Analysis and Calorimetry* 2004; 75:521–530.
45. Samms, S.R., Wasmus, S. and Savinell, R.F. Thermal stability of Nafion® in simulated fuel cell environments. *Journal of the Electrochemical Society* 1996; 143:1498–1504.
46. Aili, D., Hansen, M.K., Pan, C., Li, Q., Christensen, E., Jensen, J.O. and Bjerrum, N.J. Phosphoric acid doped membranes based on Nafion®, PBI and their blends—Membrane preparation, characterization and steam electrolysis testing. *International Journal of Hydrogen Energy* 2011; 36: 6985–6993.
47. Yang, J., Li, Q., Cleemann, L.N., Jensen, J.O., Pan, C., Bjerrum, N.J. and He, R. Crosslinked hexafluoropropylidene polybenzimidazole membranes with chloromethyl polysulfone for fuel cell applications. *Advanced Energy Materials* 2013; 3:622–630.
48. Yang, J.S., Cleemann, L.N., Steenberg, T., Terkelsen, C., Li, Q.F., Jensen, J.O., Hjuler, H.A., Bjerrum, N.J. and He, R.H. High molecular weight polybenzimidazole membranes for high temperature PEMFC. *Fuel Cells* 2014; 14:7–15.
49. Teranishi, K., Kawata, K., Tsushima, S. and Hirai, S. Degradation mechanism of PEMFC under open circuit operation. *Electrochemical and Solid-State Letters* 2006; 9: A475–A477.
50. Kundu, S., Fowler, M.W., Simon, L.C., Abouatallah, R. and Beydokhti, N. Degradation analysis and modeling of reinforced catalyst coated membranes operated under OCV conditions. *Journal of Power Sources* 2008; 183:619–628.
51. Sugawara, S., Maruyama, T., Nagahara, Y., Kocha, S.S., Shinohra, K., Tsujita, K., Mitsushima, S. and Ota, K. Performance decay of proton-exchange membrane fuel cells under open circuit conditions induced by membrane decomposition. *Journal of Power Sources* 2009; 187:324–331.
52. Kundu, S., Fowler, M., Simon, L.C. and Abouatallah, R. Reversible and irreversible degradation in fuel cells during open circuit voltage durability testing. *Journal of Power Sources* 2008; 182:254–258.
53. Eisman, G.A. The application of Dow chemical's perfluorinated membranes in proton-exchange membrane fuel cells. *Journal of Power Sources* 1990; 29:389–398.
54. Arcella, V., Troglia, C. and Ghielmi, A. Hyflon ion membranes for fuel cells. *Industrial & Engineering Chemistry Research* 2005; 44:7646–7651.
55. Aricò, A.S., Di Blasi, A., Brunaccini, G., Sergi, F., Dispenza, G., Andaloro, L., Ferraro, M., Antonucci, V., Asher, P., Buche, S., Fongalland, D., Hards, G.A., Sharman, J.D.B., Bayer, A., Heinz, G., Zandonà, N., Zuber, R., Gebert, M., Corasaniti, M.,

Ghielmi, A. and Jones, D.J. High temperature operation of a solid polymer electrolyte fuel cell stack based on a new ionomer membrane. *Fuel Cells* 2010; 10:1013–1023.
56. Ghielmi, A., Vaccarono, P., Troglia, C. and Arcella, V. Proton exchange membranes based on the short-side-chain perfluorinated ionomer. *Journal of Power Sources* 2005; 145:108–115.
57. Gebert, M., Merlo, L. and Arcella, V. Aquivion—The short-side-chain PFSA for next generation PEFCs presents D79-20BS as new stabilized low-EW dispersion grade. *ECS Transactions* 2011; 30:91–95.
58. Mohamed, H.F.M., Kobayashi, Y., Kuroda, S. and Ohira, A. Positron trapping and possible presence of so3h clusters in dry fluorinated polymer electrolyte membranes. *Chemical Physics Letters* 2012; 544:49–52.
59. Liu, Y., Horan, J.L., Schlichting, G.J., Caire, B.R., Liberatore, M.W., Hamrock, S.J., Haugen, G.M., Yandrasits, M.A., Seifert, S. and Herring, A.M. A small-angle x-ray scattering study of the development of morphology in films formed from the 3M perfluorinated sulfonic acid ionomer. *Macromolecules* 2012; 45:7495–7503.
60. Giffin, G.A., Haugen, G.M., Hamrock, S.J. and Di Noto, V. Interplay between structure and relaxations in perfluorosulfonic acid proton conducting membranes. *Journal of the American Chemical Society* 2013; 135:822–834.
61. Kocha, S.S. Polymer electrolyte membrane (PEM) fuel cells, automotive applications, in: K.-D. Kreuer (Ed.) *Fuel Cells: Selected Entries from the Encyclopedia of Sustainability Science and Technology*, Springer, New York, 2013, pp. 473–518.
62. Bae, B., Chun, B.H., Ha, H.Y., Oh, I.H. and Kim, D. Preparation and characterization of plasma treated PP composite electrolyte membranes. *Journal of Membrane Science* 2002; 202:245–252.
63. Rodgers, M.P., Berring, J., Holdcroft, S. and Shi, Z. The effect of spatial confinement of Nafion® in porous membranes on macroscopic properties of the membrane. *Journal of Membrane Science* 2008; 321:100–113.
64. Shim, J., Ha, H.Y., Hong, S.A. and Oh, I.H. Characteristics of the Nafion ionomer-impregnated composite membrane for polymer electrolyte fuel cells. *Journal of Power Sources* 2002; 109:412–417.
65. Ballengee, J.B., Haugen, G.M., Hamrock, S.J. and Pintauro, P.N. Properties and fuel cell performance of a nanofiber composite membrane with 660 equivalent weight perfluorosulfonic acid. *Journal of the Electrochemical Society* 2013; 160:F429–F435.
66. Mauritz, K.A. and Warren, R.M. Microstructural evolution of a silicon-oxide phase in a perfluorosulfonic acid ionomer by an *in situ* sol–gel reaction. 1. Infrared spectroscopic studies. *Macromolecules* 1989; 22:1730–1734.
67. Mauritz, K.A., Stefanithis, I.D., Davis, S.V., Scheetz, R.W., Pope, R.K., Wilkes, G.L. and Huang, H.H. Microstructural evolution of a silicon-oxide phase in a perfluorosulfonic acid ionomer by an *in situ* sol–gel reaction. *Journal of Applied Polymer Science* 1995; 55:181–190.
68. Chen, S.Y., Han, C.C., Tsai, C.H., Huang, J. and Chen-Yang, Y.W. Effect of morphological properties of ionic liquid-templated mesoporous anatase TiO_2 on performance of PEMFC with Nafion/TiO_2 composite membrane at elevated temperature and low relative humidity. *Journal of Power Sources* 2007; 171:363–372.
69. Apichatachutapan, W., Moore, R.B. and Mauritz, K.A. Asymmetric Nafion/(zirconium oxide) hybrid membranes via *in situ* sol–gel chemistry. *Journal of Applied Polymer Science* 1996; 62:417–426.
70. Nørgaard, C.F., Nielsen, U.G. and Skou, E.M. Preparation of Nafion 117™–SnO_2 composite membranes using an ion-exchange method. *Solid State Ionics* 2012; 213:76–82.
71. Costamagna, P., Yang, C., Bocarsly, A.B. and Srinivasan, S. Nafion® 115/zirconium phosphate composite membranes for operation of PEMFCs above 100°C. *Electrochimica Acta* 2002; 47:1023–1033.
72. Alberti, G., Casciola, M., Pica, M., Tarpanelli, T. and Sganappa, M. New preparation methods for composite membranes for medium temperature fuel cells based on precursor solutions of insoluble inorganic compounds. *Fuel Cells* 2005; 5:366–374.
73. Rodgers, M.P., Shi, Z. and Holdcroft, S. *Ex situ* characterisation of composite Nafion membranes containing zirconium hydrogen phosphate. *Fuel Cells* 2009; 9:534–546.
74. Mohajeri, N., Pearman, B.P., Rodgers, M., Agarwal, R., Slattery, D., Bonville, L., Kunz, H.R. and Fenton, J.M. 750 EW perfluorosulfonic acid composite membranes with stabilized phosphotungstic acid for high temperature/low relative humidity PEM fuel cells. *ECS Transactions* 2008; 16:1163–1171.
75. Tian, H. and Savadogo, O. Silicotungstic acid Nafion composite membrane for proton-exchange membrane fuel cell operation at high temperature. *Journal of New Materials for Electrochemical Systems* 2006; 9:61–71.
76. Alberti, G. and Casciola, M. Composite membranes for medium-temperature PEM fuel cells. *Annual Review of Materials Research* 2003; 33:129–154.
77. Savinell, R., Yeager, E., Tryk, D., Landau, U., Wainright, J., Weng, D., Lux, K., Litt, M. and Rogers, C. A polymer electrolyte for operation at temperatures up to 200°C. *Journal of the Electrochemical Society* 1994; 141:L46–L48.
78. Aili, D., Savinell, R.F., Jensen, J.O., Cleemann, L.N., Bjerrum, N.J. and Li, Q. The electrochemical behavior of phosphoric-acid-doped poly(perfluorosulfonic acid) membranes. *ChemElectroChem* 2014, doi:10.1002/celc.201402053.
79. Doyle, M., Choi, S.K. and Proulx, G. High-temperature proton conducting membranes based on perfluorinated ionomer membrane-ionic liquid composites. *Journal of the Electrochemical Society* 2000; 147:34–37.
80. Savadogo, O. Emerging membranes for electrochemical systems: (I) Solid polymer electrolyte membranes for fuel cell systems. *Journal of New Materials for Electrochemical Systems* 1998; 1:47–66.
81. Prober, M. The synthesis and polymerization of some fluorinated styrenes. *Journal of the American Chemical Society* 1953; 75:968–973.
82. Gubler, L., Gursel, S.A. and Scherer, G.G. Radiation grafted membranes for polymer electrolyte fuel cells. *Fuel Cells* 2005; 5:317–335.
83. Huslage, J., Rager, T., Schnyder, B. and Tsukada, A. Radiation-grafted membrane/electrode assemblies with improved interface. *Electrochimica Acta* 2002; 48:247–254.
84. Hickner, M.A., Ghassemi, H., Kim, Y.S., Einsla, B.R. and McGrath, J.E. Alternative polymer systems for proton exchange membranes (PEMs). *Chemical Reviews* 2004; 104:4587–4611.
85. Zhang, H.W. and Shen, P.K. Recent development of polymer electrolyte membranes for fuel cells. *Chemical Reviews* 2012; 112:2780–2832.
86. Kreuer, K.-D. On the development of proton conducting polymer membranes for hydrogen and methanol fuel cells. *Journal of Membrane Science* 2001; 185:29–39.
87. Jones, D.J. and Rozière, J. Recent advances in the functionalisation of polybenzimidazole and polyetherketone for fuel cell applications. *Journal of Membrane Science* 2001; 185:41–58.

88. Noshay, A. and Robeson, L.M. Sulfonated polysulfone. *Journal of Applied Polymer Science* 1976; 20:1885–1903.
89. Genova-Dimitrova, P., Baradie, B., Foscallo, D., Poinsignon, C. and Sanchez, J.Y. Ionomeric membranes for proton exchange membrane fuel cell (PEMFC): Sulfonated polysulfone associated with phosphatoantimonic acid. *Journal of Membrane Science* 2001; 185:59–71.
90. Dizman, C., Tasdelen, M.A. and Yagci, Y. Recent advances in the preparation of functionalized polysulfones. *Polymer International* 2013; 62:991–1007.
91. Wang, F., Hickner, M., Ji, Q., Harrison, W., Mecham, J., Zawodzinski, T.A. and McGrath, J.E. Synthesis of highly sulfonated poly(arylene ether sulfone) random (statistical) copolymers via direct polymerization. *Macromolecular Symposia* 2001; 175:387–395.
92. Wang, F., Hickner, M., Kim, Y.S., Zawodzinski, T.A. and McGrath, J.E. Direct polymerization of sulfonated poly(arylene ether sulfone) random (statistical) copolymers: Candidates for new proton exchange membranes. *Journal of Membrane Science* 2002; 197:231–242.
93. Hou, J.B., Li, J. and Madsen, L.A. Anisotropy and transport in poly(arylene ether sulfone) hydrophilic–hydrophobic block copolymers. *Macromolecules* 2010; 43:347–353.
94. Takamuku, S. and Jannasch, P. Properties and degradation of hydrocarbon fuel cell membranes: A comparative study of sulfonated poly(arylene ether sulfone)s with different positions of the acid groups. *Polymer Chemistry* 2012; 3:1202–1214.
95. Zhu, Z., Walsby, N.M., Colquhoun, H.M., Thompsett, D. and Petrucco, E. Microblock ionomers: A new concept in high temperature, swelling-resistant membranes for PEM fuel cells. *Fuel Cells* 2009; 9:305–317.
96. Miyatake, K. and Hay, A.S. Synthesis of novel phosphinic acid-containing polymers. *Journal of Polymer Science Part A: Polymer Chemistry* 2001; 39:1854–1859.
97. Miyatake, K. and Hay, A.S. New poly(arylene ether)s with pendant phosphonic acid groups. *Journal of Polymer Science Part A: Polymer Chemistry* 2001; 39:3770–3779.
98. Asano, N., Aoki, M., Suzuki, S., Miyatake, K., Uchida, H. and Watanabe, M. Aliphatic/aromatic polyimide ionomers as a proton conductive membrane for fuel cell applications. *Journal of the American Chemical Society* 2006; 128:1762–1769.
99. Genies, C., Mercier, R., Sillion, B., Petiaud, R., Cornet, N., Gebel, G. and Pineri, M. Stability study of sulfonated phthalic and naphthalenic polyimide structures in aqueous medium. *Polymer* 2001; 42:5097–5105.
100. Genies, C., Mercier, R., Sillion, B., Cornet, N., Gebel, G. and Pineri, M. Soluble sulfonated naphthalenic polyimides as materials for proton exchange membranes. *Polymer* 2001; 42:359–373.
101. Yan, J.L., Liu, C.P., Wang, Z., Xing, W. and Ding, M.M. Water resistant sulfonated polyimides based on 4,4′-binaphthyl-1,1′,8,8′-tetracarboxylic dianhydride (BNTDA) for proton exchange membranes. *Polymer* 2007; 48:6210–6214.
102. Chung, T.S. A critical review of polybenzimidazoles: Historical development and future R&D. *Journal of Macromolecular Science, Reviews in Macromolecular Chemistry and Physics* 1997; C37:277–301.
103. Linkous, C.A. and Slattery, D.K. Characterization of sulfonic acids of high-temperature polymers as membranes for water electrolysis. *Abstracts of Papers of the American Chemical Society* 1993; 205:122–123.
104. Staiti, P., Lufrano, F., Aricò, A.S., Passalacqua, E. and Antonucci, V. Sulfonated polybenzimidazole membranes—Preparation and physico-chemical characterization. *Journal of Membrane Science* 2001; 188:71–78.
105. Ariza, M.J., Jones, D.J. and Rozière, J. Role of post-sulfonation thermal treatment in conducting and thermal properties of sulfuric acid sulfonated poly(benzimidazole) membranes. *Desalination* 2002; 147:183–189.
106. Peron, J., Ruiz, E., Jones, D.J. and Rozière, J. Solution sulfonation of a novel polybenzimidazole. A proton electrolyte for fuel cell application. *Journal of Membrane Science* 2008; 314:247–256.
107. Mader, J.A. and Benicewicz, B.C. Sulfonated polybenzimidazoles for high temperature PEM fuel cells. *Macromolecules* 2010; 43:6706–6715.
108. Asensio, J.A., Borros, S. and Gomez-Romero, P. Proton-conducting polymers based on benzimidazoles and sulfonated benzimidazoles. *Journal of Polymer Science Part A: Polymer Chemistry* 2002; 40:3703–3710.
109. Qing, S., Huang, W. and Yan, D. Synthesis and characterization of thermally stable sulfonated polybenzimidazoles. *European Polymer Journal* 2005; 41:1589–1595.
110. Thomas, O.D., Peckham, T.J., Thanganathan, U., Yang, Y. and Holdcroft, S. Sulfonated polybenzimidazoles: Proton conduction and acid–base crosslinking. *Journal of Polymer Science Part A: Polymer Chemistry* 2010; 48:3640–3650.
111. Gieselman, M.B. and Reynolds, J.R. Water-soluble polybenzimidazole-based polyelectrolytes. *Macromolecules* 1992; 25: 4832–4834.
112. Glipa, X., El Haddad, M., Jones, D.J. and Rozière, J. Synthesis and characterisation of sulfonated polybenzimidazole: A highly conducting proton exchange polymer. *Solid State Ionics* 1997; 97:323–331.
113. Bae, J.M., Honma, I., Murata, M., Yamamoto, T., Rikukawa, M. and Ogata, N. Properties of selected sulfonated polymers as proton-conducting electrolytes for polymer electrolyte fuel cells. *Solid State Ionics* 2002; 147:189–194.
114. Wycisk, R. and Pintauro, P.N. Sulfonated polyphosphazene ion-exchange membranes. *Journal of Membrane Science* 1996; 119:155–160.
115. Rozière, J. and Jones, D.J. Non-fluorinated polymer materials for proton exchange membrane fuel cells. *Annual Review of Materials Research* 2003; 33:503–555.
116. Vilčiauskas, L., Tuckerman, M.E., Bester, G., Paddison, S.J. and Kreuer, K.-D. The mechanism of proton conduction in phosphoric acid. *Nature Chemistry* 2012; 4:461–466.
117. Wainright, J.S., Wang, J.T., Weng, D., Savinell, R.F. and Litt, M. Acid-doped polybenzimidazoles—A new polymer electrolyte. *Journal of the Electrochemical Society* 1995; 142:L121–L123.
118. Wang, J.T., Wainright, J.S., Savinell, R.F. and Litt, M. A direct methanol fuel cell using acid-doped polybenzimidazole as polymer electrolyte. *Journal of Applied Electrochemistry* 1996; 26:751–756.
119. Weng, D., Wainright, J.S., Landau, U. and Savinell, R.F. Electro-osmotic drag coefficient of water and methanol in polymer electrolytes at elevated temperatures. *Journal of the Electrochemical Society* 1996; 143:1260–1263.
120. Pu, H.T. and Liu, Q.Z. Methanol permeability and proton conductivity of polybenzimidazole and sulfonated polybenzimidazole. *Polymer International* 2004; 53:1512–1516.
121. Lobato, J., Cañizares, P., Rodrigo, M.A., Linares, J.J. and López-Vizcaíno, R. Performance of a vapor-fed polybenzimidazole (PBI)-based direct methanol fuel cell. *Energy & Fuels* 2008; 22:3335–3345.
122. Araya, S.S., Andreasen, S.J., Nielsen, H.V. and Kær, S.K. Investigating the effects of methanol–water vapor mixture on a PBI-based high temperature PEM fuel cell. *International Journal of Hydrogen Energy* 2012; 37:18231–18242.

123. Vogel, H. and Marvel, C.S. Polybenzimidazoles, new thermally stable polymers. *Journal of Polymer Science* 1961; 50: 511–539.
124. Neuse, E.W. Aromatic polybenzimidazoles—Synthesis, properties and applications. *Advances in Polymer Science* 1982; 47:1–42.
125. Iwakura, Y., Imai, Y. and Uno, K. Polyphenylenebenzimidazoles. *Journal of Polymer Science Part: A General Papers* 1964; 2:2605–2615.
126. Hanley, T.R., Helminiak, T.E. and Benner, C.L. Expansion of aromatic heterocyclic polymers in salt solution. *Journal of Applied Polymer Science* 1978; 22:2965–2978.
127. Li, Q., He, R.H., Berg, R.W., Hjuler, H.A. and Bjerrum, N.J. Water uptake and acid doping of polybenzimidazoles as electrolyte membranes for fuel cells. *Solid State Ionics* 2004; 168:177–185.
128. Xing, B.Z. and Savadogo, O. The effect of acid doping on the conductivity of polybenzimidazole (PBI). *Journal of New Materials for Electrochemical Systems* 1999; 2:95–101.
129. Xiao, L.X., Zhang, H.F., Scanlon, E., Ramanathan, L.S., Choe, E.W., Rogers, D., Apple, T. and Benicewicz, B.C. High-temperature polybenzimidazole fuel cell membranes via a sol–gel process. *Chemistry of Materials* 2005; 17: 5328–5333.
130. Liu, Z., Tsou, Y.-M., Calundann, G. and De Castro, E. New process for high temperature polybenzimidazole membrane production and its impact on the membrane and the membrane electrode assembly. *Journal of Power Sources* 2011; 196:1055–1060.
131. Litt, M., Ameri, R., Wang, Y., Savinell, R. and Wainright, J. Polybenzimidazoles/phosphoric acid solid polymer electrolytes: Mechanical and electrical properties, in: G.A. Nazri, C. Julien, A. Rougier (Eds.) *Solid State Ionics V*, vol. 548, Materials Research Society, Warrendale, PA, 1999, pp. 313–323.
132. Aili, D., Li, Q., Christensen, E., Jensen, J.O. and Bjerrum, N.J. Crosslinking of polybenzimidazole membranes by divinylsulfone post-treatment for high-temperature proton exchange membrane fuel cell applications. *Polymer International* 2011; 60:1201–1207.
133. Aili, D., Cleemann, L.N., Li, Q., Jensen, J.O., Christensen, E. and Bjerrum, N.J. Thermal curing of PBI membranes for high temperature PEM fuel cells. *Journal of Materials Chemistry* 2012; 22:5444–5453.
134. Bouchet, R. and Siebert, E. Proton conduction in acid doped polybenzimidazole. *Solid State Ionics* 1999; 118:287–299.
135. Hughes, C.E., Haufe, S., Angerstein, B., Kalim, R., Mahr, U., Reiche, A. and Baldus, M. Probing structure and dynamics in poly[2,2′-(*m*-phenylene)-5,5′-bibenzimidazole] fuel cells with magic-angle spinning NMR. *Journal of Physical Chemistry B* 2004; 108:13626–13631.
136. Ma, Y.L., Wainright, J.S., Litt, M.H. and Savinell, R.F. Conductivity of PBI membranes for high-temperature polymer electrolyte fuel cells. *Journal of the Electrochemical Society* 2004; 151:A8–A16.
137. He, R.H., Li, Q., Jensen, J.O. and Bjerrum, N.J. Doping phosphoric acid in polybenzimidazole membranes for high temperature proton exchange membrane fuel cells. *Journal of Polymer Science Part A: Polymer Chemistry* 2007; 45: 2989–2997.
138. Perry, A.K., More, L.K., Andrew Payzant, E., Meisner, R.A., Sumpter, B.G. and Benicewicz, B.C. A comparative study of phosphoric acid–doped *m*-PBI membranes. *Journal of Polymer Science Part B: Polymer Physics* 2014; 52:26–35.
139. Mecerreyes, D., Grande, H., Miguel, O., Ochoteco, E., Marcilla, R. and Cantero, I. Porous polybenzimidazole membranes doped with phosphoric acid: Highly proton-conducting solid electrolytes. *Chemistry of Materials* 2004; 16:604–607.
140. Jayakody, J.R.P., Chung, S.H., Durantino, L., Zhang, H., Xiao, L., Benicewicz, B.C. and Greenbaum, S.G. NMR studies of mass transport in high-acid-content fuel cell membranes based on phosphoric acid and polybenzimidazole. *Journal of the Electrochemical Society* 2007; 154:B242–B246.
141. Li, Q., Hjuler, H.A. and Bjerrum, N.J. Phosphoric acid doped polybenzimidazole membranes: Physiochemical characterization and fuel cell applications. *Journal of Applied Electrochemistry* 2001; 31:773–779.
142. He, R.H., Li, Q., Xiao, G. and Bjerrum, N.J. Proton conductivity of phosphoric acid doped polybenzimidazole and its composites with inorganic proton conductors. *Journal of Membrane Science* 2003; 226:169–184.
143. Aili, D., Allward, T., Alfaro, S.M., Hartmann-Thompson, C., Steenberg, T., Hjuler, H.A., Li, Q., Jensen, J.O. and Stark, E.J. Polybenzimidazole and sulfonated polyhedral oligosilsesquioxane composite membranes for high temperature polymer electrolyte membrane fuel cells. *Electrochimica Acta* 2014; 140:182-190.
144. Neyerlin, K.C., Singh, A. and Chu, D. Kinetic characterization of a Pt–Ni/C catalyst with a phosphoric acid doped PBI membrane in a proton exchange membrane fuel cell. *Journal of Power Sources* 2008; 176:112–117.
145. Schmidt, T.J. and Baurmeister, J. Properties of high-temperature PEFC Celtec®-P 1000 MEAs in start/stop operation mode. *Journal of Power Sources* 2008; 176:428–434.
146. Yu, S., Xiao, L. and Benicewicz, B.C. Durability studies of PBI-based high temperature PEMFCs. *Fuel Cells* 2008; 8: 165–174.
147. Asensio, J.A., Borrós, S. and Gómez-Romero, P. Polymer electrolyte fuel cells based on phosphoric acid–impregnated poly(2,5-benzimidazole) membranes. *Journal of the Electrochemical Society* 2004; 151:A304–A310.
148. Gulledge, A.L., Gu, B. and Benicewicz, B.C. A new sequence isomer of AB–polybenzimidazole for high-temperature PEM fuel cells. *Journal of Polymer Science Part A: Polymer Chemistry* 2012; 50:306–313.
149. Xiao, L., Zhang, H., Jana, T., Scanlon, E., Chen, R., Choe, E.W., Ramanathan, L.S., Yu, S. and Benicewicz, B.C. Synthesis and characterization of pyridine-based polybenzimidazoles for high temperature polymer electrolyte membrane fuel cell applications. *Fuel Cells* 2005; 5:287–295.
150. Carollo, A., Quartarone, E., Tomasi, C., Mustarelli, P., Belotti, F., Magistris, A., Maestroni, F., Parachini, M., Garlaschelli, L. and Righetti, P.P. Developments of new proton conducting membranes based on different polybenzimidazole structures for fuel cells applications. *Journal of Power Sources* 2006; 160:175–180.
151. Sannigrahi, A., Ghosh, S., Maity, S. and Jana, T. Structurally isomeric monomers directed copolymerization of polybenzimidazoles and their properties. *Polymer* 2010; 51: 5929–5941.
152. Yu, S. and Benicewicz, B.C. Synthesis and properties of functionalized polybenzimidazoles for high-temperature PEMFCs. *Macromolecules* 2009; 42:8640–8648.
153. Suarez, S., Kodiweera, N., Stallworth, P., Yu, S., Greenbaum, S.G. and Benicewicz, B.C. Multinuclear NMR study of the effect of acid concentration on ion transport in phosphoric acid doped poly(benzimidazole) membranes. *Journal of Physical Chemistry B* 2012; 116:12545–12551.

154. Sannigrahi, A., Ghosh, S., Lalnuntluanga, J. and Jana, T. How the monomer concentration of polymerization influences various properties of polybenzimidazole: A case study with poly(4,4′-diphenylether-5,5′-bibenzimidazole). *Journal of Applied Polymer Science* 2009; 111:2194–2203.
155. Kim, T.-H., Kim, S.-K., Lim, T.-W. and Lee, J.-C. Synthesis and properties of poly(aryl ether benzimidazole) copolymers for high-temperature fuel cell membranes. *Journal of Membrane Science* 2008; 323:362–370.
156. Dai, H., Zhang, H., Zhong, H., Jin, H., Li, X., Xiao, S. and Mai, Z. Properties of polymer electrolyte membranes based on poly(aryl ether benzimidazole) and sulphonated poly(aryl ether benzimidazole) for high temperature PEMFCs. *Fuel Cells* 2010; 10:754–761.
157. Li, Q., Rudbeck, H.C., Chromik, A., Jensen, J.O., Pan, C., Steenberg, T., Calverley, M., Bjerrum, N.J. and Kerres, J. Properties, degradation and high temperature fuel cell test of different types of PBI and PBI blend membranes. *Journal of Membrane Science* 2010; 347:260–270.
158. Yang, J.S., Li, Q.F., Cleemann, L.N., Xu, C.X., Jensen, J.O., Pan, C., Bjerrum, N.J. and He, R.H. Synthesis and properties of poly(aryl sulfone benzimidazole) and its copolymers for high temperature membrane electrolytes for fuel cells. *Journal of Materials Chemistry* 2012; 22:11185–11195.
159. Yang, J., Aili, D., Li, Q., Cleemann, L.N., Jensen, J.O., Bjerrum, N.J. and He, R. Covalently cross-linked sulfone polybenzimidazole membranes with poly(vinylbenzyl chloride) for fuel cell applications. *ChemSusChem* 2013; 6:275–282.
160. Chuang, S.W. and Hsu, S.L.C. Synthesis and properties of a new fluorine-containing polybenzimidazole for high-temperature fuel-cell applications. *Journal of Polymer Science Part A: Polymer Chemistry* 2006; 44:4508–4513.
161. Qian, G. and Benicewicz, B.C. Synthesis and characterization of high molecular weight hexafluoroisopropylidene-containing polybenzimidazole for high-temperature polymer electrolyte membrane fuel cells. *Journal of Polymer Science Part A: Polymer Chemistry* 2009; 47:4064–4073.
162. Yang, J., Aili, D., Li, Q., Xu, Y., Liu, P., Che, Q., Jensen, J.O., Bjerrum, N.J. and He, R. Benzimidazole grafted polybenzimidazoles for proton exchange membrane fuel cells. *Polymer Chemistry* 2013; 4:4768–4775.
163. Yang, J.S., Li, Q.F., Jensen, J.O., Pan, C., Cleemann, L.N., Bjerrum, N.J. and He, R.H. Phosphoric acid doped imidazolium polysulfone membranes for high temperature proton exchange membrane fuel cells. *Journal of Power Sources* 2012; 205:114–121.
164. Li, Q., Pan, C., Jensen, J.O., Noyé, P. and Bjerrum, N.J. Cross-linked polybenzimidazole membranes for fuel cells. *Chemistry of Materials* 2007; 19:350–352.
165. Noyé, P., Li, Q., Pan, C. and Bjerrum, N.J. Cross-linked polybenzimidazole membranes for high temperature proton exchange membrane fuel cells with dichloromethyl phosphinic acid as a cross-linker. *Polymers for Advanced Technologies* 2008; 19:1270–1275.
166. Han, M., Zhang, G., Liu, Z., Wang, S., Li, M., Zhu, J., Li, H., Zhang, Y., Lew, C.M. and Na, H. Cross-linked polybenzimidazole with enhanced stability for high temperature proton exchange membrane fuel cells. *Journal of Materials Chemistry* 2011; 21:2187–2193.
167. Kerres, J., Ullrich, A., Meier, F. and Häring, T. Synthesis and characterization of novel acid–base polymer blends for application in membrane fuel cells. *Solid State Ionics* 1999; 125:243–249.
168. Kerres, J., Ullrich, A., Häring, T., Baldauf, M., Gebhardt, U. and Preidel, W. Preparation, characterization and fuel cell application of new acid–base blend membranes. *Journal of New Materials for Electrochemical Systems* 2000; 3: 229–239.
169. Deimede, V., Voyiatzis, G.A., Kallitsis, J.K., Li, Q. and Bjerrum, N.J. Miscibility behavior of polybenzimidazole/sulfonated polysulfone blends for use in fuel cell applications. *Macromolecules* 2000; 33:7609–7617.
170. Hasiotis, C., Li, Q., Deimede, V., Kallitsis, J.K., Kontoyannis, C.G. and Bjerrum, N.J. Development and characterization of acid-doped polybenzimidazole/sulfonated polysulfone blend polymer electrolytes for fuel cells. *Journal of the Electrochemical Society* 2001; 148:A513–A519.
171. Kerres, J., Schönberger, F., Chromik, A., Häring, T., Li, Q., Jensen, J.O., Pan, C., Noyé, P. and Bjerrum, N.J. Partially fluorinated arylene polyethers and their ternary blend membranes with PBI and H_3PO_4. Part I. Synthesis and characterisation of polymers and binary blend membranes. *Fuel Cells* 2008; 8:175–187.
172. Li, Q., Jensen, J.O., Pan, C., Bandur, V., Nilsson, M.S., Schönberger, F., Chromik, A., Hein, M., Häring, T., Kerres, J. and Bjerrum, N.J. Partially fluorinated aarylene polyethers and their ternary blends with PBI and H_3PO_4. Part II. Characterisation and fuel cell tests of the ternary membranes. *Fuel Cells* 2008; 8:188–199.
173. Katzfuß, A., Krajinovic, K., Chromik, A. and Kerres, J. Partially fluorinated sulfonated poly(arylene sulfone)s blended with polybenzimidazole. *Journal of Polymer Science Part A: Polymer Chemistry* 2011; 49:1919–1927.
174. Zaidi, S.M.J. Preparation and characterization of composite membranes using blends of sPEEK/PBI with boron phosphate. *Electrochimica Acta* 2005; 50:4771–4777.
175. Zhang, H., Li, X., Zhao, C., Fu, T., Shi, Y. and Na, H. Composite membranes based on highly sulfonated PEEK and PBI: Morphology characteristics and performance. *Journal of Membrane Science* 2008; 308:66–74.
176. Wycisk, R., Lee, J.K. and Pintauro, P.N. Sulfonated polyphosphazene–polybenzimidazole membranes for DMFCs. *Journal of the Electrochemical Society* 2005; 152:A892–A898.
177. Ainla, A. and Brandell, D. Nafion®–polybenzimidazole (PBI) composite membranes for DMFC applications. *Solid State Ionics* 2007; 178:581–585.
178. Zhai, Y.F., Zhang, H.M., Zhang, Y. and Xing, D.M. A novel H_3PO_4/Nafion–PBI composite membrane for enhanced durability of high temperature PEM fuel cells. *Journal of Power Sources* 2007; 169:259–264.
179. Gubler, L., Kramer, D., Belack, J., Unsal, O., Schmidt, T.J. and Scherer, G.G. Celtec-V—A polybenzimidazole-based membrane for the direct methanol fuel cell. *Journal of the Electrochemical Society* 2007; 154:B981–B987.
180. Berber, M.R., Fujigaya, T., Sasaki, K. and Nakashima, N. Remarkably durable high temperature polymer electrolyte fuel cell based on poly(vinylphosphonic acid)-doped polybenzimidazole. *Scientific Reports* 2013; 3:2–7.
181. Yuan, S., Guo, X., Aili, D., Pan, C., Li, Q. and Fang, J. Poly(imide benzimidazole)s for high temperature polymer electrolyte membrane fuel cells. *Journal of Membrane Science* 2014; 454:351–358.
182. Kallitsis, J.K., Geormezi, M. and Neophytides, S.G. Polymer electrolyte membranes for high-temperature fuel cells based on aromatic polyethers bearing pyridine units. *Polymer International* 2009; 58:1226–1233.

183. Lobato, J., Cañizares, P., Rodrigo, M.A., Úbeda, D. and Pinar, F.J. A novel titanium PBI-based composite membrane for high temperature PEMFCs. *Journal of Membrane Science* 2011; 369:105–111.
184. Plackett, D., Siu, A., Li, Q., Pan, C., Jensen, J.O., Nielsen, S.F., Permyakova, A.A. and Bjerrum, N.J. High-temperature proton exchange membranes based on polybenzimidazole and clay composites for fuel cells. *Journal of Membrane Science* 2011; 383:78–87.
185. Kurdakova, V., Quartarone, E., Mustarelli, P., Magistris, A., Caponetti, E. and Saladino, M.L. PBI-based composite membranes for polymer fuel cells. *Journal of Power Sources* 2010; 195:7765–7769.
186. Ghosh, S., Maity, S. and Jana, T. Polybenzimidazole/silica nanocomposites: Organic–inorganic hybrid membranes for PEM fuel cell. *Journal of Materials Chemistry* 2011; 21: 14897–14906.
187. Staiti, P., Minutoli, M. and Hocevar, S. Membranes based on phosphotungstic acid and polybenzimidazole for fuel cell application. *Journal of Power Sources* 2000; 90:231–235.
188. Staiti, P. and Minutoli, M. Influence of composition and acid treatment on proton conduction of composite polybenzimidazole membranes. *Journal of Power Sources* 2001; 94:9–13.
189. Staiti, P. Proton conductive membranes based on silicotungstic acid/silica and polybenzimidazole. *Materials Letters* 2001; 47:241–246.
190. Lobato, J., Cañizares, P., Rodrigo, M.A., Úbeda, D. and Pinar, F.J. Promising $TtiOSO_4$ composite polybenzimidazole-based membranes for high temperature PEMFCs. *ChemSusChem* 2011; 4:1489–1497.
191. Di, S.Q., Yan, L.M., Han, S.Y., Yue, B.H., Feng, Q.X., Xie, L.Q., Chen, J., Zhang, D.F. and Sun, C. Enhancing the high-temperature proton conductivity of phosphoric acid doped poly(2,5-benzimidazole) by preblending boron phosphate nanoparticles to the raw materials. *Journal of Power Sources* 2012; 211:161–168.
192. Suryani and Liu, Y.L. Preparation and properties of nanocomposite membranes of polybenzimidazole/sulfonated silica nanoparticles for proton exchange membranes. *Journal of Membrane Science* 2009; 332:121–128.
193. Hu, Y., Jensen, J.O., Zhang, W., Cleemann, L.N., Xing, W., Bjerrum, N.J. and Li, Q. Hollow spheres of iron carbide nanoparticles encased in graphitic layers as oxygen reduction catalysts. *Angewandte Chemie International Edition* 2014; 53:3675–3679.
194. Lu, S.F., Pan, J., Huang, A.B., Zhuang, L. and Lu, J.T. Alkaline polymer electrolyte fuel cells completely free from noble metal catalysts. *Proceedings of the National Academy of Sciences of the United States of America* 2008; 105:20611–20614.
195. Couture, G., Alaaeddine, A., Boschet, F. and Ameduri, B. Polymeric materials as anion-exchange membranes for alkaline fuel cells. *Progress in Polymer Science* 2011; 36:1521–1557.
196. Merle, G., Wessling, M. and Nijmeijer, K. Anion exchange membranes for alkaline fuel cells: A review. *Journal of Membrane Science* 2011; 377:1–35.
197. Hickner, M.A., Herring, A.M. and Coughlin, E.B. Anion exchange membranes: Current status and moving forward. *Journal of Polymer Science Part B: Polymer Physics* 2013; 51:1727–1735.
198. Edson, J.B., Macomber, C.S., Pivovar, B.S. and Boncella, J.M. Hydroxide based decomposition pathways of alkyltrimethylammonium cations. *Journal of Membrane Science* 2012; 399:49–59.
199. Arges, C.G. and Ramani, V. Two-dimensional NMR spectroscopy reveals cation-triggered backbone degradation in polysulfone-based anion exchange membranes. *Proceedings of the National Academy of Sciences of the United States of America* 2013; 110:2490–2495.
200. Marino, M.G., Melchior, J.P., Wohlfarth, A. and Kreuer, K.-D. Hydroxide, halide and water transport in a model anion exchange membrane. *Journal of Membrane Science* 2014; 464:61–71.
201. Staudt, R. Development of polybenzimidazole-based high temperature membrane and electrode assemblies for stationary applications, in: *FY 2006 Annual Progress Report*, DOE Energy Program, 2006, pp. 718–721. (see http://www.hydrogen.energy.gov/annual_progress06.html)
202. Staiti, P., Arico, A.S., Baglio, V., Lufrano, F., Passalacqua, E. and Antonucci, V. Hybrid Nafion–silica membranes doped with heteropolyacids for application in direct methanol fuel cells. *Solid State Ionics* 2001; 145:101–107.
203. Bennett, B., Koraishy, B.M. and Meyers, J.P. Modeling and optimization of the DMFC system: Relating materials properties to system size and performance. *Journal of Power Sources* 2012; 218:268–279.
204. Pan, J., Lu, S.F., Li, Y., Huang, A.B., Zhuang, L. and Lu, J.T. High-performance alkaline polymer electrolyte for fuel cell applications. *Advanced Functional Materials* 2010; 20:312–319.
205. Xiao, G., Li, Q.F., Hjuler, H.A. and Bjerrum, N.J. Hydrogen oxidation on gas diffusion electrodes for phosphoric acid fuel cells in the presence of carbon monoxide and oxygen. *Journal of the Electrochemical Society* 1995; 142:2890–2893.
206. Debe, M.K. Novel catalysts, catalyst supports and catalyst coated membrane methods, in: W. Vielsstich, A. Lamm, H.A. Gasteiger (Eds.) *Handbook of Fuel Cells*, vol. 3, John Wiley & Sons, Chichester, 2003, pp. 576–589.
207. Song, Y., Fenton, J.M., Kunz, H.R., Bonville, L.J. and Williams, M.V. High-performance PEMFCs at elevated temperatures using Nafion 112 membranes. *Journal of the Electrochemical Society* 2005; 152:A539–A544.
208. Oetjen, H.F., Schmidt, V.M., Stimming, U. and Trila, F. Performance data of a proton exchange membrane fuel cell using H_2/CO as fuel gas. *Journal of the Electrochemical Society* 1996; 143:3838–3842.
209. Borup, R., Meyers, J., Pivovar, B., Kim, Y.S., Mukundan, R., Garland, N., Myers, D., Wilson, M., Garzon, F., Wood, D., Zelenay, P., More, K., Stroh, K., Zawodzinski, T., Boncella, J., McGrath, J.E., Inaba, M., Miyatake, K., Hori, M., Ota, K., Ogumi, Z., Miyata, S., Nishikata, A., Siroma, Z., Uchimoto, Y., Yasuda, K., Kimijima, K.I. and Iwashita, N. Scientific aspects of polymer electrolyte fuel cell durability and degradation. *Chemical Reviews* 2007; 107:3904–3951.

16 Oxygen Reduction Non-PGM Electrocatalysts for PEM Fuel Cells—Recent Advances

Surya Singh, Anil Verma, and Suddhasatwa Basu

CONTENTS

16.1 INTRODUCTION

Polymer electrolyte membrane fuel cell (PEMFC) is one of the most promising energy conversion devices in terms of commercialization for use in transportation and distributed power generation systems. PEMFC offers attractive features such as high power density, modularity, solid-state construction, low operating temperature and pressure, longer lifetime, minimal emissions, quick start-up, negligible noise, and high chemical to electrical energy conversion efficiency. PEMFC may generate power in the range of watts to kilowatts.

PEMFC exhibits good energy efficiency and higher power density per unit volume when platinum is used as the electrocatalyst. However, the high cost of the platinum catalyst, PEM, and control system, as well as the lack of efficient bipolar plate are the main constraints for the worldwide commercialization of PEMFC. A big portion of a fuel cell cost comes from the materials, namely precious metals like platinum. Platinum is being used as an electrocatalyst at the anode for hydrogen oxidation reaction (HOR) and also at the cathode for oxygen reduction reaction (ORR).

The ORR that occurs at the cathode follows a more complicated mechanism than HOR. It is well known for its sluggish kinetics, as it is the source of more than half of the voltage loss in a PEMFC. Moreover, ORR confers a major challenge for PEMFCs because the catalyst material must be stable under extremely corrosive conditions at a fuel cell cathode yet chemically active enough to reduce O_2. Owing to the difficulties of ORR, the cathode requires a higher Pt loading, typically more than several times that of the anode.

It is reported by the US Department of Energy that the cost of automotive fuel cell systems is reduced to \$47 kW^{-1} in 2012 (projected to high volume manufacturing), which is more than 35% reduction since 2008 and more than 80% reduction since 2002, and well on the way to achieve the target of \$30 kW^{-1} by 2017. It is reported that the reduced platinum content of fuel cells is achieved by more than doubling the catalyst-specific power from the 2008 baseline of 2.8 kW g^{-1} of platinum group metal (PGM) to 5.8 kW g^{-1} in 2012. Current catalyst-specific power is approaching the 2017 target of 8.0 kW g^{-1}, and it reflects a more than 80% reduction in PGM content since 2005. An average fuel cell–powered car needs about 30 g of platinum, which may cost more than \$2000 (99.99% pure bulk platinum). Moreover, it is anticipated that the availability of platinum may not be sufficient to cater the demand, as platinum is an extremely rare metal and exists at a concentration of 0.005 ppm in the crust of the earth.

Platinum and other PGMs are highly catalytically active metals. PGM is a commonly used term and collectively refers to six transition metal elements clustered together in the *d*-block (groups 8, 9, and 10, period 5 and 6) of the periodic table. The six PGMs are ruthenium, rhodium, palladium, osmium, iridium, and platinum.

Various efforts are being carried out to replace and/or minimize the use of PGMs for HOR and ORR in PEMFCs. ORR needs a significantly higher amount of platinum catalyst as compared with HOR, as mentioned above, and the research work is focused on finding a suitable non-PGM electrocatalyst for the ORR. Therefore, this chapter reviews the state of art in ORR using non-PGM electrocatalysts in PEMFCs.

TABLE 16.1
ORR Pathway, Reactions, and Corresponding Standard Electrode Potential (Acidic Electrolyte)

Reaction Pathway	ORR Reactions	Standard Electrode Potential, E^o (V versus SHE)
2+2-Electrons	$O_2 + 2H^+ + 2e^- \rightarrow H_2O_2$	0.70
	$H_2O_2 + 2H^+ + 2e^- \rightarrow 2H_2O$	1.76
4-Electrons	$O_2 + 4H^+ + 4e^- \rightarrow 2H_2O$	1.229

Source: Bard, A.J. and Faulkner, L.R., *Electrochemical Methods: Fundamentals and Applications*, Wiley, New York, 1980; Yeager, E., *J. Mol. Catal.*, 38, 5, 1986.

The chapter covers the basic mechanism of ORR, synthesis of non-PGM catalysts, and the electrocatalytic performance reported in the last 5 years.

16.2 BASIC MECHANISM OF ORR

The mechanism of ORR is rather complicated and involves many intermediates, depending on the nature of the electrode material, electrocatalyst, and electrolyte. ORR, in the acidic condition of PEMFC, generally follows a 2 + 2 or 4-electron pathway. Table 16.1 [1,2] shows the ORR for these pathways along with their standard electrode potentials. The formation of the 2 + 2 reaction pathway forms H_2O_2 as an intermediate, which is highly oxidizing in nature. The formation of H_2O_2 or peroxide radical is detrimental to the PEM used in PEMFC. Therefore, a 4-electron pathway is desired in PEMFCs for the ORR.

The performance of an electrocatalyst-like activity, stability, and selectivity is directly related to the electrocatalyst morphology and the outer-shell configuration of the constituting elements. The morphology of the electrocatalyst varies with the method of synthesis, metal precursor, support material, catalyst pretreatment, etc. A wide range of literature is available on the pure metal electrocatalysts. Surprisingly, these pure metal electrocatalysts for ORR are either noble metal catalysts or PGMs. Although platinum is used at both the anode and cathode, comparatively much higher loading is required at the cathode because of the sluggish nature of ORR. Hence, the focus is more on finding an alternative catalyst for ORR, which is preferably non-PGM and works efficiently. The use of a non-PGM catalyst is preferred because of its easy availability, which might result in the low cost of the PEMFC. In general, the studies on pure non-PGM shows that they are transition metals, which have low electrocatalytic activity along with stability issues in the fuel cell environment. The major obstacle in the search for non-PGM catalysts is their poor stability in acidic environment. Other than pure metals, metal oxides, alloys, and metal macrocycles, etc., have also been investigated for the ORR. The aim of this chapter is to provide some basic knowledge on ORR using a recently developed non-PGM-based electrocatalyst. Therefore, the chapter will primarily focus on the electrocatalysts based on metals or their alloys (especially Co, Fe, and Ni) with or without support material for the oxygen electroreduction reaction in acidic medium.

16.3 RECENT APPROACHES IN THE FIELD OF NON-PGM ELECTROCATALYSTS

Long back in 1964, Jasinski [3] found that Co–phthalocyanine exhibits good electrocatalytic activity toward ORR. Since then, a number of transition metal macrocycles have been developed for ORR. Macrocycles are important and powerful ligands present in transition metal coordination chemistry. These macrocycles are typically very stable species and can give rise to unusual structural, electronic, and electrochemical properties. Generally, transition metal macrocycles are prepared by impregnation of a transition metal N_4-chelate precursor onto a support material, followed by heat treatment under an inert environment. Bezerra et al. [4] reviewed the metal precursors and ligands and listed the various heat treatment conditions. Co, Ni, and Fe are well-investigated non-PGM electrocatalysts. However, these metals are very poor ORR electrocatalysts in their unmodified states. Hence, the catalysts including these metals also incorporate the significant use of ligands, another metals, some efficient support, or various other treatments. Figure 16.1 shows the performance of non-PGMs in fuel cell, along with a comparison with Pt–C [5]. It can be seen that the support is playing an important role in determining the overall performance of non-PGMs. Whereas non-PGMs supported onto metal organic frameworks (MOFs) show better power density and cell voltage, the activity of non-PGMs supported over carbon catalysts slows down very quickly.

Non-PGMs have been tried along with the nitrogen-doped carbon support [6,7]. Carbon support works as a conductive support on which Co was found to form Co–N_4 macrocycles. The method involving the preparation of such catalysts includes sputtering and pyrolysis. The sputtering method has also been used in case of Ni along with other metals. Ta as an assisting component with Ni helped to passivate the catalyst against corrosion, while electrocatalytic activity was found to be improved with increasing content of Ni. Recently, Lee and Kim [8] have reported the sputtering of Ni metal over carbon paper. It was found that heat treatment temperature is the most important factor for determining the catalytic activity of the electrocatalysts. An optimum heat treatment temperature is required as the ligands become denatured at higher temperature, whereas catalyst formation does not take place at low temperatures. However, the current density in case of pure

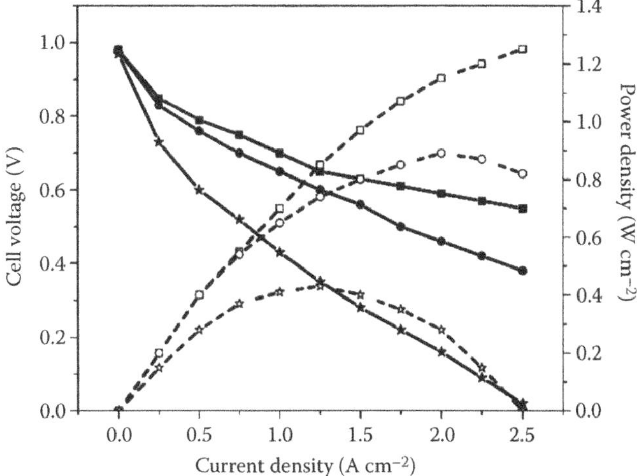

FIGURE 16.1 Cell voltage (solid lines) and power density (broken lines) curves for state-of-the-art Pt–C catalyst (square), non-platinum group metal (non-PGM) catalysts based on MOFs (circle), and non-PGM based on Black Pearl 2000 carbon support (star). (With kind permission from Springer Science+Business Media: *Electrocatalysis in Fuel Cells: A Non- and Low-Platinum Approach*, 2013, Shao, M.)

metals (whether single or in combination) was much less in comparison to commercial Pt–C catalysts. This has led the research in the direction of some other alternatives.

Transition metal macrocyclic compounds are rather interesting as they provide various pathways for reducing oxygen. They can catalyze ORR by undergoing a 2-electron or 4-electron pathway to generate H_2O_2 or H_2O, respectively, as described previously. However, catalysis through a mixed pathway of 2 and 4-electron transfer is also not uncommon. Interestingly, sometimes these complexes undergo only 1-electron O_2 reduction, thereby producing superoxide ions. The transition metals that are commonly used in macrocyclic complexes are Co, Fe, and Ni. The ligands usually incorporate phthalocyanine, porphyrin, pyridine, and salen, etc. The electrocatalytic activity of these complexes depends both on the metal center as well as the ligands. It has been seen that for Fe and Co complexes, N_4 chelating ligands are rather promising, while for Ni complex, O_4 ligands behave in a similar way.

M–N_4 complexes such as tetramethoxyphenylporphyrin, tetraphenylporphyrin, phthalocyanine, and tetraazaannulene, whose molecular structures are shown in Figure 16.2, are able to strongly adhere on the carbon supports to form the ORR catalysts, which help to enhance their durability. Among the investigated N_4 macrocycles, CoN_4 has been shown to be the best for ORR; however, still much more improvement is needed to realize their practical application in PEMFCs. Recently, Rao et al. [9] have used reduced graphene oxide as the support for cobalt(II)tetramethoxyphenylporphyrin (CoTMPP) and found that the stability and oxygen reduction activity has improved notably. Thus, the support material also plays an important role for the ORR activity.

It has also been seen that the ORR electrocatalytic activity is also influenced by the substituents on the macrocyclic rings. Electron donor substituents in M–phthalocyanine complexes (M stands for metal) are expected to show improved ORR catalytic activity, as they help increase the binding strength

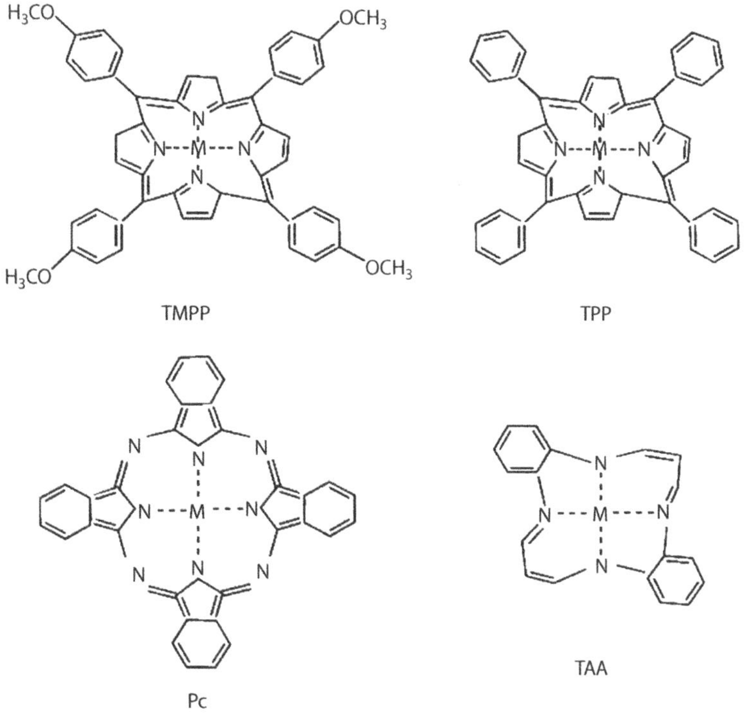

FIGURE 16.2 Various ligands used in the formation of metal complexes. TMPP, tetramethoxyphenylporphyrin; TPP, tetraphenylporphyrin; Pc, phthalocyanine; TAA, tetraazaannulene.

between O_2 and metal centers [10]. Furthermore, heat treatment in case of transition metal macrocyclic complexes is very important. Heat treatment is usually needed to stabilize the transition metal macrocyclic complexes as they are not very stable in acidic or alkaline conditions. It has been reported that thermal treatment helps stabilize the electrocatalysts along with helping improve their catalytic activity. In many of the reported works, pyrolysis, annealing, or some other type of heat treatment is given at temperatures higher than 400°C, usually in an inert atmosphere. Sometimes, heat treatment results in the disruption of the macrocyclic ring. It has also been reported that heat treatment at higher temperatures might lead to phase transition in transition metal macrocycles. In the case of Fe phthalocyanine, it was found to be in β-phase after heat treatment, in contrast to its α-phase in untreated molecule [11], which may be responsible for ORR activity.

Transition metal chalcogenides are a new class of non-PGM electrocatalysts. Usually, the non-PGM centers are partially embedded in/or bonded to chalcogen atoms, forming metastable or thermodynamically stable catalytic sites. Owing to their low cost, good abundance, and easy synthesis, chalcogenides are rather promising candidates for ORR. The research in this field had been started with the work of Baresel et al. [12] in the 1970s, where they used materials based on Co–S and Co–Ni–S. Researches have shown that the catalytic activity in transition metal chalcogenides occurs through the interaction of oxygen with the *d*-band of the transition metals. Like others, chalcogenides can catalyze ORR either by 2-electron or 4-electron transfer pathways. Tritsaris et al. [13] have used density functional theory (DFT) to understand the effect of chalcogenides over substrate metals. Modifying the surface of transition metal with chalcogenides affects the binding energy of the reaction intermediates, thus affecting the overall performance toward ORR activity. Behret et al. [14] have investigated the chalcogenides and reported that the electrocatalytic activity of these materials depend on the metal utilized as well as on the chosen chalcogenide. To be specific, the order of the activity for the metals under consideration is Co > Ni > Fe, and that for metal chalcogenides is S > O > Se > Te. According to this, the cobalt sulfides should exhibit the highest activity among other metal–chalcogenides combination. Recently, DFT studies have also shown that Co_9S_8 has the potential to replace the Pt electrocatalyst [15].

16.3.1 Cobalt as an Electrocatalyst for ORR

Cobalt as electrocatalyst has been used in various forms—as pure sputtered metal; in the form of oxides [16,17], in the form of various complexes such as porphyrin [18,19], pyrrole [20], amine complexes [21,22]; etc. Most of these catalysts undergo four electron mechanisms, which is desirable for the better performance of ORR. A reasonable degree of success has also been achieved if compared with the performance of commercial Pt–C. It is very common for these catalysts that a variety of support materials have been used, ranging from pure carbon to polyacrylonitrile, graphene films to graphene oxides, carbon nanotubes, etc., which not only act as support but also act as a co-catalyst and help in the enhancement of overall performance of the catalyst. Usually, the C–N bonding formed with the catalyst gives a chelate-like structure and helps in the binding of the catalyst with the support materials. A survey based on recent studies, on the synthesis procedures of cobalt electrocatalysts, is provided in Table 16.2. The list of Co-based catalysts, their precursors, ligands, support material, and method of preparation are given in Table 16.2.

For cobalt catalysts, a few investigators have also tried to manipulate the performance for ORR by changing the cobalt precursors (without changing the ligand moiety) such as oxalate, nitrate, acetate, and chloride. It was found that cobalt acetate exhibits the highest activity among all [19]. This is further illustrated through Figure 16.3. It can be seen in Figure 16.3a that the peak potential value in all the cases is approximately same, viz. 0.71 V versus NHE (normal hydrogen electrode); however, the highest peak current was found in case of the catalyst prepared from a cobalt acetate precursor. The onset potential was also highest in case of cobalt acetate, as can be seen in Figure 16.3b. Thus, it becomes clear that the chosen metal precursor plays an important role for the ORR activity.

Recently, a few of the researchers have reported the use of cobalt oxides such as CoO and Co_3O_4 using various preparation methods [16,23–26]. In these studies, the metal oxides were supported over highly efficient graphene or carbon nanotube supports. It was found that the synergistic effects among the active moieties, including C–N, Co–N–C, and CoO–Co, rendered the high performance of the electrocatalysts. Moreover, the use of high-surface-area-support materials must also have played a significant role by avoiding agglomeration and promoting better dispersion of the electrocatalysts. Many of these cobalt oxides, however, have also been used as ORR catalysts in alkaline medium, with reasonable success [23,24,26,27].

Cobalt porphyrins have attracted attention in the last few years. These complexes are usually formed through pyrolysis at high temperature. The porphyrins are synthesized under specific conditions of temperature, pyrolysis time, and metal concentration. The active sites in these catalysts are CoN_x sites ("$_x$" represents the number of N atoms) [28]. In this kind of catalysts, the nitrogen content at the active site plays an important role. It has also been shown that most active catalysts can be obtained by means of heat treatment at 500–700°C, resulting in a catalytic site having a metal bound to nitrogen, which are not stable. More stable catalysts may be produced by means of heat treatment at much higher temperatures, such as >900°C. Faubert et al. [29] have reported that after such a heat treatment, the metal is usually found as a cluster enveloped by graphite, which is responsible for the stability.

Cobalt chalcogenides have also been studied for ORR owing to their high stability in acidic environment especially when coordinated with metal. For the synthesis of chalcogenides, the hydrothermal or solvothermal routes have been employed [30–32], as they are simple and environmentally benign. The reaction time, reactant concentration, and the amount of the ligand greatly affect the morphologies and crystal purity.

TABLE 16.2
Different Types of Co-Based Electrocatalysts for ORR

Electrocatalysts	Metal Precursor	Ligand Precursor	Electrocatalyst Support	Method	Posttreatment	Ref.
CoO@Co/N–C	$CoCl_2·6H_2O$	1,10-Phenanthroline	Carbon black	Reflux at 95°C for 4 h	Annealing in Ar at 700°C for 1 h	[23]
CoTEPA/C	$CoCl_2·6H_2O$	Tetraethylenepentamine (TEPA)	Black Pearl 2000 modified by H_2O_2	Dried under reduced pressure	Pyrolyzed in N_2 at 800°C for 90 min	[21]
Co_3O_4	$CoC_2O_4·2H_2O$	–	–	Low-temperature degradation of precursor	–	[25]
3D crumpled graphene (CG)–CoO nanohybrids	$CoCl_2$	–	Graphene oxide	Nebulization to form aerosol particles	Heat treatment at 750°C	[24]
Self-assembled 2,9,16,23-tetrahydroxythiophenylphthalocyanato cobalt(II) (4β-Co^{II}THTPhPc) on Au surface	$CoCl_2$	4-Hydroxythiophenol and 4-nitrophthalonitrile	–	Self-assembly achieved by immersing Au electrode in 4β-Co^{II}THTPhPc	–	[33]
Carbon-supported cobalt–polypyrrole–4-toluenesulfinic acid	$Co(CH_3COO)_2·4H_2O$	Pyrrole and toluenesulfonic acid (TsOH)	Pyrolyzed BP2000 carbon powder	Stirring and sonication	Pyrolysis in Ar at 800°C	[34]
Cobalt poly(phenylenediamine)	$Co(NO_3)_2$	*o*-Phenylenediamine and *m*-phenylenediamine	Acetylene black	Oxidative polymerization	–	[35]
Co–N–rGO_x	Any salt of Co	Tetramethoxyphenylporphyrin	Reduced graphene oxide (rGO_x)	–	Heat treatment at 800°C for 2 h	[9]
Cobalt porphyrin	$CoCl_2$	5,10,15,20-Tetrakis(4-pyridyl) porphyrin (H_2TPyP)	Poly(sodium-*p*-styrenesulfonate) modified reduced graphene oxide	–	–	[36]
Pt-coated Co nanowires	Co nanowires	–	Potassium tetrachloroplatinate	Coating by partial galvanic displacement	–	[37]
Cobalt–polypyrrole–carbon	$Co(NO_3)_2$	Pyrrole, polymerized by ammonium peroxydisulfate	Carbon support (Ketjen Black EC 300)	Carbon support can be added before or after the polymerization step	Heat treatment at 900°C for 1 h	[38]
Co–CoO nanoparticles onto graphene	$CoCl_2·6H_2O$	–	Graphene films	Electrodeposition using three-electrode cell	–	[16]
Cobalt/polypyrrolenano composites	$CoCl_2·6H_2O$	Pyrrole	–	Electrochemical synthesis by potentiostatic anodic/cathodic pulse plating	–	[39]
Cobalt phthalocyanine/CB	NA	NA	Carbon black (Printex L6)	CoPC was employed as modifier	–	[40]

(*Continued*)

TABLE 16.2 (CONTINUED)
Different Types of Co-Based Electrocatalysts for ORR

Electrocatalysts	Metal Precursor	Ligand Precursor	Electrocatalyst Support	Method	Posttreatment	Ref.
Pyrolyzed carbon–supported cobalt–polypyrrole	$CoCl_2$, $Co(NO_3)_2$, $Co(CH_3COO)_2$, CoC_2O_4	Pyrrole and *p*-toluenesulfonic acid (TsOH)	Carbon black BP2000	–	Heat treated at 800°C for 2 h in Ar	[19]
CoO–nitrogen-doped carbon nanotube (NCNT)	$Co(CH_3COO)_2$	–	Nitrogen-doped oxidized CNT	–	Annealing at 400°C for 3 h in NH_3/Ar	[26]
Sputtered Co	Pure Co	–	–	RF magnetron sputtering	Heat treatment at 750°C	[6]
N-doped Co_3O_4/rmGO	$Co(CH_3COO)_2$	–	Reduced mildly oxidized graphene (rmGO)	Hydrolysis, oxidation, and hydrothermal reaction	–	[17]
Co oxyphosphide	$Co(NO_3)_2$	–	Carbon black	Incipient wetness	Impregnation for 12 h, calcination at 500°C	[32]
CoN_x	$Co(NO_3)_2$	Imidazole	–	Chelation	Heat treatment at 600–900°C	[41]
CoTETA	$CoCl_2$	Triethylenetetramine (TETA)	–	Chelation	Pyrolysis at 800°C	[22]
Carbon black–supported Co–PPy	$Co(NO_3)_2$	Pyrrole	Carbon black (black pearls)	Chemical method of polymerization synthesis	Heat treatment at 600°C, 800°C, or 1000°C for 2 h	[42]
CoTMPP/BP	$Co(CH_3COO)_2$	Tetramethoxyphenylporphyrin (TMPP)	Carbon black pearls (BP)	Dissolved in acetic acid and refluxed in Ar for 90 min	Thermal treatment at 900°C for 2h	[18]
CoEDA/CB	$Co(NO_3)_2$	Ethylenediamine (EDA)	Carbon black (CB)	–	Heat treatment at 800°C in Ar	[43]

Note: @, Core–shell electrocatalysts.

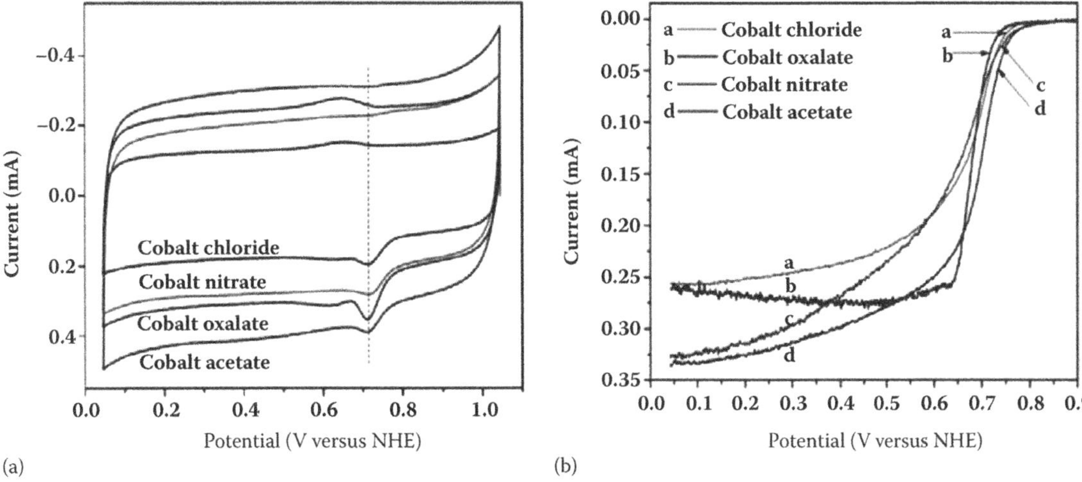

FIGURE 16.3 Co–Ppy–TsOH/C (Ppy, polypyrrole; TsOH, *p*-toluenesulfonic acid) catalysts prepared from various cobalt precursors in O_2-saturated 0.5 M H_2SO_4 solution: (a) CV curves; (b) rotating disk electrode (RDE) polarization curves. (Reproduced from Yuan, X., Hu, X.X., Ding, X.L., Kong, H.C., Sha, H.D., Lin, H., Wen, W., Shen, G., Guo, Z., Ma, Z.F., and Yang, Y., *Nanoscale Res. Lett.* 8, 478, 2013.)

Wang et al. [44] have reported that flower-like CoS nanostructures display good cycle stability and discharge capacity.

The synthesis procedures of the cobalt-based electrocatalysts is rather variable and depending upon the chosen metal precursor, ligand precursor, support material, and operating conditions, different types of electrocatalysts morphologies and activity are produced. Various methods used for the synthesis of Co-based electrocatalysts in the last 5 years are summarized in Table 16.2. It can be seen that in most of the electrocatalysts, heat treatment under an inert atmosphere is very common, as it helps in increasing the activity as well as stability. Moreover, it can be seen that catalyst support is used, which increases the surface area of the electrocatalysts.

It appears that pure cobalt metal-based nanoparticles supported over high-surface-area materials have considerable activity toward ORR in acidic medium. The graphene sheets (e.g., graphene oxide) are very common support materials owing to their extremely high electrical conductivity [45,46]. Hence, they helps achieve better performance in ORR. The electrocatalytic activity of ORR catalysts in acidic medium is summarized in Table 16.3. Most of the reported catalysts have shown to follow a 4-electron transfer pathway for ORR, which is desirable for better performance. The catalyst performance results are compared with the Pt–C (20 wt.%). It can be seen that where only pure cobalt metal is impregnated onto the support, the onset potential is comparable to that of the Pt–C. The Co tetramethoxyphenylporphyrin ORR catalyst–based fuel cell shows comparable results with PGM-based catalysts and a substantial power density.

16.3.2 Iron as an Electrocatalyst for ORR

Some of the highest oxygen reduction activities among non-Pt electrocatalysts have been shown by Faubert et al. [47] using Fe^{II} atoms supported over pyrolyzed perylenetetracarboxylic dianhydride. They reported the current density as 0.038 $A{\cdot}cm^{-2}$ at 800 mV versus reversible hydrogen electrode (RHE) at 50°C. The turnover frequency of Fe-pyrolyzed perylenetetracarboxylic dianhydride is approximately 1/15th of the turnover frequency of the commercial Pt catalyst. Thus, Fe-based catalysts seem to be a possible alternative for PGM catalysts. Many of the reported studies show that Fe is usually coordinated with the N atom in the form of an Fe–N_4 structure. The superior activity of this structure may correspond to the inductive and mesomeric effects of the central metal atom and ligand. For Fe-based N_4 catalysts, the ORR usually takes place through a 4-electron transfer pathway, as reported by Gao et al. [48]. Recently, Kamiya et al. [49] have modified graphene by Fe and N at 900°C for just 45 s, and found that Fe–N-modified graphene electrocatalyst catalyzes a 4-electron-transfer ORR. However, it was seen that the formation of H_2O_2 increases (viz. 2-electron pathway) on increasing the heat treatment period, which suggests that excess heat treatment deactivates the active sites, which help in the 4-electron ORR pathway.

On the basis of the central metal ion, oxygen reduction may proceed through a 4-electron or a 2-electron pathway. Wiesener et al. [50] have shown an order for the activity of N_4-chelates (phthalocyanins) viz. Fe > Co > Ni. Iron phthalocyanine complexes are known to favor a 4-electron reduction pathway; however, stability is an issue. Usually, Co–N_4 chelates are comparatively more stable but they can undergo only 2-electron reduction; therefore, they are not a good choice. To incorporate the catalytic activity and durability of both these complexes, a few workers have worked over mixtures of Co–Fe metalloporphyrins, and found that the activity shown is higher than that for individual Co or Fe complexes [51].

Table 16.4 lists some of the recent developments in the field of Fe-based electrocatalyst synthesis. It is noticeable that most of the catalysts incorporate the N atom, which is considered to be very important for catalytic activity. With the increasing nitrogen content, the activity of the catalysts has been found

TABLE 16.3
Performance Comparison of Co-Based ORR Catalysts

Metal/Metal Complex/Bimetallic Electrocatalysts	Electrolyte Half/ Fuel Cell	Mechanism	Cyclic Voltammetry Results (V versus RHE)	Fuel Cell Results	Ref.
Pt–C (20 wt.%)	0.5 M H_2SO_4	4-Electron	Onset potential: 0.91; peak potential: 0.78	Open circuit voltage (V): 0.96	[52]
		Pure Cobalt			
Co–N_x–C	0.5 M H_2SO_4/ Nafion-212	4-Electron	Onset potential: 0.81	Open circuit voltage (V): 0.74–0.82; maximum power density (mW cm^{-2}): 200 at 80°C	[41]
Co–N–rGO_x	0.5 M H_2SO_4	–	Onset potential: 0.84	–	[9]
		Cobalt Porphyrin			
Co tetramethoxyphenylporphyrin	0.5 M H_2SO_4/ Nafion-115	2-Electron	–	Open circuit voltage (V): 0.975; power density (mW cm^{-2}): 150 at 50°C	[18]
Pyrolyzed carbon–supported cobalt–polypyrrole	0.5 M H_2SO_4	4-Electron	Peak potential: 0.71	–	[19]
Carbon-supported cobalt polypyrrole pyrolyzed at 800°C	0.5 M H_2SO_4	4-Electron	Onset potential: 0.81; peak potential: 0.614	–	[42]
		Co with Amine Complexes			
Cobalt tetraethylenepentamine (CoTEPA/C)	0.5 M H_2SO_4	4-Electron	Peak potential: 0.752	–	[21]
Cobalt poly(phenylenediamines)	0.05 M H_2SO_4	–	Onset potential: 0.43; peak potential: 0.22	–	[35]
Co triethylenetetramine (CoTETA)	0.5 M H_2SO_4	4-Electron	Onset potential: 0.85; peak potential: 0.71	Open circuit voltage (V): 0.785; maximum power density (mW cm^{-2}): 135 at 25°C	[22]
Co ethylenediamine (CoEDA)	0.5 M H_2SO_4	4-Electron	Onset potential: 0.82; peak potential: 0.52	–	[43]

to be increasing [53]. Along with nitrogen, the metal loading over the support material also affects the performance of the catalysts. An optimum metal loading is essential, as too low or too high loading would result in lesser activity and poisoning of the catalyst surface, respectively. Apart from the metal loading, the structure of the catalyst layer also needs special attention. Elzing et al. [54] studied the effect of different preparation methods of Fe-based catalysts on the catalyst activity. They considered five different methods and found the sequence of activity to be as follows: incorporation in polypyrrole > irreversible adsorption > evaporation of the solvent > vacuum deposition method. As mentioned earlier, heat treatment has also been suggested as an important step for the stability of the electrocatalyst in ORR (Table 16.4). However, a deleterious effect may also be observed on the catalytic activity at very high temperatures. Iliev et al. [55] prepared carbon activated with CoTMPP and heat treated in argon at temperatures between 200°C and 1200°C. It was found that the temperatures between 460°C and 810°C improved the initial polarization characteristics of air electrode. The further increase in the treatment temperature decreases the catalyst activity sharply.

As in the case of cobalt-based electrocatalysts, the precursors also play an important role for Fe-based electrocatalysts. Figure 16.4 demonstrates the effect of different Fe precursors on the oxygen reduction activity. If Fe acetate is used as an Fe precursor, then the catalyst activity increases quickly but then levels off over 5000 ppm loading. When chloro-iron tetramethoxyphenylporphyrin is used as precursor, higher Fe loadings could be achieved before the ORR activity levels off. The difference is attributed to the interaction of Fe precursors with that of PTCDA (perylene-3,4,9,10-tetracarboxylic dianhydride). For instance, iron porphyrin being dispersed in a better way than Fe acetate results in better activity [56].

Gupta et al. [57] have studied the stability of iron(III) chlorotetramethoxyphenylporphyrin (Cl-FeTMPP) along with the μ-oxo-iron(III) tetramethoxyphenylporphyrin ($[FeTMPP]_2O$) and iron(III) octaethylporphyrin (FeOEP-Cl) catalysts supported over high-surface-area carbon. Along with providing better activity, it is shown that the cathodes made up of these catalysts are resistant to degradation even after 24 h.

Kong et al. [52] recently developed an innovative method to synthesize Fe complexes. They converted Fe–N_4 complexes into a three-dimensional (3D) ordered mesoporous material

TABLE 16.4
Different Fe-Based ORR Electrocatalysts

Electrocatalysts	Metal Precursor	Ligand Precursor/ Alloying Metal	Electrocatalyst Support	Method	Posttreatment	Ref.
Graphene-supported iron-based nanoparticles (Fe@N–C)	Ammonia ferric citrate ($C_6H_{11}FeNO_7$)	–	Graphene oxide	Pyrolysis and acid leaching	Pyrolysis in Ar atmosphere at 600°C	[58]
Fe–N–C electrocatalyst	$FeCl_3$	Aniline hydrochloric salt	Vulcan XC-72; EC-300	–	–	[59]
Iron tetrasulfophthalocyanine (FeTSPC) functionalized graphene	–	Tetrasulfophthalocyanine (TSPC)	Functionalized graphene nanosheets	–	–	[48]
Nitrogen-doped porous Fe/Fe_3C@C nanoboxes supported on RGO sheets (N-doped Fe/Fe_3C@C/RGO)	Prussian blue ($C_{18}Fe_7N_{18}$)	–	Reduced graphene oxide (RGO)	Pyrolysis method	–	[60]
Fe–N–C sites as electrocatalysts	Iron acetylacetonate [$Fe(C_5H_7O_2)_3$]	–	Polyacrylonitrile (PAN)	Pulverization and carbonization	Carbonization at 1000°C for 1 h in N_2 atmosphere	[61]
Iron phthalocyanine coordinated with pyridine cycloaddition of graphene sheets (PyNGs)	NA	–	Graphene with pyridine	One-step chemical modification process—refluxing PyNG and FePc mixture under Ar atmosphere	–	[62]
Fe–N-coordinated modified graphene	$FeCl_3{\cdot}6H_2O$	Pentaethylenehexamine	Graphene oxide	Short-duration heat treatment (45 s)	Heat treatment at 900°C for 45 s	[49]
Fe–N decorated hybrids of CNTs on ordered porous carbon (Fe–N–CNT–OPC)	$FeCl_2 + FeCl_3$	–	Resol and graphitic carbon nitride	–	Heat treatment at 900°C	[63]
Fe–N/carbons	$FeCl_3{\cdot}6H_2O$	Pentaethylenehexamine (PEHA)	Carbon black (Vulcan XC-72)	–	Heat treatment at 800°C for 2 h	[64]

(Continued)

TABLE 16.4 (CONTINUED)
Different Fe-Based ORR Electrocatalysts

Electrocatalysts	Metal Precursor	Ligand Precursor/ Alloying Metal	Electrocatalyst Support	Method	Posttreatment	Ref.
Fe–N-doped mesoporous tungsten carbide nanostructures	Fe-TMPP [5,10,15,20-tetrakis(4-methoxyphenyl)-21*H*,23*H*-porphyrin] iron(III)	Ammonium tungstate $[(NH_4)_{10}H_2(W_2O_7)_6]$	–	Hydrothermal and carburization process	First heat treatment in NH_3 flow for 3 h at 700°C; second heat treatment in ($CH_4 + H_2$) flow at 900°C	[65]
Supported Ni–Fe nanoparticles	Iron acetylacetonate $[Fe(C_5H_7O_2)_3]$	Nickel acetylacetonate $[Ni(C_5H_7O_2)_2]$	Carbon black (Vulcan XC-72)	Organic solution phase nanocapsule approach	–	[66]
Fe–N_x/rGO electrocatalyst	$Fe(CH_3COO)_2$	Tripyridyltriazine (TPTZ)	Reduced graphene oxide (rGO)	–	Pyrolysis at 900°C for 2 h under N_2 atmosphere	[67]
Poly-*m*-phenylenediamine–based FeN_x/C catalyst (PmPDA–FeN_x/C)	$FeCl_3$	Poly-*m*-phenylenediamine	Carbon black (Ketjen Black EC600JD)	High-temperature pyrolysis	Pyrolysis at 950°C	[68]
Fe-N–C catalysts	$Fe(NO_3)_3{\cdot}9H_2O$	–	Polyacrylonitrile (PAN) nanofibers	Electrospinning and heating method	Carburization at 800°C for 2 h under Ar atmosphere	[69]
Mesoporous Fe–porphyrin	$FeCl_3{\cdot}6H_2O$	Tetrapyridylporphyrin	SBA-15	Nanocasting procedure	Pyrolysis at 700°C for 4 h in N_2 atmosphere	[52]
Fe co-doped OMPC	FeTMPPCl (porphyrin system)	Ordered mesoporous porphyrinic carbons (OMPC)	Self-supported	Solid-state nanocasting method	Heat treatment at 600–1000°C for 3 h in N_2 flow	[70]
Fe phthalocyanins supported over Py–CNTs	NA		Pyridyl (Py) functionalized CNTs	–	–	[71]
Iron-based nitrogen-coordinated transition metal catalysts (NTMCs)	Iron tetrasulfophthalocyanine	–	Single-wall CNTs	–	–	[72]

Note: NA, not available; @, core–shell electrocatalyst.

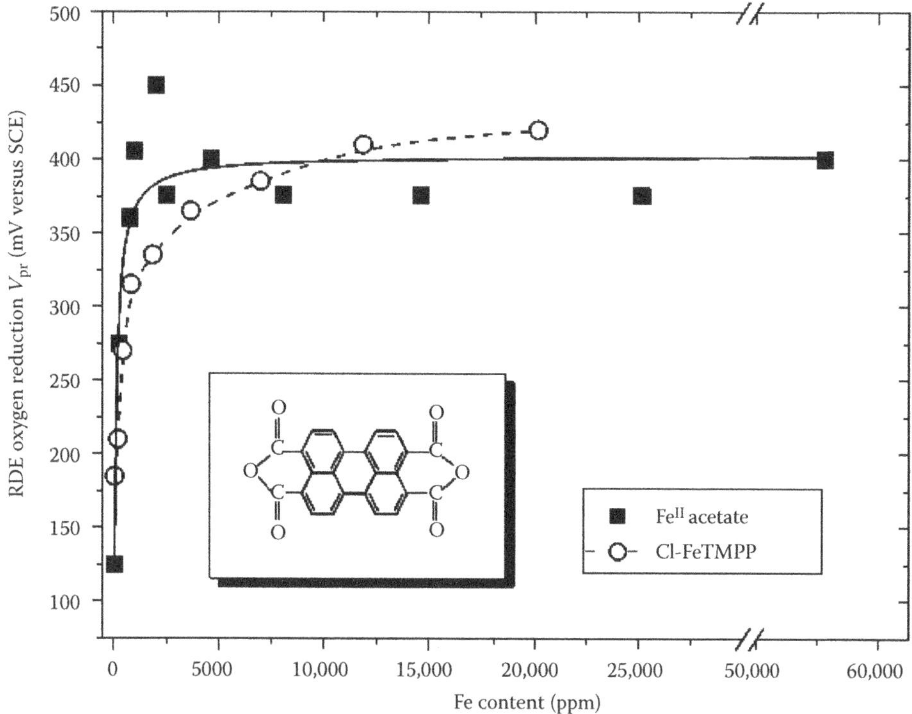

FIGURE 16.4 Catalytic activity for O_2 reduction (in O_2-saturated H_2SO_4 solution at pH 1) versus the Fe content for catalysts prepared by adsorbing either iron acetate or Cl-FeTMPP on PTCDA (perylene-3,4,9,10-tetracarboxylic dianhydride) and pyrolyzing the material at 900°C in H_2/Ar/NH_3 (1:1:2). (With kind permission from Springer Science+Business Media: *Electrocatalysis in Fuel Cells: A Non- and Low-Platinum Approach*, 2013, Shao, M.)

having high surface area. In such a structure, active Fe–N_x sites become well incorporated into mesoporous graphite backbones, and thus provide better electrical and mechanical connections between active sites and the support material. The Fe–N_4 complex showed very good performance in acidic medium, reaching a current density of up to 0.20 mA cm^{-2} with onset potential and peak potential of 0.89 and 0.75 V versus RHE, respectively, which is comparable to those of commercial Pt catalysts. The porphyrinic structure was also studied by He et al. [73], who demonstrated the catalytic activity of iron porphyrin (Fe(III)TMPyP) coordinated with Nafion®. This catalyst was found to be very active and, at the same time, its degradation rate was 2.7 mV · h^{-1} under constant current discharging at 50 mA cm^{-2}.

Yan et al. [69] synthesized Fe–N–C-based catalysts by electrospinning and heating in very low oxygen flow, and found very high efficiency with onset potential of 0.88 V versus RHE. Durability was also found to be better than Pt-based catalysts. The significance of the work was that it provided the simplest method to convert commercial carbon fibers to effective ORR catalysts, only by adding iron salt with proper heat treatment. Investigating in the same line, Liu et al. [64] tried to determine the impact of iron and nitrogen precursors on the synthesis of FeN_x–carbon, and reported that catalysts having pentaethylenehexamine and iron(III) chloride possess superior electrocatalytic property. Kamiya et al. [49] also worked on Fe–N coordination with modified graphene as an electrocatalyst support, which proceeds through a 4-electron reaction pathway, resulting in an onset potential of 0.85 V versus RHE. It was investigated that the strength of the Fe–N coordination bond depends on the time provided for heat treatment, and the bond breaks if the heat treatment period exceeds 5 min, resulting in loss of activity. Lefevre et al. [74] have developed microporous carbon-supported iron-based catalysts having an efficacy similar to that of platinum-based cathode with a loading of 0.4 mg platinum per square centimeter at a cell voltage of ≥0.9 V.

The details of the electrocatalytic activity of the Fe-based electrocatalysts are summarized in Table 16.5. On comparing the peak potential values with Pt–C catalysts, the developed materials seem to be very effective as ORR catalysts. Many of these materials have been found to be rather stable in acidic medium, especially Fe^{+2} species [75], thus proving them to be efficient alternatives to the PGM. It may be noted that although the half-cell performance of these catalyst is promising, the performance of the catalyst in fuel cells is not very encouraging. In fact, researchers have studied the half-cell performance, but the fuel cell performance needs to be evaluated as there are few results available.

16.3.3 Nickel as an Electrocatalyst for ORR

Usually, Ni metal is less studied in comparison to Co and Fe, as many of the studies show that Ni complexes are not much efficient toward ORR [50]. Nickel-based electrocatalysts have usually been studied either after doping with nitrogen or in

TABLE 16.5
Performance Comparison of Different Fe-Based ORR Catalysts

Metal/Metal Complex/Bimetallic Electrocatalysts	Electrolyte Half/ Fuel Cell	Mechanism	Cyclic Voltammetry Results (V versus RHE)	Fuel Cell Results	Ref.
Pt–C (20 wt.%)	0.5 M H_2SO_4	4-Electron	Onset potential: 0.91; peak potential: 0.78	Open circuit voltage (V): 0.96	[52]
		N-Doped Fe Nanoparticles			
H–Fe@N–C/RGO	0.1 M $HClO_4$	4-Electron	Onset potential: 0.89; half-wave potential: 0.67	–	[68]
Fe–N–C/Vu	0.1 M $HClO_4$	4-Electron	Onset potential: 089; half-wave potential: 0.67	–	[59]
Fe–N–G	0.5 M H_2SO_4	4-Electron	Onset potential: 0.85	–	[49,76]
Fe–N–C/C-3.1	0.5 M H_2SO_4	4-Electron	Onset potential: 0.83	–	[64]
Fe–N–C	0.1 M $HClO_4$	–	Onset potential: 0.88; half-wave potential: 0.76	–	[69]
		Fe Porphyrins			
Fe(III) meso-tetra(*N*-methyl-4-pyridyl) porphine (TMPyP) chloride in triflouromethanesulfonic acid (TFMSA)	0.1 M TFMSA + 0.8 mM Fe(III) TMPyP/Nafion-212	4-Electron	Peak potential: 0.73	Open circuit voltage (V): 0.712; maximum power density (mW cm^{-2}): 25.8 at 80°C	[73]
3D ordered mesoporous Fe porphyrin-like material (Fe–N–C)	0.5 M H_2SO_4	4-Electron	Onset potential: 0.89; peak potential: 0.75	–	[52]

bimetallic form with Co. The synthesis of various Ni-based ORR catalysts is summarized in Table 16.6. In contrast to the results of Co-impregnated nitrogen modified carbon (NC), the Ni-impregnated modified carbon had a negative effect, and it inhibited the ORR activity in acidic medium [77]. It has been revealed by x-ray photoelectron spectroscopic studies that it is the formation of nickel oxide that has deleterious effects on ORR activity. Ni-doped NC undergoes a 2-electron reduction of oxygen and results in the formation of H_2O_2. However, this result is in contrary to the findings of Yang et al. [78] where sputtered Ni-doped NC resulted in high activity in alkaline electrolytes. In this case, the nickel oxide layer was not formed and the active moiety was composed of nickel nanoparticles, which resulted in its activity toward ORR. The discrepancy is attributed to the synthesis procedures of both the catalysts. While the incorporation of Ni-containing *meso*-tetra(4-pyridyl) porphyrins in NC resulted in the corrosion of catalyst, in the case of sputtering, Ni in NC remained in pure metal form and exhibited activity toward ORR in alkaline medium [77].

For bimetallic Ni-based non-PGM catalysts, Co has been chosen as the alloying metal preferentially. The recent studies report the chalcogenides-incorporated Co–Ni bimetals as ORR catalysts. However, Garcia-Contreras et al. [79] reported about the mechanically alloyed Ni and Co (without any chalcogenide) at 70:30 ratios, exhibiting a synergistic effect resulting in a 4-electron mechanism. Ni–Co alloy mixed with chalcogenides (such as $NiCo_2S_4$@N/S-rGO and $NiCo_2O_4$-G-2) provide much better stability than Pt–C. The activity of these catalysts was also found to be rather impressive [80,81].

Yang et al. [78] have studied the activity of sputtered Ni metal in acidic medium for ORR. It was found that, it is required to heat above critical temperature to achieve phase transformation and activity toward ORR. Phase transformation results in conversion of sputtered films, which are in amorphous structure, to the nanoscale mixture of N-containing carbon and Ni. However, the passivation provided against corrosion was not much, and it was suggested that incorporation of Cr into the sample may improve the stability. Yang et al. [78] found that the onset potential shown by $Ni_{10.1}C_{50.5}N_{39.4}$ ORR catalyst was 0.60 V (versus RHE). Recently, lanthanum-based perovskite-type electrocatalysts have also been investigated using Ni metal for ORR; however, their activity was not seen in acidic medium [83]. Garcia et al. [83] found the 2-electron pathway using Ni complex ($\{[NiTPyP(Crphen_2Cl)_4]^{8+}/SiW_{12}O_{40}^{4-}\}n_i$). However, the catalysts are not evaluated in fuel cell, and thus there is a pressing need to evaluate the potential ORR catalysts in fuel cells.

16.4 COMPARATIVE ACCOUNT OF THE NON-PGM-BASED ELECTROCATALYSTS

Recently, Du et al. [84] have done an interesting study of Co, Fe, and Ni metals by complexing these metals with the salen ligands for ORR in alkaline medium. They avoided the post-treatment that is usually given to most of the catalysts and applied direct pyrolysis of the precursors. It was found that the process helps to generate both micropores and mesopores, which increases the surface area, thereby resulting in better ORR activity. The electrocatalyst is supported over nitrogen-doped carbon. All the catalysts were synthesized in a similar manner; however, the surface area and pore size distribution was found to be rather different. It is deduced that metal

TABLE 16.6
Different Ni-based ORR Catalysts

Electrocatalysts	Metal Precursor	Ligand Precursor/ Alloying Metal	Electrocatalyst Support	Method	Posttreatment	Ref.
Ni(salen)-derived nitrogen-doped porous nanocomposites	$Ni(NO_3)_2$	H_2(salen) [salen = *N,N'*-bis(salicylidene)-ethylenediamine]	–	Direct reaction of H_2(salen) with $Ni(NO_3)_2$	Heat treatment under Ar, acid washing, annealing in Ar for further graphitization	[83]
{[NiTPyP(Crphen$_2$Cl$_4$]$^{8+}$/ $SiW_{12}O_{40}^{4-}$ }n_i, where TpyP = 5,10,15,20-tetra(4-pyridyl)porphyrin, phen = 1,10-phenanthroline, $SiW_{12}O_{40}^{4-}$ = silicotungstate, and n_i = number of multilayers	$Ni(CH_3COO)_2$	5,10,15,20-Tetra(4-pyridyl)-21*H*,23*H*-porphine; $CrCl_3{\cdot}6H_2O$; 1,10-phenanthroline monohydrate; and $H_4SiW_{12}O_{40}$	–	Reflux of mixture of $Ni(CH_3COO)_2$ and ligand precursors for 4 h; precipitated and further dried in vacuum at 120°C	–	[84]
Lanthanum-based Ni perovskites	$Ni(NO_3)_2{\cdot}6H_2O$ and $La(NO_3)_3$	Tetrapropylammonium bromide (TPAB), tetramethylammonium hydroxide pentahydrate (TMAH)	Natural carbon or graphitic carbon	Colloidal method	Calcination at 700°C for 4 h in dehumidified air	[82]
Ni-impregnated nitrogen modified carbon catalyst (NC)	$NiCl_2$ (anhydrous)	*Meso*-tetrakis(4-pyridyl) porphyrin (TPyP) containing *meso*-tetra(4-pyridyl)porphyrins	NC prepared by pyrolysis of Vulcan/polypyrrole composite at 800°C in He for 2 h	Impregnation of NC with Ni-containing *meso*-tetra(4-pyridyl)porphyrins, followed by ultrasonication	Pyrolysis at 650°C for 2 h	[77]
Carbon-supported Ni-doped MnO_x	$Ni(NO_3)_2$	$KMnO_4$	Carbon black (CB)	Mild hydrothermal treatment	–	[85]
Mesoporous $NiCo_2O_4$ nanoplatelet and graphene hybrid	$Ni(NO_3)_2{\cdot}6H_2O$	$Co(NO_3)_2{\cdot}6H_2O$	Graphene oxide (GO)	Concurrent precipitation and hydrothermal reactions	Calcination in air at 300°C for 3 h	[80]
$NiCo_2S_4$@graphene bifunctional electrocatalyst	$Ni(CH_3COO)_2$	$Co(CH_3COO)_2$	Graphene oxide/ethylene glycol suspension	Solvothermal reaction	–	[81]
$NiCo_2O_4$ spheres (urchin like)	$Ni(NO_3)_2{\cdot}6H_2O$	$Co(NO_3)_2{\cdot}6H_2O$	–	Hydrothermal reaction	Annealing at 400°C in air for 3 h	[86]
$LaNiO_3$ perovskite-type oxide	Ni	$LaNiO_3$	–	Self-combustion method	–	[87]

(*Continued*)

TABLE 16.6 (CONTINUED)
Different Ni-based ORR Catalysts

Electrocatalysts	Metal Precursor	Ligand Precursor/ Alloying Metal	Electrocatalyst Support	Method	Posttreatment	Ref.
Sputtered Ni metal over CNT on carbon paper	Ni sputter target	–	CNT on carbon paper	Sputtering through RF magnetron	Heat treatment at 750–900°C in NH_3 atmosphere	[8]
Ni diphosphine complexes with positioned pendant amines	$[Ni(CH_3CN)_6](BF_4)_2$	1,3-bis(diphenylphosphino) propane (dppp)	–	–	–	[88]
Nanocomposites of heteroatomic polymer and Ni supported on CB	$Ni(NO_3)_2{\cdot}6H_2O$	Monomer either aniline, pyrrole or 3-methyl-thiophene	Vulcan XC-72	Oxidation–reduction reactions	–	[89]
3D α–Nickel hydroxide	$Ni(NO_3)_2{\cdot}6H_2O$	Urea	–	Microwave-assisted hydrothermal method	Calcination at 450°C for 3 h	[90]
Sputtered Ni–C–N catalysts	Commercial Ni sputtering target	Commercial C sputtering target	–	Sputtering in an Ar/N_2 mixture	–	[78]
Co–Ni alloys	Elemental Ni powder	Elemental cobalt powder	–	Mechanical alloying in Ar atmosphere	–	[79]
Anthraquinone-modified Ni electrodes	Ni rod	9,10-Anthraquinone	–	Electrochemical reduction of diazonium salts	–	[91]
Ni–MnO_x/C nanoparticles	$Ni(NO_3)_2{\cdot}6H_2O$	$KMnO_4$, $MnSO_4$	Hydrophilic CB	–	–	[92]
Sputtered Ta–Ni–C	Commercial Ni sputtering target	Commercial Ta sputtering target	C (graphite) sputtering target	Sputtering method	–	[93]
Nanostructured Ni–21% W alloy	$NiSO_4$	$Na_2WO_4{\cdot}2H_2O$	–	Electrodeposition method	–	[94]

Note: @, Core–shell electrocatalysts.

influences the nature of pore formation, where Co and Fe resulted in mesopores, whereas micropore formation is found in the case of Ni.

In this study, Co and Fe complexes were found to have better performance than the Ni complex in alkaline medium. As can be seen in Figure 16.5a, the current density for the Co–salen complex is higher than that for the Pt–C. However, the Ni–salen complex resulted in the lowest current density among the three metals. The onset potential was found to be the least in the case of Fe–salen. The Koutecky–Levich plots in Figure 16.5b present good linearity at potentials ranging from 0.3 to 0.6 V. It was found that more or less a 4-electron pathway mechanism at 0.5 V occurs for Co–N–C (n = 4.1), Fe–N–C (n = 3.96), and Pt–N–C (n = 3.6) electrocatalysts, respectively. In fact, the Tafel plots in the Figure 16.5c show close kinetic current densities, especially in the low overpotential regions, which make Co–N–C and Fe–N–C superior over the platinum electrocatalyst. Besides high activity, the metal–salen complexes also resulted in excellent durability (even over 10,000 continuous cycles) than Pt–C in alkaline medium as can be seen in the cyclic voltammetry (CV) curves shown in Figure 16.5d. The inset of Figure 16.5d represents a transmission electron microscope (TEM) image of a metal–salen electrocatalyst (Co–N–C). Thus, it may be concluded that salen–metal complexes are very effective catalysts for ORR in alkaline medium. The main reason for the excellent activity is attributed to the small particle size of the synthesized particles, resulting in high surface area of the catalyst and uniform dispersion over the high-surface-area support [83]. However, the activity and stability of these complexes is yet to be proved in acidic medium.

16.5 CONCLUDING REMARKS

The synthesis and electrochemical properties of the recently developed non-PGM electrocatalysts, e.g., Ni-/Co-/Fe-based nitrides, macrocyles, and chalcogenides, have been discussed in detail. In recent times, many developments took place in non-PGM catalysts, which are rather promising as ORR catalysts for application in PEM fuel cells. The synthesis of a large variety of catalysts has been undertaken; however, the performance of many of the catalysts in acidic medium in PEMFC condition is yet to be proven. More focus is to be given over the kinetic studies and determination of exchange current

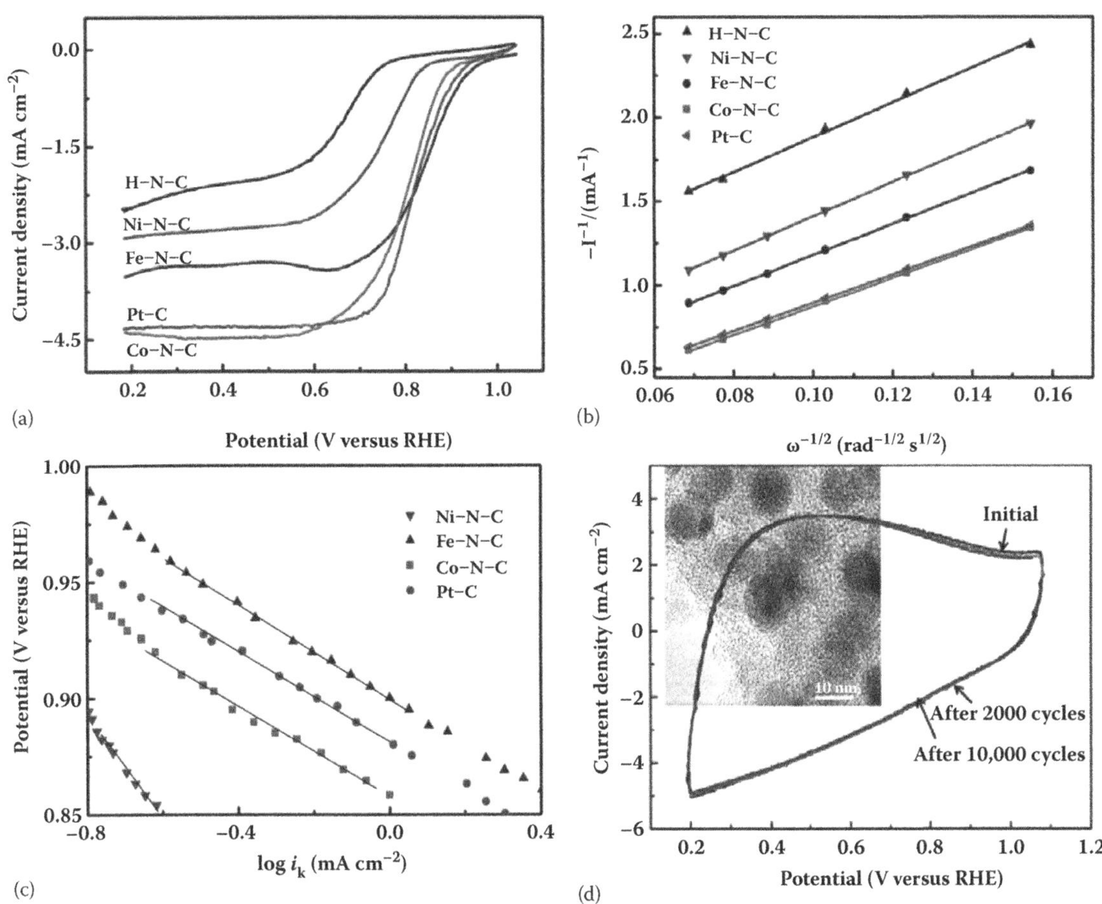

FIGURE 16.5 Comparative performance of Co–N–C, Fe–N–C, Ni–N–C, H–N–C, and Pt–C in 0.1 M KOH solution: (a) ORR polarization curves; (b) Koutecky–Levich plots at 0.5 V; and (c) Tafel plots. (d) CV and durability study of Co–N–C with TEM image after 10,000 cycles. (Reproduced by permission from Macmillan Publishers Ltd., *Nat. Sci. Rep.*, Du, J., Cheng, F., Wang, S., Zhang, T. and Chen, J., 4, 1–7, copyright 2014.)

densities, charge transfer coefficient, ORR pathways, and catalyst degradation so that easy comparison can be made with the commercial Pt–C catalyst. Moreover, the lack of studies in fuel cell also makes it difficult to understand the real performance of electrocatalysts in the operative environment, such as in high temperature conditions and the effect of redox cycling.

The approaches toward non-PGM catalysts for ORR in fuel cells is largely carried out by working on catalyst supports, e.g., multiwall carbon nanotubes (treated and untreated), graphene (functionalized), carbon–nitrogen framework, and chalcogenides along with chelated or coordinate bonded with transition metals, such as Fe, Co, and Ni. Apart from this, efforts are also being undertaken by some of the national laboratories and automobile companies to use aligned nitrogen-doped carbon nanotubes decorated with Fe. Aligned functionalized carbon nanotubes in the direction of mass transport would help reduce concentration polarization. Other approaches are use of MOFs and porous polymers as precursors to Fe, Co, or Ni non-PGM catalyst, which have been discussed in detail. The desired properties of non-PGM catalysts are higher ORR onset potential (>0.9 V), higher mass and volume activities (>0.18 A mg^{-1}), and higher active surface area (>200 m^2 g), with micropore distribution between 10 and 18 Å, low mass transport resistance, high electronic conductivity (similar to carbon), and, most important, high chemical resistance/tolerance to acidic and oxidative environments. Among all the different types of non-PGM catalysts studied, the transitional metal–organic framework and porous polymer precursor–loaded transition metals Fe and Co were found to be the most promising. However, all the properties as mentioned above are not satisfied, and hence there is a requirement for further work in this field.

REFERENCES

1. Bard, A.J. and Faulkner, L.R. *Electrochemical Methods: Fundamentals and Applications*, 1980, Wiley, New York.
2. Yeager, E. Dioxygen electrocatalysis: Mechanisms in relation to catalyst structure. *Journal of Molecular Catalysis* 1986; 38: 5–25.
3. Jasinski, R. A new fuel cell cathode catalyst. *Nature* 1964; 201: 1212–1213.
4. Bezerra, C.W.B., Zhang, L., Lee, K., Liu, H., Marques, A.L.B., Marques, E.P., Wang, H. and Zhang, J. A review of Fe–N/C and Co–N/C catalysts for the oxygen reduction reaction. *Electrochimica Acta* 2008; 53: 4937–4951.
5. Shao, M. *Electrocatalysis in Fuel Cells: A Non- and Low-Platinum Approach*, 2013, Springer, South Windsor, CT, USA.
6. Lee, K.S., Jang, C., Kim, D., Ju, H., Hong, T.W., Kim, W. and Kim, D. The catalytic activities of sputtered cobalt metal electrocatalysts for polymer electrolyte membrane fuel cells. *Solid State Ionics* 2012; 225: 395–397.
7. Wei, G., Wainright, J.S. and Savinell, R.F. Catalytic activity for oxygen reduction reaction of catalysts consisting of carbon, nitrogen and cobalt. *Journal of New Materials for Electrochemical Systems* 2000; 3: 121–129.
8. Lee, K.S. and Kim, D.M. Sputtering and heat treatment of pure Ni metal onto a carbon nanotube on carbon paper to fabricate electrocatalysts for the oxygen reduction reaction in PEMFC. *International Journal of Hydrogen Energy* 2012; 37: 6272–6276.
9. Rao, C.V., Muthukumar, K., Kumar, L.H. and Viswanathan, B. Development of cobalt based non-precious electrocatalyst for oxygen reduction reaction. *Advanced Chemistry Letters* 2014; 1: 1–7.
10. Shi, Z. and Zhang, J. Density functional theory study of transitional metal macrocyclic complexes dioxygen-binding abilities and their catalytic activities toward oxygen reduction reaction. *Journal of Physical Chemistry C* 2007; 111: 7084–7090.
11. Baranton, S., Coutanceau, C., Garnier, E. and Leger, J.M. How does α-FePc catalyst dispersed onto high surface carbon support work toward oxygen reduction reaction. *Journal of Electroanalytical Chemistry* 2006; 590: 100–110.
12. Baresel, D., Sarholz, W., Scharner, P. and Schmitz, J. Transition metal chalcogenides as oxygen catalysts for fuel cells. *Berichte der Bunsengesellschaft für Physikalische Chemie* 1974; 78: 608–611.
13. Tritsaris, G.A., Nørskov, J.K. and Rossmeisl, J. Trends in oxygen reduction and methanol activation on transition metal chalcogenides. *Electrochimica Acta* 2011; 56: 9783–9788.
14. Behret, H., Binder, H. and Sandstede, G. Electrocatalytic oxygen reduction with thiospinels and other sulphides of transition metals. *Electrochimica Acta* 1975; 20: 111–117.
15. Sidik, R.A. and Anderson, A.B. Co_9S_8 as a catalyst for electroreduction of O_2: Quantum chemistry predictions. *Journal of Physical Chemistry B* 2006; 110: 936–941.
16. Bozzini, B., Bocchetta, P., Gianoncelli, A., Mele, C. and Kiskinova, M. Electrodeposition of Co/CoO nanoparticles onto graphene for ORR electrocatalysis: A study based on micro-X-ray absorption spectroscopy and X-ray fluorescence mapping. *Acta Chimica Slovenica* 2014; 61: 263–271.
17. Liang, Y., Li, Y., Wang, H., Zhou, J., Wang, J., Regier, T. and Dai, H. Co_3O_4 nanocrystals on graphene as a synergistic catalyst for oxygen reduction reaction. *Nature Materials* 2011; 10: 780–786.
18. Ma, Z.F., Xie, X.Y., Ma, X.X., Zhang, D.Y., Ren, Q., Heβ-Mohr, N. and Schmidt, V.M. Electrochemical characteristics and performance of CoTMPP/BP oxygen reduction electrocatalysts for PEM fuel cell. *Electrochemistry Communications* 2006; 8: 389–394.
19. Yuan, X., Hu, X.X., Ding, X.L., Kong, H.C., Sha, H.D., Lin, H., Wen, W., Shen, G., Guo, Z., Ma, Z.F. and Yang, Y. Effects of cobalt precursor on pyrolyzed carbon-supported cobalt–polypyrrole as electrocatalyst toward oxygen reduction reaction. *Nanoscale Research Letters* 2013; 8: 478.
20. Bashyam, R. and Zelenay, P. A class of non-precious metal composite catalysts for fuel cells. *Nature Letters* 2006; 443: 63–66.
21. Li, X., Zhang, H.J., Li, H., Zhao, B. and Yang, J. Non-precious metal oxygen reduction electrocatalyst from pyrolyzing cobalt tetraethylenepentamine complex on carbon. *Journal of Electrochemical Society* 2014; 161: F925–F932.
22. Zhang, H.J., Yuan, X., Sun, L., Zeng, X., Jiang, Q.Z., Shao, Z. and Ma, Z.F. Pyrolyzed CoN_4-chelate as an electrocatalyst for oxygen reduction reaction in acid media. *International Journal of Hydrogen Energy* 2010; 35: 2900–2903.
23. Huang, D., Luo, Y., Li, S., Zhang, B., Shen, Y. and Wang, M. Active catalysts based on cobalt oxide@cobalt/N-C nanocomposites for oxygen reduction reaction in alkaline solutions. *Nano Research* 2014; 7: 1054–1064.
24. Mao, S., Wen, Z., Huang, T., Hou, Y. and Chen, J. High performance bi-functional electrocatalysts of 3D crumpled graphene–cobalt oxide nanohybrids for oxygen reduction and evolution reactions. *Energy and Environmental Science* 2014; 7: 609–616.

25. Menezes, P.W., Indra, A., Gonzalez-Flores, D., Ranjbar, N., Zaharieva, I., Strasser, P., Dau, H. and Driess, M. Single source precursor derived nanochains of cobalt oxide as highly efficient multifunctional catalyst for energy conversion and storage, in: *Symposium Z, E-MRS 2014 Spring Meeting*, 2014, Lille, France.
26. Liang, Y., Wang, H., Diao, P., Chang, W., Hong, G., Li, Y., Gong, M., Xie, L., Zhou, J., Wang, J., Regier, T.Z., Wei, F. and Dai, H. Oxygen reduction electrocatalyst based on strongly coupled cobalt oxide nanocrystals and carbon nanotubes. *Journal of American Chemical Society* 2012; 134: 15849–15857.
27. Liang, Y., Li, Y., Wang, H., Zhou, J., Wang, J., Regier, T. and Dai, H. Co_3O_4 nanocrystals on graphene as a synergistic catalyst for oxygen reduction reaction. *Nature Materials* 2011; 10: 780–786.
28. Wang, B. Recent development of non-platinum catalysts for oxygen reduction reaction. *Journal of Power Sources* 2005; 152: 1–15.
29. Faubert, G., Lalande, G., Cote, R., Guay, D., Dodelet, J.P., Weng, L.T., Bertrand, P. and Denes, G. Heat treated iron and cobalt tetraphenylporphyrins adsorbed on carbon black: Physical characterization and catalytic properties of these materials for the reduction of oxygen in polymer electrolyte fuel cells. *Electrochimica Acta* 1996; 41: 1689–1701.
30. Jiang, L. and Zhu, Y.J. General solvothermal route to the synthesis of CoTe, Ag_2Te/Ag, and CdTe nanostructures with varied morphologies. *European Journal of Inorganic Chemistry* 2010; 1238–1243.
31. Dong, W., Wang, X., Li, B., Wang, L., Chen, B., Li, C., Li, X., Zhang, T. and Shi, Z. Hydrothermal synthesis and structure evolution of hierarchical cobalt sulphide nanostructures. *Dalton Transactions* 2011; 40: 243–248.
32. Wang, M., Zhang, H., Zhong, H. and Ma, Y. Cobalt oxyphosphide as oxygen reduction electrocatalyst in proton exchange membrane fuel cell. *International Journal of Hydrogen Energy* 2011; 36: 720–724.
33. Jeevagan, A.J. and John, S.A. Self-assembled monolayer of 2,9,16,23-tetrahydroxythiophenylphthalocyanatocobalt(II) on gold electrode and its electrocatalytic activity towards dioxygen reduction. *Journal of Electroanalytical Chemistry* 2014; 713: 77–81.
34. Sha, H.D., Yuan, X., Li, L., Ma, Z., Ma, Z.F. and Zhang, L. Experimental identification of the active sites in pyrolyzed carbon supported cobalt–polypyrrole–4-toluenesulfinic acid as electrocatalysts for oxygen reduction reaction. *Journal of Power Sources* 2014; 255: 76–84.
35. Kurys, Y.I., Ustavytska, O.O., Koshechko, V.G. and Pokhodenko, V.D. Non-precious metal oxygen reduction nanocomposite electrocatalysts based on poly(phenylenediamines) with cobalt. *Electrocatalysis* 2015; 6(1): 117–125. doi:10.1007/s12678-014-0229-7.
36. Jiang, L., Cui, L. and He, X. Cobalt–porphyrin noncovalently functionalized graphene as nonprecious-metal electrocatalyst for oxygen reduction reaction in an alkaline medium. *Journal of Solid State Electrochemistry* 2015; 19(2): 497–506. doi:10.1007/s10008-014-2628-3.
37. Alia, S.M., Pylypenko, S., Neyerlin, K.C., Cullen, D.A., Kocha, S.S. and Pivovar, B.S. Platinum coated cobalt nanowires as oxygen reduction reaction electrocatalysts. *ACS Catalysis* 2014; 4: 2680–2686.
38. Chung, H.T., Wu, G., Li, Q. and Zelenay, P. Role of two carbon phases in oxygen reduction reaction on the Co–Ppy–C catalyst. *International Journal of Hydrogen Energy* 2014; 39: 15887–15893.
39. Bocchetta, P., Gianoncelli, A., Abyaneh, M.K., Kiskinova, M., Amati, M., Gregoratti, L., Jezersek, D., Mele, C. and Bozzini, B. Electrosynthesis of Co/Ppy nanocomposites for ORR electrocatalysis: A study based on quasi-in situ X-ray absorption, fluorescence and *in-situ* Raman spectroscopy. *Electrochimica Acta* 2014; 137: 535–545.
40. Reis, R.M., Valim, R.B., Rocha, R.S., Lima, A.S., Castro, P.S., Bertotti, M. and Lanza, M.R.V. The use of copper and cobalt phthalocyanines as electrocatalysts for the oxygen reduction reaction in acid medium. *Electrochimica Acta* 2014; 139: 1–6.
41. Ma, Y., Zhang, H., Zhong, H., Xu, T., Jin, H., Tang, Y. and Xu, Z. Cobalt based non-precious electrocatalysts for oxygen reduction reaction in proton exchange membrane fuel cells. *Electrochimica Acta* 2010; 55: 7945–7950.
42. Lee, K., Zhang, L., Lui, H., Hui, R., Shi, Z. and Zhang, J. Oxygen reduction reaction (ORR) catalyzed by carbon-supported cobalt polypyrrole (Co-Ppy/C) electrocatalysts. *Electrochimica Acta* 2009; 54: 4704–4711.
43. Subramanian, N.P., Kumaraguru, S.P., Colon-Mercado, H., Kim, H., Popov, B.N., Black, T. and Chen, D.A. Studies on Co-based catalysts supported on modified carbon substrates for PEMFC cathodes. *Journal of Power Sources* 2006; 157: 56–63.
44. Wang, Q., Jiao, L., Du, H., Peng, W., Han, Y., Song, D., Si, Y., Wang, Y. and Yuan, H. Novel flower like CoS hierarchitectures: One pot synthesis and electrochemical properties. *Journals of Materials Chemistry* 2010; 21: 327–329.
45. Liang, Y., Li, Y., Wang, H., Zhou, J., Wang, J., Regier, T. and Dai, H. Co_3O_4 nanocrystals on graphene as a synergistic catalyst for oxygen reduction reaction. *Nature Materials* 2011; 10: 780–786.
46. Chen, J.H., Jang, C., Xiao, S., Ishigami, M. and Fuhrer, M.S. Intrinsic and extrinsic performance limits of graphene devices on SiO_2. *Nature Nanotechnology* 2008; 3: 206–209.
47. Faubert, G., Cote, R., Dodelet, J.P., Lefevre, M. and Bertrand, P. Oxygen reduction catalysts for polymer electrolyte fuel cells from the pyrolysis of Fe^{II} acetate adsorbed on 3,4,9,10-perylenetetracarboxylic dianhydride. *Electrochimica Acta* 1999; 44: 2589–2603.
48. Gao, X., Wang, J., Ma, Z. and Ye, J. Iron tetrasulfophthalocyanine functionalized graphene nanosheets for oxygen reduction reaction in alkaline media. *Electrochimica Acta* 2014; 130: 543–550.
49. Kamiya, K., Koshikawa, H., Kiuchi, H., Harada, Y., Oshima, M., Hashimoto, K. and Nakanishi, S. Iron–nitrogen coordination in modified graphene catalyzes a four-electron-transfer oxygen reduction reaction. *ChemElectroChem* 2014; 1: 877–884.
50. Wiesener, K., Ohms, D., Neumann, V. and Franke, R. N_4 macrocycles as electrocatalysts for the cathodic reduction of oxygen. *Materials Chemistry and Physics* 1989; 22: 457–475.
51. Chu, D. and Jiang, R. Novel electrocatalysts for direct methanol fuel cells. *Solid State Ionics* 2002; 148: 591–599.
52. Kong, A., Dong, B., Zhu, X., Kong, Y., Zhang, J. and Shan, Y. Ordered mesoporous Fe-porphyrin like architectures as excellent cathode materials for the oxygen reduction reaction in both alkaline and acidic media. *Chemistry: A European Journal* 2013; 19: 16170–16175.
53. Jaouen, F., Marcotte, S., Dodelet, J.P. and Lindbergh, G. Oxygen reduction catalysts for polymer electrolyte membrane fuel cells from the pyrolysis of iron acetate adsorbed on various carbon supports. *Journal of Physical Chemistry B* 2003; 107: 1376–1386.
54. Elzing, A., Vander Putten, A., Visscher, W. and Barendrecht, E. The cathodic reduction of oxygen at cobalt phthalocyanine: Influence of electrode preparation on electrocatalysis. *Journal of Electroanalytical Chemistry* 1986; 200: 313–322.

55. Iliev, I., Gamburzev, S. and Kaisheva, A. Optimization of the pyrolysis temperature of active carbon–CoTMPP catalysts for air electrodes in alkaline media. *Journal of Power Sources* 1986; 17: 345–352.
56. Lefevre, M., Dodelet, J.P. and Bertrand, P. O_2 reduction in PEM fuel cells: Activity and active site structural information for catalysts obtained by the pyrolysis at high temperature of Fe precursors. *Journal of Physical Chemistry B* 2000; 104: 11238–11247.
57. Gupta, S., Tryk, D., Zecevic, S.K., Aldred, W., Guo, D. and Savinell, R.F. Methanol tolerant electrocatalysts for oxygen reduction in a polymer electrolyte membrane fuel cell. *Journal of Applied Electrochemistry* 1998; 28: 673–682.
58. Wang, J., Wang, G., Miao, S., Li, J. and Bao, X. Graphene supported iron based nanoparticles encapsulated in nitrogen doped carbon as a synergistic catalyst for hydrogen evolution and oxygen reduction reactions. *Faraday Discussions* 2014; 176: 135–151. doi:10.1039/C4FD001123K.
59. Bukola, S., Merzougui, B., Akinpelu, A., Laoui, T., Hedhili, M.N., Swain, G.M. and Shao, M. Fe-N-C electrocatalysts for oxygen reduction reaction synthesized by using aniline salt and Fe^{3+}/H_2O_2 catalytic system. *Electrochimica Acta* 2014; 146: 809–818. doi:10.1016/j.electacta.2014.08.152.
60. Hou, Y., Huang, T., Wen, Z., Mao, S., Cui, S. and Chen, J. Metal–organic framework-derived nitrogen doped core–shell structured porous $Fe/Fe_3C@C$ nanoboxes supported on graphene sheets for efficient oxygen reduction reactions. *Advanced Energy Materials* 2014; 4: 1–8.
61. Jeong, B., Shin, D., Jeon, H., Ocon, J.D., Mun, B.S., Baik, J., Shin, H.J. and Lee, J. Excavated Fe–N–C sites for enhanced electrocatalytic activity in the oxygen reduction reaction. *ChemSusChem* 2014; 7: 1289–1294.
62. Zhong, X., Liu, L., Wang, X., Yu, H., Zhuang, G., Mei, D., Li, X. and Wang, J. A radar like iron based nanohybrid as an efficient and stable electrocatalyst for oxygen reduction. *Journal of Materials Chemistry A* 2014; 2: 6703–6707.
63. Liang, J., Zhou, R.F., Chen, X.M., Tang, Y.H. and Qiao, S.Z. Fe–N decorated hybrids of CNTs grown on hierarchically porous carbon for high performance oxygen reduction. *Advanced Materials* 2014; 26: 6074–6079.
64. Liu, S.H., Wu, J.R., Pan, C.J. and Hwang, B.J. Synthesis and characterization of carbon incorporated Fe–N/carbons for methanol-tolerant oxygen reduction reaction of polymer electrolyte fuel cells. *Journal of Power Sources* 2014; 250: 279–285.
65. Moon, J.S., Lee, Y.W., Han, S.B., Kwak, D.H., Lee, K.H., Park, A.R., Sohn, J.I., Cha, S.N. and Park, K.W. Iron–nitrogen doped mesoporous tungsten carbide nanostructures as oxygen reduction electrocatalysts. *Physical Chemistry Chemical Physics* 2014; 16: 14644–14650.
66. Qiu, Y., Xin, L. and Li, W. Electrocatalytic oxygen evolution over supported small amorphous Ni–Fe nanoparticles in alkaline electrolyte. *Langmuir* 2014; 30: 7893–7901.
67. Videla, A.H.A.M., Ban, S., Specchia, S., Zhang, L. and Zhang, J. Non-noble $Fe–N_x$ electrocatalysts supported on the reduced graphene oxide for oxygen reduction reaction. *Carbon* 2014; 76: 386–400.
68. Wang, Q., Zhou, Z.Y., Lai, Y.J., You, Y., Liu, J.G., Wu, X.L., Terefe, E., Chen, C., Song, L., Rauf, M., Tian, N. and Sun, S.G. Phenylenediamine-based $Fe–N_x/C$ catalyst with high activity for oxygen reduction in acid medium and its active site probing. *Journal of American Chemical Society* 2014; 136: 10882–10885.
69. Yan, X., Gan, L., Lin, Y.C., Bai, L., Wang, T., Wang, X., Luo, J. and Zhu, J. Controllable synthesis and enhanced electrocatalysis of Iron-based catalysts derived from electrospun nanofibres. *Small* 2014; 10(20): 4072–4079. doi:10.1002/smll.201401213.
70. Cheon, J.Y., Kim, T., Choi, Y.M., Jeong, H.Y., Kim, M.G., Sa, Y.J., Kim, J., Lee, Z., Yang, T.H., Kwon, K., Terasaki, O., Park, G.G., Adzic, R.R. and Joo, S.H. Ordered mesoporous porphyrinic carbons with very high electrocatalytic activity for oxygen reduction reaction. *Nature Scientific Reports* 2013; 3: 1–8.
71. Cao, R., Thapa, R., Kim, H., Xu, X., Kim, M.G., Li, Q., Park, N., Liu, M. and Cho, J. Promotion of oxygen reduction by a bio-inspired tethered iron phthalocyanine carbon nanotube based catalyst. *Nature Communications* 2013; 4: 1–7.
72. Mo, G., Liao, S., Zhang, Y., Zhang, W. and Ye, J. Synthesis of active iron based electrocatalyst for the oxygen reduction reaction and its unique electrochemical response in alkaline medium. *Electrochimica Acta* 2012; 76: 430–439.
73. He, Q., Mugadza, T., Kang, X., Zhu, X., Chen, S., Kerr, J. and Nyokong, T. Molecular catalysis of the oxygen reduction reaction by iron porphyrin catalysts tethered into Nafion layers: An electrochemical study in solution and a membrane-electrode-assembly study in fuel cells. *Journal of Power Sources* 2012; 216: 67–75.
74. Lefevre, M., Proietti, E., Jaouen, F. and Dodelet, J.P. Iron based catalysts with improved oxygen reduction activity in polymer electrolyte fuel cells. *Science* 2009; 324: 71–74.
75. LaConti, A.B., Hamdan, M. and McDonald, R.C. Mechanisms of membrane degradation for PEMFCs, in: Vielstich, W., Gasteiger, H. and Lamm, A. (Eds.), *Handbook of Fuel Cells—Fundamentals, Technology and Applications*, vol. 3, 2003, Wiley, Chichester, UK, pp. 647.
76. Kamiya, K., Hashimoto, K. and Nakanishi, S. Instantaneous one-pot synthesis of Fe–N modified graphene as an efficient electrocatalyst for the oxygen reduction reaction in acidic solutions. *Chemical Communications* 2012; 48: 10213–10215.
77. Masa, J., Zhao, A., Xia, W., Muhler, M. and Schuhmann, W. Metal free catalysts for oxygen reduction in alkaline electrolytes: Influence of the presence of Co, Fe, Mn and Ni inclusions. *Electrochimica Acta* 2014; 128: 271–278.
78. Yang, R., Stevens, K. and Dahn, J.R. Investigation of Activity of sputtered transition-metal (TM)–C–N (TM = V, Cr, Mn, Co, Ni) catalysts for oxygen reduction reaction. *Journal of the Electrochemical Society* 2008; 155: B79–B91.
79. Garcia-Contreras, M.A., Fernandez-Valverde, S.M. and Vargas-Garcia, J.R. Oxygen reduction reaction on cobalt–nickel alloys prepared by mechanical alloying. *Journal of Alloys and Compounds* 2007; 434–435: 522–524.
80. Lee, D.U., Kim, B.J. and Chen, Z. One-pot synthesis of a mesoporous $NiCo_2O_4$ nanoplatelet and graphene hybrid and its oxygen reduction and evolution activities as an efficient bi-functional electrocatalyst. *Journal of Materials Chemistry A* 2013; 1: 4754–4762.
81. Liu, Q., Jin, J. and Zhang, J. $NiCo_2S_4$@graphene as a bifunctional electrocatalyst for oxygen reduction and evolution reactions. *ACS Applied Materials & Interfaces* 2013; 5: 5002–5008.
82. Hardin, W.G., Mefford, J.T., Slanac, D.A., Patel, B.B., Wang, X., Dai, S., Zhao, X., Ruoff, R.S., Johnston, K.P. and Stevenson, K.J. Tuning the electrocatalytic activity of perovskites through active site variation and support interactions. *Chemistry of Materials* 2014; 26: 3368–3376.

83. Garcia, C., Diaz, C., Araya, P., Isaacs, F., Ferraudi, G., Lappin, A.G., Aguirre, M.J. and Isaacs, M. Electrostatic self-assembled multilayers of tetrachromated metalloporphyrins/polyoxometalate and its electrocatalytic properties in oxygen reduction. *Electrochimica Acta* 2014; 146: 819–829. doi:10.1016/j.electacta.2014.08.117.
84. Du, J., Cheng, F., Wang, S., Zhang, T. and Chen, J. M(salen)-derived nitrogen-doped M/C (M = Fe, Co, Ni) porous nanocomposites for electrocatalytic oxygen reduction. *Nature Scientific Reports* 2014; 4: 1–7.
85. Garcia, A.C., Lima, F.H.B., Ticianelli, E.A. and Chatenet, M. Carbon supported nickel-doped manganese oxides as electrocatalysts for the oxygen reduction reaction in the presence of sodium borohydride. *Journal of Power Sources* 2013; 222: 305–312.
86. Liu, Z.Q., Xu, Q.Z., Wang, J.Y., Li, N., Guo, S.H., Su, Y.Z., Wang, H.J., Zhang, J.H. and Chen, S. Facile hydrothermal synthesis of urchin-like $NiCo_2O_4$ spheres as efficient electrocatalysts of oxygen reduction reaction. *International Journal of Hydrogen Energy* 2013; 38: 6657–6662.
87. Silva, R.A., Soares, C.O., Carvalho, M.D., Melo Jorge, M.E., Gomes, A., Rangel, C.M. and Silva, P. Redox stability and bifunctionality of $LaNiO_3$-based oxygen electrodes, in: *XVIII Meeting of the Portuguese Electrochemical Society 2012*, March 24–27, 2012, Porto, Portugal.
88. Yang, J.Y., Bullock, R.M., Dougherty, W.G., Kassel, W.S., Twamley, B., DuBois, D.L. and DuBois, M.R. Reduction of oxygen catalyzed by nickel diphosphine complexes with positioned pendant amines. *Dalton Transactions* 2010; 39: 3001–3010.
89. Millan, W.M. and Smit, M.A. Study of electrocatalysts for oxygen reduction based on electroconducting polymer and nickel. *Journal of Applied Polymer Science* 2009; 112: 2959–2967.
90. Xu, L., Ding, Y.S., Chen, C.H., Zhao, L., Rimkus, C., Joesten, R. and Suib, S.L. 3D Flowerlike α-nickel hydroxide with enhanced electrochemical activity synthesized by microwave assisted hydrothermal method. *Chemistry of Materials* 2008; 20: 308–316.
91. Kullapere, M. and Tammeveski, K. Oxygen electroreduction on anthraquinone-modified nickel electrodes in alkaline solution. *Electrochemistry Communications* 2007; 9: 1196–1201.
92. Roche, I., Chainet, E., Chatenet, M. and Vondrak, J. Carbon supported manganese oxide nanoparticles as electrocatalysts for the oxygen reduction reaction (ORR) in alkaline medium: Physical characterizations and ORR mechanism. *Journal of Physical Chemistry C* 2007; 111: 1434–1443.
93. Yang, R., Bonakdarpour, A. and Dhan, J.R. Investigation of sputtered Ta–Ni–C as an electrocatalyst for the oxygen reduction reaction. *Journal of the Electrochemical Society* 2007; 154: B1–B7.
94. Ma, M., Donepudi, V.S., Sandi, G., Sun, Y.K. and Prakash, J. Electrodeposition of nano-structured nickel–21% tungsten alloy and evaluation of oxygen reduction reaction in a 1% sodium hydroxide solution. *Electrochimica Acta* 2004; 49: 4411–4416.

17 Advanced Technologies for Proton-Exchange Membrane Fuel Cells

Sivakumar Pasupathi, Huaneng Su, Huagen Liang, and Bruno G. Pollet

CONTENTS

17.1 INTRODUCTION

When oil, one of the most important energy sources in the history of mankind, was first discovered in Pennsylvania almost 150 years ago, the fuel cell had already been known for 20 years, invented by Sir William Grove, "father of the fuel cell," in 1839. Back then, it was an idea that was far ahead of its time. Today, however, it is the most important development in the world history of decentralized energy supply. Fuel cells are now widely accepted to be efficient and nonpolluting power sources, offering much higher energy densities and energy efficiencies than any other current energy storage devices. They are therefore considered to be promising energy devices for the transport, mobile, and stationary sector, and have emerged as a vital alternative energy solution to reduce societal dependence on internal combustion engines and lead–acid batteries. Indeed, they promise significantly improved energy efficiency with zero or low greenhouse gas emissions, and they are now expected to play a key role in the so-called hydrogen economy.

17.2 FABRICATION AND MANUFACTURING OF FUEL CELLS

17.2.1 Introduction

A fuel cell is an "electrochemical" device operating at various temperatures (up to 1000°C) that transforms the chemical energy of a fuel (hydrogen, methanol, natural gas, etc.) and an oxidant (air or pure oxygen), in the presence of a catalyst, into water, heat, and electricity. Furthermore, the power generated by a fuel cell depends largely on the catalytic electrodes and

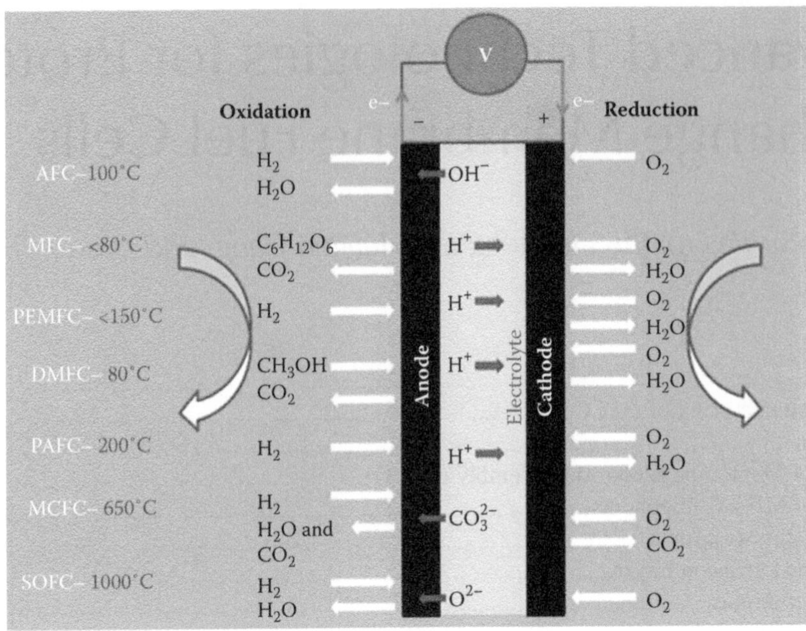

FIGURE 17.1 Schematic of a single typical PEMFC. (Copyright - Bruno G. Pollet.)

materials used. The majority of fuel cell technology is based on the simple reaction given in Equation 17.1:

$$H_2 + O_2 \rightarrow H_2O \tag{17.1}$$

There are currently six main groupings of fuel cell available (Figure 17.1): (i) proton-exchange membrane fuel cell (PEMFC), including direct methanol fuel cell (DMFC), (ii) alkaline fuel cell (AFC), (iii) phosphoric acid fuel cell (PAFC), (iv) molten carbonate fuel cell (MCFC), (v) solid oxide fuel cell (SOFC), and (vi) microbial fuel cell (MFC). PEMFC, AFC, PAFC, and MFC operate at low temperatures in the range of 50–200°C, and MCFC and SOFC at high temperatures in the range of 650–1000°C. This chapter focuses only on PEMFC.

The PEMFC is one of the most elegant types of fuel cells; it uses hydrogen (H_2) as a fuel that is oxidized at the anode, and oxygen that is reduced at the cathode. The protons released during the oxidation of hydrogen are conducted through the proton-exchange membrane (PEM) to the cathode. Since the polymer membrane is not electrically conductive, the electrons released from the hydrogen travel along the electrical detour provided, and an electrical current is generated. These reactions and pathways are shown schematically in Equation 17.1.

17.2.2 Principles of Membrane Electrode Assembly (MEA)

The MEA is the heart of PEMFCs, where the electrochemical reactions take place to generate electrical power. The MEA is pictured in the schematic of a single PEMFC shown in Figure 17.2. It is typically sandwiched by two flow-field plates that are often mirrored to make a bipolar plate when cells are stacked in series for larger cell voltages.

The MEA consists of a polymeric PEM, catalyst layers (CLs), and gas diffusion layers (GDLs). Typically, these components are fabricated individually and then pressed together at high temperatures and pressures. An ideal MEA would allow all active catalyst sites in the CL to be accessible to the reactant (H_2 or O_2), protons, and electrons, and would facilitate the effective removal of produced water from the CL and GDL.

Analogous cathode and anode reactions in the MEA for an H_2/O_2 fuel cell are as follows:

$$\text{Cathode: } \frac{1}{2}O_2 + 2H^+ + 2e^- \rightarrow H_2O \tag{17.2}$$

$$\text{Anode: } H_2 \rightarrow 2H^+ + 2e^- \tag{17.3}$$

The flow of ionic charge through the electrolyte must be balanced by the flow of electronic charge through an outside circuit, and it is this balance that produces electrical power.

Several review articles in the literature cover-specific aspects of PEMFC. The MEA fabrication technology is a key element for creating a fuel cell that will perform at a high level. Over the past several decades, great efforts have been focused on the optimization of the CL, MEA structures, and fabrication methods that have been developed. In this chapter, it is attempted to cover some MEA fabrication methods commonly used in low-temperature fuel cell laboratories, with an

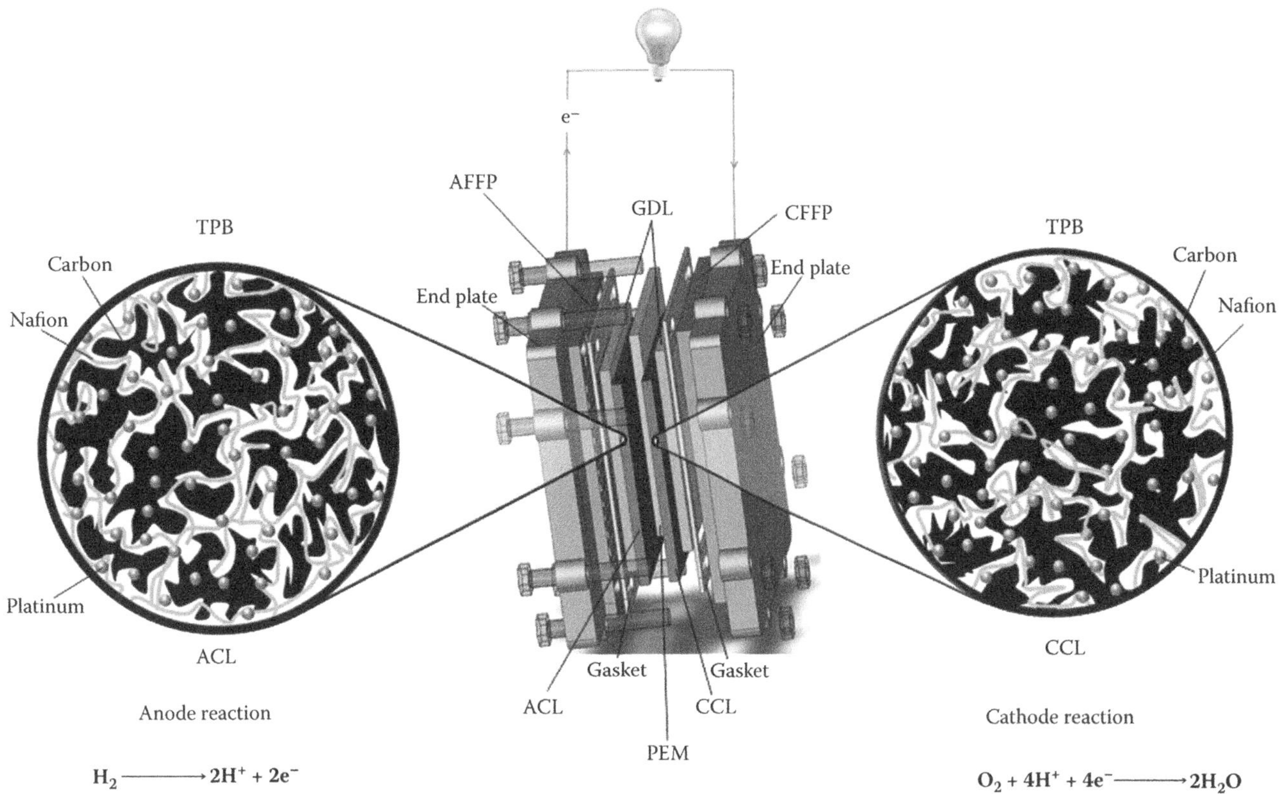

FIGURE 17.2 Single PEMFC schematic. ACL: Anode catalyst layer; AFFP: Anode flow field plate; CCL: Cathode catalyst layer; CFFP: Cathode flow field plate; GDL: Gas diffusion layer; PEM: Polymer electrode membrane; TPB: Triple-phase boundary. (Copyright - Pollet & Reddy.)

emphasis on applied aspects. Some techniques that are useful in the investigation of various aspects of fuel cells, such as the AFC and the PAFC, are mentioned.

17.2.3 Overview of MEA Components

17.2.3.1 Catalyst Layer

The CL is a key component in the GDE, as the location where electrochemical reactions take place. The CLs need to be designed to generate high rates of the desired reactions and minimize the amount of catalyst necessary for reaching the required levels of power output. An ideal CL should maximize the active surface area per unit mass of the electrocatalyst, and minimize the obstacles for reactant transport to the catalyst, for proton transport to exact positions, and for product removal from the cell; these requirements entail an extension of the three-phase boundary.

17.2.3.2 Gas Diffusion Layer

The porous GDL in PEMFCs ensures that reactants effectively diffuse to the CL. In addition, the GDL is the electrical conductor that transports electrons to and from the CL. Typically, GDLs are constructed from porous carbon paper, or carbon cloth, with a thickness in the range of 100–300 μm. The GDL also assists in water management by allowing an appropriate amount of water to reach, and be held at, the membrane for hydration. In addition, GDLs are typically wet-proofed with a polytetrafluoroethylene (PTFE) (Teflon) coating to ensure that the pores of the GDL do not become congested with liquid water.

17.2.3.3 Membrane

The main function of the membrane in PEMFCs is to transport protons from the anode to the cathode; membrane polymers have sulfonic groups that facilitate the transport of protons. The other functions include keeping the fuel and oxidant separated, which prevents mixing of the two gases and withstanding harsh conditions, including active catalysts, high temperatures or temperature fluctuations, strong oxidants, and reactive radicals. Thus, the ideal polymer must have excellent proton conductivity, chemical and thermal stability, strength, flexibility, low gas permeability, low water drag, low cost, and good availability [1]. Although the membrane and GDL is the critical part of the MEA, a review of the design and fabrication of polymer electrolyte membranes and GDL is beyond the scope of this paper.

17.2.4 Methods for MEA Fabrication

The process of MEA fabrication should be guided by both fuel cell performance and cost reduction. Table 17.1 identifies various methods available for the fabrication of MEAs. The following

TABLE 17.1
Various Methods Available for the Fabrication of MEAs

MEA Fabrication Methods	
Ink-Based Methods	***In Situ* Methods**
Spray coating, screen printing, and inkjet printing	Sputter deposition, dual ion-beam-assisted deposition, electrodeposition method, reactive spray deposition technology, and atomic layer deposition

description provides an overview of two common MEA fabrication methods: ink-based method and *in situ* method.

17.2.4.1 Ink-Based Method

The first generation of PEMFC used PTFE-bound Pt black electrocatalysts that exhibited excellent long-term performance at a prohibitively high cost [2]. However, the CL formed by this Pt black catalyst has several disadvantages, including high platinum loadings (4 mg/cm^2), large platinum agglomerates (~1 μm on average), lower electrochemical surface area (~25 m^2/g), and poor access to the catalyst surface for gas, electrons, and protons. Approaches to improve the CL construction by significantly increasing the three-phase boundary in the CL can be achieved in two ways. First, a breakthrough was made by replacing pure Pt black with supported Pt catalysts [3], which can significantly reduce the Pt loading from 4 mg/cm^2 down to 0.4 mg/cm^2 [4,5]. Second, the PTFE-bound CLs are typically impregnated with Nafion® by brushing or spraying to provide ionic transport to the catalyst site. The impregnation of an ionomer (Nafion) into these CLs was found to be extremely effective in improving the three-dimensional reaction zone for fuel cell applications. The latter can be accomplished by blending the solubilized ionomer and the platinized carbon into a homogeneous "ink" from which the thin-film CL of the electrode can be made. However, the platinum utilization in this PTFE-bound CL still remains low—in the vicinity of 20% [6,7].

Numerous efforts have been made to improve existing thin-film catalysts in order to prepare a CL with low Pt loading and high Pt utilization without sacrificing electrode performance. In thin-film CL fabrication, the most common method is to prepare catalyst ink by mixing the Pt/C agglomerates with a solubilized polymer electrolyte, such as Nafion ionomer, and then to apply this ink on a porous support or membrane using various methods. In this case, the CL always contains some inactive catalyst sites not available for fuel cell reactions because the electrochemical reaction is located only at the interface between the polymer electrolyte and the Pt catalyst where there is reactant access.

The procedure for forming a thin-film CL on the membrane, according to Wilson's 1993 patent [8], is as follows:

1. Combine a 5% solution of solubilized perfluorosulfonate ionomer (such as Nafion) and 20% wt Pt/C support catalyst in a ratio of 1:3 Nafion/catalyst.
2. Add water and glycerol to weight ratios of 1:5:20 carbon–water–glycerol.
3. Mix the solution with ultrasound until the catalyst is uniformly distributed and the mixture is adequately viscous for coating.
4. Ion exchange the Nafion membrane to the Na^+ form by soaking it in NaOH, then rinse and let dry.
5. Apply the carbon–water–glycerol ink to one side of the membrane. Two coats are typically required for adequate catalyst loading.
6. Dry the membrane in a vacuum with the temperature of approximately 160°C.
7. Repeat steps 5 and 6 for the other side of the membrane.
8. Ion exchange the assembly to the protonated form by lightly boiling the MEA in 0.1 M H_2SO_4 and rinsing in deionized water.
9. Place carbon paper/cloth against the film to produce a GDL.

Alternatively, the CL ink can be (i) applied to a PTFE blank or some other substrate, and then decal transferred onto the membrane, called the decal-transfer method [9,10]; (ii) deposited onto the diffusion layer and then hot pressed to the membrane for MEA fabrication, called the hot-pressing method [11–15]; or (iii) deposited onto Nafion membranes and then assembled with two GDLs, called the catalyst coating membrane method [16–18]. On the basis of the nature of the catalyst ink and its application method, several thin-film CL fabrication techniques have been developed, including decal transfer [9,10], brush painting [19,20], spray coating [21,22], doctor blade coating [23], screen printing [24–26], inkjet printing [27,28], and rolling [29]. Currently, screen printing and spray coating have become the standard methods for conventional CL fabrication. Inkjet printing is also showing promise for fabricating low-Pt-loading MEAs.

17.2.4.1.1 Spray Coating

The spray-coating method [30] is widely used for catalyst fabrication. Typically, the catalyst ink is coated on the GDL or cast directly on the membrane. The application apparatus can be a manual spray gun or an autospraying system with programmed *X–Y* axes, movable robotic arm, an ink reservoir and supply loop, ink atomization, and a spray nozzle with adjustable flux and pressure. To prevent distortion and swelling of the membrane, Xu et al. developed a novel catalyst-coated membrane (CCM) approach—a catalyst-sprayed membrane under irradiation, for the preparation of MEAs for low-temperature PEMFC. Catalyst ink was sprayed directly onto the membrane and an infrared light was used simultaneously to evaporate the solvents. The resultant MEAs prepared by this method yielded very high performance.

Shin et al. [31] prepared a colloidal catalyst ink with a method similar to the conventional spray-coating approach, which is called the colloidal method. A mixture of Pt/C powder and Nafion ionomer was dripped drop by drop into the normal-butyl acetate solvent to form the ionomer colloids. The ink was then treated ultrasonically to allow the colloids to absorb the Pt/C powder. The ink was then sprayed via air brushing onto the carbon paper, which was to be used as the GDL. It was stated that the colloidal method is the preferred ink for spraying methods, as it forms larger agglomerates. Small agglomerates formed by the solution method have a tendency to penetrate too far into the GDL, blocking pores needed for gas transport. In addition, it was proposed that in the colloidal method, the ionomer colloid absorbs the catalyst particles and larger Pt/C agglomerates are formed. The colloidal method is known to cast a continuous network of ionomer that enhances proton transport.

The colloidal method dramatically outperformed the solution method at high current densities. This is attributed to a significant increase in the proton conductivity, as well as a moderate enhancement of the mass transport in the CL formed with the colloidal ink. The increase in proton conductivity is due to the continuous network of ionomers in the colloidal CL. The increased mass transport is a product of the larger agglomerates of Pt/C in the colloidal CL, which translates to a higher porosity, allowing a greater flux of the reactant and product gases.

17.2.4.1.2 Screen Printing

Similar to spray coating, the screen-printing method is also widely used for CL fabrication. The major difference between the two is that the viscosity of the ink for spray coating is much lower than that for screen printing. In a typical screen-printing process, the ink slurry is first cast onto certain substrates, and then the CL is transferred to a Nafion membrane by hot pressing. The membrane is converted into Na^+ form to avoid swelling during hot pressing and decal transfer because, in this form, the Nafion membrane is mechanically strong and stable for hot pressing from 150°C to 160°C. In the direct screen-printing process [2], the ink slurry is applied to a membrane in either Na^+ or tert-butyl alcohol (TBA) form to stabilize the catalytic layer, thereby enhancing the membrane's physical strength.

The coating apparatus consists of a silk-screen mesh fixed to a frame with sufficient tension to squeeze the ink through the screen and onto the blank substrate (e.g., polyimide). The substrate is fixed on an *XY* table with adhesion tape, and the silken screen mesh is masked, with an open window in the center for screen printing. The silicon rubber squeeze is a fixed support and can be moved in both *X* and *Y* directions. A hot-air or infrared ramp is used to dry the coating for solvent removal.

The coating procedure consists of positioning the substrate layer under the silk-screen mesh, which is not masked, and using a squeegee on top of the mesh at one side. An appropriate amount of ink is micropipetted near the squeegee, and the slurry is first spread across and then pushed through the mesh to the substrate layer by rapid movements of the squeegee. Hot air is used to dry the CL coated on the substrate. The same procedure is repeated until the pipette volume of catalyst ink has been transferred to the substrate layer.

17.2.4.1.3 Inkjet Printing

Inkjet printers utilize drop-on-demand technology to deposit various materials or "inks." This is a popular deposition technique used not only in desktop printers, but also to deposit various other coating materials, such as those required for CL fabrication. Using inkjet-printing technology, a research group in the Pacific Northwest National Lab has successfully fabricated CLs for hydrogen–air PEMFCs [27]. In their study, a slightly modified commercial desktop inkjet printer was used to deposit catalyst ink directly from a print cartridge onto the Nafion membrane to form a CL. The ink cartridges were filled with the catalyst inks. The membrane was secured to a cellulose acetate sheet and fed through the printer using the original paper feed platen. Computer software was used to control the print parameters, such as electrode dimensions, thickness, and resolution. Fuel cell testing on the fabricated CLs showed power densities up to 155 mW/cm^2, with a cathode catalyst loading of 0.20 mg Pt/cm^2.

These studies demonstrate some of the advantages of inkjet printing for CL fabrication, such as varied composition layer printing, and suggest that inkjet-based fabrication technology might be the way to cost-effective and large-scale fabrication of PEMFCs in the future. However, the question still remains of whether inkjet fabrication can offer any performance advantage and still lead to more efficient utilization of the Pt catalyst.

Similarly, Taylor et al. [28] used inkjet printing to deposit catalyst materials onto GDLs. Their inkjet-printed CLs with a catalyst loading of 0.020 mg Pt/cm^2 showed high Pt utilizations. This research also demonstrated the capacity of the inkjet-printing technique to control ink volume precisely down to picoliters for ultralow catalyst loading, the flexibility of using different carbon substrates, and the functionality of gradient catalyst structure fabrication using these techniques.

17.2.4.2 *In Situ* Method

Numerous efforts have been made to develop *in situ* CL fabrication methods to lower Pt loading and increase platinum utilization without sacrificing electrode performance.

17.2.4.2.1 Sputter Deposition (Vacuum Deposition Methods)

Common vacuum deposition methods include chemical vapor deposition, physical or thermal vapor deposition, and sputter deposition. Sputtering is commonly employed to form CLs and is known for providing denser layers than the alternative evaporation methods [32]. The sputtering of CLs consists of a vacuum evaporation process that removes portions of a coating material (the target) and deposits a thin and resilient film of the target material onto an adjacent substrate. In the case of sputtered CLs, the target material is the catalyst material and the substrate can be either the GDL or the membrane. Sputtering provides a method of depositing a thin CL (onto either the membrane or the GDL) that delivers high performance combined with a low Pt loading. The entire CL is in such intimate contact with the membrane that the need for ionic conductors in the CL is resolved [33]. Moreover, platinum and its alloys are easily deposited by sputtering [34]. The success of the sputtering method on reducing platinum loading depends heavily on the reduction in the size of catalyst particles below 10 nm. State-of-the art thin-film electrodes feature Pt loading of 0.1 mg/cm^2 [33]. A 5-nm sputtered platinum film amounts to a platinum loading of 0.014 mg/cm^2. However, the performance of a fuel cell with a sputtered CL can vary by several orders of magnitude depending on the thickness of the sputtered CL [35].

The advantages of sputter deposition include

1. Precise Pt loading and thickness, as well as controlled microstructure morphology
2. Much smaller Pt particle size
3. Homogeneous distribution of the Pt particles on the support and extremely low metal loadings (down to 10 ng/cm^2)
4. A simple preparation process that is easy to scale up
5. Adaptiveness to various substrates, such as GDL and membrane

Although the sputter deposition technique can provide a cost-effective and directly controlled deposition method, the performance of PEMFCs with sputtered CLs is still inferior to that of conventional ink-based fuel cells. In addition, other issues arise related to the physical properties of sputtered CLs, such as low lateral electrical conductivity of the thin metallic films [36,37]. Furthermore, the smaller particle size of sputter-deposited Pt can hinder water transport due to the high resistance to water transport in a thick, dense, sputtered Pt layer [37]. Currently, the sputter deposition method is not considered an economically viable alternative for large-scale fuel cell electrode fabrication [14] and further research is under way to improve methods.

17.2.4.2.2 Dual Ion-Beam-Assisted Deposition

Saha et al. [38] have proposed an improved ion deposition method based on a dual ion-beam-assisted deposition (dual IBAD) method. Dual IBAD combines physical vapor deposition (PVD) with ion-beam bombardment. The unique feature of dual IBAD is that the ion bombardment can impart substantial energy to the coating and coating/substrate interface, which could be employed to control film properties such as uniformity, density, and morphology. Using the dual IABD method, an ultralow, pure Pt-based CL (0.04–0.12 mg Pt/cm^2) can be prepared on the surface of a GDL substrate, with film thicknesses in the range of 250–750 Å. The main drawback is that the fuel cell performance of such a CL is much lower than that of conventional ink-based CLs. Further improvement in catalyst utilization and the gas/liquid diffusivity of the CLs prepared by IBAD is necessary.

17.2.4.2.3 Electrodeposition Method

The first disclosure of electrodeposition of the catalytic layer in PEMFCs was in the form of Vilambi Reddy et al.'s 1992 US patent [39]. This patent detailed the fabrication of electrodes featuring low platinum loading in which the platinum was electrodeposited into the uncatalyzed carbon substrate in a commercial plating bath. The uncatalyzed carbon substrate consisted of a hydrophobic porous carbon paper that was impregnated with dispersed carbon particles and PTFE. Nafion was also impregnated onto the side of the carbon substrate that was to be catalyzed.

The Nafion-coated carbon paper was placed in a commercial platinum acid-plating bath, along with a platinum counter electrode. The face of the substrate that was not coated with Nafion was most likely masked with some form of a nonconducting film. This step would have been taken to ensure that platinum would only be deposited in regions impregnated with Nafion. Thus, when an interrupted direct current was applied to the electrodes in the plating bath, catalyst ions would pass through the Nafion to the carbon particles and successfully be deposited only where protonic and electronic conduction coexists. This method was able to produce electrodes featuring platinum loadings of 0.05 mg/cm^2. This is a significant reduction in loading from the state-of-the-art thin-film electrode.

In the following years, additional research on electrodeposition of platinum onto porous substrates was continued by Verbrugge [40]. According to Verbrugge, a distinguishing difference between his study and the aforementioned patent is the larger amount of sulfuric acid employed by Vilambi Reddy [39]. Another distinguishing feature of Verbrugge's study is the employment of a membrane instead of a Nafion-impregnated layer. Using the area provided by the deposition channel, platinum was selectively electrodeposited through the membrane and into the membrane–electrode interfacial region. Verbrugge suggested that this method has the potential to increase platinum utilization because of the concentrated platinum found at the membrane–electrode interface. However, he did not provide the results of these electrodes implemented in a functional fuel cell.

The objective of studies by Hogarth et al. [41], and later by Gloaguen et al. [42], was to improve the reaction kinetics for the oxidation of methanol using electrodes fabricated with electrodeposition. Hogarth et al. placed electrodes in a plating bath that contained 0.02 M chloroplatinic acid and exposed only 1 cm^2 of the PTFE impregnated carbon cloth electrode face by using a water seal. In this study, neither a Nafion layer nor a membrane film was applied to the carbon substrate prior to electrodeposition. The Gloaguen et al. study focused on the oxygen reduction reaction kinetics of electrodes formed with the electrodeposition of platinum on carbon supports that were bound by Nafion onto a glassy carbon stick. One of the most significant conclusions of the study was that Pt activity is less related to particle size and more to the fine structure of the platinum surface.

A noticeable increase in catalyst utilization resulted when Pt deposition took place only in the three-phase reaction zone. Because Pt electrodeposition in aqueous solution only occurs in the region of the ionic and electronic pathways, it should be possible to reduce Pt loading significantly and increase Pt utilization in the CL. Pulse electrodeposition is promising and could replace conventional methods for fabricating cost-effective, low-Pt-loading CLs. However, CL durability may be an issue if the active CL sites change with time.

17.2.4.2.4 Reactive Spray Deposition Technology

Reactive spray deposition technology (RSDT) has been developed for direct catalyst deposition onto substrates, including polymer electrolyte membranes, to form CCMs [43]. The process involves dissolving a Pt precursor in appropriate solvents, followed by spraying the solution with an expansion gas through a nozzle to produce micron-sized droplets. The droplets are burnt out in a flame, resulting in metal atoms and/or metal oxide molecules in the gas phase. At the same time, a quench gas is used to induce rapid condensation of Pt vapors into Pt particles with a size of ~5 nm. A mixture of carbon and Nafion ionomer is subsequently introduced into the gas stream. The cooling effect of the quench gas helps avoid thermal damage to the ionomer and the membrane substrate. As a result, a thin-film CL forms on the electrolyte membrane.

In the RSDT process, the steps for introducing catalyst, ionomer, and carbon into the gas mix are decoupled and can be independently controlled in such a manner that the Pt/C and ionomer/C ratios can be continuously modified during the deposition process. RSDT has the capacity and flexibility required to produce compositionally and structurally optimized CLs. The technology can be used to generate supported and unsupported platinum CLs with thicknesses from 10 to 200 nm, and with varied morphologies, including dense films, porous films (packed particles), and dendritic and island-growth structures. However, the usage efficiency of the catalyst ink is rather low in the fabrication process, which is a major drawback of this technology for fuel cell CL preparation.

17.2.4.2.5 Pulsed Laser Deposition

Cunningham et al. [44] used the pulsed laser deposition (PLD) method to deposit platinum onto E-TEK gas diffusion electrodes (GDEs) to prepare low-catalyst-loading electrodes for PEMFCs. In the PLD process, the laser beam is focused by a quartz focus lens onto a polycrystalline platinum target. The Pt target is continuously rastered across the laser beam, via a dual rotation and translation motion, to obtain a uniform ablation over the entire target surface. The chamber is evacuated by a turbomolecular pump and filled with helium at a constant pressure throughout the PLD process. The platinum loadings are controlled by the number of pulses during deposition.

This technique yields a catalyst composed entirely of metal nanoparticles or nanocrystalline thin film, and it allows for the control of size and distribution while eliminating the need for a dispersing and supporting medium. The obtained electrodes contained as little as 0.017 mg Pt/cm^2 and performed as well as standard E-TEK electrodes (Pt loading 0.4 mg/cm^2). The PLD technique may be of special interest as an alternative to the sputtering process in the production of micro fuel cells.

17.2.4.2.6 Atomic Layer Deposition

To lower the cost of PEMFCs, the catalytic activity and utilization efficiency of Pt must be increased and, therefore, the CL structure must be modified [45]. This may be achieved by the catalyst deposition method. Thin-film methods have been used to prepare a Pt catalyst for PEMFC [36,46]. Complex processes, such as reflux, centrifugation, and mixing with solvent before pasting on the gas diffusion layer, are needed in the thin-film method to prepare a supported catalyst [46,47]. However, there is still a high Pt loading (>0.1 mg/cm^2) in the state-of-the-art thin-film electrode [36,46]. Electrodeposition has been applied to increase Pt utilization efficiency by depositing the catalyst on ionic and electronic pathways [36]. However, it has relatively low productivity and high Pt loading. Sputtering can be used to directly deposit Pt on the substrate [46]; however, the Pt particles cannot penetrate into deep regions of the porous support [47]. Consequently, atomic layer deposition (ALD) is introduced to overcome these difficulties. The particle density and loading of Pt could be controlled by the substrate surface condition and cycle number. Since the ALD of Pt has the advantages of simple processing, large area, and batch production, small particle size, lower loading, and good uniformity and conformality, it is a promising technique for applications in PEMFCs, especially for micro fuel cells where both lower loading and uniform, conformal coating of Pt with controlled particle size is required.

ALD is a gas-phase thin-film deposition method that is unique because the film growth proceeds through self-limiting surface reactions [48–51]. As a consequence, ALD offers excellent large-area uniformity and conformality [52] and enables simple and accurate control of film thickness and composition at an atomic layer level. All these characteristics make ALD an important film deposition technique for future microelectronics.

Liu et al. deposited Pt nanoparticles on carbon cloths and carbon nanotubes (CNTs) for applications in PEMFCs by using the ALD method [53]. In the ALD of Pt, (methylcyclopentadienyl) trimethylplatinum (MeCpPtMe3) and air or oxygen have been used as precursors to grow a uniform and thickness-controllable particle-like Pt film [54,55]. Conformal Pt nanoparticle coating on acid-treated CNTs was achieved and good dispersion was observed on acid-treated carbon cloth and CNTs. ALD of Pt was conducted for 100 cycles to deposit 0.016 mg/cm^2 of Pt on the CNTs. In addition, the MEA prepared by Pt deposited on acid-treated CNTs using the ALD method performs better than that of the commercial E-TEK electrode. Platinum nanocatalyst deposited by ALD on CNTs has been demonstrated to have a higher utilization efficiency in PEMFCs than commercial E-TEK electrodes.

17.3 DEGRADATION MECHANISMS AND MITIGATION STRATEGIES

The performance of a PEM fuel cell or stack is affected by many internal and external factors, such as fuel cell design and assembly, degradation of materials, operational conditions, and impurities or contaminants [56–62]. Performance degradation is unavoidable, but the degradation rate can be minimized through a comprehensive understanding of degradation and failure mechanisms.

17.3.1 PEMFC Components Degradation

A schematic cross section of a single cell of a PEMFC showing individual components is shown in Figure 17.2.

An individual fuel cell consists of an anode, a cathode, and a polymer–electrolyte membrane. Each electrode has an electrocatalyst layer and a GDL. The CLs can be attached to either the membrane or at times to the GDL material (i.e., the GDE). The individual cells are electrically connected in series by bipolar plates to form a fuel cell stack, and special end plates terminate the stacks to provide the compressive forces needed for the stack structural integrity. The bipolar plates provide conducting paths for electrons between cells, distribute the reactant gases across the entire active MEA surface area (through flow channels integrated into the plates), remove waste heat (through cooling channels), and provide stack structural integrity as well as barriers to anode and cathode gases. The aim of this section is to review the mechanisms affecting the lifetime of these fuel cell components, and the mitigation strategies for these degradations.

17.3.1.1 Catalyst Layer

In CLs, the electrocatalysts are nanoparticles of platinum or platinum alloys deposited on high-surface-area carbon supports. The nanoparticles increase the active catalytic surface area per unit mass of platinum. However, they also pose material stability and interaction issues. The stability of platinum, platinum alloys, and carbon particles, and the interaction of the nanocatalyst particles with the carbon supports are of concern for fuel cell durability.

17.3.1.1.1 Platinum Dissolution and Particle Growth

Several studies have observed the growth of Pt particles and the dissolution of Pt in the membrane phase. Both phenomena were already described for the PAFC, and are closely related as will be explained below. Thermodynamics predict substantial Pt dissolution in acid media, strongly increasing with potential in the region of +0.85–+0.95 V versus the reversible hydrogen electrode (RHE), according to the reaction

$$Pt \rightarrow Pt^{2+} + 2e^- \quad E_0 = +1.188 \text{ V versus RHE} \qquad (17.4)$$

However, experiments on Pt thin film, Pt wire, and on nanosized Pt on C have shown that at 80°C in the region above +0.85 V versus RHE, Pt dissolution does not show the strong Nernstian potential dependence, and even has a maximum around +1.15 V versus RHE. This was ascribed to the formation of an oxide layer, as also predicted by a kinetic model taking into account electrochemical formation of platinum oxide and its subsequent chemical dissolution:

$$Pt + H_2O \rightarrow PtO + 2H^+ + 2e^- \quad E_0 = +0.980 \text{ V versus RHE} \qquad (17.5)$$

$$Pt^{2+} + H_2O \rightarrow PtO + 2H^+ \qquad (17.6)$$

Also, the catalyst may lose its activity because of sintering or migration of Pt particles on the carbon support. Several mechanisms have been proposed to explain the coarsening in catalyst particle size during PEMFC operation: (i) small Pt particles may dissolve in the ionomer phase and redeposit on the surface of large particles, leading to particle growth, a phenomenon known as Ostwald ripening. On the other hand, the dissolved Pt species may diffuse into the ionomer phase and subsequently precipitate in the membrane via reduction of Pt ions by the crossover hydrogen from the anode side, which dramatically decreases membrane stability and conductivity; (ii) the agglomeration of platinum particles on the carbon support may occur at the nanometer scale owing to random cluster–cluster collisions, resulting in a typical log-normal distribution of particles sizes with a maximum at smaller particle sizes and a tail toward the larger particle sizes; (iii) the growth in catalyst particles may also take place at the atomic scale by the minimization of the clusters' Gibbs free energy. In this case, the particle size distribution can be characterized by a tail toward the smaller particle sizes and a maximum at larger particle sizes.

17.3.1.1.2 Carbon Corrosion

Corrosion of the catalyst carbon support is another important issue pertaining to electrocatalyst and CL durability. Thermodynamically, carbon (graphite) can be electrochemically oxidized to CO_2 at rather low potentials:

$$C + 2H_2O \rightarrow CO_2 + 4H^+ + 4e^- \quad E_0 = +0.207 \text{ V versus RHE} \qquad (17.7)$$

Owing to the slow kinetics of these reactions, carbon can still be used in fuel cells. The presence of Pt catalyzes the subsequent oxidation to CO_2 as verified by DEMS for carbon activated with 20% Pt. The mechanism consists of the following steps:

$$C + H_2O \rightarrow CO_{surf} + 2H^+ + 2e^- \quad E > 0.3 \text{ V versus RHE} \qquad (17.8)$$

$$CO_{surf} + H_2O \rightarrow CO_2 + 2H^+ + 2e^- \quad E = +0.8 \text{ V versus RHE} \qquad (17.9)$$

At a potential higher than +0.3 V versus RHE, CO_{surf} starts to form irreversibly on the carbon particle surface. CO_{surf} is oxidized on neighboring Pt sites to CO_2 at a potential of +0.8 V versus RHE. The amount of CO_2 formed is proportional to the total length of the two-dimensional grain boundaries between the Pt particles and the carbon support. Although carbon is stable in air at temperatures as high as 195°C, a high platinum loading can lead to a loss of more than 80% of all carbon, where the time needed to reach the maximum amount of combusted carbon is determined by the temperature and the platinum loading. At platinum loadings used in commercial PEMFC catalysts, 40 wt.% and higher, the loss of carbon at the lowest experimental temperature, 125°C, amounted to 15% after 1000 h. From the relation between carbon loss and temperature, it was concluded that below 100°C, thermal oxidation of platinum-loaded carbon in air did not take place. However, the humidification of air substantially enhanced the thermal corrosion rate of carbon, by providing an additional pathway for chemical carbon oxidation through a direct reaction with water. In these studies, the corrosion rate was also found to be higher for high-surface-area carbon blacks, which was ascribed to a finer dispersion of Pt particles on this material.

17.3.1.1.3 Ionomer Degradation

The ionomer in the electrode is susceptible to many of the same degradation mechanisms that were described for the fuel cell membrane. In comparison to the membrane, the ionomer is in a state that is more soluble than the Nafion membrane, which only starts to dissolve when in water at 210°C and high pressure of 68 atm for more than 2 h. The stability of recast Nafion ionomer in CL is reported to be much lower. The trace of the soluble ionomer is found in the cathode effluent water. The dissolution becomes more severe over time. As a result of the chemical ionomer degradation, the interface between catalyst and ionomer is lost, consequently reducing the electrochemically active surface of the electrode. As the ionomer acts as a binder in the electrode, it has a strong impact on the ionic and electronic conductivity of the electrode. The structural instability due to ionomer swelling and dissolution during long-time exposure in high-humidity conditions can change the electrical properties, leading to degradation of the fuel cell.

17.3.1.1.4 Mitigation Strategies for CL Degradation

Researchers have proposed and successfully employed several strategies to enhance catalyst durability. First of all, fuel cell operating conditions play a major role in catalyst degradation. The dissolution of Pt from the carbon support is less favorable at low electrode potentials, which makes Pt catalysts more stable at the anode electrode than that at the cathode side. Second, corrosion of the carbon support due to fuel starvation can be alleviated by enhancing water retention on the anode, such as through modifications to the PTFE and/or ionomer, the addition of water-blocking components like graphite, and the use of improved preferable catalysts for water electrolysis. Third, Pt–alloy catalysts such as PtCo and Pt–Cr–Ni have shown better activity and stability than pure Pt catalysts, By strengthening the interaction between the metal particles and the carbon support, sintering and dissolution of the metal alloy catalysts can be alleviated.

17.3.1.2 Membrane

Membrane degradation can be classified into three categories: mechanical, thermal, and chemical/electrochemical. Chemical degradation is a major cause of failure of PFSA membrane. It is thought that hydroxyl (OH) or peroxy (OOH) radicals attack polymer end groups that still contain residual terminal H-groups. Characterization by x-ray photoelectron spectrometry revealed that during fuel cell operation, the interactions between carbon, fluorine, and oxygen are changing. An example is the attack of hydroxyradicals on carboxylic end groups:

$$\text{Rf–CF}_2\text{COOH} + \text{OH} \rightarrow \text{Rf–CF}_2 + \text{CO}_2 + \text{H}_2\text{O} \quad (17.10)$$

$$\text{Rf–CF}_2 + \text{OH} \rightarrow \text{Rf–CF}_2\text{OH} \rightarrow \text{Rf–COF} + \text{HF} \quad (17.11)$$

$$\text{Rf–COF} + \text{H}_2\text{O} \rightarrow \text{Rf–COOH} + \text{HF} \quad (17.12)$$

The exact mechanism of the direct radical pathway has not been established yet. Studies on chemical degradation have shown that the conditions leading to increased radical attack in PFSAs include higher temperature (especially above 90°C), low humidification, high gas pressure, use of pure hydrogen and pure oxygen, and high cell voltages.

Mechanical degradation causes early life failure due to perforations, cracks, tears, or pinholes, which may result from congenital membrane defects or from improper MEA fabrication processes. The local areas corresponding to the interface between the lands and channels of the flow field or the sealing edges in a PEMFC, which are subjected to excessive or nonuniform mechanical stresses, are also vulnerable to small perforations or tears. During fuel cell operation, the overall dimensional change due to nonhumidification, low humidification, and relative humidity (RH) cycling are also detrimental to mechanical durability. The constrained membrane in an assembled fuel cell experiences in-plane tension resulting from shrinkage under low RH and in-plane compression during swelling under wet conditions.

Several studies have addressed the issue of thermal stability and thermal degradation of PFSA membranes. The PTFE-like molecular backbone gives Nafion membranes their relative stability until beyond 150°C due to the strength of the C–F bond and the shielding effect of the electronegative fluorine atoms. At higher temperatures, Nafion begins to decompose via its side sulfonate acid groups.

17.3.1.2.1 Mitigation Strategies for Membrane Degradation

To prevent mechanical failure of the membrane, the MEA and flow field structure must be carefully designed to avoid local drying of the membrane, especially at the reactant inlet area. With respect to the chemical and electrochemical degradation of the membrane, developing membranes that are chemically stable against peroxy radicals has drawn particular attention. First, one solution is to develop novel membranes with higher chemical stability, such as a radiation-grafted FEP-*g*-polystyrene membrane, in which polystyrene was used as a sacrificial material owing to its low resistance to radicals. Free-radical stabilizers and inhibitors such as hindered amines or antioxidants also have the potential to be mingled during membrane fabrication. Second, increased chemical stability can also be realized by modifying the structure of the available membrane. It was suggested that radical attack of the residual H-containing terminal bonds of the main chain of the

PFSA membrane was the primary degradation mechanism. By eliminating the unstable end group, chemical stability was significantly enhanced. Third, the damage caused by hydrogen peroxide can be suppressed by redesigning the MEA. For example, a composite membrane, in which a thin recast Nafion membrane was bonded with a polystyrene sulfonic acid (PSSA) membrane, when positioned at the cathode of the cell could successfully prevent oxidation degradation of the PSSA membrane. Fourth, introduction of peroxide-decomposition catalysts like heteropoly acids within the membrane has proven to moderate or eliminate membrane deterioration due to peroxide. However, the advantage of this approach would be partially counteracted by a decrease in membrane stability and conductivity caused by the mixture of the catalysts. Last but not least, the development and implementation of new metal coatings with improved corrosion resistance and of catalysts that produce less hydrogen peroxide are long-term goals for membrane durability enhancement.

17.3.1.3 Gas Diffusion Layer

The GDL typically consists of two layers bonded together: a macroporous layer made of conductive carbon fibers and a microporous layer (MPL) made of carbon particles and a Teflon binder. The understanding of how a GDL degrades during operation and the effects of its degradation on fuel cell performance is based on only a few recent studies. Unfortunately, there is a large gap in the literature between studies of the GDL's physical properties and studies that relate these properties to PEMFC durability data.

Some researchers found that the loss of GDL hydrophobicity increased with operating temperature and when sparging air was used instead of nitrogen. Additionally, they concluded that changes in the GDL properties were attributed mostly to the MPL. It was found that weight loss and the MPL contact angle increased with the time of exposure, and the increase was attributed to oxidation of the carbon in the MPL. As the fuel cell operates, the PTFE and carbon composite of the GDLs are susceptible to chemical attack (i.e., OH radical as an electrochemical by-product) and electrochemical (voltage) oxidation. The loss of PTFE and carbon results in the changes in GDL physical properties, such as the decrease of GDL conductivity and hydrophobicity, which further lowers MEA performance and negatively affects the durability of the whole fuel cell. With regard to the quantitative correlation between performance loss and the changes in GDL properties, it was found that the decomposition of PTFE in the electrodes induced an approximately two times higher performance loss than that related to the agglomeration of the platinum catalyst after 1000 h of fuel cell operation.

17.3.1.3.1 Mitigation Strategies for GDL Degradation

Little information about mitigating GDL degradation is available from the literature. To improve GDL oxidative and electro-oxidative stability, it was suggested using graphitized fibers during GDL preparation. It was also proposed that higher PTFE loading could benefit the water management ability of aged GDLs.

17.3.1.4 Bipolar Plate

The bipolar plate serves as a separator between two adjacent cells. It conducts the electronic current between these two cells, separates the gases, and contains the flow patterns both for the reactants and for the coolant, in most cases a liquid coolant. With respect to the electric conductivity of bipolar plates, the contact resistance is more important than the bulk resistance, especially with regard to long-term behavior. Whereas a BOL (beginning of life) resistance of 10 mΩ cm^2 is suggested as a specification, for long-term stability a combined specification for anode and cathode side of lower than 50 mΩ cm^2 has been used as a target specification that has to be maintained up to 5000 h of fuel cell operation. The release of contaminants from the bipolar plates is a feature that can determine to a large extent the lifetime of the fuel cell. Whereas it does not so much affect the properties of the bipolar plates itself, it leads to poisoning of the membrane and the catalysts.

17.3.1.4.1 Mitigation Strategies for Bipolar Plate Degradation

For graphite and graphite composite plates, corrosion and release of contaminants under normal operation is not an issue. At normal operating conditions, the cathode potential is not high enough for oxide formation on the surface, and the contact resistance remains constant during fuel cell operation. Although not reported, under extreme conditions, corrosion of the graphite is conceivable. Especially under start/stop or fuel starvation conditions, electrodes can be exposed to potentials promoting carbon corrosion.

For metal-based bipolar plates, the stability depends on the nature of the metal, the potential, and RH. Corrosion of the bipolar plates takes place when the plate material is oxidized at the potential to which it is exposed, and the surface oxide layer is soluble in the medium it is situated in. The medium is often not well defined. When in direct contact with the electrolytic membrane, the plate–membrane contact can lead to direct exchange of cations with protons of the electrolyte. On the other hand, the water produced in the fuel cell also contains ions, emanating from the electrodes and the membranes. When using stainless-steel bipolar plates, corrosion is a generally observed phenomenon, which varies strongly between the various stainless-steel grades.

Another common concern related to bipolar plates is the possible deformation or even fracture caused by the compressive forces that are used to ensure good electric contact and reactant sealing during fuel cell operation. Some operational factors, such as thermal cycling, nonuniform current, or thermal misdistributions over the active area, can impair the mechanical properties of the bipolar plate materials. A surface treatment method is particularly promising since the treatment is a modification to the surface rather than a coating procedure, and therefore delamination is not an issue. At the cathode side, the potential is high enough to form a partly conducting passive layer on the metal surface. Although it protects the surface from corrosion, this layer results in an increase in contact resistance. Many metals show a buildup of such a passive layer under long-term testing in a fuel cell. This is a process that can take place over thousands of hours, thus finally exceeding a maximum tolerable resistance.

17.3.1.5 Seals

Sealing material is placed between the bipolar plates to prevent gas and coolant leakage and crossover. Typical sealing materials utilized in PEMFCs include fluorine caoutchouc, ethylene propylene diene monomer (EPDM), and silicone. Not only are they meant to prevent leakage of the gases and the coolants outside their containments, seals also function as electrical insulation, stack height control, and variability control. The degradation phenomena connected to seals do not only refer to the loss of the functionality of the seals themselves, but also to the leakage of seal components that could poison the MEA. In a recent study the stability of various sealing materials has been tested in a simulated environment at 60°C and 80°C. This simulated environment consisted of solutions containing HF and H_2SO_4, in two different concentrations. It was concluded that silicone S and silicone G are heavily degraded in the concentrated solutions as well as the diluted concentrations, although most of the data are collected in the concentrated solutions. Degradation reveals itself by weight loss, complete disintegration, as well as by leaching of Mg and Ca. The latter stems from magnesium oxide and calcium carbonate, which are used as fillers for obtaining the desired tensile strength, hardness, and resistance to compression.

17.3.1.5.1 Mitigation Strategies for Seals Degradation

Only in a couple of long-term experiments was seal degradation observed, and this might have been the consequence of an inappropriate materials selection. Silicone seals in direct contact with a perfluorosulfonic acid membrane suffer from degradation, at the anode as well as at the cathode. The degradation is probably caused by acidic decomposition of the sealing material, leading to coloration of the membrane and detectable amounts of silicon on the electrodes. No fuel cell performance loss or increase in gas leakage along the seal has been observed. Seal selection through *ex situ* and *in situ* screening processes should be based on the overall chemical and mechanical properties of the materials. With regard to seal material degradation, no available publications relevant to mitigation strategies have yet been found.

17.3.2 Operational Effects on Degradation

Operating conditions are known to have an impact on fuel cell durability. These effects include exposure to impurities (on both the cathode and anode sides of the fuel cell), exposure to and start-up from subfreezing conditions, potential cycling, fuel starvation, start/stop cycling, and changes in temperature and/or RH.

17.3.2.1 Impurity Effects

The effect of contaminants on fuel cells is one of the most important issues in fuel cell operation and applications. Impurities in hydrogen fuel, such as CO, H_2S, NH_3, organic sulfur–carbon, and carbon–hydrogen compounds, and in air, such as NO_x, SO_x, and small organics, are brought along with the fuel and air feed streams into the anodes and cathodes of a PEMFC stack, causing performance degradation, and sometimes permanent damage to MEAs. For the PEM, the contaminants—in particular the cations—can get into the membrane to compete with the proton for the -$SO3^-$ sites (Nafion membrane) and at the same time decrease the water content, resulting in a reduction in proton conductivity. On the other hand, metal ions such as Fe^{3+}/Fe^{2+} inside a membrane can also accelerate membrane degradation during a fuel cell operation through a peroxide formation mechanism.

17.3.2.1.1 Carbon Oxide Contamination

Both carbon monoxide and carbon dioxide have become major concerns in PEMFCs using reformate H_2-rich gas as fuel, particularly at conventional operating temperatures (<80°C). It is well documented that CO binds strongly to Pt sites, resulting in the reduction of surface-active sites available for hydrogen adsorption and oxidation. The CO poisoning effect was strongly related to the concentration of CO, the exposure time to CO, the cell operation temperature, and anode catalyst types. Normally, CO poisoning on Pt electrocatalysts becomes more severe with increases in CO concentration and exposure time. The CO impurities from fuel streams, even at a level of a few parts per million, can cause a substantial degradation in cell performance, especially at high current densities. The cell voltage losses became deeper with prolonged exposure to CO, due to its accumulation on the Pt catalyst surface over time. On the other hand, although the severity of catalyst poisoning by CO can be strongly affected by fuel cell operation temperature, it may not be sensitive to pressure. At low temperatures (<80°C), a trace amount of CO can cause a significant performance drop. The performance loss due to CO_2 contamination in anode fuel can be observed especially at higher current densities. On a Pt catalyst, CO_2 can be catalytically converted into CO, which then poisons the catalyst.

17.3.2.1.2 Hydrogen Sulfide (H_2S)

Metal catalysts, in general, have a strong chemical affinity with H_2S, and Pt catalysts are not an exception and are particularly vulnerable. The degrading effects of the presence of this impurity in the FC are significant and commensurate with H_2S concentration and time of exposure. The electrochemical desorption of sulfur requires potentials unachievable in a continuously operating H_2/air fuel cell. H_2S at 10 ppb has been shown during long operating times to have a degrading effect on fuel cell performance.

17.3.2.1.3 Ammonia (NH_3)

The presence of ammonia levels as low as 13 ppm in the fuel stream has rapid deleterious effects on performance. Higher concentrations (80, 200, and 500 ppm) of NH_3 have shown a marked decrease in performance in simulated reformate: cell performance decreases with exposure to NH_3 in reformate from 825 to 200 mA cm^{-2} at 0.6 V. Short-term exposure (<1 h) to NH_3 shows reversible effects. However, the negative effects caused by long-term exposure are irreversible, meaning that further operation on neat H_2 results only in a partial recovery. Cyclic voltammetry (CV) of the anode, after exposure, does not indicate any noticeable NH_3 adsorption onto the CL. Thus, the degradation mechanism appears to be due to protonic conductivity loss. The likely culprit is that NH_3 reacts with ionomeric H^+, generating NH^{4+} and consequently lowering the protonic activity. The negative effect gradually starts at the anode CL, the first region exposed, and continues into the membrane as the ammonia diffuses deeper and deeper.

17.3.2.1.4 Cationic Ions

The foreign cationic ions that originate from impurities in fuel cell stack component materials, fuels, and coolants can cause water management problems in fuel cells. The cationic ions such as alkali metals, alkaline earth metals, transition metals, and rare earth metals were reported to directly affect the transport properties of the electrolyte membrane. Iron ions from stainless-steel end plates resulted in severe Nafion degradation as evidenced by a massive fluoride loss. The iron contamination also led to several types of performance losses, including cathode and anode kinetic losses, ohmic loss, and mass transport loss. For example, metallic ions including Cu^{2+}, Fe^{3+}, Ni^{2+}, and Na^+ presented in sulfate salt solutions at a concentration level of 100 ppm were found to significantly decrease the ionic conductivity of a Nafion 117 membrane; among these, the ferric ions were more harmful. Researchers have discussed the critical role played by trace metallic ions in membrane degradation by reviewing the degradation modes of polymer electrolyte membranes; it was concluded that the displacement of H^+ with foreign cationic ions directly affected water flux and proton conductivity inside the membrane, leading to membrane degradation.

17.3.2.1.5 Sulfur Dioxide (SO_2)

Sulfur dioxide is a common air contaminant resulting from fossil fuel combustion and can be found in high concentrations in urban areas with heavy traffic and in close proximity to some chemical plants. The effects of SO_2 on the cathode are similar to those produced by the presence of H_2S in the anode. Performance degradation appears to be a function of SO_2 concentration in the bulk, as the performance decrease was measured to be 53% at 2.5 ppm SO_2 as compared with a 78% decrease at 5 ppm SO_2 for the same applied dosage. Performance does not improve after impurity injection is turned off; thus, is not reversible just by normal operation. The severity of the effect is due to the strong chemisorption of SO_2 (or other S-species) onto the Pt catalyst surface. Electrochemical oxidation (during a CV) of adsorbed SO_2 shows full cell performance recovery.

17.3.2.1.6 Nitrogen Oxides (NO_x)

Nitrogen oxides (NO_x) are air contaminants that mostly originate in the combustion of fossil fuels. Internal combustion engine emissions are the major source of NO_x; thus, they are abundant in urban areas. NO_2 has been shown to quickly degrade fuel cell performance, with a gradual decrease over 30 h of operation, after which degradation did not continue. The rate of poisoning of PEMFCs by NO_2 does not strongly depend on NO_2 bulk concentration. The degradation of performance of the fuel cell can reach 50%, while the cell performance completely recovers after applying neat air for 24 h. The poisoning effects of NO_2 do not appear to be a catalyst poisoning issue, since no surface species can be detected during cyclic voltammetry; the poisoning mechanism is still not understood.

17.3.2.1.7 Sodium Chloride (NaCl)

The presence of NaCl at the electrode decreases its performance. The performance loss is mostly due to a decrease of protonic conductivity as a consequence of exchange of H^+ by Na^+ at the CL and at the membrane. Large concentrations of the salt also decreased the hydrophobicity of the GDL, increased liquid water retention, and correspondingly decreased oxygen transport to the electrocatalyst at high current densities. Surprisingly, Cl^- does not appear to block adsorption on the catalyst surfaces, as revealed by CV measurements. However, chloride has a dramatic effect on the oxygen reduction kinetics of cathode electrocatalysts. Chloride has also been noted to affect GDL materials, which can lead to changes in water and gas transport.

17.3.2.1.8 Mitigation Strategies for Impurity Effects

As discussed above, the impurities in the hydrogen stream and pollutants in the air stream can contaminate the fuel cell MEA in many ways, causing performance degradation and failure. Therefore, the measures and methods for mitigation contamination have to be developed in order to minimize or eliminate its effects. For fuel, the reformed H_2-rich gas is the dominant source. As discussed above, this fuel contains appreciable amounts of CO and CO_2, which are the major fuel cell anode contaminants. There are several effective methods available to mitigate CO poisoning in PEMFCs, such as enhancing CO oxidation by pretreating reformate, introducing an anode oxidant-bleed, developing CO-tolerant catalysts, and optimizing fuel cell operating conditions. Pretreatment of reformate is one of the most popular ways to purify H_2-rich gas to reduce the CO concentration to as low as 10 ppm. Air or oxygen (or H_2O_2) bleeding has been demonstrated to be another effective way to reduce CO contamination if the fuel cell stack is operated with a reformate fuel stream. During fuel cell operation, air or oxygen will be intermittently blown into the anode. CO-tolerant catalyst development is another important mitigation method. As discussed above, the addition of a second or third metal into the Pt can form an alloying catalyst. The second or third metal can greatly help in CO oxidation. Operating a PEMFC at high temperatures (>80°C) has a significant benefit for contamination tolerance owing to the weaker adsorption and faster oxidation rate of CO on the Pt catalyst. It is worthwhile to note that for a long-term supply of hydrogen, reformate may not be an option due to the fossil hydrogen carbon shortage. It is expected that the external supply of hydrogen will rely on electrolysis and reformate from renewable biomass materials such as methanol and ethanol. For H_2 production from water electrolysis, CO fuel contamination may not be a problem. However, for short-term hydrogen supplies, reformate is still an option in terms of cost and reliability. For oxidant (air), the use of filters to purify the cathode feed stream effectively eliminated contamination from diesel and dust emission, hence improving the performance of a PEMFC operated in an underground mine.

17.3.2.2 Subfreezing Temperature

One criterion PEMFCs are required to meet for automotive applications is the ability to survive at and start-up from subfreezing temperatures. Early literature suggested that there was little degradation from freezing fuel cells to subfreezing temperatures for a limited number of cycles. However, a recent paper suggested that there could be significant degradation in the performance of PEMFCs subjected to cycling from −10°C. These apparent discrepancies (from no degradation to 5.4% degradation/cycle) in the various literature results can be attributed to several factors. The preparation method of the MEA is critical in determining the freeze/thaw durability of the PEMFC. If the electrode/electrolyte adhesion is weak, then there is a greater degradation in the performance due to ice formation resulting in delamination. This may explain some of the discrepancies in the literature results where some electrodes were sprayed on to the GDL while others were prepared directly on the membrane. Moreover, the degradation is also a function of the rate of heating/cooling in addition to the freezing temperature. This is evidenced by rapid cycling (quenching) to −80°C, leading to delamination of the electrode while normal cycling to −40°C (even for 100 cycles) shows no such delamination in an identically prepared MEA. Studies have also revealed that there is little loss in performance of even fully humidified cells using carbon cloth backing. However, the breakage of carbon fibers in carbon paper backing leads to loss in performance during freeze/thaw cycles. Therefore, in addition to the preparation method and cycling conditions, the component materials used in the fuel cell assembly will also play a vital role in determining the durability of fuel cells subjected to multiple freeze/thaw cycling.

The reason for the performance degradation at freezing temperature is thought to be the effect of freezing water on the MEA (including the Nafion membrane, CL, and GDL) properties. The conductivity of Nafion is highly dependent on the state of water in the polymer, and it has been shown to have increased activation energy at lower temperatures, where the water in the membrane is likely in the frozen state. There is only limited literature available on the durability of the CL and GDL under freeze/thaw cycling. These reveal that even a free-standing hydrated CL can be subjected to cracking and peeling while cycling from −30°C. This damage was associated with a loss in the electrochemical surface area of the catalyst that can be avoided by drying the catalyst.

17.3.2.2.1 Mitigation Strategies for Degradation Due to Subfreezing Temperatures

The mitigation strategies utilized to avoid degradation due to subfreezing temperatures can fall into three categories: (i) those that prevent ice formation either by drying out the fuel cell or by replacing the water with a nonfreezing liquid, (ii) those that prevent ice formation during start-up by providing heat, and (iii) those that keep the fuel cell warm, thus preventing ice formation. Fuel cell stacks can be kept warm by providing insulation or by providing heat either through a battery or by operating the cell intermittently in a low power mode. The freezing water can be avoided by eliminating carrying a water tank on board and by running the fuel cell at lower inlet RHs and reclaiming the exhaust water, although it is not clear if the current class of membrane materials will operate stably under such conditions. The water inside the stack can be minimized by running the cell under dry reactant gases (H_2, air) or dry nitrogen before shutdown or by vacuum drying the fuel cell. Moreover, during start-up, extra heat can be provided from a battery, by catalytically combusting hydrogen, or by preheating the reactant gases, and the heat-carrying capacity of the reactant gases is rather low.

17.3.2.3 Relative Humidity

Another aspect of fuel cell operation that is likely to affect the integrity of the cell is the changes in temperature and RH that are associated with transitions between low and high power. In general, for cells that operate at fixed stoichiometric ratios, operation at low current implies a relatively cool and wet cell; higher currents imply a hotter, drier cell. The fact that the ionomer swells with water uptake suggests that increases in water uptake as the membrane is exposed to high RH conditions can lead to compressive stresses in the membrane that then yield tensile residual stresses during drying. These stresses are suggested as a significant contributor to mechanical failures of the membrane. Another recent study suggests that drying can considerably strain the membrane–electrode assembly and that mechanical failure of membranes can result from gradual reduction in ductility combined with excessive strains induced by constrained drying of the MEA. Both temperature and RH have been shown to affect the rate of catalyst surface area loss due to platinum particle growth. These studies suggest that more needs to be learned about material properties and how they change over the course of fuel cell operation.

17.3.2.4 Fuel Starvation

Full-sized cells, on the order of several hundred square centimeters in area, will experience different conditions between the inlet and the outlet, and this can lead to current distributions that cannot easily be simulated in subscale testing. Furthermore, cells arranged in a stack configuration can experience different flows of fuel, air, and coolant resulting from imperfect manifolding. Therefore, adjacent cells in a stack can experience different conditions in terms of hydrogen and oxygen content, but they will be forced to carry the same current as their neighboring cells, as they are connected in series. It is noticed that, in the case of gross fuel starvation, cell voltages can become negative, as the anode is elevated to positive potentials and the carbon is consumed instead of the absent fuel. In the case of gross fuel starvation, for multiple cells in a stack, fuel maldistributions can lead to some cells having insufficient fuel to carry the current that is being pushed through them by adjacent cells. In the absence of a sufficient anodic current source from hydrogen, the cell potential climbs higher until oxidation occurs, in this case the oxidation of the carbon support of the CL.

Mitigation strategies: Proper reactant distribution is critical to avoid this problem, and stack developers have accordingly sought to monitor the voltage of each cell to avoid such a problem. Obviously, such an extensive monitoring system will add considerable cost and complexity to the fuel cell stack and control scheme.

17.3.2.5 Load Cycling

A fuel cell, particularly one that must meet the challenging dynamic load of an automotive application, will undergo many rapid changes in load over the course of its lifetime. As the fuel cell cycles from high to low current, its cell potential will also vary, generally between +0.6 and +1.0 V versus RHE. For cells operating with relatively pure hydrogen as a fuel, the anode will stay fairly close to the reversible hydrogen potential, owing to the facile nature of the hydrogen oxidation reaction. This implies that the cathode experiences potential swings as cell potential changes to match variable power demands. The variation of the cathode potential will change several properties of the electrode materials, notably the degree of oxide coverage of both platinum and carbon, and the hydrophobicity of the surfaces. A more subtle distinction has to do with the fact that the oxide can actually serve to protect the platinum surface from dissolution at higher potentials. When the cathode potential rises rapidly to higher values, the platinum can dissolve at a rapid rate until a passivating oxide layer is formed.

Mitigation strategies: Any attempts to develop stable catalysts for fuel cell applications must consider the stability of the catalysts not only under constant potential conditions but also under potential cycling. To design catalysts that are robust to this degradation mode, considerably more information is needed about the nature of the oxide, the kinetics of its formation, and its ability to protect the catalyst from dissolution over the entire range of potentials.

17.3.2.6 Start/Stop Cycling

Performance degradation during start/stop cycling is considered an important issue affecting the durability and lifetime of PEMFCs. Owing to the high potentials experienced by the cathode during start/stop cycling, the conventional carbon support for the cathode catalyst is prone to oxidation by reacting with oxygen or water. Both start-up and shutdown are dynamic processes that a fuel cell inevitably must confront in automobile applications. Compared with steady-state processes, start/stop cycling processes experience different profiles under operating conditions. Under conditions of a prolonged shutdown, unless the stack is continually provided with fuel, hydrogen crossover from the anode to the cathode will eventually empty out the anode chamber and result in an air-filled flow channel. In this case, the starting flow of fuel will induce a transient condition in which fuel exists at the inlet but the exit is still fuel starved. As a result, starting and stopping the fuel cell can induce considerable damage to the cell. This phenomenon has been modeled and reveals that an unprotected start can induce local potentials on the cathode in excess of +1.8 V versus RHE. On the other hand, catalyst degradation at the cathode is considered a major failure mode for PEMFCs when the catalysts are exposed to reverse current conditions during start/stop cycling.

17.3.2.6.1 Mitigation Strategies for Degradation Due to Start/Stop Cycling

The strategies for degradation due to start/stop cycling can be classified into two major categories: (i) material improvement for more stable catalyst supports; (ii) system mitigation strategies for conventional carbon black supports. In the aspect of material improvement, replacing conventional carbon supports with corrosion-resistant materials (e.g., graphitized carbon) is an important mitigation strategy. The use of graphitized carbon resulted in a reduced mass transport limitation of the GDL, in comparison with the mass transport limitation enhanced by conventional carbon oxidation at the MPL/electrode interface. In addition to graphitized carbon, other carbon materials, such as CNTs or carbon nanofibers, and carbon aerogel and xerogel, have also been considered as catalyst supports because of their more stable electrochemical behaviors. In the aspect of system mitigation strategies, it includes the following: (i) gas purge to anode before start-up and after shutdown; (ii) auxiliary load applied to consume residual oxygen at the cathode with potential control; (iii) exhaust gas recycle as purging gas or reaction gas; and (iv) electronic short to eliminate high potential at the cathode. In comparison with materials improvement by using graphitized carbon or non-carbon supports, system strategies are relatively simple and cheap to implement in real fuel cell engines.

17.4 CONCLUSIONS

MEA is the core component of PEMFCs and electrochemical reaction only takes place at the "triple-phase boundaries," where the reactant, electrolyte, and electrons are brought together. The high cost of Pt and Pt-group metals, along with low Pt utilization in the CL, is a major barrier to the commercialization of fuel cell technology. The basic technical strategies and methodologies to maximize the Three-Phase Boundary (TPB) of the CL are to increase the CL Pt utilization, and to reduce Pt loading. Thus far, researchers have found that these goals can be achieved through optimizing existing CLs with respect to composition and structure, developing novel fabrication technologies, and introducing innovative CL approaches. This report outlined major advances made in the fabrication of MEA for PEMFCs from the PTFE-bound CLs of almost 20 years ago to the present investigation of low-Pt-loading MEA. Those methods are not only used in PEMFCs, but also can be utilized for other low-temperature fuel cells such as AFCs and PAFCs.

REFERENCES

1. Zhang, J. *PEM Fuel Cell Electrocatalysts and Catalyst Layers: Fundamentals and Applications.* Canada: Springer, 2008.
2. Wilson, M. S., Valerio, J. A., and Gottesfeld, S. Low platinum loading electrodes for polymer electrolyte fuel cells fabricated using thermoplastic ionomers. *Electrochimica Acta* 1995; 40:355–363.
3. Wagner, N., Kaz, T., and Friedrich, K. A. Investigation of electrode composition of polymer fuel cells by electrochemical impedance spectroscopy. *Electrochimica Acta* 2008; 53:7475–7482.
4. Raistrick, I. D. Electrode assembly for use in a solid polymer electrolyte fuel cell. US Patent 4,876,115, 1989.
5. Raistrick, I. D. Diaphragms, separators, and ion exchange membranes. *The Electrochemical Society Proceedings Series*, Pennington, NY, 1986, 156.
6. Murphy, O. J., Hitchens, G. D., and Manko, D. J. High-power density proton exchange membrane fuel cells. *Journal of Power Sources* 1994; 47:353–368.
7. Cheng, X., Yi, B., Han, M., Zhang, J., Qiao, Y., and Yu, J. Investigation of platinum utilization and morphology in catalyst layer of polymer electrolyte fuel cells. *Journal of Power Sources* 1999; 79:75–81.
8. Wilson, M. S. Membrane catalyst layer for fuel cells. US Patent 5,234,777, 1993.
9. Wilson, M. S. and Gottesfeld, S. Thin film catalyst layers for polymer electrolyte fuel cell electrodes. *Journal of Applied Electrochemistry* 1992; 22:1–7.
10. Xie, J., Garzon, F., Zawodzinski, T., and Smith, W. Porosimetry of MEAs made by "thin film decal" method and its effect on performance of PEFCs. *Journal of the Electrochemical Society* 2004; 151:A1841–A1846.
11. Antolini, E., Giorgi, L., Pozio, A., and Passalacqua, E. Influence of Nafion loading in the catalyst layer of gas-diffusion electrodes for PEFC. *Journal of Power Sources* 1999; 77:136–142.
12. Passalacqua, E., Lufrano, F., Squadrito, G., Patti, A., and Giorgi, L. Influence of the structure in low-Pt loading electrodes for polymer electrolyte fuel cells. *Electrochimica Acta* 1998; 43:3665–3673.
13. Sasikumar, G., Ihm, J. W., and Ryu, H. Dependence of optimum Nafion content in catalyst layer on platinum loading. *Journal of Power Sources* 2004; 132:11–17.
14. Taylor, E. J., Anderson, E. B., and Vilambi, N. R. K. Preparation of high-platinum-utilization gas diffusion electrodes for proton-exchange-membrane fuel cells. *Journal of the Electrochemical Society* 1992; 139:L45–L46.
15. Paganin, V. A., Ticianelli, E. A., and Gonzalez, E. R. Development and electrochemical studies of gas diffusion electrodes for polymer electrolyte fuel cells. *Journal of Applied Electrochemistry* 1996; 26:297–304.
16. Liang, H., Zheng, L., and Liao, S. Self-humidifying membrane electrode assembly prepared by adding PVA as hygroscopic agent in anode catalyst layer. *International Journal of Hydrogen Energy* 2012; 37:12860–12867.
17. Liang, H., Dang, D., Xiong, W., Song, H., and Liao, S. High-performance self-humidifying membrane electrode assembly prepared by simultaneously adding inorganic and organic hygroscopic materials to the anode catalyst layer. *Journal of Power Sources* 2013; 241:367–372.
18. Liang, H., Su, H., Pollet, B. G., Linkov, V., and Pasupathi, S. Membrane electrode assembly with enhanced platinum utilization for high temperature proton exchange membrane fuel cell prepared by catalyst coating membrane method. *Journal of Power Sources* 2014; 266:107–113.
19. Ticianelli, E. A., Derouin, C. R., Redondo, A., and Srinivasan, S. Methods to advance technology of proton exchange membrane fuel cells. *Journal of the Electrochemical Society* 1988; 135:2209–2214.
20. Park, H. S., Cho, Y. H., Cho, Y. H., Park, I. S., Jung, N., Ahn, M., and Sung, Y. E. Modified decal method and its related study of microporous layer in PEM fuel cells. *Journal of the Electrochemical Society* 2008; 155:B455–B460.
21. Subramaniam, C. K., Rajalakshmi, N., Ramya, K., and Dhathathreyan, K. S. High-performance gas diffusion electrodes for PEMFC. *Bulletin of Electrochemistry* 2000; 16: 350–353.
22. Møller-Holst, S. Preparation and evaluation of electrodes for solid polymer fuel cells. *The Journal of the Electrochemical Society of Japan* 1996; 64:699–705.
23. Bender, G., Zawodzinski, T. A., and Saab, A. P. 2003. Fabrication of high-precision PEFC membrane electrode assemblies. *Journal of Power Sources* 2003; 124:114–117.
24. Ihm, J. W., Ryu, H., Bae, J. S., Choo, W. K., and Choi, D. K. High performance of electrode with low Pt loading prepared by simplified direct screen printing process in PEM fuel cells. *Journal of Materials Science* 2004; 39:4647–4649.
25. Kim, C. S., Chun, Y. G., Peck, D. H., and Shin, D. R. A novel process to fabricate membrane electrode assemblies for proton exchange membrane fuel cells. *International Journal of Hydrogen Energy* 1998; 23:1045–1048.
26. Rajalakshmi, N. and Dhathathreyan, K. S. Catalyst layer in PEMFC electrodes—Fabrication, characterization and analysis. *Chemical Engineering Journal* 2007; 129:31–40.
27. Towne, S., Viswanathan, V., Holbery, J., and Rieke, P. Fabrication of polymer electrolyte membrane fuel cell MEAs utilizing inkjet print technology. *Journal of Power Sources* 2007; 171:575–584.
28. Taylor, A. D., Kim, E. Y., Humes, V. P., Kizuka, J., and Thompson, L. T. Inkjet printing of carbon supported platinum 3-D catalyst layers for use in fuel cells. *Journal of Power Sources* 2007; 171:101–106.
29. Bolwin, K., Giilzow, E., Bevers, D., and Schnurnberger, W. Preparation of porous electrodes and laminated electrode–membrane structures for polymer electrolyte fuel cells (PEFCs). *Solid State Ionics* 1995; 77:324–330.
30. Kumar, G. S., Raja, M., and Parthasarathy, S. High performance electrodes with very low platinum loading for polymer electrolyte fuel cells. *Electrochimica Acta* 1995; 40:285–290.
31. Shin, S.-J., Lee, J.-K., Ha, H.-Y., Hong, S.-A., and Chun, I.-H. Effect of the catalytic ink preparation method on the performance of polymer electrolyte membrane fuel cells. *Journal of Power Sources* 2002; 106:146–152.
32. Cavalca, C. A., Arps, J. H., and Murthy, M. Fuel cell membrane electrode assemblies with improved power outputs and poison resistance. US Patent No. 6,300,000, 2001.
33. Cha, S. Y. and Lee, W. M. Performance of proton exchange membrane fuel cell electrodes prepared by direct decomposition of ultrathin platinum on the membrane surface. *Journal of the Electrochemical Society* 2008; 146:4055–4060.
34. Weber, M. F., Mamiche-Afare, S., Dignam, M. J., Pataki, L., and Venter, R. D. Sputtered fuel cell electrodes. *Journal of the Electrochemical Society* 1987; 134:1416–1419.

35. O'Hayre, R., Lee, S. J., Cha, S. W., and Prinz, F. B. A sharp peak in the performance of sputtered platinum fuel cells at ultra-low platinum loading. *Journal of Power Sources* 2002; 109:483–493.
36. Wee, J. H., Lee, K. Y., and Kim, S. H. Fabrication methods for low-Pt-loading electrocatalysts in proton exchange membrane fuel cell systems. *Journal of Power Sources* 2007; 165:667–677.
37. Kadjo, A. J. J., Brault, P., Caillard, A., Coutanceau, C., Garnier, J. P., and Martemianov, S. Improvement of proton exchange membrane fuel cell electrical performance by optimization of operating parameters and electrodes preparation. *Journal of Power Sources* 2007; 172:613–622.
38. Saha, M. S., Gullb, A. F., Allen, R. J., and Mukerjee, S. High-performance polymer electrolyte fuel cells with ultralow Pt loading electrodes prepared by dual ion-beam assisted deposition. *Electrochimica Acta* 2006; 51:4680–4692.
39. Vilambi Reddy, N. R. K., Anderson, E. B., and Taylor, E. J. High utilization supported catalytic metal-containing gas-diffusion electrode, process for making it, and cells utilizing it. US Patent No. 5,084,144, 1992.
40. Verbrugge, M. Selective electrodeposition of catalyst within membrane–electrode structures. *Journal of the Electrochemical Society* 1994; 141:46–53.
41. Hogarth, M. P., Munk, J., Shukla, A. K., and Hamnett, A. Performance of carbon-cloth bound porous-carbon electrodes containing an electrodeposited platinum catalyst towards the electrooxidation of methanol in sulphuric acid electrolyte. *Journal of Applied Electrochemistry* 1994; 24:85–88.
42. Gloaguen, F., Leger, J. M., and Lamy, C. Electrocatalytic oxidation of methanol on platinum nanoparticles electrodeposited onto porous carbon substrates. *Journal of the Electrochemical Society* 1997; 27:1052–1060.
43. Maric, R., Roller, J., and Vanderhoek, T. Reactive spray formation of coatings and powders. BC, Canada, WO Patent /2007/045089, 2007.
44. Cunningham, N., Irissou, E., Lefevre, M., Denis, M. C., Guay, D., and Dodelet, J. P. PEMFC anode with very low Pt loadings using pulsed laser deposition. *Electrochemical and Solid-State Letters* 2003; 6:A125–A128.
45. Lee, K., Zhang, J. J., Wang, H. J., and Wilkinson, D. P. Progress in the synthesis of carbon nanotube- and nanofiber-supported Pt electrocatalysts for PEM fuel cell catalysis. *Journal of Applied Electrochemistry* 2006; 36:507–522.
46. Litster, S. and McLean, G. PEM fuel cell electrodes. *Journal of Power Sources* 2004; 130:61–76.
47. Chan, K. Y., Ding, J., Ren, J. W., Cheng, S. A., and Tsang, K. Y. Supported mixed metal nanoparticles as electrocatalysts in low temperature fuel cells. *Journal of Materials Chemistry* 2004; 14:505–516.
48. Niinistö, L., Ritala, M., and Leskelä, M. Synthesis of oxide thin films and overlayers by atomic layer epitaxy for advanced applications. *Materials Science and Engineering: B* 1996; 41:23–29.
49. Ritala, M. Advanced ALE processes of amorphous and polycrystalline films. *Applied Surface Science* 1997; 112:223–230.
50. Klaus, J. W., Sneh, O., and George, S. M. Growth of SiO_2 at room temperature with the use of catalyzed sequential half-reactions. *Science* 1997; 278:1934–1936.
51. Klaus, J. W., Sneh, O., Ott, A. W., and George, S. M. Atomic layer deposition of SiO_2 using catalyzed and uncatayzed self-limiting surface reactions. *Surface Review and Letters* 1999; 6:435–448.
52. Ritala, M., Leskelä, M., Dekker, J.-P., Mutsaers, C., Soininen, P. J., and Skarp, J. Perfectly conformal TiN and Al_2O_3 films deposited by atomic layer deposition. *Chemical Vapor Deposition* 1999; 5:7–9.
53. Liu, C., Wang, C.-C., Kei, C.-C., Hsueh, Y.-C., and Perng, T.-P. Atomic layer deposition of platinum nanoparticles on carbon nanotubes for application in proton-exchange membrane fuel cells. *Small* 2009; 5:1535–1538.
54. Aaltonen, T., Ritala, M., Sajavaara, T., Keinonen, J., and Leskela, M. Atomic layer deposition of platinum thin film. *Chemistry of Materials* 2003; 15:1924–1928.
55. Aaltonen, T., Ritala, M., Tung, Y.-L., Yun Chi, Y., Arstila, K., Meinander, K., and Leskelä, M. Atomic layer deposition of noble metals: Exploration of the low limit of the deposition temperature. *Journal of Materials Research* 2004; 19:3353–3358.
56. Borup, R., Meyers, J., Pivovar, B., Kim, Y. S., Mukundan, R., Garland, N., Myers, D., Wilson, M., Garzon, F., and Wood, D. Scientific aspects of polymer electrolyte fuel cell durability and degradation. *Chemical Reviews* 2007; 107:3904–3951.
57. Cheng, X., Shi, Z., Glass, N., Zhang, L., Zhang, J. J., Song, D., Liu, Z. S., Wang, H., and Shen, J. A review of PEM hydrogen fuel cell contamination: Impacts, mechanisms, and mitigation. *Journal of Power Sources* 2007; 165:739–756.
58. Bruijn, F., Dam, V., and Janssen, G. Review: Durability and degradation issues of PEM fuel cell components. *Fuel Cells* 2008; 8:3–22.
59. Wu, J., Yuan, X., Martin, J., Wang, H., Zhang, J., Shen, J., Wu, S., and Merida, W. A review of PEM fuel cell durability: Degradation mechanisms and mitigation strategies. *Journal of Power Sources* 2008; 184:104–119.
60. Yousfi-Steiner, N., Moçotéguy, P., Candusso, D., and Hissel, D. A review on polymer electrolyte membrane fuel cell catalyst degradation and starvation issues: Causes, consequences and diagnostic for mitigation. *Journal of Power Sources* 2009; 194:130–145.
61. Yu, Y., Li, H., Wang, H., Yuan, X., Wang, G., and Pan, M. A review on performance degradation of proton exchange membrane fuel cells during startup and shutdown processes: Causes, consequences, and mitigation strategies. *Journal of Power Sources* 2012; 205:10–23.
62. Zhang, S., Yuan, X., Hin, J., Wang, H., Friedrich, K., and Schulze, M. A review of platinum-based catalyst layer degradation in proton exchange membrane fuel cells. *Journal of Power Sources* 2009; 194:588–600.

Section VI

Advanced Materials and Technologies for Supercapacitors

18 Advanced Materials for Supercapacitors

Gaind P. Pandey, Yueping Fang, and Jun Li

CONTENTS

18.1 INTRODUCTION

Electricity is one of the most convenient forms of energy from the utilization point of view. Electrical energy storage (EES) technologies have wide applications covering portable electronic devices, electrical vehicles, and grid-scale energy storage. Particularly, effective EES is critical in the utilization of renewable energy generated from intermittent energy sources such as sunlight and wind. Supercapacitors and batteries are at the forefront of EES technologies. Although both technologies have their own established markets, broader applications require the combination of the high power density property of supercapacitors with the high energy density capability of batteries. Extensive research has been focused on the development of new electrode materials that can potentially hybridize the merits of these two types of EES technologies in one system. In this chapter, we review the progress in the development of advanced materials for supercapacitors and highlight the promising research directions. However, it is impossible to cover the vast research activities in this field in one short chapter. For interested readers, it is recommended to refer to the cited comprehensive review articles on specific topics for more details.

In general, supercapacitors or ultracapacitors are common names referring to electrochemical capacitors (ECs), which store energy in the form of charges at the electrochemical interfaces between high-surface-area porous electrodes and electrolytes, as schematically illustrated in Figure 18.1. The specific capacitance of ECs is typically several orders of magnitude greater than conventional electrostatic capacitors because of the much larger specific surface area (SSA) (500–2500 $m^2\ g^{-1}$) and small ion–electrode separation (~1–2 nm), which can be approximately described by the Helmholtz model as

$$C = \frac{\varepsilon_r \varepsilon_o}{d} A \tag{18.1}$$

where ε_r is the relative electrolyte permittivity, ε_o (in F m^{-1}) is the vacuum permittivity, A (in $m^2\ g^{-1}$) is the SSA of the electrode accessible to the electrolyte ions, and d (in m) is the effective thickness of the electrical double layer (EDL).

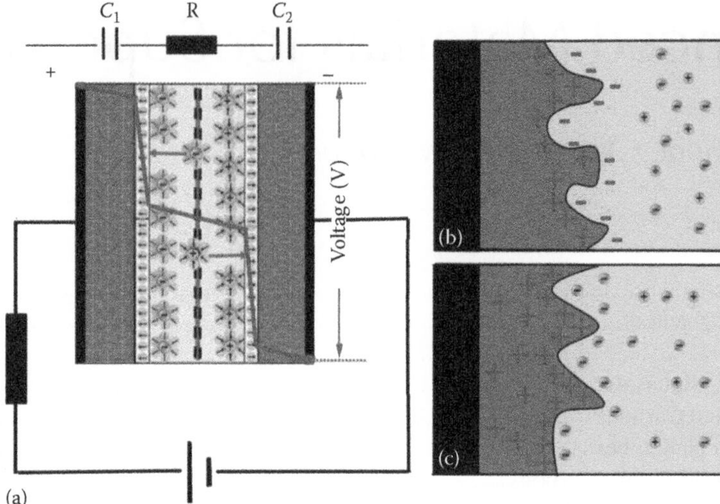

FIGURE 18.1 (a) Schematic of charge storage mechanism in a symmetric supercapacitor; (b) charge accumulation at the electrode-electrolyte interface in an EDLC; and (c) charge storage in the redox film via faradaic processes in a pseudocapacitive supercapacitor.

18.2 CLASSIFICATION OF SUPERCAPACITORS

On the basis of the charge storage mechanisms, ECs can be classified into two general categories: electrochemical (or electrical) double-layer capacitors (EDLCs), which rely on nonfaradaic processes, and pseudocapacitors, which rely on faradaic processes. EDLCs, which store charges electrostatically via reversible ion redistribution in the EDL at the electrode–electrolyte interfaces (Figure 18.1b), commonly use various forms of stable high-SSA carbon materials as the active electrode materials. In contrast, pseudocapacitors store charges through fast and reversible surface redox reactions of thin-film electroactive materials deposited on conductive current collectors (Figure 18.1c). The major difference is that charges only present at the outmost surface of the conductive electrodes in EDLCs but present across the whole thin film of redox materials with a gradient as a function of the state of charge, much like that in batteries. As a result, the specific capacitances of pseudocapacitive electrodes (typically 300–1000 F g^{-1}) are much higher than those of carbon-based EDLC electrodes (typically 100–250 F g^{-1}). Typical active pseudocapacitive materials include transition metal oxides such as RuO_2, MnO_2, Fe_3O_4, NiO_x, CoO_x, and electronically conducting polymers (ECPs) such as polyaniline (PANI), polypyrrole (PPy), and polythiophene and its derivatives. Details about these two energy storage mechanisms (EDLCs and pseudocapacitors) and related materials are discussed in the following sections.

A complete supercapacitor cell may be constructed with two identical electrodes (in the form of either EDLCs or pseudocapacitors) to form a symmetric cell, or by a hybrid of two different electrodes (an EDLC with a pseudocapacitor or two different pseudocapacitors) to form a battery-like asymmetric supercapacitor. This chapter focuses on the discussion of the principles and electrode materials of symmetric supercapacitors first in Section 18.3, and leaves the detailed discussions on asymmetric supercapacitors in Section 18.4.

The performance of ECs can be evaluated by three most commonly used techniques: (i) cyclic voltammetry (CV), (ii) galvanostatic charge–discharge, and (iii) electrochemical impedance spectroscopy (EIS). The overall specific capacitance C_0 (in F) from the CV method is

$$C_0 = \frac{dQ}{dV} \times \frac{1}{m} = \frac{I}{dV/dt} \times \frac{1}{m} = \frac{I}{\upsilon} \times \frac{1}{m} \tag{18.2}$$

where I (in A) is the instantaneous current in CV, dV/dt or υ (in V s^{-1}) is the scanning rate of voltage, and m (in g) is the total mass of the active materials.

For the galvanostatic charge–discharge method, the EC is charged or discharged at a constant current until the voltage between the working electrode and the reference electrode (in three-electrode cell) or the counter electrode (in two-electrode cell) reaches the desired value. The value for specific capacitance C_0 can be deduced from the charge–discharge curve using following equation:

$$C_0 = \frac{I \times \Delta t}{\Delta V \times m} \tag{18.3}$$

where Δt (in s) is the time period required to reach the desired potential change (ΔV) in a charge–discharge curve.

EIS is a powerful tool to characterize the structure, and electron and ion transport properties at the electrodes. Various useful electrochemical data can be extracted. The impedance response for an ideal capacitor is a straight line parallel to the imaginary axis of the complex impedance spectra, i.e., Nyquist plots. However, in practical ECs, the steep-rising capacitive impedance response is only observed in the low-frequency region ($f < 10$ Hz), while the high-frequency region typically shows semicircular features due to charge transfer across the electrode–electrolyte interface. The EIS can be fitted with proper equivalent models to derive the parameters related to the properties of the electrode materials.

In addition to the specific capacitance, energy density and power density are the two important parameters to evaluate the performance of an EC. These values are related to the specific capacitance C_0 of a two-electrode system. The maximum energy E (in W h kg^{-1}) that can be stored and the maximum power P (in W kg^{-1}) that can be delivered by an EC are calculated according to the following equations:

$$E = \frac{1}{2} C_0 V^2 \tag{18.4}$$

$$P = \frac{V^2}{4R_s} \tag{18.5}$$

where V is the operating voltage, which is determined by the thermodynamic stability of the electrolyte and the electrodes. According to the above equations, a high-performance EC must have a large specific capacitance C_0, wide operating voltage V, and minimum R_s (equivalent series resistance, ESR). These values all critically depend on the properties and structures of the electrode materials.

18.3 ELECTRODE MATERIALS

In recent years, extensive research is being undertaken to develop new materials in order to improve energy storage performance. Significant efforts are devoted to developing high-SSA carbon electrodes with defined pore size distributions for EDLCs and incorporating redox-active materials with optimized structure, electrical conductivity, and electrochemical activities for pseudocapacitors.

18.3.1 EDLC Electrode Materials

A large variety of carbons can be considered as active electrode materials in ECs [1–3]. The main materials with high SSA include activated carbons (ACs) [4], carbon nanotubes (CNTs) [5,6], carbon aerogels (CAGs) [7], templated porous carbons [8,9], activated carbon nanofibers (ACNFs) [10], and AC fabrics [11]. Several comprehensive reviews are available in the literature, which can be consulted for more detailed information [3,5,12–16]. Table 18.1 summarizes the different carbon materials as EC electrodes according to the electrochemical test methods and capacitance values. The specific capacitance largely depends on the material properties, especially the SSA, pore size distribution, pore structure, electrical conductivity, and surface wettability.

18.3.1.1 Activated Carbons

ACs are the most widely used EDLC materials because of their high Brunauer–Emmett–Teller SSA (in the range of

TABLE 18.1
Various Carbon Materials and Their Specific Capacitances

Materials	Test Method	Electrolyte	Specific Capacitance (F g^{-1})	SSA (m^2 g^{-1})	Ref.
Activated carbon (coal)	Charge–discharge	1 M $LiClO_4$/PC	220	2647	[29]
Activated carbon	CV, charge–discharge	1 M H_2SO_4	198	270	[26]
Single-walled carbon nanotubes	EIS	38 wt.% H_2SO_4	102	430	[30]
(SWCNTs)	Charge–discharge	1 M Et_4NBF_4/PC	160	1250	[31]
Multiwalled carbon nanotubes (MWCNTs)	CV, EIS	6 M KOH	137	411	[32]
Templated porous carbons (Na–Y zeolite)	Charge–discharge	Aqueous electrolyte	340	1680	[33]
Templated porous carbons	CV	1 M H_2SO_4	200–350	1094–2090	[34]
Carbide-derived carbon (CDC)	CV	1 M H_2SO_4	175	~1800	[35]
	Charge–discharge	1.5 M Et_4NBF_4/AN	130	~1500	[36]
	CV	$EMIBF_4$	100	1698	[37]
Graphene	Charge–discharge	5.5 M KOH; Et_4NBF_4/AN	135; 99	705	[38]
Graphene	Charge–discharge	30 wt.% KOH	205	320	[39]
Activated graphene	Charge–discharge	$BMIBF_4$/AN	166	3100	[40]
Carbon nano-onions (CNOs)	CV	1M H_2SO_4	70–100	~550	[41]
Oxidized carbon nano-onions	CV	Aqueous electrolyte	9.68	–	[42]
Activated carbon fibers (ACFs)	Charge–discharge	30 wt.% KOH	175	1450	[43]
	CV	5.5M H_2SO_4	150–175	~2400–2800	[44]
Carbon aerogels (CAGs)	CV	1 M Et_4NBF_4/PC	100–130	592	[45]
	CV	6 M KOH	110	–	[46]

1000–3000 $m^2 g^{-1}$), commercial availability, and relatively low cost. ACs are generally produced from physical (thermal) and/or chemical activation of various types of natural carbonaceous materials (e.g., wood, coal, nutshell, etc.). Physical activation usually refers to the treatment of carbon precursors at high temperature (from 700°C to 1200°C) in the presence of oxidizing gases such as steam, CO_2, and air [17,18]. Chemical activation is usually carried out at lower temperatures (from 400°C to 700°C) with activating agents like phosphoric acid, potassium hydroxide, sodium hydroxide, and zinc chloride [19,20]. Depending on the carbon precursors and activation methods, the produced ACs present varied physicochemical properties with the highest SSA of 3000 $m^2 g^{-1}$ [21–27]. The ACs have a broad pore size distribution consisting of well-connected micropores (<2 nm), mesopores (2–50 nm), and macropores (>50 nm). However, it has been observed that a high SSA sample (3000 $m^2 g^{-1}$) only showed a relatively small specific capacitance of ~10 μF cm^{-2}, much smaller than the theoretical EDL value (15–25 μF cm^{-2}). It is likely that not all pores were effective in charge accumulation [1,24]. Therefore, other aspects of the carbon materials such as pore size distribution, pore shape and structure, electrical conductivity, and surface functionality can also significantly influence their electrochemical performance. In most AC materials, the pore size distribution is not optimum and cannot be controlled by the activation process, thus limiting the possibility to fully utilize the whole developed surface area to form EDL. Excessive activation may create a larger accessible pore volume, but at the expense of lowering the material density and reducing the electrical conductivity, leading to lower volumetric energy density and loss of power capability.

Several studies have been reported for the relationship between the nanoporous structure of ACs and their capacitance performance in different electrolytes [22,25,28]. In general, the capacitance of ACs is higher in aqueous electrolytes (ranging from 100 to 300 F g^{-1}) than in organic electrolytes (<150 F g^{-1}). One reason for this is that the effective size of the electrolyte ions in organic solutions is larger compared with those in water, thus only being able to access fewer number of pores. The smaller interaction between the organic solvent and the carbon surface may be another reason. The exact size for optimum EDLCs is still controversial. Salitra et al. concluded that a pore size >0.4 nm is needed for EDL charging in aqueous solution [25]. Beguin and coworkers reported the optimal pore size for EDLCs is 0.7 nm in aqueous media and 0.8 nm in organic electrolytes [22].

18.3.1.2 Carbon Nanotubes

CNTs have attracted a great deal of attention as EDLC materials owing to their unique pore structure, superior electrical properties, and good mechanical and thermal stability [5,30–32,47,48]. CNTs are produced from hydrocarbons by catalytic decomposition. Depending on the synthesis parameters, single-walled CNTs (SWCNTs) and multiwalled CNTs (MWCNTs) can be prepared, both of which have been widely explored as EDLC materials, particularly for high-power applications. Moreover, their high mechanical resilience and ease in forming entangled network make them a good support for active materials. However, the specific capacitances and, hence, energy density is relatively small owing to low SSA (generally <500 $m^2 g^{-1}$) as compared with ACs. This is mainly attributed to the poorly developed microporous volume of CNTs. In the case of MWCNTs, this volume can be increased by activation of the nanotubes [49]; however, the capacitance values still remain lower than those of ACs. Surface functionalization by introducing pseudocapacitance contribution through oxidation treatments leads to significant improvement of the capacitive behavior in protic media [5]; however, the cycle life tends to decrease in this case. Currently, efforts are focused on the development of dense, nano-ordered, aligned CNT forests perpendicular to the current collector that could help increase the capacitance by fine-tuning the intertube distance [50–53]. Such CNT-based nanoarchitectured electrodes appear to be very promising, mainly for microelectronics applications [54,55].

18.3.1.3 Templated Porous Carbons

The template method offers a powerful way to produce both microporous and mesoporous carbons materials with a tunable pore size, large SSAs, and interconnected pore network, making them promising candidates for EDLC materials [56]. In disordered porous carbon, high resistance will be formed as a result of the tortuous diffusion path for the ions, especially at high current density. After the first report by Ryoo et al. [57,58] on the fabrication of ordered mesoporous carbons (OMCs) by a hard template nanocasting strategy, there has been remarkable progress in the synthesis of ordered nanostructured carbons through template techniques [59]. In general, the preparation procedure of templated carbons is infiltration of a carbon precursor into the pores of the template, followed by a carbonization treatment and finally the removal of the template to leave behind a porous carbon structure as shown by schematic representation in Figure 18.2. Lee et al. [60] first reported that the OMCs possess excellent electrochemical performance compared with commercial ACs. Xing et al. [61] reported superior EDLC performances of OMC by optimizing the ordered pore symmetries and the mesopore structure. Despite all efforts, only moderate improvement has been achieved.

18.3.1.4 Carbide-Derived Carbon

Various metal carbides were found to give highly microporous carbons through the chlorination at temperatures from 400°C to 1200°C [62]. The SSA was reported to be 1000 to 2000 $m^2 g^{-1}$. After preparation, the CDC samples are annealed at 600°C under H_2 atmosphere to clean the surface and neutralize the reactive carbon dangling bonds. This method allows the synthesis of carbons with controlled narrow pore size distribution, since porosity is issued from the leaching out of metal atoms. The pore structure of these carbons depends strongly on the precursor carbide, temperature, and thermal treatment duration. In these CDCs, micropores are mainly formed up to about 800°C, but mesopores become predominant above 800°C; as a consequence, the SSA shows

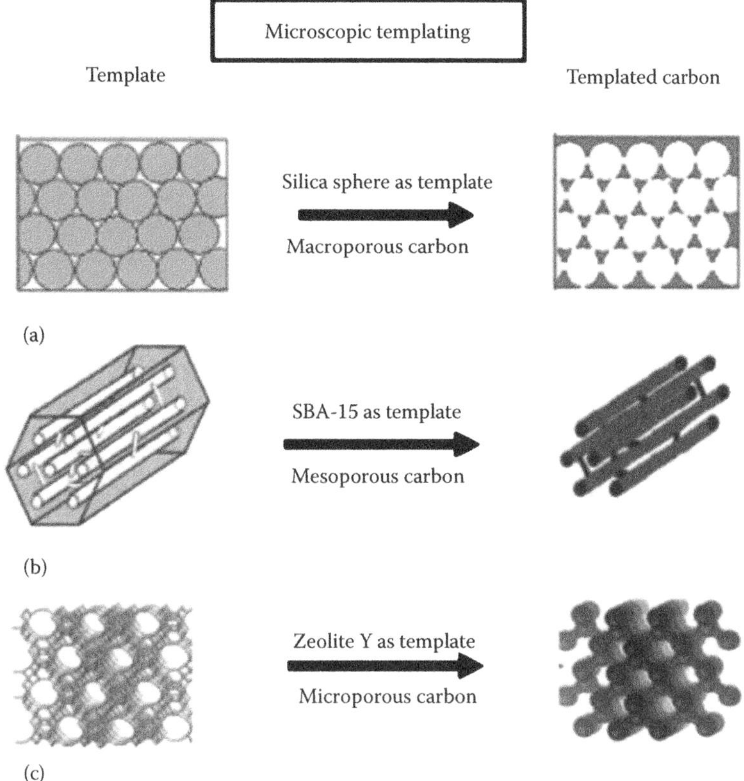

FIGURE 18.2 Microscopic templating synthesis of (a) macroporous carbons using silica spheres as template, (b) mesoporous carbons using SBA-15 as template, and (c) microporous carbons using zeolite Y as template. (From Zhang, L.L. and Zhao, X.S., *Chem. Soc. Rev.*, 38, 2520, 2009. With permission.)

a maximum at around 800°C in most of the cases [63–66]. By using CDCs with pore size of less than 1 nm, the desolvation of ions as they enter into micropores were observed, and EDLC behavior was studied in various nonaqueous solutions [67–69], an H_2SO_4 solution [35], and an ionic liquid electrolyte [37], and the effects of ion sizes of electrolytes were discussed [70,71]. Contrary to the traditional behavior, the normalized capacitance decreases with pore size until a critical value (~1 nm) is reached, and then increases. It seems that pores smaller than the size of solvated ions greatly enhance the double-layer charge storage. This was initially explained by the distortion of the ion solvation shell, leading to a closer approach of the ion to the carbon surface, which by Equation 18.1 leads to improved capacitance. References [72–76] can be consulted for additional information regarding capacitor performance and the effects of pore accessibility and porosity saturation.

18.3.1.5 Graphene

Graphene is an attractive robust sheet-like electrode material for ECs because of its high theoretical SSA (2630 $m^2\ g^{-1}$), excellent electrical conductivity, high surface-to-volume ratio, outstanding intrinsic EDL capacitance (21 $\mu F\ cm^{-2}$), and theoretical specific capacitance (550 $F\ g^{-1}$) [38,39,77–79]. A variety of graphene-based materials with different chemical structures and morphologies, such as chemically modified graphene, microwave expanded graphite oxide, laser-scribed graphene, and activated and curved graphene, have been explored as electrode materials for ECs [38,78,80–82].

Recently, a graphene-derived carbon with a SSA up to ~3100 $m^2\ g^{-1}$ has been prepared by microwave irradiation of graphite oxide (GO) followed by chemical activation with potassium hydroxide (KOH) [40]. This material has a large fraction of micro- and mesopores that provide a large and accessible surface area for charge accommodation, and thus showed improved specific capacitance in organic and ionic liquid electrolytes with a relatively high gravimetric energy density [40].

However, most graphene-derived carbons only showed EDL capacitance on a portion of the surface, presumably because they form stacked agglomerates and thus may have limited ion diffusion paths inside some narrow channels [81]. If such graphene platelets can be assembled in a hierarchical architecture so that mesopores within the interconnected macroporous scaffolds remain accessible [83,84], it could allow electrolyte ions to diffuse to the interior surface and efficiently form extensive EDL, potentially leading to more effective energy storage devices. Despite extensive efforts, the overall performance of graphene-based ECs, especially regarding the power density, has not been improved remarkably. Some scanning electron microscopic (SEM) images of various carbon materials are shown in Figure 18.3.

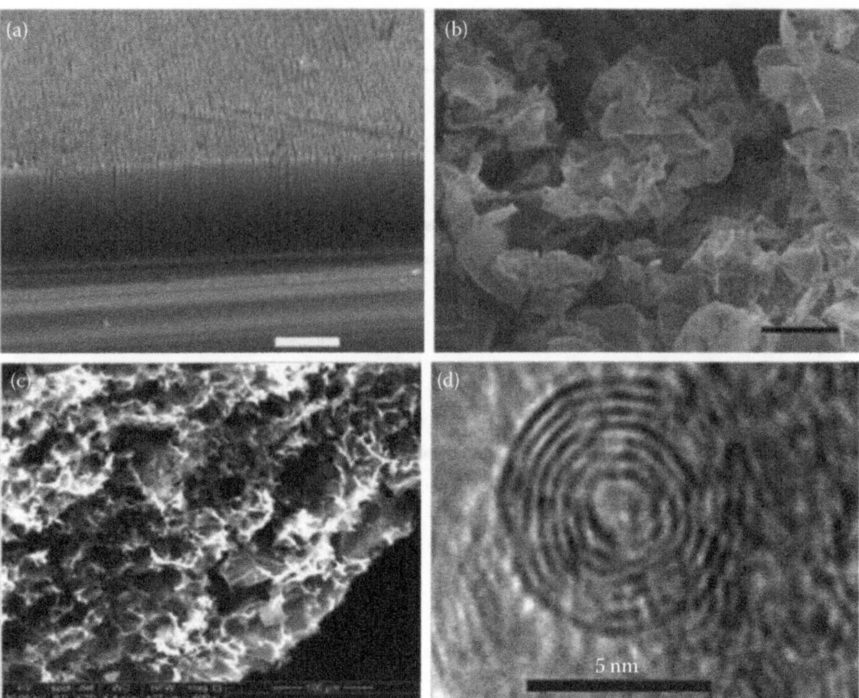

FIGURE 18.3 SEM image of (a) plasma-etched vertically aligned CNT electrode (scale bar: 100 μm) (From Lu, W. et al., *J. Power Sources*, 189, 1270, 2009. With permission.); (b) curved graphene sheets (scale bar, 10 μm) prepared by a conventional chemical route (From Liu, C. et al., *Nano Lett.*, 10, 4863, 2010. With permission.); (c) high-surface-area activated microwave-expanded graphite oxide (scale bar: 100 μm) (From Kim, T. et al., *ACS Nano*, 7, 6899, 2013. With permission.); and (d) high-resolution TEM images of CNOs from thermally annealing nanodiamonds (From Palkar, A. et al., *Chemistry—An Asian J.*, 2, 625, 2007. With permission.).

18.3.1.6 Carbon Nano-Onions

Carbon nano-onions (CNOs) were discovered in 1992 by Ugarte [85], and until quite recently they were rarely studied because of the lack of convenient methods for synthesis and separation. CNOs are spherical particles typically 4–25 nm in diameter, consisting of concentric shells of graphitic carbon, that can be also described as multishelled fullerenes (Figure 18.3d) [86–88]. Methods for CNO synthesis include vacuum annealing of nanodiamonds, arc discharging of graphite underwater, flash pyrolysis, ion implantation into copper and silver, vapor deposition, reactions in stainless-steel autoclaves, and laser ablation [86]. Most of these methods gave low yields and/or many by-products. For applications as EC electrodes, large quantities of materials with controlled physico–chemical properties are required, which is still challenging. A detailed review on CNOs as EC electrodes is available in the literature [86].

A nonporous CNO electrode containing nearly spherical particles with a concentric shell structure and a narrow particle size distribution has been reported [89]. The closed structures of high-temperature annealed CNOs enabled their use in organic electrolytes at temperatures as low as −40°C and in ionic liquid electrolytes at −50°C [90]. Pech et al. reported microsupercapacitors produced by electrophoretic deposition of a several-micrometer-thick layer of nanostructured CNOs, showing a power density comparable to electrolytic capacitors [91]. The maximum power density was close to 1 kW cm^{-3}, about 100 times higher than that of ACs [91].

18.3.1.7 Activated Carbon Fibers and Carbon Aerogels

ACFs are different from ACs owing to their pore structures. In ACFs, micropores are directly exposed to the surface of the fiber; however, they are on the interior wall of macropores and mesopores in ACs [92]. Therefore, fast ion access can be expected on ACFs. The ACFs are usually produced from carbonization of preformed fibrous carbon precursors followed by activation processes. ACFs derived from silk fibroin have high SSAs (up to 3000 m^2 g^{-1}) and large cell capacitance. However, because of low bulk density (~0.2 g mL^{-1}), it is not suitable for ECs requiring small volumes [93,94].

CAGs are another type of ultralight highly porous materials, predominated by mesopores, which may be used in EC electrodes without binding substances. CAGs are typically fabricated through a sol–gel process with subsequent pyrolysis of the organic aerogels. The unique porosity of a CAG is based on the interconnection of colloidal carbon nanoparticles. Since it is dominated by mesopores (>2 nm), the EDL capacitance is not so high in CAG-based electrode materials [45,95].

18.3.1.8 Heteroatoms Containing Carbons

The specific capacitance of EDLCs is limited by the electrochemically active surface area of ACs and the pore accessibility. Therefore, a number of approaches are under development to overcome these problems. One approach is introducing heteroatoms on the carbon surface for pseudocapacitance

[96–100]. Apart from typical electrostatic interactions in the EDL, redox reactions on the electrode–electrolyte interface can be employed to enhance the charge storage. However, due to their faradaic origin, these heterogeneous reactions exhibit a slow kinetics and a moderate cycle life. The presence of oxygen and nitrogen functionalities in the carbon network has been demonstrated to enable quick faradaic reactions by modifying the local electronic structure [101].

Oxygen-containing functional groups (such as –COOH, –(CO)–) can be introduced by activating the carbon surface in O_2 [102,103] or by chemical/electrochemical oxidation of carbon in HNO_3 solution [104–107]. However, oxidizing carbon materials have adverse effects on the reliability of ECs using nonaqueous electrolytes due to self-discharge, leakage current, and many other reactions. Substantial costs are added to remove oxygen in production of ACs for commercial EDLCs. In contrast, oxygen-containing functional groups increase the total capacitance in aqueous electrolyte solutions, mostly in H_2SO_4 solutions, by introducing pseudocapacitance.

Recently, nitrogen-enriched carbon has attracted a lot of attention because of its pseudocapacitance effect for supercapacitors [108]. Increasing capacitance by nitrogen doping is often attributed to faradaic reactions of the nitrogen-containing functional groups, similar to the case of oxygen-containing carbons. However, in most of the cases, the nitrogen content is not sufficiently high to explain the large gain in capacitance. Additional rationales are offered, such as improving the wettability of the pore walls by the formation of polar functional groups, increasing the capacitance of the space-charge layer due to the increase in carrier concentration, and others [3]. Nitrogen is often introduced onto carbon frameworks by a postprocess, such as treating it with ammonia or melamine, resulting in a tedious process and the instability of pseudocapacitance properties given by nitrogen upon cycling the capacitors. Also, only a low nitrogen content (<10%) is usually reached. Detailed discussions on pseudocapacitive contributions by heteroatoms on the carbon surface are available in the literature [3,108].

18.3.2 Pseudocapacitive Materials

Two main types of redox materials are transition metal oxides (and/or nitrides) and ECPs. Recently, the composites consisting of both of them present a new avenue.

18.3.2.1 Redox-Active Metal Oxides and/or Nitrides

Transition metal oxides (and nitrides) are generally considered the best candidate electrode materials for pseudocapacitors because they have a variety of oxidation states available for redox reactions. They not only store energy like electrostatic carbon materials but also exhibit fast electrochemical faradaic reactions between electrode materials and electrolytes within appropriate potential windows [109–112].

The general requirements for transition metal oxides in EC applications are summarized in Ref. [1], including the following: (i) the oxide should be electronically conductive; (ii) the metal can exist in two or more oxidation states that coexist over a continuous range with no phase changes involving irreversible modifications of a three-dimensional (3D) structure; and (iii) the protons (and/or other cations such as Na^+, Li^+, etc.) can freely intercalate into the oxide lattice on reduction (and out of the lattice on oxidation), allowing facile interconversion such as $O^{2-} \leftrightarrow OH^-$. The ions in electrolyte during charging (usually OH^- or H^+) diffuse from the electrolyte into the electrode–solution interface under the electric field and proceed with the following electrochemical reaction:

$$(MO_y)_{surface} + xH^+ + xe^- \leftrightarrow [MO_{y-x}(OH)_x]_{surface} \quad (18.6)$$

The common metal oxides are summarized in Table 18.2 and discussed in the following sections.

TABLE 18.2
Summary of Redox-Active Oxides and Their Electrical Conductivity and Specific Capacitance

Materials	C (F g^{-1})	σ (S cm^{-1})	Scan Rate (mV/s)	Voltage Window (V)	Electrolyte	Ref.
RuO_2	2000[a]					[113–115]
	1300	10^5	1000	0–1	1.0 M H_2SO_4	[116]
	1300	10^5	1	−0.35–1.05	0.1 M H_2SO_4	[113]
MnO_2	1380	10^{-5}–10^{-6}	5	0–0.9	0.1 M Na_2SO_4	[117]
	1380[a]					[141–143]
Co_3O_4	3560[a]					[118]
	281		5	−02–0.6	2 M KOH	[119]
NiO	3750[a]					[149,150]
	696		1	0.2–0.6	7 M KOH	[120]
α-Fe_2O_3	100		2	0–2.8	1 M $LiClO_4$ in EC/DMC	[121]
Fe_3O_4	118.2	10^2–10^3	10	−1– 0.1	0.1 M Na_2SO_3	[122]
V_2O_5	>1000[a]	10^{-5}–10^{-3}				[123]
	910	10^{-3}	10	1.5–4 versus Li/Li^+	1 M $LiClO_4$	[124]
VN	850–1340	10^4	2		1 M KOH	[125,126]

[a] Theoretical values.

18.3.2.1.1 Ruthenium Oxide

Ruthenium oxide, RuO_2 and $RuO_2 \cdot xH_2O$, is widely studied because it has very high conductivity ($\sigma = 10^5$ S cm^{-1}) and has three distinct oxidation states accessible within 1.4 V [111]. It has an ultrahigh theoretical capacitance of (~2000 F g^{-1}) and excellent chemical stability [113–115]. The pseudocapacitance behavior of RuO_2 in acidic solutions has been the focus of research in the past 35 years [1]. It can be described as a fast, reversible electron transfer together with an electro-adsorption of protons on the surface of RuO_2 particles, according to Equation 18.7, where Ru oxidation states can change from (II) up to (IV):

$$RuO_2 + xH^+ + xe^- \leftrightarrow RuO_{2-x}(OH)_x \quad \text{where } 0 \leq x \leq 2 \tag{18.7}$$

The continuous change of x during proton insertion or deinsertion occurs over a window of about 1.2 V, and leads to a capacitive behavior with ion adsorption following a Frumkin-type isotherm [1,109]. The following factors play key roles in the electrochemical behavior of $RuO_2 \cdot xH_2O$ [110]: (i) SSA [127,128], (ii) the degree of hydration [129,130], (iii) the crystallinity [131], (iv) the particle size [132], and (v) the selection of electrolytes [133]. The major problems are that ruthenium is very expensive and the 1-V voltage window limits the energy density. Some representative SEM images of various nanostructured oxide electrodes are shown in Figure 18.4.

18.3.2.1.2 Manganese Oxide

The charge storage mechanism of MnO_2 is based on surface adsorption of electrolyte cations C^+ (K^+, Na^+, ...) as well as proton incorporation according to the reaction [109,110,134–136]:

$$MnO_2 + xC^+ + yH^+ + (x + y)e^- \leftrightarrow MnOOC_xH_y \tag{18.8}$$

MnO_2 exhibits pseudocapacitive behavior owing to the surface adsorption and insertion/extraction of metal alkali

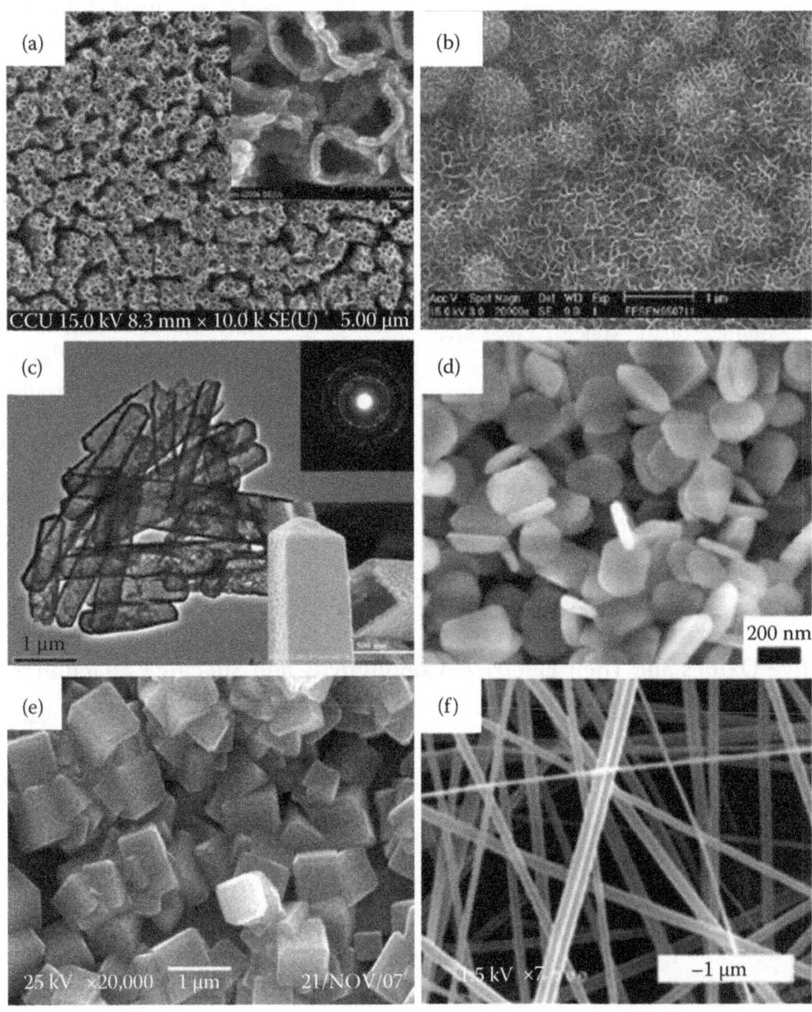

FIGURE 18.4 Summary of representative transition metal oxide pseudocapacitive materials. (a) SEM image of a hydrous RuO_2 nanotube arrayed electrode (inset: enlarged view) (From Hu, C.-C. et al., *Nano Lett.*, 6, 2690, 2006. With permission.); (b) SEM image of MnO_2 electrodes (From Chang, J.-K. et al., *J. Power Sources*, 179, 435, 2008. With permission.); (c) TEM image of Co_3O_4 boxes (inset: SEM image and selected area electron diffraction pattern) (From Du, W. et al., *J. Power Sources*, 227, 101, 2013. With permission.); (d) SEM image of the porous NiO nanocrystals (From Zhang, X. et al., *Nano Res.* 3, 643, 2010. With permission.) (e) SEM image of octadecahedron Fe_3O_4 thin films (From Chen, J. et al., *Electrochim. Acta*, 55, 1, 2009. With permission.); and (f) SEM image for as-synthesized electrospun V_2O_5 nanofibers (From Wee, G. et al., *J. Mater. Chem.*, 20, 6720, 2010. With permission.).

cations inside the open crystalline framework. The theoretical specific capacitance for MnO_2 is approximately 1380 F g^{-1}, based on one-electron redox reaction for every Mn atom [137]. Owing to the low diffusion coefficient of protons (D_{H^+} = 6 × 10^{-10} cm^2 s^{-1}) and alkali cations (D_{Li^+} = 3 × 10^{-10} cm^2 s^{-1}) to the interior of MnO_2 particles and low electrical conductivity (~10^{-5}–10^{-6} S cm^{-1}), conventional MnO_2 electrodes require binder and conductive additives, and typically show inhibited specific capacitance below 200 F g^{-1} [138]. Current efforts are focused on developing binder-free nanostructured MnO_2 using core–shell structures or composites to reduce electronic and charge diffusion distances in MnO_2 and thus to promote charge-transfer reactions [139–141].

18.3.2.1.3 Cobalt Oxide

Co_3O_4 seems to exhibit excellent reversible redox behavior, large surface area, high conductivity, long-term performance, and good corrosion stability [142,143]. Therefore, it has been considered an alternative electrode material for ECs [144] and has been assessed by many groups [143,145]. One of the difficulties in utilizing this material in pseudocapacitors is that Co_3O_4 has poor electronic conductivity and it is difficult for the electrolyte to diffuse into its structure [140]; therefore, increasing the SSA is the key to making a high-performance electrode from Co_3O_4 [146].

18.3.2.1.4 Nickel Oxide

Nickel oxide (NiO) is considered an alternative electrode material for ECs in alkaline electrolytes owing to its ease in synthesis, relatively high specific capacitance (theoretical specific capacitance of 3750 F g^{-1}) [147,148], environmental friendliness, and low cost [149]. The challenges are the poor cycle performance [150] and high resistivity [151].

18.3.2.1.5 Vanadium Oxides

Vanadium oxides, specifically vanadium pentoxide (V_2O_5), are more recently studied as potential pseudocapacitive materials [152,153]. Some characteristics of V_2O_5 even exceed those of RuO_2, such as the large number of oxidation states and higher theoretical capacitance (>1000 F g^{-1}) [123,154]. The higher theoretical values for the specific capacitance have been attributed to the lower equivalent molecular weight and the greater differences between the oxidation states. The challenge for V_2O_5 is its low conductivity (~10^{-5}–10^{-3} S cm^{-1}). Hydrogen thermal treatment has recently been demonstrated as an effective approach to increasing the conductivity of VO_2 by nearly 3 orders of magnitude, possibly due to the synergistic effects of H-doping, oxygen vacancies, and slightly reduced vanadium valence state. A specific discharge capacity of 300 F g^{-1} and a specific energy density of 17 Wh kg^{-1} was obtained at a rate of 1 A g^{-1} with good long-term cycling stability [155].

18.3.2.1.6 Iron Oxides (Fe_2O_3 and Fe_3O_4)

α-Fe_2O_3 is another candidate electrode material for ECs, offering low cost and minimal environmental impact. However, poor electronic conductivity restricts its applications in high-power storage devices. The incorporation of MWCNTs or graphene into α-Fe_2O_3 nanomaterials seems to be a way to solve this problem [121]. Fe_3O_4 is another inexpensive material exhibiting pseudocapacitance in alkali sulfite electrolytes. However, the specific capacitance is rather low. For example, an Fe_3O_4 film prepared *via* a hydrothermal process displayed a typical pseudocapacitive behavior in a 1 M Na_2SO_3 solution, with a specific capacitance of 118.2 F g^{-1} between −1 and 0.1 V, and a capacitive retention of 88.8% after 500 cycles of charging–discharging [122]. Overall, neither Fe_2O_3 nor Fe_3O_4 seem to be realistic materials for ECs.

18.3.2.1.7 Transition Metal Nitrides

In recent years, transition metal nitride materials have attracted much attention owing to their high electrical conductivity (up to ~10^6 Ω^{-1} m^{-1}) and excellent chemical stability in common acids and bases. Vanadium nitride (VN) is a good example. Its specific capacitance was found to be 161 F g^{-1} at 30 mV s^{-1}, and 70% of its original capacitance value was retained even when the potential scan rate was increased from 30 to 300 mV s^{-1} [156]. Using a nanostructured VN, a high specific capacitance of 1200–1400 F g^{-1} was obtained when tested at a scan rate of 2 mV s^{-1}, while 554 F g^{-1} was achieved at a high scan rate of 100 mV s^{-1} [126]. These properties were attributed to reversible redox reactions of a thin surface layer of vanadium oxides on highly conductive VN nanocrystals [126]. Molybdenum nitride (MoN) was also explored as an EC electrode material in sulfuric acid solution [157]. Its specific capacitance was found to be comparable to RuO_2. Unfortunately, MoN has a substantially smaller operating voltage range (only 0.7 V) due to its electrochemical decomposition.

18.3.2.2 Electronically Conducting Polymers

ECPs are an attractive class of electrode materials for pseudocapacitors. They have been intensively studied because of their capability for fast and reversible switching between the redox states, high charge density, good intrinsic conductivity (from a few S cm^{-1} to 500 S cm^{-1}), and low cost [158,159]. ECPs conduct electrons through conjugated bonds along the polymer backbone. They are typically formed either through chemical oxidation or electrochemical oxidation of the monomer. Two oxidation reactions occur simultaneously, one is the oxidation of the monomer and second is the oxidation of the polymer with the insertion of a dopant/counter ion (e.g., Cl^-) [160]. The doping level (in p-type ECP) is typically 0.3–0.5, i.e., one counter ion on each two to three monomer units. This is limited by how closely the positive charges can be spaced along the polymer chain. The polymers that are most commonly studied as pseudocapacitive electrodes are PPy, PANI, and derivatives of polythiophene [159]. The typical dopant level for these ECPs, theoretical specific capacitances and voltage ranges, and conductivities are given in Table 18.3.

Electric energy can be stored and delivered in conducting polymers as delocalized π-electrons are accepted and released during electrochemical doping/dedoping, respectively. There are two types of doping process below:

TABLE 18.3
Typical Values of Theoretical Specific Capacitances and Conductivities of Representative ECPs [159,161]

ECP	Monomer MW (g mol^{-1})	Dopant Level	Potential Window (V)	Theoretical Specific Capacitance (F g^{-1})	Conductivity (S cm^{-1})
Polyaniline (PANI or PAN)	93	0.5	0.7	750	0.1–5
Polypyrrole (PPy)	67	0.33	0.8	620	10–50
Polythiophene	84	0.33	0.8	485	300–500
Poly(3,4-ethylenedioxythiophene) (PEDOT)	142	0.33	1.2	210	300–400

Source: Snook, G. A. et al., *Journal of Power Sources*, 196, 1–12, 2011; Lota, K. et al., *Journal of Physics and Chemistry of Solids*, 65, 205–301, 2004.

p-Doping (with counter anions):

$$(\text{polymer}) + yA^- \leftrightarrow [(\text{polymer})^{y+}\, yA^-] + ye^- \quad (18.9)$$

n-Doping (with counter cations):

$$(\text{polymer}) + yC^+ + ye^- \leftrightarrow [(\text{polymer})^{y-}\, yC^+] \quad (18.10)$$

ECP-based ECs can be classified into three configurations [158]: Type I (symmetric) capacitors utilize p-doping ECPs for both positive and negative electrodes; the drawbacks are low operating potential (<1 V) and utilization of only half of the total capacity. Type II (asymmetric) capacitors utilize two different p-doping ECPs with distinct electroactivities. Thus, higher capacity and higher working voltage can be obtained. Type III (symmetric) capacitors utilize the same ECP for both electrodes but making it p-doping at the positive electrode and n-doping at the negative electrode. Type III capacitors can provide the widest operating voltage (2.5–3.0 V), about two times higher than type II, and the highest energy density [158]. They have similar discharge characteristics as batteries, where operating voltage drops very rapidly after discharge. Thus far, there is only limited success with the type III configuration because of the difficulty of the n-doping process and low chemical stability. Table 18.4 summarizes different ECP materials as EC electrodes with the electrochemical evaluation method and capacitance values.

18.3.2.2.1 Polyaniline (PANI or PAN)

PANI or PAN has been studied extensively as a pseudocapacitor electrode material [162–164]. PANI has many desirable properties for use in an EC device; it has high electroactivity, a high doping level (0.5; Table 18.3), good stability, and a high specific capacitance (400–500 F g^{-1} in an acidic medium) [165]. In addition, it has good environmental stability, controllable electrical conductivity (around 0.1–5 S cm^{-1} in the doped state), and can be easily processed. A major disadvantage of PANI is that it requires a proton to be properly charged and discharged; therefore, a protic solvent, an acidic solution, or a protic ionic liquid is required [166]. PANI has the most variable specific capacitance of all ECPs with that of electrodeposited materials higher than those by chemical polymerization.

PANI-based ECs generally show low power density and poor cycling performance. It has been reported that the cycle life of PANI for an Li-doped ($LiPF_6$) positive electrode was over 9000 cycles. However, the achieved specific capacitance (107 F g^{-1}) is rather low [163]. The low electronic conductance limits them from high power applications. This has been addressed by numerous researchers. In one attempt, the ECP was fabricated into suitable structures by optimizing deposition/preparation conditions. Stable nanostructured

TABLE 18.4
Different Conducting Polymers and Their Specific Capacitances

Materials	Test Method	Electrolyte	Specific Capacitance (F g^{-1})	Ref.
Polyaniline (PANI or PAN)	CV at 10 mV s^{-1}	1 M H_2SO_4	775	[164]
	Charge–discharge method	1 M Et_4NBF_4	107	[163]
	Charge–discharge method	1 M $HClO_4$	950	[167]
	Charge–discharge method	1 M H_2SO_4	548	[195]
Polypyrrole (PPy)	Charge–discharge method	PVdF–HFP-based gel polymer electrolyte	78–137	[172]
	CV	0.5 M H_2SO_4	400	[174]
	CV at 2 mV s^{-1}	0.5 M H_2SO_4	586	[176]
Poly(3-methylthiophene) (pMeT)	Charge–discharge method	Ionic liquid	200	[196]
	CV at 20 mV s^{-1}	1 M Et_4NBF_4 in PC	165	[180]
Poly(3,4-ethylenedioxythiophene) (PEDOT)	Charge–discharge method	1 M H_2SO_4	130	[189]
	Charge–discharge method	1 M H_2SO_4	127	[190]
	CV	0.1 M $LiClO_4$ aqueous electrolyte	124	[194]

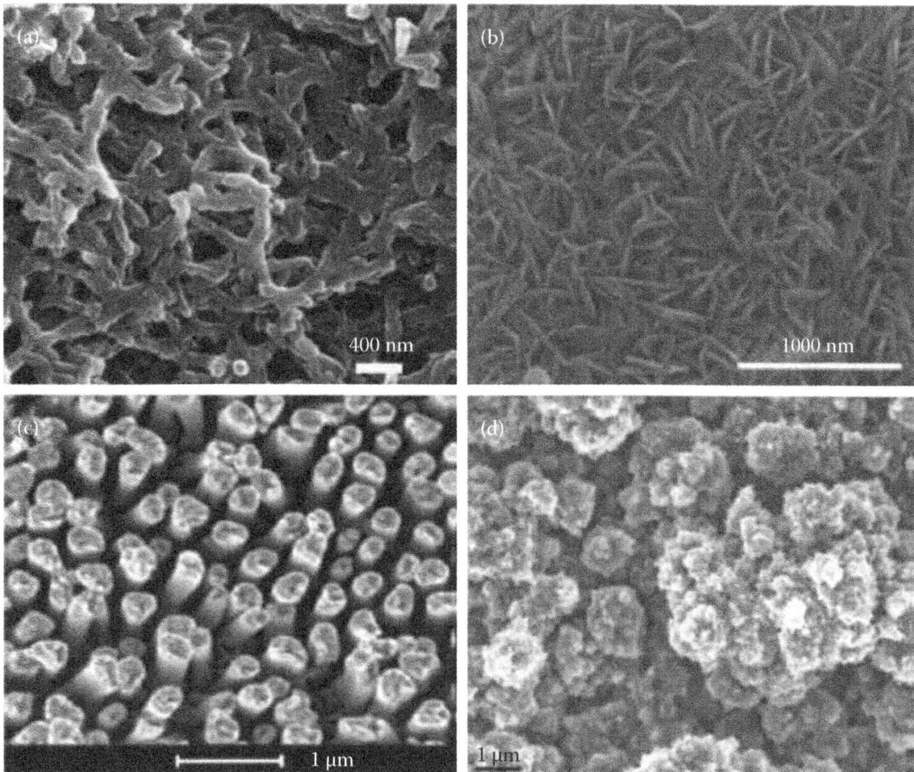

FIGURE 18.5 SEM image of (a) PANI nanofibers obtained by interfacial polymerization (From Guan, H. et al., *Electrochim. Acta*, 56, 964, 2010. With permission.); (b) nanosheets of PPy thin film (From Dubal, D.P. et al., *J. Mater. Chem.*, 22, 3044, 2012. With permission.); (c) PEDOT nanowires (From Li, C. et al., *Chem. Soc. Rev.*, 38, 2397, 2009. With permission.); and (d) highly porous PEDOT film surface (From Pandey, G.P. and Rastogi, A.C., *Electrochim. Acta*, 87, 158, 2013. With permission.).

PANI with high SSA and high porosity have been prepared and showed improved performances [164,167–170]. Some representative SEM images of various nanostructured ECP are shown in Figure 18.5.

18.3.2.2.2 Polypyrrole (PPy)

PPy is one of the most studied conducting polymers for the pseudocapacitive electrodes owing to its fast charge–discharge kinetics, high energy density, low cost, and high degree of flexibility in electrochemical processing than other conducting polymers [171–175]. PPy has a high capacitance per unit volume (400–500 F cm^{-3}), owing to its higher density [159]. One disadvantage due to dense growth is the limited access to the interior sites of the polymer by dopant ions. This reduces the capacitance per gram, especially for thicker coatings on electrodes. Wang et al. [175] reported PPy films doped by *p*-toluenesulfonic, prepared by pulse current polymerization in aqueous solutions. They characterized the PPy films in liquid electrolytes containing different cations (H^+, Li^+, Na^+, and K^+ ions). The PPy electrode shows very stable cycling performance (20,000 charge–discharge cycles at current density of 40 A g^{-1}) in 3 M HCl. Different nanostructured PPy have been prepared on stainless-steel substrates using the potentiodynamic mode by Dubal et al. [176]. Multilayer nanosheets offered a specific capacitance of 586 F g^{-1} at a scan rate of 2 mV s^{-1}.

PPy has been used as an electrode to make a type I supercapacitor, as well as combined with poly(3-methylthiophene) (PMeT) to form a type II version [177]. The type I device has a discharge capacitance of 8–15 mF cm^{-2}, which is similar to that for the type II device. The voltage range is around 0.5–1.0 V for the type I device, but is extended to 1.2 V for the type II.

18.3.2.2.3 Thiophene-Based Conducting Polymers

Thiophene-based ECPs, both p- and n-doped, were synthesized and studied for pseudocapacitors with a focus on optimizing the working voltage range of the device [178–180]. In general, the mass specific capacitance in the n-doped form is found to be lower than that of the p-doped form. It is also found that the conductivity in the n-doped form is lower. This limits the use of these materials in the n-doped form as an anode material. Most polythiophene derivatives are stable in air and moisture in both the p-doped and undoped forms [180]. Mastragostino and coworkers developed type III configuration (n/p type of polymer device) using pMeT [181,182]. The same group also developed a hybrid device based on pMeT as positive electrode and AC as negative [183]. A gel polymer electrolyte-based hybrid device using pMeT and AC has been reported by Sivaraman et al. [184]. Poly3-(4-fluorophenylthiophene), which can be both n- and p-doped, has been chemically synthesized and fabricated all-solid-state device by Kim and Chung [185]. Stenger-Smith et al. [186] reported poly(propylenedioxy) thiophene-based supercapacitor operating at low temperatures (–35°C) in a mixture of the ionic liquids 1-ethyl-3-methylimidazolium bis(trifluoromethanesulfonyl)imide (EMIBTI) and 1-ethyl-3-methylimidazolium hexafluorophosphate ($EMIPF_6$). More detailed information can be found in a recent review by Snook et al. [159].

The thiophene derivative poly(3,4-ethylenedioxythiophene) (PEDOT) is the most popular one for EC applications owing to its high conductivity and chemical stability compared with many of the thiophene derivatives. The polymer has a higher potential range of 1.2 V but the smallest specific capacitance due to a combination of the large molecular weight of the monomer unit and the low doping level [187]. PEDOT is an electron-rich polymer and consequently has a low oxidation potential together with a wide potential window [188]. PEDOT have many attractive properties such as a low band gap of 1–3 eV, being highly conducting in the p-doped state (300–500 S cm^{-1}), good thermal and chemical stability, and high charge mobility that results in fast electrochemical kinetics [161,189,190]. The polymer has also been found to have good film-forming properties and can be switched rapidly with a minimum of side reactions leading to a long cycle life [188].

However, owing to its large molecular weight and a low doping level (around 0.33), it has a relatively low specific capacitance compared with the other ECPs, e.g., PANI and PPy. Recent research is focused onto the increase in the specific capacitance by various strategies such as formation of composites with nanostructured carbon, modified microstructure, and novel dopant schemes [161,190,191]. Both dopant incorporation and microstructures are strongly influenced by the polymerization process that modifies the electrochemical and physical properties of the polymer [190]. Various techniques have been reported for PEDOT electrode preparation using the sulfonate-doped aqueous emulsions and chemical polymerization or electrooxidation of EDOT monomer to optimize its performance as pseudocapacitive electrode [192,193]. Lota et al. reported a high specific capacitance of about 150 F g^{-1} with PEDOT–CNT composite electrode [161]. Liu et al. reported electropolymerization of PEDOT using an ionic liquid as both the growth medium and the supporting electrolyte [189]. The PEDOT electrodes were showing stable cycling performance with high specific capacitance 130 F g^{-1}. Huang and Chu achieved a high specific capacitance of 124 F g^{-1} by controlling the polymerization kinetics of PEDOT growth by adding inhibitors [194].

18.3.2.3 Composite Materials

18.3.2.3.1 Carbon–Metal Oxide Composites

In recent years, composites containing metal oxides (such as RuO_2, MnO_2, V_2O_5, etc.) and carbon materials (such as AC, carbon nanofibers, CAGs, CNTs, graphene, etc.) have been intensively studied as EC materials [127,128,197–201]. Many extensive reviews are available on carbon–metal oxide composites [110,202,203]. Both materials have their own merits and shortcomings as EC electrodes. The electro-active metal oxides are known for high specific capacitances; however, the conductivity of most metal oxides except for RuO_2 is very low (Table 18.2). The high resistivity of metal oxides increases the charge transfer resistance and causes a large ESR. Thus, the rate capability and power density are poor. The strain developed in the metal oxide electrodes during the charge–discharge processes causes cracking, which leads to poor long-term cycling stability. On the other hand, carbon material–based electrodes have limited specific capacitance and, hence, lower energy density. However, the high electronic conductivity of carbon materials benefits the rate capability and power density. Thus, approaches to composite electrodes combining metal oxides with carbon offer an effective solution to overcome the shortcomings and combines the merits of both components. The high electrochemical activity of metal oxides contributes to high specific capacitance and high energy density, whereas the carbon materials serve as the physical support and faster pathway for charge transport. CNTs have been commonly used as a robust electrical percolating structure. SWCNTs have an electrical resistivity of 0.03 to 0.1 mΩ cm, and MWCNTs can reach an electrical resistivity of 5.1 μΩ cm, lower than the in-plane resistivity of graphite (0.04 mΩ cm) [204]. Graphene, a 2D aromatic monolayer of graphite, presents another attractive carbon material for energy storage. Recent studies have demonstrated that graphene combines a low electrical resistivity of 5.1 μΩ cm, a high tensile strength of 35–41.8 GPa, and a high surface area ~700–800 m^2 g^{-1} (2630 m^2 g^{-1} for a single layer of graphene) [205,206].

For RuO_2–carbon composites, the addition of carbon to Ru oxide has been demonstrated to improve the homogeneity of the electrochemical reaction, reduce the ionic resistance of the metal oxide, expand the active sites, increase electrical conductivity, and consequently increase the power and energy densities of the corresponding supercapacitors [197–199]. The combination of MnO_2 with graphene and CNTs has also been investigated extensively [137,207–212]. Graphene–MnO_2 composites (78 wt.% MnO_2), which were prepared by a redox reaction between graphene and potassium permanganate under microwave irradiation, displayed a specific capacitance as high as 310 F g^{-1} at 2 mV s^{-1}, which is almost three times higher than that of pure graphene (104 F g^{-1}) and birnessite-type MnO_2 (103 F g^{-1}). The improved electrochemical performance might be attributed to the increased electrode conductivity in the presence of graphene network, the increased effective interfacial area between MnO_2 and the electrolyte, and the contact area between MnO_2 and graphene [201]. Most of the reports on CNT composites used randomly stacked CNT networks. Vertically aligned CNTs are more attractive in reducing contact resistance, facilitating fast kinetics involving electron and ion transport, and providing open space for metal oxide deposition [213].

As mentioned above, the challenge for V_2O_5 is its low conductivity. Highly conductive materials such as carbon have been introduced into V_2O_5 to solve this problem [214–218]. For example, CNTs have been explored to form V_2O_5–CNT composites, which have a larger specific capacitance than pure V_2O_5. A $V_2O_5 \cdot xH_2O$/CNT film electrode also showed a high specific capacitance of 910 F g^{-1} at a potential scan rate of 10 mV s^{-1}, three times higher than that of a $V_2O_5 \cdot xH_2O$ thin-film electrode (300 F g^{-1}) [124].

18.3.2.3.2 Carbon–ECP Composites

ECP-based ECs are known to suffer from poor cycling stability (only a few thousand cycles) and fast capacitance decay in high-rate cycling processes, presumably because of significant volume changes during the capacitor operation and the accompanied decrease of their electrical conductivity. Electroreduction or electrooxidation of ECP generates charges, which are balanced by counterions entering or leaving the polymer film to maintain neutrality. The solvation shell of ions may be also involved. Therefore, electrochemical charging and discharging irreversibly changes the 3D structure of polymers owing to polymer swelling and shrinking [219,220]. The composite materials of carbon and ECPs show synergistic effects that combine the advantages of both materials. ECPs provide superior pseudocapacitance while carbon materials act as a framework that helps ECPs sustain the strains in charging–discharging cycling processes. Current research is focused on improving the interface between nanostructured carbons and conducting polymers, and controlling the chemical structure of conducting polymer for improved electrochemical activities. Several reviews are available in the literature for detailed information [220–222].

Various forms of carbon (e.g., ACs, carbon nanofibers, CNTs, and graphene) have been used to form composites with ECPs for EC applications. Lei et al. coated PANI on hollow carbon spheres [223]. The composite material (with 65 wt.% PANI loading) showed improved specific capacitance of 525 F g^{-1} as compared with the pristine hollow carbon sphere (268 F g^{-1}). Kovalenko et al. reported hybrid structures of PANI and detonation nanodiamond, which showed improved cycling stability along with the increased specific capacitance and rate capability [224]. Wang et al. prepared the CMK-3/PANI hybrid nanomaterials, with ordered PANI nanofibers grown on the surface of OMCs [225]. The composite material exhibited a high specific capacitance of 900 F g^{-1} at current density of 0.5 A g^{-1}. Liu et al. synthesized PANI/SWCNT hybrid films through an *in situ* electrochemical polymerization/degradation process, and they achieved a high specific capacitance of 706.7 F g^{-1} [226]. Freestanding and flexible graphene–PANI composite paper was prepared by an *in situ* anodic electropolymerization of PANI film on graphene paper by Wang et al. [227]. Graphene–PANI composite paper showed stable large electrochemical capacitance of 233 F g^{-1}. Feng et al. prepared graphene–PANI composite films by a novel one-step electrochemical codeposition synthesis of using graphite oxide and aniline as the starting materials [228]. The composite film showed a high specific capacitance of 640 F g^{-1} and good cycling stability of 90% capacitance retention after 1000 cycles. Davies et al. developed flexible, uniform graphene–PPy composite films using a pulsed electropolymerization technique for supercapacitor electrodes, which showed a high specific capacitance of 237 F g^{-1} [229]. Zhang et al. synthesized graphene–PPy composite via *in situ* polymerization of pyrrole monomer in the presence of graphene [230]. The specific capacitance of composite based on the three-electrode cell configuration is as high as 482 F g^{-1} at a current density of 0.5 A g^{-1}. Fang et al. developed a self-supported supercapacitor electrode, showing a high specific capacitance ~427 F g^{-1}, by homogeneously coating PPy on MWCNT membranes [231]. A few representative SEM images of various composite electrodes are shown in Figure 18.6.

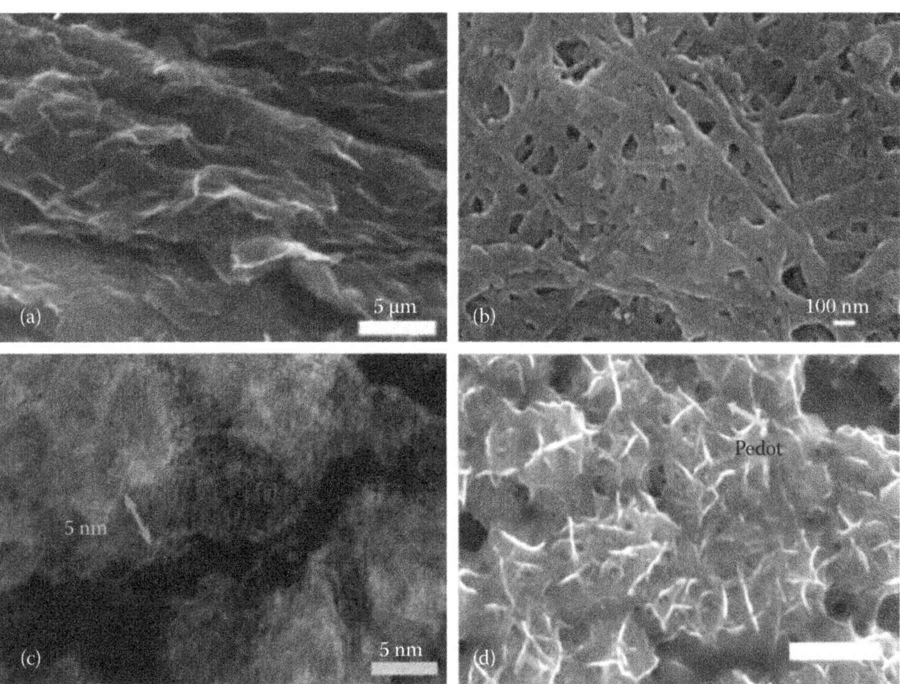

FIGURE 18.6 SEM images of (a) graphene–PANI composite paper (From Wang, D.-W. et al., *ACS Nano*, 3, 1745, 2009. With permission.); (b) SWNT–PANI composite (From Liu, J. et al., *J. Phys. Chem. C*, 114, 19614, 2010. With permission.); (c) graphene–MnO_2 composite (From Yan, J. et al., *Carbon*, 48, 3825, 2010. With permission.); and (d) graphene/MnO_2/PEDOT:PSS nanostructures (scale bar: 1 μm) (From Yu, G. et al., *Nano Lett.*, 11, 4438, 2011. With permission.).

18.3.2.3.3 MnO_2–ECP Composites

MnO_2–ECP composites may have the advantage of improving the poor electrical conductivity of MnO_2 and incorporating the high stability and mechanical flexibility of ECPs. Liu and Lee demonstrated a one-step method to synthesize MnO_2–ECP coaxial nanowires by electrodeposition in a porous alumina template [232]. Such material preserved 85% of its specific capacitance (from 210 to 185 F g^{-1}) as the current density was increased from 5 to 25 mA cm^{-2}.

18.3.2.3.4 Ternary (Carbon–Metal Oxide–ECP) Composites

In addition to using binary composites to overcome the problems associated with single-component electrodes made of metal–oxides or ECPs, ternary hybrid structures consisting of pseudocapacitive metal oxides and conducting polymers as well as carbon have been explored recently as a new design that could combine the advantages of all components. Hou and coworkers designed a ternary hybrid material composed of MnO_2, CNTs, and ECP [233]. This ternary composite electrode showed that, by 3D conductive wrapping of graphene–MnO_2 nanostructures with CNTs or ECPs, the specific capacitance of the electrodes reached as high as ~380 F g^{-1} [210]. MWCNT–PSS/PEDOT/MnO_2 nanocomposite electrodes were fabricated by Sharma and Zhai [234] with the specific capacitance as high as 375 F g^{-1}, significantly higher than the binary composite of MWCNT–PSS/MnO_2. A ternary composite of CNT–PPy–hydrous MnO_2, prepared by an *in situ* chemical method, was reported by Sivakkumar et al. [235]. The specific capacitance values of the ternary composite, CNT–MnO_2 composite, and PPy–MnO_2 composite were found to be 281, 150, and 32 F g^{-1}, respectively.

18.4 ASYMMETRIC SUPERCAPACITORS (HYBRID SYSTEMS)

As shown in Figure 18.1, a complete cell normally consists of two identical EDLCs or pseudocapacitor electrodes with equal capacitance, $C_1 = C_2 = C$, forming the so-called symmetric supercapacitors. The total cell capacitance C_{tot} of this serial circuit is

$$C_{tot} = C/2 \tag{18.11}$$

The working voltages at the two electrode are equal with $V_1 = V_2 = V$, which are limited by the irreversible electrooxidation or electroreduction of solvent and electrode materials, defined as V^0. The total cell voltage is defined as

$$V_{tot} = V_1 + V_2 = 2V \tag{18.12}$$

As a result, the total energy stored in a symmetric supercapacitor cell is

$$E = (1/2)C_{tot}V_{tot}^2 = CV^2 \tag{18.13}$$

which is simply the sum of the energy stored in each electrode (with $E_1 = E_2 = CV^2/2$). The energy storage capability is thus limited by the potential window (with V = ~1 V in aqueous solution). One approach to expand the cell voltage is using asymmetric supercapacitors (or hybrid systems) that combine a battery-like electrode (serving as energy source) with a capacitor-like electrode (power source) [236]. In most cases, it employs an EDLC negative electrode and a positive electrode based on a pseudocapacitive mechanism (such as transition metal oxides or ECPs) or lithium insertion reactions; however, some systems did use the EDLC materials as positive electrodes. A comprehensive review article by Cericolar and Kotz can be found in the literature [237].

Designing an asymmetric supercapacitor cell involves fairly more complicated factors besides the potential windows. In both symmetric and asymmetric supercapacitor cells, the amount of charge Q stored in the positive and negative electrodes has to be equal [238], i.e.

$$Q = C_pV_p = C_nV_n \tag{18.14}$$

where C_p and C_n are the capacitance, and V_p and V_n are the working voltage of the positive and negative electrodes, respectively. Thus, the working voltage at each electrode is different, depending on the ratio of their capacitance

$$V_n = C_pV_p/C_n \tag{18.15}$$

The total cell capacitance, C_{tot}, and total cell voltage can be derived as

$$C_{tot} = \frac{C_pC_n}{C_p + C_n} = \frac{C_p}{1 + C_p/C_n} \tag{18.16}$$

and

$$V_{tot} = V_p + V_n = V_p(1 + C_p/C_n) \tag{18.17}$$

Thus, the working voltage on one electrode may be larger than the other. However, the applicable electrode potential at each electrode cannot exceed the upper and lower limits of its fundamental "exploitable potential range," V_p^o or V_n^o, i.e., it has to satisfy both $V_p \le V_p^o$ and $V_n \le V_n^o$. Owing to the asymmetric nature, the voltage V at the electrode with low capacitance may be augmented (see Equation 18.15). As a result, the electrode that first reaches its voltage limit will determine the overall applicable cell voltage V_{tot}. Selecting a proper electrode pair and further tuning the value of C_p/C_n are both required to optimize the system. Peng et al. demonstrated an asymmetric supercapacitor with an EDLC electrode made of 0.3 mg Cabot Monarch 1300 pigment black (CMPB) and a pseudocapacitive positive electrode made of PANI electrodeposited onto CNTs [238]. The CVs in Figure 18.7a show that the CMPB negative electrode present 34.2 mF capacitance (114 F g^{-1}) in the range from −0.65 to 0.72 V versus Ag/AgCl, i.e., $V_n^o = 1.37$ V. On the other hand, the PAN–CNT positive electrode showed similar

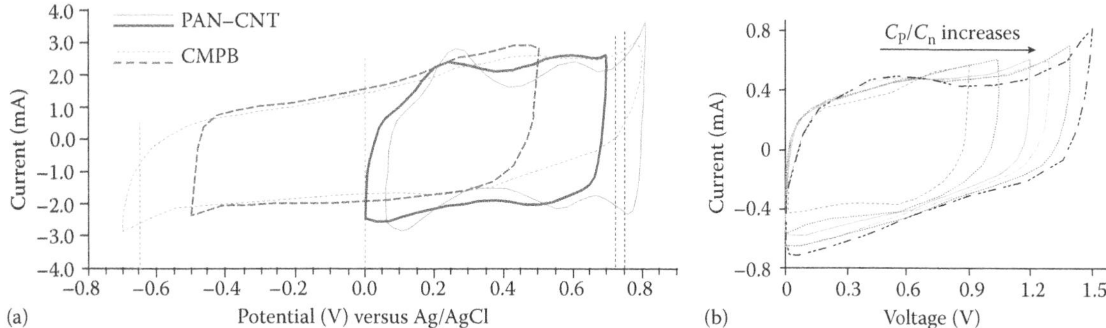

FIGURE 18.7 (a) CVs of a CMPB negative electrode (dashed lines) and a PCN–CNT positive electrode versus Ag/AgCl to demonstrate their potential limits. The scan rate was 50 mV/s. (b) CVs of the asymmetric supercapacitor cell consisting of the CMPB negative electrode and the PCN–CNT positive electrode as the C_p/C_n ratio was increased from 0.8 to 1.5. The CVs were measured within the largest potential window without inducing irreversible redox reactions. (From Peng, C. et al., *Energy Environ. Sci.*, 3, 1499, 2010. With permission.)

capacitance in a much narrower potential range from 0 to 0.75 V versus Ag/AgCl, i.e., $V_p^o = 0.75$ V. Clearly, the positive electrode is the limiting electrode. The two potential ranges overlap between 0 to 0.72 V, making it possible to construct an asymmetric suprercapacitor with $V_{tot} = 1.4$ V (from −0.65 to 0.75 V). They have elegantly controlled the amount of PAN deposition and demonstrated that the C_p/C_n value is indeed important in determining the cell voltage (see Figure 18.7b). The cell with $C_p/C_n = 1.3$ gave the highest performance with the specific energy increased by 72.6% and the specific power by 29.3% compared to the condition with $C_p/C_n = 1$.

It is noteworthy that MnO_2 only has limited interests as electrodes for symmetric supercapacitors owing to the low specific capacitance (150 F g^{-1}) and low working voltage (<1 V) in neutral aqueous electrolytes. However, it has a high overpotential for water oxidation and thus is a good pseudocapacitive positive electrode for high-voltage asymmetric cells. The combination of a MnO_2 positive electrode with an AC negative electrode has provided a 2-V working voltage in aqueous medium, and a PEDOT negative electrode gave a 1.8-V working voltage [239,240]. These asymmetric supercapacitors can produce a power density of ~120 kW kg^{-1} and an energy density of 21.0 and 13.5 Wh kg^{-1}, respectively, in aqueous electrolytes, comparable or higher than the symmetric counterparts in organic electrolytes. A recent study using an MnO_2/SWCNT hybrid film as the positive electrode and In_2O_3/SWCNT hybrid film as the negative electrode has obtained stable charging–discharging cycles in a 2-V potential window [241]. A power density of 50.3 kW kg^{-1} and over ~90% columbic efficiency were obtained.

The working voltage of an asymmetric supercapacitor, even after optimizing the C_p/C_n ratios, frequently cannot achieve the full voltage allowed by the electrolyte, due to the misalignment of the potential window of the two electrodes. One of the electrode may reach its upper or lower potential limit, while the other one has not been fully utilized. Ideally, the system will provide optimum performance if the potential window of each electrode can be shifted to match the system. The midpoint of the potential window can be represented by the open circuit potential E_{ov} before the cell is charged or discharged. However, E_{ov} is defined by the nature of the electrode materials and is not easily adjusted. Recently, Weng et al. demonstrated an elegant approach to tune the E_{ov} of individual electrodes by electrochemical charge injection [242]. Two identical graphene electrodes were separately constructed into two half-cells against an Li electrode in a nonaqueous electrolyte consisting of 1-M $LiPF_6$ in a mixture of ethylene carbonate and dimethyl carbonate (1:1 vol). Both of these two half-cells can only achieve ~3-V working voltage, although the electrolyte allows a 4.5-V operation voltage (from 0 to 4.5 versus Li/Li^+). This was because the E_{ov} of the graphene electrode was at ~3 V, largely offset to the positive end of the potential window, which limited the positive charging to only 1.5 V before reaching the upper limit. To solve this problem, the graphene electrode to be later used as the positive electrode was discharged from 3.0 to 1.16 V, while the one to be used as the negative electrode was discharged to 0.0 V and then charged back to 1.16 V. These two half-cells were then taken apart, and the two graphene electrodes were reassembled into a new full cell in the identical electrolyte. Since the electrode potentials are the same, i.e., $E_{ov} = 0$ V versus Li/Li^+, the cell voltage is 0 V. However, they carried different amounts of electrochemical charge and further charge–discharge capacities. The combination of the initial E_{ov} and the proper C_p/C_n ratio made it possible for the full cell to reach a 4.5 V working voltage, i.e., the positive electrode potential increased from 1.16 to 4.5 V versus Li/Li^+ and the negative electrode decreased from 1.16 to 0 V when the cell is fully charged. This method, in principle, can be applied to any electrodes if the charging can be stably controlled.

18.5 ELECTROLYTES FOR SUPERCAPACITORS

The performance of an EC not only depends on the electrode materials, but is also strongly affected by the properties of the electrolyte. Both power and energy density of ECs are proportional to the square of the cell operating voltage, which is limited by the electrochemical stability of the electrolyte. The current trend in EC development involves switching from aqueous electrolytes to nonaqueous media. The advantage of aqueous electrolytes such as acids (H_2SO_4) and bases (KOH) is the higher conductivity (up to ~1 S cm^{-1})

than other electrolytes. However, the stable voltage range is only ~1 V. The most recent commercial devices use organic electrolytes, e.g., tetraethylammonium tetrafluoroborate in acetonitrile and $LiBF_4$ in propylene carbonate, so that the operating voltage reaches ~2.7–2.8 V. However, there are many other issues involved in the use of organic electrolytes, such as high cost, low conductivity (leading to power deterioration), low dielectric constant (resulting in smaller capacitance), complex purification procedure, and safety concerns (high flammability and high toxicity of the organic solvent such as acetonitrile).

To achieve a high-power EC system, the internal electrolyte resistance and the structural resistance of the porous carbon electrode material should be minimized [1]. This can be achieved by an electrochemically compatible electrolyte salt or an acid or alkali that is strongly soluble in the solvent to be used. Minimum ion pairing and maximum free mobility of dissociated ions should be achieved in dissolved state. The dissociation of salt molecules in nonaqueous (organic) electrolyte solutions is significantly lower than aqueous solutions (nearly 100%). This leads to higher internal resistance (or ESR) with organic electrolytes.

18.5.1 Aqueous and Volatile Organic Liquid Electrolytes

The most studied aqueous electrolytes for supercapacitors are sulfuric acid (H_2SO_4) and KOH. Highly concentrated solutions are used to overcome the ESR factor and to increase the power capability. However, the acid solutions are highly corrosive in nature as compared with concentrated KOH, especially for current collectors. Most of the fundamental studies in KOH and H_2SO_4 have been performed using gold current collectors, and the operating voltage window in these media is less than 1 V [243]. Recently, Beguin and coworkers have shown that it is possible to enhance the operating voltage of carbon-based supercapacitors in aqueous H_2SO_4 up to 1.6 V, by different optimized carbons as positive and negative electrodes and/or by balancing the mass of the electrodes [244].

Recently, research on neutral electrolytes has started to overcome the limitations in both acidic and alkaline electrolytes [108]. Beguin and coworkers demonstrated that AC electrodes show a stability window of 2 V in Na_2SO_4 aqueous electrolyte, and a symmetric carbon/carbon cell can operate up to 1.6 V with good charge–discharge cycle life [245,246]. The migration of hydrated alkali ions in the bulk electrolyte and within the inner pores of AC increases in the order of $Li^+ < Na^+ < K^+$, and the suitability of electrolytes for capacitors should vary as $Li_2SO_4 < Na_2SO_4 < K_2SO_4$ [247]. However, the highest operating voltage is displayed in Li_2SO_4. It has been suggested that the stronger hydration of Li^+ compared with Na^+ and K^+ ions is responsible for the larger voltage in Li_2SO_4 solution [248]. The use of neutral aqueous electrolytes, like lithium sulfate, not only eliminates the disadvantages related to corrosion, but also gives an opportunity to realize high-voltage and high-energy-density supercapacitors with environmentally friendly, cost-effective, and safe materials [108]. However, further extensive research is required using these media to optimize the performance of supercapacitors.

Organic electrolytes are preferred for supercapacitors to obtain higher operating voltages, to realize higher energy densities compared with aqueous electrolytes. Tetraalkylammonium (R_4N^+) salts have been used for many years because of their good solubility in nonaqueous solvents (such as acetonitrile) and moderately good conductivity (~5×10^{-2} S cm^{-1}). However, there are some disadvantages, such as being expensive and sensitive to moisture contents and subsequent recombination reactions leading to self-discharge. An appropriate candidate in this category appears to be 1 mol·L^{-1} Et_4NBF_4 in propylene carbonate (PC) or acetonitrile. A comparison between aqueous and organic electrolytes shows that the latter allows high voltages desirable for high energy and power density, while the former allows high capacitance and lower resistance (R_s) values [249].

A few groups have investigated the use of low-viscosity linear carbonates and PC mixtures for the realization of EDLCs with an operative voltage as high as 3 V that have high performance and high cycling stability [250,251]. Recently, an operating voltage of 3.75 V has been reported for EDLCs using 0.7 mol·L^{-1} Et_4NBF_4 in adiponitrile at room temperature [252,253]. However, the conductivity of such a medium is much lower than that of conventional solutions in acetonitrile. Hence, owing to the higher ohmic loss, the real voltage window between the electrodes is not as high as claimed.

18.5.2 Ionic Liquids Electrolytes

Room-temperature ionic liquids (ILs) are under considerable research for use as next-generation electrolytes in energy storage [254–256]. They are attractive candidates for electrolytes because of their good chemical and physical properties, such as high thermal stability, high electrochemical stability window (>3 V), nontoxicity, nonflammability, a variety of combination choices of cations and anions, acceptable conductivity at elevated temperature (~10 mS cm^{-1}), etc. EDLCs based on ILs seem to be able to operate at voltages as high as 3.5–3.7 V with high cycling stability [257–260]. However, because of the relatively high viscosity and low conductivity of ILs, the ESR value at room temperature of IL-based EDLCs is considerably higher than that of conventional electrolytes. Therefore, designing an IL having a wider potential window (>4 V) together with high conductivity and low viscosity for practical applications is still challenging.

The main ILs studied for supercapacitor applications are imidazolium, pyrrolidinium, and asymmetric, aliphatic quaternary ammonium salts with anions such as tetrafluoroborate, trifluoromethanesulfonate, bis(trifluoromethanesulfonyl) imide, bis(fluorosulfonyl)imide, or hexafluorophosphate [261–263]. Room-temperature ILs are usually quaternary ammonium salts such as tetralkylammonium $[R_4N]^+$, and cyclic amines such as aromatic pyridinium, imidazolium, and saturated piperidinium, pyrrolidinium. Low-temperature molten salts based on sulfonium $[R_3S]^+$ as well as phosphonium $[R_4P]^+$ cations are also explored in the literature [255,264].

18.5.3 Solid Electrolyte

In recent years, considerable attention has been devoted to the development of solid electrolyte–based supercapacitors. Replacement of liquid electrolytes to solid electrolytes in the supercapacitors offers prevention of electrolyte leakage, flexible and thin configuration of devices, volumetric stability during the cell operation, and easy packing and handling [265–267]. Gel polymer electrolytes have been widely studied as solid electrolytes because they have preferable features as solid electrolytes for supercapacitors [184,268,269]. Gel polymer electrolytes are formed by mixing the polymer host, solvent, and salt, where the polymer host acts simply as a stiffener for the low molecular weight solvent working as a medium for the mobility of cationic and anionic species [270,271]. Ionic conductivities of gel polymer electrolytes utilizing aqueous solvents, organic solvents, and ILs for ECs range between 10 and 100 mS cm^{-1}, 1 and 10 mS cm^{-1}, and 1 and 10 mS cm^{-1}, respectively [265–268]. These values are lower than those of liquid electrolytes by an order of magnitude.

Although interesting results have been found in aqueous medium with alkaline PVA electrolytes (such as PVA–H_3PO_4, PVA–H_2SO_4, PEO–KOH–H_2O, etc.) [272–274], organic medium–based all-solid-state ECs show lower power densities. The ESR of the ECs with gel polymer electrolyte was about four times higher than that with the liquid electrolyte (5 versus 20 Ω cm) [267,268,275]. However, the ESR of the gel electrolyte–based ECs is comparable to those using IL-based electrolytes. The use of gel electrolytes for the realization of safe and high-performance devices appears an interesting strategy. The high operative voltage is also beneficial in enhancing the energy density.

18.6 CURRENT COLLECTORS AND EFFECT ON PERFORMANCE

Most of the pseudocapacitive transition-metal oxides have poor electrical conductivity (see Table 18.2). In addition, the surface redox reactions only take place within a very thin layer of the active materials. Hence, great effort has been devoted to the fabrication of nanostructured pseudocapacitive materials to enhance the SSA and shorten the electron conduction length [222]. Another approach is loading pseudocapacitive materials onto conductive nanoarchitectured current collectors with large SSA for an efficient charge-transport process. Various substrates such as carbon foam, porous carbon paper, nickel foam, AC fabric, and nanoarchitectures such as vertically aligned carbon nanofibers [213] and vertically aligned CNTs (VACNTs) [276] are utilized as templates to support a thin layer of conducting pseudocapacitive materials. Some of representative templates with EC materials coating are shown in Figure 18.8.

Carbon paper is an attractive supporting substrate for active materials because of its high porosity, large surface area, and good conductivity. Yang et al. reported hybrid electrodes prepared from carbon paper loaded with Co_3O_4 nanomaterials [277]. The Co_3O_4 deposited on the graphite fiber surface,

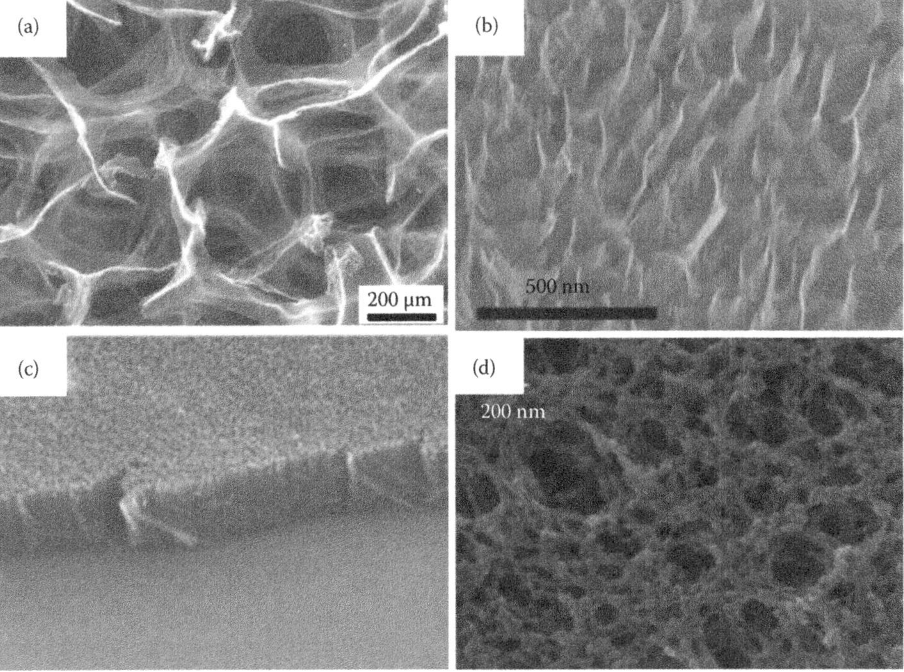

FIGURE 18.8 SEM images of (a) 3D graphene networks after removal of Ni foam (From Cao, X. et al., *Small*, 7, 3163, 2011. With permission.); (b) graphene bridges standing vertically on the nickel–foam surface (From Bo, Z. et al., *Adv. Mater.*, 25, 5799, 2013. With permission.); (c) potentiostatic-deposited MnO_2 on CNTs (From Amade, R. et al., *J. Power Sources*, 196, 5779, 2011. With permission.); (d) Toray paper–supported carbon nanofoam with macropores sized at 50–300 nm (From Lytle, J.C. et al., *Energy Environ. Sci.*, 4, 1913, 2011. With permission.).

forming a secondary network consisting of nanowires with diameters of a few tens of nanometers. The electrode showed high specific capacitance (1124 F g^{-1}) with good rate capability due to the effective hierarchical structure design, which facilitated electron/ion transport and the redox reaction. Fisher et al. utilized a carbon nanofoam substrate for MnO_2 loading through a self-limiting electroless deposition method and controlled permanganate self-decomposition under a neutral pH condition [278]. The resulting MnO_2–carbon nanofoam material maintained a highly porous structure with pore sizes in the range of 10–60 nm and showed a specific capacitance of 110 F g^{-1} (area capacitance of 1.5 F cm^{-2}). These values were further enhanced to 150 F g^{-1} (7.5 F cm^{-2}) by optimizing the pore sizes and increasing the sample thickness [279].

Cao et al. reported the preparation of novel 3D graphene networks by using Ni foam as a sacrificial template [280]. They electrochemically deposited nickel oxide (NiO) on the 3D graphene networks. The obtained NiO–graphene composite exhibited a high specific capacitance of ~816 F g^{-1} at a scan rate of 5 mV s^{-1} and a stable cycling performance. Bo and coworkers reported the growth of vertically oriented graphene bridges on a nickel-foam current collector and demonstrated the reduced constriction/spreading resistance caused by the limited contact points at the active layer–current collector interface [281]. The electrode showed high power capability. Liu et al. reported the fabrication of hybrid supercapacitor electrodes by coaxially coating of manganese oxide thin films on a vertically aligned carbon nanofiber array [213]. A maximum specific capacitance of 365 F g^{-1} has been achieved with chronopotentiometry in 0.1 M Na_2SO_4 aqueous solution with ~7.5-nm-thick manganese oxide at a scan rate of up to 2 V s^{-1}. VACNTs were also used to deposit MnO_2 [276]. A specific capacitance of 642 F g^{-1} was obtained for MnO_2–VACNTs nanocomposite electrode at a scan rate of 10 mV s^{-1}.

18.7 SUMMARY AND FUTURE PERSPECTIVES

Tremendous progress has been made in the past decades to improve the performance of EES devices driven by the new developments in materials synthesis, assembly, and characterization techniques. It is foreseen that rechargeable EES devices and systems with higher energy and higher power density as well as longer lifetime will make great impact, covering a wide range of applications spanning from portable electronics, electrical vehicles, to grid-scale energy storage. For electrochemical supercapacitors, research in the following areas are expected to continue driving this field:

1. Development of new stable, low-cost, high-SSA active electrode materials with suitable electropotential windows and redox properties for both EDLCs and pseudocapacitors
2. Development of highly conductive current collectors with high SSA and novel architectures, which can optimize electron and ion transport and provide reliable mechanical support to the active electrode materials from the macro-, micro-, to nanoscales
3. Innovation with electrolytes that have large stable potential windows, high ionic conductivities, and wide operating temperatures
4. Exploration of various hybrid systems by combining electrode materials with suitable EDLC, pseudocapacitor, and battery properties to achieve a wider working voltage and higher energy density without sacrificing the system's power density
5. Fundamental understanding of the new phenomena involving desolvation of ions in ultrasmall pores (<1 nm)

ACKNOWLEDGMENTS

We are grateful for the financial support by NSF grant CMMI-1100830, NASA grant NNX13AD42A, NSF EPSCoR Award EPS-0903806, and matching funds to these grants by the state of Kansas during the preparation of the manuscript. Yueping Fang would like to acknowledge the financial support by NSF of China grant (21173088) and China Scholarship Council (CRC) for scholarship to visit the United States.

REFERENCES

1. Conway, B. E. 1999. *Electrochemical Supercapacitors: Scientific Fundamentals and Technological Applications.* New York: Kluwer Academic/Plenum.
2. Lu, M., F. Beguin and E. Frackowiak (Eds.). 2013. *Supercapacitors: Materials, Systems and Applications.* Weinheim, Germany: John Wiley & Sons.
3. Inagakia, M., H. Konno and O. Tanaike. 2010. Carbon materials for electrochemical capacitors. *Journal of Power Sources,* 195, 7880–7903.
4. Qu, D. Y. and H. Shi. 1998. Studies of activated carbons used in double-layer capacitors. *Journal of Power Sources,* 74, 99–107.
5. Frackowiak, E. and F. Beguin, F. 2002. Electrochemical storage of energy in carbon nanotubes and nanostructured carbons. *Carbon,* 40, 1775–1787.
6. Shiraishi, S., H. Kurihara, K. Okabe, D. Hulicova and A. Oya. 2002. Electric double layer capacitance of highly pure single-walled carbon nanotubes (HiPco™ Buckytubes™) in propylene carbonate electrolytes. *Electrochemistry Communications,* 4, 593–598.
7. Lee, Y. J., J. C. Jung, J. Yi, S. H. Baeck, J. R. Yoon and I. K. Song. 2010. Preparation of carbon aerogel in ambient conditions for electrical double-layer capacitor. *Current Applied Physics,* 10, 682–686.
8. Morishita, T., Y. Soneda, T. Tsumura and M. Inagaki. 2006. Preparation of porous carbons from thermoplastic precursors and their performance for electric double layer capacitors. *Carbon,* 44, 2360–2367.
9. Lee, J., J. Kim and T. Hyeon. 2006. Recent progress in the synthesis of porous carbon materials. *Advanced Materials,* 18, 2073–2094.
10. Xu, B., F. Wu, R. Chen, G. Cao, S. Chen, Z. Zhou and Y. Yang. 2008. Highly mesoporous and high surface area carbon: A high capacitance electrode material for EDLCs with various electrolytes. *Electrochemistry Communications,* 10, 795–797.

11. Ra, E. J., E. Raymundo-Piñero, Y. H. Lee and F. Béguin. 2009. High power supercapacitors using polyacrylonitrile-based carbon nanofiber paper. *Carbon*, 47, 2984–2992.
12. Pandolfo, A. G. and A. F. Hollenkamp. 2006. Carbon properties and their role in supercapacitors. *Journal of Power Sources*, 157, 11–27.
13. Obreja, V. V. N. 2008. On the performance of supercapacitors with electrodes based on carbon nanotubes and carbon activated material—A review. *Physica E: Low-Dimensional Systems and Nanostructures*, 40, 2596–2605.
14. Noked, M., A. Soffer and D. Aurbach. 2011. The electrochemistry of activated carbonaceous materials: Past, present, and future. *Journal of Solid State Electrochemistry*, 15, 1563–1578.
15. Huang, Y., J. Liang and Y. Chen. 2012. An overview of the applications of graphene-based materials in supercapacitors. *Small*, 8, 1805–1834.
16. Simon, P. and Y. Gogotsi. 2012. Capacitive energy storage in nanostructured carbon–electrolyte systems. *Accounts of Chemical Research*, 46, 1094–1103.
17. Rodriguez-Reinoso, F. and M. Molina-Sabio. 1992. Activated carbons from lignocellulosic materials by chemical and/or physical activation: An overview. *Carbon*, 30, 1111–1118.
18. Navarro, M. V., N. A. Seaton, M. A. Mastral and R. Murillo. 2006. Analysis of the evolution of the pore size distribution and the pore network connectivity of a porous carbon during activation. *Carbon*, 44, 2281–2288.
19. Illán-Gómez, M. J., A. Garcia-Garcia, C. Salinas-Martinez de Lecea and A. Linares-Solano. 1996. Activated carbons from Spanish coals. 2. Chemical activation. *Energy & Fuels*, 10, 1108–1114.
20. Marsh, H. and F. R. Reinoso. 2006. *Activated Carbon*. London, UK: Elsevier.
21. Endo, M., T. Maeda, T. Takeda, Y. J. Kim, K. Koshiba, H. Hara and M. S. Dresselhaus. 2001. Capacitance and pore-size distribution in aqueous and nonaqueous electrolytes using various activated carbon electrodes. *Journal of the Electrochemical Society*, 148, A910–A914.
22. Raymundo-Pinero, E., K. Kierzek, J. Machnikowski and F. Béguin. 2006. Relationship between the nanoporous texture of activated carbons and their capacitance properties in different electrolytes. *Carbon*, 44, 2498–2507.
23. Barbieri, O., M. Hahn, A. Herzog and R. Kötz. 2005. Capacitance limits of high surface area activated carbons for double layer capacitors. *Carbon*, 43, 1303–1310.
24. Kierzek, K., E. Frackowiak, G. Lota, G. Gryglewicz and J. Machnikowski. 2004. Electrochemical capacitors based on highly porous carbons prepared by KOH activation. *Electrochimica Acta*, 49, 515–523.
25. Salitra, G., A. Soffer, L. Eliad, Y. Cohen and D. Aurbach. 2000. Carbon electrodes for double-layer capacitors I. Relations between ion and pore dimensions. *Journal of the Electrochemical Society*, 147, 2486–2493.
26. Raymundo-Piñero, E., F. Leroux and F. Béguin. 2006. A high-performance carbon for supercapacitors obtained by carbonization of a seaweed biopolymer. *Advanced Materials*, 18, 1877–1882.
27. Liu, P., M. Verbrugge and S. Soukiazian. 2006. Influence of temperature and electrolyte on the performance of activated-carbon supercapacitors. *Journal of Power Sources*, 156, 712–718.
28. Largeot, C., C. Portet, J. Chmiola, P.-L. Taberna, Y. Gogotsi and P. Simon. 2008. Relation between the ion size and pore size for an electric double-layer capacitor. *Journal of the American Chemical Society*, 130, 2730–2731.
29. Lozano-Castello, D., D. Cazorla-Amorós, A. Linares-Solano, S. Shiraishi, H. Kurihara and A. Oya. 2003. Influence of pore structure and surface chemistry on electric double layer capacitance in non-aqueous electrolyte. *Carbon*, 41, 1765–1775.
30. Niu, C., E. K. Sichel, R. Hoch, D. Moy and H. Tennent. 1997. High power electrochemical capacitors based on carbon nanotube electrodes. *Applied Physics Letters*, 70, 1480–1482.
31. Izadi-Najafabadi, A., S. Yasuda, K. Kobashi, T. Yamada, D. N. Futaba, H. Hatori, M. Yumura, S. Iijima and K. Hata. 2010. Extracting the full potential of single-walled carbon nanotubes as durable supercapacitor electrodes operable at 4 V with high power and energy density. *Advanced Materials*, 22, E235–E241.
32. Frackowiak, E., K. Metenier, V. Bertagna and F. Beguin. 2000. Supercapacitor electrodes from multiwalled carbon nanotubes. *Applied Physics Letters*, 77, 2421–2423.
33. Ania, C. O., V. Khomenko, E. Raymundo-Piñero, J. B. Parra and F. Beguin. 2007. The large electrochemical capacitance of microporous doped carbon obtained by using a zeolite template. *Advanced Functional Materials*, 17, 1828–1836.
34. Yamada, H., H. Nakamura, F. Nakahara, I. Moriguchi and T. Kudo. 2007. Electrochemical study of high electrochemical double layer capacitance of ordered porous carbons with both meso/macropores and micropores. *The Journal of Physical Chemistry C*, 111, 227–233.
35. Chmiola, J., G. Yushin, R. K. Dash, E. N. Hoffman, J. E. Fischer, M. W., Barsoum and Y. Gogotsi. 2005. Double-layer capacitance of carbide derived carbons in sulfuric acid. *Electrochemical and Solid-State Letters*, 8, A357–A360.
36. Dash, R., J. Chmiola, G. Yushin, Y. Gogotsi, G. Laudisio, J. Singer, J. Fischer and S. Kucheyev. 2006. Titanium carbide derived nanoporous carbon for energy-related applications. *Carbon*, 44, 2489–2497.
37. Kurig, H., A. Jänes and E. Lust. 2010. Electrochemical characteristics of carbide-derived carbon 1-ethyl-3-methylimidazolium tetrafluoroborate supercapacitor cells. *Journal of the Electrochemical Society*, 157, A272–A279.
38. Stoller, M. D., S. Park, Y. Zhu, J. An and R. S. Ruoff. 2008. Graphene-based ultracapacitors. *Nano Letters*, 8, 3498–3502.
39. Wang, Y., Z. Shi, Y. Huang, Y. Ma, C. Wang, M. Chen and Y. Chen. 2009. Supercapacitor devices based on graphene materials. *The Journal of Physical Chemistry C*, 113, 13103–13107.
40. Zhu, Y., S. Murali, M. D. Stoller, K. J. Ganesh, W. Cai, P. J. Ferreira, A. Pirkle, R. M. Wallace, K. A. Cychosz and M. Thommes. 2011. Carbon-based supercapacitors produced by activation of graphene. *Science*, 332, 1537–1541.
41. Bushueva, E. G., P. S. Galkin, A. V. Okotrub, L. G. Bulusheva, N. N. Gavrilov, V. L. Kuznetsov and S. I. Moiseekov. 2008. Double layer supercapacitor properties of onion-like carbon materials. *Physica Status Solidi (B)*, 245, 2296–2299.
42. Plonska-Brzezinska, M. E., A. Palkar, K. Winkler and L. Echegoyen. 2010. Electrochemical properties of small carbon nano-onion films. *Electrochemical and Solid-State Letters*, 13, K35–K38.
43. Kim, C., Y.-O. Choi, W.-J. Lee and K.-S. Yang. 2004. Supercapacitor performances of activated carbon fiber webs prepared by electrospinning of PMDA–ODA poly (amic acid) solutions. *Electrochimica Acta*, 50, 883–887.
44. Leitner, K., A. Lerf, M. Winter, J. O. Besenhard, S. Villar-Rodil, F. Suarez-Garcia, A. Martinez-Alonso and J. M. D. Tascon. 2006. Nomex-derived activated carbon fibers as electrode materials in carbon based supercapacitors. *Journal of Power Sources*, 153, 419–423.

45. Fang, B. and L. Binder. 2006. A modified activated carbon aerogel for high-energy storage in electric double layer capacitors. *Journal of Power Sources*, 163, 616–622.
46. Li, J., X. Wang, Q. Huang, S. Gamboa and P. J. Sebastian. 2006. Studies on preparation and performances of carbon aerogel electrodes for the application of supercapacitor. *Journal of Power Sources*, 158, 784–788.
47. Liu, C. G., M. Liu, F. Li and H. M. Cheng. Frequency response characteristic of single-walled carbon nanotubes as supercapacitor electrode material. *Applied Physics Letters*, 92, 143108-143101–143108-143103.
48. An, K. H., K. K. Jeon, J. K. Heo, S. C. Lim, D. J. Bae and Y. H. Lee. 2002. High-capacitance supercapacitor using a nanocomposite electrode of single-walled carbon nanotube and polypyrrole. *Journal of the Electrochemical Society*, 149, A1058–A1062.
49. Frackowiak, E., S. Delpeux, K. Jurewicz, K. Szostak, D. Cazorla-Amoros and F. Beguin. 2002. Enhanced capacitance of carbon nanotubes through chemical activation. *Chemical Physics Letters*, 361, 35–41.
50. Simon, P. and A. F. Burke. 2008. Nanostructured carbons: Double-layer capacitance and more. *The Electrochemical Society Interface*, 17, 38–43.
51. Futaba, D. N., K. Hata, T. Yamada, T. Hiraoka, Y. Hayamizu, Y. Kakudate, O. Tanaike, H. Hatori, M. Yumura and S. Iijima. 2006. Shape-engineerable and highly densely packed single-walled carbon nanotubes and their application as supercapacitor electrodes. *Nature Materials*, 5, 987–994.
52. Lu, W., L. Qu, K. Henry and L. Dai. 2009. High performance electrochemical capacitors from aligned carbon nanotube electrodes and ionic liquid electrolytes. *Journal of Power Sources*, 189, 1270–1277.
53. Kim, B., H. Chung and W. Kim. 2012. High-performance supercapacitors based on vertically aligned carbon nanotubes and nonaqueous electrolytes. *Nanotechnology*, 23, 155401.
54. Talapatra, S., S. Kar, S. K. Pal, R. Vajtai, L. Ci, P. Victor, M. M. Shaijumon, S. Kaur, O. Nalamasu and P. M. Ajayan. 2006. Direct growth of aligned carbon nanotubes on bulk metals. *Nature Nanotechnology*, 1, 112–116.
55. Pushparaj, V. L., M. M. Shaijumon, A. Kumar, S. Murugesan, L. Ci, R. Vajtai, R. J. Linhardt, O. Nalamasu and P. M. Ajayan. 2007. Flexible energy storage devices based on nanocomposite paper. *Proceedings of the National Academy of Sciences*, 104, 13574–13577.
56. Zhang, L. L. and X. S. Zhao. 2009. Carbon-based materials as supercapacitor electrodes. *Chemical Society Reviews*, 38, 2520–2531.
57. Ryoo, R., S. H. Joo and S. Jun. 1999. Synthesis of highly ordered carbon molecular sieves via template-mediated structural transformation. *The Journal of Physical Chemistry B*, 103, 7743–7746.
58. Ryoo, R., S. H. Joo, M. Kruk and M. Jaroniec. 2001. Ordered mesoporous carbons. *Advanced Materials*, 13, 677–681.
59. Zhao, X. S., F. Su, Q. Yan, W. Guo, X. Y. Bao, L. Lv and Z. Zhou. 2006. Templating methods for preparation of porous structures. *Journal of Materials Chemistry*, 16, 637–648.
60. Lee, J., S. Yoon, T. Hyeon, S. M. Oh and K. B. Kim. 1999. Synthesis of a new mesoporous carbon and its application to electrochemical double-layer capacitors. *Chemical Communications*, 2177–2178.
61. Xing, W., S. Z. Qiao, R. G. Ding, F. Li, G. Q. Lu, Z. F. Yan and H. M. Cheng. 2006. Superior electric double layer capacitors using ordered mesoporous carbons. *Carbon*, 44, 216–224.
62. Gogotsi, Y., A. Nikitin, H. Ye, W. Zhou, J. E. Fischer, B. Yi, H. C. Foley and M. W. Barsoum. 2003. Nanoporous carbide-derived carbon with tunable pore size. *Nature Materials*, 2, 591–594.
63. Gogotsi, Y. R. K. Dash, G. Yushin, T. Yildirim, G. Laudisio and J. E. Fischer. 2005. Tailoring of nanoscale porosity in carbide-derived carbons for hydrogen storage. *Journal of the American Chemical Society*, 127, 16006–16007.
64. Dash, R. K., G. Yushin and Y. Gogotsi. 2005. Synthesis, structure and porosity analysis of microporous and mesoporous carbon derived from zirconium carbide. *Microporous and Mesoporous Materials*, 86, 50–57.
65. Chmiola, J., G. Yushin, G., R. Dash and Y. Gogotsi. 2006. Effect of pore size and surface area of carbide derived carbons on specific capacitance. *Journal of Power Sources*, 158, 765–772.
66. Jänes, A., T. Thomberg and E. Lust. 2007. Synthesis and characterisation of nanoporous carbide-derived carbon by chlorination of vanadium carbide. *Carbon*, 45, 2717–2722.
67. Jänes, A., L. Permann, M. Arulepp and E. Lust. 2004. Electrochemical characteristics of nanoporous carbide-derived carbon materials in non-aqueous electrolyte solutions. *Electrochemistry Communications*, 6, 313–318.
68. Chmiola, J., G. Yushin, Y. Gogotsi, C. Portet, P. Simon and P.-L. Taberna. 2006. Anomalous increase in carbon capacitance at pore sizes less than 1 nanometer. *Science*, 313, 1760–1763.
69. Lin, R., P.-L. Taberna, J. Chmiola, D. Guay, Y. Gogotsi and P. Simon. 2009. Microelectrode study of pore size, ion size, and solvent effects on the charge/discharge behavior of microporous carbons for electrical double-layer capacitors. *Journal of the Electrochemical Society*, 156, A7–A12.
70. Arulepp, M., L. Permann, J. Leis, A. Perkson, K. Rumma, A. Jänes and E. Lust. 2004. Influence of the solvent properties on the characteristics of a double layer capacitor. *Journal of Power Sources*, 133, 320–328.
71. Lust, E., G. Nurk, A. Jänes, M. Arulepp, P. Nigu, P. Möller, S. Kallip and V. Sammelselg. 2003. Electrochemical properties of nanoporous carbon electrodes in various nonaqueous electrolytes. *Journal of Solid State Electrochemistry*, 7, 91–105.
72. Ania, C. O., J. Pernak, F. Stefaniak, E. Raymundo-Piñero and F. Béguin. 2009. Polarization-induced distortion of ions in the pores of carbon electrodes for electrochemical capacitors. *Carbon*, 47, 3158–3166.
73. Cachet-Vivier, C., V. Vivier, C. S. Cha, J. Y. Nedelec and L. T. Yu. 2001. Electrochemistry of powder material studied by means of the cavity microelectrode (CME). *Electrochimica Acta*, 47, 181–189.
74. Portet, C., J. Chmiola, Y. Gogotsi, S. Park and K. Lian. 2008. Electrochemical characterizations of carbon nanomaterials by the cavity microelectrode technique. *Electrochimica Acta*, 53, 7675–7680.
75. Mysyk, R., E. Raymundo-Pinero and F. Béguin. 2009. Saturation of subnanometer pores in an electric double-layer capacitor. *Electrochemistry Communications*, 11, 554–556.
76. Mysyk, R., E. Raymundo-Pinero, J. Pernak and F. Béguin. 2009. Confinement of symmetric tetraalkylammonium ions in nanoporous carbon electrodes of electric double-layer capacitors. *The Journal of Physical Chemistry C*, 113, 13443–13449.
77. Zhu, Y., S. Murali, W. Cai, X. Li, J. W. Suk, J. R. Potts and R. S. Ruoff. 2010. Graphene and graphene oxide: Synthesis, properties, and applications. *Advanced Materials*, 22, 3906–3924.

78. Liu, C., Z. Yu, D. Neff, A. Zhamu and B. Z. Jang. 2010. Graphene-based supercapacitor with an ultrahigh energy density. *Nano Letters*, 10, 4863–4868.
79. Yoo, J. J., K. Balakrishnan, J. Huang, V. Meunier, B. G. Sumpter, A. Srivastava, M. Conway, A. L. Mohana Reddy, J. Yu and R. Vajtai. 2011. Ultrathin planar graphene supercapacitors. *Nano Letters*, 11, 1423–1427.
80. Zhu, Y., S. Murali, M. D. Stoller, A. Velamakanni, R. D. Piner and R. S. Ruoff. 2010. Microwave assisted exfoliation and reduction of graphite oxide for ultracapacitors. *Carbon*, 48, 2118–2122.
81. El-Kady, M. F., V. Strong, S. Dubin and R. B. Kaner. 2012. Laser scribing of high-performance and flexible graphene-based electrochemical capacitors. *Science*, 335, 1326–1330.
82. Kim, T., G. Jung, S. Yoo, K. S. Suh and R. S. Ruoff. 2013. Activated graphene-based carbons as supercapacitor electrodes with macro-and mesopores. *ACS Nano*, 7, 6899–6905.
83. Wu, Z.-S., Y. Sun, Y.-Z. Tan, S. Yang, X. Feng and K. Mu'llen. 2012. Three-dimensional graphene-based macro- and mesoporous frameworks for high-performance electrochemical capacitive energy storage. *Journal of the American Chemical Society*, 134, 19532–19535.
84. Choi, B. G., M. Yang, W. H. Hong, J. W. Choi and Y. S. Huh. 2012. 3D macroporous graphene frameworks for supercapacitors with high energy and power densities. *ACS Nano*, 6, 4020–4028.
85. Ugarte, D. 1992. Curling and closure of graphitic networks under electron-beam irradiation. *Nature*, 359, 707–709.
86. Plonska-Brzezinska, M. E. and L. Echegoyen. 2013. Carbon nano-onions for supercapacitor electrodes: Recent developments and applications. *Journal of Materials Chemistry A*, 1, 13703–13714.
87. Nasibulin, A. G., A. Moisala, D. P. Brown and E. I. Kauppinen. 2003. Carbon nanotubes and onions from carbon monoxide using $Ni(acac)_2$ and $Cu(acac)_2$ as catalyst precursors. *Carbon*, 41, 2711–2724.
88. Palkar, A., F. Melin, C. M. Cardona, B. Elliott, A. K. Naskar, D. D. Edie, A. Kumbhar and L. Echegoyen. 2007. Reactivity Differences between Carbon Nano Onions (CNOs) Prepared by Different Methods. *Chemistry—An Asian Journal*, 2, 625–633.
89. Li, S., G. Feng, P. F. Fulvio, P. C. Hillesheim, C. Liao, S. Dai and P. T. Cummings. 2012. Molecular dynamics simulation study of the capacitive performance of a binary mixture of ionic liquids near an onion-like carbon electrode. *The Journal of Physical Chemistry Letters*, 3, 2465–2469.
90. Lin, R., P.-L. Taberna, S. Fantini, V. Presser, C. R. Pérez, F. Malbosc, N. L. Rupesinghe, K. B. K. Teo, Y. Gogotsi and P. Simon. 2011. Capacitive energy storage from −50 to 100 C using an ionic liquid electrolyte. *The Journal of Physical Chemistry Letters*, 2, 2396–2401.
91. Pech, D., M. Brunet, H. Durou, P. Huang, V. Mochalin, Y. Gogotsi, P.-L. Taberna and P. Simon. 2010. Ultrahigh-power micrometre-sized supercapacitors based on onion-like carbon. *Nature Nanotechnology*, 5, 651–654.
92. Inagaki, M. 2009. Pores in carbon materials-importance of their control. *New Carbon Materials*, 24, 193–232.
93. Kim, Y. J., Y. Abe, T. Yanagiura, K. C. Park, M. Shimizu, T. Iwazaki, S. Nakagawa, M. Endo and M. S. Dresselhaus. 2007. Easy preparation of nitrogen-enriched carbon materials from peptides of silk fibroins and their use to produce a high volumetric energy density in supercapacitors. *Carbon*, 45, 2116–2125.
94. Nakagawa, H., A. Shudo and K. Miura. 2000. High-capacity electric double-layer capacitor with high-density-activated carbon fiber electrodes. *Journal of the Electrochemical Society*, 147, 38–42.
95. Biener, J., M. Stadermann, M. Suss, M. A. Worsley, M. M. Biener, K. A. Rose and T. F. Baumann. 2011. Advanced carbon aerogels for energy applications. *Energy & Environmental Science*, 4, 656–667.
96. Frackowiak, E., G. Lota, J. Machnikowski, C. Vix-Guterl and F. Béguin. 2006. Optimisation of supercapacitors using carbons with controlled nanotexture and nitrogen content. *Electrochimica Acta*, 51, 2209–2214.
97. Hulicova-Jurcakova, D., M. Seredych, G. Q. Lu and T. J. Bandosz. 2009. Combined effect of nitrogen- and oxygen-containing functional groups of microporous activated carbon on its electrochemical performance in supercapacitors. *Advanced Functional Materials*, 19, 438–447.
98. Jeong, H. M., J. W. Lee, W. H. Shin, Y. J. Choi, H. J. Shin, J. K. Kang and J. W. Choi. 2011. Nitrogen-doped graphene for high-performance ultracapacitors and the importance of nitrogen-doped sites at basal planes. *Nano Letters*, 11, 2472–2477.
99. Kim, W., J. B. Joo, N. Kim, S. Oh, P. Kim and J. Yi. 2009. Preparation of nitrogen-doped mesoporous carbon nanopipes for the electrochemical double layer capacitor. *Carbon*, 47, 1407–1411.
100. Li, W., D. Chen, Z. Li, Y. Shi, Y. Wan, J. Huang, J. Yang, D. Zhao and Z. Jiang. 2007. Nitrogen enriched mesoporous carbon spheres obtained by a facile method and its application for electrochemical capacitor. *Electrochemistry Communications*, 9, 569–573.
101. Béguin, F. and E. Frackowiak. 2010. *Carbons for Electrochemical Energy Storage and Conversion Systems*. Boca Raton, FL: CRC Press.
102. Bleda-Martínez, M. J., J. A. Maciá-Agulló, D. Lozano-Castelló, E. Morallon, D. Cazorla-Amorós and A. Linares-Solano. 2005. Role of surface chemistry on electric double layer capacitance of carbon materials. *Carbon*, 43, 2677–2684.
103. Ruiz, V., C. Blanco, E. Raymundo-Piñero, V. Khomenko, F. Béguin and R. Santamaría. 2007. Effects of thermal treatment of activated carbon on the electrochemical behaviour in supercapacitors. *Electrochimica Acta*, 52, 4969–4973.
104. Hsieh, C.-T. and H. Teng. 2002. Influence of oxygen treatment on electric double-layer capacitance of activated carbon fabrics. *Carbon*, 40, 667–674.
105. Nian, Y.-R. and H. Teng. 2002. Nitric acid modification of activated carbon electrodes for improvement of electrochemical capacitance. *Journal of the Electrochemical Society*, 149, A1008–A1014.
106. Oda, H., A. Yamashita, S. Minoura, M. Okamoto and T. Morimoto. 2006. Modification of the oxygen-containing functional group on activated carbon fiber in electrodes of an electric double-layer capacitor. *Journal of Power Sources*, 158, 1510–1516.
107. Okajima, K., K. Ohta and M. Sudoh. 2005. Capacitance behavior of activated carbon fibers with oxygen-plasma treatment. *Electrochimica Acta*, 50, 2227–2231.
108. Frackowiak, E., Q. Abbas and F. Béguin. 2013. Carbon/carbon supercapacitors. *Journal of Energy Chemistry*, 22, 226–240.
109. Simon, P. and Y. Gogotsi. 2008. Materials for electrochemical capacitors. *Nature Materials*, 7, 845–854.
110. Wang, G., L. Zhang and J. Zhang. 2012. A review of electrode materials for electrochemical supercapacitors. *Chemical Society Reviews*, 41, 797–828.

111. Zhao, X., B. M. Sanchez, P. J. Dobson and P. S. Grant. 2011. The role of nanomaterials in redox-based supercapacitors for next generation energy storage devices. *Nanoscale*, 3, 839–855.
112. Zhao, D.-D., S.-J. Bao, W.-J. Zhou and H.-L. Li. 2007. Preparation of hexagonal nanoporous nickel hydroxide film and its application for electrochemical capacitor. *Electrochemistry Communications*, 9, 869–874.
113. Hu, C.-C. and W.-C. Chen. 2004. Effects of substrates on the capacitive performance of $RuOx{\cdot}nH_2O$ and activated carbon–RuO*x* electrodes for supercapacitors. *Electrochimica Acta*, 49, 3469–3477.
114. Conway, B. E. 1991. Transition from "supercapacitor" to "battery" behavior in electrochemical energy storage. *Journal of the Electrochemical Society*, 138, 1539–1548.
115. Zheng, J. P., P. J. Cygan and T. R. Jow. 1995. Hydrous ruthenium oxide as an electrode material for electrochemical capacitors. *Journal of the Electrochemical Society*, 142, 2699–2703.
116. Hu, C.-C., K.-H. Chang, M.-C. Lin and Y.-T. Wu. 2006. Design and tailoring of the nanotubular arrayed architecture of hydrous RuO_2 for next generation supercapacitors. *Nano Letters*, 6, 2690–2695.
117. Toupin, M., T. Brousse and D. Bélanger. 2004. Charge storage mechanism of MnO_2 electrode used in aqueous electrochemical capacitor. *Chemistry of Materials*, 16, 3184–3190.
118. Mazloumi, M., S. Shadmehr, Y. Rangom, L. F. Nazar and X. Tang. 2013. Fabrication of three-dimensional carbon nanotube and metal oxide hybrid mesoporous architectures. *ACS Nano*, 7, 4281–4288.
119. Wang, G., X. Shen, J. Horvat, B. Wang, H. Liu, D. Wexler and J. Yao. 2009. Hydrothermal synthesis and optical, magnetic, and supercapacitance properties of nanoporous cobalt oxide nanorods. *The Journal of Physical Chemistry C*, 113, 4357–4361.
120. Cheng, J., G.-P. Cao and Y.-S. Yang. 2006. Characterization of sol–gel-derived NiO*x* xerogels as supercapacitors. *Journal of Power Sources*, 159, 734–741.
121. Zhao, X., C. Johnston and P. S. Grant. 2009. A novel hybrid supercapacitor with a carbon nanotube cathode and an iron oxide/carbon nanotube composite anode. *Journal of Materials Chemistry*, 19, 8755–8760.
122. Chen, J., K. Huang and S. Liu. 2009. Hydrothermal preparation of octadecahedron Fe_3O_4 thin film for use in an electrochemical supercapacitor. *Electrochimica Acta*, 55, 1–5.
123. Do, Q. H., T. R. Fielitz, C. Zeng, O. A. Vanli, C. Zhang and J. P. Zheng. 2013. Vanadium oxide–carbon nanotube composite electrodes for energy storage by supercritical fluid deposition: Experiment design and device performance. *Nanotechnology*, 24, 315401.
124. Kim, I.-H., J.-H. Kim, B.-W. Cho, Y.-H. Lee and K.-B. Kim. 2006. Synthesis and electrochemical characterization of vanadium oxide on carbon nanotube film substrate for pseudocapacitor applications. *Journal of the Electrochemical Society*, 153, A989–A996.
125. Choi, D. and P. N. Kumta. 2005. Chemically synthesized nanostructured VN for pseudocapacitor application. *Electrochemical and Solid-State Letters*, 8, A418–A422.
126. Choi, D., G. E. Blomgren and P. N. Kumta. 2006. Fast and reversible surface redox reaction in nanocrystalline vanadium nitride supercapacitors. *Advanced Materials*, 18, 1178–1182.
127. Kim, Y.-T., K. Tadai and T. Mitani. 2005. Highly dispersed ruthenium oxide nanoparticles on carboxylated carbon nanotubes for supercapacitor electrode materials. *Journal of Materials Chemistry*, 15, 4914–4921.
128. Lee, J.-K., H. M. Pathan, K.-D. Jung and O.-S. Joo. 2006. Electrochemical capacitance of nanocomposite films formed by loading carbon nanotubes with ruthenium oxide. *Journal of Power Sources*, 159, 1527–1531.
129. Fu, R., Z. Ma and J. P. Zheng. 2002. Proton NMR and dynamic studies of hydrous ruthenium oxide. *The Journal of Physical Chemistry B*, 106, 3592–3596.
130. Sugimoto, W., K. Yokoshima, Y. Murakami and Y. Takasu. 2006. Charge storage mechanism of nanostructured anhydrous and hydrous ruthenium-based oxides. *Electrochimica Acta*, 52, 1742–1748.
131. Zheng, J. P. and Y. Xin. 2002. Characterization of $RuO_2{\cdot}xH_2O$ with various water contents. *Journal of Power Sources*, 110, 86–90.
132. Sugimoto, W., H. Iwata, Y. Yasunaga, Y. Murakami and Y. Takasu. 2003. Preparation of ruthenic acid nanosheets and utilization of its interlayer surface for electrochemical energy storage. *Angewandte Chemie International Edition*, 42, 4092–4096.
133. Egashira, M., Y. Matsuno, N. Yoshimoto and M. Morita. 2010. Pseudo-capacitance of composite electrode of ruthenium oxide with porous carbon in non-aqueous electrolyte containing imidazolium salt. *Journal of Power Sources*, 195, 3036–3040.
134. Pang, S. C., M. A. Anderson and T. W. Chapman. 2002. Novel electrode materials for thin-film ultracapacitors: Comparison of electrochemical properties of sol–gel-derived and electrodeposited manganese dioxide. *Journal of the Electrochemical Society*, 147, 444–450.
135. Hu, C.-C. and T.-W. Tsou. 2002. Ideal capacitive behavior of hydrous manganese oxide prepared by anodic deposition. *Electrochemistry Communications* 4, 105–109.
136. Ye, C., Z. M. Lin and S. Z. Hui. 2005. Electrochemical and capacitance properties of rod-shaped MnO_2 for supercapacitor. *Journal of the Electrochemical Society*, 152, A1272–A1278.
137. Lee, S. W., J. Kim, S. Chen, P. T. Hammond and Y. Shao-Horn. 2010. Carbon nanotube/manganese oxide ultrathin film electrodes for electrochemical capacitors. *ACS Nano*, 4, 3889–3896.
138. Desilvestro, J. and O. Haas. 1990. Metal oxide cathode materials for electrochemical energy storage: A review. *Journal of the Electrochemical Society*, 137, C5–C22.
139. Chang, J.-K., C.-H. Huang, W.-T. Tsai, M.-J. Deng and I.-W. Sun. 2008. Ideal pseudocapacitive performance of the Mn oxide anodized from the nanostructured and amorphous Mn thin film electrodeposited in BMP–NTf_2 ionic liquid. *Journal of Power Sources*, 179, 435–440.
140. Yu, C., L. Zhang, J. Shi, J. Zhao, J. Gao and D. Yan. 2008. A simple template-free strategy to synthesize nanoporous manganese and nickel oxides with narrow pore size distribution, and their electrochemical properties. *Advanced Functional Materials*, 18, 1544–1554.
141. Xiao, W., H. Xia, J. Y. H. Fuh and L. Lu. 2009. Growth of single-crystal α-MnO_2 nanotubes prepared by a hydrothermal route and their electrochemical properties. *Journal of Power Sources*, 193, 935–938.
142. Kulesza, P. J., S. Zamponi, M. A. Malik, M. Berrettoni, A. Wolkiewicz and R. Marassi. 1997. Spectroelectrochemical characterization of cobalt hexacyanoferrate films in potassium salt electrolyte. *Electrochimica Acta*, 43, 919–923.
143. Srinivasan, V. and J. W. Weidner. 2002. Capacitance studies of cobalt oxide films formed via electrochemical precipitation. *Journal of Power Sources*, 108, 15–20.
144. Liu, T. C., W. G. Pell and B. E. Conway. 1999. Stages in the development of thick cobalt oxide films exhibiting reversible redox behavior and pseudocapacitance. *Electrochimica Acta*, 44, 2829–2842.

145. Du, W., R. Liu, Y. Jiang, Q. Lu, Y. Fan and F. Gao. 2013. Facile synthesis of hollow Co_3O_4 boxes for high capacity supercapacitor. *Journal of Power Sources*, 227, 101–105.
146. Lin, P., Q. She, B. Hong, X. Liu, Y. Shi, Z. Shi, M. Zheng and Q. Dong. 2010. The nickel oxide/CNT composites with high capacitance for supercapacitor. *Journal of the Electrochemical Society*, 157, A818–A823.
147. Castro, E. B., S. G. Real and L. F. Pinheiro Dick. 2004. Electrochemical characterization of porous nickel–cobalt oxide electrodes. *International Journal of Hydrogen Energy*, 29, 255–261.
148. Zhang, X., W. Shi, J. Zhu, W. Zhao, J. Ma, S. Mhaisalkar, T. L. Maria, Y. Yang, H. Zhang, H. H. Hng and Q. Yan. 2010. Synthesis of porous NiO nanocrystals with controllable surface area and their application as supercapacitor electrodes. *Nano Research*, 3, 643–652.
149. Nam, K.-W., K.-H. Kim, E.-S. Lee, W.-S. Yoon, X.-Q. Yang and K.-B. Kim. 2008. Pseudocapacitive properties of electrochemically prepared nickel oxides on 3-dimensional carbon nanotube film substrates. *Journal of Power Sources*, 182, 642–652.
150. Wu, M.-S., Y.-A. Huang, C.-H. Yang and J.-J. Jow. 2007. Electrodeposition of nanoporous nickel oxide film for electrochemical capacitors. *International Journal of Hydrogen Energy*, 32, 4153–4159.
151. Nam, K.-W., E. S. Lee, J.-H. Kim, Y.-H. Lee and K.-B. Kim. 2005. Synthesis and electrochemical investigations of $Ni_{1-x}O$ thin films and $Ni_{1-x}O$ on three-dimensional carbon substrates for electrochemical capacitors. *Journal of the Electrochemical Society*, 152, A2123–A2129.
152. Wee, G., H. Z. Soh, Y. L. Cheah, S. G. Mhaisalkar and M. Srinivasan. 2010. Synthesis and electrochemical properties of electrospun V_2O_5 nanofibers as supercapacitor electrodes. *Journal of Materials Chemistry*, 20, 6720–6725.
153. Jayalakshmi, M., M. M. Rao, N. Venugopal and K.-B. Kim. 2007. Hydrothermal synthesis of SnO_2–V_2O_5 mixed oxide and electrochemical screening of carbon nano-tubes (CNT), V_2O_5, V_2O_5–CNT, and SnO_2–V_2O_5–CNT electrodes for supercapacitor applications. *Journal of Power Sources*, 166, 578–583.
154. Lee, H. Y. and J. B. Goodenough. 1999. Ideal supercapacitor behavior of amorphous $V_2O_5{\cdot}nH_2O$ in potassium chloride (KCl) aqueous solution. *Journal of Solid State Chemistry*, 148, 81–84.
155. Pan, X., Y. Zhao, G. F. Ren and Z. Y. Fan. 2013. Highly conductive VO_2 treated with hydrogen for supercapacitors. *Chemical Communications*, 49, 3943–3945.
156. Zhou, X., H. Chen, D. Shu, C. He and J. Nan. 2009. Study on the electrochemical behavior of vanadium nitride as a promising supercapacitor material. *Journal of Physics and Chemistry of Solids*, 70, 495–500.
157. Liu, T. C., W. G. Pell, B. E. Conway and S. L. Roberson. 1998. Behavior of molybdenum nitrides as materials for electrochemical capacitors: Comparison with ruthenium oxide. *Journal of the Electrochemical Society*, 145, 1882–1888.
158. Rudge, A., J. Davey, I. Raistrick, S. Gottesfeld and J. P. Ferraris. 1994. Conducting polymers as active materials in electrochemical capacitors. *Journal of Power Sources*, 47, 89–107.
159. Snook, G. A., P. Kao and A. S. Best. 2011. Conducting-polymer-based supercapacitor devices and electrodes. *Journal of Power Sources*, 196, 1–12.
160. Suematsu, S., Y. Oura, H. Tsujimoto, H. Kanno and K. Naoi. 2000. Conducting polymer films of cross-linked structure and their QCM analysis. *Electrochimica Acta*, 45, 3813–3821.
161. Lota, K., V. Khomenko and E. Frackowiak. 2004. Capacitance properties of poly (3, 4-ethylenedioxythiophene)/carbon nanotubes composites. *Journal of Physics and Chemistry of Solids*, 65, 295–301.
162. Fusalba, F., P. Gouérec, D. Villers and D. Bélanger. 2001. Electrochemical characterization of polyaniline in nonaqueous electrolyte and its evaluation as electrode material for electrochemical supercapacitors. *Journal of the Electrochemical Society*, 148, A1–A6.
163. Ryu, K. S., K. M. Kim, N.-G. Park, Y. J. Park and S. H. Chang. 2002. Symmetric redox supercapacitor with conducting polyaniline electrodes. *Journal of Power Sources*, 103, 305–309.
164. Gupta, V. and N. Miura. 2006. High performance electrochemical supercapacitor from electrochemically synthesized nanostructured polyaniline. *Materials Letters*, 60, 1466–1469.
165. Talbi, H., P. E. Just and L. H. Dao. 2003. Electropolymerization of aniline on carbonized polyacrylonitrile aerogel electrodes: Applications for supercapacitors. *Journal of Applied Electrochemistry*, 33, 465–473.
166. Wu, M., G. A. Snook, V. Gupta, M. Shaffer, D. J. Fray and G. Z. Chen. 2005. Electrochemical fabrication and capacitance of composite films of carbon nanotubes and polyaniline. *Journal of Materials Chemistry*, 15, 2297–2303.
167. Wang, K., J. Huang and Z. Wei. 2010. Conducting polyaniline nanowire arrays for high performance supercapacitors. *The Journal of Physical Chemistry C*, 114, 8062–8067.
168. Guan, H., L.-Z. Fan, H. Zhang and Q. Qu. 2010. Polyaniline nanofibers obtained by interfacial polymerization for high-rate supercapacitors. *Electrochimica Acta*, 56, 964–968.
169. Zhou, H., H. Chen, S. Luo, G. Lu, W. Wei and Y. Kuang. 2005. The effect of the polyaniline morphology on the performance of polyaniline supercapacitors. *Journal of Solid State Electrochemistry*, 9, 574–580.
170. Horng, Y.-Y., Y.-C. Lu, Y.-K. Hsu, C.-C. Chen, L.-C. Chen and K.-H. Chen. 2010. Flexible supercapacitor based on polyaniline nanowires/carbon cloth with both high gravimetric and area-normalized capacitance. *Journal of Power Sources*, 195, 4418–4422.
171. Ingram, M. D., H. Staesche and K. S. Ryder. 2004. 'Activated' polypyrrole electrodes for high-power supercapacitor applications. *Solid State Ionics*, 169, 51–57.
172. Tripathi, S. K., A. Kumar and S. A. Hashmi. 2006. Electrochemical redox supercapacitors using PVdF-HFP based gel electrolytes and polypyrrole as conducting polymer electrode. *Solid State Ionics*, 177, 2979–2985.
173. Kim, B. C., J. M. Ko and G. G. Wallace. 2008. A novel capacitor material based on Nafion-doped polypyrrole. *Journal of Power Sources*, 177, 665–668.
174. Sharma, R. K., A. C. Rastogi and S. B. Desu. 2008. Pulse polymerized polypyrrole electrodes for high energy density electrochemical supercapacitor. *Electrochemistry Communications*, 10, 268–272.
175. Wang, J., Y. Xu, J. Wang and X. Du. 2011. Toward a high specific power and high stability polypyrrole supercapacitors. *Synthetic Metals*, 161, 1141–1144.
176. Dubal, D. P., S. H. Lee, J. G. Kim, W. B. Kim and C. D. Lokhande. 2012. Porous polypyrrole clusters prepared by electropolymerization for a high performance supercapacitor. *Journal of Materials Chemistry*, 22, 3044–3052.
177. Hashmi, S. A. and H. M. Upadhyaya. 2002. Polypyrrole and poly (3-methyl thiophene)-based solid state redox supercapacitors using ion conducting polymer electrolyte. *Solid State Ionics*, 152, 883–889.

178. Laforgue, A., P. Simon, C. Sarrazin and J.-F. Fauvarque. 1999. Polythiophene-based supercapacitors. *Journal of Power Sources*, 80, 142–148.
179. Mastragostino, M., C. Arbizzani and F. Soavi. 2001. Polymer-based supercapacitors. *Journal of Power Sources*, 97, 812–815.
180. Mastragostino, M., C. Arbizzani and F. Soavi. 2002. Conducting polymers as electrode materials in supercapacitors. *Solid State Ionics*, 148, 493–498.
181. Mastragostino, M., C. Arbizzani, R. Paraventi and A. Zanelli. 2000. Polymer selection and cell design for electric-vehicle supercapacitors. *Journal of the Electrochemical Society*, 147, 407–412.
182. Mastragostino, M., R. Paraventi and A. Zanelli. 2000. Supercapacitors based on composite polymer electrodes. *Journal of the Electrochemical Society*, 147, 3167–3170.
183. Di Fabio, A., A. Giorgi, M. Mastragostino and F. Soavi. 2001. Carbon-poly (3-methylthiophene) hybrid supercapacitors. *Journal of the Electrochemical Society*, 148, A845–A850.
184. Sivaraman, P., A. Thakur, R. K. Kushwaha, D. Ratna and A. B. Samui. 2006. Poly (3-methyl thiophene)-activated carbon hybrid supercapacitor based on gel polymer electrolyte. *Electrochemical and Solid-State Letters*, 9, A435–A438.
185. Kim, J. Y. and I. J. Chung. 2002. An all-solid-state electrochemical supercapacitor based on poly3-(4-fluorophenylthiophene) composite electrodes. *Journal of the Electrochemical Society*, 149, A1376–A1380.
186. Stenger-Smith, J. D., A. Guenthner, J. Cash, J. A. Irvin and D. J. Irvin. 2010. Poly (propylenedioxy) thiophene-based supercapacitors operating at low temperatures. *Journal of the Electrochemical Society*, 157, A298–A304.
187. Snook, G. A. and G. Z. Chen. 2008. The measurement of specific capacitances of conducting polymers using the quartz crystal microbalance. *Journal of Electroanalytical Chemistry*, 612, 140–146.
188. Stenger-Smith, J. D., C. K. Webber, N. Anderson, A. P. Chafin, K. Zong and J. R. Reynolds. 2002. Poly (3, 4-alkyl enedioxythiophene)-based supercapacitors using ionic liquids as supporting electrolytes. *Journal of the Electrochemical Society*, 149, A973–A977.
189. Liu, K., Z. Hu, R. Xue, J. Zhang and J. Zhu. 2008. Electropolymerization of high stable poly (3, 4-ethylene dioxythiophene) in ionic liquids and its potential applications in electrochemical capacitor. *Journal of Power Sources*, 179, 858–862.
190. Pandey, G. P. and A. C. Rastogi. 2013. Synthesis and characterization of pulsed polymerized poly (3, 4-ethylenedioxy thiophene) electrodes for high-performance electrochemical capacitors. *Electrochimica Acta*, 87, 158–168.
191. Li, C., H. Bai and G. Shi. 2009. Conducting polymer nanomaterials: Electrosynthesis and applications. *Chemical Society Reviews*, 38, 2397–2409.
192. Bhandari, S., M. Deepa, S. Singh, G. Gupta and R. Kant. 2008. Redox behavior and optical response of nanostructured poly (3, 4-ethylenedioxythiophene) films grown in a camphorsulfonic acid based micellar solution. *Electrochimica Acta*, 53, 3189–3199.
193. Patra, S. and N. Munichandraiah. 2007. Supercapacitor studies of electrochemically deposited PEDOT on stainless steel substrate. *Journal of Applied Polymer Science*, 106, 1160–1171.
194. Huang, J.-H. and C.-W. Chu. 2011. Achieving efficient poly (3, 4-ethylenedioxythiophene)-based supercapacitors by controlling the polymerization kinetics. *Electrochimica Acta*, 56, 7228–7234.
195. Dhawale, D. S., A. Vinu and C. D. Lokhande. 2011. Stable nanostructured polyaniline electrode for supercapacitor application. *Electrochimica Acta*, 56, 9482–9487.
196. Biso, M., M. Mastragostino, M. Montanino, S. Passerini and F. Soavi. 2008. Electropolymerization of poly (3-methylthio phene) in pyrrolidinium-based ionic liquids for hybrid supercapacitors. *Electrochimica Acta*, 53, 7967–7971.
197. Sun, Z., Z. Liu, B. Han, S. Miao, J. Du and Z. Miao. 2006. Microstructural and electrochemical characterization of RuO_2/CNT composites synthesized in supercritical diethyl amine. *Carbon*, 44, 888–893.
198. Ramani, M., B. S. Haran, R. E. White, B. N. Popov and L. Arsov. 2001. Studies on activated carbon capacitor materials loaded with different amounts of ruthenium oxide. *Journal of Power Sources*, 93, 209–214.
199. Panić, V., T. Vidaković, S. Gojković, A. Dekanski, S. Milonjić and B. Nikolić. 2003. The properties of carbon-supported hydrous ruthenium oxide obtained from RuOxHy sol. *Electrochimica Acta*, 48, 3805–3813.
200. Yan, J., Z. Fan, T. Wei, J. Cheng, B. Shao, K. Wang, L. Song and M. Zhang. 2009. Carbon nanotube/MnO_2 composites synthesized by microwave-assisted method for supercapacitors with high power and energy densities. *Journal of Power Sources*, 194, 1202–1207.
201. Yan, J., Z. Fan, T. Wei, W. Qian, M. Zhang and F. Wei. 2010. Fast and reversible surface redox reaction of graphene–MnO_2 composites as supercapacitor electrodes. *Carbon*, 48, 3825–3833.
202. Zhi, M., C. Xiang, J. Li, M. Li and N. Wu. 2013. Nanostructured carbon–metal oxide composite electrodes for supercapacitors: A review. *Nanoscale*, 5, 72–88.
203. Fisher, R. A., M. R. Watt and W. J. Ready. 2013. Functionalized carbon nanotube supercapacitor electrodes: A review on pseudocapacitive materials. *ECS Journal of Solid State Science and Technology*, 2, M3170–M3177.
204. Ebbesen, T., H. Lezec, H. Hiura, J. Bennett, H. Ghaemi and T. Thio. 1996. Electrical conductivity of individual carbon nanotubes. *Nature*, 382, 54–56.
205. Dikin, D. A., S. Stankovich, E. J. Zimney, R. D. Piner, G. H. Dommett, G. Evmenenko, S. T. Nguyen and R. S. Ruoff. 2007. Preparation and characterization of graphene oxide paper. *Nature*, 448, 457–460.
206. Li, D., M. B. Mueller, S. Gilje, R. B. Kaner and G. G. Wallace. 2008. Processable aqueous dispersions of graphene nanosheets. *Nature Nanotechnology*, 3, 101–105.
207. Nam, K.-W., C.-W. Lee, X.-Q. Yang, B. W. Cho, W.-S. Yoon and K.-B. Kim. 2009. Electrodeposited manganese oxides on three-dimensional carbon nanotube substrate: Supercapacitive behaviour in aqueous and organic electrolytes. *Journal of Power Sources*, 188, 323–331.
208. Chen, S., J. Zhu, X. Wu, Q. Han and X. Wang. 2010. Graphene oxide- MnO_2 nanocomposites for supercapacitors. *ACS Nano*, 4, 2822–2830.
209. Ko, J. M. and K. M. Kim. 2009. Electrochemical properties of MnO_2/activated carbon nanotube composite as an electrode material for supercapacitor. *Materials Chemistry and Physics*, 114, 837–841.
210. Yu, G., L. Hu, N. Liu, H. Wang, M. Vosgueritchian, Y. Yang, Y. Cui and Z. Bao. 2011. Enhancing the supercapacitor performance of graphene/MnO_2 nanostructured electrodes by conductive wrapping. *Nano Letters*, 11, 4438–4442.
211. Li, Z., Y. Mi, X. Liu, S. Liu, S. Yang and J. Wang. 2011. Flexible graphene/MnO_2 composite papers for supercapacitor electrodes. *Journal of Materials Chemistry*, 21, 14706–14711.

212. Lee, H., K. Junmo, M. S. Cho, J.-B. Choic and Y. Lee. 2011. MnO_2/graphene composite electrodes for supercapacitors: The effect of graphene intercalation on capacitance. *Journal of Materials Chemistry*, 21, 18215–18219.
213. Liu, J., J. Essner and J. Li. 2010. Hybrid supercapacitor based on coaxially coated manganese oxide on vertically aligned carbon nanofiber arrays. *Chemistry of Materials*, 22, 5022–5030.
214. Passerini, S., J. Ressler, D. Le, B. Owens and W. Smyrl. 1999. High rate electrodes of V_2O_5 aerogel. *Electrochimica Acta*, 44, 2209–2217.
215. Dobley, A., K. Ngala, S. Yang, S., P. Y. Zavalij and M. S. Whittingham. 2001. Manganese vanadium oxide nanotubes: Synthesis, characterization, and electrochemistry. *Chemistry of Materials*, 13, 4382–4386.
216. Chen, Z., V. Augustyn, J. Wen, Y. Zhang, M. Shen, B. Dunn and Y. Lu. 2011. High-performance supercapacitors based on intertwined CNT/V_2O_5 nanowire nanocomposites. *Advanced Materials*, 23, 791–795.
217. Sathiya, M., A. S. Prakash, K. Ramesha, J. M. Tarascon and A. K. Shukla. 2011. V_2O_5-anchored carbon nanotubes for enhanced electrochemical energy storage. *Journal of the American Chemical Society*, 133, 16291–16299.
218. Bonso, J. S., A. Rahy, S. D. Perera, N. Nour, O. Seitz, Y. J. Chabal, K. J. Balkus Jr., J. P. Ferraris and D. J. Yang. 2012. Exfoliated graphite nanoplatelets–V_2O_5 nanotube composite electrodes for supercapacitors. *Journal of Power Sources*, 203, 227–232.
219. Frackowiak, E., V. Khomenko, K. Jurewicz, K. Lota and F. Beguin. 2006. Supercapacitors based on conducting polymers/nanotubes composites. *Journal of Power Sources*, 153, 413–418.
220. Pieta, P., I. Obraztsov, F. D'Souza and W. Kutner. 2013. Composites of conducting polymers and various carbon nanostructures for electrochemical supercapacitors. *ECS Journal of Solid State Science and Technology*, 2, M3120–M3134.
221. Peng, C., S. Zhang, D. Jewell and G. Z. Chen. 2008. Carbon nanotube and conducting polymer composites for supercapacitors. *Progress in Natural Science*, 18, 777–788.
222. Yu, G., X. Xie, L. Pan, Z. Bao and Y. Cui. 2013. Hybrid nanostructured materials for high-performance electrochemical capacitors. *Nano Energy*, 3, 213–234.
223. Lei, Z., Z. Chen and X. S. Zhao. 2010. Growth of polyaniline on hollow carbon spheres for enhancing electrocapacitance. *The Journal of Physical Chemistry C*, 114, 19867–19874.
224. Kovalenko, I., D. G. Bucknall and G. Yushin. 2010. Detonation nanodiamond and onion-like carbon-embedded polyaniline for supercapacitors. *Advanced Functional Materials*, 20, 3979–3986.
225. Wang, Y. G., H. Q. Li and Y. Y. Xia. 2006. Ordered whisker-like polyaniline grown on the surface of mesoporous carbon and its electrochemical capacitance performance. *Advanced Materials*, 18, 2619–2623.
226. Liu, J., J. Sun and L. Gao. 2010. A promising way to enhance the electrochemical behavior of flexible single-walled carbon nanotube/polyaniline composite films. *The Journal of Physical Chemistry C*, 114, 19614–19620.
227. Wang, D.-W., F. Li, J. Zhao, W. Ren, Z.-G. Chen, J. Tan, Z.-S. Wu, I. Gentle, G. Q. Lu and H.-M. Cheng. 2009. Fabrication of graphene/polyaniline composite paper via *in situ* anodic electropolymerization for high-performance flexible electrode. *ACS Nano*, 3, 1745–1752.
228. Feng, X. M., R. M. Li, Y. W. Ma, R. F. Chen, N. E. Shi, Q. L. Fan and W. Huang. 2011. One-step electrochemical synthesis of graphene/polyaniline composite film and its applications. *Advanced Functional Materials*, 21, 2989–2996.
229. Davies, A., P. Audette, B. Farrow, F. Hassan, Z. Chen, J.-Y. Choi and A. Yu. 2011. Graphene-based flexible supercapacitors: Pulse-electropolymerization of polypyrrole on free-standing graphene films. *The Journal of Physical Chemistry C*, 115, 17612–17620.
230. Zhang, D., X. Zhang, Y. Chen, P. Yu, C. Wang and Y. Ma. 2011. Enhanced capacitance and rate capability of graphene/polypyrrole composite as electrode material for supercapacitors. *Journal of Power Sources*, 196, 5990–5996.
231. Fang, Y., J. Liu, D. J. Yu, J. P. Wicksted, K. Kalkan, O. Topal, B. N. Flanders, J. Wu and J. Li. 2010. Self-supported supercapacitor membranes: Polypyrrole-coated multi-walled carbon nanotube networks enabled by pulsed electrodeposition. *Journal of Power Sources*, 195, 674–679.
232. Liu, R. and S. B. Lee. 2008. MnO_2/poly(3,4-ethylenedioxythiophene) coaxial nanowires by one-step coelectrodeposition for electrochemical energy storage. *Journal of the American Chemical Society*, 130, 2942–2943.
233. Hou, Y., Y. Cheng, T. Hobson and J. Liu. 2010. Design and synthesis of hierarchical MnO_2 nanospheres/carbon nanotubes/conducting polymer ternary composite for high performance electrochemical electrodes. *Nano Letters*, 10, 2727–2733.
234. Sharma, R. K. and L. Zhai. 2009. Multiwall carbon nanotube supported poly (3, 4-ethylenedioxythiophene)/manganese oxide nano-composite electrode for super-capacitors. *Electrochimica Acta*, 54, 7148–7155.
235. Sivakkumar, S. R., J. M. Ko, D. Y. Kim, B. C. Kim and G. G. Wallace. 2007. Performance evaluation of CNT/polypyrrole/MnO_2 composite electrodes for electrochemical capacitors. *Electrochimica Acta*, 52, 7377–7385.
236. Conway, B. E. and W. G. Pell. 2003. Double-layer and pseudocapacitance types of electrochemical capacitors and their applications to the development of hybrid devices. *Journal of Solid State Electrochemistry*, 7, 637–644.
237. Cericola, D. and R. Kötz. 2012. Hybridization of rechargeable batteries and electrochemical capacitors: Principles and limits. *Electrochimica Acta*, 72, 1–17.
238. Peng, C., S. Zhang, X. Zhou and G. Z. Chen. 2010. Unequalisation of electrode capacitances for enhanced energy capacity in asymmetrical supercapacitors. *Energy & Environmental Science*, 3, 1499–1502.
239. Khomenko, V., E. Raymundo-Pinero and F. Beguin. 2006. Optimisation of an asymmetric manganese oxide/activated carbon capacitor working at 2 V in aqueous medium. *Journal of Power Sources*, 153, 183–190.
240. Khomenko, V., E. Raymundo-Pinero, E. Frackowiak and F. Beguin. 2006. High-voltage asymmetric supercapacitors operating in aqueous electrolyte. *Applied Physics A*, 82, 567–573.
241. Chen, P.-C., G. Shen, G. Y. Shi, H. Chen and C. Zhou. 2010. Preparation and characterization of flexible asymmetric supercapacitors based on transition-metal-oxide nanowire/single-walled carbon nanotube hybrid thin-film electrodes. *ACS Nano*, 4, 4403–4411.
242. Weng, Z., F. Li, D. W. Wang, L. Wen and H. M. Cheng. 2013. Controlled electrochemical charge injection to maximize the energy density of supercapacitors. *Angewandte Chemie International Edition*, 52, 3722–3725.
243. Ruiz, V., R. Santamaría, M. Granda and C. Blanco. 2009. Long-term cycling of carbon-based supercapacitors in aqueous media. *Electrochimica Acta*, 54, 4481–4486.
244. Khomenko, V., E. Raymundo-Piñero and F. Béguin. 2010. A new type of high energy asymmetric capacitor with nanoporous carbon electrodes in aqueous electrolyte. *Journal of Power Sources*, 195, 4234–4241.

245. Demarconnay, L., E. Raymundo-Pinero and F. Béguin. 2010. A symmetric carbon/carbon supercapacitor operating at 1.6 V by using a neutral aqueous solution. *Electrochemistry Communications*, 12, 1275–1278.
246. Bichat, M. P., E. Raymundo-Piñero and F. Béguin. 2010. High voltage supercapacitor built with seaweed carbons in neutral aqueous electrolyte. *Carbon*, 48, 4351–4361.
247. Qu, Q. T., B. Wang, L. C. Yang, Y. Shi, S. Tian and Y. P. Wu. 2008. Study on electrochemical performance of activated carbon in aqueous Li_2SO_4, Na_2SO_4 and K_2SO_4 electrolytes. *Electrochemistry Communications*, 10, 1652–1655.
248. Fic, K., G. Lota, M. Meller and E. Frackowiak. 2012. Novel insight into neutral medium as electrolyte for high-voltage supercapacitors. *Energy & Environmental Science*, 5, 5842–5850.
249. Tanahashi, I., A. Yoshida and A. Nishino. 1990. Comparison of the electrochemical properties of electric double-layer capacitors with an aqueous electrolyte and with a nonaqueous electrolyte. *Bulletin of the Chemical Society of Japan*, 63, 3611–3614.
250. Jänes, A., H. Kurig, T. Romann and E. Lust. 2010. Novel doubly charged cation based electrolytes for non-aqueous supercapacitors. *Electrochemistry Communications*, 12, 535–539.
251. Naoi, K. 2010. 'Nanohybrid capacitor': The next generation electrochemical capacitors. *Fuel Cells*, 10, 825–833.
252. Brandt, A., P. Isken, A. Lex-Balducci and A. Balducci. 2012. Adiponitrile-based electrochemical double layer capacitor. *Journal of Power Sources*, 204, 213–219.
253. Brandt, A. and A. Balducci. 2012. The influence of pore structure and surface groups on the performance of high voltage electrochemical double layer capacitors containing adiponitrile-based electrolyte. *Journal of the Electrochemical Society*, 159, A2053–A2059.
254. Armand, M., F. Endres, D. R. MacFarlane, H. Ohno and B. Scrosati. 2009. Ionic-liquid materials for the electrochemical challenges of the future. *Nature Materials*, 8, 621–629.
255. Galiński, M., A. Lewandowski and I. Stępniak. 2006. Ionic liquids as electrolytes. *Electrochimica Acta*, 51, 5567–5580.
256. Wishart, J. F. 2009. Energy applications of ionic liquids. *Energy & Environmental Science*, 2, 956–961.
257. Arbizzani, C., M. Biso, D. Cericola, M. Lazzari, F. Soavi and M. Mastragostino. 2008. Safe, high-energy supercapacitors based on solvent-free ionic liquid electrolytes. *Journal of Power Sources*, 185, 1575–1579.
258. Handa, N., T. Sugimoto, M. Yamagata, M. Kikuta, M. Kono and M. Ishikawa. 2008. A neat ionic liquid electrolyte based on FSI anion for electric double layer capacitor. *Journal of Power Sources*, 185, 1585–1588.
259. Lewandowski, A., A. Olejniczak, M. Galinski and I. Stepniak. 2010. Performance of carbon–carbon supercapacitors based on organic, aqueous and ionic liquid electrolytes. *Journal of Power Sources*, 195, 5814–5819.
260. Vatamanu, J., Z. Hu, D. Bedrov, C. Perez and Y. Gogotsi. 2013. Increasing energy storage in electrochemical capacitors with ionic liquid electrolytes and nanostructured carbon electrodes. *The Journal of Physical Chemistry Letters*, 4, 2829–2837.
261. McEwen, A. B., H. L. Ngo, K. LeCompte and J. L. Goldman. 1999. Electrochemical properties of imidazolium salt electrolytes for electrochemical capacitor applications. *Journal of the Electrochemical Society*, 146, 1687–1695.
262. Sato, T., G. Masuda and K. Takagi. 2004. Electrochemical properties of novel ionic liquids for electric double layer capacitor applications. *Electrochimica Acta*, 49, 3603–3611.
263. Devarajan, T., S. Higashiya, C. Dangler, M. Rane-Fondacaro, J. Snyder and P. Haldar. 2009. Novel ionic liquid electrolyte for electrochemical double layer capacitors. *Electrochemistry Communications*, 11, 680–683.
264. Frackowiak, E., G. Lota and J. Pernak. 2005. Room-temperature phosphonium ionic liquids for supercapacitor application. *Applied Physics Letters*, 86, 164104.
265. Lewandowski, A. and A. Świderska. 2003. Electrochemical capacitors with polymer electrolytes based on ionic liquids. *Solid State Ionics*, 161, 243–249.
266. Meng, C., C. Liu, L. Chen, C. Hu and S. Fan. 2010. Highly flexible and all-solid-state paperlike polymer supercapacitors. *Nano Letters*, 10, 4025–4031.
267. Pandey, G. P., A. C. Rastogi and C. R. Westgate. 2014. All-solid-state supercapacitors with poly (3, 4-ethylenedioxythiophene)-coated carbon fiber paper electrodes and ionic liquid gel polymer electrolyte. *Journal of Power Sources*, 245, 857–865.
268. Lu, W., K. Henry, C. Turchi and J. Pellegrino. 2008. Incorporating ionic liquid electrolytes into polymer gels for solid-state ultracapacitors. *Journal of the Electrochemical Society*, 155, A361–A367.
269. Pandey, G. P., S. A. Hashmi and Y. Kumar. 2010. Multiwalled carbon nanotube electrodes for electrical double layer capacitors with ionic liquid based gel polymer electrolytes. *Journal of the Electrochemical Society*, 157, A105–A114.
270. Sung, H.-Y., Y.-Y. Wang and C.-C. Wan. 1998. Preparation and characterization of poly (vinyl chloride-co-vinyl acetate)-based gel electrolytes for Li-ion batteries. *Journal of the Electrochemical Society*, 145, 1207–1211.
271. Agrawal, R. C. and G. P. Pandey. 2008. Solid polymer electrolytes: Materials designing and all-solid-state battery applications. An overview. *Journal of Physics D: Applied Physics*, 41, 223001–223018.
272. Lewandowski, A., M. Zajder, E. Frąckowiak and F. Beguin. 2001. Supercapacitor based on activated carbon and polyethylene oxide–KOH–H_2O polymer electrolyte. *Electrochimica Acta*, 46, 2777–2780.
273. Yuan, L., X.-H. Lu, X. Xiao, T. Zhai, J. Dai, F. Zhang, B. Hu, X. Wang, L. Gong and J. Chen. 2011. Flexible solid-state supercapacitors based on carbon nanoparticles/MnO_2 nanorods hybrid structure. *ACS Nano*, 6, 656–661.
274. Yu, M., Y. Zeng, C. Zhang, X. Lu, C. Zeng, C. Yao, Y. Yang and Y. Tong. 2013. Titanium dioxide@polypyrrole core–shell nanowires for all solid-state flexible supercapacitors. *Nanoscale*, 5, 10806–10810.
275. Schroeder, M., P. Isken, M. Winter, S. Passerini, A. Lex-Balducci and A. Balducci. 2013. An investigation on the use of a methacrylate-based gel polymer electrolyte in high power devices. *Journal of the Electrochemical Society*, 160, A1753–A1758.
276. Amade, R., E. Jover, B. Caglar, T. Mutlu and E. Bertran. 2011. Optimization of MnO_2/vertically aligned carbon nanotube composite for supercapacitor application. *Journal of Power Sources*, 196, 5779–5783.
277. Yang, L., S. Cheng, Y. Ding, X. Zhu, Z. L. Wang and M. Liu. 2011. Hierarchical network architectures of carbon fiber paper supported cobalt oxide nanonet for high-capacity pseudocapacitors. *Nano Letters*, 12, 321–325.

278. Fischer, A. E., K. A. Pettigrew, D. R. Rolison, R. M. Stroud and J. W. Long. 2007. Incorporation of homogeneous, nanoscale MnO_2 within ultraporous carbon structures via self-limiting electroless deposition: Implications for electrochemical capacitors. *Nano Letters*, 7, 281–286.
279. Lytle, J. C., J. M. Wallace, M. B. Sassin, A. J. Barrow, J. W. Long, J. L. Dysart, C. H. Renninger, M. P. Saunders, N. L. Brandell and D. R. Rolison. 2011. The right kind of interior for multifunctional electrode architectures: Carbon nanofoam papers with aperiodic submicrometre pore networks interconnected in 3D. *Energy & Environmental Science*, 4, 1913–1925.
280. Cao, X., Y. Shi, W. Shi, G. Lu, X. Huang, Q. Yan, Q. Zhang and H. Zhang. 2011. Preparation of novel 3D graphene networks for supercapacitor applications. *Small*, 7, 3163–3168.
281. Bo, Z., W. Zhu, W. Ma, Z. Wen, X. Shuai, J. Chen, J. Yan, Z. Wang, K. Cen and X. Feng. 2013. Vertically oriented graphene bridging active-layer/current-collector interface for ultrahigh rate supercapacitors. *Advanced Materials*, 25, 5799–5806.

19 Advanced Technologies for Supercapacitors

Hao Liu and Li-Min Liu

CONTENTS

19.1 INTRODUCTION

Supercapacitors (SCs), also called "ultracapacitors" and "electrochemical capacitors," have been accepted as one of the most promising energy storage devices owing to their high power density and long cycling life. These excellent properties make SCs a very good complementary device for batteries in certain applications by providing backup power supplies to protect against sudden current density changes in both charging and discharging processes. Besides high power density and long cycling life, SCs also have some other unique advantages such as quick charging, high efficiency, and deep charge and discharge without decreasing the performance or lifetime. This chapter will cover the fundamentals, fabrication, performance characterization, degradation, and theoretical modeling of SCs. Furthermore, the challenge and new designs of SCs will also be discussed.

19.2 FABRICATION OF SUPERCAPACITORS

The design and fabrication of SCs are important in SC technology, since several factors must be considered to make full use of electrode materials, such as good contact between the current collector and active materials, minimization of equivalent series resistance (ESR), avoidance of electrolyte overflew during cell compression, achieving best performance with least cost and materials, etc.

19.2.1 Main Types and Structures of Supercapacitors

SCs are generally classified, based on their charge storage mechanisms, into three types: electrochemical double-layer capacitors (EDLCs), pseudocapacitors, and hybrid capacitors (see Figure 19.1). In EDLCs, the charges are stored electrostatically, or nonfaradaically, and there is no charge transfer between the electrode and the electrolyte. Distinctly,

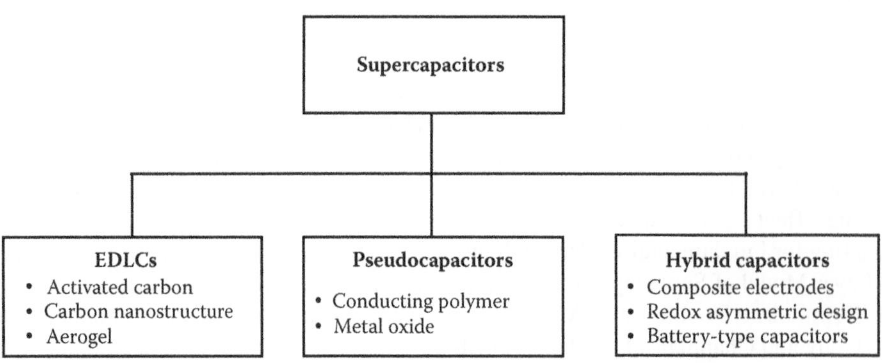

FIGURE 19.1 Classification of SC types.

pseudocapacitors use a faradaic process to store charges, in which redox reactions are commonly involved. The hybrid capacitor, as the name indicates, employs both storage mechanisms.

19.2.1.1 Electrochemical Double-Layer Capacitor

EDLCs accommodate charges similarly to dielectric capacitors, as shown in Figure 19.2. While a voltage is applied, charges accumulate on the electrode's surface. As a consequence, ions with opposite charges in the electrolyte also diffuse to the surface of electrode to maintain electroneutrality. Thus, the charges are stored at the interface between electrodes and the electrolyte, and this interface can be viewed as a capacitor. Owing to the nonfaradaic nature, charges are highly reversible, which brings about a very high cycling stability even under full charge–discharge. The capacitance C produced by this electrical double layer was first described by Helmholtz [1], according to

$$C = \frac{\varepsilon A}{d} \tag{19.1}$$

where ε is the dielectric constant, A the area of interface, and d the effective thickness of the double layer.

From Equation 19.1, the surface area of the electrode is the most important factor. The most popular electrodes thus far are fabricated from nanoscale carbon materials that have a very high surface area and high interconnected porosity, high electrical conductivity, excellent wettability toward the electrolyte, and presence of electrochemically active surface functionalities. The application of carbon nanomaterials, including nanoporous carbon, carbon nanotubes, fullerene, and graphene, in SCs has been intensively reviewed [2–4]. The highest capacitance reached by carbon-based EDLCs is 385 F g^{-1} with well-maintained capacity at increasing current density. These results justify the possibility of expanding the applications of EDLCs [5].

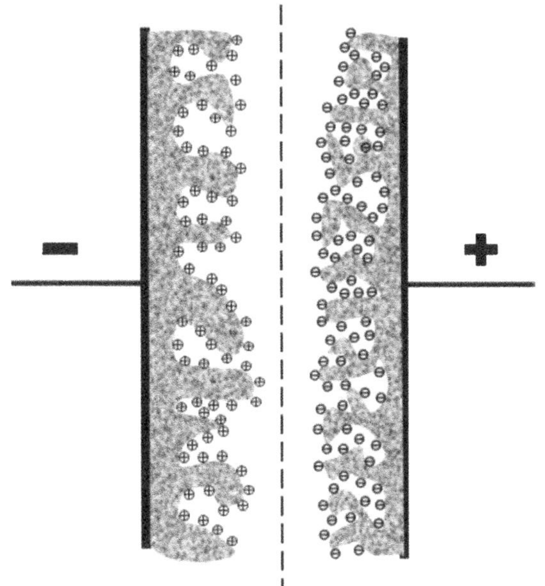

FIGURE 19.2 Charge storage mechanism of EDLCs.

19.2.1.2 Pseudocapacitor

In contrast to EDLCs, as voltage is applied to a pseudocapacitor, quick reversible faradaic reactions take place, inducing the transfer of charges across the interface like in batteries. However, different from batteries, which are based on Nernstian processes, in charging and discharging at a certain voltage, the charge transfer that takes place during these faradaic reactions is voltage dependent because of thermodynamic reasons [6]. There are three types of faradaic processes, including reversible adsorption (e.g., surface adsorption of hydrogen on the surface of electrode), redox reactions on and near the surface of electrode, and reversible electrochemical doping–dedoping in electrodes [7]. Studies demonstrate that the working voltage of a pseudocapacitor can be enlarged owing to the expansion of the charge developed in the defined potential [8]. Meanwhile, another significance of the pseudocapacitor has been recognized as the increased capacitance compared with EDLCs [7]. However, it is noticeable that these faradaic charge storage mechanisms are usually slower than the nonfaradaic processes occurring in EDLCs. As a result, pseudocapacitors normally suffer from a relatively lower power density. Moreover, similar to batteries, the redox reactions that occur at the electrode in pseudocapacitors usually lead to lack of stability during cycling.

The most popular electrode materials utilized for pseudocapacitors are metal oxides and conducting polymers. Transition metal oxides are considered attractive materials for pseudocapacitors because of their good conductivities, high rate capability, excellent electrochemical reversibility, and relatively long cycling life. MnO_2 and RuO_2 are the most popular pseudocapacitor materials. However, the use of transition metal oxide in pseudocapacitors is often limited by irreversible reactions, insufficient conductivity, dissolution (MnO_2), and/or high cost and toxicity (RuO_2). Recent progress in pseudocapacitors based on metal oxides has been reviewed by Augustyn et al. [9]. Conducting polymers are used as alternative pseudocapacitor electrode materials because of their inexpensiveness and ease of synthesis. Charges can be stored in and released from the polymer backbones through redox processes associated with the π-conjugated polymer chain [10]. Reviews of conductive polymers for pseudocapacitors show that they could reach relatively high capacitance and good conductivities but suffer from poor stability [11].

19.2.1.3 Hybrid Capacitors

As a result of the drawbacks of EDLCs and pseudocapacitors, hybrid capacitor integrates both faradaic and nonfaradaic processes to store charges to alleviate their disadvantages to achieve better performance. The most traditional hybrid capacitors are mainly assorted into two configurations: composite and asymmetric.

Composite electrodes involve combinations of carbon-based materials with either metal oxides or conducting polymers, incorporating both faradaic and nonfaradaic reactions in a single electrode. Meanwhile, carbon-based materials render a high surface area to increase the contact between pseudocapacitive materials and the electrolyte, further increasing the composite electrodes' capacity through faradaic reactions.

The asymmetric hybrid capacitor integrates an EDLC electrode (providing high power density) and a pseudocapacitor electrode (delivering high energy density) in the same capacitor cell. This trade-off between EDLCs and pseudocapacitors promises enhanced capacitances of the hybrid capacitor, greatly improved energy density compared with EDLCs, and better cycling stability than pseudocapacitors.

Numerous combinations have been investigated. Overall, the key to achieving high power and energy density with a long cycle life is to explore novel electrode materials and systems with the consideration of new materials with controllable morphology and size, new electrolyte with high voltage, ionic conductivity and stability, as well as new energy storage strategies, which will be discussed and reviewed later in Section 19.6.

19.2.2 Fabrication of Single Supercapacitor Cell

Although the fabrication technique varies based on electrode materials, electrolyte, and their applications, a single SC consists of two electrodes separated by a thin, porous, nonconducting separator. Subsequently, this composite is rolled or folded into a cylindrical or rectangular shape and sealed in a container with impregnation of the electrolyte.

19.2.2.1 Electrode Fabrication

The fabrication of commercial electrodes can be similar to that of experimental ones, although the recipes of commercial electrodes are commonly patented and not disclosed to the public. In general, an electrode is composed of a current collector and an active electrode material paste.

Considering the balance of high conductivity and relatively low cost, copper, nickel, or aluminum foil is ordinarily chosen as the current collector, with a thickness of 20–80 μm depending on cell design and applications. Before electrode preparation, current collectors are initially etched by acid or other chemicals to produce irregularities along the surface, and higher surface area leading to improved contact between the electrode and the current collector.

The electrode paste is prepared by mixing active materials (porous carbon or metal oxides), conducting additives (if applicable), and a binder agent (often a fluorine-containing polymer such as polytetrafluoroethylene, to increase durability and prevent the degradation of the collector–electrode interface over time) by using a solvent to obtain a paste or slurry material, with ball milling or ultrasonication to ensure a good dispersion and homogeneous composition. However, the amount of binder agent and conducting additive has to be carefully considered since they can seldom contribute any capacitance.

After the electrode paste preparation and current collector pretreatment, the electrode paste is attached to the current collectors followed by drying and pressing. Several strategies could be used to do this attachment; however, the adhesion between the electrode paste and the current collector has to be sufficient to minimize the internal resistance. At the very last step of electrode preparation, the electrodes are cut into the desired shape for further uses.

19.2.2.2 Electrolyte

The electrolyte is another critical component affecting the overall performance of SCs considering their functions during the energy storage mechanism. First, the working voltage of an SC must be kept within a certain range to restrict the breakdown of the electrolyte. Considering the quadratic relationship of the operating potential window to the energy density, the electrochemical stability of an electrolyte at a higher voltage is of great importance. Second, the electrolyte conductivity and subsequent dependence on operating temperature, as well as the electrode–electrolyte interface, contribute to the major part of ESR. Third, in a porous structured electrode, the size of cation and anion has to be optimized to guarantee their good accessibility to the electrode surface. Additionally, the electrolyte should have low volatility, low flammability, low corrosion potential, and low toxicity, for safety considerations.

The advantages of aqueous electrolytes include high ionic conductivity, low cost, nontoxicity, and availability. The relatively narrow potential operating range (0.8–1.0 V) avoids electrolyte decomposition and cell rupture, and metal collector corrosion due to chloride ion attack limits the wide application of aqueous electrolytes.

Organic electrolytes are currently more widely used in commercial SCs owing to their higher operation voltage (2.2–2.7 V with a short-term voltage tolerance of 3.5 V [12]). However, the specific capacitance achieved with organic electrolytes is normally lower than in an aqueous electrolyte, which negatively affects power density. Meanwhile, organic electrolytes suffer from a higher resistance. Fortunately, these reductions in power performance are balanced by the increase of potential window. Nevertheless, the raised safety issues, including toxicity and flammability, have to be noticed. The organic electrolyte is also very sensitive to moisture. Hence, the preparation of the electrolyte and the saturation of the electrode must be operated in a moisture-free environment. Any water within the electrolyte will cause gas formation at high voltage, resulting in higher internal resistance, lower power ratings, higher leakage current, disruption in cell sealants, and eventually reduced life span.

19.2.2.3 Separator

The separator allows the transfer of the charged ions but forbids the electrical contact between the electrodes. In consequence, the separator should have satisfactory ion conductivity and be insulated to electrons. Meanwhile, considering that the separator has no contribution to the capacitance of the SC, the best separator should be as thin as possible but maintain sufficient mechanical strength in order to resist the migration of charged electrode materials motivated by electrophoresis and allow the winding in the following procedure. The resistance to corrosion from electrolytes and by-products of electrode degradation should also be taken into account. Additionally, the ion flow resistance and interfacial contact resistance between the electrolyte and the separator can be alleviated by increasing the wettability.

19.2.2.4 Cell Design

The selection of the design, assembly, and packaging of SCs is ultimately dependent on the targeted market and cost-efficiency. On the basis of the stacking strategy of electrode and appearance, there are mainly three kinds of designs: coin cells, cylindrical cells, and pouch cells, with the same construction of a pair of electrodes separated by a separator and immersed in electrolyte.

Coin cells are exclusively used in applications where power and capacity are not the key factors. The schematic of a coin cell is shown in Figure 19.3a. A single-layer electrode–separator–electrode sandwich system is tightly sealed in a metal case with a plastic insulator between the top and bottom cases by high pressure. The hydrophobic insulator or gasket can also help prevent the electrolyte from leaking and the moisture from entering the cell. Limited to the volume of the cell, the capacitance of a coin cell is generally smaller than 1 F, and the utilization of the volume is important to attain higher capacitance. Meanwhile, a high-level compression is needed to achieve good contact between the sandwich system and the cases.

With the quest for high capacitance, it is necessary to increase the mass of the electrode. In cylindrical cells (Figure 19.3b), the electrode and separator are made in long strips of foil and rolled into a spiral or cylindrical shape. Cylindrical cells thus have only two electrode strips, simplifying the construction considerably. A single tab connects each electrode to its corresponding terminal placed on different sides, although high-power cells may have multiple tabs welded along the

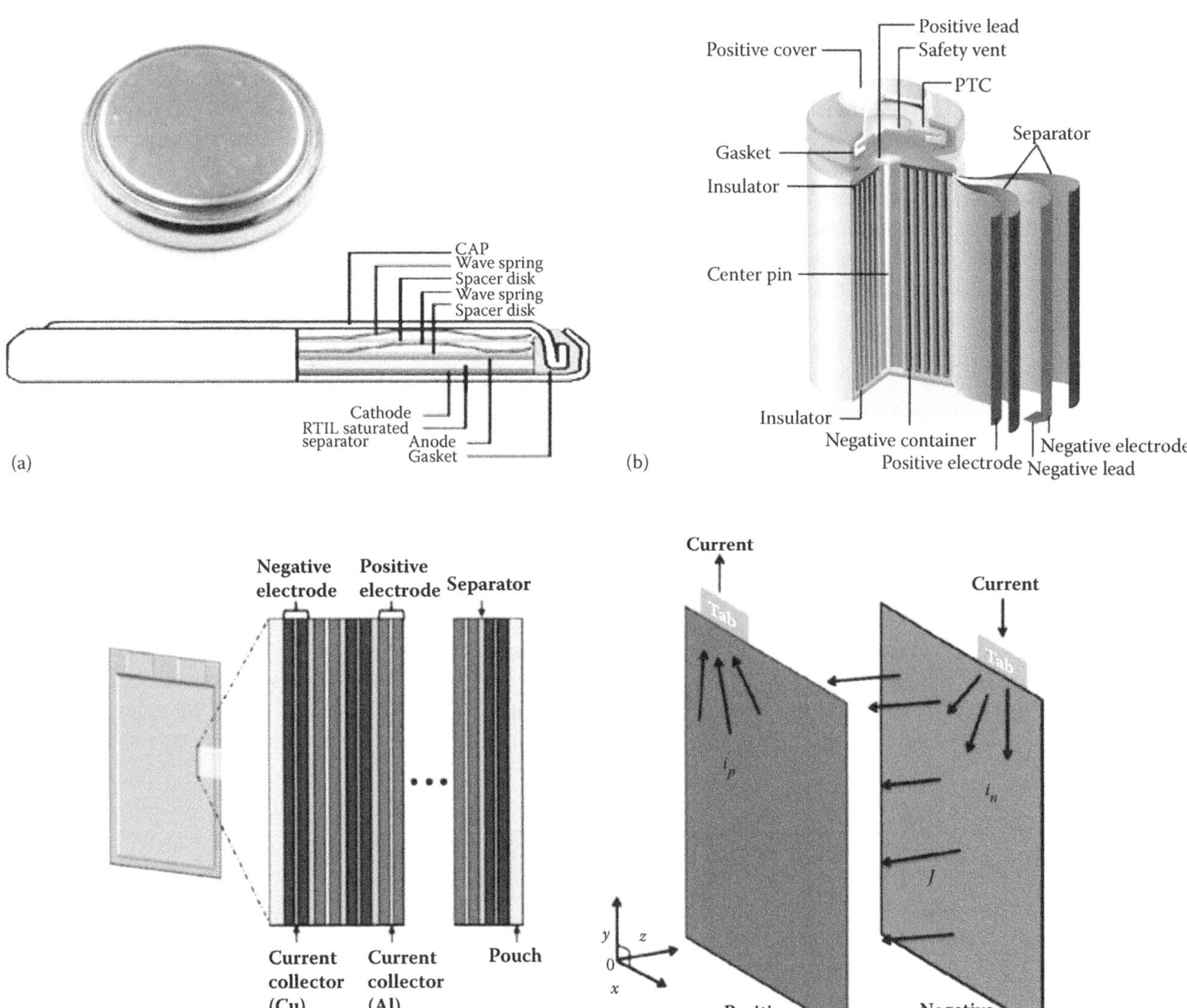

FIGURE 19.3 Schematic of (a) coin cell, (b) cylindrical cell, and (c) pouch cell. (From *J. Electrochem. Soc.*, 160, A437–443, 2013.)

edges of the current collector. Then, the roll together with the terminals is positioned in a conductive casing with an insulating gasket separating the outer casing. The electrolyte is then injected into the cell. A gas release vent, gasket, and top cap are located on the top of the can. The whole device is eventually sealed mechanically. This construction provides cells with very low internal resistance. The downside is that since the electrodes take up more space within the can, there is less room for the electrolyte and so the potential energy storage capacity of the cell is reduced. Cylindrical cells are popularly commercialized with a capacitance in the range of subfarad to thousands of farad.

To further increase the packaging efficiency, the pouch cell design (Figure 19.3c) eliminates the metal enclosure to reach maximized energy density. In pouch cells, the electrode and separators are usually stacked in layers or laminated together and enclosed in a foil envelope. Cells thus can be made very thin and offer optimal space, resulting in improved gravimetric and volumetric power and energy densities. This thin rectangular shape also guarantees a shorter interconnection between several cells in modules and lowers the pack resistance. However, pouch cells are very vulnerable to mechanical damages from both external and internal factors. Moreover, the possibility of swelling induced by temperature variation and/or gas generation during the reaction has to be considered when designing cells for some specific applications. Owing to the simplified assembling, pouch cells are not suitable for restrictive environment applications since they are not water and airtight for the long term. The durability of pouch cells might also be influenced compared with cylindrical cells.

19.2.3 Module Design

Most of the applications of SCs require much higher voltage, energy, and power density that no single SC can offer. A number of SCs could be combined together to attain those requirements. The fabrication of SC stacks is generally classified into two families: monopolar cell stacking and the bipolar cell design.

19.2.3.1 Monopolar Cell Stacking

The simplest way to construct an SC stack is to connect a series of single SC cells. There are two ways to connect single SC cells: stacking in series and stacking in parallel. High operating voltages could be attained by connecting single cells in series, while greater capacitance could be achieved by a parallel connecting design.

New components, such as connections between cells, stack terminals, external casing, and insulator between cells and external casing, have to be involved to design stacking cells. Terminals, which should possess the ability to undertake high current density, are generally welded directly to the first or last cell to minimize the resistance. On the other hand, the key problem of connections is that they have to be designed in accordance to the geometry of cells and terminal, and low resistance is also preferred. Besides providing insulation, the insulator should also assist in the heat transfer from the stack to the environment and absorb the geometrical dispersion of the stack. However, considering that only the active materials in individual cells contribute to the performance of the SC stack, minimizing the passive materials is always a big concern to maximize the power and energy densities. Manufacturers thus have to balance the safety, energy deliverability, and weight of the stacks. Meanwhile, the welding joint, long connections, and the accumulation of the resistance induced by the partial contact between tabs and current collectors bring considerable pack resistance to SC stacks.

More attention should be paid to parallel connections because the voltage of each individual SC has to be similar and the voltage difference that each SC subjected to should be smaller than the limitation that one SC could tolerate. Meanwhile, nonuniform charge distribution could also be stimulated during usage. Details of voltage and charge balancing will be discussed later in this chapter.

19.2.3.2 Bipolar Cell Design

To minimize the employment of passive materials such as connections, terminals, extra casing, etc., the bipolar design, which has been commonly incorporated in batteries and fuel cells, was then introduced to the fabrication of SCs. The schematic of a typical bipolar-designed SC is depicted in Figure 19.4. Each bipolar electrode, the key component in this design, was attached to positive and negative electrode materials each on one side. Each bipolar electrode thus works as the anode electrode of one SC unit, and functions as the cathode electrode of the adjacent unit. The electrolyte floods between the spaces existing between neighboring bipolar plates, and bipolar plates and endplates. By this means, a number of individual SC cells can be assembled in series without the addition of materials and external connections. The volume,

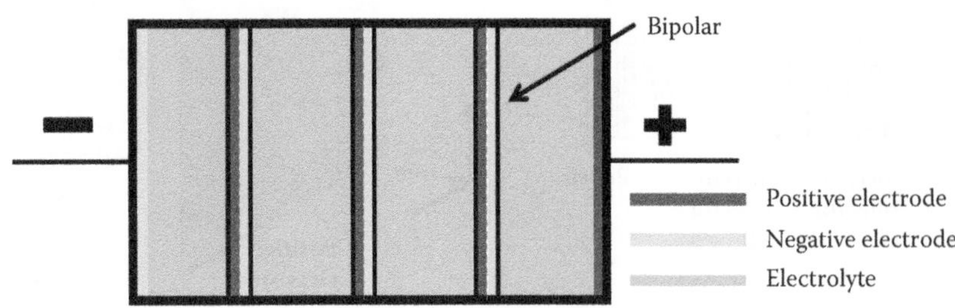

FIGURE 19.4 Schematic of bipolar design.

weight, stacking sizes, and internal resistances are extensively reduced compared with the stacking module.

Besides the above benefits, close attention should be paid to two main concerns. First, the electrolyte must be prohibited from crossing any bipolar plate in order to prevent short circuits. To ensure that each space between neighboring bipolar plates or a bipolar plate and an endplate is electrically isolated from its adjacent space, sealants are placed between bipolar electrodes and outer casing. On the other hand, heat dispersion of bipolar-designed SC is more difficult than that of stacking cells, because of the reduced surface area and resultant high volumetric efficiency of bipolar cells. Overheating could be more severe in bipolar-designed SC if the heat generated during the charge and discharge process is not effectively managed. Hence, bipolar plates demand excellent intrinsic conductivity and good contact with electrode materials to minimize internal resistance. External pressure is also optional to further improve material contact with the bipolar plate.

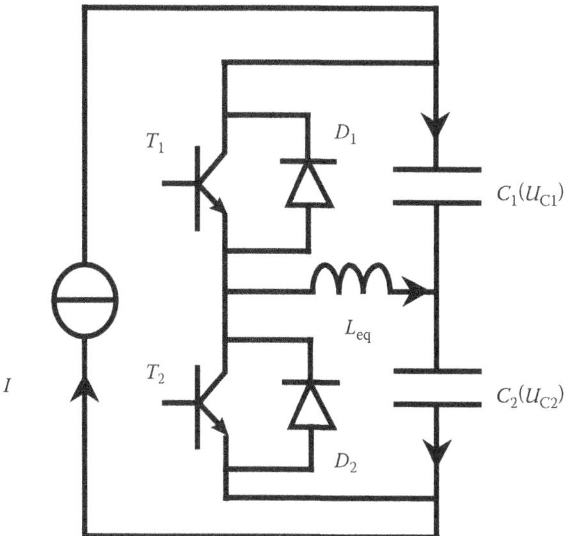

FIGURE 19.5 Active voltage balancing schematic based on buck–boost topology.

19.2.4 Other Considerations

A number of factors have to be considered, since the performance of SCs is also highly dependent on some other factors such as voltage decays and voltage balancing, besides the topics discussed above.

19.2.4.1 Voltage Decay

On the basis of the principle of SC, the voltage of an SC is linear with respect to the amount of remaining charges. As the SC discharges, the voltage will subsequently decrease. The voltage requirement of a loading has to be considered. If the load can function over a range of voltages, SCs could be directly applied. In an application that requires a constant voltage output, a DC–DC converter should be combined with SCs to maintain a constant voltage output.

With respect to the adaptability of loads to voltage, the relation between voltage and energy of an SC ($E = CV^2/2$) implies that 75% energy is released with a 50% voltage decay, while only 10% energy remains when the voltage decreases to 30% of its fully charged voltage.

19.2.4.2 Voltage Balancing of the Cells

Variations in the resistance and capacitance of individual SC make it impossible to have the same voltages across each single cell in series SC stacks during discharge and charge. In general, small capacitance leads to high local voltage and big internal resistance causes low local voltage. A higher cell voltage during the charging process might lead to excessive aging, gas evolution, and even stack rupture. Although large effort has been done to improve the consistency in the capacitance and resistance of SCs during manufacture, cell to cell voltage balancing is still required since cell parameters are never fully equal because of process and aging variation.

The simplest method of voltage balancing is achieved by connecting a resistor across each cell. This parallel resistor drives a bypass current out of the cells when voltage goes beyond a balanced voltage. The disadvantage of this solution includes waste of energy through resistors in the form of heat, and low efficiency resulting in loss of energy and a lengthy charging process [13]. An improved solution uses Zener diodes, instead of resistors, as the bypassing elements to regulate the voltage of each cell equal to that of the diodes. Energy losses are lessened by this means since current only passes the diodes when the voltage is beyond the local threshold voltage.

The passive balancing methods by drawing bypass current to normalize cell voltages are mainly suited for low-power applications. More efficient balancing methods are desired for high charging and discharging rates. An active balance schematic based on the buck–boost topology is shown in Figure 19.5. When a significant voltage difference is detected between U_{C1} and U_{C2}, an equaling current I_{eq} is generated by the coordination of the transistors until the voltage is balanced. The current density and direction is determined by U_{C1} and U_{C2}.

19.2.4.3 Polarity

Polarity is obviously very important for asymmetric SCs because two electrodes are made from different materials. Especially, when electrochemical reactions are involved in one of the electrodes, the electrode potentials might be different. Therefore, two electrodes must be distinguished and their polarity also must be labeled.

As for the symmetric SCs, two electrodes are commonly assembled with the same material. Theoretically, they thus do not have this polarity issue. However, because materials have different behaviors when positively or negatively charged, the volume or mass, and even integration, of the electrode possibly suffers a change during the preparation of different electrodes. Taking this factor into consideration, labeling of the polarity of each electrode is preferred. However, no catastrophic failure strikes when an SC is not charged according to expected polarities. There are even no obvious instant effects on cell performance when the voltage of SCs is reversed. However, the cell will show degraded performance in the subsequent cycles even after it is reconnected correctly, and the lifetime of the device will surely be reduced. Therefore, manufacturers always indicate the intended polarities for use to avoid reverse potential and resultant loss of lifetime.

19.2.4.4 Venting

Most of the commercial SCs are sealed to minimize the environmental effect and prohibit the vaporization of the electrolyte. The gas evolution is hence a concern especially in stiff metal cylindrical cells. A one-way safety vent can be integrated in SCs when the device works in a critical environment where pressure and/or heat buildup is a concern. This one-way vent releases gas as long as the internal pressure exceeds a defined stress, but prevents moisture and other hazard substances from getting into the cell.

19.2.4.5 Temperature Effects

SCs work normally in a relatively wider temperature range than batteries, especially with organic electrolytes. However, the SC performance can be affected by operating temperatures. In general, an increased operating temperature will reduce the resistance of the electrolyte resulting from the increased mobility at higher temperatures. Meanwhile, high temperature can also lead to degradation in capacitance and performance, and bring the risk of electrolyte decomposition. This temperature effect could be more severe in SC stacks because of their high volumetric efficiency.

19.3 SUPERCAPACITOR PERFORMANCE TESTING AND DIAGNOSIS

The characterizations of SCs including cyclic voltammetry (CV), charging–discharging curves, and electrochemical impedance spectroscopy, are commonly used to evaluate the performance of SCs, determine the suitability of SC to certain usage, and investigate the individual function of each component. In this section, a number of methods for testing and diagnosis of SCs will be presented.

19.3.1 Cyclic Voltammetry

CV is widely used to characterize several performances of SCs, including electrochemical kinetics, reaction reversibility, and reaction mechanism. During a CV test, a cyclic linear potential sweep between the specific upper and lower potential, which is established by the electrolyte working window, is added across the electrodes. The resultant current is measured between the working electrode and the counter electrode, which could reflect the thermodynamics and kinetics of electron transfer between electrodes and electrolyte, as well as the kinetics of the chemical reactions initiated by electron transfer in solution. In general, the current will increase as the potential reaches the reduction/oxidation potential, but then decrease as the concentration of the electrolyte near the electrode is depleted. For SCs, the recorded current can be estimated by the predetermined constant potential scan rate ($v = dV/dt$) from Equation 19.2.

$$I = vC \tag{19.2}$$

where C is the capacitance at a certain potential. Since the potential sweep rate is constant, the voltammetry can also be depicted as capacitance versus voltage as well as current versus voltage.

For ideal EDLCs with no resistance, the CV should present a rectangular shape, since the capacitance should not change at various voltages. A typical CV curve of porous carbon materials presents a nearly ideal performance, as shown in Figure 19.6a. The deviation during initial charging and discharging at corresponding voltage boundaries, which could be observed for most real EDLCs, is commonly attributed to the diffusion resistances. In pseudocapacitors, because the charging and discharging processes possess faradaic reactions, prominent peaks appear within corresponding voltages. A typical CV of pseudocapacitors is presented in Figure

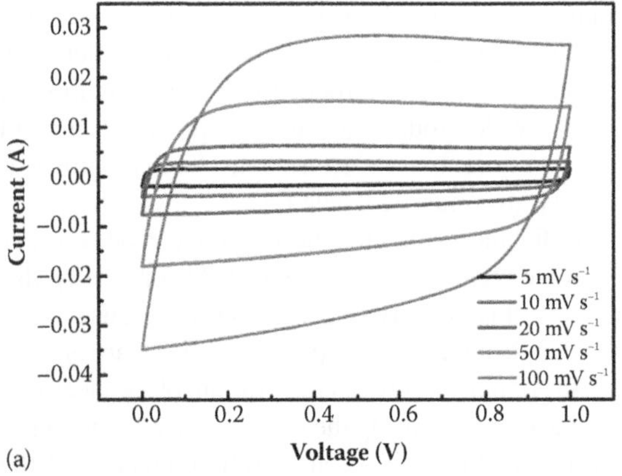

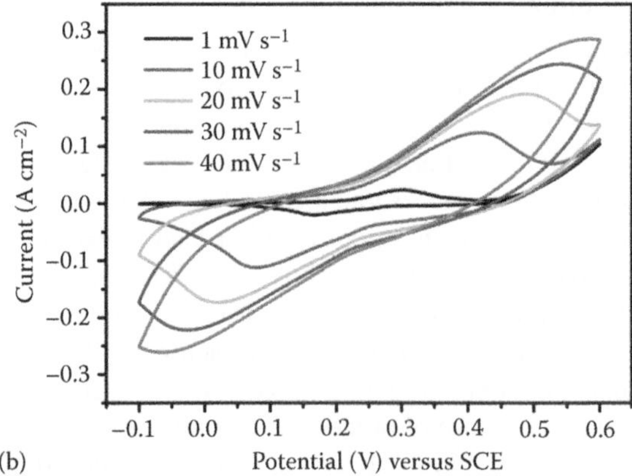

FIGURE 19.6 CV of a typical EDLC (a) and pseudocapacitor (b) at various scan rates.

19.6b. In both cases, the capacitance of an SC within the voltage boundaries can be estimated from the CV curves by integrating the current response, which could be expressed as the following equation:

$$C = \int \frac{I(V)}{v} dV \tag{19.3}$$

The rate capability of SCs can be reflected from CV as well by manipulating the potential scan rate, because a bigger scan rate means that the SC is charged or discharged in a shorter time between the limiting potentials. A systematical study of CV curves at increasing scan rate, as shown in Figure 19.6a, demonstrates the impact of power level on charging and discharging characteristics. At a lower scan rate, a CV curve shows a nearly ideal rectangular shape, while upon the increase of scan rate, the CV curves are distorted due to the increase of resistance at the higher scan rate. Calculation shows that the specific capacitance of SC decreases with increasing scan rate, which is ascribed to the sluggish ion transport from the electrolyte to the electrode surface at rapid cell voltage or current changes, and the resultant boosting inaccessible pore portions of the electrode. This capacitance decay with increasing scan rate is even worse in pseudocapacitors (Figure 19.6b) because it suffers from a slower electrochemical reaction due to the difference between faradaic and nonfaradaic reactions. This dependency of capacitance on the potential scan rate reflects the mass transfer kinetics with SCs.

The degree of reversibility of electrode reactions, which is very important for rechargeable energy storage devices, can also be revealed from CV curves. If the reaction is reversible within the potential range, the peaks present in the forward and backward scans should be in pairs, and similar in shape and position, although there might be some deviation due to polarization, and the CV curve should be closed after each cycle. More information about the CV technique can be found from Ref. [14].

19.3.2 Chronopotentiometry and Charging–Discharging Curves

Chronopotentiometry is another basic technique to determine the capacitances and round-trip efficiencies of SCs. Different from CV, a constant current is applied to the electrodes in this technique. The potential of the working electrode against the reference electrode is then recorded as a function of time from a three-electrode system, while the voltage between working electrode and the other electrode is measured from a two-electrode system. This method is also called galvanostatic cycling.

For nonfaradaic processes, the potential will change linearly with time when a constant current is applied due to the presence of an interfacial charging between the interface of the electrode and the electrolyte. A schematic diagram of representative chronopotentiometry curve on an EDLC is presented in Figure 19.7. As soon as the current is applied,

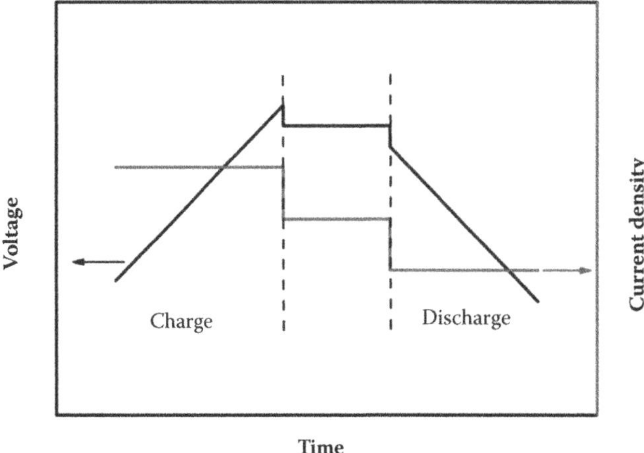

FIGURE 19.7 Charging and discharging the chronopotentiometry curve.

an instantaneous voltage jump appears. The jump is initiated from the ohmic current flow in capacitors due to the existence of series resistance. Then, the voltage increases linearly with time until the predetermined voltage boundary, followed by an instant voltage drop due to the disappearance of the current referred to as the IR drop. During the rest time between charging and discharging, voltage may drop somewhat due to self-discharge redistribution, and this phenomenon is more severe at higher current densities. The mechanism of self-discharge will be discussed in detail in Chapter 19.4. During the following discharge process, similar features are observed.

In pseudocapacitors or hybrid capacitors, the character of voltage is not absolutely linear versus time due to the existence of faradaic reactions. In a typical chronopotentiogram of the charging process, the voltage increases linearly at the beginning of charging as the double-layer capacitance at the electrode is charged until the voltage at which the nonfaradaic reaction begins is reached. Then, the increase of voltage slows down due to the pseudocapacitive chemical process until the reactants in electrolyte near the surface are exhausted and the diffusion of reactants is no longer sufficient to maintain the reaction. The chronopotentiogram in this range also presents linearity with a smaller slope if the pseudocapacitive process occurs continuously. Otherwise, the chronopotentiogram shows nonlinearity in this range. After the exhaustion of the reactant near the surface of electrode, the voltage increases more quickly again. The slope change would not be obvious as the nonfaradaic reaction occurs continuously over the whole voltage window or at large constant current densities.

The capacitance of SCs can also be calculated from charging and discharging chronopotentiometry curves. As described by $C = \Delta Q/\Delta V$, the capacitance of a device that exhibits a linear relationship of voltage versus time in a chronopotentiogram can be determined directly by constant current charge/discharge data as follows:

$$C = I\frac{\mathrm{d}t}{\mathrm{d}V} = I\frac{\Delta t}{\Delta V} \tag{19.4}$$

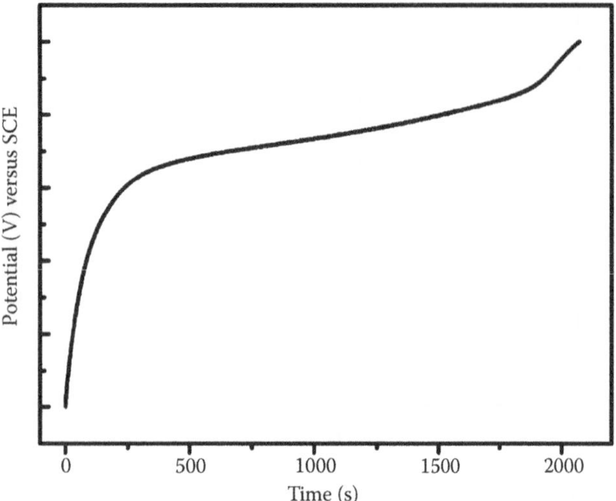

FIGURE 19.8 Schematic diagram of a chronopotentiogram for the charging process including a faradaic reaction.

where I is the constant current, ΔV is the potential difference between two boundary voltages, and Δt is the time used for voltage changes from one boundary to the other. It is also noticeable that $\Delta t/\Delta V$ can be read directly from Figure 19.8 since it is the inverse slope of the chronopotentiogram during charging or discharging. The selected voltage window should have no influence on capacitance. For pseudocapacitors, where the chronopotentiogram is not as linear as desired, or for the hybrid SCs, whose V–t curve profile is not exactly linear, capacitance can be determined by integrating the inverse slope, which can be expressed by the following equation:

$$C = I\int \frac{1}{V(t)}\mathrm{d}t \tag{19.5}$$

Owing to this nonlinearity, the capacitance is dependent to some extent on the selection of the voltage window. The selected voltage range should cover all voltages at which the nonfaradaic reaction occurs.

The resistance of a capacitor can also be determined by using a chronopotentiogram from the IR drop at the initiation of a constant current discharge, or the voltage recovery at the interruption of a discharge or charge current as indicated by Ban [15]. Meanwhile, the energy and power density can also be calculated based on the chronopotentiogram. This will be comprehensively discussed in the following section.

19.3.3 Energy Density and Power Density

Specific energy density and specific power density are two of the most important characters for energy storage devices to evaluate their performance. High energy and power density are both desired in applications, especially for those involving high loads in short times. However, increasing one of them essentially means to decrease the other owing to the polarization losses and a reduction in cell voltage resulting from the current density increase. A balancing between energy and power density is necessary when choosing a device or electrode material.

19.3.3.1 Energy Density

The energy density is defined as $Q\cdot\Delta V$. The energy density can be achieved by integrating charges as follows:

$$E = \int Q\,\mathrm{d}V = \int CV\,\mathrm{d}V \tag{19.6}$$

Assuming that the capacitance is independent of voltage, as it is in the ideal EDLCs and pseudocapacitors, the energy density could be further calculated as

$$E = C\int V\,\mathrm{d}V = \frac{1}{2}CV^2 \tag{19.7}$$

In practice, the voltage of the capacitor is restricted to the range between its maximum voltage and half of the maximum voltage. A complicated and costly circuitry must be involved to regulate the voltage at a relatively high level to satisfy the requirement of a load, if we want to utilize the remaining 25% of the stored energy. Therefore, the maximum usable specific energy in a SC is given by

$$E_{\text{usable}} = \frac{3}{8}\frac{C}{M}V_0^2 \tag{19.8}$$

where V_0 is the maximum voltage and M is the device weight. Hence, the most effective way to increase the energy density of an SC is to find an electrolyte with a high voltage window, an electrode with high capacitance, and lighter other components.

19.3.3.2 Power Density

Power density is another important performance factor of SCs indicating how quickly the energy can be stored or released, which is determined by the nature of the electrode materials, resistance within the electrolyte, and kinetics of polarization at certain potentials. The power delivered from an SC is related to the discharge current density on a certain voltage and defined as

$$P = I\cdot V \tag{19.9}$$

In general, the voltage V or ΔV decreases with the increase of current I, because the existence of internal resistance in both electrolyte and electrodes, and the kinetic polarization will lead to voltage losses at higher current densities. The maximum deliverable power density is grossly estimated as [16]

$$P_{\max} = \frac{V_C^2}{4R_{\text{esr}}} \tag{19.10}$$

where V_C is the open circuit voltage of an SC and R_{esr} is the ESR, considering the capacitance of the material in series with an ESR. However, this estimated maximum power can only be reached at a specific circumstance in which half of the energy is dissipated in the form of heat while the rest of the energy is electricity. The resultant energy efficiency (50%) is unacceptable for almost all applications.

Another method to estimate the power density of an SC is derived by Burke [17] based on pulse discharge instead of constant discharge, expressed as

$$P = \frac{9}{16}(1 - EF)\frac{V_0^2}{R} \tag{19.11}$$

where EF stands for pulse efficiency. In both forms, the power density is proportional to V_0^2/R. An SC that can work in a wider voltage window with a smaller ERS is hence desired to attain high power density.

19.3.3.3 Ragone Plot

Ragone plot is used to evaluate and compare the performance of different energy storage devices and related materials by plotting the specific energy versus specific power density in logarithmic axes [18]. It indicates the energy storage capabilities of energy storage devices at different power levels.

In the Ragone plot, the leakage current is neglected and the current only flows through the ESR (R_{esr}) and the external load R_L. The energy and power density available for the load (E_L and P_L) can hence be written as

$$E_L = \frac{R_L}{R_L + R_{esr}} E_{max} \tag{19.12}$$

$$P_L = \frac{V_L^2}{R_L} = \frac{1}{R_L}\left(\frac{R_L}{R_L + R_{esr}} V_C\right)^2 = \frac{V_C^2}{4R_{esr}} \frac{4R_L R_{esr}}{(R_L + R_{esr})^2} \tag{19.13}$$

where E_{max} is the maximum energy delivered from the SC and V_C and V_L are the open circuit voltage and the voltage across R_L, respectively. Considering the definition of P_{max} in Equation 19.10, Equation 19.13 can be further developed as

$$P_L = \frac{4R_L R_{esr}}{(R_L + R_{esr})^2} P_{max} \tag{19.14}$$

Combining Equations 19.12 and 19.14, the relationship of available energy density and power density is

$$E_L = \frac{1}{2} E_{max}\left(1 \pm \sqrt{1 - \frac{P_L}{P_{max}}}\right) \tag{19.15}$$

where "positive" should be taken when load is bigger than ESR and "negative" corresponds to the opposite case. Since the loading resistance is bigger than ESR in most of the applications considering the efficiency, the available energy density can be depicted as a function of the output power across the SC, as shown in Figure 19.9.

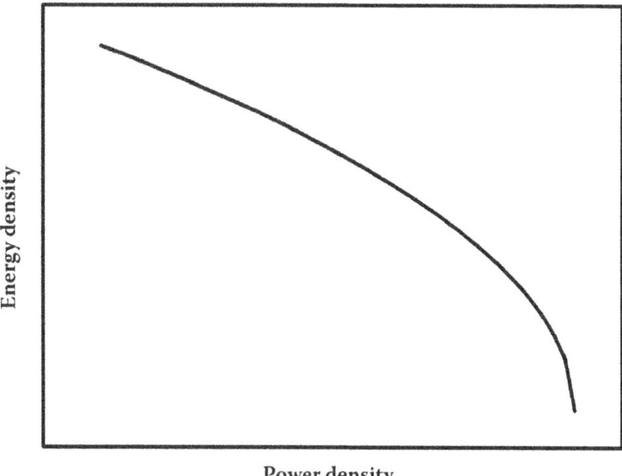

FIGURE 19.9 Ragone plot representing the available energy as a function of available power based on Equation 19.15 in the case of $R_L > R_{esr}$.

19.3.4 Cycling at Constant Power

The estimation of capacity, specific energy, and power density mentioned above based on chronopotentiometric charging and discharging does not provide the real situation because these parameters are all dependent on voltages but the voltage is changing during the measurement of these parameters.

To accurately determine the practical power and energy performance of an SC, constant power measurements can be applied. In a typical constant power measurement process, devices are discharged and charged at a fixed constant power by varying the current density with the help of a feedback loop, while the voltage changes is recorded, until a preset voltage is reached. The corresponding energy density of an SC at a certain constant power can therefore be calculated from the practical power and time.

It should be noticed that the current density becomes larger as the voltage decreases, and hence, more energy will be dissipated in the form of heat, resulting in reduced efficiency.

19.3.5 Electrochemical Impedance Spectroscopy

Electrochemical impedance spectroscopy (EIS) is a powerful technique for evaluating the electrochemical behavior of the electrode and/or the electrolyte, and their interface. Different from previous techniques, EIS applies a small periodical time-dependent electrical stimulus (typically sinusoidal voltage or current) with different pulsations to the electrodes, and analyzes the response functions of the resulting current or voltage without disturbing those measured properties. Since the stimulus signal is very small, a linear relationship exists between the stimulus signal and the measured response. To get accurate measurement, the target system or device should be under steady-state conditions. On the basis of this technique, system properties can be expressed as a function of time considering the response to the stimulus. The system can hence be simplified by the application of Fourier transformation, but changing from time domain to frequency domain. By this transformation, the impedance can be represented as follows, and the details about this transformation can be found from Ref. [19]:

$$Z(\omega) = Z_r + jZ_{im} \tag{19.16}$$

where Z_r and Z_{im} are the real and imaginary parts of the total resistance. Commonly, the EIS is expressed in the form of a Nyquist plot that plots Z_{im} versus Z_r. This technique enables the separation and determination of circuit elements of the constructed device such as ESR, double-layer capacitance, faradaic capacitance, and leakage resistance by fitting the impedance spectra with equivalent circuit (EC), which behaves similarly.

In an ideal case, EC generally contains the ESR (R_{esr}) related to the electrode, electrolyte, separator and contact resistance, the double-layer capacitance (C_{dl}) according to the charge accumulation at the interface of the electrolyte and electrode, the charge transfer resistance (R_{ct}) linked to electrochemical reactions, and the pseudocapacitance (C_F). Detailed discussions about the equivalent circuit containing all these elements have been presented [12]. The effect of each element on the shape Nyquist plots is also inspected.

In a real case, the local concentrations of reactant in the interface of the electrode and electrolyte have to be involved especially at very high charging and discharging rates, and/or when the concentration of the reactant is low, where diffusion control takes the most important role in kinetics. The Warburg impedance element is then involved to define the polarization of an electrochemical system due to diffusion limitation. For simple charge transfer reactions limited by diffusion control, the Randles equivalent circuit (as shown in Figure 19.10) is employed to represent the impedance behavior.

The Nyquist plot based on the Randles equivalent circuit is displayed in Figure 19.11, which is composed of three parts. A semicircular form occurs at high frequencies, which is similar to the real case because this semicircular form is mainly attributed to the double-layer capacitance. However, a nonvertical slope is always observed from low frequencies.

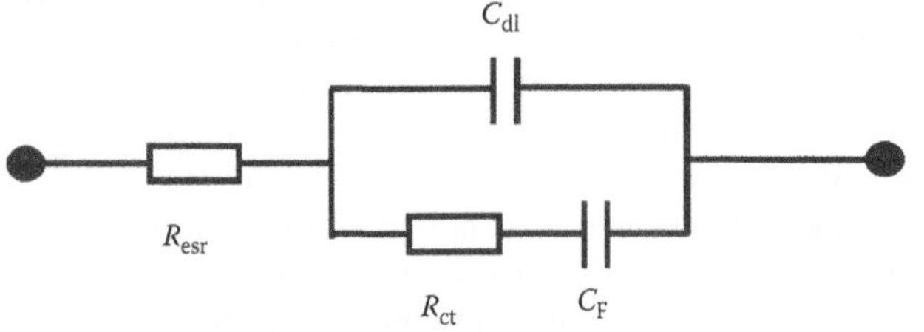

FIGURE 19.10 Randles equivalent circuit.

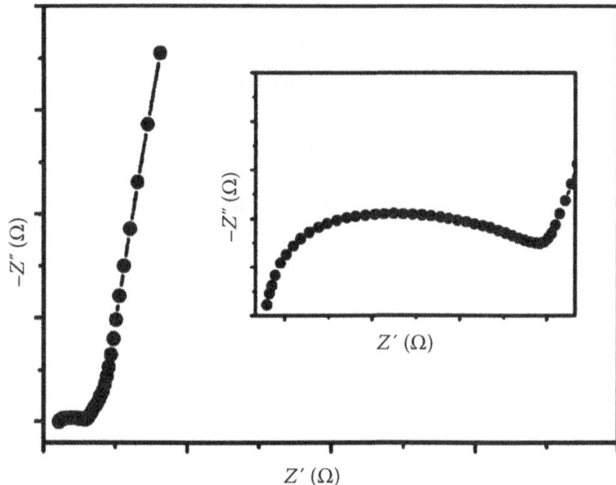

FIGURE 19.11 Typical Nyquist plot based on Randles equivalent circuit.

Meanwhile, a 45° region presents between high and low frequencies. This 45° region arises for the Warburg impedance element, which demonstrates a frequency dependence on the square root of ω^{-1} or t in both imaginary and real components of resistance. This square root relation of Warburg impedance element and ω^{-1} or t can be explained by the dynamics of particle diffusion in a concentration gradient [20]. The most reliable parameter decided from the Randles equivalent circuit is R_{esr}, which is in series with other components. The value of R_{esr} can therefore be obtained directly from the intercept of real impedance when the frequency goes very high.

Actually, real electrode behavior is more complex since other effects including electrode porosity, surface roughness, and active sites for energy storage have a big influence on the shape of the Nyquist plot. For example, small pore size prohibits the penetration of alternating current (AC), especially at high frequencies [21]. Thus, the applicable surface area increases as the frequency decreases.

19.3.6 Leakage Current

Self-discharge is a common phenomenon and often observed when charged SCs are left on open circuit, similar to batteries. However, energy stored in SCs can be lost faster than in batteries, especially at higher temperatures and voltages because of the difference in thermodynamics and kinetics of the energy storage mechanisms compared with batteries [22]. The driving force for self-discharge is the thermodynamic unstable state of higher free energy compared with the discharged state, which is the same as in the discharge process through a load. The performance of SCs is therefore diminished because of the voltage decay caused by self-discharge. As a result, the reliability of the SCs becomes an issue in device designs. The characterization of the self-discharge of an SC is required to evaluate its performance before further applications. The self-discharge of a charged SC can be triggered by different reasons, which will be discussed in Section

19.4, and the methodologies for self-discharge measurement will also be discussed.

Self-discharge can be measured in two primary ways, i.e., recording the current required to keep a constant voltage and measuring the open circuit voltage as a function of time. For the first method, a float current is applied to a charged SC to maintain the preset potential. The float current is then just equal to the self-discharge rate at the preset potential. This method, however, cannot help us to understand what occurs in the electrodes.

In the other practical procedures, the open circuit potential difference between two electrodes is recorded with respect to time. The measurement period varies from hundreds of seconds to weeks, and even months, depending on the requirements. For fundamental research, the time interval should be between milliseconds to hundreds of seconds, while it requires weeks or months to evaluate device performance. It is noticeable that the input impedance of the device used to measure the potential should be big enough to make sure that the measuring circuit itself does not significantly influence the time-dependent voltage signal by drawing charge from the capacitor, leading to an enhanced rate of self-discharge.

If the self-discharge is controlled by an activation-controlled faradaic process from overcharge or the oxidation–reduction reactions that follow the Tafel equation [23], the voltage drops logarithmically with time. The slope is determined by the Tafel slope factor. On the other hand, if the self-discharge is a diffusion control process, the voltage declines linearly with the square root of time. The most important fact determining the slope is the initial concentration of the diffusible redox species. In the case that the self-discharge is solely caused by leakage current, the logarithm of voltage then falls linearly with time. The fundamental differences between the above situations offer a facile way to distinguish the types of self-discharge mechanisms in practical cases.

For asymmetric hybrid SCs, a three-electrode arrangement can be employed, in which the potential changes of two electrodes depending on time are individually recorded against the reference. The respective performance of two electrodes can be determined separately.

19.3.7 Cycle Life Measurement

The cycle life of an SC is a principal criterion in evaluating its performance. A charge–discharge cycling test over many cycles (up to 100,000 cycles) is the most reliable method to investigate the degradation of an SC by comparing the initial capacitance and the ESR of the first cycle with those of subsequent cycles. In general, the ESR of an SC increases with prolonged charge–discharge cycles and the capacitance of a SC decreases, resulting in the degradation of both energy and power density of the device.

19.3.8 Other Test Procedures

Other test procedures are also important to evaluate the performance of SCs. Except for the constant current or power charge–discharge test, the charge or discharge test can also be carried out under a constant voltage or through a constant load. In applications where SCs experience frequent transient operations, pulse cycle tests other than the constant cycle test should be employed to evaluate the performance of these SCs. The temperature effect on the previously discussed performance can also be involved in device tests especially for commercial applications. Hence, these test procedures can be done at various temperatures based on the application requirement to estimate the undesired effects arising at high or low temperatures.

The safety test is another crucial aspect of SCs. The temperature generation especially on a high rate charge or discharge should be recorded. The resistance to overcharge and overdischarge can also be monitored. A heat-to-vent test can help decide whether the SC is safe enough even at extreme circumstances.

19.4 DEGRADATION MECHANISMS AND MITIGATION STRATEGIES

Incessant charging and discharging of an SC generally leads to decay of capacitance and increase in ESR. The lifetime of an SC is defined by the time duration from its first use in the application until the SC cannot fulfill its specified function. On the basis of industry standards, the life criteria are defined as a 20% reduction of the capacitance, at which time the accessible electrode surface and the availability of ions are reduced, and/or a 100% increase of the equivalent serial resistance, at which time the electrode adhesion on the collectors weakens.

SC may also fail in a more severe way, in which gas is generated inside the cell during the charge and/or discharge process, and increases with prolonged cycles. When the pressure is high enough to be released automatically from the cell by opening a groove on wall, the SC fails. The reasons for this failure mode will be discussed in this section.

19.4.1 Cell Aging

The failure of an SC seldom occurs suddenly, but its performance gradually degrades due to capacitance loss, deterioration in rate capabilities, and increase in ESR. These factors are diagnosed as cell aging. Aging can be induced by several reasons, including electrolyte decomposition, electrode oxidation, pore access closure, and impurity in the electrolyte.

The gas evolution resulting from the decomposition of the electrolyte in electric double-layer capacitors was first reported by Hahn et al. in 2005 [24]. In their experiments, differential electrochemical mass spectrometry was employed to examine the gas evolution from an electrolyte solution composed of 1 M $(C_2H_5)_4NBF_4$ dissolved in propylene carbonate. The study showed the formation of carbon dioxide, hydrogen, and propene as principal products of electrolyte decomposition. Further study confirmed this result and distinguished the difference of positive and negative electrodes in gas evolution [25]. The results showed that only CO_2 and CO were formed at the positive electrode due to the oxidation of electrolyte, while propylene, ethylene, CO, and a small amount of hydrogen were generated at the negative electrode resulting from the solvent reduction.

In acetonitrile-based electrolytes, the decomposition of the electrolyte has also been observed [26]. The authors suggest that the active carbon electrodes work as the catalytic materials for the decomposition of acetonitrile and the formation of heterocyclic compounds such as pyrazines in liquid phase. Meanwhile, the acetonitrile decomposed into acetamide, acetic and fluroacetic acid, and derivatives depending on the cell voltage and humidity of the electrolyte. Fluoroborate, the solute, then works as the source of fluoride and hydrogen fluoride, and boric acid derivatives. The alkylammonium cation is destroyed at elevated temperatures by the elimination of ethane. The resultant fluorination destroys the etched aluminum support material under the active carbon layer. The active carbon positive electrode, on the other hand, loses cyclic siloxanes and aromatic contaminations even at room temperature. This surface layer exfoliation is accompanied by the generation of CO and CO_2.

As mentioned above, the partially oxidized electrode with the surface functional group also contributes to the performance degradation of SCs by participating in the decomposition of electrolyte as well as by the consequential reduction of the porosity of the electrodes. The BF_4^- anion in fluoroborate decomposes to $BF_xO_y^-$ ions in the presence of an oxygen-containing group accompanied by the formation of solid products from the decomposition of the electrolyte. The amount of products is related to the nature of activated carbon and their porosities, which suggest that the surface functionalities are involved in the electrolyte decomposition on the surface of electrodes [27]. These resulting solid products can be trapped in the pores within electrodes, causing a decrease in accessibility to the porous surfaces of the electrode, resulting in the decline in capacitance. The pores within the separator can also be blocked by these solid products, leading to an increase in electrolyte resistance [28]. The positive electrode is found to be aging faster than the negative electrode because fluoride and nitrogen species, including pyridinic and amine groups, can be bonded covalently to the graphitic structure of positive electrode, which facilitates the polymerization of acetonitrile and accelerates the aging of the positive electrode [29]. The above researches also showed that the concentration and nature of the surface group play an important role in the decrease in performance with operation time.

Water impurity in the electrolyte is another reason for the aging of SCs. Similar to the effect of the oxygen-containing group, the BF_4^- anion reacts with residual water to produce hydrogen fluoride in the positive electrode. Some protons forming H_3O^+ ions are transmitted to the negative electrode and generate hydrogen gas. The remaining protons work as a catalyst during the electrolyte decomposition. Water impurity in organic electrolytes is generally considered an important reason for the fading of electrode capacitance.

As a result of all above reactions, various gases are generated and the ESR is increased due to the decrease of the active surfaces of electrodes and the blocking of separator pores. This aging phenomenon is more severe at alleviated temperatures and/or overcharging and discharging voltages, because the reaction rates of the interactions between electrodes and the electrolyte are exponentially accelerated at a temperature or an operating voltage above the cell rating limits. The details about the temperature and voltage accelerating effect will be discussed in Section 19.4.3.

On the basis of this understanding of the aging mechanisms of SCs, mitigation strategies can be developed. First, the water content of the electrolyte should be controlled at a very low level (<20 ppm in general), and the application of extensively dried separator and electrodes is crucial for SCs to present longer lifetimes with higher capacitances and lower resistances. In the meantime, the mechanical stress caused by gas generation must be considered when designing an SC cell to make sure that the SC can mechanically resist the increase of internal pressure or that the pressure can be released when accumulated to a certain level. Except for the application of thicker covers to reinforce the stress of the cell, some other strategies have also been developed to solve this gas generation problem. Valves or selected membranes can be used to evacuate the generated gases. If a closed system is requested, a getter material can be used in the cell to condense the gases, or new chemicals can be developed to decrease the generation of gases.

19.4.2 Self-Discharge

Different from the performance decay over prolonged cycles, which has been attributed to cell aging as intensively discussed above, SCs also experience voltage decay in open circuit over time. This loss of voltage in open circuit is normally caused by self-discharge. The driving force of self-discharge is the higher positive energy state after charging relative to that for the discharged state. Although the self-discharge mechanisms vary depending on the chemistry and electrochemistry of SCs, such as the purity of electrode materials and electrolytes, the environmental temperatures, and the charged state, the practical phenomena are the same. Typically, when the SC stands without any connection between two electrodes for some time, certain voltage deterioration can be detected due to the loss of charge on electrodes. The performance of SCs in terms of energy and power densities is hence reduced. The self-discharge rate of SCs is the most important consideration for applications that are designed to deliver a reliable power supply after a period of time.

Different from the self-discharge in batteries, whose open-circuit voltage is thermodynamically decided by potential-determining electrode reactions, the potential difference of two electrodes in SCs electrostatically depend on the stored charge amount without any stabilization mechanism based on thermodynamics or kinetics. Hence, the cell voltage of SCs tends to be more sensitive to adventitious depolarizing processes, and SCs suffer greater self-discharge rates compared with batteries.

The mechanisms of self-discharge in SCs must be clarified before directing research efforts to minimize the self-discharge rates of SCs.

19.4.2.1 Charge Redistribution

Redistribution of charges within the deeper pores of the electrodes is a dominant self-discharge mechanism especially at a higher charge current density. Owing to the difficulty of charge diffusion during the charging process, the surface of deep pores in the electrode materials could not be fully accessed by charges in the charging process due to lack of time. After the SC is fully charged, the charges are still trying to travel further down the pores, and additional ions will continue to enter the unoccupied surfaces based on a diffusion controlled process until the charges are uniformly distributed on all surfaces of the electrodes. This kind of self-discharge normally occurs right after the charging finished and lasts for tens of hours, depending on the microstructures of the electrode materials including the surface area, pore size, pore volume, and geometry of the pores. In general, 50 h or less is required to guarantee a uniform distribution of charges along the pore surfaces. Hence, a slow voltage decline is observed if charge redistribution occurs. A smaller charging rate is helpful to alleviate this kind of self-discharge. Moreover, an unsophisticated structure can also reduce the time required for the redistribution of charges but results in diminished capacitance.

19.4.2.2 Decomposition of Electrolyte Due to Overcharge

If the SC is charged to a voltage beyond the respective thermodynamic limitation of any component in the electrolyte, the electrolyte will be electrochemically oxidized or decomposed. This oxidation or decomposition of the electrolyte leads to spontaneous self-discharge. In aqueous electrolytes, water could be oxidized to oxygen at the positive electrode and reduced to hydrogen at the negative electrode. The overcharge potential decreases with the self-discharge until the overcharge potential reaches almost zero. The self-discharge rate in this mechanism also decreases with the overcharge potential in an exponential relationship based on a Tafel relation, since the self-discharge process is in accordance with a faradaic charge transfer reaction [30]. The self-discharge based on this mechanism stops automatically when the potential difference of two electrodes is equal to the respective reversible potential for the oxygen or hydrogen evolution reaction in an aqueous system. Self-discharge induced by decomposition of the electrolyte due to overcharge can be prevented by restricting the working voltage of the SCs. This kind of self-discharge might be a great problem in series stacks if the cells in the stack do not have equal capacitance, where overcharge would be common if the performance of the individual SCs is not consistent. Hence, the optimization of manufacturing procedures of SCs is required to ensure quality control.

19.4.2.3 Impurity-Induced Redox Reactions

Impurities in both the electrolyte and electrodes may cause self-discharge if they are reducible or oxidizable within the operating potential range of the SCs. The resultant potential-dependent faradaic charge leakage currents occur when the SC becomes nonpolarizable because of the existence of impurities.

The most common impurities present in electrode materials are the functional groups residing on the edge of carbon materials, which might be oxidized in the positive electrode and reduced in the negative electrode, forming a complete electrochemical reaction. The electrochemical and physiochemical behavior of different kinds of surface functional groups present in electrode and electrochemistry applications can be found from Ref. [31]. Self-discharge based on this mechanism does not consume considerable charges for traditional electrode materials since the distribution of functional groups on the surface of the electrode is sparse, and not all of the functional groups are redox active; however, this has to be taken into consideration as the increase of surface area of electrode materials. For high-surface-area carbon electrode materials, the dangling bonds on the surface of porous or powder carbon materials can be another source of degradation as the charging of electrochemical double-layer capacitors. The shuttle effect can also be stimulated in SCs by transition metal ions (such as Fe^{2+} or Fe^{3+}). The reductant and oxidant species diffuse between the positive and negative electrodes, become oxidized or reduced in the electrodes, and then travel back.

In the electrolyte, dissolved traceable oxygen can be reduced to peroxide or water in the negative electrode. The peroxide or water then gives rise to free radicals by surface reactions. If oxidation of other impurities occurs simultaneously in the negative electrode, a self-discharge current occurs resulting in the loss of charges. Otherwise, the peroxide or water can be then reduced or reoxidized, presenting a shuttle effect. This mechanism is probably important since oxygen can also be adsorbed on the surface of high-surface-area carbon electrode materials and introduced into SCs. To eliminate oxygen in SCs, electrode materials should be vacuumed and the electrolyte should be treated to remove the dissolved or adsorbed oxygen before cell assembly.

19.4.2.4 Short Circuit

A short circuit also could be a reason for the self-discharge of SCs. Short-circuit pathways could be induced by fibrils across separator films owing to improper electrode isolation or as a result of electrolyte leakage around the inappropriately sealed cells or passing through the bipolar electrode plates in a bipolar configuration design. In this self-discharge mechanism, the SCs release their energy slowly approximating through a load resistor.

In real cases, self-discharge seldom results from a single mechanism. For example, an accidental overcharge over an SC produces hydrogen and oxygen at the negative and positive electrode separately. Oxygen and hydrogen diffuse across the electrolyte to the opposite electrodes (oxygen to the negative electrode and hydrogen to the positive electrode). Then, oxygen is reduced to water or peroxide, and hydrogen is oxidized on the positive electrode forming a self-discharge current. If hydrogen is not easily oxidized, water or peroxide behaves as an oxidizing or reducing agent in the following depolarization reactions. The impurity-induced self-discharge will be set up.

19.4.3 Accelerating Effect

It is notable that degradation is also related to the application requirement. There is an acceleration of the degradation when the voltage is close to the electrochemical decomposition voltage of the electrodes, and/or at a higher temperature. The charging voltage and working temperature are recognized as principal factors for the aging of SCs. A study shows that the aging rate of an SC is doubled by an increase of 10 K above the rated temperature or an increase of 100 mV above the rated cell voltage [32]. The temperature and voltage effect on SCs has also been comprehensively reviewed [33]. Electrodes and/or electrolyte decomposition can be enhanced by elevating the temperature or charging voltage, resulting in the increase of cell resistance.

The lifetime of an SC related to the working temperature and voltage can be estimated on the basis of the following relationship:

$$\frac{t_1}{t_2} = \left(\frac{V_2}{V_1}\right)^n \exp\left[\frac{E_a}{k}\left(\frac{1}{T_1} - \frac{1}{T_2}\right)\right] \quad (19.17)$$

where t_1 is the lifetime at the temperature T_1 and voltage V_1, and t_2 is the lifetime at temperature T_2 and voltage V_1. k is the Boltzmann constant, and E_a and n are the activation energy and a constant, which are both determined by the experimental data.

Except for the estimation of the lifetime of an SC at elevated temperature and/or operating voltage, this relationship can also be used for accelerating the test of SCs since a normal cycling test is time and energy consuming.

19.5 THEORETICAL MODELING FOR FUNDAMENTAL UNDERSTANDING

SCs can be utilized as energy storage devices based on the understanding of their physical mechanism during the charge–discharge processes. To design well-optimized energy storage units by simulation-based analysis, an accurate theoretical model is greatly desirable, to describe the dynamic behavior of the SCs. Many researchers have devoted their studies to the theoretical models of SC [7,34]. This section covers the typical theoretical models that can be used to understand the scientific principles behind the capabilities of the SC.

19.5.1 Double-Layer Model of Supercapacitor

The double-layer model is the first proposed SC model, which was established based on the charge storage mechanism [34,35]. On the basis of the double-layer model, three different models have been developed over the years.

19.5.1.1 Helmholtz's Double-Layer Model

In 1879, Helmholtz proposed the electrical double-layer model to study the capacitive properties of the interface between solid electronic conductors and liquid ionic conductors [36]. It is considered to be the classical physical models of SCs, which can uncover the charge storage mechanism.

According to Helmholtz's model, each interface between the electrolyte and electrode is composed of ions and anions. A simplified diagram of anion accumulation on the surface of the electrode is shown in Figure 19.12. The differential capacitance proposed by Helmholtz can be described as

$$C_H = \frac{A\varepsilon}{4\pi d}$$

where ε represents the dielectric constant of the electrolyte, A is the surface area, and d is the thickness of electric double layer (EDL), which is defined as the molecular diameter of the solvent in Helmholtz's double-layer model. It can be seen that C_H is a constant value dependent only on the charge layer separation and the relative permittivity.

Since Helmholtz's double-layer method did not take the temperature and voltage dependency of the capacitance into account, it is inadequate to describe the double-layer capacitance accurately. The calculated values from Helmholtz's double-layer method are about one order of magnitude of surface capacitances larger than the experimental values experimentally (generally of 10–30 μF cm^{-2}).

19.5.1.2 Gouy and Chapman's Double-Layer Model

It is known that the double-layer capacitance is not a constant, which is dependent on the terminal voltage and the ionic concentration. To account for this phenomenon, Gouy introduced the random thermal motion in 1910, which prompted him to consider the distribution of ions to form a diffuse space charge instead of accumulating on the surface of the electrode [7]. This diffuse space is well known today as the diffuse layer, as shown in Figure 19.13.

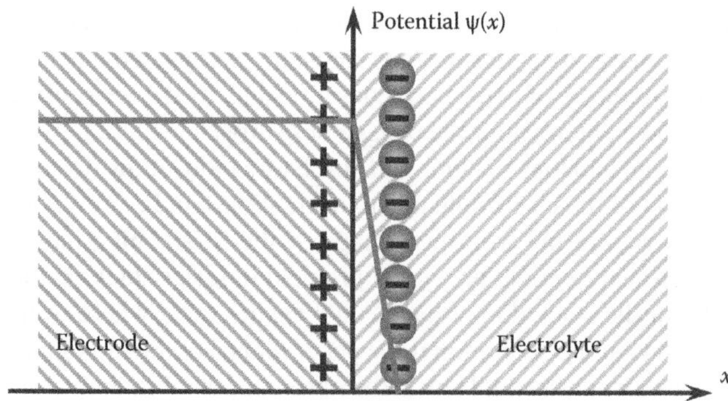

FIGURE 19.12 Helmholtz's double-layer model.

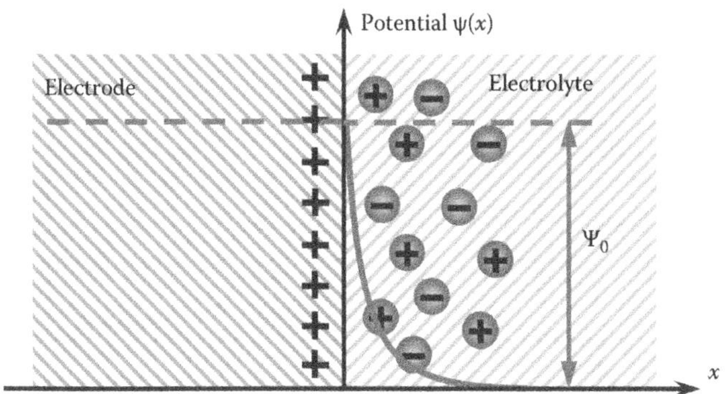

FIGURE 19.13 Gouy and Chapman's double-layer model.

Subsequently, in 1913, Chapman [37] presented a detailed mathematical formulation of Gouy's diffused layer, which was established by the integration of the Boltzmann distribution function and the Poisson equation. The double-layer capacitance C at temperature T can be described as the function of surface potential ψ_0 by the following equation:

$$C = z \cdot \sqrt{\frac{2q^2 n_0 \varepsilon}{kT}} \cdot \cosh\left(\frac{zq\psi_0}{2kT}\right) \tag{19.18}$$

where z is the valence of the ions, q is the elementary charge, n_0 is the number of ions per centimeter, ε is the dielectric constant of the electrolyte, and k is the Boltzmann constant. It was found that the double-layer capacitance was overestimated by this model, and the reason is that the ion was assumed to be a point charge that can be infinitely close to the interface of the electrode and electrolyte.

19.5.1.3 Stern and Grahame's Double-Layer Model

In 1924, Stern improved Gouy and Chapman's model by dividing the space charges in the electrolyte into two contributions: one is a compact layer (also referred to as the Helmholtz layer) constituted by a layer of adsorbed ions at the electrode surface, and the other is the diffused layer as defined in Gouy and Chapman's model, as shown in Figure 19.14. Subsequently, Grahame established the metal–solution interface model in 1951, which further subdivided the compact plane into the inner Helmholtz plane (IHP) and the outer Helmholtz plane (OHP) [35]. In this case, the total charges on the solution side are

$$Q_s = q_E = q_H + q_D$$

where q_E is the charge on the electrode. It can be seen that the total charges are comprised by the Gouy diffuse charge q_D and the Helmholtz fixed charge q_H. The total capacitance across the electrode/electrolyte is then governed by

$$\frac{1}{C} = \frac{1}{C_H} + \frac{1}{C_G}$$

where C_H is the Helmholtz capacitance and C_G is the Gouy–Chapman capacitance, which remains governed by that in Gouy–Chapman's model.

It should be noted that the above three models were primarily used to calculate the capacitance of the double-layer capacitors, while they could not effectively reflect the dynamic characteristics of the SCs during the charge–discharge process. Therefore, new types of models, which can be used to analyze the dynamic process, are desired to be further developed.

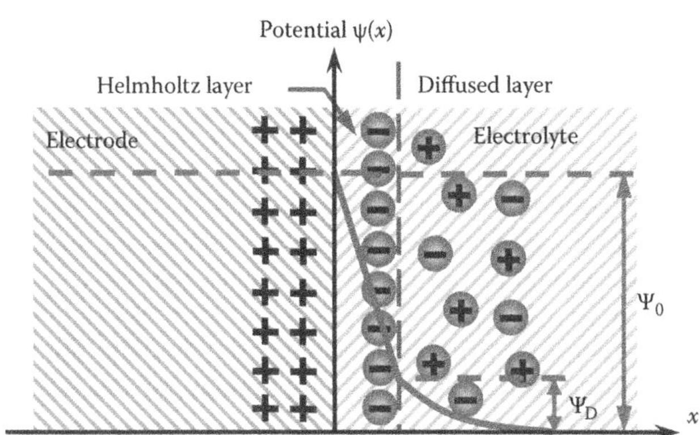

FIGURE 19.14 Stern and Grahame's double-layer model.

19.5.2 Equivalent Circuit Models

This is one of the most practical application models. It uses basic circuit components (resistor, capacitor, and inductor) to describe the external characteristics of SCs during the working process, which can easily be modeled and calculated in practice.

19.5.2.1 Classical Equivalent Circuit Model

In this model, an EDLC could be modeled by a capacitance (C_{dl}) with an equivalent parallel resistance (EPR) and an ESR, as shown in Figure 19.15. The EPR and ESR represent the effect of leakage current acting on the long-term performance and the ohmic loss during the charge–discharge process, respectively. According to these three parameters, the classical equivalent circuit can be considered as a first-order approximation to the actual capacitor behavior. Spyker et al. suggest that the EPR in a slow-discharge application can be determined by its rated voltage [38]. Since the decay is

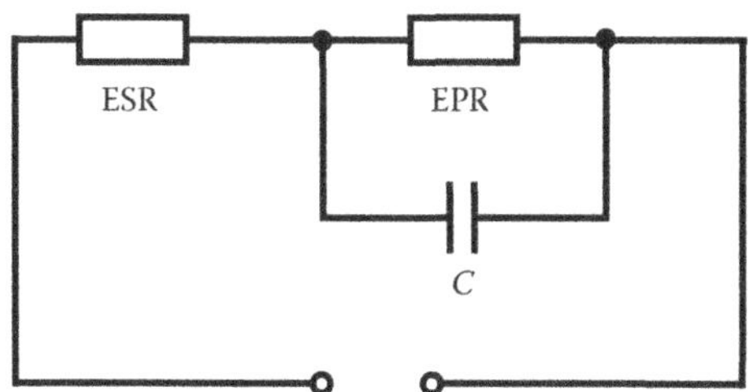

FIGURE 19.15 Classic equivalent circuit model of the SC.

exponential, the EPR can be determined by the following equation:

$$\text{EPR} = -t/(\ln(V_2/V_1)C) \tag{19.19}$$

where V_1 and V_2 are the initial and final voltage, respectively. C is assumed to be equal to the rated capacitance and t is the time. Owing to the quite large time constant of C and EPR, the EPR can usually be ignored in the case of a short discharge within few minutes.

To confirm the ESR, the equivalent circuit is considered to be composed by only the capacitance and the ESR, while the EPR is ignored. In this case, the ESR can then be calculated by the change in current, ΔI, and voltage, ΔV, during the charging process.

$$\text{ESR} = \Delta V/\Delta I \tag{19.20}$$

Using Spyker's method, the capacitance, C, can be measured by determining the change in energy, ΔE, that occurs during the charging–discharging process, which can be determined by

$$C = 2\Delta E/\left(V_1^2 - V_2^2\right) \tag{19.21}$$

since ΔE can be calculated by the integral of the instantaneous power. Thus, the capacitance C can be calculated from

$$C = 2\int_{t_1}^{t_2} v(t)i(t)\,\mathrm{d}t/\left(V_1^2 - V_2^2\right) \tag{19.22}$$

Owing to the simple topology of the classical equivalent circuit, it can be quickly simulated and easily integrated into many systems. According to the experimental tests, the performance of classical equivalent circuit has a strong correlation with that of a real device in low-frequency applications. However, it is not sufficient to reproduce the characteristics of SC at high frequencies [38]. Therefore, further improved circuit models have been established, such as the ladder circuit model [39,40], multiple branches model [41], transmission line model [42], and RC parallel branches model [43].

19.5.2.2 Ladder Circuit Model

The ladder circuit model is generally used to examine the capacitive behavior in slow discharge and pulse load applications, which demonstrates success in modeling carbon–nickel fiber electrodes in electrochemical capacitors [44,45]. The distributed nature of SCs can be modeled by the ladder circuit shown in Figure 19.16. The different ladder circuits are assigned with a two-character label, which are consistent with an uppercase L and a numeral to represent the number of capacitors, such as L_1 consisting of C_1, R_1, and R_L. As shown in Figure 19.16, the circuit could be referred to as "ladder circuit L_n." By use of the software employing various statistical techniques, the parameters for the n ladder circuits (L_1 to L_n) can be assessed.

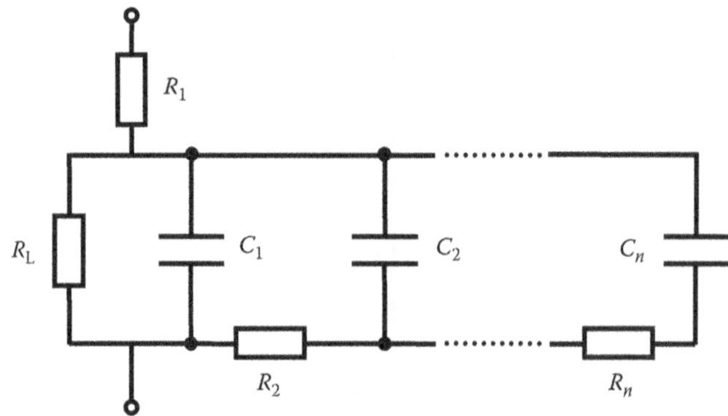

FIGURE 19.16 Ladder circuit model of the SC.

Nelms et al. [40] evaluated the ladder circuit model using a nonlinear least squares fitting technique to determine the circuit parameters from L_1 to L_n circuits. They attempted to match the AC impedance data of an ELNA 50-F, 2.5-V double-layer SC through a comparison of different developed models. In their approach, five time constant parameters of R_1–C_1 to R_5–C_5 were determined using software analysis of AC impedance data. The measurement of the leakage current was done by application of a measured current over time to maintain double-layer charging. Their calculated results suggested that ladder circuits of L_3 or higher is required to obtain a reasonable fit to the actual impedance data.

19.5.2.3 Multiple Branches Model

The multiple branches model was developed by Zubieta and Bonert [41,46] for power electronic applications. In this model, it considers the bias voltage dependency of capacitance by designing a nonlinear capacitance varying as a function of terminal voltage. In addition, the charge redistribution in pore sections through the interface between the electrode and electrolyte can be evaluated by adding RC branches. The multiple branches model is similar to the ladder circuit model. However, each branch in the equivalent circuit of the multiple branches model has a different time constant.

Theoretically, the number of RC branches should be large enough so that the model can precisely describe the characteristics of SC. However, to simplify the model, while ensuring reasonable accuracy, two or three branches are typically selected. Figure 19.17 shows the three branch equivalent model. The first branch containing R_f denoted the immediate branch, and it also includes a voltage-dependent capacitor C_{f1} that represents the voltage dependence of the capacitance in double-layer capacitors. The immediate branch has the lowest time constant and simulates the behavior of fast charge–discharge in a few seconds. The second branch, containing R_d, is denoted as the delayed branch. It is usually used to describe the characteristics of capacitance over the range of a few minutes. At last, the third branch, also referred to as the slow branch, simulates the self-discharge behavior for a time longer than 10 min. Apart from the RC branches, a leakage

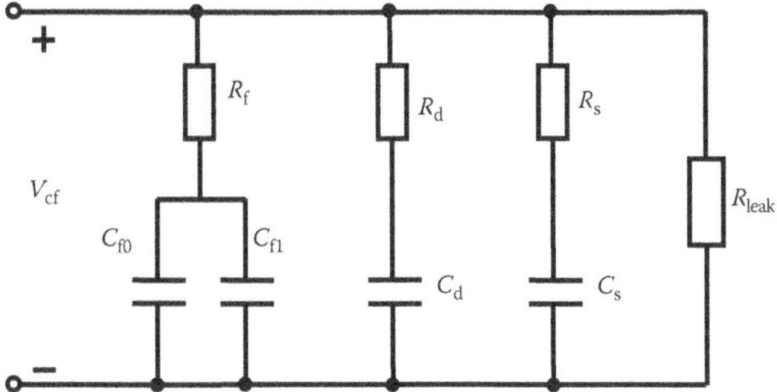

FIGURE 19.17 Three branches equivalent circuit model of SC.

resistor R_{leak} is added parallel to the RC branches so as to determine the leakage current effect.

The multiple branches model takes into account the dependency of capacitance and voltage; thus, it can better describe the allocation process of the inner charge of SCs. However, the application of this model is impeded by the assumption that only the immediate branch of the capacitor is voltage dependent, resulting in a greater error in the model's predictions at low voltages, e.g., as voltage levels increase above 40% of the rated terminal voltage, the model displays satisfactory accuracy within 30 min [41]. For a lower terminal voltage and a longer time, the simulation results cannot agree well with the measured data. For this reason, the model is not fit to predict the behavior of a capacitor after several hours or days.

19.5.3 Porous Electrode Theory

The electrodes with porous structures have been extensively analyzed because of their significant effects on the characteristics of the SC. The porous electrode theory was first introduced by de Levie to uncover the effect of porosity on the double-layer capacitance and electrolyte resistance [42]. The porous electrode theory was subsequently developed in 1963. According to this theory, all pores in the electrodes are assumed to be cylindrical, identical, and independent, and the total impedance on each porous electrode is defined to be the same. Thus, the porous behavior of electrodes can be investigated by the impedance of a single pore.

According to the porous electrode theory, two typical equivalent circuit models, the transmission line model [42,47] and the RC parallel branches model [43], are proposed.

19.5.3.1 Transmission Line Model

On the basis of the porous electrode theory, de Levie proposed that each pore in the porous electrodes can be modeled as a transmission line, as shown in Figure 19.18. It is the simplest way to integrate pore impedance Z_p into the emulators by the transmission line topology. In this case, the pore impedance Z_p can be calculated by the following equation [42]:

$$Z_p = \sqrt{\frac{R_{el}}{j \cdot \omega \cdot C_{dl}}} \coth\left(\sqrt{j \cdot \omega \cdot C_{dl} \cdot R_{el}}\right) \quad (19.23)$$

where ω is the frequency and $j \equiv \sqrt{-1}$. R_{el} and C_{dl} are related to the local electrolyte resistance r_{el} and double-layer capacitance c_{dl} per unit length of the pore, respectively.

$$R_{el} = r_{el} \cdot l$$

$$C_{dl} = c_{dl} \cdot l$$

The transmission line model has the following advantages to simulate SC compared with the above models [40]:

1. It takes both dynamic and long-time behaviors into account.
2. This model can describe the scientific principles of the SC accurately within a wide range of frequencies.
3. It can be easily used by other simulation software to find analytical and numerical solutions.

However, the transmission line has a complex analytical expression, which is not suitable for simulation and also severely hinders its practical application in the modeling of SCs.

19.5.3.2 RC Parallel Branches Model

The RC parallel branches model was established by Buller based on the porous electrode theory [43]. Figure 19.19 demonstrates the circuit schematic with n series parallel branches. Buller simply employed an alternative analytical representation of the pore impedance by the pulse response of Z_p.

$$Z_p = \sqrt{\frac{R_{el}}{j \cdot \omega \cdot C_{dl}}} \coth\left(\sqrt{j \cdot \omega \cdot C_{dl} \cdot R_{el}}\right) \Leftrightarrow \frac{1}{C_{dl}} + \frac{2}{C_{dl}} \sum_{n=1}^{\infty} e^{\frac{-n^2 \cdot \pi^2}{R_{el} \cdot C_{dl}} t} \quad (19.24)$$

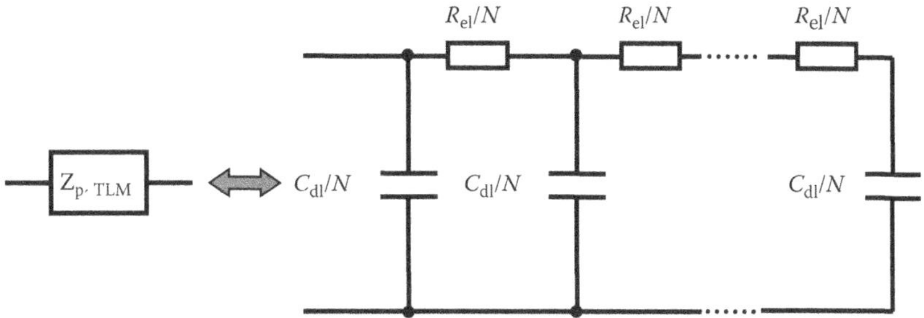

FIGURE 19.18 Transmission line model of pore impedance.

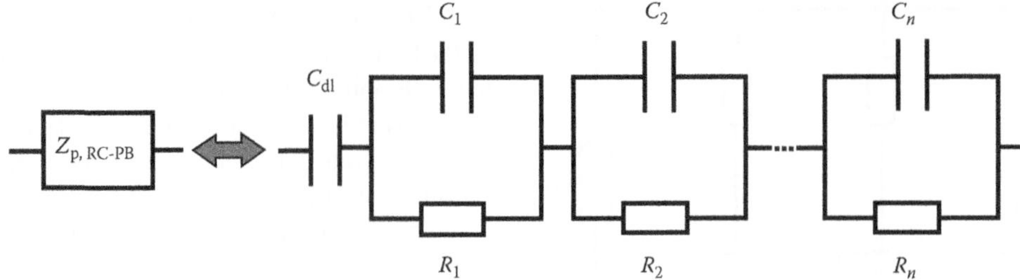

FIGURE 19.19 RC parallel branches model of pore impedance.

where the pulse response satisfies the following equation:

$$\sqrt{\frac{R_{el}}{j \cdot \omega \cdot C}} \Leftrightarrow \frac{1}{C}\sum_{n=1}^{\infty} e^{\frac{-t}{R \cdot C}} \tag{19.25}$$

The determining parameters of pore impedance can be expressed as

$$R_k = \frac{2R_{el}}{\pi^2 \cdot k^2}, \; k = 1,2 \ldots n$$

$$C_1 = C_2 = \ldots C_{dl}/2$$

Compared with the transmission line model, the RC parallel branches model displays faster convergence in simulations. The main difference between the two models is the analytical representation of pore impedance. With the increasing of the RC branches, both of these two circuit models characterize the porous nature of the SC more precisely. However, the computational cost will be greatly increased to establish the higher-order models.

Since the SCs generally operate at a low-frequency environment, as well as relatively small inductance coefficient (usually in nH level) in case of high frequency, all the models described above did not consider the inductance components, making them unable to describe the spectral characteristics of SCs accurately. Therefore, a complete theoretical model should be coupled with a series inductor.

19.5.4 Frequency Domain Model

The frequency domain model [33,48] was established to precisely describe the porous impedance characteristics of SCs over a wide range of frequency, which can be achieved by extending the pore impedance ($\underline{Z}_p$) in series with an internal resistance (R_i) as a first-order approximation for leakage current and an inductor (L) representing the stray inductance of electrodes, as shown in Figure 19.20.

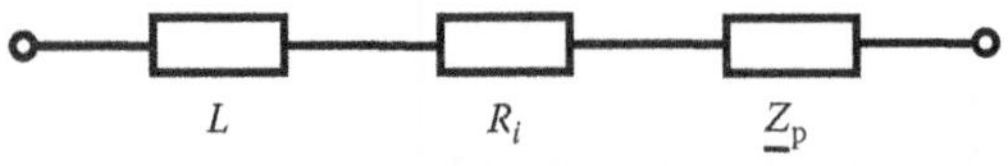

FIGURE 19.20 Frequency domain model of SCs.

The typical Nyquist plot, which reflects the capacitive behavior of SCs under different frequency regions, is shown in Figure 19.21. It is worth noting that the spectra line is almost vertical for the low frequency, which is the typical characteristic of an ideal capacitor. For the middle frequency, the impedance forms an angle of −45° with the real axis, which can be explained by the complex impedance properties of porous electrodes. On the basis of the frequency domain model, the complete impedance model for SCs, which includes the series inductor and the internal resistor, can be described by the following equation:

$$Z = R_i + j\omega L_i + \sqrt{\frac{R_{el}}{(j\omega)^{\gamma} C_{dl}}} \coth \sqrt{(j\omega)^{\gamma} R_{el} C_{dl}} \tag{19.26}$$

where R_{el} and C_{dl} are the electrolyte resistance and double-layer capacitance of the pore, respectively.

The frequency domain model can fit well to the measured spectrum of the SCs over the whole frequency range. However, the special purpose of this model is to obtain the parameters, as well as their overly complex mathematical expressions, resulting

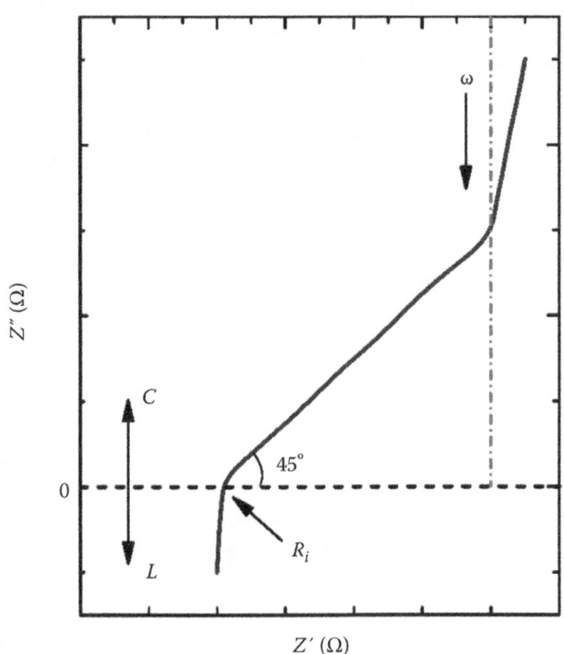

FIGURE 19.21 Typical Nyquist plot of SCs.

in the inconvenience to apply in the time-domain simulations and dynamic analyses.

19.5.5 Artificial Intelligent Model

With the fast development of artificial intelligence technology, the artificial neural network (ANN) is also applied to examine the performance of SCs. The ANN can predict the electrical and thermal behaviors of SCs; thus, it can help reflect the intrinsic, synthetic, and operating characteristics of SCs [49,50].

Compared with all other training models, the ANN can be used as a black-box model. Only the experimental data are required during the whole ANN training process; thus, the quality of training is closely related to the input data. The training may fail to produce the true characteristic of the problem if the learning samples do not meet the requirement of the applied domain.

The quality of the trainings can be estimated by calculating the average quadratic error (AQE), which is assessed in terms of their generalization capacity [49]:

$$\mathrm{AQE} = \frac{1}{N_{\mathrm{eg}}} \sum_{i=1}^{N_{\mathrm{eg}}} (O_{\mathrm{e}}(i) - O_{\mathrm{m}}(i))^2 \tag{19.27}$$

where N_{eg} is the number of examples for the training process, and $O_e(i)$ and $O_m(i)$ are the expected and simulated outputs, respectively. The artificial intelligent models can accurately describe the nonlinear characteristics of the real system. However, the extensive applications of the ANN are greatly hindered by the requirements of a large amount of training data.

19.5.6 Other Advanced Theoretical Models

The theoretical models discussed above have been proposed to describe the microstructure of the electric double-layer SCs. However, most of these theories are dependent on the assumption that the ions and dilute electrolyte are well separated, while only the electrostatic interaction between ions is taken into consideration, which is inconsistent with the actual situation. For this purpose, many theories are attempted to provide more realistic information on the structural and dynamic characteristics of SC, such as the continuum theory, first-principles calculation, and molecular dynamics (MD) simulation.

19.5.6.1 Continuum Theory

On the basis of the continuum theory, the solvent is considered to be the continuous medium fluid [51]. Therefore, the interaction between ions and solvent becomes electrostatic. To calculate the energy of the solvent–ion interaction, the initial and final states have to be considered. The interaction between solvent and ions does not exist at the initial state, while the interaction appears at the final state. To know the energy change during the mixing process, the reversible Gibbs energy, ΔG_{i-s} (J mol^{-1}), can be calculated as

$$\Delta G_{i-s} = N_{\mathrm{A}}(W_1 + W_2 + W_3) \tag{19.28}$$

where W_1 and W_2 are the reversible energies used to discharge an ion in a vacuum ($\varepsilon = 1$) and to put a discharged ion in the solvent, respectively. Under the electrostatic interaction, W_2 can be ignored and W_3 is the reversible energy to charge a discharged ion in the fluid with the dielectric constant (ε).

The continuum model is feasible for the modeling of electronic double-layer SCs. However, owing to the assumption that the electrolytes and ion distribution are continuous, it describes the interfacial properties of SCs on average, which neglects the discretization of charges in nature. Therefore, many important properties of SCs, such as the ion distribution, structure of nanosized materials, localized nonlinear material deformation, and various nanosized phenomena, could not be precisely characterized by this model. Additionally, the continuum simulations based on Poisson–Boltzmann equation seriously overestimate the concentration of electrolytes, and the further calculated capacitance, and this overestimation is mainly due to the lack of definition of the ion size in this model.

19.5.6.2 First-Principles Calculation

The charge storage and the capacitance of electrochemical SCs are closely related to the electrode materials. Thus, it is urgent to improve the performance of existing electrode materials and further develop new materials for use in SCs [52]. First-principles calculation is a powerful tool to precisely investigate the structural and electronic properties of electrodes in SC. It has been successfully used to predict the effect of electrode materials on the charge–discharge performance of SCs. In general, the electrode materials of electric SCs can be divided into three categories: carbon materials with high specific surface area, metal oxides, and conductive polymers.

Huang et al. [53] employed first-principles calculations combined with the heuristic theoretical model to study nanoporous carbon SCs, revealing the reason for the abnormal increase of the capacitance for pores less than 1 nm, as well as the slight increase of capacitance when the pore size exceeds 2 nm. Stengel et al. [54] presented the dielectric properties of realistic $SrRuO_3/SrTiO_3/SrRuO_3$ nanocapacitors by the theoretical calculations. It shows that the observed remarkable decrease of capacitance is indeed an inherent effect. Subsequently, they proposed a correspondence between the dielectric dead layer and the hardening of the collective $SrTiO_3$ zone-center polar modes, providing a practical guideline for decreasing the harmful effects of the dead layer in nanosized devices.

Moreover, conductive polymers, such as polypyrrole and polyaniline (PANI), have been widely used as electrode materials for pseudocapacitors, owing to their high capacitance value and improved cycling stability. First-principles calculations have been successfully used to study the electronic structures of the PANI-G hybrid system [55]. It is suggested that the carbonyl group in the amide plays a crucial role in forming a π-conjugated structure, which facilitates the charge transfer and ultimately improves the cycling capability and capacitance of SCs.

19.5.6.3 MD Simulation

Although the theoretical models mentioned before are useful for understanding EDLCs, they have many limitations in application. For instance, the chemical details of ions and solvent molecules, such as the charge distributions and electrode morphologies, are not easy to take into account with such models. In MD simulations, individual atoms with finite size are constructed, and the interatomic potentials and molecular shape are taken into consideration. It can clearly characterize the interfacial properties of the electrolyte and the structure of SCs, such as ion distribution and orientation [56]—properties that are difficult to characterize experimentally.

MD simulation can describe the dynamic characteristics of the system more realistically. It has been successfully used to predict the structural separation in room temperature ionic liquids (RTILs), and the obtained structure factors are consistent with the experimental results measured by small angle x-ray scattering [57]. The interfacial dynamics of RTILs simulated by the MD approach also agree well with that the values measured by nuclear magnetic resonance [58]. Moreover, MD simulations are widely used in the study of carbon electrode EDLs with RTILs in the atomic scale [59].

Moreover, the MD approach can provide detailed information of ions packing inside the SCs. For instance, it is suggested that Na^+ ions were tightly adsorbed on the electrode with a surface charge density of −0.26 C/m^2, and the nearest distance between these ions is 0.92 nm. Since the separation is slightly larger than the hydrated diameter of Na^+ ions (0.72 nm), the packing of Na^+ ions do not reach the spatial limit defined by their hydrated diameters [60].

19.6 ADVANCED DESIGNS FOR SUPERCAPACITORS

From a practical point of view, increasing the energy density of SCs is one of the most crucial problems since the energy density of SCs is far from being satisfying compared with batteries. To further increase the energy density of SCs, many efforts have been made in both aqueous and organic systems.

19.6.1 Asymmetric Electrochemical Capacitor in Aqueous Electrolyte

Since the energy density of SCs is proportional to its capacity and the square of the operating voltage as expressed in Equation 19.7, increasing the operation voltage could be a feasible way to improve the energy density of SCs. Although organic electrolytes and ionic liquids could offer a wider operating voltage, their high cost, inflammability, low conductivity, and lower capacitance compared with aqueous electrolytes [61,62] limit their usage. In addition, SCs are normally used for high-power-density applications, which always require high current density and fast cycling rates. Thus, the aqueous electrolyte SCs are more desirable since they always have better electrochemical safety than organic-based electrolytes.

The working voltage of aqueous-based SCs is theoretically limited to 1.23 V, beyond which the water will be dissociated, resulting in gas evolution. In a symmetrical design, each carbon electrode thus has to be operated within an electrochemical window smaller than 0.6 V. Compared with that measured in a three-electrode device, the carbon electrode working in a symmetrical configuration only contributes one-fourth of its capacitance [63].

To increase the energy density by enlarging the working voltage window of the carbon electrode in an aqueous electrolyte, a faradaic positive electrode (such as a pseudocapacitive electrode or a battery-like electrode) was introduced to form an asymmetric design. In this design, the faradaic electrode has much higher capacity compared with the electrical double-layer negative electrode. Thus, the faradaic electrode could work in a very narrow voltage window since only a limited capacity could be used. In addition, because of the addition of the faradaic electrode, the oxygen evolution overpotential, which is strongly dependent on the chemical nature of the electrode, also increases, leading to an overall voltage of SC even as high as 2 V. As a result, the nonfaradaic electrode can work in its full electrochemical window and the total capacitance of this new design is much higher than that of symmetrical ones. The asymmetric capacitors therefore possess the advantages of both kinds of electrodes such as rate capability, long cycle life, and higher energy densities.

19.6.1.1 Carbon/Pseudocapacitive Electrode Asymmetric Design

As the most extensively studied pseudocapacitive material, RuO_2 is the first option to build an asymmetric device. The synthesis of RuO_2 can be achieved by various strategies, including electrostatic spray deposition, chemical precipitation, oxidative process, hydrothermal method, etc. An asymmetric electrochemical capacitor can be easily built by employing activated carbon and RuO_2-based material as two electrodes [64]. However, considering the capacitance difference between activated carbon and RuO_2, the thickness difference of the two electrodes would be remarkable. To balance the capacitance of the RuO_2 electrode, the carbon electrode has to be modified. An anthraquinone-modified carbon was employed as the negative electrode, while the RuO_2 was the positive electrode. The negative potential limit could be extended from 1.0 to 1.3 V by the anthraquinone modification and a maximum energy density of 26.7 Wh kg^{-1} can be reached in 1 mol L^{-1} H_2SO_4 solution [65].

Considering the costliness and scarcity of RuO_2, MnO_2, another popular pseudocapacitive material with a lower price, has also been envisioned as a positive electrode in asymmetric aqueous SCs. The poor electronic conductivity of MnO_2 is generally improved by the reduction in its size and the formation of composites [66]. The cell voltage of carbon–MnO_2 asymmetric SC can reach as high as 2.2 V in 1 mol L^{-1} K_2SO_4 solution [67]. Similar asymmetric SC based on activated carbon nanofiber and MnO_2 nanowires@graphene electrodes exhibits an energy density of 51.1 Wh kg^{-1} [68]. The cycling lives of these SCs are mostly more than thousands of cycles as long, as they do not work at a very high power density and energy density [69].

On the basis of the same principle, several other devices, composed of a nonfaradaic electrode (including activated carbon, carbon nanotubes, and graphene) and a pseudocapacitive electrode (such as V_2O_5, MoO_3, Co_3O_4, and conducting polymer) have been proposed. However, their cycling abilities are not as good as the ones with MnO_2 and RuO_2 electrodes for unknown reasons [70].

The performance of carbon/pseudocapacitive electrode asymmetric SCs can be affected by various parameters. The interface between the electrode and the current collector is subjected to corrosion by oxygen evolution reaction, resulting in the increase of the ESR of the SC [71]. The cyclic volumetric variation of the electrode particle during cycling may lead to the mechanical failure of the pseudocapacitive electrode depending on the pristine polymorph of electrode materials [72]. The dissolution of electrode materials could also cause severe problems during the electrochemical processes, leading to a consequent capacitance fading on cycling [73]. The mass and capacitance have to be balanced to make full use of both electrodes, and to make sure that the pseudocapacitive electrode will not exceed its stable electrochemical window.

19.6.1.2 Carbon/Battery-Type Electrode Asymmetric Design

Asymmetric SC can also be made by combining a carbon nonfaradaic electrode with a battery-type electrode, such as replacing the negative electrode of lead–acid battery with activated carbon. The positive and negative electrodes reaction are described by the following equations:

$$PbO_2 + 3H^+ + HSO_4^- + 2e^- \leftrightarrow PbSO_4 + 2H_2O \quad (19.29)$$

$$C_6^{x-}(H^+)_x \leftrightarrow C_6^{(x-2)-}(H^+)_{x-2} + 2H^+ + 2e^- \quad (19.30)$$

Upon charging, H^+ accumulated on the surface of the negative electrode, and $PbSO_4$ from the electrolyte was oxidized into PbO_2. During the discharging process, H^+ absorbed on the surface of the negative electrode is transported to the positive electrode, resulting in the dissolution of PbO_2 and the formation of water. The lifespan of the electrodes within this design was maintained over 10,000 cycles with acceptable coulombic and energy efficiencies [63]. Since the redox reaction mainly occurs on the surface, efforts have been made to increase the electrochemically active surface area. Consequently, thin films as well as porous and various nanostructures of PbO_2 have been investigated [74,75]. The hybrid system exhibits excellent energy and power performance with a specific energy of 43.6 Wh kg^{-1} at a power density of 654.2 W kg^{-1} [76].

Nickel hydroxide is another typical battery type positive electrode. Asymmetric SC can also be made based on activated carbon and $Ni(OH)_2$ as negative and positive electrodes, respectively, in the presence of an alkaline electrolyte, which is very similar to those used in Ni–Cd or Ni–H batteries. The electrode reaction during charging and discharging is as follows:

$$Ni(OH)_2 + OH^- \leftrightarrow NiOOH + H_2O + e^- \quad (19.31)$$

Recent studies on this kind of asymmetric SC mostly focused on the preparation of $Ni(OH)_2$ with different morphologies and structures by various methods, such as heteroatom doping and compositing $Ni(OH)_2$ with carbon-based materials. The results showed that (i) high-surface-area porous structures or nanostructures are preferred [77]; (ii) heteroatom doping can improve the electronic conductivity, and/or enhance the active material utilization and ion-moving path [78]; and (iii) the composition with carbon materials will stimulate a good distribution of $Ni(OH)_2$ and keep the particles from aggregating [79].

On the basis of the principle of asymmetric design and related review papers [66,69,80], the key point of building a successful aqueous asymmetric SC can be summarized as that the capacity of the SC should be controlled by the nonfaradaic electrode. In this case, the faradaic-type electrode can work within a narrow voltage window, which gives the other electrode a wider operating voltage range to realize a higher capacity and assures the faradaic electrode a longer cycle life, because a narrow working voltage range guarantee a reservation of a big amount of electrode materials from those affected by electrochemical cycles, leading to limited structural changes. Meanwhile, the reserved electrode materials would be activated as some of the materials are consumed and become inactive due to mechanical or chemical changes of electrodes, which also improved the cycling performance of SCs. Otherwise, the cycle life of the faradaic electrode (especially for the battery-like electrode) will be reduced since the electrode will behave similarly to that in a battery or a symmetric pseudocapacitor. On the other hand, since the kinetics of the faradaic electrode is usually 1 or 2 orders of magnitude greater than that of nonfaradaic ones, the rate capability of asymmetric SC is limited to the nonfaradaic electrode. On this occasion, the structure of faradaic electrode has to be adapted to the rate requirement, such as the employment of nanostructured electrode.

19.6.2 Lithium-Ion Hybrid Design

As described above, the increase in the working voltage of SCs is the most effective way to increase the energy density due to the square dependence between energy density and voltage. The employment of an organic electrolyte could increase the voltage from ~1.2 to 2.5–2.7 V. However, when further increasing the working voltage of SCs, serious damage resulting from significant side effects such as gas evolution, electrolyte decomposition, and film formation on the surface of electrode could be induced [81].

Similar to the situation in aqueous electrolytes, many studies have been carried out to increase the energy density of nonaqueous electrolyte SCs. Except for the changing of electrode materials, developing hybrid SCs has also been introduced in nonaqueous SCs. Among the various hybrid capacitors, the lithium-ion capacitor (LIC) has drawn much attention [82,83]. A typical LIC is composed of an activated carbon positive electrode, a prelithiated graphite negative electrode, and a lithium-containing nonaqueous electrolyte. In other words, the LIC capacitor can be built by replacing the positive electrode of LIB with an EDLC electrode. However, different from that in LIB, the lithium intercalation and deintercalation of the negative electrode in LIC are limited to within a reasonable state of charge. On the other hand, the positive electrode in LIC performs similarly as that in EDLC, in which anions, typically BF_4^- and PF_6^-, are adsorbed or desorbed on the surface of electrode. Hence the total process in LIC is mainly capacitive instead of the battery type.

As presented in Figure 19.22, since the reaction potential of the negative electrode after prelithiation is reduced to lower than 0.1 V versus Li/Li^+, the working voltage of an LIC can reach as high as 3.8–4.0 V, which assures a high energy density of approximately 20–30 Wh kg^{-1}, while keeping an acceptable power density of 5 kW kg^{-1}. Recently, a report shows that LIC could deliver a maximum energy density of 55 Wh kg^{-1} when the cell works in the operation voltage between 3.1 and 4.1 V, and the energy density increased to 110 Wh kg^{-1} under an operation voltage between 2.0 and 4.1 V [84]. This kind of device has also been reported to process an inspiring cycle life of 100,000 cycles [85]. However, the rate capability of this device may be limited owing to the slow intercalation kinetics in graphite electrode.

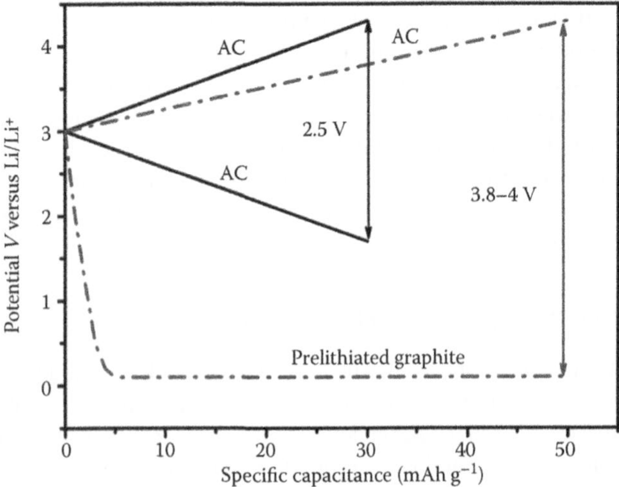

FIGURE 19.22 Comparison of the voltage profile of EDLC (activated carbon as both electrodes) and LIC (composed of an LIB negative electrode and an activated carbon positive electrode).

19.7 TECHNICAL CHALLENGES AND PERSPECTIVES

SCs have been proven to be a very promising energy storage device as a complement to batteries owing to their high power density and long cycling life. Furthermore, the electrode materials used are recyclable, providing an environmentally friendly energy storage device compared with lithium-ion batteries. However, to realize commercial expansion, the energy density has to be improved. Since the energy density of devices is greatly related to the electrode, electrolyte, and related materials, greater efforts should be paid to the following crucial factors.

19.7.1 Connection between Current Collector and Electrode

The interface between the current collector and the electrode induces a charge transfer resistance. The overall internal resistance of SCs hence increases. The increase of the contact between active materials and current collector has to be taken into consideration to decrease the charge transfer resistance. The modification of current collectors is one of the feasible methods. For example, the etching of the current collector could effectively increase the contacting area between the current collector and active materials, leading to the reduction of the internal resistance.

Another strategy to improve the contact between the current collector and the electrode is the exploration of new current collectors. The utilization of highly graphitized nanostructured carbon materials can offer significant advantages. The novel current collectors show great promise and potential to significantly improve the interface and decrease the overall resistance. Meanwhile, the adoption of novel current collector can also alleviate the corrosion of present current collector during prolonged periods of operation, the corrosion of which would inevitably result in the increase of resistance, detachment of active materials, and considerable performance loss. However, detailed studies are desired to justify the practical feasibility of these current collectors.

19.7.2 Electrode Innovation

The innovation of electrode materials, regardless of the type of SCs, is one of the most effective ways to increase the energy density of SCs. Hence, attention should be paid to electrode materials.

For EDLCs, carbon materials with tunable nanoporous structure should be designed to fit with the specific ion species of different electrolytes. The mechanism occurring at the interface between the active carbon and the electrolyte should also be further clarified. As for pseudocapacitor electrode materials, since the most predominant problems are the limited electronic conductivity of both the transition metal oxide or the conductive polymer, and the poor cyclability due to the redox nature and electrochemical instability, studies should be focused on the understanding of the performance of these materials and on developing different structures. Meanwhile, the composition of the above two kinds of electrode material has been considered as a promising approach to realize significant improvement of SCs, and the integration of their combined benefits should further be investigated.

19.7.3 Electrolyte Innovation and Selection

Increasing the operating voltage of SCs, which depends on the electrolyte, is another practical way to improve the energy density. While the resistance of the electrolyte can also limit the power density, its ion concentration and operating voltage can limit the energy density of an SC. Hence, the innovation of electrolytes is important. The development of ionic liquid electrolytes is a promising approach. On the other hand, a relationship between the pore size of the electrode and the ion size of the electrolyte has been observed. The compatibility of the electrode and the electrolyte should also be investigated.

REFERENCES

1. Helmholtz, H. V. 1879. Uber den Einfluß der elektrischen Grenzschichten bei galvanischer Spannung und der durch Wasserströmungerzeugten Potentialdiffernz. *Ann. Phys.* 29: 337.
2. Ghosh, A. 2012. Carbon-based electrochemical capacitors. *ChemSusChem* 5:480–499.
3. Bose, S. 2012. Carbon-based nanostructured materials and their composites as supercapacitor electrodes. *J. Mater. Chem.* 22:767–784.
4. Li, X. 2013. Supercapacitors based on nanostructured carbon. *Nano Energy* 2:159.
5. Fan, Z., Yan, J., Zhi, L., Zhang, Q., Wei, T., Feng, J., Zhang, M., Qian, W., Wei, F. 2010. A three-dimensional carbon nanotube/graphene sandwich and its application as electrode in supercapacitors. *Adv. Mater.* 22:3723–3728.
6. Conway, B. E. 1997. The role and utilization of pseudocapacitors, *J. Power Sources* 66:1–14.
7. Conway, B. E. 1999. *Electrochemical Supercapacitors*. Kluwer Academic/Plenum Publishing, New York.
8. Shukla, A. K. 2000. Electrochemical supercapacitors: Energy storage beyond batteries. *Curr. Sci.* 79:1656–1661.
9. Augustyn, V., Simon, P., Dunn, B. 2014. Pseudocapacitive oxide materials for high-rate electrochemical energy storage. *Energy Environ. Sci.* 7:1597–1614.
10. Frackowiak, E. 2013. Electrode materials with pseudocapacitive properties, in *Supercapacitors*, eds. F. Beguin, and E. Frackowiak, 207–238. Wiley-VCH, New York.
11. Du, B., Jiang, Q., Zhao, X. F., Huang, B., Zhao, Y. 2009. Preparation of PPy/CNT composite applications for supercapacitor electrode material. *Mater. Sci. Forum* 610–613: 502–505.
12. Yu, A., Chabot, V., Zhang, J. 2013. *Electrochemical Supercapacitors for Energy Storage and Delivery*. CRC Press, Boca Raton, FL.
13. Barrade, P. 2002. Series connection of supercapacitors: Comparative study of solutions for active equalization of voltage, in *7th International Conference on Modelling and Simulation of Electric Machines*. Montreal, Canada.
14. Gosser, D. K. Jr. 1993. *Cyclic Voltammetry*. VCH, New York.
15. Ban, S. 2013. Charging and discharging electrochemical supercapacitors in the presence of both parallel leakage process and electrochemical decomposition of solvent. *Electrochim. Acta* 90:542–549.
16. Miller, J. E. 1995. Capacitor-battery power sources: Designing for optimal performance, *5th International Seminar on Double-Layer Capacitors and Similar Energy Storage Devices*.
17. Burke, A. F. 2011. The power capability of ultracapacitors and lithium batteries for electric and hybrid vehicle applications. *J. Power Sources* 1996:514–522.
18. Ragone, D. 1968. Review of battery systems for electrically powered vehicles. *SAE Tech. Pap.* 680453.
19. Barsoukov, B., Macdonald, J. R. 2005. *Impedance Spectroscopy Theory, Experiment, and Applications*. John Wiley & Sons.
20. Huggins, R. A. 2000. Supercapacitors and electrochemical pulse sources. *Solid State Ionics* 134(1–2):179–195.
21. Song, H.-K., Hwang, H.-Y., Lee, K.-H., Dao, L. H. 2000. The effect of pore size distribution on the frequency dispersion of porous electrodes. *Eletrochim. Acta* 45:2241–2257.
22. Conway, B. E., Pell, W. G., Liu, T. C. 1997. Diagnostic analyses for mechanisms of self-discharge of electrochemical capacitors and batteries. *J. Power Sources* 65:53–59.
23. Tilak, B. V., Conway, B. E. 1976. Overpotential decay behavior—I. Complex electrode reactions involving adsorption. *Electrochim. Acta* 21:745–752.
24. Hahn, M., Wursig, A., Gallay, R., Novak, P., Kotz, R. 2005. Gas evolution in activated carbon/propylene carbonate based double-layer capacitors. *Electrochem. Commun.* 7:925–930.
25. Ishimoto, S., Asakawa, Y., Shinya, M., Naoi, K. 2009. Degradation responses of activated-carbon-based EDLCs for higher voltage operation and their factors. *J. Electrochem. Soc.* 156:A563–A571.
26. Kurzweil, P., Chwistek, M. 2008. Electrochemical stability of organic electrolytes in supercapacitors: Spectroscopy and gas analysis of decomposition products. *J. Power Sources* 176: 555–567.
27. Azais, P., Duclaux, L., Florian, P., Massiot, D., Lillo-Rodenas, M.-A., Linares-Solano, A., Peres, J.-P., Jehoulet, C., Beguin, F. 2007. Causes of supercapacitors ageing in organic electrolyte. *J. Power Sources* 171:1046–1053.
28. Ruch, P. W., Cericola, D., Foelske-Schmitz, A., Kotz, R., Wokaun, A. 2010. Aging of electrochemical double layer capacitors with acetonitrile-based electrolyte at elevated voltages. *Electrochim. Acta* 55:4412–4420.
29. Zhu, M., Weber, C. J., Yang, Y., Konuma, M., Starke, U., Kern, K., Bittner, A. M. 2008. Chemical and electrochemical ageing of carbon materials used in supercapacitor electrodes. *Carbon* 46:1829–1840.
30. Bard, A. J., Faulkner, L. R. 2001. *Electrochemical Methods Fundamentals and Application*, 2nd Ed. Wiley, New York.
31. Kinoshita, K. 1988. *Carbon: Electrochemical and Physicochemical Properties*. Wiley, New York.
32. Kotz, R., Hahn, M., Gallay, R. 2006. Temperature behavior and impedance fundamentals of supercapacitors. *J. Power Sources* 154:550–555.
33. Bohlen, O., Kowal, J., Sauer, D. U. 2007. Aging behavior of electrochemical double-layer capacitors. *J. Power Sources* 172:468–475.
34. Belhachemi, F., Rael, S., Davat, B. 2000. In *Industry Applications Conference. Conference Record of the 2000 IEEE*, vol. 5, pp. 3069–3076.
35. Matsumoto, M. 1998. Electrocapillarity and double layer structure. *Surfac. Sci. Ser.* 76:87.
36. Sparnaay, M. J. 1972. *The Electrical Double Layer*. Pty. Ltd.

37. Chapman, D. L. 1913. LI. A contribution to the theory of electrocapillarity. *Philos. Mag. Ser.* 6:475.
38. Spyker, R. L., Nelms, R. M. 2000. Classical equivalent circuit parameters for a double-layer capacitor. *IEEE Trans. Aerosp. Electron. Syst.* 36:829.
39. Nelms, R. M., Cahela, D. R., Tatarchuk, B. J. 2003. Modeling double-layer capacitor behavior using ladder circuits. *IEEE Trans. Aerosp. Electron. Syst.* 39:430.
40. Dougal, R. A., Gao, L., Liu, S. 2004. Ultracapacitor model with automatic order selection and capacity scaling for dynamic system simulation. *J. Power Sources* 126:250.
41. Zubieta, L., Bonert, R. 2000. Characterization of double-layer capacitors for power electronics applications. *IEEE Trans. Ind. Appl.* 36:199.
42. de Levie, R. 1963. On porous electrodes in electrolyte solutions: I. Capacitance effects. *Electrochim. Acta* 8:751.
43. Buller, S., Karden, E., Kok, D., de Doncker, R. W. 2002. Modeling the dynamic behavior of supercapacitors using impedance spectroscopy. *IEEE Trans. Ind. Appl.* 38:1622.
44. Cahela, D. R., Tatarchuk, B. J. 1997. in *23rd International Conference on Industrial Electronics, Control and Instrumentation, IECON 97*, vol. 3, pp. 1080–1085.
45. Cahela, D. R. 1998. Electrochemical impedance spectroscopy of metal fiber/activated carbon fiber composite materials for electrochemical capacitors. Auburn University.
46. Bonert, R., Zubieta, L. 1997. in *Industry Applications Conference, Thirty-Second IAS Annual Meeting, IAS '97. Conference Record of the 1997 IEEE*, vol. 2, p. 1097.
47. Itagaki, M., Suzuki, S., Shitanda, I., Watanabe, K., Nakazawa, H. 2007. Impedance analysis on electric double layer capacitor with transmission line model. *J. Power Sources* 164:415.
48. Rafik, F., Gualous, H., Gallay, R., Crausaz, A., Berthon, A. 2007. Frequency, thermal and voltage supercapacitor characterization and modeling. *J. Power Sources* 165:928.
49. Marie-Francoise, J. N., Gualous, H., Berthon, A. 2006. Supercapacitor thermal- and electrical-behaviour modelling using ANN. *IEE Proc. Electr. Power Appl.* 153:255.
50. Farsi, H., Gobal, F. 2007. Artificial neural network simulator for supercapacitor performance prediction. *Comput. Mater. Sci.* 39:678.
51. Butt, H. J., Graf, K., Kappl, M. 2006. *Physics and Chemistry of Interfaces.* Wiley.
52. Arico, A. S., Bruce, P., Scrosati, B., Tarascon, J.-M., van Schalkwijk, W. 2005. Nanostructured materials for advanced energy conversion and storage devices. *Nat. Mater.* 4:366.
53. Huang, J., Sumpter, B. G., Meunier, V. 2008. Theoretical model for nanoporous carbon supercapacitors. *Angew. Chem. Int. Ed.* 47:520.
54. Stengel, M., Spaldin, N. A. 2006. Origin of the dielectric dead layer in nanoscale capacitors. *Nature* 443:679.
55. An, J., Liu, J., Ma, Y., Li, M., Yu, M., Li, S. 2012. A polyaniline-grafted graphene hybrid with amide groups and its use in supercapacitors. *J. Phys. Chem. C* 116:19699–19708.
56. Shim, Y., Kim, H. J. 2010. Nanoporous carbon supercapacitors in an ionic liquid: A computer simulation study. *ACS Nano* 4:2345.
57. Li, S., Banuelos, J. L., Guo, J., Anovitz, L., Rother, G., Shaw, R. W., Hillesheim, P. C., Dai, S., Baker, G. A., Cummings, P. T. 2012. Alkyl chain length and temperature effects on structural properties of pyrrolidinium-based ionic liquids: A combined atomistic simulation and small-angle X-ray scattering study. *J. Phys. Chem. Lett.* 3:125–130.
58. Li, S., Han, K. S., Feng, G., Hagaman, E. W., Vlcek, L., Cummings, P. T. 2013. Dynamic and structural properties of room-temperature ionic liquids near silica and carbon surfaces. *Langmuir* 39:9744–9749.
59. Li, S., Feng, G., Fulvio, P. F., Hillesheim, P. C., Liao, C., Dai, S., Cummings, P. T. 2012. Molecular dynamics simulation study of the capacitive performance of a binary mixture of ionic liquids near an onion-like carbon electrode. *J. Phys. Chem. Lett.* 3:2465–2469.
60. Cagle, C., Feng, G., Qiao, R., Huang, J., Sumpter, B. G., Meunier, V. 2010. Structure and charging kinetics of electrical double layers at large electrode voltages. *Microfluid. Nanofluid.* 8:703–708.
61. Chen, P.-C., Shen, G., Shi, Y., Chen, H., Zhou, C. 2010. Preparation and characterization of flexible asymmetric supercapacitors based on transition-metal-oxide nanowire/single-walled carbon nanotube hybrid thin-film electrodes. *ACS Nano* 4:4403–4411.
62. Wu, Z.-S., Ren, W., Wang, D.-W., Li, F., Liu, B., Cheng, H.-M. 2010. High-energy MnO_2 nanowire/graphene and graphene asymmetric electrochemical capacitors. *ACS Nano* 4:5835–5842.
63. Pell, W. G., Conway, B. E. 2004. Peculiarities and requirements of asymmetric capacitor devices based on combination of capacitor and batter-type electrodes. *J. Power Sources* 136:334–345.
64. Wang, Y.-G., Wang, Z.-D., Xia, Y.-Y. 2005. An asymmetric supercapacitor using RuO_2/TiO_2 nanotube composite and activated carbon electrodes. *Electrochim. Acta* 50:5641–5646.
65. Algharaibeh, Z., Liu, X., Pickup, P. G. 2009. An asymmetric anthraquinone-modified carbon/ruthenium oxide supercapacitor. *J. Power Sources* 187:640–643.
66. Wei, W., Cui, X., Chen, W., Ivey, G. D. 2011. Manganese oxide-based materials as electrochemical supercapacitor electrodes. *Chem. Soc. Rev.* 40:1697–1721.
67. Cottineau, T., Toupin, M., Delahaye, T., Brousse, T., Belanger, D. 2006. Nanostructured transition metal oxides for aqueous hybrid electrochemical supercapacitors. *Appl. Phys. A* 82:599.
68. Fan, Z., Yan, J., Wei, T., Zhi, L., Ning, G., Li, T., Wei, F. 2011. Asymmetric supercapacitors based on graphene/MnO_2 and activated carbon nanofiber electrodes with high power and energy density. *Adv. Funct. Mater.* 21:2366–2375.
69. Wang, F., Xiao, S., Hou, Y., Hu, C., Liu, L., Wu, Y. 2013. Electrode materials for aqueous asymmetric supercapacitor. *RCS Adv.* 3:13059–13084.
70. Qu, Q. T., Shi, Y., Li, L. L., Guo, W. L., Wu, Y. P., Zhang, H. P., Guan, S. Y., Holze, R. 2009. V_2O_5 $0.6H_2O$ nanoribbons as cathode material for asymmetric supercapacitor in K_2SO_4 solution. *Electrochem. Commun.* 11:1325–1328.
71. Brousse, T., Taberna, P. L., Crosnier, O., Dugas, R., Guillemet, P., Scudeller, Y., Zhou, Y., Favier, F., Belanger, D., Simon, P. 2007. Long-term cycling behavior of asymmetric activated carbon/MnO_2 aqueous electrochemical supercapacitor. *J. Power Sources* 173:633–641.
72. Hsieh, Y. C., Lee, K. T., Lin, Y. P., Wu, N. L., Donne, S. W. 2008. Investigation on capacity fading of aqueous $MnO_2 \cdot nH_2O$ electrochemical capacitor. *J. Power Sources* 177:660–664.
73. Chang, J. K., Huang, C. H., Lee, M. T., Tsai, W. T., Deng, M. J., Sun, I. W. 2009. Physicochemical factors that affect the pseudocapacitance and cyclic stability of Mn oxide electrodes. *Electrochim. Acta* 54(12):3278–3284.

74. Dan, Y., Lin, H., Liu, X., Lu, H., Zhao, J., Shi, Z., Guo, Y. 2012. Porous quasi three-dimensional nano-Mn_3O_4+PbO_2 composite as supercapacitor electrode material. *Electrochim. Acta* 83:175–182.
75. Yu, N., Gao, L. 2009. Electrodeposited PbO_2 thin film on Ti electrode for application in hybrid supercapacitor. *Electrochem. Commun.* 11:220–222.
76. Wu, Z., Qu, Y. H., Gao, L. J. 2012. Performance of PbO_2/activated carbon hybrid supercapacitor with carbon foam substrate. *Chin. Chem. Lett.* 23:623–626.
77. Zhong, J.-H., Wang, A.-L., Li, G.-R., Wang, J.-W., Ou, Y.-N., Tong, Y.-X. 2012. Co_3O_4/$Ni(OH)_2$ composite mesoporous nanosheet networks as a promising electrode for supercapacitor application. *J. Mater. Chem.* 22:5656–5665.
78. Park, J. H., Kim, S., Park, O. O., Ko, J. M. 2006. Improved asymmetric electrochemical capacitor using Zn–Co co-doped $Ni(OH)_2$ positive electrode material. *Appl. Phys. A* 82(4):593–597.
79. Wang, Y.-G., Yu, L., Xia, Y.-Y. 2006. Electrochemical capacitance performance of hybrid supercapacitors based on $Ni(OH)_2$/carbon nanotube composites and activated carbon. *J. Electrochem. Soc.* 153(4):A743–A748.
80. Burke, A. 2000. Ultracapacitors: Why, how, and where is the technology. *J. Power Sources* 91:37–50.
81. Simon, P., Gogotsi, Y. 2008. Materials for electrochemical capacitors. *Nat. Mater.* 7:845.
82. Smith, P. H., Tran, T. N., Jiang, T. L., Chung, J. 2013. Lithium-ion capacitors: Electrochemical performance and thermal behavior. *J. Power Sources* 243:982–992.
83. Naoi, K., Naoi, W., Aoyagi, S., Miyamoto, J., Kamino, T. 2013. New generation "nanohybrid supercapacitor." *Acc. Chem. Res.* 46(5):1075–1083.
84. Sivakkumar, S. R., Pandolfo, A. G. 2012. Evaluation of lithium-ion capacitors assembled with pre-lithiated graphite anode and activated carbon cathode original research. *Electrochim. Acta* 65:280–287.
85. Yoshino, A., Tsubata, T., Shimoyamada, M., Satake, H., Okano, Y., Mori, S., Yata, S. 2004. Development of a lithium-type advanced energy storage device. *J. Electrochem. Soc.* 151:A2180–A2182.

20 Supercapacitors' Applications

Guoping Wang, Hongqing Wang, Benhe Zhong, Lei Zhang, and Jiujun Zhang

CONTENTS

20.1 INTRODUCTION

The supercapacitor is also known as the ultracapacitor or electrochemical capacitor (EC). Its capacitance originates from two ways, namely, electrical double-layer capacitance (EDLC) and faradaic pseudocapacitance (FP).

1. EDLC—Electrostatic storage has been achieved by separation of charge in a Helmholtz double layer, and no charge transfers across the electrode/electrolyte interface. The charge storage is highly reversible, allowing high cycling stabilities. The charge separation distance in a double layer is in the order of 0.3–0.8 nm. The most common EDLC electrode material is carbon. Currently, carbon-based EC accounts for approximately 85% of the worldwide revenue of EC. Active carbon represents the foremost carbon material in commercial devices owing to its precursor availability and low cost.
2. FP—Fast and reversible redox reactions take place on the surface of the electrode materials, significantly enhancing the capacitance of electrode materials. FP capacitance can be 10–100 times higher than that of EDLC [1,2]. This kind of material mainly includes transition-metal oxides (e.g., RuO_2) and conducting polymers (e.g., polyaniline, polypyrrole, polythiophene, and their corresponding derivatives). The reactions involve reversible adsorption (e.g., adsorption of hydrogen), redox reactions (e.g., RuO_2), and reversible electrochemical doping/dedoping (e.g., conductive polymer-based electrodes).

FP usually suffers from relatively lower power density than EDLC because faradic processes are normally slower than nonfaradaic processes [3]. Moreover, FP often lacks stability during cycling because of the redox reactions occurring at the electrodes.

Both EDLC and FP capacitance contribute to the total capacitance of EC. The ratio of the two varies greatly, depending on the design of the electrodes and the composition of the electrolyte.

On the basis of the design of the electrodes, EC is divided into three families.

1. EDLC capacitors—Carbon or its derivatives serve as electrodes. Commercial active carbons have a capacitance lower than 200 F g^{-1} in aqueous electrolytes and <100 F g^{-1} in organic electrolytes. The average energy density is about 5 Wh kg^{-1}.
2. FP capacitors—The electrodes are made of metal oxides or conducting polymers, which show much higher FP capacitance than EDLC capacitance. A capacitance of 600–1000 F g^{-1} is attainable.
3. Hybrid capacitors—This kind of capacitor has an asymmetrical electrode configuration (e.g., one electrode is carbon material while the other is FP material).

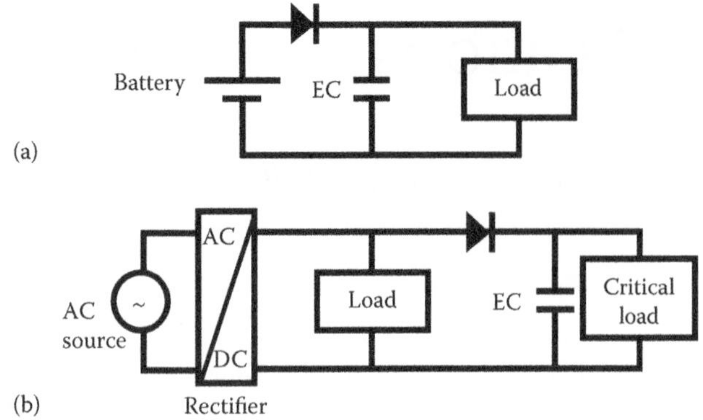

FIGURE 20.1 Use of EC in electronic circuits. (a) Battery-powered device where EC provides power backup of the load in case of disconnection of the battery. (b) Alternating current (AC) voltage–powered device involving heavy switching currents. EC protects the critical load (e.g., memory) from large voltage drops. (Reprinted from *Electrochimica Acta*, vol. 45, Kötz R. and Carlen M., Principles and applications of electrochemical capacitors, pp. 2483–2498, Copyright 2000, with permission from Elsevier.)

Electrodes exhibit both significant EDLC and FP capacitance. It has been extensively studied recently to improve overall cell voltage, energy, and power density.

The earliest EC patent was filed in 1957; however, it is still in the early commercialization stage because of its relatively low energy density and high cost. Not until 1978 did Nippon Electric Company (NEC) introduce the first commercial EC device (with H_2SO_4 as the electrolyte) into the market under license from SOHIO. NEC named this device as the supercapacitor, which was later referred to as the ultracapacitor by Pinnacle Research Institute [4,5]. At that time, EC was mainly used to provide backup power, enabling devices such as computer to retain memory in the power interruption. Those old NEC ECs possessed a capacitance of 1.0 F and a voltage of 5.5 V. Now, EC with various sizes is manufactured in several dozen companies around the world for various applications. Some of them have a capacitance of up to 9 kF or larger.

In 2010, the EC market generated a revenue of about $400 million. However, it occupied less than 1% of the world market in electrical energy storage system (batteries and EC). In spite of this fact, EC is playing a more and more important role in our daily life, and now its wide application prospect is attracting significant attention throughout the world. Figure 20.1 shows EC application on electronic circuits [6].

20.2 MARKET DRIVING FORCES FOR EC APPLICATION

1. *High power density.* EC, one of the most efficient and cost-effective devices to store energy, can be rapidly charged and discharged repeatedly. For EDLC, since the charge and discharge are dependent only on the physical movement of ions, it can store and release energy much faster (meaning more power) than a battery, which relies on slower chemical reactions instead. EC can be fully charged within seconds (~30 s), whereas batteries' charging or discharging time is on the scale of hours. The rapid charge–discharge ability makes EC an ideal energy storage device in automotive subsystems for load leveling (with discharge time of 2 min), in smart weapons for bridge power (with a discharge time of <5 s), in self-destruct input pin for pulse power (with a discharge time of milliseconds or less), and in memory protection for standby power (with a discharge time of hours at a low rate), etc.

   EC can reach a power density as high as 15 kW kg^{-1}, 10–100 times higher than that of batteries (e.g., lithium-ion battery can only reach 150 W kg^{-1}). Recently, with the special development of a tailored composite electrode, EC has achieved a maximum power density of 990 kW kg^{-1} [7]. This is a great advantage for those applications requiring high instantaneous power.
2. *Long life expectancy.* Rechargeable batteries typically wear out after hundreds to a few thousand charge–discharge cycles. On the contrary, EC's life expectancy, except for those with polymer electrodes, is much longer than that of batteries. For EDLC, they have an almost unlimited cycle life. They can withstand a large number of charge–discharge cycles, up to millions of cycles. During their cycling, no substantial drops of capacity or significant increase of internal resistance occur, because no or negligibly small chemical charge transfer reactions and phase changes are involved in the charging and discharging process. Even for FP, although fast redox reactions are involved, their cycling life is also much longer than that of batteries. As a result, minimal routine maintenance is required, meaning less influence on the total cost.
3. *Wide operating temperature range.* EC can function effectively at a wide temperature window ranging from −40°C to 70°C without significant degradation. Such a wide operating temperature range is incredible for lithium-ion battery, particularly at low temperatures. Because of overheating, lithium-ion battery cannot run at extremely high temperatures.
4. *Environmental friendliness.* The increasing focus on the environment is a key driving force for the application of EC. As they are mostly made up of carbon and its compounds, EC is easily biodegradable and its waste materials are easy to dispose. Meanwhile, packaging is carefully designed to minimize the negative impact on the environment.
5. *Significant progress in the voltage of EC.* The application of two different electrode materials, operating in their optimal potential ranges, extends the working voltage of the whole EC system [8]. Besides, efforts on the development of new electrolytes in the

past have increased the working voltage to 2.7 V by 2012. Further efforts are expected to allow the working voltage to achieve 3.0 V or higher.

6. *High efficiency*. Energy efficiency is a primary concern in renewable power generation. EC demonstrates up to 90–95% energy conversion or higher, which is much higher than that of batteries. During charging, the energy loss for EDLC is only 1%, while for a lead–acid battery, the loss reaches up to 30%.

20.3 SURVEY OF EC's MARKET

Owing to the market driving forces, new applications for EC are developed worldwide. North America is the early adopter of EC technology in research and development. Besides, Europe and Asia-Pacific have been paying great attention to searching various applications for EC, such as hybrid heavy vehicles, rail networks, and wind blades, etc.

With the increasing development of EC application, the global revenue is growing at a remarkable rate. Frost & Sullivan has forecasted that the global market of EC is likely to experience a consistent year-on-year double-digit growth during 2009–2019, and the Global Compound Annual Revenue Growth Rate will be over 20% [9]. Especially, the Asia-Pacific market is expected to grow at an annual growth rate of 22.3%. By 2015, the global revenue is expected to reach more than $500 million, indicating that the EC's market is gradually moving toward maturity. By 2019, the revenue for EC is expected to reach up to $1069 million.

The wide application and bright future have attracted more and more companies to join in EC industries. Table 20.1 [4] and Figure 20.2 [10] show the current main manufacturers of EC and their corresponding market share. Among these companies, Maxwell occupies the largest market share (25.5%), and Panasonic occupies the second largest market share (23.9%).

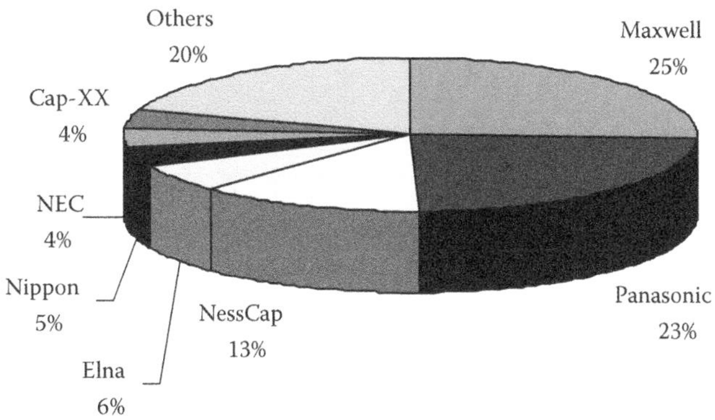

FIGURE 20.2 Total EC market: company market share by revenues (world), 2008. (Data from Frost & Sullivan, *Ultracapacitor Market Alternative Energy Storage Holds a Promising Future*, 2009. Reprinted with permission from Frost & Sullivan.)

Companies' research and development activities have facilitated the technology progress of EC. The introduction of organic electrolytes and asymmetric EC has significantly increased both the voltage and the energy density of EC. Organic electrolytes can extend the cell voltage range to 2.5–3.0 V from 1 to 1.5 V for the aqueous electrolytes. The electrolytes containing quaternary ammonium salts in propylene carbonate or acetonitrile possess an operative voltage as high as 2.7–2.8 V [11]. With asymmetric EC, ESMA Company used NiOOH as the positive electrode and active carbon as the negative electrode, increasing the specific energy from several to over 10 Wh kg^{-1}. Recently, the asymmetric design of EC has achieved an energy density of ~29 Wh kg^{-1} with the power density of 10 kW kg^{-1} [12]. Fuji Heavy Industries has achieved cell voltages as high as 3.8 V and energy density up to 25 Wh kg^{-1} by using active carbon predoped with lithium ions as a positive electrode and active carbon as a negative

TABLE 20.1
Main Manufacturers of EC

Company	Device	Country	Voltage Range (V)	Capacitance (F)
AVX	BestCap	USA	3.50–12.0	0.022–0.56
Maxwell	BoostCap	USA	2.50	1.60–2600
Panasonic	Gold Capacitor	Japan	2.30–5.50	0.10–2000
NessCap	EDLC	South Korea	2.70	10.0–5000
ELNA	Dynacap	USA	2.50–6.80	0.033–100
NEC	Supercapacitor	Japan	3.50–12.0	0.01–6.50
Cap-XX	Supercapacitor	Australia	2.25–4.50	0.090–2.80
Copper	PowerStor	USA	2.50–5.00	0.470–50.0
ESMA	Capacitor modules	Russia	12.0–52.0	100–8000
EPCOS	Ultracapacitor	USA	2.30–2.50	5.00–5000
Evans	Capattery	USA	5.50–11.0	0.01–1.50
Kold Ban	Kapower	USA	12.0	1000

Source: Reprinted from *Energy Conversion and Management*, vol. 51, Pawan S., and T.S. Bhatti, A review on electrochemical double-layer capacitors, pp. 2901–2912, Copyright 2010, with permission from Elsevier.

electrode [13]. All of these, in turn, boost the application development of EC.

20.4 EC's APPLICATION

Although EC is still considered as a relatively new energy storage device, it is playing an important role in shaping the quality of modern life and technological experience. Various EC devices, such as EDLC, pseudocapacitor, and hybrid EC, are intensively studied, and their performances have been improved greatly, bringing EC wide new applications in emerging technologies.

Generally, EC is used in many application areas based on its two characteristics [14]:

1. *Application of high power.* Thanks to its high power capability, EC finds new opportunities in power electronics, where short-time power peaks are required. Typical examples are the fast energy management in hybrid vehicles or the starting of heavy diesel engines.
2. *Application of long cycle life.* In the low-power applications, batteries suffer from maintenance problems or insufficient lifetime performance. EC can present a good solution for this issue. The uninterruptable power supply (UPS) as well as security installations are the most representative examples.

EC's applications cover transportation (electric vehicles [EV], hybrid EVs [HEVs], automotive subsystems, energy conservation, regenerative braking, and cold-starting assistance, etc.), backup (consumer electronics, digital cameras, computers, uninterruptible power supplies, and telecommunications), industrial (factory automation, and robotics), and renewable energy (solar and wind power), etc. [14–19].

Different applications correspond to different performance requirements for EC. For example, with memory backup, low power and low voltage are the performance requirements; for EV load leveling, high power and high voltage are needed; for space, medium or high power, high voltage, and reliability are necessary; for military, medium/high power and high voltage are indispensable; and for automotive subsystems, medium/high power and medium voltage are required.

Frost & Sullivan [9] has classified EC's applications into three segments, namely, the transportation, industrial, and consumer electronics.

In this chapter, the applications in transportation, backup power, and renewable energy will be discussed in detail.

20.4.1 Application in Transportation

HEVs, metro trains, and tram transportation, etc., involve frequent stops across short distances. These particularly require safe and reliable energy storage devices, which should have high power capability, long operational lifetime, and low maintenance requirements. EC is one of the devices that can meet these requirements. It is used predominantly as a backup energy storage device for engine start, regenerative braking, and short-launch boost.

Top vehicle manufacturers such as Honda, Toyota, BMW, Nissan, Ford, and GM have been pursuing EC technology as a supplemental power source for HEV. As a temporary energy storage device, EC is coupled with high-energy batteries or fuel cells to provide power for acceleration and engine starting, to balance load leveling, to increase the battery lifetime, to reduce operating cost, and to secure power for brakes and other critical functions in case of battery failure [20,21]. In addition, it can also act as a power buffer to recapture brake energy. Aircraft, airbuses, and ships have similar needs.

20.4.1.1 Transportation Market Driving Forces

1. *Need for fuel efficiency and green energy.* Currently, gasoline still remains to be the main energy source for transportation in the world. The depletion of oil reserves, instability of the Middle East, traffic congestion, the rising costs of fossil fuels, and the emissions (CO_2, sulfur and nitrogen oxides, etc.), etc., are pushing automakers to search for alternative energy storage devices. A battery-only system may operate at a 72% turnaround efficiency; however, an EC can contribute an efficiency of at least nearly 90%. By substituting the primary engine in an electric motor with EC, fuel consumption can be reduced by more than 50%. Particularly, emissions can be reduced by 90%, and nitrogen oxide emissions can be reduced by 50% [22].
2. *Relevant regulations.* With increasing concerns about the energy security and the environment, new designs and ideas especially for the transportation segment are necessary. Some increasing fuel efficiency standards have been released. For example, the corporate average fuel economy regulation of the United States requires all car companies to have a certain fleet average of miles per gallon. The current level issued by the US House of Representative is 35 miles per gallon for cars and light trucks by 2020 [22]. Besides, to encourage the development of environmentally desirable vehicles, the US government has introduced a number of federal tax incentives for hybrids and full EVs. For instance, the special tax deduction of $2000 was granted for HEV buyers. In 1991, California legislators passed the "Zero Emission Vehicles" legislation, which was enforced by allowing the car companies to earn numerical "credits" for their zero-emission cars.

   An analogous situation can be seen in other countries. For example, a voluntary agreement has been reached between the European Commission and the automakers. According to the agreement, major European car companies would reduce the average CO_2 emission from their fleets in several steps. In Japan, the government shows high concern about fuel efficiency. It provides various financial and tax

technologies with better fuel efficiency [23].

incentives to encourage the development of vehicle technologies with better fuel efficiency [23].

3. *EV's and HEV's requirements for peak power/current.* The electrical loads in today's cars, when applied simultaneously, require much more power than 2–3 kW, which can be provided by the current 14-V electrical systems. For example, even for a small car with a mean power of about 30 kW, its peak power may reach up to 60 kW during starting or acceleration. Obviously, additional "peak" power is frequently necessary for the energy storage devices of EV and HEV.

Batteries alone cannot meet the requirements. The high internal resistance limits their initial peak current. Trying to meet the peak load requirements with batteries will result in an oversized configuration because only a small fraction of the total energy is available from a battery at a start/stop. Besides, too high drain from batteries will shorten their life expectancy. Moreover, under harsh conditions (e.g., extremely low temperature, high duty cycles, and frequent deep discharging etc.), batteries cannot work efficiently.

EC has no such limitations. Although it may be unsuitable to serve as a primary energy source for EV and HEV, it is an ideal energy storage device for providing peak power and taking the main load from the battery during acceleration and braking, for capturing/storing the energy from regenerative braking, and for frequently starting the engine within 100 s, even at lower temperatures.

With the increasing requirements of EV and HEV for power and energy, it is reasonable to separate the energy and power requirements by combining a pulse power device (e.g., EC) with a primary energy storage unit (e.g., battery). This is called hybridized energy storage device. It consists of two basic energy storage types: one with high specific energy and the other with high specific power. In accelerating or climbing, ECs provide the peak load requirements, protecting batteries from degrading and overheating by rapid charge–discharge cycling. Meanwhile, in normal braking that lasts 5–10 s, ECs capture the kinetic energy to store for the next acceleration. Furthermore, ECs can guarantee a quick and reliable engine start regardless of the low-temperature environment.

20.4.1.2 Superior Performances of Transportation Tools When Using EC in Rechargeable Energy Storage Systems

When EC is in conjunction with batteries, the following superior performances can be observed in EV, HEV, cranes, etc.

1. *Remarkably reduced fuel consumption and accompanying emissions.* A battery-only system may operate at 72% turnaround efficiency, but an EC can contribute efficiency near 90% at least. EV's and HEV's energy storage system with EC can greatly improve fuel economy over conventional gasoline-only powered vehicles, saving millions of barrels of oil. When EC is used in electrical trains, trams, and subways, 25% energy savings can be obtained, corresponding to thousand liters of fuel per year. Even with a minor role in micro- to mild-hybrid cars, the energy savings can be up to 15% [24]. More attractive is the Komatsu hybrid excavator with a capacitor storage system, whose fuel savings can reach up to 30–40% [25]. Generally, the fuel consumption was 6.1 L 100 km^{-1} for those without regenerative braking and 5.3 L 100 km^{-1} for those with regenerative braking, amounting to an energy saving of 15% [26]. Meanwhile, CO_2 and other emissions can be reduced by an equivalent amount (5 g km^{-1}).

   In practice, a 5% energy saving can drastically reduce gasoline consumption during the life of the vehicle. The savings depend on the frequency of stopping and the weight. For train stations separated by 4 km, about 30% energy savings can be achieved. With 10 km station separation, around 18% energy savings can be obtained [27].

   Energy savings originate from the following: (i) It is not necessary to size the diesel engine for lifting the maximum load. As EC can take over some fraction of the power, it is possible to use a smaller and more fuel-efficient engine. (ii) EC can efficiently capture and store virtually all of the braking energy generated at each stop and the energy that is normally wasted as heat. The captured energy can be quickly fed back to the electric motor during subsequent reaccelerating, lifting, or climbing. This is a large amount of energy. For example, with a 20,000 kg bus travelling at 50 km h^{-1} and stopping in 10 s, the kinetic energy of the bus is 1929 kJ and the average power generated during braking is 193 kW. (iii) Heating can be performed immediately before the use of machines. Hence, EC makes it possible to avoid machine idling continuously, reducing the associated energy waste from idling. Meanwhile, fast engine cranking can significantly reduce the conservation of fuel and the emissions [27].

   Scientists at the Argonne National Laboratory have successfully demonstrated that an integrated system combining batteries with EC dramatically improves braking energy recuperation efficiency. In April 1997, as one of the first vehicles in the world, the "Blue-Angel" prototype vehicle was able to recuperate energy by means of an EC Module, which rated at 42 F at 400 V (286 × 3000 F cells). An overall recuperation efficiency of about 70 % can be expected [28].

2. *Lowered brake maintenance and replacement expenses.* The Cap-XX Stop–Start test has shown that battery alone failed after 44,000 starts, while the "Battery and EC" system still ran after 120,000 starts in New European Drive Cycle at 23°C. The reason is that EC have taken most of the load off batteries or mechanical brakes, extending the batteries' life and reducing brake maintenance and replacement expense.

3. *Substantially reduced weight and volume of the energy storage system.* For the battery system, to overcome the limitations of charge–discharge rate and to meet either the power or energy demands, it is typically overdimensioned. As a result, the volume, weight, and cost are added to the battery system. A hybrid power source combining EC with batteries can eliminate the need for oversizing battery, creating a smaller and lighter energy storage system. Many devices can benefit from such a configuration to address the increasing requirement for the reduction of the battery size, weight, and miniaturization of devices. About 60% weight can be reduced if the hybrid power source is selected. It has been reported that using a continental "E-Booster" system containing MXWL EC can save about 6 m of heavy and expensive copper battery cable, and can reduce the size of the battery by 30%, making it small enough to be located under the hood instead of in the trunk [6,29].
4. *Greatly extended battery life.* EC can make batteries mainly supply a smaller average load and meet short power pulses by offering the required peak power. These relieve batteries of the most severe load demands, ensuring batteries run under favorable conditions. Moreover, EC can retain and release a burst of power under extreme cold environments down to −40°C, extending the operational temperature scale. As the result, the batteries' cycling life is significantly lengthened. A doubled battery life has been reported by Ivanov et al. [30] in the starter application of diesel engine.
5. *Fast restarting engine.* EC can be charged from batteries within no more than 1 min, and can provide a short burst of power for restarting the motor. With the use of EC, engine can restart within just 400 ms, twice as fast as a manual key restart and 30% faster than those with a reinforced starter (dual mass flywheel). Such a start–stop technology is beneficial for heavy-duty vehicles. It refuses vehicles to shut down when they come to a stop at a red light, picking up/dropping off passengers, sitting in traffic, or in cold climates. Several papers about engine cranking using EC are available [31,32]. Besides, starter systems with EC eliminate morning idle heat up in cold climates and the cost of jump starting. Moreover, engine noises appear practically unnoticeable.
6. *Less need for cooling.* Thermal loading is a serious problem for batteries, particularly for advanced batteries, owing to the considerable heat generated and the limited dissipation. ECs show much lower resistance than batteries. This not only increases the efficiency of ECs but also ensures less cooling needs. They can be air-cooled far more easily than batteries that require either liquid cooling or full climate control systems.

There is something we have to watch out for. Since ECs are normally connected in parallel with the battery in these applications, they can only be charged up to the battery upper voltage level and can only be discharged down to the battery lower discharge level, leaving considerable unusable charge in EC and limiting its effective or useful energy storage capacity.

20.4.1.3 Briefing of Transportation Market

An important and genuine breakthrough in EC technology was made in the early 1990s by ESMA of Moscow, which adopted an asymmetric design. In the design, one electrode was an electric double-layer material (activated carbon), while the other used faradaic material (NiOOH). This design significantly increases the energy density and remarkably reduces the self-discharge rate. A bank of such asymmetric EC with 30 MJ (8 kWh) of energy allows the bus to run 15 km [27].

Additionally, in 1991, hybrid vehicles have been developed and realized in the department of electrical engineering of the HTA Lucerne (University of Applied Science of Central Switzerland). Mass-transit buses, fleet vehicles, long-haul trucks and other heavy-transportation vehicles (trains, light rail, trams, and metros) all benefit from the adoption of a hybrid power using EC. The issues, "long operational lifetimes, low maintenance requirements, and operating efficiently under harsh conditions," have been resolved with the use of EC modules.

In the early 1990s, Moscow has the EC-powered bus running a shorter circular route in an exhibition park. In 2007, Toyota made the history by winning the Tokashi 24-Hour Race with its hybrid race car that adopted a quick-charging system with two series connected 1200 F EC. In 2010, EC-powered buses were exhibited in Shanghai, China, where batteries and EC are integrated in the rechargeable energy storage system. They can be rapidly recharged in stations within about 20 s. Since the charge of energy storage system occurs at only several fixed locations, suspended catenary electric lines, which are seen in conventional electric bus or trolley systems, are absent. Such capacitor-powered buses not only create a visually more attractive solution to the delivery of bus power, but also increase the route flexibility. It is possible for the bus to go from one charging station to other charging stations, depending on transportation demands [27]. In addition, we can see other hybrid vehicles running on the roads, such as Toyota Prius Hybrid, Honda Civic Hybrid, the Ford Escape Hybrid, Mercury Mariner Hybrid, and the Nissan Altima Hybrid.

With the increasing interest on the use of EC in conjunction with batteries in HEV for stop-and-go operations, EC application in transport market will be boosted significantly. It can be seen from the fact that Maxwell Technology alone has received purchase orders, worth $13.5 million, from three transit bus manufacturers in China for its 48-V EC modules in April 2009. The EC modules are used to support braking energy recuperation and torque assist functions in 850 diesel–electric hybrid buses. Besides, ISE's EC-based hybrid bus drive systems power nearly 300 buses in daily revenue service.

TABLE 20.2
Transportation EC Market: Revenue Forecasts (World), 2005–2015

Year	2005	2006	2007	2008	2009	2010	2011	2012	2013	2014	2015
Revenue ($ million)	29.2	31.7	34.5	44.0	55.1	66.5	81.0	101.5	131.4	176.3	242.6
Revenue growth rate (%)	–	8.6	8.5	27.8	25.4	20.6	21.8	25.4	29.4	34.2	37.6

Source: Frost & Sullivan, *Ultracapacitor Market Alternative Energy Storage Holds a Promising Future*, 2009. Reprinted with permission from Frost & Sullivan.

As shown in Table 20.2, the transportation market of EC added up to $44.0 million in 2008. It is expected to reach $242.6 million by 2015. Various mild-, micro-, and full-hybrid vehicles, forklifts, and trucks contribute to this market. According to Frost & Sullivan, the estimated global commercial EV and HEV demand is expected to reach approximately 240,000 units by 2015 [10].

In the light of a quick return on investment, many manufacturers shift their interest to the transportation market. The main manufacturers are Panasonic Corporation, Maxwell Technologies Inc., and NessCap Co. Ltd. Other companies include Tavrima, Nippon Chemi-Con Corporation, Axion Power International Inc., and EEStor [10].

Panasonic acts as the leader in the transportation EC market and occupies approximately 40.0% of total revenues in 2008. It has developed the "UP Cap" series and introduced them to the automotive market in the early 2000. All of its transportation ECs are produced and managed in the Japanese automotive branches of the company.

Maxwell Technologies tread on the heel of Panasonic in transportation market, with 26% of total revenues. Its EC can be used in conjunction with a primary energy source (e.g., battery) for targeting toward regenerative braking. The products mainly include

1. *The 125-V Heavy Transportation Module.* It stores more energy per unit volume, delivers more power per unit volume and weight, and lasts longer. It is designed especially for vehicles requiring the highest power performance available with shock and vibration immunity.
2. *The 48-V EC Module.* It provides efficient cooling and a maximum continuous current of 90 A without compromising reliability. It is used primarily in hybrid buses and train engine starting systems.
3. *The K2 EC Cells.* It is a solution for hybrid train systems where the EC is integrated into custom modules, and the 16-V EC Module is used in engine starting systems for buses.

Now, Maxwell Technologies is trying to increase its operating profits by reducing the manufacturing costs. For example, carbon is obtained from different suppliers to cut the price of carbon down to around $10 per pound. Besides, Maxwell is considering outsourcing its manufacturing to China to reduce overall costs as well as to enter the potential Asia-Pacific market.

NessCap Inc., Ltd., based in South Korea, is particularly aiming at the transportation market, and its market share (13%) ranks the third. NessCap's EC for transportation has ultralow internal resistance, rugged design for harsh environments, and integrated EC management unit for individual cell balancing, voltage, and temperature diagnosis and optional communication interface.

It offers two product lines, namely NessCap High Power and NessCap High Energy. Besides producing the line of prismatic-type ECs from "2.3 V 1200 F" to "2.7 V 5000 F," NessCap also offers a radial-type EC of 3–120 F.

20.4.2 Application in Backup Power and Portable Consumer Electronics

Since consumer electronic devices are becoming more and more power intensive and requiring instantaneous power to operate their various features, the need for energy storage devices with high power density is increasingly growing. EC has the ability to instantly charge and discharge for numerous cycles. It is an ideal solution to the above issue, especially in the applications where peak power and pulsed currents have to be provided, or grid power disturbances/outages occur.

Typical examples are UPS systems, telecom backup power systems, digital camera flashes, music players with superior audio quality, signal transmission in mobile phones, Internet and e-mail services, wireless remote control, toys, personal digital assistants, pulse power sources for high-power lasers, clocks of measuring or control equipment, low-voltage emergency lighting, and board-mounted backup in solid-state disk drives used in enterprise computing systems [33]. In the military field, the suggested devices include rail guns, electromagnetic pulse weapons, radars, and torpedoes.

According to EC's operation time in the above-mentioned electronic devices, NEC Company has classified EC into four types—long-time backup, backup for 1 h or less, backup for 10 s or less, and backup for power assist (see Table 20.3 [34]). In these application devices, EC may play roles as improving safety and reliability of power sources, quick charging within seconds or minutes, extending battery life or eliminating batteries, and optimizing size and cost, etc.

TABLE 20.3
Backup Application Examples of EC

Intended Use	Power Supply	Application	Examples of Equipment
Long-time backup	≤500 μA	Clocks	Measuring device, control equipment, communication devices, automotive power sources, microcomputer, static digital tuning system, etc.
Backup for ≤1 h	≤50 mA	Driving motor, camera, microcomputer	Microwave oven, microcomputer, memory equipped device, printer, projector, video disk, camera
Backup for ≤10 s	≤1 A	Toys, buzzer	Toys, display device, alarm device, actuator, relay solenoid, gas igniter
Power assist	Several A	Power supply	Street sign, UPS, display light

Source: NEC Tokin. 2013. Super capacitors. Available at www.nec-tokin.com/english/product/pdf_dl/supercapacitors.pdf. Reprinted with permission from NEC Tokin Corporation.

20.4.2.1 Roles of EC in Backup Power and Portable Electronics

1. *Improving the safety, durability, and reliability of power sources.* In our daily lives, we often encounter electrical outages and electrical faults, including breaker trips, voltage sages, power surges, frequency variations, waveform distortions, spikes, and noise. For digital appliances with a memory component, a short interruption of power supply will cause critical data loss and substantial losses in industrial production lines or commercial electronics [6]. These often bring great inconvenience to users, or even worse, cripple a business for days, weeks, or longer. Typical examples are large enterprises that increasingly rely on solid-state disk drives for flash memory in applications such as credit card transaction processing.

   To make a variety of functions (e.g., memory buffers) active during a power loss or shutdown of a system, backup power systems are deployed to supply continuous power for critical mission applications, such as network computer servers, Internet data centers, telecommunications, hospitals, factories and broadcast systems, etc. UPS is such a device that can provide short-term bridge power, allowing a network to be transferred to a secondary power source or gracefully shut down in a controlled and orderly fashion.

   Batteries are one of the candidates for backup power system or UPS; however, batteries' life expectancy is limited, and high-power or pulsed currents can deteriorate battery performance. This often results in a regular replacement, boosting the price of battery appliances by 20%. Generally, more than 90% of power disturbances last for less than several seconds. Hence, it is only for very short periods of time, although high energy, voltage, and power are required [15].

   Under these situations, EC gets the opportunity to offer the most reliable and longest-life worry-free energy storage owing to its high pulse power within a short time (milliseconds to seconds), versatile temperature tolerance, and long life expectancy [4]. It has demonstrated high durability and reliability in delivering power bursts for over a million cycles, virtually eliminating site maintenance visits. System designers are increasingly turning to EC to avoid maintenance or replacement.

   The following are some examples using EC-based power/backup power for improving the safety, durability, and reliability of power sources.

   a. *Real-time clocks.* With respect to the long-time backup power, the typical example is in the real-time clocks of electronic systems, which keep track of the time in measuring device, control equipment, microcomputer, digital tuning system, communication device, and automotive power source. By using EC, continued operation can be guaranteed when the battery is being replaced or removed.
   b. *Car audio system, taxi meter, and emergency lighting.* In car audio systems and taxi meters, EC often serve as the backup for radio station memory, taxi fare programs, and accumulated fare data. In emergency lighting, EC can power emergency lighting for hours in the event of a power failure if it is matched with light-emitting diodes (LEDs).
   c. *Hospitals and medical life-sustaining equipment.* The long predictable life of EC makes it ideal for hospitals and medical equipment. UPS systems with EC can provide smooth power to shut down safely or ride through for a permanent primary power backup (typically a fuel cell or diesel generator).
   d. *Remote solar-powered installations or roadside emergency telephone.* These systems need reliable backup power to ensure continuous and safe operation in the event of power failure. Maintenance visits to monitor or replace batteries at regular intervals are difficult and expensive. EC is ideal for these mission-critical

systems. It can be charged for a million times or more and can continue performing reliably without replacement. It can provide bridge power to avoid service interruptions until primary backup power sources take over.

e. *Cameras, programmable pocket calculators, electronic agendas, and mobile phones.* In these appliances, EC can provide power for seconds to minutes during the replacement of the batteries [6]. In video recorders and TV satellite receivers, EC is needed to provide power backup for duration of hours to weeks.
f. *Flashlights.* In flashlights, EC can power LED. It can be recharged in a mere 90 s for another operation of 2 h. Meanwhile, it can be cycled for up to 50,000 times, making maintenance free and thus producing extremely low life-cycle cost. In addition, EC can work at very-low-temperature environments.

2. *Optimizing size, weight, and cost.* Portable electronics, such as cameras, cell phone, MP3 players, emergency beacons, video, and GPS navigation, etc., are now very popular. Stuffing all these features into a small package requires a lot of energy and high power.

   Using battery alone will lead to a big size and overweight of appliances, defeating the "pocket" concept. To keep up with the trend of smaller lightweight systems and meet the need of power pulses, design engineers are specifying small energy storage devices to facilitate a simpler design.

   By separately addressing the power and energy needs, that is, intelligently combining battery for energy with an EC for power, the above problem can be easily resolved.

   EC is making its way into many portable products in this way. For example, in the mobile phone market, EC is increasingly used to reduce the size and enhance the performance of mobile phones. In digital camera, EC makes smaller and lighter camera truly portable. In moving toys, EC makes toys accelerate very fast owing to their compactness and low weight.

   In addition, by using EC, UPS may also find economical interests, thanks to dispensing with an inverter. Meanwhile, EC's longer life can cut down the overall cost per unit by over $200 compared with a pure battery.

3. *Quickly charging/discharging.* Presently, most devices using battery power supplies need to be charged for a long time. With the use of EC, they can be charged/discharged very quickly without significant loss in efficiency.

   An interesting application is the portable driller and screwdriver, where EC (no batteries) powers an electric motor for driving screws. Being powered by EC, they only need to be charged for about a minute to drill a few holes or drive a few screws [27]. There is no need to wait for an hour recharge to do a 2-min job. The weekend handyman is freed to do quick jobs with quickly charging cordless tools, which is particularly popular with homeowners. In cordless copper tubing cutter, using EC for peak power can reduce the cutting time by 50%. In a solar-powered LED-illuminated sign or roadside emergency telephone, EC also allows more efficient storage since it can accept charge as quickly as sun rises or sets.

4. *Increasing batteries' life expectancy and operating time.* The short and high current bursts in portable electronic devices (e.g., laptops, camera, and mobile phones) often deteriorate battery performance. Combining EC with batteries offers an ideal solution to this problem, because EC can significantly relieve batteries of the most severe load demands by providing peak power for signal transmission, camera flash, and Internet connectivity in cell phones, etc. As the result, batteries are allowed to supply the average load, and their operating time and life expectancy are in turn extended.

   For instance, in digital cameras, a combination of EC with battery can make cameras take more pictures by 50%. In wireless communication devices, small EC can help batteries work much longer, generally by several folds. In automatic meter reading equipment that requires high current pulse power, EC is used for wireless data transmission. In robotic carriers, EC is helpful for hauling payloads ranging from semiconductor wafers to heavy equipment. Other similar applications include machines, elevators, crane recuperation, emergency lighting system backup, and emergency diesel starters, etc.

20.4.2.2 Overview of EC's Backup Power and Portable Consumer Electronics Market

The information about the consumer electronics EC market was obtained from the report of Frost & Sullivan. As shown in Table 20.4 [10], the consumer electronics EC market amounted to revenues of $31.3 million in 2008, representing approximately 30.4% of the total EC market.

The Consumer Electronics EC market is currently dominated by Panasonic, Elna, NEC Tokin, and Cap-XX Corporation. Other companies, such as Evans Capacitor, Kanthal Globar, and Cellergy, also contribute to increasing the revenues of this market. All these companies actively sell their EC to UPS and backup power vendors, who in turn incorporate them in their designs.

It is worth pointing out that Maxwell does not directly manufacture EC for this market but sells its electrode technology to the EC market. EC using Maxwell electrode technologies can be rapidly charged and discharged. Their cycle life can reach up to 1,000,000 cycles. They can store energy in a way that is energy efficient, environmentally friendly, and cost-effective. Maxwell's PC10 series (2.5 V DC, 10 F)

TABLE 20.4
Consumer Electronics EC Market: Revenue Forecasts (World), 2005–2015

Year	2005	2006	2007	2008	2009	2010	2011	2012	2013	2014	2015
Revenue ($ million)	26.7	28.4	30.3	31.3	32.4	33.5	35.5	37.9	40.7	44.2	48.2
Revenue growth rate (%)		6.4	6.7	3.2	3.5	3.6	5.8	6.8	7.5	8.5	9.0

Source: Frost & Sullivan, *Ultracapacitor Market Alternative Energy Storage Holds a Promising Future*, 2009. Reprinted with permission from Frost & Sullivan.

are ideal for consumer electronics and wireless transmission, which require more energy per unit volume, deliver more power per unit volume and weight, and perform longer. The HC series (2.7 V DC, 5–150 F) are suitable for consumer electronics and portable power tools. The 48-V Module is widely utilized in UPS for power quality and more demanding requirements. The 56-V Module provides power during dips and sags in the main power source. In longer-term outages, the modules provide transition/bridge power to a longer-term backup source such as a motor generator or fuel cell. D Cells are assembled onto printed circuit boards and packaged into a module for a variety of operating voltages and lower-power UPS [33].

Panasonic Gold EC includes various series. Stacked Coin Type SD, SG, SE, NF series (5.5 V, −25–70°C) are recommended to be used in memory backup for video and audio equipment, cameras, telephones, printers, data terminals, rice cookers, and intelligent remote controls. RG (3.6 V), RF (5.5 V) series, whose working temperature ranges from −25°C to +85°C, are used for backup of data of base station, electronic meter, and assist of rapid load change; EN, EP (3.3 V, −10–60°C), ER (2.6 V, −25–60°C) series for memory cards (power supply to hold memory), mobile phones, and digital cameras, etc.; HZ (2.5 V, −25–70°C) series for solar battery–operated circuits and backup power supplies (UPS); and HW (2.3 V, −25–60°C/70°C) for solar battery–operated circuits (road guidance flasher), quick-charging motor drives (toy car), and UPS.

Elna, a Japanese manufacturer, mainly focuses on navigation system cards, steering controls, entertainment systems inside a vehicle, vending machines, and specialized hospital beds. It offers two kinds of product: small cell (0.047–1 F, 5.5 V) is mainly for memory backup for PC to standard clocks; large cell (1–50 F, 2.5 V) is used for memory backup and as an additional power source.

NEC targets the small-sized markets. Its three major product groups are the energy device business such as condensers and secondary cells, network device business such as IC cards and optical communication devices, and functional devices such as piezoelectric components and magnetic components. It has several series. Among these, the E/SV, F/SV, and SV/Z series (2.5–35 V, −55–125°C) are recommended to be used in mobile phone, PC and PC peripheral products, audiovisual products, and measuring instruments. The PS/L and PS/G series (2.5–16 V, −55–105°C) are used for PC and game, mobile phone, storage device, noise reduction, and voltage fluctuation suppression for video/digital camera, etc. The NB/P series (2.5 V, −55–125°C) are employed in high-speed image processors for game machines.

Cap-XX, an Australian company, has revolutionized the consumer electronics EC market by introducing thinner, lighter, and smaller EC. Cap-XX's high-temperature HS and HW series EC, introduced in September 2007, are capable of withstanding temperatures up to 85°C.

20.4.3 Applications in Renewable Energy

The most prominent application of EC is the capture and reuse of waste energy. In the present industry, significant quantities of energy can be reused, provided that suitable energy storage devices are available. Since the storage processes typically occur in a very short duration (seconds to a few minutes), the storage devices should have the ability to provide high power. EC is the preferred device because of its rapid and efficient charging as well as long cycle life.

1. Solar and wind power often fluctuate with weather conditions. The weather variations occur within milliseconds or a minute. As weather varies, the power output and resulting voltage levels can vary up to 10%. This variability creates instability on the grid, resulting in poor utilization of the resources as low as 30–50%.

   With the growing use of renewable energy sources, the percentage of grid power met by renewals increases. Consequently, grid instability is increasingly becoming a serious challenge. Most countries around the world are placing particular emphasis on improving the reliability of the electricity grid. EC provides the most cost-effective and maintenance-free solution to the issue. It can absorb energy to keep voltage spikes down, and release energy to smooth voltage to the grid. Meanwhile, it can accept the challenge by providing a few seconds of reactive power in a smaller package size. Therefore, it is ideal for firming the output of renewable installations. By using EC, solar and wind power can be valuable to help ride through periods when there is little or no power generation.
2. In solar photovoltaic applications, photovoltaic power supplies cycle every day, leading to a detrimental effect on batteries. As a result, batteries have to be replaced every 3–7 years [4]. EC can be charged and discharged

quickly, and also can experience a large number of charge–discharge cycles (up to 30 years) without significant losses in performance, eliminating the frequent maintenance and saving a lot of maintenance costs.

Besides, some stations may be located in extremely cold or hot climates. If batteries are used for energy storage, the temperature has to be maintained at about room temperature with auxiliary systems. This represents additional cost and energy consumption [4]. EC can run efficiently at a wider range of temperatures, and no additional auxiliary devices are necessary.

3. Wind turbine operators need systems with zero or minimal maintenance, and require systems with reliability for years under all weather conditions. Pitch systems from 6 to 750 MW and up must be guaranteed to operate across broad temperature ranges and wind conditions.

When environmental temperature ranges from −40°C to +75°C, EC has demonstrated its durability and reliability, enabling electric pitch control systems to meet the grueling wind energy expectations under any conditions. Moreover, it can provide burst power for electric blade pitch control systems to ensure that rotor speed remains within a safe operating range and/or to optimize wind turbine output [33].

In 2007, the renewable energy market generated approximately revenues of $34.9 million (accounting for 35.0% of the total market). This figure increased to around $37.9 million in 2008, and is expected to reach $91.2 million in 2015 [10]. The compounded annual growth rate is estimated to be 13.4%. With more understanding of this technology, EC is expected to be used in conjunction with the battery as a hybrid solution.

The significant trend observed in this market is toward the wind power blade pitch systems. Companies that actively participate in renewable energy market are Maxwell Technologies, NessCap Inc., and NEC Tokin. Other companies are Evans Capacitor, Nippon Chemi-con, and Axion Power International [22].

Of these companies, Maxwell Technologies has exclusive deals with various European blade pitch manufacturers to incorporate Maxwell's Boostcap technology into their systems. Maxwell Technologies ramped up their production to keep pace with demand. Currently, this market is characterized by lack of competition because there are very few vendors.

Of Maxwell ECs for grid storage applications [33], the K2 cells allow custom-designed systems for pitch control and backup power; the 16-V Modules, cost-effective modules, can be built up to custom voltages and power levels for pitch control; and the 75-V Wind Turbine Module can be used in wind turbine pitch drive systems, and double as backup for DC–DC links.

NessCap Inc., Ltd., also has a significant presence in the renewable energy market. It offers lines of prismatic-type and radial-type supercapacitors, similar to its products applied in the transportation market. Domestic sales of its products represent 5–10% of its revenues.

The factors influencing the future revenues of renewable energy market involve the continuous demand for power quality products, cost reductions in EC manufacture, and demand for battery-free solutions.

20.5 EC's MARKET RESTRAINTS

Although EC shows potential applications in many areas, there are some factors restraining it from practical use.

1. *High price.* The key challenge for EC's application is its high price, which is estimated to be about 10 times higher than that of batteries. Current costs of large (>1000 F) EDLC cells as well as midsize EDLC cells (20–500 F) range from 3 to 6 cents F^{-1}. In terms of \$ Wh^{-1} (including the cost associated with the EDLC packs), its costs are between \$35 Wh^{-1} and \$70 Wh^{-1} [23].

The high price is due to insufficient production capabilities, high raw material costs, auxiliary devices, and slow adoption by end users. Raw materials (including electrode, aluminum foil, binder, conductive carbon, separator, and electrolyte) account for 50–65% of the total cost of ECs. At present, the most common electrode materials are carbon and RuO_2. Carbon with high specific surface area can be US\$ 50–100 kg^{-1} because it is produced in small quantities and there are only few suppliers. It goes without saying that the rare metal oxide RuO_2 will result in much higher cost. In addition, the separator and electrolyte also boost the expense. If organic electrolytes are used, the cost further increases.

In addition, EC needs to be assembled together in series or in parallel for high voltage and high power. The variations in the self-discharge of the cells can lead to imbalance of the voltages in an EC unit. Therefore, balancing cell voltages to avoid overvoltage is the main issue. To achieve cell balancing, other components are added, also boosting the price. When EC is applied in a vehicle, the total cost also involves the added electronics for each of the installed systems in addition to the additional energy storage unit. With the addition of more components, complex control mechanisms are needed to incorporate for coordinating and assisting. As the result, the total cost of the vehicle will increase remarkably (on the average, at \$15 Wh^{-1}).

Moreover, to make the current flow smoothly, large pieces of metals are installed and complex designs are necessary, boosting EC's price again. This also leads to greater weight and volume for the EC unit. Meanwhile, at the market introduction stage, the limited sales volume and manufacturing capabilities also heighten EC's price.

The key factors to reducing EC's price are to utilize lower-cost electrode materials, produce high-voltage EC, and develop assembly processes that can be automated at reasonable investments [35]. Another efficient strategy is to increase volumes and to target heavy-volume markets such as transportation and renewable energy.

2. *Low energy density.* Currently, EC cannot fully substitute batteries because of its limited energy density (about 5 Wh kg^{-1}). Particularly, commercially available ECs can only provide energy densities of 3–4 Wh kg^{-1}. Owing to their low energy density, they cannot sustain loads for a few minutes.

   Two strategies are adopted to improve the energy density, namely, enhancing the specific capacitance of the electrode materials and expanding the operating potential window. Electrolytes, especially ionic liquids, have a large range of operating voltage. Asymmetric or hybrid capacitors are also a potential candidate for improving the potential window and then energy density. For example, MnO_2/AC (active carbon) systems can operate within a voltage range of 0–2.2 V [36]. In mild Li_2SO_4 aqueous electrolyte, the $LiMn_2O_4$/AC system was operated within an operation voltage range of 0.8–1.8 V [37].

3. *Low voltage.* Generally, the potential window of the EC ranges between 1 and 2.5 V. In the case of organic electrolyte, it lies between 2 and 3 V. However, in practice, most powerful applications need much higher voltages, and some reaches up to 700 V. To achieve the required application voltages, EC must be connected in series.

   Each EC has a slight difference in capacitance value, voltage, and equivalent series resistance (ESR). These give rise to the issue of cell balancing, such as voltage and ESR balances. A higher ESR results in a higher voltage drop, leading to poor performance and product degradation. Hence, it is necessary to actively or passively balance them with a control unit or electronic equilibration. These affect the volumetric efficiency of the finished EC module. A 500-V module will be the size of a desk, which is a big challenge for the device design.

4. *Lack of standard.* When using EC as an energy storage device, a standardized form factor will enable their easy adaptability. However, till now, there is still no common standard for EC. The lack of standards has greatly affected the adoption of the market, because each EC company has to identify different design parameters for each client. Meanwhile, the lack of standards makes it difficult to compare EC products made from different manufacturers, and gives rise to an additional challenge in technology and design development, slowing down the market step of EC.

   For example, current EC does not have size and shape specifications, causing the difficulty in designing. As the result, companies are striving to reduce the footprint of their products by adopting various shapes and forms. Thus, it is essential to regulate different parameters such as size, shape, and voltage.

5. *Limited consumer awareness.* Owing to its low energy density, EC is not a direct requirement for the end users, resulting in limited consumer awareness. In the consumer electronic market, EC does not serve as the main energy storage device, but is particularly used in value-added products, such as high-tech mobile phones and digital cameras, etc. Hence, it is necessary to increase the awareness and place correct emphasis on the advantages of the technology.

6. *Leakage.* The driving force for self-discharge, leakage, is the thermodynamic instability of the charged state. EC's leakage depends on capacitance, voltage, temperature, and the chemical stability of electrode and electrolyte combination.

   With the leakage, the reliability of energy storage may become an issue in the design. Moreover, owing to leaking, there is a tendency to eventually short out, leading to immediate loss of data memory when batteries are removed. Hence, the leaking characteristics of EC are required to be evaluated and improved before their use.

7. *Long development cycles and integration concerns.* A complete EC product needs about 4 years. The detailed design process takes almost 1 year, including drawing, modeling, feasibility discussion, and consultation with engineers. The engineering phase, which usually lasts for 18–20 months, is the second step of product timeline. Immediately, the product is subjected to various tests to check for compliance with industry automotive standards before being introduced into the market. Along with this is the need to negotiate with device manufacturers to obtain approvals for incorporating EC into their devices.

   Moreover, in EC's integration, more space allocation is required. The added space to house EC is a challenge for device design engineers.

8. *Safety factor.* Although the low internal resistance of EC allows extremely rapid discharge, it may result in a much greater spark hazard than batteries when short-circuiting occurs. To guarantee safety, protection circuit and device are often installed.

20.6 SUMMARY AND APPLICATION TRENDS

EC is still in its infancy of application; however, it does have great potential application in a number of areas. It holds the most opportunities in the transportation market. This market area includes forklifts and cranes, etc., which require high power density. Meanwhile, in LED, lighting, and self-generating batteryless remote controls, EC is becoming a major energy storage device. Online gaming, Wi-Fi accessibility, camera phones, and global positioning system navigation are the upcoming trends in the consumer electronics market. Renewable energy markets such as solar or wind energies are very likely to generate great opportunities for EC adoption.

As for the application trend, there are several possibilities. (i) One of the future research areas is to maximize the obtained wind energy using EC. In this area, EC can be used in blade pitch control systems, or be applied for draining and storing the excess power for a few minutes until it can be fed back into the grid, optimizing the use of unutilized wind capacity. This will enable the grid to run at efficiency close to 100%. (ii) Hybrid energy storage system in conjunction EC with battery is another possible hot area. (iii) Asymmetric cell design may be popular in the coming years. This design combines the best features of EDLC and FP together in a unified EC. In this way can the cell voltage be increased significantly, since the two different electrodes work in their optimal potential range: the negative one with high hydrogen overpotential and the positive one with high oxygen overpotential [4].

ACKNOWLEDGMENTS

The authors would like to thank the financial support project (2014GK3091) from Hunan Provincial Science and Technology Department.

REFERENCES

1. Conway B. E., Birss V., and Wojtowicz J. 1997. The role and utilization of pseudocapacitors for energy storage by supercapacitors. *Journal of Power Sources* 66: 1–14.
2. Conway B. E. 1999. *Electrochemical Supercapacitors*. Kluwer Academic/Plenum Press, New York.
3. Ming C. C., Huang C. W., Teng H., and Ting J. M. 2010. Effects of carbon nanotube grafting on the performance of electric double layer capacitors. *Energy Fuel* 24: 6476–6482.
4. Pawan S., and Bhatti T. S. 2010. A review on electrochemical double-layer capacitors. *Energy Conversion and Management* 51: 2901–2912.
5. Namisnyk A. M. 2003. A survey of electrochemical supercapacitor technology. B.E. Thesis, University of Technology, Sydney, Australia.
6. Kötz R., and Carlen M. 2000. Principles and applications of electrochemical capacitors. *Electrochimica Acta* 45: 2483–2498.
7. Ali I., Takeo Y., Don N., Masako Y., Hideyuki T., Hiroaki H., Sumio I., and Kenji H. 2011. High-power supercapacitor electrodes from single-walled carbon nanohorn/nanotube composite. *ACS Nano* 5: 811–819.
8. Malak A., Fic K., Lota G., Vix-Guterl C., and Frackowiak E. 2010. Hybrid materials for supercapacitor application. *Journal of Solid State Electrochemistry* 14: 811–816.
9. Frost & Sullivan. 2013. *Ultracapacitor Market Alternative Energy Storage Holds a Promising Future*. Frost & Sullivan, Palo Alto, CA.
10. Frost & Sullivan. 2009. *Ultracapacitor Market Alternative Energy Storage Holds a Promising Future*. Frost & Sullivan, Palo Alto, CA.
11. Kötz R., Hahn H., and Gallay R. 2006. Temperature behavior and impedance fundamentals of supercapacitors. *Journal of Power Sources* 154: 550–555.
12. Katsuhiko N., Shuichi I., Yusaku I., and Shintaro A. 2010. High-rate nano-crystalline $Li_4Ti_5O_{12}$ attached on carbon nanofibers for hybrid supercapacitors. *Journal of Power Sources* 195: 6250–6254.
13. Katsuhiko N., and Patrice S. 2008. New materials and new configurations for advanced electrochemical capacitors. *Electrochemical Society Interface* 17: 34–37.
14. Adrian S., and Roland G. 2000. Properties and applications of supercapacitors from the state-of-the-art to future trends. In: *Proceedings of the PCIM*, 1–10.
15. Conway B. E. 1999. *Electrochemical Supercapacitors, Scientific Fundamentals and Technological Applications*. Kluwer Academic/Plenum Press, New York.
16. Guoping W., Lei Z., and Jiujun Z. 2012. A review of electrode materials for electrochemical supercapacitors. *Chemical Society Reviews* 41: 797–828.
17. Miller J. R. 2006. Electrochemical capacitor thermal management issues at high-rate cycling. *Electrochimica Acta* 52: 1703–1708.
18. Jim P. Z. 2005. Theoretical energy density for electrochemical capacitors with intercalation electrodes. *Journal of the Electrochemical Society* 152: A1864–A1869.
19. Changzhou Y., Bo G., and Xiaogang Z. 2007. Electrochemical capacitance of NiO/Ru0.35V0.65O2 asymmetric electrochemical capacitor. *Journal of Power Sources* 173: 606–612.
20. Li L. Z., and Zhao X. S. 2009. Carbon-based materials as supercapacitor electrodes. *Chemical Society Reviews* 38: 2520–2531.
21. Kötz R., and Carlen M. 2000. Principles and applications of electrochemical capacitors. *Electrochimica Acta* 45: 2483–2498.
22. Frost & Sullivan. 2007. *Ultracapacitor Market Alternative Energy Storage Holds a Promising Future*. Frost & Sullivan, Palo Alto, CA.
23. Menahem A. 2005. *The Ultracapacitor Opportunity Report—Could Ultracapacitors Become the Preferred Energy-Storage Device for Future Vehicles?* Advanced Automotive Batteries.
24. NessCap. 2014. Overview: Ultracapacitors are ten times more powerful than batteries and contain thousand times more energy than electrolytic capacitors. Available at http://www.nesscap.com/product/overview.jsp (accessed August 30, 2014).
25. Komatsu. 2008. Komatsu introduces the world's first hydraulic excavator: Hybrid evolution plan for construction equipment. Available at http:/www.komatsu.com/CompanyInfo/press/2008051315113604588.html (accessed August 30, 2014).
26. Kötz R., Bartschi M., Buchi F., Gallay R., and Dietrich P. 2002. Power—A fuel cell car boosted with supercapacitors. In: *Proceedings of the 12th International Seminar on Double Layer Capacitors and Similar Energy Storage Devices, Deerfield Beach, FL.*
27. François B., and Elzbieta F. 2013. *Supercapacitors: Materials, Systems, and Applications*. Wiley-VCH Verlag GmbH & Co. KGaA, Weinheim, Germany.
28. Kötz R., Müller S., Bärtschi M., Schnyder B., Dietrich P., Büchi F. N., Tsukada, A., Scherer G. G., Rodatz P., Garcia O., Barrade P., Hermann V. and Gallay R. 2001. Supercapacitors for peak-power demand in fuel-cell-driven cars. *Electrochemical Society Proceedings* 21: 564–575.
29. Schupback R. M., and Balda J. C. 2003. Design methodology of a combined battery–ultracapacitor energy storage unit for vehicle power management. In: *IEEE Power Electronic Specialists Conference, PEEC'03, Hyatt-Regency Hotel, Acapulco, Mexico, June 15–19.*
30. Ivanov A., Poliashov L., Gerasimov A., and Lev F. December 1995. In: *Proceedings of the 5th International Seminar on Double Layer Capacitors and Similar Energy Storage Devices*, Florida Educational Seminars.
31. Furukawa T. 2008. Engine cranking with green technology. In: *Proceedings of Advanced Capacitor World Summit 2008, Hilton San Diego Resort, San Diego, CA, July 14–16.*
32. Miller, J. R. 2005. Standards for engine-starting capacitors. In: *Proceedings of the 15th International Seminar on Double Layer Capacitors and Hybrid Energy Storage Devices, Deerfield Beach, FL, December 5–7.*

33. Maxwell Technologies. Ultracapacitor applications for uninterruptible power supplies (UPS). Available at http://www.maxwell.com/products/ultracapacitors (accessed March 10, 2014).
34. NEC Tokin. 2013. Super capacitors. Available at http://www.nec-tokin.com/english/product/pdf_dl/supercapacitors.pdf.
35. Andrew B. 2000. Supercapacitors: Why, how, and where is the technology. *Journal of Power Sources* 91: 37–50.
36. Khomenko V., Raymundo-Piñero E., and Béguin F. 2006. Optimisation of an asymmetric manganese oxide/activated carbon capacitor working at 2V in aqueous medium. *Journal of Power Sources* 153: 183–190.
37. Wang Y. G., and Xia Y. Y. 2005. A new concept hybrid electrochemical surpercapacitor: Carbon/$LiMn_2O_4$ aqueous system. *Electrochemistry Communications* 7: 1138–1142.

Section VII

Advanced Materials and Technologies for Liquid–Redox Rechargeable Batteries

21 Advanced Materials for Liquid–Redox Flow Batteries

Jianlu Zhang, Jinfeng Wu, Lei Zhang, and Jiujun Zhang

CONTENTS

21.1 INTRODUCTION

Liquid–redox flow batteries (LRFBs) are rechargeable batteries that store electrical energy in two soluble redox couples contained in two separate external electrolyte tanks. The sizes of the electrolyte tanks can be adjusted to contain different amounts of electrolyte solutions according to application requirements. Figure 21.1 shows a schematic diagram of a typical LRFB [1]. It can be seen that the aqueous liquid electrolytes containing active redox species are pumped from the storage tanks to the flow-through electrodes, where chemical energy is converted to electrical energy (discharge) or vice versa (charge). The electrolytes flowing through the cathode and anode are usually different and referred to as anolyte and catholyte, respectively. Like proton-exchange membrane fuel cells, there is a membrane between the anodic and cathodic compartments that selectively allows the cross-transport of nonactive species (e.g., H^+, Na^+, etc.) to maintain the electronic neutrality and electrolyte balance of the whole battery system. Ideally, the membrane should be impermeable to electroactive redox species to avoid the cross-contamination of electrolytes and subsequent battery capacity loss. Different from the traditional batteries that store energy in electrodes, LRFBs are more like regenerative fuel cells in which the chemical energy stored in the incoming fuels is converted into electricity at the electrodes. Therefore, the power and energy

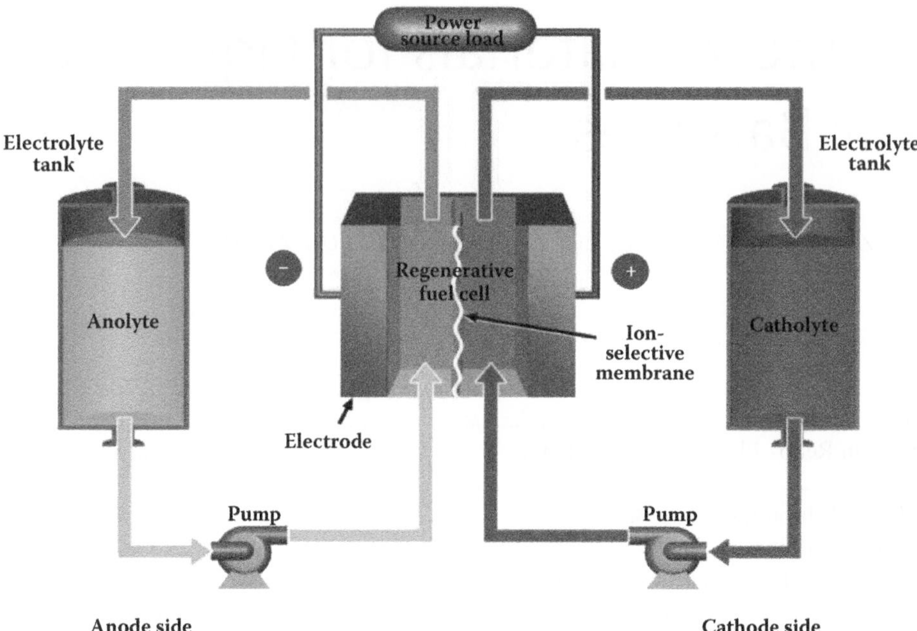

FIGURE 21.1 Schematic diagram of a typical liquid–redox rechargeable battery. (From Yang, Z., Liu, J., Baskaran, S., Imhoff, C.H., and Hollady, J.D., *JOM*, 62, 14, 2010.)

capacity of an LRFB system can be designed separately. In general, the power (kW) of the system is determined by the active area of the electrodes and the number of cells in the stack of the power system, while the energy storage capacity (kWh) is determined by the concentration of the active species and the volumes of the electrolytes. Depending on the application, the power of the system can be easily adjusted by changing the electrode active area or the number of the cells in stack. The energy storage capacity is usually increased by enlarging the volume of the electrolyte, but is limited by the concentration of the active species due to their limited solubility. Depending on the application, a charge–discharge cycle of an LRFB power system could range from minutes to hours, days, or even weeks because of the variation of the system's power and capacity. This indicates an important advantage of LRFBs for the application in renewable integration, such as backup powers, portable power supplies, and uninterruptible power supplies. Generally, LRFBs can sustain no damage to the cells when completely discharged, although an overcharge or overdischarge may need to be avoided. There is an only negligible irreversible loss due to self-discharge. The liquid electrolytes and their intimate contacts with electrodes suggest high current densities and quick response (in a matter of subseconds), which makes the LRFBs possible for utility applications. The simplicity in the cell and stack structure allows for building large systems based on module design, which is another important advantage of LRFBs for their applications in electrical grids.

21.2 TYPES OF LRFBs

For an ideal LRFB, both the negative and positive electrode reactions should be electrochemically reversible, and both the negative and positive active species must be soluble and their redox potentials should be as far apart as possible. The reactants in both negative and positive electrolytes should be cost-effective, easy to get, and stable in the LRFB working conditions. To increase the energy density, the reactants should have concentrations as high as possible. Many kinds of LRFBs have been developed in history, and they can be classified as follows according to the type of reactant species in the positive and negative electrolyte compartments: polysulfide/bromine flow batteries (PSBs) [2–5], vanadium/bromine flow batteries (VBrs) [6,7], all-vanadium redox flow batteries (VRFBs or VRBs) [8,9], vanadium/cerium flow batteries (VCes) [10], iron/chromium flow batteries (FeCrs) [11], zinc/bromine flow batteries (ZnBrs) [11], zinc/cerium flow batteries (ZnCrs), and soluble lead/acid batteries (SLAs). In the above-mentioned LRFBs, the VRBs are the ones with relatively high maturity and have been accomplished by many demonstration projects with the integration with wind/solar power. Therefore, VRB will be taken as an example and will be discussed in detail in the following sections in terms of its chemistries and key materials. The other flow batteries will also be introduced briefly in this chapter.

21.2.1 All-Vanadium Redox Flow Batteries (VRBs)

Vanadium exists in solution in four different oxidation states, V^{2+}, V^{3+}, VO^{2+}, and VO_2^+, in which V^{2+} and V^{3+} form a redox couple of V^{2+}/V^{3+} for the anode reaction, and VO^{2+} and VO_2^+ form another redox couple of VO^{2+}/VO_2^+ (sometimes expressed as V^{4+}/V^{5+}) for the cathode reaction. VRBs are designed using these two redox couples with only one active element (vanadium) involved in both negative and positive electrolytes. Thus, the cross-transport of the active

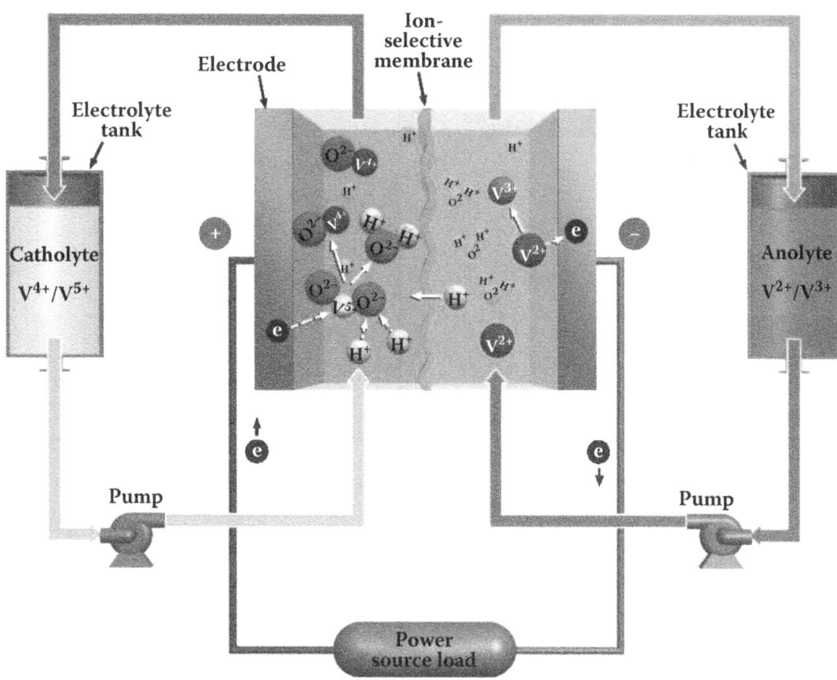

FIGURE 21.2 Working principle of VRBs. (From Yang, Z., Zhang, J., Kintner-Meyer, M.C.W., Lu, X., Choi, D., Lemmon, J.P., and Liu, J., *Chem. Rev.*, 111, 3577, 2011.)

component in the VRBs is significantly diminished. Figure 21.2 shows the working principle of VRBs. The use of vanadium redox couples as the reactive species in the flow batteries was suggested by Pissoort [12] first, and then developed by National Aeronautics and Space Administration (NASA) researchers and by Pellegri and Spaziante in 1978 [13]. The present form of VRBs (with sulfuric acid as a supporting electrolyte) was proposed by Skyllas-Kazacos and coworkers at the university of New South Wales in 1986 [8,14], and then the first successful demonstration was developed at the university of New South Wales in the 1980s [15]. Recently, the scientists at Pacific Northwest National Laboratory (PNNL) have developed the VRBs employing the mixture of sulfuric acid and hydrochloric acid as a supporting electrolyte with the vanadium component dissolved in the mixture [16,17]. This development increased the concentration of vanadium in the electrolyte (up to 2.5 M), leading to the improved energy density of VRBs. More important, the stability of VO_2^+ ion at higher temperatures (up to 50°C) was claimed to be improved as well [16,17].

The first large VRB system (50 kW/200 kWh) was built by Kashima-Kita Electric Power (a Mitsubishi subsidiary) and put in operation in 1995 by licensing the Skyllas-Kazacos' invention [13,14]. Later, Sumitomo Electric Industries constructed and demonstrated a series of VRB systems in Japan from the kW scale to the MW scale for applications in energy storage [18]. In 2005, Sumitomo successfully installed a 4.0-MW/6.0-MWh system at the 30.6-MW Tomamae wind farm in Hokkaido in northern Japan [19]. The system was used to regulate wind output, enabling the supply of firmed power to the grid. VRB Power System, a Vancouver British Columbia company (the company was purchased in 2009 by Prudence Power Inc.), bought most of the early intellectual rights to the technology, and VRB Power System targeted the VRBs in both grid and off-grid stationary applications.

In general, it has been claimed that the flow batteries could demonstrate a long-term cycle and calendar life (>6000 cycles, 80% depth of discharge, and >15 years, respectively) [20]. Recently, VRBs have received much more attention because of their promising applications in energy storage [21–27]. In March 2012, a company, UniEnergy Technologies, LLC, was founded based on PNNL's VRB technology, aiming at the commercialization of VRBs [28].

In VRBs, the energy conversions are realized via changes in vanadium valence states through the following electrode reactions:

$$\text{Cathode side: } VO^{2+} + H_2O \underset{\text{Discharge}}{\overset{\text{Charge}}{\rightleftharpoons}} VO_2^+ + 2H^+ + e^- \tag{21.1}$$

$$\text{Anode side: } V^{3+} + e^- \underset{\text{Discharge}}{\overset{\text{Charge}}{\rightleftharpoons}} V^{2+} \tag{21.2}$$

Cell reaction:

$$VO^{2+} + V^{3+} + H_2O \underset{\text{Discharge}}{\overset{\text{Charge}}{\rightleftharpoons}} VO_2^+ + V^{2+} + 2H^+ (E^\circ = 1.26\ V) \tag{21.3}$$

The overall electrochemical reaction gives a cell voltage of 1.26 V at 25°C and unit activities (i.e., standard voltage). Figure 21.3 shows typical charge–discharge profiles of a VRB

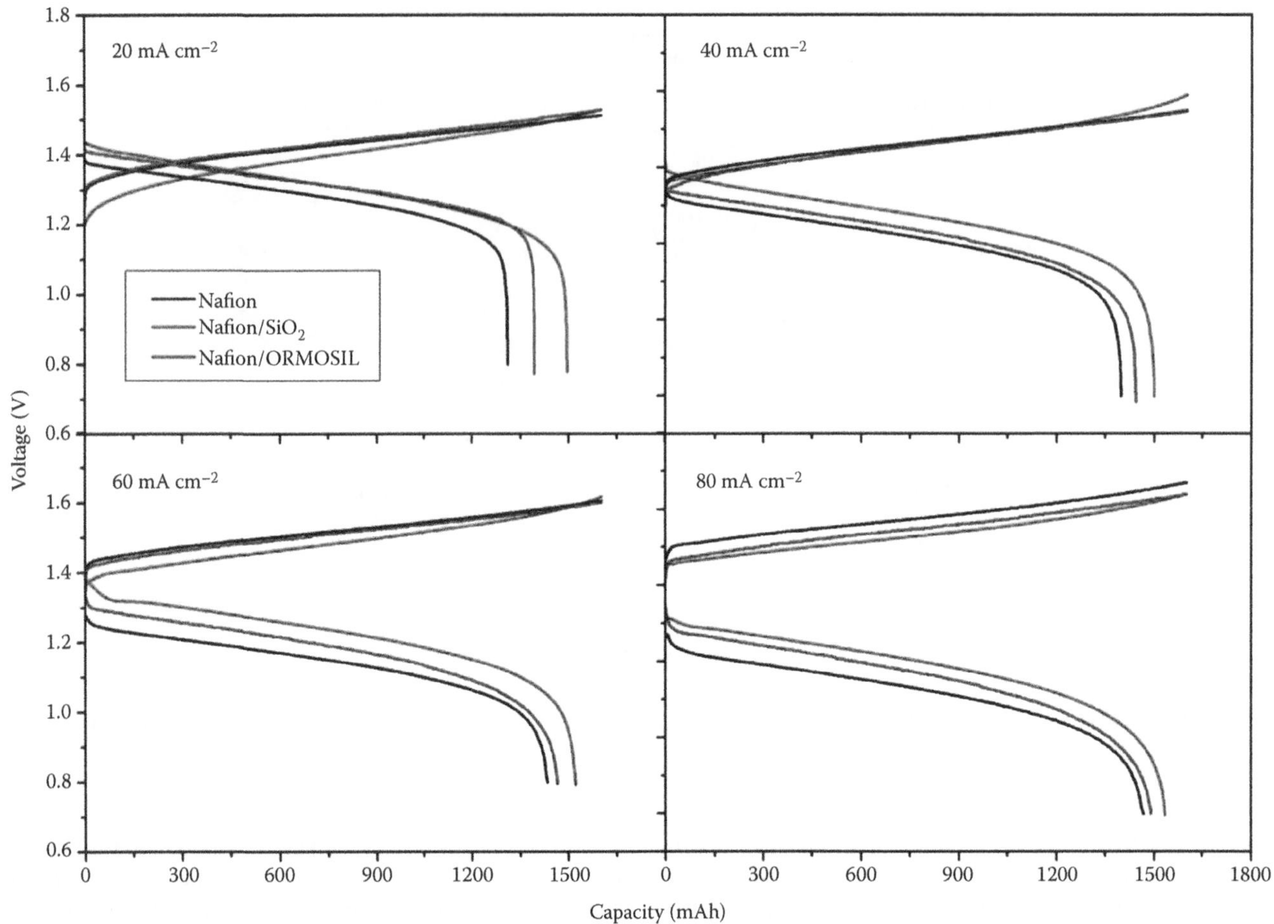

FIGURE 21.3 Charge–discharge curves of a VRB with Nafion, Nafion/SiO_2, and Nafion/ORMOSIL hybrid membranes at different current densities. Charge capacity was controlled to be 1600 mAh corresponding to a redox-couples utilization of 75%. Mixtures of 40 mL of 2-M V^{3+}/V^{4+} and 2.5-M H_2SO_4 solutions were used as the starting anolyte and catholyte. (From Teng, X.G., Zhao, Y.T., Xi, J.Y., Wu, Z.H., Qiu, X.P., and Chen, L.Q., *J. Power Sources*, 189, 1240, 2009; Xi, J.Y., Wu, Z.H., Qiu, X.P., and Chen, L.Q., *J. Power Sources*, 166, 531, 2007.)

with 2-M V^{3+}/V^{4+} in 2.5-M H_2SO_4 electrolytes, at current densities of 20–80 mA cm^{-2} [29,30].

21.2.2 Other LRFBs

In addition to the all-vanadium redox couples, there are a number of active ions that can be potentially used to design LRFBs. Figure 21.4 compiles varied redox couples and their standard potentials (except the H^+/H_2 couple that is based on the overpotential of carbon electrodes). Their combinations for a useful voltage are, however, limited by the electrochemical reaction window of hydrogen and oxygen evolutions in aqueous systems. In addition to VRBs, other flow battery chemistries that were explored include V^{2+}/V^{3+} versus $Br^-/ClBr_2^-$ [6,31], Br_2/Br^- versus S/S^{2-} [2,20,32], Fe^{3+}/Fe^{2+} versus Cr^{3+}/Cr^{2+} [33,34], Br^-/Br_2 versus Zn^{2+}/Zn [35,36], Ce^{4+}/Ce^{3+} versus V^{2+}/V^{3+} [10], Fe^{3+}/Fe^{2+} versus Br_2/Br^- [37], Mn^{2+}/Mn^{3+} versus Br_2/Br^- [38], Fe^{3+}/Fe^{2+} versus Ti^{2+}/Ti^{4+} [39], Fe^{3+}/Fe^{2+} versus V^{2+}/V^{3+} [40,41], etc. [11]. Among these LRFBs that were not all-vanadium, technologies of PSBs with Br^2/Br^- versus S/S^{2-} redox couples, ICBs with Fe^{3+}/Fe^{2+} versus Cr^{3+}/Cr^{2+}, and Zn/bromide (ZBB) with Br^-/Br^{2-} versus Zn^{2+}/Zn have received wide attention and have been demonstrated at scales up to 100 kW and even to MW levels.

21.2.2.1 Iron/Chromium Flow Batteries (ICBs)

In the early 1970s, Thaller [42,43] at NASA invented an electrochemical storage system that employed the redox couples of Fe^{3+}/Fe^{2+} and Cr^{3+}/Cr^{2+} as catholyte and anolyte in an acid medium (usually hydrochloric acid solution), respectively. This iron/chromium battery (ICB) was considered the earliest energy storage device using two fully soluble redox couples that were pumped through a battery cell. The electrode and cell reactions of ICBs are as follows:

$$\text{Anode side: } Cr^{2+} - e^- \xrightarrow{\text{Discharge}} Cr^{3+} \quad (21.4)$$

$$\text{Cathode side: } Fe^{3+} + e^- \xrightarrow{\text{Discharge}} Fe^{2+} \quad (21.5)$$

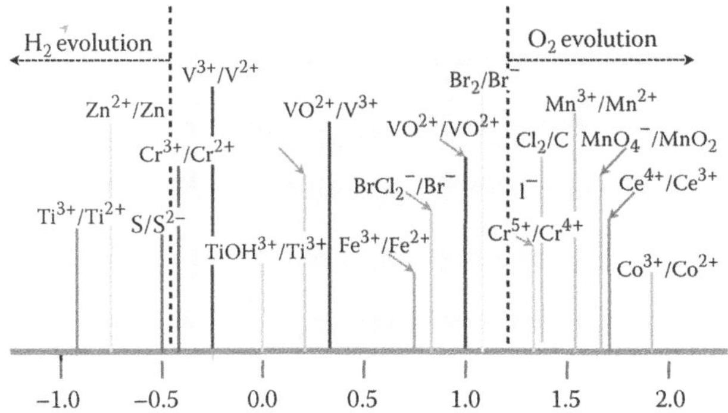

FIGURE 21.4 Standard potential (versus the standard hydrogen electrode) of redox couples, except the H_2 evolution potential that is the overpotential at carbon electrodes. (From Yang, Z., Zhang, J., Kintner-Meyer, M.C.W., Lu, X., Choi, D., Lemmon, J.P., and Liu, J., *Chem. Rev.*, 111, 3577, 2011.)

$$\text{Overall reaction: } Cr^{2+} + Fe^{3+} \xrightarrow{\text{Discharge}} Cr^{3+} + Fe^{2+} \quad (21.6)$$

The cell reaction as shown by Equation 21.6 offers a standard voltage of 1.18 V. It was reported that the ICB could operate with either a cation- or anion-exchange membrane/separator and typically employed carbon fiber, carbon felt, or graphite as electrode materials. An issue associated with the early ICBs was identified as the cross-transport of iron and chromium active species. Later on, a significant reduction in the cross-transport was achieved by using mixed electrolytes at both the cathode and anode sides [44]. The mixed electrolytes also made the use of a cost-effective microporous separator feasible.

In addition, there were some extensive attempts [33,34,45–50] to optimize and scale-up the ICB system for the applications in energy storage. In the middle of 1980s, NASA licensed its Fe/Cr flow battery technology to Sohio (Standard Oil of Ohio, Cleveland, OH), which was later bought by British Petroleum.

21.2.2.2 Polysulfide Bromide Flow Batteries (PSBs)

PSBs are a kind of flow batteries using electrolytes of sodium bromides and sodium polysulfides as the electrolytes [2,20,32]. These chemicals are abundant, reasonably low cost, and also soluble in aqueous media. A standard cell voltage of 1.36 V is given by the following electrochemical reactions:

$$\text{Anode side: } 2S_2^{2-} - 2e^- \xrightarrow{\text{Discharge}} S_4^{2-} \quad (21.7)$$

$$\text{Cathode side: } Br_3^- + 2e^- \xrightarrow{\text{Discharge}} 3Br^- \quad (21.8)$$

$$\text{Overall reaction: } 2S_2^{2-} + Br_3^- \xrightarrow{\text{Discharge}} S_4^{2-} + 3Br^- \quad E° = 1.36\text{ V} \quad (21.9)$$

According to Equations 21.7 and 21.8, during the charging cycle, the bromide ions are oxidized to bromine, which is then complexed as tribromide ions in the cathode side, and the soluble polysulfide anion is reduced to sulfide ion in the anode side. On discharge, the sulfide ion is the reducing agent, and the tribromide ion is the oxidizing agent. The open-circuit cell voltage is around 1.5 V, as shown in Equation 21.9, which is dependent on the activity of the active species. The electrolyte solutions are separated by a cation-selective membrane to prevent the sulfur anions from reacting directly with the bromine. The electrical balance can be achieved by transporting Na^+ across the membrane. Nafion® membranes were often used in PSB cells [20,51], and the electrodes were made from materials including high-surface-area carbon/graphite, a nickel form, and even sulfide nickel [4,5,52,53].

21.2.2.3 Zinc Bromide Batteries (ZBBs)

ZBBs are often classified into the traditional LRFB categories although the anode redox process contains solid zinc. The electrode reactions during discharge process are as follows:

$$\text{Anode side: } Zn - 2e^- \xrightarrow{\text{Discharge}} Zn^{2+} \quad (21.10)$$

$$\text{Cathode side: } Br_2(aq) + 2e^- \xrightarrow{\text{Discharge}} 2Br^- \quad (21.11)$$

$$\text{Cell reaction: } Zn + Br_2(aq) \xrightarrow{\text{Discharge}} 2Br^- + Zn^{2+} \quad (21.12)$$

As indicated by Equation 21.12, the cell reaction gives a standard voltage of 1.85 V. The ZBB employs an aqueous solution of zinc bromides that is added with agents [35,36]. During operation, the zinc bromide electrolyte is pumped through both the positive and negative electrode surfaces, which are separated by a microporous plastic film, or alternatively, an ionic membrane. This ionic membrane can selectively allow the transport of zinc and bromide but not the aqueous bromine, polybromide ions, or complex phase. At the positive cathode, bromide ions are converted to bromine during charging or vice versa during discharging. To reduce the serious health hazard of bromine, some complex agents are used to reduce the evolution of bromine. At the negative anode, zinc is reversibly deposited from the ions. Because of the solid Zn that is produced, ZBBs are not real redox flow batteries and are often referred to as a "hybrid" redox flow battery. Because the total available energy of ZBB is limited by the available amount of plated zinc, the power/energy relationship of a ZBB is more fixed than that of the traditional LRFBs. To facilitate the battery reactions, the electrodes are generally made from high-surface-area, carbon-based materials [54]. Table 21.1 compares the technical parameters of VRBs, PSBs, ICBs, and ZBBs [55]. As shown in Table 21.1, ZBB cells offer the highest reaction reversibility, cell voltage, and energy density among the four major LRFB technologies.

TABLE 21.1
Technical Comparison of VRB with Other Chemistries

Type	Open Circuit Voltage (V)	Specific Energy (Wh kg^{-1})	Characteristic Discharge Time (h)	Self-Discharge at 20°C (% per Month)	Cycle Life (Cycles)	Round-Trip DC Energy Efficiency
VRB	1.4	10 (29)[a]	4–12	5–10	>5000	70–80%
PSB	1.5	20 (41)	4–12	5–10	>2000	60–70%
ICB	1.18	<10	4–12		>2000	70–80%[b]
ZBB	1.8	65 (429)	2–5	12–15	>2000	65–75%

Source: Yang, Z., Zhang, J., Kintner-Meyer, M.C.W., Lu, X., Choi, D., Lemmon, J.P., and Liu, J., *Chem. Rev.*, 111, 3577, 2011; Electrical Power Research Institute, *Handbook of Energy Storage for Transmission and Distribution Applications*, Department of Energy, Washington, DC, 2003; Chen, H., Cong, T.N., Yang, W., Tan, C., Li, Y., and Ding, Y., *Progr. Nat. Sci.*, 19, 291, 2009.

[a] Theoretical specific energy.

[b] Depending on operating temperatures.

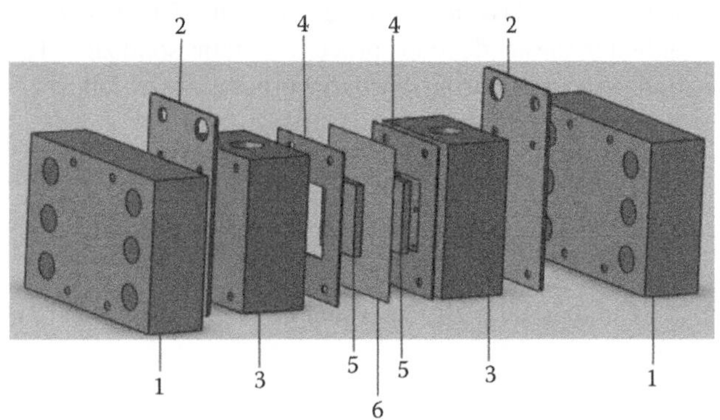

FIGURE 21.5 Hardware of a single VRB. 1—end plate, 2—current collector, 3—flow plate, 4—gasket, 5—electrode, and 6—membrane. (From Yang, Z., Zhang, J., Kintner-Meyer, M.C.W., Lu, X., Choi, D., Lemmon, J.P., and Liu, J., *Chem. Rev.*, 111, 3577, 2011.)

21.3 LRFB COMPONENTS

Regarding the structure of an LRFB, taking the VRB as an example, as shown in Figure 21.5, the components of a typical VRB hardware includes the end plates, current collectors, flow plates, gaskets, electrodes, and membrane. Among them, the electrode, membrane, and electrolyte together are the key materials for the VRBs. They will be discussed in detail in the following sections.

21.4 ELECTROLYTE CHEMISTRIES AND PROPERTIES

21.4.1 Solubility of $VOSO_4$

For VRBs, the anolyte is the solution containing the redox couple of V^{2+}/V^{3+} and the catholyte is the solution containing the redox couple of V^{4+}/V^{5+}, as shown in Figure 21.2. H_2SO_4 is used as the supporting electrolyte in both anolyte and catholyte. Usually, the concentration of vanadium is controlled at less than 2 M, and the total concentration of SO_4^{2-} is controlled at less than 5 M, respectively. This is limited by the stability of vanadium species at VRB operating conditions. The solution of V^{4+} is usually prepared by dissolving $VOSO_4$ in H_2SO_4 solutions. However, according to the literature data, the solubility of $VOSO_4$ can be decreased with increasing H_2SO_4 concentration [56], as shown in Figure 21.6 [57]. The solubility of $VOSO_4$ can be increased with increasing solution temperature; particularly, this temperature effect can be more significant when the H_2SO_4 concentration is low [56], as shown in Figure 21.7.

21.4.2 State of Dissolved V(IV) and V(V)

As identified, in the media of H_2SO_4 <4 M, the VO^{2+} ion exists in noncomplexing acidic solutions as a blue oxovanadium ion $[VO(H_2O)_5]^{2+}$ (an aquo cation of VO^{2+}) [58,59] with a radius roughly estimated to be 0.28 nm if based on the crystallographic data (V–O interatomic distance for VO^{2+} plus radius of O^{2-}) [57]. As shown in Figure 21.8, the structure can be described as a tetragonal bipyramid. This structure contains four equatorial waters having a residence time of 1.35×10^{-3} s, and an axial water being more weakly held with a residence time of 10^{-11} s [60]. However, if the H_2SO_4 concentration is higher than 5 M, VO^{2+} tends to form ion pairs with anions (such as SO_4^{2-} and HSO_4^-), resulting an increase in radius.

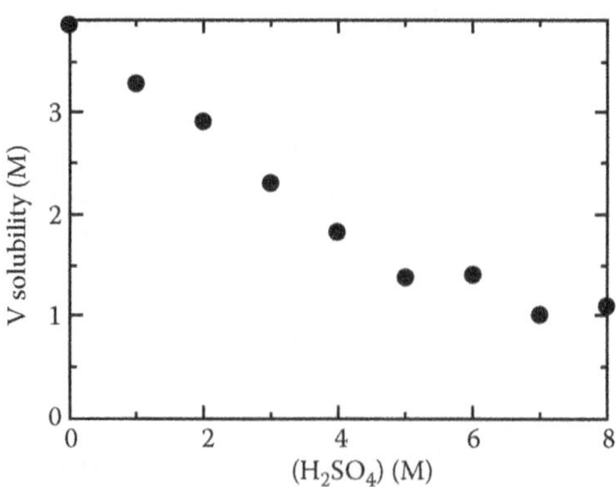

FIGURE 21.6 Dependence of the solubility of the V(IV)–H_2SO_4 solution on the H_2SO_4 concentration at 25°C. (From Oriji, G., Katayama, Y., and Miura, T., *Electrochim. Acta*, 49, 3091, 2004.)

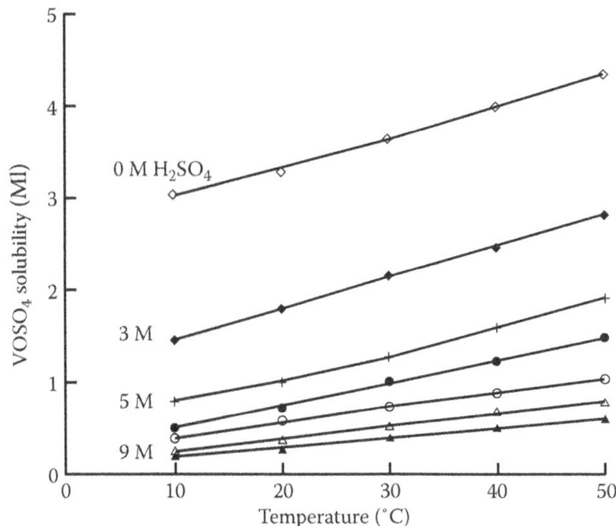

FIGURE 21.7 Effect of temperature on solubility of $VOSO_4$ in various H_2SO_4 concentrations. (From Rahman, F. and Skyllas-Kazacos, M., *J. Power Sources*, 72, 105, 1998.)

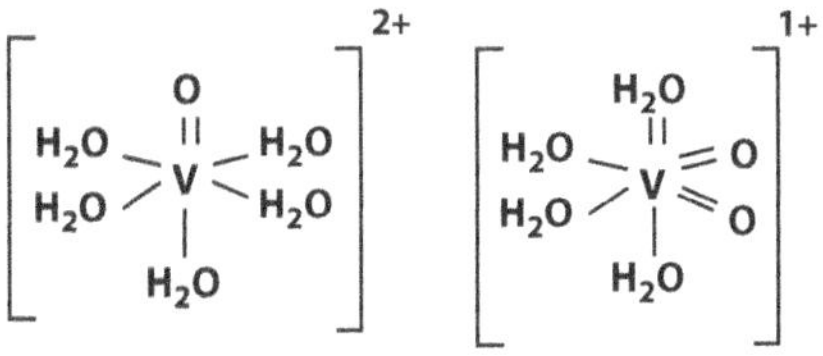

FIGURE 21.8 Structures for VO^{2+} and VO_2^+ in noncomplexing acidic solutions. (From *Advanced Inorganic Chemistry*. Cotton, F.A. and Wilkinson, G., Eds., John Wiley & Sons: New York, p. 665, 1988; Gattrell, M., Park, J., MacDougall, B., Apte, J., McCarthy, S., and Wu, C.W., *J. Electrochem. Soc.*, 151, A123, 2004; Baes, C.F. and Mesmer, R.E., *The Hydrolysis of Cations*, John Wiley & Sons: New York, 1976.)

As shown in Figure 21.8, VO_2^+ ion exists as the yellow dioxovanadium ion, *cis*-$[VO_2(H_2O)_4]^+$ in strong acid media [58–60]. However, it was reported that many kinds of V(V) species could be found in aqueous solutions, depending on the vanadium concentration and pH value [61]. At high vanadium and sulfuric acid concentrations, complex species such as $[V_2O_3]^{4+}$, $[V_2O_4]^{2+}$ [62] and complexes with sulfate and bisulfate [61,63] could be formed. In concentrated H_2SO_4 and $HClO_4$, except the dimer species, the VO_2^+ ion could polymerize into chains of vanadate octahedral [64,65].

21.4.3 Stability of Vanadium Species

In general, the concentration of vanadium species in the electrolyte determines the energy density of VRBs, that is, increasing the concentration of vanadium can increase the energy density. However, if the concentration is increased to >2 M in H_2SO_4 supporting electrolyte, a supersaturation of solutions will occur, leading to the formation of precipitates in the electrolyte at temperatures higher than 40°C for the V(V) solution and at the temperature lower than 10°C for the V(II), V(III), and V(IV) solutions [66,67]. Therefore, the extent and rate of precipitate formation depends on the temperature, concentration of vanadium, concentration of sulfuric acid, and the state-of-charge (SOC) of the electrolyte (ratio of V(V) to V(IV) ions or the ratio of V(II) to V(III)) [61,66,68]. Thus, to improve the stability of both anolyte and catholyte, the VRB operating conditions might be optimized. In this regard, some organic or inorganic chemicals were selected and added into the vanadium solutions as stabilizing agents, among which some exhibited a positive effect on the improvement in the stability of vanadium solutions [69,70]. However, it seems not feasible to further increase the vanadium concentration above 2 M for the real-world operation of VRBs with H_2SO_4 as the supporting electrolyte. The studies from PNNL showed that the vanadium concentration could be increased to >3.0 M without precipitates forming at temperatures up to 60°C when the mixture of sulfuric acid and hydrochloric acid was used as the supporting electrolyte [16,17].

21.4.4 Cyclic Voltammetry of Vanadium Species

Regarding the electrochemical activity of the vanadium species, as shown in Figure 21.9 [71], there are four peaks on the cyclic voltammograms (CVs) obtained at a glassy carbon electrode in 1 M $VOSO_4$–2 M H_2SO_4 solution. The anodic peak at about 1000–1100 mV corresponds to the oxidation of V(IV) to V(V) (i.e., $VO^{2+} + H_2O \rightarrow VO_2^+ + 2H^+ + e^-$), and the corresponding reduction peak occurs at about 700 mV during the negative scan. The anodic peak at about −400 mV and the cathodic peak at about −750 mV correspond to the oxidation and reduction of the redox couple of V(II)/V(III), respectively (i.e., $V^{3+} + e^- \rightarrow V^{2+}$). The peak potential separation (>200 mV) of the redox couple peaks in both the positive

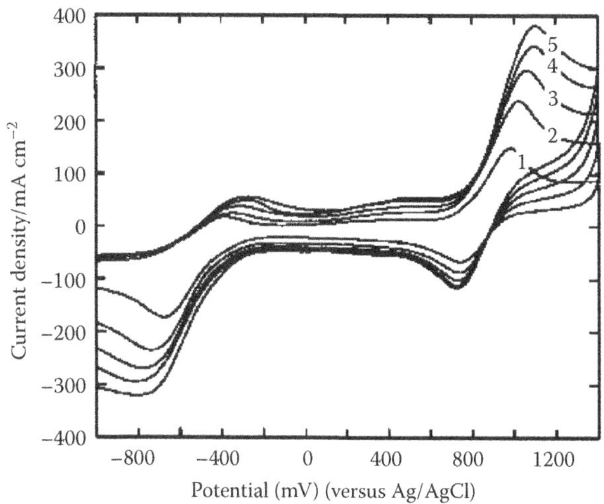

FIGURE 21.9 CV obtained at a glassy carbon electrode in 1-M $VOSO_4$–2-M H_2SO_4 solution. Scan rates: (1) 100 mV s^{-1}, (2) 200 mV s^{-1}, (3) 300 mV s^{-1}, (4) 400 mV s^{-1}, and (5) 500 mV s^{-1}. (From Fang, B., Wei, Y., Arai, T., Iwasa, S., and Kumagai, M., *J. Appl. Electrochem.*, 33, 197, 2003.)

and the negative potential range suggests a poor reversibility of the electrochemical reaction. From the literature, it can be seen that owing to the differences in electrode preparation and the operating conditions, the study results from different research groups might be different. Sum et al. [72] indicated that the redox reaction of V(V)/V(IV) at glassy carbon electrodes was electrochemically irreversible, and the reversibility of the V(III)/V(II) process was critically dependent on the surface preparation of the glassy carbon electrode. Rahman and Skayllas-Kazacos [66,67] studied the CVs of V(V) with different vanadium concentrations, and found that the peak current was increased linearly with increasing V(V) concentration from 2 to 3.5 M and then started to decrease with further increasing V(V) concentration. It was discussed that the decrease in peak current with increasing V(V) above 3.5 M might be due to the sharp increase in viscosity of V(V). In addition, the changes in the interfacial tension properties of the more concentrated solution might reduce the wettability of the glassy carbon electrode in the solution, reducing the effective surface area, leading to the decrease in peak current [66,67].

21.5 ELECTRODE MATERIALS

21.5.1 Graphite Felt

In LRFBs, VRBs for example, electrodes provide the places for electrochemical reactions, suggesting that the electrode materials must possess high surface area, suitable porosity, low electronic resistance, and high electrochemical activity toward the reactions between the active vanadium species. Actually, in acidic and strong oxidative environments of VRB, there are limited choices of materials for electrodes. In general, inert, high-surface-area graphite- or carbon-based materials in forms of porous felt and layer have been the most common materials for electrodes [30,73–89]. In general, the electrochemical reaction kinetics on the electrode may be dependent on electrode structure, surface chemistry, surface area, etc. The electric resistance of electrode assembly can also have a strong effect on the cell performance. To minimize electrical resistance then power loss, the electrodes are usually integrated with current collectors or bipolar plates to form a tight electrode assembly. One early combination was graphite felt as electrodes mechanically compressed to graphite plates. Thus far, graphite felt is the most widely used porous electrode material in VRBs. Before use, the surface modification of graphite felt is necessary to improve its electrochemical activity of the redox reactions of vanadium ions and reduce the reaction overpotentials on both negative and positive electrodes.

21.5.2 Modification of Graphite Felt

Normally, graphite felt has a large specific surface area, wide operation potential range, low cost, and good stability at both negative and positive electrolytes, which has been considered to be a promising material for VRBs. However, untreated graphite felt has a low electrochemical activity and poor kinetic reversibility toward the electrochemical reactions between the vanadium species. In recent years, much attention has been paid to modifying the graphite felt to improve its electrochemical properties [83,84,90–93]. The reported methods for modifying the graphite felt mainly include heat treatment [83,87,94], chemical treatment [84,86,90], electrochemical oxidation [91,93,95,96], and doping by depositing other metals on carbon fibers [85,92,97].

21.5.2.1 Heat Treatment

With respect to heat treatment, Sun and Skyllas-Kazacos [83] found that the electrochemical activity of graphite felt could be greatly improved by heat treatment at 400°C for 30 h. As a result, the energy efficiency of vanadium redox cell was improved from 78% to 88%. The increased activity could be attributed to the increased surface hydrophilicity, reduced resistance, and formed functional groups of C–O–H and C=O on the surface of graphite felt. They suggested that the C–O groups on the electrode surface could behave as active sites, and then catalyze the reactions of vanadium species. The proposed mechanisms for reactions are as follows [83,84]:

In the positive half-cell, the reactions occur as

$$VO^{2+} + H_2O \underset{\text{Discharge}}{\overset{\text{Charge}}{\rightleftharpoons}} VO_2^+ + 2H^+ + e^- \quad (21.13)$$

where oxygen transfer is involved in this reaction during charge and discharge processes. During the charge, the following processes could occur as follows:

1. First, VO^{2+} ions transport from solution to electrode surface and exchange protons with phenolic groups on the electrode surface, and thus bond onto the electrode surface by Equation 21.14:

$$\text{(–OH)}_2 + VO^{2+} \longrightarrow \text{(–O)}_2{>}V=O + 2H^+ \quad (21.14)$$

2. Second, electrons transfer from VO^{2+} to the electrode along the C–O–V bond, and one of the oxygen atoms in the C–O functional group transfers to the VO^{2+}, forming a surface VO_2^+ through Equation 21.15:

$$\text{(–O)}_2{>}V=O + H_2O \longrightarrow \text{–O–V(=O)(=O)} + 2H^+ + 2e^- \quad (21.15)$$

3. Last, the VO_2^+ exchanges with an H^+ from solution and diffuses back into the bulk solution by Equation 21.16:

$$\text{–O–V(=O)(=O)} + H^+ \longrightarrow \text{–O–H} + VO_2^+ \quad (21.16)$$

During discharge, the reactions are the reverse of the charge processes. The formation of C–O–V bond could facilitate the electron transfer and oxygen transfer processes, and thus reduces the activation overpotential for the V(IV)/V(V) redox process.

In the negative half-cell, the reactions occur as

$$V^{3+} + e^- \underset{\text{Discharge}}{\overset{\text{Charge}}{\rightleftharpoons}} V^{2+} \quad (21.17)$$

In Equation 21.17, the electron transfer is involved during charge and discharge processes. In the charge process, three reactions can happen. The first step is V^{3+} diffusion from the bulk solution to the surface of the electrode, and then combination with the proton of the phenol groups through Equation 21.18:

$$\text{–OH} + V^{3+} \longrightarrow \text{–O–}V^{2+} + H^+ \quad (21.18)$$

The second step is the electron transfer from the electrode surface to V^{3+} along the –C–O–V bond to form V^{2+} through Equation 21.19:

$$\text{–O–}V^{2+} + e^- \longrightarrow \text{–O–}V^{+} \quad (21.19)$$

The final step is the V^{2+} ion exchange with protons and then diffusion into the bulk solution through Equation 21.20:

$$\text{–O–}V^{+} + H^+ \longrightarrow \text{–O–H} + V^{2+} \quad (21.20)$$

In the discharge process, the reactions are reversed. The formation of C–O–V bond could facilitate the transfer of electrons and thus reduce the activation overpotential for the V(II)/V(III) process.

21.5.2.2 Chemical Treatment

Chemical treatment is another method to activate carbon materials. During the chemical treatment with NaOCl, $KMnO_4$, or $(NH_4)_2S_2O_8$ solutions, a strong acidic carboxyl group can be formed on the carbon electrode surface at room temperature. For example, Sun and Skyllas-Kazacos [84] investigated the electrochemical properties of VRB carbon felt electrodes treated with concentrated H_2SO_4 or HNO_3. A significant improvement in both coulombic and voltage efficiencies were achieved with graphite felt treated with concentrated H_2SO_4. The increased activity of acid-treated graphite felt was attributed to the increased concentrations of surface functional groups such as C=O and C–OOH that formed during the acid treatment. It was found that these functional groups could not only lead to an increase in hydrophilicity of graphite felt but also behave as some active sites for vanadium redox reactions. In addition, the chemical treatment could also decrease the resistance of graphite felt and the resistance decrease with the increase of both acid concentration and treatment time was also observed.

The mechanism for the enhanced activity by chemical treatment is generally attributed to the formation of a C–O–V bond, which can facilitate the electron transfer for the reactions on both the positive and negative electrodes as well as the oxygen transfer for reactions on the positive electrode. Li et al. [90] investigated the electrochemical behavior of diverse vanadium ions at the graphite-felt electrode modified by combining acid treatment and heat treatment. In their study, the graphite felt was first treated in 98% sulfuric acid for 5 h and then kept at 450°C for 2 h. Their results indicated that the synergistic effect between acid and heat treatments could increase the –COOH functional groups on the graphite-felt surface, and also the surface area of graphite felt could be increased from 0.31 to 0.45 $m^2\ g^{-1}$. They also observed that the electrochemical activity of the electrode was improved greatly, which was mainly ascribed to the increase of the –COOH groups that could behave as active sites, catalyzing the reactions of vanadium species and accelerating both electron and oxygen transfer processes.

21.5.2.3 Electrochemical Oxidation

In recent years, electrochemical oxidation has been employed to modify the graphite felts for improving their electrochemical activity toward the redox reactions between vanadium species in VRBs. For example, Li et al. [91] studied the characteristics of graphite felt by electrochemical oxidation for VRBs, and Tan et al. [93] studied the activation mechanism of electrochemically treated graphite felt. Their studies showed that both the surface area of graphite felt and the ratio of O to C on the felt surface could be increased owing to the formation of –COOH functional groups during the electrochemical oxidation. AC impedance data showed that the resistances of the vanadium specie reactions were also decreased. The mechanism of the improvement in electrochemical activity of graphite felt was attributed to the formation of the C–O–V bond, which could facilitate the electron transfer for the reactions on both positive and negative electrodes, and also the oxygen transfer for the reactions in the positive electrode.

21.5.2.4 Metal Doping Treatment

Doping the carbon electrode by depositing of metals is another effective way to improve electrode performance. In this regard, various metal compounds were employed to dope graphite felt, and some improvement in both the electrochemical activity of vanadium redox processes and stability in vanadium electrolyte solutions was observed [85]. Sun et al. [85] reported a chemical modification of graphite fibers by impregnation using solutions containing Pt^{4+}, Pd^{2+}, Au^{4+}, Ir^{3+}, Mn^{2+}, Te^{4+}, or In^{3+}, and found that the electrode modified by Ir^{3+} could exhibit the best electrochemical performance for vanadium redox species. Wang and Wang [92] investigated Ir-modified carbon felt as the VRB positive electrode, and some improvement in electrode activity was observed, indicated by both the reduced cell internal resistance and the overpotential for the V(IV)/V(V) redox reaction. The enhanced activity was attributed to the formation of active functional groups of Ir.

21.6 CATALYST MATERIALS

In LRFBs, both the rates of anode and cathode redox reactions play important roles in battery performance. To speed up the reactions, some electrocatalysts have to be used. For example, the Cr^{3+}/Cr^{2+} redox couple for flow batteries shows relatively slow kinetics with the electrode materials; thus, catalysts were employed to enhance its electrode kinetics [98,99]. It is required that the catalysts should have a high overpotential toward hydrogen evolution during the reducing process of redox couples. For the Cr^{3+}/Cr^{2+} couple, hydrogen evolution appeared to be its competitive reaction during charging. In this regard, some catalysts having high overpotentials toward H_2 evolution were used in the literature, such as Au [99], Pb [100], Tl [101], Bi [102], or their compounds. It was reported that deposition of Pb or Bi on the electrode surface in HCl solution could not only enhance the rate of Cr^{3+}/Cr^{2+} redox reaction but also increase the hydrogen overpotential [99]. Lopez-Atalaya et al. [99] investigated the behavior of Cr^{3+}/Cr^{2+} reaction on gold–graphite electrodes in the effort to eliminate the H_2 evolution.

21.7 MEMBRANES

The membrane in VRB systems is one of the necessary components; it can not only serve as a separator to prevent cross-mixing of the positive and negative half-cell electrolytes but also allow the transport of charge-balancing ions (such as H^+, SO_4^{2-}, and HSO_4^-) to complete the circuit. An ideal membrane for VRB should have several properties [103]: (i) low permeation rates of the vanadium ions to minimize self-discharge, (ii) low area resistance to minimize losses in internal energy, (iii) good chemical stability for long lifetime cycling, (iv) high ion conductivity for the transport of the charge-carrying ions to maintain the circuit, and (v) low cost. It seems that the major technical challenges of membrane in VRB systems are the undesired reactant crossover from one side to other side in the cell, and the low stability.

Regarding the membrane crossover, some undesired transports of species across the membrane should be given more attention. These undesired transports, including both the water and vanadium ion transports across the membrane, can decrease the capacity and energy of VRBs. The transport of vanadium ions through the cation-exchange membrane was found to be caused by an undesired exchange of H^+ ions with vanadium ions [104,105]. The vanadium ions with different oxidation states could transport from one half-cell to its opposite half-cell and react with the vanadium ions, leading to the loss of energy density and the reduction of the energy efficiency. Sun et al. [106] investigated the membrane crossover of both water and vanadium ions and found that the diffusion coefficients of the vanadium ions across the N115 membrane were in the order of $V^{2+} > VO^{2+} > VO_2^+ > V^{3+}$.

Regarding the water transfer across membrane, a number of processes can make contributions. These processes include those water molecules carried by ions moving due to the concentration gradient, those carried by the charge-balancing species, and those transported owing to the osmotic pressure difference between the positive and negative electrolyte solutions. For example, vanadium and sulfate and bisulfate ions can carry hydrated water when they are transported across the membrane, which contributes to the net water transfer across membrane. The net water transfer would cause the electrolyte in one half-cell condensed and diluted in other half-cell, resulting in the decrease of the capacity and efficiency of VRBs. As reported, the direction of preferential water transfer was dependent on the SOC of the vanadium solutions [107] and also the type of ion-exchange membrane [108]. The water was transferred toward the positive half-cell at an initial SOC of 100% and 50%, and it was reversed toward the negative half-cell from 50% to 0% SOC. It appeared that the most significant level of water transfer occurred when the cell went into overdischarge [107]. For a cation-exchange membrane, a significant amount of water could be transferred from the negative half-cell to the positive half-cell by the hydration shells of V^{2+} and V^{3+} ions, which could carry a large amount of water and easily permeate through a cation-exchange membrane. For an anion-exchange membrane, the permeation of V^{2+} and V^{3+} co-ions could be restricted. However the ions of SO_4^{2-} and HSO_4^- in the positive side might permeate through the membrane, resulting in a net water transfer from positive to negative half-cells [108].

Chemical stability is another challenge of membranes for VRB systems. This is because of both the strong acidic environment of VRB and the strong oxidative V(V) ions in the positive half-cell electrolyte. Mohammadi and Skyllas-Kazacos [109] extensively studied the chemical stability of some membranes in vanadium solution, and found that the Selemion CMV membrane showed the lowest chemical stability in VRB, whereas Nafion 112 showed an excellent stability. The composite membrane prepared by cross-linking Daramic membrane with divinylbenzene (DVB) showed good chemical stability. It was found that the degradation of membranes was associated with the oxidation of polymers by V(V) ions.

Actually, during the development of VRB, various ion-exchange membranes have been explored. Both cation and anion-exchange membranes have been tested and found to be feasible for the application of VRB systems [110].

21.7.1 Cation-Exchange Membrane

21.7.1.1 Nafion and Modified Nafion Membrane

Nafion membrane is the most commonly used proton-exchange membrane in VRB systems because of its high proton conductivity and good chemical stability in strong acid and oxidation conditions. However, the poor selectivity of Nafion membrane between vanadium ions and protons seemed to be a drawback. The vanadium ions were found to easily transport through the membrane, leading to the decrease of coulombic, voltage, and energy efficiencies of VRB systems [30,104,106,111–113]. To overcome this drawback, much effort has been done to modify the Nafion membranes to suppress the permeation of vanadium ions.

In membrane modification, Nafion/organic hybrid membranes have been explored. For example, Zeng et al. [114] modified a Nafion 117 membrane using pyrrole by electrolyte soaking, oxidation, and electrodeposition. Compared with the Nafion 117 membrane, the modified membranes showed a relatively lower V(IV) permeability and also a lower water transfer property. The V(IV) ion permeability was reduced from 2.87×10^{-6} cm^2 min^{-1} for the Nafion 117 membrane to 5.0×10^{-7} cm^2 min^{-1} for the Nafion/PPR membrane, and the water transfer was decreased from 0.72 mL (72 h cm)$^{-2}$ to 0.22 mL (72 h cm)$^{-2}$ correspondingly. Luo et al. [115] prepared a Nafion/sulfonated poly(ether ether ketone) (SPEEK) layered composite membrane (N/S membrane) consisting of a thin layer of recast Nafion membrane and a layer of SPEEK membrane by chemically cross-linking the sulfonic acid groups of different ionomer membranes. As seen in Table 21.2, compared with the Nafion 117 membrane, the SPEEK and N/S membranes showed higher ion-exchange capacities and significantly decreased VO^{2+} permeability, although the thickness was smaller. The VRB single cells with SPEEK and N/S membranes showed higher coulombic efficiencies than that with a Nafion 117 membrane, but a little lower overall energy efficiency. It was discussed that although the SPPEK membrane can be easily oxidized by the highly oxidizing pentavalent vanadium ions (VO_2^+), the layer of Nafion could effectively prevent the oxidation degradation of the SPEEK layer in the composite membrane of N/S, resulting in good chemical stability in the positive half-cell solutions. They claimed that N/S could be a promising membrane for the VRB system because of its low cost, low vanadium ion permeability, and good chemical stability although the overall energy efficiency of VRB single cell with this membrane was slightly lower [115]. Luo et al. [116] also modified the Nafion membrane using interfacial polymerization with polyethylenimine (PEI) and obtained a Nafion/PEI composite membrane. The experimental results showed that the vanadium ion permeability of the Nafion/PEI composite membrane could be reduced significantly from 36.55×10^{-7} cm^2 min^{-1} for the unmodified Nafion 117 membrane to 5.23×10^{-7} cm^2 min^{-1} for the modified membrane. Furthermore, the water transfer across the membrane could also be greatly reduced. The VRB single cell with this Nafion/PEI membrane showed a higher coulombic efficiency than that with the unmodified Nafion membrane, although the voltage efficiency was lower owing to a higher area resistance of the Nafion/PEI membrane. Thus, the overall energy efficiency of VRB with the Nafion/PEI membrane was improved.

TABLE 21.2
Properties of Nafion 117, SPEEK, and N/S membranes, and Their Performance Efficiencies in VRB Single Cells

Membrane	Nafion 117	SPEEK	N/S
Thickness (μm)	175	100	100
IEC (mmol g^{-1})	0.91	1.8	1.67
Area resistance (Ω cm^2)	1.06	1.27	1.6
Permeability of VO^{2+} ($\times 10^{-7}$ cm^2 min^{-1})	36.55	2.43	1.93
Coulombic efficiency (%)	93.8	97.1	97.6
Voltage efficiency (%)	90.7	87.3	85.3
Energy efficiency (%)	85.0	84.8	83.3

Source: Luo, Q., Zhang, H., Chen, J., You, D., Sun, C., and Zhang, Y., *J. Membr. Sci.*, 325, 553, 2008.

Using a polyelectrolyte layer-by-layer self-assembly technique, Xi et al. [113] fabricated a barrier layer onto the surface of Nafion membrane by alternate adsorption of polycation poly(diallydimethylammonium chloride) (PDDA) and polyanion poly(sodium styrene sulfonate) (PSS), and obtained the multilayer modified Nafion membrane (Nafion–[PDDA–PSS]***n***, where ***n*** is the number of multilayer). The Nafion–[PDDA–PSS]$_n$ composite membrane exhibited a significantly reduced vanadium ion permeability, and the VRB single cell assembled with this composite membrane showed improved coulombic and energy efficiencies.

Membrane modification using inorganic chemicals has also be explored. For example, Nafion/SiO_2 [30] and Nafion/organically modified silicate (Nafion/ORMOSIL) [29] hybrid membranes were synthesized via an *in situ* sol–gel method, and their performance was tested in VRB systems. The results showed that both the hybrid membrane exhibited significantly lower vanadium ion permeabilities when compared with the Nafion 117 membrane. For example, the permeability of VO^{2+} was reduced from 36.9×10^{-7} cm^2 min^{-1} for the Nafion 117 membrane to 1.85×10^{-7} cm^2 min^{-1} for the Nafion/ORMOSIL membrane. This low vanadium permeability might be attributed to the fact that the polar clusters of Nafion membrane were filled or partly filled by SiO_2 nanoparticles, leading to the reduction of vanadium ion diffusion across the membrane. Figure 21.10 shows the coulombic efficiency (CE), voltage efficiency (VE), and energy efficiency (EE) of VRB single cells with Nafion, Nafion/SiO_2, and Nafion/ORMOSIL membranes as a function of current density. As seen from Figure 21.10, the performance of VRB single cells with both two hybrid membranes is superior to that of the Nafion membrane, and the cell with the Nafion/ORMOSIL membrane exhibits the best performance. The higher performance of hybrid membrane may be attributed to their lower vanadium permeability. Compared with the Nafion/SiO_2 membrane, the Nafion/ORMOSIL membrane shows better performance owing to its unique network structure, i.e., Q and D units as shown in Figure 21.11. It is believed that both the ORMOSIL phase and the presence of relatively large $-CH_3$ groups within the hybrid membrane may be able to more effectively suppress the vanadium ion transportation across the membrane than the single Q units in the Nafion/SiO_2 membrane.

In addition, the cycle performances of VRB single cells with the hybrid Nafion/SiO_2 and Nafion/ORMOSIL membranes were also measured, as shown in Figure 21.12, from which it can be seen that both the CE and EE of VRBs with these two hybrid membranes had nearly no decay after 100 cycles at the current density of 60 mA cm^{-2}. This indicates that the hybrid membranes possess good chemical stability

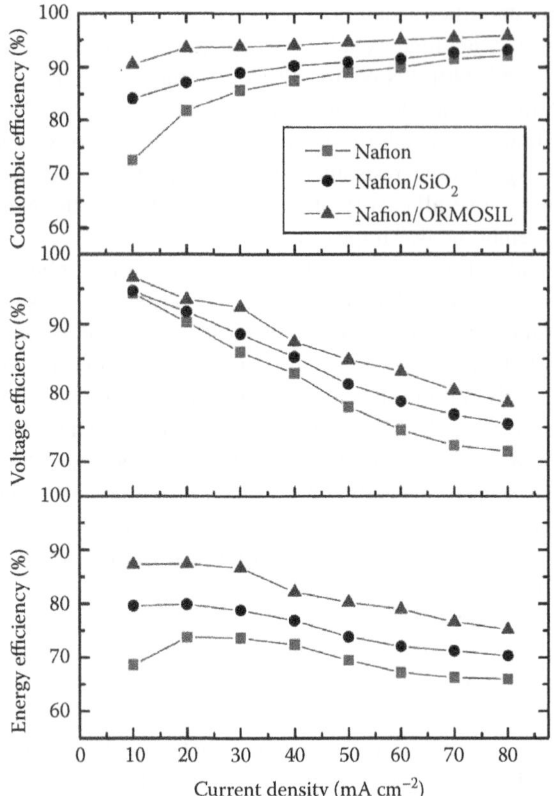

FIGURE 21.10 Change of CE, VE, and EE of VRB single cell with Nafion, Nafion/SiO_2 hybrid membrane, and Nafion/ORMOSIL hybrid membranes under different current densities. (From Teng, X.G., Zhao, Y.T., Xi, J.Y., Wu, Z.H., Qiu, X.P., and Chen, L.Q., *J. Power Sources*, 189, 1240, 2009.)

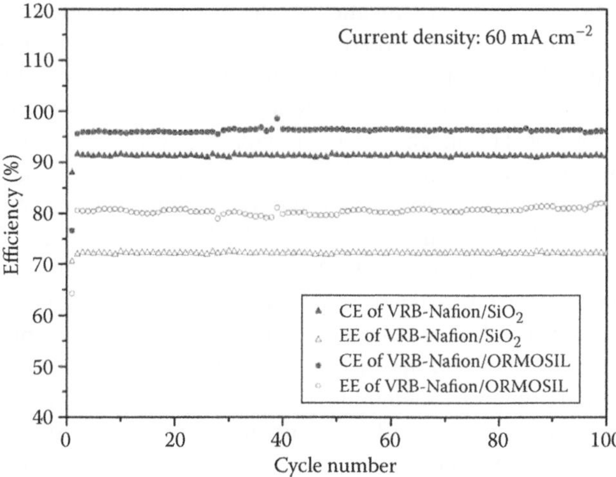

FIGURE 21.12 Cycle performance of VRB with Nafion/ORMOSIL and Nafion/SiO_2 membranes at a current density of 60 mA cm^{-2}. (From Teng, X.G., Zhao, Y.T., Xi, J.Y., Wu, Z.H., Qiu, X.P., and Chen, L.Q., *J. Power Sources*, 189, 1240, 2009.)

during long-term charge–discharge processes under strong acidic and oxidative conditions. The above results suggested that the Nafion/SiO_2 hybrid membrane, especially the Nafion/organically modified silicate hybrid membrane, could be a promising membrane material for application in VRB systems owing to its lower vanadium ion permeability and high chemical stability.

Similarly, Teng et al. [117] synthesized a Nafion/organic silica/TiO_2 composite membrane (Nafion/Si/Ti hybrid membrane) using an *in situ* sol–gel method, and evaluated its vanadium permeability and water transfer ability, as well as its corresponding VRB performance. As seen in Figures 21.13 and 21.14, both the permeability of vanadium ions and the water transfer across the Nafion/Si/Ti hybrid membrane were much lower than that of the unmodified Nafion membrane. The VRB single cell test results revealed that the VRB with this Nafion/Si/Ti hybrid membrane could give both slightly higher CE and EE than those of the Nafion 117 membrane. In addition, this Nafion/Si/Ti hybrid membrane could also exhibit a better open circuit voltage (OCV) behavior than that of the Nafion 117 membrane. Regarding the stability, the cycle performance of the VRB single cell with Nafion/Si/Ti hybrid membrane showed no decay of CE and a little decrease of EE after 100 cycles at a current density of 50 mA cm^{-2}, indicating a good stability of this composite membrane.

Sang et al. [118] synthesized some Nafion 1135/zirconium phosphate (ZrP) hybrid membranes by an impregnation method and characterized their transport properties for both H^+ and VO^{2+}. They claimed that the Nafion 1135/(ZrP) hybrid membranes had a suppressed VO^{2+} permeability when compared with the unmodified Nafion 1135 membrane, although

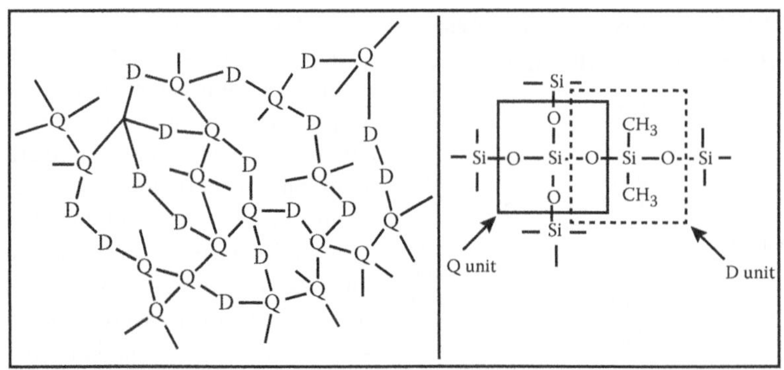

FIGURE 21.11 Hypothesized ORMOSIL phase structure consisting of copolymerized Q and D units within Nafion/ORMOSIL hybrid membranes. (From Teng, X.G., Zhao, Y.T., Xi, J.Y., Wu, Z.H., Qiu, X.P., and Chen, L.Q., *J. Power Sources*, 189, 1240, 2009.)

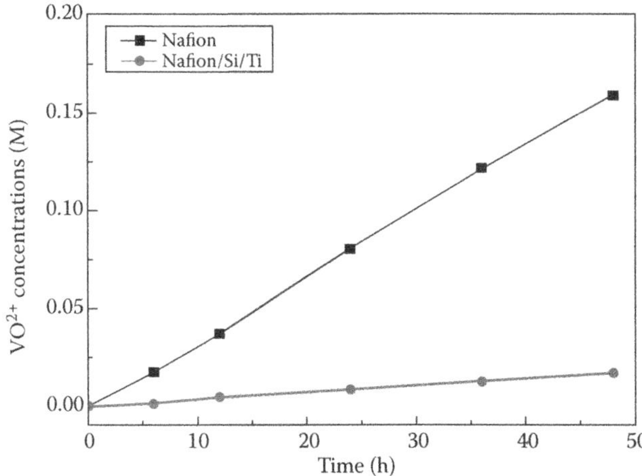

FIGURE 21.13 Concentration of VO^{2+} in the right reservoir of the cell with Nafion and Nafion/Si/Ti hybrid membrane. (From Teng, X., Zhao, Y., Xi, J., Wu, Z., Qiu, X., and Chen, L., *J. Membr. Sci.*, 341, 149, 2009.)

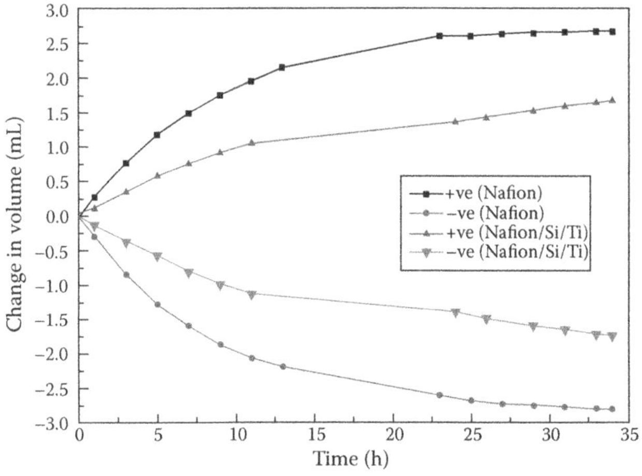

FIGURE 21.14 Transfer of water across the modified and unmodified Nafion membrane with electrolytes at an SOC of 50%. (From Teng, X., Zhao, Y., Xi, J., Wu, Z., Qiu, X., and Chen, L., *J. Membr. Sci.*, 341, 149, 2009.)

the proton transport across the membrane was also suppressed. This can be evidenced by the diffusion coefficients of $\bar{D}_{H^+}$ and $\bar{D}_{VO^{2+}}$ insert equation number in Nafion 1135 and the Nafion 1135/(ZrP) hybrid membranes, as listed in Table 21.3. This result suggests that the hybrid membranes have a higher selectivity for H^+ and VO^{2+} and a potential prospect application in VRB systems.

TABLE 21.3
Diffusion Coefficients of $\bar{D}_{H^+}$ and $\bar{D}_{VO^{2+}}$ in Nafion 1135 and the Nafion 1135/(ZrP) Composite Membranes

Membrane	$\bar{D}_{H^+}$ ($\times 10^{12}$ m^2 s^{-1}) (%)	$\bar{D}_{VO^{2+}}$ ($\times 10^{13}$ m^2 s^{-1}) (%)
Nafion 1135	3.412 ± 0.2	4.012 ± 0.06
Nafion 1135/ZrP1	3.093 ± 0.2	2.325 ± 0.08
Nafion 1135/ZrP3	2.848 ± 0.2	2.091 ± 0.09
Nafion 1135/ZrP5	2.827 ± 0.2	0.3227 ± 0.09

Source: Sang, S., Wu, Q., and Huang, K., *J. Membr. Sci.*, 305, 118, 2007.

21.7.1.2 Modified Daramic Membrane

Modified Daramic membranes have also been explored for VRB systems [109,110,119–123]. In general, Daramic membrane (Daramic, LLC) is a microporous polymeric separator consisting primarily of an ultrahigh-molecular-weight polyethylene, amorphous silica, and specially formulated hydrocarbon oils. This kind of membrane possesses excellent oxidation and puncture resistance as well as low ionic resistance. Although the Daramic membranes without modification are normally unsuitable for the VRB application owing to their high vanadium ion permeability, they can exhibit promising performance after modification [109,110,119–123]. Both the low cost and good chemical stability in vanadium electrolytes further make them candidate membranes for VRB systems.

Composite Daramic membranes made by cross-linking Daramic membrane with DVB [110,119–121] could exhibit a significantly improved VRB performance owing to the pore-blocking effect by the ion-exchange resin (Amberlite CG400) and the cross-linked DVB. Tian et al. [122] made a Daramic/Nafion composite membrane by soaking Daramic membrane in a 5 wt.% Nafion solution. Their composite membrane exhibited a low area resistance and reasonable ion-exchange capacity for the VRB. The transport of the vanadium ions across the membrane was also suppressed, and thus the self-discharge was reduced significantly.

21.7.1.3 Fluorinated or Partly Fluorinated Membrane

Fluorinated or partly fluorinated membranes have been explored as another kind of VRB membrane candidates for VRBs [124,125]. Luo et al. [111] synthesized a poly(vinylidene fluoride)-graft-poly(styrene sulfonic acid), denoted as PVDF–g–PSSA membrane, by a solution-grafting method, and characterized its properties and performance in VRB systems. Their studies showed that this PVDF–g–PSSA membrane could give a high conductivity and significantly reduced vanadium ion permeability when compared with the Nafion 117 membrane. The VRB assembled with this low-cost PVDF–g–PSSA membrane exhibited a higher performance than that with Nafion 117 at the same operating conditions. This membrane also showed an excellent chemical stability observed by the result of cycle performance of its corresponding VRB at a current density of 60 cm^2. As shown in Figure 21.15, there is no capacity decay after running more than 220 cycles. In addition, the PVDF-based [124,125], poly(tetrafluorotheylene)-based [126] and poly(ethylene-co-tetrafluoroethylene) (ETFE)-based [127] ion-exchange membranes were also synthesized by Qiu et al. recently. All of these membranes seemed to have higher conductivities and lower vanadium ion permeabilities than the Nafion 117 membrane, suggesting their promising applications in VRB systems.

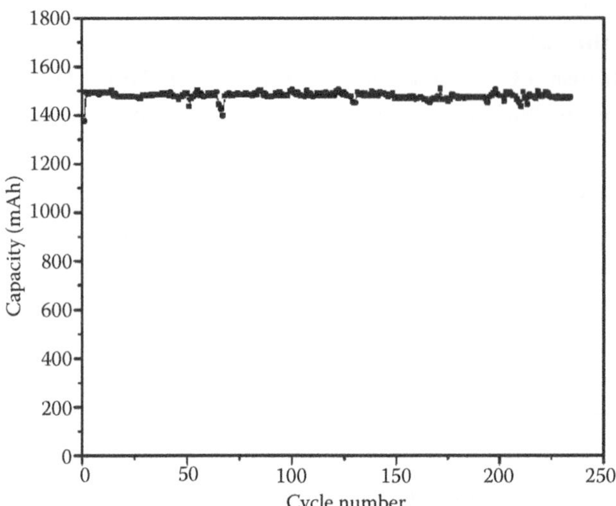

FIGURE 21.15 Discharge capacity of VRB with a PVDF–g–PSSA-22 membrane during cycling at a current density of 60 mA cm^{-2}. (From Luo, X., Lu, Z., Xi, J., Wu, Z., Zhu, W., Chen, L., and Qiu, X., *J. Phys. Chem. B*, 109, 20310, 2005.)

Besides Nafion and Daramic-based membranes, various kinds of nonfluorinated membranes have also been exploited for VRB systems in recent years [128,129]. For example, SPEEK-based membranes have received much attention in recent years owing to their low cost, adequate conductivity, good thermal stability, and mechanical strength. It was reported that a SPEEK membrane had a lower vanadium ion permeability when compared with the Nafion 117 membrane, and the VRB's OCV evaluation demonstrated that the SPEEK membrane was superior to the Nafion 117 membrane [128]. Unfortunately, the degradation of the SPEEK membrane is normally faster than that of the Nafion 117 membrane due to the oxidation by the VO_2^+ ions in the positive electrolyte.

Jia et al. [128] synthesized a novel sandwich-type SPEEK/tungstophosphoric acid (TPA)/polypropylene (PP) composite membrane that consisted of a layer of PP membrane between two layers of SPEEK/TPA membrane, and tested its VRB performance. The permeability of V(IV) across this membrane could be reduced significantly compared with the Nafion 212 membrane. The single VRB cell testing showed a better performance than the Nafion 212 membrane, and there was no decrease in voltage, coulombic, and efficiencies after 80 cycles (>350 h), as shown in Figure 21.16. This result indicates that this membrane had good chemical/electrochemical stability in strong acidic and oxidative vanadium solutions.

Hwang and Ohya [130] prepared two membranes for VRB, one was the chlorosulfonated homogeneous polyethylene (PE) dense film (PE-X) membrane, and the other was asymmetric membrane (MH-X). The VRB performances of these two membranes were compared with that of Nafion 117 membrane at the same experimental conditions. It was observed that the PE-X membrane showed the highest permeability of vanadium ions followed by the Nafion 117 membrane, and the MH-X membrane had the lowest permeability. Their study also showed that the durability of the chlorosulfonated homogeneous membrane could be improved by cross-linking.

Sulfonated aromatic polymers–based proton-exchange membranes have also been exploited for VRBs owing to their high proton conductivity. As shown in Table 21.4 [131], the proton conductivities of both sulfonated poly(arylene thioether ketone) (SPTK) and sulfonated poly(arylene thioether ketone ketone) (SPTKK) membranes are comparable to that of the Nafion 117 membrane. The VRB batteries assembled with these two membranes showed higher coulombic efficiencies than that assembled with the Nafion 117 membrane owing to their lower permeability of vanadium ions as shown in Table 21.4.

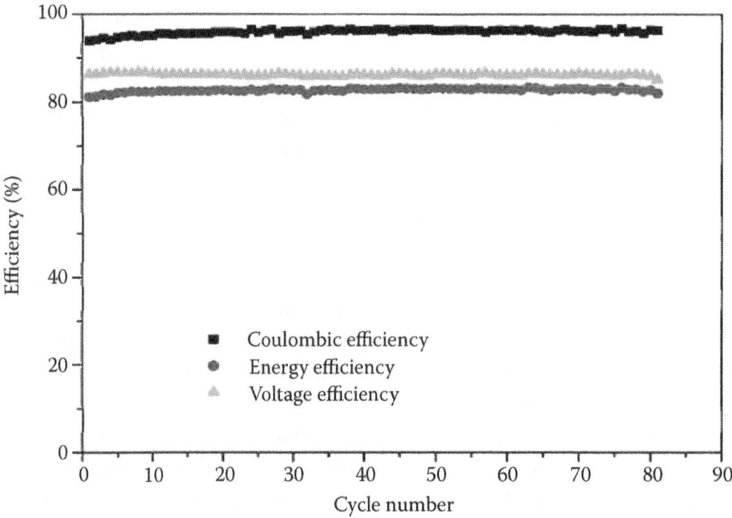

FIGURE 21.16 Cycle performance for VRB single cell with SPEEK/TPA/PP membrane at a current density of 35.7 mA cm^{-2}. (From Jia, C., Liu, J., Yan, C., *J. Power Sources*, 195, 4380, 2010.)

TABLE 21.4
Proton Conductivity and VO^{2+} Permeability of SPTK, SPTKK, and Nafion 117 Membranes at Room Temperature

Membrane	Proton Conductivity ($\times 10^{-2}$ S cm^{-1})	VO^{2+} Permeability (m^2 s^{-1})
SPTK	1.05	1.2×10^{-13}
SPTKK	1.36	3.1×10^{-13}
Nafion 117	2.03	4.9×10^{-12}

Source: Chen, D., Wang, S., Xiao, M., and Meng, Y., *Energy Environ. Sci.*, 3, 622, 2010.

21.7.2 Anion-Exchange Membrane

To decrease the permeability of vanadium ions across membranes, anion-exchange membranes have been exploited for VRB systems in the past several years [112,129,132]. For example, a polysulfone anion-exchange membrane (from Asahi Glass) was cross-linked by the accelerated electron radiation [112], and reasonable membrane stability and high cell performance were achieved for the VRB single cell assembled with this cross-linked membrane. The quaternized poly(phthalazinone ether sulfone ketone) (QPPESK) [129] and poly(phthalazinone ether sulfone ketone) (QPPES) [132] anion-exchange membranes were also synthesized for VRB. The QPPESK membrane showed good chemical stability in the VO_2^+ solutions for 20 days. The results of VRB single cell tests indicated that the cell with the QPPESK membrane had a much higher energy efficiency than that with the Nafion 117 membrane, as shown in Table 21.5 [129].

Qiu et al. [133] prepared an ETFE-based anion-exchange membrane in the effort to reduce the permeability of vanadium ions in VRBs. Experimental results showed that the ETFE-based anion-exchange membrane had a high ion-exchange capacity, lower area resistance, and much lower vanadium ion permeability when compared with the Nafion 117 membrane.

Modification of anion-exchange membranes has also been employed to reduce the vanadium ion permeability [134,135]. Tian et al. [134] modified a JAM anion-exchange membrane with sodium 4-styrenesulfonate by *in situ* polymerization. The modified JAM membrane showed very good stability, and the average efficiencies of the cell with this JAM membrane were higher than that with Nafion 117 after eight charge–discharge cycles. Mohammadi and Skyllas-Kazacos [135] found that the modified anion-exchange membranes could also reduce the water transfer across the membrane, which could reduce both the capacity loss and possible reservoir flooding during long-term continuous VRB operations.

TABLE 21.5
Performance Properties of VRB Single Cells Assembled with QPPESK and Nafion 117 Membranes

Membrane	Current Efficiency (%)	Voltage Efficiency (%)	Energy Efficiency (%)
QPPESK	96.4	91.6	88.3
Nafion 117	91.5	90.6	82.9

Source: Jian, X.G., Yan, C., Zhang, H.M., Zhang, S.H., Liu, C., and Zhao, P., *Chin. Chem. Lett.*, 18, 1269, 2007.

Generally, using anion-exchange membranes can effectively reduce the permeation of vanadium ions across the membrane; however, they will not assist the conduction of protons, which might lead to the sacrifice of voltage efficiency of the VRBs.

21.8 MATERIALS OF FLOW PLATE/BIPOLAR PLATE

Flow plates for both the anode and cathode are another necessary component in LRFBs. As shown in Figure 21.5, the electrode components of VRBs include the flow plate and the porous electrode. In general, the flow plate provides flow channels, supports the porous electrode, and collects the produced current; therefore, it is also called a solid current collector. In a flow battery stack, the solid current collector with flow channels on both sides is also called the bipolar plate. The material for flow plate must possess the characteristics of good mechanical strength, high stability at the VRB operating conditions, low bulk and contact resistance, as well as low ion permeation. With respect to this, the available materials that can be used as flow plates are limited. Currently, the materials employed for flow plates in VRBs are mainly polymer-impregnated graphite plate, conductive carbon–polymer composite [73,86], and polymer-impregnated flexible graphite [136,137]. It seems that the polymer-impregnated graphite plate is widely used owing to its low electronic resistance and machinability. However, it is relatively expensive and brittle, limiting its practical application. In recent years, conductive carbon–polymer composites have received much more attention as promising candidates for VRB bipolar plates owing to their low cost, light weight, and flexibility [73,86,88,89,138–140].

21.9 MATERIAL CHALLENGES OF LRFBs

Although some of the LRFBs such as VRBs have been well developed in recent years, and although successful demonstration of systems up to MWh levels have been performed, the LRFB technologies have not seen broad market penetration. One of the important reasons may be the high cost of the current technologies. For example, VRB is about \$500 kWh^{-1} or higher [141], which is at least three times higher than the target for broad market penetration. The high cost of LRFBs is directly dependent on the high cost of materials/components and performance parameters, including reliability, cycle/calendar life, energy efficiency, system energy capacity, etc. Therefore, some challenges in the materials must be overcome before the LRFBs can be commercialized. First, as commonly identified, the membrane material is one of the challenges—or may be the largest challenge—that contribute to the largest portion of the high cost of LRFBs. The most feasible membranes, Nafion-based cation-exchange membranes, have good chemical stability and high ion conductivity; however, their cost is high and the selectivity between protons and metal ions (e.g., vanadium ions) is low. Furthermore, the high permeability of vanadium ions across the Nafion membrane could result in the loss of energy density of VRBs, and the water transport from one half-cell to another could cause the energy density to decrease further. Although the Nafion-based composite membrane modified by incorporating organic or inorganic chemicals could give promising applications in VRBs, the cost is still high. In addition, the long-term stability of Nafion and the modified Nafion membranes is still not sufficient. Besides Nafion-based membranes, most of the hydrocarbon membranes are not stable in strong acidic and oxidative solutions of vanadium. Although some membranes could give promising performance, further work is needed to optimize their properties. Although anion-exchange membranes could effectively reduce the permeation of vanadium ions and transportation of water across the membrane, they could not assist the conduction of protons, resulting in the decrease in voltage efficiency of flow batteries.

The electrode materials need to be optimized further. Currently, porous graphite felts are widely used as the porous electrode materials in VRBs; however, the overpotentials in both the negative and positive electrodes are still too high due to the low electrochemical activities of graphite felts toward the reactions between the vanadium ions. Active and stable catalysts may need to be integrated into the electrodes to further improve the electrochemical activity.

The materials for current collectors also have some challenges. For example, the commonly used graphite plates are brittle and expensive. The conductive carbon–polymer composite plates could be the potential candidates for VRBs; however, further work is definitely needed to optimize the materials and polymers.

REFERENCES

1. Yang, Z., Liu, J., Baskaran, S., Imhoff, C. H., and Hollady, J. D. Enabling renewable energy and the future grid with advanced electricity storage. *JOM* 2010; 62: 14–23.
2. Price, A., Bartley, S., Male, S., and Cooley, G. A novel approach to utility scale energy storage. *Power Engineering Journal* 1999; 13: 122–129.
3. Zito, R. Process for energy storage and/or power delivery with means for restoring electrolyte balance. US Patent 5,612,148, 1997.
4. Zhao, P., Zhang, H., Zhou, H., and Yi, B. Nickel foam and carbon felt applications for sodium polysulfide/bromine redox flow battery electrodes. *Electrochimica Acta* 2005; 51: 1091–1098.
5. Zhou, H., Zhang, H., Zhao, P., and Yi, B. A comparative study of carbon felt and activated carbon based electrodes for sodium polysulfide/bromine redox flow battery. *Electrochimica Acta* 2006; 51: 6304–6312.
6. Skyllas-Kazacos, M., and Limantari, Y. Kinetics of the chemical dissolution of vanadium pentoxide in acidic bromide solutions. *Journal of Applied Electrochemistry* 2004; 34: 681–685.
7. Vafiadis, H., and Skyllas-Kazacos, M. Evaluation of membranes for the novel vanadium bromine redox flow cell. *Journal of Membrane Science* 2006; 279: 394–402.
8. Skallas-Kazacos, M., and Robins, R. G. All-vanadium redox battery. US Patent 4,786,567, 1988.
9. Skyllas-Kazacos, M., Rychcik, M., Robins, R. G., Fane, A. G., and Green, M. A. New all-vanadium redox flow cell. *Journal of the Electrochemical Society* 1986; 133: 1057–1058.
10. Xia, X., Liu, H. T., and Liu, Y. Studies of the feasibility of a $Ce^{4+}/Ce^{3+}–V^{2+}/V^{3+}$ redox cell. *Journal of the Electrochemical Society* 2002; 149: A426–A430.
11. Bartolozzi, M. Development of redox flow batteries. A historical bibliography. *Journal of Power Sources* 1989; 27: 219–234.
12. Pissoort, P. A. FR Patent 754,065, 1933.
13. Pellegri, A., and Spaziante, P. M. Process and accumulator for storing and releasing electrical energy. IT GB 2,030,349, 1980.
14. Skyllas-Kazacos, M., Miron, R., and Robins, R. G. All vanadium redox battery. AU Patent 5,556,286, 1987.
15. Rychcik, M., and Skyllas-Kazacos, M. Characteristics of a new all-vanadium redox flow battery. *Journal of Power Sources* 1988; 22: 59–67.
16. Li, L., Kim, S., Yang, Z., Wang, W., Nie, Z., Chen, B., Zhang, J., and Xia, G. Redox flow batteries based on supporting solutions containing chloride. US Patent 2,014,234,753, 2014.
17. Li, L., Kim, S., Wang, W., Vijayakumar, M., Nie, Z., Chen, B., Zhang, J., Xia, G., Hu, J., Graff, G., Liu, J., and Yang, Z. A stable vanadium redox-flow battery with high energy density for large-scale energy storage. *Advanced Energy Materials* 2011; 1: 394–400.
18. Sumitomo Electric Industries, Ltd. Vanadium redox-flow battery (VRB) for a variety of applications. Report.
19. CleanEnergy | ACTION PROJECT, Tomamae Wind Villa Power Plant: Vanadium Redox Flow Battery Energy Storage System. Available from http://www.cleanenergyactionproject.com/CleanEnergyActionProject/CS.Tomamae_Wind_Villa_Power_Plant___Energy_Storage_Case_Study.html.
20. de Leon, C. P., Frias-Ferrer, A., Gonzalez-Garcia, J., Szanto, D. A., and Walsh, F. C. Redox flow cells for energy conversion. *Journal of Power Sources* 2006; 160: 716–732.
21. You, D., Zhang, H., and Chen, J. A simple model for the vanadium redox battery. *Electrochimica Acta* 2009; 54: 6827–6836.
22. Skyllas-Kazacos, M. Secondary batteries—Flow systems: Vanadium redox-flow batteries. In *Encyclopedia of Electrochemical Power Sources*, Dyer, C. K., Moseley, P. T., Ogumi, Z., Rand, D. A. J., Scrosati, B., Garche, J., Eds. Elsevier: Amsterdam, 2009, pp. 444–453.
23. Huang, K. L., Li, X. G., Liu, S. Q., Tan, N., and Chen, L. Q. Research progress of vanadium redox flow battery for energy storage in China. *Renewable Energy* 2008; 33: 186–192.
24. Witmer, D. VRB flow battery demonstration. Presentation in *Wind Diesel Conference*, Girdwood, Alaska, April 24, 2008.
25. Li, Z., Huang, K., and Man, R. Research progress in the key materials of all-vanadium redox flow battery. *Battery* 2006; 36: 150–152.
26. Chen, J., Wang, Q., and Wang, B. Research progress in key materials for all vanadium redox flow battery. *Modern Chemical Industry* 2006; 26: 21–24.

27. Fabjan, C., Garche, J., Harrer, B., Jörissen, L., Kolbeck, C., Philippi, F., Tomazic, G., and Wagner, F. The vanadium redox-battery: An efficient storage unit for photovoltaic systems. *Electrochimica Acta* 2001; 47: 825–831.
28. UniEnergy Technologies, LLC. Website. Available from http://www.uetechnologies.com/.
29. Teng, X. G., Zhao, Y. T., Xi, J. Y., Wu, Z. H., Qiu, X. P., and Chen, L. Q. Nafion/organically modified silicate hybrids membrane for vanadium redox flow battery. *Journal of Power Sources* 2009; 189: 1240–1246.
30. Xi, J. Y., Wu, Z. H., Qiu, X. P., and Chen, L. Q. Nafion/SiO_2 hybrid membrane for vanadium redox flow battery. *Journal of Power Sources* 2007; 166: 531–536.
31. Skyllas-Kazacos, M. Novel vanadium chloride/polyhalide redox flow battery. *Journal of Power Sources* 2003; 124: 299–302.
32. Electrical Power Research Institute. *Handbook of Energy Storage for Transmission and Distribution Applications*. Washington, DC: Department of Energy, 2003.
33. Pawar, S., Madhale, R., Patil, P., and Lokhande, C. Studies on iron–chromium redox storage system. *Bulletin of Materials Science* 1988; 10: 367–372.
34. Lopez-Atalaya, M., Codina, G., Perez, J. R., Vazquez, J. L., and Aldaz, A. Optimization studies on a Fe/Cr redox flow battery. *Journal of Power Sources* 1992; 39: 147–154.
35. Cedzynska, K. Properties of modified electrolyte for zinc–bromine cells. *Electrochimica Acta* 1995; 40: 971–976.
36. Eustace, D. J. Bromine complexation in zinc–bromine circulating batteries. *Journal of the Electrochemical Society* 1980; 127: 528–532.
37. Wen, Y. H., Zhang, H. M., Qian, P., Zhou, H. T., Zhao, P., Yi, B. L., and Yang, Y. S. Studies on iron (Fe^{3+}/Fe^{2+})–complex/bromine (Br^{-2}/Br^{-}) redox flow cell in sodium acetate solution. *Journal of the Electrochemical Society* 2006; 153: A929–A934.
38. Xue, F. Q., Wang, Y. L., Wang, W. H., and Wang, X. D. Investigation on the electrode process of the Mn(II)/Mn(III) couple in redox flow battery. *Electrochimica Acta* 2008; 53: 6636–6642.
39. Wang, Y. Y., Lin, M. R., and Wan, C. C. A study of the discharge performance of the Ti/Fe redox flow system. *Journal of Power Sources* 1984; 13: 65–74.
40. Wang, W., Kim, S., Chen, B., Nie, Z., Zhang, J., Xia, G. G., Li, L., and Yang, Z. A new redox flow battery using Fe/V redox couples in chloride supporting electrolyte. *Energy & Environmental Science* 2011; 4: 4068–4073.
41. Wang, W., Nie, Z., Chen, B., Chen, F., Luo, Q., Wei, X., Xia, G. G., Skyllas-Kazacos, M., Li, L., and Yang, Z. A new Fe/V redox flow battery using a sulfuric/chloric mixed-acid supporting electrolyte. *Advanced Energy Materials* 2012; 2: 487–493.
42. Thaller, L. H. Electrically rechargeable redox flow cell. US Patent 3,996,064, 1976.
43. Thaller, L. H. Electrochemical cells for rebalancing redox flow system. US Patent 4,159,366, 1979.
44. Gahn, R. F., Hagedorn, N. H., and Ling, J. S. Single cell performance studies on the Fe–Cr redox energy storage systems using mixed reactant solutions at elevated temperature. NASA Report, DOE/NASA/12726-21, NASA TM-83385, National Aeronautics and Space Administration, Lewis Research Center, Cleveland, Ohio, USA, 1983.
45. Jalan, V., Morriseau, B., and Swette, L. Optimization and fabrication of porous carbon electrodes for Fe/Cr redox flow cells. NASA Report, DOE/NASA/0198-1, NASA CR-167921, National Aeronautics and Space Administration, Lewis Research Center, Cleveland, Ohio, USA, 1982.
46. Bae, C. H., Roberts, E. P. L., and Dryfe, R. A. W. Chromium redox couples for application to redox flow batteries. *Electrochimica Acta* 2002; 48: 279–287.
47. Codina, G., Perez, J. R., Lopezatalaya, M., Vazquez, J. L., and Aldaz, A. Development of a 0.1 kW power accumulation pilot-plant based on an Fe/Cr redox flow battery. 1. Considerations on flow-distribution design. *Journal of Power Sources* 1994; 48: 293–302.
48. Pupkevich, V., Glibin, V., and Karamanev, D. The effect of activation on the electrochemical behaviour of graphite felt towards the Fe^{3+}/Fe^{2+} redox electrode reaction. *Electrochemistry Communications* 2007; 9: 1924–1930.
49. Chen, Y.-W. D., Santhanam, K. S. V., and Bard, A. J. Solution redox couples for electrochemical energy storage. *Journal of the Electrochemical Society* 1981; 128: 1460–1467.
50. Wen, Y. H., Zhang, H. M., Qian, P., Zhou, H. T., Zhao, P., Yi, B. L., and Yang, Y. S. A study of the Fe(III)/Fe(II)-triethanolamine complex redox couple for redox flow battery application. *Electrochimica Acta* 2006; 51: 3769–3775.
51. Szanto, D. A. Characterization of electrochemical filter-press reactors. PhD thesis. University of Portsmouth, UK, 1999.
52. Cathro, K. J., Cedzynska, K., and Constable, D. C. Preparation and performance of plastic-bonded-carbon bromine electrodes. *Journal of Power Sources* 1987; 19: 337–356.
53. Kinoshita, K., Leach, S. C., and Ablow, C. M. Bromine reduction in a 2-phase electrolyte. *Journal of the Electrochemical Society* 1982; 129: 2397–2403.
54. Lim, H. S., Lackner, A. M., and Knechtli, R. C. Zinc–bromine secondary battery. *Journal of the Electrochemical Society* 1977; 124: 1154–1157.
55. Chen, H., Cong, T. N., Yang, W., Tan, C., Li, Y., and Ding, Y. Progress in electrical energy storage system: A critical review. *Progress in Natural Science* 2009; 19: 291–312.
56. Rahman, F., and Skyllas-Kazacos, M. Solubility of vanadyl sulfate in concentrated sulfuric acid solutions. *Journal of Power Sources* 1998; 72: 105–110.
57. Oriji, G., Katayama, Y., and Miura, T. Investigation on V(IV)/V(V) species in a vanadium redox flow battery. *Electrochimica Acta* 2004; 49: 3091–3095.
58. Cotton, F. A., and Wilkinson, G. Eds. *Advanced Inorganic Chemistry*. John Wiley & Sons: New York, 1988, p. 665.
59. Gattrell, M., Park, J., MacDougall, B., Apte, J., McCarthy, S., and Wu, C. W. Study of the mechanism of the vanadium $4^{+}/5^{+}$ redox reaction in acidic solutions. *Journal of the Electrochemical Society* 2004; 151: A123–A130.
60. Baes, C. F., and Mesmer, R. E. *The Hydrolysis of Cations*. John Wiley & Sons: New York, 1976.
61. Lu, X. Spectroscopic study of vanadium(V) precipitation in the vanadium redox cell electrolyte. *Electrochimica Acta* 2001; 46: 4281–4287.
62. Madic, C., Begun, G. M., Hahn, R. L., Launay, J. P., and Thiessend, W. E. Dimerization of aquadioxovanadium(V) ion in concentrated perchloric and sulfuric acid media. *Inorganic Chemistry* 1984; 23: 469–476.
63. Kausar, N., Howe, R., and Skyllas-Kazacos, M. Raman spectroscopy studies of concentrated vanadium redox battery positive electrolytes. *Journal of Applied Electrochemistry* 2001; 31: 1327–1332.
64. Pozarnsky, G. A., and McCormick, A. V. ^{51}V NMR and EPR study of reaction kinetics and mechanisms in V_2O_5 gelation by ion exchange of sodium metavanadate solutions. *Chemistry of Materials* 1994; 6: 380–385.

65. Pozarnsky, G. A., and McCormick, A. V. ^{17}O nuclear magnetic resonance spectroscopy of the structural evolution of vanadium pentaoxide gels. *Journal of Materials Chemistry* 1994; 4: 1749–1753.
66. Rahman, F., and Skyllas-Kazacos, M. Vanadium redox battery: Positive half-cell electrolyte studies. *Journal of Power Sources* 2009; 189: 1212–1219.
67. Skyllas-Kazacos, M., Menictas, C., and Kazacos, M. Thermal stability of concentrated V(V) electrolytes in the vanadium redox cell. *Journal of the Electrochemical Society* 1996; 143: L86–L88.
68. Kazacos, M., Cheng, M., and Skyllas-Kazacos, M. Vanadium redox cell electrolyte optimization studies. *Journal of Applied Electrochemistry* 1990; 20: 463–467.
69. Kazacos, M. S., and Kazacos, M. Stabilized vanadium electrolyte solutions for all-vanadium redox cells and batteries. US Patent 6,562,514 B1, 2003.
70. Skyllas-Kazacos, M., Peng, C., and Cheng, M. Evaluation of precipitation inhibitors for supersaturated vanadyl electrolytes for the vanadium redox battery. *Electrochemical and Solid-State Letters* 1999; 2: 121–122.
71. Fang, B., Wei, Y., Arai, T., Iwasa, S., and Kumagai, M. Development of a novel redox flow battery for electricity storage system. *Journal of Applied Electrochemistry* 2003; 33: 197–203.
72. Sum, E., Rychcik, M., and Skyllas-Kazacos, M. Investigation of the V(V)/V(IV) system for use in the positive half-cell of a redox battery. *Journal of Power Sources* 1985; 16: 85–95.
73. Kazacos, M., and Skyllas-Kazacos, M. Performance-characteristics of carbon plastic electrodes in the all-vanadium redox cell. *Journal of the Electrochemical Society* 1989; 136: 2759–2760.
74. Sum, E., and Skyllas-Kazacos, M. A study of the V(II)/V(III) redox couple for redox flow cell applications. *Journal of Power Sources* 1985; 15: 179–190.
75. Radford, G. J. W., Cox, J., Wills, R. G. A., and Walsh, F. C. Electrochemical characterisation of activated carbon particles used in redox flow battery electrodes. *Journal of Power Sources* 2008; 185: 1499–1504.
76. Hagg, C. M., and Skyllas-Kazacos, M. Novel bipolar electrodes for battery applications. *Journal of Applied Electrochemistry* 2002; 32: 1063–1069.
77. Inoue, M., Tsuzuki, Y., Iizuka, Y., and Shimada, M. Carbon-fiber electrode for redox flow battery. *Journal of the Electrochemical Society* 1987; 134: 756–757.
78. Kaneko, H., Nozaki, K., Wada, Y., Aoki, T., Negishi, A., and Kamimoto, M. Vanadium redox reactions and carbon electrodes for vanadium redox flow battery. *Electrochimica Acta* 1991; 36: 1191–1196.
79. Mohammadi, F., Timbrell, P., Zhong, S., Padeste, C., and Skyllas-Kazacos, M. Overcharge in the vanadium redox battery and changes in electrical-resistivity and surface functionality of graphite-felt electrodes. *Journal of Power Sources* 1994; 52: 61–68.
80. Rychcik, M., and Skyllas-Kazacos, M. Evaluation of electrode materials for vanadium redox cell. *Journal of Power Sources* 1987; 19: 45–54.
81. Shao, Y. Y., Zhang, S., Kou, R., Wang, X. Q., Wang, C. M., Dai, S., Viswanathan, V., Liu, J., Wang, Y., and Lin, Y. H. Noncovalently functionalized graphitic mesoporous carbon as a stable support of Pt nanoparticles for oxygen reduction. *Journal of Power Sources* 2010; 195: 1805–1811.
82. Zhong, S., and Skyllas-Kazacos, M. Electrochemical-behavior of vanadium(V) vanadium(IV) redox couple at graphite electrodes. *Journal of Power Sources* 1992; 39: 1–9.
83. Sun, B., and Sykllaskazacos, M. Modification of graphite electrode materials for vanadium redox flow battery application. 1. Thermal treatment. *Electrochimica Acta* 1992; 37: 1253–1260.
84. Sun, B., and Skyllas-Kazacos, M. Chemical modification of graphite electrode materials for vanadium redox flow battery application. 2. Acid treatments. *Electrochimica Acta* 1992; 37: 2459–2465.
85. Sun, B. T., and Skyllas-Kazacos, M. Chemical modification and electrochemical behavior of graphite fiber in acidic vanadium solution. *Electrochimica Acta* 1991; 36: 513–517.
86. Zhong, S., Kazacos, M., Burford, R. P., and Skyllas-Kazacos, M. Fabrication and activation studies of conducting plastic composite electrodes for redox cells. *Journal of Power Sources* 1991; 36: 29–43.
87. Zhong, S., Padeste, C., Kazacos, M., and Skyllas-Kazacos, M. Comparison of the physical, chemical and electrochemical properties of rayon-based and polyacrylonitrile-based graphite felt electrodes. *Journal of Power Sources* 1993; 45: 29–41.
88. Haddadiasl, V., Kazacos, M., and Skyllas-Kazacos, M. Carbon–polymer composite electrodes for redox cells. *Journal of Applied Polymer Science* 1995; 57: 1455–1463.
89. Haddadiasl, V., Kazacos, M., and Skyllas-Kazacos, M. Conductive carbon polypropylene composite electrodes for vanadium redox battery. *Journal of Applied Electrochemistry* 1995; 25: 29–33.
90. Li, X. G., Huang, K. L., Liu, S. Q., and Chen, L. Q. Electrochemical behavior of diverse vanadium ions at modified graphite felt electrode in sulphuric solution. *Journal of Central South University of Technology* 2007; 14: 51–56.
91. Li, X. G., Huang, K. L., Liu, S. Q., Tan, N., and Chen, L. Q. Characteristics of graphite felt electrode electrochemically oxidized for vanadium redox battery application. *Transactions of Nonferrous Metals Society of China* 2007; 17: 195–199.
92. Wang, W. H., and Wang, X. D. Investigation of Ir-modified carbon felt as the positive electrode of an all-vanadium redox flow battery. *Electrochimica Acta* 2007; 52: 6755–6762.
93. Tan, N., Kelong, H., Suqin, L., Xiaogang, L., and Zhifeng, C. Activation mechanism study of electrochemical treated graphite felt for Vanadium redox cell by electrochemical impedance spectrum. *Acta Chimica Sinica* 2006; 64: 584–588.
94. Lee, W. H., Lee, J. G., and Reucroft, P. J. XPS study of carbon fiber surfaces treated by thermal oxidation in a gas mixture of $O_2/(O_2+N_2)$. *Applied Surface Science* 2001; 171: 136–142.
95. Pittman, J. C. U., Jiang, W., Yue, Z. R., Gardner, S., Wang, L., Toghiani, H., and Leon, C. A. Surface properties of electrochemically oxidized carbon fibers. *Carbon* 1999; 37: 1797–1807.
96. Yue, Z. R., Jiang, W., Wang, L., Gardner, S. D., and Pittman, J. C. U. Surface characterization of electrochemically oxidized carbon fibers. *Carbon* 1999; 37: 1785–1796.
97. Bulska, E., Jedral, W., Kopysc, E., Ortner, H. M., and Flege, S. Secondary ion mass spectrometry for characterizing antimony, arsenic and selenium on graphite surfaces modified with noble metals and used for hydride generation atomic absorption spectrometry. *Spectrochimica Acta Part B: Atomic Spectroscopy* 2002; 57: 2017–2029.
98. Watt-Smith, M. J., Wills, R. G. A., and Walsh, F. C. Secondary batteries—Flow systems: Overview. In *Encyclopedia of Electrochemical Power Sources,* Dyer, C. K., Moseley, P. T., Ogumi, Z., Rand, D. A. J., Scrosati, B., Garche, J., Eds. Elsevier: Amsterdam, 2009, pp. 438–443.
99. Lopez-Atalaya, M., Codina, G., Perez, J. R., Vazquez, J. L., Aldaz, A., and Climent, M. A. Behaviour of the Cr(III)/Cr(II) reaction on gold/graphite electrodes. Application to redox flow storage cell. *Journal of Power Sources* 1991; 35: 225–234.

100. Yang, C. Y. Catalytic electrodes for the redox flow cell energy storage device. *Journal of Applied Electrochemistry* 1982; 12: 425–434.
101. Cheng, D. S., and Hollax, E. The influence of thallium on the redox reaction Cr^{3+}/Cr^{2+}. *Journal of the Electrochemical Society* 1985; 132: 269–273.
102. Wu, C. D., Scherson, D. A., Calvo, E. J., Yeager, E. B., and Reid, M. A. A bismuth-based electrocatalyst for the chromous-chromic couple in acid electrolytes. *Journal of the Electrochemical Society* 1986; 133: 2109–2112.
103. Yang, Z., Zhang, J., Kintner-Meyer, M. C. W., Lu, X., Choi, D., Lemmon, J. P., and Liu, J. Electrochemical energy storage for green grid. *Chemical Reviews* 2011; 111: 3577–3613.
104. Wiedemann, E., Heintz, A., and Lichtenthaler, R. N. Sorption isotherms of vanadium with H_3O+ ions in cation exchange membranes. *Journal of Membrane Science* 1998; 141: 207–213.
105. Wiedemann, E., Heintz, A., and Lichtenthaler, R. N. Transport properties of vanadium ions in cation exchange membranes: Determination of diffusion coefficients using a dialysis cell. *Journal of Membrane Science* 1998; 141: 215–221.
106. Sun, C., Chen, J., Zhang, H., Han, X., and Luo, Q. Investigations on transfer of water and vanadium ions across Nafion membrane in an operating vanadium redox flow battery. *Journal of Power Sources* 2010; 195: 890–897.
107. Sukkar, T., and Skyllas-Kazacos, M. Water transfer behaviour across cation exchange membranes in the vanadium redox battery. *Journal of Membrane Science* 2003; 222: 235–247.
108. Mohammadi, T., Chieng, S. C., and Skyllas Kazacos, M. Water transport study across commercial ion exchange membranes in the vanadium redox flow battery. *Journal of Membrane Science* 1997; 133: 151–159.
109. Mohammadi, T., and Skyllas-Kazacos, M. Evaluation of the chemical stability of some membranes in vanadium solution. *Journal of Applied Electrochemistry* 1997; 27: 153–160.
110. Chieng, S. C., Kazacos, M., and Skyllas-Kazacos, M. Preparation and evaluation of composite membrane for vanadium redox battery applications. *Journal of Power Sources* 1992; 39: 11–19.
111. Luo, X., Lu, Z., Xi, J., Wu, Z., Zhu, W., Chen, L., and Qiu, X. Influences of permeation of vanadium ions through PVDF–g–PSSA membranes on performances of vanadium redox flow batteries. *The Journal of Physical Chemistry B* 2005; 109: 20310–20314.
112. Hwang, G. J., and Ohya, H. Crosslinking of anion exchange membrane by accelerated electron radiation as a separator for the all-vanadium redox flow battery. *Journal of Membrane Science* 1997; 132: 55–61.
113. Xi, J., Wu, Z., Teng, X., Zhao, Y., Chen, L., and Qiu, X. Self-assembled polyelectrolyte multilayer modified Nafion membrane with suppressed vanadium ion crossover for vanadium redox flow batteries. *Journal of Materials Chemistry* 2008; 18: 1232–1238.
114. Zeng, J., Jiang, C., Wang, Y., Chen, J., Zhu, S., Zhao, B., and Wang, R. Studies on polypyrrole modified Nafion membrane for vanadium redox flow battery. *Electrochemistry Communications* 2008; 10: 372–375.
115. Luo, Q., Zhang, H., Chen, J., You, D., Sun, C., and Zhang, Y. Preparation and characterization of Nafion/SPEEK layered composite membrane and its application in vanadium redox flow battery. *Journal of Membrane Science* 2008; 325: 553–558.
116. Luo, Q., Zhang, H., Chen, J., Qian, P., and Zhai, Y. Modification of Nafion membrane using interfacial polymerization for vanadium redox flow battery applications. *Journal of Membrane Science* 2008; 311: 98–103.
117. Teng, X., Zhao, Y., Xi, J., Wu, Z., Qiu, X., and Chen, L. Nafion/organic silica modified TiO_2 composite membrane for vanadium redox flow battery via *in situ* sol–gel reactions. *Journal of Membrane Science* 2009; 341: 149–154.
118. Sang, S., Wu, Q., and Huang, K. Preparation of zirconium phosphate (ZrP)/Nafion1135 composite membrane and H^+/VO^{2+} transfer property investigation. *Journal of Membrane Science* 2007; 305: 118–124.
119. Mohammadi, T., and Skyllas-Kazacos, M. Preparation of sulfonated composite membrane for vanadium redox flow battery applications. *Journal of Membrane Science* 1995; 107: 35–45.
120. Mohammadi, T., and Skyllas-Kazacos, M. Characterisation of novel composite membrane for redox flow battery applications. *Journal of Membrane Science* 1995; 98: 77–87.
121. Chieng, S. C., Kazacos, M., and Skyllas-Kazacos, M. Modification of Daramic, microporous separator, for redox flow battery applications. *Journal of Membrane Science* 1992; 75: 81–91.
122. Tian, B., Yan, C. W., and Wang, F. H. Proton conducting composite membrane from Daramic/Nafion for vanadium redox flow battery. *Journal of Membrane Science* 2004; 234: 51–54.
123. Mohammadi, T., and Skyllas-Kazacos, M. Use of polyelectrolyte for incorporation of ion-exchange groups in composite membranes for vanadium redox flow battery applications. *Journal of Power Sources* 1995; 56: 91–96.
124. Qiu, J., Zhang, J., Chen, J., Peng, J., Xu, L., Zhai, M., Li, J., and Wei, G. Amphoteric ion exchange membrane synthesized by radiation-induced graft copolymerization of styrene and dimethylaminoethyl methacrylate into PVDF film for vanadium redox flow battery applications. *Journal of Membrane Science* 2009; 334: 9–15.
125. Qiu, J., Zhao, L., Zhai, M., Ni, J., Zhou, H., Peng, J., Li, J., and Wei, G. Pre-irradiation grafting of styrene and maleic anhydride onto PVDF membrane and subsequent sulfonation for application in vanadium redox batteries. *Journal of Power Sources* 2008; 177: 617–623.
126. Qiu, J., Ni, J., Zhai, M., Peng, J., Zhou, H., Li, J., and Wei, G. Radiation grafting of styrene and maleic anhydride onto PTFE membranes and sequent sulfonation for applications of vanadium redox battery. *Radiation Physics and Chemistry* 2007; 76: 1703–1707.
127. Qiu, J., Zhai, M., Chen, J., Wang, Y., Peng, J., Xu, L., Li, J., and Wei, G. Performance of vanadium redox flow battery with a novel amphoteric ion exchange membrane synthesized by two-step grafting method. *Journal of Membrane Science* 2009; 342: 215–220.
128. Jia, C., Liu, J., and Yan, C. A significantly improved membrane for vanadium redox flow battery. *Journal of Power Sources* 2010; 195: 4380–4383.
129. Jian, X. G., Yan, C., Zhang, H. M., Zhang, S. H., Liu, C., and Zhao, P. Synthesis and characterization of quaternized poly(phthalazinone ether sulfone ketone) for anion-exchange membrane. *Chinese Chemical Letters* 2007; 18: 1269–1272.
130. Hwang, G. J., and Ohya, H. Preparation of cation exchange membrane as a separator for the all-vanadium redox flow battery. *Journal of Membrane Science* 1996; 120: 55–67.
131. Chen, D., Wang, S., Xiao, M., and Meng, Y. Synthesis and characterization of novel sulfonated poly(arylene thioether) ionomers for vanadium redox flow battery applications. *Energy & Environmental Science* 2010; 3: 622–628.

132. Xing, D., Zhang, S., Yin, C., Zhang, B., and Jian, X. Effect of amination agent on the properties of quaternized poly(phthalazinone ether sulfone) anion exchange membrane for vanadium redox flow battery application. *Journal of Membrane Science* 2010; 354: 68–73.
133. Qiu, J., Li, M., Ni, J., Zhai, M., Peng, J., Xu, L., Zhou, H., Li, J., and Wei, G. Preparation of ETFE-based anion exchange membrane to reduce permeability of vanadium ions in vanadium redox battery. *Journal of Membrane Science* 2007; 297: 174–180.
134. Tian, B., Yan, C. W., and Wang, F. H. Modification and evaluation of membranes for vanadium redox battery applications. *Journal of Applied Electrochemistry* 2004; 34: 1205–1210.
135. Mohammadi, T., and Skyllas-Kazacos, M. Modification of anion-exchange membranes for vanadium redox flow battery applications. *Journal of Power Sources* 1996; 63: 179–186.
136. Qian, P., Zhang, H., Chen, J., Wen, Y., Luo, Q., Liu, Z., You, D., and Yi, B. A novel electrode–bipolar plate assembly for vanadium redox flow battery applications. *Journal of Power Sources* 2008; 175: 613–620.
137. Yazici, M. S., Krassowski, D., and Prakash, J. Flexible graphite as battery anode and current collector. *Journal of Power Sources* 2005; 141: 171–176.
138. Zhong, S., and Michael, K. Flexible, conducting plastic electrode and process for its preparation. WO Patent 9,406,164, 1994.
139. Brungs, A., Haddadi-Asl, V., and Skyllas-Kazacos, M. Preparation and evaluation of electrocatalytic oxide coatings on conductive carbon–polymer composite substrates for use as dimensionally stable anodes. *Journal of Applied Electrochemistry* 1996; 26: 1117–1123.
140. Zhong, S., and Kazacos, M. Flexible, conducting plastic electrode and process for its preparation. US Patent 5,665,212, 1997.
141. Johnstone, R. Making green energy work may depend on three unlikely heroes: An Australian engineer, a battery, and the element vanadium. *Discover* 2008, p. 25.

22 Advanced Technologies for Liquid–Redox Rechargeable Batteries

Yuezhong Meng, Yufei Wang, Min Xiao, and Shuanjin Wang

CONTENTS

22.1 INTRODUCTION

Resources and the environment are the decisive factor for the sustainable development of the human society. The demand for global energy is climbing rapidly due to population growth and continuing industrialization. The use of renewable energy resources, such as solar and wind, as an energy policy is being planned in larger numbers than ever before owing to concern over the energy crisis that is caused by the increasing international oil price and the exhaustion of the limited supply of fossil fuels.[1–5] These sources are, however, intermittent and often unpredictable. In this regard, as a possible route to solving the energy crisis, the development of large-scale energy-storage systems in combination with renewable resources has attracted much attention in recent years. The usage of stored energy is fundamental to the generation of electric power, whether in fuel stockpiles for fossil or nuclear power plants or the seasonal runoff and dammed waterways for hydroelectric power plants. The use of energy-storage technologies provides many advantages to electric power transmission systems, such as effective use of existing plant investment, flexibility in operation, and better response to price changes.[6] Storage could be used to defer or eliminate the need for high-cost investments in new or upgraded transmission and distribution facilities (wires, transformers, capacitors or capacitor banks, and substations). In addition, storage could also be a cost-effective option for utilities to improve power quality or service reliability for customers with high-value processes or critical operations.[7] Other benefits of energy-storage systems are use of off-peak power for pumping and/or charging and maximization of operations and flexibility for commercial buying or selling electricity during on-peak or off-peak periods.[6] Finally, broad technical, economic, and social factors also suggest a promising future for energy-storage technologies.

If electricity is to be stored, it must first be converted to some other form of energy. There are many strategies for energy storage using various energy-storage systems. Electricity storage can greatly increase the usefulness of noncontinuous renewable power sources, such as solar, wind, or tidal power. Energy is stored as compressed air and can be withdrawn by a combustion turbine generator in compressed air. This method is suitable for large plants; it is restricted by special terrain.[8] At present, large pumped hydrofacilities are the dominant means of electricity storage, primarily not only for daily load shifting but also for regulation control and spinning reserve applications. The addition of these facilities is very limited, however, because of the scarcity of further cost-effective and environmentally acceptable sites.[9] Thermal storage is used for time shifting in solar power towers for grid applications. It is suitable for the energy-management services because of its high power, high energy, and slow response.[10] Superconducting magnetic energy system, in which electricity is stored on a superconductor material and is discharged directly as dc power, can be used only under very low temperatures.[11] Electric double-layer capacitors, are which called supercapacitors, are used mainly to assist other power supplies in coping with surge power requirements, particularly in electric/hybrid vehicles. Both superconducting magnetic energy systems and electric double-layer capacitors are very expensive, even though the latter ones are maintenance free.[8] Flywheels are a technology that stores electricity into kinetic energy that can be taken back by an electrical generator. This is a newer technology that needs large maintenance costs. Flywheels and supercapacitors, which are low-energy, high-power storage systems used for power management, are not considered to have wide applications in transmission due to material limitations.[12] Redox batteries refer to the conversion of electrical energy into chemical energy that can be recovered by reversing the electrochemical reaction. Here battery-storage technologies include lead–acid, lithium-ion, sodium-based batteries and redox flow batteries (RFBs).

Conventional rechargeable batteries, such as lead–acid batteries, offer a simple and efficient way to store electricity, but development to date has largely focused on transportation systems and smaller systems for portable power or intermittent backup power; metrics relating to size and volume are far less critical for grid storage than in portable or transportation applications.[13] The disadvantages of lead–acid battery restrict its development and application, including low specific energy (watt-hours per kilogram) and specific power (watts per kilogram), short cycle life, high maintenance requirements, and environmental hazards associated with lead and sulfuric acid. RFBs for large-scale grid storage have convenient operating temperature, large capacity as well as long cycle life, high round-trip efficiency, rapid responding ability to changes in load or input, and reasonable capital costs advantages over other batteries.[14]

Ever since Thaller proposed the concept of the RFB, researchers have been exploring possible redox couples (V/V, S/Br_2, Zn/Br_2, V/Br_2, Fe/Cr, Ce/Zn, and Pb/Pb). RFBs can be classified into three groups according to the phases of the electroactive species. In the first group both redox couples have reactants/products dissolved in the liquid electrolytes. All of these batteries require ion-exchange membranes (IEMs). In the second group both redox couples involve solid species during the charge process, and chemical energy is stored in an active material on the electrode plates. In most of the systems proposed, a single electrolyte is circulated through the cell and no IEM is required. Hybrid RFBs have redox couples that involve solid species or gaseous species in one half cell during the charge process. Vanadium redox flow batteries (VRBs) and polysulfide–bromine flow battery, which belong to the first type, are all mature technologies. In this review, we give more attention on the first type.

22.2 FABRICATION/MANUFACTURING OF LIQUID–REDOX RECHARGEABLE BATTERIES

RFBs, which provide an alternative solution to the problems of balancing power generation and consumption, load leveling, and facilitating renewable energy deployment, convert and store electrical energy into chemical energy and release it in a controlled fashion when required. RFBs show greater promise for economical storage of electrical energy where the energy is stored by chemical changes to species dissolved in a working fluid, rather than being stored at the electrodes, such as in batteries. The key transport mechanisms are shown in Figure 22.1 for this generic system.[13] RFB storage systems use soluble redox couples as electroactive species that are oxidized or reduced to store or deliver energy. In the divided mode, two sides of the cell are typically separated with an IEM to control mixing of the two solutions. Electrolytes stored in two tanks are pumped and circulated through the stack where the electrochemical reaction occurs. In the charge and discharge processes, the following reactions happen:

$$A^{n+} + xe^- \xrightarrow{charge} A^{(n-x)+} \text{ and } A^{(n-x)+} \xrightarrow{discharge} A^{n+} + xe^-, \quad (22.1)$$

$$B^{m+} - ye^- \xrightarrow{charge} B^{(m+y)+} \text{ and } B^{(m+y)+} \xrightarrow{discharge} B^{m+} - ye^-, \quad (22.2)$$

for the anode (negative electrode) and cathode (positive electrode), respectively.

The system power is determined by the rate of reaction of the redox species at each electrode and the total surface area of the electrodes. The concentration of the redox species and the volume of the reservoirs determine the amount of energy stored in the system. By increasing the concentration of the electroactive species and/or the volume of the electrolytes, the energy-storage capacity of an RFB will increase, which is a great advantage over other energy-storage battery systems.

The arrangement of a typical cell stack is shown in Figure 22.2.[15] The assembly of a cell stack is a straightforward mechanical process. Electrodes, membranes, and bipolar plates are manufactured to form a stack. The RFB single cell is fabricated by sandwiching a membrane between two pieces of carbon felt electrodes and then clamped by two graphite polar plates that are carved with serpentine flow field. All these components are fixed between two stainless plates. The cell size can be determined from the desired power rate, but in practice, cell size is driven by many other factors, such as mechanical strength, thermal stability, the dimension of the components, and the shunt current.[14] The reactor in Figure 22.1 consists of a stack of an individual cell, where each cell contains the sites where electrochemical charge-transfer reactions occur as electrolyte flows through them, as well as a separator (either an electrolyte-filled gap or a selective membrane) to force the electrons through the external circuit. Frame gasket seals are normally used to compress the cell stack in order to avoid electrolyte leakages. Since the electroactive species used in RFBs are often highly oxidizing, no metallic component should be in contact with the electrolytes. Chemically resistant polymers, such as polytetrafluoroethylene (PTFE), are the typical materials for producing the battery components.[16] Practical applications tend to require high currents and voltages. To meet this,

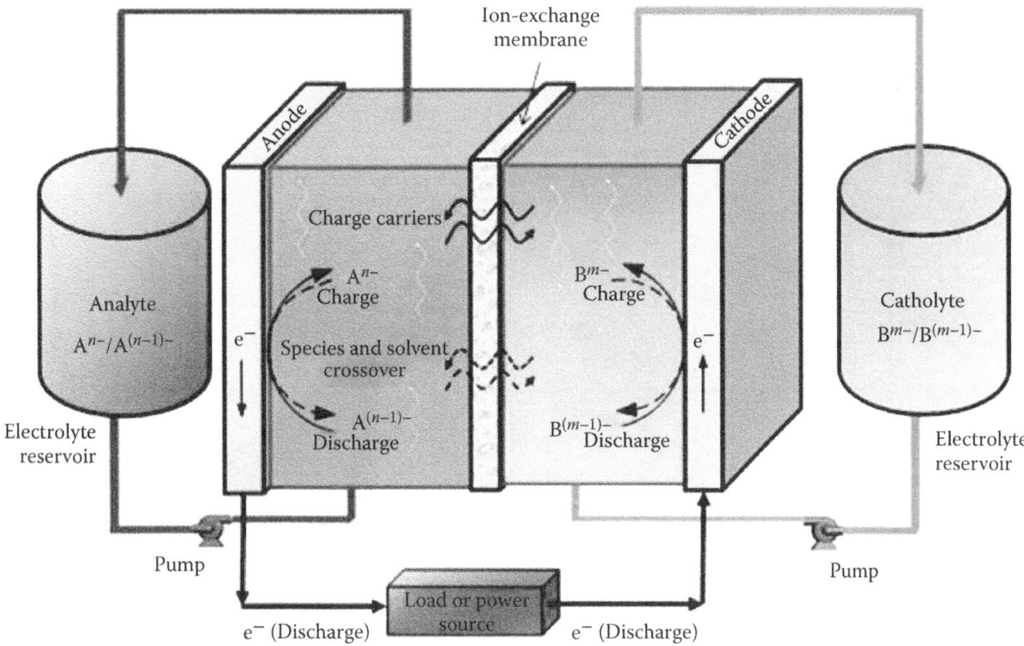

FIGURE 22.1 A schematic diagram of an RFB with electron transport in the circuit, ion transport in the electrolyte and across the membrane, active species crossover, and mass transport in the electrolyte. (With kind permission from Springer Science+Business Media: *Journal of Applied Electrochemistry*, Redox flow batteries: A review, 41, 2011, 1137–1164, Weber, A. Z., M. M. Mench, J. P. Meyers, P. N. Ross, J. T. Gostick, and Q. Liu.)

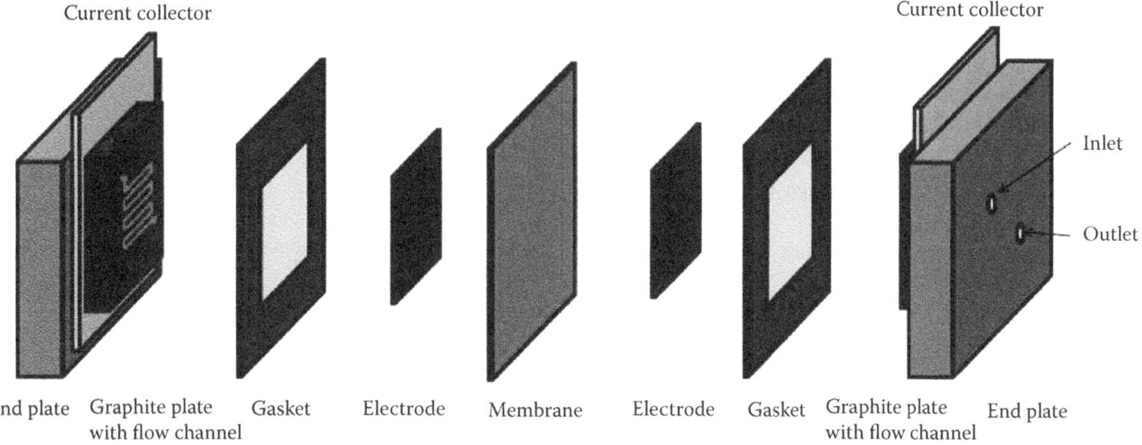

FIGURE 22.2 Schematic of VRB test setup in the laboratory. (Reprinted from *Journal of Power Sources*, 206, Aaron, D., Q. Liu, Z. Tang, G. Grim, A. Papandrew, A. Turhan, T. Zawodzinski, and M. Mench, Dramatic performance gains in vanadium redox flow batteries through modified cell architecture, 450–453, Copyright (2012), with permission from Elsevier.)

a number of unit cells in series in a bipolar configuration can be stacked to increase the overall stack voltage and the stacks can be electrically connected in parallel to yield high currents (Figure 22.3).[8] In real applications, the batteries are assembled from several single cells to form a battery stack. As presented in Figure 22.4, Zhao et al. reported that a 10 kW VRB stack system was constructed with the configuration 4 × 2 (serial × parallel) of the improved aforementioned kilowatt class stack modules.[6] And later they have described their 20 kW stack module that has been shown to operate at 80 mA cm^{-2} with an overall EE of 80%. These stack modules have been incorporated into a 260 kW subsystem (Figure 22.5).[17]

22.3 TESTING AND DIAGNOSIS OF LIQUID–REDOX RECHARGEABLE BATTERIES' PERFORMANCE

In this section, we introduce the parameters of RFB performance. Redox flow cell storage systems use two soluble redox couples as electroactive species that are oxidized or reduced to store or deliver energy. Redox energy-storage systems possess features such as flexible design, long life, and high reliability with acceptable operation and maintenance costs. One of the most important features of these batteries is that the power and energy capacity of the system can be separated.[18]

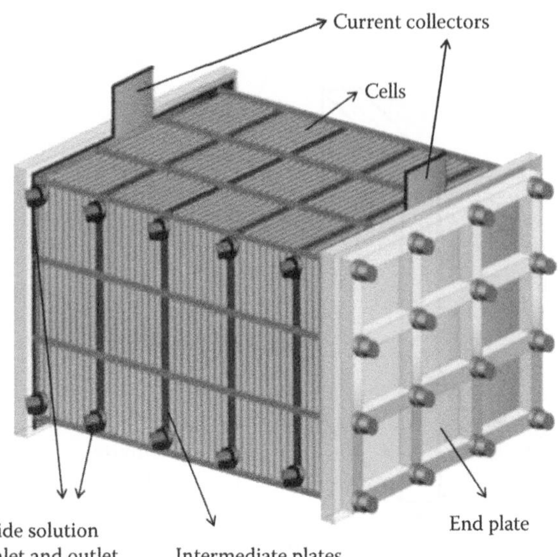

FIGURE 22.3 RFB stacks with side fluid feeding series of about 100 cells with active areas as large as 0.4 × 0.4 m^2 are usual. (Reprinted from *Renewable and Sustainable Energy Reviews*, 29, Alotto, P., M. Guarnieri, and F. Moro, Redox flow batteries for the storage of renewable energy: A review, 325–335, Copyright (2014), with permission from Elsevier.)

In an RFB, the power of the system is controlled by the size of the electrolyte reservoirs and the number of cells in the stack. The energy-storage capacity is determined by the concentration and volume of the electrolyte. The power and energy capacity of RFBs can be easily varied; hence, the flexibility of the energy storage is enhanced.

The following several main parameters can represent performance of RFBs.[19–22] The cell voltage efficiency (VE) is determined by the following equation:

$$VE = \frac{V_{discharge}}{V_{charge}} \times 100\%, \tag{22.3}$$

where $V_{discharge}$ and V_{charge} are the discharge and the charge of the cell voltages, respectively, at a certain time or state of charge during the operation of the cell.

FIGURE 22.5 Zhao et al.[6] developed 20 kW VRB stack module. (Reproduced from China Greentech Initiative, *The China Greentech Initiative Report*, http://www.china-greentech.com, last accessed: Dec. 2010. With permission.)

The coulombic efficiency (CE) is the ratio of a cell's discharge capacity ($C_{discharge}$) to the charge capacity (C_{charge}), described as the following equation:

$$CE = \frac{C_{discharge}}{C_{charge}} \times 100\%. \tag{22.4}$$

FIGURE 22.4 Photographs of 10 kW class VRB cell stack composed of eight 1 kW stack modules with the configuration 4 × 2 (serial × parallel). (Reprinted from *Journal of Power Sources*, 162, Zhao, P., H. Zhang, H. Zhou, J. Chen, S. Gao, and B. Yi, Characteristics and performance of 10 kW class all-vanadium redox-flow battery stack, 1416–1420, Copyright (2006), with permission from Elsevier.)

When the discharge current density ($I_{discharge}$) is the same as the charge current density (I_{charge}), CE can be presented as the discharge time ($t_{discharge}$) divided by charge time (t_{charge}):

$$CE = \frac{I_{discharge} t_{discharge}}{I_{charge} t_{charge}} \times 100\% = \frac{t_{discharge}}{t_{charge}} \times 100\%. \quad (22.5)$$

The energy efficiency (EE) and power efficiency (PE) of the cell are calculated by the following equations:

$$EE = \frac{E_{discharge}}{E_{charge}} \times 100\%, \quad (22.6)$$

$$PE = \frac{I_{discharge} V_{discharge}}{I_{charge} V_{charge}} \times 100\%. \quad (22.7)$$

EE is the ratio of energy between the discharge and charge processes; PE is the ratio of power between the discharge and charge processes.

It is important to refer these figures of merit to electrolyte volume, reactant conversion, and state of charge as well as to consider practical design and operational factors.

In a typical redox flow cell, positive and negative electrodes are separated by an IEM. The membrane allows ion transport and helps prevent mixing of the two half-cell electrolytes, which are stored in separate tanks and circulated through the battery by using pumps. Chemical energy is stored in the electrolyte and released by transformation or dissolution of the electrodeposits during discharge. The power is determined by the numbers of cells in the stack and by the number and size of the electrodes. The energy-storage capacity can be increased by simply using a larger volume of electrolytes or high concentration of electrolytes. All of these batteries require IEMs or separators to divide the anode and cathode compartments while allowing ion transport to maintain the electrical neutrality in the cell. The operational factors that primarily determine the RFB performance is the redox-active species, solvent, supporting electrolyte, and separator material. The VE, the CE, the EE, the PE, and the cycling performance of the RFBs are used to investigate whether the condition is suitable for the cell. VRBs that are typical examples of this type have been shown (see Figures 22.6 through 22.9).

Figure 22.6 illustrates the charge–discharge curves of the second cycle of VRBs with an S/T/P membrane and a Nafion® 212 (N212) membrane, respectively.[23] The VRB with N212 membrane demonstrated higher discharge voltage, while its capacity of discharge, CE, and EE are lower. That is due to the lower vanadium ion permeability of the S/T/P membrane, which can reduce the self-discharge of the VRB single cell. As shown in Figure 22.7, Qiu et al. gave the relationships of the CE, VE, and EE with charge–discharge current density. In all current densities, the VRB with Nafion–[poly(diallyldimethylammonium chloride) (PDDA)–poly(sodium styrene sulfonate) (PSS)]$_5$ membrane shows

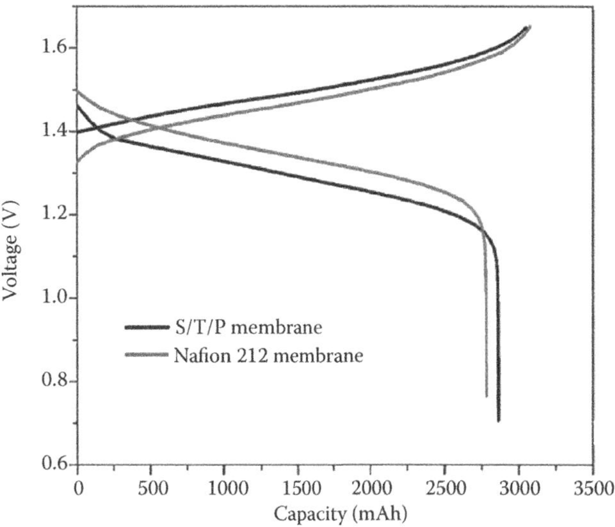

FIGURE 22.6 Charge–discharge curves of VRB single cells with an N212 membrane and an S/T/P (S: sulfonated poly(ether ether ketone) [SPEEK]; T: tungstophosphoric acid [TPA]; P: polypropylene [PP]) membrane at current density of 35.7 mA cm^{-2}. (Reprinted from *Journal of Power Sources*, 195, Jia, C., J. Liu, and C. Yan, A significantly improved membrane for vanadium redox flow battery, 4380–4383, Copyright (2010), with permission from Elsevier.)

CE, VE, and EE higher than those of the VRB with Nafion membrane. The remarkable improvement of the cell efficiency can also be attributed to the excellent performance of Nafion-[PDDA-PSS]$_5$ membrane to suppress the crossover of vanadium ion.[24] The different cyclic properties of VRBs with SFPAE-1.8 and N212 membranes are shown in Figure 22.8a. The VRB with SFPAE membrane demonstrated low capacity fade during cycling. The stability measurements are in agreement with the VRB cyclic tests shown in Figure 22.8b.[25] Even after 75 cycles, the CE, the VE, and the EE of the VRB with SFPAE membrane did not decrease. As demonstrated in Figure 22.9, a self-discharge test was conducted to investigate the transfer of vanadium ions across $N_{0.8}P_{0.2}$ and r-Nafion under "online" conditions.[26] Obviously, the capacity of the cell with $N_{0.8}P_{0.2}$ decreases much more slowly than that with r-Nafion. This result reveals that $N_{0.8}P_{0.2}$ significantly lowers the permeation rate of vanadium ions. The membrane must reduce the transport of reactive species between the anode and cathode compartments to a minimum rate and allow the transport of nonreactive species and water to maintain electroneutrality and electrolyte balance. The electrolytes must be chemically stable and easy to prepare at high concentrations.

To sum up, the RFBs' performance is valued as VE, CE, EE, and cycling performance. The composition and structure of the electrodes, as well as the reactive species' composition and concentration, has a great influence on the performance of the RFBs. In addition, the ion conductivity, the permeability, and the oxidation stability of an IEM affect the performance of RFBs. The membrane should have low electric resistivity, long life span, easy manufacture, and handling and moderate cost.

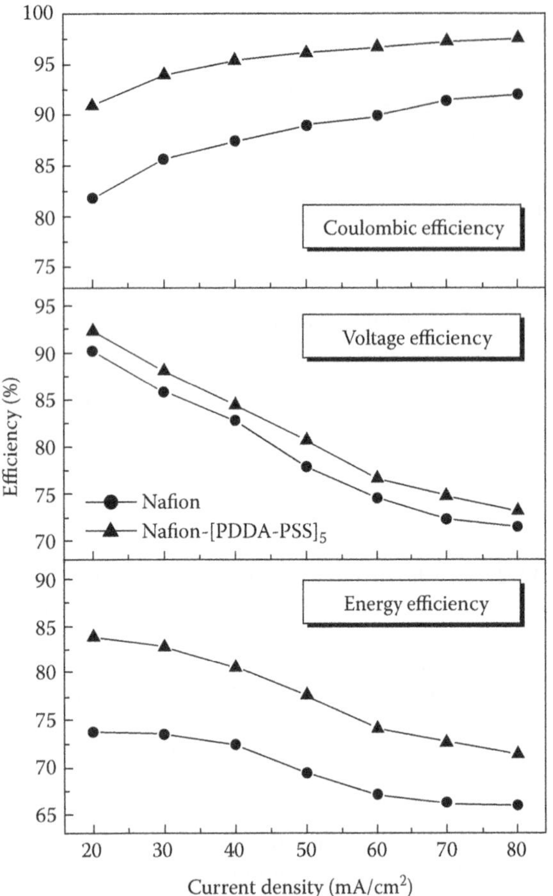

FIGURE 22.7 Effect of charge–discharge current density on CE, VE, and EE of the VRBs with Nafion and Nafion-[PDDA-PSS]$_5$ membranes. (Xi, J., Z. Wu, X. Teng, Y. Zhao, L. Chen, and X. Qiu, Self-assembled polyelectrolyte multilayer modified Nafion membrane with suppressed vanadium ion crossover for vanadium redox flow batteries, *Journal of Materials Chemistry*, 18 (11), 1232–1238, 2008. Reproduced by permission of The Royal Society of Chemistry.)

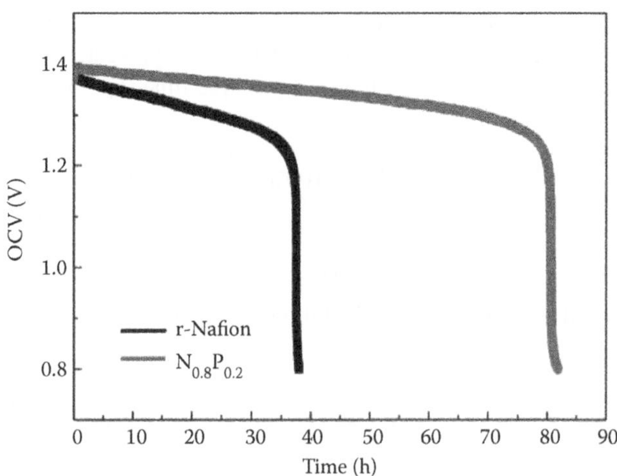

FIGURE 22.9 VRB self-discharge curves. (Reprinted from *Journal of Power Sources*, 196, Mai, Z., H. Zhang, X. Li, S. Xiao, and H. Zhang, Nafion/polyvinylidene fluoride blend membranes with improved ion selectivity for vanadium redox flow battery application, 5737–5741, Copyright (2011), with permission from Elsevier.)

22.4 DEGRADATION MECHANISMS AND MITIGATION STRATEGIES

As shown previously, a type of energy-storage device capable of providing reversible conversion between electrical and chemical energy, typically in two soluble redox couples contained in external electrolyte tanks sized in accordance with application requirements, is demonstrated in Figure 22.1. Flow batteries store energy in liquid electrolytes that contain reversible redox couples and have desirable attributes of long life, active thermal management, and independent energy and power ratings depending on their configuration.[27] Particularly the all-vanadium RFB has received much attention because of its good electrochemical reversibility, high efficiency, and absence of cross contamination between the anolyte and catholyte.[28]

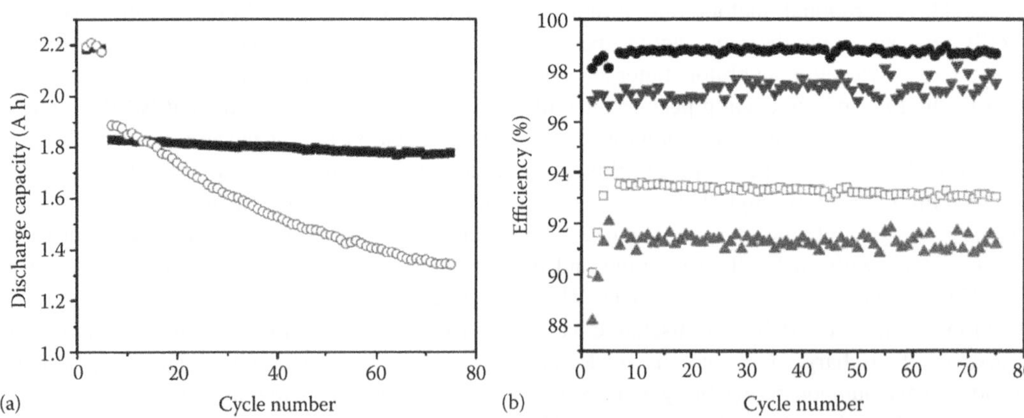

FIGURE 22.8 Cyclic performance comparison of VRBs with N212 and SFPAE-1.8 membranes at 50 mA cm^{-2}: (a) capacity as a function of cycle number (○: N212; ●: SFPAE-1.8) and (b) CE (▲: N212; □: SFPAE-1.8) and EE (▼: N212; ●: SFPAE-1.8) with cycle number. (Chen, D., S. Kim, L. Li, G. Yang, and M. A. Hickner, Stable fluorinated sulfonated poly(arylene ether) membranes for vanadium redox flow batteries, *RSC Advances*, 2 (21), 8087–8094, 2012. Reproduced by permission of The Royal Society of Chemistry.)

As we know, the fundamental performance characteristics of an RFB cell are determined primarily by the redox-active species, solvent, supporting electrolyte, and separator material. The RFB cell voltage is determined by the redox-active species. In combination with the solvent, the active species also dictate the electrochemical stability and charge capacity of the fluids.[29] Separator materials separates the positive solution (anolyte) and the negative solution (catholyte), preventing self-discharge while allowing proton transfer to complete the circuit. Because ions from the supporting electrolyte move across the separator to maintain an overall charge balance during cell operation, the separator and support also affect the PE and EE of the battery.[29] Additionally, a key factor in many of these systems is crossover of species through the separator, which is dependent on current and membrane permeability. The original RFB configurations used different active species in the anolyte and catholyte. According to the types of active species and electrochemical potential involved, a number of RFBs, including iron/chromium, iron/titanium, all-vanadium, Zn/Br_2 flow batteries, polysulfide/Br_2, and zinc/bromine hybrid system flow batteries, have been successively proposed and developed.[29–33] Following long-term operation, most RFBs require periodic electrolyte reactivation, which suffers from active-species crossover like what happens in polymer electrolyte fuel cells. Periodic rebalancing of the electrolyte solutions is required by mixing of the anolyte and catholyte solutions to recover the capacity, which increases the operational cost of the device. Dual active-species chemistries may be problematic because they can degrade irreversibly when constituents of the anolyte and catholyte mix. Among them, the all-vanadium RFB uses the same element in both half cells, which avoids cross contamination of the two half-cell electrolytes and therefore becomes the most promising technology for large-scale energy storage. The VRB suggested by Pelligri and Spaziante[34] and later realized by Skyllas-Kazacos et al.,[35] Sum et al.,[36] and Sum Skyllas-Kazacos[37] employs the redox couples VO_2^+/VO^{2+} and V^{2+}/V^{3+} in aqueous sulfuric acid solution as positive and negative half cells, respectively. The two half cells, with the reactions occurring on carbon cloth or carbon felt electrodes, are separated by an IEM. The demands for the membrane are stringent: low transport of vanadium ions, low ionic resistance, low cost, and good longevity in the presence of the electrolyte solutions. A high proton conductivity and a low vanadium ion crossover indicate high electrochemical selectivity, which can increase the VRB's CE, leading to high overall efficiency.[27] The carbon electrodes show no or very little corrosion during cycling, and the anolyte and catholyte solutions do not degrade during the lifetime of the VRB if they are kept oxygen free.[38] Thus, in single-metal RFBs, crossover causes only self-discharge, a reversible process that can occur without degrading the electrolyte materials. Crossover in single-metal RFBs has been observed to cause as much as 10% self-discharge over 72 h.[39] During discharge, electrochemical reduction and oxidation of vanadium ions occur, as described in reactions I and II below:

$$\text{I: } VO_2^+ + e \rightarrow VO^{2+}$$
$$\text{II: } V^{2+} \rightarrow V^{3+} + e$$
$$\text{III: } VO_2^+ + V^{3+} \rightarrow 2VO^{2+}$$
$$\text{IV: } 2VO_2^+ + V^{2+} \rightarrow 3VO^{2+}$$
$$\text{V: } VO^{2+} + V^{2+} \rightarrow 2V^{3+}$$

Once vanadium ions cross mix, the overall cell reaction will happen (reactions III through V) and further reduce the quantity of the active species for reactions I and II.[40,41]

At any time, the reactants entering the cell are at lower concentration than they would be if mixing has occurred, causing a gradual drop in the cell potential. On the other hand, the water transferred across the membrane by osmosis or electroosmosis changes the concentration of ionic species during the operation of the battery. Therefore, the electrolyte must be treated by a suitable method such as reverse osmosis, water evaporation, or electrodialysis to remove unwanted formed species and to maintain the redox couple concentrated and pure.[13]

In a word, the cell degradation was caused by the cross mixing of the ions, which can be controlled by the membrane performance. The performance of the membrane was measured by ion selectivity and oxidative stability. Thus, in order to ease the degradation rate, it is a good method to improve the membrane performance. As the key component of a VRB, the biggest challenge for IEMs is the balance between oxidation stability and ion selectivity under VRB operating conditions. In practice, the most often used membrane is still Nafion, which is extremely expensive and the ion selectivity of which still really needs to be further improved.[42,43] Modified Nafion membranes could be a good choice to improve selectivity, but their cost will remain a problem for further large-scale applications.[38,44] Pore-filled composite membranes could be very good candidates to replace Nafion in VRBs, because of their tunable chemical stability and ion conductivity.[44] To decrease the cost of IEMs, nonfluorinated polymer membranes could be the best choices.[19,21,45] For the cation-exchange membranes (CEMs) and pore-filled membranes, another important issue besides increasing their chemical stability is how to increase their ion conductivity under VRB conditions. For example, for the CEMs based on sulfonated aromatic polymers, the proton conductivity can be tuned by adjusting the degree of sulfonation (DS). However, with increasing DS, the chemical stability of the membranes decreases. To find the balance between DS and chemical stability is a very important issue for this kind of membranes. Because of concerns about the balance between chemical stability and ion conductivity, the detailed morphology of VRB membranes needs to be further investigated. The morphology of IEMs plays a crucial role in determining the final performance of membranes. Normally two regions coexist in IEMs, a hydrophilic and a hydrophobic one, which are, respectively, in charge of the membrane conductivity and the mechanical stability. The distribution of these two regions will greatly affect membrane properties such as ion selectivity, ion conductivity, and mechanical and chemical stability. The full understanding of the relationship between membrane morphology and membrane properties can offer very important information to guide the most appropriate research design. And still, to meet the requirements of a VRB, one of the biggest challenges is to decrease the vanadium ion permeability of CEMs with a high ion-exchange capacity (IEC) and high chemical stability. An anion-exchange membrane (AEM) separator for the VRB mitigates vanadium ion crossover, which is responsible for coulombic losses and self-discharge of the battery.[46] As believed by many researchers, a low vanadium permeability and a high ion conductivity are not likely to coexist in cation-selective membranes, between which a compromise always needs to be made. Therefore, the anion-selective membrane, which has the potential to couple both features, might be a good candidate for VRB applications.[47–49] The amphoteric ion-exchange membrane (AIEM), with both cation- and anion-exchange capabilities, exhibits lower vanadium ion permeability, as well as higher CE and EE, and maintains a rather high conductivity.[50–54] This part will be explained in detail later.

Recently, the research on the applications of IEM in VRBs has just started. And more effort need to be dedicated to design and prepare new polymers with outstanding ion selectivity with chemical stability.[28] The topic requiring future study as the systems with the greatest potential become defined is performance degradation. To enable more complete studies in these areas, a new class of RFB diagnostics will also be needed. Recently, many more studied modes of material degradation will likely be associated with transport processes that can be better optimized to promote longevity. Finally, throughout this review not much mention has been made concerning other components within the RFB system. In particular, the typical solvents and chemistries are inherently highly corrosive due to their high ionic and perhaps protonic concentrations. It is necessary for RFB systems to gain entrance to the market to find solutions to decrease the cell performance degradation.

22.5 ADVANCED DESIGNS FOR LIQUID–REDOX RECHARGEABLE BATTERIES

22.5.1 Recent Developments in Electrode Materials

Electrodes play an important role in RFBs, providing the reaction place of charge transfer of redox couples. An ideal electrode material should provide a high surface area, low electronic resistance, suitable porosity, good mechanical properties, and strong chemical resistance; be of reasonable price; and have a high electrochemical activity toward the reactions between redox species for optimized performance.[27] Typical electrodes used in RFBs are divided into carbon-based composites and inert metallic materials. Although precious metals, such as platinum and gold, have excellent chemical stability and electrical conductivity, their cost renders them impractical for routine large-scale energy storage.[16] Metallic electrodes are less commonly used in RFBs because of corrosion, dissolution, and weight issues. When metallic electrodes dissolve into the electrolyte during discharge and corrosion, it can lead to unstable redox potentials and disturb the chemistry of the RFBs. Moreover, precious metals do not necessarily show good electrochemical behavior.

Carbon-based materials, because of their wide operation potential range, high chemical stability, and reasonable cost, are widely used in RFBs; these materials include graphite felt,[55] carbon cloth,[56] and carbon fiber.[57] In recent years, many researchers pay more attention on the application of carbon-based nanomaterials, such as carbon nanotube,[58] graphite oxide,[59,60] and graphene oxide (GO) nanosheets, as the VRB electrode.[61,62] Han et al. treated GO nanoplatelets at various temperatures and demonstrated excellent electrochemical activity in a VRB.[63] Later they introduced multiwalled carbon nanotubes into GO nanosheets to form an electrocatalytic hybrid with a mixed conducting network. The hybrid not only provides rapid ion transport channel but also effectively transfers the electron produced by the redox reaction of the VO^{2+}/VO_2^+ couple.[61]

However, these common carbon-based electrodes often exhibit inadequate electrochemical activity and kinetic reversibility toward the electrochemical reactions between the vanadium species. In addition, slow oxidation due to of oxygen evolution during cell overcharge was observed to cause a slow disintegration of the carbon and graphite electrode surfaces. Therefore, much attention has been paid to the modification of the electrode materials to address these inherent problems.[64] A variety of surface treatment modification methods have been reported, such as heat treatment,[65] electrochemical oxidation,[66] nitrogenization treatment,[67] and introduction of an electrocatalyst.[57] Shao et al. prepared nitrogen-doped mesoporous carbon materials (N-MPCs) by heat-treating mesoporous carbon in NH_3.[67] On the N-MPC electrode the electrocatalytic kinetics and the reversibility of the redox couple VO^{2+}/VO_2^+ were significantly enhanced compared with that for MPC and graphite electrodes. Shao et al. thought nitrogen doping could facilitate the oxidation and reduction processes.[67] Sun and Skyllas-Kazakos reported chemical modification on graphite fiber by impregnating it with a solution containing Pt^{4+}, Pd^{2+}, Au^{4+}, Ir^{3+}, etc.[57] The cyclic voltammetry results showed that the electrode modified by Ir^{3+} exhibited the best electrochemical behavior for the various vanadium redox species.[57] However, these methods are not advantageous to commercial application for the use of noble materials, dangerous concentrated acid, or tedious treatment time.

22.5.2 Recent Developments of Novel Membrane Materials

The IEM as the key part of a VRB is responsible for transferring ions from the anode to the cathode as well as for preventing cross mixing of the positive and negative electrolytes. The properties of the IEM greatly affect the resulting performance of the RFBs. Therefore, an ideal IEM should have high ion conductivity, low permeability of active species, high selectivity, good chemical stability, and low cost.[64,68]

The high ion conductivity is required to minimize the losses in VE. The membrane must be permeable to the charge balancing ions (e.g., protons and sulfate ions) of the supporting electrolyte to complete the current circuit, but must prevent diffusion of active species in solution, which would result in self-discharge of the battery.[68,69] The ideal membrane, thus, should exhibit low permeation rates of the active species to minimize self-discharge and allow high CEs. Since some of the RFBs membranes are operated in strong acid medium in the presence of strong oxidizing solutions, excellent chemical stability is needed to ensure the battery's cycle life. The following gives the equals to characterize the membrane performance.[28]

The conductivity was calculated from the impedance plot with a computer curve-fihe co technique, the electrode area of the cell, and the thickness of the membrane, which was measured by a micrometer. The proton conductivity (σ) (in siemens per centimeter) was calculated according the following equation:

$$\sigma(\mathrm{s \cdot cm^{-1}}) = \frac{d}{RS}, \tag{22.8}$$

where S and d are the face area and thickness of the membrane, respectively, and R is the impedance. The area resistance of a membrane may indicate the internal resistance of the battery, which will finally determine the VE.

The membrane chemical stability was evaluated by immersing the membrane in 1 M VO_2^+ solution for 15 days at 25°C to measure the changes in performance of VRB single cell. However, it needs a very long time to assess. In VRB, the oxidative stability of the membrane was tested in Fenton's reagent (3 wt.% H_2O_2 + 2 parts per million $FeSO_4$) at 80°C. The membrane oxidative stability was determined by the following equation:

$$\text{weight loss}(\%) = \frac{W_0 - W_1}{W_0} \times 100\%, \tag{22.9}$$

where W_0 and W_1 are the weights of the dry membranes before and after immersed in Fenton's reagent for 1 h, respectively.

The permeability of VO^{2+} was measured at 25°C according to the method of Mohammadi and Skyllas-Kazacos,[70] as shown in Figure 22.10.[22] Forty milliliters of 1 M $VOSO_4$ in 2 M H_2SO_4 solution and 40 mL of 1 M $MgSO_4$ in 2 M H_2SO_4 solution were separately used to fill each of two reservoirs that are separated by the membrane. $MgSO_4$ was used to equalize the osmotic pressure. The membrane area exposed to electrolytes was 5.3 cm². The two solutions were continuously magnetically stirred and the $MgSO_4$ solution was sampled at

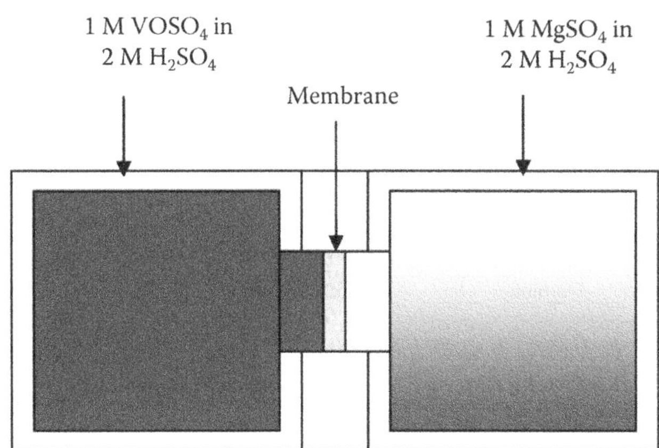

FIGURE 22.10 Equipment for vanadium permeability evaluation. (Reprinted from *Journal of Membrane Science*, 297, Qiu, J., M. Li, J. Ni, M. Zhai, J. Peng, L. Xu, H. Zhou, J. Li, and G. Wei, Preparation of ETFE-based anion exchange membrane to reduce permeability of vanadium ions in vanadium redox battery, 174–180, Copyright (2007), with permission from Elsevier.)

regular time intervals for analysis. The concentration of VO^{2+} was measured by ultraviolet–visible light spectrophotometer at 25°C. The VO^{2+} permeability in square centimeters per minute was determined by Fick's diffusion law as the following equation:

$$P(\mathrm{cm^2 \cdot min^{-1}}) = \frac{V_B d}{A(C_A - C_B(t))} \cdot \frac{dC_B(t)}{dt}. \tag{22.10}$$

where P is the permeability of VO^{2+}; V_B is the volume of $MgSO_4$ solution; d and A are the thickness and effective area of the membrane, respectively; C_A is the VO^{2+} concentration of $VOSO_4$ solution; and $C_B(t)$ is the VO^{2+} concentration of $MgSO_4$ solution as a function of time.

The selectivity (S) in siemens-minute per cubic centimeter is defined as the ratio of proton conductivity to vanadium ion permeability:

$$S(\mathrm{s \cdot min \cdot cm^{-3}}) = \frac{\sigma}{P}, \tag{22.11}$$

where σ is the ion conductivity and P is the permeability of VO^{2+}.

22.5.2.1 Cation-Exchange Membrane

Since a polymer material that is easily processed and modified has good mechanical properties, it is widely used for CEM substrate materials. By introducing acidity groups, such as sulfonic acid, phosphoric acid, and carboxylic acid, into the main chain or side chain of the polymer, the membrane made by this polymer has certain proton conductivity.

Original work done by Vafiadis and Skyllas-Kazacos established in 1984.[48] Evaluation of a number of commercially IEMs for VRB has shown that except for a few that

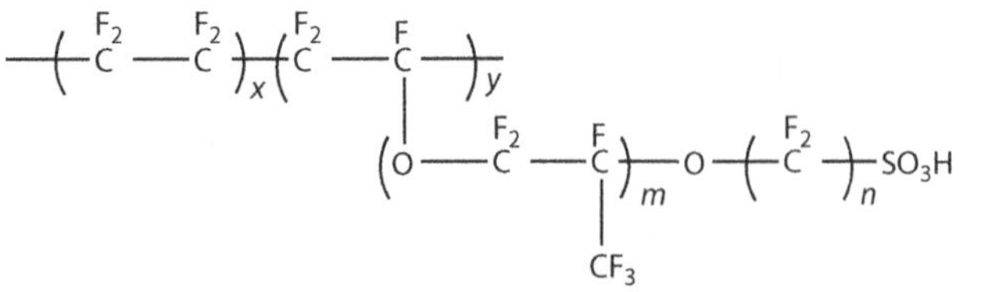

Nafion 117	$m \geq 1; n = 2; x = 5 - 13.5; y = 1000$
Flemion	$m = 0, 1; n = 1 - 5$
Aciplex	$m = 0, 3; n = 2 - 5; x = 1.5 - 14$
Dow membrane	$m = 0; n = 2; x = 3.6 - 10$

FIGURE 22.11 The proposed structure of Nafion.

are currently too expensive for commercial use, most of these IEMS are not suitable because of their poor chemical stability in the vanadium solutions; these include sulfonated polyethylene membrane, sulfonated Daramic (a microporous separator composed of polyethylene), and Selemion CMV (a sulfonated polystyrene ethylene butylene block copolymer).[71,72] All of them demonstrated instability by weight loss when immersed in VO_2^+ solutions and deteriorated when implemented in a VRB. That is to say, these sulfonated aliphatic backbone–based membranes are not stable in strongly oxidizing VO_2^+ solutions. Because of its low cost and reasonable chemical stability in VO_2^+ solutions, Daramic membrane was used in VRB. But very low CE is obtained for its high vanadium ion permeability.[73] Chieng et al.[71] and Mohammadi and Skyllas-Kazacos[72] first introduced Amberlite CG-400 into Daramic porous supports and then cross-linked it with divinylbenzene. Compared with only 40% for untreated Darmamic, VRB performance with the composite membranes showed an EE of 75% at a current density of 40 mA cm^{-2}.[71,72] In order to improve the ion conductivity and ion selectivity of pore-filled membranes, PTFE-reinforced pore-filled membranes can be a good choice.

The traditional membranes used in VRB are perfluorinated sulfonated membranes, such as Nafion, Flemion®, and Aciplex® membranes. These perfluorinated polymer electrolytes are currently the most commercially utilized electrolyte membranes for RFBs. The hydrophobic Teflon backbone provides the membranes with excellent mechanical and chemical stability, while the hydrophilic zone, originating from the assembled sulfonated groups, ensures ion conductivity.[28]

Nafion is currently the most used membrane in VRBs, showing a quite good electrochemical property and chemical stability. Figure 22.11 shows the proposed structure of Nafion. Nafion 115 shows a CE of 94% and an EE of 84% at the current density of 80 mA cm^{-2}, and Nafion 112, which is around 2.5 times thinner, shows a CE of 91% under the same conditions. Nafion membranes exhibit high proton conductivity and chemical and mechanical stabilities. However, they have limited commercial use because of their significant high cost as well as high vanadium ion permeability. Thus, there has been extensive research on the IEM. Different types of polymers, thus, have been developed and studied for VRB applications.

In recent years, considerable efforts have been devoted to the development of suitable and stable membranes for VRB application. Sulfonated aromatic polymers are well known for their excellent thermal and chemical stabilities, and the hydrophilic sulfonic acid groups in these polymers provide the membranes with high proton conductivity. Sulfonated aromatic membranes have been used in VRB systems as a candidate for Nafion membrane. Chen et al. reported sulfonated poly(arylene thioether ketone), sulfonated poly(fluorenyl ether ketone), and poly(arylene ether sulfone)-based membranes via direct synthesis of sulfonated monomer.[19,20,74] These types of membranes showed much lower VO^{2+} permeability and higher CE than those of Nafion. In Mai et al.'s work, a series of sulfonated poly(tetramethydiphenyl ether ether ketone) membranes with different DSs were prepared and characterized[40] (Figure 22.12). In VRB single-cell test, the sulfonated poly(ether ether ketone) (SPEEK) membranes exhibited comparative, even superior, performances compared with Nafion 115. In the 80-cycle charge–discharge test, the SPEEK 40 membrane exhibited stable performance and its internal structure remained dense.

In order to improve ion selectivity and chemical properties, the membranes have been modified as hybrid or composite membranes using organic and inorganic materials with different methods.

Teng et al.[75] successfully prepared Nafion/SiO_2 hybrid membranes by using the sol–gel method, which is illustrated in Figure 22.13. The results showed that the incorporation of SiO_2 into Nafion can effectively reduce the crossover of vanadium, because the polar clusters (pores) of the original Nafion were filled with SiO_2 nanoparticles during the in situ sol–gel reaction of tetraethyl orthosilicate (TEOS). The VRB with a

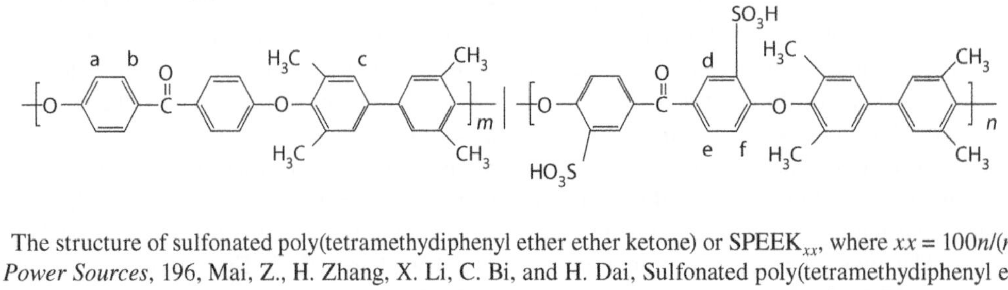

FIGURE 22.12 The structure of sulfonated poly(tetramethydiphenyl ether ether ketone) or SPEEK$_{xx}$, where $xx = 100n/(m + n)$. (Reprinted from *Journal of Power Sources*, 196, Mai, Z., H. Zhang, X. Li, C. Bi, and H. Dai, Sulfonated poly(tetramethydiphenyl ether ether ketone) membranes for vanadium redox flow battery application, 482–487, Copyright (2011), with permission from Elsevier.)

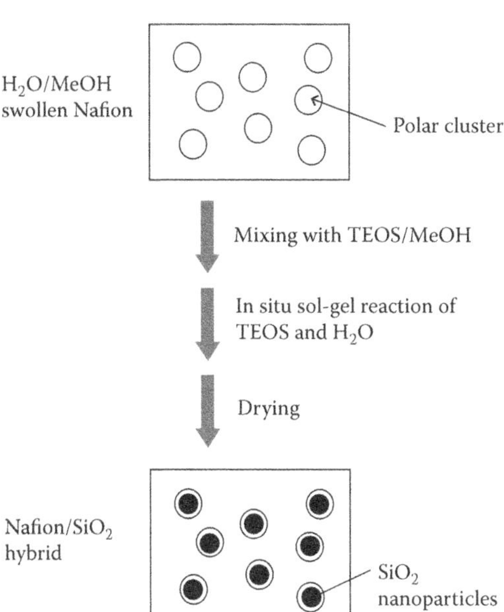

FIGURE 22.13 Schematic depiction of the preparation of Nafion/SiO_2 hybrid membrane. (Reprinted from *Journal of Membrane Science*, 341, Teng, X., Y. Zhao, J. Xi, Z. Wu, Z., X. Qiu, and L. Chen, Nafion/organic silica modified TiO_2 composite membrane for vanadium redox flow battery via in situ sol–gel reactions, 149–154, Copyright (2009), with permission from Elsevier.)

Nafion/SiO_2 hybrid membrane shows a CE and an EE than higher than those of the VRB with a Nafion membrane.[75] Later Teng et al.[75] used the same method to prepare a novel Nafion/organically modified silicate (ORMOSIL) hybrids membrane via in situ sol–gel reactions for mixtures of tetraethoxysilane and diethoxydimethylsilane. It was found that the CE, the VE, and the EE of VRB with Nafion/ORMOSIL hybrid membrane were much higher than those of the VRB with unmodified Nafion membrane and previously reported Nafion/SiO_2 membrane.[76] Chen et al.[20,74] reported novel hybrid membranes based on sulfonated poly(fluorenyl ether ketone) (SPFEK) and SiO_2 or sulfonic acid group containing SiO_2 (SiO_2–SO_3H, IEC = 1.92 milliequivalent per gram [meq g^{-1}]) particles with silica-rich layer embedded structure. The as-made membranes show dramatically higher proton selectivity when compared with Nafion 117. The VRB-SPFEK/3%SiO_2 and the VRB-SPFEK/9%SiO_2 show CEs and VEs higher than those of the VRB-SPFEK at all the tested current densities. Also, organic silica–modified TiO_2/Nafion composite membranes showed a remarkably lower crossover of vanadium ions and water transfer behavior compared with the unmodified membrane.[75] And introduction of ZrP into the Nafion 1135 membrane can decrease the permeability of VO^{2+} more than 10 times while maintaining the same level of proton conductivity.[77]

Here, Xi et al. reported a novel approach using polyelectrolyte layer-by-layer (LbL) self-assembly technique to fabricate a barrier layer onto the surface of Nafion membrane by alternate adsorption of polycation PDDA and polyanion PSS (see Figure 22.14), which can suppress the crossover of vanadium ions.[24] The LbL self-assembled polyelectrolyte multilayer modified Nafion membrane reported here is an easy and promising strategy to overcome the vanadium ions crossover in VRB, which may be extended to other proton-exchange membrane systems. Wang et al. also used PDDA and PSS to modify SPFEK, which showed lower vanadium permeability and better VRB performance compared to pure SPFEK membrane and Nafion membrane.[78] Zeng et al.[79] modified Nafion membranes successfully by electrolyte soaking, oxidation polymerization, and electrodeposition. The experimental results showed that the V(IV) ion permeability of Nafion/polypyrrole (PPR) membrane was reduced by an order of magnitude compared to that of Nafion.[79]

In order to reduce the cost of membrane used in a VRB system while keeping its chemical stability, Luo et al. reported that Nafion/SPEEK layered composite (N/S) membrane consisting of a thin layer of recast Nafion membrane and a layer of SPEEK membrane was prepared by chemically cross-linking the sulfonic acid groups of different ionomer membranes.[80] The VRB single cell employing N/S membrane exhibited a

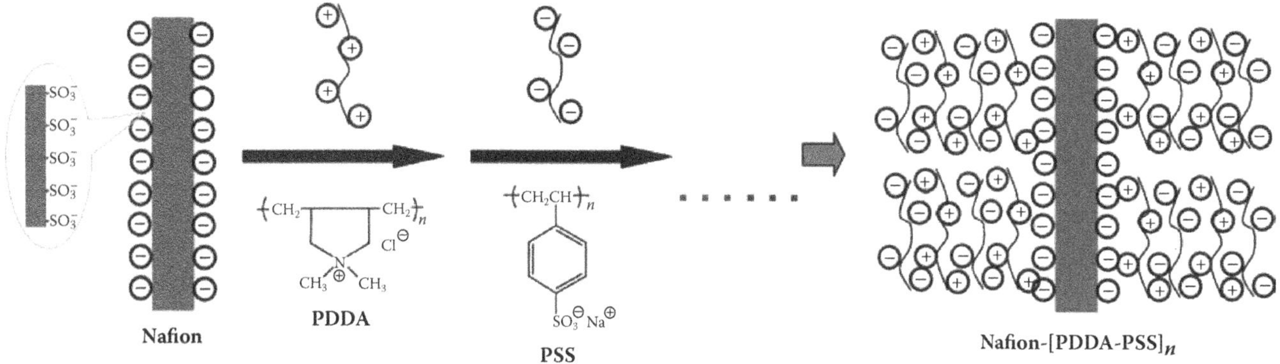

FIGURE 22.14 Schematic representation of the preparation of Nafion-[PDDA-PSS]$_n$ membrane through LbL self-assembly technique. (Xi, J., Z. Wu, X. Teng, Y. Zhao, L. Chen, and X. Qiu, Self-assembled polyelectrolyte multilayer modified Nafion membrane with suppressed vanadium ion crossover for vanadium redox flow batteries, *Journal of Materials Chemistry*, 18 (11), 1232–1238, 2008. Reproduced by permission of The Royal Society of Chemistry.)

lower VE and a higher CE compared with that employing Nafion membrane. Although the overall EE of VRB single cell with N/S membrane was a little lower, its good chemical stability and low cost make it a promising membrane used in VRB system.

22.5.2.2 Anion-Exchange Membrane

It is generally recognized that protons are the charge carriers passing through the separator of VRBs. An AEM separator for the VRB offers several advantages since it mitigates vanadium ion crossover, which is responsible for coulombic losses and self-discharge of the battery. AEMs block the transport of cations due to the Donnan repulsion effect and are widely used in electrodialysis.[81] Furthermore, minimizing the inclusion of VO_2^+ in the membrane is postulated to reduce chemical degradation of the membrane by VO_2^+.

Different polymer backbone structures, namely, aliphatic and aromatic, have been explored as AEMs.[82–84] And various cation groups have been reported, such as quaternary ammonium group,[85] imidazolium group,[86] guanidinium group,[87] quaternary phosphonium group,[88] pyridinium group,[89] and tertiary sulfonium group.[90] Limited data exist for comparison of the stabilities of all these cation groups; however, quaternary ammonium group is the commercialized and most studied cation group because of its ease of preparation and relative high stability.

There are mainly two methods for preparation of AEMs. The synthesis of AEMs generally involves halomethylation of a polymer, such as poly(sulfone) or poly(styrene), and subsequent treatment with the desired cation precursor.[91] The polymer was chloromethylated with chloromethyl methyl ether, which is carcinogenic, and then quaternized with a tertiary amine to form the quaternary ammonium cationic group ethered to an aromatic ring through a benzyl linkage. Most AEMs that are hydrocarbon based, such as polystyrene or polysulfone (PSF), use this route to synthesis. Chloromethylation quaternary amination poly(phthalazinone ether sulfone ketone) AEM was prepared by Zhang et al. and this material has been applied in VRB.[49] In Chen et al.'s work, quaternary ammonium randomly functionalized poly(arylene ether sulfone)s (Radel) with IECs from 1.7 to 2.4 meq g^{-1} were synthesized as the model AEMs for VRB.[47] Vijay Min-suk et al. reported that a PSF-based AEM containing quaternary benzyl trimethylammonium groups was determined to be chemically stable in VO_2^+ solutions and demonstrated better EE compared to N212 in a single-cell VRB.[92]

Although this is a facile, convenient route to the production of AEMs and this reagent is highly reactive for chloromethylation of the aryl rings in a polymer, some specific limitations of the chloromethylation method exist. Chloromethyl methyl ether is carcinogenic and the amount of chloromethyl groups that is potentially harmful to human health and their location along the polymer backbone could not be controlled precisely.[93] In addition, most of AEMs are not soluble in common organic solvent, which leads to difficulty of directly preparing various functional membranes. Another route for synthesizing AEMs is by direct polymerization of cationic monomers or precursors to cationic groups, which has expanded the number of approaches to new AEMs. This novel process could precisely control the amount of quaternary ammonium groups and their locations along the polymer backbone by adjusting the composition of comonomers. The resulting quaternized copolymers exhibited outstanding solubility in polar aprotic solvents and formed flexible and tough AEMs. Additionally, this more environmentally friendly route avoids the use of chloromethyl methyl ether, a toxic chemical. Wang et al. synthesized a novel bisphenol monomer containing two tertiary amine groups, 2,2′-dimethylaminemethylene-4, 4′-biphenol. With this monomer, they used different comonomers to achieve poly(arylene ether sulfone)s ionomers with pendant quaternary ammonium groups.[94,95] By using the same method, Chen et al. produced a new monomer, 9,9′-bis(3-dimethylaminemethylene-4-hydroxyphenyl) fluorene (DABPF). They reported firstly the synthesis of quaternary ammonium functionalized fluorene-containing cardo poly(arylene ether)s with a wide range of IECs by a direct polymerization method.[96] In addition, Qiu et al. group made dimethylaminoethyl methacrylate (DMAEMA), which is tertiary anion monomer grafted onto ethylene tetrafluoroethylene by gamma ray simultaneous radiation technique, and DMAEMA can be easily changed into quaternary ammonium salt.[22]

22.5.2.3 Amphoteric Ion-Exchange Membrane

Although CEMs exhibit high conductivity and higher EE, their low permeation of vanadium ions limit their use for RFBs. AEMs exhibit lower vanadium ion permeability owing to the Donnan exclusion effect, but they are simultaneously restricted by lower conductivity which results in the sacrifice of VE of the battery.[22,68] Therefore, it is necessary to exploit a novel IEM with high conductivity and low vanadium permeability. The AIEM, with both cation and anion-exchange capabilities, has potential applications in many fields since it was first proposed by Sollner in 1932. Luo et al. coated polyetherimide on Nafion membranes via interfacial polymerization.[97] Zeng et al. reported PPR modified on the surface of Nafion membranes.[79] And Xi et al. used the LbL self-assembly technique to fabricate a barrier layer onto the surface of Nafion membrane.[24] All these methods can introduce anion-exchange groups onto a Nafion membrane to reduce the permeability of vanadium ions. Hu et al.,[50] Ma et al.,[51] Pu et al.,[52] and Qiu et al.[54] prepared a series of AIEMs via a handy radiation grafting approach to obtain quaternary ammonium salt group and by sulfonation to get sulfonic group.[7] This AIEM series showed that these amphoteric membranes exhibited lower vanadium ion permeability, as well as higher CE and EE, and maintained a rather high conductivity. The facile synthesis of AIEMs with pendant quaternary ammonium groups is by one-pot direct polymerization method to decrease the vanadium ion permeability. The advantage of this method is that the amount of quaternary ammonium groups and sulfonic acid group could be controlled precisely by adjusting the feed ratio of comonomers.

22.5.2.4 Recent Advances in Electrolytes

The issue of reactant solubility in a flowing electrolyte solution can be important. The energy density of an RFB system is set by the concentration of dissolved species, but the maximum concentration in any stream is limited by the solubility of the least soluble species. Precipitation of reactants or products in a porous electrode is calamitous. Concentration limits on the electroactive species not only reduce the energy density of a system but also negatively impact the power density and cell efficiency. Lower concentrations mean reduced mass-transfer rates and current density, thus increasing concentration polarization and/or pumping power. Solubility is a function of temperature as well, which must be factored into cell design. For instance, it is observed that V_2O_5 precipitation occurs at elevated temperature, limiting the operating temperature to the range of 10 to 40°C.[3] Li et al. improved this situation with the development of a vanadium sulfate and chloride mixed electrolyte, enabling a vanadium concentration of up to 2.5 M over a temperature range of −5 to 50°C.[3] The addition of chloride acid achieves two purposes: solubility of the V(III) and V(IV) ions improved because the decrease in the sulfate ion concentration and the chloride ion improved the stability of V(V) ions at elevated temperatures. The mixed-acid all-vanadium RFB technology has been licensed for commercial scale-up production.

To increase the energy density of the VRB, several research groups used additives that are expected to enhance the stability of the supersaturated electrolytes as well as improve their electrochemical performance.[98] Zhang et al. found that polyacrylic acid was a potential stabilizing agent for the negative electrolyte; a mixture of polyacrylic acid and methane sulfonic acid appeared to be suitable for the positive electrolyte.[99] Li et al. reported that the electrochemical activity of the electrolyte improved with the addition of D-sorbitol.[100] Chang et al. observed that the addition of Coulter IIIA dispersant significantly delayed precipitate formation and achieved higher EE.[101] Peng et al. observed that the addition of trishydroxymethyl aminomethane (Tris) to the positive electrolyte improved the charge discharge behavior and exhibited a lower discharge capacity fade rate.[102] However, organic compounds such as glycerol are slowly oxidized by V(V), while inorganic compounds are more stable.[103,104] Other issues regarding concentrations include the fact that for many systems increasing the concentration of the reactants can lead to more complexing and lower diffusivities and perhaps even more viscous solutions.

22.5.3 Flow Field Designs

As a key component of flow batteries, the flow field is for distributing electrolytes on the electrode surface, for applying/collecting electric current to/from cells, for providing a structural support for the electrode material, and for facilitating heat management.[105–107] Zhu et al. investigated the effects of two different flow fields, one with a flow-pass pattern and the other with a flow-through pattern, on the performance of a VRB. The flow-through pattern increased the electrode effective area and the uniformity of electrolyte flow, increasing EE by 1.5%–5%.[108] Aaron et al.[109] and Liu et al.[110] used a fuel cell–type structure with zero gap for electrolyte flow in a VRB system. The enhancement of the cell performance was attributed to the enhanced mass transport and reduced internal resistance for the use of a serpentine flow field.[15,109,110] Xu et al.[111] proposed and applied a three-dimensional (3D) numerical model to the study of flow field designs (flow-throughs with no flow field, with serpentine flow field, and with parallel flow field, as shown in Figure 22.15) for a VRB. The VRB with the serpentine flow field shows the lowest overpotential due to the more even electrolyte distribution over the electrode surface and the enhanced convective mass transport toward the membrane. Therefore, the serpentine type appears to be a more suitable flow field design for VRBs.[111]

22.5.4 Improvements in Stack Design and Configuration

In addition to the electrolyte, electrode, and membrane materials, the design and configuration of the cell stack and the whole

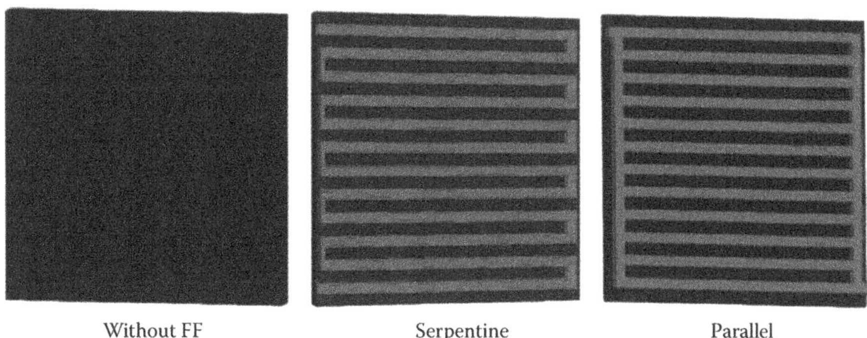

FIGURE 22.15 Schematic of the three types of flow fields in the VRB. (Reprinted from *Applied Energy*, 105, Xu, Q., T. Zhao, and P. Leung, Numerical investigations of flow field designs for vanadium redox flow batteries, 47–56, Copyright (2013), with permission from Elsevier.)

system in general is critical to improving the performance and economy of RFB technologies.[13,112] One of the challenges of stack design that must be given particular attention in RFB configurations is protection against shunt currents, which are created as a result of the voltage difference over different cells and improving other performance parameters in order to reduce the overall system cost. In a well-designed stack, there should be no current flow except directly from one cell to another in the preferred series configuration. Recently, Aaron et al. have demonstrated dramatic performance gains in VRB through modified cell architecture.[15] This study utilized an RFB that had a zero-gap configuration. The membrane, electrode, and current collectors were all in direct contact. The zero-gap flow field architecture and thin carbon paper electrodes ensured good contact between all components of the cell and reduced charge transport distances, which made the power density of a VRB increase more than fivefold compared to that of other conventional published systems. Of particular concern in flow batteries is the development of shunt currents via the liquid electrolyte. Because RFBs involve the circulation of electrolyte to each of the individual cells, there is an obvious ionic current path from one cell to another. The currents that flow in the circulating electrolyte from one cell to another via the electrolyte flow manifolds are best managed by increasing the effective resistance of the flow path, either by increasing the effective path length between cell flow inputs and outputs in the manifold or by reducing the cross-sectional area of the ports.[13] Several researchers have investigated the design implications for flow batteries for particular systems, although optimization will be required for specific electrolyte and cell configurations.

22.6 THEORETICAL MODELING FOR FUNDAMENTAL UNDERSTANDING

Modeling and simulation is indispensable and plays an important role in the development and commercialization of RFB technologies. Because a cell model once validated experimentally can be used to identify cell-limiting mechanisms and forecast cell performance for design, scale-up, and optimization. But extensive laboratory testing of different materials, components, and additives over a broad range of conditions is both time consuming and costly. Modeling can be used to systematically reduce the number of test cases and to analyze the results of tests and trials, as is already the case for fuel cells and static batteries in pilot studies.[113] Thanks in part to the advent of high-performance computers and advanced numerical algorithms, comprehensive models are frequently employed to study fuel cells and static batteries. In the past 30 years we have put a large academic effort into the development of mathematical modeling to rationalize and predict battery behavior and system performance, which is driven by the great interest in the development and applications of RFB for large-scale energy storage.[16] Mathematical modeling is often used to understand the limiting phenomena and processes in an RFB; yet, relative to the experimental and demonstration system development, analytical and computational modeling of RFBs has trailed, which may be due to the era in which they were heavily researched. Advanced modeling is needed to understand fully the various physiochemical phenomena involved to help minimize transport losses and facilitate optimized material design and architectures.

Due to early commercial interest, a number of attempts were done on the simulation of the zinc/bromine cell in the 1980s.[114–117] Several models of Zn/Br_2 have been developed to understand the physical phenomena and to determine how cell performance can be improved. Various types of models have been used to investigate the cell and stack performance, including partial differential equation models for predicting current and potential distributions, an algebraic model for shunt currents and associated energy losses, and ordinary differential equation models for predicting the EE of the cell as a function of the state of charge.[16] These models have been used on species transport, secondary electrode reactions, and chemical reactions in the bulk electrolyte, including issues during the plating of zinc. Putt developed thin diffusion-layer models.[118] These models include electrolyte convection with Butler–Volmer kinetics. Evans and White[114,115] and Simpson and White[116,117] developed a mathematical model for the cell mainly to predict performance of the cell as a function of architecture and operating conditions. They also used their model to determine the effects of the mass transfer and electrokinetics in the porous bromine electrode on the round-trip performance of the cell. It was found that the cell efficiency increases with the porous electrode thickness. Weber et al.[13] also developed models for the chlorine electrode in a Zn/Cl_2 cell. This electrode is very similar to the bromine one and they showed that the flow from the gap to the zinc electrode can impact the current density and reaction-rate distribution significantly.[13]

Scamman et al. investigated the performance of bromide/polysulfide RFB systems by using a numerical model via an equivalent-circuit approach in which the individual overpotential losses were calculated using the Butler–Volmer equation.[87] This model is able to predict the concentration and current variation along the electrode and determine various efficiencies, energy density, and power density in the charge–discharge processes. Fedkiw and Watts developed a single anode–separator–cathode Fe/Cr cell with a mathematical isothermal model.[121] The model incorporates redox reaction kinetics, mass transport, and ohmic effects as well as the parasitic hydrogen reaction that occurs in the chromium electrode. In addition, this model provides a method of determining a charge–discharge protocol that obtains the maximum chromium conversion and minimum hydrogen evolution at the same time. Codina and Aldaz[122] have reported scale-up studies of the Fe/Cr RFB based on shunt current and flow distribution analyses. A shut current calculation model for a three-stack assembly has been developed. Both shunt-current and flow distribution analysis have yielded a prototype with a 93% current efficiency with an homogenous intrastack flow distribution.[122]

Trinidad et al. developed an oxidation–reduction or redox potential model to monitor the Ce(III)/Ce(IV) couple.[123] The model fits well with the experimental results, showing that the model is useful to predict concentration versus time in a simple redox system. This model provides useful design criteria for improving cell performance.

The most extensively modeled system is the VRB.[124,125] Modeled VRBs were studied for concentration, temperature, electrolyte flow rate, electrode porosity, gas evolution, self-discharge process, and open-circuit voltage. Enomoto et al.[126] and Mingghua and Hikihara[127] both proposed a zero-dimensional transient model, which was based on the chemical reactions and circuit response, to study the relationship between the resistance in series and the state of the charge and further predict the performance of the VRB. You et al.[128] presented a two-dimensional mathematical model for a flow-through porous electrode, in which the effects of mass transfer, ohmic resistance, and electrode reaction kinetics on the performance of the porous electrode were investigated. Zhang's group described a single all-vanadium redox flow cell with a two-dimensional stationary model, based on the universal conservation laws and coupled with electrochemical reactions. The model results indicate that bulk reaction rate depends on the applied current density. Ma et al.[129] developed a 3D stationary model for a negative half cell that included an inlet/outlet channel and electrode. The effect of velocity on the distributions of concentration, overpotential, and transfer current density in the sections that are perpendicular and parallel to the applied current was investigated. Li and Hikihara developed a model considering the transient behavior in a VRB and the model was also examined based on the tests of a micro-RFB.[130] They found that the chemical reaction rate is restricted by the attached external electric circuit and the concentration change of vanadium ions depends on the chemical reactions and electrolyte flow. The most extensively and detailed modeled system study that includes concentration, temperature, electrolyte flow rate, electrode porosity, gas evolution, self-discharge process, and open-circuit voltage is referred to in the review by Ding et al.[14]

Despite this early activity, modeling of RFBs is not a well-developed area. Therefore, it is an urgent requirement to develop and validate practical stack-level models to aid the design and optimization of large-scale systems. The existing models can be used to improve the cell designs such as optimized heat management, power density, and electrolyte utilization. Meanwhile, a model including sufficient detailed mass, charge, and fluid transfers incorporated with the heat or electric circuit is necessary to control and a monitor large-scale system. Application devices and auxiliary equipment can then be incorporated to develop comprehensive control and monitoring tools.

22.7 TECHNICAL CHALLENGES AND PERSPECTIVES

There is an increasing demand for a reduction in the global use of fossil fuels for economic, domestic political, geopolitical, and environmental reasons. But global energy consumption shows no sign of abating and it is forecast to increase significantly. Existing energy resources are used much more efficiently through a combination of measures in which energy-storage technologies should feature prominently. As the energy production from renewable energy sources is usually intermittent and is not connected to the grid, energy storage is hence important. The demand for stationary energy storage has rapidly changed the worldwide landscape of energy system research, which recently brought the RFB technology into the spotlight. To be competitive with other energy-storage options, the capital cost of electrochemical electrical energy storage (EES) should be decreased and the cycle life and reliability increased. Redox energy-storage systems possess features such as flexible design, long life, and high reliability with acceptable operation and maintenance costs. Because of their modular design, their construction and maintenance costs could be the lowest. The modular nature of these batteries simplifies their maintenance, which can be done separately by individual battery modules. A major advantage is their flexibility during charge–discharge cycles; the batteries can be discharged completely without damaging the cells. Because of their large capacity and the long discharge time, flow batteries are attractive when coupling with renewable energy sources. RFB technologies have to be cost competitive for broad market penetration. High cost equally means high energy consumption. If an RFB technology stands at the high cost, it would be meaningless to use in coupling with renewable energy sources to store/release energy.[131]

The rapidly growing demand for energy generated by renewable energy sources has given rise to massive market opportunities. As already known, over the last four decades, a number of RFB systems with potential applications have been introduced and the systems, such as VRB, zinc–bromine, and polysulfide–bromine, have been tested or commercialized on a large scale. At the present, there is no "best" RFB chemistry that could be confidently declared to reach the cost level required for the absolute practical applications; development continues through industrial and academic research supported by government and industry. For example, the most developed VRB system is still expensive for broad market penetration, which is evaluated at \$500 kWh^{-1},[112] because of its dependence on expensive vanadium, IEM, and other components. That is to say, both cost competitiveness and effectiveness should be kept as the key objectives in research and development (R&D) work for RFB technology if aiming to win in practical applications over other EES technologies. The projected overall costs of renewable energy sources and batteries can be revised downward by considering the potential growth in renewable power and by considering improved EE measures, together with an associated reduction in environmental insults.

As with most battery systems that have been developed and commercialized so far, developing cost effective materials and optimizing the operational parameters for a RFB will further expand the scope of application of these systems. Several groups around the world are now focusing their research on four aspects, namely, electrolyte optimization, membrane development, electrode development, and optimization of the stack design of the RFB. Improved IEMs should have better selectivity, controlled solvent transport, low cost, and high stability. Low-cost perfluorinated membranes with excellent long-term stability are now being manufactured commercially by several companies; even greater cost reduction is expected with some of the new membrane materials currently under development. The electrode material influences the electrochemical performance of the cell; hence, developing low-cost, high-performance electrode materials with high power densities will enable considerable reduction in stack size as well as cost. Good electrode structure should have high electrocatalytic activity and high surface area and should be robust and capable of low cost and volume production. Improvements in cell architecture can also enhance system performance by several folds and this is an area that warrants greater attention in the future. With increasing interest in the VRB for large-scale energy storage, an expansion in the research efforts around the world in all the above-mentioned areas is expected to yield significant improvements in cost and performance of the VRB for stationary applications while potentially extending its use in mobile applications with the development of the vanadium–air redox fuel cell. Further investigations on the use of bipolar electrodes in a stack are necessary to scale up the battery and to increase the power density and energy outputs. Integration of electrochemical reactors with other devices and unit processes is also crucial in an RFB system. Electrochemical engineering and design in RFB development is another area that needs to be highlighted. Electrodes, IEMs, cells, and stack and system design are critical to improving the performance and economy of the RFB technologies. Unfortunately, there is less attention on environmental compatibility and energy/material sustainability of RFB technologies, which should be an environmentally attractive technology. We can do more efforts on the following specific areas: avoidance of hazardous chemicals and materials; limitation of electrical leakage currents and prevention of electrolyte escape in the large-scale systems; use of biodegradable polymers in stack construction; and consideration of problems of material degradation and corrosion of electrodes, membranes and other cell/stack components.[131]

It is clear that industrial development of prototypes and working systems has outpaced the fundamental research at this point. Inevitably, for the science to progress and the underlying fundamental problems to be resolved, a much more fundamental understanding is required. New RFB systems are often proposed and reported with promising characteristics and VE or EE. These systems are usually based on small lab cells and often the long-term cycle performance is not considered. In general, many RFB systems encountered some unaddressed problems on the road to system scale-up. In terms of transport from a generic, chemistry-agnostic perspective, much more in-depth and fundamental study and characterization are needed through combined experimental and analytical or computational modeling. In fact, the electrochemical reactions in an RFB system are usually more complicated, involving reactions at or near the electrode surface, mechanisms of charge transport and crossover in the IEM, and behavior of active species in a flowing electrolyte environment. Moreover, the fluid mechanics, transport of electrolyte through the various electrode, and cell architectures including coupled reaction rates and flow distribution for determining optimal electrode structures and properties are all the influential factors contributing to the complexity of RFB chemistry. Hence, more research work on fundamental studies of RFBs would result in a better understanding of RFBs and well-optimized RFB systems. Modeling and simulation are certainly of great importance but until now these have been a low priority in RFB technologies. Excellent modeling and simulation work would bring those benefits to RFB technology, including building up approximate and detailed mathematical models to understand the effects of variations of the entire system and ancillary equipment; examining the proposed mechanism of electrochemical and chemical reactions and their kinetics against the experimental data; avoiding working in hazardous environments as many RFB systems utilize toxic species; providing simulated long-term reliability and durability test; and reducing the R&D cost.

REFERENCES

1. Cheng, F.; Liang, J.; Tao, Z.; Chen, J. Functional materials for rechargeable batteries. *Advanced Materials* 2011; *23* (15): 1695–1715.
2. Huang, K.-L.; Li, X.-g.; Liu, S.-q.; Tan, N.; Chen, L.-q. Research progress of vanadium redox flow battery for energy storage in China. *Renewable Energy* 2008; *33* (2): 186–192.
3. Li, L.; Kim, S.; Wang, W.; Vijayakumar, M.; Nie, Z.; Chen, B.; Zhang, J.; Xia, G.; Hu, J.; Graff, G., A stable vanadium redox-flow battery with high energy density for large-scale energy storage. *Advanced Energy Materials* 2011; *1* (3): 394–400.
4. Mohamed, M. R.; Sharkh, S. M.; Walsh, F. C. Redox flow batteries for hybrid electric vehicles: Progress and challenges, Vehicle Power and Propulsion Conference, 2009 (VPPC '09). IEEE, IEEE: 2009; 551–557.
5. Shin, S.-H.; Yun, S.-H.; Moon, S.-H., A review of current developments in non-aqueous redox flow batteries: Characterization of their membranes for design perspective. *RSC Advances* 2013; *3* (24): 9095–9116.
6. Zhao, P.; Zhang, H.; Zhou, H.; Chen, J.; Gao, S.; Yi, B. Characteristics and performance of 10 kW class all-vanadium redox-flow battery stack. *Journal of Power Sources* 2006; *162* (2): 1416–1420.
7. EPRI. Energy storage technology valuation primer: Techniques for financial modeling. EPRI report 1008810, Palo Alto, California; December 2004.
8. Alotto, P.; Guarnieri, M.; Moro, F. Redox flow batteries for the storage of renewable energy: A review. *Renewable and Sustainable Energy Reviews* 2014; *29*: 325–335.
9. Bradshaw, D. T. Pumped hydroelectric storage (PHS) and compressed air energy storage (CAES). IEEE PES Meeting Energy Storage, Seattle, Washington, 2000.
10. Oró, E.; Gil, A.; de Gracia, A.; Boer, D.; Cabeza, L. F. Comparative life cycle assessment of thermal energy storage systems for solar power plants. *Renewable Energy* 2012; *44*: 166–173.
11. Rummel, T.; Risse, K.; Ehrke, G.; Rummel, K.; Jhon, A.; Mönnich, T.; Buscher, K. P.; Fietz, W. H.; Heller, R.; Neubauer, O.; Panin, A. The super conducting magnet system of the Stellarator Wendelstein 7-X. *IEEE Transactions on Plasma Science* 2012; *40* (3): 769–776.

12. Wolsky, A. M. The status and prospects for flywheels and SMES that incorporate HTS. *Physica C: Superconductivity* 2002; *372–376*: 1495–1499.
13. Weber, A. Z.; Mench, M. M.; Meyers, J. P.; Ross, P. N.; Gostick, J. T.; Liu, Q. Redox flow batteries: A review. *Journal of Applied Electrochemistry* 2011; *41* (10): 1137–1164.
14. Ding, C.; Zhang, H.; Li, X.; Liu, T.; Xing, F. Vanadium flow battery for energy storage: Prospects and challenges. *The Journal of Physical Chemistry Letters* 2013; *4* (8): 1281–1294.
15. Aaron, D.; Liu, Q.; Tang, Z.; Grim, G.; Papandrew, A.; Turhan, A.; Zawodzinski, T.; Mench, M. Dramatic performance gains in vanadium redox flow batteries through modified cell architecture. *Journal of Power Sources* 2012; *206*: 450–453.
16. Leung, P.; Li, X.; Ponce de León, C.; Berlouis, L.; Low, C. T. J.; Walsh, F. C. Progress in redox flow batteries, remaining challenges and their applications in energy storage. *RSC Advances* 2012; *2* (27): 10125.
17. China Greentech Initiative. *The China Greentech Initiative Report*, http://www.china-greentech.com, last accessed: Dec. 2010.
18. Ponce de León, C.; Frías-Ferrer, A.; González-García, J.; Szánto, D. A.; Walsh, F. C., Redox flow cells for energy conversion. *Journal of Power Sources* 2006; *160* (1): 716–732.
19. Chen, D.; Wang, S.; Xiao, M.; Han, D.; Meng, Y. Sulfonated poly(fluorenyl ether ketone) membrane with embedded silica rich layer and enhanced proton selectivity for vanadium redox flow battery. *Journal of Power Sources* 2010; *195* (22): 7701–7708.
20. Chen, D.; Wang, S.; Xiao, M.; Meng, Y. Preparation and properties of sulfonated poly(fluorenyl ether ketone) membrane for vanadium redox flow battery application. *Journal of Power Sources* 2010; *195* (7): 2089–2095.
21. Kim, S.; Yan, J.; Schwenzer, B.; Zhang, J.; Li, L.; Liu, J.; Yang, Z.; Hickner, M. A. Cycling performance and efficiency of sulfonated poly(sulfone) membranes in vanadium redox flow batteries. *Electrochemistry Communications* 2010; *12* (11): 1650–1653.
22. Qiu, J.; Li, M.; Ni, J.; Zhai, M.; Peng, J.; Xu, L.; Zhou, H.; Li, J.; Wei, G. Preparation of ETFE-based anion exchange membrane to reduce permeability of vanadium ions in vanadium redox battery. *Journal of Membrane Science* 2007; *297* (1–2): 174–180.
23. Jia, C.; Liu, J.; Yan, C. A significantly improved membrane for vanadium redox flow battery. *Journal of Power Sources* 2010; *195* (13): 4380–4383.
24. Xi, J.; Wu, Z.; Teng, X.; Zhao, Y.; Chen, L.; Qiu, X. Self-assembled polyelectrolyte multilayer modified Nafion membrane with suppressed vanadium ion crossover for vanadium redox flow batteries. *Journal of Materials Chemistry* 2008; *18* (11): 1232–1238.
25. Chen, D.; Kim, S.; Li, L.; Yang, G.; Hickner, M. A. Stable fluorinated sulfonated poly(arylene ether) membranes for vanadium redox flow batteries. *RSC Advances* 2012; *2* (21): 8087–8094.
26. Mai, Z.; Zhang, H.; Li, X.; Xiao, S.; Zhang, H. Nafion/polyvinylidene fluoride blend membranes with improved ion selectivity for vanadium redox flow battery application. *Journal of Power Sources* 2011; *196* (13): 5737–5741.
27. Wang, W.; Luo, Q.; Li, B.; Wei, X.; Li, L.; Yang, Z. Recent progress in redox flow battery research and development. *Advanced Functional Materials* 2013; *23* (8): 970–986.
28. Li, X.; Zhang, H.; Mai, Z.; Zhang, H.; Vankelecom, I. Ion exchange membranes for vanadium redox flow battery (VRB) applications. *Energy & Environmental Science* 2011; *4* (4): 1147–1160.
29. Shinkle, A. A.; Sleightholme, A. E.; Griffith, L. D.; Thompson, L. T.; Monroe, C. W. Degradation mechanisms in the non-aqueous vanadium acetylacetonate redox flow battery. *Journal of Power Sources* 2012; *206*: 490–496.
30. Ge, S.; Yi, B.; Zhang, H. Study of a high power density sodium polysulfide/bromine energy storage cell. *Journal of Applied Electrochemistry* 2004; *34* (2): 181–185.
31. Lim, H.; Lackner, A.; Knechtli, R. Zinc-bromine secondary battery. *Journal of the Electrochemical Society* 1977; *124* (8): 1154–1157.
32. Savinell, R.; Liu, C.; Galasco, R.; Chiang, S.; Coetzee, J. Discharge characteristics of a soluble iron-titanium battery system. *Journal of The Electrochemical Society* 1979; *126* (3): 357–360.
33. Zhao, P.; Zhang, H.; Zhou, H.; Yi, B. Nickel foam and carbon felt applications for sodium polysulfide/bromine redox flow battery electrodes. *Electrochimica Acta* 2005; *51* (6): 1091–1098.
34. Pelligri, A.; Spaziante, P. M. Process and accumulator for storing and releasing electrical energy, Patent GB 2030349 A, United Kingdom, 1980.
35. Skyllas-Kazacos, M.; Rychcik, M.; Robins, R. G.; Fane, A.; Green, M. New all-vanadium redox flow cell. *Journal of the Electrochemical Society* 1986; *133*: 1057.
36. Sum, E.; Rychcik, M.; Skyllas-Kazacos, M. Investigation of the V(V)/V(IV) system for use in the positive half-cell of a redox battery. *Journal of Power Sources* 1985; *16* (2): 85–95.
37. Sum, E.; Skyllas-Kazacos, M. A study of the V(II)/V(III) redox couple for redox flow cell applications. *Journal of Power Sources* 1985; *15* (2): 179–190.
38. Kim, S.; Tighe, T. B.; Schwenzer, B.; Yan, J.; Zhang, J.; Liu, J.; Yang, Z.; Hickner, M. A. Chemical and mechanical degradation of sulfonated poly(sulfone) membranes in vanadium redox flow batteries. *Journal of Applied Electrochemistry* 2011; *41* (10): 1201–1213.
39. Rychcik, M.; Skyllas-Kazacos, M. Evaluation of electrode materials for vanadium redox cell. *Journal of Power Sources* 1987; *19* (1): 45–54.
40. Mai, Z.; Zhang, H.; Li, X.; Bi, C.; Dai, H. Sulfonated poly(tetramethydiphenyl ether ether ketone) membranes for vanadium redox flow battery application. *Journal of Power Sources* 2011; *196* (1): 482–487.
41. Sun, C.; Chen, J.; Zhang, H.; Han, X.; Luo, Q. Investigations on transfer of water and vanadium ions across Nafion membrane in an operating vanadium redox flow battery. *Journal of Power Sources* 2010; *195* (3): 890–897.
42. Schwenzer, B.; Zhang, J.; Kim, S.; Li, L.; Liu, J.; Yang, Z. Membrane development for vanadium redox flow batteries. *ChemSusChem* 2011; *4* (10): 1388–1406.
43. Sukkar, T.; Skyllas-Kazacos, M. Membrane stability studies for vanadium redox cell applications. *Journal of Applied Electrochemistry* 2004; *34* (2): 137–145.
44. Zhang, H.; Zhang, H.; Li, X.; Mai, Z.; Zhang, J. Nanofiltration (NF) membranes: The next generation separators for all vanadium redox flow batteries (VRBs)? *Energy & Environmental Science* 2011; *4* (5): 1676–1679.
45. Wang, N.; Peng, S.; Li, Y.; Wang, H.; Liu, S.; Liu, Y. Sulfonated poly(phthalazinone ether sulfone) membrane as a separator of vanadium redox flow battery. *Journal of Solid State Electrochemistry* 2012; *16* (6): 2169–2177.
46. Chen, D.; Hickner, M. A.; Agar, E.; Kumbur, E. C. Optimized anion exchange membranes for vanadium redox flow batteries. *ACS Applied Materials & Interfaces* 2013; *5* (15): 7559–7566.

47. Chen, D.; Hickner, M. A.; Agar, E.; Kumbur, E. C. Selective anion exchange membranes for high coulombic efficiency vanadium redox flow batteries. *Electrochemistry Communications* 2013; *26*: 37–40.
48. Vafiadis, H.; Skyllas-Kazacos, M. Evaluation of membranes for the novel vanadium bromine redox flow cell. *Journal of Membrane Science* 2006; *279* (1): 394–402.
49. Zhang, B.; Zhang, S.; Xing, D.; Han, R.; Yin, C.; Jian, X. Quaternized poly(phthalazinone ether ketone ketone) anion exchange membrane with low permeability of vanadium ions for vanadium redox flow battery application. *Journal of Power Sources* 2012; *217*: 296–302.
50. Hu, G.; Wang, Y.; Ma, J.; Qiu, J.; Peng, J.; Li, J.; Zhai, M. A novel amphoteric ion exchange membrane synthesized by radiation-induced grafting α-methylstyrene and N,N-dimethylaminoethyl methacrylate for vanadium redox flow battery application. *Journal of Membrane Science* 2012; *407–408*: 184–192.
51. Ma, J.; Wang, Y.; Peng, J.; Qiu, J.; Xu, L.; Li, J.; Zhai, M. Designing a new process to prepare amphoteric ion exchange membrane with well-distributed grafted chains for vanadium redox flow battery. *Journal of Membrane Science* 2012; *419–420*: 1–8.
52. Pu, H.; Luo, H.; Wan, D. Synthesis and properties of amphoteric copolymer of 5-vinyltetrazole and vinylbenzyl phosphonic acid. *Journal of Polymer Science Part A: Polymer Chemistry* 2013; *51* (16): 3486–3493.
53. Qiu, J.; Zhai, M.; Chen, J.; Wang, Y.; Peng, J.; Xu, L.; Li, J.; Wei, G. Performance of vanadium redox flow battery with a novel amphoteric ion exchange membrane synthesized by two-step grafting method. *Journal of Membrane Science* 2009; *342* (1–2): 215–220.
54. Qiu, J.; Zhang, J.; Chen, J.; Peng, J.; Xu, L.; Zhai, M.; Li, J.; Wei, G. Amphoteric ion exchange membrane synthesized by radiation-induced graft copolymerization of styrene and dimethylaminoethyl methacrylate into PVDF film for vanadium redox flow battery applications. *Journal of Membrane Science* 2009; *334* (1–2): 9–15.
55. Weier, K.; Doran, J.; Power, J.; Walters, D. Denitrification and the dinitrogen/nitrous oxide ratio as affected by soil water, available carbon, and nitrate. *Soil Science Society of America Journal* 1993; *57* (1): 66–72.
56. Kaneko, H.; Nozaki, K.; Wada, Y.; Aoki, T.; Negishi, A.; Kamimoto, M. Vanadium redox reactions and carbon electrodes for vanadium redox flow battery. *Electrochimica Acta* 1991; *36* (7): 1191–1196.
57. Sun, B.; Skyllas-Kazakos, M. Chemical modification and electrochemical behaviour of graphite fibre in acidic vanadium solution. *Electrochimica Acta* 1991; *36* (3): 513–517.
58. Li, W.; Liu, J.; Yan, C. Multi-walled carbon nanotubes used as an electrode reaction catalyst for/VO^{2+} for a vanadium redox flow battery. *Carbon* 2011; *49* (11): 3463–3470.
59. González, Z.; Botas, C.; Álvarez, P.; Roldán, S.; Blanco, C.; Santamaría, R.; Granda, M.; Menéndez, R. Thermally reduced graphite oxide as positive electrode in vanadium redox flow batteries. *Carbon* 2012; *50* (3): 828–834.
60. Li, W.; Liu, J.; Yan, C. Graphite–graphite oxide composite electrode for vanadium redox flow battery. *Electrochimica Acta* 2011; *56* (14): 5290–5294.
61. Han, P.; Yue, Y.; Liu, Z.; Xu, W.; Zhang, L.; Xu, H.; Dong, S.; Cui, G. Graphene oxide nanosheets/multi-walled carbon nanotubes hybrid as an excellent electrocatalytic material towards VO_2^+/VO^{2+} redox couples for vanadium redox flow batteries. *Energy & Environmental Science* 2011; *4* (11): 4710–4717.
62. Pengxian, H.; Haibo, W.; Zhihong, L.; Chen, X.; Ma, W.; Yao, J.; Zhu, Y.; Cui, G. Graphene oxide nanoplatelets as excellent electrochemical active materials for VO^{2+}: VO_2^+ and V^{2+}: V^{3+} redox couples for a vanadium redox flow battery. *Carbon* 2011; *49* (2); 693–700.
63. Han, P.; Wang, H.; Liu, Z.; Chen, X.; Ma, W.; Yao, J.; Zhu, Y.; Cui, G. Graphene oxide nanoplatelets as excellent electrochemical active materials for VO^{2+}/VO_2^+ and V^{2+}/V^{3+} redox couples for a vanadium redox flow battery. *Carbon* 2011; *49* (2): 693–700.
64. Fabjan, C.; Garche, J.; Harrer, B.; Jörissen, L.; Kolbeck, C.; Philippi, F.; Tomazic, G.; Wagner, F. The vanadium redox-battery: An efficient storage unit for photovoltaic systems. *Electrochimica Acta* 2001; *47* (5): 825–831.
65. Sun, B.; Skyllas-Kazacos, M. Chemical modification of graphite electrode materials for vanadium redox flow battery application—Part II: Acid treatments. *Electrochimica Acta* 1992; *37* (13): 2459–2465.
66. Pittman Jr, C.; Jiang, W.; Yue, Z.; Gardner, S.; Wang, L.; Toghiani, H.; Leon y Leon, C. Surface properties of electrochemically oxidized carbon fibers. *Carbon* 1999; *37* (11): 1797–1807.
67. Shao, Y.; Wang, X.; Engelhard, M.; Wang, C.; Dai, S.; Liu, J.; Yang, Z.; Lin, Y. Nitrogen-doped mesoporous carbon for energy storage in vanadium redox flow batteries. *Journal of Power Sources* 2010; *195* (13): 4375–4379.
68. Hwang, G.-J.; Ohio, H. Crosslinking of anion exchange membrane by accelerated electron radiation as a separator for the all-vanadium redox flow battery. *Journal of Membrane Science* 1997; *132* (1): 55–61.
69. Obi, D.-G. Permeation of vanadium cations through anionic and cationic membranes. *Journal of Applied Electrochemistry* 1985; *15* (2): 231–235.
70. Mohammadi, T.; Skyllas-Kazacos, M. Preparation of sulfonated composite membrane for vanadium redox flow battery applications. *Journal of Membrane Science* 1995; *107* (1): 35–45.
71. Chieng, S.; Kazacos, M.; Skyllas-Kazacos, M. Modification of Daramic, microporous separator, for redox flow battery applications. *Journal of Membrane Science* 1992; *75* (1): 81–91.
72. Mohammadi, T.; Skyllas-Kazacos, M. Characterisation of novel composite membrane for redox flow battery applications. *Journal of Membrane Science* 1995; *98* (1): 77–87.
73. Mohammadi, T.; Skyllas-Kazacos, M. Use of polyelectrolyte for incorporation of ion-exchange groups in composite membranes for vanadium redox flow battery applications. *Journal of Power Sources* 1995; *56* (1): 91–96.
74. Chen, D.; Wang, S.; Xiao, M.; Meng, Y. Synthesis and properties of novel sulfonated poly(arylene ether sulfone) ionomers for vanadium redox flow battery. *Energy Conversion and Management* 2010; *51* (12): 2816–2824.
75. Teng, X.; Zhao, Y.; Xi, J.; Wu, Z.; Qiu, X.; Chen, L. Nafion/organic silica modified TiO_2 composite membrane for vanadium redox flow battery via *in situ* sol–gel reactions. *Journal of Membrane Science* 2009; *341* (1): 149–154.
76. Teng, X.; Zhao, Y.; Xi, J.; Wu, Z.; Qiu, X.; Chen, L. Nafion/organically modified silicate hybrids membrane for vanadium redox flow battery. *Journal of Power Sources* 2009; *189* (2): 1240–1246.
77. Sang, S.; Wu, Q.; Huang, K. Preparation of zirconium phosphate (ZrP)/Nafion1135 composite membrane and H^+/VO^{2+} transfer property investigation. *Journal of Membrane Science* 2007; *305* (1): 118–124.
78. Wang, Y.; Wang, S.; Xiao, M.; Han, D.; Hickner, M. A.; Meng, Y. Layer-by-layer self-assembly of PDDA/PSS-SPFEK composite membrane with low vanadium permeability for

vanadium redox flow battery. *RSC Advances* 2013; *3* (35): 15467–15474.

79. Zeng, J.; Jiang, C.; Wang, Y.; Chen, J.; Zhu, S.; Zhao, B.; Wang, R. Studies on polypyrrole modified Nafion membrane for vanadium redox flow battery. *Electrochemistry Communications* 2008; *10* (3): 372–375.
80. Luo, Q.; Zhang, H.; Chen, J.; You, D.; Sun, C.; Zhang, Y. Preparation and characterization of Nafion/SPEEK layered composite membrane and its application in vanadium redox flow battery. *Journal of Membrane Science* 2008; *325* (2): 553–558.
81. Sata, T. Studies on anion exchange membranes having permselectivity for specific anions in electrodialysis—Effect of hydrophilicity of anion exchange membranes on permselectivity of anions. *Journal of Membrane Science* 2000; *167* (1): 1–31.
82. Clark, T. J.; Robertson, N. J.; Kostalik IV, H. A.; Lobkovsky, E. B.; Mutolo, P. F.; Abruna, H. D.; Coates, G. W. A ring-opening metathesis polymerization route to alkaline anion exchange membranes: Development of hydroxide-conducting thin films from an ammonium-functionalized monomer. *Journal of the American Chemical Society* 2009; *131* (36): 12888–12889.
83. Robertson, N. J.; Kostalik IV, H. A.; Clark, T. J.; Mutolo, P. F.; Abruña, H. c. D.; Coates, G. W. Tunable high performance cross-linked alkaline anion exchange membranes for fuel cell applications. *Journal of the American Chemical Society* 2010; *132* (10): 3400–3404.
84. Tanaka, M.; Koike, M.; Miyatake, K.; Watanabe, M. Synthesis and properties of anion conductive ionomers containing fluorenyl groups for alkaline fuel cell applications. *Polymer Chemistry* 2011; *2* (1): 99–106.
85. Yan, J.; Hickner, M. A. Anion exchange membranes by bromination of benzylmethyl-containing poly(sulfone) s. *Macromolecules* 2010; *43* (5): 2349–2356.
86. Qiu, B.; Lin, B.; Qiu, L.; Yan, F. Alkaline imidazolium-and quaternary ammonium-functionalized anion exchange membranes for alkaline fuel cell applications. *Journal of Materials Chemistry* 2012; *22* (3): 1040–1045.
87. Wang, J.; Li, S.; Zhang, S. Novel hydroxide-conducting polyelectrolyte composed of an poly(arylene ether sulfone) containing pendant quaternary guanidinium groups for alkaline fuel cell applications. *Macromolecules* 2010; *43* (8): 3890–3896.
88. Gu, S.; Cai, R.; Luo, T.; Chen, Z.; Sun, M.; Liu, Y.; He, G.; Yan, Y. A soluble and highly conductive ionomer for high-performance hydroxide exchange membrane fuel cells. *Angewandte Chemie International Edition* 2009; *48* (35): 6499–6502.
89. Zhang, F.; Zhang, H.; Qu, C. Imidazolium functionalized polysulfone anion exchange membrane for fuel cell application. *Journal of Materials Chemistry* 2011; *21* (34): 12744–12752.
90. Dewi, E. L.; Oyaizu, K.; Nishide, H.; Tsuchida, E. Cationic polysulfonium membrane as separator in zinc–air cell. *Journal of Power Sources* 2003; *115* (1): 149–152.
91. Komkova, E.; Stamatialis, D.; Strathmann, H.; Wessling, M. Anion-exchange membranes containing diamines: Preparation and stability in alkaline solution. *Journal of Membrane Science* 2004; *244* (1): 25–34.
92. Min-suk, J. J.; Parrondo, J.; Arges, C. G.; Ramani, V. Polysulfone-based anion exchange membranes demonstrate excellent chemical stability and performance for the all-vanadium redox flow battery. *Journal of Materials Chemistry A* 2013; *1* (35): 10458–10464.
93. Wang, G.; Weng, Y.; Chu, D.; Chen, R.; Xie, D. Developing a polysulfone-based alkaline anion exchange membrane for improved ionic conductivity. *Journal of Membrane Science* 2009; *332* (1): 63–68.
94. Wang, J.; Wang, J.; Li, S.; Zhang, S. Poly(arylene ether sulfone)s ionomers with pendant quaternary ammonium groups for alkaline anion exchange membranes: Preparation and stability issues. *Journal of Membrane Science* 2011; *368* (1): 246–253.
95. Wang, J.; Zhao, Z.; Gong, F.; Li, S.; Zhang, S. Synthesis of soluble poly(arylene ether sulfone) ionomers with pendant quaternary ammonium groups for anion exchange membranes. *Macromolecules* 2009; *42* (22): 8711–8717.
96. Chen, D.; Hickner, M. A.; Wang, S.; Pan, J.; Xiao, M.; Meng, Y. Synthesis and characterization of quaternary ammonium functionalized fluorene-containing cardo polymers for potential anion exchange membrane water electrolyzer applications. *International Journal of Hydrogen Energy* 2012; *37* (21): 16168–16176.
97. Luo, Q.; Zhang, H.; Chen, J.; Qian, P.; Zhai, Y. Modification of Nafion membrane using interfacial polymerization for vanadium redox flow battery applications. *Journal of Membrane Science* 2008; *311* (1): 98–103.
98. Parasuraman, A.; Lim, T. M.; Menictas, C.; Skyllas-Kazacos, M. Review of material research and development for vanadium redox flow battery applications. *Int. J. Electrochimica Acta* 2012; *101*: 27–40.
99. Zhang, J.; Li, L.; Nie, Z.; Chen, B.; Vijayakumar, M.; Kim, S.; Wang, W.; Schwenzer, B.; Liu, J.; Yang, Z. Effects of additives on the stability of electrolytes for all-vanadium redox flow batteries. *Journal of Applied Electrochemistry* 2011; *41* (10): 1215–1221.
100. Li, S.; Huang, K.; Liu, S.; Fang, D.; Wu, X.; Lu, D.; Wu, T. Effect of organic additives on positive electrolyte for vanadium redox battery. *Electrochimica Acta* 2011; *56* (16): 5483–5487.
101. Chang, F.; Hu, C.; Liu, X.; Liu, L.; Zhang, J. Coulter dispersant as positive electrolyte additive for the vanadium redox flow battery. *Electrochimica Acta* 2012; *60*: 334–338.
102. Peng, S.; Wang, N.; Gao, C.; Lei, Y.; Liang, X.; Liu, S.; Liu, Y. Influence of trishydroxymethyl aminomethane as a positive electrolyte additive on performance of vanadium redox flow battery. *International Journal of Electrochemical Science* 2012; *7* (5): 2440–2447.
103. Kazacos, M. S.; Kazacos, M. Stabilized electrolyte solutions, methods of preparation thereof and redox cells and batteries containing stabilized electrolyte solutions. Google Patents 2000.
104. Skyllas-Kazacos, M.; Peng, C.; Cheng, M., Evaluation of precipitation inhibitors for supersaturated vanadyl electrolytes for the vanadium redox battery. *Electrochemical and Solid-State Letters* 1999; *2* (3): 121–122.
105. Hamilton, P.; Pollet, B. Polymer electrolyte membrane fuel cell (PEMFC) flow field plate: Design, materials and characterisation. *Fuel Cells* 2010; *10* (4): 489–509.
106. Li, X.; Sabir, I. Review of bipolar plates in PEM fuel cells: Flow-field designs. *International Journal of Hydrogen Energy* 2005; *30* (4): 359–371.
107. Yang, H.; Zhao, T. Effect of anode flow field design on the performance of liquid feed direct methanol fuel cells. *Electrochimica Acta* 2005; *50* (16): 3243–3252.
108. Zhu, S.-q.; Chen, J.-q.; Wang, Q.; Wang, B.-g. Influence of flow channel structure and electrolyte flow state on the performance of VRB [J]. *Battery Bimonthly* 2008; *5*: 005.

109. Aaron, D.; Tang, Z.; Papandrew, A. B.; Zawodzinski, T. A. Polarization curve analysis of all-vanadium redox flow batteries. *Journal of Applied Electrochemistry* 2011; *41* (10): 1175–1182.
110. Liu, Q.; Grim, G.; Papandrew, A.; Turhan, A.; Zawodzinski, T. A.; Mench, M. M. High performance vanadium redox flow batteries with optimized electrode configuration and membrane selection. *Journal of the Electrochemical Society* 2012; *159* (8): A1246–A1252.
111. Xu, Q.; Zhao, T.; Leung, P. Numerical investigations of flow field designs for vanadium redox flow batteries. *Applied Energy* 2013; *105*: 47–56.
112. Yang, Z.; Zhang, J.; Kintner-Meyer, M. C.; Lu, X.; Choi, D.; Lemmon, J. P.; Liu, J. Electrochemical energy storage for green grid. *Chemical Reviews* 2011; *111* (5): 3577–3613.
113. Kear, G.; Shah, A. A.; Walsh, F. C. Development of the all-vanadium redox flow battery for energy storage: A review of technological, financial and policy aspects. *International Journal of Energy Research* 2012; *36* (11): 1105–1120.
114. Evans, T.; White, R. E. A review of mathematical modeling of the zinc/bromine flow cell and battery. *Journal of the Electrochemical Society* 1987; *134* (11): 2725–2733.
115. Evans, T.; White, R. E. A mathematical model of a zinc/bromine flow cell. *Journal of the Electrochemical Society* 1987; *134* (4): 866–874.
116. Simpson, G.; White, R. E. A simple model for a zinc/bromine flow cell and associated storage tanks. *Journal of the Electrochemical Society* 1990; *137* (6): 1843–1846.
117. Simpson, G. D.; White, R. E. An algebraic model for a zinc/bromine flow cell. *Journal of the Electrochemical Society* 1989; *136* (8): 2137–2144.
118. Putt, R. Assessment of technical and economic feasibility of zinc/bromine batteries for utility load-levelling. EPRI report Em-1059, Project 635-1, Palo Alto, California, Appendix M, 1979.
119. Scamman, D. P.; Reade, G. W.; Roberts, E. P. Numerical modelling of a bromide–polysulphide redox flow battery: Part 1—Modelling approach and validation for a pilot-scale system. *Journal of Power Sources* 2009; *189* (2): 1220–1230.
120. Scamman, D. P.; Reade, G. W.; Roberts, E. P. Numerical modelling of a bromide-polysulphide redox flow battery: Part 2—Evaluation of a utility-scale system. *Journal of Power Sources* 2009; *189* (2): 1231–1239.
121. Fedkiw, P. S.; Watts, R. W. A mathematical model for the iron/chromium redox battery. *Journal of the Electrochemical Society* 1984; *131* (4): 701–709.
122. Codina, G.; Aldaz, A. Scale-up studies of an Fe/Cr redox flow battery based on shunt current analysis. *Journal of Applied Electrochemistry* 1992; *22* (7): 668–674.
123. Trinidad, P.; de León, C. P.; Walsh, F. The use of electrolyte redox potential to monitor the Ce(IV)/Ce(III) couple. *Journal of Environmental Management* 2008; *88* (4): 1417–1425.
124. (You, D.; Zhang, H.; Chen, J. A simple model for the vanadium redox battery. *Electrochimica Acta* 2009; *54* (27): 6827–6836.
125. You, D.; Zhang, H.; Sun, C.; Ma, X. Simulation of the self-discharge process in vanadium redox flow battery. *Journal of Power Sources* 2011; *196* (3): 1578–1585.
126. Enomoto, K.; Sasaki, T.; Shigematsu, T.; Deguchi, H. Evaluation study about redox flow battery response and its modeling. *Transactions of the Institute of Electrical Engineers of Japan B* 2002; *122* (4): 554–560.
127. Minghua, L.; Hikihara, T. A coupled dynamical model of redox flow battery based on chemical reaction, fluid flow, and electrical circuit. *IEICE Transactions on Fundamentals of Electronics, Communications and Computer Sciences* 2008; *91* (7): 1741–1747.
128. You, D.; Zhang, H.; Chen, J. Theoretical analysis of the effects of operational and designed parameters on the performance of a flow-through porous electrode. *Journal of Electroanalytical Chemistry* 2009; *625* (2): 165–171.
129. Ma, X.; Zhang, H.; Xing, F. A three-dimensional model for negative half cell of the vanadium redox flow battery. *Electrochimica Acta* 2011; *58*: 238–246.
130. Li, M.; Hikihara, T. A coupled dynamical model of redox flow battery based on chemical reaction, fluid flow, and electrical circuit. *IEICE Transactions on Fundamentals of Electronics Communications and Computer Science* 2008; *E91A*: 1741.
131. Leung, P.; Li, X.; de León, C. P.; Berlouis, L.; Low, C. J.; Walsh, F. C. Progress in redox flow batteries, remaining challenges and their applications in energy storage. *RSC Advances* 2012; *2* (27): 10125–10156.

23 Liquid–Redox Flow Battery Applications

Jinfeng Wu, Jianlu Zhang, and Jiujun Zhang

CONTENTS

23.1 INTRODUCTION

Over the past several decades, the continually widening gap between supply and demand in electricity grids has drawn unprecedented attention and effort around the world. The balancing between supply and demand has significant operational and cost implication. Since it is impossible to directly store energy in the form of electricity, various energy-storage technologies have been considered and employed to convert electrical energy to another form that can be stored, including kinetic energy (flywheels or compressed air), gravitational potential energy (pumped hydroelectric), electrical and magnetic fields, and electrochemical energy (batteries and supercapacitors). Different types of energy-storage technologies have their own benefits and flaws. For examples, pumped hydroelectric and compressed air energy storages are both reliable and relatively mature energy-management technologies. However, not only their application and employment require large facilities, which have a strong impact on the local environment, but also they are extremely limited by geographical conditions. In Table 23.1, the technical suitabilities of large-scale energy-storage systems (EESs) for different applications are provided [1].

Normally, direct comparison and selection are generally not appropriate. A more useful approach is to begin by identifying the application or applications that the device is to serve and then to identify the types of storage technologies that are appropriate for that application [2]. Depending on the customers' requirements, the operating times of ESSs normally range from the timescale of fractions of a second, with response times on the order of milliseconds to minutes. Meanwhile, their powers range from kilowatts for domestic utilities to multi-gigawatts for large plants.

Electrochemical energy–storage systems store electrical energy in the form of chemical energy in a rechargeable battery, which consists of a liquid, paste, or solid electrolyte together with a positive electrode and a negative electrode. During discharge, electrochemical reactions occur at the two electrodes, generating a flow of electrons through an external circuit. In order to deliver high power and store more energy, multiple electrochemical cells are generally electrically connected together in series within a common housing referred to as an electrochemical "stack." Multiple stacks electrically connected together are generally referred to as "strings." Furthermore, multiple strings electrically connected together are generally referred to as "sites."

Electrochemical energy–storage devices include lead–acid, nickel–cadmium, lithium-ion, supercapacitors, fuel cells, liquid-redox rechargeable, and so on, among which liquid-redox flow batteries are unique because they can convert electrical energy into chemical potential energy by means of a reversible electrochemical reaction between two liquid electrolyte solutions. Differing from other conventional batteries, liquid-redox flow batteries are electrochemical energy–conversion devices that exploit solution redox processes of species in fluid form. These solution redox species can be stored in external tanks and introduced into the batteries when needed. Therefore, the power and energy capacity can be independent, meaning that the storage capacity is determined by the quantity of electrolyte used and the rating power is decided by the active area as well as the cell number of the battery stacks. This feature enables liquid-redox flow batteries to span a wide range of energy capacity from kilowatt-hours to hundreds of megawatt-hours and makes them suitable for integration with renewable energy. The other features of liquid-redox flow batteries that are claimed to be appealing are their abilities to tolerate fluctuating power supplies, bear repetitive charge–discharge cycles at maximum rates without damage to the battery, and initiate charge–discharge cycling at any state of charge and their long durability, high round-trip efficiency, fast responsiveness, flexible choice of battery location, and low operational cost, as well as high level of security without fire hazards.

TABLE 23.1
Technical Suitabilities of Large-Scale EESs

Storage Application	Lead–Acid Batteries	Flow Batteries	Flywheels	Pumped Hydro-EES	Compressed Air EES
Transit and end-use ride-through	✓	✓	✓		
Uninterruptible power supply	✓	✓	✓		
Emergency backup	✓	✓			✓
Transmission and distribution stabilization and regulation	✓	✓			
Load leveling[a]	✓	✓		✓	✓
Load following[b]	✓	✓			
Peak generation	✓	✓	✓	✓	✓
Fast response spinning reserve	✓	✓	✓		
Conventional spinning reserve	✓	✓	✓	✓	✓
Allowance for renewable integration	✓	✓	✓	✓	✓
Suitability for renewable backup	✓	✓		✓	✓

Source: Poullikkas, A., A comparative overview of large-scale battery systems for electricity storage, *Renewable and Sustainable Energy Reviews*, 27, 778–788, 2013. With permission.

[a] Reducing the large fluctuations that occur in electricity demand.

[b] Adjusting power output as demand for electricity fluctuates throughout the day.

It has been expected that liquid-redox flow batteries, as a kind of efficient energy-storage technologies, should have wide applications and vast market prospects [3]. Various liquid-redox flow battery systems have been introduced in the past few decades, but only all-vanadium redox flow batteries, polysulfide bromide systems (PSBs), zinc–bromine (Zn-Br) systems, and zinc–cerium (Zn-Ce) systems have been tested or precommercialized on a large scale. The main application areas of liquid-redox rechargeable batteries include the following:

1. Integration with renewable energy sources
2. ESSs for grids
3. Uninterruptible power systems (UPSs) and emergency backup power sources
4. Electric vehicles' power batteries
5. Electric vehicle charging stations
6. Telecommunication base stations

In this chapter, these six application areas will be given a detailed discussion in the following sections.

23.2 INTEGRATION WITH RENEWABLE ENERGY SOURCES

Rising concerns about environmental pollution and climate change have dramatically accelerated the research and development of renewable energy in the past few years. The financial support by feed-in tariffs worldwide has led to a rapid increase in the installation of renewable energy capacities. For example, more than 73% of the national supply on May 11, 2014, in Germany came from renewable energy [4]. On June 16, 2013, the renewable energy accounted for 60% of power supply, in which wind energy contributed with approximately 9 GW and photovoltaics with 20 GW [5]. More importantly, it is worthy to point out that power quality has become a problematic topic, with increasing shares of renewable energy in the total energy supply [6].

As recognized, the renewable power sources, specifically solar photovoltaics and wind, are highly unreliable and have large fluctuations in power output due to weather variations, which makes them difficult to be fully integrated into the existing electrical power grids and distribution networks, although they are capable of being supplemental to the grid. Furthermore, with the increasing proportion of renewable energy in the total energy supply in the future, matching supply with demand will become increasingly difficult. Recently, connecting large-scale electrical energy storage with renewable energy has been accepted as a promising way to create renewable distributed generation and enable these intermittent sources to smooth power output, improve efficiency, and then further achieve the popularization of renewable energy. Particularly, for grid applications, the energy from these fluctuating small, medium, and large renewable energy sources, such as wind, solar, and ocean thermal, has to be adjusted before integrating in terms of power peak and frequency modulation to ensure a high standard of grid operation. Therefore, to develop an efficient energy-storage technology compatible with renewable energy generation is highly necessary for achieving a large-scale utilization of renewable energy [7,8].

In 1994, the first implementation of liquid-redox flow battery with solar energy sources was a 5 kW/12 kWh vanadium redox flow battery (VRB) system installed in Thailand by Thai Gypsum Products Co. Ltd. [9]. In recent years, Austria, the United States, Japan, China, and some European countries have integrated, one after another, liquid-redox flow batteries with wind or photovoltaic energy supplies for peak shaving of power stations [10]. For example, VRB Power Systems Inc.

in Canada (purchased by Prudent Energy Co. in 2009) set up a 250 kW/1 MWh VRB system in 2003 for application in the Australian King Island wind power system to provide a stable and reliable power supply [11]. It had four sets of 1500 kW diesel power stacks and three sets of 250 kW and two sets of 850 kW wind power stacks. In 2005, SOLON AG in Germany purchased the 10 kWh VRB system from the VRB Power Systems Inc., which was integrated with a photovoltaic system [12]. It was claimed that the VRB system played the role of stabilizing the transient power transport changes and load fluctuations in wind power generation, controlling frequency and voltage, enabling load transfer, optimizing the performance of diesel–wind hybrid power systems, and reducing the workload of the diesel power stack, thereby reducing fuel costs and waste emission. The 170 kW/1 MWh VRB system developed by Sumitomo Electric Power (SEI) in 2001 was built in Hokkaido, Japan, for smoothing power output fluctuations of wind turbines [13]. Later on, SEI installed another 4 MW/6 MWh VRB system in Hokkaido for the Subaru wind farm [14]. The 4 MW/6 MWh system consisted of four banks, each made of 24 stacks and rated 1 MW (maximum 1.5 MW). Each stack contained 108 cells, with a rated power of 45 kW. Over 3 years of operation it completed more than 270,000 cycles. A typical 45 kW/90 kWh VRB system is shown in Figure 23.1 [15]. In China, the Zhangbei national wind and solar energy storage and transmission demonstration project had a 2 MW VRB system provided by Prudent Energy [16]. In 2013, Dalian Rongke Power Co. Ltd. successfully installed a 5 MW/10 MWh VRB ESS to complement a 50 MW wind farm in Shenyang, Liaoning Province, which has the VRB ESS with the largest power and capacity in the world so far [17]. This VRB system was used for power generation tracking, wind output smoothing, frequency regulating, and voltage supporting.

Regarding other type of liquid-redox flow batteries, the Zn-Br flow battery has been explored in utility energy-storage applications [18]. However, it is still in the early stage of demonstration/commercialization, mainly limited by material corrosion, high self-discharge rate, dendrite formation, electrical shorting during zinc deposition on charge, short cycle life, and unsatisfied energy efficiency. Early installation of the Zn-Br system had a capacity of 50 kW/100 kWh, which was developed by ZBB Energy Corporation in the United States in 2002, coupled with 50 kW_{ac} rooftop solar panels in Brooklyn, New York [19]. In 2008, ZBB declared that they provided Ireland with a 500 kWh Zn-Br system for energy storage and wind energy technology. In 2011, U.S. electric utilities conducted early trials of 2.8 MWh Zn-Br liquid-redox flow battery systems coupling with 500 kW solar panels for grid support and reliability in Albuquerque [20]. Within the same year, the Sacramento Municipal Utility District tried to demonstrate a 1 MW Zn-Br system for utility grid application, which would validate the benefits of a Zn-Br storage system for load shifting, peak shaving, and support for microgrid operations, and renewable energy integration. Generally speaking, the demand for energy storage from wind farms and solar panels will be increased dramatically in the coming decades.

(a)

(b)

FIGURE 23.1 Power and energy components of an SEI VRB system rated 125 V, 500 A, 45 kW, 90 kWh: (a) stack, consisting of 100 cells in series; (b) two tanks, sized 5 m^3 and storing 4000 L. (From Alotto, P., M. Guarnieri, and F. Moro, Redox flow batteries for the storage of renewable energy: A review, *Renewable and Sustainable Energy Reviews*, 29: 325–335, 2014. With permission.)

From the above examples, it can be foreseen that liquid-redox flow batteries will be an indispensable technology for construction of dynamic and renewable energy–storage systems, including wind and solar energy. They would be able to bring significant improvements in the efficiency and profits of power plants and promote the development of renewable energy industries. The major operational and planned large-scale liquid-redox flow battery systems coupling with renewable energy sources around the world are tabulated in Table 23.2 [1–21].

TABLE 23.2
Major Liquid-Redox Rechargeable Battery Systems Coupling with Renewable Energy Sources around the World

Project	Technology	Rated Power (kW)	Duration (h)	Status	Applications	State/ Province	Country	EES Provider
Prudent Energy VRB-ESS® (Gills Onions, Calif.)	VRB	600	6	Operational	Grid-connected commercial (reliability and quality)	California	U.S.	Prudent Energy Co.
City of Painesville Municipal Power VRB Demonstration	VRB	1080	8	Contracted	Load following (tertiary balancing); electric energy time shift; transmission congestion relief; electric supply reserve capacity (spinning); voltage support	Ohio	U.S.	Ashlawn Energy LLC
Illinois Institute of Technology RDSI Perfect Power Demonstration	Zn-Br	250	2	Under construction	Black start	Illinois	U.S.	ZBB Energy Corp.
UCSD ZBB/ Sunpower Energy Storage CSI 2	Zn-Br	100	3	Operational	Electric bill management	California	U.S.	ZBB Energy Corp.
UEP CCNY Demonstration	Zinc-nickel oxide flow battery	100	2	Operational	Electric bill management	New York	U.S.	UEP
Sumitomo Densetsu Office	VRB	3000	0.27	Operational	Electric bill management	Kansai	Japan	SEI
PacifiCorp Castle Valley VRB	VRB	250	8	Decommissioned	Electric energy time shift; electric supply capacity	Utah	U.S.	VRB Power Systems
Sumba Island Microgrid Project	VRB	400	1.25	Operational	Electric energy time shift; electric supply capacity	East Nusa Tenggara	Indonesia	Prudent Energy Corp.
Powerco's RedFlow Battery Demonstration	Zn-Br	3	2.67	Operational	Electric supply capacity; grid-connected residential (reliability)	Taranaki	New Zealand	RedFlow
Samyoung VRB Project	VRB	50	2	Operational	Electric energy time shift	Chungnam	South Korea	H2 Inc.
VISA (data-processing center)	Zn-Br	25	2	Operational	On-site power	Virginia	U.S.	ZBB Energy Corp.
BPC Energy	Zn-Br	25	2	Under construction	Voltage support; on-site power	Moscow	Russia	ZBB Energy Corp.
PSE Storage Innovation Project 1	Zn-Br	500	2	Under construction	Transportable transmission/ distribution upgrade deferral; electric supply capacity; ramping	Washington	U.S.	Primus Power

(Continued)

TABLE 23.2 (CONTINUED)
Major Liquid-Redox Rechargeable Battery Systems Coupling with Renewable Energy Sources around the World

Project	Technology	Rated Power (kW)	Duration (h)	Status	Applications	State/ Province	Country	EES Provider
American Vanadium Corp. MTA CellCube Installation	VRB	90	4.33	Under construction	Grid-connected commercial (reliability and quality); electric bill management; electric supply capacity; black start; electric energy time shift	New York	U.S.	Gildemeister Energy Solutions
UET HQ BESS	VRB	600	3	Under construction	On-site power; electric energy time shift	Washington	U.S.	UET
Snohomish PUD: MESA 2 BESS	VRB	2000	2	Contracted	Electric energy time shift	Washington	U.S.	UET
Avista UET BESS	VRB	1000	3.2	Contracted	Electric energy time shift; load following (tertiary balancing); ramping; microgrid capability; voltage support	Washington	U.S.	UET
Flathead Electric ViZn Z20	Zn-Fe Flow Battery	80	2	Operational	Electric energy time shift	Montana	U.S.	ViZn Energy Systems
Zn-Br–based Transportable ESS	Zn-Br	500	5.6	Operational	Grid-connected commercial (reliability and quality)	New Mexico	U.S.	Primus Power

Source: Shibata, A. S., and K. Sato, Development of vanadium redox flow battery for electricity storage, *Power Engineering Journal*, 13, 130–135, 1999; http://www.cleanbreak.ca/2005/09/30/vrb-power-strikes-again-another-solar-deal/ (accessed November 4, 2014); V-Fuel Pty. Ltd., Status of energy storage technologies as enabling systems for renewable energy from the sun, wind, waves and tides, http://www.aph.gov.au/parliamentary_business/committees/house_of_representatives_committees?url=isr/renewables/submissions/sub21.pdf (accessed November 4, 2014); McDowell, J., International Renewable Energy Storage (IRES) Conference 2006, Gelsenkirchen, Germany, pp. 30–31; http://www.pdenergy.com/case-studies.php (accessed October 30, 2014); http://www.rongkepower.com/index.php/article/show/id/140/language/en (accessed October 30, 2014); Alotto, P., M. Guarnieri, and F. Moro, Redox flow batteries for the storage of renewable energy: A review, *Renewable and Sustainable Energy Reviews*, 29, 325–335, 2014; Luo, X., J. Wang, M. Dooner, and J. Clarke, Overview of current development in electrical energy storage technologies and the application potential in power system operation, *Applied Energy*, 137, 511–536, 2015; Norris, B. L., G. J. Ball, P. Lex, and V. Scaini, Grid-connected solar energy storage using the zinc-bromine flow battery, http://www.zbbenergy.com/files/1112/8285/0124/technicalpaper_grid.pdf (accessed October 28, 2014); Rastler, D., Electric energy storage applications in the electric enterprise: Overview of applications and U.S. demonstration activities, Electric Power Research Institute (EPRI), EDF Energy Storage Workshop, October 2011.

Note: BESS: battery energy-storage system; CCNY: City College of New York; HQ: headquarters; MESA: Modular Energy Storage Architecture; MTA: Metropolitan Transportation Authority; PSE: Puget Sound Energy; PUD: Public Utility District; UCSD: University of California, San Diego; UEP: Urban Electric Power; UET: UniEnergy Technologies.

23.3 APPLICATION AS ENERGY-STORAGE SYSTEM FOR THE GRID

In addition to facilitating the integration of renewable wind and solar energy, the incorporation of large-scale electric energy–storage systems has great benefits for current electrical grid management. With the fast development of society and economy, there is an ever-increasing demand for more power. Meanwhile, demands differ greatly between night and day, and the power plant must be able to satisfy peak demand. In order to satisfy such demands and maintain a normal operation, more and more localized power plants have been built, resulting in larger and larger differences between peak and off-peak loads and causing a low generator efficiency as well as high waste of fossil fuels during off-peak times [3]. To resolve this issue, ESSs can be used to store the excess power of large-scale power plants and avoid high-cost plants to cycle on and off frequently. For example, ESSs can store off-peak electricity for use in peak hours, which reduces both the power generation requirements at peak hours and the scale of the power plant. This is particularly important for improving the efficiency of power generation and maintaining the safety and stability of the grid. Furthermore, the stored electricity generated in off-peak hours, which is normally inexpensive, can be used as needed, significantly cutting the operational cost of manufacturing. By employing liquid-redox flow batteries as ESSs, the bottleneck within the transition system can be relieved, which is helpful in reducing power transmission losses and leading to more reliable electrical power. Therefore, ESSs such as liquid-redox flow batteries can play an important role in grid energy management for optimizing energy uses and have many positive effects, such as frequency regulation, spinning reserves, fast ramping capacity, black start capacity, possibly replacement of fossil fuel peaking systems, transmission and delivery support by increasing capability of existing assets and deferring grid upgrade investments, microgrid support, time shifting, and peak shaving and power shifting.

Regarding the application of Zn-Br redox flow batteries, as early as 1990, Kyushu Electric Power Co. and Meidensha Co. installed a 1 MW/4 MWh Zn-Br battery system in Imajuku, Japan, for electric utility application [22]. In 1996, Kashima-Kita and SEI in Japan successfully developed 200 kW/800 kWh and 450 kW/900 kWh VRB systems, respectively, for demonstration of load leveling and peak shaving of the power grid [3]. Subsequently, SEI developed a series of VRB systems with powers/capacities of (200 kW/1.6 MWh)/(500 kW/5 MWh) in 2000 and (1.5 MW/1.5 MWh)/(500 kW/2 MWh) in 2001 and 2003. These are all for the improvement of grid safety and stability [3]. In 2004, VRB Power Systems Inc. built a 250 kW/2 MWh VRB system for Castle Valley, Utah Pacific Corporation, also for load leveling and peak shaving [23]. Most recently, UniEnergy Technologies LLC (UET) in the United States announced to install three VRB systems by deploying UET's Uni.System™ grid-scale ESS, which includes (1) a 1 MW/3.2 MWh system for Avista in Pullman, Washington, to support the Washington State University's smart campus operations; (2) a 2 MW/6.4 MWh system for Snohomish County; and (3) a 500 kW system in California [24]. Uni.System system is modular, factory integrated (including power conversion), and "plug and play" and comprises five 20 ft standard containers that provide 500 kW_{ac} of power for 4 hours, with power up to 600 kW_{ac} and energy up to 2.2 MWh_{ac}, as shown in Figure 23.2 [24].

In the early 2000s, the United Kingdom's Innogy Company built the first business-scale PSB flow batteries in Little Barford, Cambridgeshire, United Kingdom, and in Columbus, Mississippi, United States, whose energy-storage capacity and output power could reach up to 15 MW/120 MWh [25]. ZBB Energy Corporation installed a 200 kWh Zn-Br battery stack in Australia's United Energies Inc., whose main function was load leveling and peak shaving [26]. For Zn-Ce flow batteries, the major installation was the 2 kW/1 MWh testing facility in Glenrothes, Scotland, installed by Plurion Inc. (United Kingdom) in 2007 [9].

From the demonstrations mentioned, it can be anticipated that liquid-redox flow batteries would play an important role in load leveling of power grids in the future. The main installations of liquid-redox flow batteries integrated with the grid as electric energy–storage systems around the world are summarized in Table 23.3 [9,21–26].

FIGURE 23.2 Uni.System grid-scale ESS string, including four battery containers and one power conversion container. (From http://www.uetechnologies.com/index.htm [accessed October 28, 2014]. With permission.)

TABLE 23.3
Main Installations of Liquid-Redox Rechargeable Batteries Integrated with the Grid around the World

Project	Technology	Rated Power (kW)	Duration (h)	Status	Applications	State/ Province	Country	EES Provider
MID Primus Power Wind Firming EnergyFarm	Zn-Br	28,000	4	Announced	Renewables capacity firming; renewables energy time shift	California	U.S.	Primus Power
Ausgrid SGSC: 20 RedFlow Systems	Zn-Br	100	2	Decommissioned	Stationary transmission/ distribution upgrade deferral; renewables energy time shift; grid-connected residential (reliability)	New South Wales	Australia	RedFlow
RedFlow, University of Queensland M90	Zn-Br	90	2	Decommissioned	Renewables energy time shift; renewables capacity firming	Queensland	Australia	RedFlow
Fort Sill Microgrid	Zn-Br	250	2	Decommissioned	Renewables capacity firming; microgrid capability	Oklahoma	U.S.	ZBB Energy Corp.
Port Hueneme Naval Facility	Zn-Br	25	2	Under construction	Renewables capacity firming; black start	California	U.S.	ZBB Energy Corp.
Pualani Manor	Zn-Br	60	2.5	Operational	Renewables capacity firming; black start; electric energy time shift	Hawaii	U.S.	ZBB Energy Corp.
San Nicolas Island Naval Facility	Zn-Br	500	2	Under construction	Renewables capacity firming; on-site renewable generation shifting	California	U.S.	ZBB Energy Corp.
St. Petersburg Solar Parks Project	Zn-Br	25	2	Under construction	Electric bill management with renewables; renewables capacity firming	Florida	U.S.	ZBB Energy Corp.
Zhangbei National Wind and Solar Energy Storage and Transmission Demonstration Project (V)	VRB	2000	4	Operational	Renewables energy time shift; renewables capacity firming; frequency regulation; ramping	Hebei	China	Prudent Energy, GE
EnStorage Technology Demonstrator	Hydrogen–bromine redox flow battery	50	2	Operational	Renewables capacity firming; renewables energy time shift; transportable transmission/ distribution upgrade deferral; transmission upgrades due to wind; on-site renewable generation shifting	Negev	Israel	
SMUD HQ Premium Power DESS	Zn-Br	500	6	Under construction	Electric supply capacity; renewables energy time shift; load following (tertiary balancing)	California	U.S.	Premium Power
National Grid DESSs Demonstration (Worcester, Mass.)	Zn-Br	500	6	Contracted	Renewables energy time shift; renewables capacity firming; electric energy time shift; voltage support; grid-connected commercial (reliability and quality)	Massachusetts	U.S.	Premium Power

(Continued)

TABLE 23.3 (CONTINUED)
Main Installations of Liquid-Redox Rechargeable Batteries Integrated with the Grid around the World

Project	Technology	Rated Power (kW)	Duration (h)	Status	Applications	State/ Province	Country	EES Provider
Yokohama Works	VRB	1000	5	Operational	On-site renewable generation shifting; renewables capacity firming; electric supply capacity	Kanagawa	Japan	SEI
DOD Marine Corps Air Station Miramar Microgrid ESS	Zn-Br	250	4	Under construction	Renewables capacity firming; electric bill management with renewables; on-site renewable generation shifting; black start; microgrid capability	California	U.S.	Primus Power
Tomamae Wind Farm	VRB	4000	1.5	Operational	Renewables capacity firming; on-site renewable generation shifting; renewables energy time shift	Hokkaido	Japan	SEI
SunCarrier Omega Net Zero Building (Bhopal)	VRB	45	6.67	Operational	On-site renewable generation shifting	Madhya Pradesh	India	SunCarrier Omega Pvt. Ltd.
CENER VRB	VRB	50	2	Operational	Electric energy time shift; renewables energy time shift; on-site power; on-site renewable generation shifting; microgrid capability	Navarra	Spain	Prudent Energy Corp.
RISO Syslab Redox Flow Battery	VRB	15	8	Operational	Renewables capacity firming; renewables energy time shift; frequency regulation; voltage support; load following (tertiary balancing)	Lyngby-Taarbæk	Denmark	Prudent Energy Corp.
PVCROPS Evora Demonstration Flow Battery Project	VRB	5	12	Operational	Renewables capacity firming; renewables energy time shift	Norte	Portugal	REDT
Global Change Institute M120	Zn-Br	120	2.5	Operational	On-site renewable generation shifting; load following (tertiary balancing); renewables energy time shift	Queensland	Australia	
Gigha Wind Farm Battery Project	VRB	100	12	Contracted	Renewables energy time shift; renewables capacity firming	Scotland	United Kingdom	REDT
Enel Livorno Test Facility: 10 kW VRB	VRB	10	10	Operational	Renewables capacity firming; renewables energy time shift; voltage support	Livorno	Italy	Cellstrom
Minami Hayakita Substation VRB	VRB	15,000	4	Contracted	Renewables capacity firming; renewables energy time shift; voltage support	Hokkaido	Japan	SEI
Prudent Energy Corp./CEPRI Vanadium Redox Battery ESS	VRB	500	2	Operational	Renewables capacity firming; renewables energy time shift; load following (tertiary balancing); voltage support	Hebei	China	Prudent Energy, GE
KIER/Juju Island Vanadium Redox Battery Project	VRB	100	2	Operational	Renewables capacity firming; renewables energy time shift; frequency regulation; voltage support	Juju	South Korea	Prudent Energy

(Continued)

TABLE 23.3 (CONTINUED)
Main Installations of Liquid-Redox Rechargeable Batteries Integrated with the Grid around the World

Project	Technology	Rated Power (kW)	Duration (h)	Status	Applications	State/ Province	Country	EES Provider
PV and EV Charging Systems at Sporting Venue	Zn-Br	25	2	Under construction	On-site renewable generation shifting; renewables capacity firming; voltage support	Arizona	U.S.	ZBB Energy Corp.
JBPHH U.S. Military Base	Zn-Br	125	3.33	Under construction	Black start; on-site renewable generation shifting	Hawaii	U.S.	ZBB Energy Corp.
Tetiaroa Brando Resort	Zn-Br	1000	2	Under construction	Renewables capacity firming; on-site power	Tahiti	French Polynesia	ZBB Energy Corp.
University of Technology Sydney	Zn-Br	25	2	Under construction	Renewables capacity firming	Sydney	Australia	ZBB Energy Corp.
CSIRO, ZBB Experimental Zinc–Bromide Flow Battery	Zn-Br	100	5	Decommissioned	Electric bill management; renewables energy time shift	New South Wales	Australia	ZBB Energy Corp.
Dundalk Institute of Technology: Coupling with 850 kW Wind Turbine	Zn-Br	125	4	Operational	Renewables capacity firming	County Louth	Ireland	ZBB Energy Corp.
SmartRegion Pellworm (VRB)	VRB	200	8	Operational	Renewables energy time shift; electric supply capacity; electric energy time shift	North Sea	Germany	Gildemeister Energy Solutions, Gustav Klein
Fotonenboer 't Spieker Dairy Farm	VRB	10	8	Operational	On-site renewable generation shifting; electric bill management with renewables; on-site power	Gelderland	Netherlands	Gildemeister Energy Solutions
Swiss Dual-Circuit Redox Flow Battery	VRB	10	6	Under construction	Renewables energy time shift	Valais	Switzerland	
SMUD Premium Power DESS, Anatolia SolarSmart Homes Project	Zn-Br	500	6	Under construction	Renewables energy time shift; electric bill management with renewables; renewables capacity firming	California	U.S.	Premium Power
NREL American Vanadium CellCube Test Site	VRB	20	4	Operational	Renewables capacity firming; grid-connected commercial (reliability and quality); microgrid capability; electric supply capacity; electric bill management	Colorado	U.S.	
Bosch Braderup ES Facility: Flow Battery	VRB	325	3.1	Under construction	Transmission congestion relief; on-site renewable generation shifting; frequency regulation; transportable transmission/distribution upgrade deferral	Schleswig-Holstein	Germany	UET, Vanadis Power, Rongke Power

(*Continued*)

TABLE 23.3 (CONTINUED)
Main Installations of Liquid-Redox Rechargeable Batteries Integrated with the Grid around the World

Project	Technology	Rated Power (kW)	Duration (h)	Status	Applications	State/ Province	Country	EES Provider
MID Primus Power Wind Energy Storage Demonstration: Renewables Firming	Zn-Br	250	4	Under construction	Renewables energy time shift	California	U.S.	Primus Power
King Island Renewable Energy Expansion VRB	VRB	200	4	Off-line/under repair	Renewables capacity firming; on-site renewable generation shifting	Tasmania	Australia	SEI
BlueSky Energy Microgrid ViZn Z20	Zinc–iron flow battery	64	2.8	Operational	Renewables capacity firming; on-site renewable generation shifting	Upper Austria	Austria	ViZn Energy Systems
Guodian Hefeng Beizhen Wind Farm: VFB	VRB	2000	2	Under construction	On-site renewable generation shifting; renewables capacity firming; frequency regulation; voltage support	Liaoning	China	Rongke Power
GuoDian LongYuan Wind Farm VFB	VRB	5000	2	Operational	On-site renewable generation shifting; renewables capacity firming; voltage support; electric supply reserve capacity (spinning)	Liaoning	China	Rongke Power
Liaoning Electric Power Research Institute Micro-grid Project	VRB	100	4	Under construction	On-site renewable generation shifting; on-site power; microgrid capability	Liaoning	China	Rongke Power
Snake Island Independent Power Supply System VFB	VRB	10	20	Operational	On-site power; on-site renewable generation shifting	Liaoning	China	Rongke Power
Dalian EV Charging Station VFB	VRB	60	10	Operational	Transportation services; renewables energy time shift; grid-connected commercial (reliability and quality)	Liaoning	China	Rongke Power
Gold Wind Smart Micro-grid VFB	VRB	200	4	Operational	On-site renewable generation shifting; microgrid capability	Beijing	China	Rongke Power
RKP R&D Center Building VFB	VRB	60	5	Operational	Renewables energy time shift	Liaoning	China	Rongke Power
CellCube Industrial Smart Grid	VRB	260	2.5	Operational	On-site renewable generation shifting; grid-connected commercial (reliability and quality); load following (tertiary balancing)	North Rhein–Westfalia	Germany	DMG Mori AG
SunCarrier Omega Net Zero Building (Bhopal)	VRB	45	6.67	Operational	On-site renewable generation shifting	Madhya Pradesh	India	SunCarrier Omega Pvt. Ltd.

(Continued)

TABLE 23.3 (CONTINUED)
Main Installations of Liquid-Redox Rechargeable Batteries Integrated with the Grid around the World

Project	Technology	Rated Power (kW)	Duration (h)	Status	Applications	State/ Province	Country	EES Provider
Terna Storage Lab 1	VRB	1000	4	Announced	Frequency regulation; transmission upgrades due to wind; transmission support; voltage support; black start	Sardinia	Italy	Tender in process
Terna Storage Lab 2	VRB	500	4	Announced	Frequency regulation; transmission upgrades due to wind; transmission support; voltage support; black start	Sicily	Italy	Tender in process

Source: Leung, P., X. Li, C. P. D. Leon, L. Berlouis, C. T. John Low, and F. C. Walsh, Progress in redox flow batteries, remaining challenges and their applications in energy storage, *RSC Advances*, 2, 10125–10156, 2012; Department of Energy's Energy Storage Database, Sandia National Laboratories, United States Department of Energy, http://www.energystorageexchange.org/ (accessed November 8, 2014); Butler, P. C., P. A. Eidler, P. G. Grimes, S. E. Klassen, and R. C. Miles, Zinc bromine batteries, In *Handbook of Batteries*, Linden, D., and T. B. Reddy (eds.), Chapter 39, McGraw-Hill, New York, 2002; Doughty, D. H., P. C. Butler, A. A. Akhil, N. H. Clark,, and J. D. Boyes, Batteries for large-scale stationary electrical energy storage, *Electrochemical Society Interface*, 49–53, Fall 2010; http://www.uetechnologies.com/index.htm (accessed October 28, 2014); Hazza, A., D. Pletcher, and R. Wills, A novel flow battery—A lead acid battery based on an electrolyte with soluble lead (II); IV. The influence of additives, *Journal of Power Sources*, 149, 103–111, 2005; Review of electrical energy storage technologies and systems and of their potential for the UK, Contract number DG/DTI/00055/00/00, URN number 04/1876, 2004, http://www.wearemichigan.com/JobsAndEnergy/documents/file15185.pdf (accessed October 30, 2014).

Note: CENER: Centro Nacional de Energías Renovables; CEPRI: China Electric Power Research Institute; CSIRO: Commonwealth Scientific and Industrial Research Organisation; DESS: distributed energy-storage system; DOD: Department of Defense; EV: electric vehicle; GE: General Electric; HQ: headquarters; JBPHH: Joint Base Pearl Harbor Hickam; KIER: Korean Institute of Energy Research; MID: Modesto Irrigation District; NREL: National Renewable Energy Laboratory; PV: photovoltaic vehicle; PVCROPS: PhotoVoltaic Cost Reduction, Reliability, Operational Performance, Prediction and Simulation; REDT: Renewable Energy Dynamics Technology Ltd.; RKP: Rongke Power; SGSC: Smart Grid, Smart City; SMUD: Sacramento Municipal Utility District; UET: UniEnergy Technologies; VFB: vanadium redox flow battery.

23.4 APPLICATION AS UNINTERRUPTIBLE POWER SUPPLY AND EMERGENCY BACKUP POWER SUPPLY

The availability of reliable and secure power supply is essential in the rapidly growing telecommunications and information technology to safeguard the vast computer networks that have been established. UPS systems have been widely used in the financial industries and telecommunication of nowadays, where reliable power supply for vital computer installations and electrical equipment is the top priority. These systems commonly rely on the sealed lead–acid batteries to provide 10 to 15 minutes of power in case of emergency power failure, enabling an orderly shutdown of computer systems or allowing switching over of smooth power feeding until generators come on line [27]. However, due to the poor performance and relatively short cycle life of lead–acid batteries under deep discharge cycling, there will always be a heavy dependence on the diesel generator for longer-term power generation in the event of power failure [27]. The main features of liquid-redox flow batteries, such as relatively low cost, scalability, easy maintenance, and especially their ability of offering efficient "instant recharge" by replacing electrolytes, should distinguish them from other batteries, such as lead–acid batteries, in emergency power supply application. In 1990s, the University of New South Wales successfully developed several kilowatt-scale VRB batteries as UPS systems with energy efficiency above 80% [27].

Besides UPS application, some other industries also require power supplies with high quality, efficiency, environmental friendliness, safety, and reliability. These industries include electrolysis, electroplating, and metallurgy and transport services (trolleybus, light railway train, and subway), all of which are intensive electricity consumers. These industries and services can use off-peak electricity to charge energy-storage batteries, which are then discharged for peak-hour use [3]. This can alleviate the workload and lower the operational cost of the power grid. Another important role for large-scale high-efficiency ESSs is to provide several hours' backup power for government operations, hospitals, transformer stations, and other crucial facilities so that they can operate properly in times of power failure. With respect to these applications, liquid-redox flow batteries are superior over lead–acid batteries in terms of energy-storage capacity, output power, safety, stability, and life span and thus have a very promising future in the area of backup power supply. For example, one 1.5 MW/1.5 MWh VRB system was developed by VRB Power Systems Inc. in 2001 for emergency backup power supply [11].

23.5 APPLICATION AS ELECTRIC VEHICLE POWER BATTERIES

The power density and energy density of liquid-redox flow batteries are relatively low compared to other batteries, making them unsuitable for mobile applications at present. To date, only limited studies and tests have been involved in the use of liquid-redox flow batteries for electric vehicle or other mobile applications due to their low specific energy compared to lithium ion battery and fuel cells, as well as their relatively narrow operating temperature range. A VRB-powered electric golf cart was field-tested in 1995–1996 by the University of New South Wales, Australia [28].

With 40 L of 1.85 M vanadium electrolyte per half-cell tank, an off-road driving range of 17 km was obtained, suggesting that the energy density of an optimized all-vanadium flow battery could approach that of lead–acid battery, with the additional advantage of rapid recharging by electrolyte replacement. Subsequent studies with a 3 M stabilized vanadium electrolyte provided a driving range of 31.5 km with partly filled electrolyte tanks and showed that up to 54 km could be achieved if the tanks were filled to their maximum capacity [28].

In the early 1980s, to compete with lead–acid batteries, a 50 kWh zinc–chloride battery was tested by Energy Development Associates for automobile power system applications, which reached an energy density of 154 Wh/kg [29]. Exxon Research and Engineering Co. and Gould Inc. were the major companies carrying out research on zinc–bromine battery system in the mid-1970s to early 1980s. Although a high level of activity was not maintained in both companies, in the mid-1989s the battery technologies were licensed to other developers, including Johnson Controls Inc. (United States), Studiengesellschaft für Energiespeicher und Antriebssysteme (SEA; Europe), Toyota Motor Corporation and Meidensha Corporation (Japan), and Sherwood Industries (Australia). Several of the Exxon licensees are still active in development toward commercial products [30].

Since 1983, SEA (now PowerCell GmbH of Boston) in Mürzzuschlag, Austria, has been developing Zn-Br flow batteries for electric vehicles and has produced batteries with capacities ranging between 5 and 45 kWh [22]. A SEA 45 kWh/216 V Zn-Br battery was installed in a Volkswagen bus for the Austrian Postal Service [9]. The battery weighed about 700 kg and the maximum speed achieved by the bus was 100 km/h. Hotzenblitz, a German company, designed an electric vehicle to be powered specifically by a 15 kWh/114 V Zn-Br battery [22]. Toyota Motor Corporation also developed 7 kWh/106 V Zn-Br batteries for the electric vehicle called the EV-30 [30]. In the 1990s electric vehicles with a 35 kWh Zn-Br battery were launched by the University of California [31]. It should be pointed out that the disadvantages of employing the Zn-Br battery as electric vehicle power normally include complexity of system, safety of operation, and low power density.

23.6 APPLICATION IN ELECTRIC VEHICLE CHARGING STATIONS

Currently, energy-efficient and environmentally friendly vehicles, represented by electrical vehicles, have become the main area of development within the automobile industry. In this regard, the construction of charging stations is required to speed up the development of electrical vehicles. However, the impact of the charging station on the electrical network and the high cost of construction and operation of charging stations are expected to be a challenge. Currently, storage batteries with large capacity are thought to represent one promising type of economical charging station [3]. They can be charged from the electrical network during off-peak hours, that is, at night, and then the stored electrical energy can be released quickly for use by electric vehicles. In this way, the energy powering the electric vehicles is not obtained directly from the electrical network so it will avoid the impact of vehicle charging on the quality and reliability of the network.

It is believed that VRB systems, with the advantages of high efficiency, long life span, independent power and capacity designs, and flexible operation, may meet the requirements of electrical vehicle charging stations [3]. They could provide an efficient way of energy storage, enhance the reliability and efficiency of the electrical network, lower the operating cost, and accelerate the development and popularization of charging stations. For example, the Austrian Cellstrom GmbH Corporation developed a VRB system in Vienna in 2008 [3]. This system, integrated with a photovoltaic system, had a power/capacity of 10 kW/100 kWh and an energy efficiency of 80%, demonstrating its application as an electric vehicle charging station [3].

To overcome the two major obstacles to the widespread adoption of electric vehicles (excessive downtime for recharging, and short-lived/high-cost batteries), RE-Fuel technology Ltd. (now Renewable Energy Dynamics Technology Ltd. [REDT], in Ireland) developed one proprietary VRB battery-refueling system [32], in which the electric vehicle powered by VRB could be instantly recharged by removing the spent electrolyte and simultaneously replacing it with charged electrolyte. The electrolyte is continually reenergized in the refueling station by wind, solar, or grid energy according to availability. In this way, off-peak base load power can be utilized and undesirable grid loads during peak times can be avoided.

23.7 APPLICATION IN TELECOMMUNICATION BASE STATIONS

In practice, communication base stations and communication machine rooms need batteries as backup power supplies for stable and reliable purposes, and the backup power supply time should not be less than several hours [33]. Currently, the inexpensive lead–acid battery is usually used to power communication base stations. However, some drawbacks, such as the insufficient performance and asset life of lead–acid batteries, could greatly limit their application in this area. For example, lead–acid batteries are limited to the number of deep discharge cycles before degradation or permanent damage occurs; that is, the operation of conventional lead–acid batteries is to limit the depth of discharge (DoD) to a maximum of 50% of the batteries fully charged state. Due to this limitation, replacement is recommended after around 1500 cycles at 50% DoD. It seems that liquid-redox flow batteries, especially VRBs, could reduce the operation and maintenance cost of the communication station, prolong the life span of diesel engines, and enhance the reliability and safety level of the station. There are some corporations, including Rongke Power Co. Ltd. in China and VRB Power Systems Inc. in Canada, Imergy Power Systems (formerly Deeya Energy) in the United States, and REDT in the United Kingdom, have demonstrated or precommercialized VRB ESS as the backup power supply for telecommunication base stations. So far, Imergy's 5 kW (optional 2.5 kW) VRB energy-storage platform was explored to replace lead–acid batteries in more than 70 off-grid or weak-grid telecom applications in remote areas of Africa and India [34]. In the United States, Deeya Energy also developed a few kilowatt-order products using the Fe-Cr system for wireless base stations [35]. It was reported that the firm RedFlow in Australia could commercialize a fully functional Zn-Br battery module product (named ZBM) for off-grid remote power and telecommunication applications, which had a capability of 3 kW (5 kW peak)/8 kWh [36].

23.8 CONCLUSION

Concerns about the environmental consequences of burning fossil fuels have led to a rapid increase in production of renewable energy sources as well as the imperative demand for advanced ESSs. The crucial advantage of flow battery ESSs is that the power size is independent on its energy capacity, providing a great flexibility and allowing systems to be designed according to specific needs of each application. As a result, the energy-storage capacity of the flow battery can be easily increased by simply using larger volumes of electrolytes. Besides that, other attractive features of liquid-redox flow batteries, such as long service life, low maintenance, easy scalability, quick response times, flexible layout, and tolerance to overcharge/overdischarge, distinguish them from other types of energy-storage technologies and make them ideally suitable for many applications, especially for large-scale energy storage in both off-grid and grid-connected applications. Liquid-redox flow battery ESSs have been successfully demonstrated or precommercialized for load balancing, peak shaving, renewable energy integration, emergency backup power supply, and microgrid or smart grid applications. A unique feature of liquid-redox flow batteries for electric vehicle applications is their ability to be both electrically recharged and mechanically refueled by simple electrolyte exchange. However, the overall power density and energy density of different liquid-redox flow batteries range from 40 to 100 W/L and from 20 to 35 Wh/L, respectively, and still remain low when compared to those of lithium-ion batteries and fuel cells, which limits their extensive applications in electric vehicles. For broad market penetration in vast telecommunication market, the cost of liquid-redox flow batteries needs to be further reduced to make them competitive with lead–acid batteries. With the development and breakthroughs in the key materials and technologies, there will be more unprecedented opportunities for liquid-redox flow batteries in energy storage and power supply markets in the future.

REFERENCES

1. Poullikkas, A. A comparative overview of large-scale battery systems for electricity storage. *Renewable and Sustainable Energy Reviews* 2013; 27: 778–788.
2. Carnegie, R., Gotham, D., Nderitu, D., and Preckel, P. Utility Scale Energy Storage Systems. State Utility Forecasting Group. 2013. http://www.purdue.edu/discoverypark/energy/assets/pdfs/SUFG/publications/SUFG%20Energy%20Storage%20Report.pdf (accessed October 30, 2014).
3. Zhang, H. M. Liquid Redox Rechargeable Batteries. In *Electrochemical Technologies for Energy Storage and Conversion*, ed. Zhang, J. J., Zhang, L., Liu, H. S. et al. pp. 277–315. Wiley, Hoboken. 2012.
4. Herbrich, V. R. Am Muttertag kam 73 Prozent des Stroms aus erneuerbaren Energien. *The Huffington Post* May 14, 2014. http://www.huffingtonpost.de/2014/05/14/rekord-erneuerbare-energien_n_5321800.html (accessed October 30, 2014).
5. Mueller, S. C., Sandner, P. G., and Welpe, I. M. Monitoring innovation in electrochemical energy storage technologies: A patent-based approach. *Applied Energy* 2015; 137: 537–544.
6. Drouineau, M., Maïzi, N., and Mazauric, V. Impacts of intermittent sources on the quality of power supply: The key role of reliability indicators. *Applied Energy* 2014; 116: 333–343.
7. Fabjan, C., Garche, J., Harrer, B., Jörissen, L., Kolbeck, C., Philippi, F., Tomazic, G., and Wagner, F. The vanadium redox-battery: An efficient storage unit for photovoltaic systems. *Electrochimica Acta* 2001; 47: 825–831.
8. Joerissen, L., Garche, J., Fabjan, C., and Tomazic, G. Possible use of vanadium redox-flow batteries for energy storage in small grids and stand-alone photovoltaic systems. *Journal of Power Sources* 2004; 127: 98–104.
9. Leung, P., Li, X., León, C. P. D., Berlouis, L., John Low, C. T., and Walsh, F. C. Progress in redox flow batteries, remaining challenges and their applications in energy storage. *RSC Advances* 2012; 2: 10125–10156.
10. Tang, A., McCann, J., Bao, J., and Skyllas-Kazacos, M. Investigation of the effect of shunt current on battery efficiency and stack temperature in vanadium redox flow battery. *Journal of Power Sources* 2013; 242: 349–356.
11. Shibata, A. S., and Sato, K. Development of vanadium redox flow battery for electricity storage. *Power Engineering Journal* 1999; 13:130–135.
12. http://www.cleanbreak.ca/2005/09/30/vrb-power-strikes-again-another-solar-deal/ (accessed November 4, 2014).
13. V-Fuel Pty. Ltd. Status of energy storage technologies as enabling systems for renewable energy from the sun, wind, waves and tides. http://www.aph.gov.au/parliamentary_business/committees/house_of_representatives_committees?url=isr/renewables/submissions/sub21.pdf (accessed November 4, 2014).
14. McDowell, J. International Renewable Energy Storage (IRES) Conference 2006, Gelsenkirchen, Germany 2006; pp. 30–31.

15. Alotto, P., Guarnieri, M., and Moro, F. Redox flow batteries for the storage of renewable energy: A review. *Renewable and Sustainable Energy Reviews* 2014; 29: 325–335.
16. http://www.pdenergy.com/case-studies.php (accessed October 30, 2014).
17. http://www.rongkepower.com/index.php/article/show/id/140/language/en (accessed October 30, 2014).
18. Luo X., Wang J., Dooner M., and Clarke J. Overview of current development in electrical energy storage technologies and the application potential in power system operation. *Applied Energy* 2015; 137: 511–536.
19. Norris, B. L., Ball, G. J., Lex, P., and Scaini, V. Grid-connected solar energy storage using the zinc-bromine flow battery. http://www.zbbenergy.com/files/1112/8285/0124/technicalpaper_grid.pdf (accessed October 28, 2014).
20. Rastler, D. Electric energy storage applications in the electric enterprise: Overview of applications and U.S. demonstration activities. Electric Power Research Institute (EPRI). EDF Energy Storage Workshop. October 2011.
21. Department of Energy's Energy Storage Database. Sandia National Laboratories, United States Department of Energy. http://www.energystorageexchange.org/ (accessed November 8, 2014).
22. Butler, P. C., Eidler, P. A., Grimes, P. G., Klassen, S. E., and Miles, R. C. Zinc Bromine Batteries. In *Handbook of Batteries*, ed. Linden, D., and Reddy, T. B. Chapter 39. McGraw-Hill, New York. 2002.
23. Doughty, D. H., Butler, P. C., Akhil, A. A., Clark, N. H., and Boyes, J. D. Batteries for large-scale stationary electrical energy storage. *Electrochemical Society Interface* Fall 2010; 49–53.
24. http://www.uetechnologies.com/index.htm (accessed October 28, 2014).
25. Hazza, A., Pletcher, D., and Wills, R. A novel flow battery—A lead acid battery based on an electrolyte with soluble lead (II); IV. The influence of additives. *Journal of Power Sources* 2005; 149: 103–111.
26. Review of electrical energy storage technologies and systems and of their potential for the UK. Contract number DG/DTI/00055/00/00, URN number 04/1876. 2004. http://www.wearemichigan.com/JobsAndEnergy/documents/file15185.pdf (accessed October 30, 2014).
27. Skyllas-Kazacos, M., and Menictas, C. The vanadium redox battery for emergency back-up application. http://www.arizonaenergy.org/Analysis/FuelCell/Vanadium%20Battery/vanadium_redox_battery_for_emerg.htm (accessed October 28, 2014).
28. Skyllas-Kazacos, M. Recent progress with the UNSW vanadium battery. http://www.arizonaenergy.org/Analysis/FuelCell/Vanadium%20Battery/recent_progress_with_the_unsw_va.htm (accessed October 28, 2014).
29. Symons, P. C., and Butler, P. C. Advanced Batteries for Electric Vehicles and Emerging Applications—Introduction. In *Handbook of Batteries*, ed. Linden, D., and Reddy, T. B. Chapter 37. McGraw-Hill, New York. 2002.
30. Crompton, T. R. *Battery Reference Book*, Third Edition. Newnes Press/Linacre House, Oxford. 2000.
31. Swan, D. H., Dickinson, B. E., Arikara, M. P., and Prabhu, M. K. Construction and performance of a high voltage zinc bromine battery in an electric vehicle. *Proceedings of the Tenth Annual Battery Conference on Applications and Advances* 1995; pp. 135–140.
32. http://www.redtenergy.com/news/electric-vehicle-applications-flow-batteries (accessed November 6, 2014).
33. Hallberg, H., and Österling, J. Method and a Radio Base Station for Handling of Data Traffic. US Patent 20120289224 A1. 2012.
34. http://www.imergypower.com/products/wireless-telecom-business-case/ (accessed October 28, 2014).
35. Shigematsu, T. Redox flow battery for energy storage, *SEI Technical Review* 2011; 73: 4–13.
36. http://redflow.com/electricity-storage/off-grid-remote-power-telcos/ (accessed October 28, 2014).

Section VIII

Advanced Materials and Technologies for Water Electrolysis Producing Hydrogen

24 Advanced Materials for Water Electrolysis

Yongjun Leng and Chao-Yang Wang

CONTENTS

24.1 INTRODUCTION

Currently our society heavily relies on fossil fuels as energy sources for electricity and transportation, causing many problems, including energy insecurity, air pollution, and enormous CO_2 emission. Because of the gradual depletion of fossil fuels and intolerance of the environment, much attention has been paid to sustainable and renewable energy. Hydrogen is an excellent energy-storage medium for sustainable and renewable energy systems.[1–3] The advantages of hydrogen as an energy carrier include (1) high efficiency, reversible conversion between hydrogen and electricity; (2) good energy density of compressed hydrogen storage; and (3) scalability of hydrogen technologies for grid-scale applications.[4] Switching fossil fuel economy to hydrogen economy could provide health, environmental, climate, and economic benefits and could reduce reliance on diminishing oil supplies.[5]

Hydrogen production is a key element of the hydrogen economy. Water electrolysis is one of the most efficient and reliable methods for producing hydrogen from renewable but intermittent energy sources, such as solar, wind, and hydropower.[6–9] On the other hand, a fuel cell provides a highly efficient and clean means with zero emission to convert hydrogen energy to

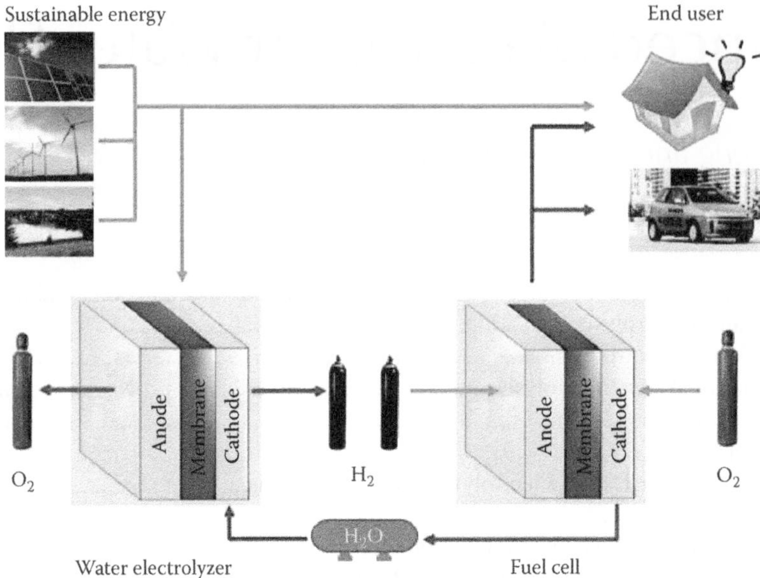

FIGURE 24.1 Combination of water electrolyzers with fuel cells for use in sustainable energy resources.

electricity. Figure 24.1 shows an example of using sustainable and renewable energies within a hydrogen economy in the future. First, hydrogen is produced by water electrolysis from sustainable but intermittent energy sources and is then stored and transported to the sites near end users, where hydrogen is converted to electricity for houses and transportation. The whole cycle produces zero emissions, causes no harm to the environment, and leaves no carbon footprint.

24.1.1 Principles of Water Electrolysis

The electrolysis of water is an electrochemical reaction that splits water into hydrogen and oxygen with electricity as the energy source, as shown in Figure 24.2. The electricity can come either from a grid or from nuclear and renewable energy. In some cases, heat may be also supplied as an additional energy source. Compared with other hydrogen-production technologies such as methanol reforming, the electrolysis of water for hydrogen production has several advantages: (1) it is one of simplest ways to produce hydrogen, since it requires no moving parts; (2) it is capable of making high-purity hydrogen; and (3) it is a clean way to produce large quantities of hydrogen without CO_2 emission if the required electricity comes from nuclear energy and carbon-free renewable energy, such as solar, wind, and hydropower.[6]

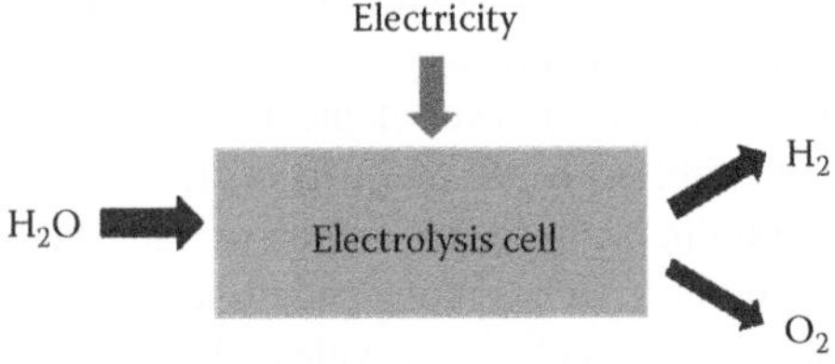

FIGURE 24.2 Principle of water electrolysis.

24.1.2 Main Types of Water Electrolysis

A water electrolysis cell (i.e., electrolyzer) consists of at least three components: a hydrogen electrode (cathode) for hydrogen evolution reaction (HER), an oxygen electrode (anode) for oxygen evolution reaction (OER), and an electrolyte for ionic transport from one electrode to another. At the cathode and anode, electrocatalysts are required for HER and OER, respectively. In most cases, ionic-conducting phases, such as ionic-conducting ionomers, are added to the electrode catalyst layers (CLs) to improve the electrode performance, since the HER/OER occurs at three-phase boundary (electronic-conducting phase/ionic-conducting phase/gas pore) reaction sites. The electrolyte can be an ionic-conducting liquid solution, a solid oxide film, or a polymer membrane. Depending on the type of electrolyte, the water electrolysis cell can be classified into four main types: alkaline liquid electrolyte, proton-exchange membrane (PEM), anion-exchange membrane (AEM), and solid oxide electrolyte. Table 24.1 shows a comparison of four types of water electrolysis technologies. In an alkaline liquid electrolyte, a PEM, an AEM, and solid oxide electrolysis cells (SOECs), a diagram/separator filled with liquid KOH solution, a proton-conducting polymer membrane, a hydroxide-conducting polymer membrane, a solid oxide film made from oxygen ion conductor are used as the electrolytes, respectively. Figure 24.3 shows the schematic diagrams of four types of water electrolysis cells.

In alkaline liquid electrolyte water electrolysis cells, water is consumed and reduced by the electrons received from an external circuit to produce hydrogen gas and hydroxide ions (OH^-) at the cathode. The OH^- ions are then transported from the cathode through the liquid KOH solution to the anode. At the anode, OH^- ions are oxidized to produce oxygen gas, water, and electrons. The half reaction of anode and cathode and the whole reaction are as follows:

TABLE 24.1
Comparison of Four Types of Water Electrolyzers

	Liquid Alkaline	PEM	AEM	SOEC
Operating temperature (°C)	25–200	25–200	25–100	600–1000
Electrolyte/conducting ion	KOH/OH^-	$Nafion/H^+$	$A201/OH^-$	YSZ/O^{2-}
Way of gas separation	Diaphragm/separator	Polymeric membrane	Polymeric membrane	Ceramic membrane
Advantages	Established; cost effective; no noble metal catalysts	High purity of hydrogen; high current density; simple cell design; high pressurization possible	High purity of hydrogen; potential for using nonnoble metal catalysts; simple cell design; high pressurization possible	Low electricity demand; no noble metal catalysts; potential for high current density
Disadvantages	Low current density; low purity of hydrogen; corrosive electrolyte	High investment costs; noble metal catalysts	Low current density; poor durability performance	Heat source required; no pressurization elaborate separation of hydrogen and vapor
Status	Commercialized	Ready for commercialization	Lab research	Ready for commercialization

Note: YSZ: Y_2O_3-stabilized ZrO_2.

$$\text{Cathode:}\quad 4H_2O + 4e^- \rightarrow 2H_2 + 4OH^- \ (E = -0.828 \text{ V}) \quad (24.1)$$

$$\text{Anode:}\quad 4OH^- \rightarrow O_2 + 2H_2O + 4e^- \ (E = 0.401 \text{ V}) \quad (24.2)$$

$$\text{Overall:}\quad 2H_2O \rightarrow 2H_2 + O_2 \ (E = 1.229 \text{ V}) \quad (24.3)$$

In alkaline liquid electrolyte water electrolysis, non–precious group metals (non-PGMs) can be used as the electrocatalysts for HER and OER.[10–12] As one of the least costly water electrolysis technologies, alkaline liquid electrolyte water electrolysis has been widely deployed for several decades in large-scale hydrogen production.[10,12] However, the alkaline liquid electrolyte used in this system, such as aqueous 10 M KOH, can react with carbonate anions formed by the adsorption of CO_2 from the air to form insoluble species, such as K_2CO_3. These insoluble carbonates can precipitate in the porous CLs and

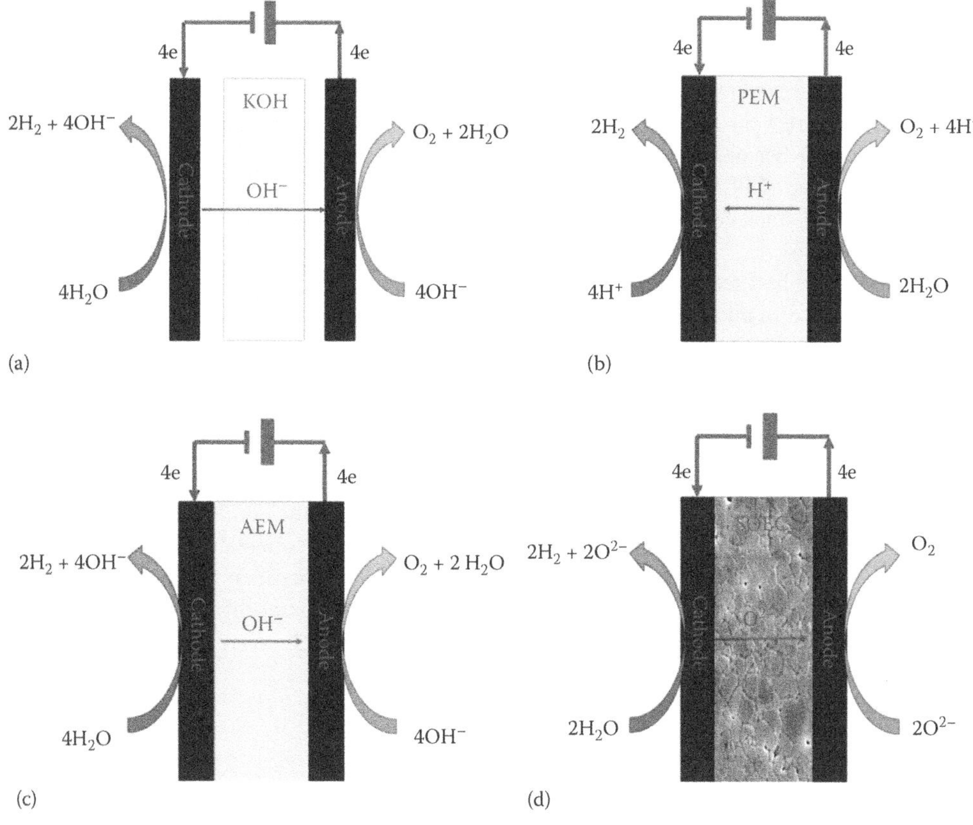

FIGURE 24.3 Four types of water electrolyzers: (a) alkaline liquid electrolyte, (b) PEM, (c) AEM, and (d) SOEC.

block the transport of products and reactants, which significantly decreases cell performance.[13] In addition, the alkaline liquid electrolyte used is very corrosive.

In PEM water electrolysis, water is consumed and oxidized at the anode to produce oxygen gas and protons (H^+) and electrons. The protons are then transported from the anode through the PEM to the cathode. At the cathode, protons are reduced by the electrons received from the external circuit at the cathode to produce hydrogen gas. The half reaction of anode and cathode and the whole reaction in PEM water electrolysis are as follows:

$$\text{Cathode:}\quad 4H^+ + 4e^- \rightarrow 2H_2\ (E = 0\ \text{V}) \tag{24.4}$$

$$\text{Anode:}\quad 2H_2O \rightarrow O_2 + 4H^+ + 4e^-\ (E = 1.229\ \text{V}) \tag{24.5}$$

$$\text{Overall:}\quad 2H_2O \rightarrow 2H_2 + O_2\ (E = 1.229\ \text{V}) \tag{24.6}$$

Compared with alkaline liquid electrolyte water electrolysis, PEM water electrolysis system offers several advantages, including higher energy efficiency, greater hydrogen production rate, and a more compact design.[14–16] These advantages are derived from the solid-state membrane electrolyte compared to a device with a free liquid electrolyte and a porous diagram/separator. However, there are several drawbacks with PEM water electrolysis, including (1) the acidic environment limits the catalysts to noble metals, which increases the cost of PEM water electrolysis; (2) cationic impurities supplied in the feed water or released from the cell components can bind to the proton-conducting sites of the PEM/ionomer such as Nafion, which reduces their conductivity and increases electrolyte and reaction resistance; and (3) the perfluorinated Nafion-based membranes used are very expensive and have limited chemical diversity for further optimizing their properties.[4,16] The distinct disadvantage of PEM water electrolysis is the high capital cost of the cell stack compared to alkaline liquid electrolyte water electrolysis.

In AEM water electrolysis, the half reactions at the anode and cathode are the same as those in alkaline liquid electrolyte water electrolysis. The difference between them is that in the case of AEM water electrolysis, OH^- are transported from the cathode through the AEM to the anode; while in the case of alkaline liquid electrolyte water electrolysis, OH^- are transported from the cathode through the alkaline liquid electrolyte to the anode. Like PEM water electrolysis, AEM water electrolysis is an electrolysis cell with a solid-state membrane electrolyte. Therefore, AEM water electrolysis combines the advantages of both alkaline liquid electrolyte and PEM water electrolysis while it avoids some of their disadvantages.[4] For example, like alkaline liquid electrolyte water electrolysis, AEM water electrolysis can also potentially employ non-precious metals as catalysts and can use cheap cell support structures to make it much less expensive than a PEM-based system. However, AEM water electrolysis is still in its infant stage. Its performance and durability still lag behind the requirements for commercialization.

In solid oxide steam electrolysis, steam is supplied into the cathode chamber. At the cathode, water (steam) is consumed and reduced by the electrons received from the external circuit to produce hydrogen gas and oxide ions (O^{2-}). The O^{2-} ions are then transported from the cathode through the electrolyte made from the solid oxide ion conductor to the anode. At the anode, O^{2-} ions are oxidized to produce oxygen gas and release electrons. The half reaction of the anode and cathode and the whole reaction are as follows:

$$\text{Cathode:}\quad 2H_2O + 4e^- \rightarrow 2H_2 + 2O^{2-} \tag{24.7}$$

$$\text{Anode:}\quad 2O^{2-} \rightarrow O_2 + 4e^- \tag{24.8}$$

$$\text{Overall:}\quad 2H_2O \rightarrow 2H_2 + O_2 \tag{24.9}$$

Compared with low-temperature water electrolyzers (such as alkaline liquid electrolyte, PEM, and AEM water electrolysis), steam electrolysis in SOECs operating at high temperatures has several advantages, including (1) less demand for electric energy because of some of the required energy being supplied by heat sources at high temperatures; (2) reduced cathodic and anodic overpotentials at high temperatures; and (3) PGM-free catalysts.[17] Moreover, the heat and power generated by nuclear power, renewable energy, and waste heat from high-temperature industrial processes can be utilized for steam electrolysis in SOECs.[18,19] However, for high-temperature solid oxide steam electrolysis cells, the additional system is needed to separate hydrogen from the mixture of hydrogen and steam.

24.1.3 Thermodynamics of Water Electrolysis

For an electrochemical system, the equilibrium condition is expressed by

$$\Delta\bar{G} = \Delta G + nF\Delta E = 0, \tag{24.10}$$

where $\bar{G}$ is the electrochemical free energy, G is the chemical free energy, n is the number of moles of electrons involved in the conversion of 1 mol of product, F is the Faraday constant, and ΔE is the potential difference measured between two electrodes. The difference between $\bar{G}$ and G is that $\bar{G}$ includes the effects from the additional large-scale electrical environment in the electrochemical system. Based on Equation 24.10, the relationship between the change in chemical free energy ΔG and the potential difference ΔE between two electrodes is described by

$$\Delta E = -\frac{\Delta G}{nF}. \tag{24.11}$$

For the water electrolysis, under standard conditions at 25°C, the change in chemical free energy ΔG is 237.18 kJ/mol and $n = 2$; therefore, the cell potential, i.e., the potential difference ΔE between hydrogen and oxygen electrodes is

$$\Delta E = \frac{237.18}{2 \times 96.485} \text{V} = 1.229 \text{ V}. \tag{24.12}$$

This is the minimum cell voltage that needs to be applied without any hydrogen production under standard conditions at 25°C.

The total energy required for water splitting and the formation of hydrogen and oxygen is the enthalpy ΔH.[20] The enthalpy ΔH can be expressed by

$$\Delta H = \Delta G + T \cdot \Delta S, \tag{24.13}$$

where T is the temperature and ΔS is the entropy. Therefore, the energy required for water splitting and the formation of hydrogen and oxygen is provided by both the form of electricity (ΔG) and the form of heat ($T \cdot \Delta S$). If the heat can be absorbed from the environment, the minimum cell voltage for water electrolysis is the thermodynamic cell potential ΔE, which is defined by Equation 24.11; while, if there is no net heat exchange between the cell and the environment, additional electric energy is required to compensate for the heat. In such a thermoneutral environment, the minimum cell voltage for water electrolysis is the thermoneutral cell potential ΔE_{tn}, which is defined by[20]

$$\Delta E_{\text{tn}} = -\frac{\Delta H}{2F}. \tag{24.14}$$

For the water electrolysis, under standard conditions at 25°C, ΔE_{tn} is equal to 1.481 V, which is a little bit higher than the thermodynamic cell potential of 1.229 V.

The thermodynamic cell voltage ΔE is independent of the pH value[21]; i.e., it is the same for alkaline liquid electrolyte (pH = 14), AEM water electrolysis (pH = 14), and PEM water electrolysis (pH = 0–1). However, the anode and cathode potentials under the equilibrium are shifted with the electrolyte pH. Temperature and pressure affect the thermodynamic cell voltage ΔE. Since more heat can provide additional energy at higher temperature, the required thermodynamic cell voltage ΔE decreases with increasing temperature. For example, at room temperature, 15% of the required energy for water splitting comes from heat and 85% from electricity; while at 1000°C, one-third and two-thirds come from heat and electricity, respectively.[21] This is why the thermodynamic cell voltage ΔE for high-temperature steam electrolysis is much lower than that for low-temperature water electrolysis, including alkaline liquid electrolyte, AEM, and PEM electrolysis. In the case of high-pressure water electrolysis, the thermodynamic cell potential ΔE increases with increasing pressure.[21] However, high-pressure water electrolysis avoids the additional system to pressurize the hydrogen.

24.1.4 Kinetic Considerations

There is no practical interest in performing water electrolysis at the thermodynamic cell potential ΔE since there is no net hydrogen production under the equilibrium. For an effective production of hydrogen at a reasonable hydrogen production rate, it is required that the applied cell voltage V be larger than thermodynamic cell potential ΔE in order to overcome all overpotentials involved with the electrochemical reaction process and ohmic drop in the cell. The overpotentials for water electrolysis include (1) the activation overpotential for HER at the cathode (η_A^c), (2) the activation overpotential for OER at the anode (η_A^a), (3) the overpotential related to the resistance of ionic transport in the electrode (anode and cathode) CLs ($\eta_{\text{ionic,T}}^c$ and $\eta_{\text{ionic,T}}^a$), and (4) bulk mass transport overpotentials due to diffusion of reactions and/or products ($\eta_{\text{bulk,T}}^c$ and $\eta_{\text{bulk,T}}^a$). The ohmic drop (iRohm) includes the ohmic drops in the electrodes, across the interface between cell components, and across the electrolyte. Therefore, the required cell voltage (V) can be expressed as

$$V = \Delta E + iR_{\text{ohm}} + \eta_A^c + \eta_{\text{ionic,T}}^c + \eta_{\text{bulk,T}}^c + \eta_A^a + \eta_{\text{ionic,T}}^a + \eta_{\text{bulk,T}}^a. \tag{24.15}$$

Clearly, the total overpotentials and the required cell voltage for water electrolysis are a function of the current density. Generally, the higher the current density, the higher the total overpotentials and the required cell voltage. In order to reduce the capital costs, it is necessary to increase the operating current density to a certain level (such as 100–1000 mA/cm^2). However, increasing the operating current density will reduce the energy efficiency. In order to improve the energy efficiency at a given operating current density, it is necessary to reduce the total overpotentials and ohmic drop as much as possible by different strategies, including developing advanced electrocatalysts with high activity, developing advanced ionomer/electrolyte with high ionic conductivity, optimizing electrode microstructure, and improving cell design and configuration.

24.1.5 Aspects of Materials

As mentioned above, the basic components for a water electrolysis cell are two electrodes (cathode and anode) and an electrolyte. For a single cell, a current collector is needed for passing the current into the cell. For a stack consisting of a number of single cells, an interconnector is needed to connect one cell with another one. Figure 24.4 shows the requirement of the electrodes, electrolyte, and current collector/interconnector for the water electrolysis cells.

For the electrode requirement, a composite consisting of electrocatalysts with high activity and ionic-conducting phases (such as ionomers or electrolyte materials) with high ionic conductivity is suitable for achieving high performance in electrode reactions. Electrodes should have a porous structure in order to reduce the resistance for the transportation of reactants and products. Chemical, thermal, and mechanical

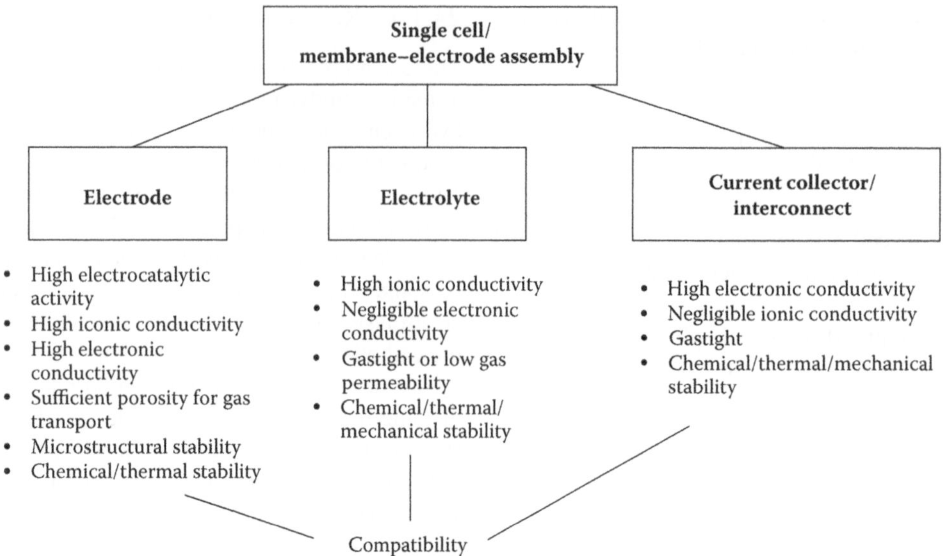

FIGURE 24.4 General criteria for selection of materials of water electrolyzers.

stability of the electrodes is a determining factor in the long-term stability of the water electrolysis cells.

Regarding the electrolyte, the main function of the electrolyte is for ionic transport from one electrode, through the electrolyte, to another electrode. Therefore, electrolyte materials need to exhibit high ionic conductivity in order to reduce the ohmic resistance related to ionic transport across the electrolyte. In order to improve the current efficiency and avoid a partial internal short circuit, the electronic conductivity of the electrolyte should be negligible. Another function of the electrolyte is to prevent the mixing of produced hydrogen and oxygen. Thus, the electrolyte should be gastight or should exhibit very low gas permeability. Chemical, thermal, and mechanical stability of the electrolyte is another determining factor in the long-term stability of the water electrolysis cells.

With regard to the current collectors and interconnectors, in order to reduce the ohmic drop for current passage, the current collectors and interconnectors need to have high electronic conductivity and negligible ionic conductivity. Also, the current collectors and interconnectors should be gastight. The stability of current collectors/interconnectors may also affect the long-term stability of the water electrolysis cells. Please note that compatibility between cell components is also important for the performance and long-term stability of water electrolysis cells.

The main challenges to the widespread use of water electrolysis technologies are performance, durability, and cost. The factors affecting cell performance, durability, and cost are summarized in Figure 24.5. The cell performance depends on many factors, including (1) the electrocatalytic activity and loading of the catalysts or electrode materials; (2) the ionic

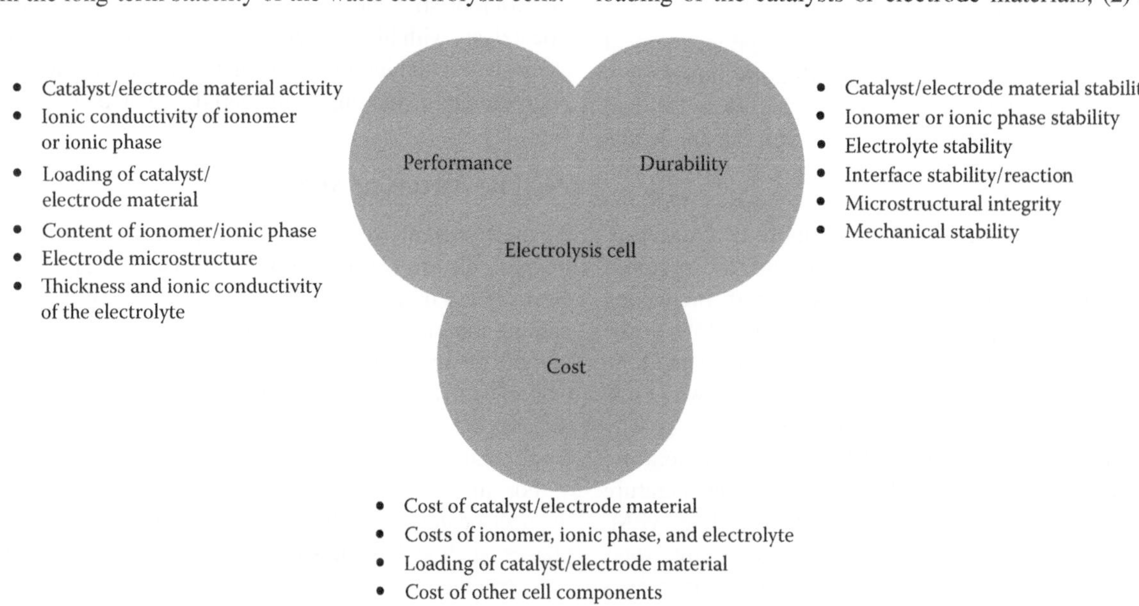

FIGURE 24.5 Factors affecting performance, durability, and cost of water electrolysis cells.

conductivity and loading of ionic-conducting phases, such as ionomer or electrolyte materials in the electrode; (3) the thickness and the ionic conductivity of the electrolyte; and (4) the electrode microstructure. The durability of water electrolysis is also determined by many factors, including (1) the stability of electrocatalysts/electrode materials; (2) the stability of ionic-conducting phases, such as ionomer or electrolyte materials in the electrode; (3) the stability of the electrolyte; (4) the stability of interface between cell components; and (5) the integrity of electrode microstructure. A number of factors determine the cost of water electrolysis, including (1) the price and loading of electrocatalysts/electrode materials; (2) the price and loading of materials used for ionic-conducting phase in the electrode; (3) the price of electrolyte materials and the thickness of the electrolyte; (4) the cost of additional other cell components; and (5) the cell performance and lifetime. It is noted that performance, durability, and cost are interrelated. For example, adding more electrocatalysts to the water electrolysis cell can boost the cell performance, but this also increases the cell cost. Similarly, using thick electrolyte membrane can extend the lifetime of the cell while at the same time this increases cell cost, by adding materials, and decreases cell performance. In considering performance, durability, and cost, there is trade-off for the selection of materials for water electrolysis.

24.2 ADVANCED MATERIALS FOR ALKALINE LIQUID ELECTROLYTE WATER ELECTROLYSIS

Alkaline liquid electrolyte water electrolysis is a mature and low-cost water electrolysis technology and has been commercialized and widely employed in large-scale hydrogen production for several decades. There are a number of reviews about alkaline water electrolysis technology in literature.[10,11,22–25] For example, Zeng and Zhang[10] reviewed recent progress in alkaline water electrolysis for hydrogen production and application. Pletcher and Li[25] reviewed recent developments in oxygen-evolving anodes, hydrogen-evolving cathodes, and hydroxide-transporting membranes related to alkaline zero-gap water electrolyzers. In this and next sections, we will briefly introduce advanced materials for alkaline liquid electrolyte water electrolysis.

24.2.1 Electrocatalysts

24.2.1.1 Electrocatalysts for Hydrogen Evolution Reaction

One of main advantages of alkaline liquid electrolyte water electrolysis is to potentially use non-PGM materials as the electrocatalysts. Ni-based materials have been considered as promising non-PGM electrocatalysts for HER in alkaline water electrolysis. There are two approaches to enhance the performance of Ni-based electrocatalysts: (1) to increase the surface area of Ni by various methods, such as Raney-type Ni,[26] or (2) to increase their intrinsic activity by combining Ni with other metals, such as Mo, to obtain alloys with optimized adsorption characteristics.[27] Raney Ni can be obtained by leeching Zn or Al from Raney Ni precursor alloys (either NiZn or $NiAl_3/Ni_2Al_3$) in alkaline solution,[11] which leads to a highly porous structure, a very small Ni particle size, and a high surface area. Thus, Raney Ni can substantially improve the electrode performance and reduce the overpotentials for HER. Raney Ni is still one of the good HER electrocatalysts because of its low cost and relatively high activity. NiMo, especially Raney-type NiMo alloy, is another promising type of Ni-based electrocatalyst for HER.[27–29] For example, Birry and Lasia[28] obtained very active and stable Raney NiMo electrodes by leeching Al from Al-rich Ni-Al-Mo alloys and found that adding Mo significantly improves catalytic activity. By combining Raney-type Ni with the Mo addition, they obtained an overpotential for HER as low as 67 mV at 250 mA cm^2 in 1 M KOH solution for the Raney-type NiMo electrode. Crnkovic et al.[29] prepared a Raney-type Ni-Fe-Mo alloy electrode by leeching Zn from Ni-Fe-Mo-Zn coating and obtained an overpotential for HER as low as 83.1 mV at 135 mA/cm^2 in 6 M KOH at 80°C. Raney Ni and NiMo electrodes showed a low overpotential in HER and very good long-term stability under continuous electrolysis, but their advantages are easily lost after intermittent operation, especially after a long period.[30,31] In order to overcome this problem, Hu[30] and Hu et al.[31] prepared composite electrocatalysts consisting of hydrogen storage alloys (such as $MmNi_{3.6}Co_{0.75}Mn_{0.4}Al_{0.27}$ alloy [Mm = misch metal], $LaNi_{4.9}Si_{0.1}$ alloy, and Ti_2Ni alloy) and Ni-Mo coatings and found that these composite electrocatalysts exhibited not only a low overpotential in HER but also excellent long-term stability under both conditions of continuous and intermittent electrolysis. Besides Ni-based materials, other materials have been explored for use as HER electrocatalysts. For example, Michishita et al.[32] found that $La_{0.6}Sr_{0.4}CoO_3$ exhibits almost the same overpotential for HER as that of Pt in water electrolysis and shows good chemical stability.

24.2.1.2 Electrocatalysts for Oxygen Evolution Reaction

The electrocatalysts for OER are more important for the improvement of alkaline liquid electrolyte water electrolysis than that for HER, since OER is much slower than HER. During the past several decades, much effort has been devoted to the development of highly active OER electrocatalysts, especially non-PGM electrocatalysts for OER.[11] OER electrocatalysts include (1) oxides with spinel structure, such as $NiCo_2O_3$,[33] and Li-doped Co_3O_4[34]; (2) oxides with perovskite structure, such as $SrFeO_3$,[35] $SrFe_{0.9}M_{0.1}O_3$ (M = Ni, Co, Ti, Mn),[36] $La_{(1-x)}Sr_xMnO_3$,[32,37] $Li_{(1-x)}Sr_xCoO_3$,[11,32,38,39] and $La_{(1-x)}Sr_xFe_{(1-y)}Co_yO_3$[39,40]; (3) mixed transition-metal oxides or hydroxides, such as $(Ni,Fe)O_2$,[41–43] $(Ni,Fe)(OH)_2$[43,44]; and (4) pyrochlore-type oxides with the general formula $(A_2[B_{(2-x)}A_x])O_{7-y}$ (A = Pb or Bi, B = Ru or Ir, $0 < x < 1$, and $0 < y < 0.5$).[45]

Singh and Lal[39] prepared a series of Co-based perovskite oxides with the molecular formula $Li_{(1-x)}Sr_xCoO_3$ ($0 \le x \le 0.5$)

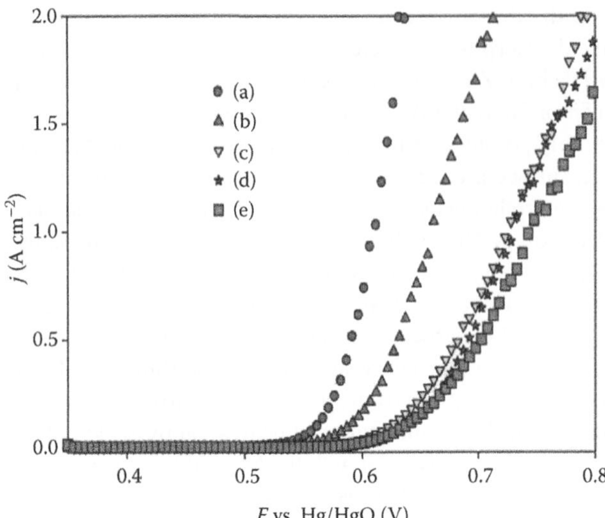

FIGURE 24.6 Steady-state polarization curves (scan rate: 1 mV/s) recorded in 1 M NaOH at 80°C. (a) a mixed $Ni/Fe(OH)_2$ layer; (b) a $NiCo_2O_4$ spinel; (c) a mesoporous $Ni(OH)_2$ layer; (d) a polished nickel microdisk; (e) a polished stainless steel microdisk. (Li, X., F. C. Walsh, and D. Pletcher, Nickel based electrocatalysts for oxygen evolution in high current density, alkaline water electrolysers, *Physical Chemistry Chemical Physics*, 13 (3), 1162–1167, 2011. Reproduced with permission of the Royal Society of Chemistry.)

and $La_{0.7}Sr_{0.3}B_{(1-y)}Co_yO_3$ (B = Cu, Fe, Ni, Cr, Mn; $0 \leq y \leq 0.2$) at 700°C, using a carbonate precipitation method, and reproduced in film forms on pretreated Ni supports by an oxide slurry painting technique. They found that in 1 M KOH at 25°C, the oxide $La_{0.7}Sr_{0.3}Fe_{0.1}Co_{0.9}O_3$ exhibits the highest apparent catalytic activity for OER; while in 30 wt.% KOH at 70°C, the oxides $La_{0.7}Sr_{0.3}Fe_{0.1}Co_{0.9}O_3$ and $La_{0.7}Sr_{0.3}CoO_3$ show similar activities.

Li et al.[44] investigated a number of Ni-based materials as potential OER catalysts under conditions close to those met in modern, high–current density alkaline water electrolyzers, as shown in Figure 24.6. They found that the activity increases in the following order: smooth Ni < mesoporous $Ni(OH)_2$ coating < spinel $NiCo_2O_4$ coating < $(Ni,Fe)(OH)_2$ coating, and the overpotential of $(Ni,Fe)(OH)_2$ coating for OER was only 265 mV at under 0.5 A/cm² in 1 M NaOH at 80°C.

24.2.2 Diaphragms/Separators/Membranes

In the early stages, asbestos was widely used as the commercial diaphragm for alkaline liquid electrolyte water electrolysis. However, because of poor chemical stability in aqueous KOH at elevated temperatures and severe harm to human health, asbestos has been gradually replaced by other materials.[10,22] Alternative porous diaphragm materials include nickel oxide, polyphenylene sulfide (such as Ryton), polysulfone (such as Udel), polysulfone-bonded ZrO_2 composite (such as Zirfon), porous Teflon films impregnated in situ with potassium titanate, and membrane fabricated from polyantimonic acid and an organic binder.[22,23]

Hung et al.[46] prepared a polyvinylidene fluoride–grafted 2-methacrylic acid 2-(bis-carboxylmethylamino)-2-hydroxylpropyl ester (PVDF-g-G-I) bipolar membrane through the plasma-induced polymerization method, and used it as the diaphragm for alkaline water electrolysis. The PVDF-g-G-I bipolar membrane was found to function very well as a diaphragm. Lu et al.[47] successfully prepared a homogeneous blend membrane from poly(ether sulfone) (PES) and poly(vinylpyrrolidone) (PVP). The PES-PVP membrane possesses combined advantages of hydrophobic and hydrophilic components, leading to good mechanical strength and excellent hydrophilicity at the same time.[47] In the work reported by Lu et al., they indicated that, when compared with the traditional asbestos diaphragm, an electrolysis cell with the PES-PVP membrane as diaphragm can achieve an energy saving of ~10%.[47]

Recently AEMs have been used as electrolyte for alkaline water electrolysis.[48–52] However, the long-term durability of current AEMs is not satisfactory, which will hamper the commercialization of AEM-based alkaline water electrolysis technology. Aili et al.[53] developed heterogeneous anion-conducting membranes based on linear and cross-linking KOH-doped polybenzimidazole (PBI) used as the electrolyte for alkaline water electrolysis. The KOH-doped PBI membranes showed high ionic conductivity in the 10^{-2} S/cm range and low gas permeability and were stable in 6 M KOH at 85°C for up to 176 days.

24.2.3 Zero-Gap Alkaline Water Electrolysis Cells

Since alkaline liquid electrolyte (such as KOH) water electrolysis is a mature and commercialized technology, the performance of the electrolysis cells, and even stacks and systems, has been well documented in literature.[10,12] Traditional alkaline water electrolysis cells are based on an aqueous alkaline electrolyte such as KOH and a porous diaphragm/separator/membrane. The maximum current density is typically ~250 mA/cm² and its energy efficiency is typically ~60%.[25] To improve cell efficiency, the compact design of zero-gap alkaline water electrolysis cells based on AEM has been proposed.[25,48,50] In AEM-based zero-gap alkaline liquid electrolyte water electrolysis cells, aqueous KOH or dilute K_2CO_3 solution is supplied into the cell to provide the conducting path for hydroxide ion transportation in the electrode and to improve the electrocatalyst utilization, while $OH^-$is transported through AEM from the cathode to the anode, similar to AEM water electrolysis.[25] AEM-based zero-gap alkaline liquid electrolyte water electrolysis cell can be considered as a transition from alkaline liquid electrolyte water electrolysis to AEM water electrolysis. Pletcher et al.[48] fabricated zero-gap alkaline water electrolyzers with NiMo and RuO_2 coating as the HER cathode, a 160 µm thick hydroxide-conducting membrane as the electrolyte, and a $NiFe(OH)_2$ coating as the OER anode. The cell can achieve a current density of 1 A/cm² with a cell voltage of ~2.1 V at 60°C.[48] Faraj et al.[50] prepared an alkaline solid polymeric electrolyte water electrolysis cell with (low density polyethylene)-g-(vinyl benzyl chloride)-(1,4-diazabicyclo[2.2.2]octane) (LDPE-g-VBC-Dabco) polymer film as the

AEM and anode and cathode electrodes consisting of proprietary non-PGM catalysts produced by Acta Spa, Italy. At 45°C, with a flowing aqueous solution of 1 wt.% K_2CO_3 at the anode side and no liquid electrolyte at the cathode side, the cell can achieve a current density of 500 mA/cm² at a cell voltage of ~2.18 V and was stable during more than 500 h operation under a current density of 460 mA/cm².

24.3 ADVANCED MATERIALS FOR PROTON-EXCHANGE MEMBRANE WATER ELECTROLYSIS

In this and next section, our emphasis will be placed on the advanced materials, especially their recent developments for two main types of low-temperature water electrolysis technologies, PEM and AEM water electrolysis. Some good reviews about PEM water electrolysis can be found in literature.[54,55] For example, Arico et al.[54] reviewed the status of PEM water electrolysis technologies and potential applications in combination with renewable energy. A comprehensive review on PEM water electrolysis was done by Carmo et al.[55]

24.3.1 Electrocatalysts

The acidic environment of PEM water electrolysis limits the catalysts to noble metals since nonnoble metals are unstable in the acidic environment due to the corrosion. Noble metals, such as Platinum (Pt) and Iridium (Ir), not only show high activity for electrode reactions but also are quite stable in acidic media. Typically, platinum black and iridium (metal or oxides) are used as the electrocatalysts for HER at the cathode and OER at the anode, respectively. The common loading of Pt or Ir is in the range of 2–5 mg/cm². Such a high noble metal loading on the electrodes significantly contributes to the cost of PEM water electrolysis technology. To reduce the cost of electrocatalysts for PEM water electrolysis, three main approaches are suggested: (1) reduce the loading of Pt and Ir with improved catalyst utilization; (2) use less expensive noble metals, such as Pd and Ru, to replace Pt and Ir at the cathode and anode, respectively; and (3) develop low-cost nonnoble metals as alternatives to noble electrocatalysts.[56] Some noble metals and non-PGM electrocatalysts reported in literature for PEM water electrolysis are summarized in Table 24.2.[57–74]

24.3.1.1 Electrocatalysts for Hydrogen Evolution Reactions

24.3.1.1.1 Pt-Based Hydrogen Evolution Reaction Electrocatalysts

To reduce the loading of Pt, the catalyst utilization and activity of Pt should be increased. This may be achieved either by reducing the particle size of Pt to nano size (2–3 nm) level or by developing Pt monolayer or Pt-rich layer (shell) onto a supported metal core substrate (i.e., core-shell structure).

When Pt particles are reduced to nano size, due to their high surface energy, Pt nanoparticles tend to agglomerate together during the preparation process or sintering during the electrochemical reaction. One effective way to prevent the agglomeration and sintering is to support the Pt nanoparticles on a support. High–surface area carbons, such as Vulcan XC-72[15,68–72] and carbon nanotube (CNT),[72] are ideal supports for Pt catalysts due to their high electronic conductivity, high chemical/electrochemical stability, capability for good dispersion of Pt nanoparticles, and low cost. The carbon support not only provides a physical surface for the dispersion of Pt nano-sized particles but also provides an effective electron-conducting network for the reaction. Several groups have adopted carbon-supported Pt to replace Pt black as the electrocatalyst for HER and demonstrated promising performance as high-loading Pt black catalysts in PEM water electrolysis cells.[15,68–72] Recently, highly dispersed carbon-supported nano-sized Pt particles have become the standard electrocatalysts for HER in PEM water electrolysis. Tungsten carbide (WC) is considered another promising support for Pt nanoparticles because it is known to exhibit good stability and to show moderate HER activity in acidic environments. Ma et al.[75] prepared WC-supported nano–Pt catalyst by an impregnation method with as-prepared WC powders and H_2PtCl_6. WC-supported nano–Pt catalyst showed higher HER activity than Pt, indicating the synergistic effect between Pt and WC.[75]

Core-shell structure is another effective way to improve the Pt catalyst utilization and activity. Brankovic et al. of Brookhaven National Laboratory in the United States are one of the pioneers in the development of core-shell catalysts for oxygen-reduction reaction (ORR) in fuel cells.[76] Since only the surface of Pt nanoparticles actually participates in the electrocatalysis, Pt needs to be only at the surface. By supporting Pt or Pt alloys monolayer on other metal core substrates, the mass-specific activity of Pt catalysts for ORR significantly increases, thus lowering the Pt loading while achieving similar performance to that of Pt nanoparticle catalysts. Moreover, a monolayer of Pt has shown higher activity for ORR compared to the bulk Pt surface.[77] Brankovic et al.,[76] Zhang et al.,[77] and Strasser et al.[78] synthesized Pt monolayer (shell) on Pd/C or Pd_9Au_1/C (core) by galvanic displacement by a Pt monolayer of a Cu monolayer deposited on the cores at underpotentials. Sasaki et al. adopted a dealloying method to develop Pt-based catalysts with core-shell structure.[79,80] Pt-based catalysts with core shell show not only high mass-specific activity of Pt for ORR but also high stability.[78–80] Similarly, core-shell–structured Pt-based catalysts can be used as the electrocatalysts for HER in PEM electrolysis to drastically decrease the Pt loading without sacrificing efficiency. Esposito et al.[58] prepared a Pt monolayer supported on low-cost transition tungsten carbides (WC and W_2C) and evaluated its HER activity in 0.5 M H_2SO_4 solution at room temperature. They found that one monolayer of Pt on WC/W_2C surface can provide HER activity that is comparable to that of bulk Pt.

Nanostructured thin-film (NSTF) catalysts with advanced structure developed by 3M Company are a promising alternative to conventional Pt black and carbon-supported Pt catalysts for PEM water electrolysis. The NSTF catalysts are formed by facial vacuum sputter deposition of catalyst alloys onto a supported monolayer of oriented crystalline organic pigment whiskers.[60,81] NSTF catalysts can be considered a

TABLE 24.2
PGM and Non-PGM Electrocatalysts for PEM Electrolysis Cells

	Electrocatalyst	Type of Catalyst	Performance	Ref.
HER	Pt/C	Supported catalyst	25°C, 0.1 M $HClO_4$ E_{onset} = 0 V vs. RHE i_0 = 0.78 mA/cm^2	57
HER	Pt monolayer/WC Pt monolayer/W_2C	Supported catalyst, core-shell structure	Close to Pt foil	58
HER	40 wt.% Pd/C, 0.7 mg Pd/cm^2	Supported catalyst	90°C, η = 50 mV at 1 A/cm^2 Close to 40 wt.% Pt/C at the same conditions	56
HER	NiMo with a loading of 3.0 mg/cm^2	Non-PGM catalyst	25°C, η = 80 mV at 20 mA/cm^2 in 0.5 M H_2SO_4	59
HER	$NiMoN_x$/C	Non-PGM catalyst Supported catalyst	Onset potential: −78 mV vs. RHE Tafel slope: 35.9 mV/decade Exchange current density: 0.24 mA/cm^2, ~1/3 of that of Pt	57
HER	3M's NSTF $Pt_{68}Co_{32}Mn_3$ or $Pt_{50}Ir_{50}$ catalysts, 0.15 mg/cm^2 Pt	NSTF	Similar to that of standard Pt blacks having loadings of >2 mg/cm^2	60
HER	MoS_2 grown on graphene	Non-PGM catalyst	E_{onset} = −0.1 V vs. RHE Tafel slope: 41 mV/decade Better performance than MoS_2	61
HER	α-$H_4SiW_{12}O_{40}$, 0.8 mg/cm^2	Non-PGM catalyst	90°C, E = 1.84 V at 1 A/cm^2 (w/ Ir anode) Slightly worse than Pt	62
HER	Boron-capped tris(glyoximato) cobalt complexes supported on Vulcan XC-72, 1 mg/cm^2	Non-PGM catalyst	90°C, E = 2.1 V at 1 A/cm^2 (w/ Ir anode) Slightly worse than Pt	62
OER	IrO_2, 2.5 mg/cm^2	PGM catalyst Sulfite-complex route	80°C, E_{cell} = 1.80 V at 1.75 A/cm^2	63
OER	3D ordered macroporous IrO_2, 0.5 mg/cm^2	3D ordered macroporous catalyst	25°C, 70 mA/cm^2 at 1.60 V Better than colloidal IrO_2	64
OER	Ir/TiC, 0.3 mg Ir/cm^2	Supported catalyst	80°C, E_{cell} = 1.80 V at 0.84 A/cm^2	65
OER	IrO_2/Ti_nO_{2n-1}, 1.0 mg IrO_2/cm^2	Supported catalyst	80°C, E_{cell} = 1.70 V at 0.50 A/cm^2	66
OER	RuO_2/Sb-doped SnO_2, 10 mg/cm^2	Supported catalyst	80°C, E_{cell} = 1.56 V at 1.0 A/cm^2 Higher activity than unsupported RuO_2	67
OER	$Ir_{0.2}Ru_{0.8}O_2$, 1.5 mg/cm^2 $Ir_{0.4}Ru_{0.6}O_2$, 1.5 mg/cm^2 $Ir_{0.4}Ru_{0.6}Mo_xO_y$, 1.5 mg/cm^2	Solid solution catalyst, modified Adams fusion method	80 °C, E_{cell} = 1.622 V at 1.0 A/cm^2; 1.646 V at 1.0 A/cm^2; 1.606 V at 1.0 A/cm^2	57
OER	$Ir_{0.6}Ru_{0.2}Ta_{0.2}O_2$, 2.0 mg/cm^2 $Ir_{0.8}Ru_{0.2}O_2$, 2.0 mg/cm^2 $Ir_{0.6}Ru_{0.4}O_2$, 2.0 mg/cm^2	Solid solution catalyst	80 °C, E_{cell} ~1.60 V at 1.0 A/cm^2; ~1.60 V at 1.0 A/cm^2; ~1.567 V at 1.0 A/cm^2	15
OER	$Ir_{0.2}Ru_{0.8}O_2$, 1.5 mg/cm^2	Solid solution catalyst	80°C, E_{cell} ~1.622 V at 1.0 A/cm^2 (with 0.5 mg/cm^2 Pt/C cathode)	57
OER	$Ir_{0.6}Sn_{0.4}O_2$, 0.77 mg Ir/cm^2	Solid solution catalyst	80°C, E_{cell} ~1.631 V at 1.0 A/cm^2; ~1.820 V at 2.0 A/cm^2	73
OER	$Ru_{0.6}Sn_{0.4}O_2$	Solid solution catalyst	Slightly better than RuO_2	74
OER	3M'sNSTF $Pt_{50}Ir_{50}$, $Pt_{50}Ir_{25}Ru_{25}$ catalysts, 0.15 mg/cm^2 Pt	NSTF	Similar to that of standard PtIr blacks having loadings of >2 mg/cm^2	60

Note: NSTF: nanostructured thin film; RHE: reversible hydrogen electrode.

special type of supported catalysts. NSTF catalysts exhibited better/comparable specific activities as well as durability in fuel cell membrane–electrode assembly (MEA) compared to conventional carbon-supported Pt catalysts.[60,81] Since the whisker supports are resistant to chemical or electrochemical dissolution/corrosion, even at a high voltage above 2 V, it is possible to use NSTF catalysts as electrocatalysts for PEM water electrolysis.[60] Recently 3M's NSTF catalysts with low Pt loading have been evaluated in PEM water electrolyzers at Proton OnSite[16,60,82] and Giner Inc.[63,83] It was found that beginning-of-life cathode performance of NSTF $Pt_{68}Co_{32}Mn_3$ and NSTF $Pt_{50}Ir_{50}$ catalysts with a loading of 0.15 mg/cm^2 Pt were similar to that of standard Pt blacks with very high Pt loadings (>2 mg/cm^2),[16,60] as shown in Figure 24.7. MEAs consisting of NSTF $Pt_{68}Co_{32}Mn_3$ or $Pt_{50}Ir_{50}$ catalysts on the cathodes and standard PtIr blacks on the anode also showed very impressive short stack durability performances under constant current with a lifetime of several thousand hours.[60]

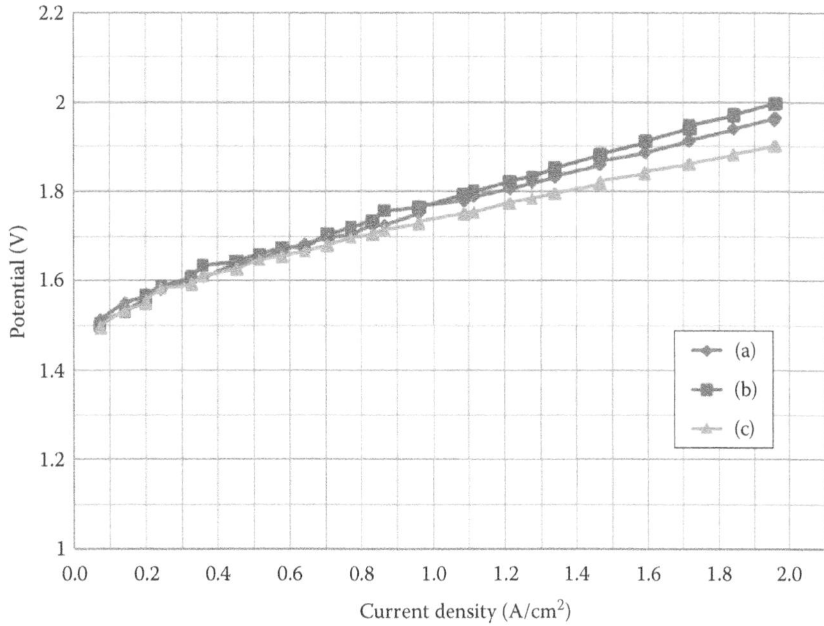

FIGURE 24.7 Polarization curves at 80°C of MEAs with Proton standard PGM anode and three different cathodes. (a) 3M's NSTF $Pt_{50}Ir_{50}$ (loading: 0.15 mg Pt/cm^2); (b) 3M's $Pt_{68}Co_{32}Mn_3$ cathode (loading: 0.15 mg Pt/cm^2); (c) Proton's Pt black cathode (loading: >2 mg Pt/cm^2). (Reproduced from Ayers, K. E., E. B. Anderson, C. B. Capuano, B. D. Carter, L. T. Dalton, G. Hanlon, J. Manco, and M. Niedzwiecki, Research advanced towards low cost, high efficiency PEM electrolysis, *ECS Transactions*, 33(1), 3–15, 2010. With permission.)

24.3.1.1.2 Pd-Based Hydrogen Evolution Reaction Electrocatalysts

Considering its abundance on earth and much lower cost than Pt (the price of Pd is about 30% of that of Pt in weight), Pd is a promising alternative to Pt for PEM water electrolysis. Grigoriev et al.[56] evaluated carbon-supported Pd for HER in PEM water electrolyzers. Their 40 wt.% Pd/Vulcan XC-72 electrocatalyst with a loading of 0.7 mg/cm^2 Pd exhibited very close performance to its counterpart–40 wt.% Pt/Vulcan XC-72 with the same loading of Pt. In their work, MEA with 40 wt.% Pd/Vulcan XC-72 electrocatalyst with a loading of 0.7 mg/cm^2 Pd on the cathode and Ir black with a loading of 2.4 mg/cm^2 Ir has achieved a cell voltage of ca. 1.68 V at a current density of 1 A/cm^2 at 90°C and has performed steadily for over 100 hours.[56] The results obtained showed that Pd can be advantageously used as an alternative electrocatalyst to Pt for the HER in PEM water electrolyzers.[56] Based on the works about Pt-based electrocatalysts with core-shell and NSTF structure, core-shell–structured or NSTF Pd-based electrocatalysts may further increase the mass activity of Pd toward HER and thus achieve high efficiency of PEM water electrolysis while further lowering the loading of Pd and significantly reducing the cost of the electrocatalysts used. Esposito et al.[58] reported that a monolayer Pd-WC and monolayer Pd-Mo_2C thin films showed the most promise as HER electrocatalysts. Although their measured HER activity was not as high as that of bulk Pt or Pd films, their relatively high activity, potential for synthesis of high–surface area structures for Mo_2C, and the lower price of Pd make these catalysts very promising.[58]

24.3.1.1.3 Non–Precious Group Metal Hydrogen Evolution Reaction Electrocatalysts

In order to significantly reduce the cost of electrocatalysts used for HER, the development of non-PGM catalysts is desired. Several non-PGM catalysts, such as transition-metal complexes,[62] carbides,[58,84] nitrides,[57] and transition-metal chalcogenides, such as MoS_2,[85,86] have been developed recently.

Millet et al.[62] explored two types of non-PGM electrocatalysts for HER in PEM water electrolyzers: cobalt clathrochelate and polyoxometalate. They found that, although the carbon-supported cobalt clathrochelate complexes were significantly less efficient than Pt for HER, a cell efficiency of ~80% was still obtained at a current density of 500 mA/cm^2 at 90°C for MEA consisting of Vulcan XC-72–supported cobalt clathrochelate complexes with a loading of 1 mg/cm^2 at the cathode and Ir black with a loading of 2.5 mg/cm^2 at the anode.[62] Very interestingly, they demonstrated that commercially available tungstosilicic acid hydrate (α-$H_4SiW_{12}O_{40}$), one of the polyoxometalates used as electrocatalyst for HER, is much more active than the cobalt clathrochelate complexes. For example, in their work, MEA consisting of α-$H_4SiW_{12}O_{40}$ adsorbed on the cathodic Ti current collector with a loading of 0.8 mg/cm^2 at the cathode and Ir black with a loading of 2.5 mg/cm^2 at the anode exhibited a cell efficiency of ~80% at a current density of 1 A/cm^2 and 90°C, slightly lower than that of MEA consisting of Pt cathode and Ir anode with similar loadings.[62] Their results offer new and interesting perspective on PEM water electrolysis. However, the long-term stability of these non-PGM electrocatalysts needs to be clarified.

Nickel and nickel-based alloys are known electrocatalysts for HER in alkaline electrolytes, as described in Section 24.2.1.1. Among them, Ni-Mo alloys exhibit high HER activity and long-term stability under alkaline conditions.[59] However, Ni-Mo catalysts are unstable in acidic conditions.[87] Incorporating nitrogen into a Ni-Mo composite to form metal nitride can enhance the HER activity and stability of Ni-Mo based catalysts under acidic conditions. Chen et al.[57] synthesized NiMo nitride nanosheets supported on carbon support ($NiMoN_x/C$), which exhibited high HER electrocatalytic activity with a small onset potential of −78 mV versus the reversible hydrogen electrode (RHE), a high exchange current density of 0.24 mA/cm^2(about one-third of that for Pt: 0.78 mA/cm^2), and a small Tafel slope of 35.9 mV/decade. $NiMoN_x/C$ catalysts also showed high long-term stability in an acidic environment with a negligible loss after 2000 cycles' potential cycling in the range of −0.3 to 0.9 V.

MoS_2 has been considered one of the promising non-PGM HER catalysts.[61,85–88] Hinnemann et al.[85] reported that MoS_2 nanoparticles supported on carbon can be used as a catalyst for HER and exhibited an overpotential in the range 0.1–0.2 V. Supporting MoS_2 on other supports may improve the HER activity of MoS_2-based catalysts.[63,87] For example, Li et al. synthesized MoS_2 nanoparticles on reduced graphene oxide (RGO) sheets via a facile solvothermal approach.[61] The MoS_2/RGO hybrid catalyst exhibited excellent HER activity in 0.5 M H_2SO_4 solution with an overpotential as small as ~150 mV at a current density of 10 mA/cm^2 and a Tafel slope as small as 40 mV/decade, which is superior to those of other MoS_2 catalysts.[61] The MoS_2/RGO hybrid catalyst also exhibited good durability with a negligible loss of the cathodic current after cycling continuously for 1000 cycles.[61] The authors attributed the high performance of their MoS_2/RGO hybrid catalyst in HER to strong chemical and electronic coupling between the graphene oxide sheets and MoS_2.[61] Chang et al.[87] prepared MoS_x grown on graphene-protected 3D Ni foam and demonstrated high activity in HER with an overpotential of ~200 mV at a current density of 45 mA/cm^2 for their catalysts, which is close to that of MoS_2/RGO hybrid catalysts reported by Li et al.[61]

24.3.1.2 Electrocatalysts for Oxygen Evolution Reactions

Due to its slower kinetics, OER in the anode is a dominant source of overpotentials in PEM water electrolysis cells, compared to HER in the cathode, similar to the case of alkaline liquid electrolyte water electrolysis cells. Therefore, much attention has been paid to the development of highly active, stable anode electrocatalysts with low cost for PEM water electrolysis.

24.3.1.2.1 Ir-Based Oxygen Evolution Reaction Electrocatalysts

Iridium (Ir) or iridium oxides (IrO_2) are the most used anode catalysts due to their high activity in OER, their high electronic conductivity (~10^4 S/cm), and their high stability in acid conditions. Typically, the Adams fusion method or its modified version was widely adopted to prepare fine noble metal oxide powders including iridium oxides.[57,89,90] During the preparation process, the calcination temperature of the precursor influences the physical properties, morphology, and electrocatalytic performance of IrO_2 for OER.[57,90] In order to further increase the OER activity and reduce the demand of noble metal oxide catalysts, the morphology and nanostructure of IrO_2 need to be improved. Alternative methods, including sol–gel,[91] modified polyol,[92] and sulfite-complex route,[63] have been proposed to increase the surface area, to reduce the particle size, and to reduce the contaminant. For example, Siracusano et al.[63] prepared nano-sized IrO_2 with a crystalline size of 1–3 nm using the sulfite-complex route. The main advantage of the sulfite-complex route is to eliminate the negative effect of chlorine contaminants as residues of chloride-based precursors that exist in other methods.[63] The template-control method is another promising way to improve the morphology and nanostructure of IrO_2. The use of hard or soft templates with a narrow size distribution as a pore-directing agent offers better control over pore size and pore connectivity of IrO_2 materials.[64,93] Ortel et al.[93] demonstrated the synthesis of mesoporous IrO_2 film by using the micelles of amphiphilic triblock-copolymer poly(ethylene oxide)–poly(butadiene)–poly(ethylene oxide) (PEO-PB-PEO) as the soft template. Template-fabricated IrO_2 films substantially reduced the overpotentials for OER compared to those of untemplated coatings.[93] Pore templating, thus, enables direct control over surface catalytic properties of iridium oxide.[93] Hu et al.[64] prepared three-dimensional ordered macroporous (3-DOM) IrO_2 materials used as OER electrocatalysts using monodispersed SiO_2 spheres as the templates. The 3-DOM IrO_2 exhibited a very large surface area of 128.7 m^2/g, three times higher than that of the IrO_2 prepared by colloidal method, thus providing much higher electrochemical surface area for OER. Considering its similar intrinsic activity to the colloidal IrO_2, the 3-DOM structure significantly increased the mass-specific activity of IrO_2 compared with the conventional nanostructure prepared by colloidal method.

In order to ensure sufficient performance and required stability, the IrO_2 used as OER electrocatalysts in PEM water electrolysis is often used in unsupported form with a loading of several milligrams per square centimeter. Unsupported IrO_2 catalysts exhibit large particle size with agglomeration, low surface area, and poor catalyst utilization. In order to reduce the loading of IrO_2 electrocatalysts, it is essential to decrease the particle size, to increase the surface area of the catalysts, and to improve their electrocatalytic activity and utilization. This may be achieved by supporting nano-sized IrO_2 on a high–surface area support. High–surface area carbon, the support commonly used for Pt- or Pd-based HER catalysts, cannot be used for the Ir-based OER catalysts, since carbon is unstable at high OER potential due to its corrosion. It is desirable to develop stable supports for OER catalysts with high surface area, high electronic conductivity, and capability for good dispersion of IrO_2 nanoparticles, as exhibited by carbon supports for HER catalysts. Several metal oxides, such as TiO_x,[66] SnO_2,[94] and TiC,[62,95] have been found to be more stable at high potentials and can be used as supports for OER electrocatalysts.

Siracusano et al.[66] prepared titanium suboxides (TiO_x) as conductive supports for IrO_2 electrocatalysts and found that IrO_2 catalysts supported on TiO_x prepared in-house showed superior electrocatalytic activity compared to those supported on commercial supports—Ebonex®. Ma et al.[95] prepared TiC-supported Ir (Ir/TiC) electrocatalysts with an Ir particle size of 10–40 nm for OER by chemical reduction and deposition. Based on mass-specific activity of Ir, the Ir/TiC catalyst was found to be more active than unsupported Ir catalyst.[65,95]

Another effective approach to reduce the loading of Ir for the Ir-based catalysts is to form a solid solution of IrO_2 with less expensive transition-metal oxides, such as SnO_2,[73,92,96–98] and Ta_2O_5,[97,99–101] as well as RuO_2.[15,57,99–101] Marshall et al.[97] used a modified polyol method to prepare nanocrystalline oxide powders of $Ir_xSn_{1-x}O_2$ ($0.2 \leq x \leq 1$). They found that the addition of SnO_2 to IrO_2 particles for OER electrocatalysts showed no beneficial effect other than dilution of the more expensive IrO_2.[97] On the other hand, Li et al.[73] used a polymer-assisted method to synthesize Ir-Sn oxide electrocatalysts. In the polymer-assisted method, the triblock polymer with repeating units of PEO–poly(propylene oxide) (PPO)–PEO was used as an efficient stabilizer and structure-directing agent. Li et al. reported that the $Ir_{0.6}Sn_{0.4}O_2$ catalysts showed an OER activity comparable to that of IrO_2 prepared using the same method and preparative process while there was an ~23% reduction in Ir.[73] Since RuO_2 is more active than IrO_2, the addition of RuO_2 into IrO_2 can enhance the activity of IrO_2 while reducing the amount of Ir.[15,57,70,99] Cheng et al.[51] prepared $Ir_xRu_{1-x}O_2$ electrocatalysts using the Adams fusion method and examined the catalysts' performance. They confirmed that the $Ir_xRu_{1-x}O_2$ ($x = 0.2, 0.4, 0.6$) compounds are more active than pure IrO_2 and are more stable than pure RuO_2. Marshall et al.[15] synthesized $Ir_xRu_yTa_zO_2$ electrocatalysts with different compositions using an aqueous hydrolysis method, followed by thermal oxidation, and examined the catalysts' performance. They found that the performance of the PEM water electrolysis cell with different anode catalysts and the same 20 wt.% Pt/C cathode increased in the order $IrO_2 < Ir_{0.6}Ru_{0.2}Ta_{0.2}O_2 \approx Ir_{0.8}Ru_{0.2}O_2 < Ir_{0.6}Ru_{0.4}O_2$,[15] as shown in Figure 24.8. Cheng et al.[57] used a modified Adams fusion method to prepare $Ir_{0.4}Ru_{0.6}Mo_xO_y$ electrocatalysts and demonstrated that $Ir_{0.4}Ru_{0.6}Mo_xO_y$ had a higher performance compared to $Ir_{0.4}Ru_{0.6}O_2$. The main reason for the higher performance of the former is that doping with a small amount of Mo can help reduce the particle size and increase the surface area of the catalysts.[57]

Ir-based catalysts with NSTF and core-shell structure may be good alternatives to Ir black or IrO_2 catalysts for OER. Like Pt-based NSTF catalysts for HER, Ir-based NSTF catalysts may be promising for OER. Debe et al.[60] reported that in single-cell tests, beginning-of-life anode performances of NSTF-$Pt_{50}Ir_{50}$ and $Pt_{50}Ir_{25}Ru_{25}$ treated with a proprietary surface energy treatment process were close to that of standard PtIr blacks having loadings an order of magnitude higher. Short stack durability tests at constant current with these NSTF catalysts on the anodes and standard Pt black on the cathodes were found to be stable for >2000 hours in multiple tests.[60] As mentioned in Section 24.3.1.1.1, core-shell structure is an effective way to improve the utilization and activity of Pt-based HER catalysts. Similarly, Ir-based catalysts with core-shell structure may improve their catalyst utilization and activity for OER.

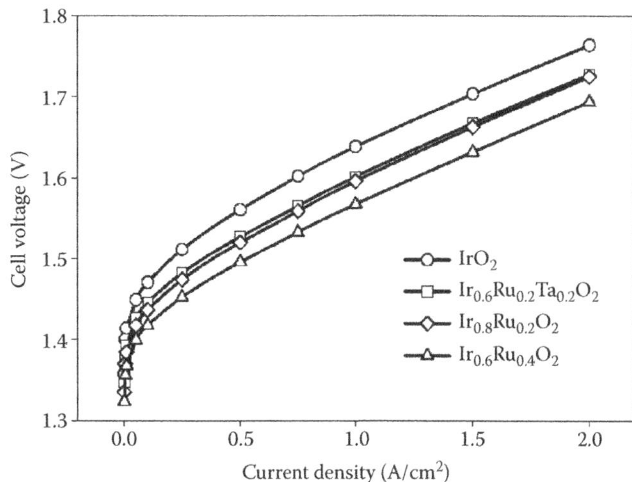

FIGURE 24.8 Polarization curves of a PEM water electrolysis cell at 80°C using different anode electrocatalysts and 20% Pt/C cathodes and Nafion 115 electrolyte membranes. (Reproduced from Marshall, A. T., S. Sunde, A. Tsypkin, and R. Tunold, Performance of a PEM water electrolysis cell using $Ir_xRu_yTa_zO_2$ electrocatalysts for the oxygen evolution electrode, *International Journal of Hydrogen Energy*, 32 (13), 2320–2324, 2007. With permission.)

24.3.1.2.2 Ru-Based Oxygen Evolution Reaction Electrocatalysts

Ruthenium oxide (RuO_2) is one of the most active electrocatalysts for OER, even more active than IrO_2. Moreover, RuO_2 is several times cheaper than IrO_2. However, it suffers from instability in acid solution and at high operating potential. The stability and activity of RuO_2 might be improved by forming a solid solution with SnO_2 or by supporting RuO_2 on SnO_2 support.[67,74,102] Wu et al.[74] prepared nanocrystalline powders of $Ru_xSn_{1-x}O_2$ ($1.0 \geq x \geq 0.2$) and used them as the electrocatalysts for OER in PEM water electrolyzers. They found that the catalyst $Ru_{0.6}Sn_{0.4}O_2$ demonstrated the best performance in general due to its smaller particle size and greater ratio of outer active surface area and had better stability compared to pure RuO_2. Since SnO_2 is a semiconductor, the electrochemical property of RuO_2 supported on SnO_2 can be improved by enhancing the electronic conductivity of the catalyst support.[67] Due to a higher electronic conductivity of Sb(V)-doped SnO_2 (antimony tin oxide or ATO) than that of SnO_2, ATO could be a promising support for the electrocatalysts. Wu et al.[67] deposited RuO_2 (10–15 nm) on ATO nanoparticles (30–40 nm) and found that RuO_2 supported on ATO exhibited higher activity than unsupported RuO_2 for OER.

24.3.2 Membrane

The function of the PEM in PEM water electrolysis is to transport protons from the anode to the cathode and to prevent the produced gases from mixing. The benchmark electrolytes for

PEM water electrolysis are perfluorosulfonic membranes, such as Nafion. A Nafion membrane meets the criteria for a number of key physical properties required for the operation in PEM water electrolysis: (1) high ionic conductivity (>10 mS/cm), (2) negligible electronic conductivity, (3) good chemical and mechanical stability, (4) high thermal conductivity, and (5) limited permeability to hydrogen and oxygen.[103] Ito et al.[104] reviewed the properties of Nafion membranes under PEM water electrolysis conditions. In their review, they pointed out that the hydration state of the membrane under electrolysis operation is quite different from that under fuel cell operation.[104] During PEM fuel cell operation, the membrane is humidified by humidified gases and equilibrated with water vapor, while during the electrolysis operation, the electrolyte membrane is exposed to liquid water and fully hydrated with liquid water.[104] Therefore, the properties of a Nafion membrane used for PEM water electrolysis, including water uptake, swelling behavior, proton conductivity, and electroosmotic drag coefficient, need to be evaluated under the equilibration with liquid water. Thickness is considered one of the important parameters for a Nafion membrane used for PEM water electrolysis. If the membrane is too thick, high ohmic resistance of the membrane will limit the cell efficiency; if the membrane is too thin, the mechanical strength may not be high enough for cell operation, especially in the case of high-pressure water electrolysis. In addition, a thin membrane will lead to high gas crossover through the membrane, thus reducing the cell current efficiency and the purity of the gas obtained. Therefore, there is a trade-off between mechanical strength, gas crossover, and ohmic resistance for the selection of the membrane thickness. The gas permeability of the membrane is another key parameter to be considered for the electrolysis operation, especially in the case of high-pressure operation. During high-pressure electrolysis operation, hydrogen and oxygen produced at the cathode and anode permeate through the PEM and then mix at the respective counterelectrode compartment. This kind of gas crossover increases the risk of gas explosions, reduces the purity of the produced gas, and decreases the current efficiency of the cell.

Besides the high loading of noble metals as electrocatalysts on the electrodes, the use of expensive perfluorosulfonic membranes also significantly contributes to the cost of PEM water electrolysis technology. The development of alternative low-cost membranes with enhanced mechanical properties for high-pressure applications, high conductivity, and reduced gas crossover is desired. Sawada et al.[105] prepared a highly proton-conductive PEM for PEM water electrolysis by γ ray–induced postgrafting of styrene into a cross-linked polytetrafluoroethylene film and subsequent sulfonation. Wei et al.[106] blended sulfonated poly(ether ether ketone) (SPEEK) with PES to prepare a PEM for water electrolysis. They reported that an electrolytic current of 1.65 A/cm^2 was obtained at 2.0 V and 80°C for the MEA with SPEEK/PES-blended membrane.[106] Although the performance of a SPEEK/PES-blended membrane–based MEA is still lower than that of a Nafion-based one, a SPEEK/PES-blended membrane is much cheaper than Nafion.

An increase in the operating temperature of a water electrolyzer can improve the electrode kinetics and increase membrane conductivity, thus enhancing MEA performance and cell efficiency. However, Nafion membrane cannot tolerate high temperatures above 100°C due to its dehydration and a subsequent loss of proton conductivity at such high temperatures. Baglio et al.[107] manufactured a composite Nafion-TiO_2 membrane by a recast procedure using in-house–prepared TiO_2 as a filler, which allowed an increase in the operating temperature of the electrolyzer to up to 120°C. They demonstrated higher performance for MEA based on Nafion-TiO_2 composite membrane at 120°C with respect to a commercial Nafion 115 membrane, and attributed the effect to the water retention properties of the TiO_2 filler.[107] However, the MEA based on Nafion-TiO_2 composite membrane at 120°C still showed worse performance than that based on Nafion 115 membrane at 80°C in the region of high current densities.

24.3.3 Ionomer

In state-of-the-art PEM water electrolyzers, commercially available Nafion dispersions with high proton conductivity and high stability are used to produce MEAs with high output and good long-term operational stability. Ionomers with high proton conductivity can facilitate an efficient three-phase boundary for the desired electrochemical reactions that significantly improve the catalyst utilization, and can reduce the ionic transport resistance in the electrocatalyst layer. Highly stable ionomers can greatly extend the lifetime of MEAs and allow for optimization of MEA and cell design for high performance. The ionomer also acts as a binder to maintain the integrated electrode structure. Unfortunately, there are only a few reports about alternatives to Nafion dispersions used as ionomers for PEM water electrolysis. Wei et al.[106] used Nafion and different SPEEK ionomers as ionic conductors in the catalyst layer (CL) for the SPEEK/PES-blended membrane and compared their MEA performances. They found that SPEEK/PES-based MEA with Nafion ionomers in the catalysts showed extremely poor performance, much lower than the performance of those with SPEEK ionomers. The authors attributed the poor performance of MEA with Nafion ionomer to the additional resistance generated at the interface between SPEEK/PES-blended membrane and the CL with Nafion ionomer since different ionomer materials were used in the membrane and in the CL.[106]

24.3.4 Membrane–Electrode Assembly

The heart of the PEM water electrolysis cell is the membrane-electrode assembly (MEA), which consists of anode gas-diffusion media, anode CL, electrolyte membrane, cathode CL, and cathode gas-diffusion media. Performance, durability, cost, and safety are four key considerations for the development of MEA for PEM water electrolysis. The MEA performance depends on many factors, including electrocatalyst activity and loading, electrolyte membrane thickness and proton conductivity, ionomer content and proton conductivity,

TABLE 24.3
MEA Specifications and Performances of PEM Electrolysis Cells Reported in Literature

Anode	Cathode	Membrane	Performance	Ref.
IrO_2-RuO_2 (1:1), 2.0 mg/cm^2	30 wt.% Pt/C, 2.0 mg/cm^2	Unknown	90°C, pressure: 30 bars E = 1.68 V at 1.0 A/cm^2	14
IrO_2, 2 mg/cm^2	10% Pt/Vulcan XC-72, 0.4 mg/cm^2	Nafion	80°C E = 1.65 V at 1.0 A/cm^2	57
$Ir_{0.6}Ru_{0.4}O_2$, 2 mg/cm^2	20 wt.% Pt/C, 0.4 mg/cm^2	Nafion 115	80°C E = 1.567 V at 1.0 A/cm^2	15
$Ir_{0.2}Ru_{0.8}O_2$, 1.5 mg/cm^2	28 wt.% Pt/C, 0.5 mg/cm^2	Nafion 1035	80°C E = 1.622 V at 1.0 A/cm^2	57
$Ir_{0.4}Ru_{0.6}Mo_xO_y$, 1.5 mg/cm^2	28 wt.% Pt/C, 0.5 mg/cm^2	Nafion 1035	80°C E = 1.606 V at 1.0 A/cm^2	57
IrO_2, 3.0 mg/cm^2	28 wt.% Pt/C, 0.5 mg/cm^2	Nafion 112	80°C E = 1.63 V at 1.0 A/cm^2	57
Ir/TiC, 0.3 mg/cm^2	Pt black, 1.0 mg/cm^2	Nafion 112	80°C E = 1.8 V at 0.84 A/cm^2	65
Ir black, 3.8 mg/cm^2	Pt black, 2.3 mg/cm^2	Nafion 112	60°C E = 1.8 V at 680 mA/cm^2	108
Ir black, 2.4 mg/cm^2	40 wt.% Pd/C, 0.7 mg/cm^2 Pd	Nafion 115	90°C E = 1.70 V at 1.0 A/cm^2	56

operating temperature and pressure, and MEA configuration and fabrication method. Table 24.3 summarizes the MEA performances and corresponding MEA specifications reported in the literature. In most cases, the cell voltage required for hydrogen production in PEM water electrolysis is in the range of 1.60–1.80 V at a current density of 1 A/cm^2 at 80–90°C. Figure 24.9 shows the typical performances of PEM water electrolysis with different anode and cathode electrocatalysts, which were reported by Grigoriev et al.[14] The durability of MEA depends on the catalyst stability, ionomer and membrane stability, and electrode structure integration. State-of-the art PEM water electrolysis cells/stack can last for several thousands of hours, up to 20,000 hours during long-term operation.[16,60,72] The cost of MEA depends on the catalyst price and loading, the cost of ionomer and electrolyte membrane cost, and the cost of other components, including the gas-diffusion media and the bipolar plate. Besides the MEA, the flow field and the separator are other major contributors to the cost of PEM water electrolysis stacks. Currently Ti-based flow fields and bipolar plates are commonly used for PEM water electrolysis cells and stacks. In order to reduce the stack cost, it is necessary to replace Ti-based diffusion media and bipolar plates with appropriate and cost-effective materials, such as stainless steel with enhanced resilience to chemical and electrochemical corrosion. For example, the company Proton OnSite reported that solid Ti parts were replaced with carbon-based composite parts, coating the carbon on the oxygen side of the cell with Ti or other protective coating to avoid the oxidation and corrosion.[16]

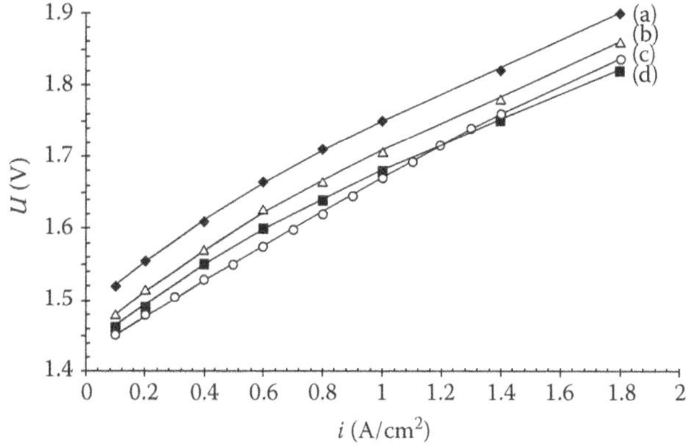

FIGURE 24.9 Polarization curves of PEM water electrolysis cell with different anode/cathode catalysts at 90°C. (a) Cathode: 2.0 mg/cm^2 Pt30/C, anode: 2.4 mg/cm^2 Ir; (b) cathode: 2.0 mg/cm^2 Pt30/C, anode: 2.0 mg/cm^2 RuO_2(30%)-IrO_2(32%)-SnO_2(38%); (c) cathode: 2.4 mg/cm^2 Pd40/C, anode: 2.4 mg/cm^2 Ir; (d) cathode: 2.0 mg/cm^2 Pt30/C, anode: 2.0 mg/cm^2 RuO_2(50%)-IrO_2(50%) (Pt30/C—30 wt.% of Pt supported on carbon, etc.). (Reproduced from Grigoriev, S. A., V. I. Porembsky, and V. N. Fateev, Pure hydrogen production by PEM electrolysis for hydrogen energy, *International Journal of Hydrogen Energy*, 31 (2), 171–175, 2006. With permission.)

24.4 ADVANCED MATERIALS FOR ANION-EXCHANGE MEMBRANE WATER ELECTROLYSIS

Figure 24.10 shows the schematic diagram of an AEM water electrolysis cell. In general, an alkaline membrane water electrolysis cell comprises an anode for OER (including anode gas-diffusion layer [GDL] and anode CL), a cathode for HER (including cathode GDL and cathode CL), an anode bipolar plate, a cathode bipolar plate, and an AEM. The key element for AEM water electrolysis cell is the MEA, which consists of anode GDL, anode CL, AEM, cathode CL, and

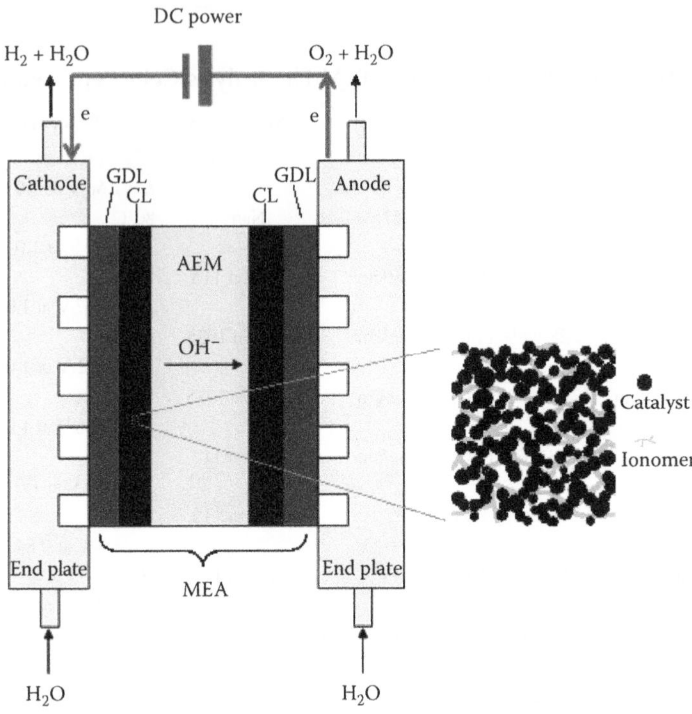

FIGURE 24.10 Schematic diagram of an AEM water electrolyzer. (Reprinted with permission from Ref. 4, 9054–9057. Copyright 2012 American Chemical Society.)

cathode GDL. The inset of Figure 24.10 shows the porous microstructure of the electrode CLs, where the catalyst particles are in intimate contact with the ionomer. Catalyst particles contact each other to form an electron-conducting network, while the ionomer in the CLs forms a network for the hydroxide ion–conducting path. Electrode reactions for HER/OER take place at the electrochemical active sites, i.e., triple-phase boundaries among catalysts, ionomer, and gas pore. Since ionomers in the electrodes and the AEM provide the conducting path for OH^- transport, no supply of KOH or K_2CO_3 is needed; therefore, pure water can be fed into the cell, which is quite different from the alkaline liquid electrolyte and AEM-based zero-gap alkaline water electrolysis cells. In AEM water electrolysis, water is generally fed into the cathode (i.e., water cathode-feed mode), since water is consumed for the HER to produce hydroxide ions. However, the cathode-feed mode increases the system complexity for the separation of hydrogen and water. It is possible to feed water to the anode (i.e., water anode-feed mode) since water can transport from the anode through the AEM to the cathode in the case of anode-feed mode.

24.4.1 Electrocatalysts

One of main advantages of AEM water electrolysis is the potential to use non-PGM catalysts for HER and OER, similar to alkaline liquid electrolyte water electrolysis. Non-PGM electrocatalysts for alkaline membrane electrolysis cells are summarized in Table 24.4.

24.4.1.1 Non–Precious Group Metal Electrocatalysts for Hydrogen Evolution Reaction

Ni-based materials and steel are generally used as electrodes and/or electrocatalysts for alkaline electrolysis. Among them, NiMo alloys exhibit high activity for HER and long-term stability under an alkaline environment. However, their performance is still not sufficiently high for an AEM water electrolysis cell. Hence, high catalyst loading is necessary for achieving high performance. In a liquid alkaline electrolyzer, since the liquid alkaline electrolyte solution can penetrate into the electrode and its ionic conductivity is quite high, the loading of Ni alloys can be very high (with a loading of >40 mg/cm^2 and >200 μm thick electrode layer) to reduce the cathode overpotential for HER without the punishment of ionic transport resistance in the electrode. However, if such high loading of Ni alloys were applied for HER in AEM water electrolysis, due to the limited ionic conductivity of anion exchange ionomer (10–50 mS/cm) and the very thick electrode CL, the penalty of ionic transport resistance in the electrode CL would be much more significant, leading to very large electrode overpotentials. Fabrication of as-thin-as-possible Ni-Mo cathode with high loading is very challenging. Xiao et al.[109] prepared a Ni-Mo cathode by filling a stainless steel fiber felt skeleton with Ni-Mo–containing precipitates produced by a codeposition procedure, followed by reducing the thus-obtained Ni-Mo substrate under a hydrogen atmosphere at 500°C. In their work, the overpotential of the ionomer-impregnated Ni-Mo cathode with an optimal metal loading of 40 mg/cm^2 was found to be only ~0.11 V in 1 M KOH solution at 400 mA/cm^2

TABLE 24.4
Non-PGM Electrocatalysts for AEM Water Electrolysis Cells

	Electrocatalyst	Activity	Ref.
HER	Ni-Mo filled in stainless steel skeleton, codeposition method, 40 mg/cm^2 NiMo	40°C, 1 M KOH η = 0.11 V at 0.4 A/cm^2	109
OER	NiFe spread into porous Ni	40°C, 1 M KOH η = 0.35 V at 0.4 A/cm^2	109
OER	$Cu_{0.7}Co_{2.3}O_4$, 3 mg/cm^2	25°C, 1 M KOH η < 0.57 V at 1 A/cm^2 [a]	49
OER	$Li_{0.721}Co_{2.79}O_4$, 2.5 mg/cm^2	More active than Co_3O_4 45°C, AEM water electrolysis cell (with Ni cathode): E = 2.05 V at 300 mA/cm^2	110
OER	$Ba_{0.5}Sr_{0.5}Co_{0.8}Fe_{0.2}O_{3-\delta}$	25°C, 1 M KOH Higher than IrO_2 at the same condition	111
OER	$Ni_{0.9}Fe_{0.1}O_x$, solution-cast thin-film method	25°C, 1 M KOH An order of magnitude higher than IrO_2 at the same condition	112
OER	NiFe (5:1) layered double hydroxide-CNT hybrid	25 °C, 1 M KOH Higher than Ir/C at the same condition	113
OER	3030 OER catalysts: $CuCoO_x$ (Acta Spa, Italy), 36 mg/cm^2	AEM water electrolysis with 1 wt.% K_2CO_3 electrolyte, Acta's 4030 HER catalyst and Acta's 3030 OER catalyst	114
HER	4030 HER catalysts: NiM/CeO_2-La_2O_3/C (Acta Spa, Italy), 7.4 mg/cm^2	T = 43°C, E = 1.89 V at 470 mA/cm^2	114
HER	NiMo from $NiMoO_4$ precursor/Ni foam, 30 mg/cm^2 NiMo	70°C, 5 M KOH η = 0.15 V at 0.7 A/cm^2	115
HER	Ni-Mo nanopowders, two-step precipitation/reduction process, 13.4 mg/cm^2 NiMo	25°C, 2 M KOH η = 0.10 V at 0.13 A/cm^2	59

[a] Estimated from the electrolysis cell performance reported by Wu and Scott.[49]

and 40°C.[109] Later, Tang et al.[115] prepared a Ni-Mo cathode using a modified method where $NiMoO_4$ powder was synthesized and used as precursor. Their new Ni-Mo cathode with a loading of 30 mg NiMo/cm^2 exhibited a current density of 700 mA/cm^2 at an overpotential of 150 mV in 5 M KOH solution at 70°C.

24.4.1.2 Non–Precious Group Metal Electrocatalysts for Oxygen Evolution Reaction

Iridium oxide (IrO_2) has been considered as one of most active catalysts for OER, in either acid or base medium.[64] Much effort has been devoted to searching for highly active, durable, and cost-effective alternatives. Transition-metal oxides, such as Co_3O_4 and doped Co_3O_4 of spinel structure; and binary mixed oxides of La with Mn or Co, of perovskite structure, such as $La_{1-x}Sr_xMnO_3$,[32,37] $Li_{1-x}Sr_xCoO_3$,[11,32,38,39] and $La_{1-x}Sr_xFe_{(1-y)}Co_yO_3$,[39,40] have been extensively investigated.

Co_3O_4 and doped Co_3O_4 are two of the active non-PGM catalysts for OER in base media. Clearly, the activity of Co_3O_4 nanoparticle anodes is dependent on the particle size. Esswein et al.[116] reported that anodes loaded with 1 mg/cm^2 of small (5.9 nm), medium (21.1 nm), and large (46.9 nm) Co_3O_4 nanoparticles showed overpotentials of 328, 363, and 382 mV at a current density of 10 mA/cm^2 in 1.0 M KOH solution, respectively. The activity increase with decreasing particle size is mainly due to the increase of accessible catalyst surface area.[116] Doping is an effective way to improve the electrocatalytic activity of Co_3O_4. Nikolov et al.[117] performed a comparative study of the electrochemical activity of binary cobalt oxides with spinel structure $M_xCo_{3-x}O_4$ (M = Li, Ni, Cu) in OER and found that the activity of the spinels increases relative to that of Co_3O_4 in the order $Co_3O_4 < Ni_xCo_{3-x}O_4 \ll Cu_xCo_{3-x}O_4 < Li_xCo_{3-x}O_4$. The authors believed that the enhanced activity is mainly determined by two interrelated factors: the cation distribution in the two types of spinel site (8a) and (16d) and the number and energetics of the active sites formed on the surface in the anodic peak potential range preceding the oxygen evolution potential.[117] Singh et al.[118] obtained Co_3O_4 and La-doped Co_3O_4 thin films on Ni using microwave-assisted synthesis and found that the overpotential for OER was ~224 mV at 100 mA/cm^2 in 1 M KOH at 25°C for La-doped Co_3O_4 electrode with an oxide loading of ~2.8 mg/cm^2, which is slightly smaller than that for Co_3O_4 electrode with an oxide loading of 3.4 mg/cm^2. Recently, Wu and Scott prepared Cu-doped[49] and Li-doped[110] Co_3O_4 nanoparticles by using a thermal decomposition method and used these nanoparticles as OER electrocatalysts for alkaline AEM water electrolyzers. In their work, alkaline AEM water electrolysis cell with Ni cathode (2 mg Ni/cm^2) and $Li_{0.21}Co_{2.79}O_4$ anode (2.5 mg Li-doped Co_3O_4/cm^2) exhibited a current density of 300 mA/cm^2 at a voltage of 2.2 to 2.05 V at temperatures in the range of 20 to 45°C.

Suntivich et al.[111] reported a distinct OER activity design principle, namely, that a near-unity occupancy of the e_g orbital of surface transition-metal ions and high covalence in bonding to oxygen can enhance the intrinsic OER activity of ABO_3 perovskite oxides (A is rare-earth or alkaline-earth metal and B is a transition metal) in alkaline solution. Their studies showed that the specific activities of the perovskite oxides exhibited a volcano shape as a function of the e_g-filling of surface B-site cation. Based on this design principle, they predicted and found a highly active oxide catalyst for OER, $Ba_{0.5}Sr_{0.5}Co_{0.8}Fe_{0.2}O_3$ (BSCF), one of high-performance cathode materials for solid oxide fuel cells (SOFCs).[119] BSCF has an intrinsic activity that is at least an order of magnitude higher than the state-of-the-art IrO_2 for OER in alkaline media.[111] However, the surface area of their as-prepared and ball-milled BSCF powders were quite low, only 0.2 and 3.9 m^2/g, respectively, because of the large particle size (about several microns to submicrons), which was much lower than that of IrO_2 (~71 m^2/g). Therefore, the mass activity of their BSCF catalysts was still lower than that of IrO_2.[111]

Ni-Fe alloy, oxides, and hydroxides are considered the promising non-PGM catalysts for the OER in alkaline media. Xiao et al.[109] prepared a Ni-Fe anode by spraying the ethanol solution of the Ni and Fe nitrate mixture onto the porous Ni electrode substrate (a Ni-foam skeleton filled with Ni powder) on a hot plate with a temperature of 60°C, followed by electrochemically reducing the thus-prepared Ni-Fe precursor. In their work, under a current density of 400 mA/cm^2, the overpotential of their ionomer-impregnated Ni-Fe cathode for OER was only ~0.35 V in 1 M KOH solution at 40°C.[109] Trotochaud et al.[112] adopted solution-cast method to prepare metal oxide thin-film electrocatalysts for OER and found that $Ni_{0.9}Fe_{0.1}O_x$ is the most active water oxidation catalyst in basic media with an overpotential of 336 mV at 10 mA/cm^2 and a Tafel slope of 30 mV/decade, as shown in Figure 24.11. Moreover, the OER activity of $Ni_{0.9}Fe_{0.1}O_x$ thin-film electrocatalysts was found to

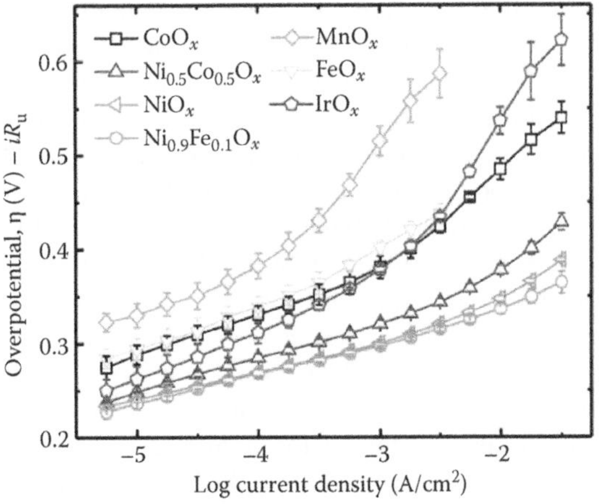

FIGURE 24.11 Comparison of steady-state polarization curves for ultrathin films with different OER catalysts in 1 M KOH. (Reprinted with permission from Ref. 112, 17253–17261. Copyright 2012 American Chemical Society.)

be roughly an order of magnitude higher than the OER activity of IrO_x control films and similar to that of the best known OER catalysts in basic media.[112] Gong et al.[113] synthesized ultrathin Ni-Fe layered double hydroxide (NiFe-LDH) nanoplates on mildly oxidized multiwalled CNTs and found that the resulting NiFe-LDH/CNT complex exhibits higher electrocatalytic activity and stability for OER than commercial Ir metal catalysts, as shown in Figure 24.12.

Up to now, several catalysts for OER were found to have an intrinsic activity better than that of IrO_2.[111–113] However, those tests were performed in alkaline liquid electrolyte solution using thin-film rotating disk electrode (RDE) techniques. The actual environment in an AEM water electrolyzer is very different from the alkaline liquid electrolyte solution (typically KOH). In some cases, test results from the RDE system in alkaline liquid electrolyte solution may not be directly transferred to the AEM water electrolyzer since the RDE test eliminates the effects of a lot of factors, including catalyst loading, catalyst morphology, the interaction between catalyst and AEM ionomer, and mass transport. It is necessary to incorporate advanced non-PGM OER catalysts into an AEM water electrolysis cell in order to evaluate their real performance and long-term stability in actual electrolysis cell. On the other hand, it is a challenge to find a better non-PGM HER catalyst than Pt. Ni-based alloys, especially NiMo, are considered to be state-of-the-art non-PGM catalysts for HER.

24.4.2 Anion-Exchange Membrane/Ionomer

Besides electrocatalysts, the AEM and ionomer should be a key technology for AEM water electrolyzers in order to achieve practical performance. The functions of the AEM in AEM water electrolysis are to transport anions (typically hydroxide ion or OH^-) from the cathode to the anode and to prevent the produced hydrogen and oxygen from mixing. The function of the anion exchange ionomer is to build an efficient three-phase boundary for electrochemical reaction (i.e., HER and OER), thus improving catalyst utilization and reducing the ionic transport resistance inside the electrocatalyst layer. The A201 membrane and AS-4 ionomer produced in Tokuyama, Japan, are one of the commercially available AEMs and anion-exchange ionomers, respectively.[120] The thickness of the A201 membrane is 28 μm, and the ionic conductivity is as high as 42 mS/cm at 23°C. The A201 membrane is thermally stable in either water or methanol at 80°C for more than 2000 hours. The ionic conductivity of the AS-4 ionomer is 13 mS/cm, which is much lower than that of A201. The main reason is that the ion-exchange capacity (IEC) of AS-4 is much lower than that of A201 (IEC: 1.3 for AS-4 vs. 1.7 for A201). AS-4 is not soluble in water, methanol, or ethanol, but it is soluble in 1-propanol, which makes the MEA fabrication process simple and cost effective. Leng et al.[4] adopted A201 and AS-4 as the AEM and ionomer in their AEM water electrolysis cells, respectively. Other AEMs and ionomers have been developed and used for AEM water electrolyzers, such as animated Radel (A-Radel) poly(sulfone) ionomers,[4] self–cross-linking

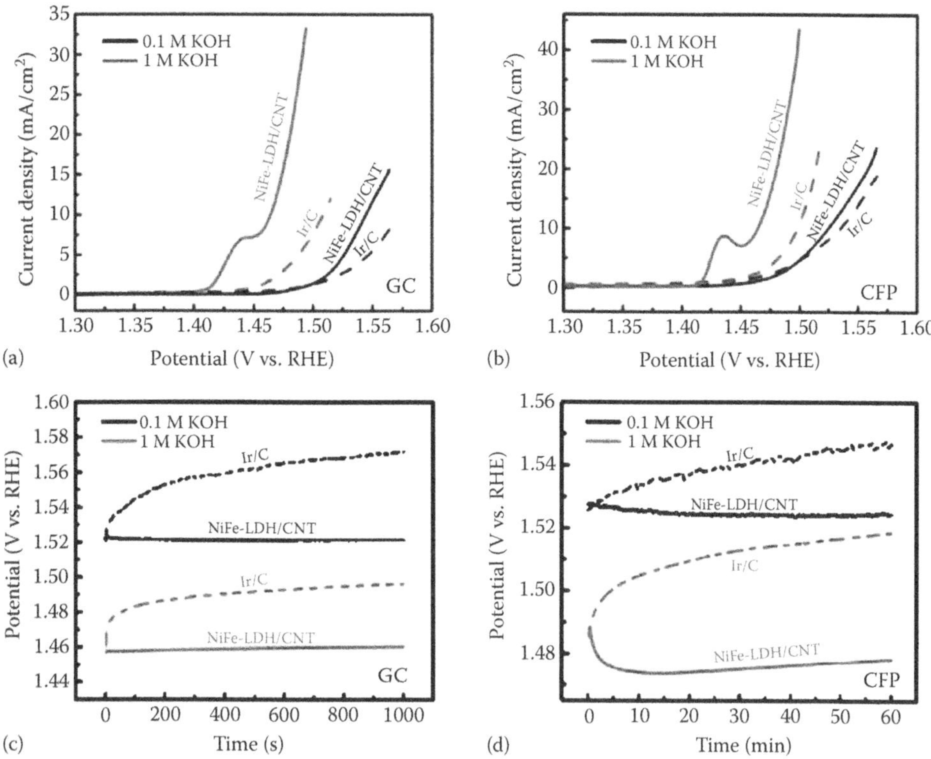

FIGURE 24.12 Electrochemical performance of NiFe-LDH/CNT hybrid OER catalyst. (a) *iR*-corrected polarization curves of NiFe-LDH/CNT hybrid (loading: 0.2 mg/cm^2) and Ir/C (loading: 0.2 mg/cm^2) catalyst on glassy carbon (GC) electrode in KOH, measured at a continuous electrode rotating speed of 1600 rotations per minute; (b) *iR*-corrected polarization curves of NiFe-LDH/CNT hybrid (loading: 0.25 mg/cm^2) and Ir/C catalysts (loading: 0.25 mg/cm^2) on carbon fiber paper (CFP); (c) chronopotentiometry curves of NiFe-LDH/CNT hybrid and Ir/C catalyst on GC electrode at a constant current density of 2.5 mA/cm^2; (d) chronopotentiometry curves of NiFe-LDH/CNT hybrid and Ir/C catalyst on CFP at a constant current density of 5 mA/cm^2. (Reproduced with permission from Ref. 113, 8452–8455. Copyright 2013 American Chemical Society.)

quaternary ammonium polysulfone (*x*QAPS) membrane/ionomers,[109,121] polymethacrylate-based (quaternary ammonium OH^- ionomers (quaternary poly(DMAEMA-co-TFEMA-co-BMA) or QPDTB),[122] and methylated melamine grafted poly(vinylbenyl chloride) (mm-qPVB/Cl^-) membranes.[123] The chemical structures of these membranes and/or ionomers are shown in Figure 24.13, and their ionic conductivities are summarized in Table 24.5. The ionic conductivity of A201, AS-4, and other membranes/ionomers is about 4–10 times lower than that of 1 M KOH solution. The low ionic conductivity of the state-of-the-art AEMs and ionomers leads to low catalyst utilization, high ohmic resistance, and high ionic transport resistance in the CL, which makes the current performance of AEM water electrolyzers much lower than that of the alkaline liquid electrolyte. Advanced AEMs and ionomers are desired for further improvement of the energy efficiency of AEM water electrolyzers. Innovative designs and novel concepts are necessary for the development of advanced AEMs and ionomers for AEM water electrolyzers.[124–126]

A critical concern for AEM water electrolysis technology is durability. The long-term stability of AEMs and ionomers is a key to the durability of AEM water electrolysis technology. It has been well documented that most AEMs suffer from poor chemical stability.[128,129] The main reason for the chemical degradation of AEMs is reported to be the nucleophilic attack on the cation fixed charged sites by OH^-.[130–131] This kind of degradation leads to a loss in the number of anion-exchange group and thus a decrease in OH^- conductivity. Since the ionomer used in the CL is in intimate contact with the catalysts, the chemical or electrochemical degradation of the ionomer may be more severe than that of AEM. However, there are a few reports on the stability of AEMs and ionomers used in AEM water electrolyzers.[4,132] Leng et al.[4] found that the lifetime of MEA used in AEM water electrolysis with AS-4 ionomer was only 27–100 hours and revealed that the degradation of MEA was mainly due to the degradation of the ionomer and/or membrane–electrode interface. Replacing AS-4 ionomer with A-Radel polysulfone ionomer can extend the MEA lifetime to more than 500 hours.[4] Leng et al. also observed pinhole formations in AEM occurred at longer cell operation times.[4] Parrondo et al.[132] demonstrated that long-term deterioration of AEM water electrolyzer over time was because of irreversible AEM polymer (especially the backbone) degradation. More effort should focus on designing alkaline stable AEMs and ionomers with higher ionic conductivity for developing advanced AEM water electrolysis competitive with PEM and alkaline liquid electrolyte water electrolysis technologies.

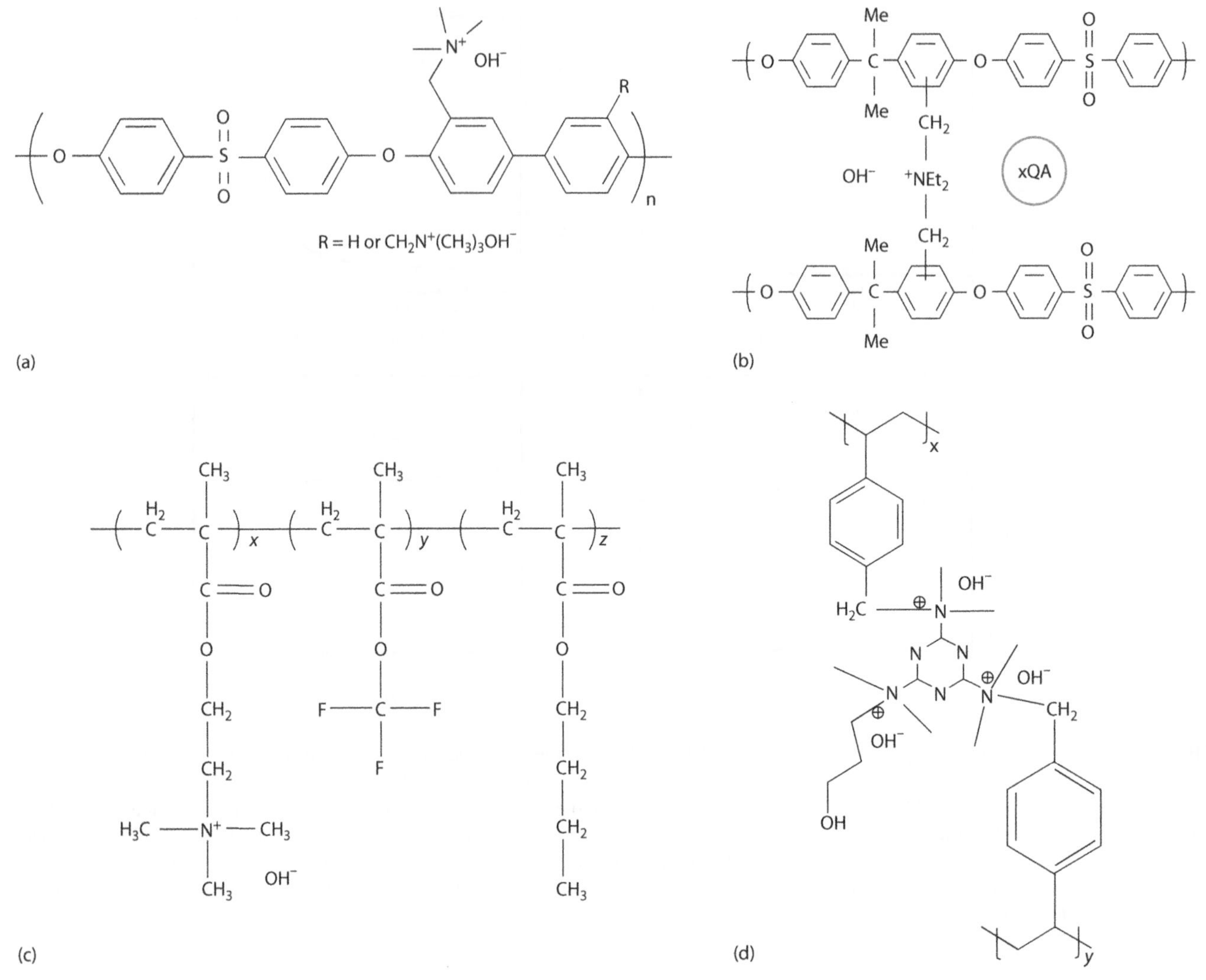

(a)

(b)

(c)

(d)

FIGURE 24.13 Chemical structures of several AEMs and ionomers: (a) A-Radel polysulfone ionomer (Reprinted with permission from Ref. 4, 9054–9057. Copyright 2012 American Chemical Society.); (b) *x*QAPS ionomer/membrane (Xiao, L., S. Zhang, J. Pan, C. X. Yang, M. L. He, L. Zhuang, and J. T. Lu, First implementation of alkaline polymer electrolyte water electrolysis working only with pure water, *Energy & Environmental Science*, 5 (7), 7869–7871; Pan, J., Y. Li, L. Zhuang, and J. T. Lu, Self-crosslinked alkaline polymer electrolyte exceptionally stable at 90 degrees C, *Chemical Communications*, 46 (45), 8597–8599. Reproduced by permission of The Royal Society of Chemistry.); (c) QPDTB ionomer (Reprinted from Wu, X., and K. Scott, A polymethacrylate-based quaternary ammonium OH^- ionomer binder for non-precious metal alkaline anion exchange membrane water electrolysers, *Journal of Power Sources*, 214, 124–129. Copyright 2012, with permission from Elsevier.); and (d) mm-qPVB-Cl membrane. (Reproduced from Cao, Y.-C., X. Wu, and K. Scott, A quaternary ammonium grafted poly vinyl benzyl chloride membrane for alkaline anion exchange membrane water electrolysers with no-noble-metal catalysts, *International Journal of Hydrogen Energy*, 37 (12), 9524–9528, 2012. With permission.)

24.4.3 Membrane–Electrode Assembly

Similarly as with the PEM water electrolysis cell, the heart of the AEM water electrolysis cell is also the MEA, which consists of an anode GDL, an anode CL, an AEM membrane, a cathode CL, and a cathode GDL. Typically, porous Ti foam is used as the anode GDL, while carbon paper is used as the GDL. There are two methods to fabricate MEAs, namely, the catalyst-coated membrane (CCM) method and the catalyst-coated substrate (CCS) method,[4] as shown in Figure 24.14. In the CCM method, the catalysts (such as Pt black or IrO_2) are mixed with water, organic solvents, and ionomer binder to obtain well-dispersed ink, and then the as-prepared ink consisting of anode and cathode catalysts is coated onto both sides of the AEM (such as A201) using hand spray and other coating techniques, respectively, to obtain CCM; finally, to form a full MEA, the Ti foam anode GDL and the carbon paper cathode GDL are mechanically pressed against the CCM during assembly in the cell hardware. In the CCS method, to obtain a CCS, the as-prepared ink consisting of anode and cathode catalysts is coated onto the surface of one side of anode and cathode GDLs, respectively; finally to obtain a full MEA, the anode CCS, the AEM, and the cathode CCS are assembled in the cell hardware. Generally, the performance of MEAs made

TABLE 24.5
Membranes and Ionomers for AEM Water Electrolysis Cells

	Chemical Structure	Conductivity	Other Properties	Ref.
Membrane	A201	42 mS/cm at 23°C	IEC = 1.8 Thickness: 28 μm Water content: 25%	4, 120
Ionomer	AS-4	13 mS/cm at 25°C in HCO_3^- form	IEC = 1.4 Soluble in 1-propanol	4, 120
Membrane	*x*QAPS	15 mS/cm at 25°C 43 mS/cm at 90°C	IEC = 1.34–1.49 Swelling degree: < 3% even at 90°C	109, 121
Ionomer	*x*QAPS	15 mS/cm at 25°C 43 mS/cm at 90°C		109, 121
Ionomer	QPDTB ionomer	59 mS/cm at 50°C	IEC = 1.275	122
Membrane	mm-qPVB/Cl⁻	16 mS/cm at 25°C 27 mS/cm at 60°C		123
Membrane	LDPE-g-VBC-Dabco in HCO_3^-/CO_3^{2-} form	14 mS/cm at 30°C 25 mS/cm at 60°C	IEC = 1.5 water uptake = 81–94%	50
Ionomer/membrane	1 M KOH/water	194 mS/cm		114
Ionomer/membrane	1 wt.% K_2CO_3/water	15.8 mS/cm		114

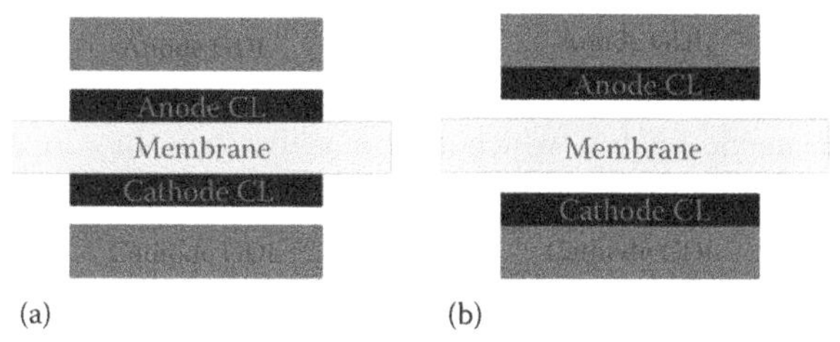

FIGURE 24.14 Fabrication methods for the MEAs of AEM water electrolyzers: (a) CCM and (b) CCS.

using the CCM method exhibits better performance,[4] while the CCM method requires high-quality AEM and catalyst ink and needs a more complex fabrication procedure.

The performance of MEA used in AEM water electrolyzers depends on the fabrication method, the ink formula, the catalyst type and loading, the property of AEM, and the ionic conductivity and loading of ionomer. Table 24.6 summarizes the MEA performances and corresponding MEA specifications reported in the literature. The best performance of MEAs with PGM catalysts was obtained by Leng et al.[4] As shown in Figure 24.15, they reported that for an MEA with Pt black as the cathode catalyst and IrO_2 as the anode catalyst, the A201 membrane and AS-4 ionomer can achieve a current density of ~400 mA/cm² at 1.80V at 50°C.[4] Recently Parrondo et al.[132]

TABLE 24.6
MEA Specifications and Performances of AEM Water Electrolysis Cells

Anode	Cathode	Membrane	Ionomer	Performance	Ref.
IrO_2, 2.9 mg/cm²	Pt black, 3.2 mg/cm²	A201 (28 μm) (Tokuyama, Japan)	AS-4 ionomer	50°C, 399 mA/cm² at 1.8 V	4
IrO_2, 2.6 mg/cm²	Pt black, 2.4 mg/cm²	A201 (28 μm) (Tokuyama, Japan)	A-Radel ionomer	50°C, 192 mA/cm² at 1.8 V	4
$Pb_2Ru_2O_{6.5}$, 2.5 mg/cm²	Pt black, 2.5 mg/cm²	PSF-TMA⁺OH⁻ membrane	PSF-TMA⁺OH⁻ ionomer	50°C, 400 mA/cm² at 1.8 V	132
Ni-Mo, 40 mg/cm²	Ni-Fe, unknown	*x*QAPS	*x*QAPS ionomer	70°C, 1.8–1.85 V at 400 mA/cm²	109
Ni, 2 mg/cm²	$Cu_{0.7}Co_{2.3}O_4$, 3 mg/cm²	Cranfield membrane	QPDTB ionomer	22°C, 1.9 V at 100 mA/cm²	122
Ni, 2 mg/cm²	$Cu_{0.7}Co_{2.3}O_4$, 3 mg/cm²	Methylated melamine grafted poly(vinylbenzyl hydroxide) (70 μm)	*q*PVB/OH⁻ ionomer	25°C, 2.19 V at 100 mA/cm² 55°C, 1.99 V at 100 mA/cm²	123
Non-PGM (Acta Spa)	Non-PGM (Acta Spa)	LDPE-g-VBC-Dabco (60 μm)	1% K_2CO_3 in water	45°C, 20 bars pressure 1% K_2CO_3 in water, 2.03 V at 200 mA/cm²	50
4030 HER catalysts: NiM/CeO_2-La_2O_3/C, 7.4 mg/cm² (Acta Spa)	3030 OER catalysts: $CuCoO_x$, 36 mg/cm² (Acta Spa)	A201 (28 μm) (Tokuyama, Japan)	1% K_2CO_3 in water	43°C, 0.1 MPa pressure 1% K_2CO_3 in water, 1.89 V at 470 mA/cm²	114

Note: PSF: poly(sulfone); TMA: Trimethylammonium.

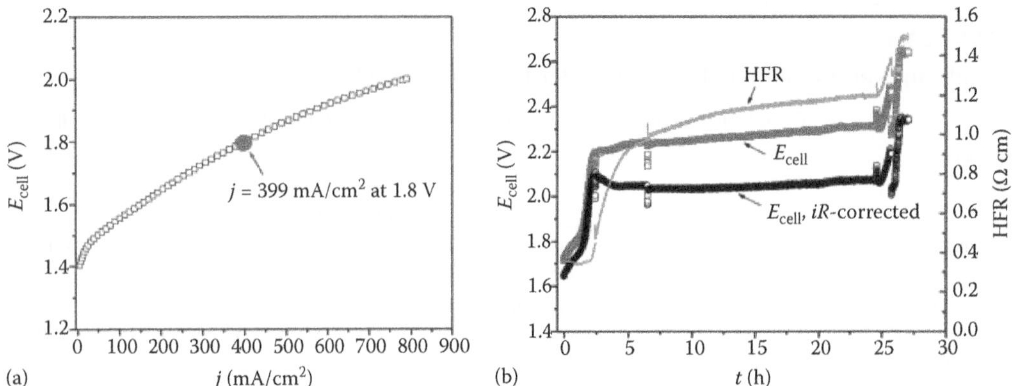

FIGURE 24.15 (a) Initial polarization curve and (b) cell voltage and high-frequency resistance (HFR) under a discharge current density of 100 mA/cm² as a function of test time for MEA fabricated with CCM method. Operation conditions: T = 50°C and water cathode-feed mode. (Reprinted with permission from Ref. 4, 9054–9057. Copyright 2012 American Chemical Society.)

obtained similar performance for an MEA with Pt black as the cathode catalyst, $Pb_2Ru_2O_{6.5}$ as the anode catalyst, and a quaternary ammonium poly(sulfone) membrane and ionomer. However, it is necessary and important to use non-PGM catalysts for improving the MEA performance while reducing the cost. Several excellent works have been reported on exploring non-PGM catalysts used for AEM water electrolyzers.[109,110,122,123] For example, Xiao et al.[109] reported the first AEM water electrolyzers with completely PGM-free catalysts, where Ni-Fe was used as the anode catalyst and Ni-Mo was used as the cathode catalyst. As shown in Figure 24.16, they demonstrated a promising cell performance with a cell voltage of about 1.80–1.85 V under a current density of 400 mA/cm² at 70°C for their AEM water electrolyzers with non-PGM catalysts and xQAPS membrane/ionomer, which is comparable to that with PGM catalysts at 50°C reported by Leng et al.[4]

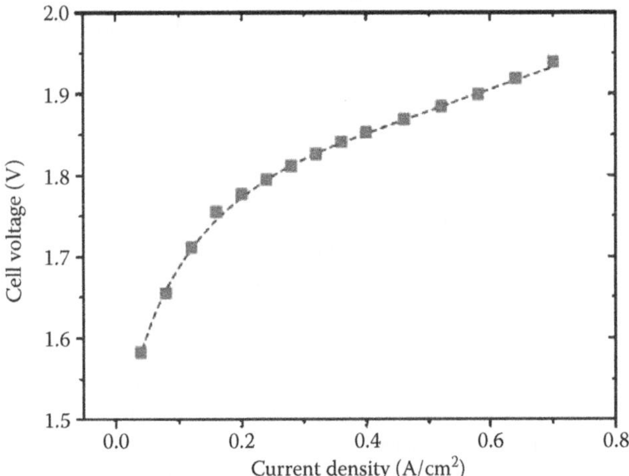

FIGURE 24.16 Steady-state cell performance of an AEM water electrolysis system using Ni-Fe anode and Ni-Mo cathode and working only with pure water at 70°C. (Xiao, L., S. Zhang, J. Pan, C. X. Yang, M. L. He, L. Zhuang, and J. T. Lu, First implementation of alkaline polymer electrolyte water electrolysis working only with pure water, *Energy & Environmental Science*, 5 (7), 7869–7871, 2012. Reproduced by permission of The Royal Society of Chemistry.)

This indicates that non-PGM catalysts are potentially used for AEM water electrolyzers to replace PGM catalysts without sacrificing cell performance while reducing the cost. Wu and Scott[122] prepared an MEA by using Ni as the cathode catalyst, $Cu_{0.7}Co_{2.3}O_4$ as the anode catalyst, and a polymethacrylate-based quaternary ammonium OH^- ionomer for the AEM water electrolysis cell, which exhibited a current density of 100 mA/cm² at a cell voltage of 1.9 V at 22°C. Later Wu and Scott[110] improved MEA performance by using Li-doped Co_3O_4 as the anode catalyst, which is more active than $Cu_{0.7}Co_{2.3}O_4$. Recently Pavel et al.[114] reported an MEA with high efficiency based on low-cost transition-metal catalysts and A201 membrane for AEM water electrolyzers. They used a material consisting of $CuCoO_x$ as the anode catalyst with a loading of 36 mg/cm² and a Ni-based nanostructured material with transition-metal deposited on CeO_2-La_2O_3/carbon support as the cathode catalyst with a loading of 0.6–7.4 mg/cm² in AEM water electrolyzers, and demonstrated a current density of 470 mA/cm² at a cell voltage of 1.89 V and 43°C, which is close to that of MEA with PGM catalysts reported by Leng et al.[4] and that of MEA with NiFe anode and NiMo cathode reported by Xiao et al.[109] However, in the work reported by Pavel et al.,[114] such promising performance was achieved by using diluted carbonate/bicarbonate aqueous solution, such as 1 wt.% K_2CO_3 or 1 wt.% K_2CO_3/$KHCO_3$, instead of using an anion-exchange ionomer as the ionic transport medium. This may reduce the purity of hydrogen and oxygen produced via water electrolyzers. Pavel et al.'s electrolysis cell is actually a zero-gap alkaline liquid electrolyte water electrolyzer using diluted carbonate/bicarbonate aqueous solution instead of KOH solution, rather than a solid-state AEM water electrolyzer, such as that reported by Leng et al.[4] and Xiao et al.[109]

The durability of MEAs used in AEM water electrolyzers is one of the key issues of their commercialization. However, because of the instability of the AEM and ionomer, most AEM water electrolyzers reported in the literature showed a lifetime of less than 600 hours. Leng et al.[4] demonstrated a lifetime of about 534 hours for AEM-based water electrolysis cells with PGM catalysts, A201 membrane, and A-Radel poly(sulfone) ionomer, as shown in Figure 24.17. For

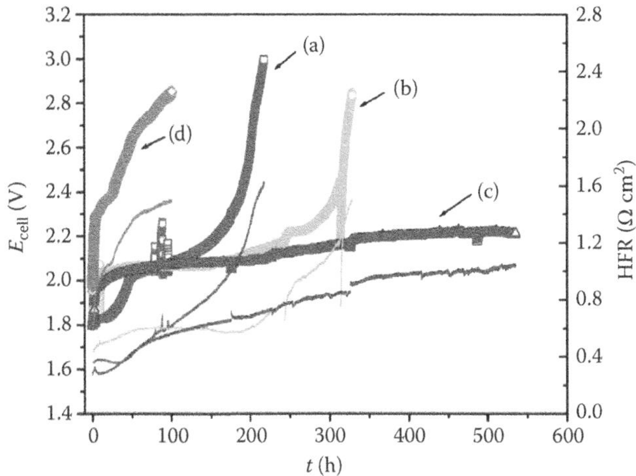

FIGURE 24.17 Cell voltage (symbol curve) and HFR (line curve) as a function of test time at 50°C for four MEAs fabricated with CCS method. (a) MEA with A-Radel ionomer, water cathode-feed mode. (b) MEA with A-Radel ionomer, water anode-feed mode. (c) MEA with A-Radel ionomer; running MEA in water cathode-feed mode for initial 2 h, then switching to water anode-feed mode. (d) MEA with AS-4 ionomer, water anode-feed mode. (Reprinted with permission from Ref. 4, 9054–9057. Copyright 2012 American Chemical Society.)

AEM-based water electrolysis cells with non-PGM catalysts, AEMs, and ionomers, there is no report about an operating time of >100 hours.[109,110,122] This may be due to more severe degradation of ionomers in AEM-based water electrolyzers with non-PGM catalysts caused by the possible interaction of non-PGM catalysts and AEM/ionomers. Replacing ionomers with 1 wt.% K_2CO_3 or 1 wt.% $K_2CO_3/KHCO_3$ as the ionic transport medium, as shown in Figure 24.18, Pavel et al.[114] demonstrated

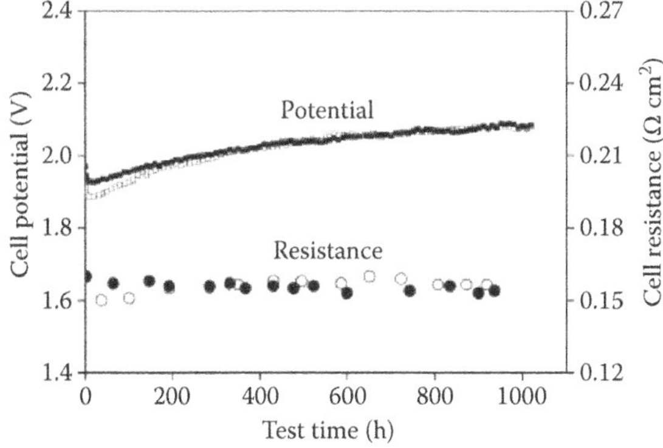

FIGURE 24.18 Long-term performance and AC resistance (1 kHz) of AEM water electrolysis cells pressured at 3 MPa operating with 1 wt.% $K_2CO_3/KHCO_3$ (solid symbols) and 1wt. % K_2CO_3 (open symbols) electrolyte solutions. Tests were performed on cells with an HER catalyst loading of 7.4 mg cm^{-2}, at constant current density of 470 mA cm^{-2} and 43°C. (Pavel, C. C., F. Cecconi, C. Emiliani, S. Santiccioli, A. Scaffidi, S. Catanorchi, and M. Comotti: Highly efficient platinum group metal free based membrane-electrode assembly for anion exchange membrane water electrolysis. *Angewandte Chemie International Edition.* 2014. 53. 1378–1481. Copyright Wiley-VCH Verlag GmbH & Co. KGaA. Reproduced with permission.)

a lifetime of about 1000 hours for an MEA with non-PGM catalysts and commercial AEM (i.e., A201 from Tokuyama) operating at a constant current density of 470 mA/cm² and 43°C. This indicates that an AEM can survive for more than 1000 hours in a less aggressive middle alkaline (pH = 10–11) environment. Clearly, the current durability performance of AEM water electrolyzers is far below the requirements of their commercialization. Fundamental understanding of MEA degradation mechanism and development of highly durable AEMs and ionomers are necessary for improving the long-term stability of AEM water electrolyzers.

AEM water electrolyzers are still in their infant stage. The performance and durability of AEM water electrolyzers needs to be improved significantly, especially for those with non-PGM catalysts. In order to develop highly efficient, durable, and low-cost AEM water electrolyzers, the focus in the near future should be placed on (1) development of highly active and durable non-PGM catalysts with low cost; (2) development of anion-exchange ionomers with high ionic conductivity, high alkaline stability, and low cost; (3) development of low-cost AEM with high ionic conductivity, high alkaline stability, high mechanical stability, and low gas crossover; and (4) fundamental understanding of degradation mechanisms of AEMs and ionomers used in the MEAs for AEM water electrolyzers.

24.5 ADVANCED MATERIALS FOR SOLID OXIDE ELECTROLYSIS CELL

For high-/intermediate-temperature SOECs, there are several excellent reviews.[133–135] For example, Hauch et al.[134] emphasized their review of state-of-the-art electrode materials and development of new materials for SOECs. Laguna-Bercero[135] reviewed the recent advances in high-temperature water electrolysis using SOECs, including the current status of electrolyte and electrode materials, and the materials' degradation issues. In this section, we will briefly introduce the advanced materials developed for hydrogen electrodes, air electrodes, and electrolytes in SOECs. Benchmarks and alternative materials for the electrolytes, steam/hydrogen electrodes, and oxygen electrode are summarized in Table 24.7.

24.5.1 Electrolyte

The most commonly used electrolyte material for SOECs is 8 wt.% Y_2O_3-stabilized ZrO_2 (YSZ) is because of its high ionic conductivity, high thermal/chemical stability, and high mechanical strength at high operation temperatures (800–1000°C).[17,136] In order to further decrease the electrolyte resistance, other doped ZrO_2 with ionic conductivity higher than that of YSZ, such as the scandia-stabilized zirconia[138–141] or ScSZ and scandia and ceria–costabilized zirconia[142] or (Sc,Ce)SZ was proposed for use as an electrolyte alternative to YSZ. However, ScSZ and (SC,Ce) SZ are much more expensive than YSZ, which limits their use as electrolyte materials for high-temperature SOECs.

Although high-temperature SOECs can improve the energy efficiency, not only does the high-temperature operation increase the interfacial reaction between cell and stack

TABLE 24.7
Materials for SOECs

	Benchmark	Examples of Alternative Materials
Steam/hydrogen electrode	Ni-YSZ	Ni/SDC $(La_{0.75}Sr_{0.25})_xCr_{0.5}Mn_{0.5}O_3$ (x = 0.9 or 1.0) $La_{0.75}Sr_{0.25}Cr_{0.5}Mn_{0.5}O_3$-YSZ LSCM-SDC-Fe $La_{0.6}Sr_{0.4}VO_{3-\delta}$-YSZ $La_{0.35}Sr_{0.65}TiO_3/Ce_{0.5}La_{0.5}O_{1.75}$ $La_{0.4}Sr_{0.4}TiO_{3-\delta}$-SDC-Ni $Pr_{0.8}Sr_{1.2}(Co,Fe)_{0.8}Nb_{0.2}O_{4+\delta}$-CoFe
Oxygen electrode	LSM-YSZ	LSM-(Gd,Ce)O_2 Sr-doped $LaFeO_3$-YSZ Sr-doped $LaCoO_3$-YSZ $La_{0.58}Sr_{0.4}Co_{0.2}Fe_{0.8}O_3$ $Ba_{0.5}Sr_{0.5}Co_{0.8}Fe_{0.2}O_3$ $La_{0.6}Sr_{0.4}Co_{0.2}Fe_{0.8}O_3$-(Gd,Ce)$O_2$ $Sm_{0.5}Sr_{0.5}O_3$-SDC
Electrolyte	YSZ	Scandia-stabilized ZrO_2 Scandia and ceria–costabilized ZrO_2 ((Sc,Ce)SZ) $La_{0.9}Sr_{0.1}Ga_{0.8}Mg_{0.1}O_3$

Note: LSCM: $(La_{0.75}Sr_{0.25})_xCr_{0.5}Mn_{0.5}O_3$ (x = 0.9 or 1.0); LSM: Sr-doped $LaMnO_3$; SDC: Sm_2O_3-doped ceria ($Ce_{0.8}Sm_{0.2}O_{2-\delta}$); YSZ: Y_2O_3-stabilized ZrO_2.

components, leading to severe degradation of SOECs and shorter lifetime, but it also limits the selection of materials and confines the application of SOECs to high-temperature processes, such as those in a high-temperature nuclear reactor used as a heat source. In order to overcome these problems, it is desirable to reduce the operation temperature of SOECs from high temperature to intermediate temperature (600–800°C). At intermediate temperatures, cheap stainless steel materials can be used as interconnectors. Moreover, at intermediate temperatures, many heat sources in the industry such as steelmaking processes or various combustors could be utilized for SOECs.[143] Unfortunately, the ionic conductivity of commonly used YSZ electrolytes is not sufficient at intermediate temperatures; hence, the performance of thick YSZ electrolyte-supported SOECs is hindered by their high electrolyte resistance. There are two approaches to solving this problem: one is to still use a YSZ electrolyte but decrease its thickness and change its cell configuration from thick YSZ (100–300 μm) electrolyte to an electrode-supported structure with thin YSZ electrolyte (10–20 μm).[144,145] Another one is to adopt new electrolyte materials with high ionic conductivity at an intermediate temperature, such as doped ceria and doped lanthanum gallate ($LaGaO_3$).[143,146]

Doped ceria, doped with either Gd_2O_3 (GDC) or Sm_2O_3 (SDC), is considered a promising oxide ion conductor for the intermediate-temperature SOFCs and SOECs since it exhibits high oxide ion conductivity at intermediate temperatures. However, partial reduction of Ce^{4+} to Ce^{3+} in the hydrogen environment leads to high electronic conductivity.[137] Moreover, the mechanical strength of doped ceria electrolyte is poor. These drawbacks limit the direct use of doped ceria as the electrolyte material for SOECs at high/intermediate temperatures. Alternatively, doped ceria has been used as the electrolyte for SOECs together with a barrier layer of YSZ between the steam/hydrogen electrode and the doped ceria electrolyte. In the case of YSZ electrolyte–based SOECs, doped ceria has also been used as a barrier layer between the oxygen electrode and the YSZ electrolyte in order to avoid the reaction between the oxygen electrode material and YSZ.[145,148–150]

Doped $LaGaO_3$, usually doped with Sr at the La site and Mg at the Ga site, is also considered as one of the most promising oxide ion conductors for the intermediate-temperature SOFCs. With suitable hydrogen/steam and oxygen electrodes, doped $LaGaO_3$ electrolyte can be used for intermediate temperature SOECs due to its high oxide ion conductivity at intermediate temperatures. For example, Ishihara et al.[143] prepared an electrolysis cell supported by a $La_{0.9}Sr_{0.1}Ga_{0.8}Mg_{0.1}O_3$ (LSGM) electrolyte (0.2 mm) with Ni-Fe (9:1) bimetallic cathode and $Ba_{0.6}Sr_{0.4}CoO_3$ anode and demonstrated a hydrogen formation rate as high as 180 μmol/cm^2 min for their electrolysis cell at 1.60 V and 600°C. Recently, they found that the H_2 formation rate at 1.60 V for LSGM-based electrolysis cells was much improved by a combination of oxygen ion–conducting oxide with Ni-Fe (9:11) bimetallic alloy as hydrogen/steam electrode.

Besides oxygen ion–conducting electrolytes, such as YSZ, doped ceria, and doped $LaGaO_3$, proton-conducting conductors have also been explored as the electrolyte materials for SOECs. The main advantage of the proton-conducting type of SOECs is that pure hydrogen is produced and not diluted with water vapor. Some perovskite-type oxides based on $SrCeO_3$, $BaCeO_3$, $BaZrO_3$, and $SrZrO_3$ exhibit high proton conductivity at intermediate temperature and have been reported to be used as electrolyte materials for SOECs including $SrCe_{0.95}Yb_{0.05}O_{3-\delta}$,[151] $SrZr_{0.9}Yb_{0.5}O_{3-\delta}$,[152] $SrZr_{0.9}Y_{0.1}O_{3-\delta}$,[153] $SrZr_{0.5}Ce_{0.4}Y_{0.1}O_{3-\delta}$,[153] $BaCe_{0.9}Y_{0.1}O_{3-\delta}$,[154] $BaZr_{0.9}Y_{0.1}O_{3-\delta}$,[154] $BaCe_{0.5}Zr_{0.3}Y_{0.2}O_{3-\delta}$,[155] and $BaCe_{0.5}Zr_{0.3}Y_{0.16}Zn_{0.04}O_{3-\delta}$.[156,157]

24.5.2 Oxygen Electrode

24.5.2.1 Sr-Doped $LaMnO_3$–Based Oxygen Electrode

For oxygen electrodes, the Sr-doped $LaMnO_3$ (LSM)–YSZ composite is the most commonly used electrode material. Since LSM is a predominantly electronic conductor with negligible ionic conductivity and oxygen mobility, the oxygen electrode reaction for a pure LSM electrode occurs mainly at a small region near the interface between the electrode and the electrolyte. The addition of ionic-conducting phases, such as YSZ,[158–160] can extend the reaction sites from the interface to the triple-phase boundaries inside the bulk oxygen electrode, thus substantially improving electrode activity. YSZ addition also improves the adhesion between the LSM-YSZ electrode and the YSZ electrolyte. The performance of LSM-YSZ composite electrodes was found to depend on the fabrication method. Yang et al.[158] prepared a LSM-YSZ composite electrode by infiltration of LSM nanoparticles into ion-conducting YSZ skeleton structure and found that the LSM-infiltrated cell had lower area-specific resistance than that of the cell with the

LSM-YSZ oxygen electrode made by a conventional mixing method. They demonstrated a stable performance under electrolysis operation with 0.33 A/cm^2 and 50 vol.% AH at 800°C for the electrolysis cell with the LSM-infiltrated oxygen electrode. One of major issues with the LSM-based electrode is the electrode delamination caused by high oxygen partial pressure at the electrode/electrolyte interface during steam electrolysis.[161,162] Interestingly, the addition of other ionic-conducting phases, such as $(Gd,Ce)O_2$ (GDC),[163] not only enhances the electrocatalytic activity significantly for the oxygen oxidation reaction but also improves the stability of the LSM-based electrode. For example, Chen et al.[163] showed that the electrode polarization resistance decreased from 8.03 Ω cm^2 for pure LSM electrode to 0.024 Ω cm^2 for GDC-impregnated/GDC-infiltrated LSM electrode, and found that the impregnation of GDC nanoparticles effectively inhibits the electrode delamination at the LSM anode/YSZ electrolyte interface.

24.5.2.2 Mixed Ionic and Electronic Conductor Electrode

At intermediate temperatures, because of the low catalytic activity of LSM and the low ionic conductivity of YSZ, the LSM-YSZ composite is not suitable to use for oxygen electrode material for intermediate-temperature SOECs. Several mixed ionic- and electronic-conducting oxides have been explored as oxygen electrode materials, such as $La_{0.58-0.60}Sr_{0.4}Co_{0.2}Fe_{0.8}O_3$ (LSCF),[142,144,150] $La_{0.6}Sr_{0.4}CoO_3$,[150] and BSCF.[164] However, LSCF and BSCF oxygen electrodes can react with YSZ electrolyte at a high temperature and cannot directly be applied onto the YSZ electrolyte. In addition, a barrier layer made of doped ceria, such as $Gd_xCe_{1-x}O_{2-\delta}$ (x = 0.1, 0.2) (GDC), between the oxygen electrodes and the YSZ electrolyte is required to avoid the interfacial reaction between the oxygen electrode materials and YSZ.[144,150,165]

24.5.2.3 Mixed Ionic and Electronic Conductor–Based Composite Electrode

The addition of ionic-conducting phases such as YSZ[166,167] and doped ceria[19,165,168] into mixed ionic and electronic conductor (MIEC) electrodes can improve the catalytic activity of the oxygen electrode and/or enhance the electrode stability. Kong et al.[166] found that $La_{0.8}Fe_{0.2}O_3$-YSZ composites exhibit an electrochemical catalytic behavior than that of $La_{0.8}Mn_{0.2}O_3$-YSZ composites. Jiang et al.[168] prepared a series of $Sm_{0.5}Sr_{0.5}O_3$-$Sm_{0.2}Ce_{0.8}O_{1.9}$ (SSC-SDC) composite as an oxygen electrode of SOEC and found that the electrode polarization resistance of SSC-SDC with a weight ratio of 7:3 was only 0.03 Ω cm^2 at 850°C, which was much smaller than that of the pure SSC electrode (i.e., 0.14 Ω cm^2). Choi et al.[165] found that, although the $La_{0.1}Sr_{0.9}Co_{0.8}Fe_{0.2}O_3$-$Gd_{0.1}Ce_{0.9}O_{1.9}$ (LSCF-GDC) composite oxygen electrode shows worse performance compared to the LSCF oxygen electrode, the LSCF-GDC composite electrode is more durable during reversible SOFC/SOEC operations.

The electrode materials currently used for the oxygen electrodes of SOECs were borrowed from that of SOFCs. It is worthwhile to note that the requirement of electrocatalytic materials for oxygen oxidation reaction in SOECs may be quite different from that for ORR in SOFCs. Some promising electrode materials for SOFCs may not show good performance for SOECs. For example, BSCF, one of best cathode materials for SOFCs,[119,169] has been found too easy to peel off from the electrolyte in SOEC and is therefore not suitable for SOEC application.[164]

24.5.3 Steam/Hydrogen Electrode

The most commonly used steam/hydrogen electrode material for high-temperature SOECs is Ni-YSZ cermet.[170] Ni-YSZ cermet exhibits good catalytic activity at high operation temperature, good electronic conductivity, and a thermal expansion behavior closely matched to that of the electrolyte.[171] Ni-1Ce10ScSZ cermet[172] was also used as the hydrogen electrode for SOECs. However, the drawbacks of Ni-based electrodes include (1) the tendency of Ni in the cermet to agglomerate after long-term operation, leading to reduced reaction sites and increased polarization resistance; and (2) the easy oxidation of Ni in the environment with high steam concentration, causing a loss of electronic conductivity and leading to poor redox cycling stability of Ni-YSZ cermet electrodes. $(La_{0.75}Sr_{0.25})_xCr_{0.5}Mn_{0.5}O_3$ (x = 0.9 or 1.0) (LSCM)[173,174] with perovskite structure has been proposed as an alternative to Ni used as electrode material since LSCM is electrochemically active and redox stable. For example, Jin et al.[174] reported an SOEC with LSCM-YSZ steam/hydrogen electrode and demonstrated a relatively stable performance in the electrolysis operations. Other alternative materials for hydrogen electrodes include LSCM-$Ce_{0.8}Sm_{0.2}O_{2-\delta}$-Fe,[175] $La_{0.6}Sr_{0.4}VO_{3-\delta}$-YSZ composite,[176] $La_{0.4}Sr_{0.4}TiO_{3-\delta}$-$Ce_{0.8}Sm_{0.2}O_{2-\delta}$-Ni composite,[177] $Pr_{0.8}Sr_{1.2}(Co,Fe)_{0.8}Nb_{0.2}O_{4+\delta}$-CoFe composite.[171]

24.5.4 Single Cell and Stack

Two basic cell configurations for YSZ electrolyte-based SOECs were suggested in literature, as shown in Figure 24.19: one is electrolyte-supported SOEC and is supported on thick YSZ electrolyte with a thickness of 100–300 μm.[17,140] Another is electrode-supported SOEC with thin YSZ electrolyte (10–20 μm) and is supported on Ni-YSZ cermet (0.5–1.5 mm).[144,145,178] In the electrolyte-supported SOECs, due to thick YSZ electrolyte, the electrolyte resistance is very high, which limits the overall cell performance. Such cells are usually suitable for high-temperature SOECs. At intermediate temperatures,

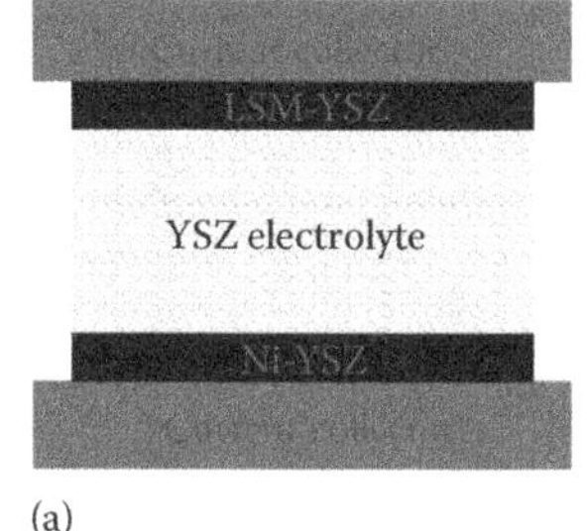

(a)

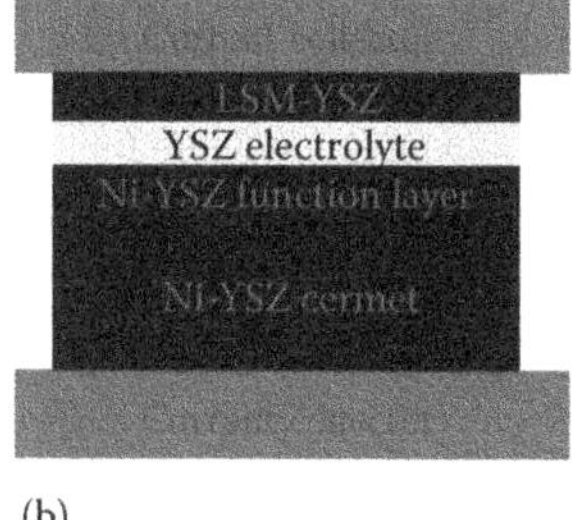

(b)

FIGURE 24.19 Two types of cell configurations for SOECs: (a) electrolyte supported and (b) electrode supported.

TABLE 24.8
Specifications, Performances, and Durabilities of Single SOECs and Stacks

Cell/Stack Configuration	Cathode	Anode	Electrolyte	Performance	Stability	Ref.
Electrolyte-supported, single cell	Ni-Fe (9:1) with Ni-SDC interlayer	$Ba_{0.6}La_{0.4}CoO_3$	$La_{0.9}Sr_{0.1}Ga_{0.8}Mg_{0.2}O_3$ (LSGM), 0.5 mm	600°C, 260 mA/cm^2 at 1.80 V	Unknown	143
Microtubular, electrode supported	NiO-YSZ tube with outer/inner diameter of 1.5/1.1 mm	$(La_{0.75}Sr_{0.25})_{0.95}MnO_3$-YSZ (50:50), ~25 μm	YSZ, ~13 μm	850°C, 0.9 A/cm^2 at 1.60 V for 60% AH	Not reported	184
Electrolyte supported, planar stack	Triple layer of Ni-ceria-gadolina, ~40 μm	$La_{0.6}Sr_{0.4}Co_{0.2}Fe_{0.8}O_3$ (LSCF)	3YSZ, 90 μm with electrolyte/LSCF interlayer $Y_{0.2}Ce_{0.8}O_{1.9}$ between	820°C, E = 1.20–1.23 V at 0.4 A/cm^2	820°C, j = 0.4 A/cm^2 4055 h	185
Electrode supported, single cell	Ni-YSZ cermet (8 μm) on Ni/YSZ support (1.5mm)	$La_{0.58}Sr_{0.4}Co_{0.2}Fe_{0.8}O_3$ (LSCF), 35–40 μm	YSZ (10 μm) with electrolyte/LSCF barrier layer of $Ce_{0.8}Gd_{0.2}O_{1.9}$ (5 μm)		780°C, 80% AH j = 1 A/cm^2 9000 h	144
Electrode supported, 30-cell stack	Ni-YSZ (400 μm) as electrode and support	YSZ (10 μm)	LSM-YSZ (30–40 μm)		800°C, 80% AH j = 0.15 A/cm^2 >1100 h	145
Electrode supported, 6-cell short stack	Ni-YSZ	YSZ with $(Gd,Ce)O_2$ barrier layer	$La_{0.6}Sr_{0.4}Co_{0.2}Fe_{0.8}O_3$ (LSCF)–$(Gd,Ce)O_2$ $La_{0.6}Sr_{0.4}CoO_3$ (LSC)	700°C LSCF-based stack: 1.60 V at 1 A/cm^2 LSC-based stack: 1.25 V at 0.8 A/cm^2	LSCF-based stack: 2400 h LSC-based stack: 1500 h	150
Electrode supported, 5-cell short stack	Ni-YSZ (470–560 μm) as electrode and support	YSZ (5 μm)	LSCF-YDC (33–55 μm)	800°C, 1.2–1.3 V at 1 A/cm^2	800°C, j = 0.5 A/cm^2 2700 h	183

it is necessary to reduce the thickness of YSZ electrolyte to 10–20 μm in order to decrease the YSZ electrolyte resistance. However, the mechanical strength of thin YSZ electrolyte is not sufficient to support the cell. Instead, SOEC is supported on electrodes (such as Ni-YSZ cermet). Therefore, electrode-supported structures are more suitable for intermediate temperature SOECs. These two cell configurations can also be applied to SOECs with other electrolytes, such as LSGM[179] and doped ceria. The selection of a suitable cell configuration depends on the operation temperature, the mechanical strength of the electrolyte, and the ionic conductivity of the electrolyte material. The cell design of an SOEC can be either tubular or planar. In the early stages of SOECs, the tests by the Hot Elly project[180] and Westinghouse Electric Co.[181] were mainly conducted on tubular-form SOECs with a porous oxygen electrode tubes or substrate tubes as cell supports. The tubular design makes the sealing easier but exhibits lower performance. Recently planar design, either electrolyte-supported[182] or electrode-supported cells,[145,183] has become more popular because of easier manufacturing process and superior cell performance.

Much effort has been devoted to SOEC material development and the electrolysis performance/durability of single SOECs and stacks. The specifications, performances, and durabilities of single SOECs and stacks reported in literature are summarized in Table 24.8. The performance of single SOECs depends on many factors, including cell design and configuration, the thickness and ionic conductivity of the electrolyte, the electrocatalytic activity of the electrode materials, the composition and microstructure of the electrodes, and the operating conditions, such as temperature and absolute humidity (AH). Durability is another important issue in SOECs. In order to become competitive with other hydrogen-production technologies, high-/intermediate-temperature SOECs need to exhibit high durability (~25,000 hours) in addition to high performance (high hydrogen-production rates, i.e., ~40 mg H_2/cm²/h, or 1 A/cm²).[183] However, only few works have been performed on the durability of SOECs.[144,149,183,186] Schefold et al.[144] demonstrated 9000 hours of operation of a solid oxide cell in steam electrolysis mode for a cell consisting of YSZ with a thickness of 10 μm as electrolyte, a 5 μm thick $Ce_{0.8}Gd_{0.2}O_{1.9}$ as a diffusion barrier layer between the electrolyte and the oxygen electrode, an oxygen electrode made of $La_{0.58}Sr_{0.4}Co_{0.2}Fe_{0.8}O_3$ with a thickness of 35–40 μm, and a steam/hydrogen electrode of Ni-YSZ cermet (8 μm thickness) supported on 1.5 mm Ni-YSZ substrate. The use of $Ce_{0.8}Gd_{0.2}O_{1.9}$ diffusion barrier layer is to avoid the reaction between YSZ and LSCF. In Schefold et al.'s work,[144] as shown in Figure 24.20, the voltage loss under constant current operation over the entire test was only 40 mV/1000 hours. The degradation of SOECs may be due to (1) decreasing ionic conduction in the electrolyte, (2) decreasing electronic conduction in the electrodes, (3) electrode deactivation, and (4) the interfacial reaction between two cell components and electrolyte–electrode delamination.[149,186] Detailed degradation mechanism of SOECs can be found in the recent review by Mocoteguy and Brisse.[187]

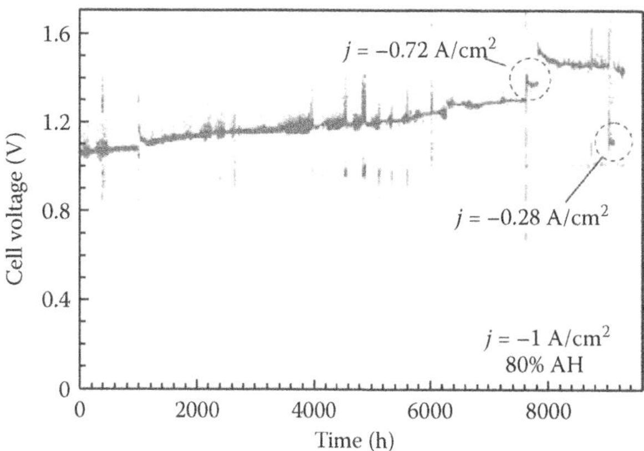

FIGURE 24.20 Long-term operation (~9000 hours) of electrode-supported SOECs under a current density of 1 A/cm², 80% AH and 780°C. (Reproduced from Schefold, J., A. Brisse, and F. Tietz, Nine thousand hours of operation of a solid oxide cell in steam electrolysis mode, *Journal of the Electrochemical Society*, 159 (2), A137, 2012. With permission.)

Several groups have developed SOEC short stacks and stack modules for large-scale hydrogen production.[18,19,145,150,178,183,188,189] For example, the Idaho National Laboratory[18,140] investigated large-scale hydrogen production using multicell stacks consisting of tens and hundreds of single cells and multistack systems. In their work, the cells are electrolyte supported, with ScSZ electrolyte (~140 μm); a graded air electrode consisting of an inner layer of manganite/zirconia (~13 μm) immediately adjacent to the electrolyte, a middle layer of manganite (~18 μm), and an outer bond layer of cobaltite; and a graded steam/hydrogen electrode with a nickel cermet layer (~13 μm) immediately adjacent to the electrolyte and a pure nickel outer layer (~10 μm). Petitjean et al.[183] investigated the performance and durability of short stacks consisting of five single electrode-supported planar SOECs with Ni-YSZ cermet (470–560 μm) as a hydrogen electrode and support, YSZ (5 μm) as an electrolyte, and YDC-LSCF composite (35–55 μm) as an air electrode. They obtained very high performances (−1 A/cm² at 1.2–1.3 V at 800°C) at the stack level and demonstrated a durability test of 2700 hours with degradation rates around 7%–13%/1000 hours under a current density of −0.5 A/cm² and a steam conversion rate of 25%. The Ningbo Institute of Material Technology and Engineering in China[145] manufactured 30-cell Ni-YSZ hydrogen electrode–supported planar solid oxide electrolyzer (SOE) stack modules and tested these in steam electrolysis mode for hydrogen production. They demonstrated up to 1100 hours of operation at 800°C under a constant current density of 0.15 A/cm² and obtained a steam conversion rate of 62% and a current efficiency of 87.4%.[99b] They also found that the main factors responsible for the degradation of the SOE stack modules were the delamination of the LSM-YSZ oxygen electrode and the agglomeration of Ni and reduction in the porosity of the Ni-YSZ steam/hydrogen electrode.[145]

24.6 CONCLUSIONS AND OUTLOOK

Hydrogen production is a key element of a hydrogen economy. For large-scale hydrogen production, alkaline liquid electrolyte water electrolysis has been employed for several decades due to its low cost. However, alkaline liquid electrolyte water electrolysis exhibits low hydrogen production rate, low energy efficiency, and a problem with the blockage of electrode pore structure by carbonate precipitation. PEM water electrolysis significantly improves the hydrogen production rate and the energy efficiency; however, its widespread use is hindered by its high capital cost due to the use of PGM catalysts and expensive perfluorosulfonic membrane and ionomer. AEM water electrolysis combines the advantages of both alkaline liquid electrolyte and PEM water electrolysis. However, its state-of-the-art performance and durability are still poor because of the limitation of AEM membrane and ionomer. High-/intermediate-temperature SOEC demands less electricity for hydrogen production, uses non-PGM catalysts, and exhibits high hydrogen-production rate and high energy efficiency. For large-scale hydrogen production, alkaline liquid electrolyte water electrolysis may be challenged by high-/intermediate-temperature SOEC, while for intermediate-/low-scale hydrogen production, such as on-site hydrogen production, alkaline liquid electrolyte water electrolysis may be challenged by PEM water electrolysis.

Performance, durability, and cost are three main factors to be considered for the widespread use of water electrolysis for hydrogen production. Currently no type of water electrolysis technology meets the three requirements: high performance, long lifetime, and low cost. Alkaline liquid electrolyte water electrolysis exhibits long lifetime and low cost but has low energy efficiency. PEM water electrolysis demonstrates high hydrogen-production rate, high energy efficiency, and relatively long lifetime while being high cost. AEM water electrolysis has a potential to meet the three requirements; however, it is still in its early stage, and its performance and durability need a significant improvement. High-/intermediate-temperature SOEC may be promising for stationary, large-scale hydrogen production because of its high energy efficiency, long lifetime, and relatively low cost.

The performance, stability, and cost of the materials for water electrolysis cell components, especially the electrocatalysts, membrane, and ionomer, are key factors for determining the performance, durability, and cost of water electrolysis. The main challenges with advanced materials for water electrolysis can be summarized as follows:

- Reducing the loading of electrocatalysts, especially PGM-based electrocatalysts, for example, Pt-based HER and Ir-based OER catalysts for PEM water electrolysis
- Increasing the utilization for the electrocatalysts for all types of water electrolysis
- Improving the electrocatalytic performance of non-PGM electrocatalysts for all types of water electrolysis
- Improving the ionic conductivity and long-term stability of AEMs for AEM water electrolysis
- Improving the ionic conductivity and long-term stability of anion-exchange ionomers for AEM water electrolysis
- Reducing the cost of PEMs and ionomers for PEM water electrolysis
- Reducing the cost of GDLs and current collectors for PEM and AEM water electrolysis
- Improving the electrocatalytic performance and long-term stability of electrode materials for high-/intermediate-temperature SOECs
- Improving the ionic conductivity and long-term stability of electrolyte materials for high-/intermediate-temperature SOECs

During the past several decades, much effort has been made to address these challenges and great progress with advanced materials for water electrolysis has been made, such as the following:

- The loading of noble metal electrocatalysts for PEM water electrolysis has been significantly reduced by supporting noble metal catalysts on supports, forming a solid-state solution with less expensive metal, developing electrocatalysts with core-shell structure, or developing NSTF-structured electrocatalysts.
- The activity of electrocatalysts has been improved by developing alloy catalysts or forming a solid-state solution with second metals.
- Several advanced non-PGM electrocatalysts have been explored, such as NiMo used as HER electrocatalysts and NiFe oxides, NiFe layered double hydroxides, and $Ba_{0.5}Sr_{0.5}Co_{0.8}Fe_{0.2}O_3$ perovskite used as OER electrocatalysts for alkaline liquid electrolyte and AEM water electrolysis.
- A number of advanced AEMs and ionomers with high ionic conductivity have been developed for AEM water electrolysis.
- Alternative low-cost PEMs with improved mechanical, electrochemical, and chemical properties have been developed for PEM water electrolysis, such as hydrocarbon PEMs and composite membranes.
- Alternative low-cost GDLs and current collectors have been developed for PEM water electrolysis.
- A number of advanced electrode and electrolyte materials used in SOFCs have been borrowed for high-/intermediate-temperature SOECs.

In order to meet the requirement of performance, durability, and cost for the widespread use of water electrolysis technologies, with regard to the materials used in water electrolysis, future focus should be placed on

- Development of advanced non-PGM catalysts with high activity, high stability, and low cost for zero-gap alkaline liquid electrolyte, AEM, and PEM water electrolysis;
- Development of innovative AEMs and ionomers with high ionic conductivity and enhanced long-term stability for zero-gap alkaline liquid electrolyte and AEM water electrolysis;

- Development of alternative proton-conducting membranes and ionomers with high ionic conductivity and enhanced long-term stability for PEM water electrolysis
- Development of advanced highly active electrode materials and highly conductive electrolyte materials for high-/intermediate-temperature SOECs;
- Fundamental understanding of the degradation mechanism of materials including electrocatalysts, polymer membranes, ionomers, electrode materials, and electrolyte materials used in water electrolysis; and
- Investigation of the relationship among the performance, property, and structure of materials used in water electrolysis.

REFERENCES

1. Turner, J. A., Sustainable hydrogen production. *Science* **2004**, *305* (5686), 972–974.
2. Cortright, R. D.; Davda, R. R.; Dumesic, J. A., Hydrogen from catalytic reforming of biomass-derived hydrocarbons in liquid water. *Nature* **2002**, *418* (6901), 964–967.
3. Zou, Z. G.; Ye, J. H.; Sayama, K.; Arakawa, H., Direct splitting of water under visible light irradiation with an oxide semiconductor photocatalyst. *Nature* **2001**, *414* (6864), 625–627.
4. Leng, Y. J.; Chen, G.; Mendoza, A. J.; Tighe, T. B.; Hickner, M. A.; Wang, C. Y., Solid-state water electrolysis with an alkaline membrane. *Journal of the American Chemical Society* **2012**, *134* (22), 9054–9057.
5. Jacobson, M. Z.; Colella, W. G.; Golden, D. M., Cleaning the air and improving health with hydrogen fuel-cell vehicles. *Science* **2005**, *308* (5730), 1901–1905.
6. Turner, J.; Sverdrup, G.; Mann, M. K.; Maness, P. C.; Kroposki, B.; Ghirardi, M.; Evans, R. J.; Blake, D., Renewable hydrogen production. *International Journal of Energy Research* **2008**, *32* (5), 379–407.
7. Gandia, L. M.; Oroz, R.; Ursua, A.; Sanchis, P.; Dieguez, P. M., Renewable hydrogen production: Performance of an alkaline water electrolyzer working under emulated wind conditions. *Energy & Fuels* **2007**, *21* (3), 1699–1706.
8. Sherif, S. A.; Barbir, F.; Veziroglu, T. N., Wind energy and the hydrogen economy—Review of the technology. *Solar Energy* **2005**, *78* (5), 647–660.
9. Barbir, F., PEM electrolysis for production of hydrogen from renewable energy sources. *Solar Energy* **2005**, *78* (5), 661–669.
10. Zeng, K.; Zhang, D. K., Recent progress in alkaline water electrolysis for hydrogen production and applications. *Progress in Energy and Combustion Science* **2010**, *36* (3), 307–326.
11. Wendt, H.; Hofmann, H.; Plzak, V., Materials research and development of electrocatalysts for alkaline water electrolysis. *Materials Chemistry and Physics* **1989**, *22* (1–2), 27–49.
12. Schiller, G.; Henne, R.; Mohr, P.; Peinecke, V., High performance electrodes for an advanced intermittently operated 10-kW alkaline water electrolyzer. *International Journal of Hydrogen Energy* **1998**, *23* (9), 761–765.
13. Naughton, M. S.; Brushett, F. R.; Kenis, P. J. A., Carbonate resilience of flowing electrolyte-based alkaline fuel cells. *Journal of Power Sources* **2011**, *196* (4), 1762–1768.
14. Grigoriev, S. A.; Porembsky, V. I.; Fateev, V. N., Pure hydrogen production by PEM electrolysis for hydrogen energy. *International Journal of Hydrogen Energy* **2006**, *31* (2), 171–175.
15. Marshall, A. T.; Sunde, S.; Tsypkin, A.; Tunold, R., Performance of a PEM water electrolysis cell using $Ir_xRu_yTa_zO_2$ electrocatalysts for the oxygen evolution electrode. *International Journal of Hydrogen Energy* **2007**, *32* (13), 2320–2324.
16. Ayers, K. E.; Anderson, E. B.; Capuano, C. B.; Carter, B. D.; Dalton, L. T.; Hanlon, G.; Manco, J.; Niedzwiecki, M., Research advanced towards low cost, high efficiency PEM electrolysis. *ECS Transactions* **2010**, *33* (1), 3–15.
17. Hino, R.; Haga, K.; Aita, H.; Sekita, K., R & D on hydrogen production by high-temperature electrolysis of steam. *Nuclear Engineering and Design* **2004**, *233* (1–3), 363–375.
18. Stoots, C. M.; O'Brien, J. E.; Condie, K. G.; Hartvigsen, J. J., High-temperature electrolysis for large-scale hydrogen production from nuclear energy—Experimental investigations. *International Journal of Hydrogen Energy* **2010**, *35* (10), 4861–4870.
19. Yu, B.; Zhang, W. Q.; Xu, J. M.; Chen, J., Status and research of highly efficient hydrogen production through high temperature steam electrolysis at INET. *International Journal of Hydrogen Energy* **2010**, *35* (7), 2829–2835.
20. Cuerrini, E., Chapter 7. Electrocatalysis in water electrolysis. In *Catalysis for Sustainable Energy Production.* Edited by Barbaro, P.; Bianchini, C. Wiley-VCH Verlag GmbH & Co. KGaA, Weinheim **2009**, 235–269.
21. Millet, P., Chapter 9. Water electrolysis for hydrogen generation. In *Electrochemical Technologies for Energy Storage and Conversion.* Edited by Liu, R.; Zhang, L.; Su, X.; Liu, H.; Zhang, J. Wiley-VCH Verlag GmbH & Co. KGaA **2012**, *1 & 2*, 383–423.
22. Renaud, R.; Leroy, R. L., Separator materials for use in alkaline water electrolyzers. *International Journal of Hydrogen Energy* **1982**, *7* (2), 155–166.
23. Bowen, C. T.; Davis, H. J.; Henshaw, B. F.; Lachance, R.; Leroy, R. L.; Renaud, R., Developments in advanced alkaline water electrolysis. *International Journal of Hydrogen Energy* **1984**, *9* (1–2), 59–66.
24. Giz, M. J.; Silva, J. C. P.; Ferreira, M.; Machado, S. A. S.; Ticianelli, E. A.; Avaca, L. A.; Gonzalez, E. R., Progress on the development of activated cathodes for water electrolysis. *International Journal of Hydrogen Energy* **1992**, *17* (9), 725–729.
25. Pletcher, D.; Li, X., Prospects for alkaline zero gap water electrolysers for hydrogen production. *International Journal of Hydrogen Energy* **2011**, *36* (23), 15089–15104.
26. Rami, A.; Lasia, A., Kinetics of hydrogen evolution on Ni-Al alloy electrodes. *Journal of Applied Electrochemistry* **1992**, *22* (4), 376–382.
27. Jaksic, J. M.; Vojnovic, M. V.; Krstajic, N. V., Kinetic analysis of hydrogen evolution at Ni-Mo alloy electrodes. *Electrochimica Acta* **2000**, *45* (25–26), 4151–4158.
28. Birry, L.; Lasia, A., Studies of the hydrogen evolution reaction on Raney nickel-molybdenum electrodes. *Journal of Applied Electrochemistry* **2004**, *34* (7), 735–749.
29. Crnkovic, F. C.; Machado, S. A. S.; Avaca, L. A., Electrochemical and morphological studies of electrodeposited Ni-Fe-Mo-Zn alloys tailored for water electrolysis. *International Journal of Hydrogen Energy* **2004**, *29* (3), 249–254.
30. Hu, W. K., Electrocatalytic properties of new electrocatalysts for hydrogen evolution in alkaline water electrolysis. *International Journal of Hydrogen Energy* **2000**, *25* (2), 111–118.

31. Hu, W. K.; Cao, X. J.; Wang, F. P.; Zhang, Y. S., A novel cathode for alkaline water electrolysis. *International Journal of Hydrogen Energy* **1997**, *22* (6), 621–623.
32. Michishita, H.; Misumi, Y.; Haruta, D.; Masaki, T.; Yamamoto, N.; Matsumoto, H.; Ishihara, T., Cathodic performance of $La_{0.6}Sr_{0.4}CoO_3$ perovskite oxide for platinum-free alkaline water electrolysis cell. *Journal of the Electrochemical Society* **2008**, *155* (9), B969–B971.
33. Rasiyah, P.; Tseung, A. C. C., A mechanistic study of oxygen evolution on $NiCo_2O_4$; 2. Electrochemical kinetics. *Journal of the Electrochemical Society* **1983**, *130* (12), 2384–2386.
34. Rasiyah, P.; Tseung, A. C. C., A mechanistic study of oxygen evolution on Li-doped Co_3O_4. *Journal of the Electrochemical Society* **1983**, *130* (2), 365–368.
35. Matsumoto, Y.; Kurimoto, J.; Sato, E., Oxygen evolution on $SrFeO_3$ electrode. *Journal of Electroanalytical Chemistry* **1979**, *102* (1), 77–83.
36. Matsumoto, Y.; Kurimoto, J.; Sato, E., Anodic characteristics of $SrFe_{0.9}M_{0.1}O_3$ (M-Ni, Co, Ti, Mn) electrodes. *Electrochimica Acta* **1980**, *25* (5), 539–543.
37. Matsumoto, Y.; Sato, E., Oxygen evolution on $La_{1-x}Sr_xMnO_3$ electrodes in alkaline-solutions. *Electrochimica Acta* **1979**, *24* (4), 421–423.
38. Matsumoto, Y.; Manabe, H.; Sato, E., Oxygen evolution on $La_{1-x}Sr_xCoO_3$ electrodes in alkaline-solutions. *Journal of the Electrochemical Society* **1980**, *127* (4), 811–814.
39. Singh, R. N.; Lal, B., High surface area lanthanum cobaltate and its A and B sites substituted derivatives for electrocatalysis of O_2 evolution in alkaline solution. *International Journal of Hydrogen Energy* **2002**, *27* (1), 45–55.
40. Matsumoto, Y.; Yamada, S.; Nishida, T.; Sato, E., Oxygen evolution on $La_{1-x}Sr_xFe_{1-y}Co_yO_3$ series oxides. *Journal of the Electrochemical Society* **1980**, *127* (11), 2360–2364.
41. Miller, E. L.; Rocheleau, R. E., Electrochemical behavior of reactively sputtered iron-doped nickel oxide. *Journal of the Electrochemical Society* **1997**, *144* (9), 3072–3077.
42. Merrill, M. D.; Dougherty, R. C., Metal oxide catalysts for the evolution of O_2 from H_2O. *Journal of Physical Chemistry C* **2008**, *112* (10), 3655–3666.
43. Corrigan, D. A., The catalysis of the oxygen evolution reaction by iron impurities in thin film Ni-oxide electrodes. *Journal of the Electrochemical Society* **1987**, *134* (2), 377–384.
44. Li, X.; Walsh, F. C.; Pletcher, D., Nickel based electrocatalysts for oxygen evolution in high current density, alkaline water electrolysers. *Physical Chemistry Chemical Physics* **2011**, *13* (3), 1162–1167.
45. Horowitz, H. S.; Longo, J. M.; Horowitz, H. H., Oxygen electrocatalysis on some oxide pyrochlores. *Journal of the Electrochemical Society* **1983**, *130* (9), 1851–1859.
46. Hung, C. Y.; Li, S. D.; Wang, C. C.; Chen, C. Y., Influences of a bipolar membrane and an ultrasonic field on alkaline water electrolysis. *Journal of Membrane Science* **2012**, *389*, 197–204.
47. Lu, S.; Zhuang, L.; Lu, J., Homogeneous blend membrane made from poly(ether sulphone) and poly(vinylpyrrolidone) and its application to water electrolysis. *Journal of Membrane Science* **2007**, *300* (1–2), 205–210.
48. Pletcher, D.; Li, X.; Wang, S., A comparison of cathodes for zero gap alkaline water electrolysers for hydrogen production. *International Journal of Hydrogen Energy* **2012**, *37* (9), 7429–7435.
49. Wu, X.; Scott, K., $Cu_xCo_{3-x}O_4$ ($0 <= x <= 1$) nanoparticles for oxygen evolution in high performance alkaline exchange membrane water electrolysers. *Journal of Materials Chemistry* **2011**, *21* (33), 12344–12351.
50. Faraj, M.; Boccia, M.; Miller, H.; Martini, F.; Borsacchi, S.; Geppi, M.; Pucci, A., New LDPE based anion-exchange membranes for alkaline solid polymeric electrolyte water electrolysis. *International Journal of Hydrogen Energy* **2012**, *37* (20), 14992–15002.
51. Hnat, J.; Paidar, M.; Schauer, J.; Zitka, J.; Bouzek, K., Polymer anion selective membranes for electrolytic splitting of water; Part I: Stability of ion-exchange groups and impact of the polymer binder. *Journal of Applied Electrochemistry* **2011**, *41* (9), 1043–1052.
52. Hnat, J.; Paidar, M.; Schauer, J.; Zitka, J.; Bouzek, K., Polymer anion-selective membranes for electrolytic splitting of water; Part II: Enhancement of ionic conductivity and performance under conditions of alkaline water electrolysis. *Journal of Applied Electrochemistry* **2012**, *42* (8), 545–554.
53. Aili, D.; Hansen, M. K.; Renzaho, R. F.; Li, Q. F.; Christensen, E.; Jensen, J. O.; Bjerrum, N. J., Heterogeneous anion conducting membranes based on linear and crosslinked KOH doped polybenzimidazole for alkaline water electrolysis. *Journal of Membrane Science* **2013**, *447*, 424–432.
54. Arico, A. S.; Siracusano, S.; Briguglio, N.; Baglio, V.; Di Blasi, A.; Antonucci, V., Polymer electrolyte membrane water electrolysis: Status of technologies and potential applications in combination with renewable power sources. *Journal of Applied Electrochemistry* **2013**, *43* (2), 107–118.
55. Carmo, M.; Fritz, D. L.; Mergel, J.; Stolten, D., A comprehensive review on PEM water electrolysis. *International Journal of Hydrogen Energy* **2013**, *38* (12), 4901–4934.
56. Grigoriev, S. A.; Millet, P.; Fateev, V. N., Evaluation of carbon-supported Pt and Pd nanoparticles for the hydrogen evolution reaction in PEM water electrolysers. *Journal of Power Sources* **2008**, *177* (2), 281–285.
57. Chen, W. F.; Sasaki, K.; Ma, C.; Frenkel, A. I.; Marinkovic, N.; Muckerman, J. T.; Zhu, Y. M.; Adzic, R. R., Hydrogen-evolution catalysts based on non-noble metal nickel-molybdenum nitride nanosheets. *Angewandte Chemie International Edition* **2012**, *51* (25), 6131–6135.
58. Esposito, D. V.; Hunt, S. T.; Kimmel, Y. C.; Chen, J. G. G., A new class of electrocatalysts for hydrogen production from water electrolysis: Metal monolayers supported on low-cost transition metal carbides. *Journal of the American Chemical Society* **2012**, *134* (6), 3025–3033.
59. McKone, J. R.; Sadtler, B. F.; Werlang, C. A.; Lewis, N. S.; Gray, H. B., Ni–Mo nanopowders for efficient electrochemical hydrogen evolution. *ACS Catalysis* **2013**, *3* (2), 166–169.
60. Debe, M. K.; Hendricks, S. M.; Vernstrom, G. D.; Meyers, M.; Brostrom, M.; Stephens, M.; Chan, Q.; Willey, J.; Hamden, M.; Mittelsteadt, C. K.; Capuano, C. B.; Ayers, K. E.; Anderson, E. B., Initial performance and durability of ultra-low loaded NSTF electrodes for PEM electrolyzers. *Journal of the Electrochemical Society* **2012**, *159* (6), K165–K176.
61. Li, Y. G.; Wang, H. L.; Xie, L. M.; Liang, Y. Y.; Hong, G. S.; Dai, H. J., MoS_2 nanoparticles grown on graphene: An advanced catalyst for the hydrogen Evolution Reaction. *Journal of the American Chemical Society* **2011**, *133* (19), 7296–7299.
62. Millet, P.; Ngameni, R.; Grigoriev, S. A.; Mbemba, N.; Brisset, F.; Ranjbari, A.; Etievant, C., PEM water electrolyzers: From electrocatalysis to stack development. *International Journal of Hydrogen Energy* **2010**, *35* (10), 5043–5052.
63. Siracusano, S.; Baglio, V.; Stassi, A.; Ornelas, R.; Antonucci, V.; Arico, A. S., Investigation of IrO_2 electrocatalysts prepared by a sulfite-complex route for the O_2 evolution reaction in solid polymer electrolyte water electrolyzers. *International Journal of Hydrogen Energy* **2011**, *36* (13), 7822–7831.

64. Hu, W.; Wang, Y. Q.; Hu, X. H.; Zhou, Y. Q.; Chen, S. L., Three-dimensional ordered macroporous IrO_2 as electrocatalyst for oxygen evolution reaction in acidic medium. *Journal of Materials Chemistry* **2012**, *22* (13), 6010–6016.
65. Ma, L. R.; Sui, S.; Zhai, Y. C., Investigations on high performance proton exchange membrane water electrolyzer. *International Journal of Hydrogen Energy* **2009**, *34* (2), 678–684.
66. Siracusano, S.; Baglio, V.; D'Urso, C.; Antonucci, V.; Arico, A. S., Preparation and characterization of titanium suboxides as conductive supports of IrO_2 electrocatalysts for application in SPE electrolysers. *Electrochimica Acta* **2009**, *54* (26), 6292–6299.
67. Wu, X.; Scott, K., RuO_2 supported on Sb-doped SnO_2 nanoparticles for polymer electrolyte membrane water electrolysers. *International Journal of Hydrogen Energy* **2011**, *36* (10), 5806–5810.
68. Rasten, E.; Hagen, G.; Tunold, R., Electrocatalysis in water electrolysis with solid polymer electrolyte. *Electrochimica Acta* **2003**, *48* (25–26), 3945–3952.
69. Cheng, J.; Zhang, H.; Chen, G.; Zhang, Y., Study of $Ir_xRu_{1-x}O_2$ oxides as anodic electrocatalysts for solid polymer electrolyte water electrolysis. *Electrochimica Acta* **2009**, *54* (26), 6250–6256.
70. Cheng, J.; Zhang, H.; Ma, H.; Zhong, H.; Zou, Y., Preparation of $Ir_{0.4}Ru_{0.6}Mo_xO_y$ for oxygen evolution by modified Adams' fusion method. *International Journal of Hydrogen Energy* **2009**, *34* (16), 6609–6613.
71. Song, S. D.; Zhang, H. M.; Ma, X. P.; Shao, Z. G.; Baker, R. T.; Yi, B. L., Electrochemical investigation of electrocatalysts for the oxygen evolution reaction in PEM water electrolyzers. *International Journal of Hydrogen Energy* **2008**, *33* (19), 4955–4961.
72. Millet, P.; Dragoe, D.; Grigoriev, S.; Fateev, V.; Etievant, C., GenHyPEM: A research program on PEM water electrolysis supported by the European Commission. *International Journal of Hydrogen Energy* **2009**, *34* (11), 4974–4982.
73. Li, G.; Yu, H.; Wang, X.; Yang, D.; Li, Y.; Shao, Z.; Yi, B., Triblock polymer mediated synthesis of Ir–Sn oxide electrocatalysts for oxygen evolution reaction. *Journal of Power Sources* **2014**, *249*, 175–184.
74. Wu, X.; Tayal, J.; Basu, S.; Scott, K., Nano-crystalline $Ru_xSn_{1-x}O_2$ powder catalysts for oxygen evolution reaction in proton exchange membrane water electrolysers. *International Journal of Hydrogen Energy* **2011**, *36* (22), 14796–14804.
75. Ma, C.; Sheng, J.; Brandon, N.; Zhang, C.; Li, G., Preparation of tungsten carbide-supported nano platinum catalyst and its electrocatalytic activity for hydrogen evolution. *International Journal of Hydrogen Energy* **2007**, *32* (14), 2824–2829.
76. Brankovic, S. R.; Wang, J. X.; Adzic, R. R., Metal monolayer deposition by replacement of metal adlayers on electrode surfaces. *Surface Science* **2001**, *474* (1–3), L173–L179.
77. Zhang, J.; Mo, Y.; Vukmirovic, M. B.; Klie, R.; Sasaki, K.; Adzic, R. R., Platinum monolayer electrocatalysts for O_2 reduction: Pt monolayer on Pd(111) and on carbon-supported Pd nanoparticles. *Journal of Physical Chemistry B* **2004**, *108* (30), 10955–10964.
78. Sasaki, K.; Naohara, H.; Cai, Y.; Choi, Y. M.; Liu, P.; Vukmirovic, M. B.; Wang, J. X.; Adzic, R. R., Core-protected platinum monolayer shell high-stability electrocatalysts for fuel-cell cathodes. *Angewandte Chemie International Edition* **2010**, *49* (46), 8602–8607.
79. Srivastava, R.; Mani, P.; Hahn, N.; Strasser, P., Efficient oxygen reduction fuel cell electrocatalysis on voltammetrically dealloyed Pt-Cu-Co nanoparticles. *Angewandte Chemie International Edition* **2007**, *46* (47), 8988–8991.
80. Mani, P.; Srivastava, R.; Strasser, P., Dealloyed Pt-Cu core-shell nanoparticle electrocatalysts for use in PEM fuel cell cathodes. *Journal of Physical Chemistry C* **2008**, *112* (7), 2770–2778.
81. Debe, M. K., Tutorial on the fundamental characteristics and practical properties of nanostructured thin film (NSTF) catalysts. *Journal of the Electrochemical Society* **2013**, *160* (6), F522–F534.
82. Ayers, K., High performance, low cost H_2 generation from renewable energy. *DOE Hydrogen Program Review* **2011**, May 11, Washington, DC.
83. Hamdan, M., PEM electrolyzer incorporating an advanced low cost membrane. *DOE Hydrogen Program Review* **2011**, May 11, Washington, DC.
84. Nikolov, I.; Petrov, K.; Vitanov, T.; Guschev, A., Tungsten carbide cathodes for electrolysis of sulfuric-acid solutions. *International Journal of Hydrogen Energy* **1983**, *8* (6), 437–440.
85. Hinnemann, B.; Moses, P. G.; Bonde, J.; Jorgensen, K. P.; Nielsen, J. H.; Horch, S.; Chorkendorff, I.; Norskov, J. K., Biomimetic hydrogen evolution: MoS_2 nanoparticles as catalyst for hydrogen evolution. *Journal of the American Chemical Society* **2005**, *127* (15), 5308–5309.
86. Benck, J. D.; Chen, Z. B.; Kuritzky, L. Y.; Forman, A. J.; Jaramillo, T. F., Amorphous molybdenum sulfide catalysts for electrochemical hydrogen production: Insights into the origin of their catalytic activity. *ACS Catalysis* **2012**, *2* (9), 1916–1923.
87. Chang, Y. H.; Lin, C. T.; Chen, T. Y.; Hsu, C. L.; Lee, Y. H.; Zhang, W. J.; Wei, K. H.; Li, L. J., Highly efficient electrocatalytic hydrogen production by MoS_x grown on graphene-protected 3D Ni foams. *Advanced Materials* **2013**, *25* (5), 756760.
88. Jaramillo, T. F.; Jorgensen, K. P.; Bonde, J.; Nielsen, J. H.; Horch, S.; Chorkendorff, I., Identification of active edge sites for electrochemical H_2 evolution from MoS_2 nanocatalysts. *Science* **2007**, *317* (5834), 100–102.
89. Adams, R.; Shriner, R. L., Platinum oxide as a catalyst in the reduction of organic compounds; III. Preparation and properties of the oxide of platinum obtained by the fusion of ceiloroplatinic acid with sodium nitrate. *Journal of the American Chemical Society* **1923**, *45*, 2171–2179.
90. Xu, J. Y.; Wang, M.; Liu, G. Y.; Li, J. L.; Wang, X. D., The physical-chemical properties and electrocatalytic performance of iridium oxide in oxygen evolution. *Electrochimica Acta* **2011**, *56* (27), 10223–10230.
91. Mattos-Costa, F. I.; de Lima-Neto, P.; Machado, S. A. S.; Avaca, L. A., Characterisation of surfaces modified by sol-gel derived $Ru_xIr_{1-x}O_2$ coatings for oxygen evolution in acid medium. *Electrochimica Acta* **1998**, *44* (8–9), 1515–1523.
92. Marshall, A.; Borresen, B.; Hagen, G.; Tsypkin, M.; Tunold, R., Preparation and characterisation of nanocrystalline $Ir_xSn_{1-x}O_2$ electrocatalytic powders. *Materials Chemistry and Physics* **2005**, *94* (2–3), 226–232.
93. Ortel, E.; Reier, T.; Strasser, P.; Kraehnert, R., Mesoporous IrO_2 films templated by PEO-PB-PEO block-copolymers: Self-assembly, crystallization behavior, and electrocatalytic performance. *Chemistry of Materials* **2011**, *23* (13), 3201–3209.
94. Marshall, A. T.; Haverkamp, R. G., Electrocatalytic activity of IrO_2-RuO_2 supported on Sb-doped SnO_2 nanoparticles. *Electrochimica Acta* **2010**, *55* (6), 1978–1984.

95. Ma, L. R.; Sui, S.; Zhai, Y. C., Preparation and characterization of Ir/TiC catalyst for oxygen evolution. *Journal of Power Sources* **2008**, *177* (2), 470–477.
96. Marshall, A.; Tsypkin, M.; Borresen, B.; Hagen, G.; Tunold, R., Nanocrystalline $Ir_xSn_{(1-x)}O_2$ electrocatalysts for oxygen evolution in water electrolysis with polymer electrolyte—Effect of heat treatment. *Journal of New Materials for Electrochemical Systems* **2004**, *7* (3), 197–204.
97. Marshall, A.; Borresen, B.; Hagen, G.; Tsypkin, M.; Tunold, R., Electrochemical characterisation of $Ir_xSn_{1-x}O_2$ powders as oxygen evolution electrocatalysts. *Electrochimica Acta* **2006**, *51* (15), 3161–3167.
98. Mayousse, E.; Maillard, F.; Fouda-Onana, F.; Sicardy, O.; Guillet, N., Synthesis and characterization of electrocatalysts for the oxygen evolution in PEM water electrolysis. *International Journal of Hydrogen Energy* **2011**, *36* (17), 10474–10481.
99. Marshall, A.; Borresen, B.; Hagen, G.; Tsypkin, M.; Tunold, R., Hydrogen production by advanced proton exchange membrane (PEM) water electrolysers—Reduced energy consumption by improved electrocatalysis. *Energy* **2007**, *32* (4), 431–436.
100. Hu, J. M.; Meng, H. M.; Zhang, J. Q.; Cao, C. N., Degradation mechanism of long service life Ti/IrO_2-Ta_2O_5 oxide anodes in sulphuric acid. *Corrosion Science* **2002**, *44* (8), 1655–1668.
101. Di Blasi, A.; D'Urso, C.; Baglio, V.; Antonucci, V.; Arico, A. S.; Ornelas, R.; Matteucci, F.; Orozco, G.; Beltran, D.; Meas, Y.; Arriaga, L. G., Preparation and evaluation of RuO_2-IrO_2, IrO_2-Pt and IrO_2-Ta_2O_5 catalysts for the oxygen evolution reaction in an SPE electrolyzer. *Journal of Applied Electrochemistry* **2009**, *39* (2), 191–196.
102. Gaudet, J.; Tavares, A. C.; Trasatti, S.; Guay, D., Physicochemical characterization of mixed RuO_2-SnO_2 solid solutions. *Chemistry of Materials* **2005**, *17* (6), 1570–1579.
103. Grigoriev, S. A.; Porembskiy, V. I.; Korobtsev, S. V.; Fateev, V. N.; Aupretre, F.; Millet, P., High-pressure PEM water electrolysis and corresponding safety issues. *International Journal of Hydrogen Energy* **2011**, *36* (3), 2721–2728.
104. Ito, H.; Maeda, T.; Nakano, A.; Takenaka, H., Properties of Nafion membranes under PEM water electrolysis conditions. *International Journal of Hydrogen Energy* **2011**, *36* (17), 10527–10540.
105. Sawada, S.; Yamaki, T.; Maeno, T.; Asano, M.; Suzuki, A.; Terai, T.; Maekawa, Y., Solid polymer electrolyte water electrolysis systems for hydrogen production based on our newly developed membranes; Part I: Analysis of voltage-current characteristics. *Progress in Nuclear Energy* **2008**, *50* (2–6), 443–448.
106. Wei, G. Q.; Xu, L.; Huang, C. D.; Wang, Y. X., SPE water electrolysis with SPEEK/PES blend membrane. *International Journal of Hydrogen Energy* **2010**, *35* (15), 7778–7783.
107. Baglio, V.; Ornelas, R.; Matteucci, F.; Martina, F.; Ciccarella, G.; Zama, I.; Arriaga, L. G.; Antonucci, V.; Arico, A. S., Solid polymer electrolyte water electrolyser based on Nafion-TiO_2 composite membrane for high temperature operation. *Fuel Cells* **2009**, *9* (3), 247–252.
108. Wei, G. Q.; Wang, Y. X.; Huang, C. D.; Gao, Q. J.; Wang, Z. T.; Xu, L., The stability of MEA in SPE water electrolysis for hydrogen production. *International Journal of Hydrogen Energy* **2010**, *35* (9), 3951–3957.
109. Xiao, L.; Zhang, S.; Pan, J.; Yang, C. X.; He, M. L.; Zhuang, L.; Lu, J. T., First implementation of alkaline polymer electrolyte water electrolysis working only with pure water. *Energy & Environmental Science* **2012**, *5* (7), 7869–7871.
110. Wu, X.; Scott, K., A Li-doped Co_3O_4 oxygen evolution catalyst for non-precious metal alkaline anion exchange membrane water electrolysers. *International Journal of Hydrogen Energy* **2013**, *38* (8), 3123–3129.
111. Suntivich, J.; May, K. J.; Gasteiger, H. A.; Goodenough, J. B.; Shao-Horn, Y., A perovskite oxide optimized for oxygen evolution catalysis from molecular orbital principles. *Science* **2011**, *334* (6061), 1383–1385.
112. Trotochaud, L.; Ranney, J. K.; Williams, K. N.; Boettcher, S. W., Solution-cast metal oxide thin film electrocatalysts for oxygen evolution. *Journal of the American Chemical Society* **2012**, *134* (41), 17253–17261.
113. Gong, M.; Li, Y.; Wang, H.; Liang, Y.; Wu, J. Z.; Zhou, J.; Wang, J.; Regier, T.; Wei, F.; Dai, H., An advanced Ni-Fe layered double hydroxide electrocatalyst for water oxidation. *Journal of the American Chemical Society* **2013**, *135* (23), 8452–8455.
114. Pavel, C. C.; Cecconi, F.; Emiliani, C.; Santiccioli, S.; Scaffidi, A.; Catanorchi, S.; Comotti, M., Highly efficient platinum group metal free based membrane-electrode assembly for anion exchange membrane water electrolysis. *Angewandte Chemie International Edition* **2014**, *53* (5), 1378–1381.
115. Tang, X.; Xiao, L.; Yang, C.; Lu, J.; Zhuang, L., Noble fabrication of Ni–Mo cathode for alkaline water electrolysis and alkaline polymer electrolyte water electrolysis. *International Journal of Hydrogen Energy* **2014**, *39* (7), 3055–3060.
116. Esswein, A. J.; McMurdo, M. J.; Ross, P. N.; Bell, A. T.; Tilley, T. D., Size-dependent activity of Co_3O_4 nanoparticle anodes for alkaline water electrolysis. *Journal of Physical Chemistry C* **2009**, *113* (33), 15068–15072.
117. Nikolov, I.; Darkaoui, R.; Zhecheva, E.; Stoyanova, R.; Dimitrov, N.; Vitanov, T., Electrocatalytic activity of spinel related cobaltites $M_xCo_{3-x}O_4$ (M = Li, Ni, Cu) in the oxygen evolution reaction. *Journal of Electroanalytical Chemistry* **1997**, *429* (1–2), 157–168.
118. Singh, R. N.; Mishra, D.; Anindita; Sinha, A. S. K.; Singh, A., Novel electrocatalysts for generating oxygen from alkaline water electrolysis. *Electrochemistry Communications* **2007**, *9* (6), 1369–1373.
119. Shao, Z. P.; Haile, S. M., A high-performance cathode for the next generation of solid-oxide fuel cells. *Nature* **2004**, *431* (7005), 170–173.
120. Fukuta, H. Y. a. K., Anion exchange membrane and ionomer for alkaline membrane fuel cells (AMFCs). *ECS Transactions* **2008**, *16* (2), 257–262.
121. Pan, J.; Li, Y.; Zhuang, L.; Lu, J. T., Self-crosslinked alkaline polymer electrolyte exceptionally stable at 90 degrees C. *Chemical Communications* **2010**, *46* (45), 8597–8599.
122. Wu, X.; Scott, K., A polymethacrylate-based quaternary ammonium OH^- ionomer binder for non-precious metal alkaline anion exchange membrane water electrolysers. *Journal of Power Sources* **2012**, *214*, 124–129.
123. Cao, Y.-C.; Wu, X.; Scott, K., A quaternary ammonium grafted poly vinyl benzyl chloride membrane for alkaline anion exchange membrane water electrolysers with no-noble-metal catalysts. *International Journal of Hydrogen Energy* **2012**, *37* (12), 9524–9528.
124. Pan, J.; Li, Y.; Han, J. J.; Li, G. W.; Tan, L. S.; Chen, C.; Lu, J. T.; Zhuang, L., A strategy for disentangling the conductivity-stability dilemma in alkaline polymer electrolytes. *Energy & Environmental Science* **2013**, *6* (10), 2912–2915.

125. Pan, J.; Chen, C.; Li, Y.; Wang, L.; Tan, L.; Li, G.; Tang, X.; Xiao, L.; Lu, J.; Zhuang, L., Constructing ionic highway in alkaline polymer electrolytes. *Energy & Environmental Science* **2014**, *7* (1), 354.
126. Robertson, N. J.; Kostalik, H. A.; Clark, T. J.; Mutolo, P. F.; Abruna, H. D.; Coates, G. W., Tunable high performance cross-linked alkaline anion exchange membranes for fuel cell applications. *Journal of the American Chemical Society* **2010**, *132* (10), 3400–3404.
127. Pan, J.; Chen, C.; Zhuang, L.; Lu, J. T., Designing advanced alkaline polymer electrolytes for fuel cell applications. *Accounts of Chemical Research* **2012**, *45* (3), 473–481.
128. Varcoe, J. R.; Slade, R. C. T., Prospects for alkaline anion-exchange membranes in low temperature fuel cells. *Fuel Cells* **2005**, *5* (2), 187–200.
129. Bauer, B.; Strathmann, H.; Effenberger, F., Anion-exchange membranes with improved alkaline stability. *Desalination* **1990**, *79* (2–3), 125–144.
130. Arges, C. G.; Ramani, V., Investigation of cation degradation in anion exchange membranes using multi-dimensional NMR spectroscopy. *Journal of the Electrochemical Society* **2013**, *160* (9), F1006–F1021.
131. Arges, C. G.; Ramani, V., Two-dimensional NMR spectroscopy reveals cation-triggered backbone degradation in polysulfone-based anion exchange membranes. *Proceedings of the National Academy of Sciences of the United States of America* **2013**, *110* (7), 2490–2495.
132. Parrondo, J.; Arges, C. G.; Niedzwiecki, M.; Anderson, E. B.; Ayers, K. E.; Ramani, V., Degradation of anion exchange membranes used for hydrogen production by ultrapure water electrolysis. *RSC Advances* **2014**, *4* (19), 9875–9879.
133. Ni, M.; Leung, M. K. H.; Leung, D. Y. C., Technological development of hydrogen production by solid oxide electrolyzer cell (SOEC). *International Journal of Hydrogen Energy* **2008**, *33* (9), 2337–2354.
134. Hauch, A.; Ebbesen, S. D.; Jensen, S. H.; Mogensen, M., Highly efficient high temperature electrolysis. *Journal of Materials Chemistry* **2008**, *18* (20), 2331–2340.
135. Laguna-Bercero, M. A., Recent advances in high temperature electrolysis using solid oxide fuel cells: A review. *Journal of Power Sources* **2012**, *203*, 4–16.
136. Brisse, A.; Schefold, J.; Zahid, M., High temperature water electrolysis in solid oxide cells. *International Journal of Hydrogen Energy* **2008**, *33* (20), 5375–5382.
137. Hauch, A.; Ebbesen, S. D.; Jensen, S. H.; Mogensen, M., Solid oxide electrolysis cells: Microstructure and degradation of the Ni/yttria-stabilized zirconia electrode. *Journal of the Electrochemical Society* **2008**, *155* (11), B1184–B1193.
138. O'Brien, J. E.; Stoots, C. M.; Herring, J. S.; Hartvigsen, J., Hydrogen production performance of a 10-cell planar solid-oxide electrolysis stack. *Journal of Fuel Cell Science and Technology* **2006**, *3* (2), 213–219.
139. Laguna-Bercero, M. A.; Skinner, S. J.; Kilner, J. A., Performance of solid oxide electrolysis cells based on scandia stabilised zirconia. *Journal of Power Sources* **2009**, *192* (1), 126–133.
140. Herring, J. S.; O'Brien, J. E.; Stoots, C. M.; Hawkes, G. L.; Hartvigsen, J. J.; Shahnam, M., Progress in high-temperature electrolysis for hydrogen production using planar SOFC technology. *International Journal of Hydrogen Energy* **2007**, *32* (4), 440–450.
141. Shao, L.; Wang, S. R.; Qian, J. Q.; Ye, X. F.; Wen, T. L., Optimization of the electrode-supported tubular solid oxide cells for application on fuel cell and steam electrolysis. *International Journal of Hydrogen Energy* **2013**, *38* (11), 4272–4280.
142. Laguna-Bercero, M. A.; Kilner, J. A.; Skinner, S. J., Performance and characterization of (La, Sr)MnO_3/YSZ and $La_{0.6}Sr_{0.4}Co_{0.2}Fe_{0.8}O_3$ electrodes for solid oxide electrolysis cells. *Chemistry of Materials* **2010**, *22* (3), 1134–1141.
143. Ishihara, T.; Jirathiwathanakul, N.; Zhong, H., Intermediate temperature solid oxide electrolysis cell using $LaGaO_3$ based perovskite electrolyte. *Energy & Environmental Science* **2010**, *3* (5), 665–672.
144. Schefold, J.; Brisse, A.; Tietz, F., Nine thousand hours of operation of a solid oxide cell in steam electrolysis mode. *Journal of the Electrochemical Society* **2012**, *159* (2), A137.
145. Zheng, Y. F.; Li, Q. S.; Guan, W. B.; Xu, C.; Wu, W.; Wang, W. G., Investigation of 30-cell solid oxide electrolyzer stack modules for hydrogen production. *Ceramics International* **2014**, *40* (4), 5801–5809.
146. Ishihara, T.; Matsushita, S.; Sakai, T.; Matsumoto, H., Intermediate temperature solid oxide electrolysis cell using LaGaO3-base oxide. *Solid State Ionics* **2012**, *225*, 77–80.
147. Eguchi, K.; Hatagishi, T.; Arai, H., Power generation and steam electrolysis characteristics of an electrochemical cell with a zirconia- or ceria-based electrolyte. *Solid State Ionics* **1996**, *86–88*, 1245–1249.
148. Lee, S. I.; Kim, J.; Son, J. W.; Lee, J. H.; Kim, B. K.; Je, H. J.; Lee, H. W.; Song, H.; Yoon, K. J., High performance air electrode for solid oxide regenerative fuel cells fabricated by infiltration of nano-catalysts. *Journal of Power Sources* **2014**, *250*, 15–20.
149. Tietz, F.; Sebold, D.; Brisse, A.; Schefold, J., Degradation phenomena in a solid oxide electrolysis cell after 9000 h of operation. *Journal of Power Sources* **2013**, *223*, 129–135.
150. Diethelm, S.; Van Herle, J.; Montinaro, D.; Bucheli, O., Electrolysis and co-electrolysis performance of SOE short stacks. *Fuel Cells* **2013**, *13* (4), 631–637.
151. Matsumoto, H.; Okubo, M.; Hamajima, S.; Katahira, K.; Iwahara, H., Extraction and production of hydrogen using high-temperature proton conductor. *Solid State Ionics* **2002**, *152*, 715–720.
152. Kobayashi, T.; Abe, K.; Ukyo, Y.; Iwahara, H., Performance of electrolysis cells with proton and oxide-ion conducting electrolyte for reducing nitrogen oxide. *Solid State Ionics* **2002**, *154*, 699–705.
153. Sakai, T.; Matsushita, S.; Matsumoto, H.; Okada, S.; Hashimoto, S.; Ishihara, T., Intermediate temperature steam electrolysis using strontium zirconate-based protonic conductors. *International Journal of Hydrogen Energy* **2009**, *34* (1), 56–63.
154. Stuart, P. A.; Unno, T.; Kilner, J. A.; Skinner, S. J., Solid oxide proton conducting steam electrolysers. *Solid State Ionics* **2008**, *179* (21–26), 1120–1124.
155. He, F.; Song, D.; Peng, R. R.; Meng, G. Y.; Yang, S. F., Electrode performance and analysis of reversible solid oxide fuel cells with proton conducting electrolyte of $BaCe_{0.5}Zr_{0.3}Y_{0.2}O_{3-\delta}$. *Journal of Power Sources* **2010**, *195* (11), 3359–3364.
156. Li, S. S.; Xie, K., Composite oxygen electrode based on LSCF and BSCF for steam electrolysis in a proton-conducting solid oxide electrolyzer. *Journal of the Electrochemical Society* **2013**, *160* (2), F224–F233.
157. Li, S. S.; Yan, R. Q.; Wu, G. J.; Xie, K.; Cheng, J. G., Composite oxygen electrode LSM-BCZYZ impregnated with Co_3O_4 nanoparticles for steam electrolysis in a proton-conducting solid oxide electrolyzer. *International Journal of Hydrogen Energy* **2013**, *38* (35), 14943–14951.
158. Yang, C. H.; Jin, C.; Coffin, A.; Chen, F. L., Characterization of infiltrated $(La_{0.75}Sr_{0.25})_{0.95}MnO_3$ as oxygen electrode for solid oxide electrolysis cells. *International Journal of Hydrogen Energy* **2010**, *35* (11), 5187–5193.

159. Yang, C. H.; Coffin, A.; Chen, F. L., High temperature solid oxide electrolysis cell employing porous structured $(La_{0.75}Sr_{0.25})_{0.95}MnO_3$ with enhanced oxygen electrode performance. *International Journal of Hydrogen Energy* **2010**, *35* (8), 3221–3226.
160. Liang, M. D.; Yu, B.; Wen, M. F.; Chen, J.; Xu, J. M.; Zhai, Y. C., Preparation of LSM-YSZ composite powder for anode of solid oxide electrolysis cell and its activation mechanism. *Journal of Power Sources* **2009**, *190* (2), 341–345.
161. Chen, K. F.; Jiang, S. P., Failure mechanism of (La,Sr) MnO_3 oxygen electrodes of solid oxide electrolysis cells. *International Journal of Hydrogen Energy* **2011**, *36* (17), 10541–10549.
162. Keane, M.; Mahapatra, M. K.; Verma, A.; Singh, P., LSM-YSZ interactions and anode delamination in solid oxide electrolysis cells. *International Journal of Hydrogen Energy* **2012**, *37* (22), 16776–16785.
163. Chen, K. F.; Ai, N.; Jiang, S. P., Development of (Gd,Ce) O_2-impregnated $(La,Sr)MnO_3$ anodes of high temperature solid oxide electrolysis cells. *Journal of the Electrochemical Society* **2010**, *157* (11), P89–P94.
164. Kim-Lohsoontorn, P.; Brett, D. J. L.; Laosiripojana, N.; Kim, Y. M.; Bae, J. M., Performance of solid oxide electrolysis cells based on composite $La_{0.8}Sr_{0.2}MnO_{3\text{-delta}}$-yttria stabilized zirconia and $Ba_{0.5}Sr_{0.5}Co_{0.8}Fe_{0.2}O_{3\text{-delta}}$ oxygen electrodes. *International Journal of Hydrogen Energy* **2010**, *35* (9), 3958–3966.
165. Choi, M. B.; Singh, B.; Wachsman, E. D.; Song, S. J., Performance of $La_{0.1}Sr_{0.9}Co_{0.8}Fe_{0.2}O_{3-\delta}$ and $La_{0.1}Sr_{0.9}Co_{0.8}Fe_{0.2}O_{3-\delta}$-$Ce_{0.9}Gd_{0.1}O_2$ oxygen electrodes with $Ce_{0.9}Gd_{0.1}O_2$ barrier layer in reversible solid oxide fuel cells. *Journal of Power Sources* **2013**, *239*, 361–373.
166. Kong, J. R.; Zhang, Y.; Deng, C. S.; Xu, J. M., Synthesis and electrochemical properties of LSM and LSF perovskites as anode materials for high temperature steam electrolysis. *Journal of Power Sources* **2009**, *186* (2), 485–489.
167. Wang, W. S.; Huang, Y. Y.; Jung, S. W.; Vohs, J. M.; Gorte, R. J., A comparison of LSM, LSF, and LSCo for solid oxide electrolyzer anodes. *Journal of the Electrochemical Society* **2006**, *153* (11), A2066–A2070.
168. Jiang, W.; Lu, Z.; Wei, B.; Wang, Z. H.; Zhu, X. B.; Tian, Y. T.; Huang, X. Q.; Su, W. H., $Sm_{0.5}Sr_{0.5}CoO_3$-$Sm_{0.2}Ce_{0.8}O_{1.9}$ composite oxygen electrodes for solid oxide electrolysis cells. *Fuel Cells* **2014**, *14* (1), 76–82.
169. Liu, Q. L.; Khor, K. A.; Chan, S. H., High-performance low-temperature solid oxide fuel cell with novel BSCF cathode. *Journal of Power Sources* **2006**, *161* (1), 123–128.
170. Liang, M. D.; Yu, B.; Wen, M. F.; Chen, J.; Xu, J. M.; Zhai, Y. C., Preparation of NiO-YSZ composite powder by a combustion method and its application for cathode of SOEC. *International Journal of Hydrogen Energy* **2010**, *35* (7), 2852–2857.
171. Yang, C. H.; Yang, Z. B.; Jin, C.; Liu, M. L.; Chen, F. L., High performance solid oxide electrolysis cells using $Pr_{0.8}Sr_{1.2}(Co,Fe)_{0.8}Nb_{0.2}O_{4+\delta}$-CoFe alloy hydrogen electrodes. *International Journal of Hydrogen Energy* **2013**, *38* (26), 11202–11208.
172. Patro, P. K.; Delahaye, T.; Bouyer, E.; Sinha, P. K., Microstructural development of Ni-1Ce10ScSZ cermet electrode for solid oxide electrolysis cell (SOEC) application. *International Journal of Hydrogen Energy* **2012**, *37* (4), 3865–3873.
173. Yang, X.; Irvine, J. T. S., $(La_{0.75}Sr_{0.25})_{0.95}Mn_{0.5}Cr_{0.5}O_3$ as the cathode of solid oxide electrolysis cells for high temperature hydrogen production from steam. *Journal of Materials Chemistry* **2008**, *18* (20), 2349–2354.
174. Jin, C.; Yang, C.; Zhao, F.; Cui, D.; Chen, F., $La_{0.75}Sr_{0.25}Cr_{0.5}Mn_{0.5}O_3$ as hydrogen electrode for solid oxide electrolysis cells. *International Journal of Hydrogen Energy* **2011,** *36* (5), 3340–3346.
175. Xu, S. S.; Chen, S. G.; Li, M.; Xie, K.; Wang, Y.; Wu, Y. C., Composite cathode based on Fe-loaded LSCM for steam electrolysis in an oxide-ion-conducting solid oxide electrolyser. *Journal of Power Sources* **2013**, *239*, 332–340.
176. Ge, X. M.; Zhang, L.; Fang, Y. N.; Zeng, J.; Chan, S. H., Robust solid oxide cells for alternate power generation and carbon conversion. *RSC Advances* **2011**, *1* (4), 715–724.
177. Gan, Y.; Qin, Q. Q.; Chen, S. G.; Wang, Y.; Dong, D. H.; Xie, K.; Wu, Y. C., Composite cathode $La_{0.6}Sr_{0.4}TiO_{3-\delta}$-$Ce_{0.8}Sm_{0.2}O_2$-delta impregnated with Ni for high-temperature steam electrolysis. *Journal of Power Sources* **2014**, *245*, 245–255.
178. Mougin, J.; Mansuy, A.; Chatroux, A.; Gousseau, G.; Petitjean, M.; Reytier, M.; Mauvy, F., Enhanced performance and durability of a high temperature steam electrolysis stack. *Fuel Cells* **2013**, *13* (4), 623–630.
179. Wang, S. J.; Inoishi, A.; Hong, J.; Ju, Y.; Hagiwara, H.; Ida, S.; Ishihara, T., Ni-Fe bimetallic cathodes for intermediate temperature CO_2 electrolyzers using a $La_{0.9}Sr_{0.1}Ga_{0.8}Mg_{0.2}O_3$ electrolyte. *Journal of Materials Chemistry A* **2013**, *1* (40), 12455–12461.
180. Donitz, W.; Dietrich, G.; Erdle, E.; Streicher, R., Electrochemical high-temperature technology for hydrogen production or direct electricity-generation. *International Journal of Hydrogen Energy* **1988**, *13* (5), 283–287.
181. Isenberg, A. O., Energy-conversion via solid oxide electrolyte electrochemical-cells at high temperatures. *Solid State Ionics* **1981**, *3–4* (AUG), 431–437.
182. Zhang, X. Y.; O'Brien, J. E.; O'Brien, R. C.; Hartvigsen, J. J.; Tao, G.; Housley, G. K., Improved durability of SOEC stacks for high temperature electrolysis. *International Journal of Hydrogen Energy* **2013**, *38* (1), 20–28.
183. M. Petitjean, M. R., A. Chatroux, L. Bruguière, A. Mansuy, H. Sassoulas; S. Di Iorio, B. M., J. Mougin, Performance and durability of high temperature steam electrolysis: from single cell to short-stack scale. *ECS Transactions* **2011**, *35* (1), 2905–2913.
184. Jin, C.; Yang, C. H.; Chen, F. L., Novel micro-tubular high temperature solid oxide electrolysis cells. In *Solid Oxide Fuel Cells 12*, Edited by Singhal, S. C.; Eguchi, K. **2011**; *35*, 2987–2995.
185. Schefold, J.; Brisse, A.; Zahid, M.; Ouweltjes, J. P.; Nielsen, J. U., Long term testing of short stacks with solid oxide cells for water electrolysis. In *Solid Oxide Fuel Cells 12*, Edited by Singhal, S. C.; Eguchi, K. **2011**, *35*, 2915–2927.
186. Zhang, X. Y.; O'Brien, J. E.; O'Brien, R. C.; Housley, G. K., Durability evaluation of reversible solid oxide cells. *Journal of Power Sources* **2013**, *242*, 566–574.
187. Mocoteguy, P.; Brisse, A., A review and comprehensive analysis of degradation mechanisms of solid oxide electrolysis cells. *International Journal of Hydrogen Energy* **2013**, *38* (36), 15887–15902.
188. Nguyen, V. N.; Fang, Q. P.; Packbier, U.; Blum, L., Long-term tests of a Julich planar short stack with reversible solid oxide cells in both fuel cell and electrolysis modes. *International Journal of Hydrogen Energy* **2013**, *38* (11), 4281–4290.
189. Kim, S. D.; Yu, J. H.; Seo, D. W.; Han, I. S.; Woo, S. K., Hydrogen production performance of 3-cell flat-tubular solid oxide electrolysis stack. *International Journal of Hydrogen Energy* **2012**, *37* (1), 78–83.

25 Advanced Practical Technologies for Water Electrolysis

Zetian Tao, Lei Zhang, and Jiujun Zhang

CONTENTS

25.1 INTRODUCTION

25.1.1 Brief Introduction to Water Electrolysis

Water electrolysis utilizing electricity from renewable energy sources, such as solar energy and wind energy, to produce hydrogen without the emission of pollutants has been recognized as a clean and sustainable option in global energy strategy. Recently, water electrolysis has attracted much more attention because of the urgent demand for clean energy and its many advantages, such as producing hydrogen with high purity, producing no pollution, being a simple process, and having plenty of electricity sources.

One of the major advantages of electrolysis is that it can be performed at a great range of scales [1]. For example, it can be carried out on small scales by using off-grid or localized electricity sources, including renewables, such as wind and solar power, to produce hydrogen, which can then be transported and converted into electricity for smart, flexible, distributed, and small-scale energy applications. It can also be performed on some large scales to produce a large quantity of hydrogen for renewable energy storage and for complementing or replacing fossil fuels.

The other main advantage is that water electrolysis can utilize unstable renewable solar and wind power to produce stable hydrogen energy medium, which then can be converted and used whenever it is needed later [2]. In general, solar and wind power can provide a variable output, which will be difficult for the electricity grid to accept due to its instability, which places a real and fundamental limit on how much of this energy can currently be incorporated into the supply. It is believed that this limit can be circumvented if the renewable energy can be stored at times of excess production to buffer the effect of the variability on the grid and provide a more predictable supply. With respect to this, using clean electricity from solar and wind power to drive water electrolysis and produce hydrogen in large quantities as an energy-storage medium has been considered to be one of the most viable options available to us.

Regarding hydrogen as a clean energy carrier, there are several advantages that can link all forms of energy use, allowing for greater integration, greater flexibility, and greater efficiency overall. As recognized, hydrogen can be produced by water electrolysis driven by either distributed renewables or grid electricity and then stored (in small or terawatt-hour–scale quantities and in a variety of ways), then fuel on-demand electricity. Hydrogen could also be used in other ways: for example, as a vehicle fuel, as industry commodity or feedstock, and as a chemical reactant to combine with carbon to produce synthetic hydrocarbon fuels.

25.1.2 Recent Developments in Water Electrolysis

In fact, water electrolysis is a very old technology that can be traced back to about 200 years ago. The first time water was decomposed into hydrogen and oxygen by electrolysis was reported by the English scientists William Nicholson (1753–1815) and Anthony Carlisle (1768–1842) [3]. After that, Johann Ritter repeated the experiment and successfully collected the produced hydrogen and oxygen. Then the law of water electrolysis, later called Faraday's law, was first studied and discovered in 1833–1834 by Michael Faraday, who was very famous in the field of electrochemistry [4]. In 1890, an industrial product of a water electrolysis unit for generating hydrogen was developed and used in French military airships. All of these discoveries contributed to the future development of water electrolysis. However, the industry application of water electrolysis seems very slow, probably because of the high cost compared with other traditional hydrogen-production techniques, such as steam reforming. It has taken about 100 years to develop electrolyzers to industrial scale. Different types of water electrolyzers were developed for commercial application to produce hydrogen. In 1927, the first water-alkaline electrolyzer for the production of ammonia was developed by a Norwegian company, in which the first 10,000 N m^3/h H_2 plant was in operation in 1939 for supplying hydrogen gas [5]. Then different engineered systems for industrial applications were researched and applied by several companies, including Bamag, Demag, Norsk Hydro, Stuart Energy, Electrolyzer Corp., and Boveri&Cie [6]. The first pressurized water electrolyzer was built and operated by Zdansky/Lonza in 1948 and then the first solid polymer electrolyte electrolzser was built by General Electric Co. in 1966. In 1972 and 1978, solid oxide water electrolysis fuel cells (or solid oxide electrolysis cells [SOECs]) and advanced alkaline water electrolysis started to develop and be applied, respectively [7].

Currently, the most advanced water electrolysis technologies can be classified into three: (1) alkaline electrolysis; (2) proton-exchange membrane (PEM) water electrolysis; and (3) solid oxide electrolysis fuel cells. All of these three technologies have been considered as the key processes that can be used to produce hydrogen with high purity from water and renewable energy sources. It is predicted that water electrolysis technologies may occupy an increasing prominent place in the hydrogen-production industry and surely obtain rapid expansion.

25.1.3 Fundamentals of Water Electrolysis

25.1.3.1 Scheme Structure of Water Electrolysis

To discuss the water electrolysis process, Figure 25.1 shows a schematic of water electrolysis in an acidic electrolyte solution. Water is fed to the oxygen (or positive) electrode (or anode), where the water is decomposed to oxygen, protons, and electrons, which are transported to the hydrogen (or negative) electrode (or cathode) through the external circuit. The protons are conducted through the proton-conducting electrolyte and combined with the transported electrons to hydrogen at the cathode electrode.

25.1.3.2 Electrochemical Process of Water Electrolysis

As shown in Figure 25.1, the overall electrochemical process can be expressed as

$$H_2O(l) \leftrightarrow \frac{1}{2}O_2(g) + H_2(g). \tag{25.1}$$

According to Equation 25.1, the Gibbs free energy of the cell reaction can be expressed as the following equation according to the second principle of thermodynamics:

$$\Delta G_R = \Delta H_R - T\Delta S_R, \tag{25.2}$$

where ΔH_R in joules per mole is the enthalpy change associated with Equation 25.1, T is the thermodynamic temperature, ΔS_R is the entropy of the reaction, and ΔG_R is defined as the Gibbs free energy of the cell reaction. The cell voltage of the electrolysis (E_{cell}) can be expressed as

$$E_{cell} = \frac{-\Delta G_R}{nF}, \tag{25.3}$$

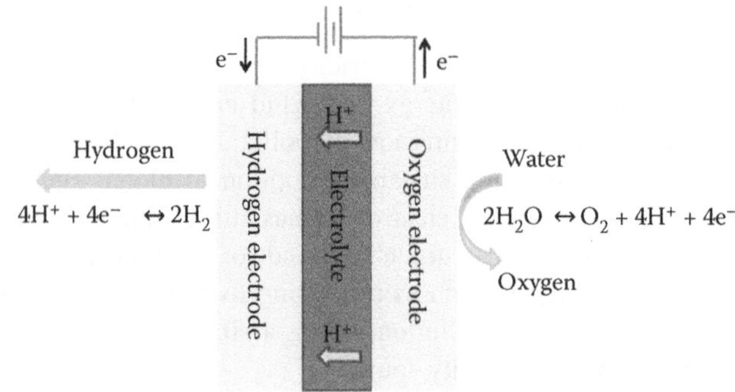

FIGURE 25.1 Schematic of water electrolysis process in acidic electrolyte solution.

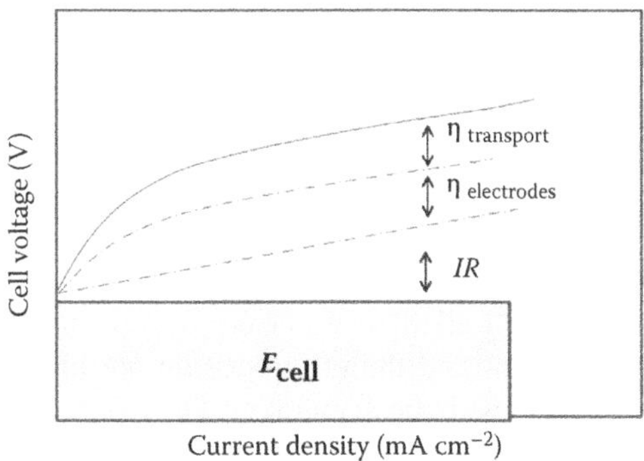

FIGURE 25.2 Schematic of the electrolytic cell voltage as a function of current density.

where n is the number of electrons, which is 2, and F is the Faraday constant (96,485C mol^{-1}). Equation 25.3 is the minimum water decomposition voltage. At the standard conditions of temperature (298.15 K) and pressure (101.325 kPa), ΔH_R (285.83 kJ mol^{-1}), ΔS_R (163.09 J mol^{-1} K^{-1}), and ΔG_R (237.23 kJ mol^{-1}) [7], the calculated cell voltage is about 1.229 V. Note that this electrolysis cell voltage is the thermodynamic or reversible one, not necessary to represent the real cell voltage when the cell has current flow.

As discussed above, although the theoretical reversible cell voltage is 1.229 V, the real cell voltage of water electrolysis is obviously higher than this value because of the slow reaction kinetics of both the oxygen evolution reaction (OER) at the oxygen electrode and the hydrogen evolution (HER) at the hydrogen electrode, which can cause overpotentials at both electrodes, mass transfer impedance, and internal and external electric resistances. The real cell voltage can be expressed as the following equation when an electrolytic current (I) passes through the cell:

$$E_{real} = E_{cell} + IR + \sum \eta, \tag{25.4}$$

where $\sum \eta$ is the sum of the overpotentials, including the electrode activation overpotentials of the two electrodes and the concentration overpotentials caused by the gas transport from the surface to the interior of the electrodes, and R is the total resistance of the electrolysis cell, including the cell internal resistance and the resistance of external circuit. Figure 25.2 shows the balance energy of the water electrolysis.

25.2 ALKALINE ELECTROLYSIS

Theoretically, water electrolysis can be carried out in acidic, neutral, or alkaline aqueous electrolyte solution. However, alkaline water electrolysis, the technology for which has been commercially available on the market, is one of the common methods of hydrogen production. The advantages of alkaline water electrolysis over other water electrolysis processes can be summarized as follows: (1) relatively lower capital expenses due to the use of cheap cell materials (electrodes and separators); (2) proven technology with well-established operational costs; (3) demonstrated large capacity units; and (4) direct use of raw water in the process without the need for specific purification procedures.

25.2.1 Working Principles

The general schematic diagram of alkaline water electrolysis cells are presented in Figure 25.3.

An alkaline electrolysis cell uses an aqueous alkaline solution, such as KOH or NaOH, for transporting ions. The electrolyte solution concentration is usually controlled at 20–40 wt.% to provide maximum electrical conductivity at temperatures of up to 90°C. The half-cell reactions can be expressed as follows:

$$2H_2O + 2e^- \rightarrow H_2 + 2OH^-, \tag{25.5}$$

$$2OH^- \rightarrow \frac{1}{2}O_2 + H_2O + 2e^-. \tag{25.6}$$

The overall cell reaction is just the same as that described in Equation 25.1, in which only water is consumed. Normally, the typical operation temperatures for alkaline electrolysis cell are between 343 and 363 K. At such operating temperatures, water vapor and traces of electrolyte can also be carried away by gas products. The operating pressure is mostly atmospheric. As stated earlier, alkaline electrolysis is a mature technology that can obtain hydrogen with a high purity of 99.9% on a dry basis [8]. Today nearly all of hydrogen that is produced through electrolysis comes from the alkaline electrolyzers.

The theoretical cell voltage (or thermodynamic or minimum electrolysis cell voltage), E_{cell}, is strongly dependent

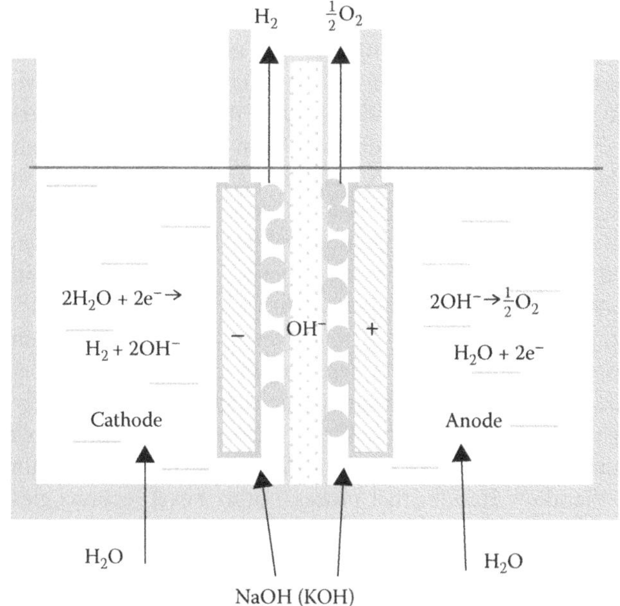

FIGURE 25.3 Schematic of alkaline water electrolysis cell structure and processes.

on the pressures of both hydrogen and oxygen. According to Equation 25.1, the dependence can be expressed as the Nernst form according to Equation 25.1:

$$E_{\text{cell}} = E^{o}_{\text{cell}} + \frac{RT}{2F}\ln\left(\frac{(P_{O_2})^{\frac{1}{2}}P_{H_2}}{a_{H_2O}}\right), \tag{25.7}$$

where E^{o}_{cell} is the cell voltage at standard conditions with both oxygen pressure (P_{O_2}) and hydrogen pressure (P_{H_2}) being 1.0 atm, R is the universal gas constant, and T and F have the same meanings as those in Equations 25.2 and 25.3.

From Equation 25.7, it can be seen that with increasing oxygen or hydrogen pressure, the electrolysis (or water decomposition) cell voltage will be increased, leading to higher energy input in order to make water decomposition.

25.2.2 Cell Components

As can be seen from the schematic shown in Figure 25.3, the two gas sides of an alkaline electrolysis cell are separated by a diaphragm (usually a porous and electrolyte-impregnated material) to avoid the possible spontaneous back recombination of H^2 and O^2 into water. In the past, the standard porous diaphragm was made of asbestos. However, this material has been abandoned now by health regulations because the inhalation of asbestos fibers can cause lung cancer. Afterward some alternative harmless materials have been developed from a large variety of alternative materials, such as polymer-based composite materials [9] or ceramic materials that are impermeable to oxygen and hydrogen, ion conductive, and stable in a serious environment at both the operating temperatures of at least 373 K and elevated pressures.

The electrodes of alkaline electrolysis cell are often made of steel grids, iron, nickel, nickel-plated iron, or materials activated by nickel sulfide [10]. In order to reduce the electrode overpotential, electrocatalysts are usually employed to cover the surface of the electrodes. Among these electrode materials, Raney nickel electrodes have been identified as the cathode materials with high performance and used for decades in alkaline electrolyzers. This kind of electrodes is often prepared by rolling or hot pressing a Ni-Al alloy to a substrate and then sintering. Then, the sintering layer is treated with concentrated sodium or potassium hydroxide at an elevated temperature of 343–363 K to activate most of the aluminum dissolved out of the alloy, resulting in a porous nickel structure with a large specific surface area and high catalytic activity toward HER. The durability of such an activated electrode can be further enhanced by adding cobalt or molybdenum to the alloy. Besides that process, other novel technologies, such as vacuum plasma spraying and plasma laser deposition, have also been used to deposit Ni-Zn, Ni-Co-Zn, or Fe-Zn alloys on the electrode support (perforated plates) [11]. Recently, several combinations of transition metals, such as Pt_2Mo, Hf_2Fe, and TiPt, have been used as cathode materials and have shown significantly higher catalysis activity than state-of-the-art electrodes [12]. The anode overpotential can also be optimized by some mixed oxides, such as ruthenium oxide (RuO_2), lanthanum nickel oxide ($LaNiO_3$), or a spinel structure (e.g., nickel cobaltite [$NiCo_2O_4$]). Among these mixed oxides, cobalt spinel cobalt oxide (Co_3O_4) presents both high activity and good stability toward oxygen evolution reaction.

Most single-cell alkaline water electrolyzer uses a filter-press configuration and the electrodes do not allow having any direct contact with the diaphragm. The electrolyzer stack, which is the core component of an electrolysis system, is made by a series connection of up to several hundred single cells [13]. In the stack, a single cell shares a common metallic plate (so-called bipolar plates) with its neighboring cell. The individual electrode is made as a perforated sheet, which is fixed on the cell support at a distance of a few millimeters from the diaphragm. During the operation, the electrolyte circulates separately in the anode and cathode compartments.

25.2.3 Cell Performance

Energy efficiency is one of the major criteria for evaluating the performance of a water electrolysis system. It is commonly defined as the percentage share of the energy output in the total energy input. Generally, energy efficiency is normally a function of the operating cell voltage and current density. Higher operating current densities are required to reduce capital costs, but at the same time lower current densities are required to reduce operational costs. Therefore, a compromise of current density is necessary and may be a great challenge.

Cell energy efficiency (η_{EE} [%]) can be expressed as the product of voltage efficiency (η_{VE} [%]) and Faraday efficiency (η_{FE} [%]):

$$\eta_{EE}(\%) = \eta_{VE}(\%)\eta_{FE}(\%) \tag{25.8}$$

Cell voltage efficiency of electrolysis is a function of practical operating current density, which can be expressed as the ratio between thermodynamic cell voltage (E_{cell}) and the practical operating cell voltage (E_{POCV}) at a certain current density:

$$\eta_{VE}(\%) = \frac{E_{\text{cell}}}{E_{\text{POCV}}} \times 100\%. \tag{25.9}$$

Due to the needed electrolysis cell voltage (E_{POCV}), for example, at 0.45 A cm^{-2} the cell voltage is normally 1.8–2.1 V, always higher than the thermodynamic one (E_{cell} = 1.229 V at standard conditions when both H_2 and O_2 pressures are all at 1.0 atm); voltage efficiency is always lower than 100%.

Faraday efficiency (also called faradaic efficiency, faradaic yield, coulombic efficiency, or current efficiency) is the efficiency with which electrons are transferred in an electrochemical system to facilitate an electrochemical reaction. It is the ratio between the mole quantity of product such as hydrogen $\left(M^{\text{act}}_{\text{prod}}\right)$ produced at a practical operating current density

(I_{cell}) and at a certain period of electrolysis time (t) and the mole quantity of product such as hydrogen $\left(M_{prod}^{theo}\right)$ calculated theoretically using I_{cell} and t:

$$\eta_{FE}(\%) = \frac{M_{prod}^{act}}{M_{prod}^{theo}} \times 100\%. \quad (25.10)$$

In practice, the value of M_{prod}^{act} can be measured, and M_{prod}^{theo} can be calculated according to Faraday's law using both values of I_{cell} and t. Inserting both Equations 25.9 and 25.10 into Equation 25.8, the energy efficiency can be expressed alternatively as

$$\eta_{EE}(\%) = \frac{E_{cell}}{E_{POCV}} \frac{M_{prod}^{act}}{M_{prod}^{theo}} \times 100\%. \quad (25.11)$$

25.2.4 Developments and Perspectives

Until today, alkaline water electrolysis is still known as the main process for the water electrolysis reaction for hydrogen production. Particularly, military applications related to the use of hydrogen isotopes have been boosting the development of the technology. For example, the first plants for the electrolysis of heavy water and the production of deuterium were built in Norway. By 1980, Aswan installed 144 electrolyzers of 162 MW which have a hydrogen generation capacity of 32,400 $m^3\ h^{-1}$ [14].

Currently, a number of companies in the world are manufacturing alkaline electrolyzers for the production of hydrogen with electrolytic grade. For example, NEL Hydrogen (a former department of Norsk Hydro Co., Norway) obtained 99.9% purity hydrogen with a high current efficiency of 98.5% by using 25% KOH as electrolyte at 80°C, while De Nora S. A. P obtained 99.9% purity hydrogen by using 29% KOH as electrolyte at 80°C [15]. There are also other companies manufacturing novel configurations for alkaline electrolyzers: Hydrogenics Corp. (which acquired Stuart Energy Systems Corp. in 2005); Teledyne Energy Systems Inc. (a subsidiary of Teledyne Technologies Inc. [Maryland, United States]); the Russian company Uralkhimmash; and De Nora (Italy), whose main products are electrolytic cells for chlorine production. The production capacity of industrial systems is usually in the 5∑500 N $m^3\ H_2\ h^{-1}$. The electrodes of these electrolyzers are often made of profiled steel coated with nickel and the diaphragms are still made of asbestos.

Alkaline electrolyzers with a high speed have been developed and also attracted a wide range of markets for the urgent demand of clean energy and hydrogen, which opens a new way to applications such as management of smart grids for more energy flexibility; chemical storage of renewable energy sources; and hydrogen-refueling stations for automotive applications.

Unfortunately, as the massive hydrocarbon energy is increasingly applied to the industry, the economic advantage of water electrolysis seems to be gradually faded as coal gasification and natural gas reforming are able to produce hydrogen on large scales at much lower costs. Although alkaline electrolysis is a commercial technology for hydrogen production, the technology still faces some challenges, such as relatively low energy efficiency and the limited market at present. The low energy efficiency is mainly caused by both insufficient voltage efficiency and Faraday efficiency. Therefore, developing new, efficient, and stable catalysts, such as some transition-metal macrocycle catalysts for both H_2 and O_2 evolution reactions to reduce electrode overpotentials, is the major effort of the current state of technology. In addition, development of advanced diaphragms with adapted electrodes/catalysts to replace poisonous asbestos is also a necessary approach. With respect to this, composite ceramic/polymer diaphragms have been proposed and technically validated but there is still room for improvement.

25.3 PROTON EXCHANGE MEMBRANE WATER ELECTROLYSIS

In 1960s, the first water electrolyzers based on PEM were developed by Grubb [16,17]. Normally, PEM water electrolysis systems can offer several advantages over traditional alkaline water electrolysis technologies, including greater energy efficiency, higher hydrogen purity, higher production rates, and more compact design [18,19]. However, there are several disadvantages of PEM electrolysis. For example, PEM electrolyzers have more special requirements on the components, including expensive polymer membranes, porous electrodes, and current collectors [20].

25.3.1 Working Principles

The working principles of PEM water electrolysis cell are presented in Figure 25.4. The proton-conducting Nafion polymer is used as both the electrolyte and the

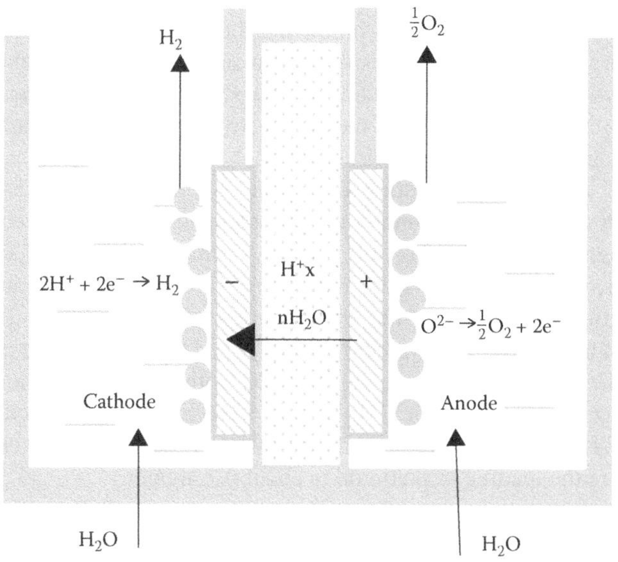

FIGURE 25.4 Schematic of PEM water electrolyzer and the electrochemical process.

separator. Two electrodes with high catalytic ability are pressed together with the electrolyte membrane, thus forming a so-called membrane–electrode assembly (MEA). The MEA is immersed in impure water. The water electrolysis reaction happens at the anode and decomposes into proton and oxygen. The mobile protons are transported through the polymer membrane to the other electrode, where they meet with protons to form hydrogen. The detailed reaction equations can be described as follows:

$$\text{Anode reaction: } H_2O \rightarrow \frac{1}{2}O_2 + 2e^- + 2H^+ \quad (E^0 = 1.299\text{ V at }25°\text{C, 1.0 atm}) \tag{25.12}$$

$$\text{Cathode reaction: } H^+ + 2e^- \rightarrow H_2 \quad (E^0 = 0.000\text{ V at }25°\text{C, 1.0 atm}) \tag{25.13}$$

The overall reaction is exactly like that described in Equation 25.1.

25.3.2 Cell Components

From Figure 25.4, it can be seen that the electrolysis cell is consisted with anode, cathode, and electrolyte membrane. The important components of the two electrodes are the electrocatalysts, which are often noble metals, and the proton-conducting electrolyte membrane.

25.3.2.1 Electrocatalysts for Proton-Exchange Membrane Water Electrolysis Cell

Because the PEM has an acidic environment, the PEM water electrolysis is actually carried out in an acidic electrolyte solution. Therefore, the requirement of catalysts for catalyst stability and activity is normally much higher than that used in alkaline medium. In practice, the catalysts used in PEM electrolyzers are noble metal–based materials. Because of the high cost of the noble catalysts, research focuses on reducing the loading through improving the catalysis ability and/or substituting the expensive noble materials with cheap catalysts. Actually, development of catalysts for water electrolysis is carried out for both of the two electrode reactions of hydrogen evolution and oxygen evolution.

At the current state of technology, carbon-supported Pt catalysts are normally used in cathode. It has been recognized that supporting strategy using highly conductive, high–surface area, and stable carbon materials can effectively improve the Pt utilization efficiency where Pt nanoparticles disperse in the catalysts. The specific area of the carbon black is about 250 m^2/g with an average grain size of about 30 nm [21]. However, the cost of the cathode catalysts are still high and need further reduction by lowering the loading of platinum to about 0.2 mg/cm^2.

Some other approaches to replacing expensive Pt have also been attempted. For example, MoS_2 and carbon nanotubes (CNTs) were used to replace platinum in some researches [22–24]. In these works, MoS_2 showed a good catalytic ability for HER but with significantly lower current densities when compared to conventional Pt cathodes. CNTs, which have higher electron conductivities and corrosion resistance when compared to conventional carbon black, could improve the catalysis activity up to 20% compared to a single Pt cathode. Pd was also used to replace Pt to form Pd/CNTs [25], but no obvious results could be found. In spite of that, some other metals, such as Co and Ni glyoximes, were also evaluated as cathode materials [26]. The catalytic activities of these metals remained stable but the performance was still not comparable to that of Pt.

For the anode, catalysts can be used to promote OER. Earlier, Damjanov et al. [27] evaluated the anode kinetics of OER on Rh, Ir, and Pt-Rh alloys using a liquid electrolyte, showing activities in the following order: Pt < Pt-Rh < Rh <Ir. Later on, some important and valuable understandings related to electrocatalysis for OER were obtained by Burke and Moynihan [28] and Burke and and O'Meara [29]. They found that the formation of oxide film on the Pt metal surface followed Elovich kinetics. In addition, the catalysis ability of Ir and RuO_2 was also evaluated [28,30,31]. In the following years, researches on PEM water electrolysis were focused on the exploration of electrocatalysts with a strong desire to have highly stable catalysts in the harsh oxidative environment and at the same time to solve the drawback of the OER irreversibility and slowness. When using single metals as anode materials, it was found that the catalysis activity strongly depended on the characters of noble metals and the activity order was Ir > Ru > Pd > Rh > Pt > Au > Nb. Based on this order, many researchers studied the potential application of Ir, Ru, and their oxides, such as RuO_2, for anode catalysis. Among the catalysts studied, RuO_2 could give the lowest overpotential for OER [32]. However, RuO_2 was limited for its unstable property, as it could be corroded at an appreciable rate with oxygen evolution by forming RuO_4 in acid electrolytes [33]. In order to solve this problem of RuO_2, many researchers attempted to reduce the erosion rate of Ru by doping other stable element or oxides into the Ru structure. Because of the high catalytic activity of Ir, introduction of this metal and its oxide into the Ru structure attracted much attention. The results demonstrated that the admixture of IrO_2 with RuO_2 could significantly improve the stability of RuO_2 during OER; even the content of IrO_2 (20%) in the catalyst was not significant [34].

Although the early researches of PEM electrolysis cells concentrated mainly on the usage and understanding of Ru and Ir catalysts and their alloys for the anode due to their high OER catalytic activity, some works were also done in exploring different catalyst alternatives for anode for improving energy efficiency and catalyst stability and reducing costs. One of the important approaches was to dilute the noble metal content by mixing some cheap metals or oxides such as SnO_2, Ta_2O_5, Sb_2O_5, Nb_2O_5 or their mixtures into the noble materials [35–37]. The noble metal Ir or Ru was coated on the surface of these oxides to form supported catalysts to obtain high utilization efficiency and chemical stability.

In order to further improve PEM electrolysis technology, many analogous catalyst materials were also explored, including core-shell catalysts, bulk metallic glasses, and

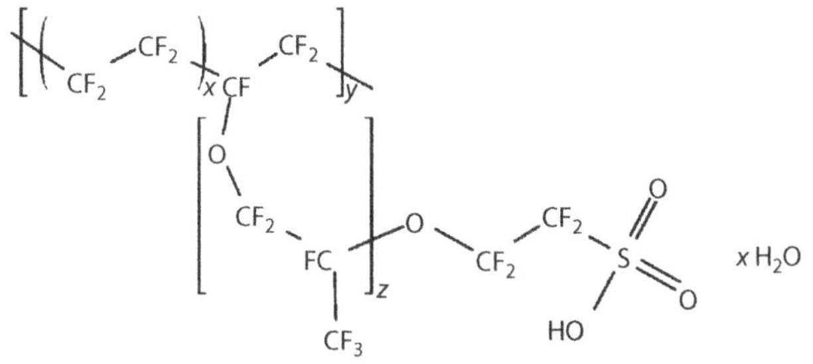

FIGURE 25.5 Chemical structure of Nafion membrane.

nanostructured thin films. All these novel materials could decrease the noble metal loadings and increase the activity and utilization of precious metal catalysts.

25.3.2.2 Polymer Membranes for Proton-Exchange Membrane Water Electrolysis Cell

The first Nafion proton-conducting membrane was developed by Dupont Corp. in 1962 and applied to fuel cells in 1966. This Nafion membrane, which contains a perfluorinated polymer with functional sulfonic acid end groups (Figure 25.5), has been commercially available for many years due to its excellent chemical and thermal stability, mechanical strength, and high proton conductivity. When the Nafion membrane is used in PEM electrolysis cell, it can be fully hydrated when it is exposed to the liquid phase of water, giving sufficient proton conductivity. There are different types of Nafion membranes for different applications. The main characters of these membranes are summarized in Table 25.1.

It should be noted that Nafion membranes are normally expensive and have some instability issue when it exposed to harsh electrochemical reactions. To overcome these drawbacks, many studies already tried to seek for hydrocarbon membranes to apply on PEM water electrolysis. Some alternative polymer materials, such as polybenzimidazoles, poly(ether ether ketones), poly(ether sulfones), and sulfonated polyphenylquinoxaline, were selected for sulfonation into ionomers/membranes to be used for PEM electrolysis [38,39]. However, these alternative membranes showed rather lower performance compared to standard Nafion membranes. Besides that, these substitute membranes face another problem, which is their low proton conductivity under operation temperature. Further works are definitely needed to improve both the conductivity and the stability of these alternative membranes.

25.3.2.3 Preparation of Proton-Exchange Membrane Electrolysis Cells

In fabricating a PEM electrolysis cells, one key step is to make MEAs, which are used to assemble the single cells then a stack of PEM electrolysis. The detailed process can be summarized as follows: (1) Pretreatment of the membrane—The aim of the pretreatment is to clear all the impurities on the membrane. The membrane is firstly treated with 3%–5% H_2O_2 at 80°C to eliminate the organic impurities. Then it is treated with diluted sulfuric acid at 80°C to clean away the inorganic impurities, impregnate protons, and washed with deionized water. (2) The prepared porous electrodes are immerged or coated with perfluorinated resin solution (ionomer). The loading of the solution is controlled at 0.6–1.2 mg/cm^2. After that, the electrodes are heated at 60–80°C. (3) The treated Nafion membrane is sandwiched between the anode and cathode to form the MEA. Then the MEA is stuck in the middle of two stainless steel plates and put inside a thermocompressor. (4) The cell is hot pressed inside the thermocompressor at a temperature of 130–150°C and a press pressure of 6–9 MPa for about 60–90 s.

In order to promote the contact between the electrode and electrolyte, the Nafion membrane can be pretreated before the hot-press process. The hot-press temperature should be improved to 150–160°C when the membrane is converted to sodium ionic and to 195°C while the membrane is converted to quaternary ammonium. The MEA can be treated with dilute sulfuric acid to reconvert after hot-press process.

25.3.3 Proton-Exchange Membrane Water Electrolysis Systems and Applications

Actually, a PEM electrolyzer has a layout similar to that of an alkaline electrolysis system consisting of an electrolysis stack, gas/liquid separators for hydrogen and oxygen, a circulation loop for the feed water at the anode, pressure control valves, a transformer, a rectifier for power conditioning, and a control system including safety devices. In the fabrication of a PEM electrolyzer, some techniques have to be applied

TABLE 25.1
Characters of Different Proton-Conducting Membranes

Corporation	Types	Thickness (μm)	Exchange Capacity (meg/g)	Moisture Content (%)
Dupont	Nafion 117	50	0.91	–
	Nafion 112	175	0.91	33
Asahi Glass	Flemion	50	1.0	–
		120	1.0	–
Asahi Chemical Industry	Aciplex S1004	100	1.0	38
	Aciplex S1004H	100	1.0	47

and developed. Among these techniques, seal technique is the main one to obtain water electrolysis system with high efficiency.

Seal technique has two types according to the technical principle. One type is the single seal technique as described in the patent of Ballard Corp. in Canada. Another type of seal technique is introduced by a patent in China which is bi-seal technique. The single-seal technique can separate the hydrogen and oxygen with high efficiency, but the utilization efficiency of membrane is only about 60%. On the contrary, the bi-seal technique can improve the utilization efficiency of expensive membrane to 90%–95%.

Currently, PEM electrolyzers have been commercially available only for small scale of hydrogen production, and there are only few manufacturers offering PEM electrolyzers with a hydrogen production rate of up to 10 Nm^3/h. The main corporations are Hamilton Sundstrand (United tatesA), Proton OnSite (United States), and Yara (Norway). These companies have created electrolyzers operating under pressures up to 2.8 MPa with capacity up to 26 m^3/h, and it is possible to combine electrolysis installations with capacity of 260 m^3/h, as shown in Figures 25.6 and 25.7 [7]. The German company H-tec can produce some small demonstration samples of water PEM electrolyzers for educational purposes. The information on the power consumption of PEM electrolysis stacks and systems as a function of the hydrogen production rate is presented in Figure 25.8 [6]. From Figure 25.8, we can find that the stacks of different size exhibit the same efficiency and the power consumption decreases only slightly with higher production rates.

FIGURE 25.7 PEM electrolyzer by Yara (Norsk Hydro Electrolysers). Productivity: 10 N m^3 H_2/h; electric power consumption: 4.4 kWh/m^3; output pressure: 3.0 MPa; hydrogen purity: 99.9% (O_2 as main impurity). (Reprinted from *Water Electrolysis Technologies*, Millet, P., and S. Grigoriev, 19–41, Copyright 2013, with permission from Elsevier.)

FIGURE 25.6 PEM electrolyzer HOGEN S Series by Proton OnSite. Productivity: 1 m^3H_2/h; electric power consumption: 5.6–9.0 kWh/m^3; electrolysis voltage: 2.3–3.8 V; output pressure: 1.4 MPa; size: 97 × 78 × 106 cm^3; weight: 215 kg. (Reprinted from *Water Electrolysis Technologies*, Millet, P., and S. Grigoriev, 19–41, Copyright 2013, with permission from Elsevier.)

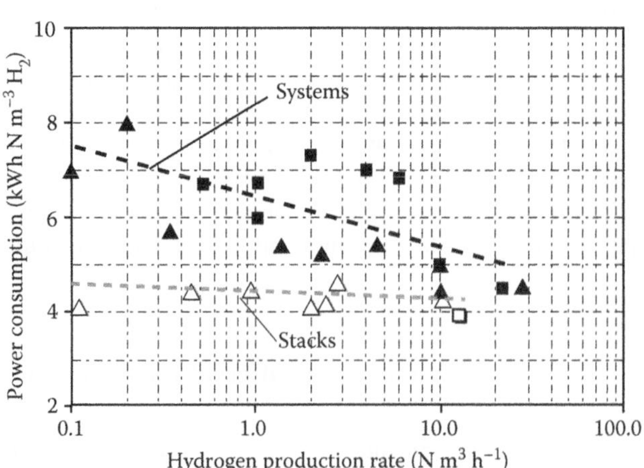

FIGURE 25.8 Power consumption of different PEM electrolysis stacks and systems as a function of the hydrogen-production rate (■: commercial systems; ▲: systems under development; □: commercial stack; △: stacks under development). (Reprinted from *Encyclopedia of Electrochemical Power Sources*, Smolinka, T., and F.-H. Production, Water electrolysis, 394–413, Copyright 2009, with permission from Elsevier.)

Researches and developments of PEM electrolyzers have been conducted and applied in many countries including Japan, France, Germany, Russia, and India. WE-NET program of Japan developed and successfully tested a cell with surface area of 2500 cm^2, operating voltage of 1.556 V at 80°C, current density of 1 A/cm^2, and an energy conversion efficiency of 95.1% [40]. In Europe, development of advanced PEM electrolyzers with elevated pressure (up to 5.0 MPa) was successfully implemented in the project GenHy PEM of 6th Framework European Program [41]. The objectives of this project were development of new gastight membranes, high-performance nanocatalysts (including nonplatinum), and bi-porous current collectors to improve the efficiency of mass transfer processes. In Russia, research and development of water PEM electrolysis systems have been carried out for more than 20 years in the National Research Center Kurchatov Institute [42].

Currently, PEM electrolyzers have been developed with capacities from a few milliliters to several cubic meters of hydrogen per hour for various purposes. It is expected that the market applications of PEM water electrolyzers should be similar to those of their alkaline counterparts, although currently commercially available systems have lower production capacities than the alkaline systems. Therefore, PEM water electrolyzers are considered to be an important component in a future energy industry where renewable energy makes a major contribution to the power supply although there is still a long way to go in the future.

25.4 SOLID OXIDE ELECTROLYSIS

In the 1980s, Doenitz and Erdle firstly reported the results from a solid oxide electrolyzer (SOECs) within the Hot Elly project at Dornier System GmbH using a supported tubular electrolyte [43]. Since then, SOECs have attracted a great interest due to the favorite reaction kinetics and thermodynamics at high-temperature electrolysis. It was demonstrated that the SOECs could convert electrical energy into chemical energy, producing hydrogen with high efficiency.

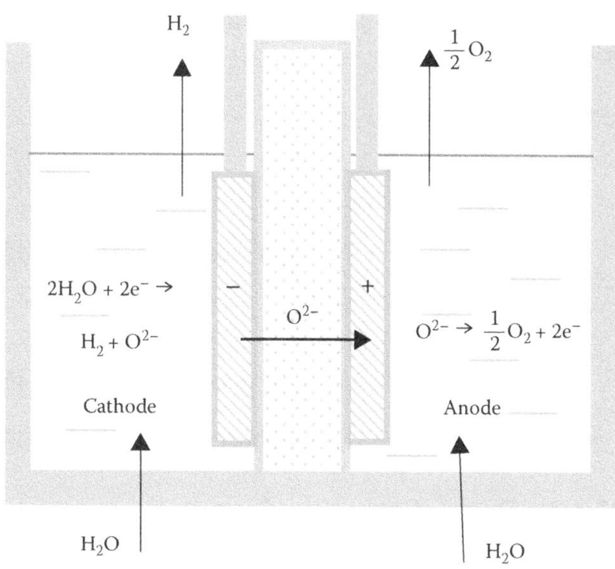

FIGURE 25.9 Schematic diagram of solid oxide-ion electrolysis cell.

25.4.1 WORKING PRINCIPLES

Traditionally, there are two types of SOECs, which can be distinguished by the electrolyte used. The electrolyte can be oxygen ion–conducting (O-SOEC) or proton-conducting (H-SOEC) materials. In the electrolysis cell, anode and cathode are painted onto the two sites of electrolyte, respectively, to form the assembly similar to membrane electrode assembly. The general principles of O-SOECs are presented in Figure 25.9 and the schematic diagram of H-SOECs is shown in Figure 25.10 [44]. According to the different conducting electrolytes, the water electrolysis reactions are different. When the electrolyte is oxygen ion conducting, the water at cathode is decomposed into hydrogen and oxygen ion which is then conducted through the electrolyte to the anode. For the proton-conducting SOECs, the water is decomposed at the anode and then the proton transports from anode to cathode

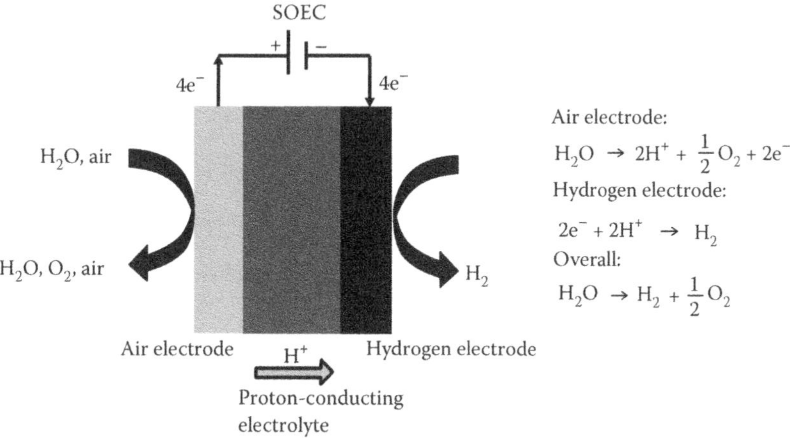

FIGURE 25.10 Schematic diagram of H-SOECs. (Reprinted from *Progress in Chemistry*, He, F., R. Peng, and S. Yang, Reversible solid oxide cell with proton conducting electrolyte: Materials and reaction mechanism, 23 (2), 477–486, 2011. With permission.)

through electrolyte. The detailed reaction equations are listed below:

O-SOECs:

$$H_2O(g) + 2e^- \rightarrow H_2(g) + O^{2-} \text{ (cathode reaction)} \quad (25.14)$$

$$O^{2-} - 2e^- \rightarrow O_2(g) \text{ (anode reaction)} \quad (25.15)$$

H-SOECs:

$$H_2O(g) - 2e^- \rightarrow \frac{1}{2} O_2(g) + 2H^+ \text{ (anode reaction)} \quad (25.16)$$

$$2H^+ + 2e^- \rightarrow H_2(g) \text{ (cathode reaction)} \quad (25.17)$$

25.4.2 Oxygen-Conducting Solid Oxide Electrolysis Cells

Over the past few decades, research and development have been focused on the advancement of solid oxide fuel cells (SOFCs) and resulted in considerable progress in materials science and technology. The development of O-SOECs has benefitted from such achievement in SOFCs and obtained quickly improvement accordingly. Recently, there has been a revival of interest in SOEC technology and different research groups in Europe, America, and Asia are working on this field to promote the applications.

An O-SOEC consists of oxygen ion–conducting electrolyte, anode, and cathode. The operation temperature of O-SOEC is highly dependent on the electrolyte materials. Many works have been done to reduce the temperature through choosing electrolyte with high ion conductivity and low activation energy [45].

25.4.2.1 Electrolyte Materials

An electrolyte material of O-SOEC conducts the oxygen ions and is sandwiched between anode and cathode. It should have high ion conductivity andhigh chemical stability and be highly compatible with the electrode materials. In addition, its sintering activity should also be good to obtain dense membrane.

Regarding the SOEC electrolyte materials, doped zirconia is the most commonly used material. Proved to be one of the most appropriate electrolytes is 8% yttrium-doped zirconia (8YSZ). Apart from the Hot Elly project [43], many researchers have also used this material as electrolyte. Normally, doped zirconia has a fluorite structure, which is highly chemically stable under reducing and oxidization atmospheres. However, SOECs with doped zirconia as electrolyte are often operated at high temperature of above 800°C because of the low ionic conductivity (~0.02 S cm^{-1} at 880°C). It has been identified that the high temperature can limit the commercial applications of SOECs due to the high costs.

To reduce the operation temperature of SOECs, some other electrolytes have also been explored to improve both the ionic conductivity and the chemical stability. Among those novel electrolytes, ceria electrolytes are probably the most promising electrolytes for intermediate temperature SOECs. Cerium oxide is usually doped with Gd_2O_3 (GDC) or Sm_2O_3 to produce the ionic conductivity, which is much higher than YSZ. Unfortunately, ceria electrolyte cannot be used as single electrolyte for SOECs due to the reduction of Ce^{4+} to Ce^{3+}, deteriorating the ionic transference number. To improve that, a bilayer structure consisting of doped zirconia and ceria has been explored. The doped zirconia can separate the ceria from the reducing atmosphere to avoid the reduction of cerium.The SOECs with the bilayer electrolyte exhibited a significantly higher performance than the single-layer SOECs [46].

$LaGaO_3$-based oxide electrolytes, usually doped with Sr on the La site and Mg on the Ga site (LSGM), have also been considered one of the most promising oxide ion conductors for intermediate temperature SOFCs and SOECs. The SOECs with LSGM as electrolyte have been studied by many researchers. For example, Burke and Meara [29] studied a SOEC using $La_{0.8}Sr_{0.2}CoO_{3-\delta}$ for oxygen electrode and $Ni_{1-x}Mg_xO$–ceria composite for fuel electrode [47].

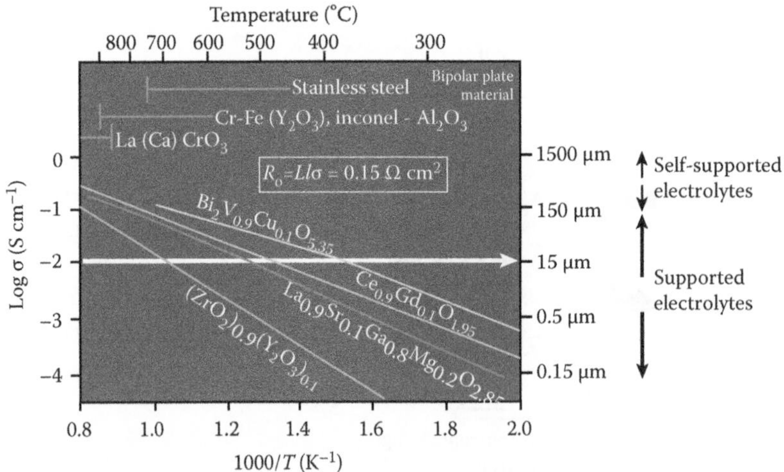

FIGURE 25.11 Ion conductivities of different electrolyte materials under different temperatures. (Reprinted by permission from Macmillan Publishers Ltd. *Nature*, Steele, B.C., and A. Heinzel, Materials for fuel-cell technologies, 414 (6861), 345–352, Copyright 2001.)

As discussed above, many electrolytes can be applied to the SOECs and the important selection criterion is the ion conductivity. Figure 25.11 [48] presents the ion conductivities of different electrolytes at elevated temperatures.

25.4.2.2 Electrode Materials

Normally, cathode of the SOECs can provide the area for water absorption, decomposition, oxygen-ion diffusion, and hydrogen production. According to the operation environment of cathode, a cathode material should meet the following requirements: (1) high catalysis activity for water decomposition; (2) enough porosity (>40%); (3) high stability under reducing atmosphere and water steam; and (4) high electron conductivity and good ion conductivity.

Traditionally, the most commonly used cathode is the composite nickel cathode with certain content of pore formers. Both the high catalytic activity and electric conductivities of nickel can certainly reduce the polarization resistances. Electrolyte material can provide the ion conductivity to transport oxygen ions and improve the compatibility with the electrolyte. In recent years, in order to improve the catalysis activity, nanoscale nickel has also been introduced onto the surface of the cathode due to their effective enhancement of the reaction rate by increasing the active reaction sites and lowering the electronic resistance. However, nickel can be oxidized, resulting in the change of the lattice constants and then mechanical stress, leading to damaged cathode structure.

In literature, lanthanum-substituted strontium titanate/ceria composite, $(La_{0.75}Sr_{0.25})_{0.95}Mn_{0.5}Cr_{0.5}O_3$ and other similar ceramic oxides have also been explored for cathode materials of SOECs [49]. The performances of these materials were tested and identified that further work would be necessary to improve electrode microstructure and current collection as well as to explore the partitioning of processes related to the conditioning of electrodes.

Regarding the anodes of SOECs, during steam electrolysis process, high oxygen partial pressure occurs at the anode/electrolyte interface; thus, delamination of anode has been observed as one of the major degradation issues [50]. Therefore, developing novel anode materials are required in order to improve their stability of both material phase and the electrode/electrolyte interface under high oxygen partial pressures during SOEC operation. According to the reaction happened at the anode, anode materials should have both high oxygen ion conductivity and electron conductivity. The chemical stability under oxidation ambient is another important factor need to be considered.

Mixed oxides with a perovskite structure, such as strontium-doped lanthanum manganite (LSM), represent the state-of-the-art for anode materials when using as YSZ electrolyte in SOECs. In order to enhance the electrochemical activity for the oxygen evolution reaction, nanoscaled Gd-doped ceria (GDC) was impregnated on LSM anode [51]. It was observed that the addition of the GDC nanoparticles could inhibit the delamination of the anode from the YSZ electrolyte.

Many other ion–electron mixed conductors such as $La_{0.8}Sr_{0.2}FeO_3$, lanthanum strontium cobalt ferrite, $Sr_2Fe_{1.5}Mo_{0.5}O_{6-\delta}$, have also used as anode materials for SOECs [52,53]. Among these novel anode materials developed, $Ba_{0.5}Sr_{0.5}Co_{0.8}Fe_{0.2}O_{3-\delta}$ (BSCF) was proved as an excellent oxygen electrode for SOECs, which had an ASR as low as 0.077 Ωcm^2 at 850°C [54]. It also exhibited a much better performance than traditional LSM anode. However, the unstable structure of BSCF could affect the long-term operation of the SOECs, limiting commercial application.

25.4.3 Proton-Conducting Solid Oxide Electrolysis Cells

H-SOECs consist of dense proton-conducting electrolyte, porous anode and porous cathode. In 1981, the first H-SOEC was developed by Iwahara et al. based on $SrCeO_3$ electrolyte to obtain hydrogen through the water electrolysis process [55]. After that, many researches put efforts on improving the conductivity and chemical stability of the electrolyte. Recently, H-SOECs have attracted more and more attention for the urgent demand of clean energy. Compared with O-SOECs, H-SOECs have many congenital structure advantages: (1) hydrogen is the carrier for water under the electrolysis process and thus can obtain pure hydrogen without purification; and (2) the proton-conducting electrolyte has low activation energy and is suitable for application under intermediate temperature, which can reduce the system costs.

At the current state of technology, the electrolysis efficiency of H-SOEC is very high due to high operation temperature. As seen in Figure 25.12 [56], ΔH is almost the same with the temperature, while $T\Delta S$ increases with the elevated temperatures, which leads to a decrease in ΔG. Therefore, the water electrolysis power decreases with the increase in temperature to improve the electrolysis efficiency. Kobayashi et al. [57] used $SrZr_{0.9}Yb_{0.1}O_{3-\delta}$ material to construct an H-SOEC, and

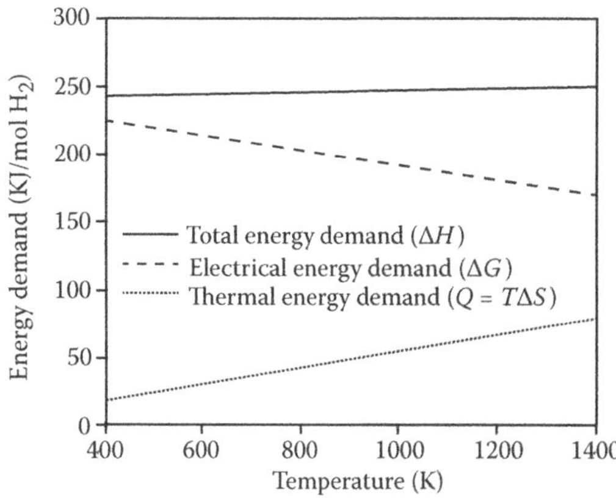

FIGURE 25.12 Energy demands for electrolytic H_2 with temperature. (Reprinted from *International Journal of Hydrogen Energy*, 33, Ni, M., M. K. Leung, and D. Y. Leung, Technological development of hydrogen production by solid oxide electrolyzer cell [SOEC], 2337–2354, Copyright 2008, with permission from Elsevier.)

obtained a high electrolysis efficiency of 94% at the current density of 2.4 mA cm^{-2} and temperature of 600°C.

25.4.3.1 Cell Components

25.4.3.1.1 Electrolyte Materials

Among those materials for H-SOECs, the electrolyte is the most critical component that plays a necessary role in conducting protons, blocking electrons and separating hydrogen from oxygen. The cell performance is strongly dependent on the properties of electrolyte, such as conductivity, chemical stability, and thermal expansion coefficient. The electrolyte materials should meet the following requirements: (1) high proton conductivity and insignificant electron conductivity; (2) high chemical stability under oxidizing and reducing atmosphere; (3) dense membrane under operation conditions to separate the gas of the two electrodes; and (4) good chemical compatibility with two electrode materials. According to these requirements, some materials with perovskite structure such as $BaCeO_3$ and $BaZrO_3$ are commonly used as electrolyte.

The first explored electrolyte was Yb-doped $SrCeO_3$, which had a high conductivity of 0.01 S cm^{-1} at 800°C [55]. After that, many studies were focused on doping this type of materials with novel elements and found that low valence elements doped $SrCeO_3$, $BaCeO_3$, $CaZrO_3$, and $BaZrO_3$ had considerable conductivities under humidified atmosphere. Among these materials, doped $BaCeO_3$ had almost the highest proton conductivity which could reach up to 0.018 S cm^{-1} at 600°C. However, the chemical stability of $BaCeO_3$ was found to not be very high because of its reactions with H_2O and CO_2. Improving the chemical stability of these materials will be a large challenge for the future study.

$BaZrO_3$ was found to be another attractive type of electrolyte due to its high chemical stability under CO_2 atmosphere which could maintain its structure after treated with saturated steam at 85°C for 3000 hours [58]. Although the chemical stability could meet the demands of commercial use for water electrolysis cell, the proton conductivity was one order lower than $BaCeO_3$ due to the high resistance. In addition, $BaZrO_3$ still faces the problem of high sintering temperature and low mechanical strength.

In order to solve the issue for $BaCeO_3$, a compromise material of $BaCeO_3$-$BaZrO_3$ solid solution was explored as electrolyte. $BaCe_{0.7}Y_{0.2}Zr_{0.1}O_3$ was studied and proved to be an ideal material because it has a good proton conductivity of 0.001 S cm^{-1} at 500°C and a high chemical stability [45]. Besides Zr in the electrolyte materials, some other elements such as Ta, In, and Sn were also explored to replace Ce to obtain high chemical stability and good conductivity [59–61].

Moreover, there are some other materials showing both good proton conductivity and chemical stability. For example, as reported by Du and Nowick [62], $Ba_3Ca_{1.18}Nb_{1.82}O_{9-\delta}$ with composite perovskite structure had a high proton conductivity of 0.01 S cm^{-1} at 600°C and maintained a stable performance under CO_2 atmosphere for 100 hours. Pyrochlore structure materials also showed some proton conductivity and high chemical stability, but the low conductivity limited their practical applications. $La_2Ce_2O_7$ series materials with fluorite structure were proved to be proton conductors through hydrogen permeation test and their chemical stability were also tested by Tao et al. [63]. It was believed that the reasonable chemical stability and enough conductivity could make these materials being potential for application in water electrolysis although the detailed proton conducting principle of these materials was still not clear.

25.4.3.1.2 Electrode Materials

The electrodes of H-SOECs are also the necessary components determining the performance. Normally, the polarization resistances (or overpotential) of anode is much higher than that of cathode in H-SOECs. Many reactions such as decomposition of water, conversion from oxygen to oxygen ion, and surface and bulk diffusion could happen at the anode. To reduce the anode overpotential, the anode materials should meet the following requirements: (1) good chemical stability; (2) favorite porous structure; (3) high electron conductivity and enough ion conductivity; (4) chemical compatible with electrolyte; and (5) high catalytic activity for oxygen evolution reaction.

At present, the anode materials of H-SOECs are almost the same as those used for conventional proton-conducting SOFCs (H-SOFCs). The anode materials can be classified into two kinds, one is the electron conducing material, and the other is the ion–electron mixed conductor. For example, Pt and other noble metals are electron-conducting materials and used for anode materials for many years. But these materials are gradually replaced by ion–electron mixed conductors to reduce the cost and increase diffusion rate at high operation temperature. Ion-electron mixed conductors normally have an ABO_3 perovskite structure and can obtain large triple-phase boundary to promote reaction rate of anode. In order to add the triple-phase boundary further, some composites anodes constituted of a proton conductor and an oxygen ion-electron mixed conductor have been explored to reduce the polarization resistances. For example, Fabbri et al. [64] used $La_{0.6}Sr_{0.4}Fe_{0.8}Co_{0.2}O_{3-\delta}$ (LSCF)/$BaCe_{0.9}Yb_{0.1}O_{3-\delta}$ as a composite air electrode for H-SOFC, and a reduced polarization resistance for the cells of approximately 0.14 Ω cm^2 at 700°C was achieved. Unfortunately, a larger polarization resistance was observed using composite anode electrode in H-SOEC. It was found that the large polarization resistance of composite cathode might result from the low oxygen partial pressure in anode side. Exploring new anode materials with an improved electrochemical reaction rate will be the key to further improve the performance of an H-SOEC using a thin film electrolyte.

The cathode of an H-SOEC can provide the area for hydrogen absorption, diffusion, and conversion. According to the operation environment of cathode, cathode materials should meet the following requirements: (1) high catalytic activity toward hydrogen evolution reaction; (2) enough porosity (> 40%); (3) high stability under reducing atmosphere; and (4) high electron conductivity and good ion conductivity.

With respect to cathode materials, composite cathode materials of nickel–electrolyte seem to be the most economical and affordable cathode materials currently. Compared to high polarization resistance of anode, cathode polarization resistance is normally much smaller, only occupies about 10% of the total polarization resistance of H-SOECs. Additionally, some oxides such as $LaTiO_3$ and $LaCrO_3$ were also used as the cathode materials [49]. However, the application of these materials is greatly limited for their low electron conductivities.

25.4.3.2 Performance and Perspectives

At present, H-SOECs are mainly adopting an electrolyte-support structure. $BaCe_{0.9}Y_{0.1}O_{3-\delta}$ supported SOEC with Pt as electrode is tested by Stuart et al. under different temperatures as shown in Figure 25.13 [65]. It can be seen that at a SOEC mode, current density is 45 mA cm^{-2} and resistance is about 9 Ω cm^2 when the potential is controlled at 1.5 V. The performance of the electrolysis cell is low due to the large resistance caused by the thick electrolyte.

In order to improve the cell performance, electrolyte-support structure could be changed into cathode-support structure to reduce the thickness of electrolyte. The change of structure could obviously improve the cell performance. In addition, some novel techniques such as impregnation were also used to modify the anode structure to reduce the polarization resistances.

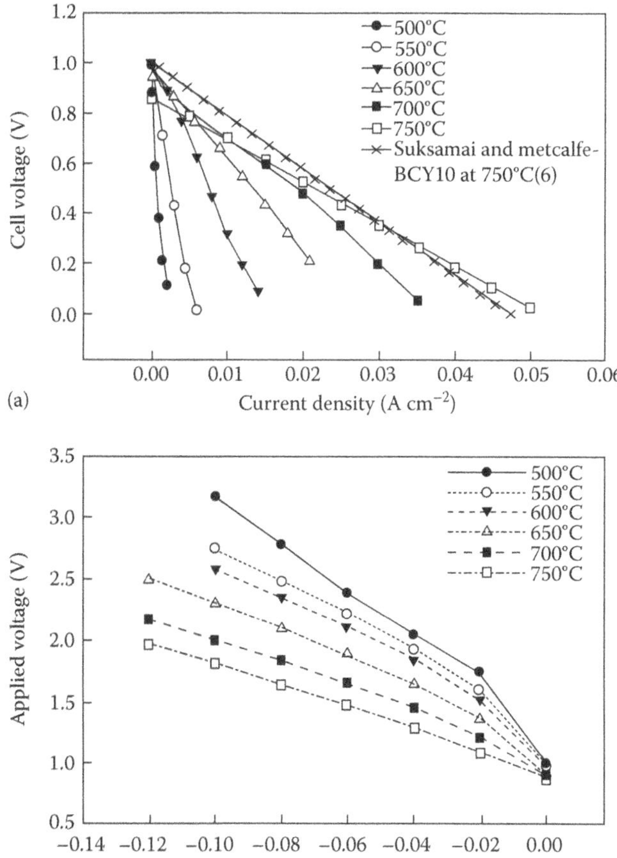

FIGURE 25.13 *I–V* curves for a cell with as electrolyte at different temperature: (a) SOFC mode and (b) SOEC mode. (Reprinted from *Solid State Ionics*, 179, Stuart, P. A. et al., Solid oxide proton conducting steam electrolysers, 1120–1124, Copyright 2008, with permission from Elsevier.)

H-SOECs have been proved to be a potential energy conversion technology. However, the improvement is still necessary. In the future, improvements of both electrolytes and anode materials may attract the most attentions.

25.4.4 Technology Developments and Applications

As discussed above, SOECs have been approved to be the promising alternatives to the existing water electrolysis methods for hydrogen production. Furthermore, due to the chemical flexibility of those devices, they can also be used for the electrolysis of CO_2 to CO, and for the coelectrolysis of H_2O/CO_2 to H_2/CO (syngas) [66]. Currently, SOECs are still under development but research has grown exponentially in the last decade, which can be singled by the great interesting from companies, research centers, and universities around the world. For example, Siemens-Westinghouse (United States; Germany) and the Institute of the High-Temperature Electrochemistry of Ural Branch of Russian Academy of Sciences (Russia) have developed some products of SOECs. These preliminary lab-scale studies are focused on the development of novel, low-cost, and highly durable materials, the inherent manufacturing processes, and the system integration for efficient and durable SOECs.

Because of the cell components all exist as solid phase, SOECs can be easily assembled and have various cell geometries. The main cell geometries are planar (similar to cells of other electrolytic cell types), tubular (the electrolyte is in the form of a thin-walled tube), and even more complex configurations such as "honeycomb" shapes.

Planar geometry has the advantages of simple structure, simple preparation technology, low cost, and high production rates. With respect to this, research and development programs were carried out by a group of U.S. national laboratories (supported by the U.S. Department of Energy) in 2006 [67]. A demonstration electrolyzer containing 25 planar cells with a hydrogen production capacity of 160 L/h (800°C) has been developed through these programs. Although the SOEC electrolyzer could be operated for 1000 hours, the seal seemed to be an issue, and the performance was not very stable under the thermal cycle. Recently, these issues have been partially solved by some research groups and the further development will accelerate the commercialization process of these planar SOECs.

Tubular SOECs can be easily assembled and grouped to large-scale cell system by either parallel connection or series connection. But the preparation of single cell seems to be very complex and expensive. Nowadays there are only a few companies in the world including USA Westinghouse [67].

25.5 CHAPTER CONCLUSION

This chapter reviews the most feasible technologies of water electrolysis for hydrogen generation, including alkaline electrolysis,

PEM electrolysis, and solid oxide electrolysis. Although there are many achievements and breakthroughs in the last several decades, there are still several challenges hindering their competition ability in commercialization. As discussed, the major challenge is the fast degradation, that is, the rate of performance degradation under high steam contents still remains too high for practical applications in the state-of-the-art technology. To solve this issue, exploring and developing innovative, stable, and highly active electrode materials including electrocatalysts for anode OER and cathode HER, as well as electrolyte materials including PEMs and solid oxide membranes should be the research directions for future approaches. With respect to these, innovative material synthesis, characterization, performance, and validation as well as fundamental understanding through both experimental and theoretical approaches are necessary.

REFERENCES

1. Zeng, K., and D. Zhang, Recent progress in alkaline water electrolysis for hydrogen production and applications. *Progress in Energy and Combustion Science*, 2010. **36**(3): pp. 307–326.
2. Ni, M. et al., A review and recent developments in photocatalytic water-splitting using TiO_2 for hydrogen production. *Renewable and Sustainable Energy Reviews*, 2007. **11**(3): pp. 401–425.
3. Andújar, J., and F. Segura, Fuel cells: History and updating; A walk along two centuries. *Renewable and Sustainable Energy Reviews*, 2009. **13**(9): pp. 2309–2322.
4. Faraday, M., *On a New Law of Electric Conduction; On Conducting Power Generally*. 1833: Royal Society.
5. Kreuter, W., and H. Hofmann, Electrolysis: The important energy transformer in a world of sustainable energy. *International Journal of Hydrogen Energy*, 1998. **23**(8): pp. 661–666.
6. Smolinka, T., and F.-H. Production, Water electrolysis. *Encyclopedia of Electrochemical Power Sources*, Elsevier, 2009: pp. 394–413.
7. Millet, P., and S. Grigoriev, *Water Electrolysis Technologies*, Elsevier. 2013: pp. 19–41.
8. Bergner, D., Membrane cells for chlor-alkali electrolysis. *Journal of Applied Electrochemistry*, 1982. **12**(6): pp. 631–644.
9. [Anon.], SOFC system passes 50% "Efficiency Threshold." *American Ceramic Society Bulletin*, 2008. **87**(1): p. 18.
10. Vandenborre, H., P. Vermeiren, and R. Leysen, Hydrogen evolution at nickel sulphide cathodes in alkaline medium. *Electrochimica Acta*, 1984. **29**(3): pp. 297–301.
11. Schiller, G., and V. Borck, Vacuum plasma sprayed electrodes for advanced alkaline water electrolysis. *International Journal of Hydrogen Energy*, 1992. **17**(4): pp. 261–273.
12. Stojić, D.L. et al., Intermetallics as cathode materials in the electrolytic hydrogen production. *International Journal of Hydrogen Energy*, 2005. **30**(1): pp. 21–28.
13. Ulleberg, Ø., Modeling of advanced alkaline electrolyzers: A system simulation approach. *International Journal of Hydrogen Energy*, 2003. **28**(1): pp. 21–33.
14. Bernard, M. et al., Application of electrocatalysis to the electrolysis of water at high temperature and high current density. In *Hydrogen as an Energy Vector*. 1980. Springer. pp. 283–294.
15. Borup, R. et al., Scientific aspects of polymer electrolyte fuel cell durability and degradation. *Chemical Reviews*, 2007. **107**(10): pp. 3904–3951.
16. Grubb, W., Ionic migration in ion-exchange membranes. *The Journal of Physical Chemistry*, 1959. **63**(1): pp. 55–58.
17. Grubb, W., Batteries with solid ion exchange electrolytes; I. Secondary cells employing metal electrodes. *Journal of the Electrochemical Society*, 1959. **106**(4): pp. 275–278.
18. Grigoriev, S., V. Porembsky, and V. Fateev, Pure hydrogen production by PEM electrolysis for hydrogen energy. *International Journal of Hydrogen Energy*, 2006. **31**(2): pp. 171–175.
19. Marshall, A. et al., Hydrogen production by advanced proton exchange membrane (PEM) water electrolysers—Reduced energy consumption by improved electrocatalysis. *Energy*, 2007. **32**(4): pp. 431–436.
20. Barbir, F., PEM electrolysis for production of hydrogen from renewable energy sources. *Solar Energy*, 2005. **78**(5): pp. 661–669.
21. Palmer, M.B., and M. Vannice, The effect of preparation variables on the dispersion of supported platinum catalysts. *Journal of Chemical Technology and Biotechnology*, 1980. **30**(1): pp. 205–216.
22. Hinnemann, B. et al., Biomimetic hydrogen evolution: MoS_2 nanoparticles as catalyst for hydrogen evolution. *Journal of the American Chemical Society*, 2005. **127**(15): pp. 5308–5309.
23. Li, Y. et al., MoS_2 nanoparticles grown on graphene: An advanced catalyst for the hydrogen evolution reaction. *Journal of the American Chemical Society*, 2011. **133**(19): pp. 7296–7299.
24. Xu, W. et al., A novel hybrid based on carbon nanotubes and heteropolyanions as effective catalyst for hydrogen evolution. *Electrochemistry Communications*, 2007. **9**(1): pp. 180–184.
25. Grigoriev, S., P. Millet, and V. Fateev, Evaluation of carbon-supported Pt and Pd nanoparticles for the hydrogen evolution reaction in PEM water electrolysers. *Journal of Power Sources*, 2008. **177**(2): pp. 281–285.
26. Pantani, O. et al., Electroactivity of cobalt and nickel glyoximes with regard to the electro-reduction of protons into molecular hydrogen in acidic media. *Electrochemistry Communications*, 2007. **9**(1): pp. 54–58.
27. Damjanovic, A., A. Dey, and J.M. Bockris, Electrode Kinetics of Oxygen Evolution and Dissolution on Rh, Ir, and Pt - Rh Alloy Electrodes. *Journal of The Electrochemical Society*, 1966. **113**(7): pp. 739–746.
28. Burke, L., and A. Moynihan, Oxygen electrode reaction; Part 1—Nature of the inhibition process. *Transactions of the Faraday Society*, 1971. **67**: pp. 3550–3557.
29. Burke, L., and T. O'Meara, Oxygen electrode reactionl Part 2—Behaviour at ruthenium black electrodes. *Journal of the Chemical Society, Faraday Transactions 1: Physical Chemistry in Condensed Phases*, 1972. **68**: pp. 839–848.
30. Buckley, D., and L. Burke, The oxygen electrode; Part 4. Lowering of the overvoltage for oxygen evolution at noble metal electrodes in the presence of ruthenium salts. *Journal of Electroanalytical Chemistry and Interfacial Electrochemistry*, 1974. **52**(3): pp. 433–442.
31. Buckley, D.N., and L.D. Burke, The oxygen electrode; Part 6—Oxygen evolution and corrosion at iridium anodes. *Journal of the Chemical Society, Faraday Transactions 1: Physical Chemistry in Condensed Phases*, 1976. **72**: pp. 2431–2440.
32. Miles, M. et al., The oxygen evolution reaction on platinum, iridium, ruthenium and their alloys at 80°C in acid solutions. *Electrochimica Acta*, 1978. **23**(6): pp. 521–526.

33. Iwakura, C., K. Hirao, and H. Tamura, Anodic evolution of oxygen on ruthenium in acidic solutions. *Electrochimica Acta*, 1977. **22**(4): pp. 329–334.
34. Andolfatto, F. et al., Solid polymer electrolyte water electrolysis: electrocatalysis and long-term stability. *International Journal of Hydrogen Energy*, 1994. **19**(5): pp. 421–427.
35. Ardizzone, S. et al., Composite ternary SnO_2–IrO_2–Ta_2O_5 oxide electrocatalysts. *Journal of Electroanalytical Chemistry*, 2006. **589**(1): pp. 160–166.
36. Morimitsu, M., R. Otogawa, and M. Matsunaga, Effects of cathodizing on the morphology and composition of IrO_2-Ta_2O_5/Ti anodes. *Electrochimica Acta*, 2000. **46**(2): pp. 401–406.
37. Terezo, A.J. et al., Separation of transport, charge storage and reaction processes of porous electrocatalytic IrO_2 and IrO/Nb_2O_5 electrodes. *Journal of Electroanalytical Chemistry*, 2001. **508**(1): pp. 59–69.
38. Jang, I.-Y. et al., Application of polysulfone (PSf)–and polyether ether ketone (PEEK)–tungstophosphoric acid (TPA) composite membranes for water electrolysis. *Journal of Membrane Science*, 2008. **322**(1): pp. 154–161.
39. Linkous, C. et al., Development of new proton exchange membrane electrolytes for water electrolysis at higher temperatures. *International Journal of Hydrogen Energy*, 1998. **23**(7): pp. 525–529.
40. Yamaguchi, M. et al., Development of 2500 cm^2 five-cell stack water electrolyzer in WE-NET. In *Environmental Aspects of Electrochemical Technology: Proceedings of the International Symposium*. 2000. The Electrochemical Society.
41. Millet, P. et al., GenHyPEM: A research program on PEM water electrolysis supported by the European Commission. *International Journal of Hydrogen Energy*, 2009. **34**(11): pp. 4974–4982.
42. Fateev, V. et al., Electrolysis of water in systems with solid polymer electrolyte. *Russian Journal Electrochemistry*, 1993. **29**(4): pp. 551–557.
43. Doenitz, W., and E. Erdle, High-temperature electrolysis of water vapor—Status of development and perspectives for application. *International Journal of Hydrogen Energy*, 1985. **10**(5): pp. 291–295.
44. He, F., R. Peng, and S. Yang, Reversible solid oxide cell with proton conducting electrolyte: Materials and reaction mechanism. *Progress in Chemistry*, 2011. **23**(2): pp. 477–486
45. Zuo, C. et al., Ba $(Zr_{0.1}Ce_{0.7}Y_{0.2})$ $O_{3-\delta}$ as an electrolyte for low-temperature solid-oxide Fuel Cells. *Advanced Materials*, 2006. **18**(24): pp. 3318–3320.
46. Kim-Lohsoontorn, P., N. Laosiripojana, and J. Bae, Performance of solid oxide electrolysis cell having bi-layered electrolyte during steam electrolysis and carbon dioxide electrolysis. *Current Applied Physics*, 2011. **11**(1): pp. S223–S228.
47. Elangovan, S., J.J. Hartvigsen, and L.J. Frost, Intermediate temperature reversible fuel cells. *International Journal of Applied Ceramic Technology*, 2007. **4**(2): pp. 109–118.
48. Steele, B.C., and A. Heinzel, Materials for fuel-cell technologies. *Nature*, 2001. **414**(6861): pp. 345–352.
49. Yang, X., and J.T. Irvine, $(La_{0.75}\,Sr_{0.25})_{0.95}Mn_{0.5}Cr_{0.5}\,O_3$ as the cathode of solid oxide electrolysis cells for high temperature hydrogen production from steam. *Journal of Materials Chemistry*, 2008. **18**(20): pp. 2349–2354.
50. Sharma, V.I., and B. Yildiz, Degradation mechanism in $La_{0.8}Sr_{0.2}CoO_3$ as contact layer on the solid oxide electrolysis cell anode. *Journal of the Electrochemical Society*, 2010. **157**(3): pp. B441–B448.
51. Chen, K., N. Ai, and S.P. Jiang, Development of (Gd, Ce) O_2-impregnated (La, Sr) MnO_3 anodes of high temperature solid oxide electrolysis cells. *Journal of the Electrochemical Society*, 2010. **157**(11): pp. P89–P94.
52. Liu, Q. et al., Perovskite $Sr_2Fe_{1.5}Mo_{0.5}O_{6-\delta}$ as electrode materials for symmetrical solid oxide electrolysis cells. *Int. J. Hydrogen Energy*, 2010. **35**: pp. 10039–44.
53. Wang, W. et al., A comparison of LSM, LSF, and LSCo for solid oxide electrolyzer anodes. *Journal of the Electrochemical Society*, 2006. **153**(11): pp. A2066–A2070.
54. Bo, Y. et al., Microstructural characterization and electrochemical properties of $Ba_{0.5}Sr_{0.5}Go_{0.8}Fe_{0.2}O_{3-\delta}$ and its application for anode of SOEC. *International Journal of Hydrogen Energy*, 2008. **33**(23): pp. 6873–6877.
55. Iwahara, H. et al., Proton conduction in sintered oxides and its application to steam electrolysis for hydrogen production. *Solid State Ionics*, 1981. **3**: pp. 359–363.
56. Ni, M., M.K. Leung, and D.Y. Leung, Technological development of hydrogen production by solid oxide electrolyzer cell (SOEC). *International Journal of Hydrogen Energy*, 2008. **33**(9): pp. 2337–2354.
57. Kobayashi, T. et al., Study on current efficiency of steam electrolysis using a partial protonic conductor $SrZr_{0.9}Yb_{0.1}O_{3-\alpha}$. *Solid State Ionics*, 2001. **138**(3): pp. 243–251.
58. Serra, J.M., and W.A. Meulenberg, Thin-film proton $BaZr_{0.85}Y_{0.15}O_3$ conducting electrolytes: Toward an intermediate-temperature solid oxide fuel cell alternative. *Journal of the American Ceramic Society*, 2007. **90**(7): pp. 2082–2089.
59. Bi, L. et al., A novel anode supported $BaCe_{0.7}Ta_{0.1}Y_{0.2}O_{3-\delta}$ electrolyte membrane for proton-conducting solid oxide fuel cell. *Electrochemistry communications*, 2008. **10**(10): pp. 1598–1601.
60. Bi, L. et al., Indium as an ideal functional dopant for a proton-conducting solid oxide fuel cell. *International Journal of Hydrogen Energy*, 2009. **34**(5): pp. 2421–2425.
61. Xie, K., R. Yan, and X. Liu, The chemical stability and conductivity of $BaCe_{0.9-x}Y_xSn_{0.1}O_{3-\delta}$ solid proton conductor for SOFC. *Journal of Alloys and Compounds*, 2009. **479**(1): pp. L36–L39.
62. Du, Y., and A. Nowick, Galvanic cell measurements on a fast proton conducting complex perovskite electrolyte. *Solid State Ionics*, 1996. **91**(1): pp. 85–91.
63. Tao, Z. et al., A stable $La_{1.95}\,Ca_{0.05}Ce_2O_{7-\delta}$ as the electrolyte for intermediate-temperature solid oxide fuel cells. *Journal of Power Sources*, 2011. **196**(14): pp. 5840–5843.
64. Fabbri, E. et al., Composite cathodes for proton conducting electrolytes. *Fuel Cells*, 2009. **9**(2): pp. 128–138.
65. Stuart, P.A. et al., Solid oxide proton conducting steam electrolysers. *Solid State Ionics*, 2008. **179**(21): pp. 1120–1124.
66. Xu, S. et al., Direct electrolysis of $C0_2$ using an oxygen-ion conducting solid oxide electrolyzer based on $La_{0.75}Sr_{0.25}Cr_{0.5}Mn_{0.5}O_{3-\delta}$ electrode. *Journal of Power Sources*, 2013. **230**: pp. 115–121.
67. Hauch, A. et al., Performance and durability of solid oxide electrolysis cells. *Journal of the Electrochemical Society*, 2006. **153**(9): pp. A1741–A1747.

Index

Page numbers followed by f and t indicate figures and tables, respectively.

D

H

I

J

K

L

N

O

P

Q

R

X

Y

Z

Printed and bound by CPI Group (UK) Ltd, Croydon, CR0 4YY
22/10/2024
01777611-0019